Basic Structural Analysis

Third Edition

About the Author

C S Reddy served as Principal, K S R M College of Engineering, Kadapa, for about twenty years before retiring from service. Earlier, he had been a faculty of IIT Kharagpur for over 20 years, before joining as Professor at N B K R Institute of Science and Technology, Vidyanagar, Nellore. He has been a Visiting Professor at the University of Basrah, Iraq, during 1974–77. Having obtained a PhD degree from IIT Kharagpur, he has guided several MTech and PhD students. He is a Fellow of the Institution of Engineers (India), Member of Senate and Chairman, Board of Studies at Sri Venkateswara University, Tirupati, Tamil Nadu. Dr Reddy can be contacted at *chintasreddy@yahoo.co.in*.

Basic Structural Analysis

Third Edition

C S Reddy

Principal (Retd.)
KSRM College of Engineering
Kadapu

McGraw Hill Education (India) Private Limited

Published by McGraw Hill Education (India) Private Limited
Registered Office: Anjana Complex No: 5/90 A, Butt Road, St. Thomas Mount, Chennai – 600016

Basic Structural Analysis, 3/e

25th reprint 2023

ISBN-13: 978-0-07-070276-9
ISBN-10: 0-07-070276-4

26 27 28 29 30 SAI 27 26 25 24 23

Managing Director: *Lalit Singh*
Head—Production: *Pankaj Manohar*
Sr. Production Manager: *Reji Kumar*

Typeset at Text-O-Graphics, B-1/56, Arawali Apartment, Sector 34, Noida, and,
Printed and Bound in India at Sai Printo Pack Pvt. Ltd., A-102/4, Okhla Industrial Area,
Phase – II, New Delhi 110020

Visit us at: www.mheducation.co.in
Write to us at: info.india@mheducation.com
CIN: U80302TN2010PTC111532
Toll free Number: 1800 103 5875

Preface

Brief Overview

The use of computers for structural analysis has completely altered the method of presentation of structural theory. While the student is expected to be familiar with this presentation, it is far more important that he understands the basic principles of structural analysis.

The book endeavours to present in one volume, the classical as well as matrix methods of structural analysis. It is expected that for some time to come, the student will be required to study both these approaches, for the matrix methods are not very different from classical methods—the only difference is in the emphasis laid in formulating them so as to be suitable for computer programming. An understanding of the basic principles in both these methods necessarily requires the solving of simple problems using hand computations.

This book is intended for a course in structural analysis following the usual course in mechanics of solid or, as it is more commonly called, strength of materials. It aims to provide a smooth transition from the classical approaches that are based on physical behaviour of structures in terms of their deflected shapes to a formal treatment of a general class of structures by means of matrix formulation. This book can be used by undergraduate students, professionals as well as those preparing for competitive examinations.

Rationale behind the Third Edition

Encouraged by the tremendous response to the first two editions, this book has been revised keeping in mind the valuable suggestions received from the reviewers, publishers, readers and colleagues. The second edition of the text came out in 1996, i.e., 14 years back and since then has undergone 23 reprints! Since this book is prescribed as a textbook and as a reference in many major universities of India, to be in tandem with the changing course requirements, revision of the text assumed prime importance. Also, to uphold the competitive edge, I felt it was necessary to include certain pedagogical features like step-by-step approach for the solved examples, objective-type questions and a solution manual.

Changes in the Third Edition

It was indeed a challenging task to undertake the revision of the textbook for its third edition! Keeping the basic approach of the first two editions intact, the third edition has been written to make the book broad-based and gain wider acceptance amongst teachers and students.

Though arches are not included under a separate chapter, the three-hinged arches are dealt with elaborately under cables and arches in **Chapter 2**, and two-hinged arches under indeterminate structures in **Chapter 10** dealing with consistent displacements. The ILD for three-hinged arches are covered in **Chapter 7**.

The scope of fixed beams is enlarged by including a large number of worked-out examples covering point loads, uniform and varying loads, applied couples and effect of sinking and rotation of supports.

Tension coefficient method is now included in the analysis of plane trusses in **Chapter 3** and space trusses in **Chapter 4**.

Organization of the Book

Chapters 1 and 2 deal with basic principles of structural analysis of simple structures using only equilibrium equations. **Chapters** 3 and 4 deal with the analysis of plane trusses and space trusses respectively. **Chapters 5 and 6** deal with displacement calculations by geometric and energy methods respectively. **Chapter** 7 discusses the analysis for rolling loads by influence lines, while cables and suspension bridges are discussed in detail in **Chapter 8. Chapter 9** is devoted to the approximate analysis of statically indeterminate structures.

Chapters 10 to 12 are devoted to the analysis of statically indeterminate structures using classical methods, such as consistent displacement, slope-deflection and moment distribution. Kani's method is presented in some detail in **Chapter 13**. Column analogy is covered in **Chapter 14. Chapters 15 and 16** discuss the preliminaries required for the formulation matrix methods of structural analysis. The flexibility and stiffness methods of analysis are presented in **chapters 17 and 18**. Simple examples needing only hand computations have been included in these chapters. However, the matrix formulation of the problems and computation techniques employed are suitable for computer programs. Finally, **Chapter 19** discusses plastic analysis of steel structures.

Four appendices are given at the end of the book which cover topics like theory of vectors and matrices, and tables on product integrals, fixed end moments in a prismatic beam and force displacement relationship in a prismatic member.

Web Supplements

This title includes web supplements for both students and instructors.

Acknowledgements

A book such as this, devoted to the basic aspects of structural analysis cannot claim to contain any original work, and only material collected over the years is presented. I gratefully acknowledge the sources I have consulted. I sincerely thank all my colleagues and students who helped me in writing this text. I am grateful to my wife for her understanding and forbearing during the long hours I spent working on the manuscript. A word of appreciation is also due to my children who refrained from disturbing me during that period.

Although this is the third edition of this book, I would still like to place on record the contribution of Usharanjan Bhattacharjee and K Subba Reddy in typing the manuscripts of the first and second editions respectively and S P Hazra for making the final diagrams. A number of experts took pains to provide valuable feedback about the book. My heartfelt gratitude goes out to those whose names are given below:

S B Singh
Birla Institute of Technology and Science (BITS) Pilani, Rajasthan
A K Dutta
National Institute of Technology (NIT) Durgapur, West Bengal
Nityanand Singh Maurya
National Institute of Technology (NIT) Patna, Bihar
R P Sharma
National Institute of Technology (NIT) Agartala, Tripura
Atanu Kumar Dutta
Jorhat Engineering College, Jorhat, Assam
R K Ingle
Visvesvarya National Institute of Technology (VNIT) Nagpur, Maharashtra
K Rajagopalan
Government College of Technology, Chennai, Tamil Nadu
K C Pazhani
Anna University, Chennai, Tamil Nadu
K Madan Mohan Reddy
Sri Kalahasteeswara Institute of Technology, Chittoor, Andhra Pradesh

I wish to offer my profound thanks and appreciation to the publishers for their guidance and skilful incorporation of the changes in the revised edition. Their initiative and interest in bringing out this work in a short span of time is really commendable. Further suggestions for the improvement of the book are welcome.

C S Reddy

Feedback

Tata McGraw-Hill invites comments, views and suggestions from readers, all of which can be sent to *tmh.mechfeedback@gmail.com*, mentioning the title and author's names in the subject line. Feel free to report any piracy of the book spotted by you.

SI Units for Structural Engineers

The international system of units (System Internationale d'Unites), commonly called SI, is being adopted all over the world as a uniform measurement system. While the complete transition from customary units to the SI system may take years, the use of SI units in the fields of engineering and science is proceeding rather rapidly, and it will soon become necessary for the modern civil engineer to gain experience in using the SI system. Fortunately, the changeover from the now common MKS units to SI units is quite simple, unlike the changeover from FPS to MKS units. In this book, SI units have been used throughout, with only minor modifications, to suit the requirements of the engineering world.

The basic and derived units for various categories of measurement are discussed in the following sections.

TYPICAL BASIC UNITS

Geometry

The basic unit of length is the metre (m), which together with the millimetre (mm) is used exclusively for geometrical quantities. Although the centimetre (cm) is a convenient quantity, its use is generally avoided in the SI system. The use of mm for section modulus and moment of inertia involves large numbers for the majority of common flexural members. This problem is met by listing steel sections properties as section modulus $\times 10^3$ mm^3 and moment of inertia $\times$ 10^6 mm^4. Very small sections, such as light gauge steel shapes may be listed as section modulus x mm^3 and moment of inertia x 10 mm^4.

Mass and Density

Mass is a basic quantity in the system. The base unit of mass is the kilogram (kg). The use of kg should not be confused with the old metric force called kgf.

Material quantities are measured in mass units rather than in weight or force units. Thus, the mass per length of a steel beam is expressed in kg/m, gravity floor loading in kg/m^2 and the mass of an object in kg. Mass density is given in kg/m^3. In contrast to weight units, these quantities do not depend upon the acceleration due to gravity. Weight is not used directly in the SI system, but force is obviously caused by gravity acting on mass.

Force, Moment and Stress

The unit of force is the newton (N), which is the force required to give 1 kg mass 1 m/s^2 acceleration. Thus 1 N is 1 kg.m/s^2. The newton is a derived unit that is independent of the acceleration due to gravity. A kilo-newton (1000 newtons) or kN, which is about 100 kgf, is a convenient quantity in structural analysis and design. Approximating the acceleration due to gravity as 9.81 m/s^2, a kg of mass exerts a force of 9.81 N on its support point.

The stress unit is newton per square metre (N/m^2) called pascal (Pa). This is a very small unit (1 kg/cm^2 approximates to 98100 Pa) and becomes practical only when used with a prefix (k or M). The most convenient SI stress unit for structures is 1,000,000 Pa, the mega pascal or MPa, which is identical to MN/m^2 or N/mm^2. The modulus of steel is about 200,000 MPa in SI units.

Surface loadings and allowable soil pressures have the units of pressure or stress and thus may be expressed in Pascals, but common usage will dictate their expression in kN/m^2 or similar units. Surface loads in particular are well expressed in kN/m^2 because their effects must be converted into kN during structural analysis.

Moment is expressed in N.m or kN.m. These units are convenient since 1 N.m is close to 10 kg.cm and 1 kN.m is close to 1/10 t.m.

Angle, Temperature, Energy and Power

Plane angles are measured in radians (rad), but degrees are also used. Temperature in the SI system should be expressed in Kelvin (K) but the use of degrees Celsius (*C), formerly called centigrade, is also permissible. Kelvin and Celsius are equal for temperature changes since an increment of 1°C equals an increment of 1 K. Energy is expressed in jonles (J), where 1 J is 1 N.m. The unit of power is the watt (W) which is equal to one joule per second (J/s).

Some Simple Rules to be Observed in Using Si Units

Prefixes are to be selected from the following table, in which each prefix is a multiple of 1000.

Prefix	*Symbol*	*Multiplying factor*
giga	G	10^9
mega	M	10^6
kilo	k	10^3
milli	m	10^{-3}
micro	μ	10^{-6}
nano	n	10^{-9}

Compound units, such as for moments, are written with a dot to indicate multiplication, such as kN.m (kilonewton-metre).

CONVERSION FACTORS FOR SI UNITS

(Standard Gravitational Acceleration = 9.80665 m/s^2)

MKS To SI Units

	Quantity	MKS	SI
1.	Force/Load/Weight	1 kgf (kg)	= 9.80665 N
		1 tonne (t)	= 9.80665 kN
2.	Force/Load/Weight per Unit Length	1 kgf/m	= 9.80665 N/m
		1 tf/m	= 9.80665 kN/m
3.	Unit Weight	1 kgf/m^3	= 9.80665 N/m-*
4.	Stress/Pressure/ Modulus of Elasticity	1 kgf/m^2	= 9.80665 N/m^2
			= 9.80665 Pa
		1 kgf/cm^2	= 98066.5 N/m^2
			= 98066.5 Pa
			= 98.0665 kN/m^2
5.	Moment of Force/ Bending moment/ Torque	1 kgf.m	= 9.80665 N.m
		1 kgf.cm	= 98.0665 × 10^{-3} N.m
		1 tf.m	= 9.80665 kN.m

1 Introduction to Structural Analysis

1.1 FORMS OF STRUCTURES

Any civil engineering structure is conceived keeping in mind its intended use, the materials available, cost and aesthetic considerations. The structural analyst encounters a great variety of structures and these are briefly reviewed here.

One of the simplest structures is a simply supported beam, supported on a pin at one end and a roller at the other (Fig. 1.1a). Such a beam, it may be recalled from the fundamentals of strength of materials, is quite stable and statically determinate, and transmits the external loads to the supports mainly through shear and moment. The other types of beams which are more complicated from the point of view of analysis are those with fixed ends and those that are continuous over supports (Figs. 1.1*b* and *c*). As we shall see later, such beams are statically indeterminate and cannot be solved using equations of static equilibrium alone.

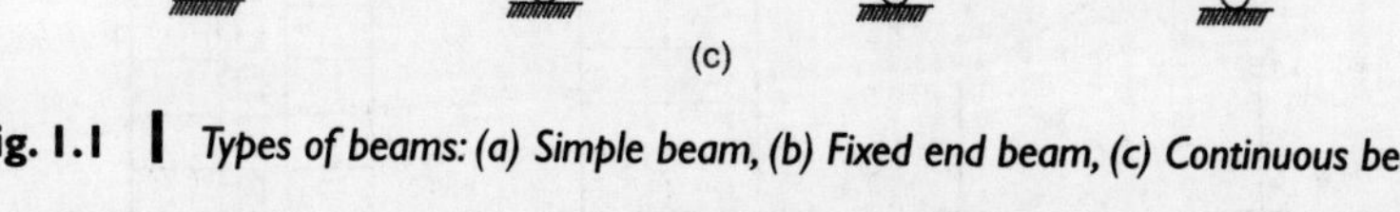

Fig. 1.1 | *Types of beams: (a) Simple beam, (b) Fixed end beam, (c) Continuous beam*

For longer spans, a truss may be employed in place of a beam. Unlike a beam in which the loads are resisted by shear and moment, the truss members transmit the load primarily by axial forces in the members. The structural action of a truss may be compared with that of a simply supported beam. For a truss under vertical loading, the top chord members of the truss are subjected to axial compressive forces and the bottom chord members to axial tensile forces. Under similar conditions, the top fibres of a beam are subjected to compressive stresses and the bottom fibres to tensile stresses. Trusses are mainly built up of prismatic members forming various structural shapes out of basic triangular elements. A typical bridge truss is shown in Fig. 1.2*a*. The truss is known as a plane truss since all the members lie in one plane. Three-dimensional trusses, known as space trusses, are also sometimes used.

Another type of structure used for long spans is the arch. From the structural point of view, arches are characterised by high axial thrust and relatively low bending moment which result from its distinguished shape as well as the horizontal reactions that develop at the support points. Almost similar in structural behaviour and equally efficient in transmitting forces is the cable structure. However, in this the forces are in tension instead of compression as in the arches. An arch and a cable structure are shown in Figs 1.2*b* and *c* respectively.

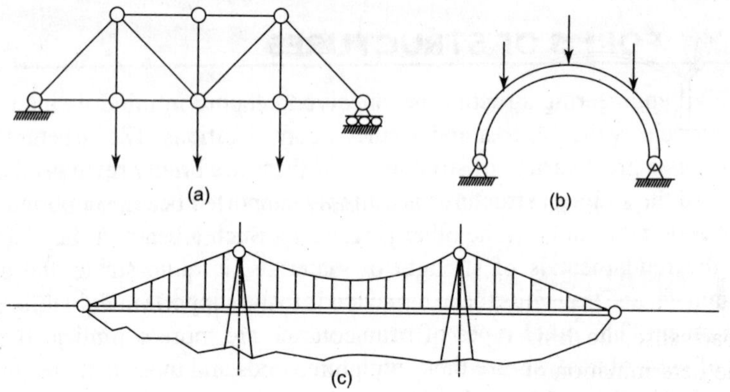

Fig. 1.2 | *Types of axial force structures: (a) Truss, (b) Arch, (c) Cable structure*

A type of structure commonly used in industrial or residential buildings is a frame. Typical frames are shown in Fig. 1.3.

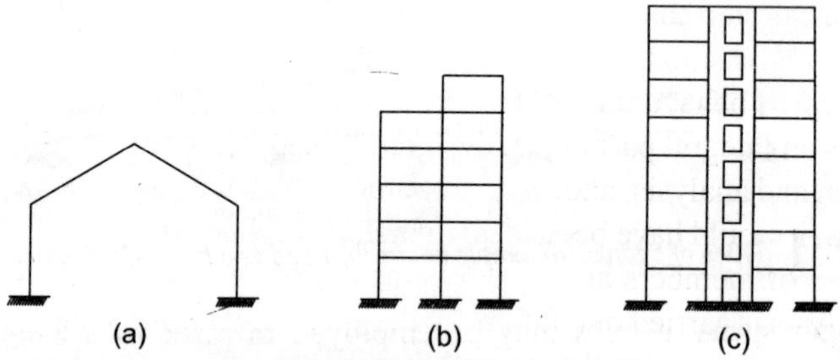

Fig. 1.3 | *Types of frames: (a) Industrial frame, (b) Multi-storey building frame, (c) Building frame with shear wall*

Frames are characterised by moment resisting members at some or all the joints. The resulting structure is rigid and, from the analytical point of view, highly statically indeterminate. As in the case of trusses, frames can be three-dimensional. However, owing to the complications of three-dimensional analysis, frames are generally treated as planar frames in two directions.

In addition to being assembled by discrete straight elements, structures such as shells can be made up of continuous surfaces. Like arches, shells derive their strength mainly from their respective shapes. The analysis of shells is generally complicated because of this surface geometry and the three-directional interaction of material. Two typical shell structures are shown in Fig. 1.4. The analysis of shell structures forms a separate topic and hence has not been included in this book.

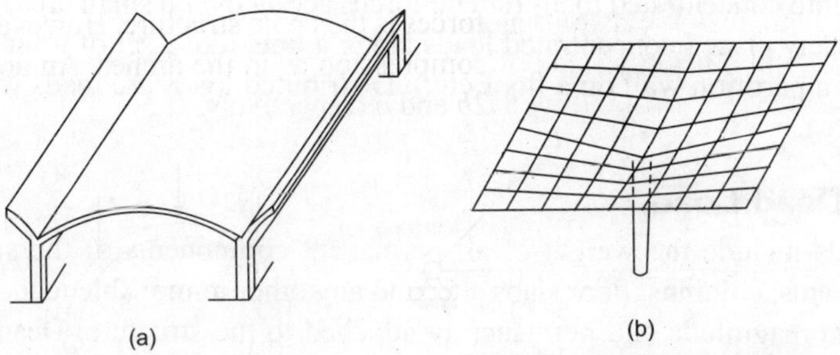

Fig. 1.4 | *Shell structures: (a) Cylindrical shell, (b) Hyperbolic parabolid*

1.2 | ANALYSIS AND DESIGN

In a broad sense, the design of a structure consists of two parts: the first part deals with the determination of forces at any point or member of the given structure and the second part deals with the selection and design of suitable sections to resist these forces so that the stresses and deformations developed in the structure due to these forces are within permissible limits. The first part can be termed as "structural analysis" and the second part as "proportioning" or "dimensioning" of members.

Before we can start the analysis, we shall require the entire details of the structure, loading and sectional properties. To proportion a structure, we must first know how it will behave under loading. Therefore, the process of analysis and design forms an integral part of any design. There is a definite advantage in combining design and analysis, and were it not for the fact that such a textbook would be enormous, it would have been ideal to include both in one volume. In practice, the properties of members are so chosen as to obtain a specified structure, and then the analysis is carried out. Often the designer may have to readjust his initial dimensions in order to get the desired response from the structure. Therefore, the intended purpose of any analysis is to know how the structure responds to a given loading and thereby evaluate the stresses and deformations.

The ultimate aim in learning the methods of analysis is to help design efficient, elegant and economical structures. Analysis helps the designer to choose the right type of sections consistent with economy and safety of the structure. The purpose of structural analysis is to determine the reactions, internal forces, such as axial, shear, bending and torsional, and deformations at any point of a given structure caused by the applied loads and forces.

1.3 LOADS AND FORCES

Although we are mainly concerned with the analysis of structures, it is desirable to give some attention to the loads and forces that are expected to come on a structure.

Loads and forces are usually classified into two broad groups: dead load and imposed loads and forces. For the purpose of structural analysis, any load can be idealised into concentrated loads (single forces acting over a small area) and line loads (closely placed concentrated loads along a line, like a set of train loads or weight of a partition wall on a floor etc.). Distributed loads are loads which act over an area.

1.3.1 Dead Load

Dead loads include the weight of all permanent components of the structure, such as beams, columns, floor slabs, etc. and any other immovable loads that are constant in magnitude and permanently attached to the structure. Dead load is perhaps the simplest of all loading types, since it can be readily computed from given dimensions and known unit weight of materials. However, exact structural dimensions are not known during the initial design phase and assumptions must first be made which may be subject to changes later as the structural proportions are developed. In some structures, such as plate girders and trusses, dead weight assumptions can be expressed by general formulae. Obviously such formulae are derived from known weights of previously built structures. The Indian Standard schedule of unit weights of building materials (first revision) (IS: 1911-1967) gives the average unit weight of materials for the purpose of dead load calculations.

1.3.2 Imposed Loads and Forces

Imposed loads are the forces that act on a structure in the use of the building or structure due to the nature of use, activities due to people, machinery installations, external natural forces, etc.

These are: (1) live load, (2) wind load, (3) seismic force, (4) snow load, (5) loads imposed by rain, (6) soil and hydrostatic forces, (7) erection loads and (8) other forces.

Live Load

Live load is categorised as: (1) live load on buildings, and (2) live load on bridges.

Live Load on Buildings The character of use of occupancy of a structure together with the detail of any specific installations would suggest the live load on the structure. In buildings, these loads include any external loads imposed upon the structure during its service, such as the weights of stored materials, furniture and people. The estimation of live loads based on any rational basis is still not possible. To aid the designer, codes usually describe uniformly distributed live loads) or equivalent concentrated loads that represent the minimum loads for that category of use. IS: 875–1964 provides conservatively superimposed loads on floors and roofs.

Live Load on Bridges Another type of live load is that of moving vehicles on highways and railway bridges. As in the case of buildings, these are the minimum specified values to be used for the design of bridges.

The live loads on a highway bridge are prescribed in the Indian Roads Congress Standard Specifications and Codes of Practice for Road Bridges: Section II. The loadings have been classified as class AA, class A and class B. The code also specifies hypothetical vehicles with wheel loads and wheel bases for the classification of vehicles and road bridges. The code also specifies the impact factor, centrifugal forces, longitudinal forces due to the tractive effect of vehicles or due to braking.

Similar information is available for loadings on a railway bridge. The nature and magnitude of the loads to be taken for railway bridges in India are given in the Bridge Rules of the Ministry of Railways, Government of India.

In moving live loads such as those on bridges and in crane gantries, the critical positions of moving vehicles or wheel loads that produce maximum forces at various points of the structure have to be determined. This is usually done with the help of influence lines discussed elsewhere in the book.

Wind Loads Wind loads are very important in the case of tall structures and also low level light structures in coastal areas. Wind forces are based upon the maximum wind velocity, which in turn depends upon the region and location. It also depends upon the shape of the structure. In the absence of any meteorological data, the wind pressure may be taken from IS: 875–1964. The code gives two basic wind maps of India; one giving the maximum wind pressure including winds of short duration as in squalls, and the other excluding winds of short duration. The code recommends the same wind pressure for all heights up to 30 m and thereafter gives values at intervals of 5 m up to 150 m. The code recommends the use of only the map giving the maximum pressure for squall conditions. But the allowable stresses can be increased by 33 to 50% depending upon the ratio of the wind pressures given by both maps for any particular area.

Earthquake Forces Earthquake forces should be considered for the design of structures in areas of seismic activity. The highly irregular or random shaking of the ground transmits acceleration to structures and the mass of the structure resists the motion due to inertia effects. The total inertia force (usually equal to the horizontal shear at the base of the structure) ranges from about 0.02 to 0.12 W or more for most buildings, where W is the total weight of the structure.

The Indian Standard Recommendations Criteria for Earthquake Resistant Design of Structures (third revision) (IS: 1893-1975) divides the whole country into five seismic zones depending on past experience and the probability of the future occurrence of earthquakes. The inertia force based on the seismic coefficient as appropriate for seismic zones depends on the type of soils and foundation system—a smaller value for hard soils and a larger value for soft soils. Buildings provided for accommodating essential services which are of post-earthquake importance, such as emergency relief stores, food grain storage structures, water works and power stations should be designed taking into account the "importance factor"

Snow and Rain Loads Snow and rain loads affect the design of roofs. The design loads corresponding to the highest accumulation of snow can be found in IS: 875-1964 and other forms of design information. These values are based on past weather records maintained by the Meteorological Department.

If storm water is drained properly, rain does not contribute to any load on the structure. However, structural failures have occurred when rain water got accumulated on roofs due to choked storm water drains. The accumulation of water causes additional load and hence deflection which permits more water to accumulate. This progressive deflection and accumulation of water may continue, leading to structural failure.

Soil and Hydrostatic Forces Structures below the ground, such as foundation walls, retaining walls or tunnels are subjected to forces due to soil pressure. The pressures may be estimated according to established theories.

The force exerted by a fluid is normal to the surface of the retaining structure. The magnitude of the force depends on the hydrostatic pressure which is taken as $\rho = vh$ where v is the unit weight of the fluid and a is the height of the fluid retained. This linear pressure distribution occurs in tanks, vessels and other structures under fluids.

Erection Loads All loads required to be carried by a structure or any part of it due to the placing or storage of construction materials and erection equipment, including all loads due to the operation of such equipment, shall be considered as erection loads.

Other Forces Impact, vibrations, temperature effects, shrinkage, creep, settlement of foundations and other such phenomena produce effects on structures, some of which may be similar to those caused by external loads and forces. These forces may sometimes be surprisingly large and should be taken into consideration while designing.

1.3.3 Load Combinations

Engineering judgement must be exercised when determining critical load combinations. It is not necessary to superpose all maximum loads. For example, a simultaneous occurrence of an earthquake and high velocity winds will have negligible statistical probability. Critical load combinations are usually specified by codes.

1.4 IDEALIZATION OF STRUCTURES

To carry out practical analysis it becomes necessary to idealize a structure. The members are normally represented by their centroidal axes. This naturally does not consider the dimensions of members or depth of joints, and hence there may be a considerable difference between clear spans and centre to centre spans ordinarily used in analysis. These differences can be ignored unless the cross-sectional dimensions of members are sufficiently large to influence the results or when the forces are applied such that these dimensions become significant. Usually, the centroidal axes or the edges of members are represented by a single line. Sometimes two lines are drawn to indicate the depth of members, and unless the depth of member is specified it is disregarded in analysis. Supports and connections are represented in a simplified form. The conventional representation of supports and connections are given in Sec. 1.5. The idealized or simplified form of the structure in Fig. 1.5a is represented in Fig. 1.5b.

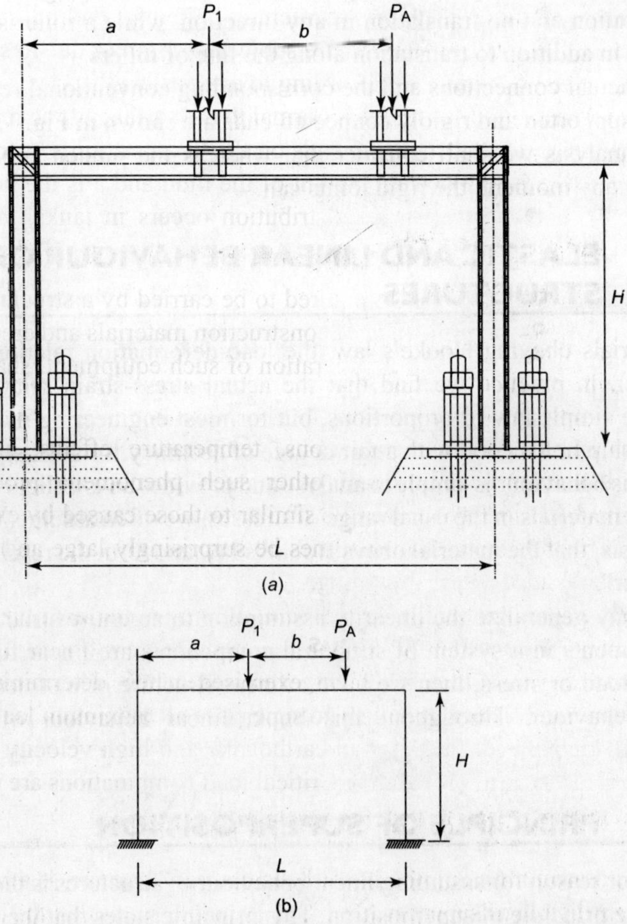

Fig. 1.5 | *Idealization of structure: (a) Actual structure, (b) Idealized structure*

1.5 SUPPORTS AND CONNECTIONS—CONVENTIONAL REPRESENTATION

Most structures are either partly or completely restrained so that they cannot move freely in space. Such restrictions on the movement of a structure are called restraints and are supplied by supports that connect the structure to some stationary body. Thus an essential part of analysis is to determine the manner in which the supports react. The reactive forces of the supports on the structure depend on the type of support condition used. As a first step in determining reactions, it is essential to understand the interacting forces between that part of the structure at the support and the supporting device

Various types of supports are used in structures. Figure 1.6 gives the commonly employed support conditions and reaction components that can be transmitted to the structure by such supports. In addition to knowing the forces that each type of support can transmit, the student should be able to recognize the type of displacement that is permitted by each. For example, a hinged support permits only rotation and no translation in any direction, while a roller support permits rotation in addition to translation along the line of rollers.

The actual connections and the corresponding conventional representation of simply supported and rigidly connected ends are shown in Fig. 1.7.

For analysis we shall consider that whereas the pinned connection cannot transmit any moment, the rigid joints can.

1.6 ELASTIC AND LINEAR BEHAVIOUR OF STRUCTURES

In materials obeying Hooke's law, the load-deformation relationship is linear. However, in practice we find that the actual stress-strain relationship differs from the simple law of proportions, but for most engineering materials a linear relationship holds good with a fair degree of accuracy for at least lower stresses. Since this behaviour is simple to analyse and provides an excellent approximation for most materials in the usual range of stresses, we often assume, for the purpose of analysis, that the material obeys Hooke's law and term the resulting behaviour as "linear".

We may generalize the linearity assumption to an entire structure. When the displacements in a system of structural components are linear functions of the applied load or stress, then we have a linear structure or a structure exhibiting linear behaviour. Throughout this book, linear behaviour of structures is assumed.

1.7 PRINCIPLE OF SUPERPOSITION

The major reason for assuming linear behaviour of structures is that it allows the use of the principle of superposition. This principle states that the displacements

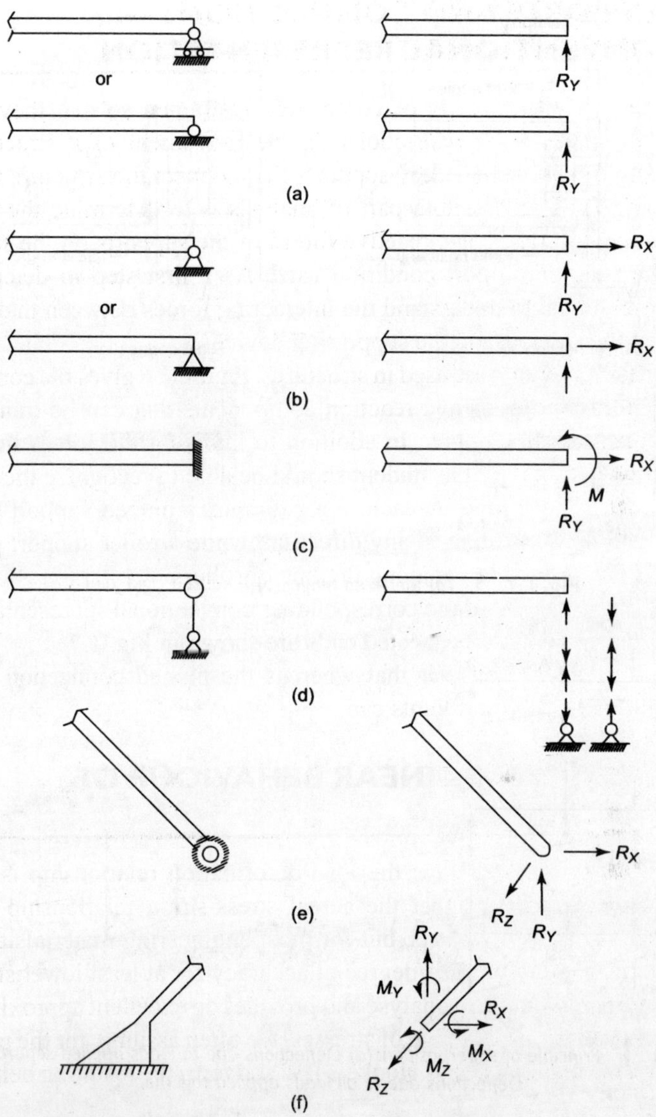

Fig. 1.6 | *Types of supports: (a) Roller support, (b) Hinged support, (c) Fixed support, (d) Link support, (e) Ball and socket, (f) Rigid support in space*

resulting from each of a number of forces may be added to obtain the displacements resulting from the sum of forces. Superposition also implies the converse, that is, the forces that correspond to a number of displacements may be added to yield the force that corresponds to the sum of displacements.

As an example, consider the cantilever beam given in Fig. 1.8. The deflections caused by the three separate loads are shown in Fig. 1.8*a*. The same final deflections would result if all the three loads are applied together as shown in

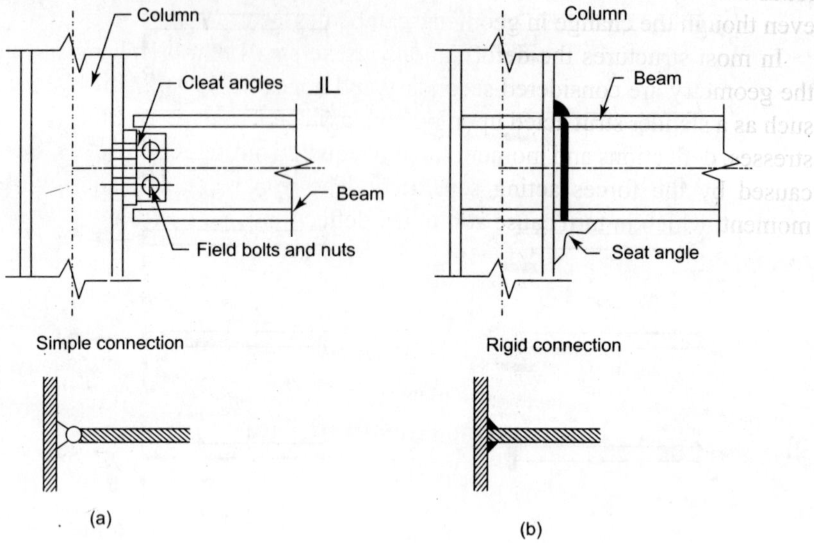

Fig. 1.7 | *(a) Idealized hinge, (b) Idealized rigid joint*

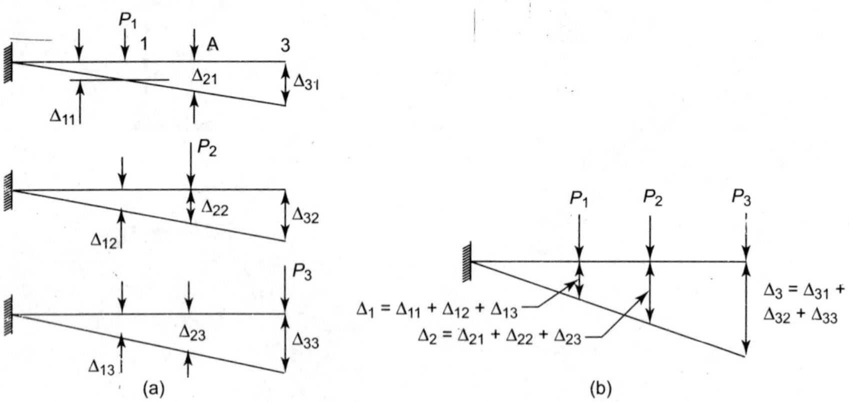

Fig. 1.8 | *Principle of superposition: (a) Deflections due to loads applied separately, (b) Deflections due to all loads applied together*

Fig. 1.8*b*. This is true even if the sequence of loading is altered. It is important to note that this useful result would not occur if the deflection was not a linear function of load.

Superposition thus allows us to separate the loads in any desired way, analyse the structure for a separate set of loads and find the result for the sum of loads by adding individual load effects. Superposition applies equally to forces, stresses, strains and displacements.

The superposition principle, however, is not valid for two important cases: (1) when the geometry of the structure changes appreciably during the application of loads and (2) when the load-deformation relationship of a structure is not linear even though the change in geometry can be neglected.

In most structures the deformations are so small that the changes caused in the geometry are considered secondary and hence neglected. However, in cases such as a slender strut acted upon by both axial and transverse loads, ie resulting stresses, deflections and moments are not equal to the algebraic sum of the values caused by the forces acting separately. The transverse deflections affect the moment, which in turn cause additional deflections.

2 Statics of Structures

2.1 EQUATIONS OF EQUILIBRIUM

Consider any stationary structure or body acted upon by a system of Forces which include external loads, reactions and body forces caused by the weight of the elements. The conditions of equilibrium are best established with reference to coordinate axes *X*, *Y* and *Z*. It is usually convenient to replace all forces by their components along the chosen reference axes. The condition of equilibrium in *X* direction expresses the fact that there is no net of unbalanced force acting in that direction which would accelerate the structure or body. Thus, for static equilibrium, the algebraic sum of all the forces along coordinate axis *X* must be zero, or mathematically, $\Sigma Fx = 0$. Similar equations hold good along coordinate axes *Y* and *Z*. Three additional equations of equilibrium state the fact that the structure or element does not spin or rotate about any of the three axes due to unbalanced moments. The satisfaction of three force equations and three moment equations establishes that the structure is in equilibrium. The six equations of equilibrium are:

$$\begin{aligned} \Sigma F_X &= 0 \qquad \Sigma M_X = 0 \\ \Sigma F_Y &= 0 \qquad \Sigma M_Y = 0 \\ \Sigma F_Z &= 0 \qquad \Sigma M_Z = 0 \end{aligned} \tag{2.1}$$

This can be expressed in vector form as:

$$\mathbf{F}_R = \text{and } \mathbf{M}_R = 0 \tag{2.2a}$$

or

$$\mathbf{F}_R = \mathbf{F}_x\mathbf{i} + \mathbf{F}_Y\mathbf{j} + \mathbf{F}_z\mathbf{k} = 0 \tag{2.2b}$$

and

$$\mathbf{M}_R = \mathbf{M}_X\mathbf{i} + \mathbf{M}_Y\mathbf{j} + \mathbf{M}_Z\mathbf{k} = 0 \tag{2.2c}$$

The primary use of equilibrium analysis is to evaluate the reactions and internal forces by forming a series of free-body diagrams. If a force or a moment acts in an arbitrary direction with respect to the coordinate axes, we replace the force or moment with its components along the three coordinate axes. In the case of a general three-dimensional structure all the six equilibrium equations are needed. However, many three-dimensional structures are idealized as series of two-dimensional components with the loading lying in one plane.

For a planar structure in the XY plane there can be no force acting in the Z direction nor any moments about X and Y directions. Moment M_Z then represents the moment about Z axis on any point in the plane. Thus, for a planar structure, we have only three equations of equilibrium:

$$\Sigma F_X = 0 \; \Sigma F_Y = 0 \; \Sigma M_Z = 0 \tag{2.3}$$

If all forces acting on a two-dimensional structure are parallel, say parallel to the coordinate axis Y, then the term $\Sigma F_X = 0$ contains no terms. Thus, there are only two effective equations of equilibrium, viz. $\Sigma F_Y = 0$ and $\Sigma M_Z = 0$, for this type of loading. Similarly, if all forces located in a plane pass through a point, the summation of moments about this point would not contain any terms and only two equations of equilibrium are available.

2.2 FREE-BODY DIAGRAMS

The analysis of all structures is based on the fact that the structure is in equilibrium under the action of external loads and reactions. The magnitudes of the reactions are such that the applied loads are exactly counteracted according to Newton's third law. Further, any part of the structure is in equilibrium along with the structure as a whole. This fact is used to determine the internal forces in a structure by drawing what are known as *free-body diagrams* for parts of a structure. Free-body diagrams are so useful in studying structural analysis that their importance cannot be over-emphasized.

The correct depiction of a free-body diagram is of extreme importance. The following steps may be followed for constructing a free-body diagram. Remove the body under consideration from its original state. To do this, cut it hypothetically or disengage some connections and supports. A drawing of the free-body diagram is then made.

On the drawing of the free-body, denote all the possible forces in the structure at the cuts and disengaged connections by appropriate force vectors. At this stage, it is neither known nor is it necessary to know the correct direction of forces. We can fix them as acting either in the positive or in the negative direction. Once the values of these quantities are ascertained by methods of statics, the proper sense for each component can be established. All external forces acting on the body in its original state must also be included on the diagram. Clearly label the forces on free-body to facilitate the writing of equilibrium equations.

For a structure that is broken down into a number of free-body diagrams, the procedure for each diagram is the same. However, in dealing with forces acting on the free-bodies, the internal forces common to two free-bodies are denoted as equal but oppositely directed force vectors. The application of this procedure and the usefulness of free-body diagrams are illustrated in the following examples.

Example 2.1 *Draw the free-body diagrams of parts of the structure shown in Fig. 2.1 by making cuts to the right of points B and C and show the forces acting an these cuts.*

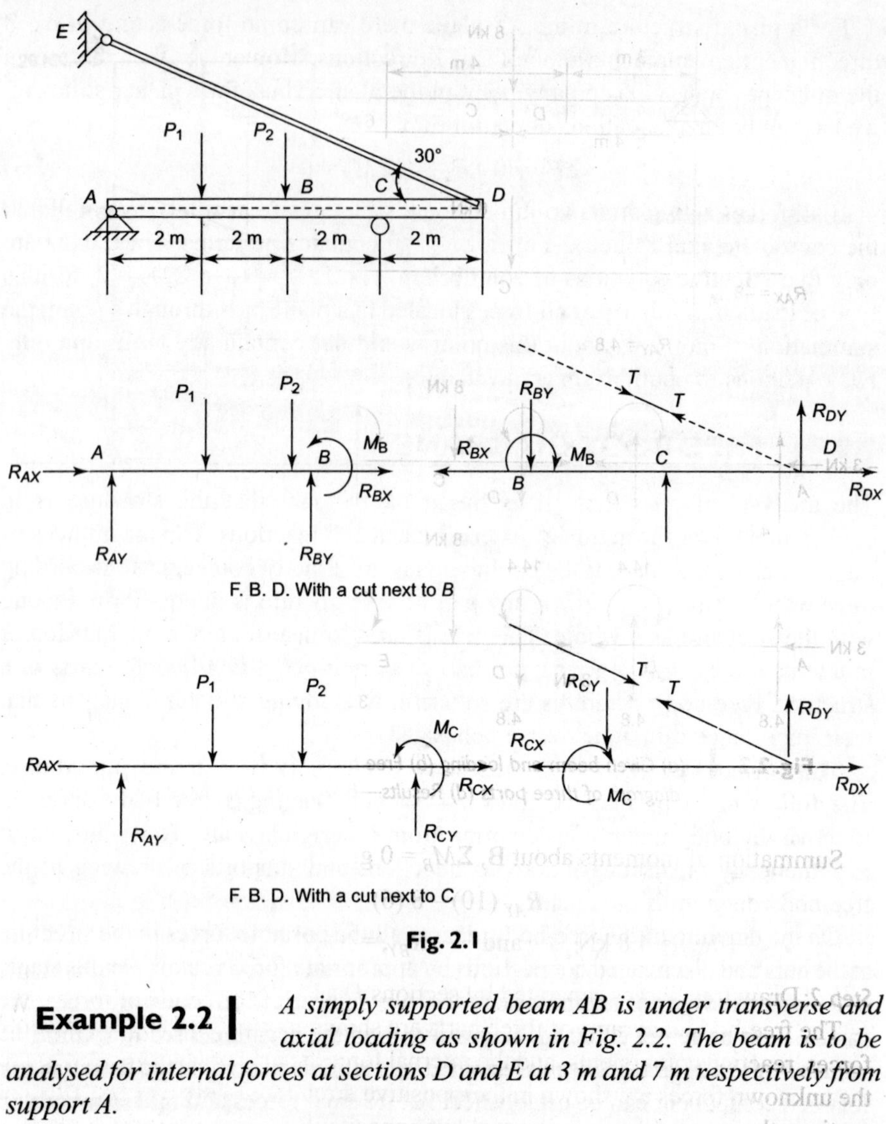

Fig. 2.1

Example 2.2 | *A simply supported beam AB is under transverse and axial loading as shown in Fig. 2.2. The beam is to be analysed for internal forces at sections D and E at 3 m and 7 m respectively from support A.*

Step 1: Evaluate reaction components

The beam is statically determinate. The reaction components can be evaluated using equations of equilibrium

The free body of the entire beam is shown in Fig. 2.2*b*. Summing up horizontal forces $\Sigma F_X = 0$

$$R_{AX} + 3.0 = 0 \quad \text{or} \quad R_{AX} = -3.0 \text{ kW}$$

The negative sign implies that the direction of R_{AX} is opposite to the direction assumed.

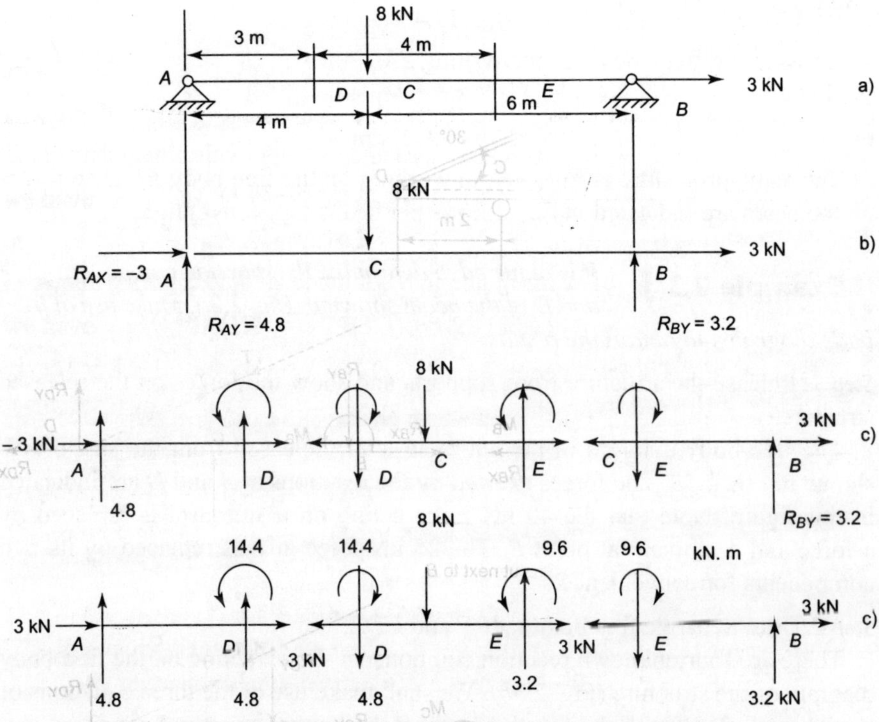

Fig. 2.2 | *(a) Given beam and loading (b) Free body of entire beam (c) Free body diagram of three parts (d) Results—Beam Sign Convention*

Summation of moments about B, $\Sigma M_B = 0$ gives

$$R_{AY}(10) - 8\,(6) = 0$$

$$R_{AY} = 4.8 \text{ kN} \quad \text{and} \quad R_{BY}, = 8.0 - 4.8 = 3.2 \text{ kN}$$

Step 2: Draw free bodies separated by sections *D* and *E*.

The free-body diagrams of three parts are shown in Fig. 2.2*c*. All the external forces, reaction components, and the internal forces at the cut sections are shown; the unknown forces are shown in their positive directions. Note that at all the cut sections the internal forces are equal but opposite in direction. All the three free bodies are under equilibrium.

Step 3: Consider each free body for equilibrium

Considering free body *AD* the summation vertical forces $\Sigma F_Y = 0$ gives

$$4.8 + V_D = 0$$

$$\therefore \quad V_D = -4.8$$

The negative sign indicates that the shear force is downwards.

Again summation of Moments $\Sigma M_D = 0$

We have $-4.8 \times 3 + M_D = 0$

or $$M_D = 14.4 \text{ kN.m}$$

Considering free body *DE* and writing $\Sigma M_E = 0$

We have $$-4.8 \times 4 - 14.4 + 8 \times 3 + M_E = 0$$

or $$M_D = 9.6 \text{ kN.m}$$

The same procedure is followed in working on the free body *EB*. The forces on the beam are indicated in Fig. 2.2d as per beam sign convention.

Example 2.3 *It is required to determine the reaction components at A and D of the beam shown in Fig. 2.3a. Make use of free-body diagrams to obtain the results.*

Step 1: Release the structure from supports and show the forces on the released structure.

The free-body diagram of the entire structure released from the supports is shown in Fig. 2.3*b*. The forces exerted by the reactions at *A* and *D* are indicated in this figure. Note that the 40 kN force acting on a stub arm is replaced by a force and a moment at point *B*. The 25 kN force at *E* is replaced by its two components for convenience.

Step 2: Draw separate free bodies *ABC* and *CDE*.

There are four unknown reaction components shown acting on the free-body diagram of the structure (Fig. 2.3*b*). We shall make use of the three equations of equilibrium along with the fourth one from the known structural condition that at hinge point *C* on the beam, the moment is zero. The free-body diagrams of the two parts separated by the hinge point are shown in Fig. 2.3c. Note that equal and oppositely directed internal forces are represented at *C*. The reactions can be determined by considering either of the two free-body diagrams. However, the consideration of free-body *CDE* directly gives R_{DY} and the internal forces at *C*.

Step 3: Consider the free body *CDE*.

Summation of horizontal forces $\Sigma_{FX} = 0$ gives

$$-R_{CX} - 15 = 0$$

or $$R_{CX} = -15 \text{ kN}$$

The negative sign implies that the direction of force R_{CX} is opposite to the direction assumed.

Summation of moments about *C*, that is $\Sigma M_C = 0$

gives $$R_{DY}(3) - 20(4.5) - 30(4.5)(4.5/2) = 0$$

or $$R_{DY} = 131.25 \text{ kN}$$

Summing up forces in the vertical direction and writing $\Sigma F_Y = 0$ we get

$$131.25 - 20 - 30(4.5) - R_{CY} = 0$$

or $$R_{CY} = -23.75 \text{ kN}$$

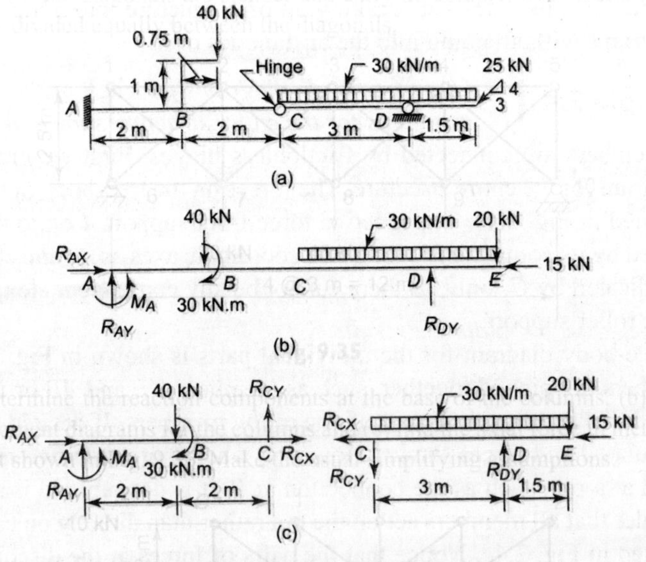

Fig. 2.3 | *(a) Given beam and loading, (b) Free-body diagram of the entire beam, (c) Free-body diagrams of two parts separated at the hinge point*

Step 4 Consider the free body *CDE*

With all the forces known on free-body *CDE*, the internal forces at any other point can be evaluated by only using statics.

Next, considering free-body *ABC*, the internal forces at hinge point *C*, R_{CX} and R_{CY} are known. The values are numerically equal but opposite to the forces determined on free-body *CDE*.

Applying condition $\Sigma F_X - 0$ for free-body *ABC* we get

$$R_{AX} + R_{CX} = 0$$

or

$$R_{AX} = 15 \text{ kN}$$

Summing up forces in *Y* direction, $\Sigma F_Y = 0$ gives

$$R_{AY} - 40 + R_{CY} = 0$$

or

$$R_{AY} - 40 - 23.75 = 0$$

or

$$R_{AY} = 63.75 \text{ kN}$$

Finally summing up moments of all the forces about support point *A*

$$M_A - 40(2) - 30 + R_{CY}(4) = 0$$

or

$$M_A - 80 - 30 - (23.75)(4) = 0$$

yields

$$M_A = 205 \text{ kNm}$$

The positive sign indicates that the direction of moment M_A assumed is in the true direction.

Thus, all the reaction components are evaluated. The internal forces at any other point along the length of the beam can be evaluated using statics.

Example 2.4 illustrates the breaking up of a structure into a number of free-body diagrams without going into the arithmetics of it.

Example 2.4 | *Draw free-body diagrams for each of the components as well as for the entire structure shown in Fig. 2.4a.*

The members are connected by frictionless hinges. First we draw the free-body diagram for the entire structure. The forces on the free-body of the structure are indicated in Fig. 2.4*b*. The reactive force from support *A* on to the structure is indicated by its components along two coordinate axes as A_X and A_Y. The force at *C* is indicated by C_Y only, as there cannot be any component along the *X* axis due to the roller support.

The free-body diagram for the individual parts is shown in Fig. 2.4*c*. When two members are pinned together, such as members *DE* and *AB* or *ED* and *BC*, it is considered that the pin is part of one of the members. If desired, the pin can also be isolated and forces shown. However, when more than two members are connected at a pin, such as the connection at *B*, it is desirable to isolate the pin and consider that all members act on the pin rather than directly on each other as is illustrated in Fig. 2.4*c*. Notice that the pairs of forces at the disconnected pin are shown oppositely directed at the points of joining.

Free-body diagrams can also be drawn for the parts of a structure hypothetically cut by a section. For example, Fig. 2.4*d* shows the free-body diagram of the assembly to the left of section 1-1 in Fig. 2.4*a*. It may be noticed that at each cut, three forces were introduced. The number of unknown force components is much more than in the previous free-body diagram. For this reason we must carefully choose the free-body diagram that is suitable for our purpose.

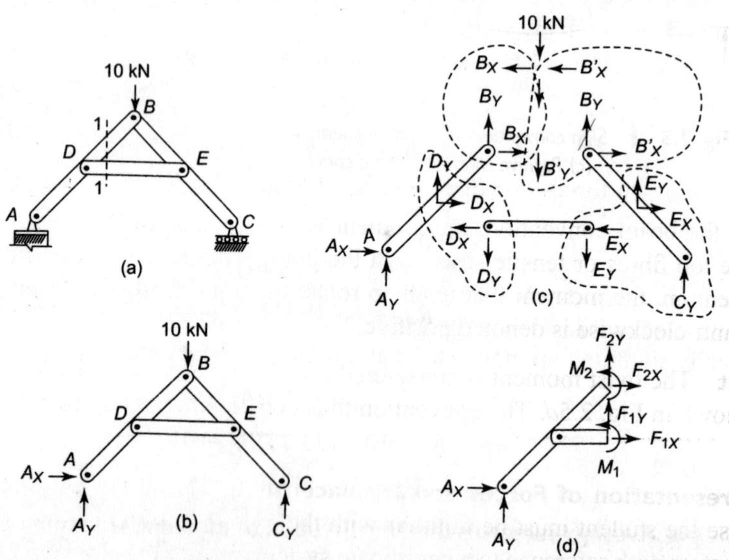

Fig. 2.4 | *(a) Structure arrangement and loading, (b) Free-body diagram of entire structure, (c) Free-body diagrams of individual parts, (d) Free-body diagram of assembly to left of section 1-1*

We shall be making use of the technique of constructing free-body diagrams in analysing various types of structures in the subsequent chapters.

2.3 SIGN CONVENTION

An essential part of structural analysis is the adoption of an appropriate sign convention for representing forces and displacements. It will become clear with the development of different methods of analyses that there are advantages in not following the same sign convention.

In this text the following sign convention for representing various forces and displacements will be followed:

Axial Force An axial force is considered positive when it produces tension in the member. A compressive force is, therefore, negative.

Shear Force Shear force which tends to shear the member as shown in Fig. 2.5*a* is considered positive. Notice that the positive shear force forms a clockwise couple on a segment.

Bending Moment There are two conventions used for bending moment: (1) the beam convention based on the nature of stress the moment produces, and (2) the static sign convention based on the direction the moment tends to rotate the joint or end of a member. The positive sense of the moments in both conventions is represented in Figs. 2.5*b* and *c*.

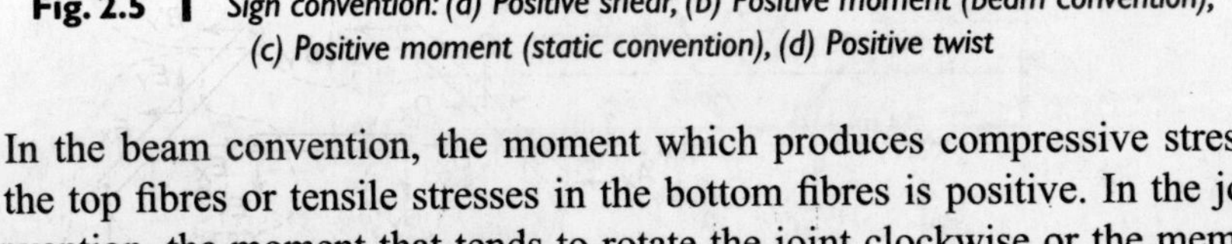

Fig. 2.5 | *Sign convention: (a) Positive shear, (b) Positive moment (beam convention), (c) Positive moment (static convention), (d) Positive twist*

In the beam convention, the moment which produces compressive stresses in the top fibres or tensile stresses in the bottom fibres is positive. In the joint convention, the moment that tends to rotate the joint clockwise or the member end anti-clockwise is denoted positive.

Twist The twist moment is considered positive when it acts on a member end as shown in Fig. 2.5*d*. The convention thus corresponds to the right-hand screw rule.

Representation of Forces and Displacements From the basic mechanics course the student must be familiar with the representation of forces by means of vectors with reference to a coordinate system. One of the common coordinate systems used is the orthogonal *X*, *Y* coordinates to describe the stresses, moments, deflections, etc.

For some analyses it is convenient to adopt a sign convention in terms of the structure or general global coordinate system. For an X, Y and Z coordinate system as shown in Fig. 2.6, the positive direction of the forces coincides with the direction of coordinate axes and the moments follow the right-hand screw rule. The moments and twists are represented by vectors with double arrow heads as in Fig. 2.6*a* or by moment vectors as shown in Fig. 2.6*b*. The same sign convention is also used for denoting deflections or rotations.

Quite often the analysis is carried out using the joint sign convention but the moment diagram is drawn based on the beam sign convention. The student should be familiar with the interpretation of sign conventions adopted in the two systems. The following example is intended to illustrate the point.

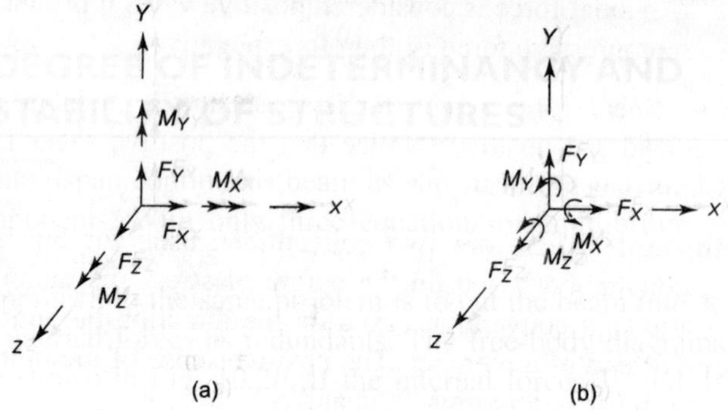

Fig. 2.6 | *General coordinate system*

Example 2.5 | *An analysis of a continuous beam shown in Fig. 2.7a has resulted in the following beam end moments in accordance with the joint sign convention.*

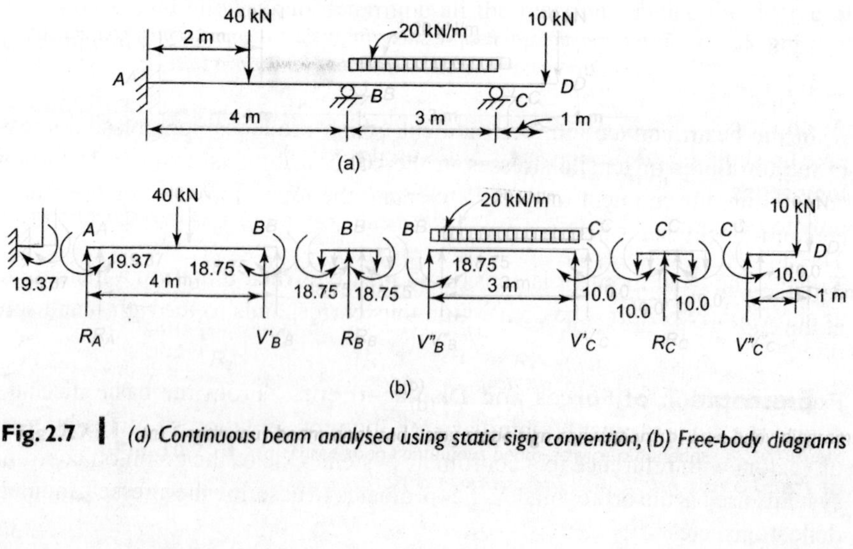

Fig. 2.7 | *(a) Continuous beam analysed using static sign convention, (b) Free-body diagrams*

$$M_{AB} = +19.37 \text{ kNm}, M_{BA} = 18.75 \text{ kNm}, M_{BC} = +18.75 \text{ kNm},$$
$$\overline{M}_{BC} = -10.00 \text{ kNm}, M_{CD} = +10.0 \text{ kNm}.$$

Construct free-body diagrams for segments *AB*, *BC* and *CD* and evaluate the reactions.

The three segments of the beam are separated from the joints and the free-body diagrams are drawn as shown in Fig. 2.7*b*. All the force components are indicated on the diagrams.

According to the beam sign convention, all the beam end moments are negative causing tension at the top. The values of the reactions are obtained by considering free-body diagrams of the segments. The reaction at *A* can be obtained by taking the summation of moments about *B* as

$$-4R_A + 40(2) + 19.37 - 18.75 = 0$$

or $$R_A = 20.16 \text{ kN}$$

The value of interior support reaction R_B is obtained by first determining the values of shear in the beams to the left and right of reaction R_B. The value of shear to the left of R_B is obtained by taking the summation of vertical forces, $\Sigma F_Y = 0$ on the free-body diagram of the segment *AB*. The shear to the right of R_B is obtained by taking the summation of moments about *C* of the forces on the free-body diagram of segment *BC* from Fig. 2.7*b*.

$$V'_B = V_B - \text{left} = 40 - 20.16 = 19.84 \text{ kN}$$

$$V''_B = V_B - \text{right} = \frac{18.75 + 20(3)(1.5) - 10}{3} = 32.92 \text{ kN}$$

Therefore, $$R_B = 19.84 + 32.92 = 52.76 \text{ kN}$$

Similarly, reaction R_C can be evaluated. Its value is 37.08 kN.

2.4 SIMPLE CABLE AND ARCH STRUCTURES

2.4.1 Cables

As an introduction to the analysis of simple determinate structures, we shall first take up simple cable structures. Cables are frequently used to support loads over long spans such as in suspension bridges and roofs of large open buildings. The only force in a cable is direct tension, since cables are too flexible to carry moment. The analysis of cables involves the straightforward application of equilibrium equations to various free-bodies. We shall first consider a cable whose supports at the ends are at the same level.

Example 2.6 *Consider a suspension cable shown in Fig. 2.8a. The loads are applied vertically downwards by the suspension cables carrying the bridge deck. Determine the reaction components at 1 and 5 and tension in the cable in different segments.*

The forces in the cable segments depend upon the geometry assumed by the cable at points 2, 3 and 4. For a given sag at any point, the shape of the cable is uniquely determined from equilibrium conditions. Knowing one coordinate, such as sag at point 2, the sag at any other .point can be calculated.

There are apparently four unknown reaction components with only three equations of equilibrium available. However, a fourth equation can be formed from the fact that the moment at any point on the cable is zero. The first equilibrium equation can be written as $\Sigma M_5 = 0$. Taking anticlockwise moments as positive, we have

$$\Sigma M_5 = -V_1(80) + 30(60) + 40(40) + 50(20) = 0$$

or

$$V_1 = 55.0 \text{ kN}$$

Next H_1 is evaluated by isolating a part of a cable by making a cut at any point on the cable. The free-body diagram of part of the cable just to the left of point 2 is shown in Fig. 2.8*b*. Writing summation of moments $\Sigma M_2 = 0$ we have

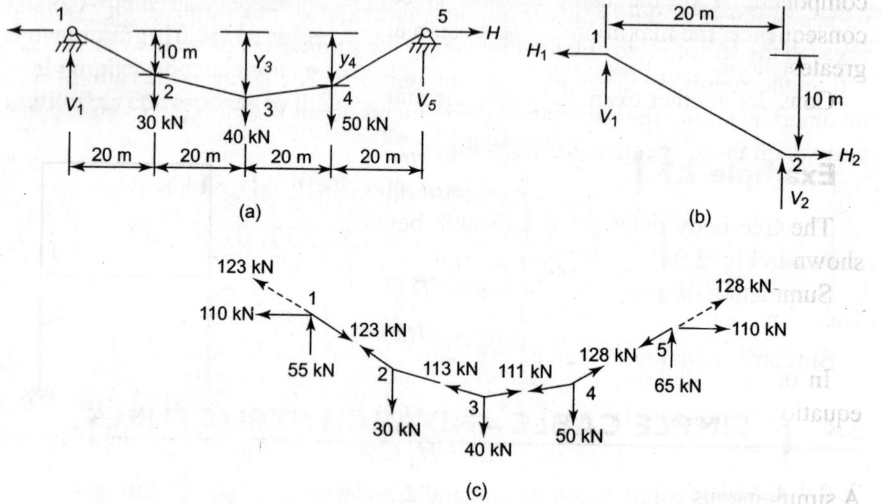

Fig. 2.8 | *(a) Cable under load, (b) Free-body diagram of cable to the left of point 2, (c) Results of analysis*

$$\Sigma M_2 = -V_1 (20) + H_1 (10) = 0$$

or

$$H_1 = 110 \text{ kN}$$

From a consideration of forces on the cable in X direction,

$$\Sigma F_X = 0 \text{ gives } H_5 = H_1 = 110 \text{ kN}$$

Again by using $\Sigma F_Y = 0$, we have

$$V_1 - 30 - 40 - 50 + V_5 = 0$$

or

$$V_5 = 65.0 \text{ kN}$$

The same result could have been obtained by summing up moments of all forces about point 1.

Having determined the four reaction components, we can evaluate the forces in the cable segments. The forces in the cable depend on the geometry of the cable segments. For this we first need to determine the sag at points 3 and 4. Considering the free-body to the left of point 3 and summing up moments about point 3 we have,

$$M_3 = -55(40) + 30(20) + 110(y_3) = 0$$

or $\quad y_3 = 14.56$ m (sag at point 3).

Similarly, taking summation of moments about point 4 yields

$$y_4 = 11.82 \text{ m (sag point 4).}$$

At each of the points 2, 3 and 4, the magnitude and direction of the applied load as well as the direction of forces in the cable are known. Using two equilibrium conditions $\Sigma F_X = 0$ and $\Sigma F_Y = 0$, the unknown forces in the cable are determined. The results are shown in Fig. 2.8c.

The horizontal equilibrium of any part of the free-body shows that horizontal component H of the cable tension is constant throughout the cable. As a consequence, the maximum cable tension always occurs in the segment with the greatest slope.

Consider another example where the end supports are not at the same level.

Example 2.7 | *For the cable structure shown in Fig. 2.9a determine the reactions at A, D and E and cable tensions.*

The free-body diagram of the cable between the support points A and D is shown in Fig. 2.9*b*.

Summation of moments about point D gives

$$M_D = -V_A(36) + 27(24) + 14(12) - H_A(3.6) = 0 \qquad (2.4)$$

In order to determine V_A and H_A, it is necessary to write another moment equation about point B where the sag is known. For the left portion of the cable

$$M_B = -V_A(12) + H_A(2.4) = 0 \qquad (2.5)$$

A simultaneous solution of Eqs. 2.4 and 2.5 yields

$$H_A = 75 \text{ kN and } V_A = 15.11 \text{ kN}$$

The sag of the cables at C, shown as *yc* in Fig. 2.9*c*, can be obtained by considering the free-body diagram of the cable to the left of point C. We can write

$$M_C = +H_A\{y_C - 3.6(24/36)\} - V_A(24) + 27(12) = 0$$

Substituting for H_A and V_A we get $y_C = 2.91$ m.

From the given dimensions of the cable structure the desired distance $d_c = 2.91 + 1.20 = 4.11$ m. With the coordinates of points B and C known, we can find the forces in the segments of the cable as in the previous example.

The tension in cable DE is obtained by considering equilibrium conditions at point D. Because the tower is pinned at both ends, we know that the horizontal component of forces in DC and DE must be equal. The results are shown in Fig. 2.9*d*.

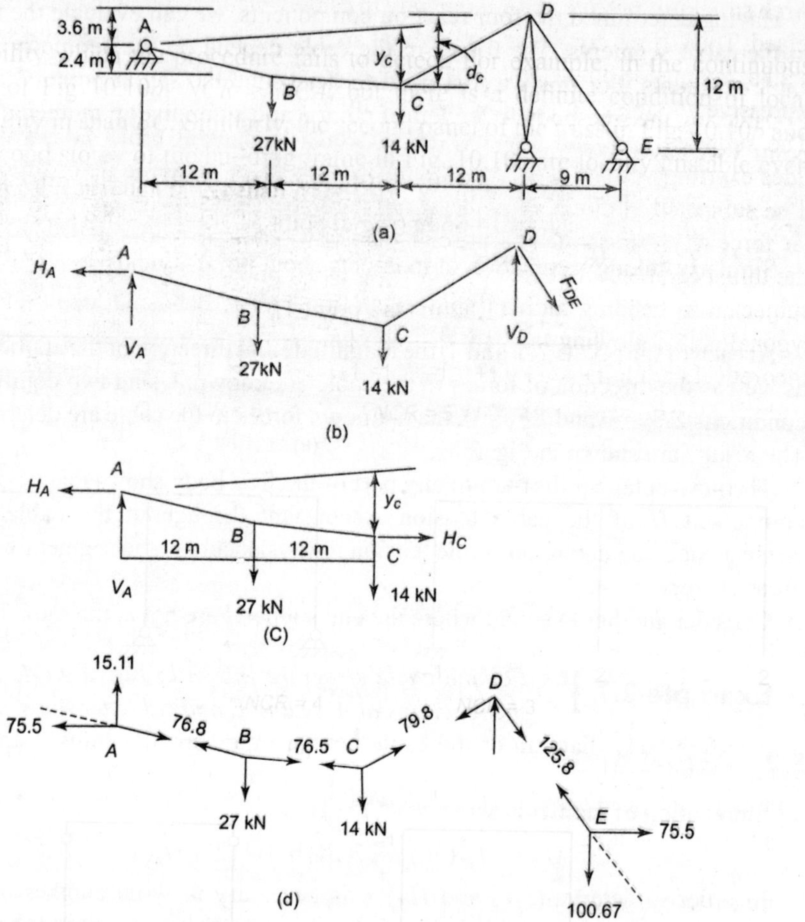

Fig. 2.9 | *(a) Cable and loading, (b) Free-body diagram of cable between A and D, (c) Free-body diagram of cable to the left of point C, (d) Results of analysis*

2.5 ARCHES

The arch is one of the oldest structures. The Romans developed the semicircular true masonry arch, which they used extensively in both bridges and aqueducts. Quite a few of the early Indian railway and highway bridges used masonry arches. They were constructed with brick or stone masonry with lime or cement mortar as the binding material. Arches are also used in buildings to carry loads over doorways, windows etc., as well as to add an aesthetic touch to the building.

2.5.1 Theoretical Arch or Line of Thrust

We have seen in the previous section that a cable can support a given set of loads by developing tensions in various segments. The shape of the cable will correspond to the funicular polygon for the given system of loads.

Arch structures behave in a similar way to cable structures but with the actions reversed. Thus, if we construct a polygonal arch similar to the cable profile as in Fig. 2.8a, but upside down as shown in Fig. 2.10, the stresses in each link will be compressive and the arch is subjected to truly axial compression.

The supports at 1 and 5 will exert an equal horizontal thrust H inwards' besides exerting vertical reactions V_1 and V_5. The polygonal arch if constructed will be subjected to direct axial thrust only and there is no bending moment or shear force anywhere. Such an arrangement will prove to be more economical as the thrust can be transmitted by a smaller cross-section, unlike a beam which is subjected to bending moment and shear force under the same load. Such a polygonal arch following the path of true compression is known as linear arch or theoretical arch.

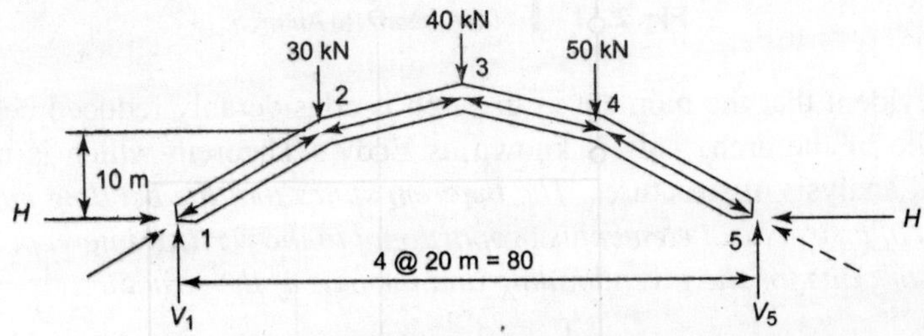

Fig. 2.10 | *Polygonal arch*

2.5.2 Actual Arch

In practice, the position and magnitude of the loading over a structure goes on changing. It is therefore neither advisable nor possible to construct an arch according to its theoretical shape. In practice, the arch can be of circular, parabolic or elliptical shape for easy construction and aesthetic appearance. Obviously such an arch is subjected to a certain amount of bending moment and radial shear.

2.5.3 Eddy's Theorem

Consider a beam and an arch of same span and under same loading as shown in Fig. 2.11.

We can draw a funicular polygon for the forces passing through two support points (see. Section 2.6.6). The polygon is the pressure line or thrust line. The ordinate between the thrust line and the axis of the beams or arch at any section (f1f2) represents B.M. at that section to some scale. The actual pending moment is obtained as under

$M_x \propto f_1 f_2$ at any Section X

or $$M_x = (f_1 f_2)\,(f)\,(s)\,(p)$$

In which f – force scale in the polygon

s – Space diagram scale

p = polar distance

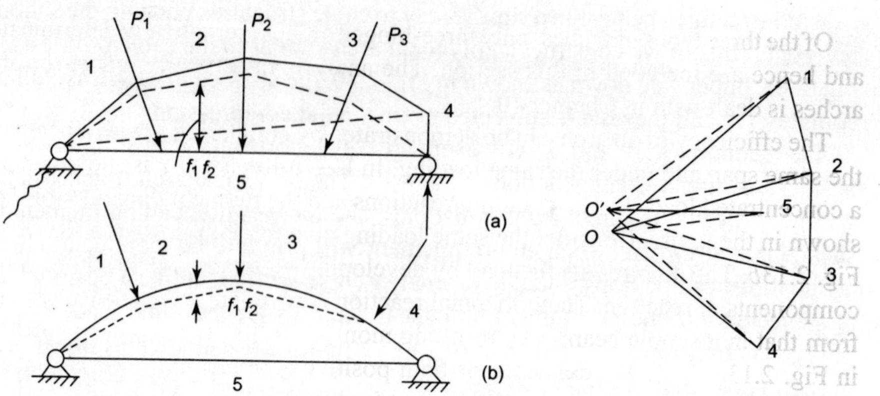

Fig. 2.11 | *(a) A beam (b) An arch*

It is evident that the moment in the arch is considerably reduced because of the profile of the arch. This is known as Eddy's Theorem which is useful in graphical analysis of structures. *The theorem states that the bending moment at a section of a structural element is proportional to the vertical intercept between the pressure line for the given loading and the axis of the structure.*

2.5.4 Types of Arches

Arches may be classified, of course, on the basis of the materials of which they are built; steel and reinforced concrete is the most common of all materials. From the point of view of structural behaviour, arches are conveniently classified as three-hinged, two-hinged and hinge less (also known as fixed) arches. On the basis of form, arches may be further classified as parabolic, circular, elliptical, etc. A number of arch forms are indicated in Fig. 2.12 which vary in the manner they are supported and in the structural arrangement of the arch ribs.

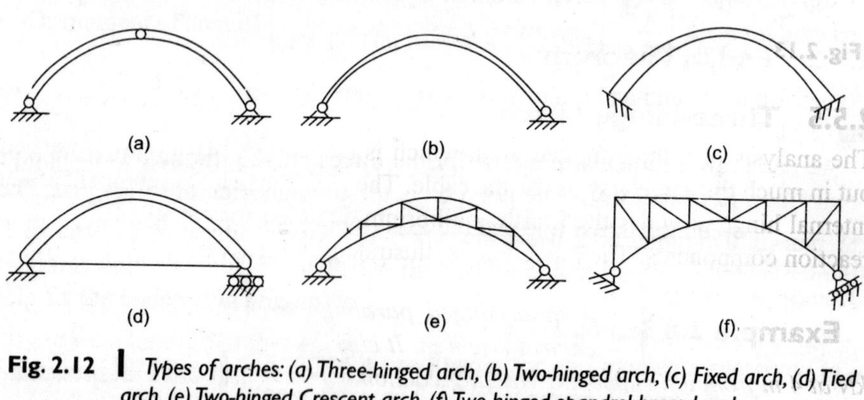

Fig. 2.12 | *Types of arches: (a) Three-hinged arch, (b) Two-hinged arch, (c) Fixed arch, (d) Tied arch, (e) Two-hinged Crescent arch, (f) Two-hinged spandrel braced arch*

Open web arch ribs, though they resemble trusses, are considered as arches because of the manner in which the loads are transmitted.

Of the three types of arches, only three-hinged arches are statically determinate and hence are included in this section. The analysis of statically indeterminate arches is dealt with in Chapter 10.

The efficiency of an arch can be demonstrated by comparing it with a beam of the same span and under the same loading. In Fig. 2.13*a* a beam is shown under a concentrated load, *P*. The resulting reactions and the moment diagram are also shown in the figure. Consider the same loading on a three-hinged arch shown in Fig. 2.13*b*. The arch resists the load by developing vertical as well as horizontal components of reaction. The horizontal reaction component reduces the moment from that in a simple beam. The resulting moment in the arch is shown hatched in Fig. 2.13*b*. Note the existence of both positive and negative moment in the arch. Thus, we see that owing to its geometric shape and proper supports, an arch supports loading with much less moment than a corresponding straight beam. It must be remembered that the reduction in moment is achieved at the expense of large axial compression in the arch rib and also horizontal reaction components at the springing.

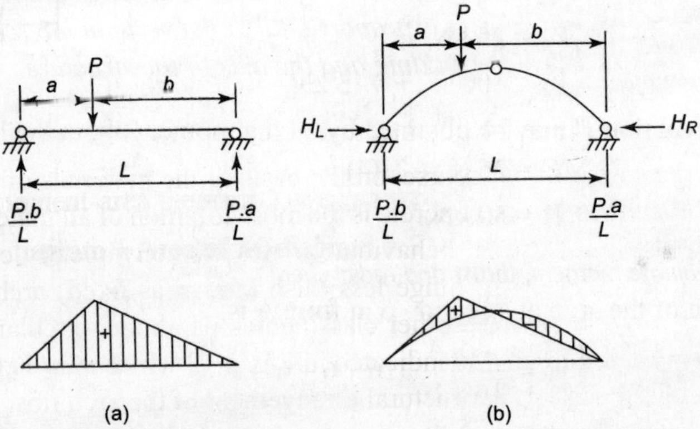

Fig. 2.13 | *(a) Beam and the moment diagram, (b) Three-hinged arch and moment diagram*

2.5.5 Three-Hinged Arch

The analysis of a three-hinged arch, which is statically determinate, is carried out in much the same way as for the cable. The condition of zero moment at the internal hinge provides the fourth equilibrium equation for calculating the four reaction components. The procedure is illustrated in the examples that follow:

Example 2.8 | *A three-hinged parabolic arch has span 16 m and central rise 4 m. It carries a concentrated load of 100 kN at 4 m from left support. Evaluate reaction components, moment, thrust and radial shear at a section 6 m from left support. Take the equation of the arch y = 4 h x (l – x) with left-hand support as origin. Draw bending moment diagram.*

Step I: To evaluate reaction components

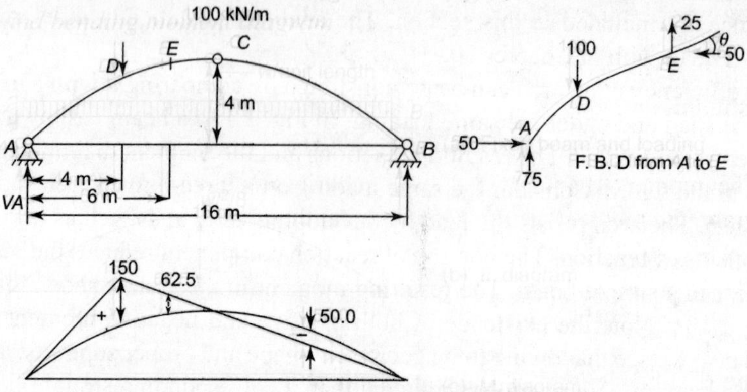

Fig. 2.14 | *(a) Given arch and loading (b) B.M diagram*

Taking moment about A,

$$M_A = V_B\,(16) - 100\,(4) = 0$$

or $$V_B = 25 \text{ kN}$$

and $$V_A = 100 - 25 = 75 \text{ kN}$$

Horizontal reaction H may be obtained by taking moment about the hinge at C

$$MC = 25\,(8) - H(4)$$

This give $$H = 50 \text{ kN.}$$

Step 2: To evaluate moment, thrust and radial shear

The ordinate of the arch at a section 6 m from A is,

$$y = \frac{4 \times 4}{16 \times 16}(16 \times 6 - 6^2)$$

or $$y = 3.75 \text{ m}$$

Moment at E,

$$M_E = 75\,(6) - 50 \times 3.75 - 100\,(2)$$

Gives $$M_E = 62.5 \text{ kN.m}$$

Maximum +ve moment occurs under the load point,

$$MD = 75 \times 4 - 50(3)$$

Gives maximum +ve moment $M_D = 150$ kN.m.

Maximum –ve moment occurs at a section in between support B and hinge C. Let it be at a distance x from B.

Writing $$M_x = 25x - 50y_x \tag{2.6}$$

$$= 25x - 50 \times \frac{4 \times 4}{16 \times 16}(16x - x^2)$$

$$= -25x + \frac{50}{16}x^2$$

Seting $\frac{dM_e}{dx} = 0$, we have $-25 + \frac{100}{16}x = 0$

or $x = 4$ m

Substituting in moment equation, maximum –ve moment

$$M_{max} = 25(4) - 50(3) = -50 \text{ kN.m}$$

The moment diagram is drawn, superimposing on simple beam moment diagram, the moment caused by horizontal thrust as shown in Fig. 2.14

Step 3: To evaluate normal thrust

From the free body of the arch from A to E as shown in Fig. 2.14 normal thrust

$$N_E = H\cos\theta - V\sin\theta$$

where θ is the inclination of the arch axis at E. Writing equation of parabola

$$y = 50 \times \frac{4h}{l^2}(lx - x^2)$$

$$\frac{dy}{dx} = \frac{4h}{l^2}(l - 2x)$$

Substituting for $x = 6$ m, $h = 4$ m and $l = 16$ m

We have $\frac{dy}{dx} = \tan\theta = 0.25$, $\theta = 14°3'$

θ is the inclination of arch axis with the horizontal

Substituting $\sin\theta = 0.24$ and $\cos\theta = 0.97$ in eqn. 2.6, we have

$$N_E = 50(0.97) - 25(0.24)$$
$$= 42.5 \text{ kN. (compression)}$$

Similarly radial shear

$$V_r = V_E\cos\theta + H\sin\theta$$
$$= 25(0.97) + 50(0.24)$$
$$= 36.25 \text{ kN}$$ –ve following the shear force sign convention

Example 2.9 | *A three-hinged segmental arch has a span of 50 m and a rise of 8m. A 100 kN load is acting at a point 15 m measured horizontally from the right-hand support.*

Find (a) the horizontal thrust H, developed at the supports, and (b) the moment, normal thrust and radial shear at a section 15 m from the left-hand support.

Step 1: To evaluate radius of arch R and reaction components

A three-hinged segmental arch as given is shown in Fig. 2.15*a*. The radius of segmental arch *R* is established using the relationship

$$(2R - y_C) = \left(\frac{L}{2}\right)^2$$

On substituting for y_C and L

$$R = 43.06 \text{ m.}$$

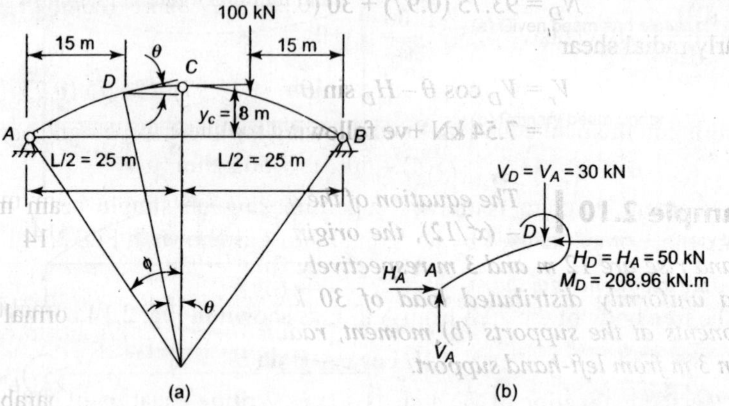

Fig. 2.15 | *(a) Arch and loading, (b) Free-body diagram of arch between A and D*

Vertical reaction component V_A may be obtained as in a simple beam by taking moment about support B.

$$V_A(50) - 100(15) = 0$$

or $$V_A = 30 \text{ kN.}$$

and $$V_B = 100 - 30 = 70 \text{ kN.}$$

Horizontal reaction H may be obtained by taking moments about hinge point C and equating to zero, that is

$$M_C = 30(25) - H(8) = 0$$

This gives $H = 93.75$ kN inwards as shown.

The inclination of the arch at 15 m from the left-hand support, or angle θ subtended at the centre as shown in Fig. 2.5*a* is given by

$$\sin\theta = \frac{10}{43.06} = 13.43°$$

Ordinate at D of arch axis, $Y_D = R\cos\theta - R\cos\varphi$ where φ is the angle subtended at the centre by segment AC of the arch, that is

$$\sin\phi = \frac{25}{43.06} = 35.49°$$

Therefore, $$Y_D = 43.06(0.97 - 0.81)$$

or $$Y_D = 6.89 \text{ m}$$

Step 2: To evaluate moment and shear

Moment at section 15 m from left-hand support

$$M_D = 30(15) - 93.75(6.89) = -195.94 \text{ kN.m}$$

From the free-body diagram in Fig. 2.15*b*, normal thrust

$$N_D = H_D\cos\theta + V_D\sin\theta$$

where θ is the inclination of arch axis at point D

$$N_D = 93.75\,(0.97) + 30\,(0.23) = 97.84 \text{ kN (compression)}$$

Similarly radial shear

$$V_r = V_D \cos\theta - H_D \sin\theta = 30\,(0.97) - 93.75\,(0.23)$$
$$= 7.54 \text{ kN +ve following the shear force sign convention.}$$

Example 2.10 | *The equation of the axis of a three-hinged arch is $y = x - (x^2/12)$, the origin being the left-hand support. The span and rise are 12 m and 3 m respectively. The let half of the arch is loaded with a uniformly distributed load of 30 kN/m. Evaluate: (a) the reaction components at the supports (b) moment, radial shear and normal thrust at a section 3 m from left-hand support.*

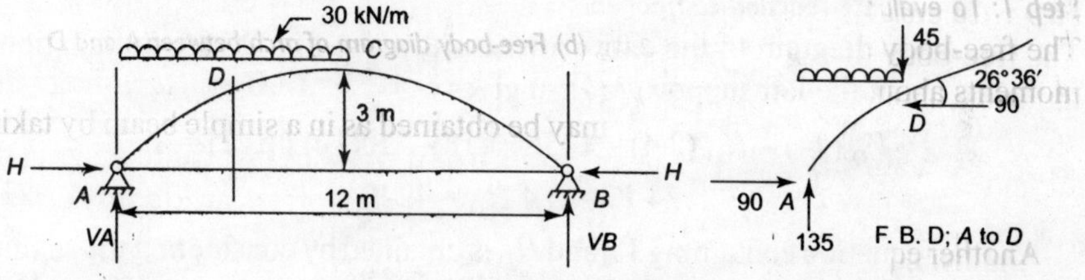

Fig. 2.16 | *Arch under load*

Step 1: To evaluate reaction components

Taking moments about the hinged support A

$$M_A = V_B\,(12) - 30 \times 6 \times 3 = 0$$

Gives $\quad V_B = 45.0 \text{ kN}$

$$V_A = 180 - 45 = 135 \text{ kN}$$

To evaluate H we take moment about hinge point C

$$M_C = V_B\,(6) - H(3) = 0$$

Which gives $H = 90$ kN

Step 2: Evaluate moment at a section 3 m from A.

Writing $M_D = 135\,(3) - 90(2.25) - \dfrac{30}{3}(3)^2 = 29.5$ kN

From Free body of the arch in Fig. 2.16

Normal thrust at D, $N_D = H\cos\theta + V_D \sin\theta$

From the arch equation inclination of arch axis at 3 m from support A, θ = 26° 36′

Substituting the values, $N_D = 90(0.8942) + 45\,(0.4478)$

$$= 100.62 \text{ kN (compression)}$$

Again radial shear $\quad V_{(r)} = H \text{ sing } \theta - V\cos\theta$

$$= 90(0.4478) - 45\,(0.8942) = 0$$

One may draw bending moment diagram by super imposing the moment diagram by horizontal thrust H over the simply supported beam bending moment diagram as shown in Fig. 2.17. The reader may verify, as an exercise, the maximum +ve and –ve moments and their sections as shown in the diagram.

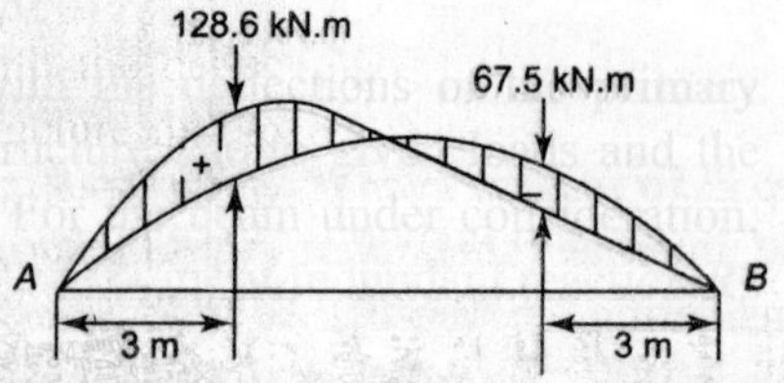

Fig. 2.17 | *B.M. Diagram*

Example 2.11 | *It is required to determine the reaction components at supports A and D and the internal forces just to the right of point C for a three-hinged arch shown in Fig. 2.18a.*

Step 1: To evaluate reaction components

The free-body diagram of the entire arch is shown in Fig. 2.12*b*. Summation of moments about the left support, $M_A = 0$ gives

$$V_D(24) - H_D(2.4) - 145680(18) - 20(10)(10/2) = 0$$

$$24\,V_D - 2.4\,H_D = 4240 \qquad (2.7)$$

Another equation containing V_D and H_D is obtained by considering the segment of the arch between B and D. Taking summation of moments about B gives

$$V_D(14) - H_D(4.9) - 180(8) = 0.$$

$$14\,V_D - 4.9\,H_D = 1440 \qquad (2.8)$$

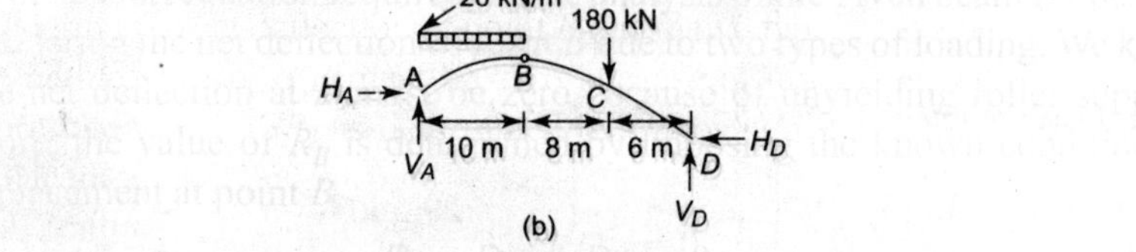

Fig. 2.18 | *(a) Arch and loading, (b) Free-body diagram of entire arch, (c) Free-body diagram of arch between C and D*

A simultaneous solution of Eqs. 2.7 and 2.8 gives

$$H_D = 295.24 \text{ kN}$$

and $\quad V_D = 206.19$ kN

For maintaining $\Sigma F_H = 0$, $H_A = H_D = 295.24$ kN

V_A is found by summing forces in the vertical direction

$$V_A - 200 - 180 + 206.19 = 0$$

or $$V_A = 173.81 \text{ kN}$$

Step 2: To evaluate internal forces just to right of C

The internal forces at a point just to the right of *C* can be determined by considering the free-body diagram of the arch between that point and support *D* (Fig. 2.18*c*).

Equilibrium condition $\Sigma F_H = 0$ and $\Sigma_{Fy} = 0$ give

$$H_C = H_D = 295.24 \text{ kN}$$

and $$V_C = V_D = 206.19 \text{ kN}$$

and summing up moments about *C*

$$M_C + 206.19(6) - 295.24\,(3.3) = 0$$

gives $$M_C = -262.85 \text{ kN m}$$

One more example is presented to illustrate the procedure involved in the case of a parabolic arch.

Example 2.12 | *The equation of a three-hinged parabolic arch with origin x^2 at its left support is $y = x - (x^2/40)$, The span of the arch is 48 m. Find the normal thrust and radial shear force at a section 6 m from the left support, when the arch is carrying a uniformly distributed load of 20 kN/m over the left half of the span. Also find the section at which the maximum positive or negative bending moment will occur and the magnitude of the same anywhere on the arch.*

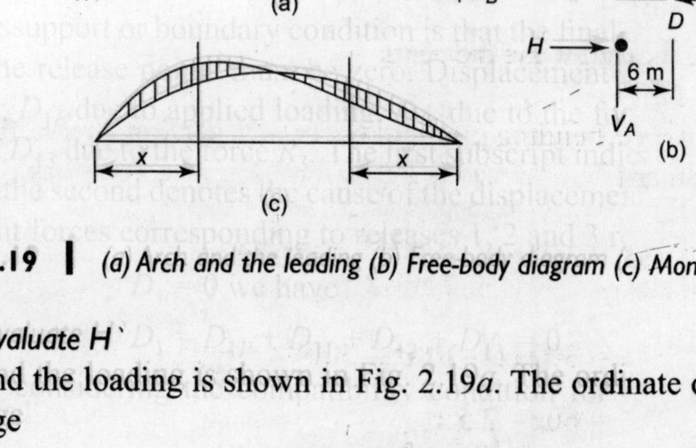

Fig. 2.19 | *(a) Arch and the leading (b) Free-body diagram (c) Moment diagram*

Step 1: To evaluate H`

The arch and the loading is shown in Fig. 2.19*a*. The ordinate of the arch at the central hinge

$$Y_C = 24 - \frac{24^2}{40} = 9.6 \text{ m.}$$

Reaction V_B is obtained by taking moments about A

$$V_B(48) - 24 \times 20 \times 12 = 0$$

$$\therefore \quad H = \frac{24 \times 20 \times 12}{48} = 120 \text{ kN.}$$

$$V_A = 24 \times 20 - 120 = 360 \text{ kN.}$$

Horizontal reaction H is obtained by taking moments about C

$$M_C = V_B(24) - H(9.6) = 0$$

$$\therefore \quad H = \frac{120 \times 24}{9.6} = 300 \text{ kN.}$$

Step 2: To evaluate normal thrust and radial shear

Let θ be the inclination of the arch axis at section D

$$\tan\theta = \frac{dy}{dx} = 1 - \frac{2x}{40}$$

Substituting for $x = 6$ m

$$\tan\theta = 1 - \frac{2 \times 6}{40} = 0.7$$

and $\theta = 35°$

From the Fig. 2.19*b*, normal thrust

$$N_D = H\cos\theta + V_D\sin\theta$$

in which V_D is the shear force at section D. Shear force $V_D = V_A - 20 \times 6$

$$V_D = 360 - 120 = 240 \text{ kN}$$

and $H = 300$ kN.

Normal thrust $N_D = 300 \times \cos 35° + 240 \sin 35°$

$= 383.41$ kN (compression)

Radial shear $V_r = V_D\cos\theta - H\sin\theta$

$= 240 \times \cos 35° - 300 \sin 35° = 24.53$ kN.

Step 3: To evaluate maximum ±ve moments

Maximum +ve B.M.

Let the maximum +ve bending moment occur at a section X, x *m* from the left end. Then the moment

$$M_x = 360(x) - 20\frac{(x)^2}{2} - 300\,Y_x$$

$$= 300(x) - 10x^2 - 300\left(x - \frac{x^2}{40}\right)$$

$$= 60x - 2.5x^2 \tag{2.9}$$

We set $\frac{dM_x}{dx} = 0$ to obtain the value for x

$$\frac{dM_x}{dx} = 60 - 5x = 0$$

$$\therefore \qquad x = \frac{60}{5} = 12$$

Maximum +ve B.M will be obtained by substituting for $x = 12$ m in Eqn. 2.9

$$M_{max} = 60(12) - 2.5\,(12)^2 = 360 \text{ kN.m}$$

Maximum –ve B.M.

It is clear from the B.M. diagram in Fig. 2.19*c* that the maximum –ve B.M. will occur in the region *C* to *B*.

Again taking a section *X*, *x m* from support *B*, moment

$$M_x = 120\,(x) - 300\left(x - \frac{x^2}{40}\right)$$

$$= 7.5\,x^2 - 180\,x \qquad (2.10)$$

To obtain maximum –ve B.M. we set

$$\frac{dM_x}{dx} = 0$$

$$\therefore \qquad \frac{dM_x}{dx} = 15\,x - 180 = 0$$

$$x = \frac{180}{15} = 12 \text{ m}$$

Substituting in Eqn. 2.10 the maximum –ve B.M.

$$M_{max} = 7.5\,(12)^2 - 180\,(12)$$

$$= 7.5\,(12)^2 - 18 = -1080 \text{ kN.m.}$$

2.6 GRAPHIC STATICS

2.6.1 General

Numerous graphical methods are available for determining the forces in the members of a truss, deflection of truss joints, analysis of cable structures and arches. We shall only review some of the common procedures in this section. The graphical method which is concerned with the visual representation of forces greatly clarifies the interaction of the force system.

A force may be represented graphically by a line drawn towards or away from the point of application and having a length that indicates the numerical size of the force to a certain scale. The slope of this line indicates the direction of the force, while the arrow head the sense in which the force acts along this line. For example, a force of 50 kN can be represented by a length of a line 25 mm if a scale of 1 mm = 2 kN is chosen (Fig. 2.20a).

2.6.2 Resultant of Two Concurrent Forces

The resultant of two concurrent forces can be obtained in accordance with the law of parallelogram of forces. Thus, to determine the resultant of two forces $\mathbf{F}_1$ and $\mathbf{F}_2$ represented by vectors **OA** and **OB**, a parallelogram is constructed as shown in Fig. 2.20*b*. The direction and magnitude of resultant $\mathbf{R}_{12}$ is obtained by diagonal vector **OC**. The same result could have been obtained by drawing either of the force triangles *OAC* or *OBC* instead of the parallelogram (Figs 2.20*c* and *d*). In constructing these triangles, either force may be drawn first and the other force laid out from the end of the first vector. The result is then obtained, in magnitude and direction, from the closing vector of the triangle drawn from the beginning of the first vector to the end of the second.

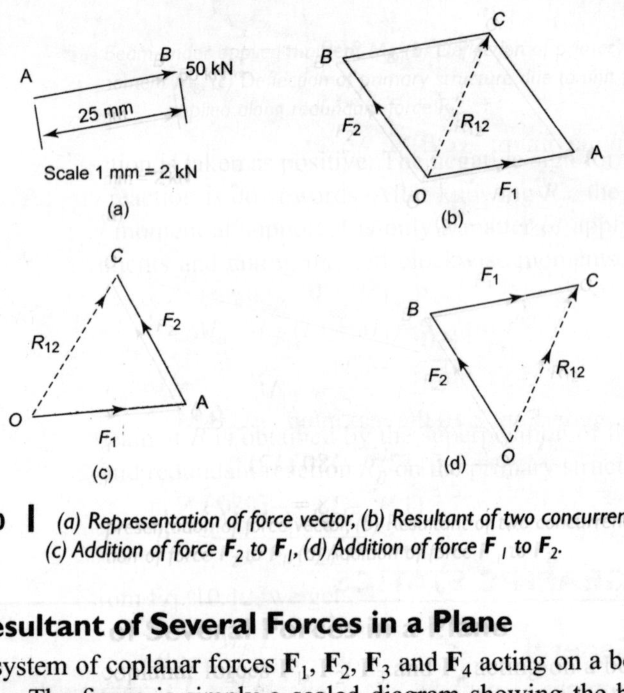

Fig. 2.20 | *(a) Representation of force vector, (b) Resultant of two concurrent forces (c) Addition of force $\mathbf{F}_2$ to $\mathbf{F}_1$, (d) Addition of force $\mathbf{F}_1$ to $\mathbf{F}_2$.*

2.6.3 Resultant of Several Forces in a Plane

Consider a system of coplanar forces $\mathbf{F}_1$, $\mathbf{F}_2$, $\mathbf{F}_3$ and $\mathbf{F}_4$ acting on a body shown in Fig. 2.21*a*. The figure is simply a scaled diagram showing the body, point of application and line of action of the forces. This diagram is known as a *space diagram*. Suppose the resultant of the forces is required to be determined graphically. As described in the previous section, resultant $\mathbf{R}_{12}$ of forces $\mathbf{F}_1$ and $\mathbf{F}_2$ may be obtained from force triangle *OAB* (Fig. 2.21*b*). The line of action of this resultant is drawn parallel to vector **OB** and through the intersection of lines of actions of forces $\mathbf{F}_1$ and $\mathbf{F}_2$ on the space diagram. In the same manner, resultant $\mathbf{R}_{123}$ is obtained by combining $\mathbf{R}_{12}$ and $\mathbf{F}_3$, and resultant $\mathbf{R}_{1234}$ by combining $\mathbf{R}_{123}$ and $\mathbf{F}_4$. The resultant of all the forces is, thus, $\mathbf{R}_{1234}$. The line of action of the resultant in the space diagram is obtained by fixing successively the line of action of the intermediate resultants. For example, the line of action of resultant $\mathbf{R}_{123}$ is obtained at the intersection point of resultant $\mathbf{R}_{12}$ and $\mathbf{F}_3$. Similarly the line of action of $\mathbf{R}_{1234}$ passes through the point of intersection of resultant R_{123} and F4. The construction involved is shown in Fig. 2.21*a*.

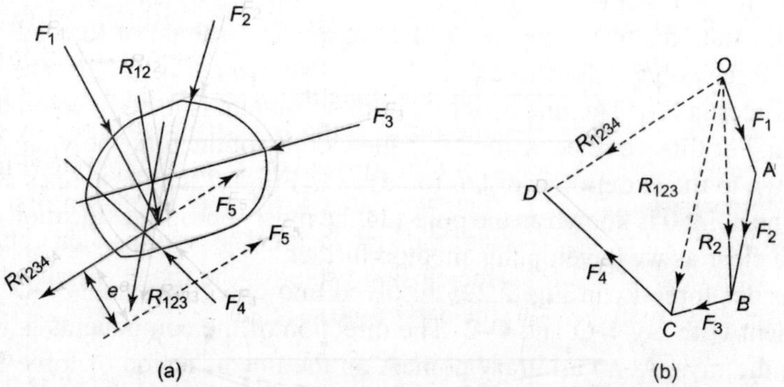

Fig. 2.21 | *(a) Space diagram, (b) Force diagram*

The same resultant could have been obtained by placing forces $\mathbf{F}_1$, $\mathbf{F}_2$, $\mathbf{F}_3$ and $\mathbf{F}_4$ tip to tail in the order in which they are encountered in going round the rigid body. This force diagram *OABCD* as indicated in Fig. 2.21*b* is called the *force polygon.* The magnitude and direction of the resultant will be given by the vector drawn from the initial to the final point of the force polygon, in this case by vector **OD**. The line of action of this resultant in the space diagram, of course, must be established as described above.

2.6.4 Equilibriant

Suppose we apply to the force system discussed above a force $\mathbf{F}_5$ which is equal in magnitude but opposite in direction to resultant $\mathbf{R}_{1234}$ or vector **OD**. If the line of action of $\mathbf{F}_5$ coincides with the line of action of $\mathbf{R}_{1234}$, then force $\mathbf{F}_5$ in effect holds the other forces in equilibrium. In such a case force $\mathbf{F}_5$ is called the *equilibriant* of the force system. The force polygon now closes thereby indicating that the resultant force on the body is zero. Suppose the line of action of force $\mathbf{F}_5$ does not coincide with $\mathbf{R}_{1234}$ but is shifted laterally by an amount e as indicated in Fig. 2.15*a*. The force polygon, of course, closes thus satisfying $\Sigma F_X = 0$ and $\Sigma F_Y = 0$. But in the space diagram, the equal and opposite forces, $\mathbf{R}_{1234}$ and $\mathbf{F}_5$ are parallel and the resultant is a moment couple. Thus, the body is not in equilibrium since $\Sigma_i M \neq 0$. Therefore, in the case of non-concurrent forces, the closure of a force polygon is a necessary but not a sufficient condition to show that the system is in equilibrium. In addition to this condition, it is necessary to show that in the space diagram the system is not equivalent to a couple. Of course, in the case of a concurrent force system, it is enough to show that the force polygon closes as a necessary condition for equilibrium.

2.6.5 Funicular Polygon

If the directions of forces in a system are parallel or nearly parallel as shown in Fig. 2.22*a*, the intersection points of the forces do not fall within the paper and, therefore, the resultant cannot be obtained. To get over this difficulty, we shall develop a general method using a *funicular polygon* which is applicable to any coplanar force system.

To illustrate the technique, let us determine the resultant of forces $\mathbf{F}_1$, $\mathbf{F}_2$ and $\mathbf{F}_3$. The force polygon for these three forces is shown in Fig. 2.22*b*. The resultant of the forces is $\mathbf{R}_{14}$. The line of action of this resultant is to be fixed on the space diagram. For this we select a point 0 in the vicinity of the force polygon and lines are drawn to the extremities of the forces as in Fig. 2.22*b*. These lines are called *rays*. The point 0 is known as the pole and the most appropriate location of it will become clear as we develop this method further.

Consider force $\mathbf{F}_1$ in Fig. 2.22*b* resolved into two components **1-O** and **O-2** coincident with ray **1-O** and **O-2**. The direction of the components is indicated next to the rays. At an arbitrary point A on the line of action of force $\mathbf{F}_1$ in the space diagram, the direction of components **1-O** and **O-2** are constructed. The line representing **O-3** is drawn through point *B*. The location of point *B* having been established by **O-2** is extended to intersect force $\mathbf{F}_2$ at **B**.

Force $\mathbf{F}_2$ is next considered and resolved into components **2-O** and **O-3** as shown in Fig. 2.22*b*. The direction of component **O-3** is drown through point *B*. The location of point *B* having been established by **O-2**, **O-3** is extended to intersect force $\mathbf{F}_3$ at *C*. In the same manner, force $\mathbf{F}_3$ is resolved into **3-O** and **O-4**, and **O-4** is drawn through point *C* as shown. Now the original force system of $\mathbf{F}_1$, $\mathbf{F}_2$ and $\mathbf{F}_3$ has been replaced by six components **1-O** and **O-2**, **2-O** and **O-3**, and **3-O** and **O-4**. Of these six components, pairs **O-2** and **2-O**, and **O-3** and **3-O** which are equal but oppositely directed balance each other. We have, in effect, replaced forces $\mathbf{F}_1$, $\mathbf{F}_2$ and $\mathbf{F}_3$ by two components **1-O** and **O-4**. Therefore, the intersection of lines of **1-O** and **O-4** at point *D* on the space diagram locates the line of action of $\mathbf{R}_{14}$ on the rigid body.

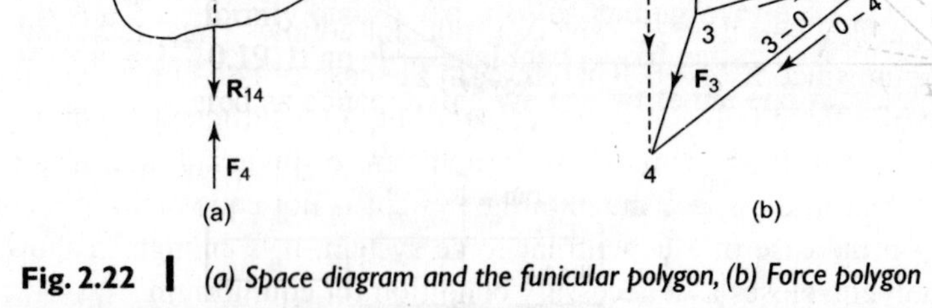

Fig. 2.22 | *(a) Space diagram and the funicular polygon, (b) Force polygon*

The polygon *ABCDA* formed on the space diagram is referred to as the *funicular polygon*. The sides of this polygon drawn between the forces are called *strings*. Note that the funicular polygon shown is not unique, as the starting point

A chosen on force $\mathbf{F}_1$ is arbitrary as also the location of pole point O. It should be apparent now that the location of pole point O is made so that the strings of the funicular polygon will intersect the lines of action of the given forces at near right angles. Thus less space is required for the diagram and hence greater accuracy can be attained.

For equilibrium of the body, a force equal and opposite to resultant $\mathbf{R}_{14}$ must be applied to the body; it must be applied though point D. If this force was to be represented as $\mathbf{F}_4$ and included in the force system, then the force polygon and the funicular polygon drawn to these forces close. Thus, for a system of non-concurrent forces to be in equilibrium, it is necessary that the force polygon as well as the funicular polygon must close.

The principle of the funicular and force polygons can be used to determine the reactions of a statically determinate structure.

Consider the beam in Fig. 2.23a. The reactions at A and B are to be determined. In this case the point of application and direction of the right-hand side reaction and only the point of application of the left-hand side reaction are known. The unknowns are the magnitudes of both the reactions and the direction of the left-hand side reaction. These three unknowns may be found by using the condition that both the force and funicular polygons must close if the combined system of loads and reactions is to be in equilibrium.

The force polygon for the given forces $\mathbf{F}_1$ and $\mathbf{F}_2$ is constructed in Fig. 2.18*b*. We select a pole point O and draw the rays to points 1, 2 and 3 as shown. The construction of a funicular polygon begins at a particular point in Fig. 2.23*a*. Although we do not know the magnitude and direction of reaction at A, we do know that it passes through point A. Therefore, we begin the construction of the funicular polygon at this point. A string parallel to ray **O-1** is drawn though point A and extended to intersect the line of action of force $\mathbf{F}_1$ at C. Note that this string represents the common component to the unknown reaction $\mathbf{R}_A$ and the force $\mathbf{F}_1$. This can be verified with reference to the developments in Fig. 2.23*b*.

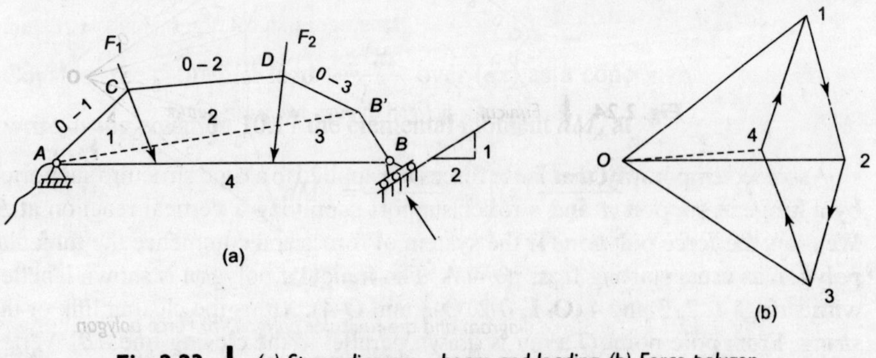

Fig. 2.23 | *(a) Space diagram—beam and loading, (b) Force polygon*

A line parallel to ray **O-2** is drawn from point C to intersect the line of force $\mathbf{F}_2$ at D. Similarly, a line **O-3** is drawn through point D until it intersects the line of action of the reaction at B at point B'. It is at this point that the known direction of reaction R_B is made use of. A closing line of the funicular polygon is drawn from A to B'. It may be pointed out that string AB' represents a common component of reactions R_A and R_B. This closing line is transferred to the force diagram by drawing a ray parallel to the closing line and passing through pole point O. From point 3 a force vector is drawn parallel to reaction R_B and extended to intersect the line just drawn through point O at 4. **3-O** and **O-4** represent the components of reaction R_B. The vector 3-4 gives the magnitude of reaction R_B while vector 4-1, the closing vector of the force polygon, gives the magnitude and direction of reaction R_A. **4-O** and **O-1** represent the components of reaction R_A.

2.6.6 Funicular Polygon through Two Points

When we consider the procedure for drawing a funicular polygon for a given system of forces, it becomes apparent that it is possible to draw an infinite number of funicular polygons for that system of forces. Similarly an infinite number of points can be chosen for pole O. Sometimes, however, it becomes necessary to draw the funicular polygon to pass through certain specific points in the space diagram.

Consider a system of forces as shown in Fig. 2.24. Suppose it is required to construct a funicular polygon passing through two points A and B.

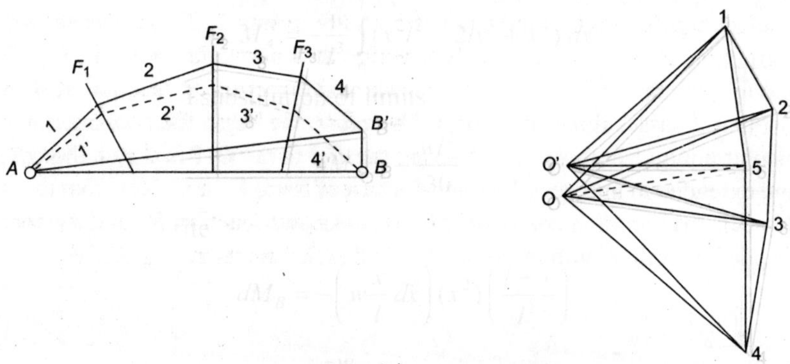

Fig. 2.24 | *Funicular polygon through two given points*

Assume, temporarily, that these forces are applied to a rigid structure supported by a hinge at support A and a roller support supplying a vertical reaction at B. We draw the force polygon for the system of forces and commence the funicular polygon as usual starting from point A. The funicular polygon is shown labelled with strings 1, 2, 3 and 4 (**O-1**, 0-2, **O-3** and **O-4**). AB' is the closing link or the string. From pole point O a ray is drawn parallel to the closing line AB'. Vertex point 5 is fixed by drawing a vertical vector through point 4 and locating the intersection point on the ray just drawn from point O. The value of reactions R_B and R_A as represented by vectors **4–5** and **5–1** are independent of the location of pole point O and thus the location of point 5 is unique.

The object was to draw a funicular polygon that passes through two points A and B. In other words the closing line of the funicular polygon must coincide with the line AB. This is easily achieved by choosing *a* pole point O' anywhere on a ray drawn parallel to AB and passing througl vertex 5. The new funicular polygon with the strings labelled 1′, 2′, 3′ and 4′ (O'-1, O'-2, O'-3 and O'-4) passes through the given points A and B.

The graphical approach is well suited for determining forces in cables carrying concentrated loads. A single construction gives the shape of the cable, reaction components of cable supports and also the tension in the cable. The following example illustrates the procedure.

Example 2.13 | *For a cable supported at end points A and B and carrying loads shown in Fig. 2.25a determine the cable shape and end reactions by a graphical construction.*

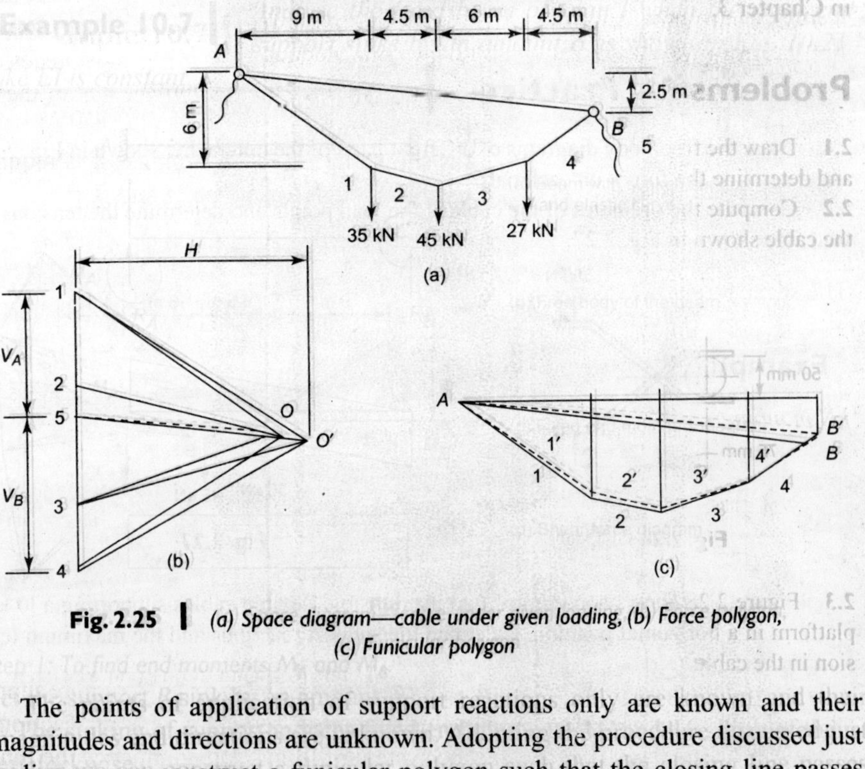

Fig. 2.25 | *(a) Space diagram—cable under given loading, (b) Force polygon, (c) Funicular polygon*

The points of application of support reactions only are known and their magnitudes and directions are unknown. Adopting the procedure discussed just earlier we can construct a funicular polygon such that the closing line passes through the chord AB in the space diagram. An arbitrary pole point O gives the closing string AB' shown in Fig. 2.20c. The corresponding ray in the force polygon is shown as O-5. As pointed out earlier, point 5 uniquely fixes up the vertical component of reactions R_B and R_A. They are independent of location of pole point O. Now a new pole point O' is chosen anywhere on the line passing through point 5 and parallel to the chord joining support points A and B. The funicular polygon drawn thus passes through support points A and B.

The sags at each load point on the cable may be scaled from Fig. 2.25*c* measuring from the chord line joining the support points. The value of the horizontal force component *H* in the cable for these particular sag values is the horizontal distance from pole point *O*′ to the vertical load vector line in the force polygon. Since the product of *H* and sag is constant for any given loading and span length, this solution defines all possible cable profiles. If the desired sag is 75% of the measured values, all sags are multiplied by 3/4 and the horizontal force *H* is increased by 4/3. Thus the profile of the cable after the sag values are adjusted gives the true profile of the cable.

It may be noted that the vertical reactions determined from the force polygon will not be true vertical reactions on the cable foundations. Actual reactions will be the forces V_A and V_B as obtained from the construction of Fig. 2.25*b* plus the vertical components of the inclined closing line represented by ray O-5.

The analysis for forces in truss members by the graphical method is discussed in Chapter 3.

Problems for Practice

2.1 Draw the free-body diagrams of different parts of the nutcracker shown in Fig. 2.26 and determine the forces on each part.

2.2 Compute the ordinates of the cable at the load points and determine the tensions in the cable shown in Fig. 2.27.

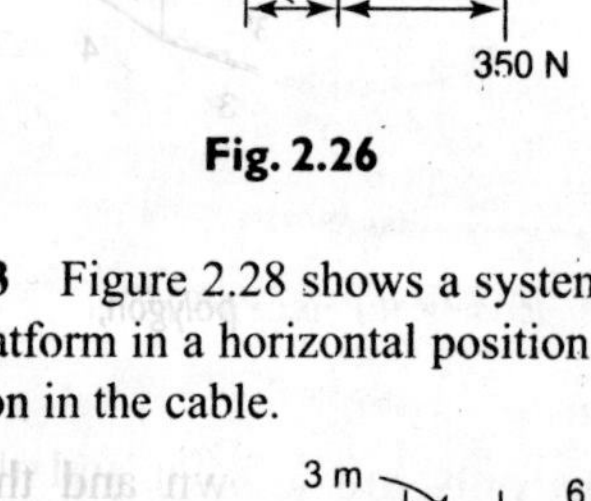

Fig. 2.26

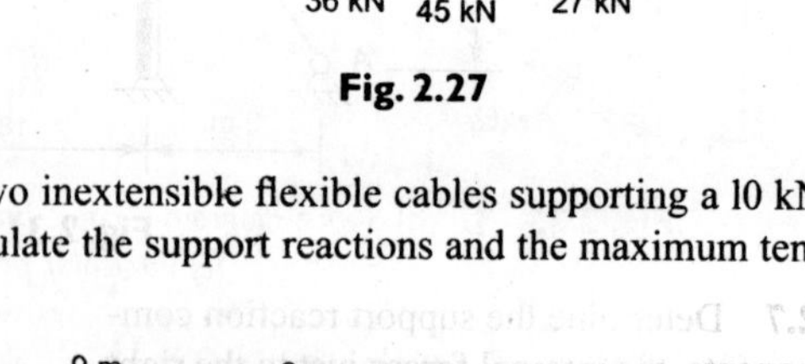

Fig. 2.27

2.3 Figure 2.28 shows a system of two inextensible flexible cables supporting a 10 kN platform in a horizontal position. Calculate the support reactions and the maximum tension in the cable.

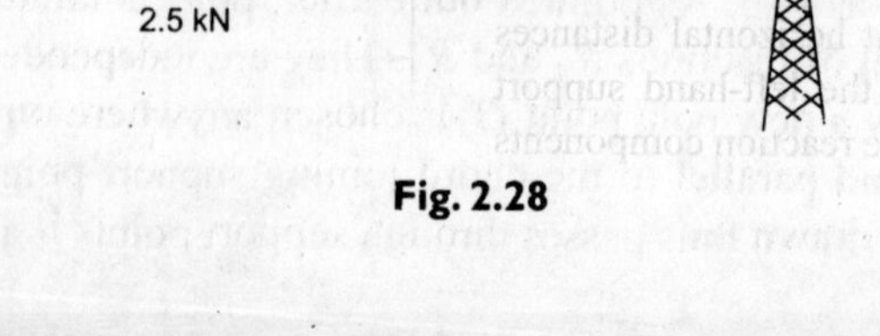

Fig. 2.28

2.4 For the cable structure shown in Fig. 2.29 determine (a) the ordinate Y_D, (b) the maximum tension in cable *BCDE*, (c) tension in cable *EF*, and (d) the value of reaction at *H*.

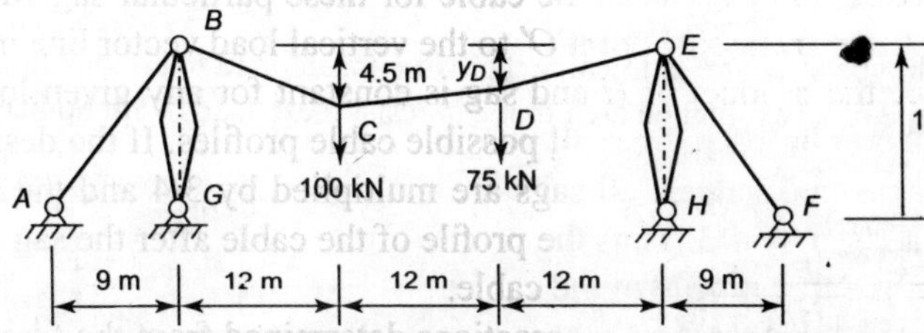

Fig. 2.29

2.5 Determine the reactions and maximum tension in the cable shown in Fig. 2.30.

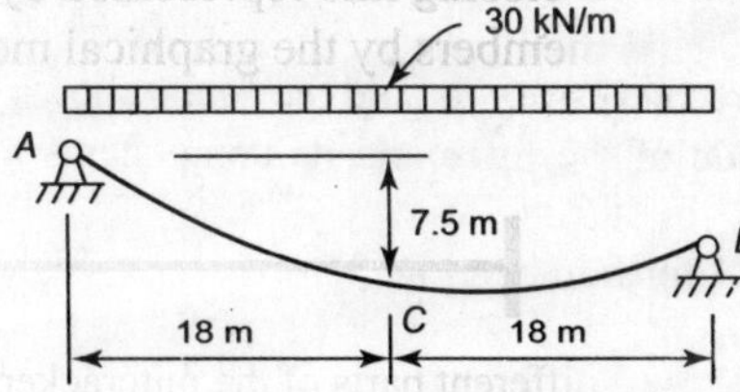

Fig. 2.30

2.6 For the cable structure shown in Fig. 2.31 determine (a) maximum tension in cable *BC* and (b) tension in cable *AB*. *D* is a point of known elevation on *BC*.

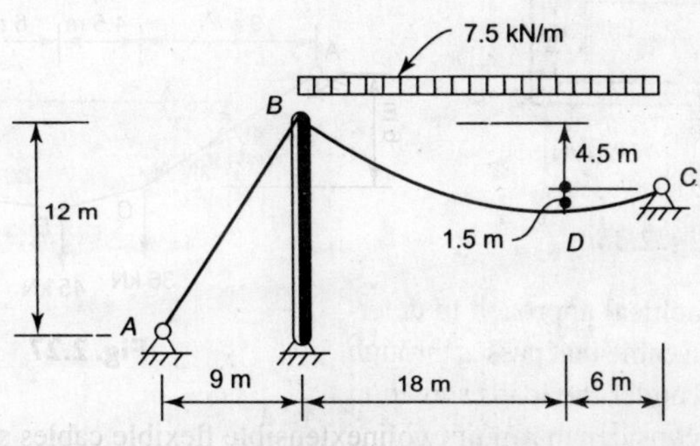

Fig. 2.31

2.7 Determine the support reaction components, the internal forces just to the right of point *D* for the three-hinged arch shown in Fig. 2.32.

2.8 A circular arched rib, span 50 m and rise 10 m, is hinged at the crown and springings and carries two vertical loads of 60 and 100 kN at horizontal distances 12 and 30 m from the left-hand support respectively. Find the reaction components

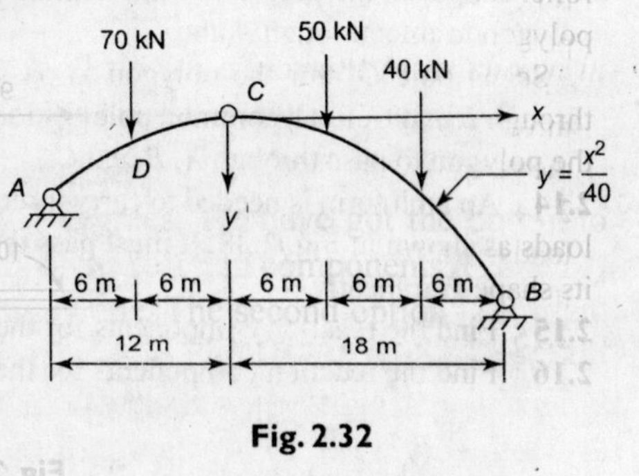

Fig. 2.32

at the springings and moment, normal thrust and radial shear at a section 10 m from the left support.

2.9 Calculate the reactions by graphical construction for the structure shown in Fig. 2.33.

2.10 An arch type structure is to carry the two concentrated loads shown in Fig. 2.34. Define the shape of a two-hinged arch that can resist these loads with no bending action in the arch.

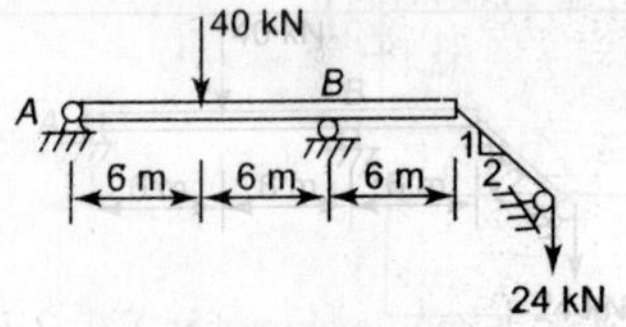

Fig. 2.33

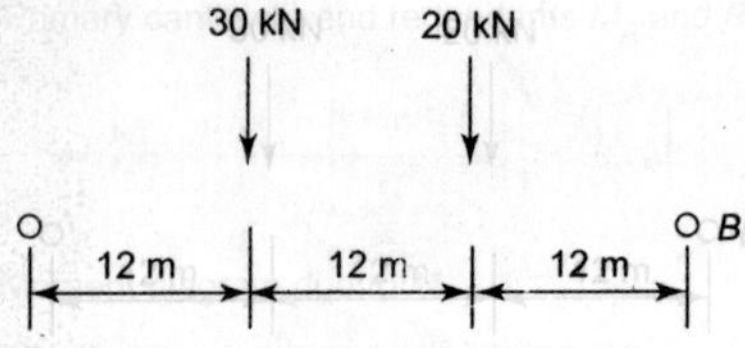

Fig. 2.34

2.11 Find graphically the force necessary to hold the frame shown in Fig. 2.35 in equilibrium. Indicate the magnitude of the force and its components along horizontal and vertical directions.

2.12 Repeat the Problem 2.11 for frame in Fig. 2.36.

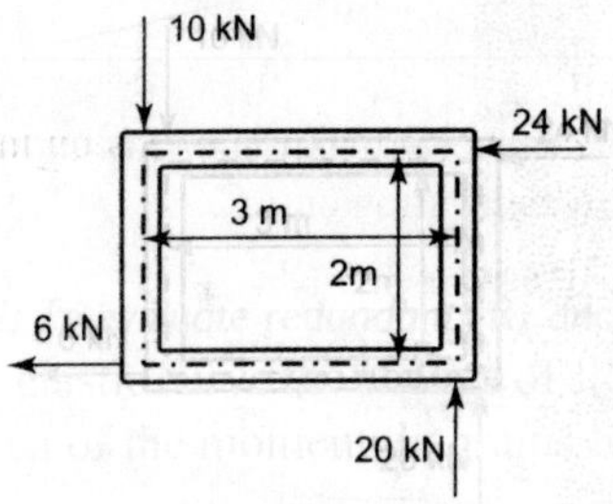

Fig. 2.35

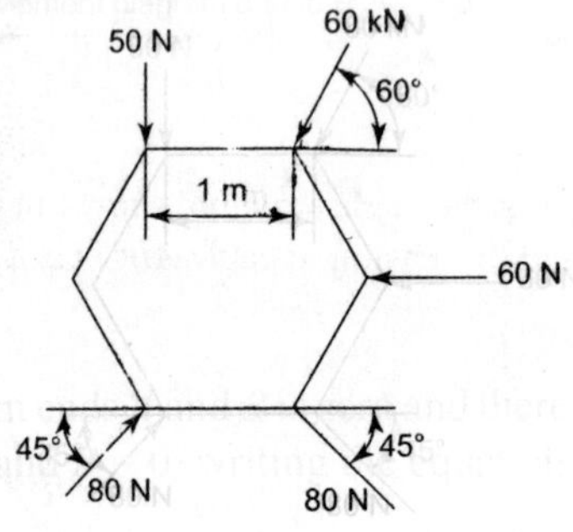

Fig. 2.36

2.13 Use the graphical approach to determine the shape of a cable that passes through points *A*, *B* and *C* under the loads shown in Fig. 2.37. (Hint: consider that point *C* is a roller support. Find pole *O′* for the funicular polygon to pass through *A* and *C*.

Select pole *O″* for the polygon to pass through *B* and *C*. Fix a common pole *O″* for the polygon to pass through *A*, *B* and *C*.)

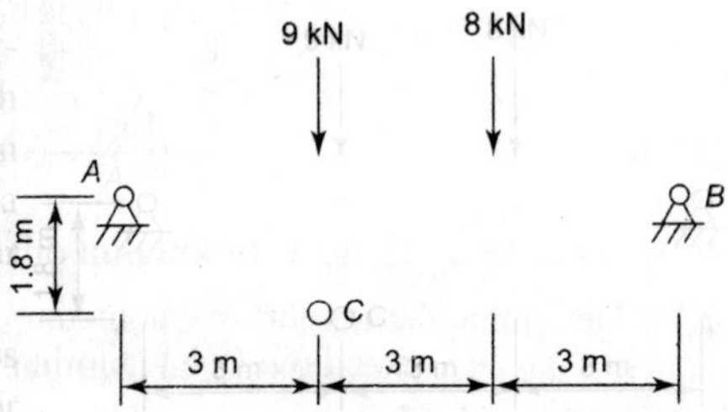

Fig. 2.37

2.14 An arch form is needed to carry three loads as shown in Fig. 2.38. It must pass through the three points *A*, *B* and *C*. Determine its shape graphically.

2.15 Find the reaction components for the structure shown in Fig. 2.39.

2.16 Find the reaction components for the structure shown in Fig. 2.40.

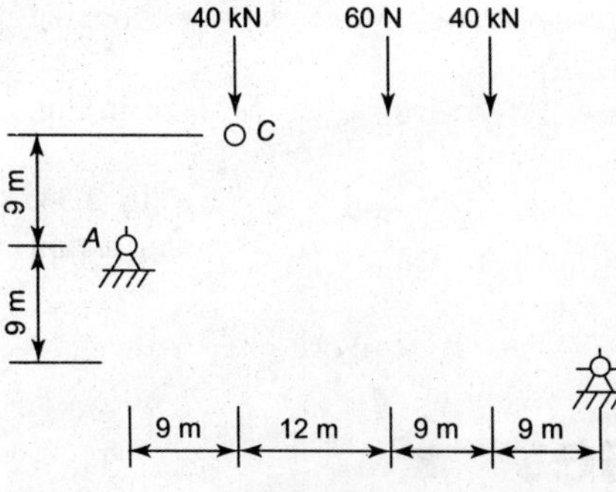

Fig. 2.38

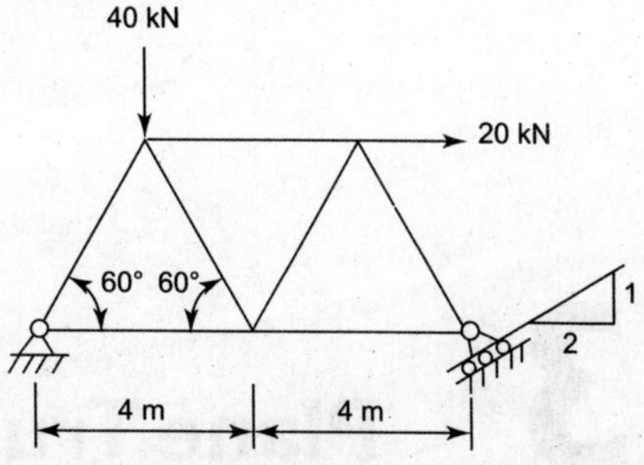

Fig. 2.39

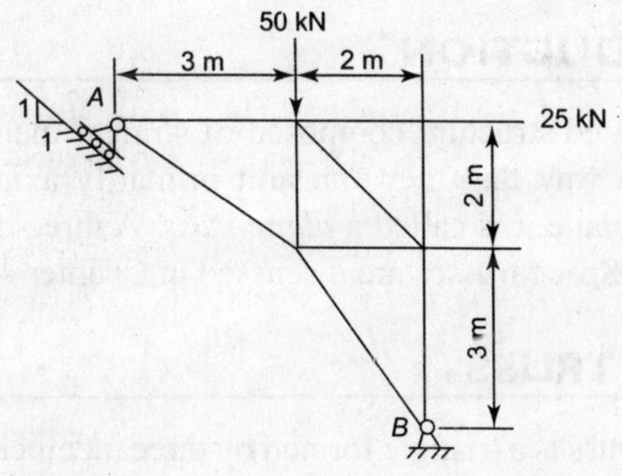

Fig. 2.40

3 Plane Trusses

3.1 INTRODUCTION

A truss is an articulated structure composed of straight members arranged and connected in such a way that they transmit primarily axial forces. If all the members lie in one plane it is called a *plane truss*. A three-dimensional truss is called a *space truss*. Space trusses are discussed in Chapter 4.

3.2 PLANE TRUSS

The basic form of a truss is a triangle formed by three members joined together at their common ends forming three joints. Such a triangle is clearly rigid. Another two members connected to two of the joints with their far ends connected to form another joint forms a stable system of two triangles. If the whole structure is built up in this way it must be internally rigid. Such a truss if supported suitably will be stable. For example, the truss has to be supported in general by three reaction components, all of which are neither parallel nor concurrent. Such a truss is called a simple truss. Various combinations of basic triangular elements produce general truss structures shown in Fig. 3.1. These trusses are stable and statically determinate.

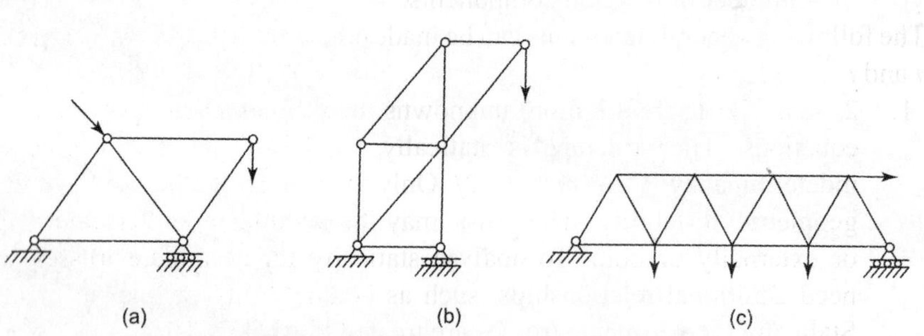

Fig. 3.1 *Trusses from triangular elements*

Several common types of trusses are shown in Fig. 3.2. Trusses given in Fig. 3.2a, *b* and *c* are roof trusses and are used up to 30 m span. The other types of trusses are commonly used in bridges.

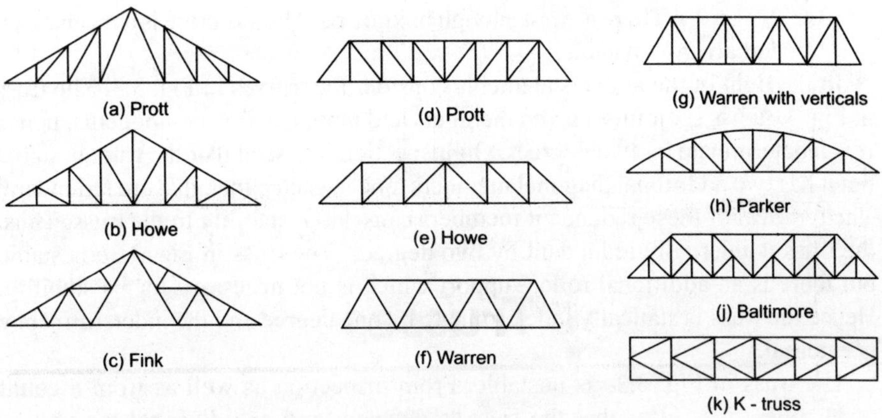

(a) Prott (b) Howe (c) Fink (d) Prott (e) Howe (f) Warren (g) Warren with verticals (h) Parker (j) Baltimore (k) K - truss

Fig. 3.2 | *Common types of trusses*

3.3 GEOMETRIC STABILITY AND STATIC DETERMINANCY OF TRUSSES

A truss which possesses just sufficient number of members or bars to maintain its stability and equilibrium under any system of forces applied at joints is called a statically determinate and stable truss.

A planar truss may be thought of as a structural device having j joints in a plane. The forces that act on the joint are the member forces, the external loads and the reactions. Since all the joints are in equilibrium, we can write two equilibrium equations, $\Sigma F_X = 0$ and $\Sigma F_Y = 0$ for each joint. Thus, for the entire truss we can write $2j$ equations. The unknowns are the member forces and the reaction components. Therefore, if the structure is statically determinate, we can write the relation

$$2j = m + r \tag{3.1}$$

where m = number of members

r = number of reaction components.

The following general statements can be made concerning the relation between j, m and r.

1. $2j < m + r$. There are more unknowns than the number of equilibrium equations. The structure is statically indeterminate. The degree of indeterminancy is $n = m + r - 2j$. Only inspection can be used to study geometric instability. The truss may be redundant either internally or externally or both. To analyse statically indeterminate trusses we need additional relationships, such as compatibility of displacements. Statically indeterminate trusses are treated in Chapter 10.
2. $2j = m + r$. The structure is statically determinate and the unknowns can be obtained from $2j$ equations. The degree of indeterminancy $n = 0$. Apart from inspection there are several ways of detecting instability.

3. $2j > m + r$. There are not enough unknowns. The structure is a mechanism and always unstable.

In the light of the above statements consider the trusses in Fig. 3.3. The truss in Fig. 3.3a has six joints, eleven members and three reaction components; hence it is indeterminate by two degrees. On inspection it is seen that the truss is stable but it has two additional diagonal members, one in each panel, that are redundant. The removal of these redundant members cause no instability to the truss. Thus, the truss is internally redundant by two degrees. The truss in Fig. 3.3*b* is stable but there is an additional roller support which is not necessary for its stability. Hence the truss is statically indeterminate by one degree and the indeterminacy is external.

The truss in Fig. 3.3*c* is unstable. From inspection as well as from a count of members it is clear that the truss is deficient and one diagonal member is necessary to make the truss rigid and stable. Consider the truss in Fig. *3.3d.* It has more members than just required. But on inspection it is clear that the end panels are made over rigid by providing diagonal members both ways and the central panel is deficient thereby making the truss unstable. It may be noted that the truss is unstable due to improper distribution of members.

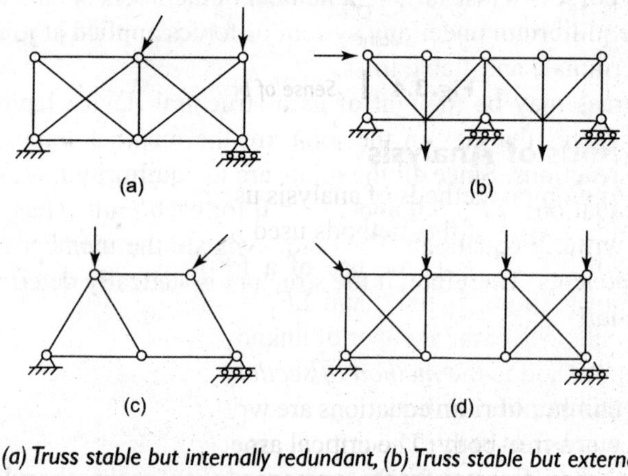

Fig. 3.3 | *(a) Truss stable but internally redundant, (b) Truss stable but externally redundant, (c) Truss unstable due to deficient member, (d) Truss unstable due to improper arrangement of members*

3.4 ANALYSIS OF TRUSSES

3.4.1 Assumptions

In analysing the trusses the following assumptions are made:

1. the members of a truss are pin-jointed at their ends on frictionless joints,
2. the loads lie in the plane of the truss and are applied only at the joints, and
3. the centroidal axes of various members framing into a joint will intersect at a common point.

Of the three, assumption I is seldom completely satisfied in practice. For example, the welded or riveted gusset plates commonly used to join the member ends do not really represent pinned connections. However, in many cases, the members are long and slender and very little moment is transmitted by the members. Hence the assumed pin connections give acceptable results. Assumptions 2 and 3 are normally satisfied. Assumption 2 implies that all truss members receive forces only through the joints at either ends and, therefore, these two end forces must be colinear and opposite to each other for equilibrium, making each a simple tension or compression member. Thus, the direction of forces away from the joint indicates, tension, and direction towards the joint indicates compression in the bars as shown in Fig. 3.4.

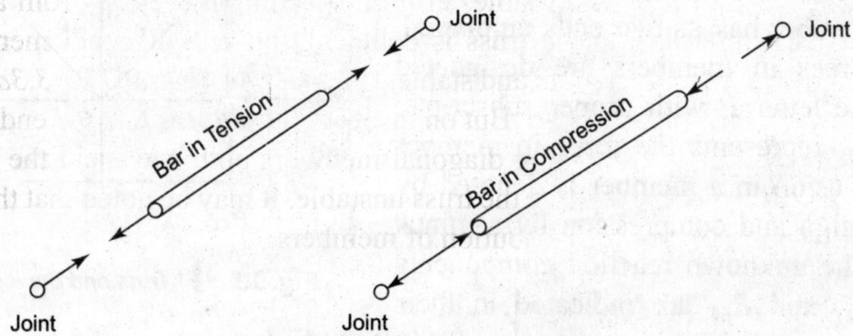

Fig. 3.4 | *Sense of bar force*

3.4.2 Methods of Analysis

There are three common methods of analysis used in calculating the forces in the members of a truss. One of the methods used in analysing a truss is the *method of joints*. This method entails the use of a free-body diagram of joints with the equilibrium equations $\Sigma F_X = 0$ and $\Sigma F_Y = 0$. Inspection of joints generally indicates the joints where the number of unknowns are two or less than two.

The second method is the *method of sections*. In this method the truss is cut into two parts and equilibrium equations are written for one of the parts of the cut truss treating it as a free-body. The critical aspect of this method is the choice of the proper free-body diagram for the purpose.

The method of joints is effective if we want to calculate forces in all members of the truss but the method of sections is obviously superior if we seek forces only in certain members. In such a case, sections can be made only through the selected members, whereas the method of joints would require the analysis of joints from one end of the structure progressively up to the particular member. The third method is known as the *tension coefficient method*. This method, formulated by Prof. Southwell is a systematic procedure of method of joints. This method is more useful for solving space trusses.

3.4.3 Method of Joints

In a determinate structure, the reaction components can be worked out using equations of equilibrium. The analysis can be commenced from a joint where there are only two unknown member forces by utilizing the equilibrium equations $\Sigma F_x = 0$ and $\Sigma F_y = 0$. We can go toe the next joint, where again, there are only two unknown member forces. The process is repeated till al the desired member forces are obtained. The procedure is illustrated by the following examples. The graphical method for the analysis of trusses is discussed in section 3.6.

Example 3.1 | *It is required to determine the forces in members of the truss shown in Fig. 3.5.*

The first step in the analysis is to assign appropriate notation to the joints. The joints are designated as 1,2, 3, 4, 5 and 6 as indicted. With the numbering of joints each member has its two ends numbered. The forces in members are designated with the letter P with proper subscripts; thus P_{15} represents the force in member 1–5. Tension in a member is denoted by a plus sign and compression by a minus sign. The unknown reaction components R_{4X}, R_{4Y} and R_{6Y} are indicated in their positive sense.

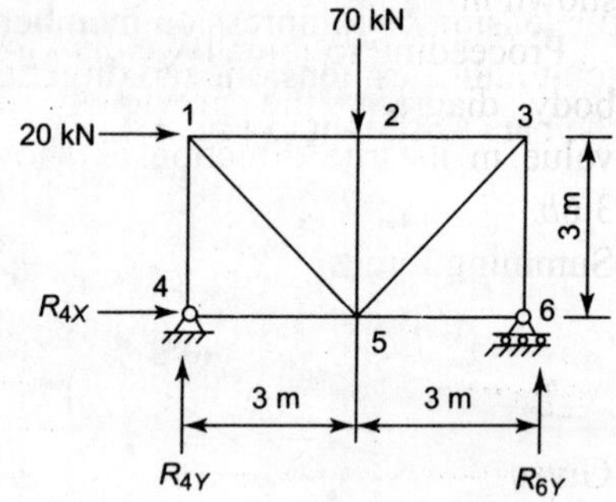

Fig. 3.5 | *Truss and the loading*

From the equilibrium equations applied to the entire truss $\Sigma F_X = 0$

we get $\quad R_{4X} = -20$ kN

Again writing summation of moments about joint 6 we have

$$M_6 = R_{4Y}(6) + 20(3) - 70(3) = 0$$

or $\quad \Sigma_{4Y} = 25$ kN

and applying $\quad \Sigma F_Y = 0$

$$R_{6Y} = 70 - 25 = 45 \text{ kN}$$

The next important step is to study the structure and decide which path can be taken up so that the joints will have two or less than two unknowns.

One path that will work is the order of joints 4-1-2-5 and 3 and check at 6. We can also start from joint 6 first and proceed in the order of 6-3-2-5 and 1 with a check at 4.

A free-body diagram of joint 4 is shown in Fig. 3.6*a*.

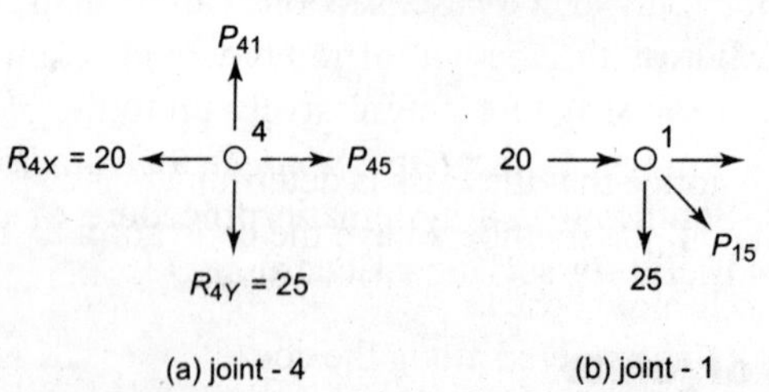

Fig. 3.6 | *Free-body diagrams of joints 4 and 1*

While drawing the free-body diagram, it may be convenient to show the unknown bar forces as tensile forces. From the equilibrium of forces if it turns out to be negative, the member is in compression. From Fig. 3.6*a* $\Sigma F_X = P_{45} - 20 = 0$ or $P_{45} = 20$ kN (tension).

Again, $\Sigma F_Y = P_{41} + 25 = 0$ or $P_{41} = -25$ kN (compression). The results obtained at each joint can be recorded on a truss diagram shown in Fig. 3.7.

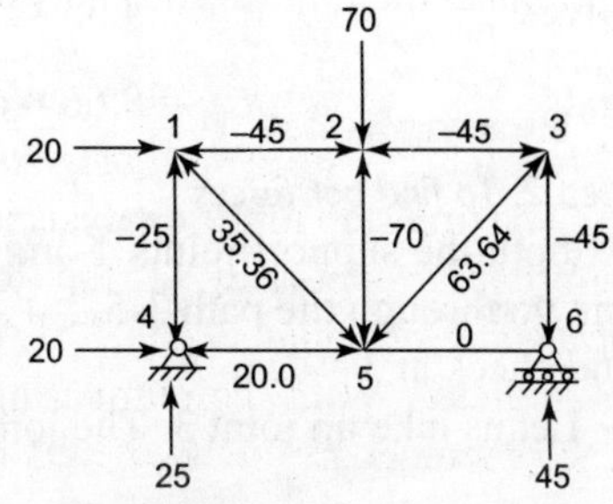

Fig. 3.7 | *Results of analysis*

Proceeding to joint 1 we draw on the free-body diagram, the previously determined value in its true direction as shown in Fig. 3.6*b*.

Summing forces

$$\Sigma F_Y = 25 - P_{15}\left(\frac{1}{\sqrt{2}}\right) = 0$$

Gives $\qquad P_{15} = 35.36$ kN (tension)

Similarly $\qquad \Sigma F_X = 20 + P_{12} + P_{15}\left(\frac{1}{\sqrt{2}}\right) = 0$

Gives $\qquad P_{12} = -45.0$ kN (compression)

The remaining bar forces are determined by the same procedure with the joints taken in the order suggested above. The results are summarised in Fig. 3.7.

Example 3.2 | *Analyse the truss shown in Fig. 3.8 for the bar forces. The 10 kN load at the joint is 30° inclined to the vertical.*

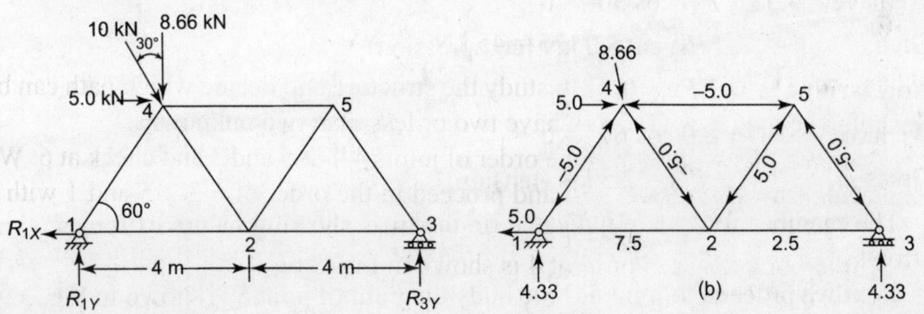

Fig. 3.8

On inspection, we notice that the truss is determinate and stable. All the joints are numbered. So that all the members have the ends numbered.

Step I: To evaluate reaction components

The applied load 10 kN is resolved along the coordinate axes so that $F_X = 5.0$ kN and $F_Y = 8.66$ kN

On summing forces in the horizontal direction

$$R_{1x} = 5.0 \text{ kN from right to left}$$

Again summing up moments about support 3

$$M_3 = R_1y\,(8) - 8.66\,(6) + 5 \times 3.464 = 0$$

Gives $\quad R_1y = 4.33$ kN

and $\quad R_3y = 8.66 - 4.33 = 4.33$ kN

Step 2: To find bar forces

At both the support joints 1 and 3, there are only two unknown bar forces. We can go through the path 3-5-2-4 and check at 1. Or, we can take the path 1-4-2-5 and check at 3

Let us take up joint 3. The joint is shown as a free body in Fig. 3.9(a)

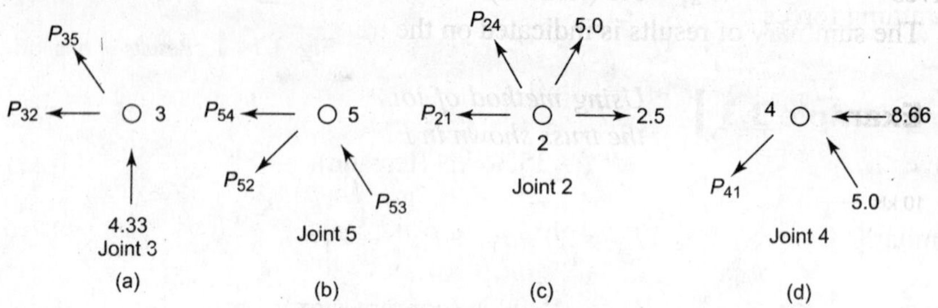

Fig. 3.9 | *Free bodies of joints*

The bar forces are shown as tensile. In the analysis, if the result turns out to be negative, the member will be under compression

Writing $\quad \Sigma F_y = 0$

We have $\quad 4.33 + P_{35} \cos 30° = 0$

Give $\quad P_{35} = -5.0$ kN (compression)

Now writing $\quad \Sigma Fx = 0$

We have $\quad -P_{32} + 5.0 \cos 60° = 0$

Gives $\quad P_{32} = 2.5$ kN (tension)

The member forces are indicated on the truss showing nature of forces with arrows.

Next we proceed to joint 5. Free body diagram of joint 5 is shown in Fig. 3.9 (b)

Writing $\quad \Sigma F_y = 0$

$$5 \cos 30° - P_{52} \cos 30° = 0$$

gives $\quad P_{52} = 5.0$ (tension)

Again writing $\quad \Sigma F_x = 0$

$$-5 \cos 60° - 5 \cos 60° - P_{54} = 0$$

gives $\quad P_{54} = -8.66$ (compression)

Again considering free body diagram of joints and writing $\Sigma Fy = 0$

We have $P_{24} \cos 30° + 5.0 \cos 30° = 0$

Gives $P_{24} = -5.0$ (compression)

Similarly writing $\Sigma F_x = 0$

$$-P_{21} + 2.5 + 2.5 + 2.5 = 0$$

gives $P_{21} = 7.5$ (tension)

Finally proceeding to joint 4 and considering the forces on the free body we can write

$$\Sigma F_y = 0$$

$$-P_{41} \cos 30° + 5 \cos 30° = 0$$

gives $P_{41} = 5.0$ (tension)

The summary of results is indicated on the truss diagram

Example 3.3 | *Using method of joints, find the forces in members of the truss shown in Fig. 3.10.*

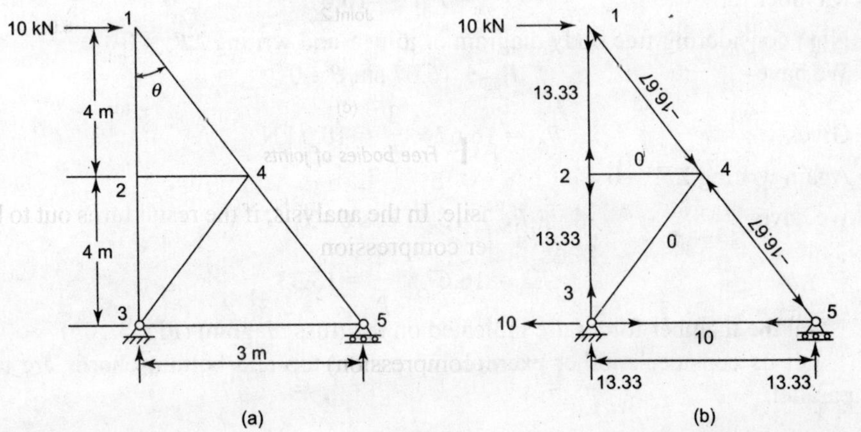

Fig. 3.10

We need not evaluate the reaction components first in this case. One can straightaway analyse the truss starting from joint 1 and following the path 1-2-4-5 with a check at 3.

Considering joint 1 and writing $\Sigma F_x = 0$ from Fig. 3.11 (a) we have $10 + P_{14} \sin \theta = 0$

$$P_{14} = \frac{-10}{\sin \theta} = -\frac{10}{3} \times 5 = -16.67 \text{ kN (compression)}$$

Againwriting $\Sigma F_y = 0$

$$P_{12} - 16.67 \cos \theta = 0$$

$$P_{12} = 16.67 \times \frac{4}{5} = 13.33$$

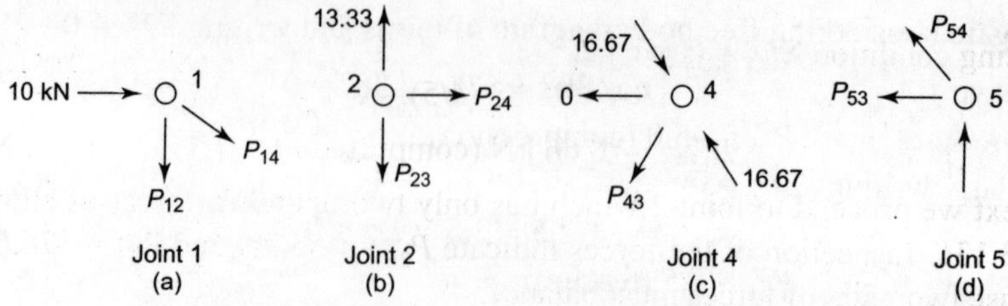

Fig. 3.11 | *Free body diagrams joints 1, 2, 4, 5*

Next considering joint 2, we notice from the free-body diagram $P_{24} = 0$

and $$P_{23} = 13.33 \text{ kN}$$

Proceeding to joint 4, member forces P_{43} and P_{45} can be evaluated by inspection

$P_{43} = 0$ since any force in member 4-3 will have component normal to 1-4-5 which cannot be balanced

Member force $$P_{45} = P_{41} = -16.67$$

Next considering free body diagram of joint 5 and writing $\Sigma F_x = 0$

We have $$P_{53} + 16.67 \sin \theta = 0$$

Gives $$P_{53} = 16.67 \times \frac{3}{5} = 10.0 \text{ kN}$$

Again writing $\Sigma F_y = 0$

We have $$R_{5y} - 16.67 \cos \theta = 0$$

$$R_{5y} = 16.67 \times \frac{4}{5} = 13.33$$

All the member forces are indicated on the truss diagram (Fig. 3.10*b*)

Let us consider another example where the top and bottom chords are not parallel.

Example 3.4 | *It is required to determine the member forces using the method of joints for the truss in Fig. 3.12.*

Since the truss and loading are symmetrical, it is enough to solve half the truss starting from one end. Each of the reactions R_Y is equal to half the total load on the truss. That is

$$R_{1Y} = R_{5Y} = 35 + \frac{45}{2} = 57.5 \text{ kN}$$

On examination we find that we can start either from joint 1 or 5 where there are only two unknowns.

Writing the forces at joint 1 as in Fig. 3.13*a* and applying first $\Sigma F_Y = 0$, we get

$$57.5 - P_{16}\,(3.6/6) = 0$$

or $$P_{16} = 95.83 \text{ kN}$$

Utilizing condition $\Sigma F_X = 0$, we get

$$P_{12} + 95.83\ (4/5) = 0$$

or $$P_{12} = -76.66 \text{ kN (compression)}$$

Next we proceed to joint 2 which has only two unknown forces as shown in Fig. 3.13*b*. Inspection of the forces indicate $P_{26} = -35$ kN and $P_{23} = -76.66$ kN because two pairs of forces must balance.

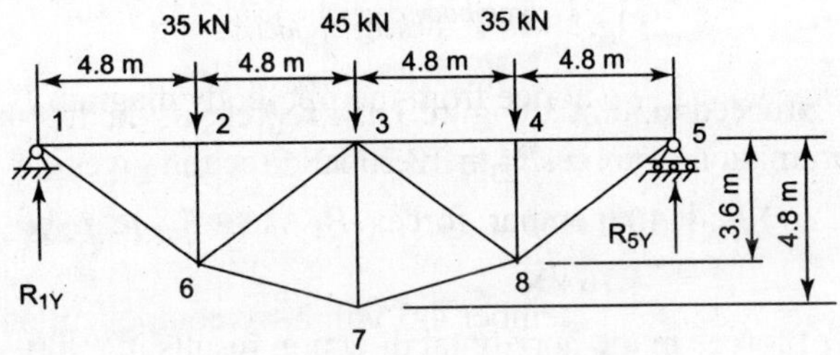

Fig. 3.12 | *Truss for analysis*

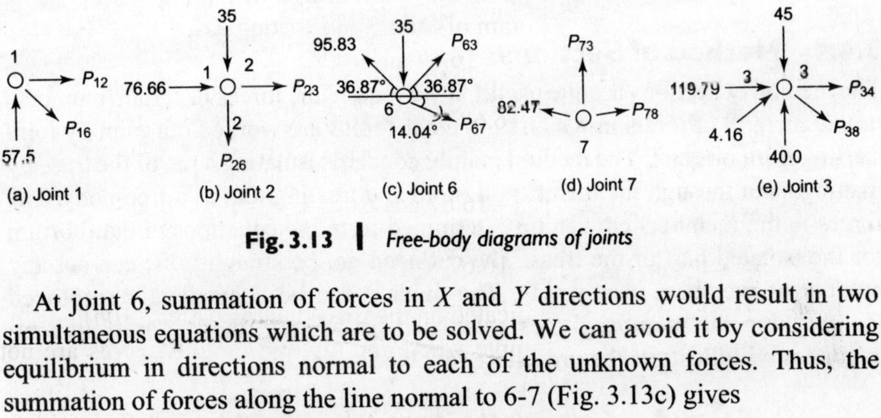

Fig. 3.13 | *Free-body diagrams of joints*

At joint 6, summation of forces in *X* and *Y* directions would result in two simultaneous equations which are to be solved. We can avoid it by considering equilibrium in directions normal to each of the unknown forces. Thus, the summation of forces along the line normal to 6-7 (Fig. 3.13c) gives

$$P_{63}\ (\cos 39.09°) - 35\ (\cos 14.04°) + 95.83\ (\cos 67.17°) = 0$$

This results in

$$P_{63} = -4.16 \text{ kN}$$

Similarly, equilibrium of forces perpendicular to member 6-3 would give directly P_{67}. Alternatively, summation of forces $\Sigma F_X = 0$, gives

$$-95.83\ (\cos 36.87°) - 4.16\ (\cos 36.87°) + P_{67}\ (\cos 14.04°) = 0$$

or $$F_{67} = 82.47 \text{ kN}$$

Next we shall proceed to joint 7. The free-body diagram of joint 7 is shown in Fig. 3.13*d*. Consideration of forces in the horizontal direction indicate P_{78} = 82.47 kN. Again summing up of forces in the vertical direction gives,

$$\Sigma F_Y = P_{73} + (82.47)(2)(\cos 75.96°) = 0$$

$$P_{73} = -40.0 \text{ kN}$$

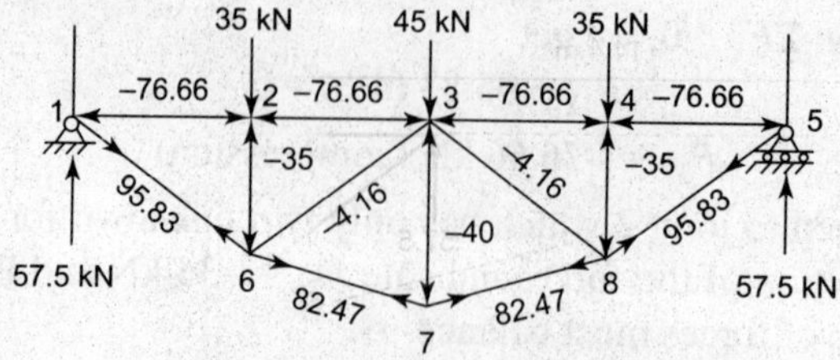

Fig. 3.14 | *Results of analysis*

Now we can proceed to joint 3. Figure 3.13*e* represents the free-body diagram of the joint. Summation of forces in the vertical direction gives

$$\Sigma F_Y = 40.0 - 45.0 + 4.16\,(\cos 53.13°) - P_{38}(\cos 53.13°) = 0$$

or $\quad P_{38} = -4.16$ kN

Summation of forces in the horizontal direction results in $P_{34} = -119.79$ kN. The values of forces P_{34}, P_{38} and P_{78} could have been anticipated because of symmetry. The summary of results of the analysis is shown in Fig. 3.14.

3.4.4 Method of Sections

The method of sections is quite useful, if we require the forces in certain members unlike method of joints in which all the bar forces are worked out joint by joint starting from one end. The method mainly concerns isolating a pat of the truss by making a cut through the members desired and treating that as a free-body. The forces in the members cut can be determined from the equations of equilibrium for the isolated part of the truss. The cut need not be straight; one can cut any number of members in any way. The method can be conveniently employed for checking the member forces calculated by other methods. The following examples will illustrate the procedure involved.

Example 3.5 | *Calculate the forces only in members 2-3, 6-3 and 6-7 of the truss solved in Example 3.4.*

Since only a few bar forces are required the method of sections is convenient. The reactions are computed as in the previous example. The truss is cut into two parts by taking a section as shown in Fig. 3.15. Consider the free-body diagram of the left part of the truss. Assume that the tensile forces are positive and the directions of unknown forces are shown accordingly. There are three unknown forces acting on this free-body which can be evaluated from the three equations of equilibrium. However, if we write these equations in a certain order, each unknown will occur in only one of the equations. For example, to evaluate P_{67} we take moments about joint 3, the point of intersection of the other two bar forces. The normal distance from joint 3 to the bar force P_{67} is

$$(6)\,(\sin 50.91°) = 4.66 \text{ m}$$

Hence writing $M_3 = (57.5)\,(9.6) - (35)\,(4.8) - P_{67}\,(4.66) = 0$ we get $P_{67} = 82.47$ kN (tension).

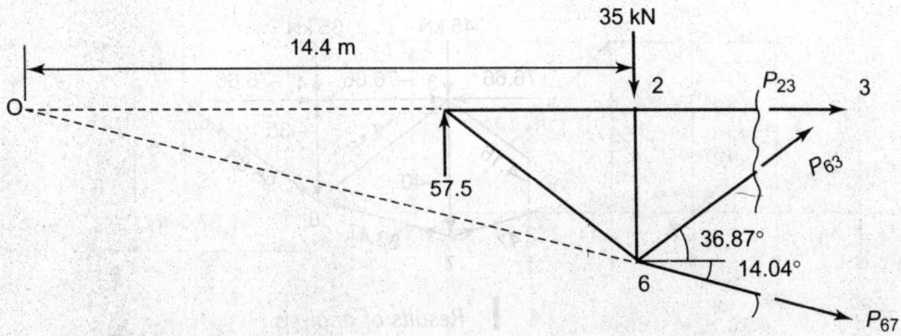

Fig. 3.15 | *Free-body diagram of part of truss*

In a similar manner, summation of moments about joint 6, the point of interaction of P_{63} and P_{67}, yields P_{23} directly. Thus

$$M_6 = 57.5(4.8) + P_{23}(3.6) = 0$$

or $$P_{23} = -76.67 \text{ kN (compression).}$$

We can follow the same approach and calculate the force in diagonal member 6-3 by taking moments about a point of intersection of the top and bottom chords denoted by O in Fig. 3.15. Point O is 19.2 m to the left of joint 3 (14.4 m from joint 2) since the slope of bottom chord is 1 : 4. The normal distance from point O to the extension of member 3–6 is 19.2 (sin 36.87°) = 11.52 m.

Writing $$M_0 = -57.5\,(9.6) + 35\,(14.4) - P_{63}\,(11.52) = 0$$

we get $$P_{63} = -4.17 \text{ (compression).}$$

Instead of working for P_{63} in this manner, it would be easier to sum up the horizontal components of these three forces and solve for P_{63}. Thus we have

$$P_{67}(\cos 14.04°) - P_{23} + P_{63}(\cos 36.87°) = 0$$

This gives $$P_{63} = -4.17 \text{ kN; same as the previous result.}$$

3.4.5 Subdivided Truss

The panels in long and deep trusses are often subdivided with more than a single web member in order to reduce the web member lengths and the distance between panel points. Two popular subdivision schemes are shown in Fig. 3.16*a* where interior panels have K-shaped web member arrangement and the end panels have the so-called Baltimore truss subdivisions. Example 3.6 illustrates the procedure for finding forces in the members meeting at a joint in a subdivided truss.

Example 3.6 | *Compute the forces in members meeting at joint 5 of the truss of Fig. 3.16a.*

In the analysis, we adopt both the methods of analysis to our advantage. A free-body diagram of joint 5 (Fig. 3.16*b*) indicates that both the members 5–7 and 5–8 are equally inclined to the horizontal and hence $P_{58} = -P_{57}$ as there can be no resultant horizontal component on the joint. A section taken through the third panel gives a free-body diagram of the left part of the truss as in Fig. 4.16c. By using

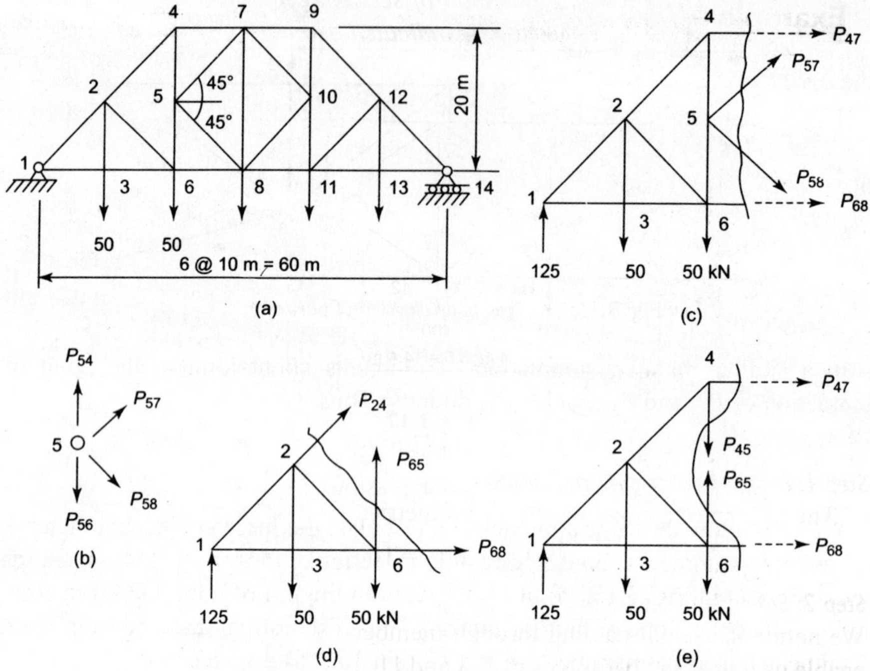

Fig. 3.16 | *Subdivided truss for analysis*

$$\Sigma F_Y = 125 - 100 + P_{57} \sin 45° - P_{58} \sin 45° = 0$$

and substituting $P_{58} = -P_{57}$ and solving we get

$$P_{58} = 17.7 \text{ kN (tension) and}$$

$$P_{57} = -17.7 \text{ kN (compression)}$$

It may be noted that the pair of members 5-7 and 5-8 carry the total shear in the panel. The division of shear between the two is done in such a manner that ensures zero horizontal resultant on joint 5.

Another section, taken through members 6-8, 6-5 and 2-4, results in a free-body diagram of Fig. 3.16*d*. Summing moments about point 1

$$\Sigma M_1 = 50\ (10) + 50\ (20) - P_{65}\ (20) = 0$$

we get $P_{65} = 75.0$ kN

Summation of forces in the vertical direction at joint 5 (Fig. 3.16*b*) gives,

$$\Sigma F_Y = P_{45} - 75 + P_{57} (\sin 45°) - P_{58} (\sin 45°) = 0$$

$$P_{45} = 100.00 \text{ kN (tension)}$$

A section which results in a free-body diagram of Fig. 3.12*e* can also be used to determine the force in member 4-5. Summing up forces in the vertical direction

$$\Sigma F_Y = -P_{45} + 125 - 100 + 75 = 0$$

we get $P_{45} = 100.0$ kN; same result as earlier.

Example 3.7 | *Using method of sections determine the bar forces in the members indicated of the truss shown in Fig. 3.17.*

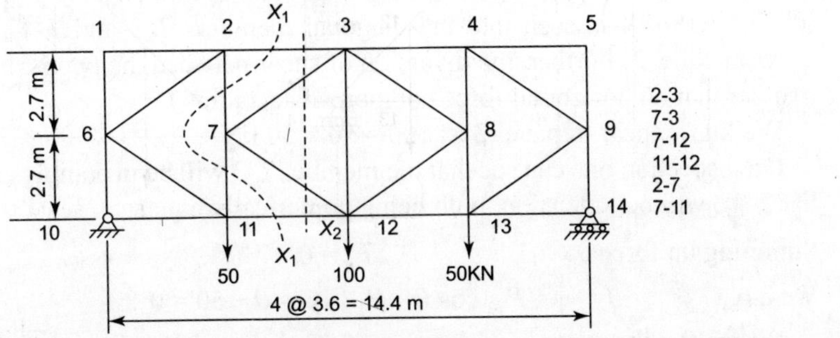

Fig. 3.17

Step 1: Evaluate reaction Components

The truss and the loading being symmetrical

$$R_{10y} = R_{14y} = 100 \text{ kN}$$

Step 2: Selection of path to the cut

We notice that a cut passing through members 11–12, 11–7, 7–2 and 2–3 will enable us to find the bar forces in 2–3 and 11–12. The left part of the truss cut by the path X_1X_1 is shown isolated as a free-body in Fig. 3.19 (a)

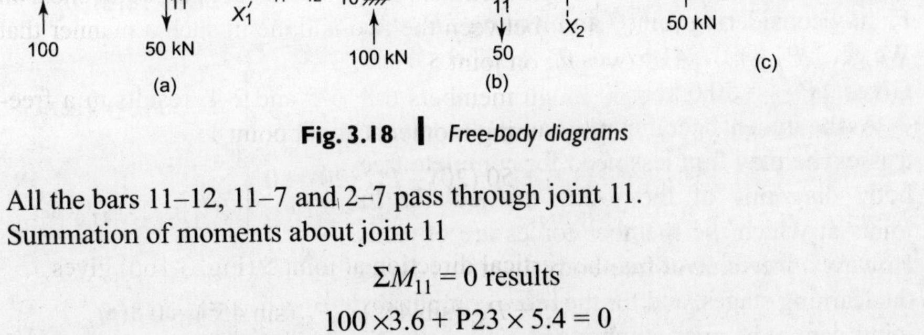

Fig. 3.18 | *Free-body diagrams*

All the bars 11–12, 11–7 and 2–7 pass through joint 11.

Summation of moments about joint 11

$$\Sigma M_{11} = 0 \text{ results}$$

$$100 \times 3.6 + \text{P23} \times 5.4 = 0$$

gives $$P_{23} = -66.67 \text{ kN (compression)}$$

Similarly summing up momens about joint 2

$$\Sigma M_2 = 0 \text{ results}$$

$$100 \times 3.6 - P_{11-12}\,(5.4) = 0$$

gives $$P_{11-12} = 66.67 \text{ (tension)}$$

Step 3: Forces in diagonal members 7–3 and 7–12

To find the forces in diagonal member, we make a cut in the second panel across members 11–12, 2–3, 7–3 and 7–12. The free body of the isolated part is shown in Fig. 3.18*b*. It is seen that the diagonal members 7–3 and 7–12 carry total shear in panel 2. Further, the diversion of shear between the two is due in such a manner that no horizontal force component acts joint 7.

We know shear in panel 2 is; 100 – 50 = 50 kN

On inspection one can see that the member 7–3 will be in compression and the bar 7–12 will be in tension, both being numerically equal.

Summing up forces $\Sigma F_y = 0$

We have $-P_{73} \cos \theta - P_{7\text{-}12} \cos \theta + 50 = 0$

Where θ is the angle of inclination of the diagonal members with the vertical, and that is 53°6.

This gives, $P_{7\text{-}12} = 41.67$ kN (tension)

and $P_{T\text{-}3} = 41.67$ kN (compression)

Step 4: Evaluation of forces in members 2–7 and 7–11

We cannot directly evaluate the member forces as there will be more than two unknowns in whatever manner we make the cut. We need to find first the forces in diagonal members in panel1, 2–6 and 6–11. We know the shear in panel is 100 kN and the bar forces are 83.34 kN: member 2–6 in compression and member 6–11 in tension following the procedure as earlier.

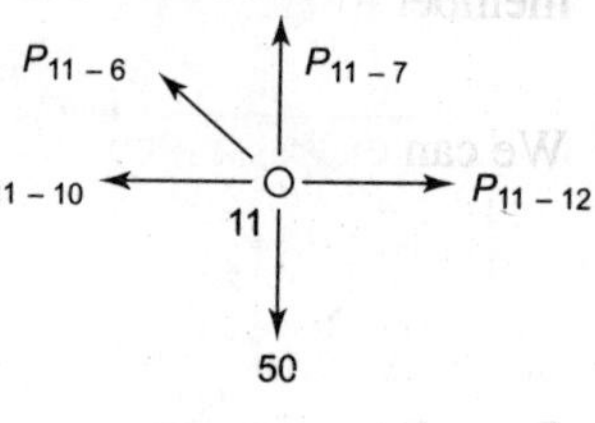

Fig. 3.18(d)

Now, using method of joints, summing up forces at joint 11, $\Sigma Fy = 0$

Gives $P_{6\text{-}11} \cos \theta + P_{11\text{-}7} - 50 = 0$

$P_{11\text{-}7} = 50 - 83.34 \times 0.6 = 0$

Finally considering joint 7 and writing $\Sigma F_y = 0$

We get $P_{7\text{-}2} + 0 - 41.67 \cos \theta - 41.67 \cos \theta = 0$

Gives $P_{7\text{-}2} = 50.0$ kN

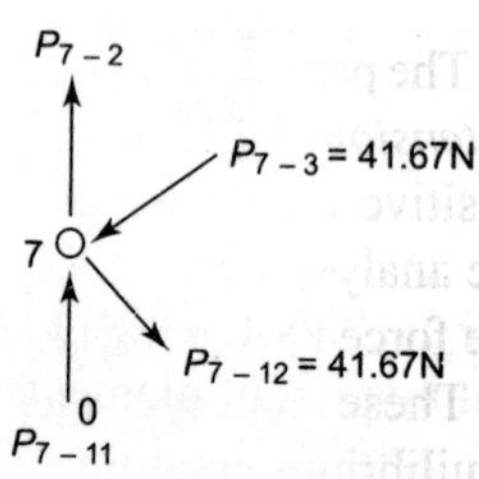

Fig. 3.18(e)

As the student becomes more adept at analysing trusses, he may find less need for complete free-body diagrams of the joints, particularly for joints at which the member forces are obvious. However, the value of free-body diagrams during the learning stages, and for the more complicated joints cannot be over-emphasized.

These examples show that for a rapid analysis, the method of sections can be combined with the method of joints. Occasionally it may be necessary to solve simultaneous equilibrium equations of the free-body, and sometimes we must solve first for some other member force before the desired member force can be found.

3.4.6 Method of Tension Coefficients

R.V. Southwell proposed in 1929 a systematic procedure based on equilibrium condition of joints in a pin-jointed truss. He noted a relationship between the components of member forces along the coordinate axes and the member coordinates at the ends of members. The analysis results in linear simultaneous equations for the components of member forces along the coordinate axes at each joint. This method gives member forces per unit length rather than member forces directly. The force per unit length of member is known as tension coefficient of the member. This concept and application of this method to plane trusses is discussed in this section. The application of this method to space frames is discussed in the next Chapter.

Tension Coefficients Consider a member ij whose coordinates are (x_i, y_i) and (x_j, y_j) as indicated in Fig. 3.19. The member force T_{ij} is considered to be tensile.

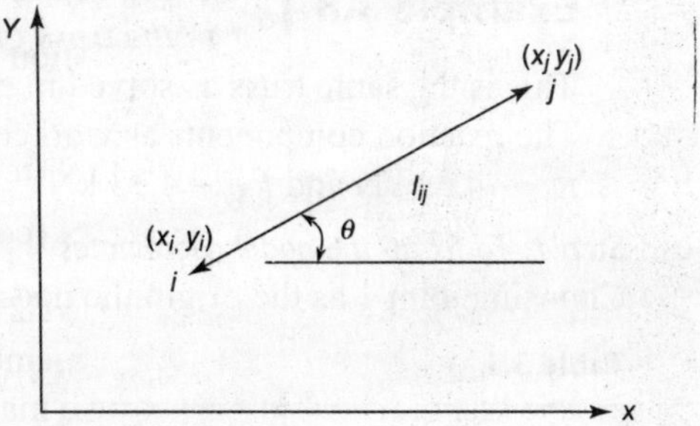

Fig. 3.19

The components of T_{ij} along the coordinate axes are

$T_x = T_{ij} \cos \theta$, $T_y = T_{ij} \sin \theta$](3.2)

Where θ is the inclination of the member to the x axis

$$\text{We can express,} \quad \cos\theta = \left(\frac{x_j - x_i}{l_{ij}}\right) = \frac{x_{ij}}{l_{ij}} \quad \text{and} \quad \sin\theta = \left(\frac{y_j - y_i}{l_{ij}}\right) = \frac{y_{ij}}{l_{ij}}$$

$$l_{ij} = (x^2_{ij} + y^2_{ij})^{1/2}$$

$$\text{From Eqn (3.2)} \ T_x = \left(\frac{T_{ij}}{l_{ij}}\right) x_{ij} = t_{ij}\, x_{ij} \quad \text{and} \quad T_y = \left(\frac{T_{ij}}{l_{ij}}\right) y_{ij} = t_{ij}\, y_{ij}$$

The parameter $t = T/l$ represents the force/unit length of member and it is known as *tension coefficient* of the member. The tension coefficients are all assumed to be positive indicating the force in member is tensile. In the analysis, if the coefficient turns out to be negative, the force in that member to be taken as compression

These concepts can be applied to formulate the equilibrium condition to a truss joint i connected to truss joints j, k, m. as shown in Fig. 3.20.

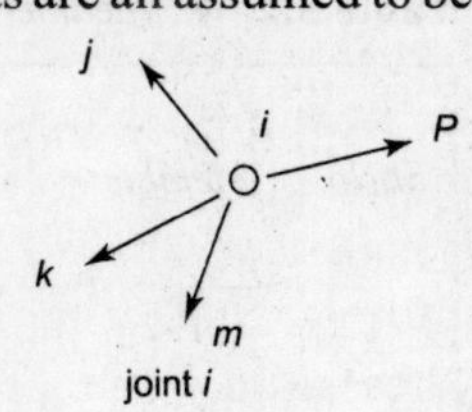

Fig. 3.20

Writing down $\Sigma F_x = 0$ and $\Sigma F_y = 0$ at the joint

$$\Sigma F_x = t_{ij}\, x_{ij} + t_{ik}\, x_{ik} + t_{im}\, x_{im} + p_x = 0 \tag{3.3}$$

and
$$\Sigma F_y = t_{ij}\, y_{ij} + t_{ik}\, y_{ik} + t_{im'}\, y_{im} + P_y = 0 \tag{3.4}$$

writing the above equations in compact form

$$\Sigma t_{ij} x_{ij} + p_x = 0 \tag{3.5}$$

and $$\Sigma t_{ij} y_{ij} + p_y = 0 \tag{3.6}$$

Such equations are formulated at each of the joints and solved for tension coefficients. As in method of joints, the sequence of joints are so chosen that only two member forces are unknown at that joint. This method is amenable to computer applications as the equilibrium equations can be set up for all the joints instead of joint by joint. Member forces are computed by multiplying tension coefficients with the corresponding member lengths. The procedure involved is illustrated by the following examples.

Example 3.8 | *Using method of tension coefficients find the bar forces in the truss given in Fig. 3.21*

This is the same truss as solved in example 3.2 using method of joints.
The reaction components as worked out earlier are
$R_{1y} = 4.33$ kN and $R_{3y} = 4.33$ kN.

Step 1: To fix up the nodal coordinates

Choosing joint 1 as the origin the nodal coordinates are given in Table 3.1

Table 3.1

Sl.No.	Node	Coordinates x	y
1	1	0	0
2	2	4.0	0
3	3	8.0	0
4	4	2.0	3.464
5	5	6.0	3.464

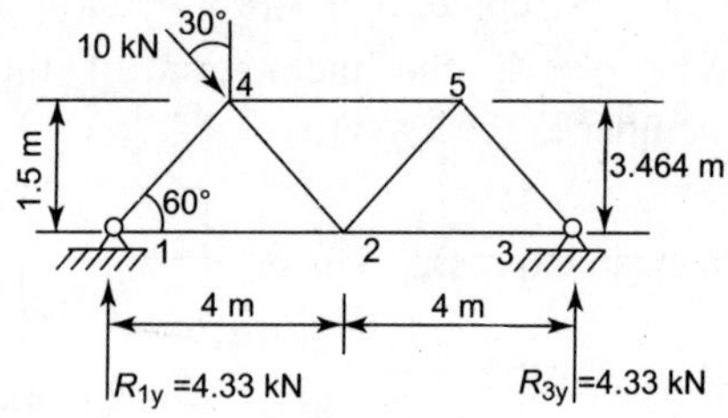

Fig. 3.21

Step 2: The member parameters x_{ij}, y_{ij} and l_{ij} are computed from the coordinates above and tabulated in Table 3.2.

Table 3.2 | *Member parameters*

Sl.No.	Member	$x_{ij} = x_i - x_i$	$y_{ij} = y_j - y_i$	$\frac{l_{ij}}{\sqrt{x_{ij}^2 + y_{ij}^2}}$	t_{ij}	T_{ij}
1	1–2	4.0	0	4.0	1.875	7.5
2	2–3	4.0	0	4.0	0.625	2.5
3	4–5	4.0	0	4.0	−1.25	−5.0
4	1–4	2.0	3.464	4.0	−1.25	-5.0
5	2–4	−2.0	3.464	4.0	−1.25	−5.0
6	2–5	2.0	3.464	4.0	+1.25	5.0
7	3–5	−2.0	3.464	4.0	−1.25	−5.0

Step 3: Resolving the external force p

$$P_x = 5.0, p_y = -8.66 \text{ kN}.$$

Step 4: As in method of joints, we proceed from joint to joint with two unknowns. We can start either from joint 1 or joint 3.

Considering joint 3 and writing

$$\Sigma F_y = t_{3-2}\, y_{3-2} + t_{3-5}\, y_{3-5} + 4.33 = 0$$

Substituting for y values, we get

$$t_{32}\,(0) + t_{35}\,(3.464) + 4.33 = 0$$

or $$t_{35} = -1.25 \text{ kN}$$

Again writing $\Sigma F_x = 0$

$$t_{3-2}\,(x_{32}) + t_{35}\, x_{35} = 0$$

or $$t_{3-2}\,(-4.0) - 1.25\,(-2.0) = 0$$

$$-4\, t_{32} = -2.50$$

$$t_{32} = +\,0.625$$

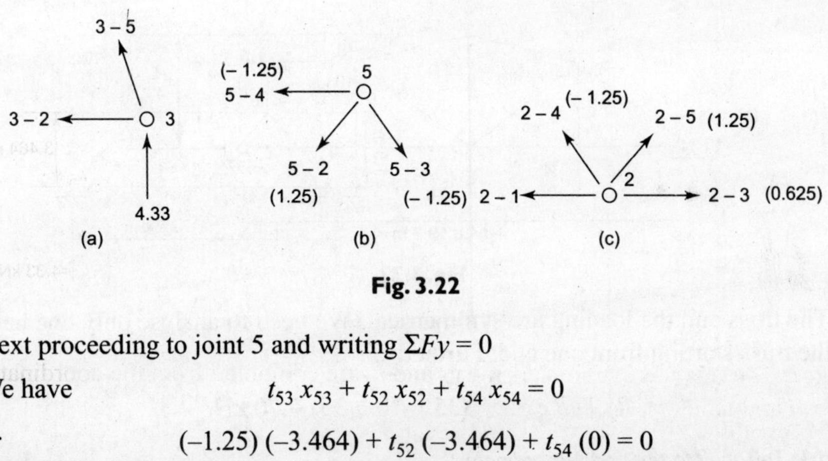

Fig. 3.22

Next proceeding to joint 5 and writing $\Sigma Fy = 0$

We have $$t_{53}\, x_{53} + t_{52}\, x_{52} + t_{54}\, x_{54} = 0$$

or $$(-1.25)\,(-3.464) + t_{52}\,(-3.464) + t_{54}\,(0) = 0$$

Gives $$t_{52} = +\,1.25$$

Writing $\Sigma F_x = 0$ we hve

$$t_{54}\, x_{54} + t_{53}\, x_{52} + t_{53}\, x_{53} = 0$$

$$t_{54}\,(-4.0) + 1.25\,(-2.0) - 1.25\,(+2.0) = 0$$

gives $$t_{54} = -1.25$$

Next proceeding to joint 2 and writing $\Sigma F_y = 0$

$$t_{24}\, y_{24} + t_{25}\, y_{25} = 0$$

$$t_{24}(3.464) + (1.25)\,(3.464) = 0$$

Gives $$t_{24} = -1.25$$

Next proceeding to joint 1 and writing $\Sigma F_y = 0$

$$t_{14}\,y_{14} + 4.33 = 0$$

$$t_{14}(3.464) = -4.33$$

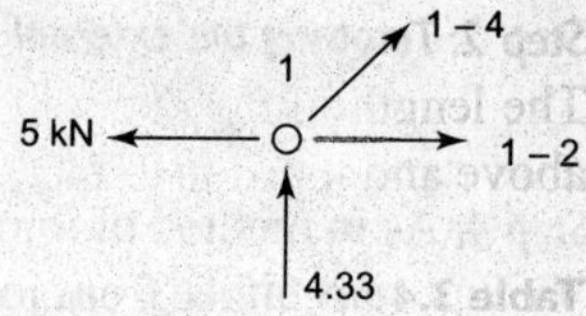

Fig. 3.22 (d)

gives $t_{14} = -1.25$ kN

A check at joint 4 satisfies $\Sigma F_x = 0$ and $\Sigma F_y = 0$

$$\Sigma F_y = -8.66 + (-1.25)(-3.464) + (-1.25)(-3.464)$$

$$= -8.66 + 4.33 + 4.33 = 0$$

Also $\Sigma F_x = 5.0 + (-1.25)(4) + (-1.25)(2)(-1.25)(-2)$

$$= 5.0 - 5.0 - 2.5 + 2.5 = 0$$

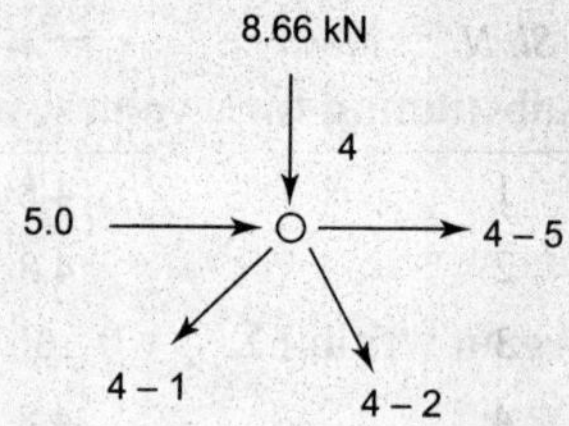

Fig. 3.22 (e)

Let us take up one more example of a truss having non parallel top ad bottom chords to further reinforce the procedure.

Example 3.9 | *Analyse the truss in Fig. 3.23 for the member forces using tension coefficients method.*

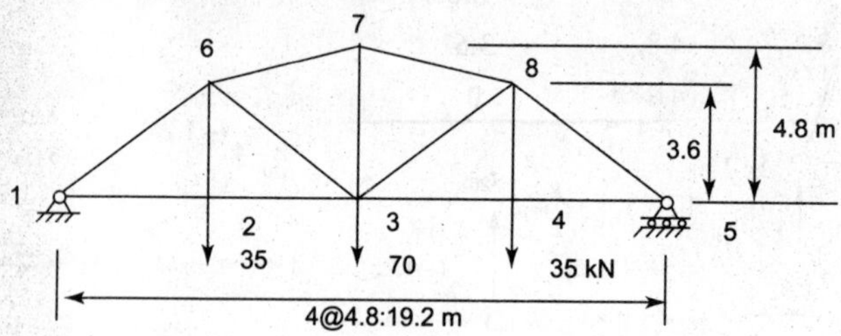

Fig. 3.23

The truss and the loading are symmetrical. We need to analyse only one half of the truss starting from one end. Further.

$$R_{1y} = R_{5y} = \frac{1}{2}(35 + 70 + 35) = 70 \text{ kN}$$

Step 1: To tabulate the nodal coordinates.

The joints are numbered. The nodes and the coordinates are tabulated taking the origin at joint 1.

Table 3.3

Sl No.	*Node*	*x*	*y*
1	1	0	0
2	2	4.8	0
3	3	9.6	0
4	4	14.4	0
5	6	4.8	3.6
6	7	9.6	4.8
7	8	14.4	3.6

Step 2: To evaluate member parameters

The length of members of the truss are computed from the coordinates in Table above and listed in Table 3.4

Table 3.4 | *Member parameters*

Sl. N.	*Member*	$x_{ij} = x_j - x_i$	$y_{ij} = y_j - y_i$	$\frac{l_{ij}}{\sqrt{x_{ij}^2 + y_{ij}^2}}$	t_{ij}	$I = t_{ij} \times l_{ij}$
1	1–2	4.8	0	4.8	15.97	76.66
2	1–6	4.8	3.6	6.0	–15.97	–95.83
3	2–6	0	3.6	3.6	9.72	35.0
4	2–3	4.8	0	4.8	15.97	76.66
5	6–7	4.8	1.2	4.95	–7.29	–36.0
6	6–3	4.8	–3.6	6.0	–8.68	–43.74
7	7–8	4.8	–1.2	4.95	–7.29	–36.0
8	7–3	0	–4.8	4.8	3.645	17.50
9	3–8	4.8	3.6	6.0	–	–
10	3–4	4.8	0	4.8	–	–

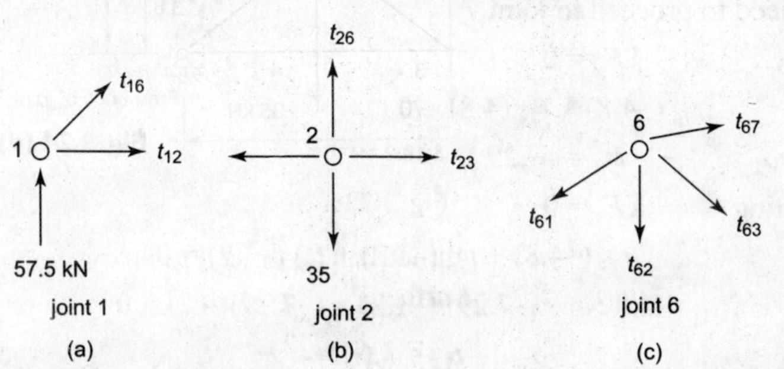

Fig. 3.24 | *Free-body diagrams*

Step 3: To consider equilibrium of joints

We can consider joint 1 where there are only two unknown bar forces. Writing $\Sigma F_y = 0$

$$t_{16}(3.6) + 57.5 = 0$$

Gives $\quad t_{16} = -15.97$

Again writing $\quad \Sigma F_x = 0$

$$t_{16}\,(4.8) + t_{12}\,(4.8) = 0$$

Gives $\quad t_{12} = -t_{16} = 15.97$

Next taking up joint 2 with two unknown bar forces

We can write $\quad \Sigma F_y = 0$

$$t_{26}\,(3.6) - 35 = 0$$

gives $\qquad t_{26} = 9.72$

Again writing $\qquad \Sigma F_x = 0$

$$t_{23}(4.8) + t_{21}\,(-4.8) = 0$$

$$t_{23} = t_{21} = 15.97$$

Proceeding to joint 6 next, we write $\Sigma F_x = 0$

$$t_{67}(4.8) + t_{63}(4.8) + t_{62}(0) + t_{61}(-4.8) = 0$$

gives $\qquad 4.8\,t_{67} + 4.8\,t_{63} = -76.66 \qquad$ (a)

We can formulate another equation by writing $\Sigma F_y = 0$

$$t_{67}(1.2) + t_{63}\,(-3\text{–}6) + t_{62}\,(-3.6) + t_{61}\,(-3.6) = 0$$

Substituting for $t_{62} = 9.72$ and $t_{61} = -15.97$

We get $\qquad 1.2\,t_{67} - 9.72$ and $t_{61} = -15.97$

We get $\qquad 1.2\,t_{67} - 3.6\,t_{63} = 22.50 \qquad$ (b)

Solving simultaneous equations (a) and (b)

We get $\qquad t_{63} = -8.68$

and $\qquad t_{67} = -7.29$

Next we need to proceed to joint 7

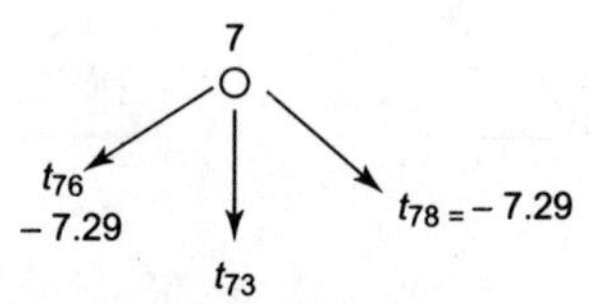

Free body of joint 7

Fig. 3.24 (d)

Writing $\qquad \Sigma F_x = 0$

We have $\qquad t_{76}\,(-4.8) + t_{78}\,(4.8) = 0$

Gives $\qquad t_{78} = -7.29$

Again writing $\qquad \Sigma F_y = 0$

$$t_{73}\,(-4.8) + t_{76}\,(-1.2) + t_{78}\,(-1.2) = 0$$

$$-4.8\,t_{73} + (-7.29)\,(-1.2) + (-7.29)(-.2) = 0$$

$$t_{73} = 3.645$$

All the tension coefficients are entered in Table 3.4 and the bars forces are worked out

3.5 COMPOUND AND COMPLEX TRUSSES

We shall now consider the stability considerations of other types of trusses which are statically determinate on the basis of criteria established in Sec. 3.3.

Simple trusses (composed of triangular panels) are always stable if supported in a suitable manner. If two simple trusses are connected with a set of bars or pin connections which provide non-concurrent, non-parallel reactive components to each simple truss, then the system is stable. Such a system is termed as a *compound truss.* Its identification is best performed by identifying the simple trusses as individual units and then identifying the bars that provide the proper connections. The reaction components must of course be non-concurrent and non-parallel. Figure 3.25 shows a compound truss.

Another type of truss which cannot be classified either as simple or compound is the *complex truss.* The truss shown in Fig. 3.26 is a complex truss. One identifying mark of a complex truss is that there is no joint where only two bars meet although the truss is statically determinate. Complex trusses are not often used. A more general method is needed to verify the stability of such trusses. Complex trusses for which $n = 0$ may be analysed for the presence of unstable or critical forms by the *zero load test.*

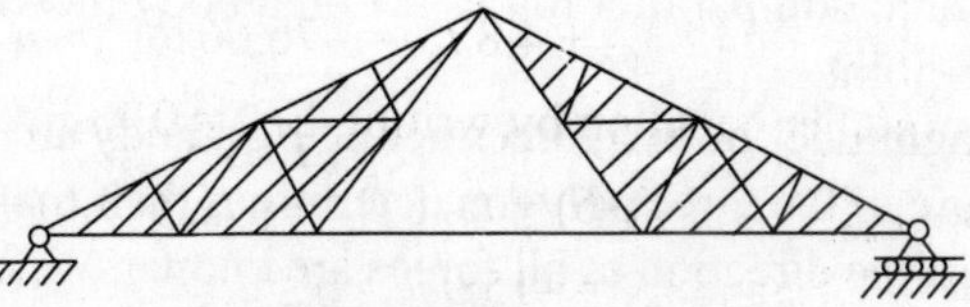

Fig. 3.25 | *Compound truss*

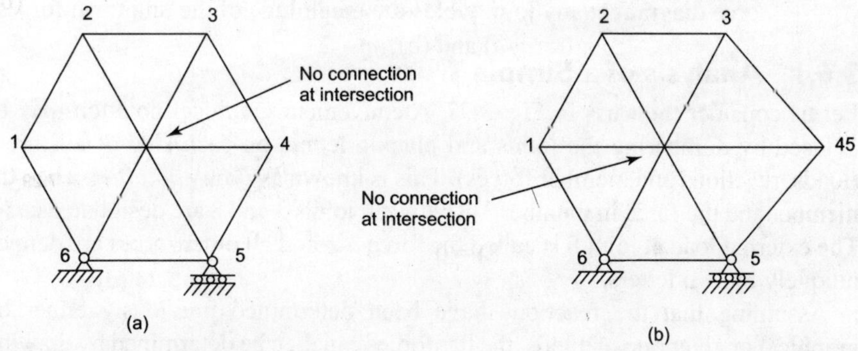

Fig. 3.26 | *Complex trusses, (a) Unstable, (b) Stable*

Zero Load Test The zero load test is simple in application. Consider the structure with no applied loads. Assume a force in a member caused by a turnbuckle arrangement and apply the rules of equilibrium to successive joints. If equilibrium can be established without developing any external reactions we have obtained a non-zero solution. It means that this set of internal forces obtained from zero load condition can be multiplied by an arbitrary constant which gives another equilibrium solution. The existence of more than one solution indicates that the structure is unstable. For example, we shall apply the zero load test for truss in Fig. 3.26a. For zero external loads the reaction components at 5 and 6 are zero. Assume now 1 kN tensile force in member 1-4. Equilibrium of joint 1 indicates 1 kN compressive force in each of the members 1-2 and 1-6. Working on joint 2 it will be seen that member 2-5 will have 1 kN tensile force, and member 2-3 will have 1 kN compressive force. Working further on joints 3, 4, 5 and 6, we find that every joint is in equilibrium with unit tension and the internal bars and force in member 1-4 would have satisfied the equilibrium condition of the structure without developing any external reaction. Since the truss is statically determinate and there exist more than one solution, we conclude that the truss is unstable. It may be of interest to note that the truss of Fig. 3.26*b* is a stable one.

3.6 GRAPHICAL ANALYSIS OF TRUSSES

The member forces in a statically determinate truss can be determined by graphical analysis. Graphical analysis is based on two facts:

1. If only three non-parallel forces act on a body they must pass through a common point.
2. If the magnitudes of two forces acting on a body are the only unknowns, the closure of the force polygon determines their magnitudes. In the case of trusses, the direction of all forces are known. We examine free-bodies of joints that have not more than two unknown forces acting on them, as in the analytical application of the method of joints. Completion of the force diagram at any joint yields the magnitude of the unknown forces.

3.6.1 Analysis of a Simple Truss

Let us consider the truss in Fig. 3.27. A convenient graphical notation may be devised by numbering the joints and placing letters on each side of all forces (loads, reactions and member forces). This is known as *Bow's notation.* Then the member and the force in member 3-8 between joints 3 and 8 are designated as *i-j.* The external load at joint 8 is called the force *a-b* and all other forces are defined uniquely by two letters.

Assuming that the reactions have been determined previously either by graphical or algebraic methods, the bar forces can then be determined by drawing a series of force polygons, one for each joint. The solution begins at joint 1 where there are only two unknown forces.

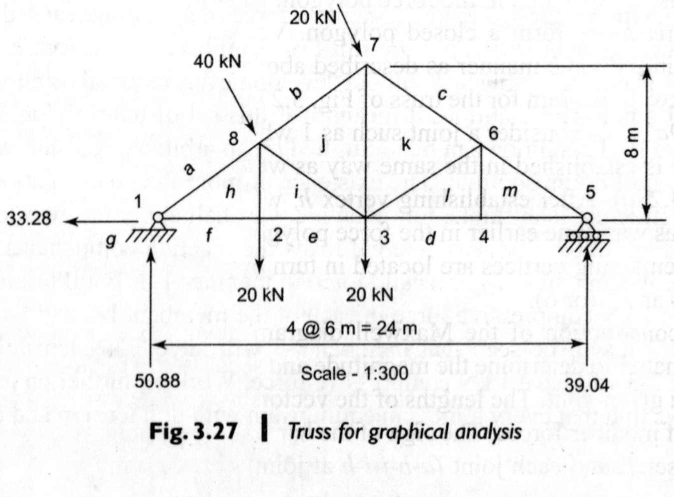

Fig. 3.27 | *Truss for graphical analysis*

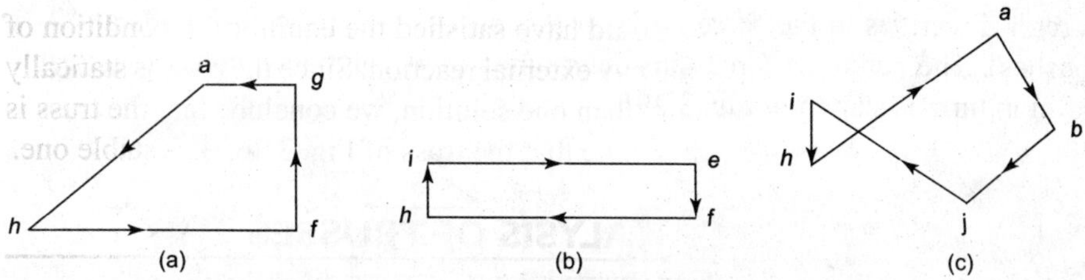

Fig. 3.28 | *Force polygons: (a) Joint 1, (b) Joint 2, (c) Joint 8*

The known reactions *f-g* and *g-a* are drawn first as shown in Fig. 3.28*a*. The directions of forces *a-h* and *f-h* are known and it is a simple matter to plot their directions to locate point *h* and thus obtain the magnitude of two forces. Note that we have proceeded clockwise around joint 1 in plotting the forces.

The forces in members *a-h* and *h-f* are measured by the vector *a-h* and *h-f* in the force polygon. The force *a-h* pushes the joint and force *h-f* pulls the joint indicating that the nature of forces in *a-h* is compression and that in *h-f* is tension. Joint 2 should be analysed next as there are more than two unknowns at joint 8. The force polygon for joint 2 is shown in Fig. 3.28*b*. Having found the force in *i-h* and *h-a* it is now possible to proceed to joint 8. The force polygon for joint 8 is shown in Fig. 3.28*c*.

Considering the remaining joints in turn, the analysis of the truss can be completed. It may be noticed that several bar forces such *as f-h, h-a, i-h,* etc. are plotted twice in Fig. 3.28. This is because we use previously determined bar forces in the successive construction of force polygons. Instead of drawing separate force polygons for each joint, it is convenient to combine all the force polygons into a single construction known as the *Maxwell diagram.*

To construct the Maxwell diagram, first draw a force polygon for all the external forces, laying out the vectors in the same order as the forces are encountered in going round the structure, say, in a clockwise direction. The reaction components should also be included in the force polygon. Remember that the external forces and the reactions form a closed polygon. Vertices of this polygon should be labelled in the same manner as described above for the force polygon of joints. The Maxwell diagram for the truss of Fig. 3.27 drawn in this manner is shown in Fig. 3.29*a*. Now consider a joint such as 1 where there are only two unknowns. Vertex *h* is established in the same way as we did in the force polygon of joint 1 (Fig. 3.28*a*). After establishing vertex *h,* we proceed to joint 2 and establish vertex *i* as was done earlier in the force polygon of joint 2.

The remaining vertices are located in turn by considering successively joints 8, 7, 3, 4 and 5 (or 6).

The construction of the Maxwell diagram having been completed, it is a simple matter to determine the magnitude and sense of the force with which a bar acts on a given joint. The lengths of the vectors in the Maxwell diagram give the values of member forces. The sign of the bar forces is determined by proceeding clockwise around each joint *(a-b-j-i-h* at joint 8) and noting the corresponding vector directions in the force polygon *(b-j* pushes, *j-i* pushes, *i-h* pulls and *h-a* pushes). The nature of force thus determined in all members, considering each joint in turn, is shown in Fig. 3.29*b*.

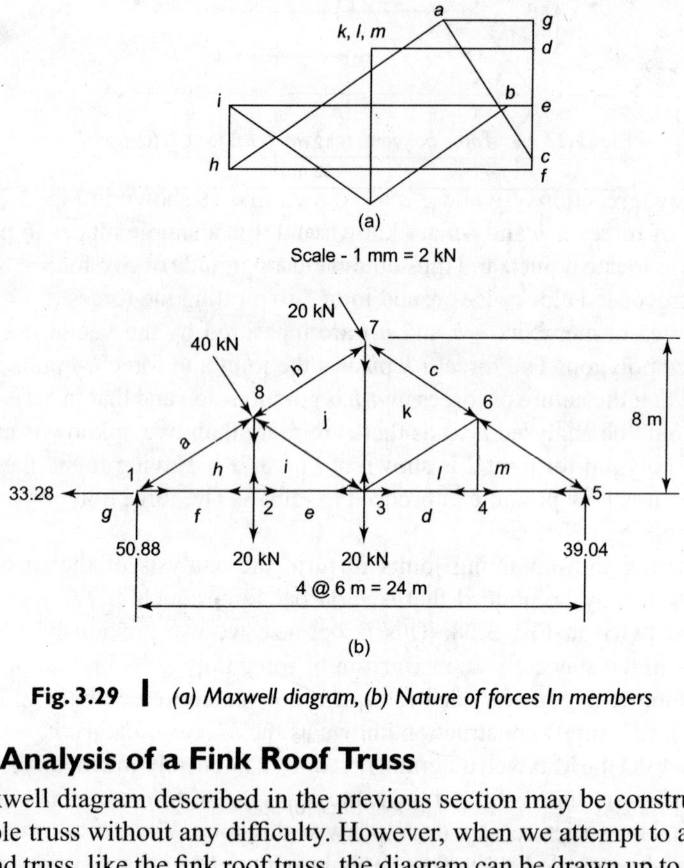

Fig. 3.29 | *(a) Maxwell diagram, (b) Nature of forces In members*

3.6.2 Analysis of a Fink Roof Truss

The Maxwell diagram described in the previous section may be constructed for any simple truss without any difficulty. However, when we attempt to analyse a compound truss, like the fink roof truss, the diagram can be drawn up to a certain point and we cannot proceed further since at each of the remaining joints there exist more than two unknowns. Consider a fink roof truss shown in Fig. 3.30a. After finding the reactions either analytically or graphically, a force polygon for external forces may be laid out and the Maxwell diagram started in the conventional manner. Starting from joint *A* one can proceed to joint *B* and then to joint *C*. Thereafter, we find, the joints at *D* or *E* contain three unknowns and, therefore, it is impossible to continue with the Maxwell diagram. We face the same problem even if we commence from the right-hand support.

We may now consider one of the methods available to overcome this difficulty. In this method, we temporarily replace bars *D-F* and *F-G* (bars 4-5 and 5-6) by a substitute bar *E-G* indicated by a dotted line. We may designate the space enclosed by triangle *EDG* by 4′ and the space enclosed by triangle EGH by 6′. This substitution does not affect the stability of the truss, or alter the forces in the

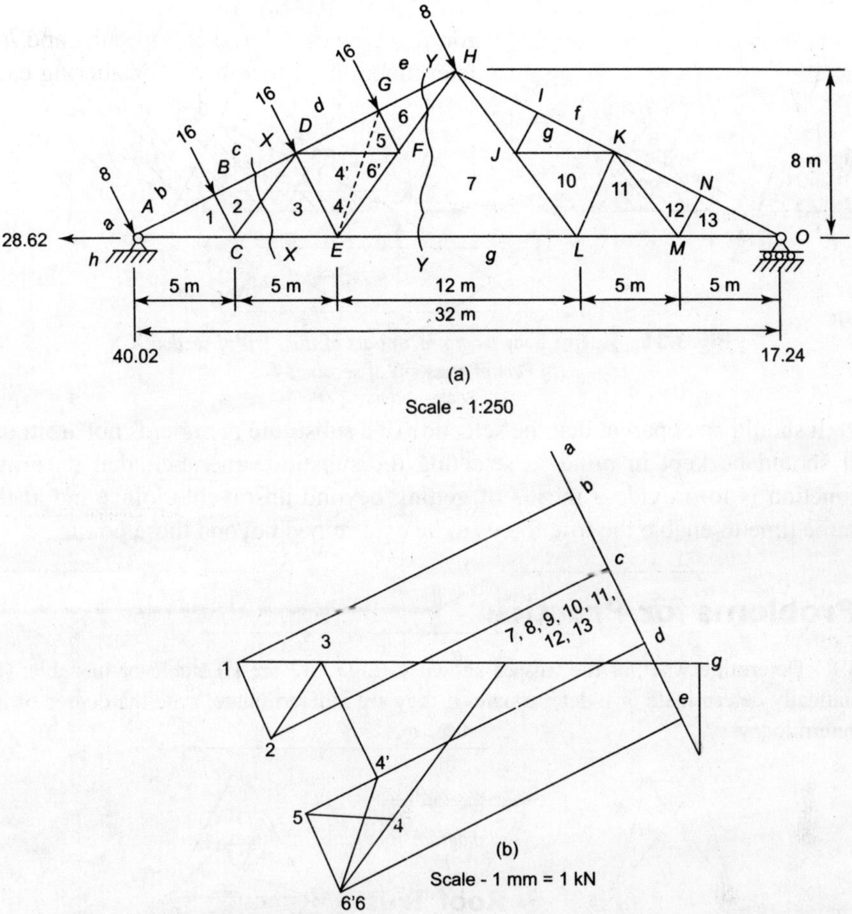

Fig. 3.30 | *(a) Analysis of fink roof truss; (a) Space diagram, (b) Maxwell diagram*

members outside the panel *DGEF*. This is evident when we take a section along *X-X* (Fig. 3.30*a*) and consider the free-body diagram of the left part of the truss for computation of forces in members *B-D*, *C-D* and *C-E*. This is also true of forces in members *G-H*, *F-H* and *E-L* when a section is taken along *Y-Y* (Fig. 3.30*a*) and a free-body diagram of the left part of the truss (Fig. 3.31) is considered. The locations of vertices 1, 2 and 3 in the Maxwell diagram, therefore, remain the same as for the original truss. It is thus possible to locate vertex 4′ of the substitute bar truss by considering joint *D* and then proceed to joint *G* to locate vertex 6′. The location of 6′ so determined for the substitute bar truss coincides with 6 of original truss, since in either case the forces in bars *G-H* and *F-H* are the same. It is now possible to return to the original truss by considering joints *G* and *D* in turn to locate the correct positions of vertices 5 and 4. It is now easy to proceed in the usual manner and locate the remaining vertices. This procedure is illustrated in the Maxwell diagram given in Fig. 3.30*b*.

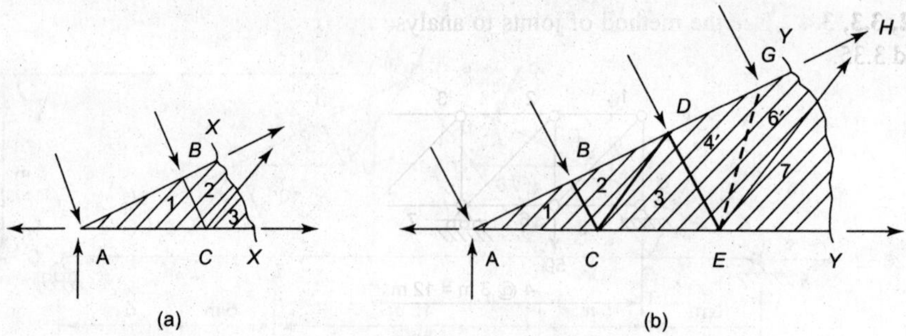

Fig. 3.31 | *Free-body diagram: (a) part of truss left of section X-X, (b) Part of truss left of section Y-Y*

It should be apparent that the selection of a substitute member is not arbitrary. It should be kept in mind in selecting the substitute member, that its prime function is to provide a means of getting beyond unsolvable joints and at the same time to enable the true forces to be determined beyond these points.

Problems for Practice

3.1 Determine whether the trusses shown in Fig. 3.32 are (i) stable or unstable; (ii) statically determinate or indeterminate. If they are indeterminate, state the degree of indeterminacy.

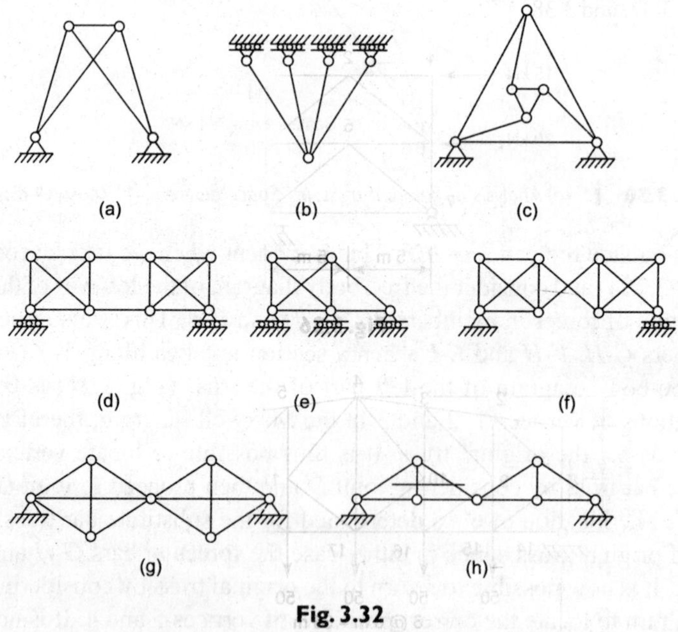

Fig. 3.32

3.2, 3.3, 3.4 Use the method of joints to analyse the trusses shown in Figs. 3.33, 3.34 and 3.35.

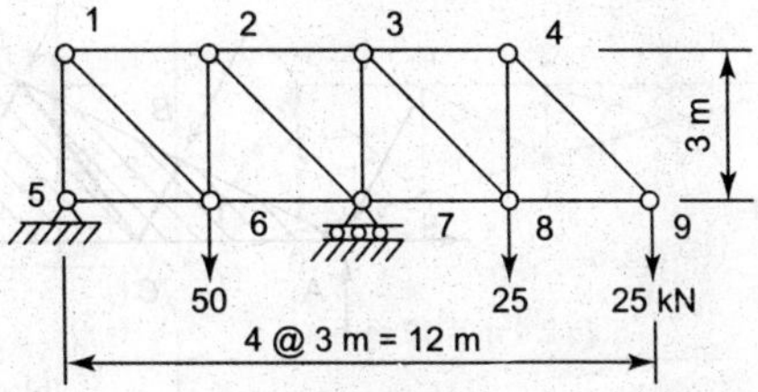

Fig. 3.33

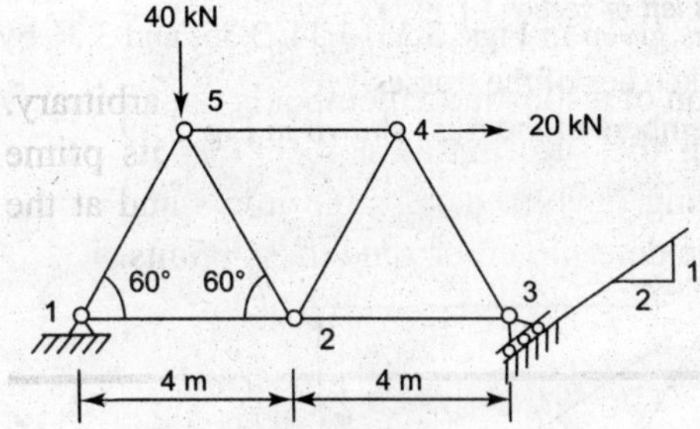

Fig. 3.34

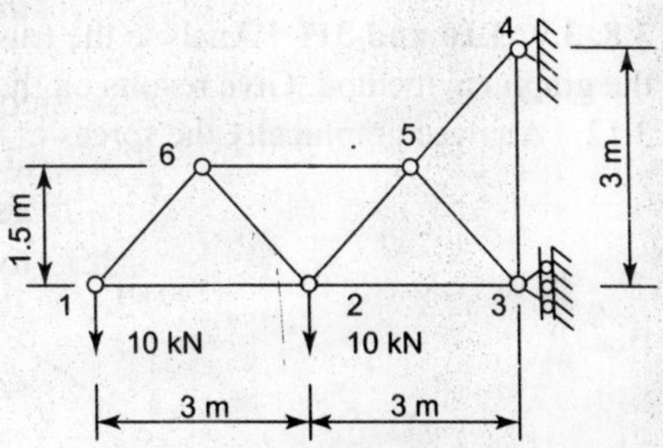

Fig. 3.35

3.5, 3.6, 3.7 Determine the bar forces in the indicated members of the trusses shown in Figs. 3.36, 3.37, and 3.38.

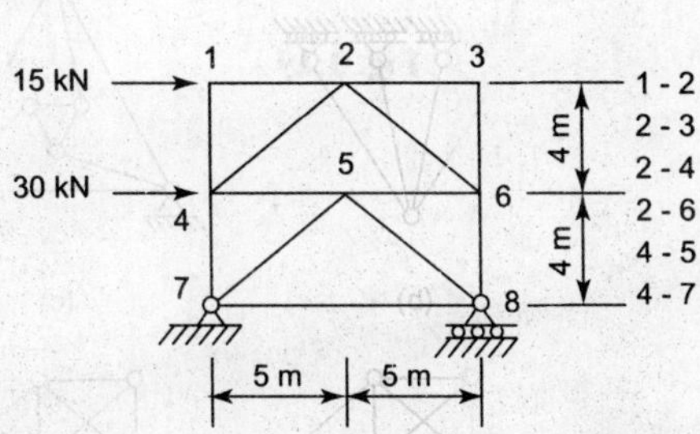

Fig. 3.36

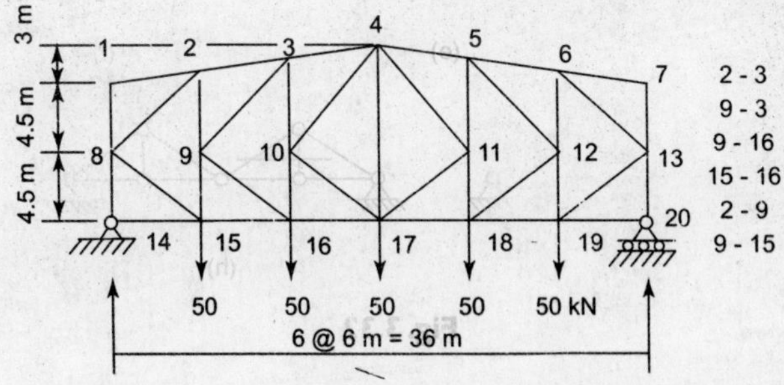

Fig. 3.37

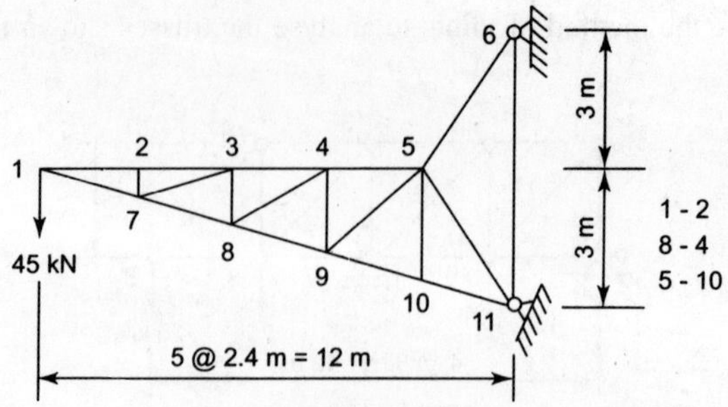

Fig. 3.38

3.8, 3.9, 3.10 and 311 Analyse the trusses given in Figs. 3,33, 3.34, 3.35, and 3.36 by the graphical method. Give results on the sketches of the trusses.

3.12 Analyse graphically the forces in members of the truss shown in Fig. 3.39.

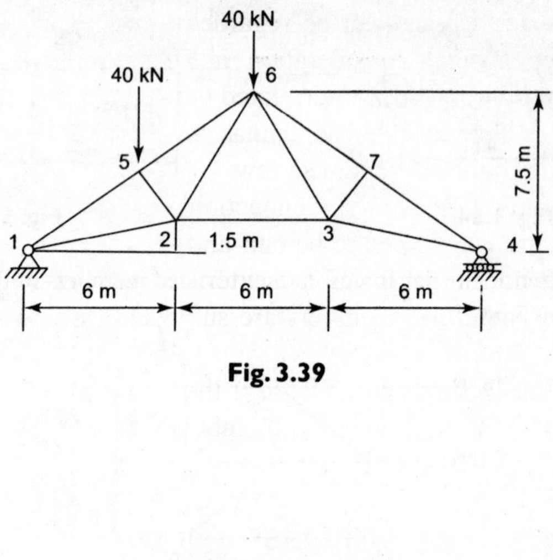

Fig. 3.39

4 Space Trusses

4.1 INTRODUCTION

The analysis of space trusses can be regarded as an extension of plane truss analysis. However, special consideration is necessary in view of the third dimension involved. Stresses are interrelated between members not lying in one plane. Space trusses include towers, antennas, guyed masts, derricks, framing for domes, aircraft framing, etc., to name a few.

To simplify computations, the connections of the members in a three-dimensional truss are considered to be ball-and-socket joints. It is assumed that the joints transmit no moments and the external loads on the truss are applied only at the joints. Therefore, members are subjected only to axial forces as in planar trusses.

An idealized truss is termed *just rigid* if the removal of any of its members destroys its rigidity. If the removal of a member does not destroy its rigidity, the truss is said to be *over rigid*.

4.2 SIMPLE SPACE TRUSS

The basic element of a space truss which is just rigid is a tetrahedron with four joints. An additional joint can be formed by adding three more members as shown in Fig. 4.1. Any number of rigid joints can be formed by adding three members for each additional joint. Such trusses, if suitably supported, are internally rigid and statically determinate.

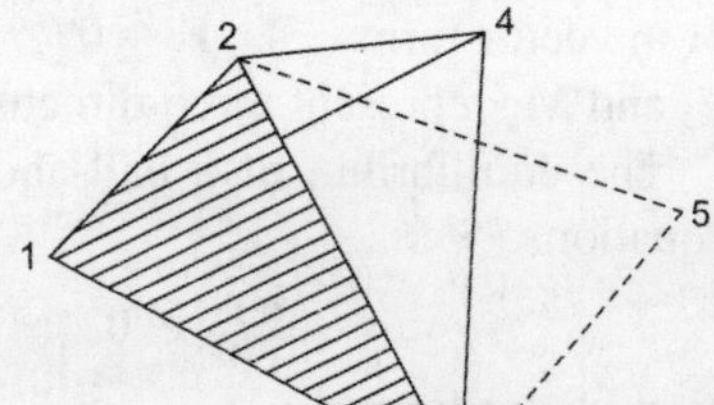

Fig. 4.1 | *Rigid space truss*

4.3 TYPES OF SUPPORTS

There are various forms of supports for space trusses. Three common types of supports and the corresponding components of reaction that are possible are shown in Fig. 4.2. The ball-and-socket support prevents the movement of the support in each of the three directions. The three component reactions are shown in Fig. 4.2*a* in their positive direction. The roller support in Fig. 4.2*b* prevents the movement of the support in two directions; movement is permitted in the third direction. The support condition in the *YZ* plane is comparable to a pinned support in a planar truss and the support condition in the *XY* plane is comparable to a roller support. The roller lies in the *XZ* plane with the axis parallel to *Z*. Other combinations of support are also possible for a roller support. For example, R_{iz} can be zero (axis of roller parallel to *X* axis) but R_{ix} and R_{iy} may exist.

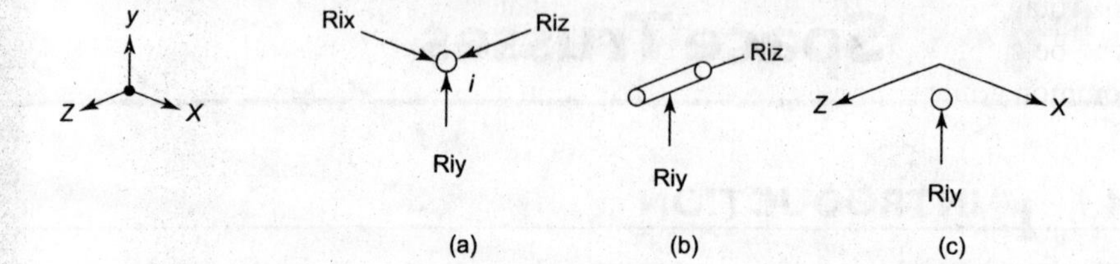

Fig. 4.2 | *Types of supports: (a) Ball-and-socket support, (b) Roller support, (c) Ball support*

The ball support in Fig. 4.2*c* prevents movement only in one direction, that is, the *Y* direction for the support shown. Movement is permitted in the other two directions, that is, the *X* and *Z* directions.

4.4 EQUILIBRIUM AND STABILITY CONDITIONS

The equilibrium of an entire space truss or sections of a space truss is described by the six scalar equations given below.

$$\Sigma F_X = 0 \qquad \Sigma M_X = 0$$
$$\Sigma F_Y = 0 \qquad \Sigma M_Y = 0$$
$$\Sigma F_Z = 0 \qquad \Sigma M_Z = 0$$

or in vector form $\qquad \mathbf{F}_R = 0 \qquad \mathbf{M}_R = 0 \qquad (4.1)$

$\mathbf{F}_R$ and $\mathbf{M}_R$ represent three-dimensional force and moment vectors.

The equilibrium of a ball-and-socket joint is described by the three scalar equations

$$\Sigma F_X = 0 \qquad \Sigma F_Y = 0 \qquad \Sigma F_Z = 0$$

or in vector from

$$\mathbf{F}_R = 0 \tag{4.2}$$

The space truss may be thought of as a structural device which contains *j* ball-and-socket joints connected in space. The forces that act on the joints are the member forces, external forces and reaction components. Thus, from the three equations of equilibrium at each joint, we can write 3*j* equilibrium equations for the entire truss. Denoting the number of members as *m,* and the number of reaction components as *r,* the necessary condition for the space truss to be statically determinate is

$$3j = m + r \tag{4.3}$$

If $3j < m + r$, the truss is statically indeterminate, that is, the number of unknowns is more than the number of equations of equilibrium.

If $3j > m + r$, the structure is statically unstable.

However, it should be remembered that a simple count alone does not prove its stability. While satisfying the count, the structure may still be statically or geometrically unstable if the bars of reaction components are not properly arranged. The reactions must be so placed that they can resist translation along and rotation about each of the three coordinate axes, if a three-dimensional truss is to be stable. For example, the reaction components in Fig. 4.3 pass through a common point *o;* they cannot resist rotation about the *Y* axis passing through *o*.

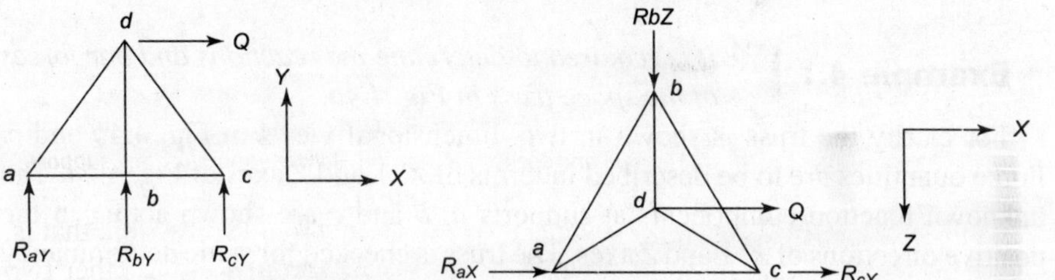

Fig. 4.3 | *Unstable support conditions*

4.5 ANALYSIS OF SPACE TRUSSES

The analysis of space trusses can be carried out by using a combination of the method of sections and method of joints.

Certain generalized theorems can also be utilized in finding bar forces. However, the analysis by these methods is tedious because of the three dimensional nature of the problem. The method of tension coefficients, being systematic, is better suited to space truss analysis than other methods. The concept of tension coefficient method to space trusses is extended by incorporating the force components in the *z* direction. The equilibrium equations in the *z* direction can be written as $\Sigma t_{ij} z_{ij} + z_i = 0$. A couple of examples at the end are presented to illustrate the procedure involved

Where it is possible, the reactions may be determined from a consideration of the equilibrium of the entire structure. Once the reactions are known, any joint with three or less unknown bar forces can be determined using the three equilibrium equations of the joint. Likewise, the joints can be solved in succession.

Sometimes it may not be possible to determine the reactions beforehand. In such cases, some of the bar forces can be determined first and then the reaction components. Often, member forces and reaction components for a given space truss can be determined with less computational effort if conditions of forces at a joint are recognised. For certain conditions of loading at a joint in a space truss, and for certain arrangement of members meeting at a joint, the forces in the bars can be determined by observation.

The following two theorems are important as their application often results in an appreciable saving in computations.

THEOREM 1 *If all the bars meeting at a joint with the exception of one bar, n, lie in a plane, the component normal to that plane of the force in the bar n is equal to the component normal to that plane of any external load applied at that joint. If no external load is applied at the joint, the force in bar n is zero.*

THEOREM 2 *If all but two bars at a joint have no bar forces and these two are not collinear, and if no external load acts at that joint, the bar force in each of these two bars is zero.*

These two theorems can be justified from a consideration of the equilibrium of a joint under the conditions stated. The usefulness of the two theorems will be seen in the examples illustrated below.

Example 4.1 | *It is required to determine the reactions and bar forces of the space truss in Fig. 4.4a.*

For clarity the truss is shown in two-dimensional views in Fig. 4.4*b* and *c*. Force quantities are to be described in terms of *X*, *Y* and *Z* axes of Fig. 4.4*a*. The unknown reaction components at supports *a*, *b* and *c* are shown acting in the positive directions of *X*, *Y* and *Z* axes. The truss is checked for static determinancy by Eq. 4.3. For this truss $j = 4$, $m = 6$ and $r = 6$, which indicates that the truss is statically determinate.

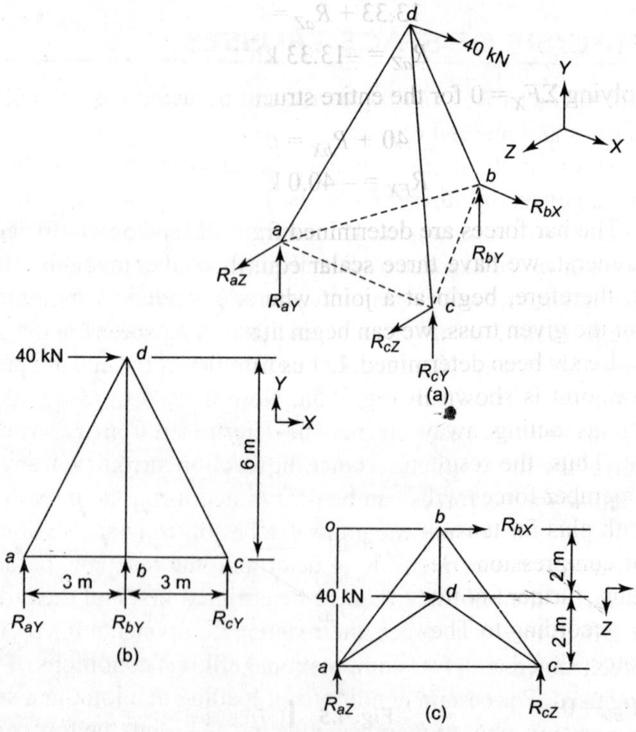

Fig. 4.4 | *Truss for analysis*

External Reactions In this case the external reactions are only six in number and can be obtained without determining the bar forces. In obtaining the reaction components, we look for an equilibrium equation that involves only one unknown at a time. Let us determine the vertical reactions first.

On applying $\Sigma M_X = 0$ about axis *ac,* we notice that R_{bY} is the only external force that could have a moment and hence $R_{bY} = 0$.

On applying $\Sigma M_z = 0$ about the line of action of R_{az}

$$-40(6) + R_{cY}(6) = 0$$

or

$$R_{CY} = 40.0 \text{ kN}$$

On applying $\Sigma F_Y = 0$ to the entire structure

$$\Sigma_{aY} + R_{cY} = 0 \; \Sigma R_{aY} = -40.0 \text{ kN}$$

To determine the horizontal reactions, if we apply $\Sigma M_Y = 0$ about the Y axis passing through the intersection of any two of the horizontal reactions, the third horizontal reaction will be the only unknown occurring in the resulting equation. Taking moments, for example, about point 0, the point of intersection of R_{az} and R_{bx} *(see* Fig. 4.4c), we have

$$40(2) - R_{eZ}(6) = 0$$

or

$$R_{cZ} = 13.33 \text{ kN}$$

Applying $\Sigma F_Z = 0$ again for the entire structure, we have

$$13.33 + R_{aZ} = 0$$

or

$$R_{aZ} = -13.33 \text{ kN}$$

Finally applying $\Sigma F_X = 0$ for the entire structure, we get

$$40 + R_{bX} = 0$$

or

$$R_{FX} = -40.0 \text{ kN}$$

Bar Forces The bar forces are determined from the equilibrium consideration of joints. In general, we have three scalar equilibrium equations for each joint and we must, therefore, begin at a joint where there are not more than three unknowns. For the given truss, we can begin at any joint since the reaction components have already been determined. Let us first analyse joint *d.* The free-body diagram of the joint is shown in Fig. 4.5a. Note that the unknown bar forces are all shown as acting away from the joint. Thus, the resulting signs for member forces will correspond with plus for tension and minus for compression. Before proceeding further, notice that at joint *d,* according to Theorem 1, bar force

$$P_{db} = 0$$

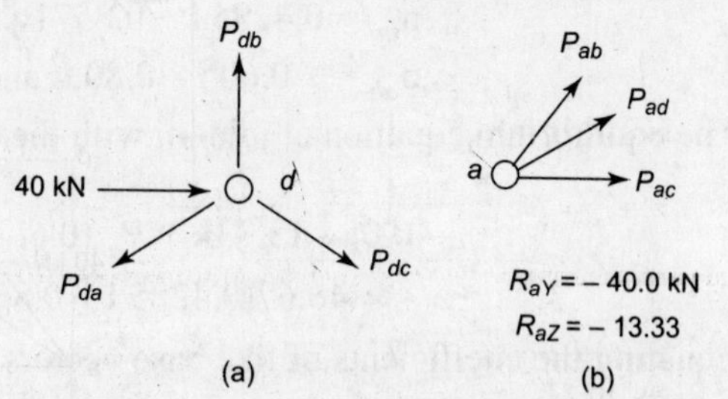

Fig. 4.5 | *Free-body diagrams: (a) Joint d, (b) Joint a*

The unit vectors along d_a and d_c are:

$$n_{da} = \frac{-3\mathbf{i}-6\mathbf{j}+2\mathbf{k}}{\sqrt{3^2+6^2+2^2}} = -0.4285\mathbf{i} - 0.8571\,\mathbf{j} + 0.2857\mathbf{k}$$

$$n_{dc} = \frac{3\mathbf{i}-6\mathbf{j}+2\mathbf{k}}{\sqrt{3^2+6^2+2^2}} = 04285\mathbf{i} - 08571\,\mathbf{j} + 0.2857\,\mathbf{k}$$

where **i**, **j** and **k** represent the unit base vectors in X, Y and Z directions respectively, (see Appendix-A for theory of vectors.) The vector equation for equilibrium of joint d is

$$40\mathbf{i} + P_{da}(-0.4285\mathbf{i} - 0.8571\,\mathbf{j} + 0.2857\,\mathbf{k})$$
$$+ P_{dc}(+0.4285\,\mathbf{i} - 0.8571\,\mathbf{j} + 0.2857\,\mathbf{k}) = 0$$

Equating the coefficients of base vectors **i** and **j** to zero, we have

$$40.0 - 0.4285\,P_{da} + 0.4285\,P_{dc} = 0$$
$$-0.8571\,P_{da} - 0.8571\,P_{dc} = 0$$

Solving the two equations simultaneously we obtain

$$P_{da} = +46.67 \text{ kN (tension)}$$
$$P_{dc} = -46.67 \text{ kN (compression)}$$

In this particular structure, because we know the vertical reactions at a, b and c, the bar forces P_{ad}, P_{bd} and P_{cd} can be computed more easily by considering $\Sigma F_Y = 0$ at each joint. For example, at joint a, the vertical reaction $R_{aY} = -40.0$ kN or, expressing this in the vector form, $R_{aY} = -40j$. The unit vector,

$$\mathbf{n}_{ad} = 0.4285\,\mathbf{i} + 0.8571\,\mathbf{j} - 0.2851\,\mathbf{k}$$

Equating the coefficients of **j** we have

$$-40\,\mathbf{j} + P_{ad}\,(0.8571) = 0$$

or $\quad P_{ad} = 46.67$ kN; same as the previous result.

Next consider joint a. The free-body diagram of joint a is shown in Fig. 4.5b. The unit vectors are

$$\mathbf{n}_{ac} = 1.0\,\mathbf{i}$$
$$\mathbf{n}_{ad} = 0.4285\,\mathbf{i} + 0.8571\,\mathbf{j} - 0.2851\,\mathbf{k}$$
$$\mathbf{n}_{ab} = +0.60\,\mathbf{i} - 0.80\,\mathbf{k} \text{ and } R_aY = -40.0\,\mathbf{j}\; R_aZ = -13.33\,\mathbf{k}$$

The equilibrium equation at joint a, with the previously determined value of P_{ad} is

$$-40.0\mathbf{j} - 13.33\mathbf{k} + P_{ab}(0.6\mathbf{i} - 0.8\mathbf{k}) + P_{ac}(1.0\mathbf{i})$$
$$+ 46.67(0.4285\,\mathbf{i} + 0.8571\,\mathbf{j} - 0.2851\,\mathbf{k}) = 0$$

Equating the coefficients of like base vectors to zero, we obtain

$$P_{ab} = -33.3 \text{ kN}$$
$$P_{ac} = 0$$

Similarly, writing the equilibrium equation for joint *c* and solving for bar force P_{cb}, we get

$$P_{cb} = 33.33 \text{ kN (tension)}$$

The complete reaction components and the bar forces are shown in Fig. 4.6.

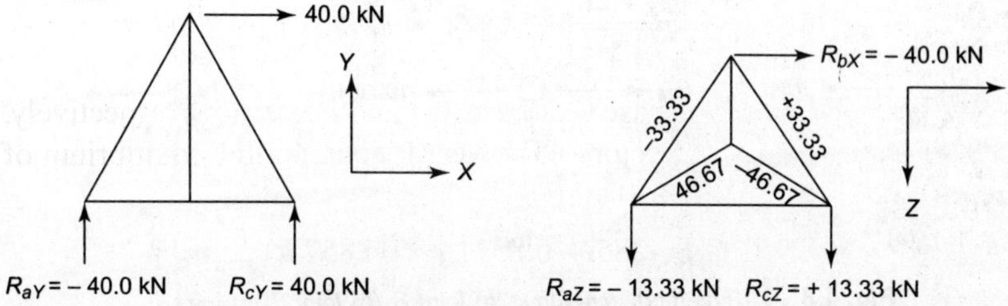

Fig. 4.6 | *Results of analysis*

Example 4.2 | *Determine the bar forces and reaction components of the space truss in Fig. 4.7. Make use of Theorems 1 and 2 wherever applicable to fix up the bar forces.*

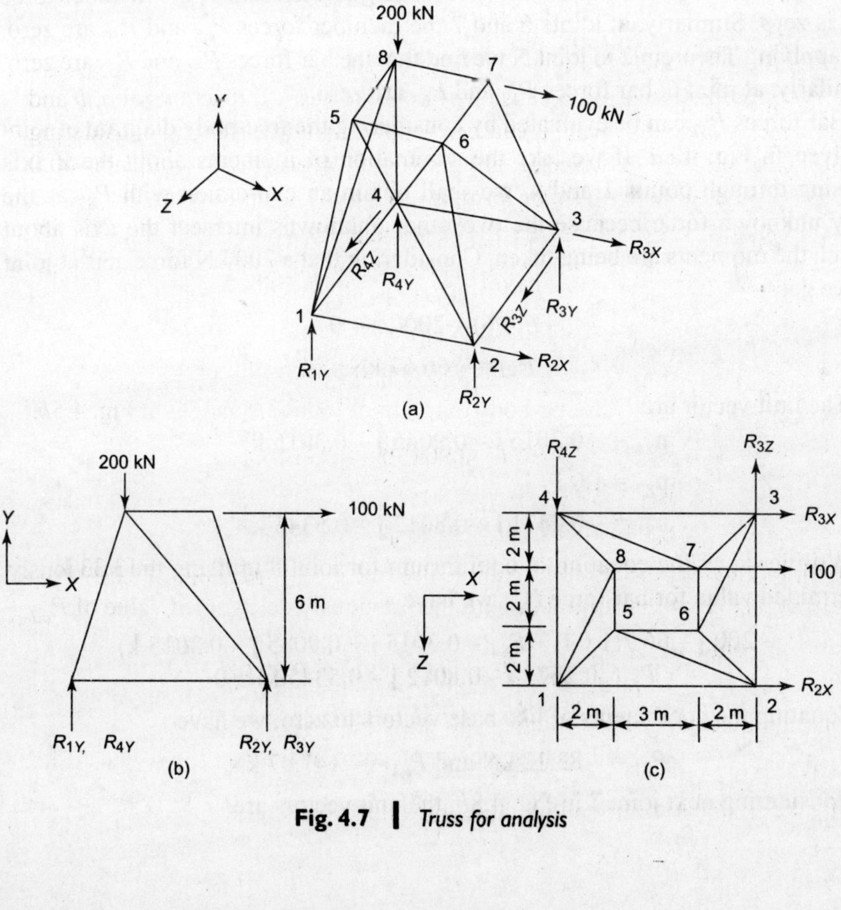

Fig. 4.7 | *Truss for analysis*

Figure 4.7*b* represents an elevation on the *XY* plane and Fig. 4.7*c* shows the plan on the *XZ* plane. The *X*, *Y* and *Z* reference axes are oriented as shown in Fig. 4.7*a*.

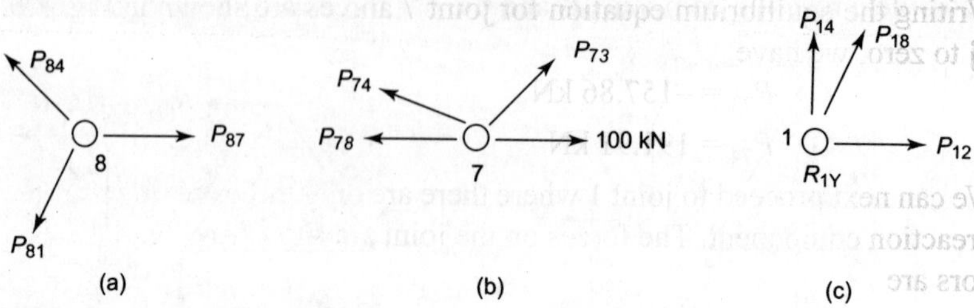

Fig. 4.8 | *Free-body diagrams: (a) Joint 8, (b) Joint 7, (c) Joint 1*

From inspection it is seen that $j = 8$, $m = 16$ and $r = 8$ satisfy Eq. 4.3. Thus, the truss is statically determinate. However, it may be noted that there are eight unknown reactions and all cannot be determined first. At joint 5 all members, except member 5-8, lie in the same plane and there are no external loads applied at the joint. Therefore, according to Theorem 1, the member force P_{58} is zero. Similarly, at joints 6 and 7 the member forces P_{65} and P_{76} are zero. By applying Theorem 2 to joint 5, we find that the bar forces P_{51} and P_{52} are zero. Similarly, at joint 6, bar forces P_{62} and P_{63} are zero.

Bar forces P_{67} can be evaluated by considering the free-body diagram of joint 8 given in Fig. 4.8*a*. If we take the summation of moments about the *Z* axis passing through points 1 and 4, we shall obtain an expression with P_{87} as the only unknown force because the two other unknowns intersect the axis about which the moments are being taken. Considering that a 200 kN force acts at joint 8, we get,

$$-P_{87}(6) - 200(2) = 0$$

or

$$P_{87} = -66.67 \text{ kN}$$

The unit vector are

$$\mathbf{n}_{84} = -0.3015\,\mathbf{i} - 0.9045\,\mathbf{j} - 0.3015\,\mathbf{k}$$

$$\mathbf{n}_{87} = 1.0\,\mathbf{i}$$

$$\mathbf{n}_{81} = -0.2671\,\mathbf{i} - 0.8012\,\mathbf{j} - 0.5345\,\mathbf{k}$$

Writing down the equation of equilibrium for joint 8 utilising the previously determined value for bar force P_{87}, we have

$$-200\,\mathbf{j} - 66.7(1.0\,\mathbf{i}) + P_{84}(-0.3015\,\mathbf{i} + 0.9045\,\mathbf{j} - 0.3015\,\mathbf{k})$$
$$+ P_{81}(-0.2671\,\mathbf{i} - 0.8012\,\mathbf{j} - 0.5345\,\mathbf{k}) = 0$$

Equating the coefficients of like base vectors to zero, we have

$$P_{81} = -83.12 \text{ kN and } P_{84} = -147.47 \text{ kN}$$

Considering next joint 7 in Fig. 4.8*b*, the unit vectors are

$$\mathbf{n}_{74} = -0.6202\,\mathbf{i} - 0.7442\,\mathbf{j} - 0.2481\,\mathbf{k}$$
$$\mathbf{n}_{73} = +0.3015\,\mathbf{j} - 0.9045\,\mathbf{j} - 0.3015\,\mathbf{k}$$
$$\mathbf{n}_{78} = +1.0\,\mathbf{i}$$

Writing the equilibrium equation for joint 7 and equating the coefficients of **i** and **j** to zero, we have

$$P_{73} = -157.86 \text{ kN}$$

and $$P_{74} = 191.91 \text{ kN}$$

We can next proceed to joint 1 where there are only three unknowns including one reaction component. The forces on the joint are shown in Fig. 4.8c. The unit vectors are

$$\mathbf{n}_{14} = -1.0\,\mathbf{k}$$
$$\mathbf{n}_{12} = +1.0\,\mathbf{i}$$
$$\mathbf{n}_{18} = 0.2671\,\mathbf{i} + 0.8012\,\mathbf{j} - 0.5345\,\mathbf{k}$$

On writing the equilibrium equation for joint 1 and equating the coefficients of like base vectors to zero, we have

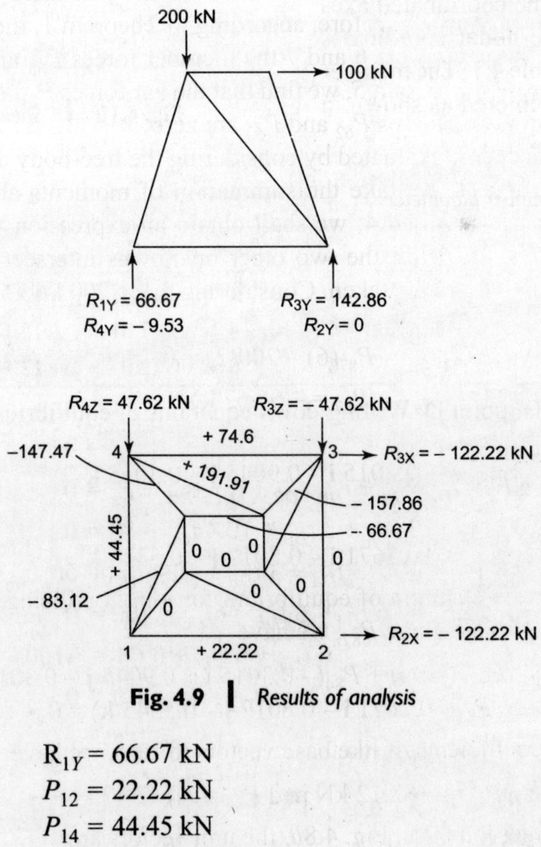

Fig. 4.9 | *Results of analysis*

$$R_{1Y} = 66.67 \text{ kN}$$
$$P_{12} = 22.22 \text{ kN}$$
$$P_{14} = 44.45 \text{ kN}$$

and $$P_{18} = -83.12 \text{ kN}$$

The remaining bar forces and reaction components can be determined from the equilibrium expressions for joints 2, 3 and 4. The results are shown in Fig. 4.9.

Example 4.3 | *Using tension coefficient method analyse the member forces in sheer legs as shown in Fig. 4.10*

Table 4.1 | *Coordinates*

Sl. No	*Node*	*x*	*y*	*z*
1	*D*	0	0	0
2	*A*	3	–6	4
3	*B*	3	–6	–4
4	*C*	10.4	–6	0

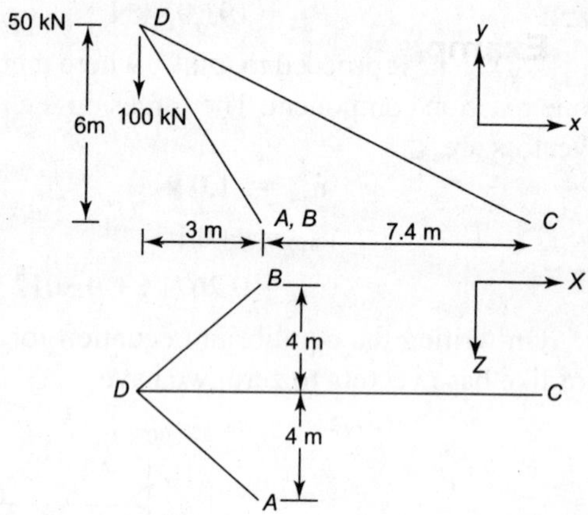

Fig. 4.10 | *Sheer legs*

The forces in sheer leg bars can be analysed without going into reaction components. Taking *D* as the origin and the coordinated axes as indicated the nodal coordinates are shown in Table 4.1. The member parameters are entered as shown in Table 4.2.

Table 4.2 | *Member parameters*

Sl.No	*Member*	x_{ij}	y_{ij}	z_{ij}	l_{ij}	t_{ij}	T_{ij}
1	DA	3	–6	4	7.81	–15.1	–117.93
2	DB	3	–6	–4	7.81	–15.1	–117.93
3	DC	10.4	–6	0	12.0	13.5	162.0

Consider nodal point D. Writing down equations of equilibrium $\Sigma F_x = 0$, $\Sigma F_y = 0$, $\Sigma F_z = 0$

we have $$\Sigma F_x = t_{DA}\, x_{DA} + t_{DB}\, x_{DB} + t_{DC} x_{DC} + P_x = 0$$

or $$3\, t_{DA} + 3\, t_{DB} + 10.4\, t_{DC} - 50 = 0$$

or $$3\, t_{DA} + 3\, t_{DB} + 10.4\, t_{DC} = 50 \qquad (4.4)$$

$$\Sigma F_y = -6\, t_{DA} - 6\, t_{DB} - 6 t_{DC} - 100 = 0$$

or $$6\, t_{DA} + 6\, t_{DB} + 6\, t_{DC} = -100 \qquad (4.5)$$

Again $$\Sigma F_z = 4 t_{DA} - 4 t_{DB} + 0 = 0 \qquad (4.6)$$

From Eqn (4.6) $\quad t_{DA} = t_{DB}$

Rewriting Eqn (4.4) $$6 t_{DB} + 10.4\, t_{DC} = 50 \qquad (4.7)$$

Rewriting Eqn. (4.5) $$12\, t_{DB} + 6\, t_{DC} = -100 \qquad (4.8)$$

Solving simultaneous equations (4.7) and (4.8)

We get $t_{DC} = 13.51$

And $t_{DB} = -15.1$

These coefficients are entered in Table 4.2 and the member forces are calculated.

Example 4.4 | *Analyse the member forces in the three bar truss as shown in Fig. 4.11 using tension coefficient method.*

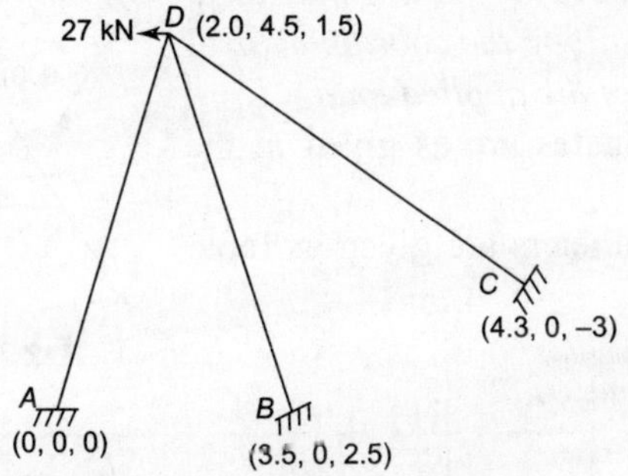

Fig. 4.11 | *Space truss*

Step 1: To fix up bar parameters

The coordinates of the nodal joints are indicated. The bar parameters are tabulated in Table 4.3.

Table 4.3 | *Bar parameters*

Sl. No	*Member*	x_{ij}	y_{ij}	z_{ij}	l_{ij}	t_{ij}	T_{ij}
1	DA	–2.0	–4.5	–1.5	5.15	–6.99	36.00
2	DB	1.5	–4.5	1.0	4.85	3.80	18.43
3	DC	2.3	–4.5	–4.5	6.77	3.18	21.50

Step 2: Writing of equilibrium equations

$\Sigma F_x = 0$ $\quad t_{DA}(-2.0) + t_{DB}(1.5) + t_{DC}(2.3) - 27 = 0$

or $\quad -2.0\, t_{DA} + 1.5\, t_{DB} + 2.3\, t_{DC} = 27$ (4.9)

$\Sigma F_y = 0$ $\quad -4.5\, t_{DA} - 4.5\, t_{DB} - 4.5\, t_{DC} = 0$ (4.10)

and SFz = 0 $\quad -1.5\, t_{AD} + 1.0\, t_{DB} - 4.5\, t_{DC} = 0$ (4.11)

Taking equation (4.9) and (4.10) t_{DB} can be eliminated by multiplying eqation (4.9) with 3 and adding Eqn. (4.10)

We get $\quad -10.5\, t_{DA} + 2.4\, t_{DC} = 81$ (4.12)

Again considering equations (4.10) and (4.11)
t_{DB} can be eliminated by multiplying Eqn (4.11)
by 4.5 and adding Eqn. (4.10)

We get $\quad -11.25\, t_{DA} - 24.75\, t_{DC} = 0$ (4.13)

Solving equations (4.12) and (4.13)

We get $\qquad t_{DC} = 3.18 \quad \text{and} \quad t_{DA} = -6.99$

Finally, we have $t_{DB} = 3.80$ from Eqn (4.13)

The values are entered and the final bar forces are worked out as in Table 4.3).

Example 4.5 | *Calculate the member forces in the truss shown in Fig. 4.12, due to the loading indicated. The forces indicated along the coordinate axes are the components of the applied load.*

The nodal coordinates are as given in the diagram.

The member parameters are given in Table 4.40.

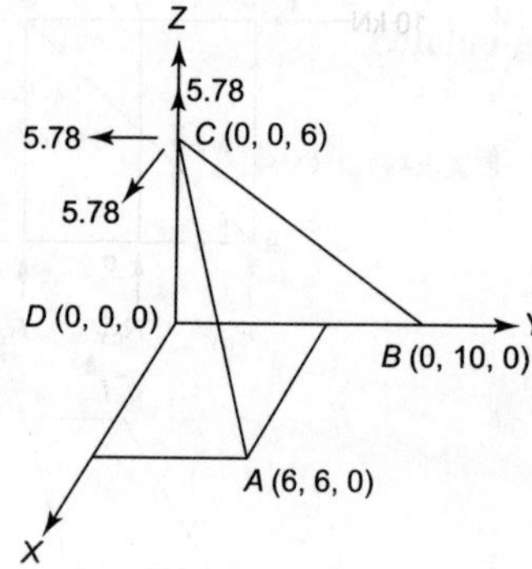

Fig. 4.12 | *Space truss*

Table 4.4 | *Bar parameters*

Sl. No.	*Member*	x_{ij}	y_{ij}	z_{ij}	l_{ij}	t_{ij}	T_{ij}
1	*CA*	6	6	–6	10.39	–0.9633	–10.0
2	*CB*	0	10	–6	11.66	1.156	13.48
3	*CD*	0	0	–6	6.0	0.77	4.62

Joint *C* is taken for the enalysis of forces.

Writing summation of forces $\Sigma F_x = 0$

we have $\qquad t_{CA}\, x_{CA} + t_{CB}\, x_{CB} + t_{CD}\, x_{CD} + 5.78 = 0$

or $\qquad 6\, t_{CA} + t_{CB}\,(0) + t_{CD}\,(0) = -5.78 = 0 \qquad (4.14)$

writing $\qquad \Sigma F_y = 6\, t_{CA} + 10\, t_{CB} + t_{CD}\,(0) - 5.78 = 0$

or $\qquad 6\, t_{CA} + 10\, t_{CB} = 5.78 \qquad (4.15)$

Again writing $\qquad \Sigma F_z = -6\, t_{CA} - 6\, t_{CB} - 6\, t_{CO} + 5.78 = 0$

or $\qquad 6\, t_{CA} + 6\, t_{CB} + 6\, t_{CO} = 5.78 \qquad (4.16)$

From Eqn. (4.14)

$$6\, t_{CA} = -5.78$$

$$t_{CA} = -0.9633$$

Substituting in Eqn. (4.15)

$$6(-.9633) + 10\, t_{CB} = 5.78$$

gives $\quad t_{CB} = 1.156$

Substituting in Eqn. (4.16)

$$(+6)\,(-0.9633) + (+6)\,(1.156) + (+6)\, t_{CO} = +\,5.78$$

gives $\quad t_{CO} = 0.77$

The tension coefficients and the bar forces are entered in Table 4.4.

Problems for Practice

4.1 Find the reaction components and the bar forces of the space truss shown in Fig. 4.13.

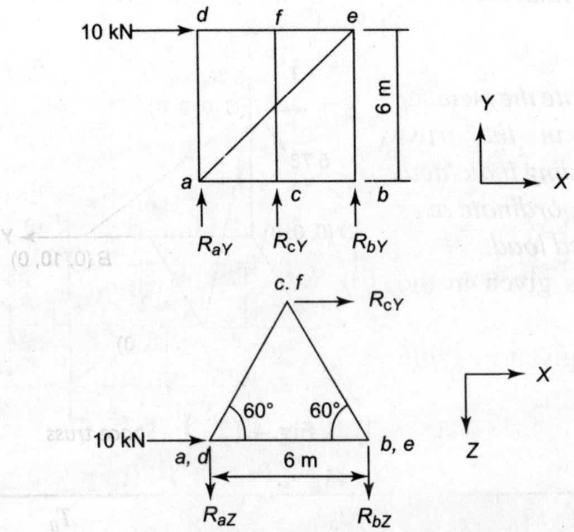

Fig. 4.13

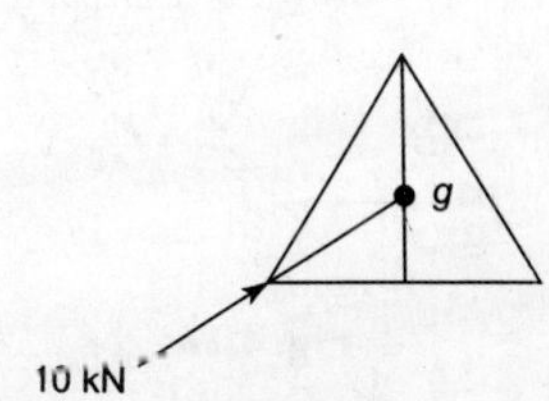

Fig. 4.14

4.2 Find the reaction components and bar forces of the truss in Problem 4.1 if a load of 10 kN is applied at joint *d,* but with a direction such that it passes through point *g,* the centre of the equilateral triangle, as shown in Fig. 4.14.

4.3 Find the reaction components and the bar forces of the truss given in Fig. 4.15.

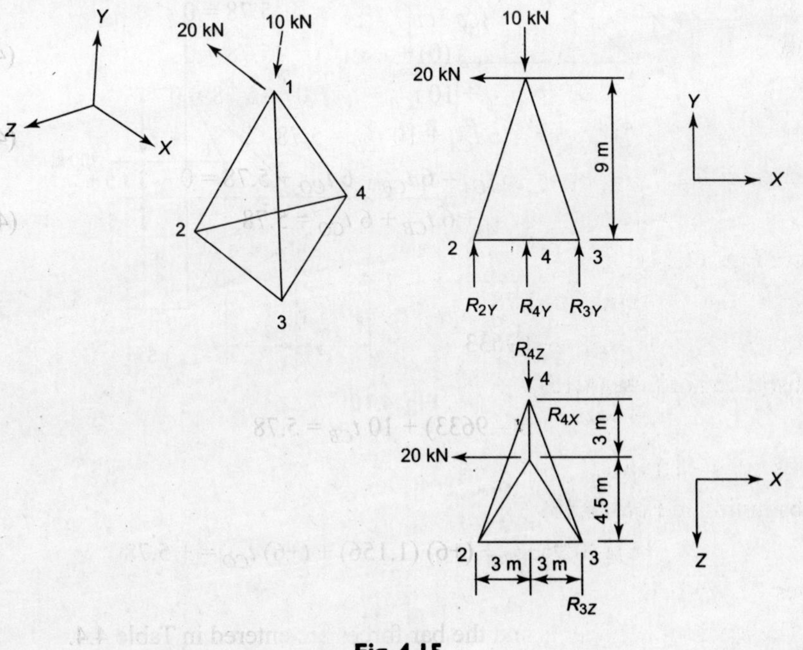

Fig. 4.15

4.4 In the truss shown in Fig. 4.16 (this is not a simple space truss), the plane of all ball sockets CDEH is in the ZY plane while the plane of FGDE is parallel to the XZ plane. Determine the forces in all the joints and then determine the supporting forces.

4.5 Find the reaction components and the bar forces of the truss given in Fig. 4.17.

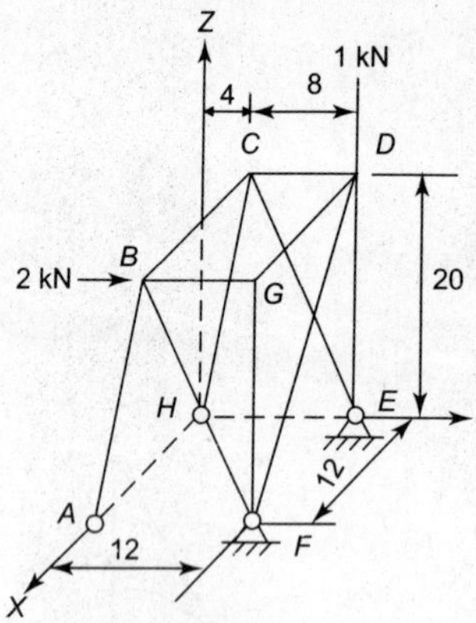

Fig. 4.16

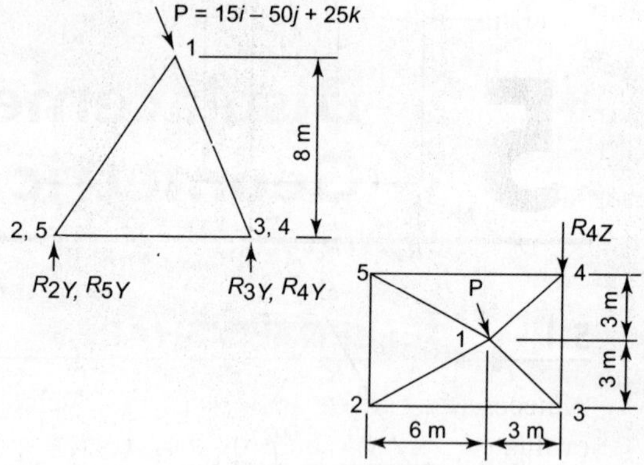

Fig. 4.17

4.6 For the space frame shown in Fig. 4.18 determine: (a) the values of reaction components, (b) the members that obviously have zero forces and (c) the forces in other members.

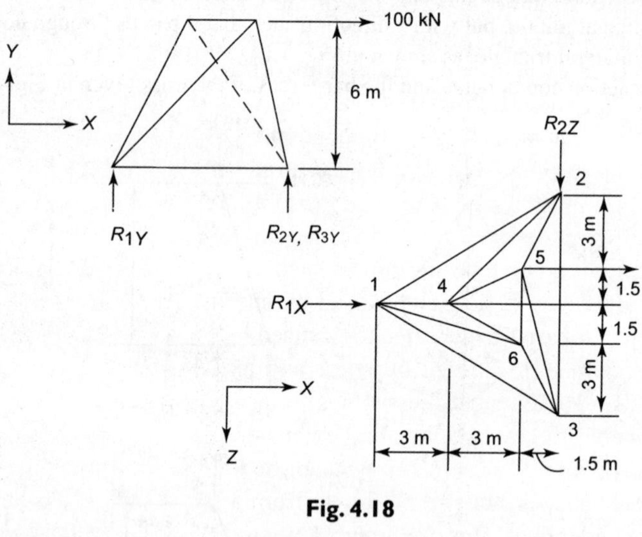

Fig. 4.18

5 Displacements—Geometric Methods

5.1 DEFLECTED SHAPES

A necessary part of analysis is the evaluation of deflections. Deflections are evaluated not only to check that they do not exceed the design limitations, but also to use them in the analysis of statically indeterminate structures. In the analysis of statically indeterminate structures, the static equilibrium equations alone are not sufficient to evaluate the unknowns. It becomes necessary to utilise additional conditions of compatibility or consistent displacements.

In this chapter, we shall discuss the geometric methods for obtaining deflections in a structure. The energy or the virtual work methods are discussed in the next chapter. The intention in both these chapters is to develop concepts and methods of evaluation of deflections. Their application in the analysis of statically indeterminate structures will be dealt with in later chapters.

As a first step in discussing deflections, it will be useful to develop an understanding of the manner in which a structure deflects under external loads. The sketching of deflected shapes helps to a large extent in understanding the response of the structure to external loads. It may be recalled that deflections are caused by various types of forces such as axial, shear and moments in the members. However, deflections caused by shear and axial forces are generally small when compared with deflection caused by moments in beams and frames. Therefore, deflections caused by shear and axial forces are generally ignored and only deflections caused by moments are evaluated. In trusses, however, deflections are primarily due to axial extensions or shortenings of the members. Normally, the forces in members will not be known quantitatively beforehand. The deflected shapes can still be drawn from a rough idea of how the structure will deflect under load. This can be substantiated by visualising the general nature of resisting moments and the condition of supports.

Let us consider an elementary case of a propped cantilever beam under a distributed load as shown in Fig. 5.1. The deflected shape is shown by a broken line. In drawing deflected shapes certain

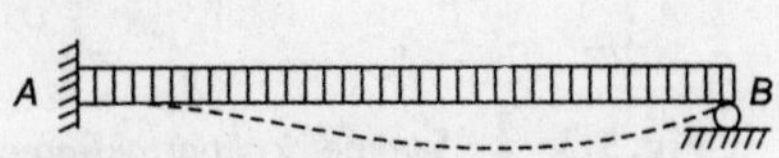

Fig. 5.1 | *Propped cantilever beam*

basic requirements must be met. For example, a deflected shape representing a member is a continuous smooth curve. It must also be such that it satisfies the support conditions and the connection between members. With regard to the propped cantilever beam of Fig. 5.1 the elastic line is a continuous smooth curve, curved downwards for some length and upwards afterwards. The fixed end at left hand support A does not allow the member to either rotate or translate; support B does not permit any translation. Deflections are very small when compared with the length of the members. While drawing deflected shapes these are highly exaggerated for a better understanding of the deflected shapes.

Now consider the continuous beam of Fig. 5.2. The deflected shape or the elastic line is shown by a dotted line for the given loading. At support points *A* and *B*, the elastic line should not show any deflections but may have some slopes. At point *C*, the fixed end, zero slope and zero deflection condition is satisfied. Note that the basic requirement that the elastic line is a smooth and continuous curve is also satisfied. The curve deflects downwards in the span *A-B* and upwards in the span *B-C* for the loading indicated. At certain points on the curve the curvature changes sign; these points are referred to as *inflection points.* Inflection points represent points of zero moment in the beam. The inflection points are indicated by asterisk marks (*) on the deflected shape.

Consider a portal frame of Fig. 5.3. The frame is symmetrical and the loading also is symmetrical. The axial deformations are considered negligible when compared with those due to bending. Therefore, points *B* and *C* will remain in the same position. However, connections at *B* and *C* do rotate. In drawing the deflected shape at *B* and *C*, it should be remembered that at the point of connection, the members rotate by the same amount so that the angular orientation of the members is maintained in the deflected shape.

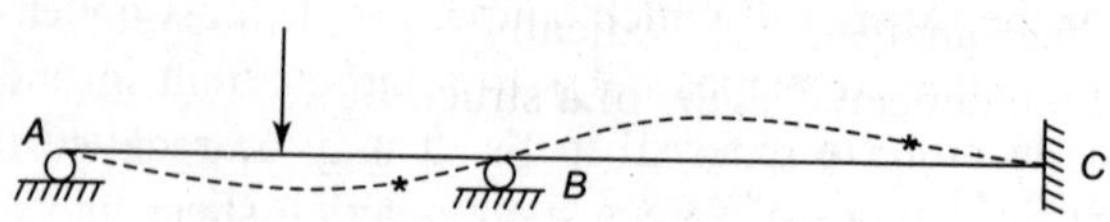

Fig. 5.2 | *Two-span continuous beam*

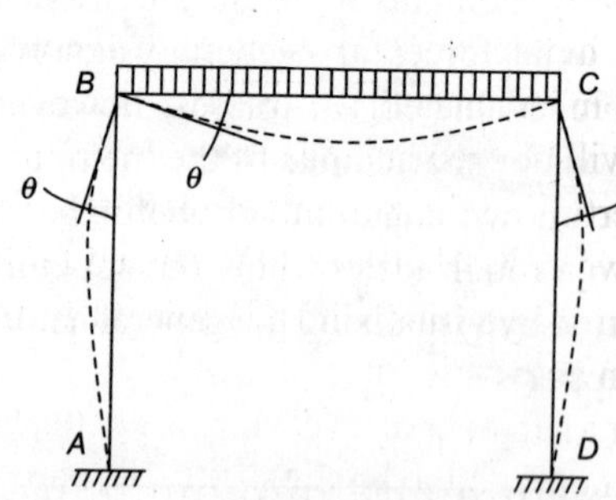

Fig. 5.3 | *Symmetrical portal frame under symmetrical loading*

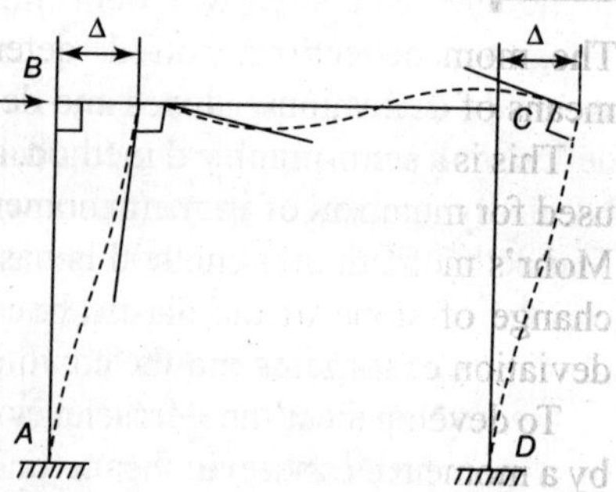

Fig. 5.4 | *Portal frame under lateral load*

Consider the same frame subjected to lateral loading as shown in Fig. 5.4. Again neglecting axial deformations, the lateral displacement of points B and C must be same. Points B and C do rotate but the angular orientation of members meeting at points B and C must be maintained even in the deflected shape.

As a last example, consider the frame of Fig. 5.5 with one leg inclined. If axial deformations are neglected, point C must move along a line perpendicular to DC at point C. Strictly speaking, this displacement is along an arc of a circle drawn with centre D and radius DC, but we are concerned with small displacements, and therefore, an arc is replaced by a straight line normal to DC. The point B moves laterally by ΔH, same as the horizontal displacement component of C since the axial deformation in member BC is neglected. No vertical displacement of point B is possible. However, both points B and C rotate. The members meeting at joints B and C will retain the same angular orientation to each other at the points of connection.

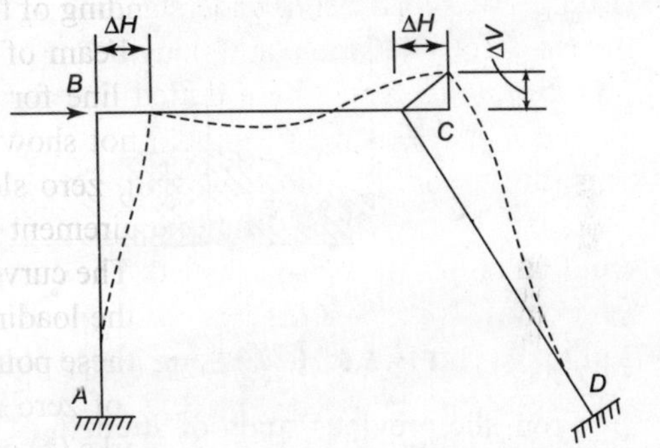

Fig. 5.5 | *Frame with one inclined leg under lateral loading*

Learning to draw the deflected shapes, qualitatively, will provide a through understanding of the methods of analysis of such structures. For example, while considering the conditions of consistent displacement to acquire additional equations for the analysis of statically indeterminate structures, a clear visualization of the deflected shape of a structure will be highly meaningful and useful.

5.2 | MOMENT-AREA METHOD

The moment-area method of determining deflections provides a convenient means of determining slopes and defections in beams and frames.

This is a semi-graphical method, and, as will be seen later, can be conveniently used for members of varying moment of inertia and discontinuous loadings. The Mohr's moment-area method is based on two basic theorems: (1) related to the change of slope of the elastic line between two points, and (2) related to the deviation of tangents drawn at two points on an elastic line.

To develop these theorems, let us consider a segment of an elastic line deflected by a moment as shown in Fig. 5.6*a*. The moment producing the deflected shape is shown in Fig. 5.6*b* in the form of moment diagram ordinates divided by flexural rigidity EI of the beam. The diagram is commonly known as the *M/EI diagram* or *curvature diagram.*

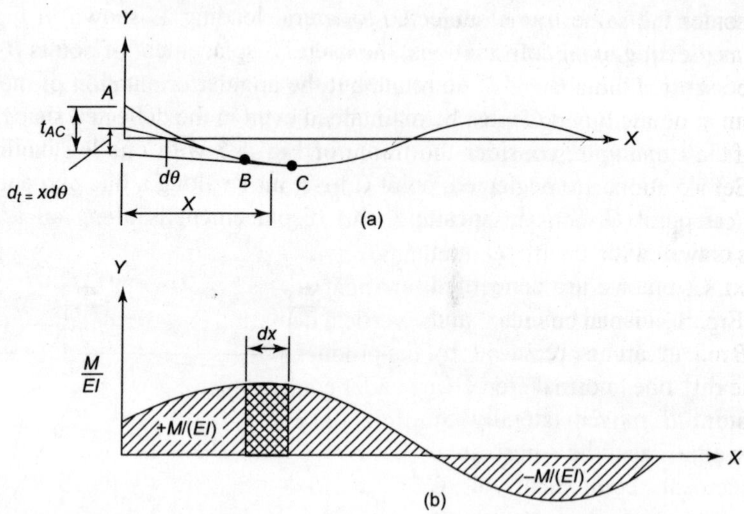

Fig. 5.6 | *(a) Elastic line, (b) M/EI diagram (curvature diagram)*

From the previous study of strength of materials it may be recalled that the differential equation of such an elastic curve is

$$\frac{d^2y}{dx^2} = \frac{M}{EI} \tag{5.1}$$

The sign of the bending moment corresponds to the beam convention of Fig. 2.3*b* and the upward direction of *y* is positive. The slope of the elastic line can be expressed as

$$\theta = \frac{dy}{dx} \tag{5.2}$$

Therefore, Eq. 5.1 can be written as

$$\frac{d\theta}{dx} = \frac{M}{EI} \tag{5.3}$$

or

$$d\theta = \frac{Mdx}{EI} \tag{5.4}$$

Considering a differential length of the beam between point *B* and *C*, (see Fig. 5.6*a*) we see that the change of slope is denoted by $d\theta$, that is the angle between the tangents drawn from points *B* and *C*. We see from Eq. 5.4 that the value of change in slope between points *B* and *C* is equal to the area of the *M/EI* diagram between those points. The change of slope between *A* and *C* may be obtained by integrating Eq. 5.4 to give.

$$\int_A^C d\theta = \theta_C - \theta_A = \Delta\theta_{CA} = \int_A^C \frac{Mdx}{EI} \tag{5.5}$$

From the results of Eq. 5.5 the first moment-area theorem related to the change of slope may be stated as follows.

THEOREM 1 *The change of slope between the two points on an elastic line is equal to the area of the M/EI diagram between those points.*

The sign for the change of slope directly results from the area of the *M/EI* diagram evaluated using Eq. 5.5. As seen from the elastic line of Fig. 5.6*a*, the sign of the slope at *A* is negative. From *A* to *C* the slope increases as can be seen from Eq. 5.5 and the *M/EI* diagram. The slope from *A* to *C* becomes less negative. Using consistent dimensions for *M, E* and *I* and the length of the member, the results obtained for θ will be in radians.

Next, let us consider the use of the moment-area concept to evaluate deflections. From Fig. 5.6a it can be seen that the vertical deviation of the two tangents drawn from *B* and *C* can be obtained from the product of the angle between the tangents and the distance to the reference line under consideration. Thus, at point *A* which is located at a distance *x* from the differential element BC, the vertical deviation *dt* between tangents *B* and *C* is

$$dt = (x)\,(d\theta) \tag{5.6}$$

It may be remembered that we are concerned with flat elastic curves with small deflections and slopes and hence the above relationship

The deviation of the tangent at *A* from the tangent at *C* denoted by t_{AC} can be obtained by repeating the procedure for each differential element between *A* and *C* and summing the resulting values of *dt.* This, of course, can be accomplished by integrating Eq. 5.6, which results in

$$t_{AC} = \int_A^C (x)(d\theta) \tag{5.7}$$

Substituting for $d\theta$ for Eq. 6.4

$$t_{AC} = \int_A^C \frac{Mxdx}{EI} \tag{5.8}$$

From this equation the second moment-area theorem can be stated as given below.

THEOREM 2 *The tangential deviation of A from a tangent to the elastic curve at C is equal to the static moment of the area of the M/EI diagram between A and C taken about point A.*

The deviation is obtained on a vertical line passing through *A*. If the area of *M/EI* diagram between *A* and *C* is denoted by A_1, Eq. 5.8 can be written as

$$t_{AC} = A_1 x_1 \tag{5.9}$$

where x_1 represents the horizontal distance of the centroid of the area from point *A* (see Fig. 5.7).

By the same reasoning the deviation of point *C* from a tangent through *A* is

$$t_{CA} = A_1 x_1' \tag{5.10}$$

where the same area of *M/EI* diagram is used but x_1' is measured from the vertical line through point C. In-Eqs. 5.9 and 5.10, the distances x_1, x'_1, are taken as positive and, as *E* and *I* are positive quantities, the sign of the tangential deviation depends on the sign of the bending moments. A positive value for the tangential deviation indicates that a given point lies above a tangent to the elastic curve drawn through the other point and vice versa.

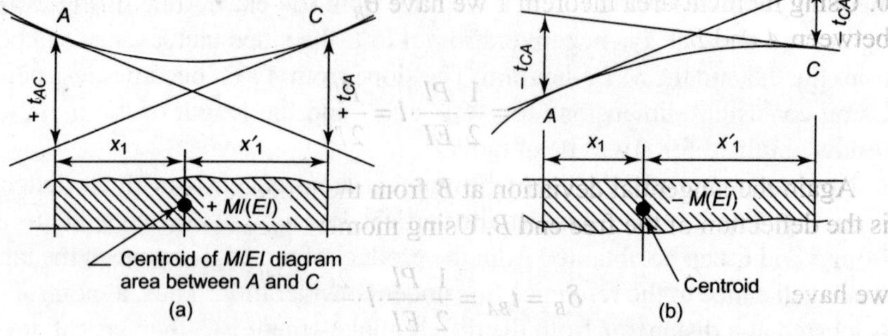

Fig. 5.7 | *Sign convention for tangential deviation*

This is shown clearly in Fig. 5.7. The order of the subscript used in the deviation at A and C may be carefully noted; the point for which the deviation is being determined is written first.

The above two theorems can be applied between any two points on a continuous elastic curve of any beam for any loading. However, it must be emphasised that only relative rotation of the tangents and only tangential deviations are obtained directly. A further consideration of the geometry of the elastic curve at the supports to include boundary conditions is necessary in every case to determine deflections. This aspect is illustrated in the following examples.

Example 5.1 | *Find the slope and deflection at the free end of a cantilever beam subjected to a concentrated load P at the free end.*

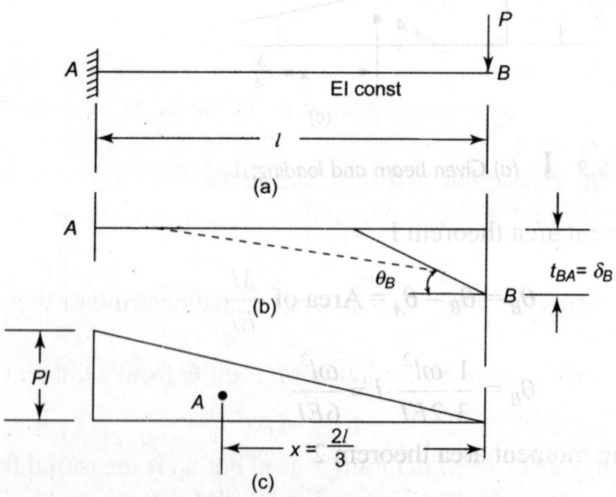

Fig. 5.8 | *(a) Beam and loading; (b) Elastic line (c) M-diagram*

The elastic curve or the deflection curve of the cantilever beam is shown in Fig. 5.8(b). Moment diagram is shown in Fig. 5.8(c). The slope of beam at the free end B can be taken as the change in slope from A to B since the slope at A = 0. Using moment-area theorem 1 we have $\theta_B = \theta_B - \theta_A$ = Area of the M diagram between A and B.

$$= \frac{1}{2}\frac{Pl}{EI}.l = \frac{Pl^2}{2EI}$$

Again the tangential deviation at B from the tangent drawn at A is t_{BA}, which is the deflection of the free end B. Using moment area theorem I.

we have, $$\delta_B = t_{BA} = \frac{1}{2}\frac{Pl}{EI}.l.\frac{2l}{3} = \frac{Pl^3}{2EI}$$

Example 5.2 | *Find the slope and deflection of the end B of a cantilever beam under a uniformly distributed load w/unit length as shown in Fig. 5.9.*

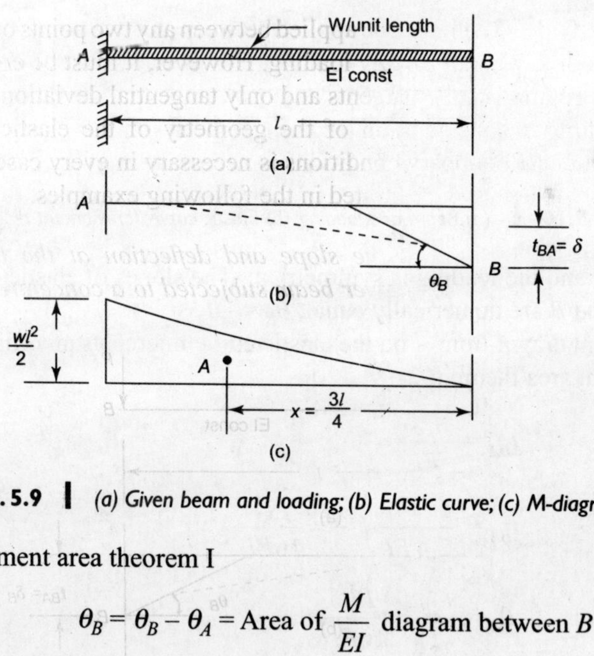

Fig. 5.9 | *(a) Given beam and loading; (b) Elastic curve; (c) M-diagram*

Using moment area theorem I

$$\theta_B = \theta_B - \theta_A = \text{Area of } \frac{M}{EI} \text{ diagram between } B \text{ and } A.$$

$$\theta_B = \frac{1}{3}\frac{\omega l^2}{2EI}.l = \frac{\omega l^3}{6EI}$$

Again using moment area theorem 2

$$\delta_B = t_{BA} = \text{Moment of the } \frac{M}{EI} \text{ diagram}$$

between A and B taken about B.

$$\therefore \qquad \delta_B = \frac{1}{3}\frac{\omega l^2}{2EI}\cdot l\cdot\frac{3}{4}l = \frac{\omega l^4}{8EI}$$

Example 5.3 | *Analyse a simply supported beam subjected to a concentrated load P at centre for its end slopes and deflection at the centre. Use moment-area method in the analysis.*

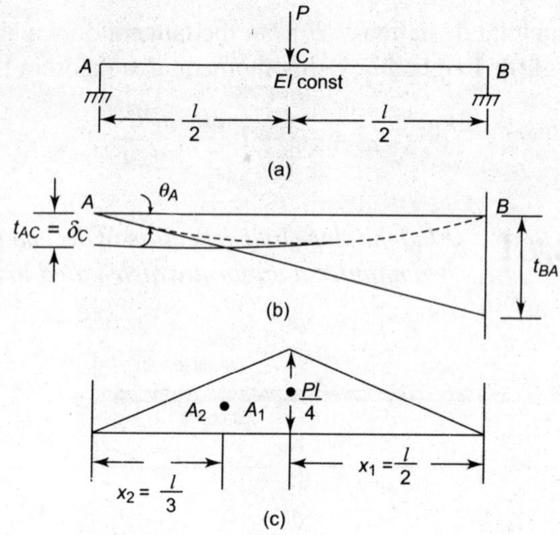

Fig. 5.10 | *(a) Beam and loading; (b) Elastic curve; (c) Moment diagram*

The beam and the loading is symmetrical. The slopes of elastic curve at both the ends A and B are numerically equal, $\theta_B = -\theta_A$

The tangent drawn from A on the elastic curve intercepts an ordinate t_{BA} at B. using moment area theorem-2

$$t_{BA} = \frac{A_i\, x_i}{EI}$$

or

$$t_{BA} = \frac{1}{2}\frac{Pl}{4EI}\cdot l\cdot\frac{l}{2} = \frac{Pl^3}{16EI}$$

$\therefore$ Then we have

$$\theta_A = \frac{t_{AB}}{l} = \frac{Pl^2}{16EI}$$

The deflection at centre of span C can be obtained indirectly. The elastic curve has zero slope at centre under load P due to symmetry. The tangent drawn from C on the elastic curve intercepts an ordinate at A equal to t_{AC} which is the deflection δ_C desired. We can also draw tangent from C to obtain t_{BC} at B according to moment area theorem-2.

$$t_{AC} = \frac{A_2 x_2}{EI}$$

or $$t_{AC} = \frac{1}{2}\frac{Pl}{4EI}\cdot\frac{l}{2}\cdot\frac{l}{3} = \frac{Pl^3}{48EI}$$

$\therefore$ $$\delta_C = t_{AC} = \frac{Pl^3}{48EI}$$

Example 5.4 | *Find slopes at the ends and deflection at centre of a simply supported beam subjected to uniformly distributed load as shown in Fig. 5.11.*

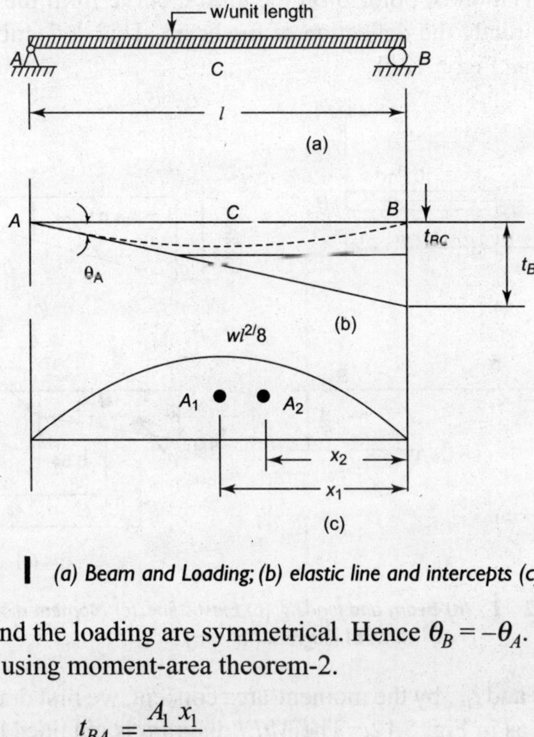

Fig. 5.11 | *(a) Beam and Loading; (b) elastic line and intercepts (c) M-diagram*

The beam and the loading are symmetrical. Hence $\theta_B = -\theta_A$. The intercept t_{BA} is obtained by using moment-area theorem-2.

$$t_{BA} = \frac{A_1\, x_1}{EI}$$

or $$t_{BA} = \frac{2}{3}\,l\cdot\frac{\omega l^2}{8EI}\cdot\frac{l}{2} = \frac{\omega l^3}{24EI}$$

$$\theta_A = \frac{t_{BA}}{l} = \frac{\omega l^2}{24EI}$$

The deflection at centre of span C is equal to t_{BC} as the tangent drawn from C and the elastic line is horizontal.

$$\delta_C = t_{BC} = \frac{A_2 x_2}{EI} = \frac{2}{3}\frac{l}{2}\frac{\omega l^2}{8EI}\cdot\frac{5}{8}\frac{l}{2} = \frac{5}{384}\frac{\omega l4}{EI}$$

Example 5.5 | *A cantilever beam having stepped moment of inertia is subjected to a 4 kN load at the free end as shown in Fig. 5.12a. Using moment-area theorems, evaluate the slope and deflection at the free end.*

$$E = 204 \times 10^6 \text{ kN/m}^2 \text{ (204,000 Mpa).}$$

The elastic curve for the deflected shape is shown in Fig. 5.12. The slope at B can be taken as the change in slope from A to B since the slope at A is zero. The slope at B is denoted by θ_B. Similarly, because the slope of the elastic curve at A is zero, the deviation of point B on the elastic curve from the tangent drawn at A represents directly the deflection of the beam. This deflection at point B is denoted by t_{BA} (*see* Fig. 5.12*b*).

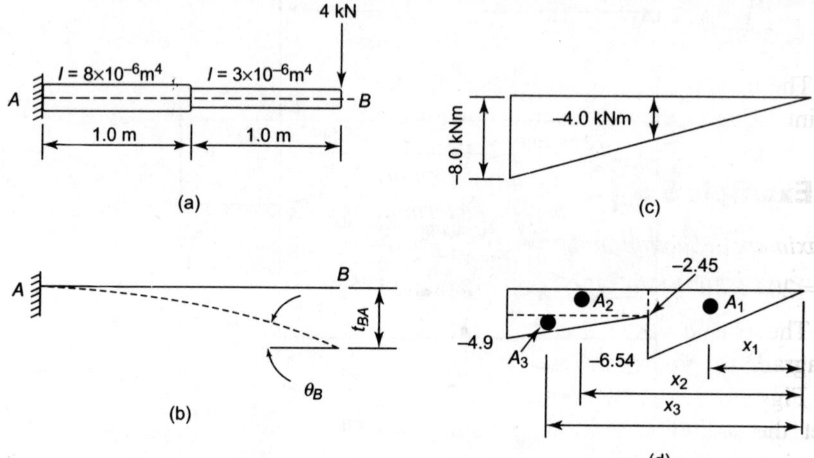

Fig. 5.12 | *(a) Beam and loading, (b) Elastic line, (c) Moment diagram, (d) M/EI diagram ($m^{-1} \times 10^{-3}$)*

To evaluate θ_B and t_{BA} by the moment-area concept, we first draw the bending moment diagram as in Fig. 5.12*c*. The *M/EI* diagram is obtained by dividing all the ordinates of the *M* diagram by *EI*. The *M/EI* diagram is shown in Fig. 5.12*d*.

The slope at B is determined according to Theorem 1 from Eq. 5.5. Since $\theta_A = 0$, we can write

$$\theta_B = \theta_B - \theta_A = \Delta\theta_{BA} = \int_A^B \frac{M}{EI}\,dx \qquad (5.11)$$

The area under the *M/EI* diagram can be conveniently broken into two triangles and one rectangle. The technique of subdividing the *M/EI* diagram into figures for which the areas and locations of centroids are commonly known is particularly useful in evaluating deflections later. The value of θ_B is, therefore, given by

$$\theta_B = A_1 + A_2 + A_3$$

From Fig. 5.12*d* we have

$$\theta_B = \left[\frac{(1)(-6.54)}{2} + (1)(-2.45) + (1)\frac{(-2.45)}{2}\right]10^{-3} = -0.0069 \text{ radians}$$

The deflection at B is determined using moment-area Theorem 2. Using areas A_1, A_2 and A_3 and taking static moments of these areas about point B, we obtain for the deflection at B

$$t_{BA} = A_{1x1} + A_{2x2} + A_{3x3}$$

where x_1, x_2 and x_3, are the horizontal distances from B to the centroids of the respective areas. From Fig. 5.12*d*, the value of the deflection is found to be

$$t_{BA} = [(-3.27)(2/3) + (-2.45)(1.5) + (-1.225)(5/3)]10^{-3}$$

$$= -7.895 \times 10^{-3}\ m = -7.9 \text{ mm}$$

The negative sign indicates that point B is below the tangent drawn through point A that is, the deflection is downwards.

Example 5.6 | *A simply supported beam is loaded as shown in Fig. 5.13a. Determine the location and magnitude of the maximum deflection in the beam. EI for the beam is constant.*

$E = 204 \times 10^6$ kN/m^2 (204,000 mPa) and I = 50×10^{-6} m^4 (50×10^6 mm^4).

The elastic curve and moment diagram for the beam are shown in Figs. 5.13*b* and *c* respectively. Let the unknown point at which maximum deflection will occur be at *D*, located at a distance *x* from the right hand support. The maximum deflection occurs where the tangent to the elastic curve is horizontal.

It is easy to visualise that the maximum deflection occurs in region *CB*. First we determine slope θ_B at support *B*. Then we determine the value of *x* for which the value of the slope changes by θ_B. From Fig. 5.13*b* we can see that θ_B can be obtained by dividing deviation t_{AB} by the distance between points *A* and *B*.

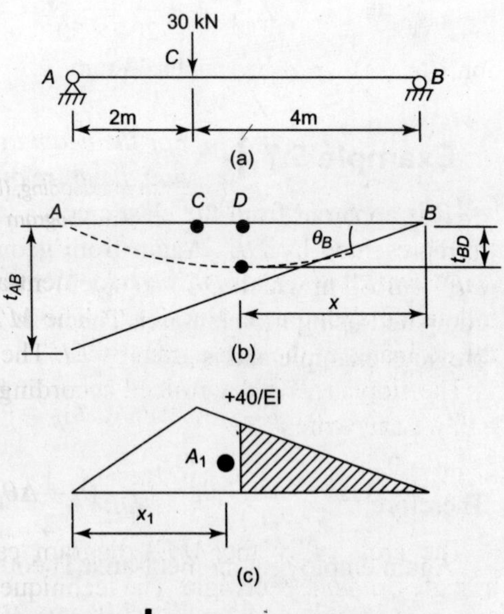

Fig. 5.13 | *(a) Beam and loading, (b) Elastic line, (c) M/EI diagram*

The *M/EI* diagram for the beam is shown in Fig. 5.13*c* in terms of *E* and *I*. The units used are kN and *m*. To evaluate t_{AB}, we take the static moment of the *M/EI* diagram area about point *A*. Thus,

$$t_{AB} = A_1 x_1 = \frac{1}{EI}(1/2)(6)(40)(8.3) = \frac{320.4}{EI} \text{ m}$$

and $$\theta_B = \frac{t_{AB}}{6} = \frac{53.4}{EI} \text{ radians}$$

The desired value for *x* is, therefore, the length in which the area of the *M/EI* diagram changes by this value of θ_B That is,

$$\frac{1}{2}(x)\left(\frac{40}{EI}\right)\left(\frac{x}{4}\right) = \frac{53.4}{EI}$$

This yields $$x = 3.27 \text{ m}$$

The magnitude of maximum deflection is found by evaluating tangential deviation t_{BD} which is equal to the static moment of the portion of *M/EI* diagram between *D* and *B* about *B*. Therefore,

$$t_{BD} = \frac{1}{2}(3.27)\left(\frac{40}{EI}\right)\left(\frac{3.27}{4}\right)\left(\frac{2}{3}\right) \tag{5.12}$$

$$= \frac{116.55}{204 \times 50} \text{ m (point } B \text{ is above the tangent drawn through point } D)$$

Substituting for *EI*, we get

$$\Delta_{max} = -\frac{116.55}{204 \times 50} = -11.43 \times 10^{-3} \text{ m}$$

or $$= 11.43 \text{ mm}$$

Example 5.7 | *For the beam in Example 5.6, find the slope and deflection at centre of beam.*

It is apparent from the elastic curve (Fig. 5.14*b*) that the required deflection is represented by *DE′*. Again from geometry or kinetic considerations, $DE' = DE'' - E'E''$ in which $DE'' = (\theta_B)(3)$ and $E'E''$ is the deviation t_{DB} which can be computed using moment-area Theorem 2. In this case t_{AB} is the same as in the previous example and is

$$t_{AB} = \frac{320.4}{EI}$$

Therefore, $$DE'' = \frac{1}{2} t_{AB} = \frac{160.2}{EI}$$

Again employing moment-area Theorem 2 we have

$$t_{DB} = \frac{1}{2}(3)\left(\frac{30}{EI}\right)(1) = \frac{45.0}{EI}$$

It may be noted that the shaded portion of *M/EI* diagram in Fig. 5.14*c* is considered and the *x* distance is measured from *D*. Therefore, the required deflection is,

$$DE' = \frac{160.2}{EI} - \frac{45.0}{EI} = \frac{115.2}{EI}$$

Substituting numerical values for *E* and *I*

$$DE' = \frac{115.2}{204 \times 50} = 11.29 \times 10^{-3} \text{ m}$$

or $DE' = 11.29$ mm

which is not much different from the value of maximum deflection of 11.43 mm.

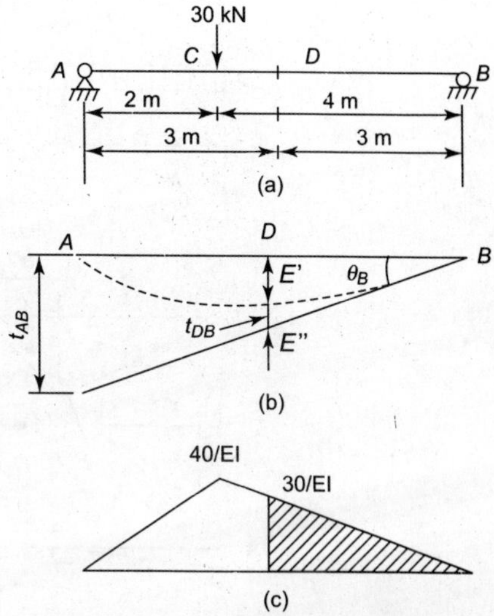

Fig. 5.14 | *(a) Beam and loading, (b) Elastic line, (c) M/EI diagram*

The positive signs of t_{AB} and t_{DB} indicate that the points *A* and *D* lie above the tangent through *B*. As may be seen from Fig. 5.14*b*, the deflection at centre of beam is in the downward direction.

The slope of the elastic curve at *D* can be found from the known slope at one of the ends and employing Eq. 5.5. For point *B* of the right hand support

$$\theta_B = \theta_D + \Delta\theta_{DB}$$

or $$\theta_D = \theta_D + \Delta\theta_{DB}$$

or $$\theta_D = \frac{t_{AB}}{6} - \frac{1}{2}(3)\left(\frac{30}{EI}\right)$$

or $$\theta_D = \frac{53.4}{EI} - \frac{45.0}{EI} = \frac{8.4}{EI} \text{ radians (counter-clockwise)}$$

The above procedure for finding the slope and deflection at a point on the elastic curve is generally applicable. The following example illustrates the application of moment-area theorems to overhanging beams.

Example 5.8 | *It is required to evaluate the deflection at free end A of the overhanging beam shown in Fig. 5.15 caused by applied loads. EI is constant.*

The moment diagram is shown in Fig. 5.15*b*. It may be seen that the point of contraflexure is at *all* from support *B*. At this point an inflection in the elastic curve takes place. The assumed profile of the elastic curve is shown in Fig. 5.15*c*.

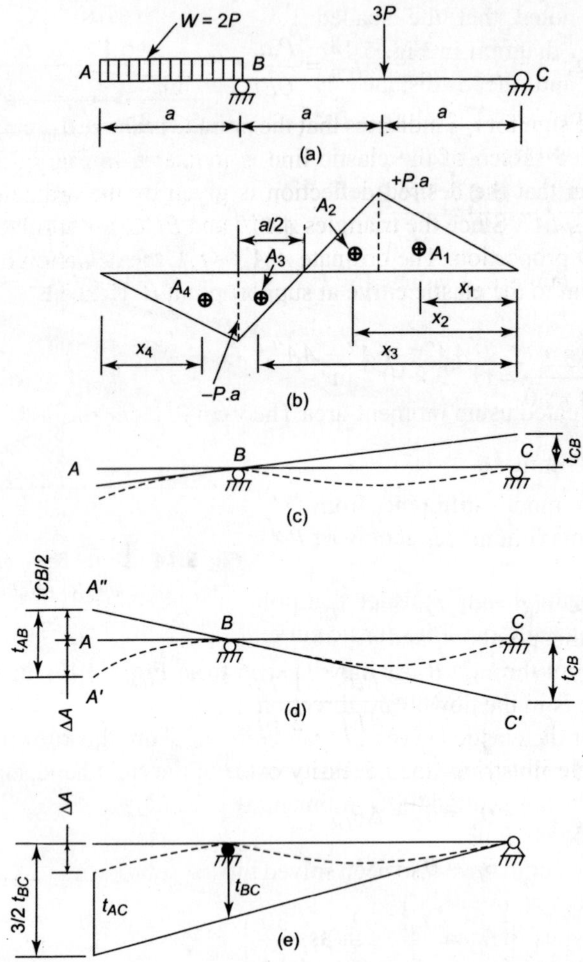

Fig 5.15 | *(a) Beam and loading, (b) Moment diagram, Elastic line—slope at B assumed positive, Elastic line—slope at B assumed negative, An alternative approach*

To start with it is not known whether the slope of the elastic line over support B is positive or negative. We shall find t_{CB} by taking moment of the M/EI diagram area between B and C about point C. This gives

$$t_{CB} = A_1x_1 + A_2x_2 + A_3x_3$$

$$= \frac{1}{EI}\left[\frac{1}{2}(a)(P.a)\left(\frac{2}{3}a\right) + \frac{1}{2}(a/2)(P.a)\left(a + \frac{a}{6}\right) + \frac{1}{2}(a/2)(-P.a)\left(\frac{3}{2}a + \frac{a}{3}\right)\right]$$

On simplifying, $$t_{CB} = \frac{P.a^3}{6EI}$$

The positive sign for t_{CB} indicates that the point C is above the tangent through B. The corrected sketch of the elastic line is indicated in Fig. 5.15d. From the figure it is seen that the desired deflection is given by the ordinate AA' and is equal to $A'A'' - AA''$. Since the triangles $AA''B$ and BCC' are similar the ordinate $AA'' = t_{CB}/2$ by proportion. The ordinate $A'A'' = t_{AB}$, the deviation of the point A from the tangent to the elastic curve at support point B. Hence,

$$AA' = A'A'' - AA'' = t_{AB} - \frac{t_{CB}}{2}$$

t_{AB} can be evaluated using moment-area Theorem 2. Hence from Fig. 5.15b

$$t_{AB} = A_4 x_4$$

or $$t_{AB} = \frac{1}{EI}\left[\frac{1}{3}(a)(-p.a)\left(\frac{3}{4}a\right)\right] = -\frac{Pa^3}{4EI}$$

Here the negative sign indicates that point A is below the tangent through B. Now from the geometry of the elastic curve

$$\Delta_A = \frac{Pa^3}{4EI} - \frac{1}{2}\frac{Pa^3}{6EI} = \frac{Pa^3}{6EI}$$

This example illustrates the necessity of watching for the sign of quantities computed by the application of the moment-area method, although in most cases the signs will be obvious.

The problem could have also been solved by first establishing t_{AC}. This scheme of analysis is shown in Fig. 5.15e.

For some types of loading, such as combination of concentrated and distributed loads, or for varying loads, the moment-area method can become complicated if directly used. The complications arise because the areas and centroids of M/EI diagrams are difficult to evaluate. This difficulty can be overcome in most cases by considering the effects of loads separately and superposing the individual results. For example, the effects of uniformly distributed load w/unit length and concentrated load P on the beam shown in Fig. 5.16 can be considered to act separately.

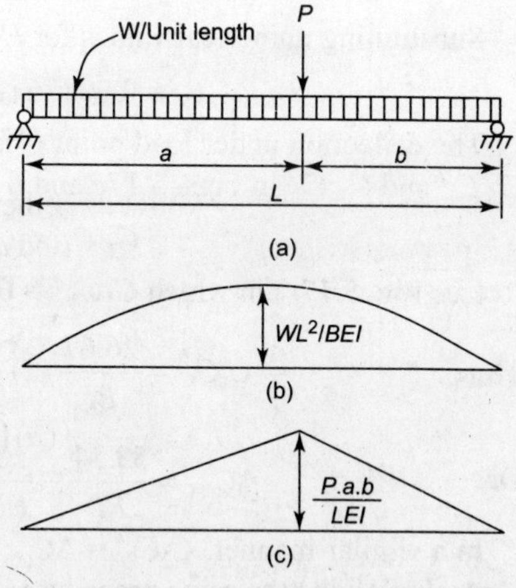

Fig. 5.16 | *(a) Combined loading, (b) M/EI diagram for distributed load, (c) M/EI diagram for concentrated load*

This results in the simpler M/EI diagrams for uniformly distributed load

(Fig. 5.16*b*) and concentrated load (Fig. 5.16*c*) for which the areas and centroids can be conveniently determined. The results from each of the diagrams are then combined for obtaining the total effect for the given loading. The following example further clarifies the point.

Example 5.9 *It is required to evaluate the slope at left hand support A and deflections under load point C and D for the beam shown in Fig. 5.17. E = 205 × 10⁻⁶ kN/m² (205,000 MPa) and I = 25 × 10⁻⁶ m⁴ (25 × 10⁶ mm⁴).*

The elastic line and the *M/EI* diagram for uniform load are shown in Figs. 5.17c and *d* and for concentrated loads in Figs. 5.17*f* and *g*. The slope of the beam at support *A* is seen to be equal to

$$\theta_A = \theta_{A(1)} + \theta_{A(2)}$$

$$\theta_A = \frac{t_{BA(1)}}{4} + \frac{t_{BA(2)}}{4}$$

Using moment-area Theorem 2,

$$t_{AB(1)} = \frac{1}{EI}\left[\frac{2}{3}(4)\,(20)(2)\right] = \frac{106.67}{EI}$$

$$t_{BA(2)} = \frac{1}{EI}\left[\frac{1}{2}(10)\,(2.4)\left(1.6 + \frac{2.8}{3}\right) + \frac{1}{2}(1.6)(-40)\left(\frac{1.6}{3}\right)\right] = \frac{13.34}{EI}$$

Therefore, the actual slope at *A*,

$$\theta_A = \frac{106.67 + 13.34}{4EI} = \frac{30.0}{EI}$$

Substituting numerical values for *E* and *I*, we have

$$\theta_A = 0.00586 \text{ radians (clockwise)}$$

The deflection under load point *C* is seen to be equal to the sum of ordinates $C_1\,C_1'$ and $C_2\,C_2'$ in Figs. 5.17*c* and *f*. We know for the case of uniform loading

$$C_1\,C_1' = C_1\,C_1'' - t_{CA(1)}$$

(see Fig. 5.17*c*) in which $C_1C_1'' = \theta_{A(1)}\,(2)$

Thus, $$C_1C_1' = \frac{26.67}{EI}(2) - \frac{2}{3}(2)\left(\frac{20}{EI}\right)\left(\frac{3}{8}\right)(2)$$

Or $$\Delta C_{(1)} = \frac{53.34}{EI} - \frac{20}{EI} = \frac{33.34}{EI} \text{ downwards.}$$

In a similar manner, $C_2C_2' = \Delta C_{(2)}$ is evaluated for the case of concentrated loads. Following the same procedure

$$C_2C_2' = C_2C_2'' - t_{CA(2)}$$

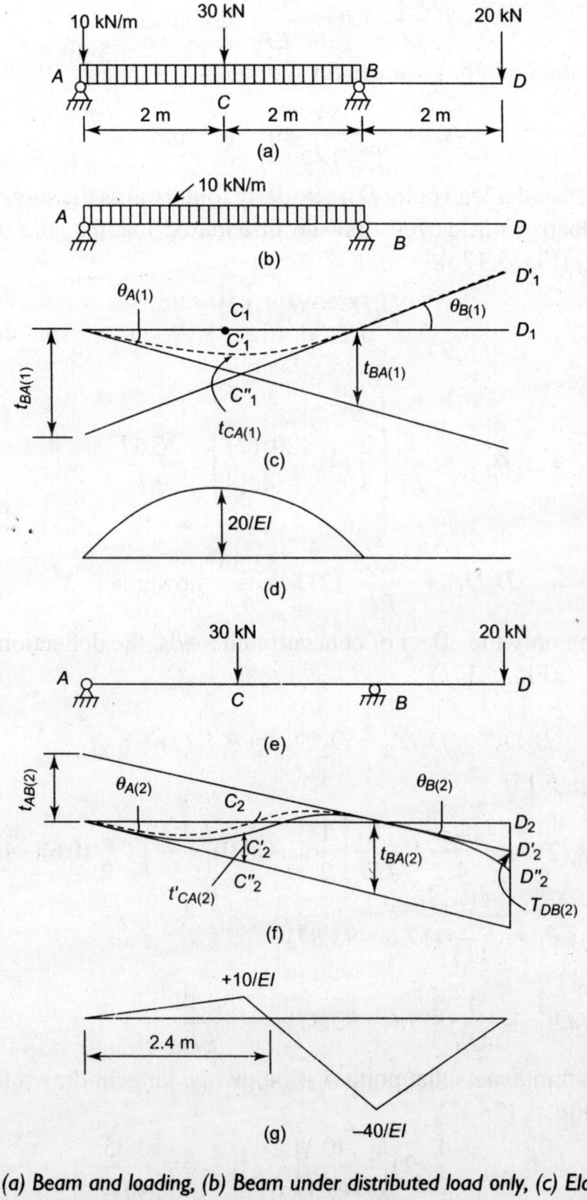

Fig. 5.17 | *(a) Beam and loading, (b) Beam under distributed load only, (c) Elastic curve due to distributed load only, (d) M/EI diagram due to distributed load only, (e) Beam under concentrated loads only, (f) Elastic curve due to concetnrated loads only, (g) M/EI diagrum due to concentrated loads only*

or
$$\Delta_{C(2)} = \frac{3.34}{EI}(2) - \frac{1}{2}(2)\left(\frac{10}{EI}\right)\left(\frac{2}{3}\right) = 0$$

The elastic curve in Fig. 5.17*f* is accordingly corrected and shown in dotted line. The deflection under load point C is therefore,

$$\Delta_C = \frac{33.34}{EI} + 0 = \frac{33.34}{EI} \text{ downwards.}$$

Substituting the numerical values for E and I

$$\Delta_C = \frac{33.34}{205 \times 25} \text{ m} = 6.5 \text{ mm}$$

The deflection under load point D is similarly found out as the sum of individual effects of the load. Considering first the distributed loading, the deflection of point D is D_1D_1 (Fig. 5.17c).

Hence,
$$D_1D'_1 = \theta_{B(1)}\,(2)$$

Again
$$\theta_{B(1)} = \frac{t_{AB(1)}}{4}$$

or
$$\theta_{B(1)} = \frac{1}{EI}\left[\frac{2}{3}(4)\frac{(20)(2)}{4}\right] = \frac{26.67}{EI}$$

Therefore,

$$D_1D_1' = \frac{26.67}{EI}(2) = \frac{53.34}{4} \text{ upwards}$$

Next considering only the effect of concentrated loads, the deflection under load point D is D_2D_2'' (Fig. 5.17*f*).

Thus,

$$D_2D_2'' = D_2D'_2 + D_2'D_2'' = \theta_{B(2)}(2) + t_{DB(2)}$$

We see from Fig. 5.17*f*

$$\theta_B(2) = \frac{t_{AB(2)}}{4} = \frac{1}{4EI}\left[\frac{1}{2}(2.4)(10)\left(\frac{4.4}{3}\right) + \frac{1}{2}(1.6)(-40)\left(\frac{8.8}{3}\right)\right]$$

$$\theta_2 = -\frac{1}{4EI}(17.6 - 93.87)$$

or
$$D_2D_2' = \frac{1}{4EI}(17.6 - 93.87) = -\frac{38.14}{EI}$$

The negative sign indicates that point A is below the tangent drawn from B.

Again from Fig. 5.17*f*

$$t_{DB(2)} = \frac{1}{2}(2)\left(-\frac{40}{EI}\right)\left(\frac{2}{3}\right)(2) = -\frac{53.33}{EI}$$

Thus the deflection under load point D is given by

$$\Delta_D = D_2D_2' + D_2'D_2''$$

$$\Delta_D = \frac{-38.14 - 53.33}{EI} = -\frac{91.47}{EI}$$

Substituting the numerical values for E and I

$$\Delta_D = -17.8 \text{ mm}$$

It may be remembered that it is necessary to keep track of the sign of the evaluated quantities and interpret to get the desired slopes and deflections as has been done in the above example.

The moment-area method can also be employed to determine the slopes and deflections in a frame. Its use on frames requires a detailed consideration of the deflected shape. The following examples will illustrate this point.

Example 5.10 | *The slope and the horizontal and vertical deflections are to be determined at point C for the frame shown in Fig. 5.18a.*

$E = 205 \times = 10^6$ kN/m^2 (205,000 MPa) and $I = 10 \times 10^{-6}$ m^4 (10×10^6 mm^4).

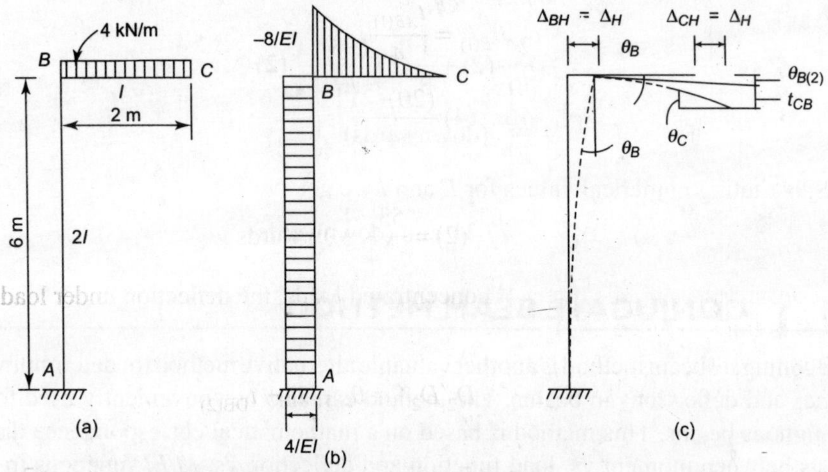

Fig. 5.18 | *(a) Frame and loading, (b) M/EI diagram, (c) Elastic line*

The *M/EI* diagrams for the frame is shown in Fig. 5.18*b*. Beam sign convention is adopted for the moment diagram for each member. End *A* is considered to be the left end of member *AB* and end *B* is considered as the left end of member *BC*.

The deflected shape of the frame is shown in Fig. 5.18*c*. The desired slope at *C* is seen to be equal to the slope at *B* plus the change in slope between *B* and *C*. If bending deflections only are considered, neglecting axial deformations, the horizontal deflection at $C = \Delta_{CH}$ will be equal to the horizontal deflection at $B = \Delta_{BH}$. The vertical deflection at *C* is due to the slope at B as well as due to deviation t_{CB}. Therefore, the vertical deflection at *C* is equal to the product of θ_B and length *BC* plus the deviation t_{CB}. From moment-area Theorem 1, θ_C is found to be

$$\theta_C = \theta_B + \Delta\theta_{BC}$$

or

$$\theta_C = \frac{6(-4)}{EI} + \frac{1}{3}(2)\frac{(-8)}{EI} = -\frac{29.33}{EI}$$

The horizontal deflection at C which is also equal to the horizontal deflection of point B is

$$t_{BA} = 6\frac{(-4)}{EI}(3) = -\frac{72.0}{EI} \text{ (to the right)}$$

Substituting the numerical values for E and I

$$\Delta_{BH} = \Delta_{CH} = 35.12 \text{ mm.}$$

Generally the direction of deflections in frames can be fixed from observation.

The vertical deflection at C is found from the expression

$$\Delta_{CV} = \theta_B(2) + t_{CB}$$

$$= -\frac{24}{EI}(2) + \frac{1}{3}(2)\frac{(-8)}{EI}\frac{3}{4}(2)$$

$$= -\frac{56}{EI} \text{ (downwards)}$$

Substituting numerical values for E and I we get

$$\Delta_{CV} = -27.31 \text{ mm (downwards)}$$

5.3 CONJUGATE BEAM METHOD

The conjugate beam method is another valuable alternative method for determining slopes and deflections in beams. The method can also be conveniently used for continuous beams. This method is based on a mathematical correspondence that exists between moment vs. load function and deflection vs. *M/EI* functions in a beam. If the deflected shape of the beam is described by the function *y(x)*, the following general relationships exist:

y = deflection ordinates of the elastic curve

$$\frac{dy}{dx} = \theta = \text{slope of the elastic curve} \quad (5.13)$$

$$\frac{d\theta}{dx} = \frac{d^2y}{dx^2} = \frac{M_x}{EI} \quad (5.14)$$

$$\frac{d^3y}{dx^3} = \frac{dM_x}{EIdx} = \frac{V_x}{EI} \quad (5.15)$$

$$\frac{d^4y}{dx^4} = \frac{dV_x}{EIdx} = \frac{w_x}{EI} \quad (5.16)$$

The validity of these relationships depends on the sign convention used for various quantities. The coordinate system that satisfies these relationships are: y upwards and x to the right are positive, shear and moment sign convention is the same as given in Figs. 2.5*a* and *b* and the load is positive when it acts upwards.

In the beam given in Fig. 5.19*a* the uniform load acting upwards is positive. Figures 5.19*b*, *c*, *d* and *e* show the corresponding shear, moment, slope and deflection quantities. The differential relationships are shown on the left side of the figures. As pointed out earlier, from Fig. 5.19 it is seen that there is a mathematical correspondence between the moment vs. load function, the deflection vs. *M/EI* function and also the slope vs. shear function. For example, the moment function can be obtained by successively integrating twice the load function. Similarly, the deflection function can be obtained by successively integrating twice the *M/EI* function. These relationships lead to the conjugate beam concept for evaluating deflection and slopes. If the *M/EI* diagram for a given beam and loading is considered to be the loading on an imaginary beam known as *conjugate beam,* the following two principles of conjugate beams can be stated.

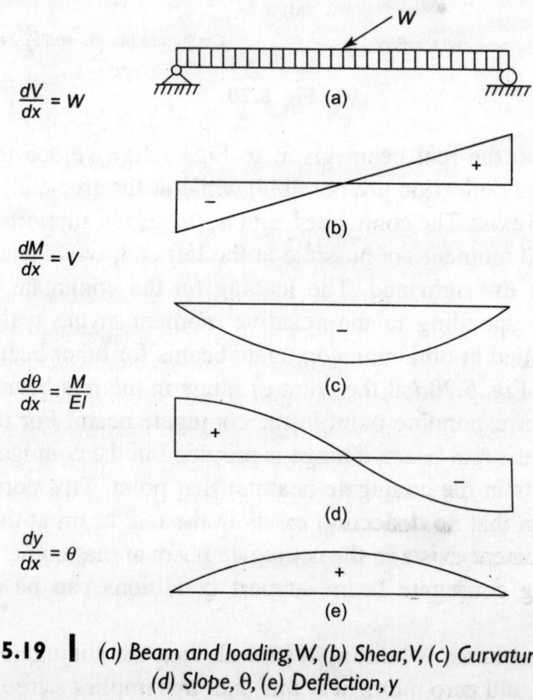

Fig. 5.19 | *(a) Beam and loading, W, (b) Shear, V, (c) Curvature, MEI, (d) Slope, θ, (e) Deflection, y*

THEOREM 1 *The shear at any point on the conjugate beam is equal (in sign and value) to the slope at the corresponding point on the real beam.*

THEOREM 2 *The moment at any point on the conjugate beam is equal (in sign and value) to the deflection at the corresponding point on the real beam.*

The supports of the conjugate beam are such that the shear and moment that are obtained in the conjugate beam are consistent with the slopes and deflections in the real beam. The conjugate beams with *(M/EI)* loading for various beams and loadings are shown in Fig. 5.20.

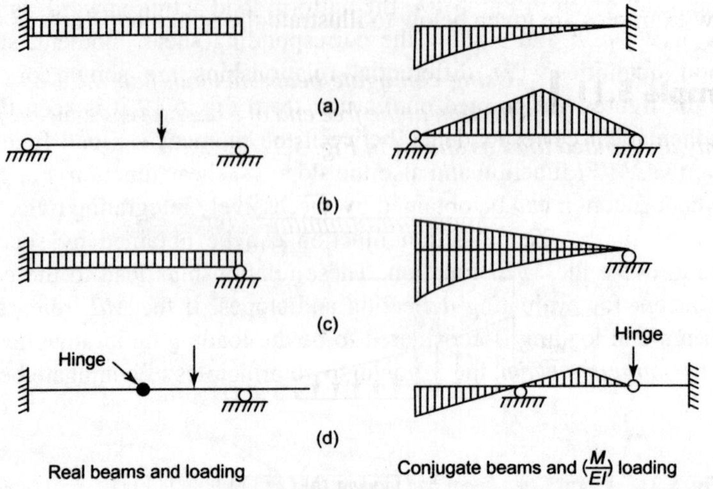

Fig. 5.20

With regard to the real beam given in Fig. 5.20*a* we see that at the fixed end, no slope and deflection are possible, while at the free end, both slope and deflection would exist. The conjugate beam is, therefore, supported in such a way that no shear and moment are possible at the left end, while shear and moment are generated at the right end. The loading on the conjugate beam is acting downwards corresponding to the negative moment in the real beam. Similar reasoning is applied in obtaining conjugate beams for other beams in Fig. 5.20. For example, in Fig. 5.20*d* at the point of hinge in the real beam, there exists a support at the corresponding point in the conjugate beam. For the intermediate support point in the real beam, a hinge is provided in the conjugate beam so that no moment exists in the conjugate beam at that point. This corresponds to the support condition that no deflection exists in the real beam at the support point and hence no moment exists in the conjugate beam at that point.

The foregoing conjugate beam support conditions can be summed up as follows:

1. If the real beam is built in at its end, then the conjugate beam has zero reaction and zero moment at this end; this implies a free end.
2. If the real beam has a free end, then the conjugate beam has a reaction and a moment at this end; this implies a fixed end.
3. If the real beam is continuous, then the conjugate beam has zero reactions at all the real beam interior support points. An exception to this occurs only when the real beam is hinged at an intermediate point. The hinge introduces an abrupt change of slope in the real beam and, therefore, a concentrated force at that point of the conjugate beam (*see* Fig. 5.20*d*).
4. The moment of the conjugate beam at all the real beam interior support points is equal to the deflection of these real supports. In the case of unyielding supports the conjugate beam moment is zero.

A few examples are given below to illustrate the principles involved.

Example 5.11 *Using conjugate beam method, find the slope and deflection at the free end of a cantilever beam subjected to uniformly distributed load as shown in Fig. 5.21a.*

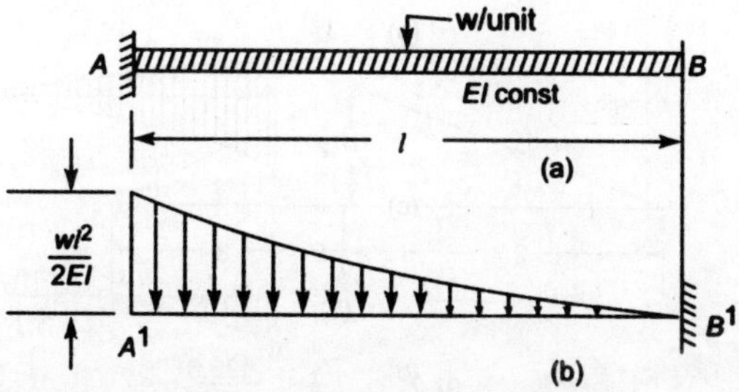

Fig. 5.21 *(a) Given beam and loading; (b) Conjugate beam under M/EI loading*

The conjugate beamand the loading on the beam is shown in Fig. 5.21*b*. In the conjugate beam, beam end A becomes free end *A′* and free end *B* becomes fixed end *B′* as shown. At end *B′* in the conjugate beam there exists shear as well as moment which relates to slope and deflection in the given beam

The shear at *B′* is equal to area of the *M*/*EI* diagram between *A′* and *B′*.

Therefore $$V'_B = \frac{1}{3}\cdot l \cdot \frac{\omega l^2}{2EI} = \frac{\omega l^3}{6EI}$$

That is $$\theta_B = \frac{\omega l^3}{6EI}$$

The deflection at *B* of the real beam is determined by evaluating M'_B in the conjugate beam.

M'_B = moment of the area of the *M*/*EI* diagram taken about *B′*

That is $$M'_B = \frac{1}{3}\cdot l \cdot \frac{\omega l^2}{2EI}\cdot\frac{3l}{4} = \frac{\omega l4}{8EI}$$

or $$\delta_B = \frac{\omega l4}{8EI}$$

Example 5.12 *Find the deflection of the beam shown in Fig. 5.22 under the point load. Take $E = 210 \times 10^6$ kN/m², $I_1 = 160 \times 10^{-6}$ m⁴ and $I_2 = 120 \times 10^{-6}$ m⁴*

Step I: To draw the M/EI diagram

The given beam is shown in Fig. 5.22*a*. the moment diagram is given 5.22*b*. The conjugate beam and the loading (*M*/*EI* diagram) is shown in Fig. 5.22*c*. The conjugate beam is also a sinply supported beam lke the given one.

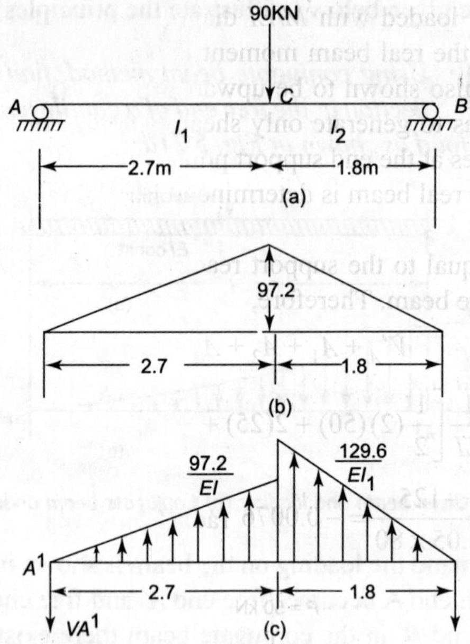

Fig. 5.22 | *(a) Given beam and loading; (b) M-diagram; (c) M/EI-diagram*

Step 2: To evaluate reaction component V'_A

$$\text{Writing } \Sigma M_{B'} = -V_{A'}(4.5) + \frac{1}{2}(2.7)\left(\frac{(97.2)}{EI_1}\right)(2.7) + \frac{1}{2}(1.8)\left(\frac{(129.6)}{EI_1}\right)(1.2) = 0$$

$$\text{Gives } V_{A'} = \frac{109.84}{EI_1}$$

Step 3: Deflection under load point

The deflection under load point *C* can be obtained by evaluating the moment in the conjugate beam at *C′*.

$$M_{C1} = \frac{109.84}{EI_1}(2.7) - \frac{1}{2}(2.7)\left(\frac{97.2}{EI}\right)\left(\frac{2.7}{3}\right) = \frac{178.47}{EI_1}$$

Substituting for *E* and *I*, values

$$\delta_C = \frac{178.47}{160 \times 210} = 0.00531 \text{ m}$$

or

$$5.31 \text{ mm}$$

Example 5.13 | *It is required to determine the deflection at centre point C and slopes at ends A and B of the beam shown in Fig. 5.23a by the conjugate beam method. $E = 205 \times 10^6$ kN/m^2 (205,000 MPa) and $I = 80 \times 10^{-6}$ m^4 (80×10^6 mm^4)*

The conjugate beam loaded with M/EI diagram of the real beam is shown in Fig. 5.23*b*. Because the real beam moment is positive, the M/EI loading on the conjugate beam is also shown to be upwards in the positive direction. The end supports are such as to generate only shearing force but no moment. This arrangement gives slopes at the end support points but no deflection.

The slope at A in the real beam is determined by evaluating the shear at A' in the conjugate beam.

The shear at A' is equal to the support reaction V'_A and is equal to half the loading on the conjugate beam. Therefore,

$$V'_A + A_1 + A_2 + A_3 = 0$$

or
$$V'_A = -\frac{1}{EI}\left[\frac{1}{2}(2)(50) + 2(25) + \frac{1}{2}(2)(25)\right] = -\frac{125}{EI}$$

that is
$$\theta_A = \frac{-125}{2.05 \times 80} = -0.0076 \text{ radians (clockwise)}$$

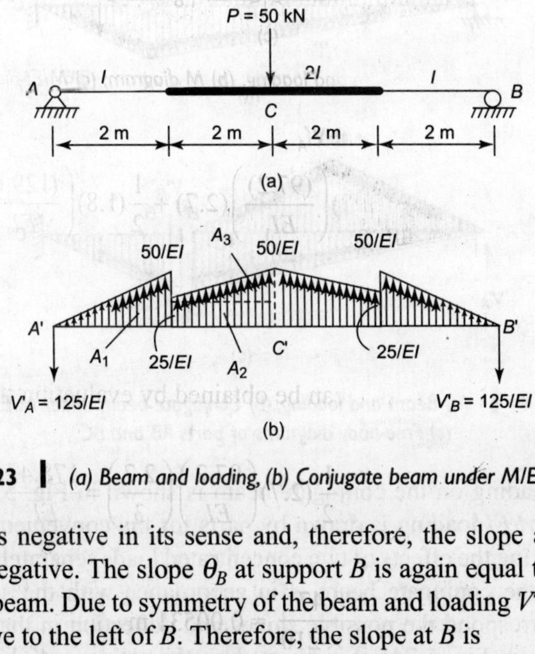

Fig. 5.23 | *(a) Beam and loading, (b) Conjugate beam under M/EI loading*

The shear is negative in its sense and, therefore, the slope at A in the real beam is also negative. The slope θ_B at support B is again equal to shear at B' in the conjugate beam. Due to symmetry of the beam and loading $V'_A = V'_{B'}$, but the shear is positive to the left of B. Therefore, the slope at B is

$$\theta_B = V'_B{}' = +0.0076 \text{ radians (anti-clockwise)}$$

The deflection at C in the real beam is determined by evaluating M'_C in the conjugate beam. The value of M'_C is obtained by summing up of moments at point C' on the conjugate beam, that is,

$$M'_C = -\frac{125}{EI}(4) + \frac{50}{EI}\left(2 + \frac{2}{3}\right) + \frac{50}{EI}(1) + \frac{25}{EI}\left(\frac{2}{3}\right) = -\frac{300}{EI}$$

The moment in the conjugate beam at C' is negative. The deflection at C in the real beam is also negative, that is, the deflection is downwards and its value is

$$\Delta_C = \frac{-300}{205 \times 80} \text{ m or } 18.29 \text{ mm.}$$

Example 5.14 | *An overhanging beam has the dimensions and loading as shown in Fig. 5.24a. Using the conjugate beam method, find the slopes at A and B, and the deflection at point C. EI is constant.*

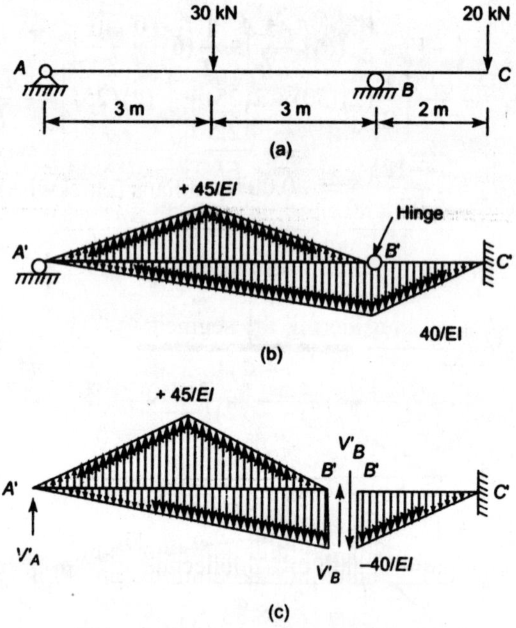

Fig. 5.24 | *(a) Beam and loading, (b) Conjugate beam under M/EI loading, (c) Free-body diagrams of parts AB and BC*

The *M/EI* loading on the conjugate beam is shown in Fig. 5.24*b*. It may be noted that the *M/EI* loading is drawn by parts for the convenience of areas and centroids by taking the effects of two concentrated loads separately. The direction of loading on the conjugate beam is in accordance with the sign convention adopted. To correspond the possible slopes and deflection in the real beam, the conjugate beam in Fig. 5.24*b* is supported on the roller at A', hinged at B' and fixed at C' The pin at B' results in zero moment, which is consistent with zero deflection at B in the real beam. The slopes at point A and B in the real beam are determined by evaluating the shear at A' and B' in the conjugate beam. The shear forces can be conveniently evaluated by considering the free-body diagram of the conjugate beam shown in Fig. 5.24*c*. Assuming V'_A and V'_B as acting upwards and summing up moments about point B', we have

$$V'_A(6) + \frac{1}{2}(6)\left(\frac{45}{EI}\right)(3) + \frac{1}{2}(6)\left(-\frac{40}{EI}\right)\left(\frac{6}{3}\right) = 0$$

$$V'_A = -\frac{27.5}{EI}$$

The sign for shear is negative and so the slope at A is also negative, that is

$$\theta_A = -\frac{27.5}{EI}\text{(clockwise)}$$

Again equating all the transverse forces on segment $A'B'$ to zero, we have

$$V'_A + V'_B + \frac{1}{2}(6)\left(\frac{45}{EI}\right) + \frac{1}{2}(6)\left(-\frac{40}{EI}\right) = 0$$

that is,

$$V'_B = \frac{12.5}{EI}$$

Thus, shear at B' is positive and so the slope at B is also positive. Therefore,

$$\theta_B = \frac{12.5}{EI}\text{ (anti-clockwise).}$$

M'_C is found now by considering the segment $B'C'$ of the conjugate beam. Thus, we have

$$M'_C = V'_B(2) + \frac{1}{2}(2)\left(-\frac{40}{EI}\right)\left(\frac{4}{3}\right)$$

$$= \frac{1}{EI}[-25.0 - 53.33] = -\frac{78.33}{EI}$$

Because this moment is negative, the deflection in the real beam at C is

$$\Delta_C = M'_C = -\frac{78.33}{EI}\text{(downwards)}$$

Note that the slope and deflection at any point on the real beam is readily obtained from the free-body diagram of the conjugate beam in Fig. 5.24*c*.

5.4 DEFLECTION OF TRUSSES—GRAPHICAL METHOD

5.4.1 Williot-Mohr diagram

A graphical method presented in the following paragraphs provides a means for obtaining the deflections of statically determinate truss structures. The deflections in a truss arise from axial extensions or shortenings of members. The two common sources of deformations are the axial forces in members due to applied loads or temperature changes. It may be recalled from studies of strength of materials that the axial deformation in a prismatic member subjected to an axial force P may be expressed

$$e = \frac{PL}{AE} \tag{5.16}$$

where L, A and E are the length, area of cross-section and Young's modulus respectively. Similarly, the deformations due to temperature change ΔT can be expressed as

$$e = \alpha L \Delta T \tag{5.17}$$

where α is the coefficient of thermal expansion. Following the same sign convention as for forces, the extension is considered as a positive quantity and shortening a negative quantity.

Let us now consider a small truss (Fig. 5.25*a*) to develop a method for determining deflections due to axial deformations in members. Under loading, let member AC undergo an extension e_{AC}, and the member BC shorten by an amount E_{BC}. The resulting displacement of joint C can be determined as shown in (Fig. 5.25*b*). The amount of extension in member

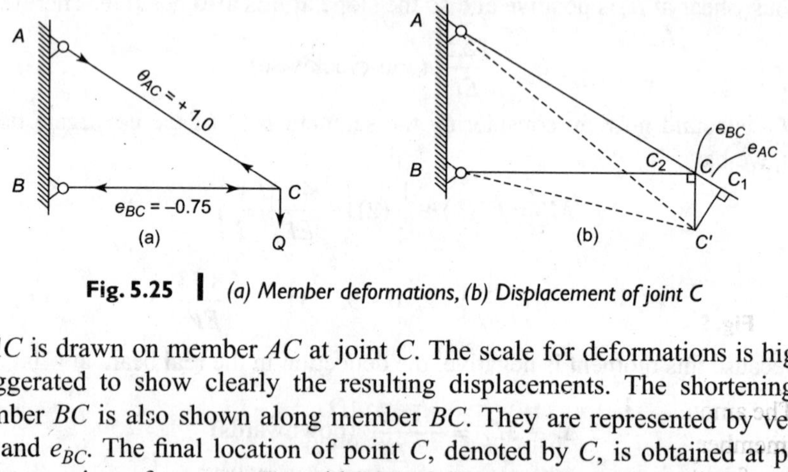

Fig. 5.25 | *(a) Member deformations, (b) Displacement of joint C*

AC is drawn on member AC at joint C. The scale for deformations is highly exaggerated to show clearly the resulting displacements. The shortening of member BC is also shown along member BC. They are represented by vector e_{AC} and e_{BC}. The final location of point C, denoted by C, is obtained at point of intersection of arcs swung with A and B as centres and extended length of AC (AC_1) and shortened length of BC(BC_2) as radii respectively. Because member deformations are small in comparison with their lengths, it is sufficient to represent the arcs by tangents or straight lines perpendicular to the original direction of the members. The resulting location of joint C is found to be at C' as shown in Fig. 5.25*b*.

Apparently it appears that the procedure can be extended to large trusses. Let us see what happens in the process. As an example, let us consider the truss of Fig. 5.26*a*. The member deformations to be considered are shown next to each member. To analyse the structure, it must be remembered that the two points A and B are fixed in space. We can determine the displacement of joint C' as we did in the previous example. The displaced position of joint C is shown as C in Fig. 5.26*b*. The displacement of D is found with respect to points A and C which can serve as fixed points. The left end of member CD now takes position C'. The amount of deformation in member CD is then constructed at its right end.

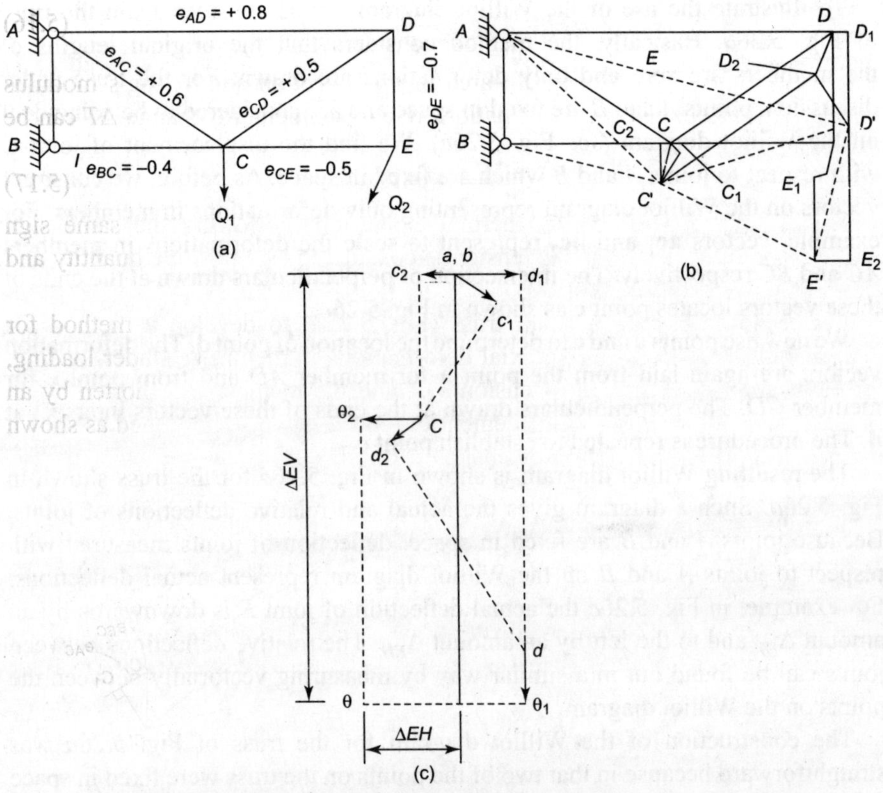

Fig. 5.26 | *(a) Truss and member deformations, (b) Graphical analysis for truss deflections, (c) Williot diagram for truss deflections*

The amount of deformation in member *AD* is constructed at the right end *D* of the member *AD*. Then the location of *D*′ is found at the intersection of the normals drawn from the ends of the deformation vectors as shown in Fig. 5.26*b*.

The location of *E*′ is found by considering *C* and *D*′ as fixed points. The member *CE* and *DE* are moved so that ends *C* and *D* coincide with ends *C* and *D*′ respectively and the member deformations are constructed at the ends in the original direction of members *CE* and *DE*. The intersection of the lines drawn perpendicular to the deformed lengths locates point *E*′. The construction is shown in Fig. 5.26*b*.

It may be noticed that in this construction the length of members and the deformations must be drawn to the same scale. For obtaining a correct solution, enormous drawing space is needed, which is neither possible nor desirable. To overcome this difficulty, we can draw a diagram of member deformations only. Such a diagram is known as *Williot diagram* named after the engineer who originated it. The Williot diagram affords an accurate graphical means of determining joint displacements without using a large scale drawing of a truss.

To illustrate the use of the Williot diagram, let us consider again the truss of Fig. 5.26*a*. Basically, the method considers that the original lengths of the members are zero and only deformations are drawn. For the truss under discussion, points *A* and *B* are fixed in space and are considered to be coincident on the Williot diagram *(see* Fig. 5.26*c*). We find the displacement of joint *C* with respect to joints *A* and *B* which are fixed in space. As before, we construct vectors on the Williot diagram representing only deformations in members. For example, vectors $\mathbf{ac}_1$ and $\mathbf{bc}_2$ represent to scale the deformations in members *AC* and *BC* respectively. The intersection of perpendiculars drawn at the ends of these vectors locates point **c** as shown in Fig. 5.26*c*.

We now use points **a** and **c** to determine the location of point **d**. The deformation vectors are again laid from the point **a** for member *AD* and from point **c** for member *CD*. The perpendiculars drawn at the ends of these vectors intersect at **d**. The procedure is repeated to establish point *e*.

The resulting Williot diagram is shown in Fig. 5.26*c* for the truss shown in Fig. 5.26*a*. Such a diagram gives the actual and relative deflections of joints. Because joints *A* and *B* are fixed in space, deflection of joints measured with respect to joints *A* and *B* on the Williot diagram represent actual deflections. For example, in Fig. 5.26*c* the actual deflection of joint *E* is downwards by an amount Δ_{EV} and to the left by an amount Δ_{EH}. The relative deflections between joints can be found out in a similar way by measuring vectorially between the points on the Williot diagram.

The construction of the Williot diagram for the truss of Fig. 5.26*a* was straightforward because in that two of the points on the truss were fixed in space and they could serve as a starting point. Note that in the construction of the Williot diagram, the location of the two adjacent points must be known in order to fix the displaced position of any other point. In the truss given in Fig. 5.27*a* no such condition exists for starting the Williot diagram. For example, joint *A* is fixed in position but joint *B* or *E* will displace. To start the construction of the Williot diagram, we need either another fixed point or a fixed direction. However, we can temporarily assume the direction of a member or, if necessary, the location of a point fixed to construct the Williot diagram which can be corrected later to yield the correct solution.

As an illustration, consider the truss in Fig. 5.27*a* and assume that the direction of member *AB* is fixed. Point **a** in the Williot diagram is fixed first. Since the direction of member *AB* is assumed fixed, point **b** is located in the direction of *AB* at a distance equal to the extension of member *AB*. Then point **e** is fixed at the intersection of the perpendiculars drawn on the deformation vectors laid for members *BE* and *AE*. The remaining Williot diagram is completed as earlier and is shown in Fig. 5.27*b*. This results in a deflected shape of the truss as shown in Fig. 5.27*c*. The indicated vertical deflection of joint *D* in Fig. 5.27*c* violates the support condition for which only horizontal displacement is possible. The inaccuracy has arisen as a result of the assumption made that member *AB* will remain fixed in direction. Therefore, it is necessary to correct so that the point

D is brought back to its proper position. For this, we rotate the whole truss by a small angle $d\theta$ in the clockwise direction about joint A until the vertical deflection at support D is eliminated. Such a rotation is equivalent to applying a correction to each joint by an amount

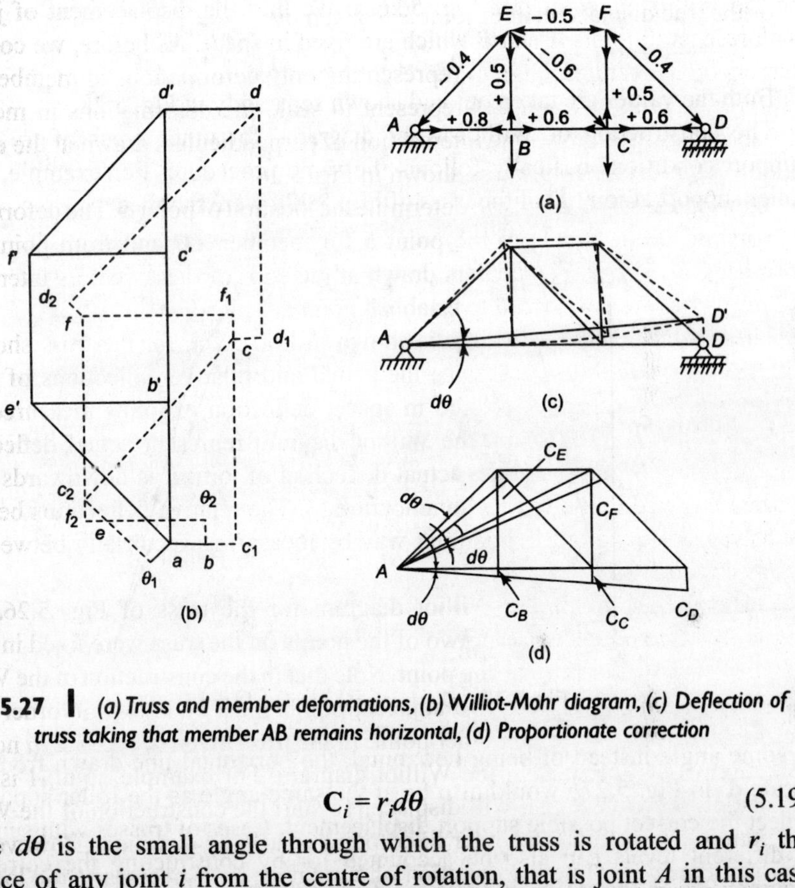

Fig. 5.27 | *(a) Truss and member deformations, (b) Williot-Mohr diagram, (c) Deflection of truss taking that member AB remains horizontal, (d) Proportionate correction*

$$\mathbf{C}_i = r_i d\theta \tag{5.19}$$

where $d\theta$ is the small angle through which the truss is rotated and r_i the distance of any joint i from the centre of rotation, that is joint A in this case (*see* Fig. 5.21*d*). Since correction C_i is proportional to radius r_i, we can draw a convenient correction diagram on the Williot diagram itself. The basis for such a correction is shown in Fig. 5.27*d* where the corrections should be perpendicular to the respective radii and magnitudes proportional to the radii.

In Fig. 5.27*b* vector **ad** represents the displacement of joint D with respect to joint A fixed in space. Vector **ad** can be resolved into vertical component **ad′** and horizontal component **d′d**. It is the vertical component that has to be eliminated. The correction can, therefore, be effected by constructing a scaled diagram of the truss on **ad′** as shown in Fig. 5.27*b*. The points on the correction diagram are denoted by primed alphabets.

The above correction is known as *Mohr's correction diagram* as it was originated by Mohr. The resulting diagram is known as the *Williot-Mohr diagram*. The true displacement of any truss joint is obtained by the vector measured from the primed alphabets to the corresponding non-primed ones. To verify this consider the displacement of joint D with respect to joint A fixed in space; **d′d** gives the displacement, that is,

$$\Delta_{DH} = C_D + \mathbf{ad} \tag{5.20}$$

or the true displacement of joint F is given by

$$\Delta_F = \mathbf{f'f} = C_F + \mathbf{af} \tag{5.21}$$

Both the values are taken out and shown separately in Fig. 5.28.

The construction of Williot-Mohr diagrams for other types of trusses or support conditions basically follows the same procedure. For example, if the roller support at joint D of the truss of Fig. 5.27*a* were inclined

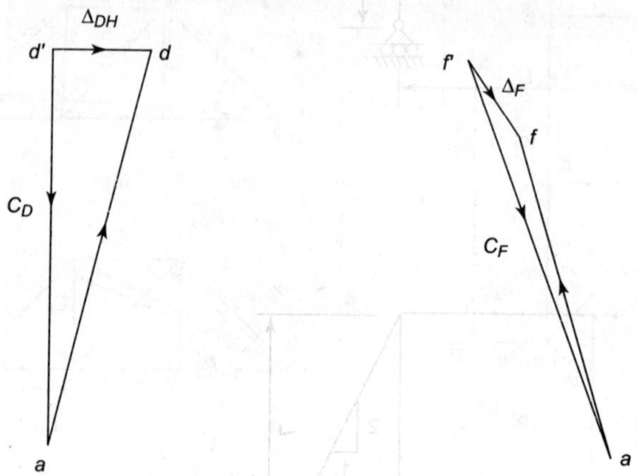

Fig. 5.28 *Displacement of Joints D and F*

at some angle instead of being horizontal, the horizontal line drawn from d to locate **d′** in Fig. 5.27*b* would also be at the same angle as the roller support to reflect the correct possible support displacement. Cases of trusses with supports at different levels can also be accounted for by constructing the correction diagram 90° to the given orientation of the truss and obtaining its scale from support considerations.

Problems for Practice

5.1 Draw the deflected shapes of the structures shown in Fig. 5.29. Neglect axial deformation of members. Indicate the possible location of points of contraflexure.

Use the moment-area method to solve problems 5.2 to 5.10.

5.2 Find the deflection and angular rotation of the free end of an aluminium cantilever beam shown in Fig. 5.30.

$E = 70 \times 10^6$ kN/m^2 (70,000 MPa)

$I_1 = 2 \times 10^{-6}$ m^4 (2×10^6 mm^4)

$I_2 = 0.4 \times 10^{-6}$ m^4 (0.4×10^6 mm^4)

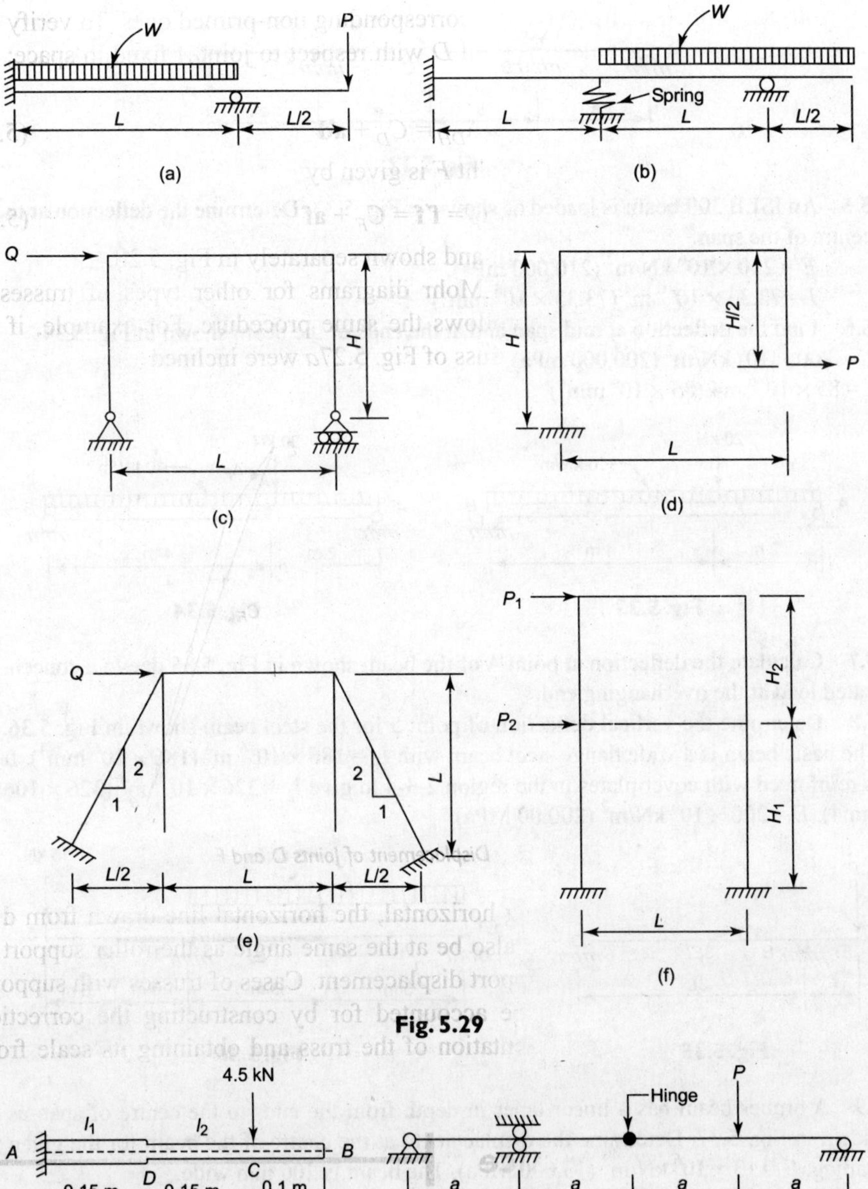

Fig. 5.29

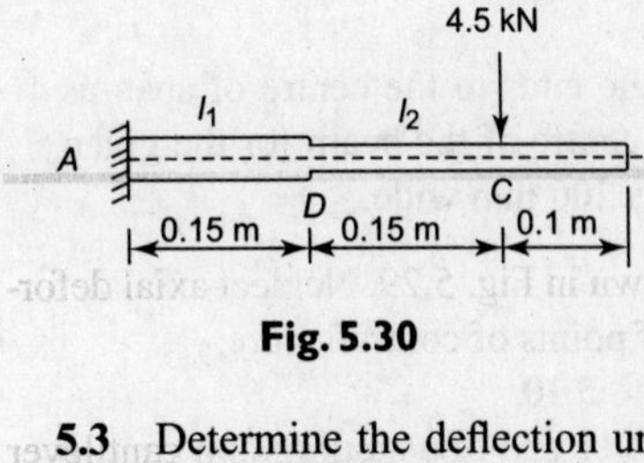

Fig. 5.30

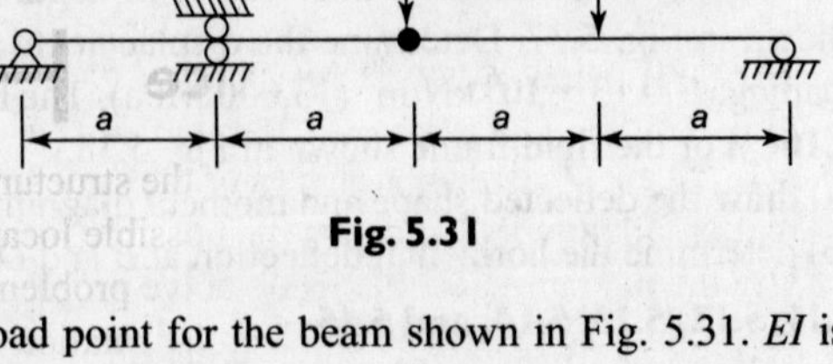

Fig. 5.31

5.3 Determine the deflection under load point for the beam shown in Fig. 5.31. *EI* is constant.

5.4 A compound beam *AE* consisting of two identical portions *AC* and *CE* hinged together at *C* is supported and loaded as shown in Fig. 5.32. Find the vertical deflection of point *E*. *El* is constant.

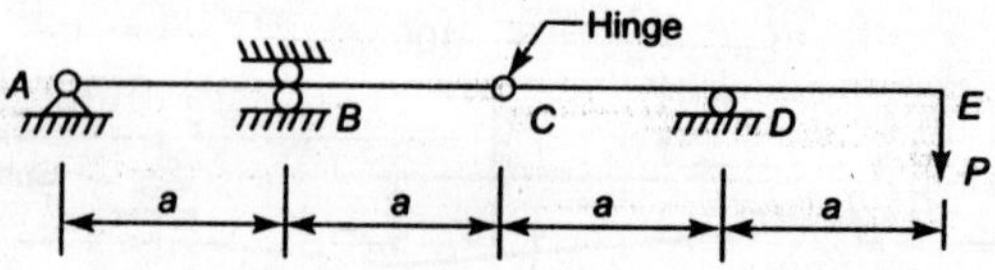

Fig. 5.32

5.5 An ISLB 300 beam is loaded as shown in Fig. 5.33. Determine the deflection at the centre of the span.

$E = 210 \times 10^6$ kN/m^2 (210,000 mPa),

$I = 73.33 \times 10^{-6}$ m^4 (73.33×10^6 mm^4)

5.6 Find the deflection at mid span and at the ends of the beam shown in Fig. 5.34.
$E = 200 \times 10^6$ kN/m^2 (200,000 mPa),
I = 85×10^{-6} m4 (85×10^6 mm^4)

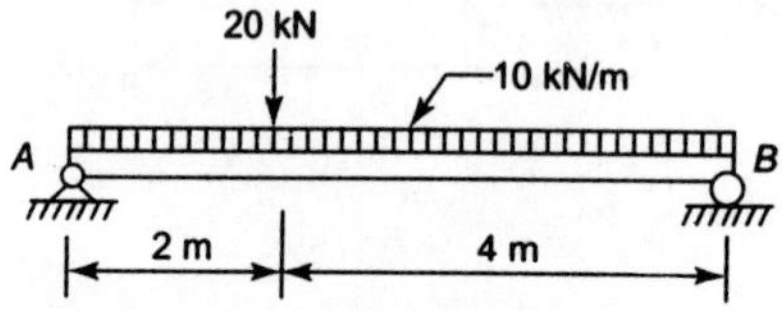

Fig. 5.33

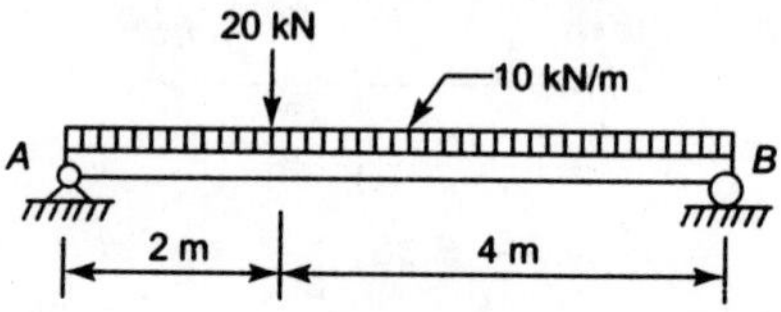

Fig. 5.34

5.7 Calculate the deflection at point A of the beam shown in Fig. 5.35 due to a concentrated load at the overhanging end.

5.8 Determine the vertical deflection of point 5 for the steel beam shown in Fig. 5.36. The basic beam is a wide flange steel beam with $I_1 = 186 \times 10^{-6}$ m^4 (186×10^6 mm^4). It is reinforced with cover plates in the region 2-3-4 to give $I_2 = 326 \times 10^{-6}$ m^4 (326 × 106 mm4). $E = 200 \times 10^6$ kN/m^2 (200,00 MPa).

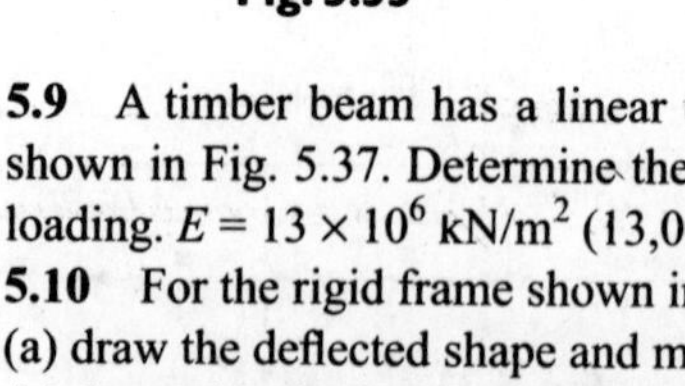

Fig. 5.35

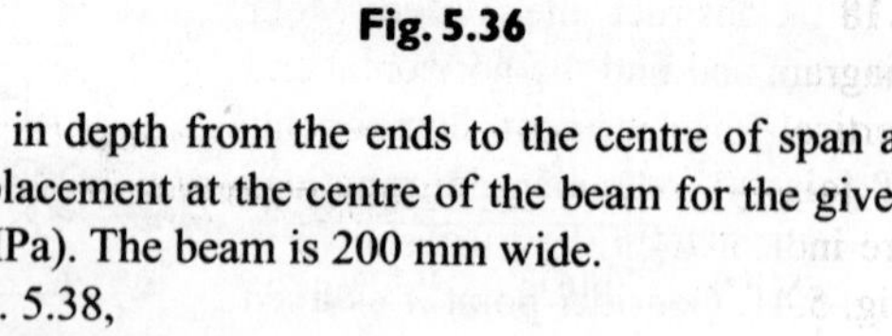

Fig. 5.36

5.9 A timber beam has a linear taper in depth from the ends to the centre of span as shown in Fig. 5.37. Determine the displacement at the centre of the beam for the given loading. $E = 13 \times 10^6$ kN/m^2 (13,000 MPa). The beam is 200 mm wide.

5.10 For the rigid frame shown in Fig. 5.38,
(a) draw the deflected shape and moment diagram,
(b) determine the horizontal deflection at *B* and *D*.

5.11, 5.12, 5.13, 5.14, and 5.15

Using the conjugate beam method solve problems 5.5, 5.6, 5.7 and 5.8

5.16 Construct the Williot diagram for the truss shown in Fig. 5.39 and find the displacement of joint *D* as caused by the applied loads. $E - 200 \times 10^6$ kN/m^2 and area of cross-section of each member = 1000 mm^2.

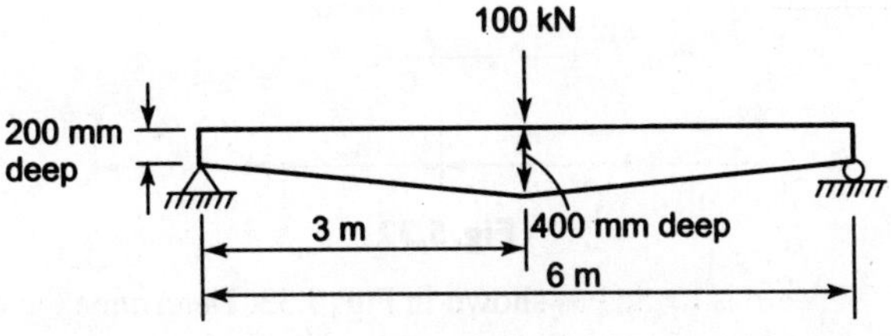

Fig. 5.37

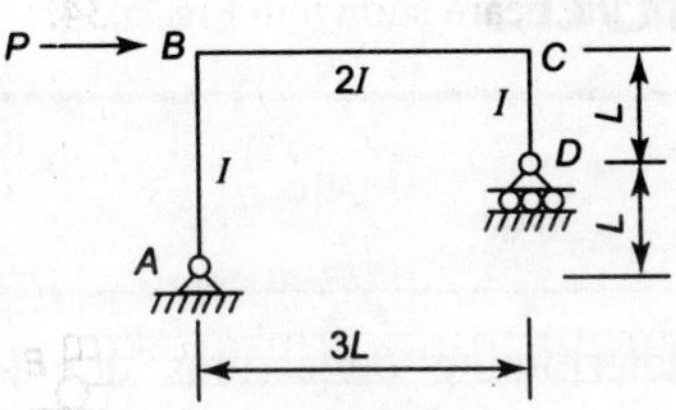

Fig. 5.38

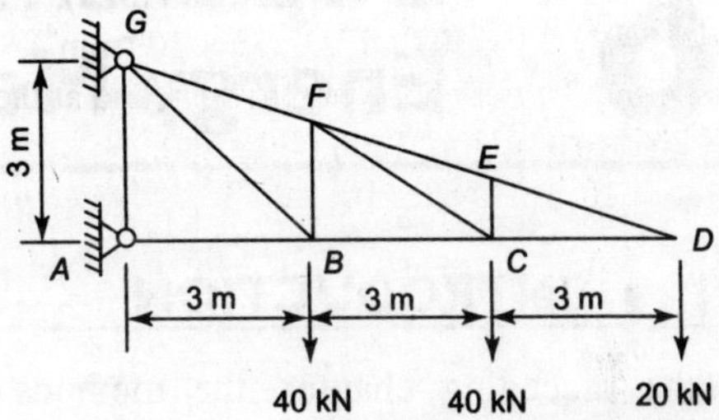

Fig. 5.39

5.17 Construct the Williot-Mohr diagram for the truss shown in Fig. 5.40. Determine the displacement of joint C along the plane of rollers and also the vertical and horizontal components of the deflection of joint B. $E = 200 \times 10^6$ kN/m^2 (200,000 MPa).

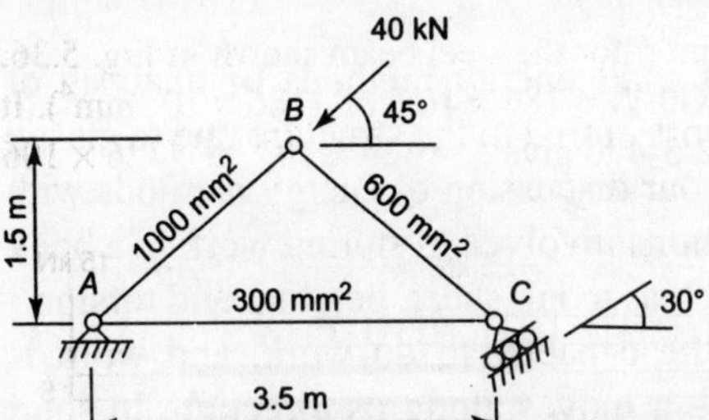

Fig. 5.40

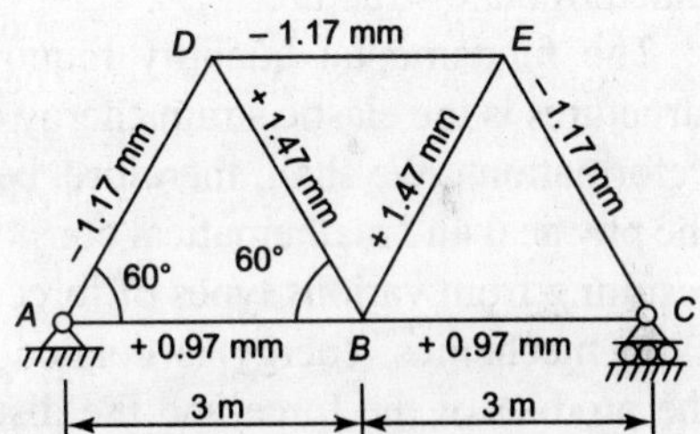

Fig. 5.41

5.18 Construct the Williot-Mohr diagram and find the horizontal and vertical displacement components of joint B. Member deformations are indicated on the truss shown in Fig. 5.41. Consider point A as fixed in position and member AD as fixed in direction.

5.19 Construct the Williot-Mohr diagram for the truss shown in Fig. 5.42 and determine the vertical and horizontal components of deflection of point D. Area of cross-section = 400 mm^2 each and $E = 200 \times 10^6$ kN/m^2 (200,000 MPa).

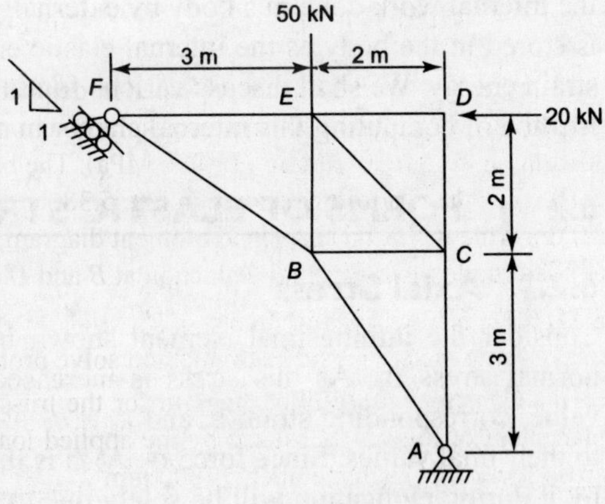

Fig. 5.42

6 Displacements—Energy Methods

6.1 INTRODUCTION

In the preceding chapter, the methods for determining deflections due to bending were based upon the geometrical interpretation of the mathematical relationship that exists between the applied moment, the curvature and flexural rigidity (*EI*), of a member. Another group of methods can be developed from energy considerations. These energy methods are a powerful tool in obtaining the numerical solutions for deflection problems and also for analysing statically indeterminate structures.

The fundamental quantity required for all energy methods of analysis of structures is the elastic strain energy or work stored in the structure due to elastic deformations. We shall, therefore, begin our discussion of energy methods with the physical and mathematical considerations involved in storing work in a body resulting from various types of forces such as axial, shear, bending and torsion.

In mechanics, energy is defined as the capacity to do work, and work is the product of the force and the distance it moves along its direction. In solid deformable bodies the stresses multiplied by the respective areas are the forces, and the deformations are the distances. The product of these two quantities is the internal work done in a body by externally applied forces. The internal work is stored in the body as the internal elastic energy of deformation or the elastic strain energy. We shall discuss various forms of the elastic strain energy and the method of computing this internal energy in structural members.

6.2 FORMS OF ELASTIC STRAIN ENERGY

6.2.1 Axial Stress

Consider the infinitesimal element shown in Fig 6.1 (a) acted upon only by normal stress σ_x. As this stress is increased gradually from zero to its final value, corresponding strain ε_x and $\varepsilon_y = \varepsilon_z = -\mu\varepsilon_x$ undergo a change from zero to their final values. Since force $\sigma_x.dy.dz$ is the only force acting, the work done by it during elongation will be solely due to the elongation of the element in *x* direction. Thus, for infinitesimal elongation $d\varepsilon'_x$, the work done by force $\sigma'_x .dy. dz$ is

$$(\sigma'_x .dy.dz)(d\varepsilon'_x)(dx) = (\sigma'_x)(d\varepsilon'_x)dV \tag{6.1}$$

where dV = volume of element.

The term $(\sigma'_z)(d\varepsilon'_z)$ is represented by the shaded area in Figs. 6.1.*b* and *c* valid for the stress-strain curve linearly or curvilinearly related. Therefore, strain energy dU_0 stored in an element of $dV = dxdydz$ in its final deformed shape is

$$dU_0 = \int_0^{\varepsilon_x} (\sigma'_x\, d\varepsilon'_x)\, dV \tag{6.2}$$

In the case of linearly elastic materials (Fig. 6.1*b*) $\sigma'_x = E\varepsilon'_x$. On substituting in Eq. 6.2 and evaluating the integral, we get

$$dU = T\frac{\varepsilon_x^2}{2}dV$$

or

$$dU = \frac{\sigma_x^2}{2E}dV = \frac{\sigma_x \varepsilon_x}{2}dV \tag{6.3}$$

The strain energy stored per unit volume of the materials or its strain energy density is

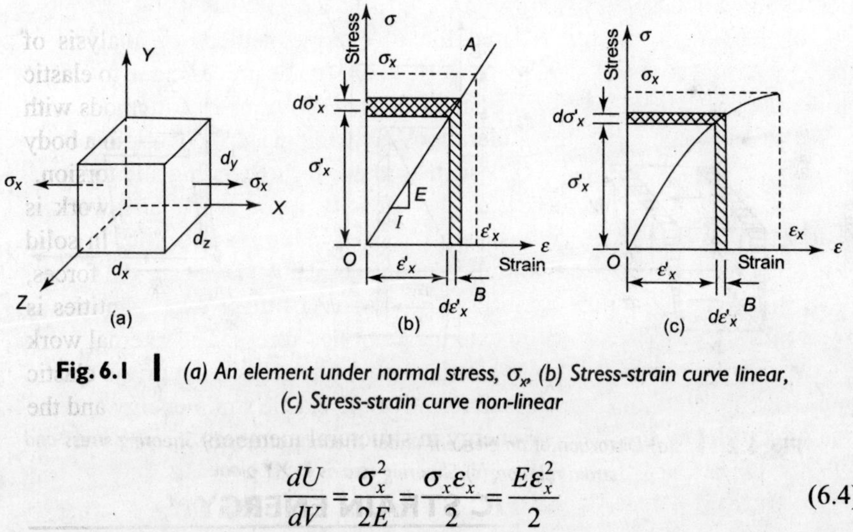

Fig. 6.1 | *(a) An element under normal stress, σ_x, (b) Stress-strain curve linear, (c) Stress-strain curve non-linear*

$$\frac{dU}{dV} = \frac{\sigma_x^2}{2E} = \frac{\sigma_x \varepsilon_x}{2} = \frac{E\varepsilon_x^2}{2} \tag{6.4}$$

This expression may be graphically interpreted as the area of triangle *OAB* in Fig. 6.16*b*.

Expressions analogous to Eq. 6.4 apply to normal stresses σ_y and σ_z and corresponding linear strains ε_y and ε_z. By the use of the stress-strain relations, strain energy can be represented either as a function of stress components or only as a function of strain components.

The complementary strain energy is defined as

$$dU_c = \int_0^{\sigma_x} (\varepsilon'_x \, d\sigma_x) \, dV \tag{6.5}$$

or
$$dU_c = \varepsilon_x \sigma_x dV - dU \tag{6.6}$$

which represents the area between the curve and the stress axis. As can be seen from Eq. 6.5 and Fig. 6.1*b*, for a linearly elastic body the strain energy and complementary strain energies are equal.

6.2.2 Shearing Stress

An expression for the elastic strain energy for an infinitesimal element under pure shear may be established in a manner analogous to uniaxial stress. Consider an element in a state of pure shear as shown in Fig. 6.2*a*. The corresponding stress-strain diagram is shown in Fig. 6.2*b*.

For simplicity the element is assumed fixed in the *XZ* plane, and gradual application of $\tau_{xy} = \tau_{yx}$ will distort the element as illustrated in Fig. 6.2*a*. Thus, for infinitesimal displacement $d\gamma_{xy}$ during deformation, the work done by force τ'_{xy} $dzdx$ acting on the parallel plane is $(\tau'_{xy}\, dzdx)\,(d\gamma_{xy},\, dy)$. Forces $\tau_{xy}\, dydz$ on the planes perpendicular to the *X* axis do no work, since for small deformation, the displacement may be assumed to be perpendicular to these forces. It follows that the shear strain energy dU_τ stored in the element in its final deformed shape is

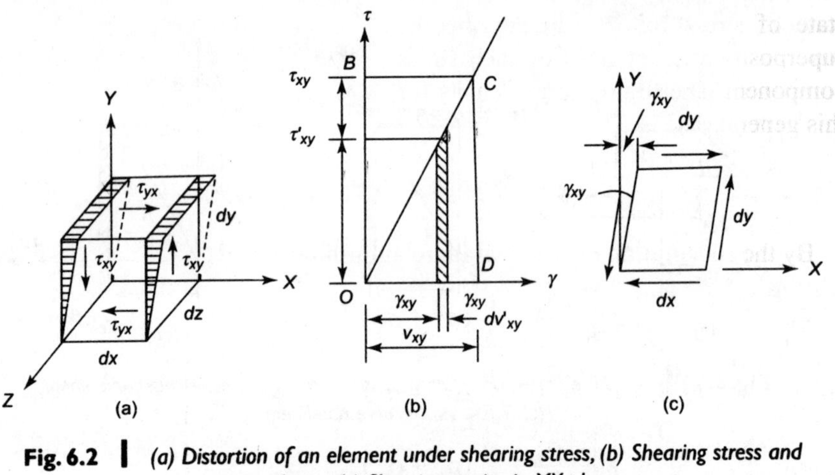

Fig. 6.2 | *(a) Distortion of an element under shearing stress, (b) Shearing stress and strain relation, (c) Shearing strains in XY plane*

$$dU_\tau = \int_0^{\gamma_{xy}} (\tau'_{xy} \, d\gamma_{xy}) \, dV \tag{6.7}$$

For linearly elastic material $\tau'_{xy} = \gamma_{xy}/G$. Substituting for τ'_{xy} in Eq. 6.7 and evaluating the integral, we get

$$dU_\tau = \frac{G\gamma_{xy}^2}{2} dV = \frac{\tau_{xy}^2}{2G} dV = \frac{\tau_{xy}\,\gamma_{xy}}{2} dV \tag{6.8}$$

Here the term $\frac{\tau_{xy}\,\gamma_{xy}}{2}$ represents the area of triangle *OCD* in Fig. 6.2*b*.

The strain energy density for shear becomes

$$\left(\frac{dU}{dV}\right)_\tau = \frac{1}{2}\tau_{xy}\,\gamma_{xy} \tag{6.9}$$

Analogous expressions apply for shearing stress τ_{xy} and τ_{zx} responding shearing strains γ_{yz} and γ_{zx}.

6.2.3 Multi-Axial State of Stress

Consider an infinitesimal element with stresses acting on its faces as shown in Fig. 6.3. During gradual loading of the body these stresses reach their final values starting from zero and increasing gradually. At the same time the element undergoes deformations gradually before it reaches its final deformed shape. The strain energy expression for this state of stress follows directly by the superposition of energies of each stress component. The strain energy density for this general case is

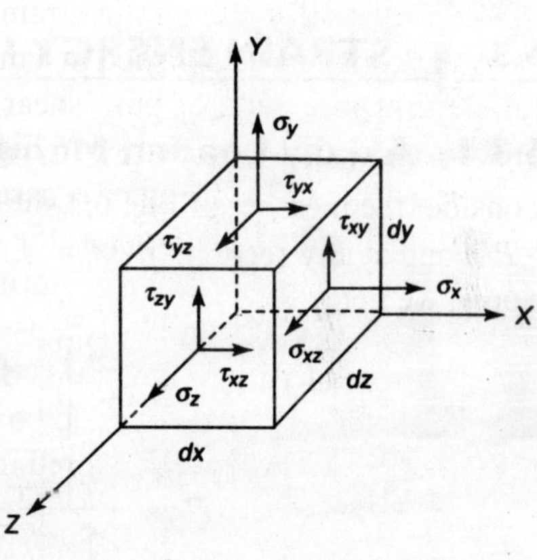

Fig. 6.3 | *Element under multi-axial state of stress*

$$\frac{dU}{dV}\left(\frac{1}{2}\right)(\sigma_x\varepsilon_x + \sigma_y\varepsilon_y + \sigma_z\varepsilon_z + \tau_{xy}\gamma_{xy} + \tau_{yz}\gamma_{yz} + \tau_{xz}\gamma_{xz}) \tag{6.10}$$

By the substitution of stress-strain relationships, we get

$$\left.\begin{aligned}
\varepsilon_x &= \frac{\sigma_x}{E} - \frac{\mu}{E}(\sigma_y + \sigma_x)\\
\varepsilon_y &= \frac{\sigma_y}{E} - \frac{\mu}{E}(\sigma_x + \sigma_z)\\
\varepsilon_z &= \frac{\sigma_z}{E} - \frac{\mu}{E}(\sigma_x + \sigma_y)\\
\gamma_{xy} &= \frac{\tau_{xy}}{G};\ \gamma_{yz} = \frac{\tau_{yz}}{G};\ \text{and}\ \gamma_{zx} = \frac{\tau_{zx}}{G}
\end{aligned}\right\} \tag{6.11}$$

In Eq. (6.10) we obtain the total strain energy per unit volume

$$dU = \frac{1}{2E}(\sigma_x^2 + \sigma_y^2 + \sigma_z^2) - \frac{\mu}{E}(\sigma_x\sigma_y + \sigma_y\sigma_z + \sigma_x\sigma_z) + \frac{1}{2G}(\sigma_{xy}^2 + \sigma_{yz}^2 + \sigma_{xz}^2) \tag{6.12}$$

The integration of Eq. (6.10) over volume *V* yields the total strain energy of the body

$$U_{total} = \frac{1}{2}\iiint_V dU\, dxdydz \tag{6.13}$$

For linearly elastic material, under uniaxial stress σ_x and shearing stress τ_{xy}, the strain energy

$$U = \iiint_V \frac{\sigma_x^2}{2E}\, dxdydz + \iiint_V \frac{\tau_{xy}^2}{2G}\, dxdydz \tag{6.14}$$

Several useful expressions can be developed from Eq. 6.14 by reducing the triple integral to single ones.

6.3 STRAIN ENERGY IN MEMBERS

6.3.1 Axially Loaded Members

Consider the member of Fig. 6.4 subjected to axial force *P*. In such situations, $\sigma_x = P/A$ and at any section, $\iint dA = A$. Therefore, since σ_x and *A* can be functions, *x* only, we get

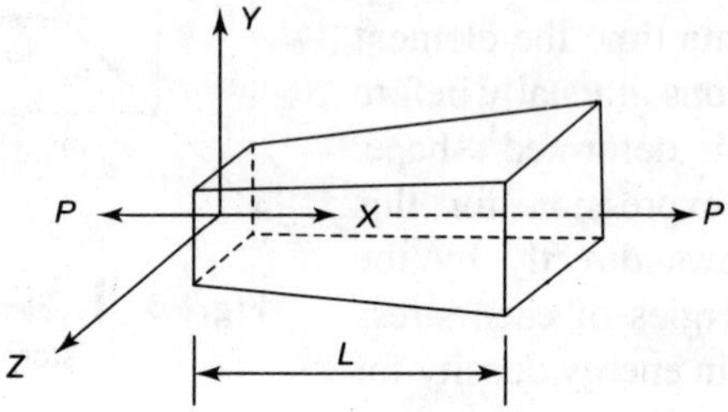

Fig. 6.4

$$U = \iiint_V \frac{\sigma_x^2}{2E}\, dxdydz = \iiint_V \frac{P^2}{2A^2E}\, dxdydz$$

$$= \int_L \frac{P^2}{2A^2E}\left[\iint_A dydz\right] dx = \int_L \frac{P^2}{2AE}\, dz \tag{6.15}$$

where a single integration along length *L* gives the required quantity.

6.3.2 Members Under Bending

In this case $\sigma_x = \dfrac{M}{I}y$. This relation must be substituted into the first right hand term in Eq. 6.14. Then noting that *M* and *I* are functions of *x* only and that by definition

$$\iint dydzy^2 = I_{XX}$$

and
$$U = \iiint_V \frac{\sigma_x^2}{2E}\, dxdydz = \iiint_V \frac{1}{2E}\left(\frac{M}{I}y\right)^2 dxdydz$$

$$= \int_L \frac{M^2}{2EI^2}\left[\iint_A y^2\, dydz\right] dx$$

$$U = \int_L \frac{M^2}{2EI} dx \qquad (6.16)$$

6.3.3 Members Under Shearing

The shear strain energy is obtained using the second term on the right hand side of Eq. 6.14.

$$U = \iiint_V \frac{\tau_{xy}^2}{2G} dxdydz$$

Substituting $\tau_{xy} = \dfrac{VQ}{It}$ in the above equation, we have

$$U = \iiint_V \frac{1}{2G}\left[\frac{VQ}{It}\right]^2 dxdydz \qquad (6.17)$$

This can be simplified if regular sections, such as rectangular, circular, elliptical or triangular cross sections are considered. Considering a rectangular cross section of breadth b and depth d (Fig. 6.5). For any value of y,

$$Q = \int_{y_1}^{d/2} bydy \qquad (6.18)$$

or

$$Q = \frac{b}{2}\left(\frac{d^2}{4} - y_1^2\right) \qquad (6.19)$$

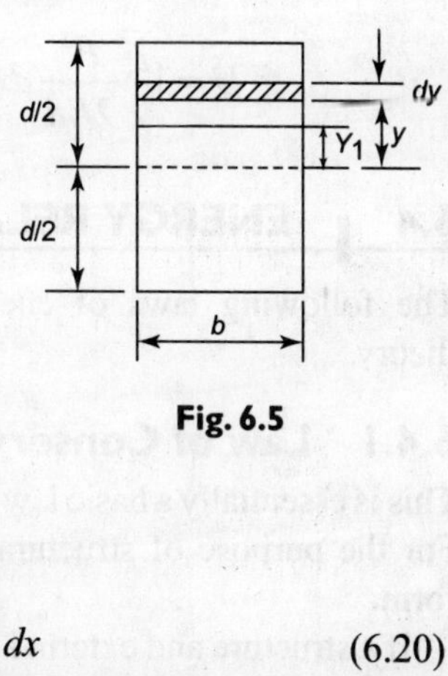

Fig. 6.5

Equation 6.17 can be written as

$$\int_L \frac{V^2}{2GI^2}\left[\iint_A \left(\frac{Q}{t}\right)^2 dydz\right] dx \qquad (6.20)$$

For a rectangular cross section, the term within the bracket

$$\iint_A \left(\frac{Q}{b}\right)^2 dydz = \int_{-d.2}^{d/2} \frac{b}{4}\left(\frac{d^2}{4} - y^2\right)^2 dy = \frac{bd^5}{120} \qquad (6.21)$$

Since $I = \dfrac{1}{12}bd^3$ and $A = bd$, Eq. 6.20 can be written as

$$U = \int_L (1.2)\frac{V^2}{2GA} dx \qquad (6.22)$$

In general, we can write the strain energy due to shear

$$U = K\int_0^L \frac{V^2}{2GA} dx \qquad (6.23)$$

where K *is a.* constant whose value is dependent on the shape of the cross section. As noted above, the value of K for a rectangle is 1.2.

6.3.4 Circular Members in Torsion

The basic expression for torsional strain energy is analogous to the last term in Eq. 6.14. Substituting into such an equation, torsional shear stress $\tau = \dfrac{T\rho}{J}$ we get

$$U = \int_L \frac{T^2}{2GJ^2}\left[\iint \rho^2\, dydz\right] dx \tag{6.24}$$

$$= \int_L \frac{T^2 dx}{2GJ} \tag{6.25}$$

Since $$\iint_A \rho^2\, dydz = J$$

Thus, for a prismatic bar of length L the total strain energy due to axial moment, shear and torsional forces becomes

$$U = \int_0^L \frac{P^2}{2EA}\, dx + \int_0^L \frac{M^2 dx}{2EI} + K\int_0^L \frac{V^2 dx}{2GA} + \int_0^L \frac{T^2 dx}{2GJ} \tag{6.26}$$

6.4 ENERGY RELATIONS IN STRUCTURAL THEORY

The following laws of energy are of fundamental importance in structural theory.

6.4.1 Law of Conservation of Energy

This is essentially a basic law of physics—energy is neither created nor destroyed. For the purpose of structural analysis, the law can be stated in the following form:

If a structure and external loads acting on it are isolated, such that these neither receive nor give out energy, then the total energy of the system remains constant.

A typical application of the law of conservation of energy can be illustrated by referring to a bar subjected to an axial pull P gradually applied as shown in Fig. 6.6. When equilibrium is reached, it will be found that the bar has extended by an amount δ. Considering that the process is adiabatic (heat is neither supplied nor taken out), according to the law of conservation of energy, loss of poten-tial energy, $P\delta/2$, must appear elsewhere. In this case, it is found in the form of strain energy stored in the bar. This is given by

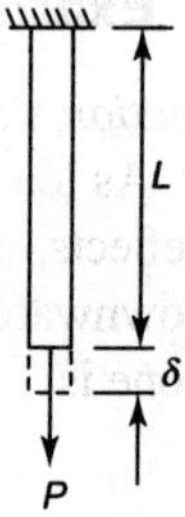

Fig. 6.6

$$W_i = \int_0^L \frac{P^2 dx}{2EA} \tag{6.27}$$

where W_i = strain energy stored in the body or internal work.

Let

$$W_e = \frac{1}{2} P\delta \tag{6.28}$$

where W_e is the external work done.

Therefore, $$W_e + W_i = -\frac{1}{2} P\delta + \int_0^L \frac{P^2 dx}{2EA} = 0 \tag{6.29}$$

The minus sign for external work is to take into account the loss of potential energy

or $$-W_e + W_i = 0 \tag{6.30}$$

or $$W_e = W_i \tag{6.31}$$

that is, the external work done is equal to the internal strain energy.

In the case of rigid bodies the real work done by all the forces including the reactions must be zero, since the internal strain energy stored in the body is zero. This shall be made clear in the following examples.

Example 6.1 *It is desired to determine reaction R_B of the rigid beam AB shown in Fig. 6.7. Use real work equation.*

A small but real displacement Δ, is given to the beam at end B as shown. Since no internal strain energy is stored in the rigid beam the total work done must be equal to zero, that is

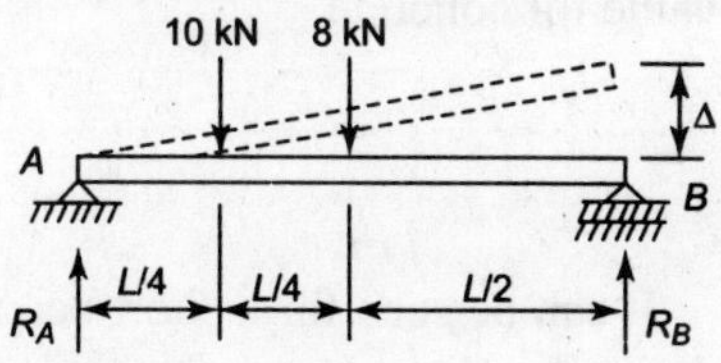

Fig. 6.7

$$R_B\,(\Delta) - 10\left(\frac{\Delta}{4}\right) - 8\left(\frac{\Delta}{2}\right) = 0$$

or $R_B = 6.5$ kN

Example 6.2 *Find the maximum deflection due to force P applied at the end of the elastic cantilever of a rectangular cross section shown in Fig. 6.8. Consider flexural and shearing deformations.*

As load P is applied, the beam deflects, say by an amount, Δ, downwards. The external work done is

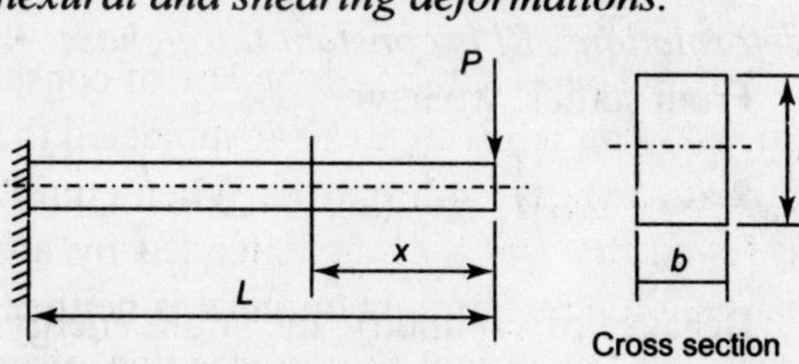

Fig. 6.8

$$W_e = \frac{1}{2} P(\Delta)$$

Internal strain energy is caused by: (1) bending stresses and (2) shearing stresses. These strain energies can be superimposed according to Eqns. 6.16 and 6.23.

$$U_{\text{bending}} = \int_0^L \frac{M_x^2 dx}{2EI} = \int_0^L \frac{(-Px)^2 dx}{2EI} = \frac{P^2 L^3}{6EI}$$

$$U_{\text{shear}} = K \int_0^L \frac{V_x^2}{2GA}\,dx = K \int_0^L \frac{P^2 dx}{2GA} = 1.2\frac{P^2 L}{2GA}$$

$$U_{\text{total}} = U_{\text{bending}} + U_{\text{shear}}$$

Equating the external work to internal strain energy

$$\frac{1}{2}P\Delta = \frac{P^2L^3}{6EI} + \frac{6}{10}\frac{P^2L}{GA}$$

$$\Delta = \frac{PL^3}{3EI} + \frac{6}{5}\frac{PL}{GA}$$

The first term in the result, $\frac{PL^3}{3EI}$, is the deflection of the beam due to flexure; the second term is the deflection due to shear.

The total deflection Δ, may be recast as

$$\Delta = \frac{PL^3}{3EI}\left(1 + \frac{3}{10}\frac{E}{G}\frac{h^2}{L^2}\right)$$

To gain further insight into this problem, if we replace $E/G = 2.4$, a typical value for concrete,

$$\Delta_{\text{total}} = \left(1 + 0.72\frac{h^2}{L^2}\right)\Delta_{\text{bending}}$$

It can be seen for a short beam $L/h = 1$, the total deflection is 1.72 times that due to bending. Hence deflections caused by shear are important. On the other hand, for $L/h = 10$, the deflection due to shear is less than 0.75 per cent. Thus deflections due to shear are small in ordinary and slender beams and are usually neglected.

Example 6.3 | *It is desired to find the deflection under the load point for the beam shown in Fig. 6.9. Consider only bending deformations. EI is constant throughout.*

From statics, we have

$$M_x = \frac{P}{2}(x), \quad 0 < x < \frac{L}{2}$$

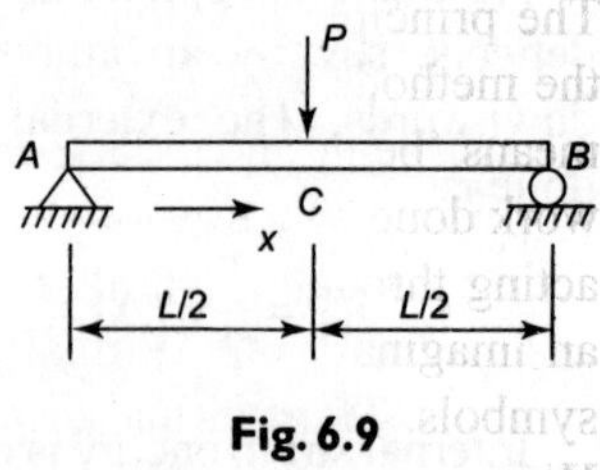

Fig. 6.9

Because of symmetry, the strain energy can be expressed as twice that of the left half of beam AC. Equating external work to internal strain energy, we have

$$\frac{1}{2}P\Delta = 2\int_0^{L/2} \frac{M_x^2\,dx}{2EI} = \int_0^{L/2}\left[\left(\frac{P}{2}\right)(x)\right]^2 \frac{dx}{2EI}$$

where Δ = deflection under load point. On integration we find

$$\frac{1}{2}P\Delta = \frac{P^2L^3}{96EI}$$

$$\Delta = \frac{PL^3}{48EI}$$

This approach can be applied to more complex structures such as the one shown in Fig. 6.10.

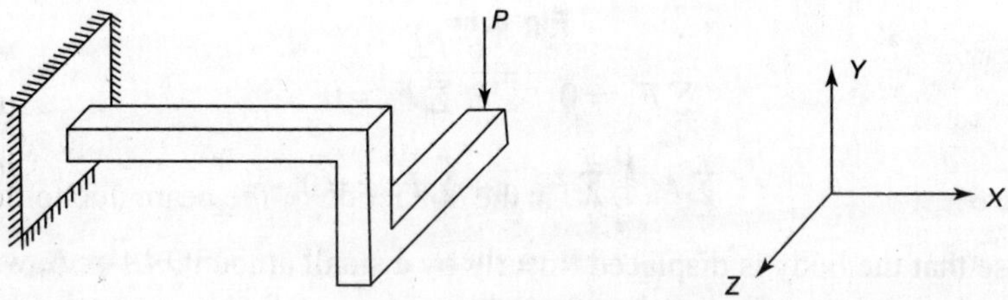

Fig. 6.10 | *Frame in three dimensions*

The strain energy for each member is evaluated separately and the results are summed up to give the total strain energy for the entire structure. Computational work can be simplified by taking only the predominant strains which contribute to the deflection. In this case, for example, the deflection under load P is a function of flexural and torsional strains only.

The approach illustrated in the above examples is limited to the determination of the deflection caused by a single force at the point of application. If more than one load is applied to the structure, more than one unknown value of deflection will appear in the expression for external work and, therefore, the resulting equation cannot be solved. Owing to these limitations, the method of real work is not widely used for deflection analysis. However, it can be used in limited cases of loading and also serves as a means for developing additional principles of deflections.

6.5 | VIRTUAL WORK

The principle developed by Johann Bernoulli in 1717 is the most versatile of the methods available for computing deflections of structures. The term virtual means 'being in essence or effect but not in fact'. The virtual work means the work done by a real force acting through a virtual displacement or a virtual force acting through a real displacement. The virtual work is not a real quantity but an imaginary one. Virtual quantities shall be denoted by bold faced letters or symbols. Thus, virtual displacement shall be denoted by A and virtual work by **W**.

6.5.1 Virtual Work on a Rigid Body

Consider a rigid body supported and loaded as shown in Fig. 6.11*a*; the body is in equilibrium and satisfies the static equilibrium conditions

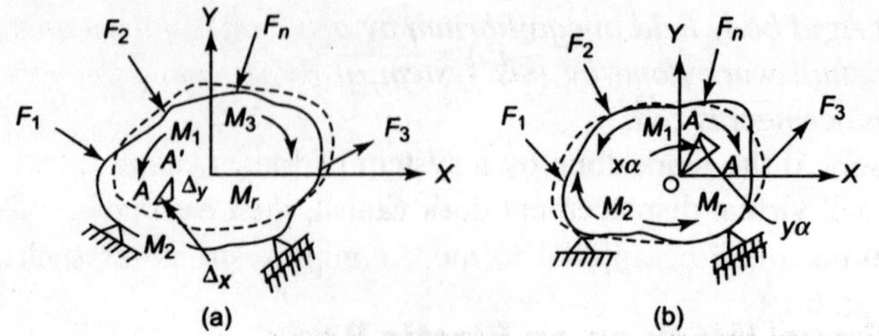

Fig. 6.11

$$\Sigma F_X = 0 \qquad \Sigma F_Y = 0 \tag{6.32}$$

and

$$\sum_1^r M + \sum_1^n F_X\ y + \sum_1^n F_Y\ x = 0 \tag{6.33}$$

Suppose that the body is displaced linearly by a small amount, $AA' = \Delta$, without any rotation due to effects other than the system of forces, and takes a new position as shown.

The two components of displacement parallel to coordinate axes X and Y are Δ_x and Δ_y. Since the translation is very small the forces are not altered in their magnitude or direction and the body remains in equilibrium at all the time. Then the virtual work done (product of real forces and virtual displacement) is

$$W_e = \sum_1^n F_X\ \Delta_x + \sum_1^n F_Y\ \Delta_y \tag{6.34}$$

Since Δ_x and Δ_y are constants for all forces, this becomes

$$W_e = \Delta_x \sum_1^n F_X + \Delta_y \sum_1^n F_y \tag{6.35}$$

This is equal to zero since $\sum_1^n F_X$ and $\sum_1^n F_Y$ are zero from Eq. 6.32.
Therefore,

$$W_e = 0 \tag{6.36}$$

If it is now assumed that the rigid body under the loading system F is rotated by a small angle α about the origin O (which could be any point) the component of displacement parallel to the X axis will be $y.\alpha$ and parallel to the Y axis will be $x.\alpha$ (*see* Fig. 6.11*b*). The total work done by the components of forces F and couples M is

$$\sum_1^r M\alpha + \sum_1^n F_{XY} \cdot \alpha + \sum_1^n F_Y \cdot x\alpha \tag{6.37}$$

or

$$\alpha\left(\sum_1^r M + \sum_1^n F_{X \cdot y} + \sum_1^n F_{Y \cdot x}\right) \tag{6.38}$$

This, by Eq. 6.33, is equal to zero.

Since any small displacement of the rigid body can be represented as the sum of a translation and a rotation about some point and since the work of F system of forces and couples has been shown to be zero in either of the displacements, Johann Bernoulli's principle of virtual work can be stated as follows.

Given a rigid body held in equilibrium by a system of forces and/or couples, the total virtual work done by this system of forces and/or couples during a virtual displacement is zero.

Conversely, if the work done by a system of forces acting on a body (rigid) during a small virtual displacement does vanish, then the system of forces is in equilibrium.

6.5.2 Virtual Work on an Elastic Body

The principle can also be applied to an elastic body. For example, suppose that the elastic body shown in Fig. 6.12 is subjected to a set of F forces, and rests in

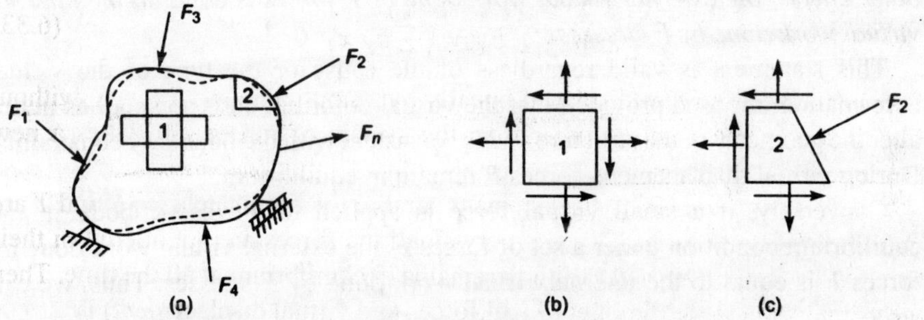

Fig. 6.12 | *(a) Elastic body under system of forces F, (b) Free-body diagram of elements 1 and 2*

equilibrium in its deformed shape. Now suppose that the body is deformed to another shape due to another set of forces, temperature, etc., while the F system of forces is present. In other words, the elastic body is given a small virtual displacement which satisfy the boundary constraints. The virtual displacement, in effect, gives a ride to the F system of forces. Certainly, during this displacement any infinitesimal element, such as 1 (interior) and 2 (exterior) will be displaced and the stresses on its boundaries will do some work. We shall designate this work as $d\mathbf{W}_s$. A part of this virtual work is done due to the rigid body movement of the element, and another part due to the change in the shape of the element. Since the change in the shape of the element is normally referred to as the deformation of the element, the work done by the stresses, due to the F system of forces on the boundaries, is designated as $d\mathbf{W}_d$. Consequently, the remaining part of the work $(d\mathbf{W}_x - d\mathbf{W}_d)$ is done by F stresses during the rigid body movement of the element. However, every element is in equilibrium under the stresses in boundaries, and the work done by them during rigid body movement is equal to zero.

Hence $$d\mathbf{W}_s - d\mathbf{W}_d = 0 \tag{6.39}$$

and for the entire body it becomes

$$\mathbf{W}_s - \mathbf{W}_d = 0 \tag{6.40}$$

or $$\mathbf{W}_s = \mathbf{W}_d \tag{6.41}$$

It should be understood here that $\mathbf{W}_s$. represents the sum of the virtual work done by F stresses on the boundaries of every element in the body. However, each element has common boundaries with an adjacent element where the stresses are equal and opposite to each other. Certainly, the work done by equal but opposite stresses during the same displacement is equal to zero. As a result of this, the work done by the F stresses on all the interior boundaries adds up to zero. Hence $\mathbf{W}_s$, consists of the work done by external F forces applied on the external boundaries only (for example F_2 on element 2). Thus, the law of virtual work can be stated. *If a system of forces F acting on a deformable body is under equilibrium, as the body is subjected to a small deformation caused by some other effects, the external virtual work done by F forces is equal to the internal virtual work done by F stresses.*

This statement is valid regardless of the cause or the type of the virtual deformation imposed provided that the virtual deformation is so small as not to alter the geometry of the structure and is consistent with the boundary constraints. During virtual displacement, forces F remain in equilibrium.

Conversely, if a small virtual force is applied to an elastic body in the equilibrium condition under a set of forces F, the external virtual work done by forces F is equal to the internal virtual work done by F stresses. Thus, we can write

Real displacement compatible

(Virtual force × real displacement) = Virtual internal forces × real internal displacements)

Virtual forces in equilibrium

Virtual displacement compatible

(Virtual displacement × real force) = (Virtual internal displacements × real internal forces

Real forces in equilibrium

6.6 BETTI'S AND MAXWELL'S LAWS OF RECIPROCAL DEFLECTIONS

Maxwell's law of reciprocal deflections is a special case of Betti's law of reciprocal work. We shall first derive Betti's law of reciprocal work. Suppose a linearly elastic structure shown in Fig. 6.13 is in equilibrium under two separate and independent systems of forces—system of forces P_m and system of forces P_n. Consider that the P_m system of forces is gradually applied first. Let the deflections at the point and direction of forces P_m be represented by

$$\Delta_m = \Delta_{m1}, \Delta_{m2} \ldots, \Delta_{mm} \tag{6.42}$$

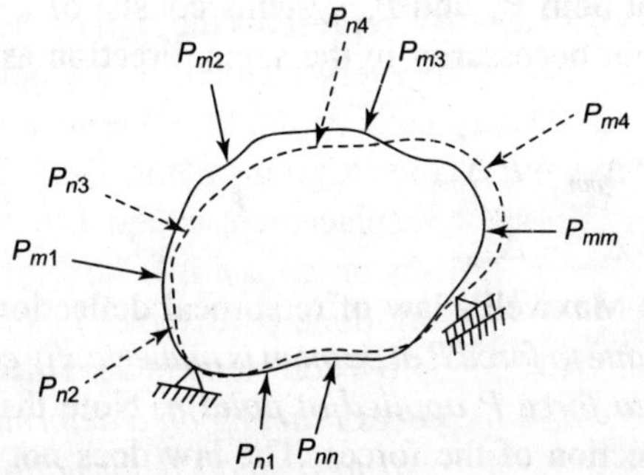

Fig. 6.13 | *Elastic body under system of forces P_m and P_n*

If the second system of forces P_n is applied on the structure while P_m forces are present, the structure will deform once more and will rest in equilibrium in its final deformed shape as shown. Let $\Delta_n = \Delta_{n1}, \Delta_{n2}, \ldots, \Delta_{nn}$, represent the deformations at the point and in the direction of forces P_n. The total external work done by these forces would then be

$$W_e = \Sigma \frac{1}{2} P_m \Delta_{mm} + \Sigma\, P_m \Delta_{mn} + \frac{1}{2} \Sigma P_n \Delta_{nn} \tag{6.43}$$

where Δ_{mn} represents the deflection of the point of application of one of the P_m forces caused by the P_n force system. The first and last terms in Eq. 6.43 represent the work done by P_m and P_n forces respectively as they are gradually applied. The middle term, however, represents the work done by P_m forces riding along the deflections caused by P_n forces.

Suppose now the loading sequence is reversed, that is, P_n forces are first applied and P_m forces later. P_n forces in full ride along the deflections caused by P_m forces. The final deformed shape of the structure will be the same as before which is shown in dotted line in Fig. 6.13. The total external work done in this case will be

$$W_e = \Sigma \frac{1}{2} P_n \Delta_{nn} + \Sigma\, P_n \Delta_{nm} + \Sigma \frac{1}{2} P_m \Delta_{mm} \tag{6.44}$$

where Δ_{nm} represents the deflection of the point of application of one of the P_n forces caused by P_m forces.

According to the principle of superposition, the total external work done in either of the sequences of loading should be same. Hence equating Eqs. 6.43 and 6.44, we get

$$\Sigma\, P_m\ \Delta_{mn} = \Sigma\, P_n\ \Delta_{nm} \tag{6.45}$$

This is known as a Betti's theorem and may be stated as follows.

For a linearly elastic structure, the work done by a set of external forces P_m acting through displacements Δ_{mn} produced by another set of forces P_n is equal, to the work done by the second set of external forces P_n acting through displacements Δ_{nm} produced by forces P_m.

Suppose now that both P_m and P_n systems consist of a single load P having the magnitude but not necessarily in the same direction as shown in Fig. 6.14, then, from Eq. 6.45,

$$P\,\Delta_{mn} = P\,\Delta_{nm}$$

or

$$\Delta_{mn} = \Delta_{nm} \tag{6.46}$$

This is known as Maxwell's law of reciprocal deflection and states that: *the deflection of point n due to force P at point m is numerically equal to the deflection of the point m due to force P applied at point n.* Note that the deflections are measured in the direction of the forces. The law does not differentiate normal force from a moment nor a linear displacement from a rotation. Such a generality is shown in Fig. 6.15.

For example, the rotation (in radians) Δ_{12} at 1 due to a unit load (*N*) at 2 is numerically equal to the deflection (in *m*) at 2 due to a unit couple (*N. m*) at 1, that is

$$\Delta_{12} = \Delta_{21}$$

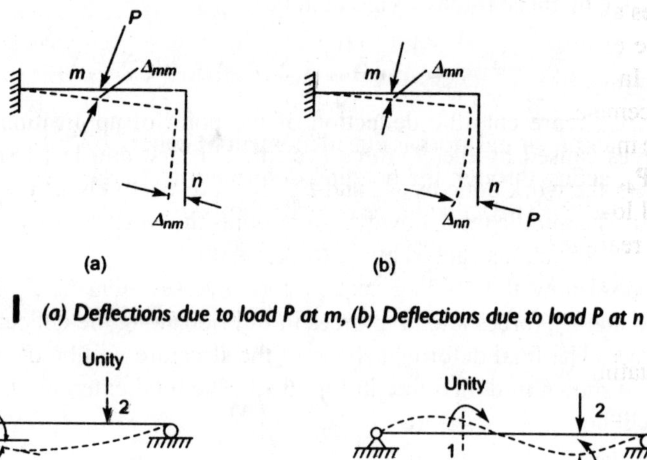

Fig. 6.14 | *(a) Deflections due to load P at m, (b) Deflections due to load P at n*

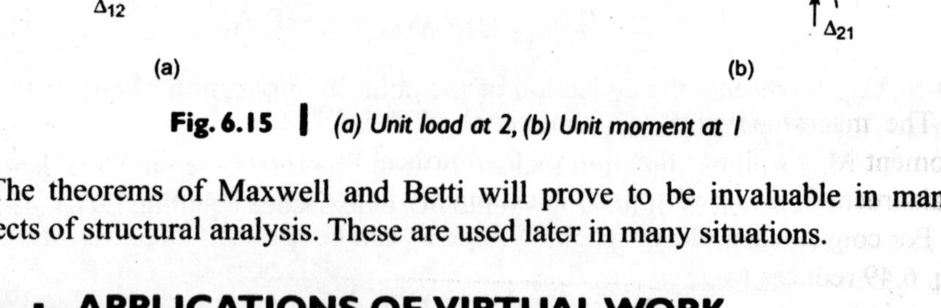

Fig. 6.15 | *(a) Unit load at 2, (b) Unit moment at 1*

The theorems of Maxwell and Betti will prove to be invaluable in many aspects of structural analysis. These are used later in many situations.

6.7 APPLICATIONS OF VIRTUAL WORK

An extremely useful general relationship can be developed by making use of the conceptual virtual work relationship. Consider a simple beam loaded as shown in Fig. 6.16*a*. For clarity only one load P_3 is considered as acting on the beam. It is required to determine the vertical downward deflection at 4. The real work method fails in this case as pointed out in Sec. 6.4. The principle of virtual work may be utilized to solve this problem.

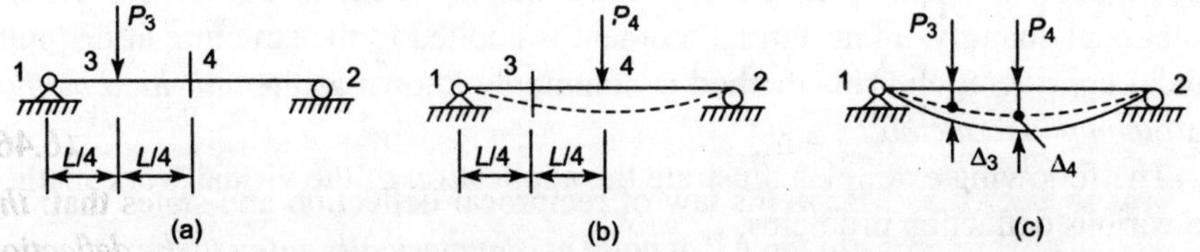

Fig. 6.16 *(a) Beam and loading, (b) Displacements due to virtual load P_4, (c) Displacement, due to loads P_3 and P_4*

We begin the analysis by temporarily removing load P_3, from the beam and placing virtual load $\mathbf{P}_4$ of unspecified magnitude at the point and in the direction of the desired deflection (Fig. 6.16*b*). The displacements caused by $\mathbf{P}_4$ are shown in Fig. 6.16*b* but we are not concerned with them as we shall see later.

We now load the beam with the real load, P_3 producing additional displacement Δ_3 and Δ_4 under loads P_3, and $\mathbf{P}_4$ as shown in Fig. 6.16*c*. The unspecified load $\mathbf{P}_4$ rides along in full during deformations caused by P_3,. Applying the principle that the external virtual work is equal to the internal virtual work we have $\mathbf{W}_e = \mathbf{W}i$. In this $\mathbf{W}_e$, $= \mathbf{P}_4.\Delta_4$, that is, the product of virtual force $\mathbf{P}_4$ and the real displacement Δ_4.

The internal virtual work is due to moment **M** caused by the unspecified virtual force $\mathbf{P}_4$, acting through the bending deformations (angle change $d\phi$) produced by real load P_3. The value of $d\phi$ is defined by $d\phi = Mdx/EI$ where moment M is due to real load P_3. The internal strain energy,

$$\mathbf{W}_i = \int_0^L \mathbf{M}d\phi = \int_0^L \frac{\mathbf{M}Mdx}{EI} \tag{6.47}$$

Equating $\mathbf{W}_e = \mathbf{W}_i$, we have

$$\mathbf{P}_4 \cdot \Delta_4 = \int_0^L \frac{\mathbf{M}Mdx}{EI} \tag{6.48}$$

or

$$\Delta_4 = \frac{1}{\mathbf{P}_4} \int_0^L \frac{\mathbf{M}Mdx}{EI} \tag{6.49}$$

The magnitude of $\mathbf{P}_4$ is immaterial; it drops out of the expression since moment **M** is a linear function of $\mathbf{P}_4$. Further, $\mathbf{P}_4$ is not actually applied to the real structure; it is applied only conceptually in the analysis.

For convenience, the unspecified force $\mathbf{P}_4$ can be replaced by a unit load. The Eq. 6.49 reduces to

$$\Delta_4 = \int_0^L \frac{\mathbf{m}Mdx}{EI} \tag{6.50}$$

where **m** is the moment caused by a unit load applied in place of $\mathbf{P}_4$. The quantity of the right side of Eq. 6.50 represents the internal virtual work. Thus, by using a unit value of external virtual force, we directly obtain the value of external displacement Δ. The unit external force can be in the form of either a force or a moment depending upon the form of external displacements that are to be determined. For example, if the external displacement to be determined is a rotational quantity, a unit virtual moment is applied to the structure at the point under consideration. This method is commonly known as the *unit load method* or *dummy load method.*

The following examples illustrate the application of the virtual work method to various deflection problems.

Example 6.4 | *Using the method of virtual work, determine the vertical deflection under the load point and at centre of cantilever beam shown in Fig. 6.17a. Consider deformations due to bending only. EI is constant.*

The moment diagram corresponding to external load P is shown in Fig. 6.17*b*. The moment diagram due to a virtual unit load at 1 is shown in Fig. 6.17*d*. Taking origin for x at the free end

$$Mx = -P(x)$$

$$\mathbf{m}_x = -(x)$$

Using Eq. 6.50

$$\Delta_1 = \int_0^L \frac{\mathbf{m}_x M dx}{EI} = \int_0^L (-x)(-P.x)\frac{dx}{EI} = \frac{PL^3}{3EI}$$

To find the deflection at 2 we apply a unit load at 2. The resulting moment diagram is shown in Fig. 6.17*f* The limits for integration for $L/2$ to L. We then have

$$M_x = -P(x)$$

$$\mathbf{m}_x = -\left(x - \frac{L}{2}\right) \quad for\ L/2 \le x \le L$$

$$\Delta_2 = \int_{L/2}^{L} P(x)\left(x - \frac{L}{2}\right)\frac{dx}{EI}$$

or

$$\Delta_2 = \frac{5}{48}\frac{PL^3}{EI}$$

The positive sign for both the deflections indicates that they are in the direction of virtual unit forces applied (in this case downward).

Example 6.5 | *It is required to determine rotations at A and B due to an applied moment M_B on the beam as shown in Fig. 6.18a. Use the method of virtual work.*

To find rotation at A apply a unit couple as shown in Fig. 6.18*c*. The origin for x is chosen at end A. Then

$$M_x = M_B\,(x/L)$$

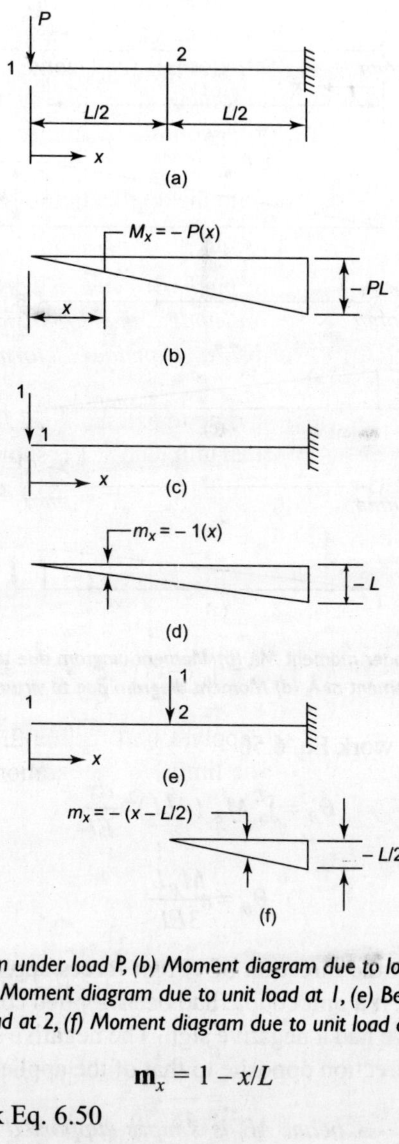

Fig. 6.17 | *(a) Beam under load P, (b) Moment diagram due to load P, (c) Beam under unit load at 1, (d) Moment diagram due to unit load at 1, (e) Beam under unit load at 2, (f) Moment diagram due to unit load at 2*

and $\quad \mathbf{m}_x = 1 - x/L$

Using virtual work Eq. 6.50

$$\theta_A = \int_0^L (1 - x/L)\, M_B\, (x/L) \frac{dx}{EI}$$

On evaluation
$$\theta_A = \frac{M_B L}{6EI}$$

The rotation at end B is found by applying a unit couple to the beam at B as shown in Fig. 6.18*d*. Again taking origin for x at end A

$$M_x = M_B\,(x/L)$$

and $\quad \mathbf{m}_x = x/L$

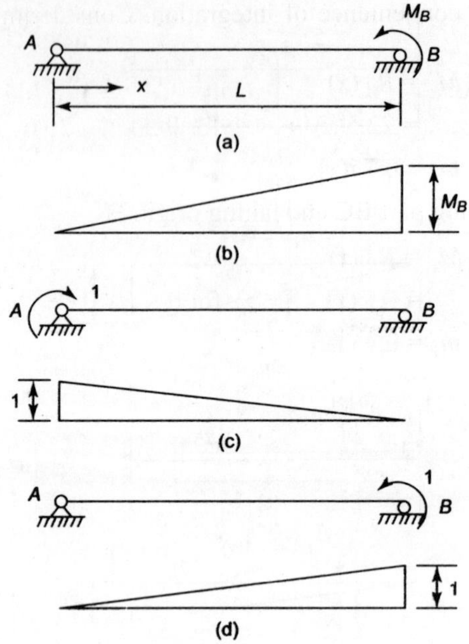

Fig. 6.18 | *(a) Beam under moment M_B, (b) Moment diagram due to M_B, (c) Moment diagram due to virtual unit moment at A, (d) Moment diagram due to virtual unit moment at B*

On applying virtual work Eq. 6.50

$$\theta_B = \int_0^L M_B\,(x/L)^2\,\frac{dx}{EI}$$

$$\theta_B = \frac{M_B L}{3EI}$$

It may be noted that the direction of virtual forces applied to the structure is arbitrary. For example, if a unit couple had been applied counter-clockwise at *A*, the result θ_A would have had a negative sign. The negative sign would mean that the rotation was in a direction opposite to that of the applied virtual couple.

Example 6.6 | *A beam AB is simply supported over a span 5 m in length. A concentrated load of 30 kN is acting at a section 1.25 m from support A. Calculate the deflection under the load point. Take E = 200 × 10⁶ kN/m² (200,000 MPa) and I = 13.0 × 10⁻⁶ m⁴ (13.0 × 10⁶ mm⁴)*

The moment diagram due to applied load is shown in Fig. 6.19*b*. To obtain the deflection under the load, we apply a virtual unit load at C.

Using equation 6.50 the deflection $\delta_C = \int_0^L M_x \dfrac{m_x dx}{EI}$. The moment due to unit load is shown in Fig. 6.19*c*. The beam length is divided into two parts, part AC and part BC, for the convenience of integration. Considering the part AC and taking origin at A

$$M_x = R_A\,(x)$$
$$= 22.5(x) \qquad \text{for } 0 \le x \le 1.25 \text{ m}$$

and
$$m_x = 0.75(x)$$

Similarly considering part BC and taking origin B

$$M_x = R_B\,(x)$$
$$= 7.5\,(x) \qquad \text{for } 0 \le x \le 3.75 \text{ m}$$

and
$$m_x = 0.25\,(x)$$

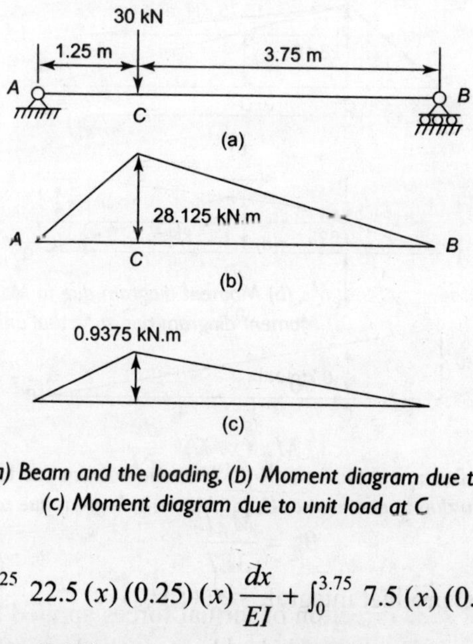

Fig. 6.19 | *(a) Beam and the loading, (b) Moment diagram due to applied loading, (c) Moment diagram due to unit load at C*

$$\therefore \qquad \delta_c = \int_0^{1.25} 22.5\,(x)\,(0.25)\,(x)\,\frac{dx}{EI} + \int_0^{3.75} 7.5\,(x)\,(0.25)\,(x)\,\frac{dx}{EI}$$

or
$$\delta_c = 22.5\,\frac{(0.25)}{3EI}\,(1.25)^3 + (7.5)\,\frac{(0.25)}{3EI}\,(3.75)^3$$

or
$$\delta_c = \frac{43.9}{3EI} = \frac{43.9 \times (1000)}{3 \times 2.6 \times 10^3} = 16.9 \text{ mm.}$$

In the above examples the product integrals have to be evaluated which are routine and time-taking. However, ready-made tables are available which can be utilized in evaluating the product integrals. The table in Appendix *B* gives the product integrals.

Example 6.7 | *Using product integrals from Appendix B, determine the deflection at the centre of the beam and slope at end A for the beam in Example 6.6.*

Deflection at Centre In order to obtain deflection at centre of beam, we apply a virtual unit load at D and draw the moment diagram as in Fig. 6.20c. The moment diagram due to applied loading is again shown in Fig. 6.20b. The beam is divided into three parts AC, CD and DB and the product integrals for these parts are obtained from Appendix B.

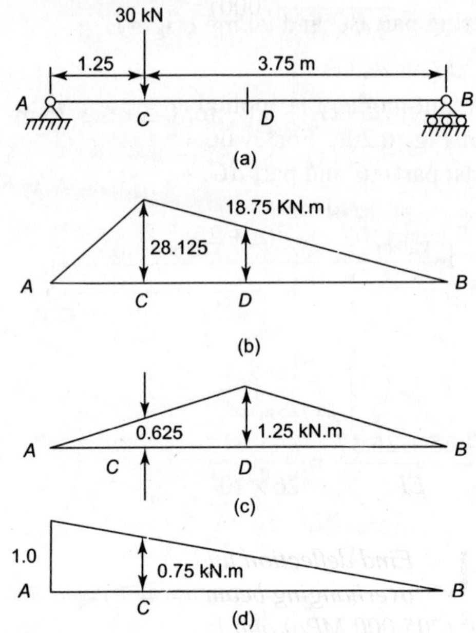

Fig. 6.20 | *(a) Beam and the loading, (b) Moment diagram due to applied loading, (a) Moment diagram due to unit load at D, (d) Moment diagram due to unit couple at A*

For the length AC product integral $= \frac{1}{3}(L)(a)(c)$

$$= \frac{1}{3}(1.25)(28.125)(0.625) = 7.32$$

For the length BD product integral $= \frac{1}{3}(L)(a)(c)$

$$= \frac{1}{3}(2.5)(18.75)(1.25) = 19.53$$

For the length CD product integral $= \frac{L}{6}\{a(2c+d) + b(2d+c)\}$

Taking $a = 28.125$

$b = 18.75$

$c = 0.625$

$d = 1.25$

and $\qquad L = 1.25$

Substituting in the relation above we have the product integral value = 26.85

$$\therefore \qquad \delta_D = \frac{7.32 + 19.53 + 26.85}{EI} = \frac{53.7}{EI}\ m$$

$$= \frac{53.7}{2.6 \times 10^3}(1000) = 20.65 \text{ mm}$$

Slope at End A A unit moment is applied at end *A* and the resulting moment diagram is shown in Fig. 6.20*c*. For evaluating product integrals the beam is divided into two parts: part *AC* and part *BC*.

$$\text{Product integral for part } AC = \frac{L}{6}(a + 2b)\,c$$

$$= \frac{1}{6}(1.25)(1 + 1.5)(28.125) = 14.65$$

$$\text{for part } BC = \frac{1}{3}Lac = \frac{1}{3}(3.75)(0.75)(28.125) = 26.37$$

$$\therefore \qquad \theta_A = \frac{14.65 + 26.37}{EI} = \frac{41.02}{26 \times 10^3} = 1.578 \times 10^{-3} \text{ radians.}$$

Example 6.8 *Find deflection under load at C the centre of span of a overhanging beam loaded as shown in Fig. 6.21a. Take $E = 205 \times 10^6$ kN/m^2 (205,000 MPa) and $I = 25 \times 10^{-6}$ m^4 (25×10^6 mm^4)*

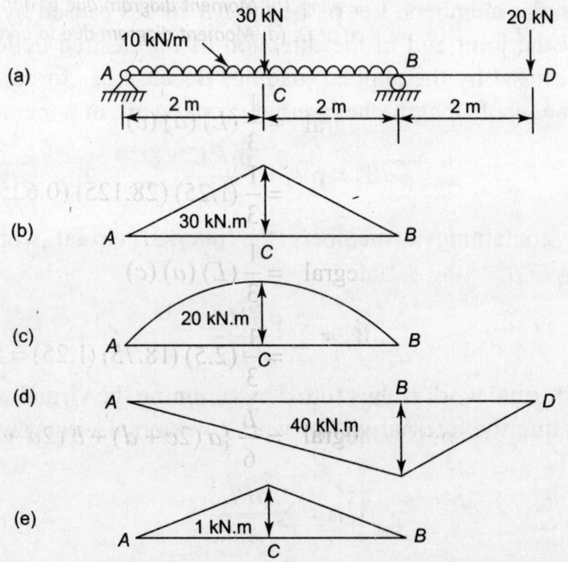

Fig. 6.21 *(a) Overhanging beam and the loading, (b) B.M.D. due to concentrated load at C, (c) B.M.D. on AB due to u.d.l, (d) B.M.D. due to concentrated load at D, (e) B.M.D. due to unit load at C*

The deflection under load at C can be obtained by superimposing the effects of individual loads. The moment diagrams for the individual loads are shown in Fig. 6.21 *b, c* and *d.* The moment diagram due to virtual unit load at C is shown in Fig. 6.21*e*. Using product integral values from Appendix *B*,

$$\delta_C \text{ due to concentrated load at } C = \frac{1}{3}\frac{Lac}{EI}$$

$$= \frac{1}{3}\frac{(4)(30)(1)}{EI} = \frac{40}{EI}$$

$$\delta_C \text{ due to u.d.l.} = \frac{5}{12}\, L\, a\, c = \frac{5}{12}\frac{(4)(30)(1)}{EI} = \frac{33.33}{EI}$$

$$\delta_C \text{ due to conc. Load at } D = \frac{1}{4}\, L\, a\, c = \frac{1}{4}(4)(-40)\,(1) = -\frac{40}{EI}$$

$$\text{Net deflection } \delta_c \text{ downward} = \frac{40}{EI} + \frac{33.33}{EI} - \frac{40}{EI}$$

$$= \frac{33.33}{EI} = \frac{33.33}{205 \times 25}(1000)\text{ mm} = 6.5\text{ mm}$$

6.8 DEFLECTION OF TRUSSES AND FRAMES

Let us now examine the virtual work method for determining the deflections of pin connected trusses. The deflections to be considered are those due to axial deformations in the members. Let $\mathbf{p}_i$, be the bar forces caused by a unit virtual load applied at the joint and in the direction of the desired deflection and P_i, the bar forces caused by the applied loading. Because bar forces $\mathbf{p}_i$ and P_i are constant over the whole length, the internal virtual work of a member is

$$dW_i = \mathbf{p}_i e_i = \frac{\mathbf{p}_i P_i L_i}{A_i E_i} \tag{6.52}$$

For a truss containing n members the internal virtual work expression becomes

$$W_i = \sum_{i=1}^{n} \frac{\mathbf{p}_i P_i L_i}{A_i E_i} \tag{6.53}$$

The internal virtual work is thus found by summing the virtual work of all the members. Equating the internal virtual work to external virtual work, we have in general

$$1 \cdot \Delta = \sum_{1}^{n} \frac{\mathbf{p}PL}{AE} \tag{6.54}$$

$$\Delta = \sum_{1}^{n} \frac{\mathbf{p}PL}{AE} \tag{6.55}$$

The following example illustrates the steps involved in evaluating deflections of truss joints.

Example 6.9 | *The vertical and horizontal deflection components of joint 4 are to be determined for the truss of Fig. 6.22a. $L = 3$ m, $A = 500 \times 10^{-6}$ m^4 (500×10^6 mm^4) and $E = 200 \times 10^6$ kN/m^2 (200,000 MPa) are constant for all members.*

To determine the vertical deflection of joint 4 denoted as Δ4V we apply a unit virtual force (1 kN) to the truss in the downward direction as shown in Fig. 6.22*b*. Member forces *P* from the applied loading are calculated using any one of the methods of truss analysis. The virtual forces in members due to unit virtual force applied are also separately evaluated. It is convenient to tabulate the results as shown in Table 6.1.

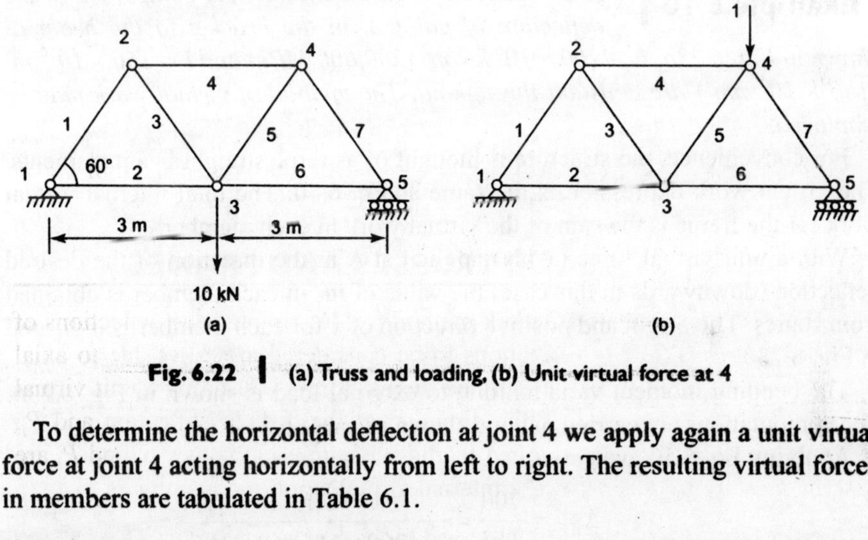

Fig. 6.22 | **(a) Truss and loading, (b) Unit virtual force at 4**

To determine the horizontal deflection at joint 4 we apply again a unit virtual force at joint 4 acting horizontally from left to right. The resulting virtual forces in members are tabulated in Table 6.1.

Table 6.1 | *Internal Virtual Work Computations for Truss of Fig. 6.22*

Member	*P* (kN)	p_v (kN)	p_h (kN)	p_vP	p_hP
1	–5.77	–0.29	+0.50	+1.55	–2.89
2	+2.89	+0.14	+0.75	+0.42	+2.17
3	+5.77	+0.29	–0.50	+1.67	–2.89
4	–5.77	–0.29	+0.50	+1.67	–2.89
5	+5.77	–0.29	+0.50	–1.67	+2.89
6	+2.89	+0.43	+0.25	+1.25	+0.72
7	–5.77	–0.87	–0.50	+5.00	+2.89
				Σ + 9.88	Σ 0

Then, the vertical displacement $\Delta_{4V} = \dfrac{1}{AE}\sum_{1}^{7} \mathbf{p}_v PL$. Similarly, the horizontal displacement $\Delta_{4V} = \dfrac{1}{AE}\sum_{1}^{7} \mathbf{p}_h PL$

$$\Delta_{4V} = \frac{9.88 \times 3}{500 \times 10^{-6} \times 200 \times 10^{6}} = 2.96 \times 10^{-3} \text{ m}$$

or $\quad \Delta_{4V} = 2.96$ mm

and $\quad \Delta_{4H} = 0$ as expected because of symmetry of structure and loading.

The following example illustrates the application of the virtual work method for frames as well.

Example 6.10 | *It is desired to compute the vertical component of the deflection of point A on the bracket to the beam as shown in Fig. 6.23a. E = 200 × 10⁶ kN/m² (200,000 MPa) and I = 160 × 10⁻⁶ m⁴ (160 × 10⁶ mm⁴) are constant throughout. The method of virtual work may be employed.*

For convenience, the structure is thought of as three straight beam elements. The virtual work expression is the same as Eq. 6.50. The total internal virtual work for the frame is the sum of the virtual work in each member.

With a unit virtual force (1 kN) applied at A in the direction of the desired deflection (downwards in this case) the value of $\mathbf{m}_x$ in each member is obtained from statics. The origin and positive direction of x for each member is indicated in Fig. 6.23*b*.

The bending moment variation due to external load is shown in Fig. 6.23*b*. The computations are carried out and shown in Table 6.2.

Applying Eq. 6.50, we get

$$\Delta_{AV} = \frac{700}{3EI} = \frac{700}{3 \times 200 \times 10^{6}\ 160 \times 10^{-6}}$$

$$= 7.28 \times 10^{-3} \text{ m}$$

or $\quad \Delta_{AV} = 7.29$ mm.

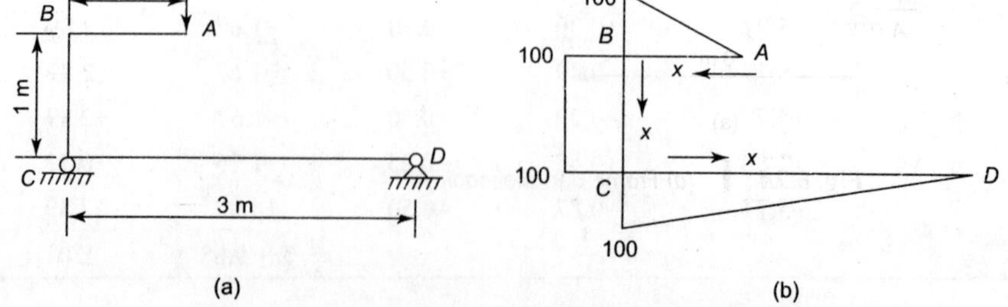

Fig. 6.23 | *(a) Structure and loading, (b) Moment diagram*

Table 6.2 | *Internal Virtual Work Computations for Structure of Fig. 6.23*

Section	Limits-For x	M_x	$\mathbf{m}_x$	$\int_0^L \frac{\mathbf{m}_x M_x dx}{EI}$
AB	0–1 m	$-100x$	$-x$	$\int_0^1 \frac{100_x^2 dx}{EI} = \frac{100}{3EI}$
BC	0–1 m	-100	-1	$\int_0^1 \frac{100\,dx}{EI} = \frac{100}{EI}$
CD	0–3 m	$-100\left(1-\frac{x}{3}\right)$	$-\left(1-\frac{x}{3}\right)$	$\int_0^3 100\left(1-\frac{x}{3}\right)^2 \frac{dx}{EI} = \frac{100}{EI}$
				$\Sigma = \frac{700}{3EI}$

Example 6.11 | *The horizontal displacement at support D is to be determined for the frame shown in Fig. 6.24a. Relative I values are indicated along the members. E = 200 × 10^6 kN/m^2 (200,000 MPa), I = 300 × 10^{-6} m^4 (300 × 10^6 mm^4).*

The deflection analysis is the same as that followed in Example 6.10. The total internal virtual work for the frame is the sum of the internal virtual work of all the sections of the frame.

The unit force (1 kN) is applied at support D in the direction of the desired displacement (Fig. 6.24*b*). The origin and the positive direction of x for each section is indicated in Fig. 6.24*b*.

The expressions for $\mathbf{m}_x$ and M_x for each member are determined as usual from statics. The resulting expressions for $\mathbf{m}_x$ and M_x for all sections are tabulated in Table 6.3.

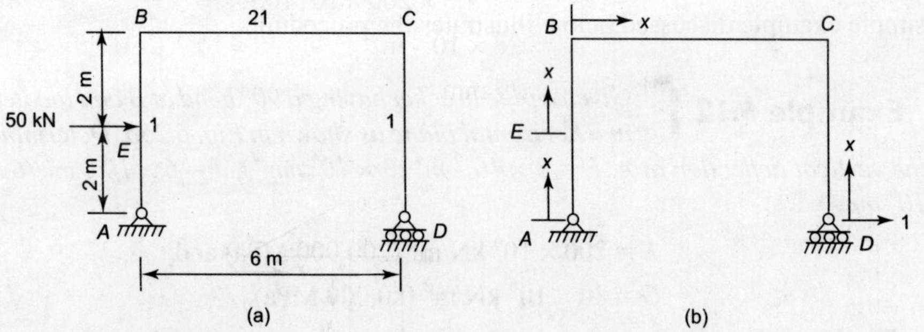

Fig. 6.24 | *(a) Frame and the loading, (b) Unit virtual force at D*

Table 6.3 | *Internal Virtual Work Computations for Frame of Fig. 6.24*

Section	*Origin for x*	*Limits for x*	M_x	$\mathbf{m}_x$	$\int_0^L \mathbf{m}_x \frac{Mdx}{EI}$
AE	*A*	0–2 m	+50x	+x	$\frac{1}{EI}\int_0^2 50x^2 dx = 400/3EI$
EB	*E*	0–2 m	+100	$(2+x)$	$\frac{1}{EI}\int_0^2 100(2+x)dx = 600/EI$
BC	*B*	0–6 m	$-\frac{50}{3}x+100$	+4	$\frac{1}{EI}\int_0^6 4\left(100-\frac{50}{3}x\right)dx = 600/EI$
DC	*D*	0–4 m	0	+x	$=0$
					$\Sigma = \frac{4000}{3EI}$

Therefore,

$$\int \frac{\mathbf{m}\,Mdx}{EI} = \frac{4000}{3EI}$$

Applying Eq. 6.50, we get

$$\Delta_{DH} = \frac{4000}{3\times 200\times 10^6 \times 300\times 10^{-6}} = 22.2\times 10^{-3}\text{ m}$$

or $\Delta_{DH} = 22.2$ mm

The deflection at any other point or direction can be found out in a similar manner by applying a unit virtual force at the point and in the direction of the desired deflection. The rotation at any point of the frame can be found by applying a unit couple at the point the rotation is desired.

The virtual work method can also be extended to three-dimensional frames taking precautions to correctly identify the direction of forces and moments. A simple example discussed below illustrates the procedure.

Example 6.12 | *Given a pipe bracket having a 90° bend at B and located in a horizontal plane as shown in Fig. 6.25a. Determine the vertical deflection at A. I. = 3×10^{-6} m^4 (3×10^6 mm^4), J = 6×10^{-6} m^4 (6×10^6 mm^4),*

$$E = 200\times 10^6\text{ kN/m}^2\text{ (200,000 MPa) and}$$
$$G = 80\times 10^6\text{ kN/m}^2\text{ (80,000 MPa).}$$

The bending and twisting strains are considered in evaluating the deflections. Because of the third dimension, it will be helpful to define positive forces on a section of a member with reference to the member co-ordinate system. The positive quantities of real and virtual forces on a section of member 1 are shown in Fig. 6.25*b* and member 2 in Fig. 6.25*c*. The double superscripts on the moment quantities indicate the axis about which the moment acts. Note that the moments and twist vectors are drawn in accordance with the right-hand screw rule. The origin for member co-ordinates are located at the ends of the members.

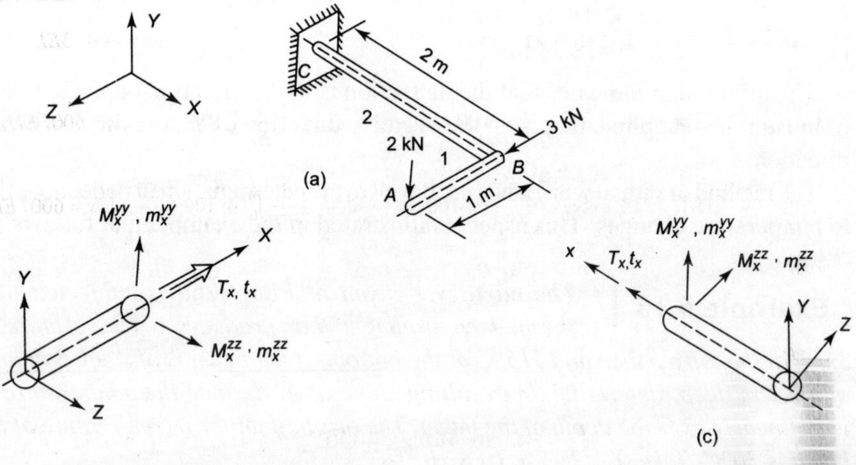

Fig. 6.25 | *(a) Bracket and loading, (b) Coordinates for member 1 (c) Coordinates for member 2*

Earlier, the expressions for virtual work were developed with the moment considered as acting only about Z axis. However, this expression also holds good for internal virtual work for moments acting about Y axis. The general expression equating the external virtual work to internal virtual work for the effects under consideration can be written as

$$W_e = W_i = \int_0^L \mathbf{m}_z^{YY} \frac{M_x^{YY}\,dx}{EI} + \int_0^L \mathbf{m}_x^{ZZ} \frac{M_x^{ZZ}\,dx}{EI} + \int_0^L \frac{t_x T_x dx}{GJ} \tag{6.55}$$

To evaluate deflection at A in the vertical direction, apply a unit force (1 kN) at A along the positive direction of Y. The resulting expressions for M_x, T_x along with $\mathbf{m}_x$ and $\mathbf{t}_x$ are given in Table 6.4.

Table 6.4 | *Computations for internal Virtual Work of Bracket in Fig. 6.25*

Member	$\mathbf{m}_x^{YY}$	M_x^{YY}	$\mathbf{m}_x^{ZZ}$	M_x^{ZZ}	$\mathbf{t}_x$	T_x
1	0	0	–x	+2x	0	0
2	0	–3x	–x	+2x	+1	–2

Applying Eq. 6.50, we get

$$\Delta_{AV} = \int_0^L \mathbf{m}_x^{YY} \frac{M_x^{YY}\,dx}{EI} + \int_0^L \mathbf{m}_x^{ZZ} \frac{M_x^{ZZ}\,dx}{EI} + \int_0^L \frac{\mathbf{t}_x T_x\,dx}{GJ} \tag{6.56}$$

Substituting values from Table 6.4 in Eq. (6.56)

$$\Delta_{AV} = \int_0^1 \left[\frac{0 + x(-2x) + 0}{EI}\right] dx + \int_0^2 \left[\frac{0 + x(-2x)}{EI} + \frac{(-1)(2)}{GJ}\right] dx = \frac{-6}{EI} - \frac{4}{GJ}$$

Substituting values for E, I, G and J

$$\Delta_{AV} = \frac{-6}{200\times10^6\times3\times10^{-6}} - \frac{4}{80\times10^6\times6\times10^{-6}}$$

$$= -18.33\times10^3 \text{ m}$$

or $\quad \Delta_{AV} = -18.33$ mm.

The minus sign indicates that the deflection is in the direction opposite to that of the unit load applied, that is, in the negative direction of Y or in the downward direction.

The method of virtual work lends itself well to the determination of deflections due to temperature changes. This aspect is illustrated in the example that follows.

Example 6.13 | *The cantilever beam in Fig. 6.26a is subjected to a thermal environment that produces a temperature of 35°C at the top surface and 115°C at the bottom. If the beam is of steel, 3 m long and 0.2 m deep, determine the resulting slope at A. Assume the temperature to vary linearly over the depth of the beam. The original uniform temperature of the beam is 30°C. Take $\alpha = 11.7\times10\times10^{-6}$/°C.*

Here we again use the condition of equality between the external virtual work and internal virtual work. The external virtual work is equal to the product of the virtual unit moment and real rotation caused by thermal effects. The internal virtual work is equal to the internal virtual force (moment **m**) multiplied by the real internal displacement.

The real internal displacement results from: (1) the average beam temperature of 75°C which is 45°C above that of the original temperature and (2) the temperature gradient of 80°C across the depth of the beam.

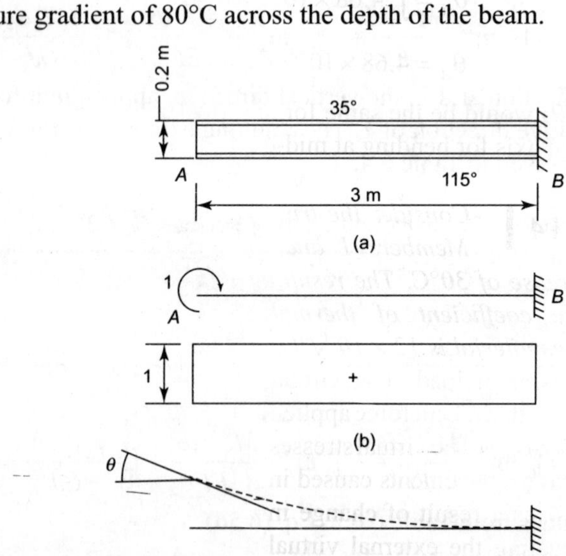

Fig. 6.26 | *(a) Beam subjected to temperature gradient, (b) Moment due to virtual unit moment at A, (c) Elastic line*

The first part of the thermal effect produces only a lengthening of the beam and does not enter into the work equation since the virtual unit moment produces no resultant axial force corresponding to an axial change along the length of the beam. The virtual moment diagram and the elastic line are shown in Fig. 6.26*b* and *c* respectively. However, the thermal gradient of the second part produces rotation $d\phi$ and the corresponding internal virtual work term is

$$W_i = \int_0^L \text{m}d\phi \tag{6.57}$$

We shall determine $d\phi$ by considering the strains at the extreme fibre caused by thermal gradient.

Considering a differential length dx of the beam (Fig. 6.27), the deformation at the top and bottom faces is given by

$$e = \alpha \Delta T dx$$

Therefore, the angle of rotation in a length $dx = d\phi = \dfrac{e}{0.1}$

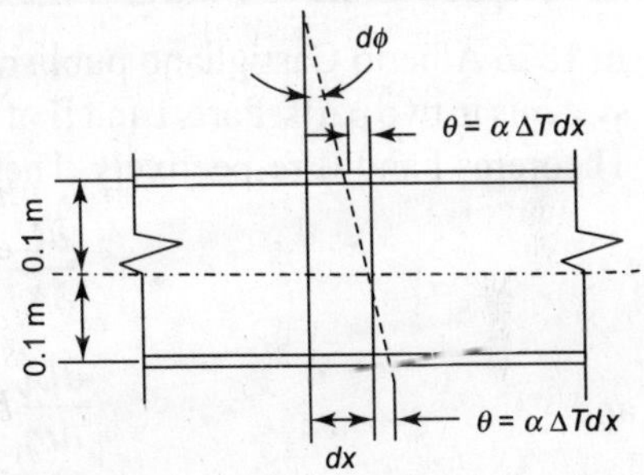

Fig. 6.27 | *Rotation caused by thermal gradient*

or $\quad d\phi = 10\alpha\Delta T dx$

$= 10 \times 11.7 \times 10^{-6} \times 40dx$

$= 4.68 \times 10^{3}\ dx$

Using Eq. 6.50, we get

$$\theta_A = \int_0^3 4.68 \times 10^{-3}\ dx$$

or
$$\theta_A = 4.68 \times 10^{-3} \times 3 = 14.04 \times 10^{-3} \text{ rad.}$$

The value of θ_A would be the same for any shape of 0.2 m depth steel beam that has its neutral axis for bending at mid-depth.

Example 6.14 | *Consider the truss given in Example 6.9 (Fig. 6.22a). Members 1 and 7 in the truss are subjected to a temperature increase of 30°C. The resulting vertical deflection at 3 has to be determined. The coefficient of thermal expansion for the material is 12×10^{-6}/°C.*

There is no external load. The virtual stresses in members due to unit force applied at 3 are given in Fig. 6.28. The virtual stresses ride along the real displacements caused in members 1 and 7 as a result of change in temperature. Equating the external virtual work to internal virtual work, we get

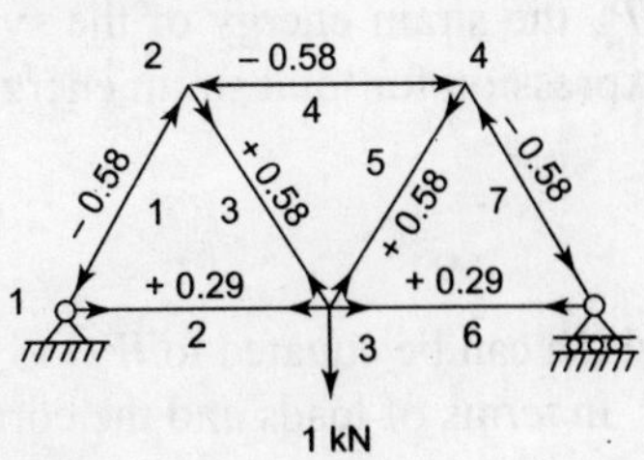

Fig. 6.28 | *Stresses due to unit virtual force at 3*

$$1 \cdot \Delta_{3V} = 2(-0.58)\alpha L dT$$

or $$\Delta_{3V} = 2(-0.58)12 \times 10^{-6} \times 3 \times 30$$
$$= -1.252 \times 10^{-3} \text{ m}$$
$$\Delta_{3V} = -1.25 \text{ mm}$$

The negative sign indicates that the deflection is upwards opposite to the direction of unit virtual load applied. The assumed direction is always identical to the direction of the applied virtual force.

6.9 CASTIGLIANO'S THEOREMS

In 1876 Alberto Castigliano published his work on the variation of strain energy systems in two parts. Parts I and II of his work are often referred to as Castigliano's Theorems I and II respectively. They are known as

$$\frac{\partial U}{\partial \Delta_i} = P_i \quad \text{Part I} \tag{6.58}$$

and
$$\frac{\partial U}{\partial P_i} = \Delta_i \quad \text{Part II} \tag{6.59}$$

where

U = Strain energy of the system

P_i = External loads applied point i.

and Δ_i = deflection of point i in the direction of P_i

The strain energy and virtual work principles play an important role in the derivation of Castigliano's theorems.

Consider for example a simple beam as shown in Fig. 6.29 subjected to a system of loads *P.* Suppose load system *P* is applied gradually. The beam undergoes deformations as shown in Fig. 6.29*a*.

Here the strain energy is a function of external loads and is equal to the external work done. Therefore,

$$U = W_e = \phi(P_1 \, P_2 \dots P_n) \tag{6.60}$$

or
$$U = \sum_{i=1}^{n} \frac{1}{2} P_i \Delta_i \tag{6.61}$$

Now, if any one of the loads, say P_i, is increased by a differential amount, say dP_i, the strain energy of the system will change by an amount $\left(\frac{\partial U}{\partial P_i}\right) dP_i$. The expression for total strain energy becomes

$$U' = U + \frac{\partial U}{\partial P_i} dP_i \tag{6.62}$$

which can be equated to W_e.

In terms of loads and the corresponding displacements shown in Fig. 6.29*a*,

$$U' = W_e' = U + \sum_{i=1}^{n} P_i d\Delta_i + \frac{1}{2} dP_i d\Delta_i \tag{6.63}$$

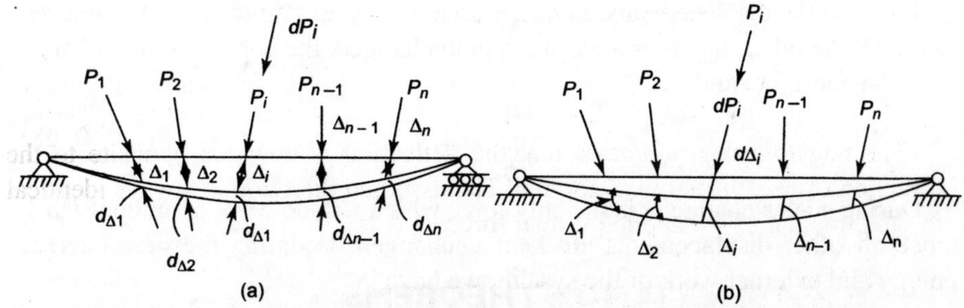

Fig. 6.29 | *(a) Deflections due to load system P applied first and then dP_i (b) Deflection due to dpi applied first and then load system P*

It may be noted that the forces, *P,* ride in full during the displacements caused by dP_i. The coefficient (1/2) in the last term is obvious. Neglecting the last term in Eq. 6.63 as being the product of two differential values, we have

$$U' = W_e' = U + \sum_{i=1}^{n} P_i d\Delta_i \tag{6.64}$$

Now, suppose, the sequence of loading is reversed, that is dP_i is applied first and then the system *P.* The corresponding displacements are shown in Fig. 6.29*b*. The application of dP_i first produces an infinitesimal displacement $d\Delta_i$. The external work of (1/2) (dP_i) $(d\Delta_i)$ can be neglected since the value is of second order. Further, the external work done by *P* system of forces is unaffected by the presence of dP_i. On the other hand, during the application of these forces, dP_i will act like a virtual load and will do (dP_i) (Δ_i) amount of work. Consequently, the total work done is

$$W'_e = U' = U + dP_i\Delta_i \tag{6.65}$$

Since the order of the application of loads is immaterial, the total work done or the total internal strain energy in both the loading cases must be equal. Therefore, equating the right-hand side quantities of Eqs. 6.62 and 6.65, we have

$$U = \frac{\partial U}{\partial P_i} dP_i = U + dP_i\Delta_i \tag{6.66}$$

or

$$\frac{\partial U}{\partial P_i} = \Delta_i \tag{6.67}$$

This quantity is known as Castigliano's second theorem in which P_i and Δ_i can also be the moment and the angular rotation respectively. The theorem can be stated as follows: *If a linearly elastic structure is subjected to a set of loads, the displacement of any load in its direction is equal to the partial derivative of the total strain energy with respect to that load.*

Castigliano'a first theorem can also be obtained in a similar manner. For example, in the beam of Fig. 6.29 loaded by *P* system of forces, we may term the external work as *W* as the internal strain energy as *U*.

If one of the displacements, say Δ_i, is changed by an infinitesimal amount $d\Delta_i$ while all the other displacements are kept unchanged, the corresponding change in strain energy would be

$$\frac{\partial U}{\partial \Delta_i} \cdot d\Delta_i \tag{6.68}$$

During such a change P_i is the only force which will do work of amount $P_i d\Delta_i$ since all other displacements are kept unchanged. Equating the new internal energy and external work of the system, we have

$$U + \frac{\partial U}{\partial \Delta_i} \cdot d\Delta_i = W + dP_i\Delta_i \tag{6.69}$$

This gives

$$\frac{\partial U}{\partial \Delta_i} = P_i \tag{6.70}$$

This is known as *Castigliano's first theorem.* Here we can see that the partial derivative of the strain energy with respect to any one of the displacements of applied loads is equal to the load. It should be noted that this theorem is not dependent upon the assumption of an elastic system and of a linear relation between the loads and displacements.

To apply Castigliano's second theorem for determining deflections, we must express the internal strain energy in terms of external loads. The expression developed in Sec. 6.3 can be utilised for this purpose. For example, in evaluating deflections due to bending strains, the internal strain energy due to bending is given as (Eq. 6.16)

$$U = \int_0^L \frac{M_x^2 dx}{2EI}$$

From Eq. 6.67 the expression for deflection can be written as

$$\Delta_i = \frac{\partial}{\partial P_i} \int_0^L \frac{M_x^2 dx}{2EI} \tag{6.71}$$

If the indicated operation were to be performed, it is necessary to square the various expressions for M, integrate and then evaluate the partial derivative. It is much easier to first differentiate the quantity under the integral sign and then evaluate the integral, that is,

$$\Delta_i = \int_0^L M_x \frac{\partial M_x}{\partial P_i} \frac{dx}{EI} \tag{6.72}$$

Similar expressions can be written for other types of strains. For example, in the case of trusses where axial strains only will be considered, the expression for the deflection of a truss joint is

$$\Delta_i = \sum P_x \frac{\partial P_x}{\partial P_i} \frac{L_x}{AE} \tag{6.73}$$

The use of Eq. 6.73 requires that the axial forces in the members be expressed in terms of the external loading.

It may be noted that if a deflection component is required at a point where no action is applied, or if an action exists at that point but not in the direction of the desired deflection, then an imaginary action is applied at that point in that direction until the partial derivative for the total strain energy has been found. The imaginary action is then reduced to zero.

The application of Castigliano's second theorem to deflection calculations is illustrated in the following examples.

Example 6.15 *It is required to determine the deflection under the load point for the beam shown in Fig. 6.30. EI is constant.*

The required deflection, denoted as Δ, can be obtained from Eq 6.72,

$$\Delta = \int_0^L M_x \frac{\partial M_x}{\partial P}\frac{dx}{EI}$$

where
$$M_x = \frac{P}{2}(x) \text{ for } 0 \le x \le \frac{L}{2}$$

$$\frac{\partial M_x}{\partial P} = \frac{x}{2}$$

Because of symmetry, the deflection can be obtained by taking twice the value of the integral for the left half of the beam, that is,

$$\Delta = 2\int_0^{L/2} \frac{P}{2}(x)\,(x/2)\frac{dx}{EI} \quad \text{or} \quad \Delta = \frac{PL^3}{48\,EI}$$

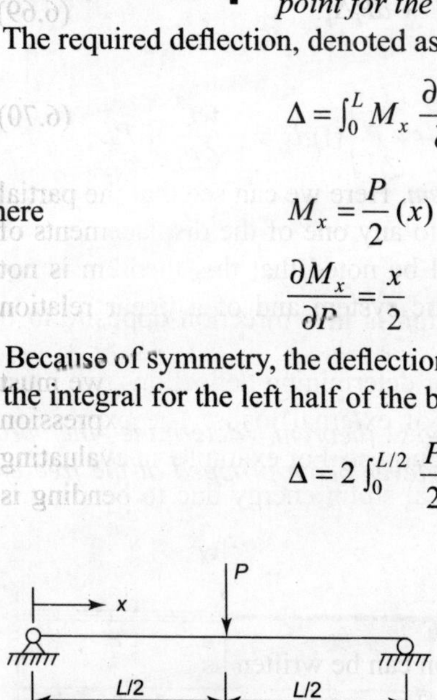

Fig. 6.30

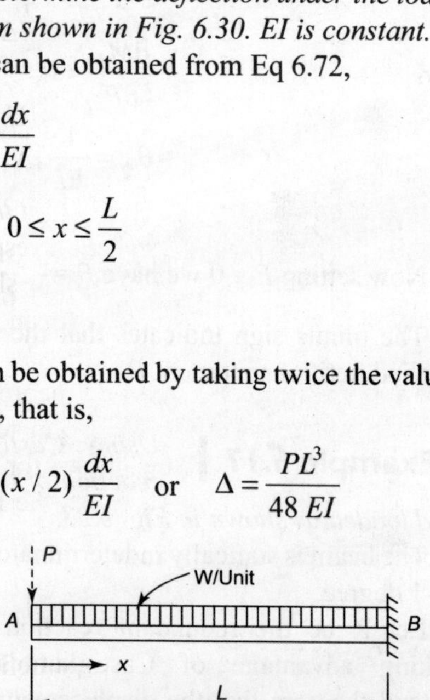

Fig. 6.31

Example 6.16 *A cantilever beam is loaded as shown in Fig. 6.31. It is required to determine the vertical deflection and rotation at free end A. EI is constant.*

The vertical deflection at the free end is obtained by applying a fictitious load *P* at *A* vertically downward. Then

$$M_x = \frac{-w_x^2}{2} - Px \quad \text{and} \quad \frac{\partial M_x}{\partial P} = -x$$

Therefore,
$$\Delta = \frac{1}{EI}\int_0^L -\left(\frac{wx^2}{2} + Px\right)(-x)dx = \frac{wL^4}{8EI} + \frac{PL^3}{3EI}$$

Now setting fictitious force $P = 0$ the desired deflection

$$\Delta_{AP} = \frac{wL^4}{8EI}$$

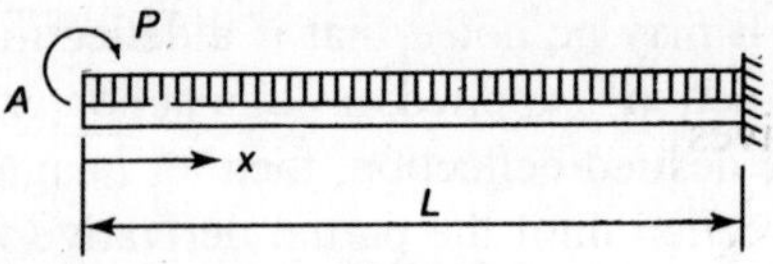

Fig. 6.32

The rotation at free end A is obtained by applying a fictitious moment force P to the beam as shown in Fig. 6.32.

Here

$$M_x = -\frac{wx^2}{2} + P$$

and
$$\frac{\partial M_x}{\partial P} = 1$$

$$\theta_A = \frac{1}{EI}\int_0^L \left(-\frac{wx^2}{2} + P\right)(1)dx = -\frac{wL^3}{6EI} + PL$$

Now setting $P = 0$ we have $\theta = -\dfrac{wL^3}{6EI}$

The minus sign indicates that the rotation is in a direction opposite to the applied fictitious moment P.

Example 6.17 | *Using Castigliano's theorem, determine the prop reaction of a centilever beam propped at the free end and loaded as shown in Fig. 6.33.*

The beam is statically indeterminate by 1 degree.

Let R be the redundant reaction. Taking advantage of Castigliano's second theorem that the displacement of reaction R in its direction is equal to the partial derivative of the total strain energy with respect to that load. We know the displacement of R = 0

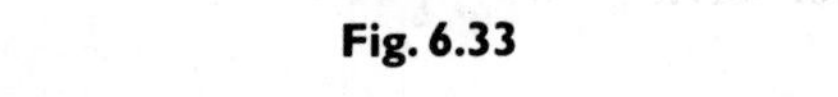

Fig. 6.33

Then we have
$$\frac{\partial U}{\partial R} = 0$$

The total strain energy U is calculated by the sum of strain energy in region AB and region BC.

For the Region A to B we have

$$Mx = R.x \text{ and } \frac{\partial MK}{\partial R} = x$$

$$\partial U_{AB} = \frac{1}{EI}\int_0^{l/2} Mx\frac{\partial Mx}{\partial R}\,dx = \frac{1}{EI}\int_0^{l/2} Rx^2 dx$$

gives $$\frac{R}{EI}\left[\frac{x^3}{3}\right]_0^{l/2} = \frac{Rl^3}{24\,EI}$$

For the region BC we have, taking x from B

$$M_x = R\left(\frac{l}{2}+x\right) - Wx$$

$$\frac{\partial Mx}{\partial R} = \left(\frac{l}{2}+x\right)$$

$$\frac{\partial U_{BC}}{\partial R} = \frac{1}{2EI}\int_0^{l/2}\left[R\left(\frac{l}{2}+x\right) - Wx\right]\left[\frac{l}{2}+x\right]dx$$

or $$\frac{1}{2EI}\int_0^{l/2}\left[R\left(\frac{l^2}{4}+lx+x^2\right) - W\left(\frac{lx}{2}+x^2\right)\right]dx$$

Integrating and substituting limits

$$\frac{1}{2EI}\left(\frac{7}{24}Rl^3\right) + \frac{W}{2EI}\left[\frac{-5}{48}Wl^3\right]$$

Now $$\frac{\partial U_{AC}}{\partial R} = \frac{Rl^3}{24\,EI} + \frac{7}{48}\frac{Rl^3}{EI} - \frac{5}{2\times 48}\frac{W\,l^3}{EI} = 0$$

or $$\frac{9}{48}Rl^3 = \frac{5}{48\times 2}Wl^3$$

or $$R = \frac{5}{48\times 2}\times\frac{48}{9}W = \frac{5}{18}W$$

This appears to be cumbersome when compared with other methods developed for solving indeterminate beams.

Example 6.18 *The sign board in Fig. 6.34 weighing 2.2 kN is supported by a cantilevered steel pipe whose axis is bent to a circular arc of 7 m radius. Taking $I = 50 \times 10^{-6}$ m^4 (50×10^6 mm^4) and $E = 200 \times 10^6$ kN/m^2 (200,00 MPa) determine the vertical displacement of the centre of the sign board.*

Since the radius of curvature is large in comparison with the cross-sectional dimension, ordinary beam deflection formulae are used replacing dx by ds. In this case $ds = Rd\theta$. Applying a fictitious load P downwards we have

$$M_\theta = -\,2.2\,R\sin\theta - PR\,\text{sing}\,\theta$$

$$\frac{\partial M_\theta}{\partial P} = -R\sin\theta$$

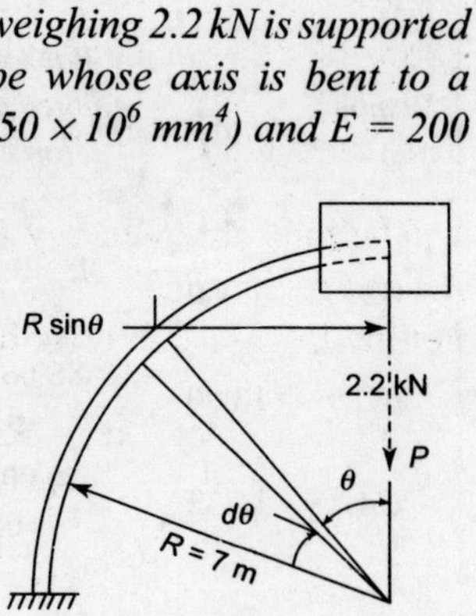

Fig. 6.34 | Sign board

Therefore, $\Delta = \dfrac{1}{EI} \int_0^{LM} \dfrac{\partial M_\theta}{\partial P}\, ds$

or $$\Delta = \frac{1}{EI} \int_0^{\pi/2} -(2.2\, R \sin \theta + PR \sin \theta)(-R \sin \theta)\, R\, d\theta$$

$$= \frac{1}{EI} \int_0^{\pi/2} -2.2\, R^3 \sin^2 \theta\, d\theta + \frac{1}{EI} \int_0^{\pi/2} PR^2 \sin^2 \theta\, d\theta$$

It is enough to evaluate the first term in the above expression as the second term reduces to zero since P is a fictitious force and is to be equated to zero.

$$\Delta = \frac{2.2\, R^3}{EI} \int_0^{\pi/2} \sin^2 \theta\, d\theta$$

On evaluating the integral and substituting values for *E, I* and *R*

$$\Delta = 59.27 \text{ mm}$$

Castigliano's theorem can also be employed to evaluate deflections in trusses. The example that follows illustrates the procedure.

Example 6.19 | *It is required to determine the vertical and horizontal deflectioncomponents of joint C of the truss in Fig. 6.35. E = 200 × 10⁶ kN/m² (200,000 MPa) and sectional area of each bar A = 100 × 10⁻⁶ m² (100 mm²).*

To find the vertical deflection component of joint *C*, it is necessary to apply a fictitious force P_v in the vertical direction. Similarly a fictitious force P_h is applied at *C* to evaluate the horizontal deflection component of joint *C*. Eq. 6.73 is made use of in evaluating the required deflections.

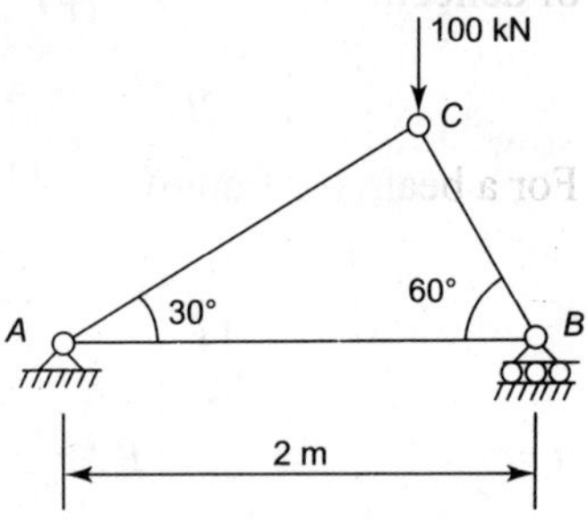

Fig. 6.35

Table 6.5 | *Computations for truss deflections*

Member	*length L(m)*	*Bar forces due to Applied and Fictitious forces: P (kN)*	$\dfrac{\partial P}{\partial P_v}$	$\dfrac{\partial P}{\partial P_h}$	$P\dfrac{\partial PL}{\partial P_v AE}$	$P\dfrac{\partial PL}{\partial P_h AE}$
1	*2*	*3*	*4*	*5*	*6*	*7*
AB	2.000	$43.30 + 0.433\, P_v + 0.250\, P_h$	+0.433	+0.250	37.24/AE	21.65/AE
BC	1.000	$-86.60 + 0.866\, P_v - 0.500\, P_h$	+ 0.866	–0.500	75.00/AE	43.30/AE
CA	1.732	$-50.00 - 0.500\, P_v + 0.866\, P_h$	–0.500	+0.866	43.30/AE	–75.00/AE
					$\Sigma \dfrac{155.54}{AE}$	$\Sigma \dfrac{-10.05}{AE}$

The complete computations involved are given in Table 6.5. Therefore,

$$\Delta_{CV} = \frac{155.54}{AE} = 7.75 \text{ mm}$$

and

$$\Delta_{CH} = \frac{-10.05}{AE} = -0.5 \text{ mm}$$

It may be noted that the value for *AE* is constant for all members and hence substituted at the end.

The negative sign for horizontal deflection indicates that the deflection is opposite to the direction of P_h, applied, that is, to the left.

It may be noted that in the last two columns of Table 6.5, P_v and P_h, terms were omitted since $P_v = P_h = 0$.

Unit Load or Dummy Load Method The unit load or dummy load method for evaluating deflections was developed in Sec. 6.7 by employing the principle of virtual work. The same result can also be obtained from a consideration of Castigliano's theorem II.

Let an elastic body shown in Fig. 6.36*a* be in equilibrium under loads P_1, P_2 ..., P_n and a load *Q* applied at point *K*. By Castigliano's theorem, the component of deflection at *K* in the direction of applied force *Q* is

$$\Delta_{KQ} = \frac{\partial U}{\partial Q} \tag{6.74}$$

For a beam (or frame)

$$\Delta_{KQ} = \int M \frac{\partial M}{\partial Q}\frac{dx}{EI} \tag{6.75}$$

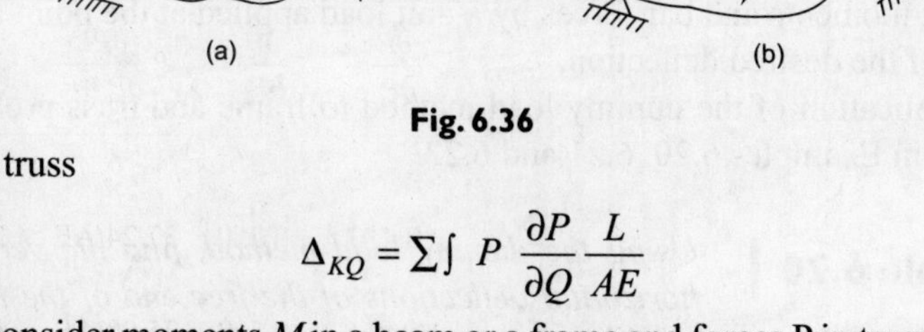

Fig. 6.36

and for a truss

$$\Delta_{KQ} = \sum \int P \frac{\partial P}{\partial Q}\frac{L}{AE} \tag{6.76}$$

Now consider moments *M* in a beam or a frame and forces P in truss members. They are necessarily functions of load *Q* as well as forces P_1, P_2, ..., P_n. Suppose a unit load is placed at *K* in the place of *Q* (see (Fig. 6.32*b*). Let the moment produced be *m* in a beam and the bar force be *p* in a truss. Therefore, the moment or bar force produced by a force *Q* will be

$$M_Q = Q \, . \, m \tag{6.77}$$

and $$P_Q = Q \, . \, p$$

Now in the body acted on by forces P_1, P_2, ...,P_n and also Q (Fig. 6.35*a*) the moment M and bar forces P can be obtained by superposition as

$$M = M_P + M_Q = M_P + Q. \, m \tag{6.78}$$

$$P = P_P + P_Q = P_P + Q. \, p \tag{6.79}$$

where M_P and P_P are the moments and bar forces respectively produced by forces $P_1, P_2, ..., P_n$ only.

Then $$\frac{\partial M}{\partial Q} = m \tag{6.80}$$

And $$\frac{\partial P}{\partial Q} = P \tag{6.81}$$

Substituting Eq. 6.80 in Eq. 6.75 for beams

$$\Delta_{KQ} = \int \frac{Mmdx}{EI} \tag{6.82}$$

Again substituting Eq. 6.81 in Eq. 6.76 for trusses

$$\Delta_{KQ} = \Sigma \frac{P \cdot p \cdot L}{AE} \tag{6.83}$$

Eqs. 6.82 and 6.83 are the same as Eqs. 6.50 and 6.54 respectively except that **m** and **p** are replaced by m and p in the expressions.

It may be pointed out that in the above derivation, the deflection point K due to P system of forces alone can be found by setting forces $Q = 0$. Then the moment and bare forces are M_P and P_P respectively. Therefore, it is not necessary to actually apply the load Q, to a body in order to find the deflection of a point. The dummy load method is a numerical method of finding the partial derivative of the moments or the bar forces with respect to Q. The derivative can be found by computing moments and bar forces by a unit load applied at the point and in the direction of the desired deflection.

The application of the dummy load method to frame and truss problems are illustrated in Examples 6.20, 6.21 and 6.22.

Example 6.20 | *Using the dummy load method, find the vertical and horizontal deflections of the free end of the lamp post shown in Fig. 6.34.*

$E = 200 \times 10^6$ kN/m^2 (200,000 MPa),

$I_1 = 2/2 = 80 \times 10^{-6}$ m^4 $(80 \times 10^6$ mm$^4)$.

The vertical and horizontal deflection components at point C can be obtained by applying each time a unit force at point C in the direction of the desired deflection. Moment M, m_v and m_h, as caused by applied loading, unit vertical force and unit horizontal force respectively are determined using statics only. The origin for x for each of the two members is shown in Fig. 6.37. The integral in Eq. 6.82 is carried out for the entire frame. The complete solution is shown in Table 6.6.

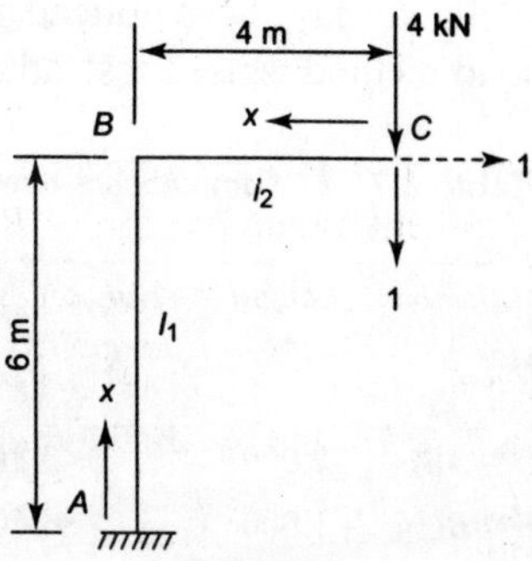

Fig. 6.37

Table 6.6 | *Computations for evaluation of* $\int \frac{Mmdx}{EI}$

Member	Limits for x	Moment due to applied load, M kN.m	Moment due to unit vertical load, m_v kN.m	Moment due to unit horizontal load, m_h kN.m	$\int \frac{Mm_v dx}{EI}$	$\int \frac{Mm_h dx}{EI}$
CB	0–4 m	–4(x)	–1.(x)	0	$\int_0^4 \frac{-4x(-x)dx}{EI_2}$	0
AB	0–6 m	–4(4)	–1.(4)	–1.x	$\int_0^6 \frac{-16(-4)dx}{EI_1}$	$\int_0^6 \frac{-16(-x)dx}{EI_1}$

On evaluation,

$$\int \frac{Mm_v dx}{EI} = \frac{1664}{3EI_1} \quad \text{and} \quad \int \frac{Mm_h dx}{EI} = \frac{288}{EI_1}$$

On substitution of numerical values for E and I_1

$$\Delta_{CV} = 34.7 \text{ mm}$$

and

$$\Delta_{CH} = 18.0 \text{ mm}$$

Example 6.21 | *Using the dummy load method, evaluate the vertical and horizontal deflection components of joint C of the truss given in Example 6.19 (see Fig. 6.35).*

Bar forces P due to the applied load are obtained as usual. To obtain vertical and horizontal deflection components of joint C, it is necessary to apply unit forces, in turn, in the direction of the deflections desired. Bar forces p_v and p_h are determined using any one of the methods.

The complete solution is worked out in Table 6.7.

Therefore

$$\Delta_{CV} = \frac{155.54}{AE} \quad \text{and} \quad \Delta_{CH} = \frac{-10.05}{AE}$$

Substituting numerical values for A and E

$$\Delta_{CV} = 7.75 \text{ mm} \quad \text{and} \quad \Delta_{CH} = -0.5 \text{ mm}$$

It may have been noticed by now that considerable effort is saved if the dummy load method is used instead of Castigliano's theorem.

Table 6.7 | *Computations to evaluate* $\Sigma \frac{P \cdot p.L}{AE}$

Member	*Length M*	*Bar forces due to applied load: P kN*	$p_v kN$	$p_h kN$	$\frac{Pp_vL}{AE}$	$\frac{Pp_hL}{AE}$
AB	2.000	+43.30	+0.430	0.250	37.24/AE	21.65/AE
BC	1.000	–86.60	–0.860	–0.500	75.00/AE	43.30/AE
CA	1.732	–50.00	–0.500	+0.866	43.30/AE	–75.00/AE
					$\sum\frac{155.54}{AE}$	$\sum\frac{-10.05}{AE}$

Example 6.22 | *For the truss shown in Fig. 6.38. Calculate the change in length of diagonal BE due to the applied loading. The areas of upper and lower chords = 400 mm² and web members = 300 mm2. Take E = 200 × 10³ N/mm².*

In order to obtain the changes in length in diagonal BE, we apply a unit force along BE as shown. The forces in all the members due to applied loading and also the forces in members due to applications of unit force are worked out and tabulated in Table 6.8.

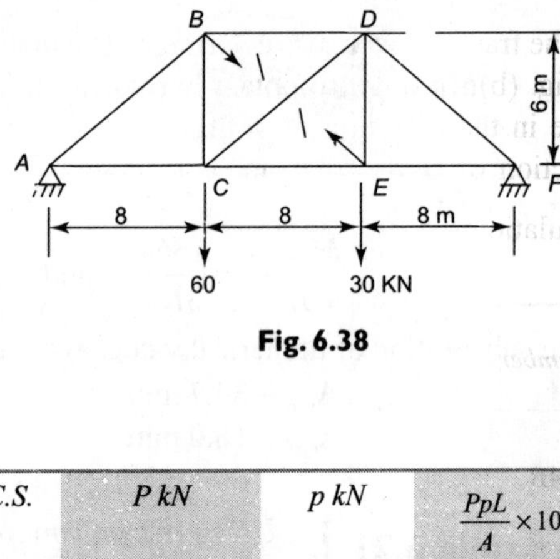

Fig. 6.38

Table 6.8

Member	*Length m*	*Area of mm² C.S.*	*P kN*	*p kN*	$\frac{PpL}{A} \times 10^6$
AC	8	400	200/3	0	0
CE	8	400	160/3	–4/5	–0.8533
EE	8	400	160/3	0	0
BD	8	400	–200/3	–4/5	1.0666
AB	10	300	–250/3	0	0
BC	6	300	50	–3/5	–0.6000
CD	10	300	50/3	1	0.5556
DE	6	300	30	–3/5	–0.3600
DF	10	300	–200/3	0	0

$$\sum -\frac{18\times10^6}{E} = -\frac{0.18\times10^6}{200\times10^6}\text{ m} = -0.9\times10^{-3}\text{ m} = -0.9\text{ mm}$$

The negative sign indicates that the diagonal BE moves away

Example 6.23 | *Using an energy method, determine the horizontal deflection of point D for the structure shown in Fig. 6.39 due to application of the force H. EI is constant for the entire frame.*

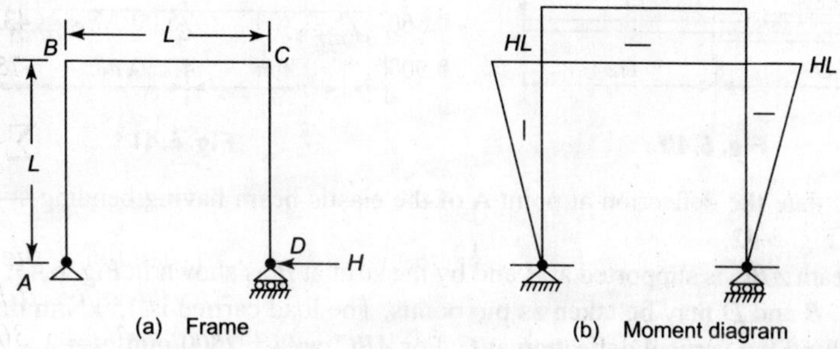

Fig. 6.39

The frame is statically determine. The moment due to applied load H is shown in Fig. (b). In order to obtain horizontal deflection of point D, we apply a unit force in the direction of deflection desired. Let the unit load is applied in the direction of H and the moment diagram is same as for H taking $H = 1$. The calculations $\int \frac{Mm\,dx}{EI}$ are tabulated below.

Member	*Limits for x*	*Moment M*	*Moment m*	$\int \frac{Mm\,dx}{EI}$
AB	$O-L$	$(-H.x)$	$(-X)$	$\int_0^L \frac{Hx^2\,dx}{EI} = \frac{HL^3}{3EI}$
BC	$O-L$	$(-H.L)$	$(-L)$	$\int_0^L \frac{HL^2\,dx}{EI} = \frac{HL^3}{EI}$
	$O-L$	$(-H.x)$	$(-x)$	$\int_0^L \frac{H\cdot x^2\,dx}{EI} = \frac{HL^3}{3EI}$
				$\sum \frac{5}{3}\frac{HL^3}{EI}$

$$\text{The deflection } D_H = \frac{5}{3}\frac{HL^3}{EI}$$

The +ve sign indicates that the deflection is inwards in the direction of H.

Problems for Practice

6.1 Calculate the deflection and slope at the free end of the cantilever beam shown in Fig. 6.40.

6.2 A uniform beam 10 m long is supported at 1 m from each end as shown in Fig. 6.41. It carries loads of 30 kN at each end and 120 kN at the centre. Find deflections under the loads and slopes at the supports.

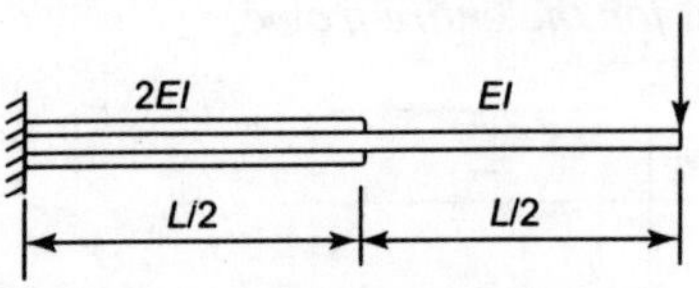

Fig. 6.40

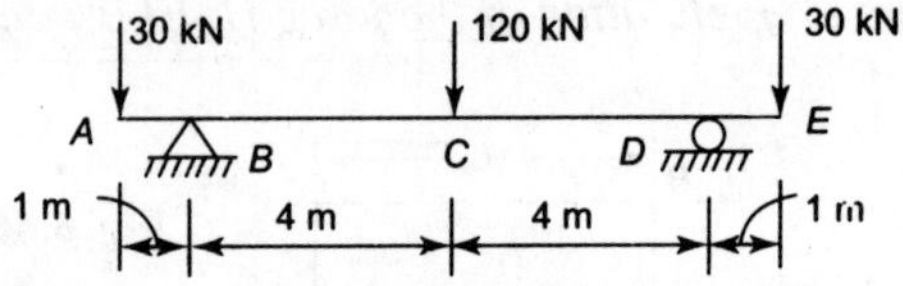

Fig. 6.41

6.3 Calculate the deflection at point A of the elastic beam having bending stiffness as shown in Fig. 6.42.

6.4 A beam *ABC* is supported at *A* and by the strut at *B* as shown in Fig. 6.43. Connections at *A*, *B* and *D* may be taken as pin points. The load carried is 15 kN/m distributed over *AB*. Find the vertical deflection at *C*. For *ABC*, area = 2500 mm^2, / = 2 × 10^6 mm^4; for BD, area = 1500 mm^2, E = 200,000 N/mra^2, (200,000 MPa).

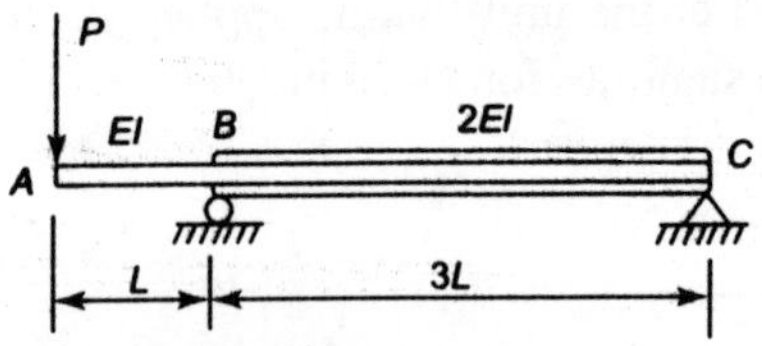

Fig. 6.42

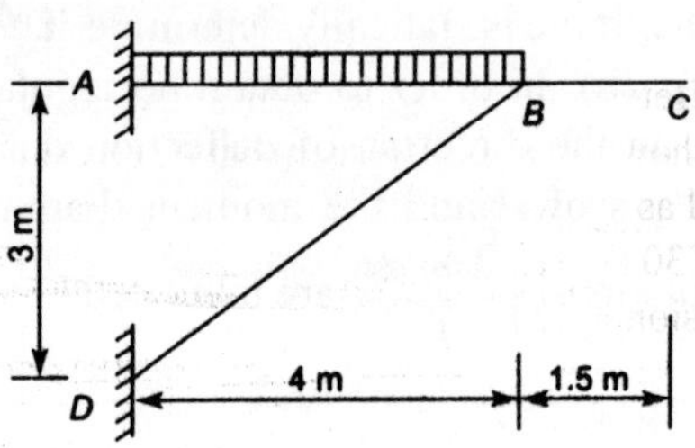

Fig. 6.43

6.5 The vertical bent cantilever shown in Fig. 6.44 carries a vertical load at free end C. If the flexural rigidity *EI* is constant throughout, estimate the vertical displacement of point C.

6.6 Determine the horizontal deflection for the simple elastic frame shown in Fig. 6.45 Consider only the deflection caused by bending. The flexural rigidity *EI* of both members is equal and constant.

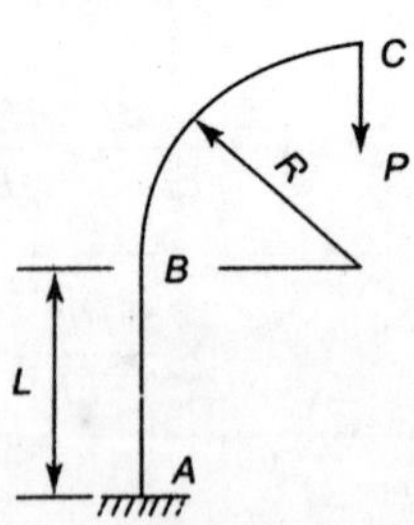

Fig. 6.44

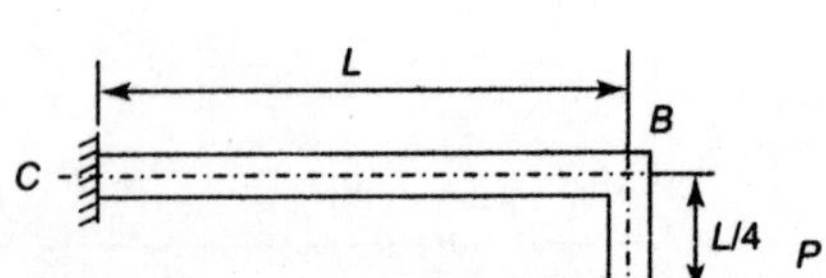

Fig. 6.45

6.7 Find the downward deflection of end C caused by the applied force of 5 kN in the structure shown in Fig. 6.46. Consider deflections only due to bending.

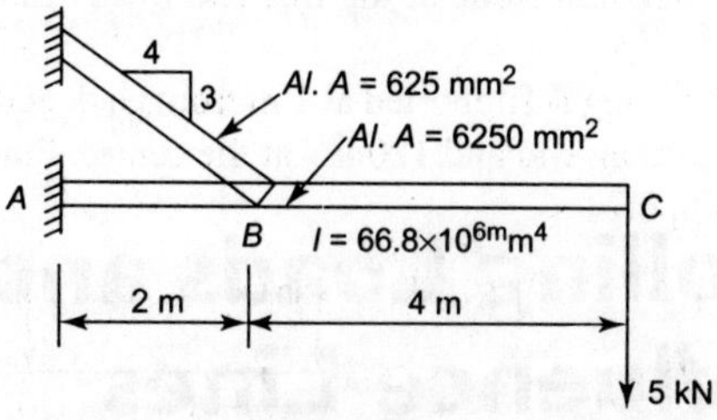

Fig. 6.46

6.8 A simply supported beam shown in Fig. 6.47 is subjected to a temperature gradient AT through its depth, the coefficient of thermal expansion is a/K. Determine the maximum deflection. (Hint: Curvature diagram is constant with ordinate $\propto \Delta T/d$).

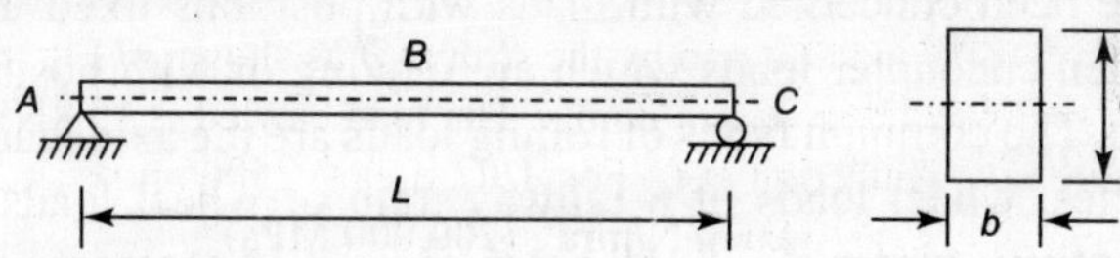

Fig. 6.47

6.9 Determine the adjustment required in the lengths of members EF and FG of the truss shown in Fig. 6.48 such that A, C, D and B all lie at the same level. The structure is loaded as shown and all members except the lower chord are subjected to a temperature rise of 30 K. AH areas are 1000 mm^2, E is 200×10^2 N/mm^2 and the coefficient of thermal expansion is $12 \times 1\,0^{-6}$/K.

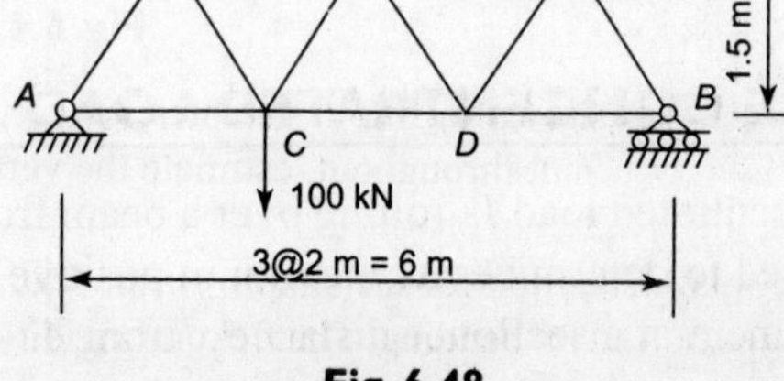

Fig. 6.48

7 Rolling Loads and Influence Lines

7.1 INTRODUCTION

So far, we have been concerned with loads with positions fixed. But in actual practice, we often encounter loads which are moving or with positions that are liable to change. The common types of rolling loads are the axle loads of moving trucks or vehicles, wheel loads of a railway train or wheel loads of a gantry assembly on a gantry girder etc. In all these cases it is necessary to determine the maximum S.F. and B.M. at different sections as the loads traverse from one end to the other. In the following sections, we shall discuss the following cases of rolling loads:

1. A single concentrated load
2. A uniformly distributed load longer than the span
3. A uniformly distributed load shorter than the span
4. Two concentrated loads spaced at some distance apart
5. A series of concentrated loads

7.2 A SINGLE CONCENTRATED LOAD

Consider a single concentrated load W rolling over a beam from A to B as shown in Fig. 7.1. It is required to determine the maximum positive and negative shear force and bending moment at a section X distance x from A.

Negative Shear Force Let us consider an instant when the load W is between A and X at a distance z from A. The shear force at section X is negative and is equal to reaction R_B'.

$$V_x = -R_B = -\frac{Wz}{l} \tag{7.1}$$

The negative shear force increases as the load advances towards the section X, reaching a maximum when the load is just to the left of section $V_x\,(\text{maxm.}) = \frac{Wx}{l}$. If we want to draw the maximum –ve shear force diagram for all the sections we vary the value for x from $x = 0$ to $x = l$. It is seen that the diagram varies linearly with X having shear force $V = 0$ at $x = 0$ and $V = W$ at $x = l$. The negative S.F. diagram is shown in Fig. 7.1*b* below the base line *AB*.

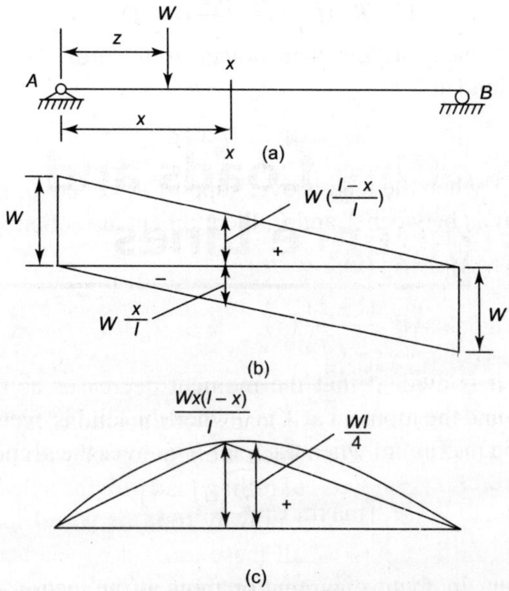

Fig. 7.1 | *(a) Beam under loading, (b) Maximum S.F. at a section X, (c) Maximum B.M. at a section*

Positive Shear Force Let us consider again an instant when the load is between X and B. We know that the shear force at section X is positive and is equal to

$$V_x = R_A = W\frac{(l-z)}{l} \tag{7.2}$$

It is evident from Eqn. 7.2 that the shear force V_x decreases as the load moves towards support B and at $z = l$ the shear force = 0. The positive shear force is at its maximum when the load is just to the right of section X. That is when $z = x$ and is equal to

$$V_x\,(\text{maxm.}) = R_A\,\frac{W\,(l-z)}{l} \tag{7.3}$$

If we want to plot the maximum positive shear force diagram for all the sections, we vary the value of x from $x = 0$ to $x = l$. It is seen that the diagram varies linearly with x having a shear force $V = W$ at $x = 0$ to $V = 0$ at $x = l$. The maximum +ve shear force diagram is shown in Fig. 7.1*b* above the base line *AB*.

From the above discussion it is clear that the maximum positive or negative S.F. at a section X occurs when the load is on the section itself. Thus the shear force diagram will consist of two parallel straight lines, one for positive and the other for negative, having end ordinates as shown in Fig. 7.1*b*.

Bending Moment The bending moment at section X for load W between A and X is equal to

$$M = R_B (l - x) = \frac{Wz}{l}(l - x) \tag{7.4}$$

The moment increases as the load advances towards X and the value of moment when the load at X is

$$M_x = W\frac{x}{l}(l - x) \tag{7.5}$$

The moment at X when the load is over support A, i.e. z = 0, is $M_x = 0$.

When the load W is between X and B, the moment at section X is

$$M_x = R_A (x)$$

$$= W\frac{(1 - z)}{l}(x) \tag{7.6}$$

From Eqn. 7.6 it is evident that the moment decreases as the load moves towards support B, and the moment at X is 0 when the load is over support B. The moment at X will be maximum when the load W is over the section, that is

$$M_x\,(\text{maxm.}) = \frac{W\,x(1 - x)}{l} \tag{7.7}$$

as before.

Thus, the bending moment diagram for moment at section X is a triangle having zero ordinates at the ends and an ordinate $\frac{W\,x(l - x)}{l}$ at the section X.

The maximum bending moment at other sections can be obtained by giving different values for x in Eqn. 7.7.

This is a second degree equation, the equation of a parabola. The section at which the absolute maximum value for M_x is obtained by differentiating M_x with respect to x and equating it to zero.

$$\frac{dM_x}{dx} = 0 \text{ or } \frac{d}{dx}\left(Wx - \frac{Wx^2}{l}\right) = 0$$

which gives $x = 1/2$.

Therefore, the absolute maximum bending moment will occur at the centre of the beam and is equal to

$$M(\text{maxm.}) = W\frac{l}{2} - \frac{Wl}{4} = \frac{Wl}{4}$$

The parabola shown in Fig. 7.1*c* envelops the maximum B.M. values.

7.3 UNIFORMLY DISTRIBUTED LOAD LONGER THAN THE SPAN

Negative Shear Force Consider a uniformly distributed load of w/unit length longer than the span l rolling from A to B, as shown in Fig. 7.2.

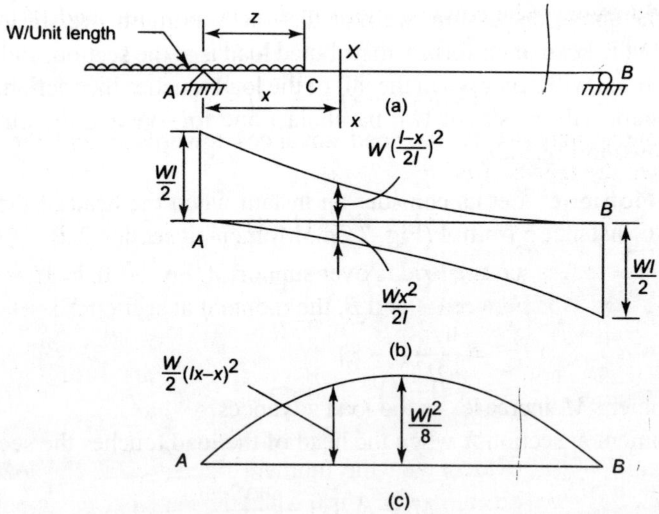

Fig. 7.2 | *(a) Beam under rolling u.d.l. longer than span, (b) Max..num S.F. diagram, (c) Maximum B.M. at different sections*

Let us consider an instant when the head of the load is at any point C at a distance z from A. The shear force at section X is negative and is equal to

$$V_x = R_B = -\frac{wz^2}{2l} \tag{7.8}$$

The shear force reaches maximum when the head of the load touches section X and is equal to

$$V_x = -R_B = -\frac{Wx^2}{2l} \tag{7.9}$$

If we want to plot the maximum –ve S.F. diagram for other sections, the value of x has to be varied in Eqn. 7.9 which represents a second degree parabola. At $x = 0$ S.F., $V = 0$ and at $x = l$, the S.F., $V = -wl/2$. The –ve S.F. diagram is shown in Fig. 7.2*b* below the base line AB.

Positive Shear Force We know that the positive shear force at section X is equal to the reaction R_A minus any load between A and X. A little consideration will show that the positive shear force is maximum when the tail of the load is at X and occupies from X to B. The maximum positive shear force for this loading position is equal to

$$V_x = R_A = w\frac{(l-x)^2}{2l} \tag{7.10}$$

If we want to plot the maximum shear force at other sect'ons the value of x has to be varied from $x = 0$ to $x = l$ in Eqn. 7.10. Eqn. (7.10) represents a second degree curve having ordinates $\frac{wl}{2}$ at $x = 0$ and zero at $x = l$. The maximum +ve shear force diagram is shown in Fig. 7.2*b* above base line AB.

From the above discussion, we find that the maximum –ve shear force will occur when the head of uniformly distributed load is at the section, and maximum +ve shear force will occur when the tail of the load touches the section. The shear force diagram will consist of two parabolas, one for –ve and the other for +ve S.F. as shown in Fig. 7.2*b*.

Bending Moment Let us consider an instant when the head of the load is at any point *C* distance *z* from *A* (Fig. 7.2*a*). Moment at section *X* is,

$$M_x = R_B\,(l-x)$$

$$= \frac{wz^2}{2l}(l-x) \tag{7.11}$$

The moment M_x increases as the load advances.

The moment at section *X* when the head of the load touches the section is

$$M_x = R_A\,(x) - \frac{wx^2}{2} \tag{7.12}$$

The moment continues to increase as the load advances since more and more load is added on to the span. The maximum moment at section *X* occurs when the load fully occupies the span from *A* to *B*. Moment at section *X* when the span is fully loaded is

$$M_x = \frac{wlx}{2} - \frac{wx^2}{2}$$

$$= \frac{w}{2}(lx - x^2) \tag{7.13}$$

It is obvious that in a simply supported beam, the maximum bending moment will occur at centre of span when the span is fully loaded and is equal to $M_{\text{max.}} = \dfrac{wl^2}{8}$

The maximum bending moment diagram for different sections is shown in Fig. 7.2*c*.

7.4 UNIFORMLY DISTRIBUTED LOAD SHORTER THAN SPAN

Negative Shear Force In Fig. 7.3*a* a u.d.l. of intensity w/unit length spreading over a length *a* crossing a simply supported beam *AB* is shown. Consider an instant when the head of the load enters the span and occupies a distance $z \le a$ from *A*. The shear force at section is –ve and is equal to reaction R_B,

$$V_x = -R_B = \frac{-wz^2}{2l} \tag{7.14}$$

It is evident that the shear force varies parabolically and reaches a value $\frac{wa^2}{2l}$ when the tail of the load just enters the span. As the load moves forward the S.F. increases as R_B increases. When the head of the load reaches section X, the S.F. is again equal to reaction R_B.

$$V_x = -R_B z - \frac{wa}{l}\left(x - \frac{a}{2}\right) \tag{7.15}$$

If we want to plot the S.F. diagram for different sections from $x = 0$ to $x = l$ it is evident that the diagram is parabolic from $x - 0$ to $x = a$

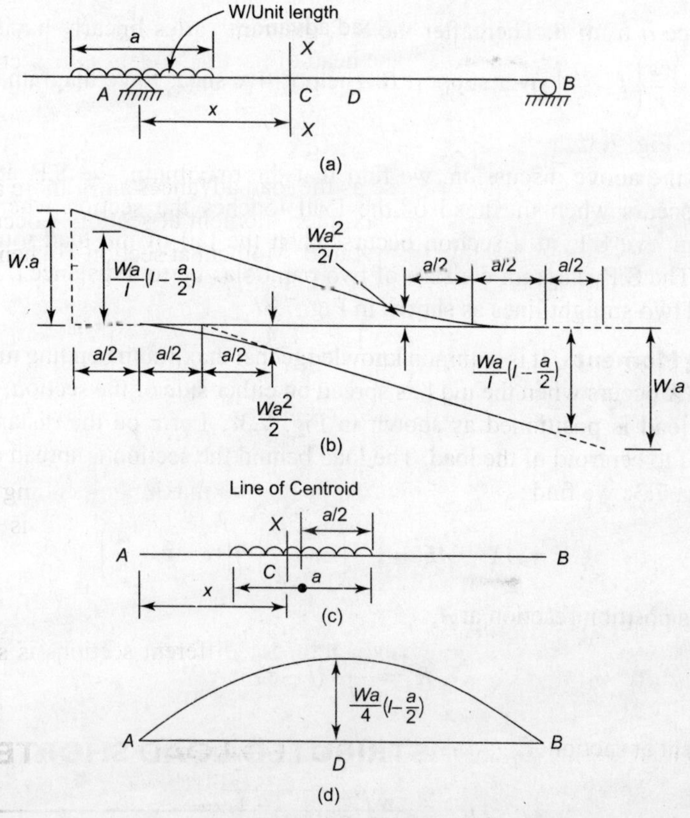

Fig. 7.3 | *(a) Beam under u.d.l. shorter than span, (b) Maximum –ve and +ve S.F. at different sections, (c) Positioning of load for maximum B.M. at a section, (d) Maximum B.M. diagram for different sections*

and varies linearly from $x = a$ to $x = l$. It is interesting to note that the straight line has its ordinates zero at $x = \frac{a}{2}$ and wa at $x = \frac{1 + a}{2}$. The –ve S.F. diagram is shown plotted in Fig. 7.3*b* below the base line *AB*.

Positive Shear Force The shear force at section X will be positive when the load lies between X to B. The positive shear force reaches maximum when the tail of the load just crosses section X and is equal to reaction R_A

$$V_x = R_A = \frac{wa}{l}\left(l - x - \frac{a}{2}\right) \tag{7.16}$$

If we want to plot maximum S.F. diagram for different sections the value of x has to be varied in Eqn. 7.16 which is linear. The S.F. ordinate $V_x = \frac{wa}{l}\left(l - \frac{1}{2}\right)$ at $x = 0$ and $V_x = 0$ at $x = \left(l - \frac{a}{2}\right)$ Thus, the S.F. diagram is a parabola for over a distance a from B having zero ordinate at B and $\frac{wa^2}{2l}$ at distance a from B. Thereafter the S.F. diagram varies linearly having a S.F. ordinate $\frac{wa}{l}\left(l - \frac{a}{2}\right)$ over support B. The positive shear force diagram is shown plotted in Fig. 7.3*b*.

From the above discussion, we find that the maximum –ve S.F. at a given section occurs when the head of the load touches the section where as the maximum +ve S.F. at a section occurs when the tail of the load touches the section. The S.F. diagram consists of two parabolas up to a distance a from the ends and two straight lines as shown in Fig. 7.3*b*.

Bending Moment It is common knowledge that maximum bending moment at a section X occurs when the u.d.l. is spread on either side of the section. Suppose that the load is positioned as shown in Fig. 7.3*c*. Let z be the distance from support A to centroid of the load. The load behind the section is spread over CX. From Fig. 7.3c we find

$$CX = AX - AC = x - \left(z - \frac{a}{2}\right) = \left(x - z + \frac{a}{2}\right)$$

At this position reaction at A,

$$R_A = \frac{w.a}{l}(l - z)$$

Moment at section X,

$$M_x = R_A\, x - \frac{w}{2}\left(x - z + \frac{a}{2}\right)^2$$

$$= \left[\frac{w.a.}{l}(l - z) - \frac{w}{2}\left(x - z + \frac{a}{2}\right)^2\right] \tag{7.17}$$

The moment M_x will be maximum for the value of z, when $\frac{dM_x}{dz} = 0$. Differentiating Eqn. 7.17 w.r.t. z and equating it to zero.

$$\frac{dM_x}{dz} = -\frac{ax}{l} + \left(x - z + \frac{a}{2}\right) = 0$$

That is
$$\frac{x}{l} = \frac{x - z + \frac{a}{2}}{a}$$

$$\frac{x}{l} = \frac{CX}{CD} \tag{7.18}$$

It means that the B.M. at the section X is maximum when the position of the load is such that the section X divides the span and the load in the same ratio. This is a very important relation which is useful later for point loads also.

It is easy to visualise that the absolute maximum bending moment will occur at the centre of span when the load is spread equally on either side of centre of span. For this position of load the absolute maximum B.M.

$$M_{\max} = R_A \frac{l}{2} - w.\frac{a}{2}.\frac{a}{4}$$

$$= \frac{wa}{2}\frac{l}{2} - \frac{wa^2}{8} = \frac{wa}{4}\left(l - \frac{a}{2}\right) \tag{7.19}$$

The maximum bending moment diagram for different section is shown in Fig. 7.3*d*.

7.5 TWO CONCENTRATED LOADS

Consider two concentrated loads W_1. and W_2 spaced d apart rolling over a beam AB as shown in Fig. 7.4*a*. Let the leading load W_1 be lighter than the trailing load W_2

Negative Shear Force Consider an instant when the loads are in the region AX and $d \leq x$. The shear force at section X is negative and is equal to R_B. The shear force will increase as the loads move to the right as R_B increases. The S.F. at section X will be maximum under one of the following two load positions,

i. When the leading load W_1 is at section X and the trailing load is between A and X.
ii. When the leading load is between X and B and the trailing load W_2 is on the section X.

Shear force at section X under load position (i) is

$$V_x = -R_B = -\left\{\frac{W_1\, x + W_2\,(x - d)}{l}\right\} \tag{7.20}$$

Shear force at section X under load position (ii) is

$$V_x = -\{R_B - W_1\}$$

$$= -\left\{\frac{W_1\,(x + d) + W_2\,(x)}{l} - W_1\right\} \tag{7.21}$$

It is obvious that the maximum –ve S.F. at section X is the larger of the two values obtained in Eqns. 7.20 and 7.21 dependent on W_1, W_2, d, x and l. If the S.F. obtained in Eqn. 7.20 were to be greater than that obtained in Eqn. 7.21 then,

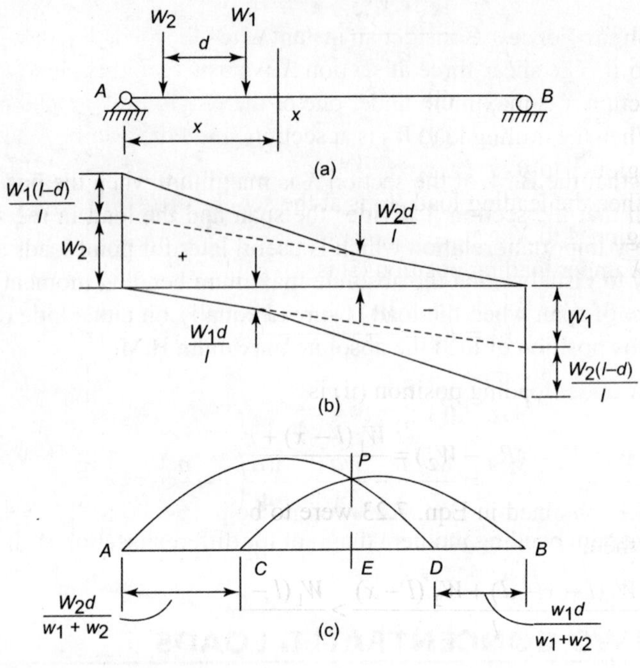

Fig. 7.4 | *(a) Beam and the two concentrated loads, (b) Maximum shear force diagram, (c) Maximum B.M. diagram at different sections*

$$\frac{W_1\,x + W_2\,(x-d)}{l} > \frac{W_1\,(x+d) + W_2\,x}{l} - W_1$$

or $$W_1x + W_2x - W_2d > W_1x + W_1d + W_2\,x - W_1l$$

This reduces to $W_1d + W_2d < W_1l$

or $$d(W_1 + W_2) < W_1l$$

$\therefore$ $$d < \frac{W_1}{W_1 + W_2}\,l \tag{7.22}$$

If $d > \dfrac{W_1}{W_1 + W_2}\,l$ the maximum S.F. at section X occurs under loading position (ii) discussed earlier.

Now if we want to plot the –ve S.F. diagram for different sections from $x = 0$ to $x = l$ the beam has to be divided into two sections: one from $x = 0$ to $x = d$ and the second from $x = d$ to $x = l$. In the first section from $x = 0$ to $x = d$ only one load W_1 will be in the span and the trailing load lies outside the beam. Substituting $x = 0$ for support section, $V = 0$.

At $x = d$, $V = \dfrac{W_1 d}{l}$ and at $x = l$, $V = W_1 + W_2 \dfrac{(l-d)}{l}$. The –ve S.F. diagram is shown plotted in Fig. 7.4*b* below base line *AB*.

Positive Shear Force Consider an instant when the loads W_1 and W_2 are in the region *X* to *B*. The shear force at section *X* is positive and is equal to R_A. Shear force at section *X* is maximum under one of the two following load positions.

i. When the trailing load W_2 is at section *X* and the leading load W_1 is in the region *X* to *B*.
ii. When the leading load W_1 is at the section and the trailing load is in the region *A* to *X*.

S.F. at *X* under loading position (i) is

$$V_x = R_A = \frac{W_1 (l-x-d) + W_2 (l-x)}{l} \tag{7.23}$$

S.F. at *X* under loading position (ii) is

$$V_x = (R_A - W_2) = \frac{W_1 (l-x) + W_2 (l-x+d)}{l} - W_2 \tag{7.24}$$

If the S.F. obtained in Eqn. 7.23 were to be greater than the S.F. obtained in Eqn. 7.24 then,

$$\frac{W_1 (l-x-d) + W_2 (l-x)}{l} > \frac{W_1 (l-x) + W_2 (l-x+d)}{l} - W_2 \tag{7.25}$$

or $W_1 l - W_1 x - W_1 d + W_2 l - W_2 x > W_1 l - W_1 x + W_2 l - W_2 x + W_2 d - W_2 l$

This reduces to $\quad W_1 d + W_2 d < W_2 l$

or
$$d < \frac{W_2}{W_1 + W_2} l \tag{7.26}$$

If $d > W_2/W_1 + W_2 l$ the S.F. at section *X* will be maximum under loading position (ii).

The +ve S.F. diagram is shown plotted in Fig. 7.4*b*. dividing the beam into two sections; one from $x = 0$ to $x = (l - a)$ and the second from $x = (l - a)$ to $x = l$. For the section from $x = (l - a)$ to $x = l$ the leading load W_1 lies outside the span and only the trailing load lies in the span.

Bending Moment The maximum bending moment at section *X* occurs when one of the two loads lie on the section.

Consider first that the leading load W_1 is at section *X*. The moment at section *X* is

$$M_x (1) = R_B (l - x)$$

$$= \left\{ \frac{W_1 x + W_2 (x-d)}{l} \right\} (l - x) \tag{7.27}$$

When the trailing load W_2 is at section X, the moment at the section is

$$M_x(2) = R_A(x)$$

$$= \left\{ W_1 \frac{(l-x-d) + W_2(l-x)}{l} \right\}(x) \quad (7.28)$$

It is obvious that the maximum bending moment at section X is the greater of the two values obtained from Eqns 7.27 and 7.28 depending upon W_1, W_2, l, d and x. If we plot the maximum moment diagrams represented by Eqns. 7.27 and 7.28 for different sections we find that these diagrams are parabolas. The first B.M. diagram $M_x(1)$ will have zero ordinates at $x = \frac{W_2 d}{W_1 + W_2}$ and at $x = l$. The second B.M. diagram $M_x(2)$ will have zero ordinates at $x = 0$ and at $x = l - \frac{W_1 d}{W_1 + W_2}$, that is at a distance $\frac{W_1 d}{W_1 + W_2}$ from end B. The two moment diagrams are shown plotted in Fig. 7.4*c*. We also notice that $M_x(1) = M_x(2)$ at section E. Equating the moments $M_x(1) = M_x(2)$ we have

$$\left\{ \frac{W_1 x + W_2(x-d)}{l} \right\}(l-x) = \left\{ \frac{W_1(l-x-d) + W_2(l-x)}{l} \right\} x \quad (7.29)$$

Simplifying, we get

$$\frac{x}{l} = \frac{W_2}{W_1 + W_2} \quad (7.30)$$

It is obvious that the average loading on the left of the section is equal to the average loading on the right of the section. It is seen that for all sections from A to E the maximum moment is given by Eqn. 7.28 in which W_2 is on the section and W_1 is ahead of it, and, for all sections from E to B, the maximum moment is given by Eqn 7.27 in which W_1 is on the section and W_2 is behind it.

7.6 SERIES OF CONCENTRATED LOADS

Consider a series of concentrated loads W_1, W_2 .... W_5 rolling from A to B over span l as in Fig. 7.5.

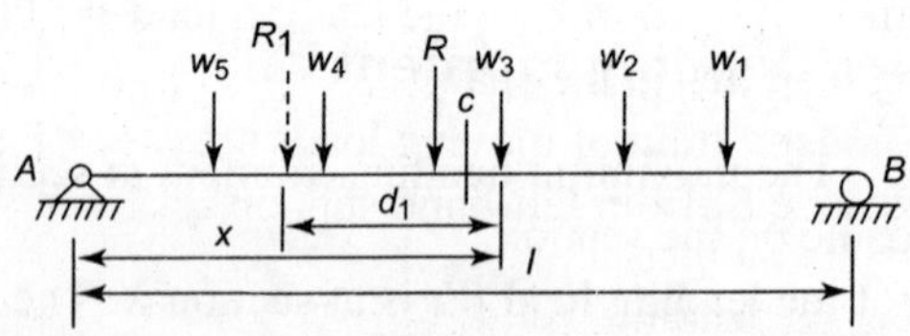

Fig. 7.5

7.6.1 Maximum S.F. at a Section

The maximum –ve or +ve S.F. at a given section occurs when one of the loads is at the section itself. From an inspection of the magnitude and disposition of the loads and making a few trials one may be able to fix up a particular load which should be placed over the section. The calculations are simple and hence the trials can be carried out quickly.

7.6.2 Maximum Bending Moment Under a Given Load

Consider the load system as in Fig. 7.5 moving from left to right. Let W_3 be the load the position of which is to be fixed for obtaining maximum bending moment under it. Let

x = Distance of load W_3 from support A.

R = Resultant of all the loads on the span and located at distance d form W_3

R_1 = Resultant of all the loads to the left of W_3 and located at distance d_1 from W_3

For the loading position indicated reaction at support A,

$$R_A = \frac{R}{l}(l - x + d_1)$$

Moment under load W_3 is,

$$M_x = R_A(x) - R_1(d_1)$$

or
$$M_x = \frac{R}{l}(lx - x^2 + xd_1) - R_1 d_1 \tag{7.31}$$

We can obtain the value for x at which M_x will be maximum by setting $\frac{dM_x}{dx} = 0$

$$\therefore \quad \frac{dM_x}{dx} = \frac{R}{l}(l - 2x + d_1) = 0$$

or
$$x = \frac{l}{2} + \frac{d_1}{2} \tag{7.32}$$

From the above we can state that the maximum bending moment under any load occurs when that load and the resultant of all the loads are located equidistant from the centre of the span; or in other words, when the centre of the beam lies midway between the resultant R and the load under consideration.

7.6.3 Maximum Bending Moment at a Given Section

Consider the beam under a train of moving loads shown in Fig. 7.6. Let section K be located at a distance L_1 from left-hand support and at a distance

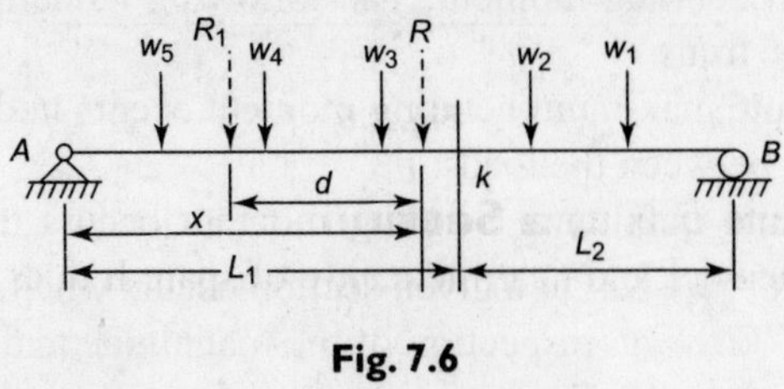

Fig. 7.6

L_2 measured from right-hand support. Let R represent the resultant of all the loads on the span and be located at a distance x from the left-hand support. Again let R_1 represent the resultant of loads to the left of section K and be located at distance d from the resultant R. The expression for the moment at K is,

$$M_K = R_A L_1 - R_1 (L_1 - x + a) \tag{7.33}$$

substituting for $R_A = \frac{R}{L}(L - x)$ and differentiating with respect to x we get,

$$\frac{dM_K}{dx} = -R\frac{L_1}{L} + R_1 \tag{7.34}$$

For a maximum value of M_K, $\frac{dM_K}{dx}$ passes from a positive value to zero and then to a negative value, that is,

$$\frac{dM_K}{dx} \text{ is positive for } \frac{R_1}{L_1} > \frac{R}{L}$$

and

$$\frac{dM_K}{dx} \text{ is negative for } \frac{R_1}{L_1} < R_L$$

This means that the maximum bending moment at any section K occurs when a particular load is on the section which changes the ratio $\frac{R_1}{L_1} > \frac{R}{L}$ to $\frac{R_1}{L_1} < \frac{R}{L}$ as the load passes over the section from left to right.

7.6.4 Absolute Maximum Shear and Moment in Beams

The absolute maximum shear in a simply supported beam subjected to a series of moving concentrated loads needs little discussion. It will occur next to one of the support sections, and therefore the solution entails only the positioning of loads such that the maximum value of reaction is obtained.

The theoretical calculations to determine the absolute maximum bending moment at any section and the curves showing these absolute maximum values are highly involved. However, the criteria arrived at for a maximum bending moment under a given loading or at a section may be used for finding out the absolute maximum bending moment anywhere on the span. This is done by first selecting a wheel load and then arranging suitably the load system as concluded in 7.6.2. The maximum bending moment under that wheel load is obtained. After that another wheel load is selected and the same procedure is adopted to arrive at the maximum bending moment under that load. Two or three trials give the absolute maximum bending moment. The following guidelines may be kept in mind in making the trials.

1. The absolute maximum bending moment occurs under one of the loads and not in between the loads.
2. The absolute maximum bending moment occurs under a wheel load which is heavier and near the centre of span. It does not occur at centre

of span unless the resultant of the load system coincides with the heavier load.

A few numerical examples that follow on the topics covered will make the discussions clear.

Example 7.1 | *A uniformly distributed load of 15 kN/m covering a length of 3 m crosses a girder of span 10 m. Find the maximum shear force and bending moment at a section 4 m from L.H. support.*

Step 1: To fix position of moving loads for maximum shear.

Maximum Negative Shear Force

We know that the maximum –ve shear force at the required section will take place when the head of the load is on the section as shown in Fig. 7.7*b*. Therefore

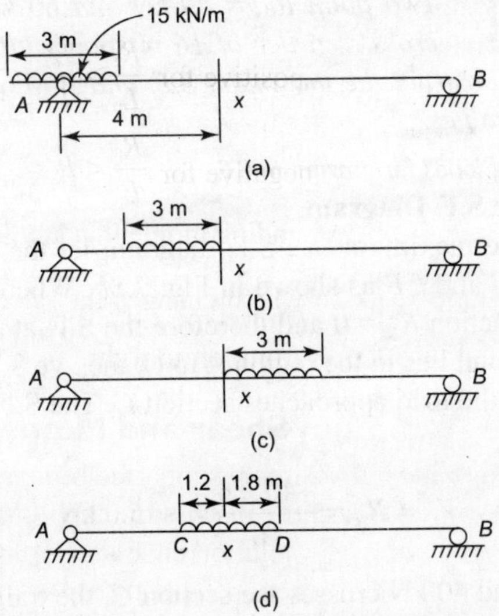

Fig. 7.7 | *(a) Beam undr u.d.l. shorter than span, (b) Position of load for maximum –ve S.F., (c) Position of load for maximum +ve S.F., (d) Position of load for maximum B.M.*

$$V_{\max} = -R_B = -\frac{15 \times 3 \times 2.5}{10} + 11.25\,\text{kN}$$

Maximum Positive Shear Force

The maximum +ve S.F. at the required section will take place when the tail of the load is on the section as shown in Fig. 7.7*c*. Therefore,

$$V_{\max} = R_A = \frac{15 \times 3 \times 4.5}{10} = 20.25\,\text{kN}$$

Step 2: To fix position of loads for maximum B.M.

Maximum Bending Moment

The maximum bending moment at the section will occur when the position of the load is such that the section X divides that load in the same ratio as it divides the span. In Fig. 7.7*d*

$$\frac{CX}{CD} = \frac{x}{l}$$

or
$$\frac{CX}{3} = \frac{4}{10}$$

∴
$$CX = 1.2 \text{ m}$$

Now bending moment
$$M_x = R_A (4) - 15 (1.2) (0.6)$$
$$= \frac{15(3)(5.7)(4)}{10} - 15(1.2)(0.6)$$
$$= 102.6 - 10.8 = 91.8 \text{ kN.m}$$

Example 7.2 | *Two point loads 40 kN and 60 kN spaced 6 m apart cross a girder of 16 m span with 40 kN load leading from left to right. Construct the maximum S.F. and B.M. diagrams stating the absolute maximum values.*

Step 1:To fix position of loads for maximum S.F.

Maximum Negative S.F. Diagram

Now for drawing the maximum –ve S.F. diagram, let the span *AB* be divided into two sections *AC* and *CB* as shown in Fig. 7.8*b*. When the leading load 40 kN is over *A*, the reaction $R_B = 0$ and therefore the S.F. at $A = 0$. As the 40 kN load enters the span and lies in the region *A* to *C*, the –ve S.F. is equal to R_B and increases linearly as the load approaches section *C*. The S.F. at *C* when the load is at section *C* is,

$$V_C = R_B = \frac{-40 \times 6}{16} = -15.0 \text{ kN}$$

When the leading load 40 kN crosses the section *C*, the trailing load also enters the span. The negative S.F. is equal to R_B for any section from the leading load to support B. The reaction R_B increases and so also the S.F. as the loads move further towards support *B*. The maximum –ve S.F. will develop when the leading load just reaches support *B*. The maximum –ve S.F. next to support *B* is,

$$V_B = -\frac{(40 \times 16 + 60 \times 10)}{16} = -77.5 \text{ kN}$$

The –ve S.F. diagram is shown in Fig. 7.8*b*.

Maximum Positive S.F. Diagram

For drawing the +ve S.F. diagram let us divide the span *AB* into two sections *AD* and *DB* so that $DB = 6$ m, the spacing between the two loads. The reaction R_A and hence the +ve S.F. is maximum at *A* when the trailing load is over support *A*. Therefore, the maximum +ve S.F. is

$$V_A = 60 + \frac{40 \times 10}{16} = 85 \text{ kN}$$

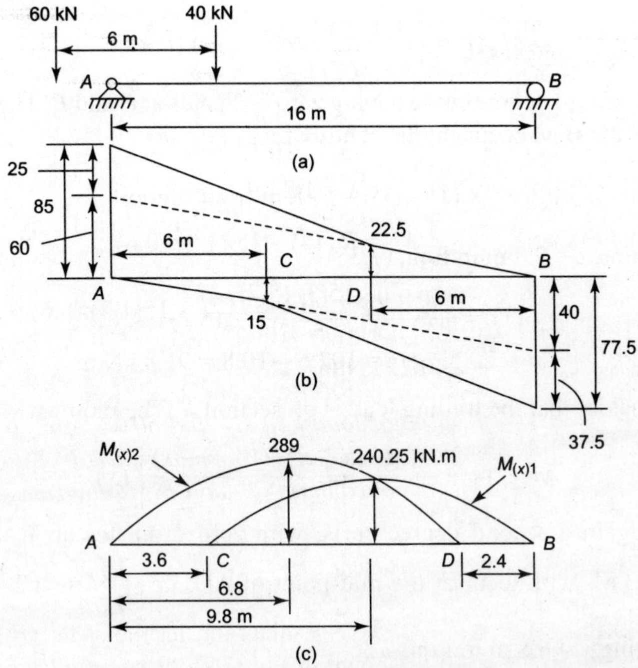

Fig. 7.8 | *(a) Beam and the rolling loads, (b) Maximum S.F. diagram, (c) Maximum B.M. diagram*

As the loads advance towards support B the reaction R_A and hence the S.F. decreases till the leading load reaches support B. At this instance the trailing load is over section D. Therefore the +ve S.F. at section D is

$$V_D = R_A = \frac{60 \times 10}{16} = 22.5 \text{ kN}$$

The maximum +ve S.F. diagram for the entire girder is shown in Fig. 7.8b.

Step 2: To fix sections C and D.

Maximum B.M. Diagram

For drawing the B.M. diagram we divide the span AB into three sections AC, CD and DB as shown in Fig. 7.8c such that

$$AC = \frac{W_2 d}{W_1 + W_2} = \frac{60(6)}{100} = 3.6 \text{ m}$$

and $$BD = \frac{W_1 d}{W_1 + W_2} = \frac{40(6)}{100} = 2.4 \text{ m}$$

Step 3: Calculation of maximum moments

The maximum B.M. at any section X at a distance x from A occurs when one of the loads is at that section. Consider first that the leading load is at the section X. The moment

$$M_x(1) = \left\{\frac{W_1(x) + W_2(x-d)}{l}\right\}(l-x) \tag{7.35}$$

This is a second degree curve having zero ordinates at C and B. The maximum ordinate for the B.M occurs at the centre of CB. i.e.,

$$3.6 + \frac{1}{2}(16 - 3.6) = 9.8 \text{ m from support } A.$$

Substituting $x = 9.8$ m in Eqn. (7.35)

$$M_{max}(1) = \frac{(40 \times 9.8 + 60 \times 3.8)}{16}(6.2)$$
$$= 240.25 \text{ kN.m.}$$

Next consider that the trailing load is on section X. The moment

$$M_x(2) = \left\{\frac{W_1(l-x-d) + W_2(l-x)}{l}\right\}(x) \tag{7.36}$$

This is again a second degree curve with zero ordinates at A and D. The maximum B.M. will occur at the mid-point of AD, i.e. $\frac{1}{2}(16 - 2.4) = 6.8$ m from A.

Substituting $x = 6.8$ m in Eqn. 7.36

$$M_{max}(2) = \left\{\frac{40(16-6.8-6) + 60(16-6.8)}{16}\right\}6.8$$
$$= 289.0 \text{ kN.m}$$

The complete B.M. diagram is shown in Fig. 7.8c.

Example 7.3 | *The load system shown in Fig. 7.9 crosses a girder 25 m span with the 120 kN load leading. Determine the value of (i) Maximum B.M. at a section 8 m from the left end of the girder and (ii) Absolute maximum B.M. on the girder.*

Step 1: To fix the position of loads for maximum B.M.

Maximum Moment at the Section

The maximum moment at section X, 8 m from support A occurs when one of the loads is on the section. Further, the load which tilts the average loading in the regions AX and XB as it passes over the section is the one which should be placed over the section for maximum B.M.

Let us consider 160 kN next to 80 kN just to the left of Section X.

Average load on $\quad AX = \frac{240}{8} = 30$

on $\quad XB = \frac{280}{17} = 16.47$

Average load on $AX >$ Average load on XB.

Consider next that the same 160 kN load is just to the right of section X.

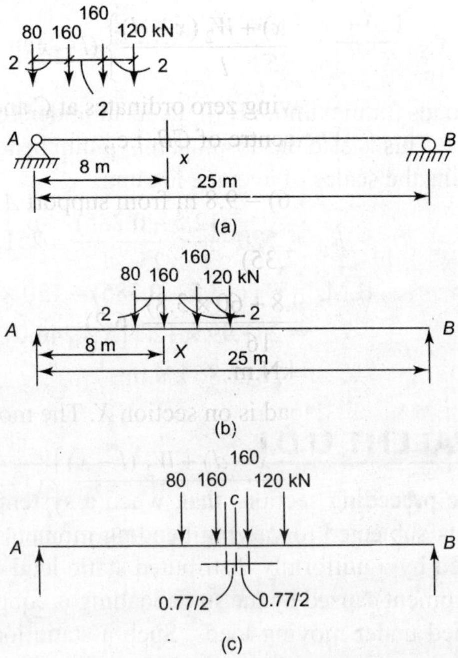

Fig. 7.9 | *(a) Beam and the rolling loads, (b) Position of rolling loads for maximum B.M. at section X, (c) Position of loads for absolute maximum B.M.*

Average load on $AX = \dfrac{80}{8} = 10$

on $XB = \dfrac{440}{17} = 25.88$

$\therefore$ Average load on $AX <$ Average load on XB.

Therefore, the maximum B.M. at section X will occur when a 160 kN load next to an 80 kN load is on the section.

The position of loads for maximum B.M. at section X is shown in Fig. 7.9*b*.

Step 2: Calculation of moment at X

The moment $M_x = R_A\,(8) - 80\,(2)$

$$R_A = \frac{120(13) + 160(15) + 160(17) + 80(19)}{25} = 328 \text{ kN}$$

$\therefore$ $M_x = 328(8) - 80(2) = 2464$ kN.m

Step 3: To fix position of loads for absolute Maximum B.M.

Absolute Maximum B.M

Let us first fix up the centroid of the loads. Taking moment about the 80 kN load

$$\bar{x} = \frac{120 + 6 \ + 160 \times 4 + 16 \times 2}{520} = 3.23 \text{ m}$$

The position of loads for maximum B.M. to occur under 160 kN next to 60 kN is shown in Fig. 7.9*c*. This is also the absolute maximum bending moment on the girder as this load tilts the scales of average loading.

Reaction $\qquad R_B = 520 \dfrac{(12.5 - 0.285)}{25} = 251.99 \text{ kN}$

Absolute Maximum $\qquad$ B.M. $= R_B (12.5 - 0.385) - 120 \times 2$

$$= 251.99 \times 12.115 - 240.0$$

$$= 2812.86 \text{ kN.m.}$$

7.7 EQUIVALENT U.D.L

We have seen in the preceding sections that, when a system of loads roll over a girder, the girder is subjected to varying bending moment. Such a system of loads can be replaced by a uniformly distributed static load covering the entire span such that the moment caused by the static loading is equal to or greater than the moments obtained under moving loads. Such a static loading is called the equivalent u.d.l.

We know that the equivalent u.d.l. produces a moment diagram which is parabolic with a maximum ordinate at the centre of the beam. This bending moment diagram should envelop the bending moment diagram under actual rolling loads. The intensity of equivalent u.d.l. which envelops the B.M. diagram produced by the actual system of loads can be evaluated by any one of the methods available. However, the simple method which is acceptable and preferred by practising engineers is the one in which the absolute maximum B.M. produced by the system of loads is equaled to the maximum B.M. produced by the equivalent u.d.l. at the centre of span. That is, if $M_{\max}$ is the absolute maximum B.M. produced by the system of loads and w' unit length is the equivalent u.d.l. then

$$\frac{w'l^2}{8} = M_{\max}$$

from which
$$w' = \frac{8 M_{\max}}{l^2} \tag{7.37}$$

Example 7.4 | *A uniformly distributed load of 20 kN/m and 3 m long rolls over a girder of 12 m span. Find the equivalent u.d.l.*

Absolute maximum B.M. at centre of span will occur when the load is placed symmetrical to the centre of span, that is 1.5 m length on either side of centre of span.

$$M_{\max} = R_A (6) - 20 \ \frac{(1.5)}{2} (1.5)$$

$$= \frac{20 \times 3 \times 6}{2} - 22.5$$

Let w' be the equivalent u.d.l., then from Eqn. 7.37 ,

$$w' = \frac{8 \times 157.5}{12 \times 12} = 8.75 \text{ kN/m}$$

Example 7.5 *Determine the maximum bending moment developed anywhere on the girder of span 15 m due to two rolling loads 150 kN and 100 kN spaced 4 m apart with the 100 kN load leading passing over the girder. Find the equivalent u.d.l. to give the same maximum bending moment.*

The centroid of the loads is found taking moment about the 100 kN load.

Then $$\bar{x} = \frac{150 \times 4}{250} = 2.4 \text{ m}$$

The absolute maximum B.M. will occur under 150 kN load when that load and the centroid of loads are at equal distantces from the centre of the girder as shown in Fig. 7,10.

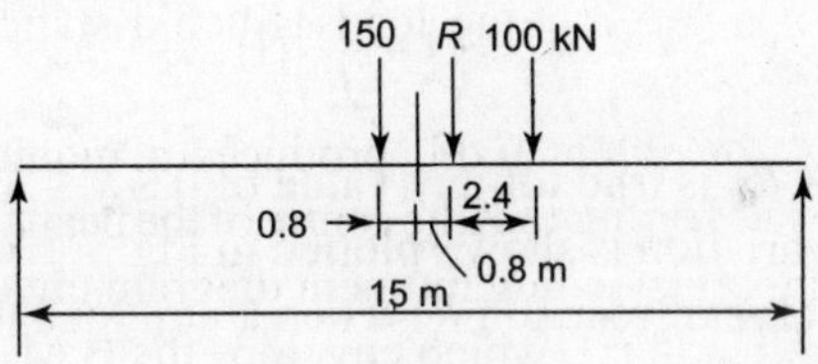

Fig. 7.10

$$M_{max} = R_A\,(7.5 - 0.8)$$

$$R_A = 250\,\frac{(7.5 + 0.8)}{15}$$

Substituting for R_A, $M_{max} = \frac{250 \times 8.3}{15} \times 6.7 = 926.83$ kN/m

If w' is the equivalent u.d.l., using Eqn. 7.37

$$w' = \frac{926.83 \times 8}{15 \times 15} = 32.95 \text{ kN/m}$$

7.8 INFLUENCE LINES

7.8.1 Introduction

The steps involved in determining S.F. and B.M. at different sections of a beam as the rolling loads move from one end to the other are rather cumbersome. Influence lines are interesting and are a very useful tool in dealing with rolling loads. In the following sections the common methods of determining influence lines are covered first and later these influence lines are utilised to determine the maximum S.F. and B.M. in beams and forces in the members of the bridge trusses.

7.8.2 Definition and Basic Concept

An influence line is defined as a function whose value at a point represents some structural quantity as a unit load is placed at that point.

The structural quantities often encountered are: reactions, shear forces, moments, deflections at specified points or member forces as in the case of trusses.

The basic concept of an influence line (I. L.) can be developed by considering the simply supported beam of Fig. 7.11.

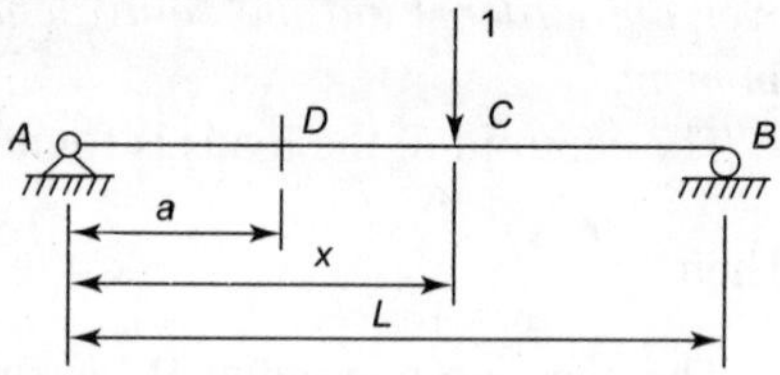

Fig. 7.11

Suppose that it is required to know the variation of left-hand support reaction R_A as a unit load moves from end A to end B. If the unit load is at a distance x from support A, the left support reaction is

$$R_A = \left(1 - \frac{x}{L}\right)(1) \tag{7.38}$$

This expression for R_A is true for any value of $0 \le x \le L$ and is positive when acting upwards. The variation is shown plotted in Fig. 7.12*a*. The ordinate at any point represents the value of reaction R_A when a unit load is at that point.

In a similar manner we can express the variation of the shear force or the moment at a section as the unit load moves along the beam. For example, the shear force at say section D, is positive (following the sign convention as in Fig. 2.3) for the unit load position shown in Fig. 7.11. The value of shear

$$V_1 = R_A = \left(1 - \frac{x}{L}\right)(1), \text{ for } a \le x \le L \tag{7.39}$$

and moment

$$M_1 = R_A(a) = \left(1 - \frac{x}{L}\right)(a) \text{ for } a \le x \le L \tag{7.40}$$

If the unit load were to be on the left of section D.

$$\text{shear force } V_1 = -R_B = \left(-\frac{x}{L}\right)(1), \qquad [0 \le x \le a] \tag{7.41}$$

$$\text{and moment } M_1 = R_B(L - a) = (x/L)(L - a), \qquad [0 \le x \le a] \tag{7.42}$$

The variations of shear and moment are shown plotted in Figs. 7.12*b* and *c* respectively.

The diagrams in Fig. 7.12 are known as influence lines. They are constructed using general equations for the structural quantity under consideration. Influence

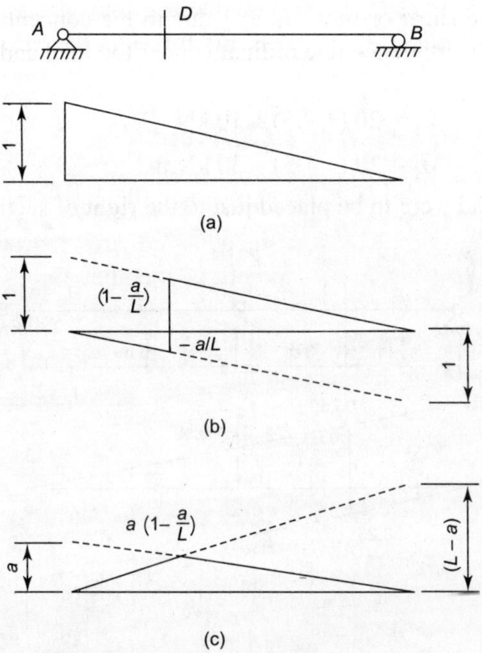

Fig. 7.12 | *(a) Variation of reaction R_A, (b) Variation of shear force at section D (c) Variation of moment at section D*

lines can also be constructed by placing a unit load at a specific number of points, evaluating the required quantity under consideration, plotting the results and joining them by a smooth curve. The general expressions above help to clarify the meaning of an influence line. The structural quantity is a function of x which reflects the position of the load.

7.8.3 Uses of Influence Lines

Once we have developed an influence line for a structural quantity, we can use it to evaluate that structural quantity for any type of moving loads.

The Influence Lines diagrams for S.F. and B.M. have been utilised to study the following cases of rolling loads covered earlier through numerical examples.

1. Single concentrated load
2. A u.d.l. longer than the span
3. A u.d.l. shorter than the span
4. Two concentrated loads
5. A series of concentrated loads.

Example 7.6 | *Consider the simple beam of Fig. 7.13a for which influence lines for shear and moment are developed for section 1 located 3 m from the left support. If a 20 kN concentrated load is placed at the mid-point of the beam, evaluate the moment and shear at section 1.*

The value of the shear or moment at 1 due to the concentrated load is equal to the product of the influence line ordinate under the load and value of the load. Thus,

$$V_1 = 20\,(+\,0.5) = 10 \text{ kN}$$

and $$M_1 = 20\,(+\,1.5) = 30 \text{ kN.m}$$

Again, if the load were to be placed just to the right of section 1, we have

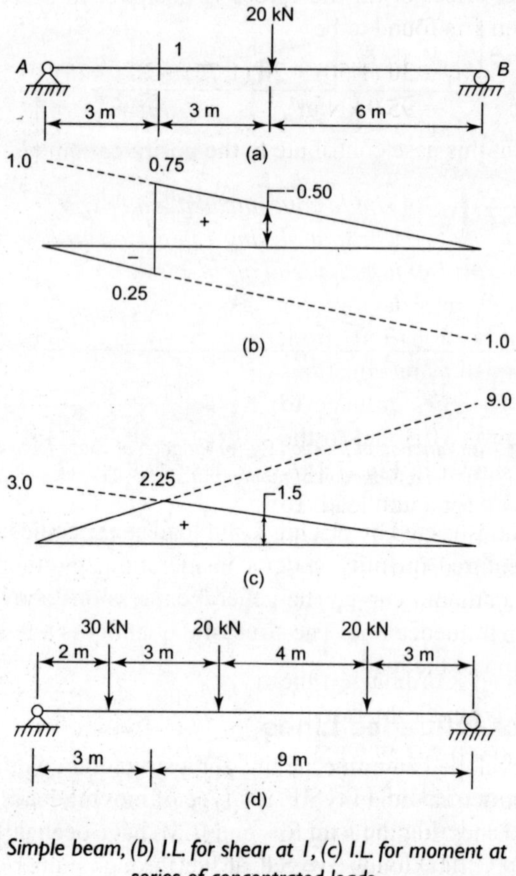

Fig. 7.13 | *(a) Simple beam, (b) I.L. for shear at 1, (c) I.L. for moment at 1, (d) Beam under series of concentrated loads*

$$V_1 = 20\,(+\,0.75) = 15 \text{ kN}$$
$$M_1 = 20\,(+\,2.25) = 45 \text{ kN.m}$$

If the load is placed just to the left of section 1, we have

$$V_1 = 20\,(-\,0.25) = 5.0 \text{ kN}$$
$$M_1 = 20\,(+\,2.25) = 45 \text{ kN.m}$$

Now let us consider the same beam subjected to a series of loads as shown in Fig. 7.13*d*. The resulting shear or moment at 1 due to the given loading can be obtained with the help of the influence lines of Figs. 7.13*b* and *c*.

The shear or moment at section 1 is obtained by summing up the effect of individual concentrated loads as calculated above. Thus,

$$V_1 = 30\,(-0.167) + 20\,(+0.58) + 20\,(0.25)$$
$$= 11.65 \text{ kN}$$

It may be noted that the influence line ordinate under 30 kN load is negative. Therefore, the contribution of 30 kN load to the shear at section 1 is negative. However, the net effect of all the forces is positive. In a similar manner, the moment at section 1 is found to be

$$M_1 = 30\,(1.50) + 20\,(1.75) + 20\,(0.75)$$
$$= 95.0 \text{ kN.m}$$

All the loads in this case contribute to the positive moment.

Example 7.7 | *A single concentrated load of 60 kN crosses a girder of 10 m span. Using I.L. diagrams find the maximum S.F. and B.M. at a section 3 m from left end of the girder.*

The Influence Line Diagram for shear force at a section 3 m from support A is obtained using equations 7.39 and 7.41. The S.F. ordinate to the left of section is –0.3 and to the right is +0.7 as shown in Fig. 7.14*b*. These ordinates are for a unit load. To obtain the maximum +ve S.F. the 60 kN load has to be placed to the right of section C. Then the maximum +ve S.F. at section C is

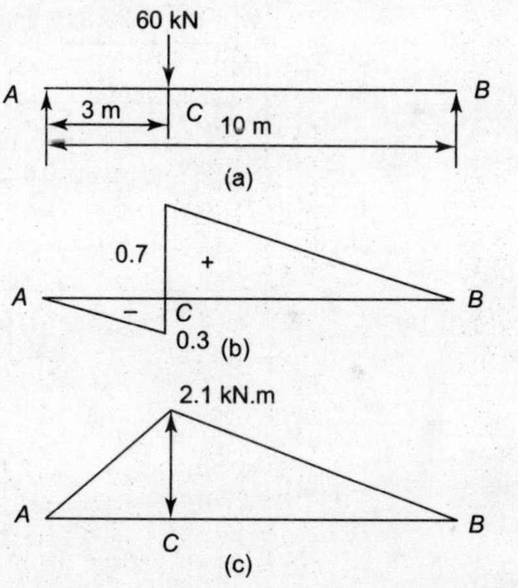

Fig. 7.14 | *(a) Beam and the rolling load (b) I.L. diagram for S.F. (c) I.L. diagram for B.M.*

V_C = load × ordinate of the I.L. under the load

$= 60 \times 0.7 = 42.0$ kN.

Similarly the maximum –ve S.F. is obtained by considering the load to the left of section C. Maximum –ve S.F. at section C is

V_C = load × ordinate of I.L. under the load 60

$= 60 \times (-0.3) = -18.0$ kN

The I.L. for moment at a section 3 m from support A is obtained using Eqn. 7.40 and 7.42 as shown in Fig. 7.14*c*. The I.L. ordinate at the section is

$$M_C = \frac{a(l-a)}{l} = \frac{3 \times 7}{10} \; 2.1 \text{ kN.m}$$

To obtain the maximum bending moment the load has to be placed over the section. The maximum B.M. is

$$M_{C\,\max} = \text{load} \times \text{ordinate of the I.L. diagram under the load}$$
$$M_{C\,(\max)} = 60 \times 2.1 = 126.0 \text{ kN.m}$$

Example 7.8 | *It is required to determine the maximum possible shear force and moment at point C of the beam shown in Fig. 7.15a due to a series of concentrated loads shown moving on the beam from right to left.*

The problem is essentially one of determining the position of loads for which the shear or moment at C is maximum. The shear and moment influence lines for point C are shown in Figs. 7.15*b* and *c*. For the purpose of discussion we consider that the loads are moving from right to left with the 20 kN load in the lead.

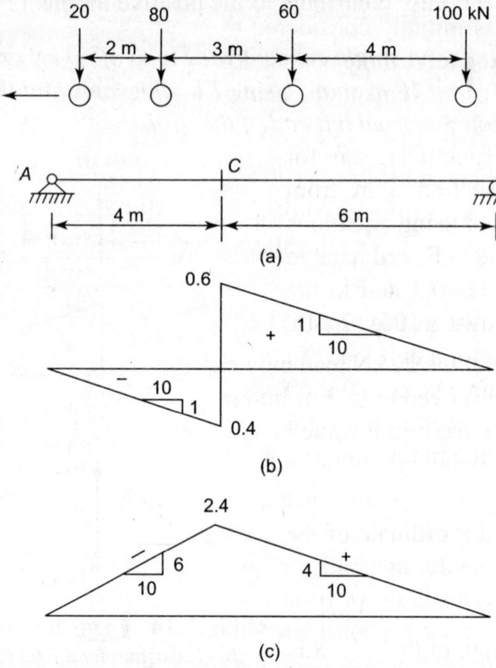

Fig. 7.15 | *(a) Beam and rolling loads, (b) I.L for shear at C, (c) I.L. for moment at C*

Let us first attempt to determine the position of loads for maximum shear. This is done by moving the loads in successive steps across the beam and observing the manner in which the shear force at C changes. Consider that the 20 kN load enters the beam first. We see that for an advance of 1 m to the left, the value of shear changes from zero to 20(0.1) = 2.0 kN. The slope of the influence line can be used to determine the change of shear force. The slope of the I.L. on either side of the section is 1 in 10. It may be noted that the slope from C to A will be considered positive because, for the movement of loads from right to left in this region, the shear at C becomes less negative.

As the loads advance further to the left, the shear at C increases continuously until the leading load 20 kN reaches C. When the 20 kN load goes past C, there is an abrupt decrease in the shear at C. However, as the loads continue to move to the left, the shear force at C again increases gradually until the next load, that is, the 80 kN load reaches C. Thus, it is seen that point C experiences a series of increases and abrupt decreases in the value of shear as each load reaches C and goes past it. The maximum shear is obtained when one of the loads is just to the right of C. The maximum value of shear can be obtained by examining the change in shear at C during each movement. If the change is positive, it implies that the maximum shear value has not yet been obtained. When the change is observed to be negative after a particular move, the location of loads just prior to that move gives the maximum positive shear force at C.

The 20 kN load is initially considered to be just to the right of C. Now the loads are moved to the left till the next 80 kN load occupies a position just to the right of C. During this movement, the change in shear force at C is

$$\Delta V_C = 20\,(-1.0) + (20 + 80 + 60)\,(2 \times 0.1) = 12.0 \text{ kN}$$

It may be noted that the last load (100 kN) did not enter the span. The present movement indicates that the shear force at C is increased by 12 kN.

The procedure is repeated with the 80 kN load being moved past C and the 60 kN being brought just to the right of C. The resulting change in the shear at C is

$$V_C = -80 + (20 + 100)\,(2 \times 0.1) + (80 + 60)\,(3 \times 0.1) = -14 \text{ kN}$$

It may be noted that the 20 kN load moved out of the span and the 100 kN load entered the span but moved only 2 m into the span. In this movement the shear force decreases. The maximum value of shear is, therefore, obtained with the 80 kN load placed just to the right of point C. With the position of loads determined, the value of shear force is obtained using the I.L. of Fig. 7.15*b*.

$$V_{C\,(\max)} = 20\,(-0.2) + 80\,(0.6) + 60\,(0.3) = 62.0 \text{ kN}$$

The maximum moment at C can be obtained in a similar way from the I.L. of Fig. 7.15*c*. As the loads move from B to C, the moment at C increases. The slope in this section may be considered positive. Loads moving from C to A cause a decrease in the moment at C, so the slope in this section may be taken as negative. The maximum moment at C will occur when one of the loads is at C.

There is a continuous increase in the moment at C as the loads move from right to left till the 20 kN load is at C. As the 20 kN load moves past C, its contribution becomes negative but the contribution of loads in the region C to B is positive. To start with, consider that the 20 kN load is just over point C. The change in the moment at C as the loads are moved to the left till the next 80 kN load reaches point C is

$$\Delta M_C = -20\,(2 \times 0.6) + (80 + 60)\,(2 \times 0.4) = 88.0 \text{ kN.m}$$

This move increases the moment at C. The loads are again moved to the left until the 60 kN load occupies point C. The change in moment for this move is

$$\Delta M_C = 20(2)\,(-0.6) + 80(3)(-0.6) + 60(3)(0.4) + 100(2)(0.4) = -16.0 \text{ kN.m}$$

It may be noted that in the last move, the 28 kN load went out of the span and the 100 kN load entered the span but moved only 2 m inside the span.

It is clear that the last move decreased the moment at C and, therefore, the position of loads which gives the maximum moment is when the 80 kN load is on point C. Thus, the maximum moment is

$$M_{C\,(\max)} = 20\,(1.2) + 80\,(2.4) + 60\,(1.2) = 288 \text{ kN.m}$$

In the example above the movement of loads was considered only from right to left. However, in the case of moving vehicles, the movement of loads can be in either direction with the front axle load leading. This aspect is illustrated in the following example.

Example 7.9 | *The maximum moment and shear force at C for the beam of Fig. 7.16a is to be computed. The loading is due to axle loads of IRC class A driving vehicle as shown on top of the beam. Assume that the vehicle can move in either direction.*

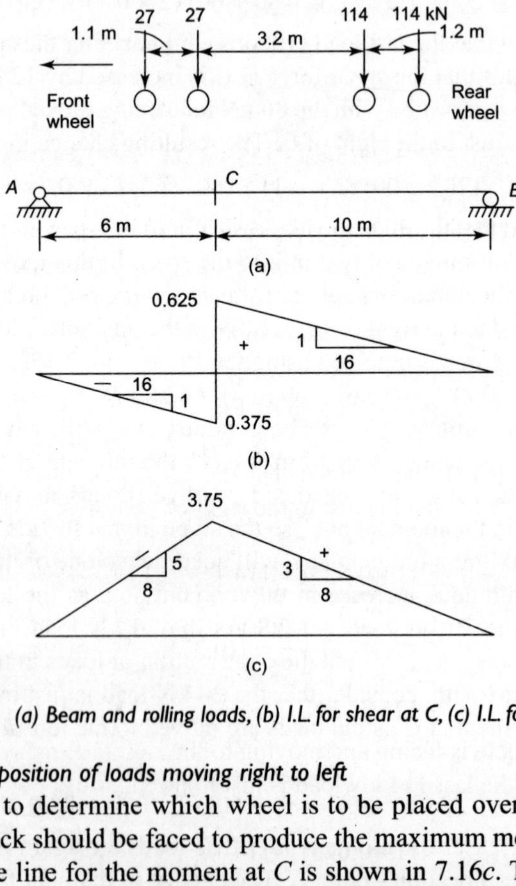

Fig. 7.16 | *(a) Beam and rolling loads, (b) I.L. for shear at C, (c) I.L. for moment at C*

Step I: To fix up position of loads moving right to left

The problem is to determine which wheel is to be placed over C and in which direction the truck should be faced to produce the maximum moment at C.

The influence line for the moment at C is shown in 7.16c. The slopes of the influence line are shown for the vehicle moving from right to left. From an inspection of the wheel loads, it is obvious that one of the two rear wheel loads must be over point C. We shall consider that initially the inner 114 kN load is over point C. If the loads are moved to the left until the rear wheel 114 kN load is over C, the change in moment is

$$\Delta M_C = (27 + 27 + 114)\,(-0.625)\,(1.2) + 114\,(0.375)\,(1.2) = -74.7 \text{ kN.M}$$

The initial position of loads, therefore, produces a maximum moment at C. The value of the moment is obtained using the influence line of Fig. 7.16c.

Step 2: Calculation of maximum moment

$$M_{C(\max)} = 27(1.7)(0.625) + 27(2.8)(0.625) + 114(6)(0.625) + 114(8.8)(0.375)$$
$$= 879.64 \text{ kN.m}$$

Step 3: To fix position of loads moving left to right

Next, let us consider that the vehicle is facing right. Initially, let the inner 114 kN load be over point C. If the loads are now moved to the right till the last 114 kN load occupies point C, the change in moment at C is

$$\Delta M_C = (27 + 27 + 114)(1.2)(-3.375) + 114(1.2)(0.625) = 9.9 \text{ kN.m}$$

The change is positive. Therefore, the maximum moment at C occurs when the last 114 kN load is over point C. For this position of loads, the moment is obtained using the I.L. of Fig. 7.16c.

$$M_{C(\max)} = 27(4.5)(0.375) + 27(5.6)(3.75) + 114(8.8)\,(0.375) + 114(10)(0.375)$$
$$= 905.96 \text{ kN.m}$$

This value is higher than the previously determined 879.64 kN.m. The maximum moment value at C is, therefore, 905.96 kN.m. It should be noted that the positioning of wheel loads is not necessarily the same as this for other sections for determining the maximum moments.

Step 4: Evaluate maximum S.F.

The I.L. for the shear at C is shown in Fig. 7.16b. We shall first evaluate the maximum positive shear force when the vehicle is moving from right to left. From inspection it may be decided that the maximum shear force will occur when the inner 114 kN load is just to the right of point C.

$$V_{C(\max)} = 114(0.625) + 114\left(\frac{8.8}{10}\right)(0.625) - 27\left(\frac{1.7}{6}\right)(0.375)$$
$$- 27\left(\frac{2.8}{6}\right)(0.375)$$
$$= 126.35 \text{ kN}$$

When the vehicle is facing and moving to the right, the maximum shear force will occur when the last 114 kN load is just to the right of point C.

$$V_{C(\max)} = 114(0.625) + 114\left(\frac{8.8}{10}\right)(0.625) + 27\left(\frac{5.6}{10}\right)(0.625)$$
$$+ 27\left(\frac{4.5}{10}\right)(0.625)$$
$$= 151.00 \text{ kN}$$

This value of the shear force is greater than the previously determined value of 126.35 kN. The maximum negative shear force can be investigated in a similar manner.

This method of determining the maximum values of structural quantities is also applicable to other types of influence lines. It may be noticed that in certain cases it is not necessary to move the loads through all possible position. Often, by inspecting the magnitude of loads and their relative position, the trials can be reduced to examining only a few alternatives. In some cases where there are only a few loads, the location can be determined simply by inspection. However, when in doubt, the general procedure should be used.

7.8.4 Distributed Loads

Influence lines can be used for distributed loads as well. Consider a segment of a beam loaded from *a* to *b* as shown in Fig. 7.17*a*. The influence line for a structural quantity is also shown in Fig. 7.17*b*. The ordinates of this influence line are indicated by a function $f(x)$. We see from Fig. 7.16 that the distributed load $w_{(x)}$ acting on a differential length of the beam dx is equivalent to a concentrated load of magnitude $w_{(x)}dx$. The effect of this load on a structural quantity F, for which the influence line is drawn is

$$dF = w_{(x)} \cdot dx \cdot f(x) \tag{7.34a}$$

that is the value of structural quantity is equal to the product of the concentrated load and the I.L. ordinate at that point. For the total distributed load from a to *b*, Eq. 7.43*a* is integrated over the length of the beam on which the load acts. Thus, the structural quantity

$$F = \int_a^b w_{(x)} \cdot f_{(x)} \cdot dx \tag{7.43b}$$

If the distributed load is uniform, then $w_{(x)} = w$ constant, and Eq. 7.43*b* can be written as

$$F = w\int_a^b f_{(x)} \cdot dx \tag{7.44}$$

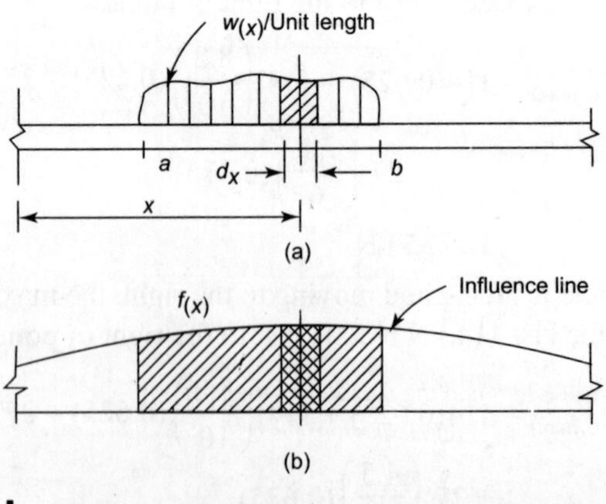

Fig. 7.17 | *(a) Beam under distributed load from a to b, (b) Influence line* $f_{(x)}$

From Eq. 7.44 we can state that the value of a structural quantity due to a uniform load is equal to the product of the magnitude of the uniform load and the area of the influence line diagram under the distributed load. For example, if the beam of Fig. 7.13*a* is loaded with a uniformly distributed load of 10 kN/m over an entire span, the moment at 1 from Fig. 7.13*c* is

$$M_1 = (10)\left(\frac{1}{2} \times 2.25 \times 12\right) = 135.0 \text{ kN.m}$$

The value of shear at 1 for the same load is found to be

$$V_1 = -(10)\left(\frac{0.25 \times 3}{2}\right) + 10\left(\frac{0.75 \times 9}{2}\right) = 30.0 \text{ kN}$$

It may be noted that the distributed load to the left of point 1 has actually reduced the shear force at 1. If the load were to be placed on the right 9 m length only, the value of shear at 1 would have been

$$V_1 = 10\left(\frac{0.75 \times 9}{2}\right) = 33.75 \text{ kN}$$

The influence lines can be used in this manner to fix up the placement of loads to obtain the maximum effects.

Example 7.10 | *A uniformly distributed load of 40 kN/m longer than the span rolls over a girder of 30 m span. Using I.L. diagram for S.F. and B.M. determine the maximum S.F. and B.M. at a section 12 m from left-hand support A.*

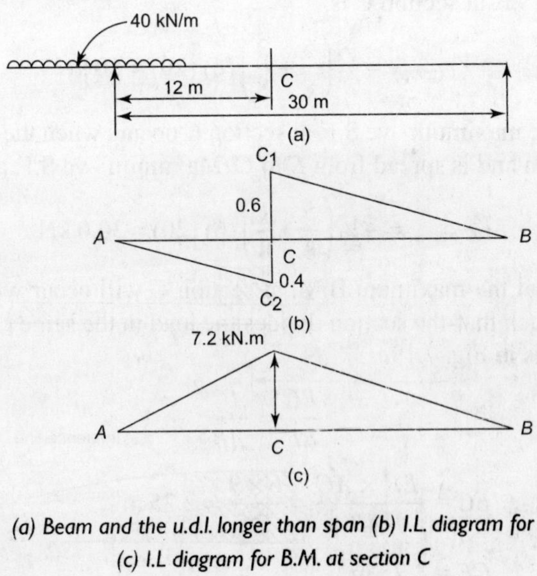

Fig. 7.18 | *(a) Beam and the u.d.l. longer than span (b) I.L. diagram for S.F. at section C (c) I.L diagram for B.M. at section C*

The I.L. diagram for shear force at section *C* is shown in Fig. 7.18*b*, the ordinates being worked out as earlier. It is evident from the diagram that the maximum +ve S.F. will occur when the tail of the load is on section *C* and occupies from *C* to *B*. The maximum +ve S.F. at section *C* is given by

$$V_C = \text{Intensity of loading} \times \text{Area of the triangle } CC_1B$$
$$= 40 \times 1/2 \times 18 \times 0.6 = 216 \text{ kN.}$$

Similarly the maximum –ve S.F. at section *C* occurs when the head of the load is on section *C* and occupies from *A* to *C*. The maxm. –ve S.F. at section *C* is

$$V_{C(\text{max.})} = \text{Intensity of loading} \times \text{Area of the triangle } ACC_2$$
$$= 40 \times 1/2 \times 12\,(-0.4) = -96.0 \text{ kN.}$$

From the I.L. diagram for B.M. shown in Fig. 7.18c, it is clear that the maximum B.M. at section *C* occurs when the load occupies the entire span from *A* to *B*. The maximum moment at section *C* is

$$M_{C\,(\text{max.})} = \text{Intensity of loading} \times \text{Area of triangle } ABC,$$

or $$M_{C(\text{max.})} = 40 \times 1/2 \times 30 \times 7.2 = 4320 \text{ kN.m}$$

Example 7.11 | *A girder simply supported has a span of 24 m. A u.d.l. of intensity 20 kN/m and 6 m long crosses the girder. Using I.L. diagrams find the maximum S.F. and B.M. at a section 9 m from the left support.*

The I.L. diagram for shear force at section C is shown in Fig. 7.19b.

From the diagram it is clear that the maximum +ve S.F. at section C occurs when the tail of the load is at C and the load is spread from C to F. Therefore, the maximum +ve S.F. at section C is,

$$V_{C(\text{max.})} = \tfrac{1}{2}\left(\frac{5}{8} + \frac{3}{8}\right)(6)\,(20) = 60 \text{ kN}$$

Similarly the maximum –ve S.F. at section *C* occurs when the head of the load is on the section and is spread from *E* to *C*. Maximum –ve S.F. at section *C* is

$$V_{C(\text{max.})} = -\tfrac{1}{2}\left(\frac{5}{8} + \frac{1}{8}\right)(6)\,(20) = 30.0 \text{ kN}$$

We know that the maximum B.M. at section *C* will occur when the position of the load is such that-the section divides the load in the same ratio as it divides the span. That is in Fig. 7.19*c*

$$\frac{EC}{EF} = \frac{AC}{AB}$$

$$\therefore \quad EC = \frac{EF \times AC}{AB} = \frac{6 \times 9}{24} = 2.25 \text{ m}$$

and $$CE = 3.75 \text{ m.}$$

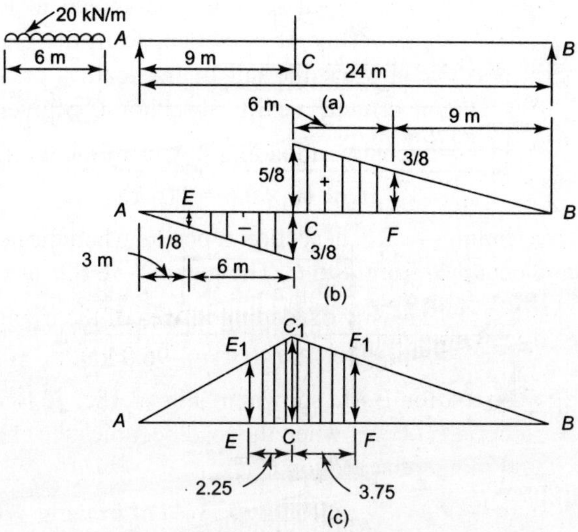

Fig. 7.19 | *(a) Beam under rolling u.d.l. shorter than span, (b) I.L. diagram for S.F. at section C, (c) I.L. diagram for B.M. at section C*

For the position of load indicated the ordinate of the I.L. diagram at E and F is 4.22 kN.m. The maximum B.M. at section C is,

$$M_{C(\text{max.})} = \text{Intensity of loading} \times \text{area of the diagram } EE_1F_1F.$$
$$= 20 \times 1/2\ (5.625 + 4.22)\ (2.25 + 3.75)$$
$$= 590.7 \text{ kN.m.}$$

Example 7.12 | *Determine the maximum shear force and moment at section C for the beam shown in Fig. 7.20a. The beam is traversed by a uniformly distributed load of intensity 20 kN/m extending over a length of 4 m. Indicate the sections that experience the absolute maximum shear and maximum moment.*

Step 1: To evaluate maximum +ve and –ve S.F. at section C

From the I.L. diagram for shear (Fig. 7.20*b)* it is obvicus that the load should cover the left of section C for maximum negative shear. Therefore,

$$V_{(\text{max})} = \frac{0.375}{2}\ (3.75)\ (20 = 14.06 \text{ kN } (-\text{ve})$$

Similarly, the maximum positive shear force will occur when the load is placed to the right of section C, that is,

$$V_{(\text{max})} = 1/2\ (0.625 + 0.225)\ (4)\ (20) = 68.0 \text{ kN } (+\text{ve})$$

Step 2: To fix up load position for maximum moment at C

For the maximum moment at section C, the load position is worked out using Eq. 7.43 and is shown in Fig. 7.20*c*. Therefore

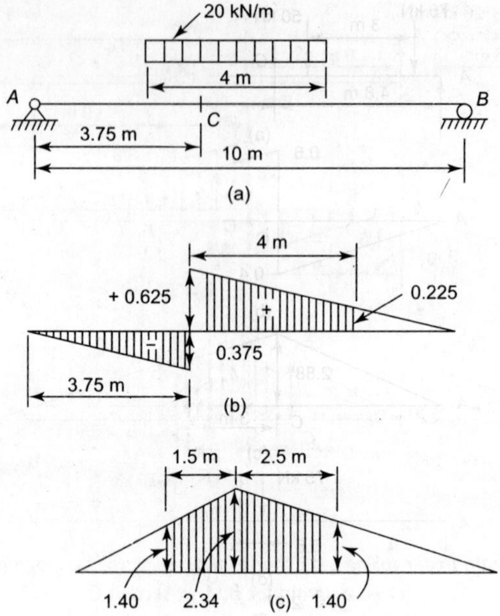

Fig. 7.20 | *(a) Beam and moving load, (b) I.L. for shear at section C (c) I.L. for moment at section C*

$$M_{C(\max)} = V_1\,(1.4 + 2.34)\,(1.5)\,(20) + 1/2(1.4 + 2.34)\,(2.5)\,(20)$$
$$= 149.6 \text{ kN.m}$$

From a knowledge of the influence lines for shear and moment, it can be said that the absolute maximum shear force will occur next to the support points and the absolute maximum moment occurs at centre of span. The values given may be verified.

$$V_{(\max)} = \pm\, 64.0 \text{ kN and } M_{(\max)} = 160.0 \text{ kN.m}$$

Example 7.13 | *Two point loads of 50 kN and 75 kN spaced 3 m apart with the 50 kN load leading passes over a simply supported span of 12 m from left to right. Using I.L. diagrams calculate the maximum S.F. and B.M. at a section 4.8 m from the left-hand support. Also find out the section and the magnitude of the absolute maximum B.M. that may occur anywhere on the beam.*

Step 1: To fix up load position for maximum S.F.

Maximum Positive S.F.

It is clear from the I.L. diagram for shear at section C that the maximum +ve S.F. at section *C* occurs when the 75 kN is on the section and the leading load is in the region *CB*. Therefore, the maximum +ve S.F. at *C* is

$$V_{C(\max.)} = 75 \times 0.6 + 50 \times \frac{0.6 \times 4.2}{7.2} = 62.5 \text{ kN.}$$

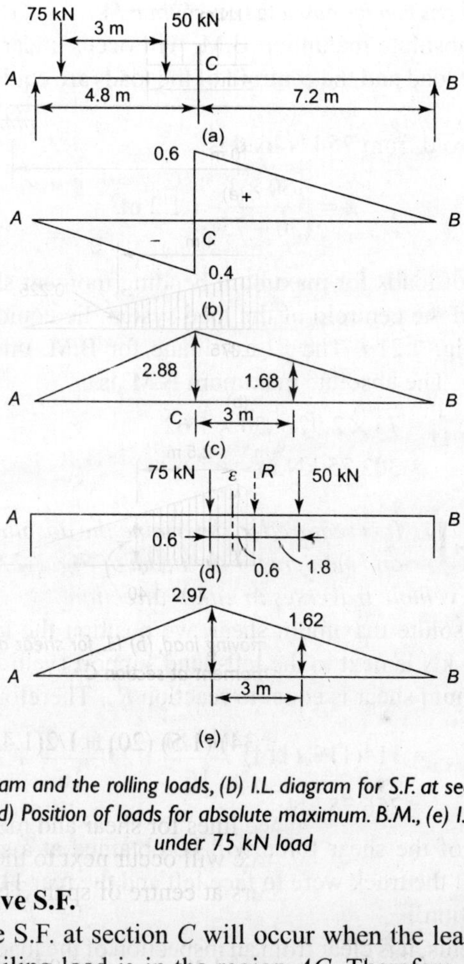

Fig. 7.21 | *(a) Beam and the rolling loads, (b) I.L. diagram for S.F. at section C, (c) I.L. diagram for B.M. at section C, (d) Position of loads for absolute maximum. B.M., (e) I.L. diagram for moment under 75 kN load*

Maximum Negative S.F.

The maximum –ve S.F. at section C will occur when the leading load is on the section and the trailing load is in the region AC. Therefore, the maximum –ve S.F. at C is

$$V_{C(\text{max.})} = -\left(50 \times 0.4 + 75 \times \frac{0.4 \times 1.8}{4.8}\right)$$
$$= -31.25 \text{ kN.}$$

Step 2: To fix up load position for maximum B.M.

Maximum Bending Moment

Maximum B.M. at the given section occurs when one of the loads is on the section. From the inspection of loads it is clear that the maximum bending moment at section C will occur when the trailing load is on the section and the leading load is in the region CB. Moment at section C is

$$M_{C(\text{max.})} = 75 \times 2.88 + 50 \times 1.68$$
$$= 300.0 \text{ kN.m.}$$

Step 3: To fix up load position for absolute maximum B.M.

We know that the absolute maximum B.M. will occur under the heavier of the two loads when that load and the centroid of the loads are equidistant from centre of span.

Distance of centroid from 75 kN load is

$$\bar{x} = \frac{50 \times 3}{(50 + 75)} = 1.2 \text{ m.}$$

The disposition of loads for maximum bending moment should be such that the trailing load and the centroid of the load system lie equidistant from centre line as shown in Fig. 7.21*d*. The I.L. ordinate for B.M. under 75 kN load is shown in Fig. 7.21*e*. The absolute maximum B.M. is,

$$M_{(\text{Absomax.})} = 75 \times 2.97 + 50 \times 1.62$$
$$= 303.75 \text{ kN.m}$$

Example 7.14 | *It is required to determine the absolute maximum shear and moment for the beam of Fig. 7.22a when a standard IRC class A driving vehicle traverses in either direction.*

To obtain the absolute maximum shear, we position the loads such that the rear wheel load 114 kN is next to the left-hand support (vehicle is facing right). The absolute maximum shear is equal to reaction R_A. Therefore

$$V_{A(\text{max.})} = 114(1) + (114)\frac{(14.8)}{16} + 27\frac{(11.6)}{16} + 27\frac{(10.5)}{16}$$
$$= 256.75 \text{ kN}$$

The same value of the shear force will be obtained at a section next to the right-hand support if the truck were to face left and the rear 114 kN load were to be placed at the section.

As regards moments, it is clear from an inspection of the loads that the absolute maximum moment occurs under one of the 114 kN loads. We shall investigate both alternatives. The location- of the resultant of the loads is found to be 1.43 m from the last wheel load as shown in Fig. 7.22*b*. First, we try under the interior 114 kN load. The loads are so positioned, that the centre line of the beam is midway between the resultant and the load under consideration. The resulting load position is shown in Fig. 7.22*c*. The value of the absolute maximum moment under the load is found to be

$$M = R_B\,(7.885) - 114(1.2)$$
$$= \frac{282}{16}(7.885)\,(7.885) - 114\,(1.2)$$
$$= 959 \text{ kN.m}$$

The positioning of loads for obtaining the maximum possible moment under the rear load 114 kN is shown in Fig. 7.22*d*. The moment under the last load is

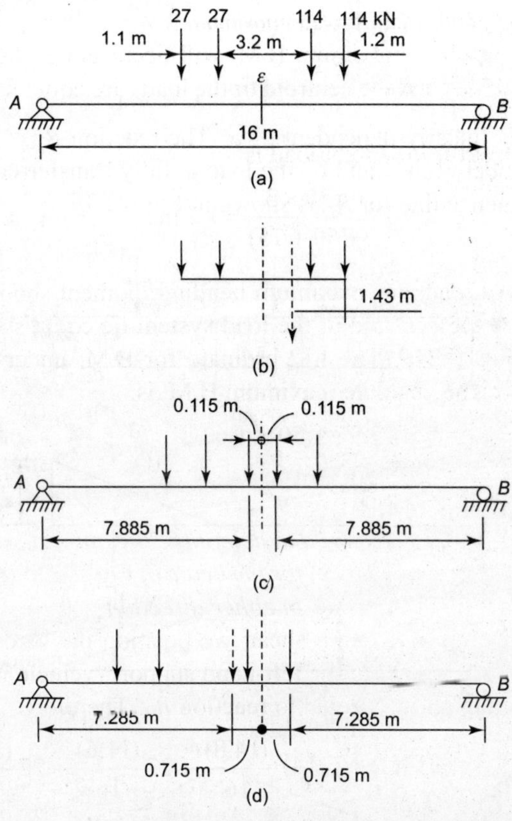

Fig. 7.22 | *(a) Beam and moving loads, (b) Resultant of load system, (c) Position of loads for maximum moment under inner 114 kN load, (d) Position of loads for maximum moment under outer 114 kN load*

$$M = R_B\,(7.285)\ \frac{282}{16} = (7.285)\,(7.285) = 935.38 \text{ kN.m}$$

Therefore, the absolute maximum moment occurs under the interior 114 kN load when placed at 0.115 m from the centre line of the beam.

7.8.5 Influence Lines for Statically Determinate Frames and Beams with Hinges

The construction of influence lines for other statically determinate beams and frames is illustrated through the following simple examples.

Example 7.15 | *Consider the beam of Fig. 7.23. It is required to construct influence lines for the reaction at D, shear at B and moments at A and B.*

To construct the influence line for reaction R_D, consider the unit load between C and D at a distance x from C. Taking the summation of the moments about hinge point C, we have

$$R_D\,(3) - (1)\,(x) = 0$$

or
$$R_D = \frac{x}{3}$$

Reaction R_D is linearly dependent on x. The reaction $R_D = 0$ for a unit load placed anywhere between A and C; the load is fully transferred to the fixed end support. The influence line for R_D is shown in Fig. 7.23*b*.

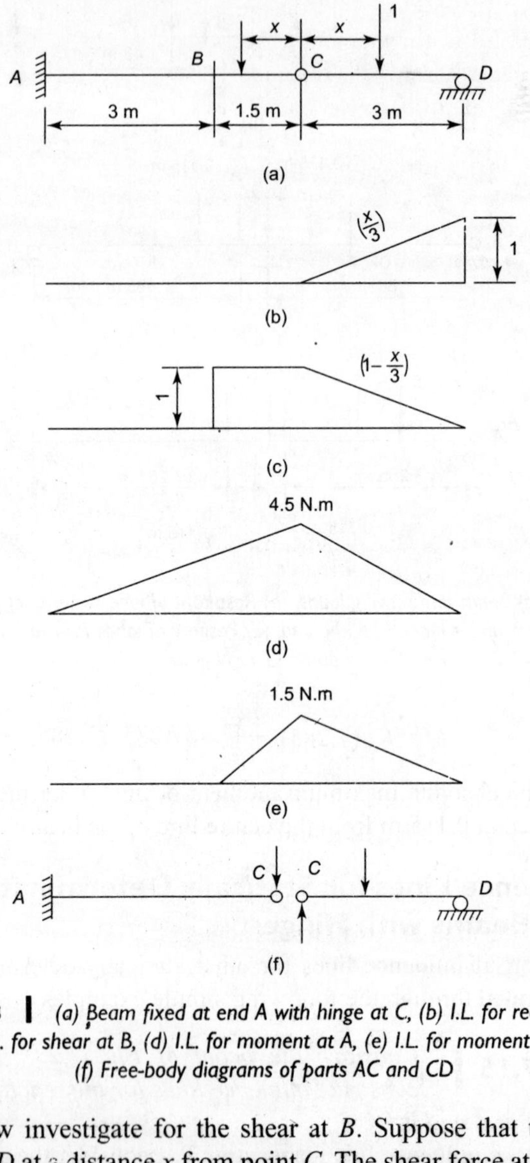

Fig. 7.23 | *(a) Beam fixed at end A with hinge at C, (b) I.L. for reaction R_D. (c) I.L. for shear at B, (d) I.L. for moment at A, (e) I.L. for moment at B, (f) Free-body diagrams of parts AC and CD*

We shall now investigate for the shear at B. Suppose that the unit load is between C and D at a distance x from point C. The shear force at B will be equal to the algebraic sum of the unit load downward and reaction R_D upward, that is,

$$V_B = \left(1 - \frac{x}{3}\right) \tag{7.45}$$

and is positive. Consider now the load between B and C. For all positions of load between B and C, the shear force at B is unity and is positive. However, for the unit load between A and B, the shear force at B is zero; the load is directly transmitted to the fixed end A as in a cantilever beam, The I.L. for shear at B is shown in Fig. 7.23*c*.

To construct the I.L. for moment at sections A and B, consider again the unit load placed between C and D at a distance x from C as shown in Fig. 7.23*a*. From the free-body diagram of parts AB and DC (Fig. 7.23*f*), we evaluate

$$M_A = \text{(reaction at the hinge) (4.5)} = \left(1 - \frac{x}{3}\right) \tag{7.46}$$

and
$$M_B = \text{(reaction at the hinge) (1.5)} = \left(1 - \frac{x}{3}\right)(1.5) \tag{7.47}$$

when the unit load is placed between B and C at a distance x from C (Fig. 7.23*a*), we obtain,

$$M_A = (1)(4.5 - x), \qquad 0 \le x \le 4.5 \text{ m} \tag{7.48}$$

$$M_B = (1)(1.5 - x), \qquad 0 \le x \le 1.5 \text{ m} \tag{7.49}$$

When the unit load is between A and B the moment at B is zero. The I.L. for moments at A and B are shown plotted in Figs. 7.23*d* and *e* respectively.

The same procedure can also be used for constructing influence lines for structural quantities of frames. The procedure is illustrated by the following example.

Example 7.16 | *Consider the frame of Fig. 7.24a. It is required to construct the I.L. for shear and moment at a point midway in column BD as the unit load moves from A to C across the horizontal member.*

The position of the unit load is described by distance x measured from end A. To develop the influence line for the shear force and moment at F, it is necessary to know the value of R_{DH} (horizontal reaction component at D) as the unit load moves from A to C.

Summing up the moment of all forces about a point 0 (not shown), the point of intersection of the two column lines extended, we have, from Fig. 7.24*b*

$$R_{DH}(12) - (1)(x - 2) = 0$$

or
$$R_{DH} = \frac{(x-2)}{12}$$

Therefore, the value of shear at F is

$$V_F = -R_{DH} = -\frac{(x-2)}{12}, \; 0 \le x \le 8 \text{ m} \tag{7.50}$$

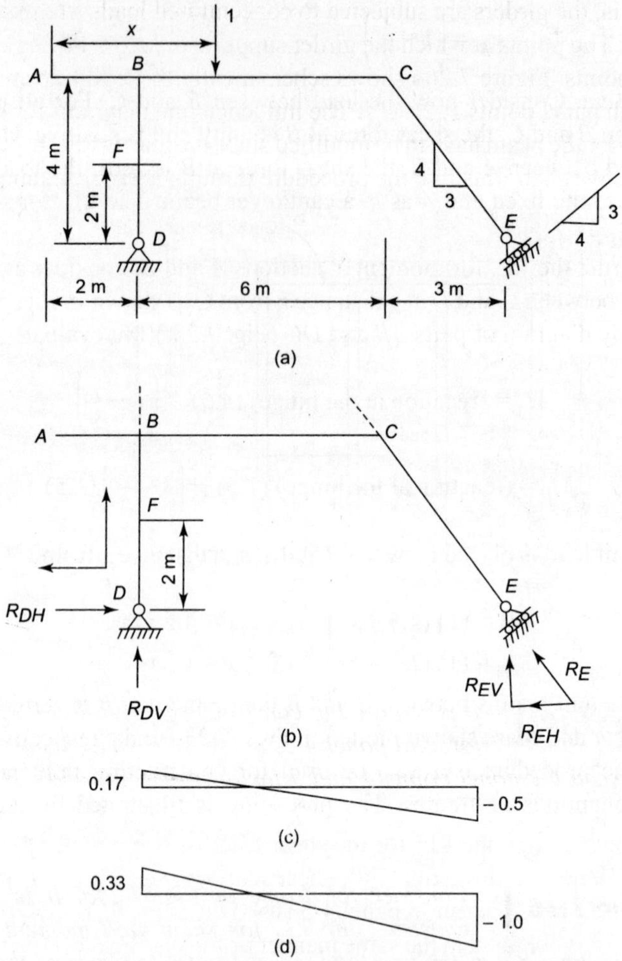

Fig. 7.24 | *(a) Frame and type of supports, (b) Free-body diagram of entire frame, (c) I.L. for shear at F, (d) I.L. for moment at F*

The resulting influence line for shear at F is shown in Fig. 7.24*c*. The moment at the same point F is evaluated as

$$M_F = - R_{DH}(2) = -\frac{(x-2)}{12}(2) = -\frac{(x-2)}{6} \quad (7.51)$$

A negative sign is given to satisfy the sign convention for the reference axes shown on column D_B. The I.L. for moment at F is shown in Fig. 7.24*d*.

7.9 INFLUENCE LINES FOR PANELLED BEAMS

Influence lines for floor beams can also be developed following the procedure outlined above. Consider the floor system of Fig. 7.25 in which the roof slab transmits the load to the cross beams and through the cross beams to the main girders, that is, the girders are subjected to concentrated loads transmitted by the cross beams. The points at which the girder supports the cross beams are referred to as panel points. Figure 7.26*a* shows schematically the load transmitted to the beam through panel points 1, 2, ..., 7. The influence line diagrams for shear force and moment for the beam have to be modified since the load is transmitted through panel points. We shall illustrate the procedure through a simple example.

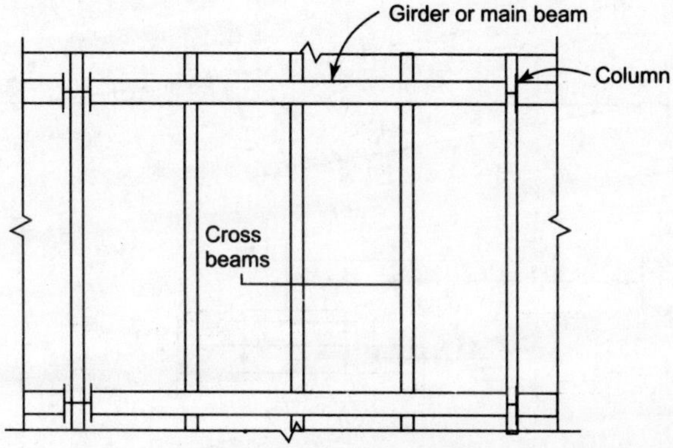

Fig. 7.25 | *Floor system*

Example 7.17 | *It is required to construct influence lines for the shear force in panel 4-5, moment at panel point 4 and also midway between the panel points 4 and 5 for the panelled beam given in Fig. 7.26.*

It may be noted that the I.L. for the shear force at any point within the panel is the same. Hence, we investigate for shear with reference to panel and not any section. The I.L. for the shear in panel 4-5 is shown in Fig. 7.26*b*. The portion of I.L. from 1 to 4 is obtained in the same manner as for a simple beam. Even though the load is applied through the joists, the shear in panel 4–5 is independent of the distribution of the unit load to any two joists between panels 1 and 4. This can also be seen from the fact that the shear in panel 4-5 is numerically equal to the reaction at the right-hand support. The right-hand support reaction is independent of the manner in which the unit load is transferred. However, when a unit load is located between panel points 4 and 5, the shear in panel 4-5 is dependent on the ratio of joists loads at 4 and 5. Suppose the unit load (N) is at a distance x from the left-hand support so that 6 m $\le x \le$ 8 m, the shear in panel 4-5 is equal to the reaction at the left support minus the reaction at panel point 4. Thus, from Fig. 7.26*c*

$$V_{4\text{-}5} = R_A - \left(\frac{8-x}{2}\right), \quad 6\text{ m} \le x \le 8\text{ m} \tag{7.52}$$

or

$$V_{4\text{-}5} = \left(\frac{12-x}{12}\right) - \left(4 - \frac{x}{2}\right) = \left(\frac{5}{12}x - 3\right)\text{ N} \tag{7.53}$$

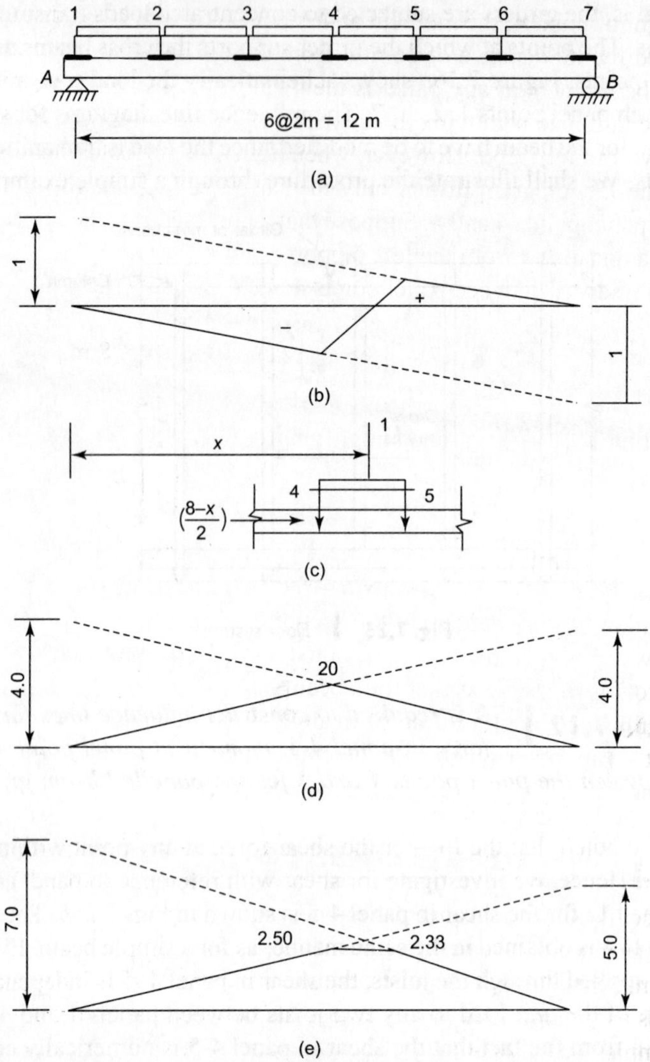

Fig. 7.26 | *(a) Panelled beam, (b) I.L. for shear in panel 4–5, (c) Unit load over panel 4–5, (d) I.L. for moment at panel point 4, (e) I.L. for moment at mid-point of panel 4–5*

The variation is linear. When $x = 6$ m, $V_{4-5} = -0.5$ N and when $x = 8$ m, $V_{4-5} = +0.33$ N. It is also apparent that at $x = 7.2$ m shear in panel 4–5 is zero.

The construction of the I.L. between panels 5 and 7 is the same at that for a sample beam. The completed I.L. diagram is shown in Fig. 7.26*b*.

Consider now the I.L. for the moment at panel point 4. When the unit load is to the left of point 4, the moment at 4 is equal to the right-hand support reaction multiplied by the distance of panel point 4 from the right-hand support. Since the right-hand support reaction is independent of the manner in which the unit load is applied, the I.L. is same as that for a simple beam. When the unit load is to the right of point 4, the moment at 4 is equal to the left-hand support reaction multiplied by the distance of point 4 from the left-hand support. Therefore, the I.L. for the moment at any panel is identical to the I.L. diagram for a beam on which the unit load is moving directly. The I.L. diagram for the moment at point 4 is shown in Fig. 7.26*d*. However, when a section is located between the panel points, the I.L. diagram differs only in that panel. Consider a section midway between panel points 4 and 5. Suppose that a unit load is placed between points 4 and 5 at a distance x from the left support point as in Fig. 7.26c. The moment at a section midway between panel points 4 and 5 can be written as

$$M = R_A\,(7) - \left(4 - \frac{x}{2}\right)(1), \quad 6\text{ m} \le x \le 8\text{ m} \tag{7.54}$$

Substituting $R_A\left(\frac{12-x}{12}\right)$ and simplifying, we get

$$M = \left(\frac{36-x}{12}\right)\text{N.m} \tag{7.55}$$

which results in $M = 2.5$ N.m when $x = 6$ m and $M = 2.33$ N.m when $x = 8$ m. The I.L. diagram for the moment at the section midway between panel points 4 and 5 is shown in Fig. 7.26*e*. It may be noted that the diagram from 1 to 4 and again from 5 to 7 is the same as that for a beam directly under a moving load.

The influence line for the shear in any other panel or the influence line for the moment at any other section can be constructed employing the same procedure as outlined above.

7.10 INFLUENCE LINES FOR TRUSS MEMBERS

Influence lines for forces in truss members can be constructed in much the same manner as those for the panelled beam above. The basic principle of an influence line is made use of, which indicates the variation of the force in any member of the truss as the unit load moves across the truss. The procedure is illustrated by an example.

Example 7.18 *It is required to develop influence lines for the forces in members 1, 2 and 3 of the truss shown in Fig. 7.27a. Consider the unit load to be moving at the level of the lower chord.*

The force in member 1 is always equal to the shear force in panel *CD*. This can be verified by making a cut along section *XX* and considering the equilibrium of one of the two parts of the truss. The free-body diagram of the left part of the truss is shown in Fig. 7.27*b*. Therefore, the I.L. diagram for the force in member 1 is same as the I.L. for shear force in panel C-D as shown in Fig. 7.27*c*. For the unit load between *a* and *c* (Fig. 7.27*c*), the force in member 1 is tensile in nature, while for any position of load between *c* to *e* the nature of force in member 1 is compressive.

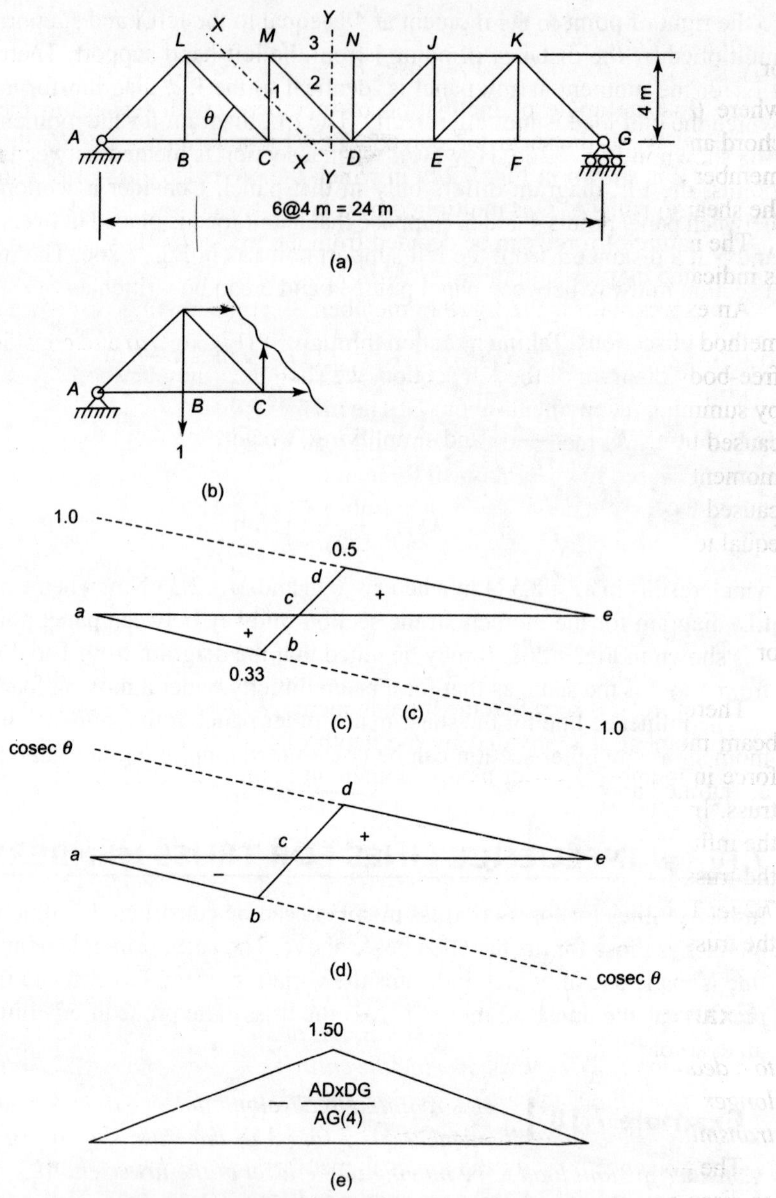

Fig. 7.27 | *(a) Truss, (b) Free-body diagram of part of truss, (c) I.L. for force in member 1, (d) I.L. for force in member 2, (e) I.L. for force in member 3*

The influence line for force in member 2 is obtained by considering the equilibrium of forces on a cut part of the truss. Imagine that the truss is cut into two parts along *YY* (Fig. 7.27*a*). Considering the equilibrium of forces in the vertical direction to the left of *YY*, we observe that the vertical component of the force in member 2 must balance the shear force in panel C-D. Writing this in the form of an expression, we have

$$P_2 \sin \theta = V_{CD}$$

or $$P_2 = V_{CD} \operatorname{cosec} \theta \tag{7.56}$$

where θ is the angle of inclination of member 2 with respect to the bottom chord and V_{CD} is the shear force in panel *C-D*. The influence line for the force in member 2 is shown in Fig. 7.27*d* in which the ordinates of the I.L. diagram for the shear in panel *C-D* is multiplied by a factor, cosec θ.

The nature of force can be decided from the free-body diagram. Tensile force is indicated by plus sign in Fig. 7.27*d*.

An expression for the force in member, 3 can be obtained by employing the method of sections. Taking a section through *YY* (Fig. 7.27*a*) and considering the free-body diagram of the left section, we shall determine the force in member 3 by summing the moments about *D*. The moment about *D* consists of the moment caused by a unit load and the left-hand reaction as in the panelled beam and the moment caused by the unknown force in member 3. In other words, the moment caused by the unknown force in member 3 about point *D* must be numerically equal to the panelled beam moment at *D* that is,

$$P_3 \,(4) = M_D$$

or $$P_3 = \frac{M_D}{4} \tag{7.57}$$

Therefore, it is seen that the force in member 3 is always equal to the panelled beam moment at *D* divided by the depth of the truss. For a moving load, the force in member 3 varies as the moment at point *D* divided by the depth of the truss. In other words, the I.L. for the force in member 3 can be obtained from the influence line for the moment at *D* by dividing the ordinates by the depth of the truss. The resulting I.L. diagram for the force in member 3 is shown in Fig. 7.27*e*. The force is obviously compressive in nature for a unit load anywhere on the truss.

Example 7.19 | *It is required to determine the maximum forces in members CE, DE and DF of the truss of Fig. 7.28a due to a dead load of 10 kN/m covering the entire span and a moving load of 20 kN/m longer than the span passing over the truss. Consider that the loads are transmitted through the lower chord.*

The influence lines for the forces in members *CE*, *DE* and *DF* are constructed as discussed in Sec. 7.10 and are shown in Fig. 7.28*b*, *c* and *d*.

From Fig. 7.28*b* it is obvious that the moving load should cover the entire span to obtain the maximum force in member *CE*. The resulting force in member *CE* due to dead and moving loads is

$$P_{CE} = (10 + 20)\left\{\frac{1}{2}(0.81)(5) + \frac{1}{2}(0.81) + 1.04)(5) + \frac{1}{2}(10.4)(15)\right\}$$

$$= 433.5 \text{ kN (tension)}$$

For the maximum force in member DE, it is obvious from the influence line in Fig. 7.28c that the moving load should occupy from the right-hand support to a point where the I.L. ordinate is zero. It may be noted that in all cases the dead load is taken as occupying the span throughout. The maximum tensile force in *DE* is

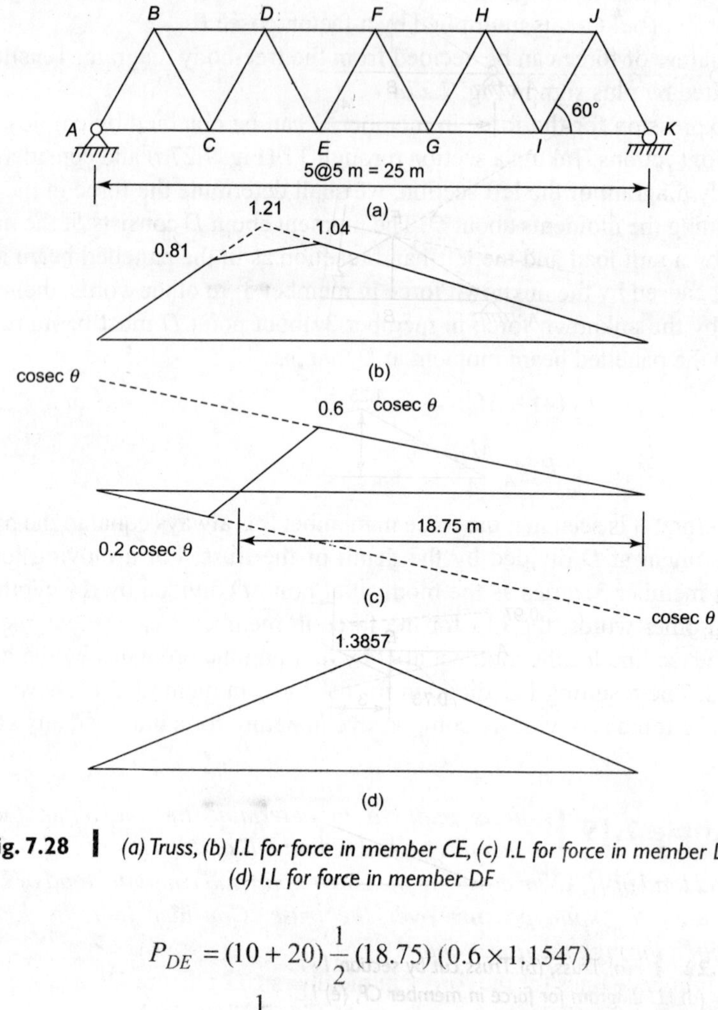

Fig. 7.28 | *(a) Truss, (b) I.L for force in member CE, (c) I.L for force in member DE, (d) I.L for force in member DF*

$$P_{DE} = (10 + 20)\frac{1}{2}(18.75)\,(0.6 \times 1.1547)$$

$$-\frac{1}{2}(10)\,(6.25)\,(0.2 \times 1.1547)$$

$$= 187.64 \text{ kN}$$

The force in member *DF* can be determined by taking the moving load to occupy the entire span. The maximum compressive force is

$$P_{DF} = (10 + 20)\ \frac{1}{2}\ (25)\,(1.3857)$$

$$= 519.64 \text{ kN}$$

Example 7.20 | *It is required to determine the maximum forces in members BC, CF and FG of the truss in Fig. 7.29 due to a live load of 25 kN/m longer than the span passing over the truss*

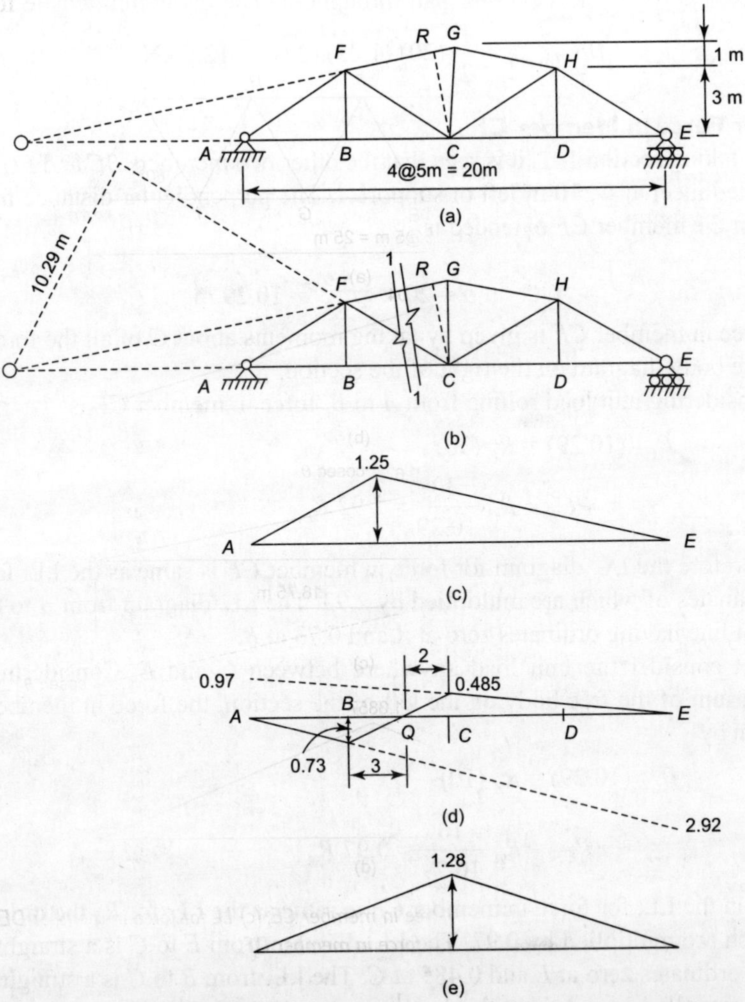

Fig. 7.29 | (a) Truss, (b) Truss cut by section 1–1, (c) I.L. diagram for force in member BC, (d) I.L. diagram for force in member CF, (e) I.L. diagram for force in member FG

Influence Lines for Force in Member BC

To construct I.L. for force in member *BC* take section 1-1 and consider the equilibrium of the free body on the left of the section. Taking moments of all the forces to the left of section, about *F* we have

$$P_{BC}(3) = R_A(AB) = M_F$$

or
$$P_{BC} = \frac{M_F}{3}$$

Therefore the I.L. for the force in member BC is obtained by drawing the I.L. for moment at F divided by a factor 3 as shown in Fig. 7.29c.

The maximum force in member BC occurs when the live load occupies the whole span.

$$\therefore \qquad PB_{C(\text{max.})} = \frac{1}{2}(20)(1.25)(25) = 312.5 \text{ kN.}$$

I.L. for Force in Member CF

Again taking section 1–1, it is seen that the other two members BC and FG when extended meet at C, 10 m left of support A. The perpendicular distance from O to P on the member CF extended is

$$O\text{₵} \sin \alpha = 20 \times \frac{3}{5.83} = 10.29 \text{ m.}$$

Force in member CF is given by taking moments about O of all the forces on the free body diagram on the right of the section.

Considering unit load rolling from A to B. force in member CF is

$$P_{CF}(10.29) = R_E(30)$$

$$\therefore \qquad PC_F = R_E \frac{30}{10.29} = 2.92 \, R_E$$

Therefore the I.L. diagram for force in member CF is same as the I.L. for R_E, the ordinates of which are multiplied by 2.92. The I.L. diagram from A to B is a straight line having ordinates zero at A and 0.73 at B.

Next consider the unit load anywhere between C and E. Considering the equilibrium of the free body on the left of the section, the force in member CF is given by

$$P_{CF}(10.29) = R_A(10)$$

$$\therefore \qquad P_{CF} = R_A \frac{10}{10.29} = 0.97 \, R_A$$

Again the I.L. for force in member CF is same as the I.L. for R_A the ordinates of which are multiplied by 0.97. The I.L. diagram from E to C is a straight line having ordinates zero at E and 0.485 at C. The I.L. from B to C is a straight line as in a panelled beam since the loads are transmitted through the joints. The complete I.L. diagram is shown in Fig. 7.29d.

Maximum tension in member CF occurs when the L.L. is in the region E to Q

$$P_{CF(\text{max.})} = \frac{1}{2}(0.485)(12)(25)$$
$$= 72.75 \text{ kN tension}$$

Maximum compression in member CF occurs when the load occupies from A to Q.

$$P_{CF\,(\text{max.})} = \frac{1}{2}(0.73)(8)(25) = 73.0 \text{ kN}$$

I.L. for Force in Member FG

From the equilibrium consideration of the free body diagram to the left of section 1-1 it is seen that the force in member *FG* is obtained by the relationship

$$P_{FG}\,(\perp \text{ distance } CR) = M_C \text{ (moment at } C)$$

Let θ be the inclination of member *FG* to the horizontal

$$\tan\theta = \frac{3}{15} \quad \therefore \quad \theta = 11.31°$$

perpendicular distance $CR = OC \sin\theta = 20 \times 0.196 = 3.92$ m

$$\therefore \qquad P_{FG} = \frac{M_C}{3.92}$$

The I.L. for force in member *FG* is obtained by drawing the I.L. for moment *C* and dividing the ordinates by 3.92. The I.L. diagram is shown in Fig. 7.29*e*.

Maximum force in member *FG* occurs when the L.L. is over the entire span.

$$P_{FG}(\text{max.}) = \frac{1}{2}\,(1.28)\,(20)\,(25) = 329 \text{ kN.m (comp.)}$$

7.11 INFLUENCE LINES FOR THREE-HINGED ARCHES

Making use of the basic principles discussed so far, influence lines for three-hinged arches, that are statically determinate may be constructed.

7.11.1 Influence Line for Horizontal Reaction H

Consider the three-hinged arch shown in Fig. 7.30*a*. Suppose it is required to construct the I.L. for horizontal reaction H, at the supports and the moment at a point *D* described by horizontal distance *x* from the left-hand support. Let a unit load be acting anywhere in the region *A* to *C* described by distance *nL*. The vertical reaction components are

$$V_A = (1 - n)$$

and
$$V_B = (n) \tag{7.58}$$

To evaluate *H,* we take moments about hinge point *C* and equate it to zero. This gives

$$(n)\frac{L}{2} - H(h) = 0 \quad \text{for } 0 \le n \le \frac{L}{2}$$

or
$$H = \frac{nL}{2h} \tag{7.59}$$

Similarly, if the unit load is in the region from *C* to *B*, we can evaluate *H* again by taking the moment about hinge point *C* and equating it to zero. This gives

$$(1 - n)\,\frac{L}{2} - H(h) = 0 \text{ for } \frac{L}{2} \le n \le L$$

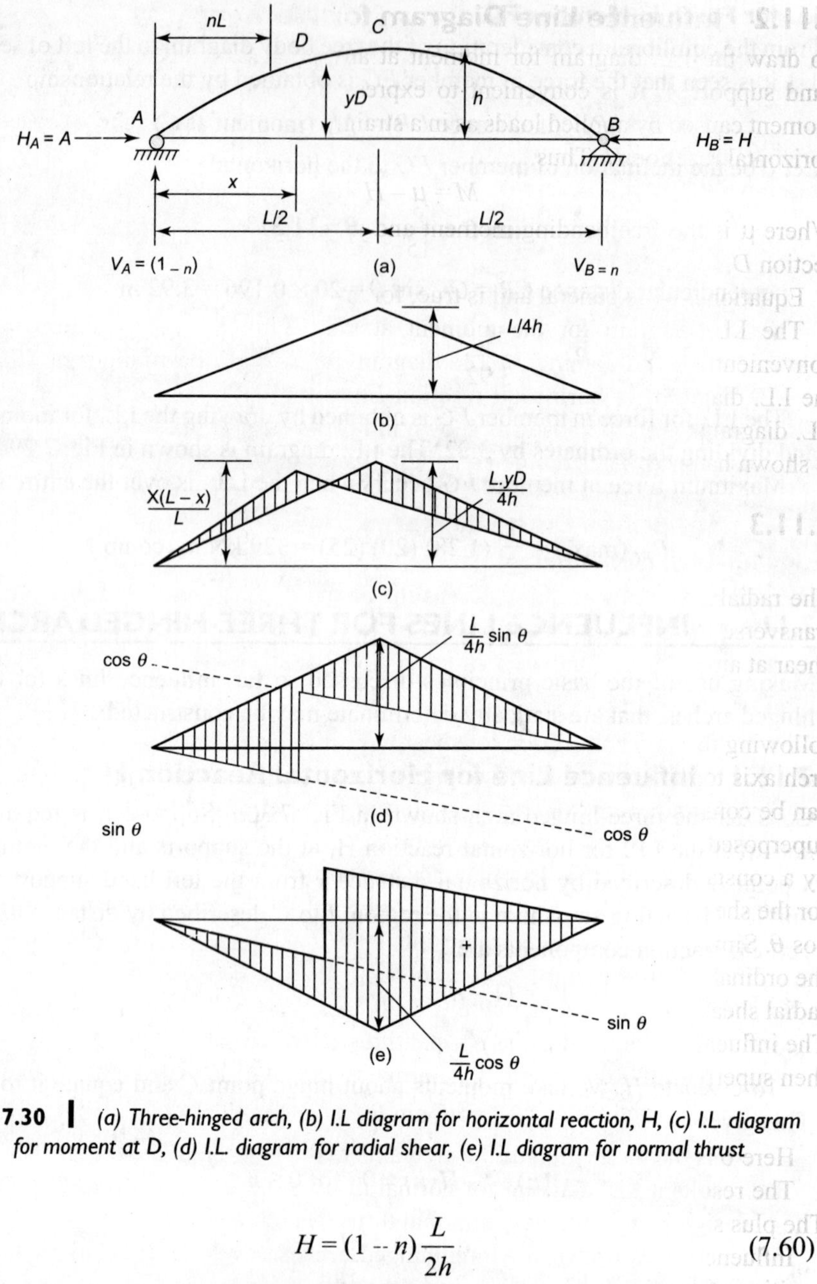

Fig. 7.30 | *(a) Three-hinged arch, (b) I.L diagram for horizontal reaction, H, (c) I.L. diagram for moment at D, (d) I.L. diagram for radial shear, (e) I.L diagram for normal thrust*

or

$$H = (1 - n)\frac{L}{2h} \tag{7.60}$$

It is seen from Eqs. 7.59 and 7.60 that horizontal reaction H is directly related to the position of the unit load described by distance nL. The variation of H for a unit load moving from A to B is shown in Fig. 7.30*b*. This diagram, by definition, is the I.L. for horizontal reaction H. It is apparent that the value of H reaches a maximum when the unit load is at the crown and its value is L/4h.

7.11.2 Influence Line Diagram for Moment

To draw the I.L. diagram for moment at any point D at distance x from left-hand support A, it is convenient to express the moment by parts, that is the moment caused by applied loads as in a straight beam and the moment caused by horizontal reaction H. Thus,

$$M = \mu - H \cdot y_D \tag{7.61}$$

Where μ is the free bending moment and y_D is the ordinate of the arch axis at section D.

Equation 7.61 is general and is true, for any section D desired.

The I.L. diagram for the moment at the section D is, therefore, drawn conveniently by combining the I.L. diagram for a simple beam moment M_f, and the I.L. diagram for horizontal reaction H, multiplied by ordinate y_D. The two I.L. diagrams are superposed as shown in Fig. 7.30*c*. The resultant I.L. diagram is shown hatched. The common area left blank cancels out.

7.11.3 Influence Line Diagrams for Radial Shear and Normal Thrust

The radial shear can be expressed as the algebraic sum of the shear caused by transverse loads and the shear caused by horizontal reaction H. Thus, the radial shear at any section D can be written as

$$V_{r(D)} = V_A \cos\theta - H \sin\theta \tag{7.62}$$

following the sign convention for shear in beams. Here θ is the inclination of the arch axis to the horizontal at section D. Hence the I.L. diagram for radial shear can be constructed in two parts representing the two terms in Eq. 7.62 and then superposed. The first term actually represents shear in a simple beam multiplied by a constant, cos θ. The I.L. diagram for this is, therefore, the same as the I.L. for the shear in a simple beam, but the ordinates are all multiplied by a constant, cos θ. Similarly, the second term results in an I.L. the same as the I.L. for H, but the ordinates are all multiplied by a constant, sin θ. The resultant I.L. diagram for radial shear is shown hatched in Fig. 7.30*d* The common blank area cancels out. The influence line diagram for normal thrust can also be drawn in two parts and then superposed. For example, the normal thrust at section D can be written as

$$N_D = V_A \sin\theta + H \cos\theta \tag{7.63}$$

Here θ is the inclination of the arch axis to the horizontal at section D.

The resultant I.L. diagram for normal thrust is shown hatched in Fig. 7.30c. The plus sign indicates that the normal thrust is compression.

Influence lines drawn for three-hinged arches can be utilised to fix up the position of moving loads for obtaining the maximum forces at any section of the arch. The example that follows illustrates the procedure.

Example 7.21 | *A three-hinged parabolic arch has a span of 25 m with a central rise of 5 m. A load of 150 kN rolls over the arch from left to right. Find the maximum shear force and B.M. at a section 8 m from the left-hand hinge.*

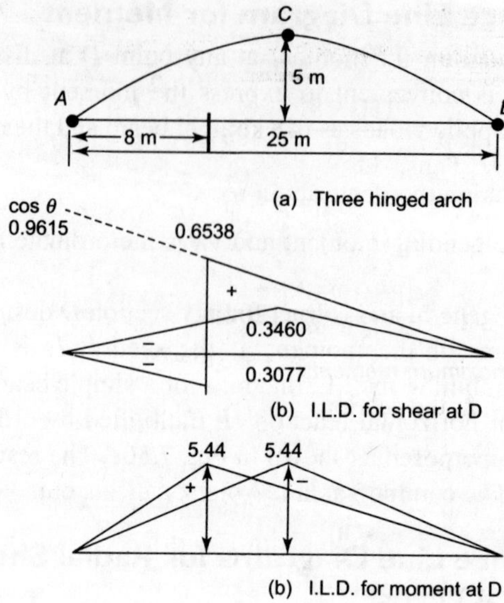

Fig. 7.31

Step 1: The influence line diagrams for shear force at a section 8 m from the left end is shown in Fig. 7.31*b*.

It is clear that the maximum –ve S.F. will occur when the load is to the left of section.

$$\text{–ve S.F., } V = \left(0.9615 \times \frac{8}{25} + 0.3460 \times \frac{8}{12.5}\right) 150$$
$$= (0.3077 + 0.2214)\ 150$$
$$= 79.37 \text{ kN}$$

$$\text{Similarly, +ve S.F., } V = \left(0.9615 \times \frac{17}{25} - 0.2214\right) 150$$
$$= (0.6538 - 0.2214)\ 150$$
$$= 64.86 \text{ kN.}$$

Step 2: To fix up I.L.D. for moment at section D

The I.L. diagram for the B.M. at a section 8 m from left end is shown in Fig. 7.31*c*.

Maximum +ve B.M. occurs when the load is on the section. The maximum –ve B.M. will occur when the load is on the crown.

Step 3: To evaluate H

$$\text{Ordinate at the section, } y = \frac{4 \times 5}{25 \times 25}\ (8)\ (17)$$
$$= 4.352 \text{ m}$$

Taking moments about A

$$V_B\,(25) = 150 \times 8$$

$$V_B = \frac{150 \times 8}{25}\ 48\ \text{kN.}$$

H is evaluated taking moments about C

$$M_C = V_B\,(12.5) - H(5) = 0$$

$$H = \frac{48 \times 12.5}{5} = 120\ \text{kN.}$$

Step 4: To evaluate maximum moments

Maximum +ve B.M. at the section

$$\begin{aligned} M_{max} &= V_A\,(8) - H\,y \\ &= 102(8) - 120 \times 4.352 \\ &= 293.76\ \text{kN.m} \end{aligned}$$

Maximum –ve B.M. when the load is on the crown is obtained as follows.

$$V_A = V_B = \frac{150}{2} = 75\ \text{kN.}$$

H is evaluated taking moments about C

$$M_C - 75 \times 12.5 - H(5) = 0$$

$$H = \frac{75 \times 12.5}{5} = 187.5\ \text{kN.}$$

$$\begin{aligned} \text{Maxm. –ve B.M.} &= V_A\,(8) - H \times 4.352 \\ &= 75 \times 8 - 187.5 \times 4.352 = -216.0\ \text{kN.m} \end{aligned}$$

Example 7.22 | *A three-hinged circular arch has a span 50 m and a rise of 10 m. A load of 200 kN crosses the arch from one end to the other. Determine (i) the maximum horizontal thrust, and (ii) the maximum +ve and –ve B.M. at a section 15 m from left-hand hinge.*

Step 1: To evaluate R and y_d

Radius of the arch using the relation

$$10\,(2R - 10) = 25 \times 25$$

$$R = 36.25\ \text{m.}$$

At the section 15 m from left end y_d is obtained from

$$(36.25 - 10 + y_d)^2 + 10^2 = 36.25^2$$

$$y_d = 8.59$$

Step 2: To fix up load position for maximum B.M.

From the I.L. diagram for B.M. at the required section (Fig. 7.32) the load is to be positioned on the section itself for obtaining maximum +ve B.M. Taking moments about A

$$V_B (5) - 200 (15) = 0$$

$$V_B = \frac{200 \times 15}{50} = 60 \text{ kN}$$

$$V_A = 200 - 60 = 140 \text{ kN}$$

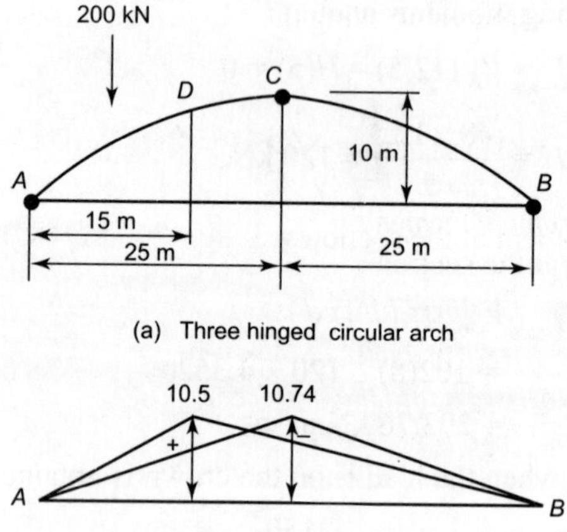

(a) Three hinged circular arch

(b) I.L.D. for moment at *D*

Fig. 7.32

Horizontal thrust H is evaluated taking moments about C

$$M_C = V_B (25) - H . y_C = 0$$

$$= 60 \times 25 - H(10) = 0$$

$$\therefore \quad H = \frac{60 \times 25}{10} = 150 \text{ kN}$$

Moment under load

$$M_D = V_B (35) - H . y_D$$

$$= 60 (35) - 150 (8.59)$$

$$= 811.5 \text{ kN.m}$$

We can also obtain the same value using the I.L.D. The +ve B.M. ordinate under load point

$$= \left(10.5 - \frac{10.74}{25} \times 15\right) = 4.056 \text{ kN.m}$$

Moment under load $M_D = 200 \times 4.056 = 811.2$ kN.m

The maximum –ve B.M. will develop when the load is placed on the crown. Horizontal thrust H can be evaluated taking moments about C

$$M_C = V_B (25) - H . y_C = 0$$

$$= 100 \times 25 - H(10) = 0$$

$$\therefore \quad H = 250 \text{ kN.}$$

Moment at the section

$$M_D = V_A(15) - H(8.59)$$
$$= 100 \times 15 - 250 \times 8.59$$
$$= -648.0 \text{ kN.m.}$$

We can also obtain the same value using the I.L. diagram. The –ve B.M. ordinate under load point

$$= \left(10.74 - \frac{10.5}{35} \times 25\right) = 3.24 \text{ kN.m}$$

Maximum –ve moment at the section = 3.24 × 200 = 648.0 kN.m

Example 7.23 *A three-hinged parabolic arch has a span 40 m and a rise of 4 m (Fig. 7.33a). Using influence lines, determine maximum horizontal thrust H, and the moment at the quarter span point from the left hand support, when two loads 100 kN and 150 kN at 3 m centres move from left to right with the 100 kN load in the lead.*

Step 1: To fix up I.L.D for H and moment M

The I.L. diagram for horizontal thrust H is the same irrespective of section and is shown in Fig. 7.33*b*. From inspection it is seen that maximum horizontal thrust H occurs when the 150 kN load is on the central hinge and the 100 kN load is 3 m to the right of the hinge. For this position of loading

$$H_{(\max)} = 150(2.5) + 100(2.5)\frac{17}{20}$$
$$= 587.5 \text{ kN.}$$

The influence line diagram for the moment at 1/4 span, that is, 10 m from the left-hand support, is shown in Fig. 7.33*c*. For convenience, the resultant influence line diagram is shown on a horizontal base in Fig. 7.33*d*.

Step 2: To fix up load position for maximum B.M.

Let point E at which the I.L. ordinate is zero, be at a distance x from the left-hand support. From triangles ACC_1 and AEE_1, we have

$$\frac{e}{x} = \frac{7.5}{20}$$

Similarly, from triangles BDD_1 and BEE_1, we have

$$\frac{e}{(40 - x)} = \frac{7.5}{30}$$

Eliminating e, we get

$$x = 16 \text{ m}$$

The maximum positive moment will occur when 150 kN is at the section itself and the 100 kN load is to the right of the section. Therefore,

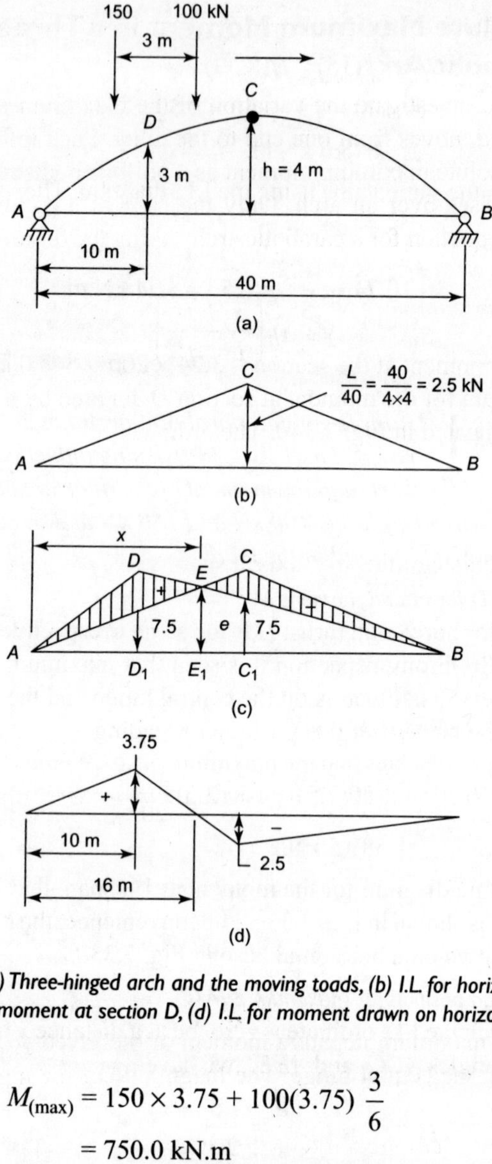

Fig. 7.33 | *(a) Three-hinged arch and the moving toads, (b) I.L. for horizontal reaction H, (c) I.L. for moment at section D, (d) I.L. for moment drawn on horizontal base*

$$M_{(\max)} = 150 \times 3.75 + 100(3.75)\ \frac{3}{6}$$

$$= 750.0 \text{ kN.m}$$

Similarly, the maximum negative moment will occur when 150 kN load is at hinge point C and the 100 kN load is to its right. For the loading position indicated

$$M_{(\max)} = 150 \times 2.5 + 100(2.5)\ \frac{17}{20}$$

$$= 587.5 \text{ kN.m}$$

7.11.4 Absolute Maximum Moment in a Three-Hinged Parabolic Arch

It is interesting to investigate the variation of the maximum moment along the arch as a unit load moves from one end to the other. Such information is useful in finding the absolute maximum moment as a uniformly distributed load longer than the span passes over an arch. Only the case of a parabolic arch will be considered. The equation for a parabolic arch, taking the origin at the left support is (Fig. 7.34*a*)

$$Y = 4h\left(\frac{x}{L} - \frac{x^2}{L^2}\right) \tag{7.64}$$

The I.L. diagram for die moment at section D denoted by distance x from the left support is indicated in Fig. 7.34*b*. The ordinate

$$cc_1 = \frac{Ly}{4h}$$

where y is the ordinate of the arch axis at section D.

Substituting for y from Eq. 7.64

$$cc_1 = \frac{x}{L}(L - x) \tag{7.65}$$

This is same as ordinate dd_1.

The I.L. diagram indicates that the maximum positive moment occurs when the unit load is at the section itself. The positive moment is given by the ordinate

$$dd_2 = dd_1 - d_1 d_2$$

$$= \frac{x(L-x)}{L} - \frac{x(L-x)}{L}\frac{x}{L/2}$$

$$= \frac{x(L-x)}{L^2}(L - 2x) \tag{7.66}$$

Similarly, the maximum negative moment at the given section occurs when the unit load is at the central hinge. The magnitude of the moment is indicated by the ordinate

$$cc_2 = cc_1 - c_1 c_2$$

$$= \frac{x(L-x)}{L} - \frac{x(L-x)}{L}\frac{L/2}{(L-x)}$$

$$= \frac{x(L-2x)}{2L} \tag{7.67}$$

In order to obtain the absolute maximum positive moment anywhere along the arch, consider that the section D denoted by distance x is variable and therefore differentiate Eq. 7.66 with respect to x and equate to zero. This results in a quadratic equation which when solved gives two values for x that is $x_1 = 0.211L$ and $x_2 = 0.789$L.

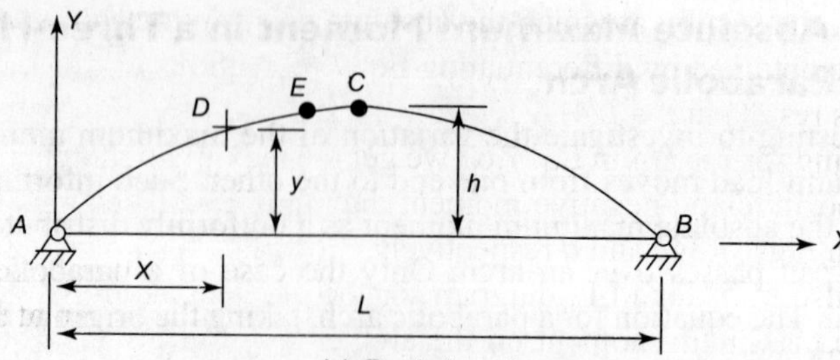

(a) Parabolic arch

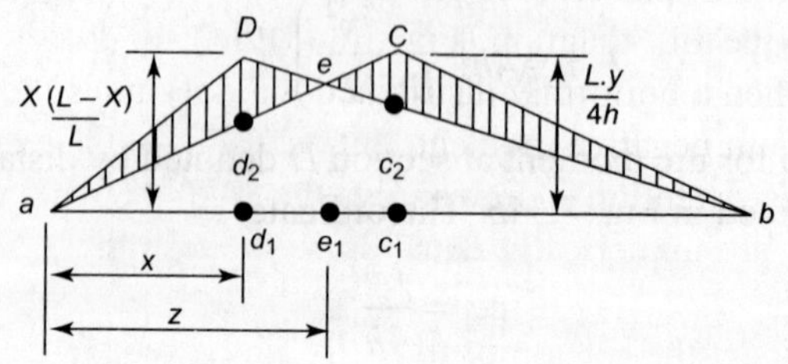

(b) I.L.D for moment at section D

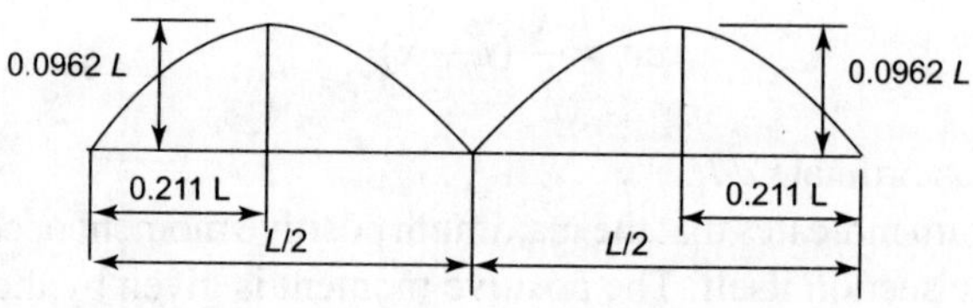

(c) Maxm.+Ve moment diagram

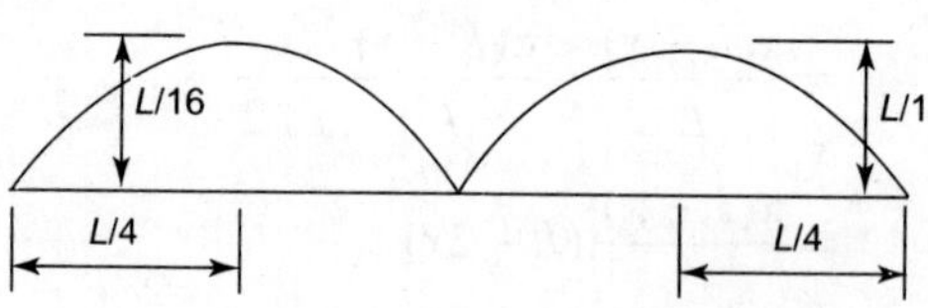

(d) Maxm-Ve moment diagram

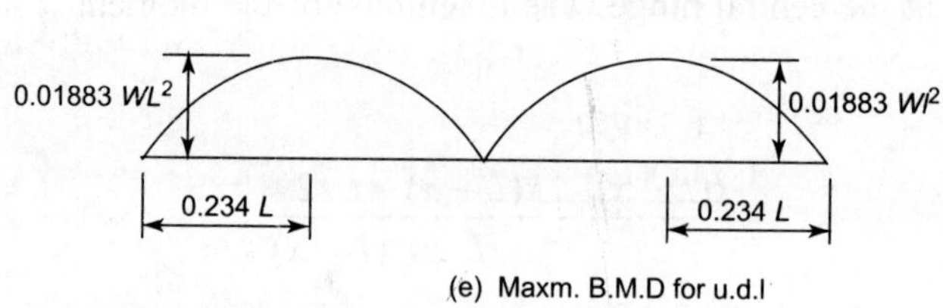

(e) Maxm. B.M.D for u.d.l

Fig. 7.34

Substituting for $x = 0.211L$ in Eq. 7.66

Maximum moment = 0.0962 L

Similarly, the section at which the absolute maximum negative moment will occur can be obtained by differentiating Eq. 7.67 with respect to x and equating to zero. This results in $x = L/4$.

Substituting for $x = L/4$ in Eq. 7.67 we get the maximum moment = $L/16$. The maximum positive and negative moment diagrams for all sections from A to B are shown in Figs. 7.34*c* and *d* respectively.

We can also utilise the I.L. diagram for the moment at any section to obtain the absolute maximum moment on the arch as a uniformly distributed load w/unit length longer than the span crosses the arch.

It is obvious from the I.L. diagram (Fig. 7.34*b*) that the maximum positive moment will occur when a uniformly distributed load occupies the region from A to E and the maximum negative moment will occur when the distributed load occupies region E to B. Attention is drawn to the fact that the maximum positive or negative moments are numerically equal since the triangular areas *ade* and *bce* are same.

In the I.L. diagram, let the section of the zero ordinate be at distance z from support A. From similar triangles, we can write

$$\frac{ee_1}{cc_1} = \frac{ae_1}{ac_1} = \frac{z}{L/2}$$

and
$$\frac{ee_1}{dd_1} = \frac{be_1}{bd_1} = \frac{(L-z)}{(L-x)} \tag{7.68}$$

Since cc_1 and dd_1 are the same, we can write

$$\frac{z}{L/2} = \frac{(L-z)}{(L-x)}$$

or
$$z = \frac{L^2}{(3L-2x)}$$

Substituting for cc_1 and z in the first of Eq. 7.68

$$ee_1 = \frac{2x(L-x)}{(3L-2x)} \tag{7.69}$$

Therefore, the area of triangle *ade* = area of triangle *adb* – area of triangle *aeb*

$$= \frac{1}{2}\frac{L(x)(L-x)}{L} - \frac{1}{2}\frac{L2x(L-x)}{(3L-2x)}$$

$$= \frac{x(L-x)(L-2x)}{2(3L-2x)} \tag{7.70}$$

Therefore,

$$M_x = \frac{wx(L-x)(L-2x)}{2(3L-2x)} \tag{7.71}$$

To obtain the absolute maximum positive moment anywhere on the arch, it is necessary to differentiate Eq. 7.71 with respect to x and equate it to zero. This gives

$$x = 0.234\ L$$

as an appropriate root of a cubic equation.

Substituting for $x - 0.234\ L$ Eq. 7.71, we get the absolute positive or negative moment $= \pm\ 0.01883\ wL^2$.

The variation of the maximum positive or negative moment at any section from A to B is shown in Fig. 7.34*e*.

Example 7.24 *A three-hinged parabolic arch has span 20 m and rise 4 m. A concentrated load of 150 kN rolls from left to right. Calculate the maximum +ve and –ve moments at a section 5 m from the left end support. Also calculate the absolute maximum B.M. that may occur anywhere in the arch.*

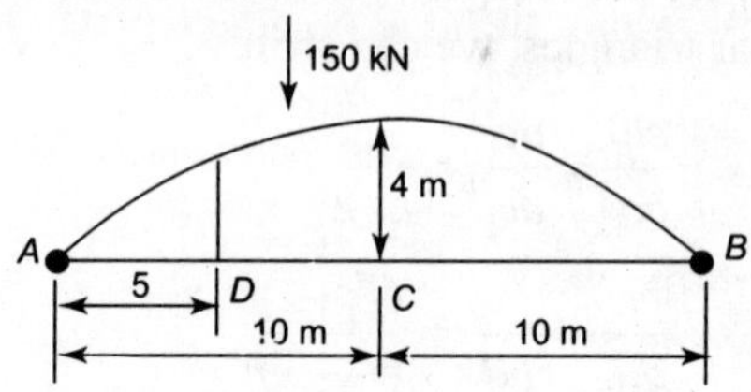

Fig. 7.35

Step 1: To evaluate H for maximum +ve moment

Using the equation for parabola the rise of the arch at section $D = 3$ m. From a diagram of I.L. for moment at D (Fig. 7.34*b*) it is clear that maximum B.M. will occur when the load is at the section itself.

Reaction R_B when the load is at the section is

$$R_B = \frac{150 \times 5}{20} = 37.5 \text{ kN}$$

Taking moments about the hinge at the crown

$$R_B\,(10) = H_B\,(4)$$

$$\therefore \qquad H_B = \frac{37.5 \times 10}{4} = 93.75 \text{ kN}$$

$$\text{Moment } M_D = 37.5 \times 15 = 93.75 \text{ kN}$$

$$= 281.25 \text{ kN.m}$$

Step 2: To evaluate H for maximum -ve moment

Again from the I.L. diagram it is seen that the maximum –ve moment will occur when the load is at the centre of the span.

$$R_A = R_B = 75 \text{ kN}.$$

Taking moments about the crown

$$H(4) = 75 \times 10$$

$$H = \frac{75 \times 10}{4} = 187.5 \text{ kN}$$

Moment $M_D = 75 \times 5 - 187.5 \times 3 = -187.5$ kN.m.

From the maximum +ve B.M. curves in Fig. 7.34*c* the absolute maximum +ve B.M. will occur at sections 0.211 l or 4.22 m from supports at either end and the magnitude of the absolute moment = 0.0962 $l \times W$

$$= 0.0962 \times 20 \times 150 = 288.6 \text{ kN.m}$$

Again from the maximum -ve moment curves in Fig. 7.33*d* the maximum –ve moment will occur at sections 0.25 l, i.e., 0.25 × 20 = 5 m from the ends and the magnitude of the moment is $\frac{-l}{16} \times w = \frac{-20}{16} \times 150 = 187.5$ kN.m.

Example 7.25 *A three-hinged parabolic arch has a span 25 m and central rise 5 m as shown in Fig. 7.36. A u.d.l. of intensity 10 kN.m longer than the span rolls over the arch. Determine the maximum positive and negative moment at a section 7.5 m from the left end. Also find out the section and the magnitude, of the absolute maximum moment that may occur anywhere on the arch.*

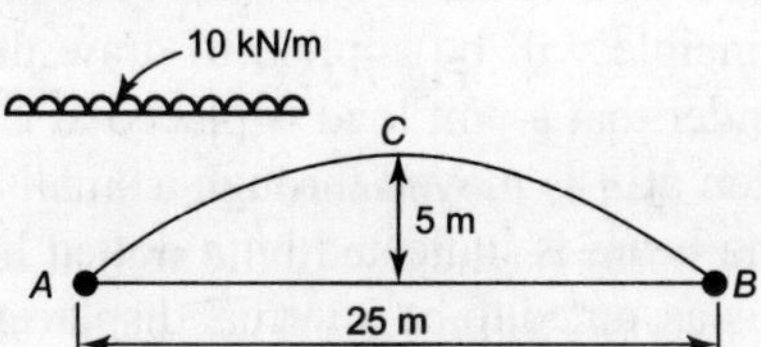

Fig. 7.36

The maximum +ve or –ve B.M. curves as the u.d.l. longer than the span rolls over the arch span are shown in Fig. 7.34*e*. It is clear that the section at which the absolute maximum +ve or –ve B.M. will occur is located at 0.234 l i.e. 0.234 × 25 = 5.85 m from either end. The magnitude of the maximum B.M. = 0.0188 wl^2 i.e., = 0.0188 × 10 × 25^2 = 117.5 kN.m.

Maximum positive B.M.

Let $M_{x(\max)}$ be the moment at the section.

Using the relation in Eqn. 7.71

$$M_{x(\max)} = \frac{Wx\,(L - x)\,(L - 2x)}{2(3L - 2x)}$$

Substituting $x = 7.5$ m

$$M_{x(\max)} = \frac{10 \times 7.5\,(25 - 7.5)\,(25 - 15)}{2(3 \times 25 - 15)} = 109.38 \text{ kN.m}$$

Maximum –ve moment

We know that the maximum –ve moment is equal to that of the maximum +ve B.M.

$$M_{max} = -109.38 \text{ kN.m}$$

7.12 INFLUENCE LINES FROM DEFLECTED SHAPES

Influence lines for structural quantities can be developed from the deflected shape of the structure. This method of constructing influence lines, as we shall see later, is particularly useful for continuous structures. This approach is mainly based on the Müller-Brcslau principle.

7.12.1 Müller-Breslau Principle

The Müller-Breslau principle states that if a reaction (or internal force) acts though an imposed displacement, the corresponding displaced shape (elastic curve) of the structure is, to some scale the influence line for the particular reaction (or internal force). The force and displacement, of course, can be replaced by moment and rotation respectively.

It must be pointed out here that the Müller-Breslau principle applies to the construction of influence lines for force quantities only.

Let us apply this principle to construct influence lines for a simple beam of Fig. 7.37*a* for which the influence lines were previously developed in Sec. 7.2. The Müller-Breslau principle will be applied to draw the I.L. for the support reaction at end A. Consider that a unit load is placed at C, an arbitrary point on the beam and the reaction at A is moved through a small displacement, Δ_A. The displaced position of the beam is indicated by a dotted line (Fig. 7.37*b*). If the small displacement Δ_A is assumed to be a virtual displacement, the virtual work done on the beam can be expressed as

$$R_A (\Delta_A) - (1) (\Delta_C) = 0$$

or
$$R_A = \frac{\Delta_C(1)}{\Delta_A} \tag{7.72}$$

Because the magnitude of the imposed displacement Δ_A is arbitrary, we can for convenience assume a value of unity. Then the displaced position represents to scale the influence line for reaction R_A.

Let us now consider the Müller-Breslau principle to develop the I.L. for shear, say at section D.

Section D is located at a distance a from left-hand support A. A unit load is placed at C, an arbitrary point on the beam. The displacement to be given to the beam is the displacement in the direction of the force quantity under consideration, in this case, the shear force. The desired displacement can be given if a pinned-bar connection is introduction at section D. This pinned-bar connection is schematically shown in Fig. 7.37*c*. This connection facilitates the required translation but prevents angular displacement between the two portions of the beam. The direction of displacement is same as the positive shear force.

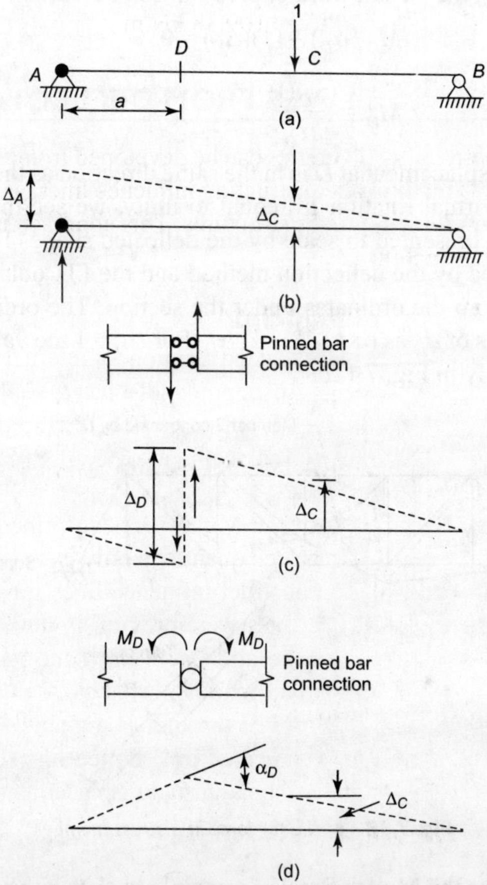

Fig. 7.37 | *(a) Beam, (b) I.L. for reaction R_A (c) I.L. for shear at D, (d) I.L. for moment at D*

The displaced beam diagram is shown in Fig. 7.37*c*. It may be noted that the two ends of the beam at section *D* must have the same slope representing continuity in transferring moment. With the displacements considered to be virtual, the virtual work equation is

$$V_D \,.\, \Delta_D - (1)(\Delta_C) = 0$$

or
$$V_D = \frac{\Delta_C(1)}{\Delta_D} \tag{7.73}$$

As before, if Δ_D is taken to be unity, the defected shape directly represents to scale the influence line for shear.

Influence line for the moment at section *D* can be obtained by inserting a pinned connection at the section under consideration and imposing rotations to the ends by applying a couple. Thus, if a pinned connection is introduced at section D as shown in Fig. 7.37d and the beam is rotated through an angle α_D at D by a couple M_D, the virtual work expression can be written as

$$M_D(\alpha_D) - (1)(\Delta_C) = 0$$

or

$$M_D = \frac{\Delta_C(1)}{\alpha_D} \tag{7.74}$$

The rotation displacement at D is in the same direction as the positive moment at D. Assuming virtual rotation α_D equal to unity, we see that the I.L. for the moment at D is represented to scale by the deflected shape. The equivalence of the values obtained by the deflection method and the I.L. obtained in Fig. 7.12c can be verified from the ordinates under the section. The ordinate under D can be written in terms of d_D as $(a/L)(L - a)\,\alpha_D$. For $\alpha_D = 1$ the value is same as that obtained previously in Fig. 7.12c.

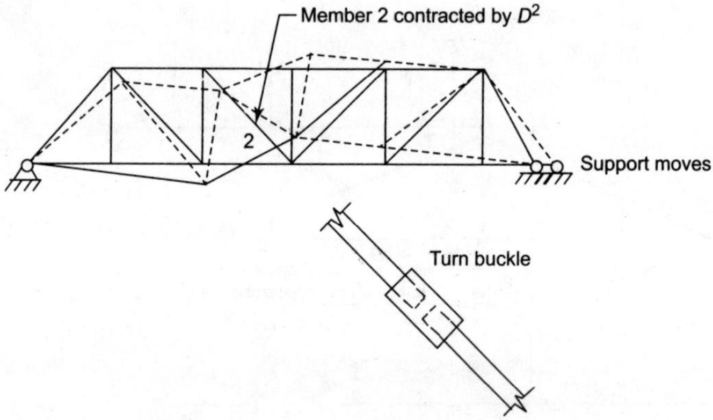

Fig. 7.38 | *I.L. for force in a struss member*

We may also use the Muller-Breslau principle to sketch the I.L. for bar forces in trusses. Consider the truss of Fig. 7.27 shown again in Fig. 7.38. The I.L. for the force in member 2 is determined by giving a displacement Δ_2 to the force in the member.

The desired displacement to the force in the member can be given by introducing a turn buckle connection somewhere on member 2. A consideration of the virtual work expression shows that for a unit displacement along member 2, the deflected shape of the bottom chord would represent to scale the I.L. for the force in member 2. A similar procedure can be followed for constructing influence lines for forces in other members. The reader may notice that the application of the Müller-Breslau principle to an ordinary truss is of no special advantage. However, influence lines can be constructed for a wide variety of structures such as space trusses, frames and continuous beams.

The use of deflected shapes for constructing influence lines for some complicated structures is even more helpful. We shall illustrate the procedure by solving a numerical example.

Example 7.26 | *It is required to construct influence lines for reaction R_A and R_B and for shear and moment at sections 1 and 2 for the balanced cantilever beam shown in Fig. 7.39a.*

Applying the Müller-Breslau principle, we obtain the influence line for the reaction at A by allowing the reaction to displace by a unit distance

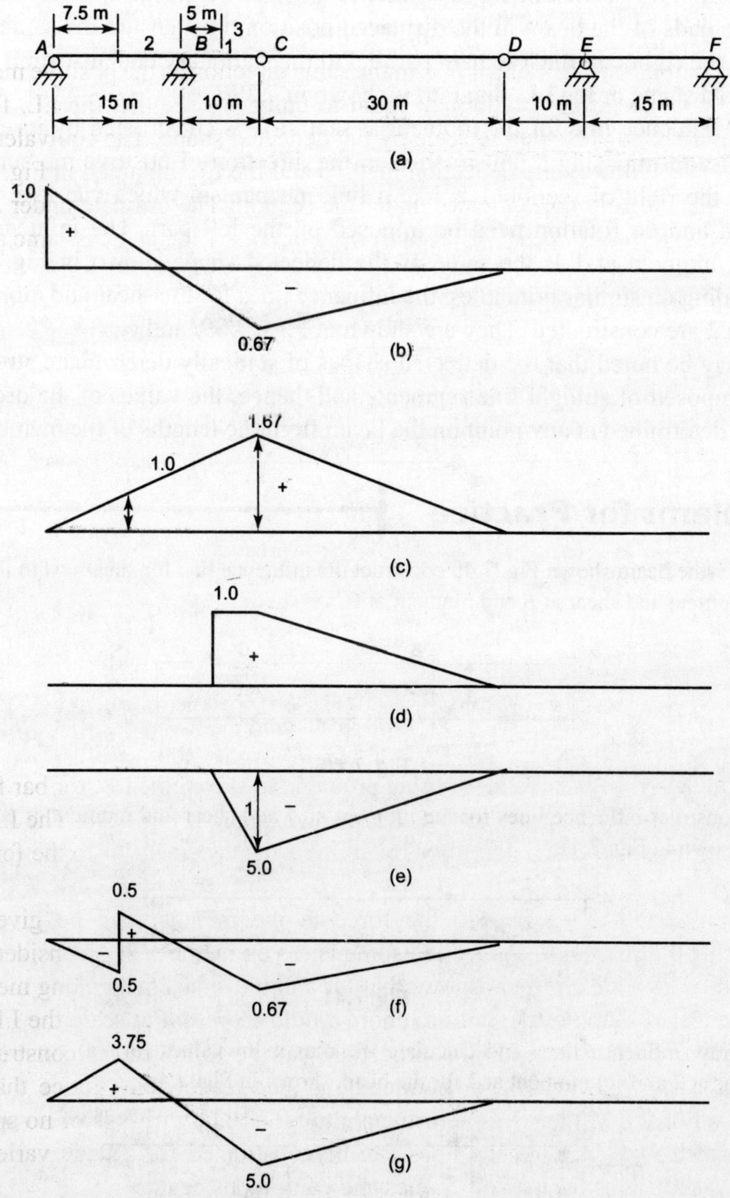

Fig. 7.39 | *(a) Balanced cantilever beam, (b) I.L. for reaction R_A, (c) I.L for reaction R_B, (d) I.L. for shear at 1, (e) I.L. for moment at 1, (f) I.L. for shear at 2, (g) I.L. for moment at 2*

The resulting deflection shape shown in Fig. 7.39*b* represents to scale the I.L. for the reaction at *A*. Proceeding in a similar manner, the I.L. for the reaction at *B* is constructed and is shown in Fig. 7.39*c*. The influence line for the shear at 1 is obtained by allowing positive shear at 1 to relatively displace a total of unity. The entire displacement is shown given to the right part of the section as the portion to the left of the section offers resistance to displacement. Again, it may be noted that the ends of the beam in the displaced position must ensure the same slope, that is, the right end must remain parallel to the stationary end at section 1. The displaced shape or the I.L. diagram is shown in 7.39*d*.

The influence line for the moment at section 1 is constructed by inserting a pin at 1 and imposing a unit rotation in the direction of positive moment. The part to the right of section 1 is like a link mechanism which can take all the rotation and no rotation need be imposed on the left part. The influence line for the moment at 1 is the same as the deflected shape shown in Fig. 7.39*e*. Proceeding on similar principles, the influence lines for the shear and moment at section 2 are constructed. They are shown in Figs. 7.39*f* and *g*.

It may be noted that the deflected shapes of statically determinate structures are composed of straight line segments and, hence, the values of the ordinates can be determined at any point on the beam from the lengths of the members.

Problems for Practice

7.1 For the beam shown Fig. 7.40 construct the influence line for shear just to the right of *A*, moment and shear at *B* and moment at *C*.

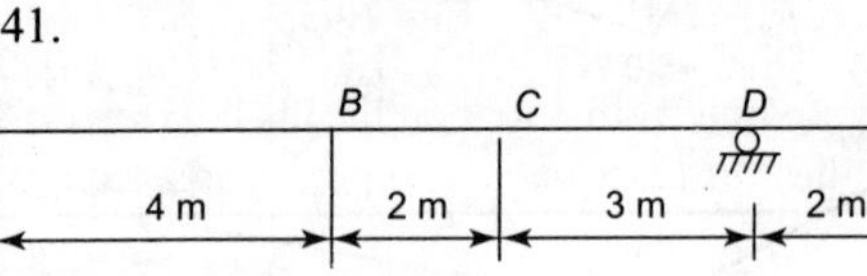

Fig. 7.40

7.2 Construct influence lines for the moment at *A* and shear and moment at *B* for the beam shown in Fig. 7.41.

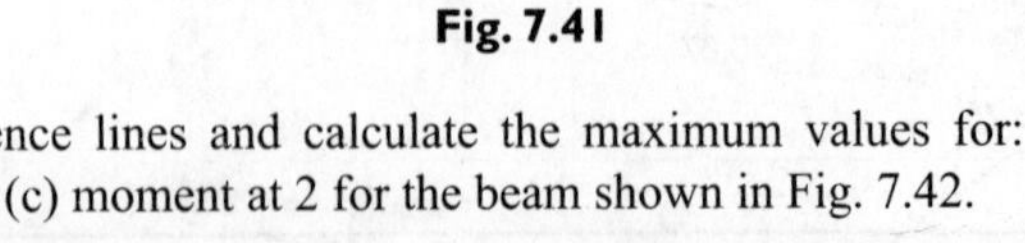

Fig. 7.41

7.3 Draw influence lines and calculate the maximum values for: (a) reaction at 1, (b) shear at 2 and (c) moment at 2 for the beam shown in Fig. 7.42.

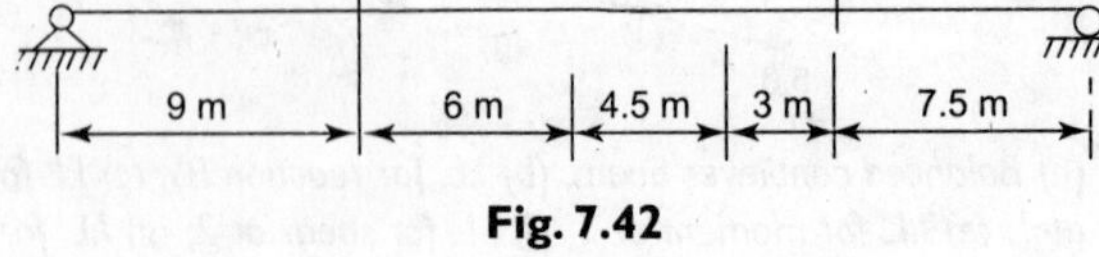

Fig. 7.42

7.4 For a unit load moving for A to C in Fig. 7.43 construct influence lines for reaction R_A, moment at support B and shear just to the left of B.

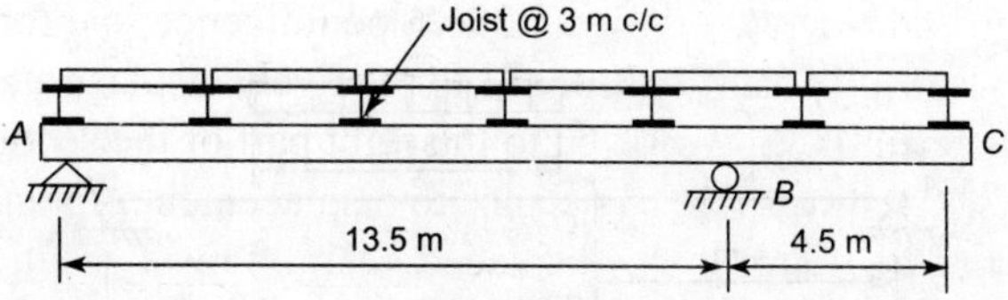

Fig. 7.43

7.5 Draw the influence line for the moment at A of the structure shown in Fig. 7.44 as a unit load moves from B to C.

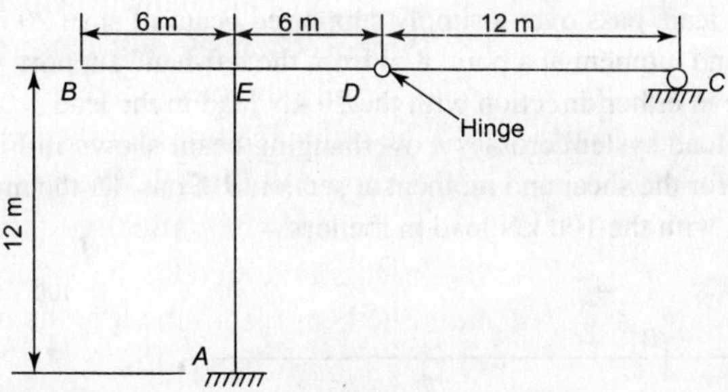

Fig. 7.44

7.6 For a unit load moving from A to E on the truss shown in Fig. 7.45 construct influence lines for members BC, CH and HJ.

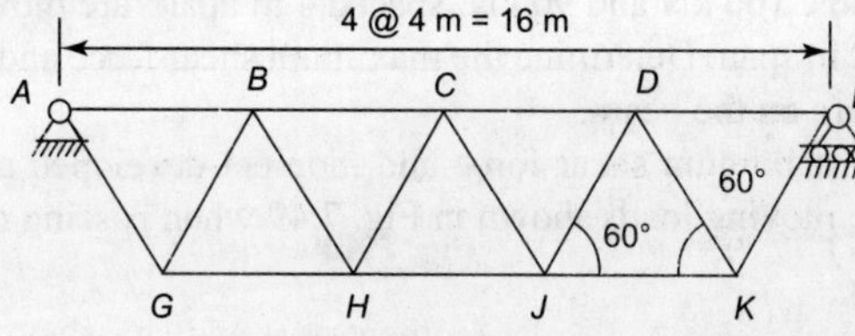

Fig. 7.45

7.7 For a unit load moving from 1 to 9 on the beam-truss structure shown in Fig. 7.46, construct influence lines for reaction at 1, moment at 2, and for forces in members 5–6, 5–12 and 11–12.

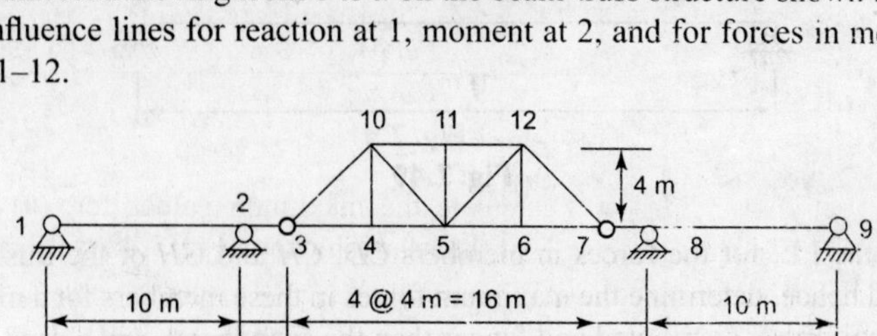

Fig. 7.46

7.8 Determine the maximum shear and moment at C of beam AB shown in Fig. 7.47. The intensity of a uniformly distributed load is 25 kN/m extending over a length of 3 m.

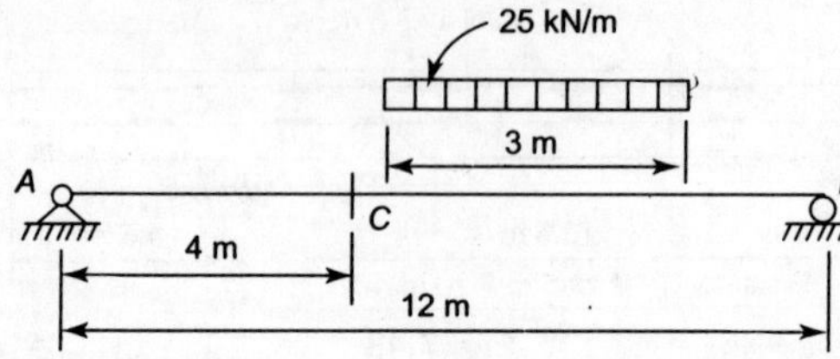

Fig. 7.47

7.9 Three wheel loads, 20, 80 and 80 kN, spaced 4 m apart from each other, with the 20 kN load in the lead, pass over a simply supported beam of span 20 m. Determine the maximum shear and moment at a point 8 m from the left-hand support. Consider that the loading can move in either direction with the 20 kN load in the lead.

7.10 The given load system crosses a overhanging beam shown in Fig. 7.48. Find the maximum values for the shear and moment at section B. Consider the movement of loads in either direction with the 100 kN load in the lead.

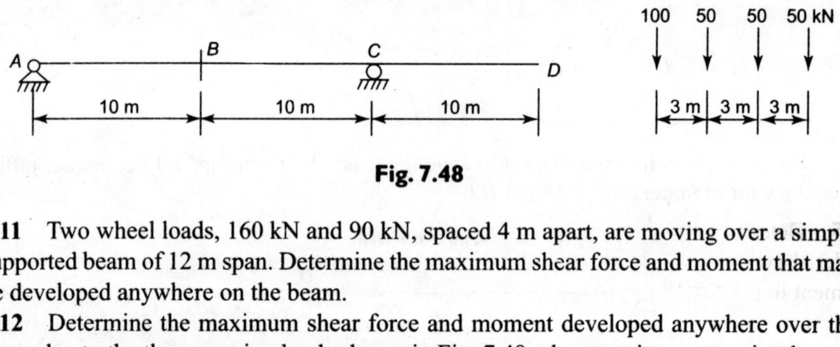

Fig. 7.48

7.11 Two wheel loads, 160 kN and 90 kN, spaced 4 m apart, are moving over a simply supported beam of 12 m span. Determine the maximum shear force and moment that may be developed anywhere on the beam.

7.12 Determine the maximum shear force and moment developed anywhere over the beam due to the three moving loads shown in Fig. 7.49 when passing over a simply supported beam.

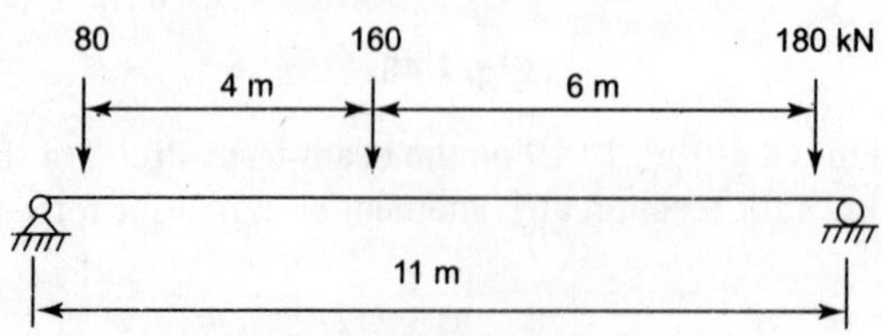

Fig. 7.49

7.13 Draw the I.L. for the forces in members CD, CH and GH of the truss shown in Fig. 7.50, and hence, determine the maximum forces in these members for a moving load of 40 kN/m uniformly distributed and longer than the span length and a dead load of 20 kN/m, covering the entire span.

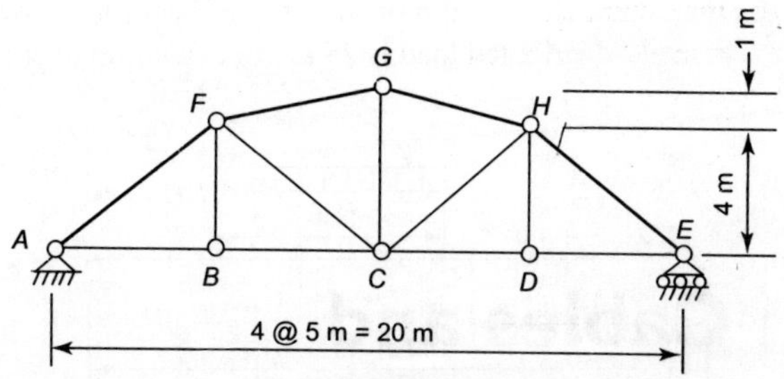

Fig. 7.50

7.14 Determine the maximum shear force and moment at a section 30 m from the left-hand support of a simply supported beam shown in Fig. 7.51. The loading system consists of four wheel loads followed by a distributed load of 15 kN/m extending over 10 m as shown. The load is reversible in direction.

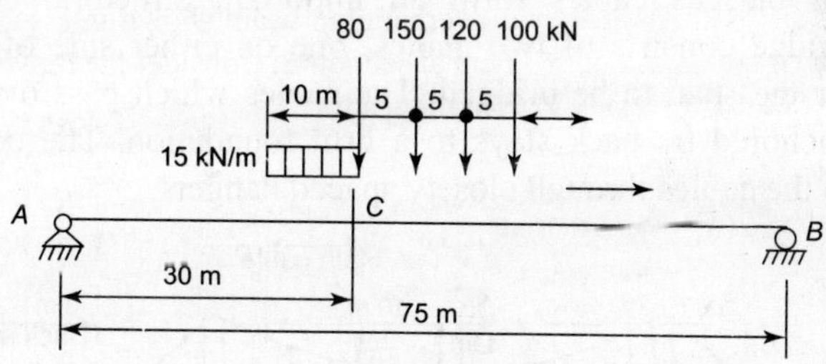

Fig. 7.51

7.15 Determine the maximum shear and moment developed at point C in the bridge girder shown in Fig. 7.52 by the truck loading indicated. What is the absolute maximum moment in the girder and where does it occur?

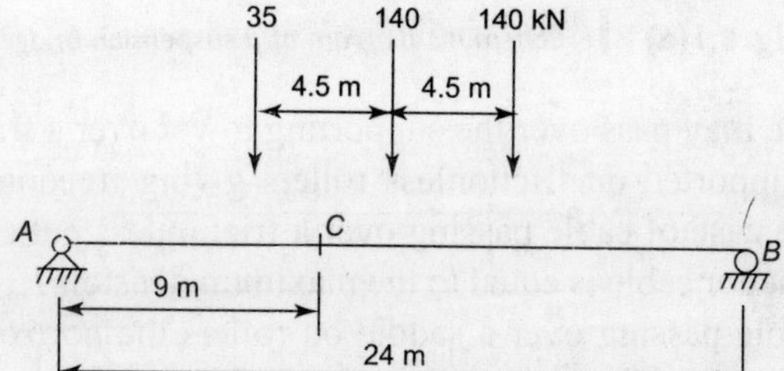

Fig. 7.52

7.16 A three-hinged circular arch has a span of 40 m and a rise of 5 m. Two point loads 160 kN and 80 kN spaced 5 m apart roll over the arch from left to right. Find the horizontal thrust and B.M. at a section 12 m from left-hand support when the 80 kN load is on the section.

8 Cables and Suspension Bridges

8.1 INTRODUCTION

In suspension bridges cables form an important structural component. A suspension bridge consists of two cables, one on either side of the roadway stretched over the span to be bridged. The cables which pass over supporting towers are anchored by back stays to a firm foundation. The deck loads are transmitted to the cables through closely spaced hangers.

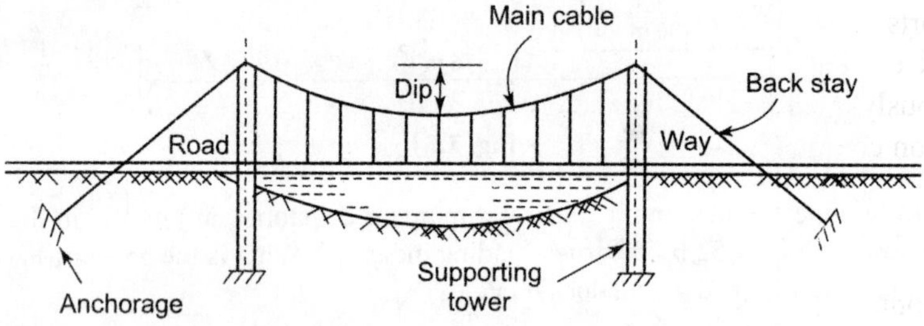

Fig. 8.1(a) | *Schematic diagram of a suspension bridge*

The main cable may pass over the supporting tower over a frictionless pulley or over saddle supported on frictionless rollers giving freedom for horizontal movement. In the case of cable passing over a frictionless pulley the maximum tension T_{max} in anchor cable is equal to the maximum tension T_{max} in main cable. In the case of cable passing over a saddle on rollers the horizontal component of cable tension is same in main and anchor cables. A schematic diagram of a suspension bridge is shown in Fig. 8.1(a) and the anchoring of cable over supporting piers is shown in Fig. 8.1 (b)

It has been shown in section 2.4 that a cable under the given loading takes the shape of a funicular polygon which represents to some scale the B.M. diagram of a simple beam under the same loading. If the number of hangers is large the load transmitted to the cables can be approximated to a uniformly distributed load for which the cable assumes the shape of a parabola similar to a B.M. diagram of a simple beam under u.d.l.

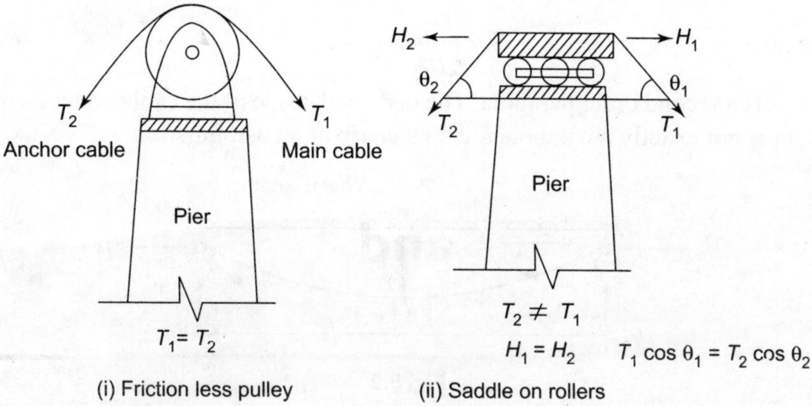

Fig. 8.1(b) | *Schematic diagram of a cable over piers*

8.2 | CABLES

8.2.1 Equation of the Cable

Let us now consider a cable subjected to a uniformly distributed loading. The loading determines the profile of the cable. A cable of span 1 suspended from supports A and B at the same level is shown in Fig. 8.2.

Let C be the lowest point of the cable and the sag of the cable at C be y_C. Obviously the point C is located midway between supports A and B. The vertical reaction components at supports A and B are,

$$V_A = V_B = \frac{wl}{2}$$

Horizontal reaction components $H_A = H_B = H$ can be obtained by taking moments about C and setting $M_C = 0$

$$M_C = V_A \frac{l}{2} - w\frac{l}{2}\frac{l}{4} - H\, y_c = 0$$

$$= \frac{wl^2}{4} - \frac{wl^2}{8} - H\, y_c = 0$$

$$\therefore \quad H = \frac{wl^2}{8y_C} \tag{8.1}$$

In order to find the shape of the cable, we write the equation for moment about a point P with coordinates x and y as shown in Fig. 8.2.

$$M_P = V_A\,(x) - \frac{wx^2}{2} - H \cdot y = 0$$

or

$$\frac{wl}{2}x - \frac{wx^2}{2} = \frac{wl^2}{8y_C}y$$

This gives $$y = \frac{4y_C}{l^2} x(l - x) \tag{8.2}$$

This is a second order parabola. The deflected shape of the cable under its own weight is not exactly a parabola but a catenary or a cosh function.

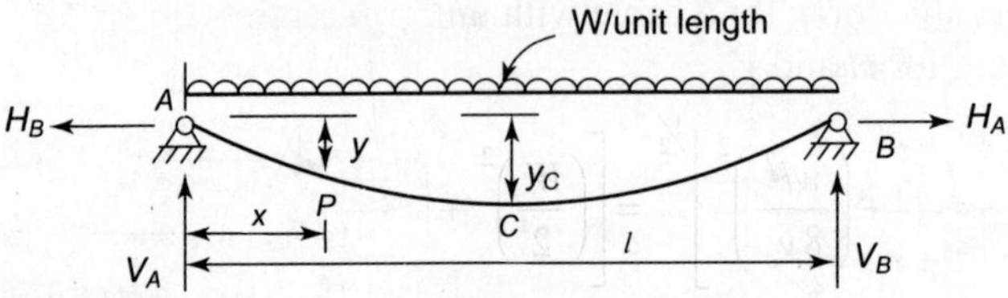

Fig. 8.2

However, in practice, the self weight of the cable is much smaller than the superimposed load and it is usually lumped together.

8.2.2 Horizontal Tension in the Cable

Considering again the cable in Fig. 8.2 subjected to a uniformly distributed load w/unit length, we can calculate horizontal tension in the cable H by taking moments about a point P as earlier. Hence,

$$M_P = \frac{wl}{2} x - \frac{wx^2}{2} - H\,y = 0$$

The first two terms in the above equation represent B.M. in a simply supported beam. The last term represents the moment caused by horizontal reaction H.

Therefore, $$M_P = \mu_P - H\,y = 0$$

or $$H = \frac{\mu_P}{y} \tag{8.3}$$

For a point C the lowest point in the cable

$$M_C = \mu_C - H\,y_C = 0$$

$$= \frac{wl^2}{8} - H\,y_C = 0$$

$$\therefore \quad H = \frac{wl^2}{8y_C} \tag{8.4}$$

From Eqn. 8.4 we see that the horizontal reaction H is inversely proportional to sag y_C. Thus for a flat cable structure, a large amount of horizontal reaction component is exerted on the supports. Further, when the cable is unusually flat, the change in sag resulting from a change in its cable length must also be considered.

The tension in the cable varies along its length. The maximum cable tension occurs at supports since H is constant all along and vertical reaction component V_A or V_B is maximum at the supports.

Thus $$T_A = \sqrt{V_A^2 + H^2} \text{ or } T_B = \sqrt{V_B^2 + H^2}$$

is the design cable force for a cable with uniform section.

Substituting for H and V_A

$$T_{\max} = \left[\left(\frac{wl}{2}\right)^2 + \left(\frac{wl^2}{8y_c}\right)^2\right]^{1/2} = \left[\left(\frac{W}{2}\right)^2 + \left(\frac{W}{8y_c}\right)^2\right]^{1/2} = \frac{W}{2}\left[\left(1 + \frac{l^2}{16y_c^2}\right)\right]^{1/2}$$

Example 8.1 *A suspension cable 140 m span and 14 m central dip carries a load of 1 kN/m. Calculate the maximum and minimum tension in the cable. Find the horizontal and vertical forces in each pier under the following conditions:*

(a) If the cable passes over a frictionless rollers on top of the piers

(b) If the cable is firmly clamped to saddles carried on frictionless rollers on top of the piers.

In each case the back stay is inclined at 30° with the horizontal.

Using the relation $$H = \frac{\omega l^2}{8y_c} = \frac{1 \times 140 \times 140}{8 \times 14} = 175 \text{ kN}$$

Vertical reaction $$V = \frac{\omega l}{2} = \frac{1 \times 140}{2} = 70 \text{ kN}$$

Maximum Tension $$T_{\max} = \sqrt{175^2 + 70^2} = 188.48 \text{ kN}$$

Shape of the cable is a parabola, and the equation taking the origin at top of pier is

$$y = \frac{4y_c}{l^2} x(l - x)$$

The slope of the cable $$= \frac{dy}{dx} = 4\frac{y_c}{l^2}(l - 2x)$$

Slope of the cable at support $$\frac{dy}{dx} = 4\frac{y_c}{l} \quad \text{at } x = 0$$

$$\tan\theta = \frac{4 \times 14}{140} = 0.40$$

$$\theta = 21° 48'$$

(a) Horizontal pull on the pier

$$H = T_{\max}(\cos 21° 48' - \cos 30°)$$

$$= 188.48\,(0.9285 - 0.8660)$$

$$H = 11.49 \text{ kN}$$

Vertical pressure on pier

$$V = T_{max}\,(\sin 21°\,48' + \sin 30°)$$
$$= 188.48\,(0.3697 + 0.5000)$$
$$V = 163.98 \text{ kN}$$

(b) In the case of saddle on rollers $H_1 = H_2$

or $$T_1 \cos 21°\,48' = T_2 \cos 30°$$

$$T_2 = \frac{T_1 \cos 21°\,48'}{\cos 30°}$$
$$= 201.97 \text{ kN}$$
$$H_1 = H_2 = 175.0 \text{ kN}$$

Vertical pressure on pier $= T_1 \sin 21°\,48' + T_2 \sin 30°$

$$= 188.48\,(0.3697) + 201.97\,(0.50)$$
$$= 170.73 \text{ kN}$$

8.2.3 Tension in Cable Supported at Different Levels

Consider cable ACB stretched between two supports A and B at different levels and subjected to a uniformly distributed load as shown in Fig. 8.3. Let C be the lowest point on the cable at distance l_1 and l_2 from supports A and B respectively. As pointed out earlier, the horizontal component H of the tension in the cable is same everywhere as it is self-balancing in the absence of any horizontal loads.

Taking moments about C, we can write

$$H_A = \frac{w(2l_1)^2}{8y_C} \quad \text{and} \quad H_B = \frac{w(2l_2)^2}{8(y_C + d)}$$

Equating $H_A = H_B$ we get

$$H = \frac{wl_1^2}{2y_C} = \frac{wl_2^2}{2(y_C + d)}$$

or $$\frac{l_1^2}{l_2^2} = \frac{y_C}{y_C + d} \tag{8.5}$$

This relationship locates the lowest point C on the cable from support A or B.

Then we can write

$$H = \frac{w(2l_1)^2}{8y_C} = \frac{w(2l_2)^2}{8(y_C + d)} \tag{8.6}$$

The vertical rections V_A or V_B can be calculated by taking moment about B or A. Taking moment about B,

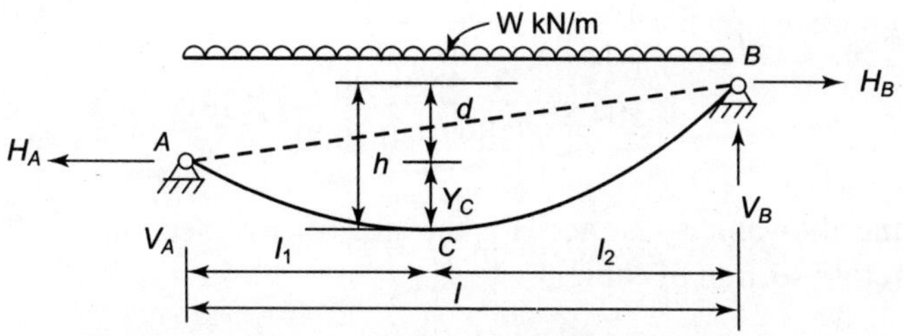

Fig. 8.3 | *Cable supported at ends at different levels*

$$V_A(l) - wl\frac{l}{2} + H\,d = 0$$

$$\therefore \quad V_A = \frac{wl}{2} - H\frac{d}{l} \tag{8.7}$$

Similarly, taking moment about *A*

$$V_B(l) - \frac{wl^2}{2} - H \cdot d = 0$$

$$\therefore \quad V_B = \frac{wl}{2} + H\frac{d}{l} \tag{8.8}$$

Tension in the cable at A,

$$T_A = \sqrt{V_A^2 + H_A^2}$$

And tension in the cable at *B*,

$$T_B = \sqrt{V_B^2 + H_B^2}$$

Example 8.2 | *A foot bridge of width 3 m and span 50 m is carried by two cables of uniform section having a central dip of 5 m. If the platform load is 5 kN/m² calculate the maximum pull in the cables. Find the necessary section area required if the allowable stress is 120 N/mm².*

Total load on each cable $W = \frac{1}{2}$ (3) (5) (50)

$$= 375 \text{ kN}$$

using the relation $\quad H = \frac{wl^2}{8y_C} = \frac{Wl}{8y_C}$

$$H = \frac{375 \times 50}{8 \times 5} = 468.75 \text{ kN}$$

Vertical reaction $\quad V_A = V_B = \frac{W}{2} = \frac{375}{2} = 187.5 \text{ kN.}$

Now using the relation $\quad T_{max} = \sqrt{V^2 + H^2}$

$$T_{max} = \sqrt{187.5^2 + 468.75^2}$$
$$= 504.86 \text{ kN.}$$

Equating $\quad T_{max} = fA$

where f is the allowable stress and A is the area of cross section,

Area of cross section of cable,

$$A = \frac{T_{max}}{f} = \frac{504.86 \times 1000}{120} = 4207.17 \text{ mm}^2$$

Example 8.3 | *A cable of uniform cross-sectional area is stretched between two supports 100 m apart with one end 4 m above the other as shown in Fig. 8.4. The cable is loaded with a uniformly distributed load of 10 kN/m and the sag of the cable measured from the higher end is 6 m. Find the horizotal tension in the cable. Also determine the maximum tension in the cable.*

Span $\quad l = 100$ m

$d = 4$ m

and $\quad h = 6$ m

We know $\quad \dfrac{l_1}{l_2} = \left(\dfrac{y_C + d}{y_C}\right)^{1/2} = \left(\dfrac{6}{2}\right)^{1/2} = 1.732$

$\therefore \quad l_1 = 1.732\, l_2$

Substituting in $\quad l_1 + l_2 = 100$

$1.732\, l_2 + l_2 = 100$

$$l_2 = \frac{100}{2.732} = 36.6 \text{ m}$$

$\therefore \quad l_1 = 63.4$ m

Using the relation $\quad H = \dfrac{w l_1^2}{2(y_C + d)}$

$$H = \frac{10 \times 63.4^2}{2 \times 6} = 335.0 \text{ kN}$$

Maximum tension in the cable

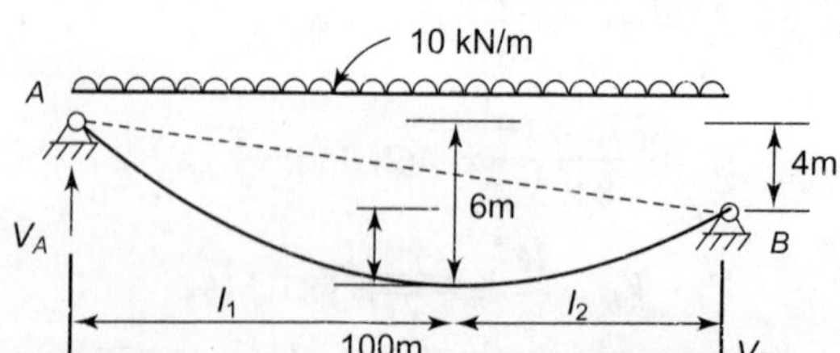

Fig. 8.4

$$V_A = \frac{wl}{2} + \frac{Hd}{l}$$

$$= \frac{10 \times 100}{2} + \frac{335 \times 4}{100} = 513.4 \text{ kN.}$$

$$T_{\max} = \sqrt{V_A^2 + H^2}$$

$$= \sqrt{513.4^2 + 335.0^2} = 613.0 \text{ kN.}$$

8.2.4 Length of the Cable

(i) Cable Supports at the Same Level

Consider a cable *ACB* supported at *A* and *B* at the same level and carrying a u.d.l. as shown in Fig. 8.5.

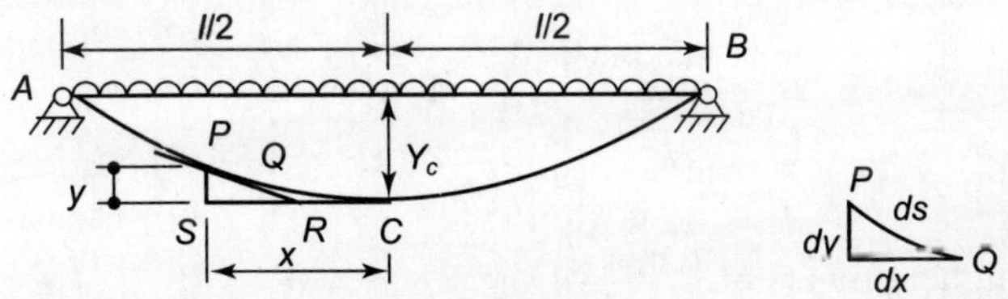

Fig. 8.5

Let *C* be the lowest point on the cable. Taking *C* as the origin, consider a point *P* (x, y) on the cable. Draw a tangent to the cable at *P* meeting the horizontal line *CS* at *R*. From the geometry of the curve, we know

$$CR = RS = \frac{x}{2}$$

A little consideration will show that the part CP of the cable is in equilibrium under the forces,

(1) Horizontal pull, *H*

(2) Downward load, w/unit length

(3) Tension in the cable, *T*

Triangle *PSR* represents to some scale the triangular forces which are in equilibrium. We can write

$$\frac{PS}{wx} = \frac{RS}{H} = \frac{RP}{T}$$

or

$$\frac{y}{wx} = \frac{x}{2H}$$

$\therefore$

$$y = \frac{wx^2}{2H} \tag{8.9}$$

Consider now an elemental length of curve *ds* between two points *P* and *Q*. Taking the length of the arc *PQ* equal to the length of chord *PQ* as it is a flat curve, we can write

$$ds = \sqrt{(dx)^2 + (dy)^2}$$

$$= dx\sqrt{1+\left(\frac{dy}{dx}\right)^2} \tag{8.10}$$

From Eqn. 8.9 we get $\frac{dy}{dx} = \frac{wx}{H}$

Substituting for $\frac{dy}{dx}$ in Eqn. 8.10

$$d_s = dx\sqrt{1+\left(\frac{wx}{H}\right)^2}$$

If $\frac{wx}{H}$ is a fraction, expanding by the binomial theorem, we can writ

$$\left(1+\frac{w^2x^2}{H^2}\right)^{1/2} = 1+\frac{1}{2}\frac{w^2x^2}{H^2}+\cdots$$

Neglecting higher powers of $\left(\frac{w^2x^2}{H^2}\right)$

$$ds = \left(1+\frac{w^2x^2}{H^2}\right)dx$$

Integrating over a range $x = 0$ to $x = l/2$

$$\int_0^{l/2} ds = \int_0^{l/2}\left(1+\frac{w^2x^2}{2H^2}\right)dx$$

$$\therefore \qquad \frac{L}{2} = \left[x+\frac{w^2x^3}{6H^2}\right]^{l/2}$$

$$= \frac{l}{2}+\frac{w^2l^3}{48H^2}$$

Total length $\qquad L = \left[l+\frac{w^2l^3}{24H^2}\right]$

Substituting for $H = \frac{wl^2}{8y_C}$

$$L = l+\frac{8}{3}\frac{y_C^2}{l} \tag{8.11}$$

(ii) Cable Supports at Different Levels

The length of a cable supported at different levels can be obtained by locating the lowest point C on the cable as earlier. Consider the cable supported at A and B at different levels having span length l as shown in Fig. 8.6.

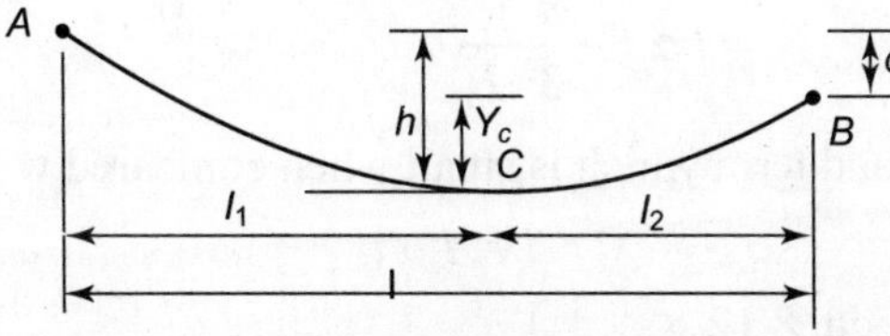

Fig. 8.6

Then length of cable $AC = \frac{1}{2}$ (the length of the cable having sag y_C and span $2l_1$)

Similarly length of cable $CB = \frac{1}{2}$ (the length of the cable having sag y_C and span $2l_2$)

Total length of the cable $AB = AC + CB$

$$L = \frac{1}{2}\left(2\,l_1 + \frac{8}{3}\frac{h^2}{2l_1}\right) + \frac{1}{2}\left(2l_2 + \frac{8}{3}\frac{y_c^2}{2l_2}\right)$$

$$L = l_1 + l_2 + \frac{4}{3}\left(\frac{h^2}{l_1} + \frac{y_c^2}{l_2}\right)$$

$$\therefore \quad L = l + \frac{4}{3}\left(\frac{h^2}{l_1} + \frac{y_c^2}{l_2}\right) \tag{8.12}$$

8.2.5 Effect on Cable Due to Change of Temperature

Consider a cable stretched between two supports A and B at the same level having sag y_C at the centre. When there is a rise in temperature, the length of the cable increases and so also the sag as the supports do not undergo any displacement. The change of the sag can be evaluated easily. Since the length of the cable,

$$L = l + \frac{8}{3}\frac{y_c^2}{l}$$

$$\delta L = \frac{16}{3}\frac{yc}{l}\,\delta\, y_c$$

or

$$\delta y_c = \frac{3}{16}\frac{l}{y_c}\,\delta\, L \tag{8.13}$$

If α is the coefficient of expansion and t°C is the rise in temperature, the length of the cable L changes to L_1, where

$$L_1 = L + L\,\alpha\, t = L\,(1 + \alpha\, t)$$

change in length $= L_1 - L = L\,\alpha\,t$

$$= (l + \frac{8}{3}\frac{y_c^2}{l})\,\alpha\,t$$

$$= l\,\alpha\,t + \frac{8}{3}\frac{y_c^2}{l}\,\alpha\,t$$

Neglecting the second term which is small when compared with $l\,\alpha.\,t$

$$L = l\,\alpha\,t$$

Substituting in Eqn. 8.12

$$y_c = \frac{3}{16}\frac{l}{y_c}(l\,\alpha\,t) = \frac{3}{16}\frac{l^2}{8\,y_c}\,\alpha\,t$$

Similarly a drop in temperature results in decrease in cable length L and hence the dip.

we know that the horizontal tension.

$$H = \frac{wl^2}{8\,y_c}$$

or $$H \propto \frac{1}{y_c}$$

and $$\frac{\delta H}{H} = -\frac{\delta\,y_c}{y_c}$$

The minus sign indicates that when dip increases the horizontal tension H decreases.

The maximum tension induced in a cable can be calculated from $T_{max} = \sqrt{V^2 + H^2}$

If the dip is small V will be small when compared with H.

We can write, $T_{max} = H$.

The stress in the cable,

$$f = \frac{T_{max}}{A} = \frac{H}{A}$$

We know the value H varies inversely with y_c

$$f \propto H \propto \frac{1}{y_c} \quad \text{or} \quad f \propto \frac{1}{y_c}$$

Again $$\delta f \propto -\frac{\delta\,y_c}{y_c^2} \quad \therefore \frac{\delta f}{f} = -\frac{\delta\,y_c}{y_c}$$

The minus sign again indicates that with an increase in cable dip the stress in the cable decreases.

Example 8.4 *A cable is suspended and loaded as shown in Fig. 8.7.*

(a) Compute the length of the cable.

(b) Compute the horizontal component of tension H, in the cable.

(c) Determine the magnitude and the position of maximum tension in the cable.

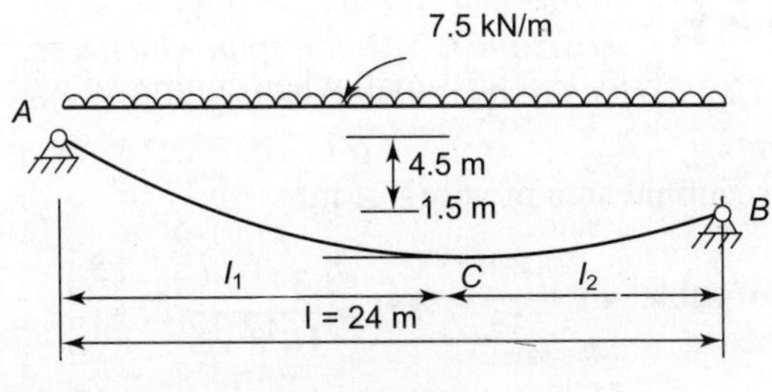

Fig. 8.7

Let C be the lowest point on the cable
L be the length of the cable ACB
l_1 be the horizontal distance from A to C
l_2 be the horizontal distance from C to B

Then using the relation

$$\frac{l_1}{l_2} = \left(\frac{y_c + d}{y_c}\right)^{1/2} = \left(\frac{6}{1.5}\right)^{1/2}$$

$$\therefore \quad \frac{l_1}{l_2} = 2.$$

We know $l_1 + l_2 = 24$

$$2l_2 + l_2 = 24.$$

$$l_2 = 8 \text{ m}$$

and $\quad l_1 = 16$ m

Now using the Eqn. $\quad L = l + \frac{4}{3}\frac{(y_c + d)^2}{l_1} + \frac{4}{3}\frac{y_c^2}{l_2}$

$$= 24 + \frac{4}{3}\frac{(6)^2}{16} + \frac{4}{3}\frac{(1.5)^2}{8} = 27.38 \text{ m}$$

Horizontal tension $\quad H = \frac{wl_1^2}{2(y_c + d)} = \frac{7.5 \times 16^2}{2 \times 6} = 160 \text{ kN}.$

We know that the vertical reaction at A, the higher support, is

$$V_A = \frac{wl}{2} = \frac{Hd}{l} = \frac{7.5 \times 24}{2} + \frac{160 \times 4.5}{24} = 120.0 \text{ kN}.$$

Now using the relation $T_{max} = \sqrt{H^2 + V_A^2}$

$$= \sqrt{160^2 + 120^2} = 200.0 \text{ kN.}$$

Example 8.5 *If the central dip is limited to 1/12 span find the maximum horizontal span which a steel wire of uniform cross section may have with the stress not exceeding 120 N/mm². Take unit weight of steel = 78 kN/m³.*

Let the cross-sectional area of wire = A m²

$$\text{Length of cable } = l + \frac{8}{3}\frac{y_c^2}{l} = l + \frac{8}{3} \times \frac{1}{12}\frac{l}{12} = \frac{55}{54} l$$

$$\text{Total weight of cable: } A\frac{55l}{54} \times 78 = 79.44\ \text{A}l = W$$

$$\text{Horizontal tension } H = \frac{wl^2}{8y_c} = \frac{Wl}{8y_c} = \frac{W}{8} \times 12 = \frac{3}{2}W$$

We also know that the vertical reaction $V_A = V_B = \dfrac{W}{2}$

Substituting in the Eqn. $T_{max} = \sqrt{V^2 + H^2}$

$$= \left(\frac{W^2}{4} + \frac{9}{4}W^2\right)^{1/2} = 1.581 \text{ W}$$

Let f be the stress in the wire

Then $f.A = 1.581\ W$

$f.A = 1.581 \times 79.44\ Al$

Substituting for $f = 120 \times 10^6$ kN/m², $l = \dfrac{120 \times 10^6}{1.581 \times 79.44} = 955.45$ m

Example 8.6 *A steel cable of 10 mm diameter is stretched across two poles 75 m apart. If the central dip is 1 m at a temperature of 15°C calculate the stress intensity in the cable. Also calculate the fall in temperature necessary to raise the stress to 70 N/mm². Take unit weight of steel $v = 78$ kN/m³ and $\alpha = 12 \times 10^{-6}$ °C*

Step 1: To find the weight of cable W

Span of cable $l = 75$ m

dip $y_c = 1$ m

Weight of cable $W = AL\,v$

in which $A = 78.54 \times 10^{-6}\ \text{m}^2$

$$L = \left(75 + \frac{8}{3} \times \frac{l^2}{75}\right) = 75.04\ \text{m}$$

and $v = 78.0\ \text{kN/m}^3$

$\therefore$ $W = 78.54 \times 10^{-6} \times 75.04 \times 78$

$$= 0.4597\ \text{kN}.$$

Step 2: T evaluate H and T

We calculate

$$H = \frac{W\,l}{8\,y_c} = W \times \frac{75}{8} = 9.375\ W$$

$$V = \frac{W}{2}$$

$$T_{\text{max}} = \sqrt{H^2 + V^2} = W\sqrt{9.375^2 + (1/2)^2}$$

$$= 9.3883\ W$$

Maximum stress $f = \dfrac{9.3883 \times 0.4597}{78.54 \times 10^{-6}}\ \text{kN/m}^2 = 54.94\ \text{N/mm}^2$

Step 3: To find change in dip

Let t°C be the fall in temperature.

Change in dip $\delta y_c = \dfrac{3}{16}\dfrac{l^2}{y_c}\alpha\,t$

$$= \frac{3}{16} \times \frac{75 \times 75}{1}(12 \times 10^{-6})\,t = 12656 \times 10^{-6}\,t$$

we know $\dfrac{\delta f}{f} = \dfrac{\delta y_c}{y_c}$

$$\left(\frac{70.0 - 54.95}{54.95}\right) = \frac{12656 \times 10^{-6}\,t}{(1)}$$

$\therefore$ $t = 21.64°\text{C}$

8.3 STIFFENING GIRDERS

We have seen that the cable takes the shape of a bending moment diagram in a simply supported beam under the given system of loads. When rolling loads pass across the bridge, the bending moment diagram changes with the position of loads and hence the profile of the cable. The bridge deck suspended from such cables will swing up and down and will not be useful. In order to make the cable retain its parabolic shape throughout the passage of loads, it is necessary to transmit the moving loads to the suspension cable as uniformly distributed load.

This is achieved by providing stiffening girders on either side of the roadway. The girders may be of three-hinged or two-hinged and are suspended from the cables through hanger-cables. The roadway is then provided on the stiffening girder.

The uniformly distributed dead load of the roadway and the stiffening girders is transmitted to the cables through hanger cables and is taken up entirely by the tension in the cables. The stiffening girders do not suffer any S.F. or B.M. under dead load as the girders are supported by closely spaced hanger cables throughout. Any live load on the bridge will be transmitted to the girders as point loads. The stiffening girders transmit the live load to the cable as uniformly distributed load. While doing so the stiffening girders will be subjected to S.F. and B.M. throughout their length.

8.4 THREE-HINGED STIFFENING GIRDER

A suspension bridge with three-hinged stiffening girder is shown in Fig. 8.8. The roadway is carried by two stiffening girders each hinged at the ends and at the centre of span.

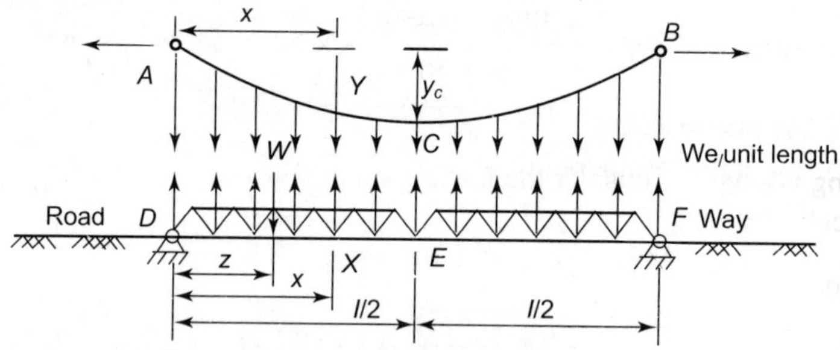

Fig. 8.8 | *Three-hinged stiffening girder*

Dead Load As already mentioned, the weight of girders and the roadway is transmitted to the cables through the hanger cables and the load is entirely taken by the tension in the cable. The stiffening girder which is being held all along by the suspenders will not be susjected to any S.F. or B.M. at any section on account of uniformly distributed dead load.

Live Load Any rolling load that crosses the bridge will be transmitted to the cables through the girders and hanger cable as uniformly distributed load so that the cables will retain their parabolic shape while the girder will have to resist a certain amount of S.F. and B.M. due to rolling loads. Let us consider the following types of rolling loads.

8.4.1 Single Concentrated Load

Let us consider a concentrated load W moving from left to right as in Fig. 8.8. ACB is the suspension cable, and DEF is the three-hinged stiffening girder.

At a particular instant let the load W be at a distance z from support D. Let us consider a section X at a distance x from A or D.

Cable Taking A as the origin the equation of the cable profile may be written as

$$y = \frac{4\,y_c}{l^2}\,x\,(l - x)$$

Let w_e be the equivalent u.d.l. on the cable. The vertical reaction at A or B, $V = w_e l/2$. The horizontal reaction component is given by,

$$H = \frac{w_e l^2}{8\,y_c}$$

Maximum tension in the cable occur at A or B.

$$T_{\max} = \sqrt{V^2 + H^2}$$

Now, consider a point $P(x, y)$ on the cable. Taking moments about P we can write

$$M_P = \frac{w_e\,l}{2}(x) - \frac{w_e\,x^2}{2} - H\,y = 0$$

$$\therefore \qquad H\,y = \frac{w_e\,lx}{2} - \frac{w_e\,x^2}{2} \qquad (8.14)$$

Stiffening Girder Consider the loading on the girder.

Vertical reaction at D,

Due to load W, $$V_D = \frac{W\,(l - z)}{l}$$

Due to upward w_e, $$V_D = \frac{W_e\,l}{2}$$

Similarly vertical reaction at F,

due to W $$V_F = \frac{Wz}{l}$$

due to w_e $$V_F = \frac{W_e l}{2}$$

B.M. at section X taking moment of all the forces to the left of section.

$$M_X = \{V_D\,x - W\,(x - z)\} - \left\{\frac{w_e l}{2}(x) - \frac{w_e x^2}{2}\right\} \quad \text{for } z \le x \qquad (8.15)$$

$$M_X = (V_D x) - \left\{w_e\,\frac{l}{2}(x) - w_e\,\frac{x^2}{2}\right\} \quad \text{for } z \ge x \qquad (8.16)$$

The first term in the above equations is evidently the equation for B.M. at section X in a simply supported beam subjected to loading W. The second term in both the equations is equal to Hy as in Eqn. 8.14

$\therefore \qquad M_x = \mu_x - H\,y$

in which

μ_x = Moment due to W at section X in a simply supported beam.
H = Horizontal tension in the cable
y = Ordinate of the cable profile at section X.

Evaluation of *H* in Terms of (*W*) We can evaluate H taking moments about hinge point at centre E of the girder.

$$M_E = \mu_E - H\,y_c = 0$$

$$\therefore \qquad H = \frac{\mu_E}{y_c}$$

Writing down equation for moment,

$$\mu_E = V_F\,\frac{l}{2} = \frac{Wz}{l}\,\frac{l}{2} = \frac{Wz}{2} \quad \text{for} \quad 0 \le z \le x$$

and

$$\mu_E = V_D\,\frac{l}{2} = \frac{W(l-z)}{l}\,\frac{l}{2} = \frac{W(l-z)}{2} \quad \text{for} \quad x \le z \le l$$

Substituting for μ_E in Eqn. (8.16)

$$H = \frac{Wz}{2\,y_c} \quad \text{for} \quad 0 \le z \le x \tag{8.17}$$

or

$$H = \frac{W(l-z)}{2\,y_c} \quad \text{for} \quad x \le z \le l \tag{8.18}$$

We can evaluate w_e by equating

$$H = \frac{w_e\,l^2}{8\,y_c} = \frac{\mu_E}{y_c} = \frac{W\,z}{2y_c} \quad \text{for} \quad 0 \le z \le l \tag{8.19}$$

$$\therefore \qquad w_e = \frac{4\,Wz}{l^2} \tag{8. 20}$$

and

$$w_e = \frac{4\,W(l-z)^2}{l^2} \quad \text{for} \quad x \le z \le l \tag{8.21}$$

8.4.2 Influence Line for *H*

The value of H varies with the position of the load. For the load position in the region D to E the variation of H is linear as in Eqn. 8.17 and again the variation of H is linear for the load position in the region E to F as in Eqn. 8.18. At $z = l/2$ the value of

$$H = \frac{W\,l}{4\,y_c} \tag{8.22}$$

It is seen that the value of H reaches maximum when the rolling load is at the centre. The I.L. diagram for H is shown in Fig. 8.9*b*.

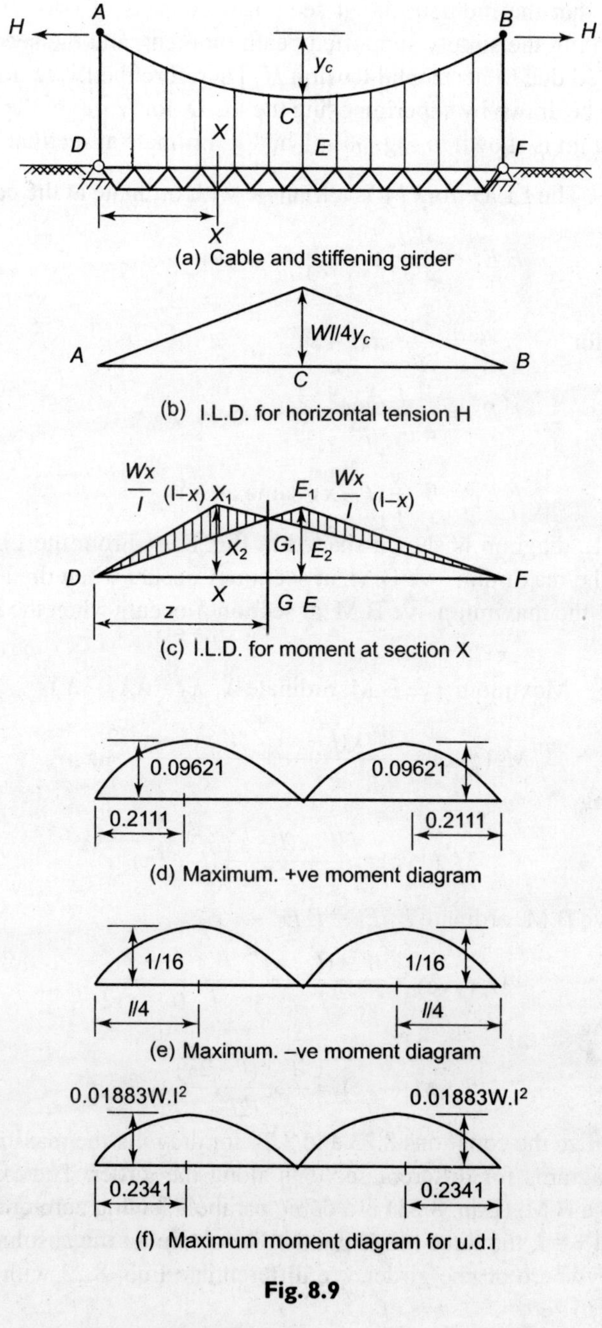

Fig. 8.9

8.4.3 I.L. for B.M. at Section *Y*

The moment expression for B.M. at section *X* is

$$M_x = \mu_x - H\,y$$

It is clear that the moment M_x at section X consists of two terms, the first term representing the simply supported beam moment, and the second term the moment caused due to horizontal tension H. Therefore the I.L.D. for moment at section X can be drawn by superimposing the I.L.D. for $H{\cdot}y$ over the I.L.D. for μ_x. The I.L. diagram is shown in Fig. 8.9c. The I.L. ordinate at section X for μ_x, XX_1 $= W\dfrac{x(l-x)}{l}$. The I.L.D. for $H{\cdot}y$ is a triangle with ordinate at the centre,

$$E E_1 = \frac{W l}{4 y_c} y$$

Substituting for $\quad y = \dfrac{4 y_c}{l^2} x(l-x)$

$$E E_1 = \frac{W l}{4 y_c} \frac{4 y_c}{l^2} x(l-x)$$

$$E E_1 = W \frac{x}{l}(l-x) \text{ same as } XX_1$$

The net I.L. diagram is shown shaded in Fig. 8.9c. From the I.L. diagram it is clear that the maximum +ve B.M. at section X occurs when the load is on the section itself; the maximum –ve B.M. at section X occurs when the load is at the centre.

$\therefore \quad$ Maximum +ve B.M. ordinate $X_1 X_2 = XX_1 - XX_2$

or
$$M_x\,(+\text{ve}) = \frac{W x(l-x)}{l} - \frac{W x(l-x)}{l} \frac{2}{l} x$$

Simplifying

$$M_x\,(+\text{ve}) = \frac{W x(l-x)}{l^2}(l-2x) \tag{8.23}$$

Maximum –ve B.M. ordinate $E_1 E_2 = E E_1 - E E_2$

or
$$M_x\,(-\text{ve}) = \frac{W x(l-x)}{l} - \frac{Wx}{l} \frac{(l-x)l}{(l-x)2}$$

Simplifying

$$M_x\,(-\text{ve}) = \frac{W x(l-2x)}{2l} \tag{8.24}$$

We can utilize the equations 8.23 and 8.24 for drawing the maximum +ve and –ve B.M. diagrams for different sections along the girder. The expression for maximum +ve B.M. (Eqn. 8.23) is a cubic parabola having zero ordinates at $x = 0$, $x = l/2$ and $x = l$, the three hinge points. For obtaining the absolute maximum +ve B.M. anywhere on the girder, we differentiate Eqn. 8.23 with respect to x and equate it to zero.

$$\frac{d M_x}{d x} = 6x^2 - 6l\,x + l^2 = 0$$

$\therefore$
$$x = \frac{l}{2} \pm \frac{l}{2\sqrt{3}}$$

That is, $x = 0.211\, l$ or $0.789\, l$ (8.25)

Substituting the value for x in Eqn. 8.23 the absolute maximum B.M. M_{max} (absolute) = 0.0962 Wl. The maximum +ve B.M. diagram is shown in Fig. 8.9d.

Similarly, the expression for maximum –ve moment at section (Eqn. 8.24) is a second degree parabola having ordinates zero, at the three hinge points. For obtaining the section at which the absolute maximum –ve B.M. will occur, we differentiate Eqn. 8.24 w.r.t. x and equate it to zero.

$$\frac{d\,M_x}{d\,x} = (l - 2x) - 2\,x = 0$$

$$\therefore \qquad x = \frac{l}{4}$$

Substituting for x in Eqn. 8.24 the absolute maximum –ve B.M.

$$M_{maxm.}\text{ (absolute)} = \frac{Wl}{16} \tag{8.26}$$

The maximum –ve B.M. diagram is shown in Fig. 8.9*e*.

The reader may notice the identity of equations in a three-hinged arch and the equations 8.23 and 8.24 in a three-hinged stiffening girder.

8.4.4 Maximum B.M. Under U.D.L. Longer than Span

We can also utilise the I.L. diagram for B.M. (Fig. 8.9) at any section X to obtain the absolute maximum B.M. anywhere on the girder as a u.d.l. w/unit length longer than the span crosses the girder.

It is obvious from th I.L. diagram (Fig. 8.9) that the maximum +ve moment will occur when a u.d.l. occupies fiom D to G and the maximum –ve moment will occur when the distributed load occupies from G to F. Attention is drawn to the fact that the maxm. +ve or –ve moments are numerically equal since the triangular areas DX_1G_1 and FE_1G_1 are same.

In the I.L. diagram, let the section of zero ordinate be at a distance z from D. From similar triangles we can write.

$$\frac{G\,G_1}{E\,E_1} = \frac{D.G}{D.E} = \frac{z}{\tfrac{1}{2}} \tag{8.27}$$

and $$\frac{GG_1}{X\,X_1} = \frac{F\,G}{F\,X} = \frac{(l - z)}{(l - x)}$$

since $$EE_1 = XX_1$$

$$\frac{z}{\tfrac{1}{2}} = \frac{(l - z)}{(l - x)}$$

solving $$z = \frac{l^2}{(3l - 2x)} \tag{8.28}$$

Substituting for $E E_1 = \frac{x}{l}(l-x)$ and $z = \frac{l^2}{(3l-2x)}$ in Eqn. 8.27

$$G G_1 = \frac{2x(l-x)}{(3l-2x)}$$

$\therefore$ The area of the triangle $D X_1 G_1 =$

(Area of the triangle DX_1F) – (Area of triangle $D G_1 F$)

$$= \frac{1}{2}\frac{l\,x(l-x)}{l} - \frac{1}{2}l\frac{2x(l-x)}{(3l-2x)}$$

Area of the triangle $$DX_1\,G_1 = \frac{x(l-x)(l-2x)}{2(3l-2x)} \qquad (8.29)$$

Therefore $$M_{x\,\text{maxm.}} = w\frac{x(l-x)(l-2x)}{2(3l-2x)} \qquad (8.30)$$

To obtain the absolute maximum +ve or –ve moment any where on the girder we differentiate Eqn. 8.30 with respect to x and equate it to zero. This gives

$$x = 0.234\,l$$

as the appropriate root of a cubic equation.

Substituting for $x = 0.234\,l$ in Eqn. 8.24

$$M_{\pm\text{maxm.}}\text{ (absolute)} = \pm\,0.01883\,wl^2 \qquad (8.31)$$

The maximum +ve or +ve moment diagram is shown in Fig. 8.9 for the girder from D to F.

8.5 INFLUENCE LINES FOR STIFFENING GIRDER

8.5.1 Influence Line for Shear Force

Consider a three-hinged stiffening girder as a rolling load moves from left to right. When the load W is in the region D to X, shear force at a section X distance x from D is –ve and is

$$V_x = -V_F + \frac{w_e\,l}{2} - w_e\,x \quad \text{for} \quad 0 \le z \le x$$

$$= -\frac{W\,z}{l} + \frac{w_e l}{2} - w_e x$$

$$= -\frac{W\,z}{l} + \frac{w_e}{2}(l-2x) \qquad (8.32)$$

When the load is in the region X to F the shear force at section X is positive and is

$$Vx = V_D - \frac{w_e l}{2} + w_e x \quad \text{for} \quad x \le z \le l$$

$$= \frac{W(l-z)}{l} - \frac{w_e}{2}(l-2x) \tag{8.33}$$

The S.F. at section X as denoted by Eqns. 8.32 and 8.33 is in two parts; the first part is similar to S.F. in a simply supported beam under load W and the second part represents shear due to w_e the equivalent u.d.l. on the cable and the girder. It can be shown that the second part is same as $H \tan \theta$, the vertical component of cable tension T at section X. We know

$$y = 4\frac{y_c}{l^2} x(l-x)$$

and $\tan \theta = \dfrac{dy}{dx} = \dfrac{4y_c}{l^2}(1-2x)$

$$\therefore \qquad H \tan \theta = \frac{w_e l^2}{8 y_c} \frac{4 y_c}{l^2}(l-2x) = \frac{w_e}{2}(l \cdot 2x)$$

Therefore S.F. at section X can be written as

$$V_x = -(v_x + H \tan \theta) \qquad \text{for } 0 \le z \le x \tag{8.34}$$

or

$$V_x = v_x - H \tan \theta \qquad \text{for } x \le z \le l \tag{8.35}$$

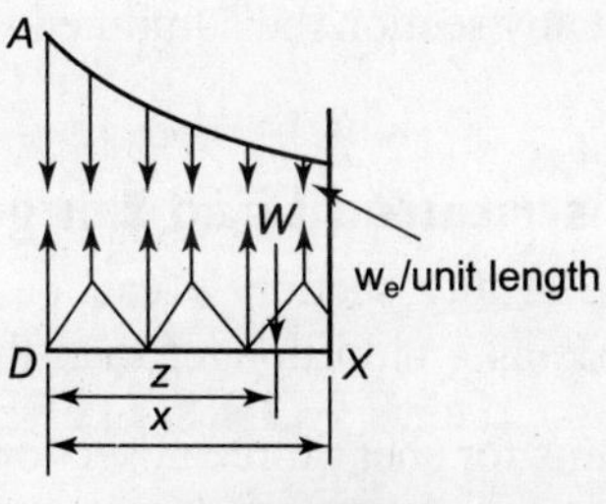

Fig. 8.10

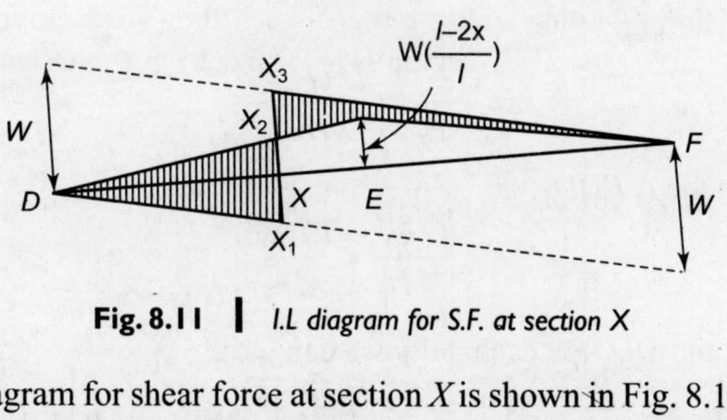

Fig. 8.11 | *I.L diagram for S.F. at section X*

The I.L. diagram for shear force at section X is shown in Fig. 8.11 Since v_x and $H \tan \theta$ are additive in the region $0 \le z \le x$, the I.L. diagrams for the two terms are shown on either side of base line DF.

From the I.L. diagram we find that the maximum –ve S.F. occurs when the load W is just to the left of section X. The ordinate $X_1 X_2$ represents the maximum –ve S.F. That is,

$$V_{x\,\text{maxm.}} = -\left\{\frac{W\,x}{l} + \frac{w_e}{l}(l-2x)\frac{2x}{l}\right\}$$

$$= -\left\{\frac{W\,x}{l} + 2\frac{w_e\,x}{l} - 4w_e\frac{x^2}{l^2}\right\}$$

The I.L. diagram for maximum –ve S.F. at different sections $(x = 0, \frac{l}{8}, \frac{l}{4}, \frac{3}{8}l, \frac{l}{2}, \frac{5}{8}l, \frac{3}{4}l, \frac{7}{8}l,$ and $l)$ is shown in Fig. 8.12.

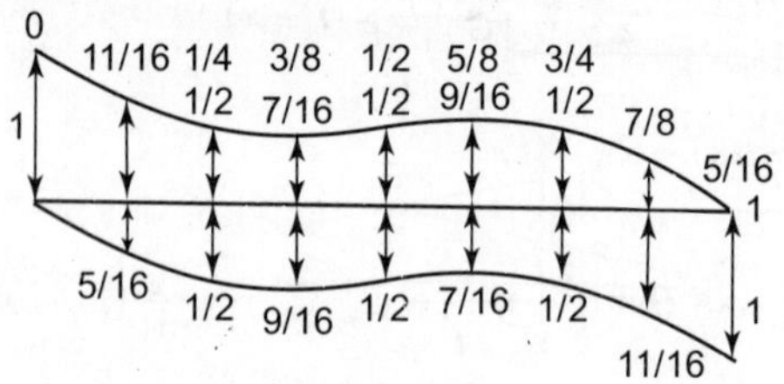

Fig. 8.12

Similarly we notice that the maximum +ve S.F. will occur when the load W is just to the right of section X. The ordinate X_2X_3 represents the maximum +ve S.F. It may be noted that at any section the combined value of maximum –ve and +ve S F. is equal to W.

8.5.2 Uniformly Distributed Load Longer than Span

The I.L. diagram for S.F. at any section X can be utilised to determine the maximum –ve and +ve S.F. as a u.d.l. longer than the girder span passes over the bridge. The I.L. diagrams for shear force at sections $x = 0$, $x = \frac{l}{4}$ and $x = \frac{l}{2}$ are shown in Fig. 8.13 a, b and c respectively At section $x = 0$, considering the triangle DEE_1 we can write

$$\frac{G\,G_1}{E\,E_1} = \frac{D\,G}{D\,E} = \frac{Z}{½}$$

Again from the $\Delta\ FDD_1$

$$\frac{G\,G_1}{D\,D_1} = \frac{l-z}{l}$$

Since $E\,E_1$ and DD_1 are each unity we can write

$$\frac{z}{½} = \frac{l-z}{l}$$

Solving, $z = \dfrac{l}{3}$

It is clear from Fig. 8.13*a* the maximum +ve S.F. will develop when the uniformly distributed load *w*/unit length occupies region *D* to *G* and maximum –ve S.F. will develop when the u.d.l. occupies the region *G* to *F*. The maximum +ve S.F. is numerically equal to maximum –ve S.F. as the areas of the triangles DD_1G and FE_1G_1 are same. Therefore, the

$$\text{maxm. +ve or –ve S.F. is } V_{max} = \frac{1}{2}\left(\frac{l}{3}\right)(1)\, w = \frac{wl}{6} \tag{8.36}$$

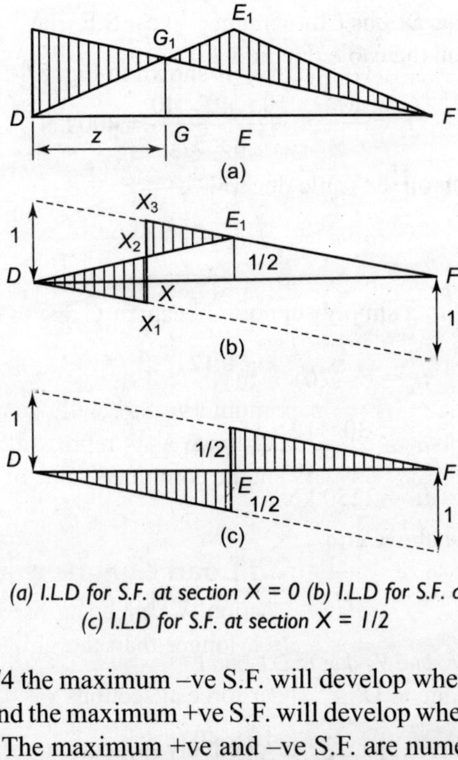

Fig. 8.13 | *(a) I.L.D for S.F. at section X = 0 (b) I.L.D for S.F. at section x = l/4 (c) I.L.D for S.F. at section X = l/2*

At section $x = l/4$ the maximum –ve S.F. will develop when the u.d.l. occupies the region *D* to *X* and the maximum +ve S.F. will develop when the u.d.l. occupies the region *X* to *E*. The maximum +ve and –ve S.F. are numerically equal as the areas of triangles DX_1X_2 and $X_2X_3E_1$, are same. Therefore the maximum +ve or –ve S.F. is

$$V_{max} = \frac{1}{2}\left(\frac{l}{4}\right)\left(\frac{1}{2}\right) w = \frac{wl}{16} \tag{8.37}$$

At section $x = l/2$ the maximum –ve S.F. will occur when the u.d.l. occupies the region *D* to *E* and the maximum +ve S.F. will develop when the u.d.l. occupies the region *E* to *F*. It is obvious from the diagram Fig. 8.13c that the maximum –ve or +ve S.F. is

$$V_{max} = \frac{1}{2}\left(\frac{l}{2}\right)\left(\frac{1}{2}\right) w = \frac{wl}{8} \tag{8.38}$$

Example 8.7 *The cables of a suspension bridge have a span of 40 m and a dip of 5 m. Each cable is stiffened by a girder hinged at the ends and at mid span to enable the cable to maintain its parabolic shape. There is a uniform dead load of 10 kN/m over the whole of the span and in addition a live load of 30 kN/m over 10 m length. Determine the maximum cable tension when the head of the live load is on the central hinge. Calculate the maximum S.F. and B.M. at a section 10 m from the left end when the live load rolls over.*

Step 1: To find H due to DL and L.L.

Horizontal tension on the cable due to D.L.

$$H_l = \frac{wl^2}{8y_c} = \frac{10 \times 40 \times 40}{8 \times 5} = 400 \text{ kN}$$

Horizontal tension on the cable due to L.L.

$$H_2 = \frac{\mu_c}{y_c}$$

μ_c is the moment in a simply supported beam at C due to the given live load position

$$\mu_c = V_F\,(20)$$

or
$$\mu_c = \frac{30 \times 10 \times 15}{40}(20)$$

$\therefore$
$$\mu_c = 2250 \text{ kN.m}$$

Substituting in the above eqn.

$$H_2 = \frac{2250}{5} = 450 \text{ kN}$$

Step 2: To determine V_A and V_B due to D.L and L.L

Vertical reaction V_A due to D.L.

$$V_{A1} = \frac{wl}{2} = \frac{10 \times 40}{2} = 200 \text{ kN}$$

If w_e is the u.d.l. on the cable due to L.L.

$$\frac{w_e\, l^2}{8\, y_c} = H_2$$

Substituting for H_2

$$w = \frac{450 \times 8 \times 5}{40 \times 40} = 11.25 \text{ kN/m}$$

Vertical reaction due to L.L

$$V_{A2} = \frac{w_e l}{2} = \frac{11.25 \times 40}{2} = 225 \text{ kN}$$

Step 3: To evaluate total V_A and H

Total vertical reaction V_A

$$V_A = V_{A1} + V_{A2} = 200 + 225 = 425 \text{ kN}$$

Total $$H = H_1 + H_2 = 400 + 450 = 850 \text{ kN}$$

$$T_{max} = \sqrt{V_A^2 + H^2}$$

or $$T_{max} = \sqrt{425^2 + 850^2}$$

$$T_{max} = 950.3 \text{ kN}$$

Step 4: To find maximum S.F.

The I.L. diagram for S.F. at a section 10 from the left end is shown in Fig. 8.14*a*. Maxm. –ve S.F. will occur when the load occupies from d to *x*. Similarly maxm. +ve S.F. will occur when the load occupies the region *x* to *e*. The maxm. –ve or +ve S.F.

$$V_X = \pm \frac{1}{2} (0.5)(10)\, 30 = 75.0 \text{ kN}$$

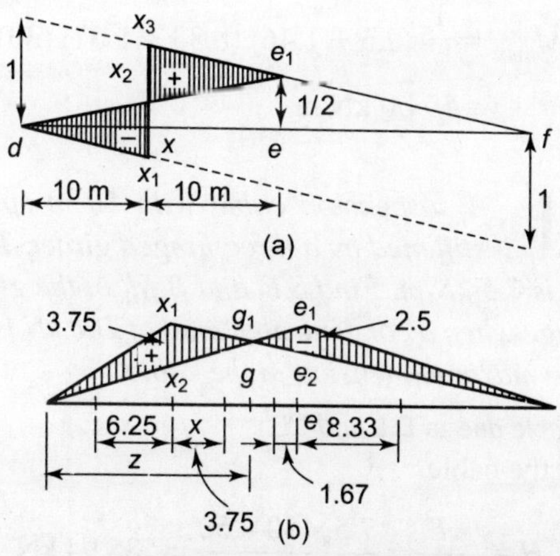

Fig. 8.14 | *(a) I.LD. for S.F. at a section 10 m from end D (b) I.LD for B.M. at a section 10 m from end D*

Step 5: To find maximum moment

The I.L. diagram for moment at section 10 m from the left end is shown in Fig. 8.14*b*.

The distance *z* from end *d* at which the I.L. ordinate is zero is determined using similar triangles.

$$\frac{Z}{gg_1} = \frac{½}{ee_1} \therefore gg_1 = \frac{zee_1}{½}$$

Again $$\frac{l - z}{gg_1} = \frac{l - x}{xx_1} \therefore gg_1 = \frac{(l - z)\, xx_1}{1 - x}$$

Substituting for $$ee_1 = xx_1 = \frac{10 \times 30}{40} = 7.5$$

$$\frac{7.5\,z}{20} = \frac{(40 - z)\,(7.5)}{30}$$

$$z = 16 \text{ m}$$

The position of load for getting maxm. B.M. is shown.

The head of the load is 3.75 m to the right of section and the tail end 6.25 m to the left of section.

For that position of the load, the maxm. +ve B.M.

$$M_{\text{max.}} = \frac{1}{2}\,(3.75 + 1.41)\,(3.75 + 6.25)\,(30)$$

$$= 774.0 \text{ kN.m}$$

Similarly the position of load for obtaining the maxm. –ve B.M. is shown ir. Fig. 8.14*b*. The head of the load is 8.33 m to the right of centre of girder and the tail end is 1.67 m to the left of centre of girder. For the loading position indicated the maxim, –ve B.M.,

$$M_{\text{max}} = \frac{1}{2}\,(2.5 + 1.46)\,(8.33 + 1.67)\,(30)$$

$$= 594.0 \text{ kN.m}$$

Example 8.8 | *A suspension cable with 50 m span and 4 m dip is stiffened by a three-hinged girder. The dead load of the girder and the deck is 7.5 kN/m. Find S.F. and B.M. in the girder at a section 10 m from left hand hinge when a concentrated load of 100 kN is placed at 8 m from the left end. Find the maximum tension in the cable.*

Step 1: To find H in cable due to D.L and L.L.

Let us consider first the cable.

Due to D.L $$H = \frac{wl^2}{8\,y_c} = \frac{7.5 \times 50 \times 50}{8 \times 4} = 585.94 \text{ kN}$$

$$V_A = \frac{wl}{2} = \frac{7.5 \times 50}{2} = 187.5 \text{ kN}$$

Due to conc, load $W = 100$ kN, the value of H can be obtained from knowing the value of w on the cable.

Taking moments about the section 10 m from support A

$$M_x = \frac{w_e l}{2}(10) - \frac{w(10)^2}{2} - H\,y = 0$$

$$\therefore \quad H\,y = w\frac{50}{2}(10) - 50\,w$$

$$H\,y = 200w$$

$$y = 2.56 \quad \text{m at the section 10 m from } A$$

We know from Eqn. 8.20 $w_e = \dfrac{4wz}{l^2} = \dfrac{4 \times 100 \times 8}{50 \times 50} = 1.28$ kN/m

$$\therefore \qquad H = \frac{200 \times 1.28}{2.56} = 100 \text{ kN}$$

Step 2: To find T

Vertical reaction due to 100 kN load $= \dfrac{wl}{2} = \dfrac{1.28 \times 50}{2} = 32$ kN

Total horizontal pull H due to dead and live load

$$= 584.94 + 100 = 685.94 \text{ kN}$$

$$\text{Total } V_A = 187.5 + 32 = 219.5 \text{ kN.}$$

$$T = \sqrt{685.94^2 + 219.5^2}$$

or $$T = 720.26 \text{ kN}$$

Now consider the girder.

Uniformly distributed dead load does not cause any shear or moment on the girder.

Step 3: To find maximum tension in cable

S.F. at section X, using Eqn. 8.34

$$V_x = -(v_x + H \tan \theta)$$

$$= -\left\{\frac{100(8)}{50} + 100\,(0.192)\right\}$$

$$= -(16 + 19.2) = -35.2 \text{ kN}$$

B.M. at section X,

$$M_x = \mu_x - Hy$$

where y is the dip of the cable above the section

$$\therefore \qquad M_x = \frac{100(8)}{50}(40) - 100 \times 2.56$$

or $$M_x = 640 - 256 = 384 \text{ kN.m}$$

Example 8.9 | *The towers of a 150 m span suspension bridge are of unequal height. One tower is 18 m and the other 6 m above the lowest point of the cable which is immediately above the inner hinge of a three-hinged stiffening girder (Fig. 8.15). Find the maximum tension in the cable due to a point load W rolling over the bridge.*

Step 1: Location of lowest point in the cable

First let us locate the position of lowest point C on the cable. Let it be at distance l_1 from end A and l_2 from end B.

Then $$\frac{l_1}{l_2} = \left(\frac{18}{6}\right)^{1/2}$$

or $$\frac{l_1}{l_2} = 1.732$$

$$l_1 = 1.732\, l_2.$$

We know $l_1 + l_2 = 150$

$$1.732\, l_2 + l_2 = 150$$

$$l_2 = \frac{150}{2.732} = 54.90$$

and $$l_1 = 95.10 \text{ m}$$

Step 2: To find H and ω

Taking moments about C of all the forces to the right of C

$$M_C = \mu_c - H{\cdot}y - \frac{Hd}{150}\,(54.9) = 0$$

The last term in the above equation is due to the towers being at different levels.

$$\therefore \quad Mc = W\frac{(54.9)(95.1)}{150} - H(6) - H\frac{(12)(54.9)}{150} = 0$$

$$= 34.81\,W - 10.39\,H = 0$$

$$\therefore \quad H = 3.35\,W$$

we know $H = \dfrac{wl_2^2}{2\,y_c}$

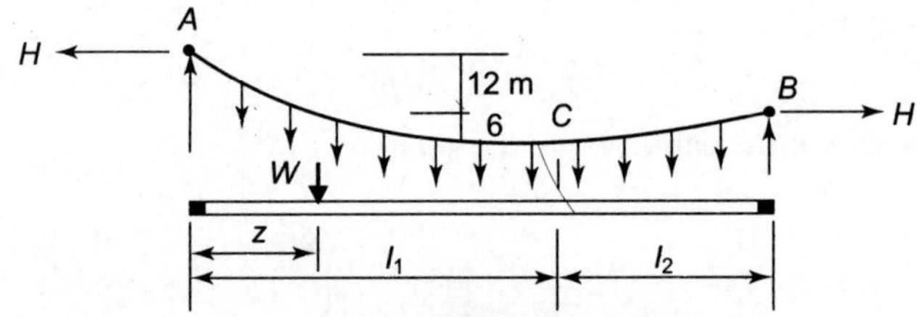

Fig. 8.15

$$\therefore \quad w = \frac{2\,H\,y_c}{l_2^2} = 2 \times \frac{3.35\,W \times 6}{54.9^2} = 0.0133W$$

Step 3: To find T_{max}

Vertical reaction at end A of the cable is obtained by taking moments about C

$$V_A\,(95.1) - 3.35\,W\,(18) - 0.0133\,\frac{(95.1)^2}{2}\,W = 0$$

$$\therefore \qquad V_A = 1.266\ W$$

$$T_{\text{maxm.}} = \sqrt{H^2 + V_A^2}$$

$$= W\sqrt{3.35^2 + 1.266^2}$$

$$= 3.58\ W.$$

Example 8.10 | *A suspension bridge with a three-hinged stiffening girder has a span of 100 m, a cenral dip of cable 8 m and weighs 2500 kN. It has to carry a live load of 50 kN/m. Calculate the sectional area of cables required and the sectional modulus for each girder if the permissible stress is 120 N/mm². The live load may cover all or any part of the span.*

Step 1: To find H and V due to D.L. and L.L.

Horizontal pull H due to D.L.

$$H_1 = \frac{Wl}{8y_c} = \frac{2500 \times 100}{8 \times 8} = 3906 \text{ kN}.$$

Vertical reaction V_A due to D.L.

$$V_{A1} = \frac{2500}{2} = 1250 \text{ kN}.$$

Horizontal pull H due to L.L.

$$H_2 = \frac{50 \times 100 \times 100}{8 \times 8} = 7813 \text{ kN}.$$

Vertical reaction V_A due to L.L.

$$V_{A2} = \frac{50 \times 100}{2} = 2500$$

Combined horizontal pull on each cable,

$$H = \frac{1}{2}(H_1 + H_2) = \frac{1}{2}(3906 + 7813) = 5860 \text{ kN}.$$

and vertical reaction $V = \frac{1}{2}(1250 + 2500) = 1875$ kN

Step 2: To find T_{max}

$$T_{\max} = \sqrt{H^2 + V^2}$$

$$= \sqrt{5860^2 + 1875^2}$$

$$= 6153 \text{ kN}$$

Step 3: To find area of cross section

We know $f.\,A = T$

$$\therefore \qquad A = \frac{6153 \times 1000}{120} = 51275 \text{ mm}^2$$

Step 4: To find modulus of section for girder

Under D.L. the stiffening girder suffers no S.F. or B.M.

Maxm. +ve or –ve B.M. occurs at section 0.234*l* from either end when the load occupies from *D* to *G* or from *G* to *F* as the case may be and is equal to 0.01883 wl^2.

$$\therefore \quad M_{\text{maxm.}} = \pm\, 0.01883 \times \frac{50}{2}(100)^2 \text{ on each girder}$$

$$= \pm\, 4708 \text{ kN.m}$$

$$\text{Modulus of section required} = \frac{M_{\max}}{f}$$

$$= \frac{4708 \times 10^6}{120} = 39.23 \times 10^6 \text{ mm}^3$$

Example 8.11 | *A suspension bridge of 100 m span has two three-hinged stiffening girders supported by two cables having central dip 10 m. The width of the road way is 8m. The roadway carries a dead load of 1 kN /m² extending over the whole span and a live load of 2 kN/m² extending over the left half of the bridge. Find B.M. and S.F. at a section 25 m and 80 m from the left hinge. Also calculate the maximum tension in the cable (see Fig. 8.16)*

Step 1: To evaluate H due to D.L. and L.L.

Let us consider first the cable. The horizontal tension in the cable is caused by dead and live loads.

$$\text{D.L. } \omega_d = 4 \times 1 = 4 \text{ kN/m}$$

$$\text{L.L. } \omega_l = 4 \times 2 = 8 \text{ kN/m}$$

Horizontal tension on the cable due to D.L.

$$H_1 = \frac{\omega_d\, l^2}{8\, y_c} = \frac{4 \times 100 \times 100}{8 \times 10} = 500 \text{ kN}$$

Horizontal tension on the cable due to L.L.

$$H_2 = \frac{\mu_c}{y_c}$$

μ_c is the moment in simply supported beam at *C* due to the given L.L. position

$$H_2 = 50 \times \frac{8}{100} \times 25 \times \frac{50}{10} = 500 \text{ kN}$$

Step 2: To evaluate T_{max}

Vertical reaction V_A dut to D.L.

$$V_{A1} = \frac{\omega_d \cdot l}{2} = \frac{4 \times 100}{2} = 200 \text{ kN}$$

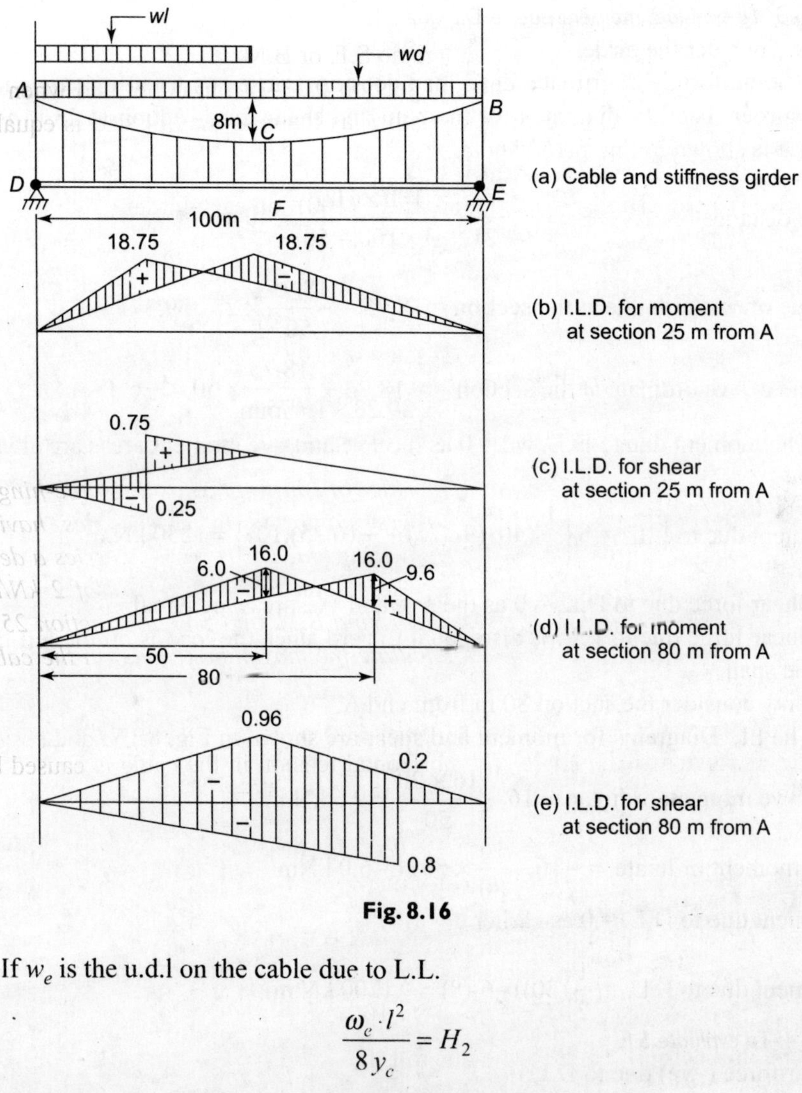

Fig. 8.16

If w_e is the u.d.l on the cable due to L.L.

$$\frac{\omega_e \cdot l^2}{8\,y_c} = H_2$$

or
$$w_e = \frac{500 \times 8 \times 10}{100 \times 100} = 4 \text{ kN/m}$$

Hence,
$$V_{A_2} = \frac{\omega_e l}{2} = \frac{4 \times 100}{2} = 200 \text{ kN}$$

Total
$$H = H_1 + H_2 = 500 + 500 = 1000 \text{ kN.}$$
$$V_A = V_{A1} + V_{A2} = 200 + 200 = 400 \text{ kN}$$

Maximum tension $T_{\max} = \sqrt{H^2 + V_A^2} = \sqrt{1000^2 + 400^2} = 1077 \text{ kN}$

Step 3: To evaluate moment due to D.L and L.L.

Now consider the girder:

The uniformly distributed dead load does not cause any shear or moment on the girder. The I.L. diagrams for moment and shear for a-section at 25 m from end *A* is shown in Fig. 8.16*b* and *c*.

The distance $z = \dfrac{l^2}{3l - 2x} = \dfrac{100 \times 100}{3 \times 100 - 50} = 40 \text{ m}$

Value of +ve ordinate at the section $= 18.75 - \dfrac{18.75}{50} \times 25 = 9.375$

Value of –ve ordinate at the section $= -18.75 + \dfrac{18.75}{75} \times 50 = -6.25$

The moment due to D.L. wd = 0 as the +ve and –ve moment areas are always equal

Moment due to L.L. $= 8\left\{\dfrac{1}{2}(40)(9.375) - \dfrac{1}{2}(6.25)(10)\right\} = 1250 \text{ kN.m}$

Shear force due to D.L. = 0 as the +ve and –ve areas are equal.

Shear force due to L.L. is also equal to zero since the load is on the left half of the span

Now consider the section 80 m from end A.

The I.L. Diagrams for moment and shear are shown in Fig. 8.16*d* and *e*.

Net +ve moment ordinate $= 16 - \dfrac{16 \times 20}{50} = 9.6 \text{ kN.m}$

–ve moment ordinate $= -16 + \dfrac{16}{80} \times 50 = -6.0 \text{ kNm}$

Moment due to D.L. = 0 as earlier

Moment due to L.L. $= \dfrac{1}{2}(50)(-6)(8) = -1200 \text{ kN.m.}$

Step 4: To evaluate S.F.

Shear force (–ve) due to D.L.

$$= 4\left\{\frac{1}{2}(-0.96)(100) + \frac{1}{2}(80)(-0.8) + \frac{1}{2}(20((0.2)\right\}$$

$$= 4(-48.0 - 32.0 + 20) = -312.0 \text{ kN}$$

Shear force (+ve) due to L.L.

$$= 8\left\{\frac{1}{2}(-0.96 - 0.5)(50)\right\}$$

$$= 8\left\{\frac{(-1.46)}{2}(50)\right\} = -292 \text{ kN}$$

Total –ve shear force = –312 – 292 = –604 kN

8.6 TWO-HINGED STIFFENING GIRDER

As already mentioned, stiffening girders can be two-hinged, having hinges at the ends only and no hinge at the centre. Such a structure is statically indeterminate and the forces in cable and stiffening girder may be obtained approximately using energy methods. However, if the girder is assumed to be rigid and the load, irrespective of its position, is transmitted as a u.d.l. to to the cable, the forces in the girder and cable may be worked out as under.

Consider a single rolling load W at a distance z from D as in Fig. 8.17a. The load is assumed to be transmitted to the cable as a u.d.l. irrespective of the load position, we have

$$W = w_e\, l \tag{8.39}$$

where w_e is the equivalent u.d.l. per unit length. The horizontal pull H is given by

$$H = \frac{w_e\, l^2}{8\, y_c} = \frac{W\, l}{8\, y_c} \tag{8.40}$$

It is thus obvious that the magnitude of horizontal pull will be constant and independent of load position.

8.6.1 Influence Lines for a Single Concentrated Load Rolling Over the Girder

The dead load of the girder and the roadway, as in a three-hinged girder, is transmitted to the cable by hangers as a u.d.l. The cable takes up the u.d.l. entirely by tension in the cable. The girder suffers no S.F. or B.M. under dead loads.

The live load is assumed to be transmitted as a u.d.l. by the hangers to the cables. Consider a suspension bridge of span l and a central dip y_c with a two-hinged stiffening girder as in Fig. 8.17. We shall draw the I.L. diagrams for horizontal pull H, shear force V and bending moment M.

I.L. Diagram for Horizontal Pull *H*

We have already seen that the equivalent u.d.l., w_e, irrespective of load position, is $w_e = \dfrac{W}{l}$

We know that in a cable subjected to a u.d.l. w_e /unit length, the horizontal pull is given by

$$H = \frac{w_e l^2}{8\, y_c} = \frac{W\, l}{8\, y_c} \tag{8.41}$$

We see, from the above equation, that the magnitude of horizontal pull H is constant and is independent of load position. Therefore, the I.L. for H is a straight line with constant ordinate as shown in Fig. 8.17b.

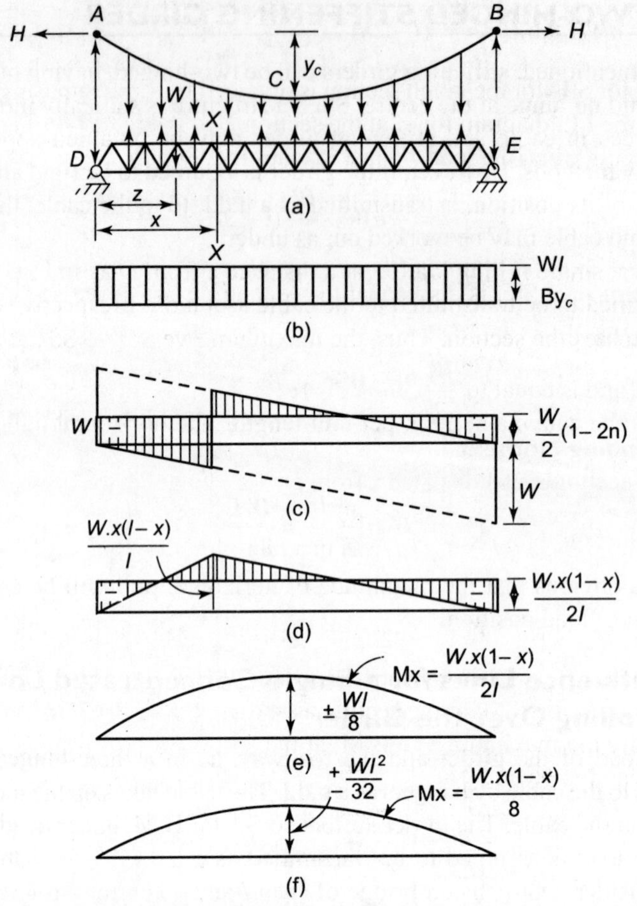

Fig. 8.17 | *(a) Cable and the two-hinged stiffening girder (b) I.L.D. for H (c) I.L.D for S.F. at section X (d) I.L.D. for B.M. due to rolling load W (e) Absolute Maxm. B.M. due to rolling load W (f) Absolute Maxm. B.M. due to u.d.l.*

I.L. for Shear Force

Consider a section X at distance x. from D. We know that the S.F. at X

$$V_x = -\{v_x + H\tan\theta\} \text{ for } 0 \le z \le x \tag{8.42}$$

or

$$V_x = v_x - H\tan\theta \text{ for } x \le z \le l \tag{8.43}$$

as in a three-hinged stiffening girder in which

v_x = shear force at section X in a simply supported beam.

$H\tan\theta$ = the vertical component of tension in the cable at section X.

We have seen $H\tan\theta = \frac{W}{2l}(l - 2x)$

Writing $x = nl$, $H \tan\theta = \frac{W}{2}(l - 2n)$

The I.L. for S.F. for the given section is obtained by superimposing the I.L. for $H \tan\theta$ on the I.L. diagram for v_x at the section X as in Fig. 8.17*c*. The maximum –ve S.F. occurs when the load W, just reaches the section.

$$V_{x_{\max m.}} = -nW - \frac{W}{2}(1 - 2n) = \frac{-W}{2} \tag{8.44}$$

The maxm. +ve S.F. is also equal to $\frac{W}{2}$ and occurs at the section when the load just crosses the section. Thus, the maximum +ve or –ve S.F. at any section is the same and is equal to $\frac{W}{2}$.

I.L. for Bending Moment

Consider a section X at a distance x from D. We know that the moment at section X is

$$M_x = \mu_x - Hy \text{ as in a three-hinged girder.}$$

or
$$M_x = \mu_x - \frac{W}{2l}x(l - x)$$

The I.L. for B.M. is drawn by superimposing the I.L. for $Hy = Wx(l - x)/2l$ on the I.L. diagram for B.M. at the section as shown in Fig. 8.17*d*.

Obviously the maximum +ve B.M. at the section occurs when the load is on the section and is given by

$$M_{\text{maxm.}} = W\frac{x}{l}(l - x) - \frac{Wx}{2l}(l - x) = \frac{Wx(l - x)}{2l}$$

The maximum –ve B.M. at the section X occurs when the load is at either end on the hinge points.

$$\therefore \quad M_{\text{maxm.}} = -\frac{Wx}{2l}(l - x)$$

Thus the maximum +ve or –ve B.M. at section X is numerically equal and is

$$M_{\text{maxm.}} = \pm\frac{W_x(1 - x)}{2l} \tag{8.45}$$

The section at which the absolute maximum B.M. occurs is obtained by differentiating $M_{x\,\text{maxm}}$ with respect to x and equating to zero.

$$\frac{d}{dx}\left\{\frac{Wx}{2l}(l - x)\right\} = 0$$

$$\frac{d}{dx}\left\{\frac{Wx}{2} - \frac{Wx^2}{2l}\right\} = 0$$

That is $$\frac{W}{2} - \frac{Wx}{l} = 0$$

$\therefore$ $$l - 2x = 0$$

or $$x = l/2$$

Substituting for $x = l/2$ in Eqn. 8.45

$$M_{\text{maxm.}} \text{ (absolute)} = \pm \frac{Wl}{8} \tag{8.46}$$

The moment diagram is shown in Fig. 8.17*e*.

8.6.2 Uniformly Distributed Load Longer than Span

Maximum Shear Force If a u.d.l. longer than the span were to traverse the girder, the maximum +ve S.F. would occur when the load occupies the middle +ve S.F. region in Fig. 8.17c extending over length $l/2$.

$$\therefore \quad V_{\text{maxm.}} = \frac{1}{2}\frac{w}{2}\frac{l}{2} = \frac{wl}{8}$$

The maximum –ve S.F. at the section occurs when the u.d.l. occupies the end parts with no load on the middle +ve S.F. region. The maximum –ve S.F. is

$$V_{\text{maxm.}} = -\frac{wl}{8}$$

$$\therefore \quad V_{\text{maxm.}} \text{ for any section} = \pm \frac{wl}{8} \tag{8.47}$$

Maximum Bending Moment The maximum +ve B.M. at a section *X* occurs when the u.d.l. occupies the +ve B.M. region as shown in Fig. 8.17*a*.

$$\therefore \quad \text{Maximum +ve B.M.} = \frac{1}{2}\frac{wx(l-x)}{2l}\frac{l}{2} = \frac{wx}{8}(l-x)$$

This is the equation of a parabola of second degree reaching a maximum value at $x = \frac{l}{2}$ and the value of the absolute maximum moment is

$$M_{\max} \text{ (absolute)} = \frac{w}{8}\frac{l}{2}\left(l - \frac{l}{2}\right)$$

$$= \frac{wl^2}{32} \tag{8.48}$$

The maximum –ve moment at the section *X* occurs when the u.d.l. occupies the two ends with –ve ordinates and no load in the +ve ordinates region in Fig. 8.17*d*

$$M_{x(\text{maxm.})} = -\frac{1}{2}\frac{wx}{2l}(l-x)\frac{l}{2} = -\frac{wx}{8}(l-x)$$

This is numerically equal to +ve $M_{x(\text{maxm})}$ derived earlier.

Therefore, the absolute maximum +ve or –ve B.M. occurs at the centre and is given by

$$M_{\text{maxm.}}\ (\text{absolute}) = \pm \frac{wl^2}{32} \tag{8.49}$$

The moment diagram is shown in Fig. 8.17*f*.

Example 8.12 | *A suspension cable of 100 m span has a dip of 10 m. It is stiffened by a two-hinged girder whose weight is 20 kN/m. Determine the maximum tension in the cable if a point load of 500 kN rolls over the girder. Find also the maximum positive and negative B.M. on the girder.*

Step 1: To find w on the cable

We know that in a two-hinged stiffening girder the live load is transmitted to the cable as equivalent u.d.l. over the entire span. Therefore total load on the cable per metre length

$$w = \frac{\text{Total (D.L. + L.L)}}{\text{span}}$$

$$= \frac{20 \times 100 + 500}{100} = 25 \text{ kN/m}$$

Step 2: To find H, the horizontal pull

$$H = \frac{wl^2}{8\,y_c} = \frac{25 \times 100 \times 100}{8 \times 10} = 3125 \text{ kN}$$

Vertical reaction at supports

$$V_A = V_B = \frac{25 \times 100}{2} = 1250 \text{ kN.}$$

Maximum tension in the cable

$$T_{\max} = \sqrt{H^2 + V^2} = \sqrt{3125^2 + 1250^2} = 3365.73 \text{ kN.}$$

Step 3: To find maximum +ve and –ve B.M.

The stiffening girder suffers no B.M. due to dead load. The maximum +ve or –ve B.M. will occur when the L.L. is at the centre of span. The I.L. diagram for moment at centre of span is shown in Fig. 8.18. The maximum +ve B.M. will occur when the load is over the central section and is

$$+ve\ M_{\max} = \frac{Wl}{8} = \frac{500 \times 100}{8} = 6250 \text{ kN.m}$$

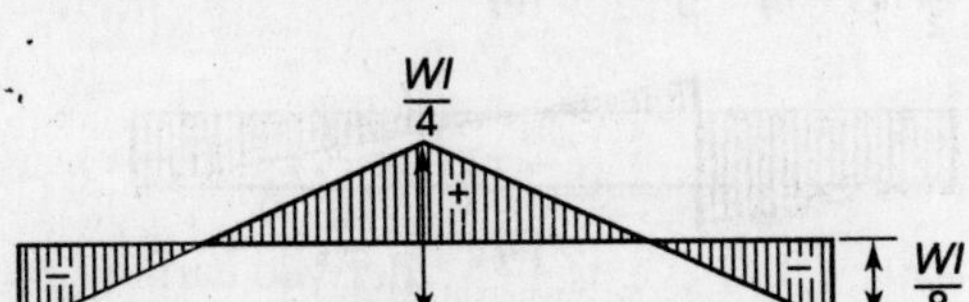

Fig. 8.18 | *I.L.D. for moment at centre of span*

The maximum –ve B.M. occurs at the central section when the load is positioned at the hinge points.

$$M_{max} = \frac{-WL}{8} = \frac{-500 \times 100}{8} = -6250 \text{ kN.m}$$

Example 8.13 | *A two-hinged stiffening girder has a span 30 m as in Fig. 8.19a. Determine: (a) the maximum bending moment on the girder when (i) a concentrated load of 100 kN rolls over the girder and (ii) a u.d.l. of 3 kN/m rolls over the girder; (b) the bending moment at 1/8 span section from either pier when a 100 kN load is at 1/4 span point from left pier; (c) the shear force at 1/4 span section due to loads as in (a) (i) and (ii).*

(a) (i) Maximum +ve and –ve B.M.

B.M. at a section $x = nl$ from C is

$$M_x = \mu_x - Hy$$

$$= \mu_x - \frac{Wl}{8\,y_c} \frac{4\,y_c}{l^2} x\,(l - x)$$

$$= \mu_x - \frac{Wl}{2} n\,(l - n)$$

The I.L.D. for moment at section X is shown in Fig. 8.19*b*. It is obvious from the I.L.D. that the maximum +ve B.M. occurs at centre when the 100 kN load is at the centre

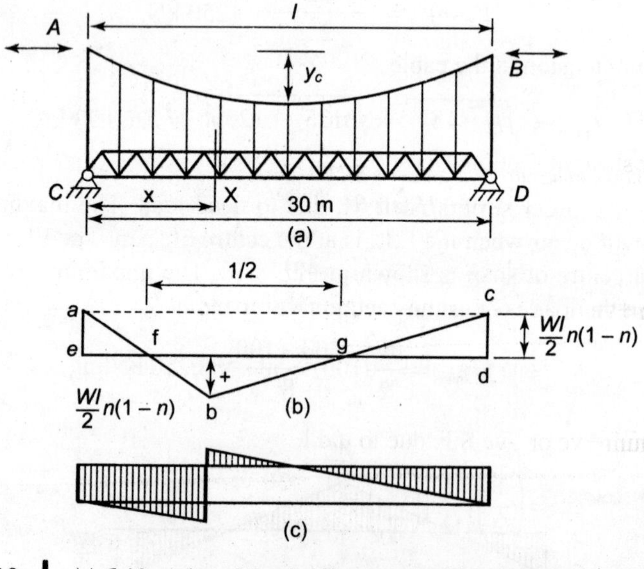

Fig. 8.19 | *(a) Cable and two-hinged stiffening girder (b) I.L.D for B.M. at section X (c) I.L.D. for S.F. at section X*

$$M_{max} = \frac{1}{2}(100)(30)\left(\frac{1}{2}\right)\left(\frac{1}{2}\right) = 375.\text{ kN.m}$$

Maximum –ve B.M. occurs when the load is at the ends and is equal to –375 kN.m.

(ii) Under u.d.l. the maximum +ve B.M. occurs when the load occupies from f to g in Fig. 8.19*b*.

$$\therefore \text{ Maximum +ve B.M. } = \left(\frac{1}{2}\right)\left(\frac{1}{2}\right)\frac{wl}{2}n(1-n)$$

$$= \frac{1}{2}\times\frac{30}{2}\times\frac{3\times 30}{2}\left(\frac{1}{2}\right)\left(\frac{1}{2}\right)$$

$$= 84.375 \text{ kN.m}$$

Maximum –ve B.M. occurs when the u.d.l. occupies e to f and g to d and is equal to –84.375 kN.m.

(b) When a 100 kN load is at $\frac{1}{4}$ span the moment at $\frac{1}{8}$ span = 0 corresponding to point f. Length $gd = \frac{3}{8}l$

$$\text{B.M. at 1/8 span from } D = \frac{2}{3}\text{ (cd)}$$

$$= \frac{2}{3}\frac{Wl}{2}n(1-n)$$

$$= \frac{2}{3}\times\frac{1}{2}\times 100\times 30\times\frac{1}{4}\times\frac{3}{4}$$

$$= 187.5 \text{ kN.m}$$

(c) I.L. for shear at section X is shown in Fig. 8.19*c*.

$$V_x = v + H\tan\theta$$

$$= v + H\,\frac{W}{2}(1-2n)$$

Maximum +ve or –ve S.F. at any section due to a 100 kN load is

$$= \frac{1}{2}(100) = 50 \text{ kN.}$$

Maximum +ve or –ve S.F. due to u.d.l.

$$= \left(\frac{1}{2}\right)\left(\frac{l}{2}\right)\left(\frac{w}{2}\right) = \frac{1}{2}\times\frac{30}{2}\times\frac{3}{2} = 11.25 \text{ kN.}$$

for all values of n.

Example 8.14 *A two-hinged stiffering girder of a suspension bridge has a span of 80 m. The dip of the supporting cable is 8 m. Two girders support a bridge deck. Two point loads of 400 kN and 600 kN at 16 and 32 m act on the deck, half of which comes on to one stiffening girder. Find S.F. and B.M. at 25 m from left hand end. Find also the maximum tension in the cable (Fig. 8.20)*

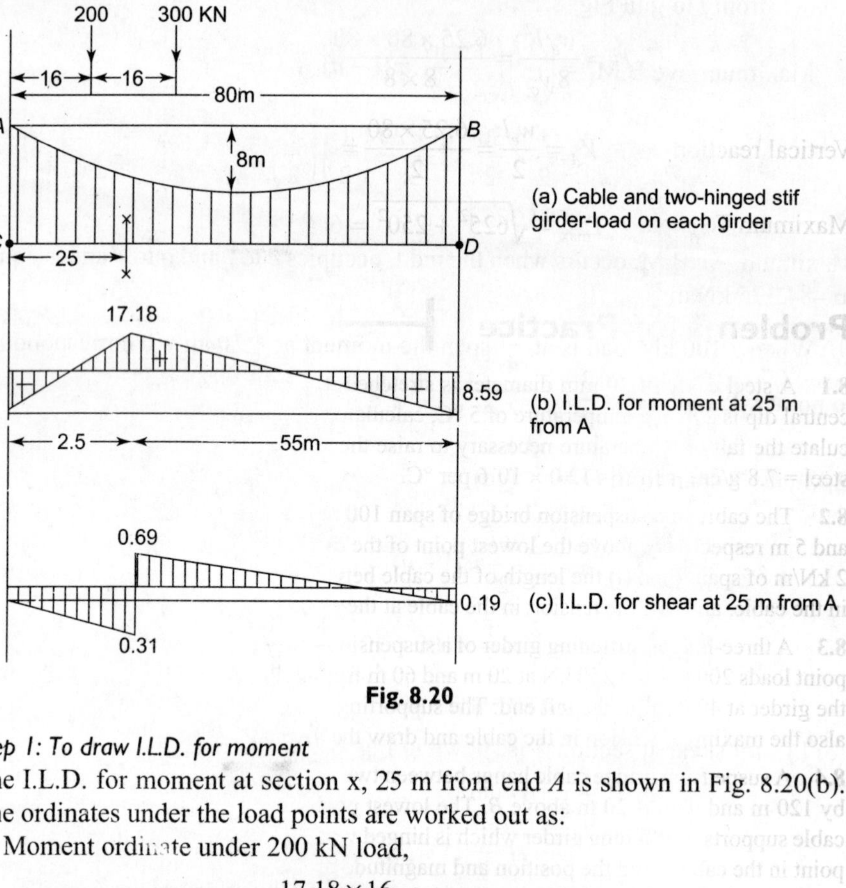

Fig. 8.20

Step 1: To draw I.L.D. for moment

The I.L.D. for moment at section x, 25 m from end A is shown in Fig. 8.20(b). The ordinates under the load points are worked out as:

Moment ordinate under 200 kN load,

$$= \frac{17.18 \times 16}{25} - 8.59 = 2.405$$

Moment ordinate under 300 kN load.

$$= \frac{17.18}{55} \times 48 - 8.59 = 6.40$$

+ve moment at section x = 200 (2.405) + 300(6.40) = 2.401 kN.m

Step 2: To draw I.L.D for S.F.

The S.F. I.L.D. is shown in Fig. 8.20(c)

Net ordinate under 200 kN. load $= -\dfrac{0.31}{25}(16) - 0.019 = -0.39$

Net ordinate under 300 kN load $= +\dfrac{0.69(48)}{55} - 0.019 = 0.41$

Therefore, S.F. = 200(–0.39) + 300(0.41) = 45.0 kN

Step 3: To evaluate tension in cable

Equivalent u.d.l. on the cable $\omega_e = \dfrac{300+200}{80} = 6.25$ kN/m

$$H = \frac{w_e\, l^2}{8\, y_c} = \frac{6.25 \times 80 \times 80}{8 \times 8} = 625.0 \text{ kN}$$

Vertical reaction $V_A = \dfrac{w_e l}{2} = \dfrac{6.25 \times 80}{2} = 250.0$ kN

Maximum Tension, $T_{max} = \sqrt{625^2 + 250^2} = 673.15$ kN.

Problems for Practice

8.1 A steel cable of 20 mm diameter is stretched across two poles 100 m apart. If the central dip is 2 m at a temperature of 5 °C, calculate the stress intensity in the cable. Calculate the fall of temperature necessary to raise the stress to 55 N/mm2. Take weight of steel = 7.8 g/cm^3 and α = 12.0 × 10¯6 per °C.

8.2 The cable of a suspension bridge of span 100 m is hung from piers which are 10 m and 5 m respectively above the lowest point of the cable. The load carried by the cable is 2 kN/m of span. Find (i) the length of the cable between the piers, (ii) the horizontal pull in the cable, and (iii) the tension in the cable at the piers.

8.3 A three-hinged stiffening girder of a suspension bridge of span 100 m is to carry two point loads 200 kN and 250 kN at 20 m and 60 m from left end. Find the S.F. and B.M. on the girder at 40 m from the left end. The supporting cable has a central dip of 10 m. Find also the maximum tension in the cable and draw the moment diagram for the girder.

8.4 A suspension bridge cable hangs between two points *A* and *B* separated horizontally by 120 m and with *A* 20 m above *B*. The lowest point on the cable is 5 m below *B*. The cable supports a stiffening girder which is hinged vertically below *A* and *B* and the lowest point in the cable. Find the position and magnitude of the largest bending moment which a point load of 20 kN can induce in the girder together with the position of the load.

8.5 A suspension cable, stiffened with a three-hinged girder has 100 m span and 10 m dip. The girder carries a dead load of 1 kN/m extending over the whole span. A live load of 25 kN rolls from left to right. Determine (i) the maximum B.M. and S.F. anywhere on the girder, (ii) the maximum tension in the cable.

8.6 A suspension bridge with two three-hinged stiffening girders has a span of 120 m and the cable has a central dip of 10 m. It carries a D.L. of 3 kN/m. It is to be designed for a rolling L.L. of 100 kN. The loads (dead and live load) can be assumed to be equally divided between the stiffening girders and corresponding cables. Determine the sectional area required for the cable if the permissible stress is 150 N/mm^2. Also find the maximum bending moment on the stiffening girder. The cable profile can be assumed to be parabolic.

8.7 A suspension cable of 60 m span having a central dip of 6 m is strengthened by stiffening girders hinged at both ends. Two girders support a bridge deck. Two point loads 500 kN and 600 kN at 16 m and 32 m respectively act on the deck, half of which comes to one girder. Find the S.F. and B.M. at 25 m from the left hand end. Find also the maximum tension in the cable.

8.8 A suspension cable, the ends of which are supported at the same level, has a span of 96 m and a central dip of 10 m. The bridge is stiffened by a stiffening girder hinged at the ends. The girder carries a single concentrated load of 10 kN at a point 24 m from left end. Assuming equal tension in the suspension hangers, calculate (i) the horizontal tension in the cable, (ii) the maximum positive and negative bending moments, and (iii) the value of absolute maximum B.M. and S.F. and where they will occur, if the 10 kN load rolls from left to right.

9 Approximate Analysis of Statically Indeterminate Structures

9.1 INTRODUCTION

The analysis of indeterminate structures, as such, will be discussed in detail in Chapters 10 to 14. However, it may be pointed out that the analysis depends on a knowledge of member proportions which are unknown at the time the design-analysis is begun. Therefore it becomes necessary to perform some approximate analysis quickly to arrive at an estimate of member sizes. Approximate analysis is also performed for checking the more elaborate computations involved in an exact analysis.

This chapter is concerned with the approximate analysis of a number of structural types that are statically indeterminate and are assumed to behave elastically. Approximate methods of analysis are discussed before exact methods of analysis with a view to making the reader familiar with a broad range of structures and their behaviour and at the same time enabling him to gain further insight into equilibrium conditions.

9.2 METHODS OF ANALYSIS

9.2.1 General

In approximate analysis, the statically indeterminate structure is reduced to a statically determinate structure, by making appropriate assumptions, and then analysed for member forces and reactions using statics. Some of the commonly used approximate methods of analysis are discussed in this chapter. It must be remembered that the results obtained are approximate and their nearness to the true values is dependent upon how good the assumptions are.

The study of approximate methods is best performed by a series of examples, since these methods are often specially related to a particular type of structure.

9.2.2 Indeterminate Trusses

Let us consider the statically indeterminate truss of Fig. 9.1 and find an approximate analysis for determining the forces in its members. The truss is statically indeterminate by four degrees because of the redundant diagonal members in four panels. If the diagonal members of the truss are considered to be long and slender, we can assume that the compression members carry negligible forces or do not take part. Therefore, the diagonals transmitting compressive forces are removed, which results in a statically determinate truss. The truss, after removing the compression diagonals for the given loading, is shown in Fig. 9.2. The identification of the diagonal members which are in tension is done by considering that the shear in each panel is carried by the diagonal members in tension.

If the diagonal members in truss of Fig. 9.1 are assumed to have considerable stiffness, we can perform an approximate analysis by assuming a certain distribution of shear in each panel between the two diagonals. It may be remembered that one of the diagonals will be in tension and the other in compression. For convenience, the shear is assumed to be distributed equally between the two diagonals. In either of these approximate methods, the number of assumptions is just equal to the degree of in-determinancy. The following example illustrates the point.

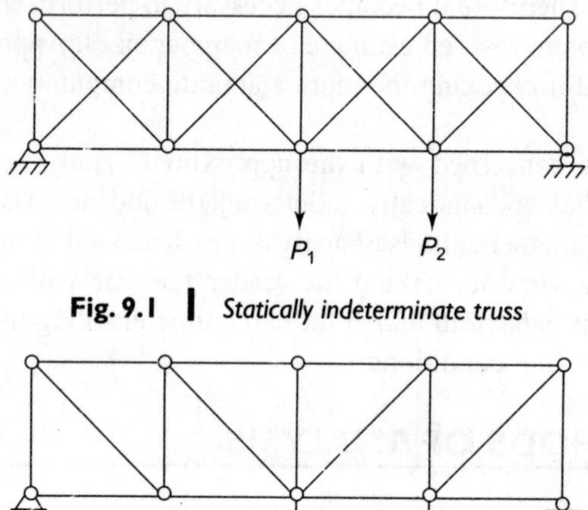

Fig. 9.1 | *Statically indeterminate truss*

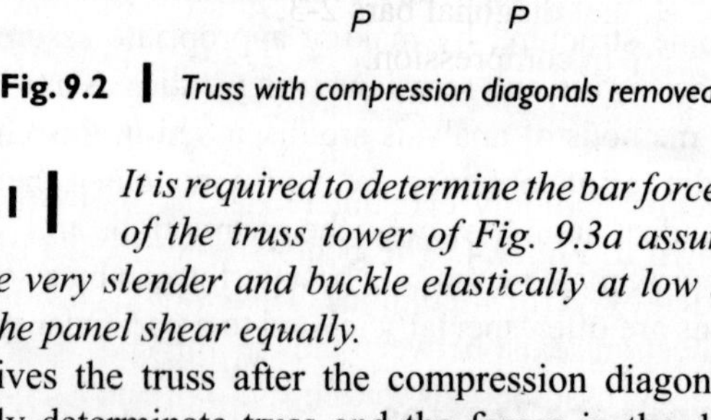

Fig. 9.2 | *Truss with compression diagonals removed*

Example 9.1 | *It is required to determine the bar forces in the diagonals of the truss tower of Fig. 9.3a assuming that (a) the diagonal bars are very slender and buckle elastically at low loads and (b) the diagonals share the panel shear equally.*

Figure 9.3*b* gives the truss after the compression diagonals are removed. This is a statically determinate truss and the forces in the diagonal members are obtained by equating the horizontal component of the forces in the diagonal members to the shear force in the respective panels. Therefore, considering the top panel, we have

$$P_{23} \cos \theta = 20$$

or $$P_{23} = 20\sqrt{2} \text{ kN}$$

Similarly, $P_{45} = 40\sqrt{2}$ kN and $P_{67} = 60\sqrt{2}$ kN

Considering that the diagonal members share the panel shear equally, we find from a cut made in the top panel (Fig. 9.3*c*)

$$P_{23} \cos \theta + P_{14} \cos \theta = 20$$

$$P_{23} = P_{14} = 10\sqrt{2} \text{ kN}$$

In a similar way, the forces in other diagonals are evaluated. They are,

$$P_{36} = P_{54} = 20\sqrt{2} \text{ kN}$$

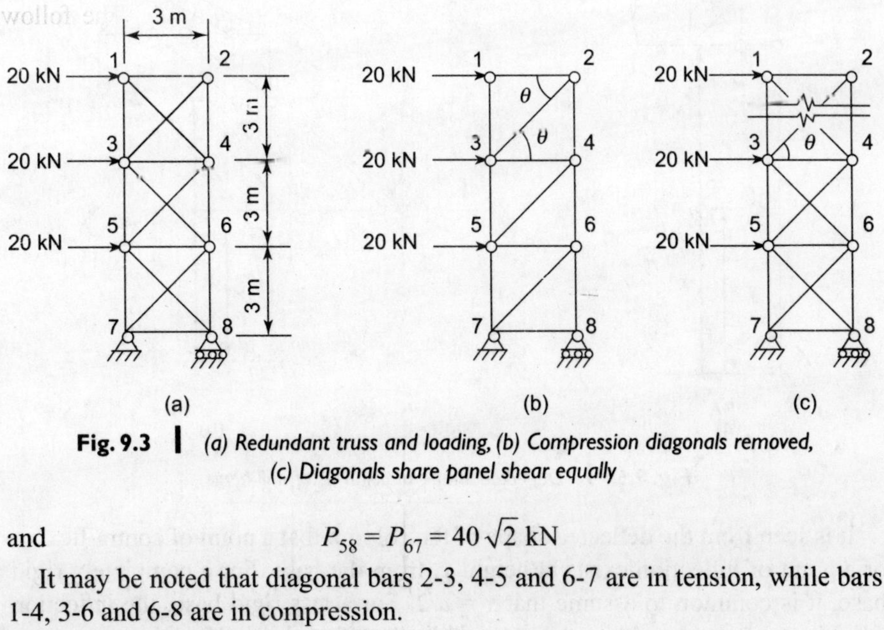

Fig. 9.3 | *(a) Redundant truss and loading, (b) Compression diagonals removed, (c) Diagonals share panel shear equally*

and $$P_{58} = P_{67} = 40\sqrt{2} \text{ kN}$$

It may be noted that diagonal bars 2-3, 4-5 and 6-7 are in tension, while bars 1-4, 3-6 and 6-8 are in compression.

9.2.3 Mill Bents

Another problem commonly encountered is the mill bent subjected to lateral loading as shown in Fig. 9.4. The bents are composed of trusses for the roof section supported by vertical columns. The columns run to the top of the bents and hinge connections exist between the column and truss.

To make rational assumptions for the approximate analysis of such structures, we must consider the manner in which the bents deflect under lateral loading. One possible deflected shape of the left hand side column of the bent in Fig. 9.4a is shown in Fig. 9.5*a*.

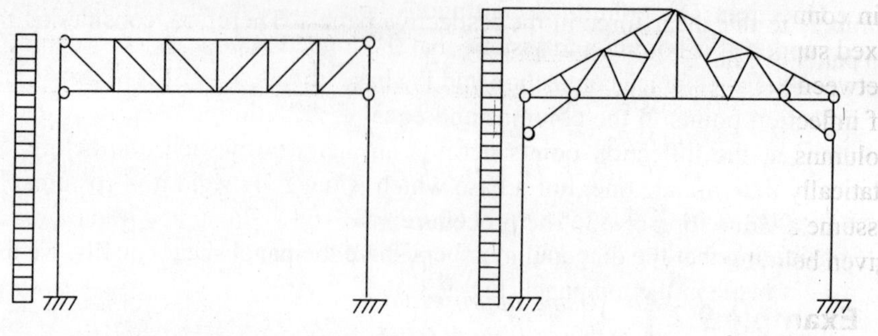

(a) (b)

Fig. 9.4 | *Mill bents: (a) Bent with flat roof, (b) Bent with sloping roof*

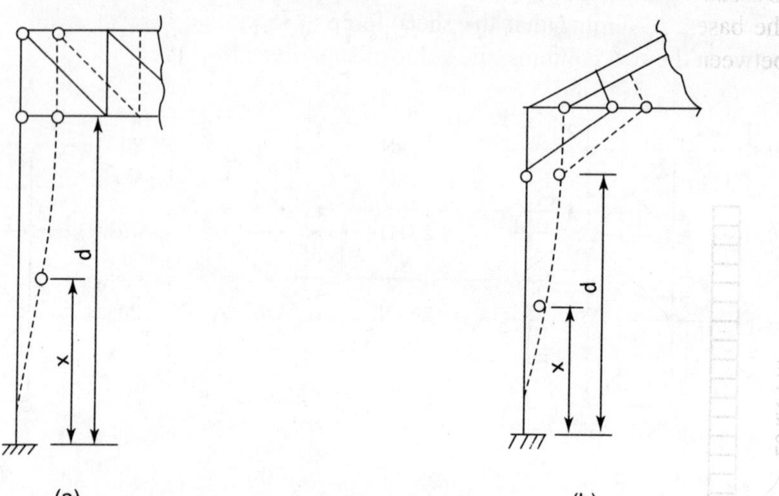

(a) (b)

Fig. 9.5 | *Deflected shape of columns of mill bents*

It is seen from the deflected shape of the column that a point of contra-flexure or a point of inflection exists at height x, from the base. For a completely rigid base, it is common to assume that $x = d/2$. For a less rigid base, the inflection point is at a lower level coinciding with the base for a hinged base.

The same reasoning is applied in fixing the contra-flexure point on the right hand side column. Since the structure is statically indeterminate by three degrees, one more assumption is necessary to make the structure a determinate one. A common third assumption is that the shear is equally shared by the columns at the inflection points. The shear in columns is equal to the summation of the horizontal forces above the level under consideration.

For the mill bent of Fig. 9.4*b*, the left hand side column will deflect as shown in Fig. 9.5*b*. The knee brace between the columns and the truss is considered to have been connected to the column at one end and the truss at the other by a pin connection. The column is continuous up to the bottom of the truss. For a fixed support it is common to assume that the inflection point is located midway between the knee brace connection and the base, that is, $x = d/2$. The assumption of inflection points in the columns and equal distribution of shear between the columns at the inflection points permits an analysis of such a structure as a statically determinate one. For a base which is not fully rigid it is customary to assume a value for $x = d/3$. The procedure involved is illustrated in the example given below.

Example 9.2 | *Using appropriate assumptions, determine for the mill bent of Fig. 9.6 the components of reaction at the bases and sketch the moment diagram for the windward and leeward columns.*

Because the columns are fixed at the bases, the inflection point in each column is assumed to be located at E and F as shown in Fig. 9.7 at a height of 4 m from the bases. Assuming that the shear force at the hinge level is divided equally between the two columns, the value of shear is

$$V_E = V_F = \frac{1}{2}(20)(6) = 60 \text{ kN}$$

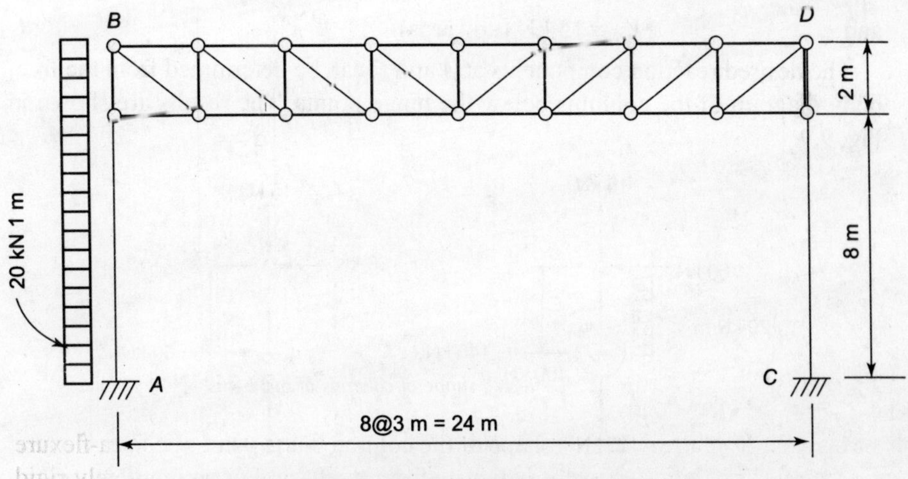

Fig. 9.6 | *Mill bent under lateral loading*

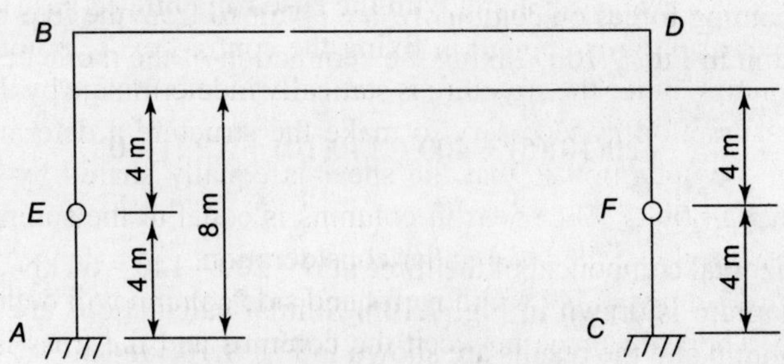

Fig. 9.7 | *Assumed position of hinges*

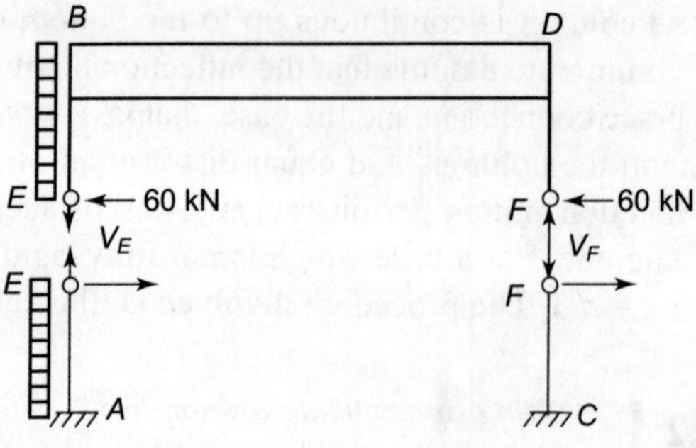

Fig. 9.8 | *Free-body diagram above hinge points*

The axial forces in the columns at the level of the hinges are evaluated by considering the free-body diagram of structure above the hinge points as shown in Fig. 9.8. Writing down the summation of the moments about F and equating to zero we have

$$-V_E(24) + 20\,(6)(3) = 0$$

or $$V_E = +\,15 \text{ kN (downwards)}$$

and $$V_F = 15 \text{ kN (upwards)}$$

The desired reaction components at A and C can be determined from the free-body diagram of the columns below the hinge points. The results are shown in Fig. 9.9

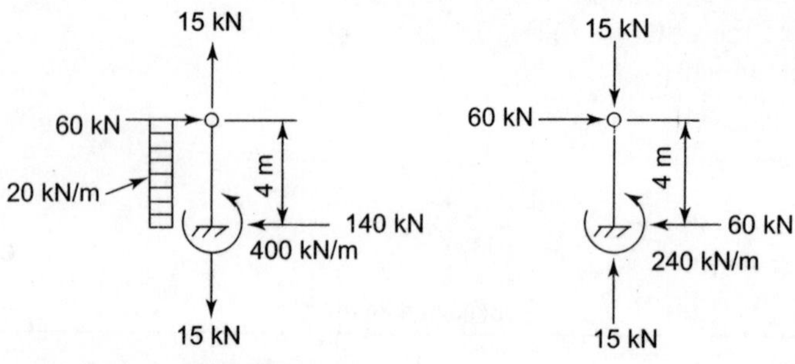

Fig. 9.9 | *Free-body diagram of columns below hinge points*

The remaining forces on column AB are obtained from the free-body diagram of the column in Fig. 9.10a. Taking the summation of the moments about B, we obtain

$$20(10)(5) + 400 - 140(10) + X_1(2) = 0$$

from which $X_1 = 0$

The horizontal component of the force at $B = 200 - 140 = 60$ kN. The resulting moment diagram is drawn in Fig. 9.10b. Similar calculations are made for the leeward column and the results are shown in Fig. 9.10c and d.

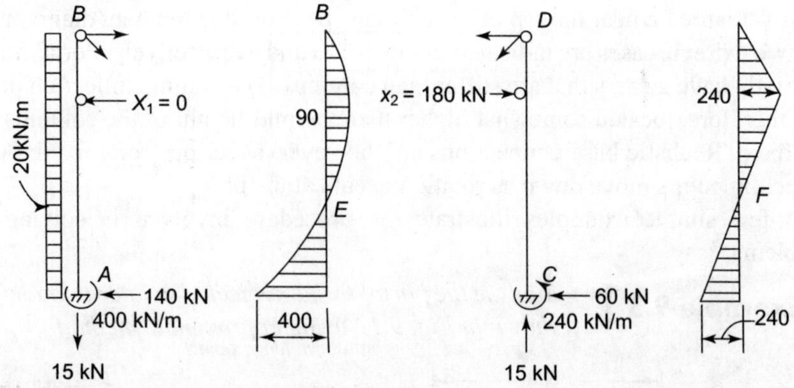

Fig. 9.10 | *(a) Forces on windward column, (b) Moment diagram for windward column, (c) forces on leeward column, (d) Moment diagram for leeward column*

9.2.4 Portal Frames

Laterally loaded portal frames can also be analysed by the approximations employed in Sec. 9.2.3 for mill bents. Consider, for example, a fixed base portal frame of Fig. 9.11*a*. Note that the deflected shape of a portal frame depends on the relative stiffness of columns and girder. Two extreme cases are considered. In Fig. 9.11*b* the deflected shape of the frame when the girder is very stiff in comparison with the columns is given. The points of contra-flexure lie at about mid-height of the columns. The deflected shape of the frame when the girder is flexible in comparison with the columns is given in Fig. 9.11*c*. The points of

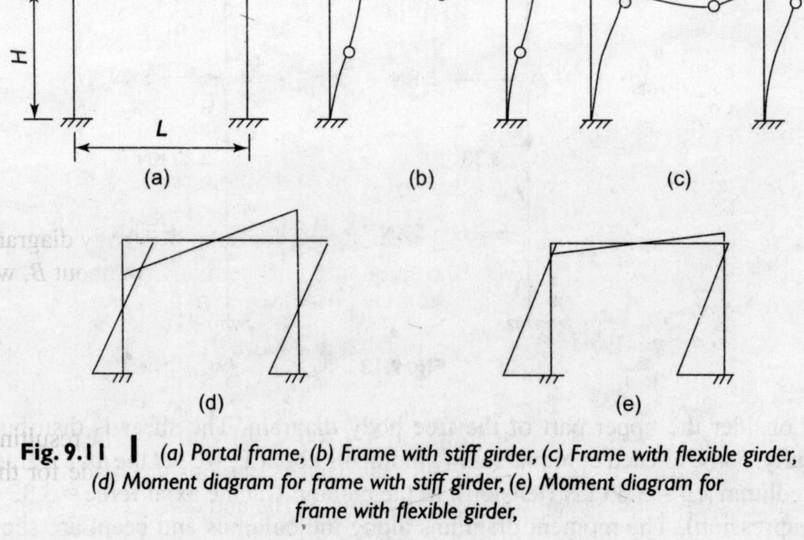

Fig. 9.11 | *(a) Portal frame, (b) Frame with stiff girder, (c) Frame with flexible girder, (d) Moment diagram for frame with stiff girder, (e) Moment diagram for frame with flexible girder,*

contra-flexure lie near the top of the column. The possible moment diagrams for the two extreme cases are indicated in Fig. 9.11*d* and *e* respectively. Portal frames normally have girders that are stiffer than columns. The column inflection points are, therefore, located somewhat higher than the mid-height of the columns that are fixed. Realistic base connections are, however, never perfectly fixed; hence inflection points move down as rotation occurs at the base.

A few simple examples illustrate the procedure involved in solving the problems

Example 9.3 | *Analyse the portal frame subjected to lateral loading as shown in Fig. 9.12 using approximate method.*

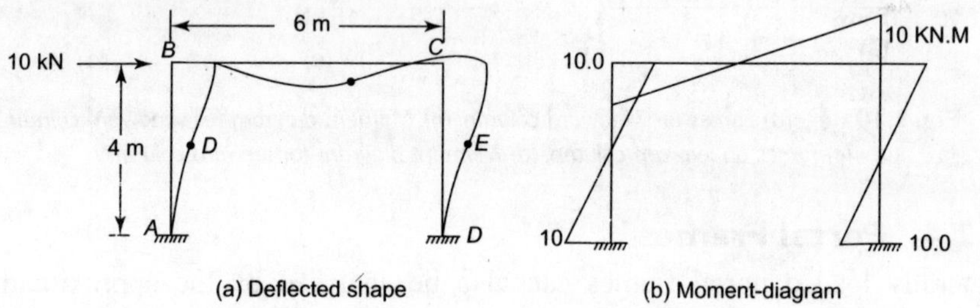

Fig. 9.12 | *Portal frame under lateral loading*

The elastic curve or the deflected shape is shown in Fig. 9.12. For the approximate analyses, it is logical to assume points of conflexure located at mid height of columns and mid span of beam. Now the frame is statically determinate and joint moments can be evaluated using equations of equilibrium. The free body diagram is shown 9.13.

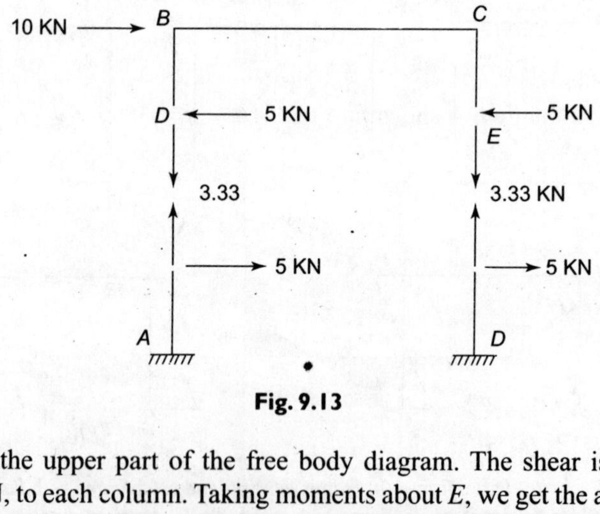

Fig. 9.13

Consider the upper part of the free body diagram. The shear is distributed equally, 5 kN, to each column. Taking moments about *E*, we get the axial force in the column *AB* = 3.33 kN (tension). In the column *CD* the axial force = 3.33 kN (compression). The moment diagrams along the columns and beam are shown in Fig. 9.12*b*.

Example 9.4 | *For the structure shown in Fig. 9.14a sketch the deflected shape; mark all points of contraflexure and construct approximate moment diagrams.*

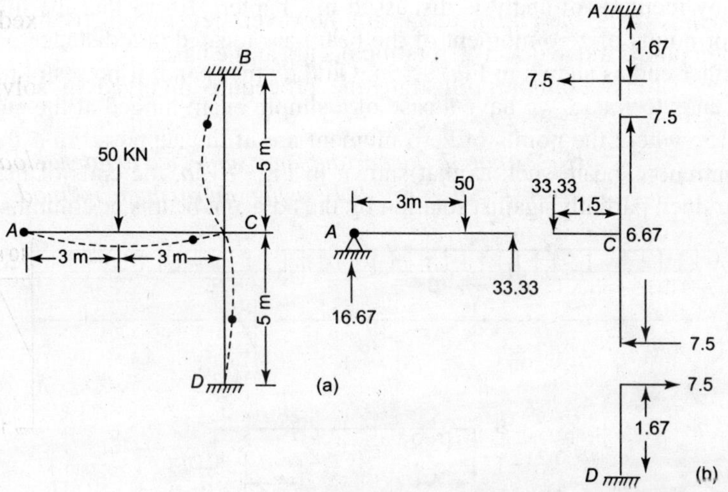

Fig. 9.14

The deflected shape indicates the approximate positions of points of contraflexure or hinge points. Assuming hinge points at 1.67 m from ends *B* and *D* and 1.5 m from C, free-body diagrams are shown in Fig. 9.14*b*. From the free-body diagrams the moments are calculated.

$$M_B = 7.5 \times 1.67 = 12.5 \text{ kN.m}$$

$$M_D = 7.5 \times 1.67 = 12.5 \text{ kN.m}$$

$$M_{CA} = 33.3 \times 1.5 = 50.0 \text{ kN.m}$$

$$M_{DC} = 7.5 \times 1.67 = 12.50 \text{ kN.m}$$

The moment diagram is shown in Fig. 9.14*c*.

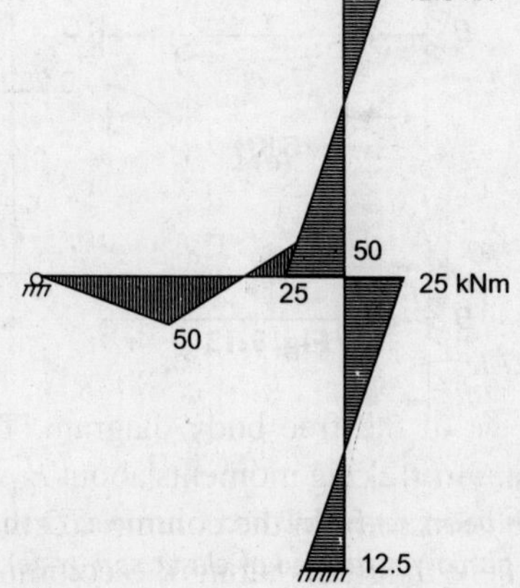

Fig. 9.14 (c) | *Moment diagram*

9.2.5 Continuous Beams and Building Frames

Consider the continuous beam shown in Fig. 9.15*a* under the distributed loading. If the ends of span *BC* do not undergo rotation as in a fixed end beam, it can be shown by methods of analysis discussed in Chapters 10–13 that the inflection points or points of zero moment of the beam are located at a distance of $0.21L$ from either end as shown in Fig. 9.15*b*. On the other hand, if no restraints exist for the ends to rotate, we have a case of a simple beam hinged at the end as in Fig. 9.15*c* where the points of zero moment are at the supports. In a frame or in a continuous beam such as that shown in Fig. 9.15*a*, the ends of the beam are restrained partially against rotation by the adjacent beams or columns which

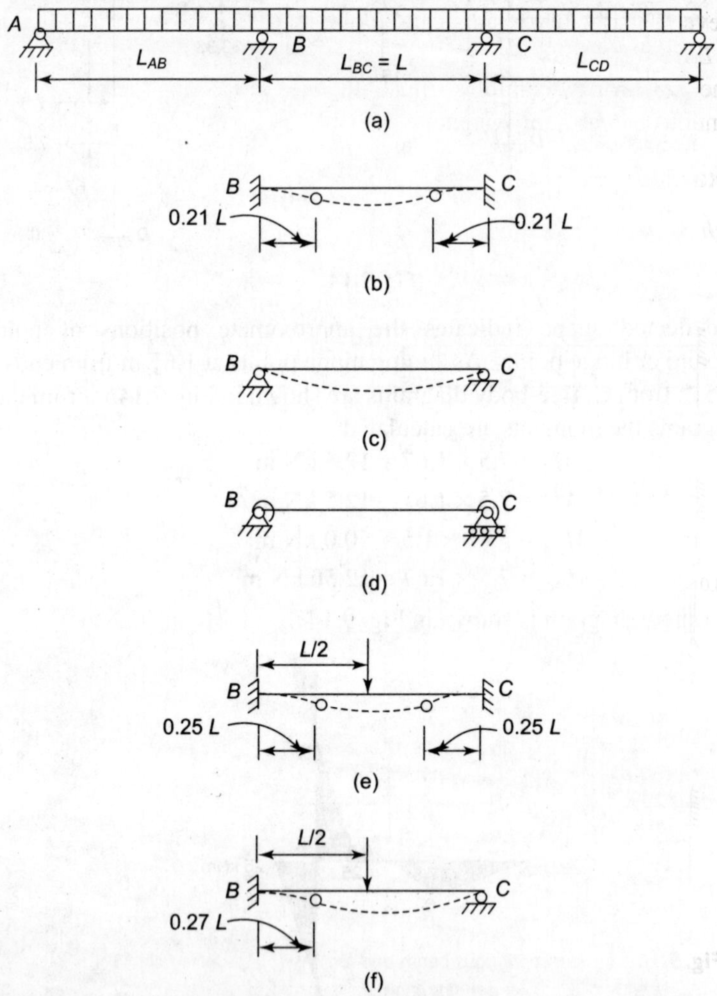

Fig. 9.15 | *(a) Continuous beam, (b) Ends B and C assumed fixed, (c) Ends B and C assumed hinged, (d) Ends B and C partially restrained by elastic springs, (e) Fixed end beam under concentrated load, (f) One end fixed and the other hinged*

in effect serve as elastic restraints. The elastic restraints can be represented by springs as in Fig. 9.15*d*. The location of inflection points for such an elastically restrained beam depends upon the stiffness of springs, but it must lie somewhere between the support points and $0.21L$ from the end as in Fig. 9.15*b*. A generally assumed location for points of zero moment is $0.1L$ from the support points. In case of beams subjected to central concentrated loads, the location of points of contra-flexure are at a distance of 0.25L from the ends (Fig. 9.15*e*). The position of the inflection point for a beam fixed at one end and hinged at the other is shown in Fig. 9.15*f*.

The positions of contra-flexure points do vary from $0.21L$ to $0.27L$ from the supports depending upon the type of loading and the restraints at the ends. However, satisfactory results can be obtained by selecting the inflection points at $0.1L$ from the ends.

The following examples illustrate the procedure followed in showing continuous beams approximately

Example 9.5 | *For the continuous beam shown in Fig. 9.16. Sketch the deflected shape and mark all points of centraflexure. Sketch the moment diagram.*

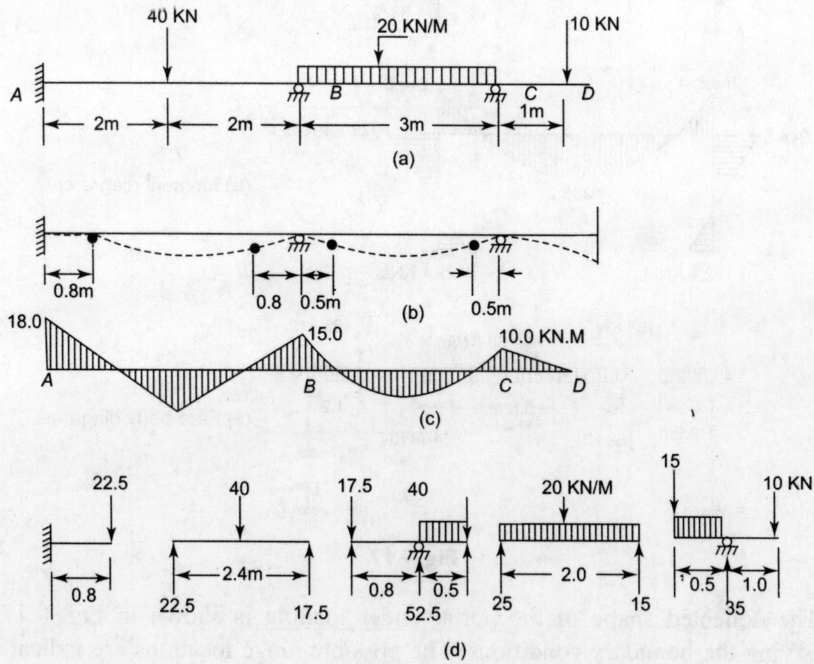

Fig. 9.16 | *(a) Continuous beam and loading (b) Deflected shape and hinge positions (c) Bending moment diagram (d) Free bodies*

The deflected shape of the continuous beam is drawn satisfying the boundary conditions. The approximate locations of points of contraflexure are decided on the basis of support conditions and loading on the span. The hinge points are shown in Fig. 9.16*b* making the distances from the adjacent supports.

The free-body diagrams are drawn extending from hinge point to hinge point. The support reactions and moments are worked out commencing from right end using equilibrium equation $\Sigma F_Y = 0$. It may be noted that the moments and reactions arrived at satisfy only $\Sigma F_Y = 0$ and not the moment $\Sigma M_B = 0$. This is expected as we have obtained moments on the basis of assumed hinge positions.

The moment diagram is drawn in Fig. 9.16*c* which is super imposition of span moment diagrams over support moment diagram.

Example 9.6 | *Using approximate method of analysis, analyse the fixed portal frame shown in Fig. 9.17. Sketch the deflected shape and moment diagram.*

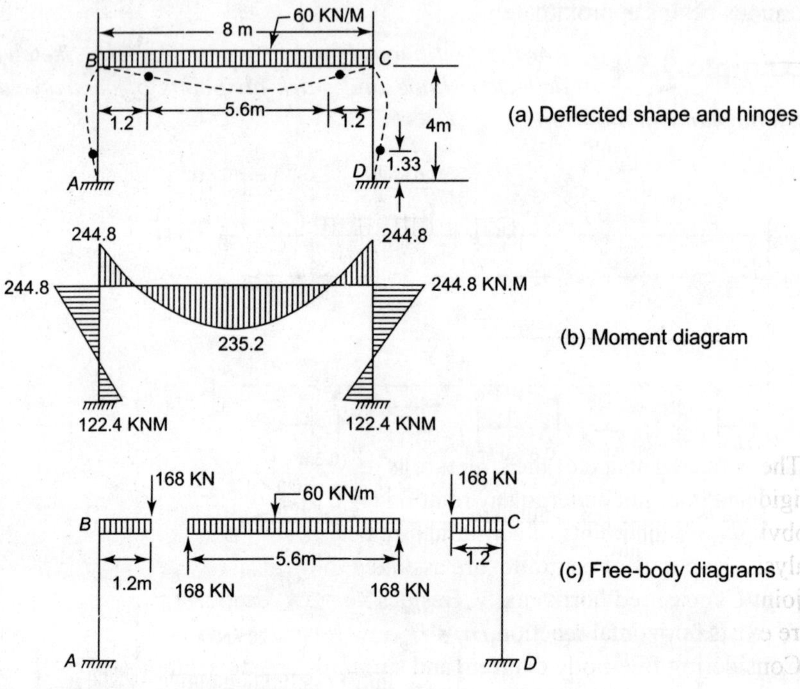

Fig. 9.17

The deflected shape of the portal under loading is shown in Fig. 9.17 a satisfying the boundary conditions. The possible hinge locations are indicated. For the beam the hinges are assumed to be located at 0.15 *l* from ends *B* and *C*. The hinges in the columns are assumed to be located at 1.33 m from the base.

The free body diagram of the three parts are shown in Fig. 9.17*c*.

$$\text{Moment at } B = 168 + \frac{60}{2}(1.2)^2 = 244.8 \text{ kN.m}$$

Moment at C = Moment at B = 244.8 kN.m

Moment at $A = \dfrac{244.8}{2} = 122.4$ kN.m

Moment at D = 122.4 kN.m

The moment diagram is shown in Fig. 9.17*b*.

Example 9.7 | *For the structure shown in Fig. 9.18. Sketch the deflected shape and mark all points of contraflexure and construct approximate moment diagram.*

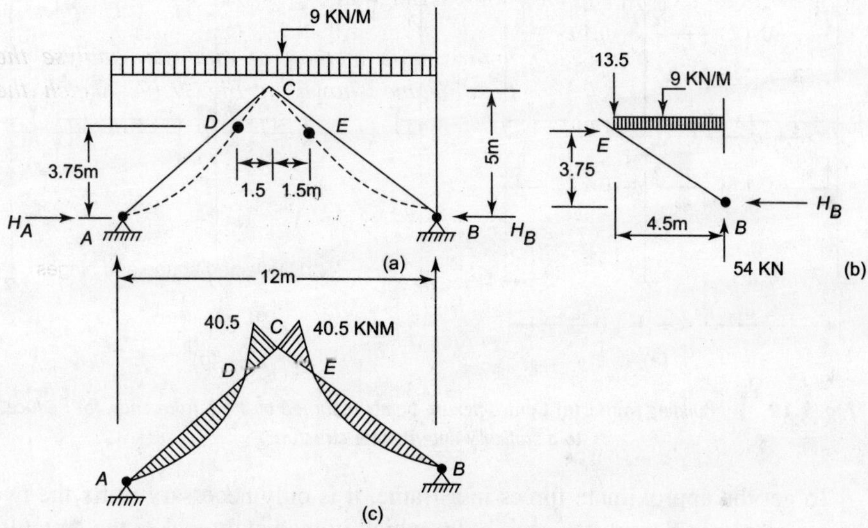

Fig. 9.18

The deflected shape of the structure is drawn keeping in mind that the joint C is rigid and does not undergo any rotation (Fig. 9.18*a*). Hence the deflected shape is obvious and the points of contraflexure occurs at D and E. In the approximate analysis the points of contraflexure assumed to be located at 1.5 m on either side of joint C measured horizontally. Besides vertical reactions at supports A and B there exists horizontal reaction $H_A = H_B$ inwards.

Considering free-body diagram and summing moments about E.

$$M_E = 54\,(4.5) - \frac{9}{2}(4.5)^2 - H_B\,(3.75) = 0$$

Gives $\quad H_B = 40.5$ kN

Due to symmetry $\quad H_A = H_B = 40.5$ kN.

Moment at C

$$M_C = 54(6) - \frac{9}{2}(6)^2 - 40.5\,(5)$$

$$= -\,40.5 \text{ kN.m.}$$

The moment diagram is shown in Fig. 9.18*c*

It is often useful to be able to approximately analyse a building frame such as the one shown in Fig. 9.19a subjected only to gravity loading.

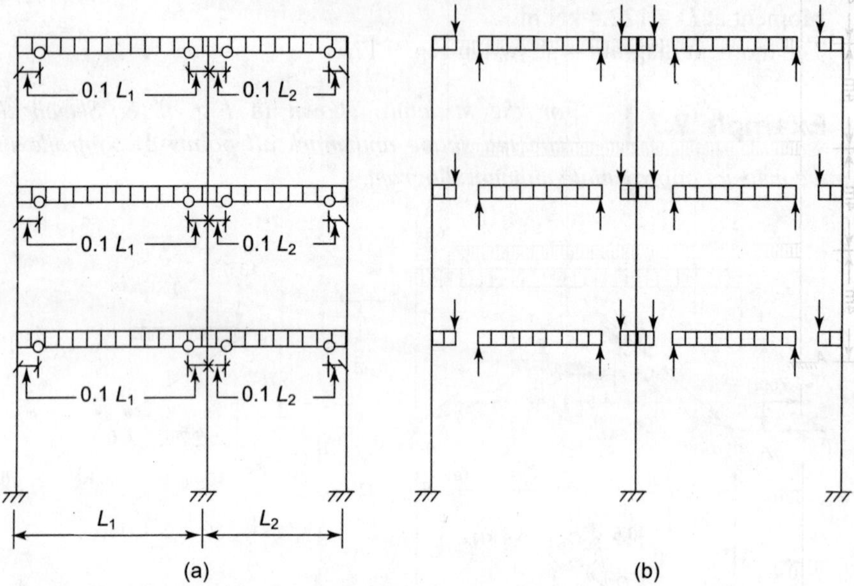

Fig. 9.19 | *Building frame: (a) Contra-flexure points assumed at 0.1 L from ends, (b) Reduced to a statically determinate structure*

To get the approximate forces in a frame, it is only necessary to fix the two points of contra-flexure for each beam which immediately makes the structure statically determinate. Figure 9.19*b* shows such a determinate structure consisting of simple beams and cantilevered columns. However, it may be noted that the determinate structure offers very little resistance to horizontal loading.

The procedure is illustrated by the following example.

Example 9.8 | *Using approximate method, analyse the building frame consisting of three bays and four storeys as shown (Fig. 9.20a). Sketch the deflected shape and moment diagram approximately.*

The deflected shape of the frame under gravity loads is shown in Fig. 9.20b. It is seen that in all the beams there are two inflexion points in each bay and one inflexion point in each of the columns on all floors. In the approximate analysis, the inflexion points are assumed located at 0.1 *l* from each end of the beam. The inflexion points in the columns are assumed at mid height in each floor. The inflexion points are shown on the deflected shape of the frame. The moments at the end of the beams and columns are worked out commencing from left end. In Fig. 9.20 c the reduced statically determinate structure is shown, Fig. 9.20*d* gives the conceptual transmission of forces in a typical bay.

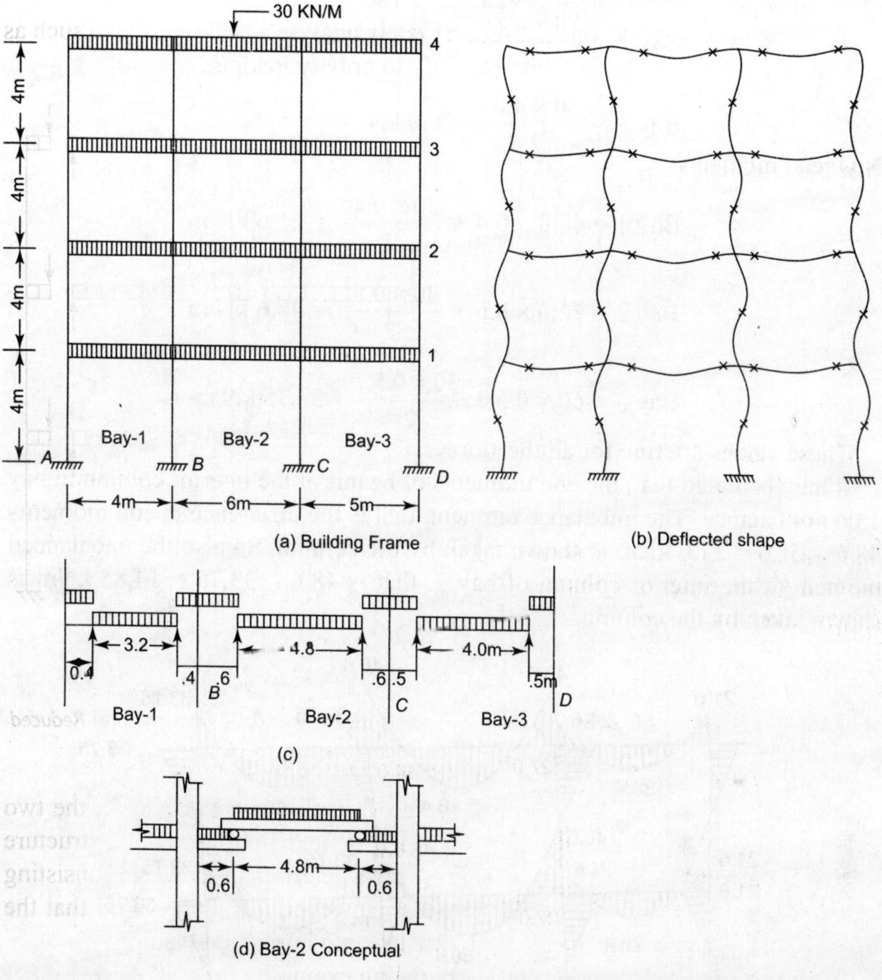

Fig. 9.20

Maximum +ve moment in beams

$$\text{Bay } 1 = \frac{30 \times 3.2^2}{8} = 38.4 \text{ kN.m}$$

$$\text{Bay } 2 = \frac{30 \times 4.8^2}{8} = 86.4 \text{ kN.m}$$

$$\text{Bay } 3 = \frac{30 \times 4.0^2}{8} = 60.0 \text{ kN.m}$$

End reactions

$$\text{Bay } 1 = \frac{30 \times 3.2}{2} = 48.0 \text{ kN}$$

$$\text{Bay } 2 = \frac{30 \times 4.8}{2} = 72.0 \text{ kN}$$

$$\text{Bay } 3 = \frac{30 \times 4.0}{2} = 60.0 \text{ kN}$$

Now end moments

$$\text{Bay } 1 = 48.0 \times 0.4 + \frac{30 \times 0.4^2}{2} = 21.6 \text{ kN.m}$$

$$\text{Bay } 2 = 72.0 \times 0.6 + \frac{30 \times 0.6^2}{2} = 48.6 \text{ kN.m}$$

$$\text{Bay } 3 = 60 \times 0.5 + \frac{30 \times 0.5^2}{2} = 33.75 \text{ kN.m}$$

These values are true for all the storeys.

It may be noted that, the end moments of beams at the interior column in bay 1 do not balance. The imbalance moment, that is the difference in end moments 48.6 – 21.6 = 27.0 kNm is shown taken by the column. So also the unbalanced moment at the interior column of bay 3, that is 48.6 – 33.75 = 14.85 kNm is shown taken by the column.

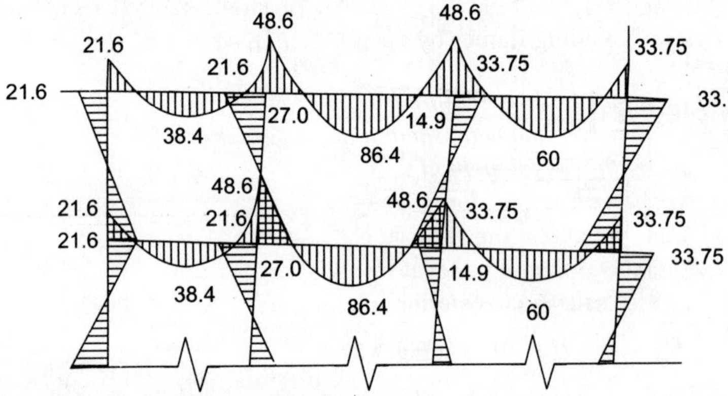

Fig. 9.21 | *Moment diagram in top storey*

The moment diagram is sketched in Fig. 9.21 giving the approximate values of moments. The moment diagram is repetitive for all the storeys below.

In addition to the approximate procedures discussed above, there are other methods for the approximate analysis of building frames subjected to lateral loading. The two common methods widely used are the *cantilever method* and the *portal method* which are discussed below.

9.3 PORTAL METHOD

The portal method is an approximate analysis used for analysing building frames subjected to lateral loading such as the one shown in Fig. 9.22*a*. This method is more appropriate for low rise (say height is less than width) building frames. In the analysis the following assumptions are made:

1. An inflection point is located at mid-height of each column;
2. An inflection point is located at the centre of each beam; and
3. The horizontal shear is divided among all the columns on the basis that each interior column takes twice as much as the exterior columns.

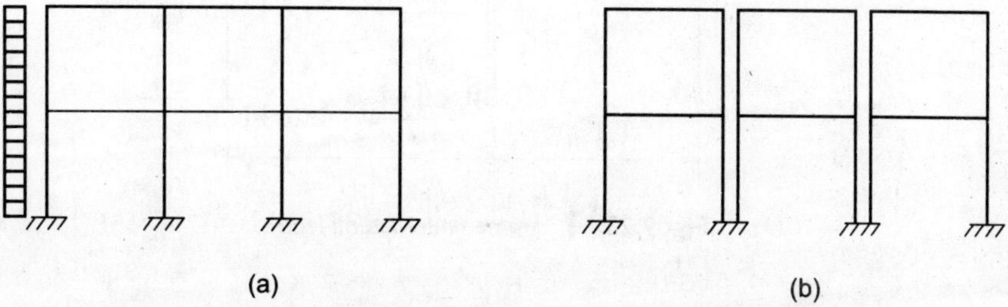

Fig. 9.22 | *(a) Building frame under lateral loading, (b) Equivalent portals*

The basis for the last assumption stems from the reasoning that the frame is composed of. individual portals as in Fig. 9.22*b*.

Obviously an interior column is in effect resisting the shear of two columns of the individual portals. The following example illustrates the procedure involved in the analysis of building frames by the portal method.

Example 9.9 | *It is required to determine the approximate values of moment, shear and axial force in each member of frame in Fig. 9.23 using the portal method.*

Considering first the upper storey, inflection points are assumed at mid-height on each column. We obtain the shear in each column from a free-body diagram of the structure above the hinge level by assigning shear to the interior column equal to twice the shear in the exterior column as shown in Fig. 9.24*a*.

$$H + 2H + H = 20$$

or

$$H = 5 \text{ kN}$$

Inflection points are also assumed at the centre of beams *GH* and *HK*. The member forces in the upper part of the frame can be evaluated from the free-body diagram of the parts shown in Fig. 9.24*b* beginning either with *G* or from *K* and working across. The resulting forces must check with the free-body diagram at the opposite end. The resulting forces are indicated on the diagram.

Again inflection points are assumed at mid-height of lower storey columns and the shear is distributed as in the upper storey. Thus, in the lower storey

$$H + 2H + H = 60$$

$$H = 15 \text{ kN}$$

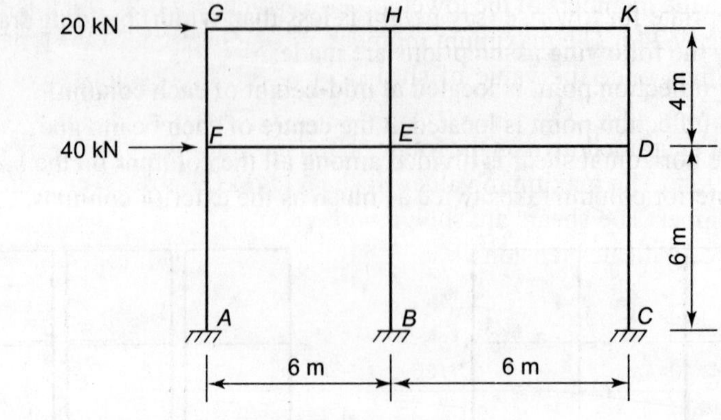

Fig. 9.23 | *Frame under lateral load*

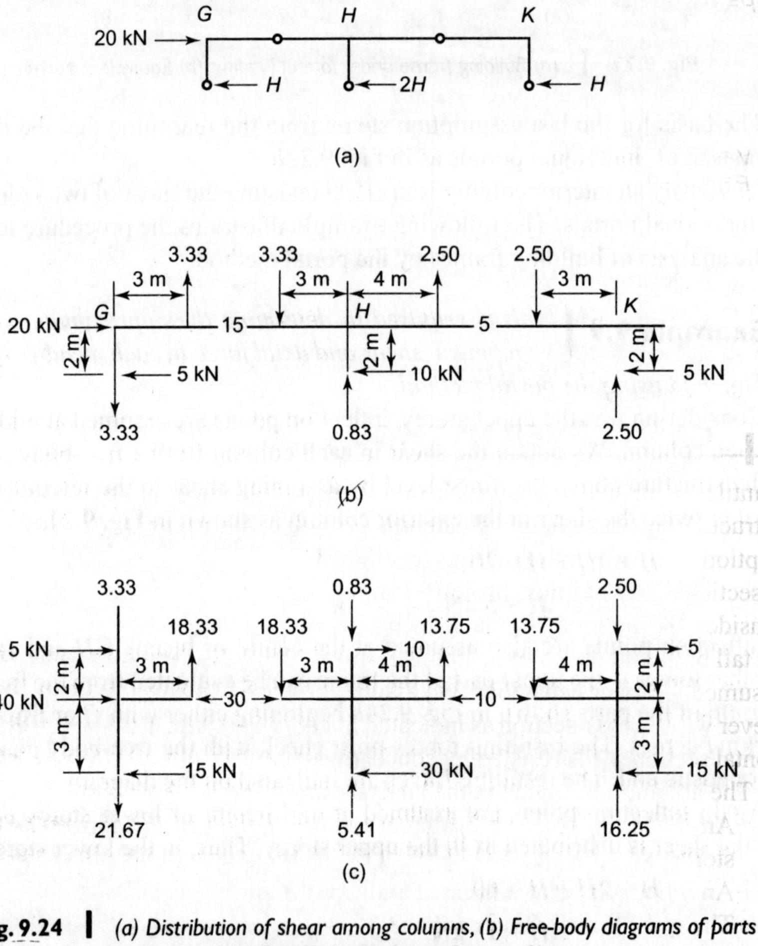

Fig. 9.24 | *(a) Distribution of shear among columns, (b) Free-body diagrams of parts of upper storey, (c) Free-body diagrams of parts of lower storey*

The forces in the members of the lower storey are obtained from the free-body diagrams of Fig. 9.24*c*. The maximum moment in each member of the structure is readily obtained once the value of shear at the inflection points have been determined.

The moment diagram drawn on the frame on the tension side of the members is shown in Fig. 9.25. The maximum values of shear and axial forces are indicated along each member. The shears are shown without any sign. The positive value of the axial force indicates tension.

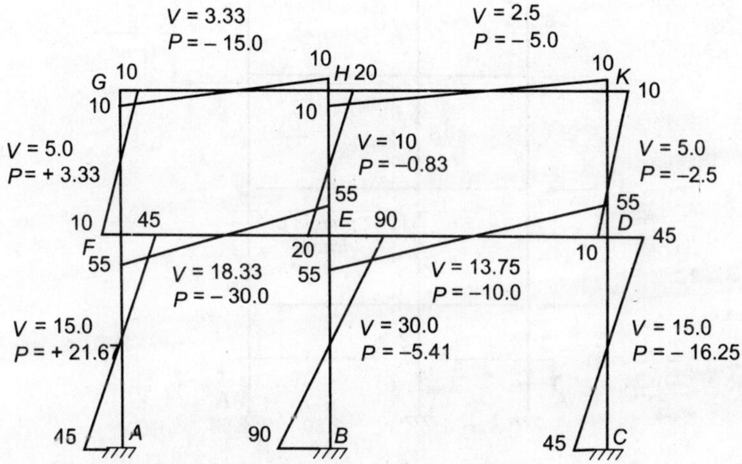

Fig. 9.25 | *Forces in frame members*

9.4 CANTILEVER METHOD

The cantilever method of analysis is more appropriate for a tall structure, that is, for a structure that has a height greater than its width. This method is based on the assumption that the building frame acts like a cantilever beam with the column cross-sectional areas as the fibres in a beam.

Consider the building frame loaded laterally as shown in Fig. 9.26*a*. For such a tall building, the column strains resulting from the overall bending action are assumed to affect behaviour. We assume that the frame is a laterally loaded cantilever with a cross-section as indicated in Fig. 8.26*b*. The moment at a typical horizontal section *AA* is resisted by concentrated column forces as shown in Fig. 9.26*c*. The assumptions made in the analysis are:

1. An inflection point is located at the mid-height of the column in each storey;
2. An inflection point is located at the mid-point of each beam; and
3. The axial force in each column is proportional to its distance from the centroid of the areas of the column group at that level.

The first two assumptions are the same as in the portal method. The third assumption gives the distribution of the axial forces in the columns instead of the distribution of the shear force among the columns as in the portal method. The last assumption enables one to include the effects of columns having different cross-sectional areas. The example that follows illustrates the procedure to be followed in analysing a frame by the cantilever method.

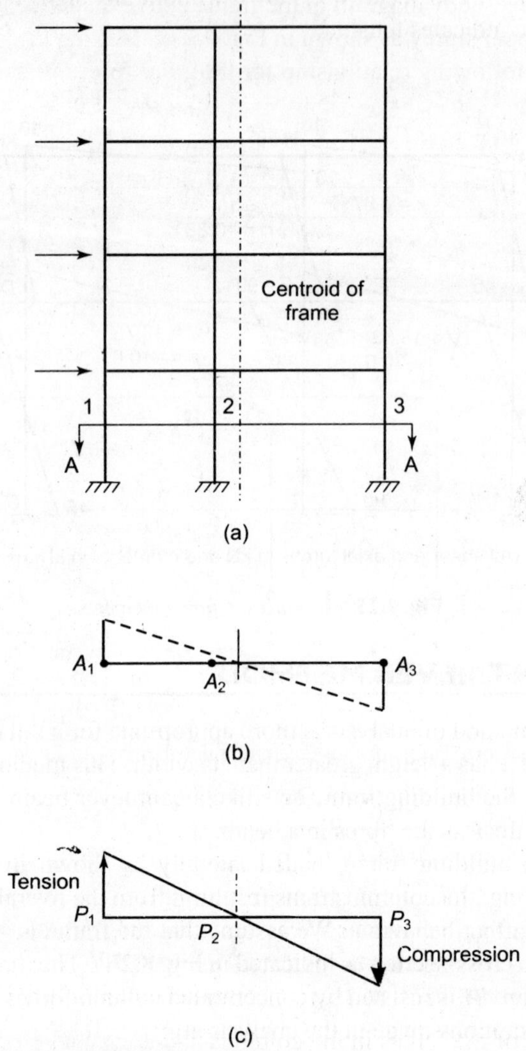

Fig. 9.26 | *(a) Building frame, (b) Cross-section of frame, (c) Axial forces in columns*

Example 9.10 | *Use the cantilever method to perform an approximate analysis of the frame analysed in Example 9.9. The cross-sectional areas of the columns are all assumed to be equal.*

The location of the centroid of the column areas is to be determined first. Taking the summation of the moments of areas of columns about the left end column, we have

$$x = \frac{6+14}{3} = 6.67 \text{ m}$$

Consider the free-body diagram of the frame above the inflection points of the columns in the upper storey as shown in Fig. 9.27*a*. Making use of assumption 3 we can write the following relationship for the axial forces in the columns:

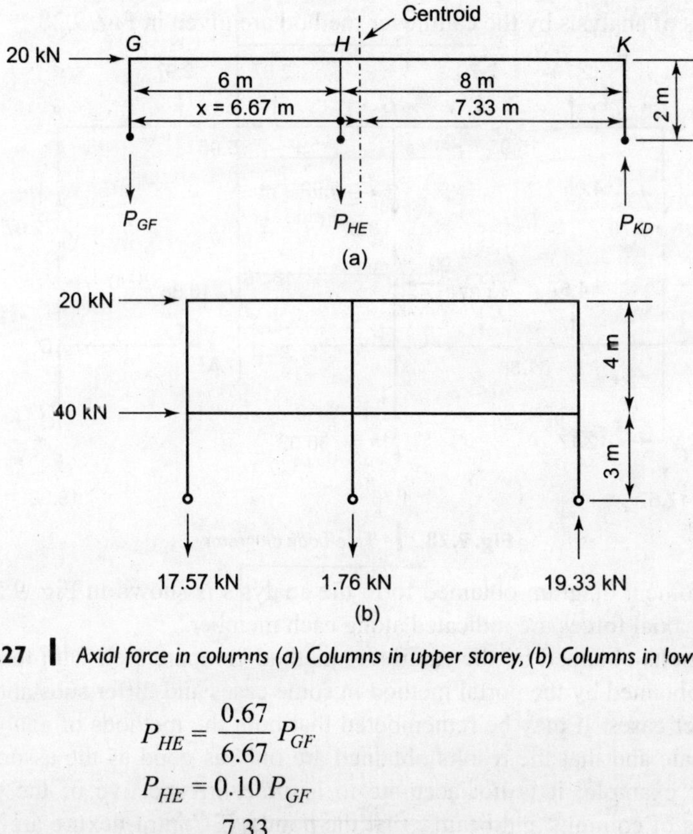

Fig. 9.27 | *Axial force in columns (a) Columns in upper storey, (b) Columns in lower storey*

$$P_{HE} = \frac{0.67}{6.67} P_{GF}$$

or
$$P_{HE} = 0.10\, P_{GF}$$

and
$$P_{KD} = \frac{7.33}{6.67} P_{GF}$$

or
$$P_{KD} = 1.10\, P_{GF}$$

Taking moments of the forces in the columns about the hinge point on the right end column, we have

$$20(2) - P_{GF}(14) - P_{HE}(8) = 0$$

Simplifying, we get

$$P_{GF} = 2.70 \text{ kN (tension)}$$
$$P_{HE} = 0.27 \text{ kN (tension)}$$

and $\qquad P_{KD} = -2.97$ kN (compression).

The columns to the left of the centroid are in tension and the column on the right is in compression.

Having determined the axial forces in the columns, the member forces are determined from the free-body diagrams as in the portal method.

The column forces in the lower storey are obtained by the same procedure used for the upper storey. While taking moments about the inflection point on the right end column, both the external loads are to be considered. The column forces are shown in Fig. 9.27*b*. The free-body diagrams for both stories showing the results of analysis by the cantilever method are given in Fig. 9.28.

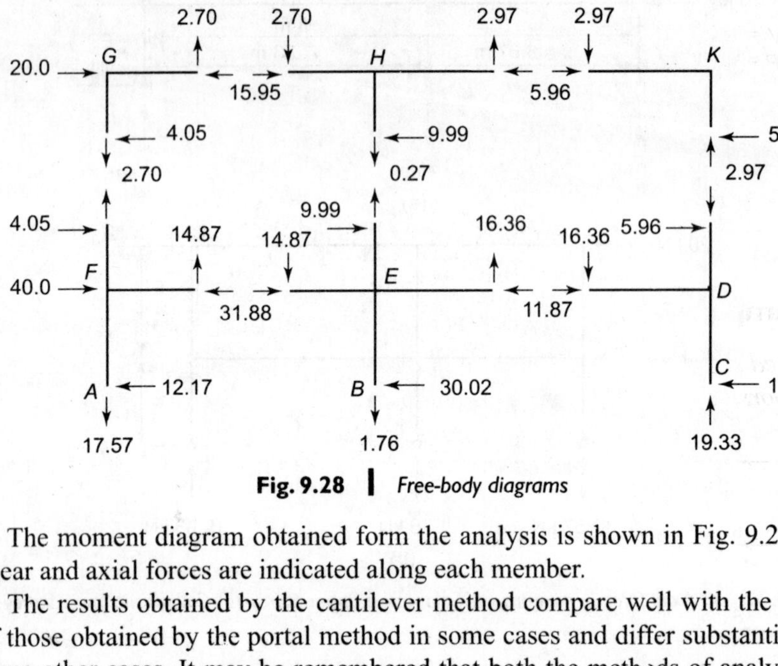

Fig. 9.28 | *Free-body diagrams*

The moment diagram obtained form the analysis is shown in Fig. 9.29. The shear and axial forces are indicated along each member.

The results obtained by the cantilever method compare well with the values of those obtained by the portal method in some cases and differ substantially in some other cases. It may be remembered that both the methods of analysis are approximate and that the results obtained are only as good as the assumptions used. For example, it is not accurate to assume, irrespective of the relative stiffnesses of columns and beams, that the points of Contra-flexure are located at mid-height of the storey on the columns and at mid-span points on the beams. It is also seen that a change in the cross-sectional area of columns can influence the location of the centroid of the column areas and, thus, appreciably alter the results, a fact not accounted for in the portal method.

Discussions on the accuracy of portal and cantilever methods as well as presentation of other approximate methods can be found in Norris and Wilbur.* The analyses presented for statically indeterminate structures in Chapters 10-14

---

*Norris, C.H. and Wilbur, J.B., Elementary Structure Analysis, McGraw-Hill Book Co., New York, 1960.

can be used to obtain exact results for the laterally loaded frame discussed above.

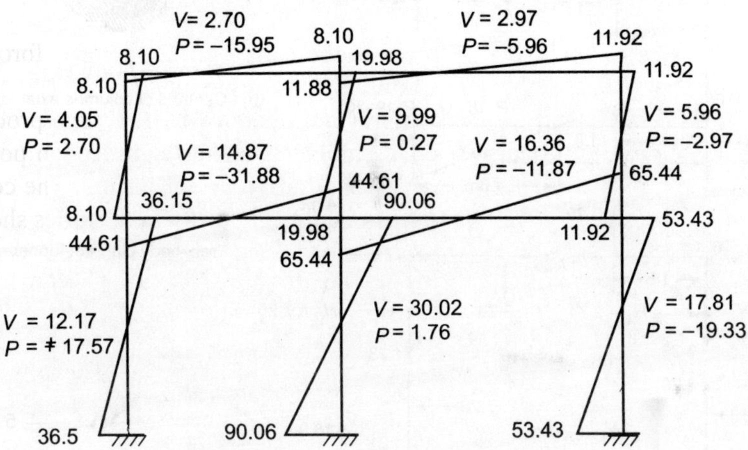

Fig. 9.29 | *Forces in frame members*

Example 9.11 | *Use cantilever method to perform an approximate analysis for the frame shown in Fig. 9.30. Draw deflected shape and sketch moment diagram. Carry out the analysis for the top two floors.*

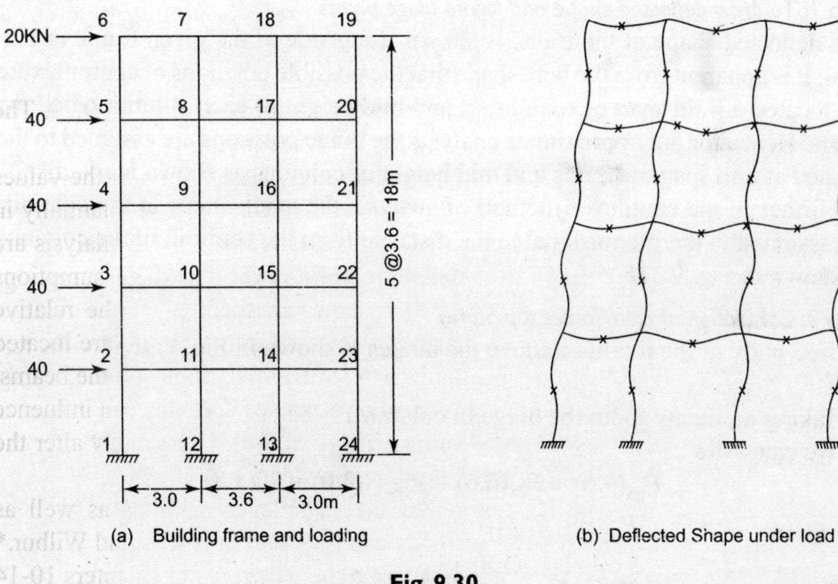

Fig. 9.30

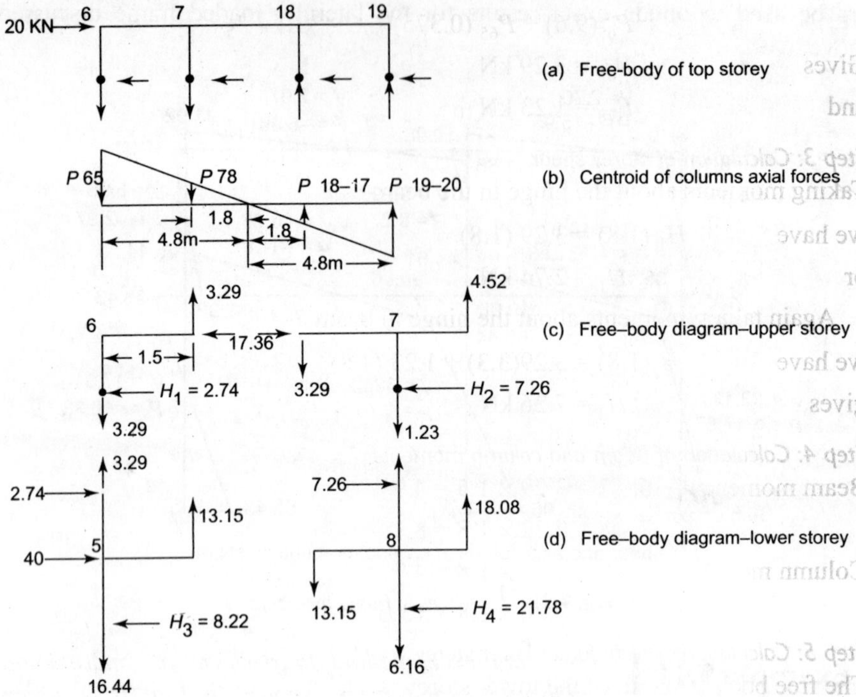

Fig. 9.31

Step 1: To draw deflected shape and locate hinge points

The deflected shape of the frame is shown along side of the given frame in Fig. 9.30. It is apparent from the bent shape that the possible positions of centraflexure are located at mid span of each beam and mid height of each column on all the floors. Hence for our approximate analysis the hinge positions are assumed to the located at mid span of beams and mid height of columns as shown marked.

Further in the cantilever method of analysis the axial forces in the columns are assumed to the proportional to the distance from the centroid of the columns as shown in Fig. 9.31*b*.

Step 2: Calculation of axial forces top Storey

Te free-body of the top floor above the hinges is shown in Fig. 9.31*c* for the left half.

Taking moments about the hinge in column 19–20

We can write

$$P_{65}(9.6) + P_{78}(6.6) - P_{18-17}(3.0) = 20(1.8)$$

We know $\qquad P_{78} = -P_{18-17}$

and $\qquad P_{78} = P_{65}\,\dfrac{(1.8)}{4.8} = 0.375\ P_{65}$

Substituting in the above equation

$$P_{65}(9.6) + P_{65}(0.375)(6.6 - 3.0) = 36$$

Gives $P_{65} = 3.29$ kN

and $P_{78} = 1.23$ kN

Step 3: Calculation of storey shear

Taking moments about the hinge in the beam 6–7

we have $H_1(1.8) = 3.29(1.8)$

or $H_1 = 2.74$ kN

Again taking moments about the hinge in beam 7–18

we have $H_2(1.8) = 3.29(3.3) + 1.23(1.8)$

gives $H_2 = 7.26$ kN

Step 4: Calculation of Beam and column moments

Beam moments $6 - 7 = 3.29 \times 1.5 = 4.94$ kN.m

$7 - 8 = 4.52 \times 1.8 = 8.14$ kN.m

Column moments $6 - 5 = 2.74 \times 1.8 = 4.93$ kN.m

$7 - 8 = 7.26 \times 1.8 = 13.07$ kM.m

Step 5: Calculation of axial forces lower storey

The free body diagram of the lower storey is shown in Fig. 9.31*d*

Taking moments about the hinge in column 19 – 20

$$P_{54}(9.6) + P_{89}(6.6) - P_{17-16}(3.0) = 20(5.4) + 40(1.8)$$

We know $P_{89} = -P_{17-16}$

and $P_{89} = P_{54}\dfrac{(1.8)}{4.8} = 0.375\,P_{54}$

Substituting in the above equation

We have $P_{54}(9.6) + P_{54}(0.375)(6.6 - 3.0) = 180$

Gives $P_{54} = 16.44$ kN

and $P_{89} = 0.375(16.44) = 6.17$ kN

Step 6: Calculation of storey shear in the lower storey

Taking moments about the hinge in beam 5–8

We have $H_3(1.8) + 2.74(1.8) - (16.44 - 3.29)(1.5) = 0$

Gives $H_3 = 8.22$ kN

Again taking moments about the hinge in beam 8–17

We have $H_4(1.8) - (6.16 - 1.23)(1.8) - 13.15(3.3) + 7.26(1.8) = 0$

Gives $H_4 = 21.78$ kN.

Step 7: Calculation of beam and column moments

Beam moments $5 - 8 = 13.15(1.5) = 19.73$ kN.m

$8 - 17 = 18.08(1.8) = 32.54$ kN.m

Column moments $\quad 5–4 = 8.22\,(1.8) = 14.80$ kN.m

$8–9 = 21.78\,(1.8) = 39.20$ kN.m

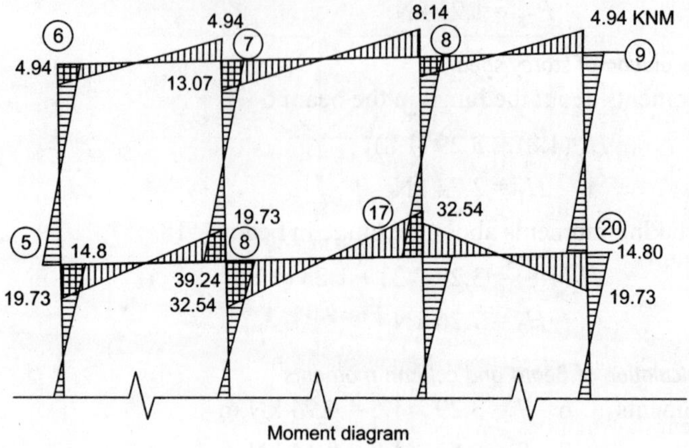

Fig. 9.32

The moments in the other half of the frame are identical due to symmetry. The procedure can be extended to lower floors on similar lines

Problems for Practice

9.1, 9.2 For the structures shown in Figs 9.33 and 9.34, sketch the deflected shape, mark all points of Contra-flexure and construct approximate moment diagrams.

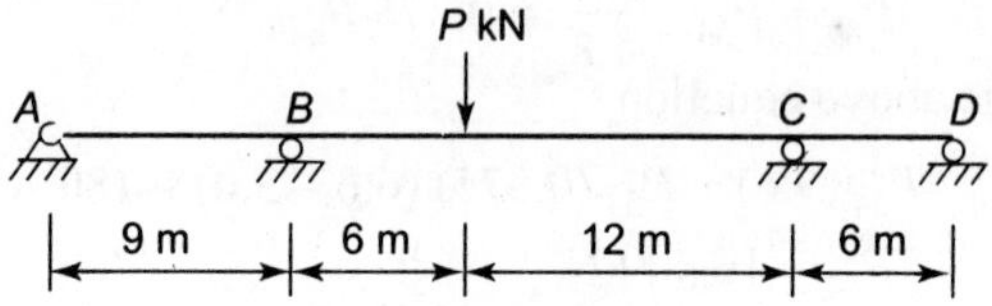

Fig. 9.33

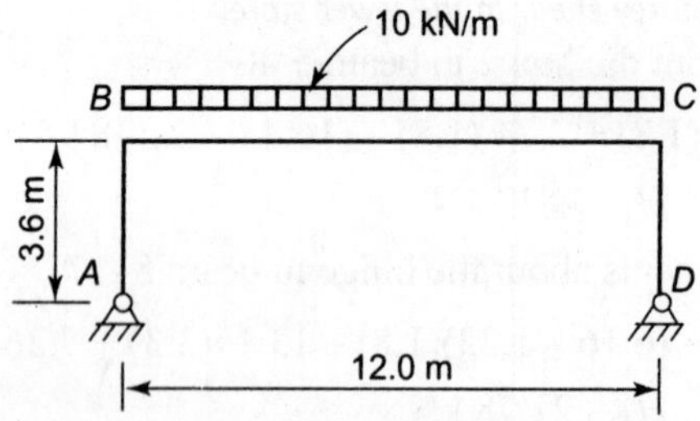

Fig. 9.34

9.3 Determine the axial forces in diagonal members of the truss shown in Fig. 9.35 assuming that: (a) the diagonals are slender and carry no compressive forces; (b) shear in each panel is divided equally between the diagonals.

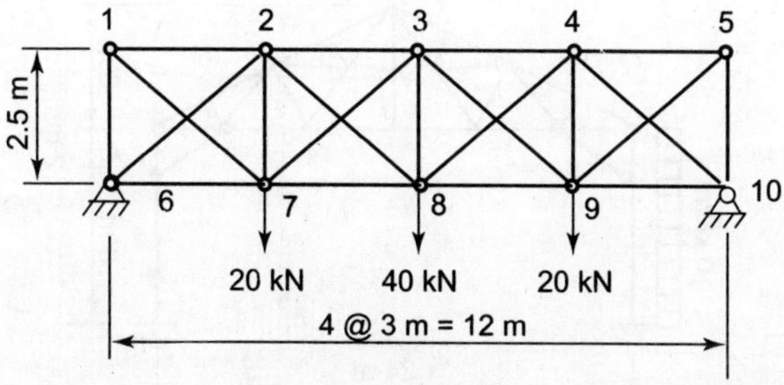

Fig. 9.35

9.4 (a) Determine the reaction components at the base of the columns, (b) sketch the shear and moment diagrams for the columns and (c) find the axial force in member 1-2 of the mill bent shown in Fig. 9.36. Make the usual simplifying assumptions.

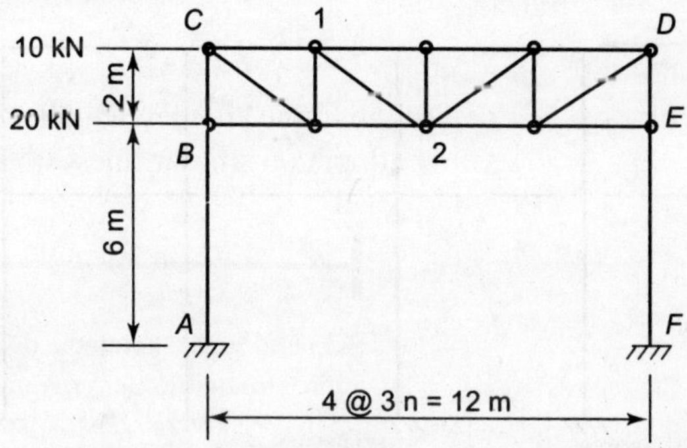

Fig. 9.36

9.5 For the mill bent shown in Fig. 9.37, sketch the moment diagram for the columns and find the forces in the members of the truss.

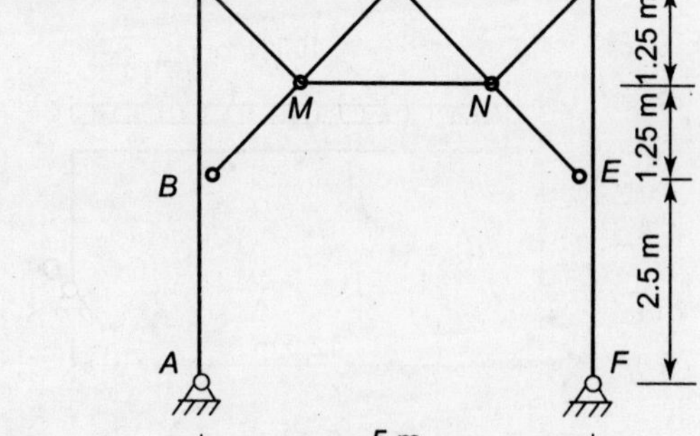

Fig. 9.37

9.6 Analyse the mills bent shown in Fig 9.38 for the reaction components at the bases and sketch the moment diagram for the columns.

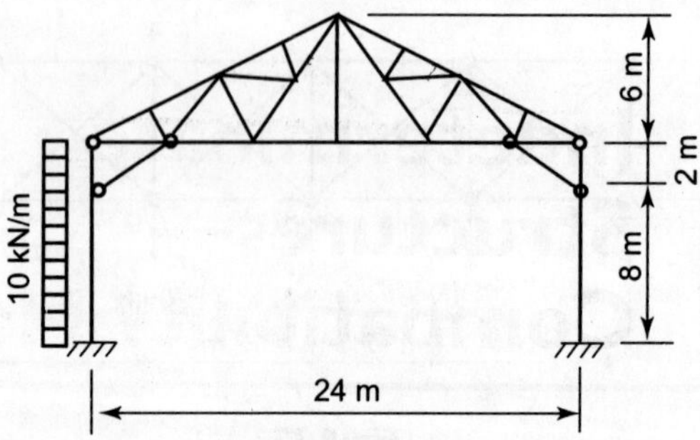

Fig. 9.38

9.7 Use the portal method to perform an approximate analysis for the frame in Fig. 9.39. Show the results by drawing the moment diagram for the entire frame.

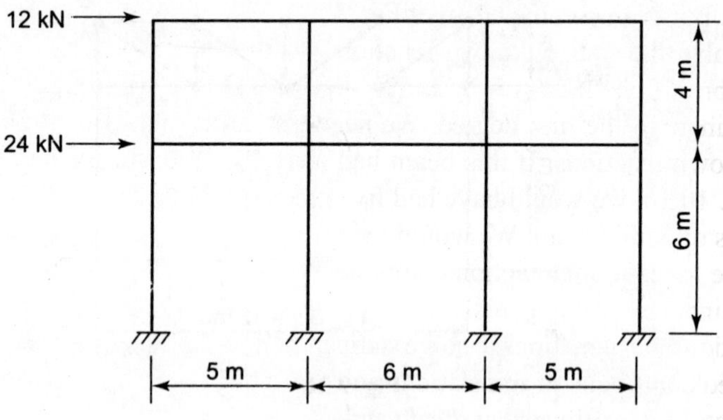

Fig. 9.39

9.8 Solve problem 9.7 by the cantilever method. Assume that the columns have equal areas.

10 Indeterminate Structures—Compatibility Methods

10.1 INTRODUCTION

Any structure whose reaction components or internal stresses cannot be established by using the equations of static equilibrium alone, is a statically indeterminate structure. For example, the beam of Fig. 10.1*a* has four reaction components. We cannot solve the four unknown reactions using only three available equations of equilibrium, viz., $\Sigma F_X = 0$, $\Sigma F_Y = 0$. and $\Sigma M_Z = 0$. Hence, it is statically indeterminate to the first degree. We need one additional equation to solve for the unknown reactions. If this beam had a fixed support at the left end support as in Fig. 10.1*b*, we would have had five unknown reactions, but still only three equations of equilibrium. We would then need two additional equations in order to be able to solve the reaction components. The beam would then be statically indeterminate by two degrees.

The additional equations to solve statically indeterminate structures come from prescribed conditions of translations and rotations, commonly called conditions of *compatibility* or *consistent displacements*.

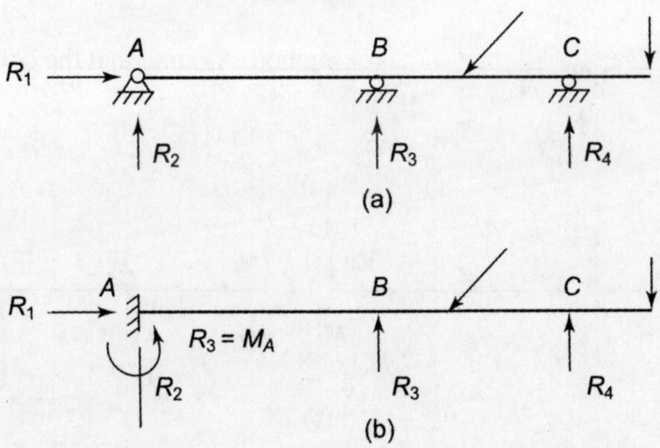

Fig. 10.1 | *(a) Support A hinged—degree of indeterminancy = 1, (b) Support A fixed—degree of indeterminancy = 2*

A statically indeterminate structure is also termed a redundant structure because of redundant reaction components or redundant members in a truss which are not necessary for stability considerations. A statically determinate structure possesses no redundants. A statically indeterminate structure of the first degree can possess one more additional reaction or member, the removal of which does not cause statical instability. For instance, in Fig. 10.1*a* any one of the roller supports can be removed without the structure becoming unstable. However, the removal of both roller supports makes the structure unstable. It follows that in a structure which is statically indeterminate to the second degree, if two redundants are removed simultaneously, the remaining structure will be statically determinate and stable. This can be verified with respect to the structure in Fig. 10.1*b*.

10.2 DEGREE OF INDETERMINANCY AND STABILITY OF STRUCTURES

Consider a three-span continuous beam as shown in Fig. 10.2*a*. There are five reaction components. With only three equations of equilibrium available the degree of indeterminancy is 5 – 3 = 2.

Another approach to the same problem is to cut the beam into segments and consider the internal forces as redundants. The free-body diagrams of the two segments are shown in Fig. 10.2*b*. If the internal forces V_6 and M_7 at the cut section are known then the external reaction components can be determined. For the left part we have five unknowns and three equilibrium equations; for the right part we have four unknowns and two equations of statics ($\Sigma F_Y = 0$ and $\Sigma M_Z = 0$). Therefore, the total number of unknown reactions exceeds the total number of equilibrium equations by two; which again establishes the degree of indeterminancy. Alternatively, a knowledge of the two internal forces at the cut section would allow us to determine all the reactions, hence the degree of indeterminancy is two.

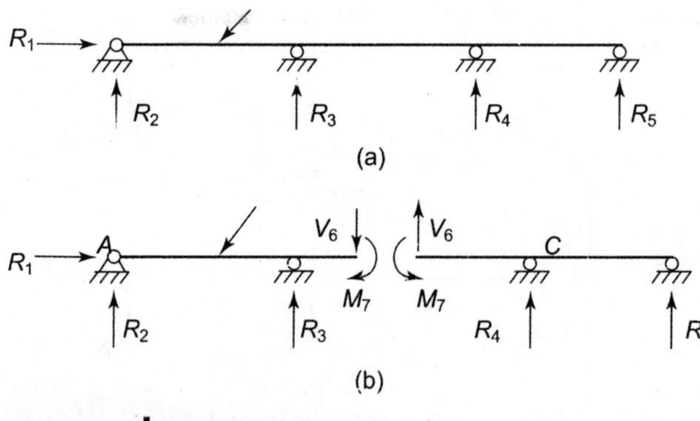

Fig. 10.2 | *(a) Three-span continuous beam—twice redundant, (b) Free-body diagram of segments*

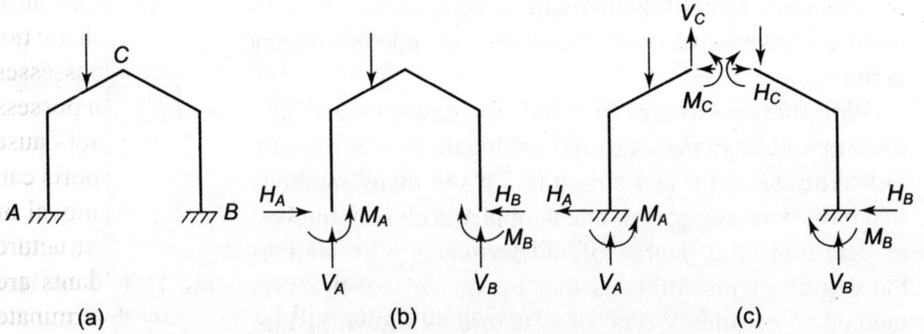

Fig. 10.3 | *(a) Gable frame, (b) Free-body diagram of entire frame, (c) Frame is cut at C and internal forces introduced*

The rigid frame of Fig. 10.3*a* provides another example. The free-body diagram in Fig. 10.3*b* indicates six unknowns. The degree of indeterminancy is, therefore, 6 – 3 = 3. Alternatively, the frame is cut at C and the three internal forces are introduced as shown in Fig. 10.3*c*. Combining these with the six reactions at the bases there are totally nine unknowns. With only six equations of equilibrium, the structure is statically indeterminate to the third degree. A consideration of each of the free-body diagrams also gives the same result.

As a last example, let us consider the two-storey frame of Fig. 10.4*a*. From the free-body diagram of the structure shown in Fig. 10.4*b* it is possible to determine the three reaction components using the three equations of equilibrium. The forces in members *AD* and *BC* can be found using statics only. However, the remainder of the structure presents a difficult situation. It is impossible to determine the forces in the members of closed loop *CDEF* from the known external reactions. Thus, we find that although the structure is determinate externally, it is statically indeterminate internally.

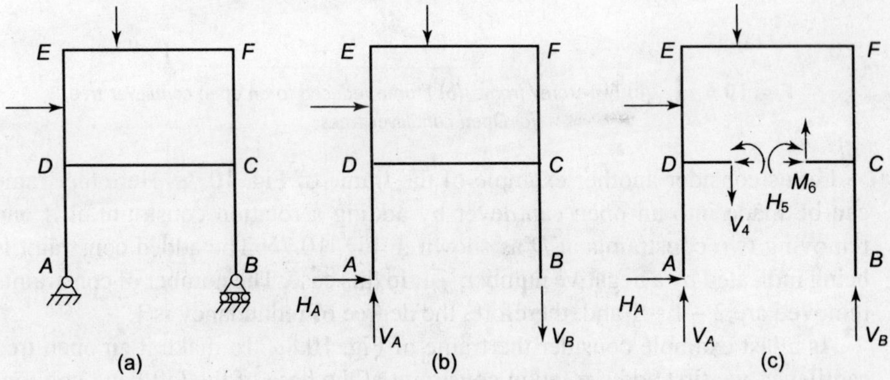

Fig. 10.4 | *(a) Two-storey frame, (b) Free-body diagram of structure, (c) Internal member is cut and forces introduced*

The degree of indeterminancy of the frame is established by taking a cut through beam *CD* and introducing three internal forces as in Fig. 10.4*c*. These three internal forces together with three external reactions yield six unknown forces. The degree of indeterminancy is three.

With these examples in mind, we now proceed to a more general statement on establishing the degree of static indeterminancy for any structure. Of the many methods that have been suggested, one simple and elegant method to determine the degree of indeterminancy as well as the degree of instability is the 'open tree' construction method. A cantilever type of structure as shown in Fig. 10.5, which is open without closed rings, is statically determinate and stable. When a structure such as the frame in Fig. 10.6 is not an open cantilever, it is a simple matter to cut into the cantilever of the open tree type. When a cut is made/we are removing constraints which are equal in number to the appropriate number of internal forces at the cut. The number of constraints removed to make the structure an open cantilever is shown in Fig. 10.6*b*. An alternative method of reducing the structure into open cantilevers is shown in Fig. 10.6*c*. In either case the number of constraints removed is 6. The total number of constraints removed to make the structure an open cantilever or cantilevers corresponds to the degree of indeterminancy.

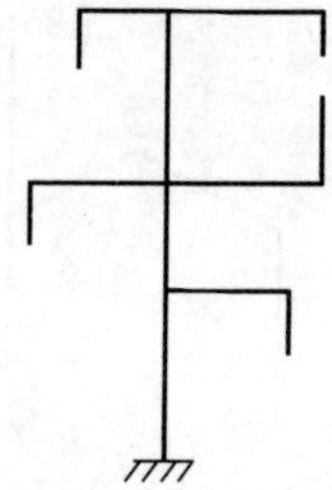

Fig. 10.5 | *An open cantilever tree*

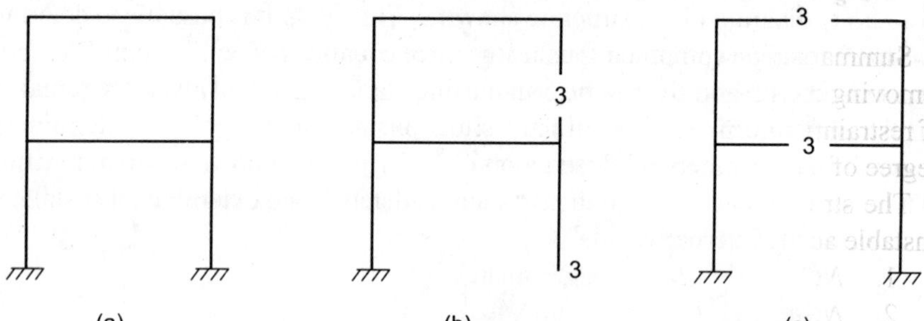

Fig. 10.6 | *(a) Two-storey frame, (b) Frame reduced to an open cantilever tree, (c) Open cantilever trees*

Let us consider another example of the frame of Fig. 10.7*a*. Here the frame can be made into an open cantilever by adding a rotation constraint at *A* and removing two constraints at *D* as shown in Fig. 10.7*b*. The added constraint is being indicated by a negative number, – 1, in this case. The number of constraints removed are 2 – 1 = 1 and, therefore, the degree of redundancy is 1.

As a last example consider the frame of Fig. 10.8*a*. To make it an open tree cantilever, we first add a rotation constraint at the base of the left hand column, remove two constraints at the base of the right hand column and two restraints at hinge 1 as shown in Fig. 10.8*b*. At this point the structure becomes unstable unless we add a rotation restraint at hinge 2. The net number of constraints removed is 4 – 2 = 2 and hence the degree of indeterminancy of the structure is 2.

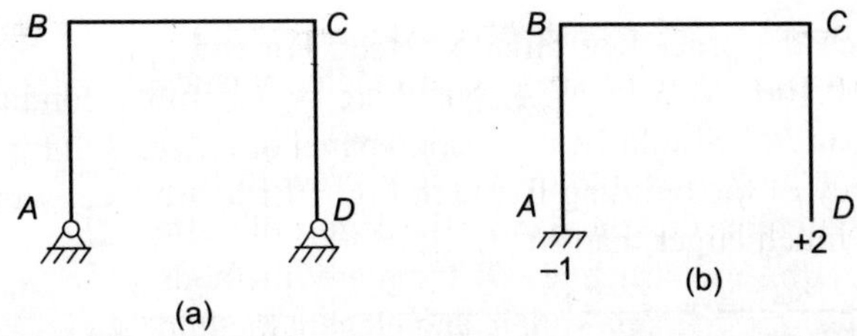

Fig. 10.7 | *(a) Frame hinged at base, (b) Number of restraints added or removed*

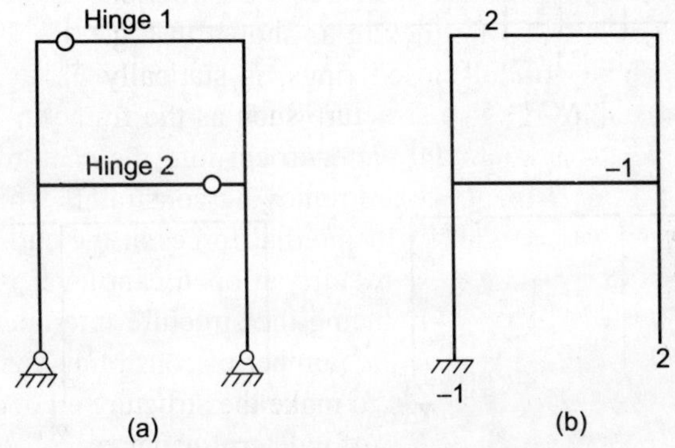

Fig. 10.8 | *(a) Structure with internal hinges, (b) Constraints added or removed*

Summarising, we reduce the structure to an open tree or trees by adding or removing constraints so that no unstable branches exist. Then we find the number of restraints added or removed; the net number of constraints removed gives the degree of indeterminancy of the structure.

The structure can be classified as statically indeterminate or determinate or unstable according to the following conditions:

1. $NCR > NCA$ Indeterminate
2. $NCR = NCA$ Determinate
3. $NCR < NCA$ Unstable

where

NCR stands for the number of constraints removed, and

NCA for the number of constraints added.

$$\text{The degree of indeterminancy} = NCR - NCA \quad (NCR > NCA)$$

$$\text{and the degree of instability} = NCA - NCR \quad (NCA > NCR)$$

The degree of instability indicates the number of constraints that must be added to a structure in order to make it stable in a statically determinate manner. The examples illustrated below in Figs. 10.9*a*, *b* and *c* will clarify the point even more. In applying this criterion to determine the degree of indeterminancy or stability of a structure, we must be careful not to overlook the conditions of local instability which the procedure fails to detect. For example, in the continuous beam of Fig 10.10*a*, $NCR > NCA$, but there is a definite condition of local instability in span *BC*. Similarly, the second panel of the truss in Fig. 10.10*b* and the second storey of the building frame in Fig. 10.10*c* are locally unstable even though *NCR* is much larger than *NCA*.

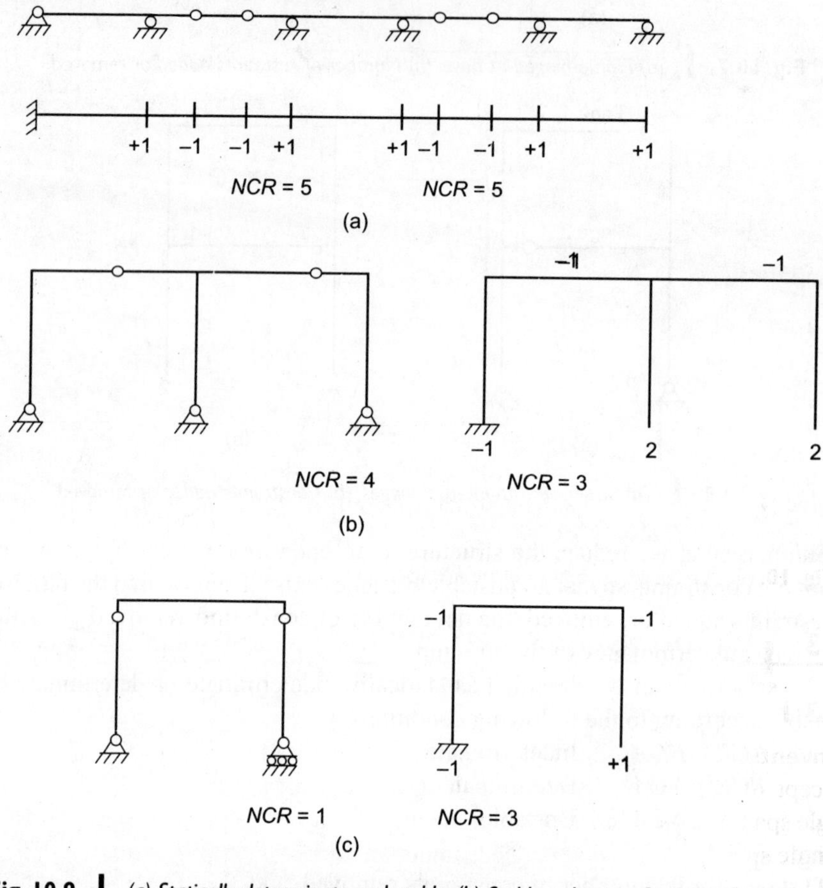

Fig. 10.9 | *(a) Statically determinate and stable, (b) Stable—degree of indeterminancy = 1, (c) Unstable—degree of instability = 2*

The discussions of this section also apply to space frames. The open tree in space will have at its base 6 restraints and to open up a closed ring requires the removal of 6 constraints—three translations and three rotations. The kinematic indeterminancy of a structure system is discussed later in Chapter 14.

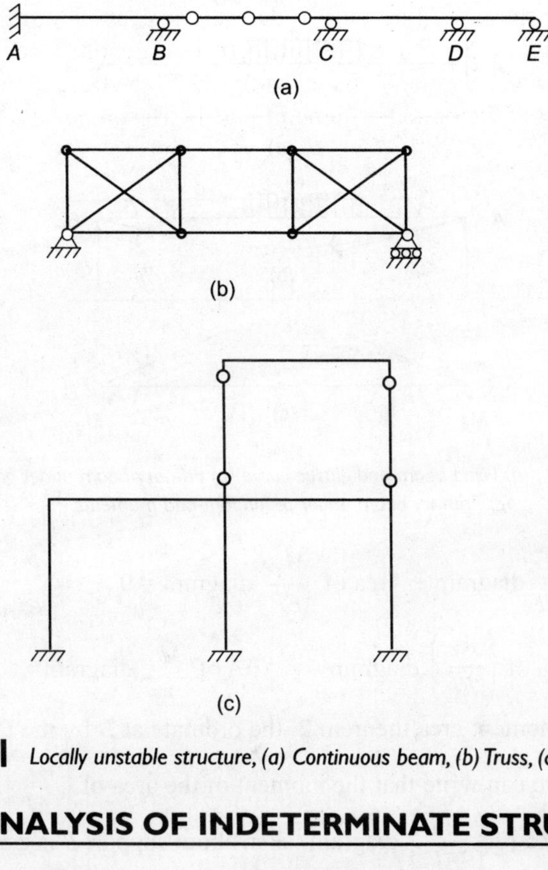

Fig. 10.10 | *Locally unstable structure, (a) Continuous beam, (b) Truss, (c) Building frame*

10.3 ANALYSIS OF INDETERMINATE STRUCTURES

10.3.1 Fixed Beams

Conventional Method The conventional method of analysis utilizes the concept of consistent displacements. The method has limited application to single span beams and makes use of the moment-area theorem 1 and 2. Consider a single span beam fixed at the ends under an arbitrary loading as shown in Fig. 10.11.

The beam is statically indeterminate by 2 degrees. The beam is made statically determinate by releasing the fixed end moments M_A and M_B. The resulting beam is known as the primary beam.

The moment at any section of the beam due to given loading can be expressed as the algebraic sum of two components: One (μ_f) due to applied loads and the other (M_f) due to redundant moments. The elastic curve under the two types of moments are indicated separately in Fig. 10.11 b and c. The combined elastic curve having zero slope and zero deflection at the supports is shown in Fig. 10.11a. We can write the resultant moment at any section X as $M_x = \mu f + M_f$

The change of slope on the elastic curve from end supports A and B is zero. Using moment area theorem 1 we can write

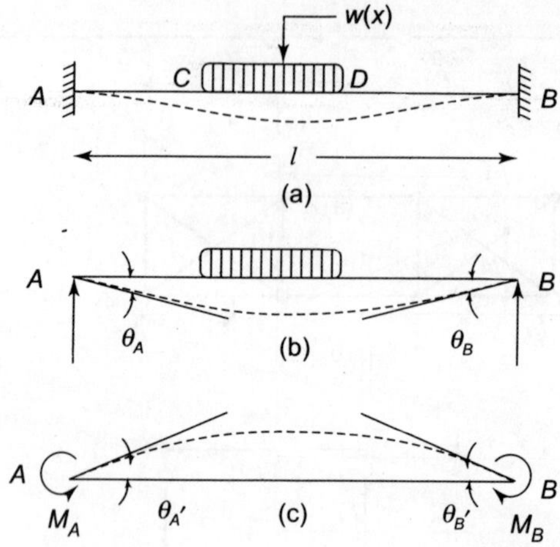

Fig. 10.11 | *(a) Fixed beam and elastic curve (b) Primary beam under given loading (c) Primary beam under redundant and moments*

$$\text{Area of the } \frac{\mu_f}{EI} \text{ diagram} + \text{Area of } \frac{M_f}{EI} \text{ diagram} = 0$$

$$\text{Therefore Area of } \left(\frac{\mu_f}{EI}\right) \text{ diagram} = -\text{Area of } \frac{M_f}{EI} \text{ diagram}$$

Again using moment area theorem 2, the ordinate at B by the tangent drawn from A is zero, we can write that the moment of the area of $\left(\frac{\mu_f}{EI}\right)$ diagram and the moment of area of $\left(\frac{M_f}{EI}\right)$ diagram taken about support B is zero

Or moment of area of $\left(\frac{\mu_f}{EI}\right)$ and moment of area of $\left(\frac{M_f}{EI}\right)$ diagram between A and B taken about $B = 0$ or moment of area of $\left(\frac{\mu_f}{EI}\right)$ taken about B = – moment of area of $\left(\frac{M_f}{EI}\right)$ taken about B.

In case of prismatic beams, EI being constant, it is enough if μ_f and Mf diagrams only are considered. A few illustrative examples presented below will help fix the underlying principles.

Example 10.1 | *Analyse a fixed beam for end moments subjected to a concentrated load W at centre of span.*

Step 1: To release end moments M_A and M_B

The beam is reduced to a primary beam by releasing the fixed end moments M_A and M_B. The μ_f diagram and M_f diagram are shown separately. Note $M_A = M_B$ due to symmetry of beam and loading but are unknown

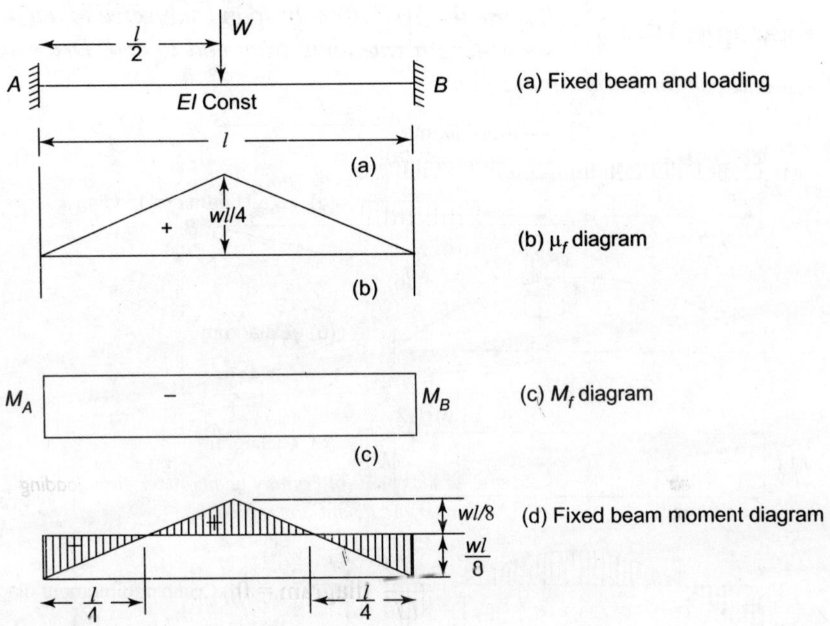

Fig. 10.12

Using moment-area theorem 1, we can write

Area of μ_f diagram + Area of M_f diagram = 0

We get
$$\frac{1}{2}(l)\left(\frac{wl}{4}\right) + M_A \cdot l = 0$$

$\therefore$
$$M_A = -\frac{wl}{8}$$

and
$$M_B = -\frac{wl}{8}$$

Step 2: To draw moment diagram

The combined moment diagram is shown in Fig. 10.12*d*. The support reactions are each $W/2$ due to symmetry. The shear force diagram is same as for the simply supported beam. Equal end moments do not contribute any thing towards shear.

Note there are two points at $l/4$ from each end at which the moment is zero. There points are known as *points of inflection* or *points of contraflexure*. At these points the elastic curve changes its curvature from hogging to sagging.

The student may, as an exercise, verify that the deflection at centre of span is $\frac{Wl^3}{192\,EI}$ which is 25% of the deflection in a simply supported beam.

Example 10.2 *Analyse the fixed beam of span l subjected to a u.d.l. of w/unit length extending from end to end. Draw shear force and bending moment diagram*

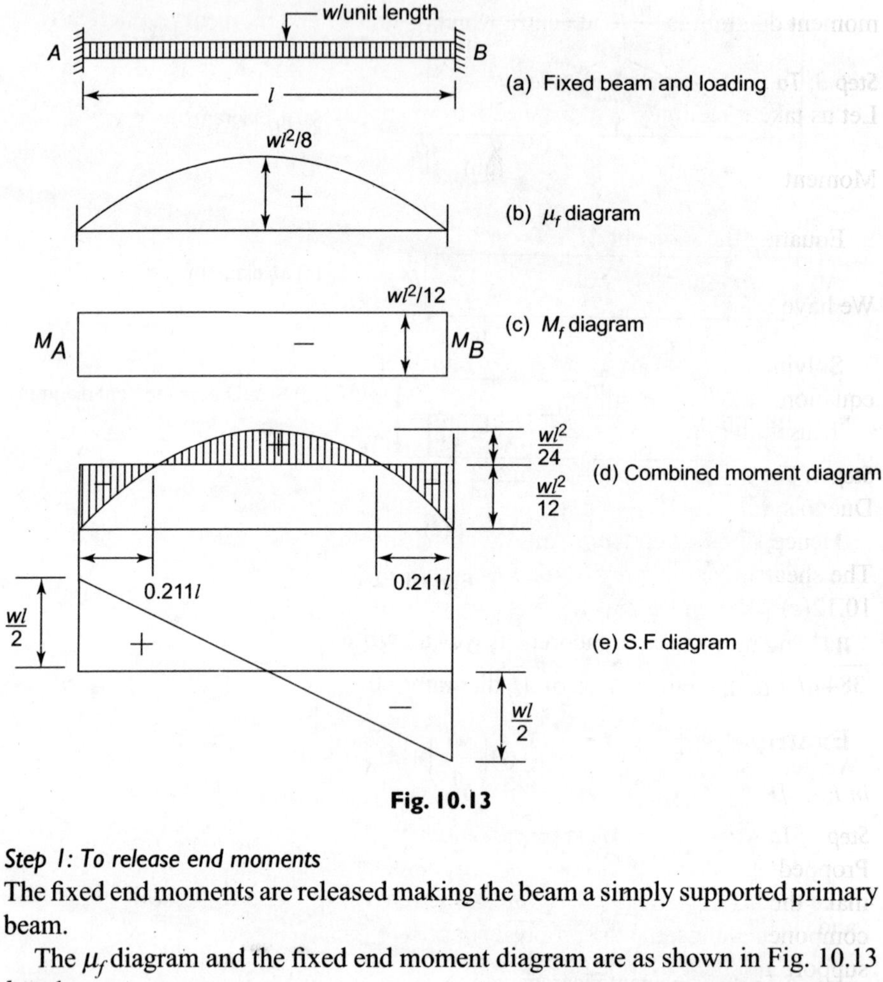

Fig. 10.13

Step 1: To release end moments

The fixed end moments are released making the beam a simply supported primary beam.

The μ_f diagram and the fixed end moment diagram are as shown in Fig. 10.13 *b* and *c*.

The fixed end moments M_A and M_B are unknown but $M_A = M_B$ due to symmetry

Using moment-area theorem 1

$$\frac{2}{3}\left(\frac{wl^2}{8}\right)l + M_A \cdot l = 0$$

or
$$M_A = -\frac{wl^2}{12}$$

and
$$M_B = -\frac{wl^2}{12}$$

Step 2: To draw combined B.M. diagram

The combined B.M. diagram is drawn by super imposing the fixed moment diagram over μ_f diagram as shown in Fig. 10.13*c*. Note that the maximum +ve moment diagram is $\frac{wl^2}{24}$ at centre which is half of end moments numerically.

Step 3: To fix up points of contraflexure

Let us take a section *x* at a distance *x* from support *A*

Moment $$M_x = \frac{wl}{2}x - \frac{wx^2}{2} - \frac{wl^2}{12}$$

Equating the moment $M_x = 0$

We have $$x^2 - l_x + \frac{l^2}{6} = 0$$

Solving for *x*, we have $x = 0.211\ l$ and $0.789\ l$ the two roots of the quadratic equation.

Thus the points of contraflexure are located at $0.211\ l$ from either end

Step 4: To draw shear force diagram

Due to symmetry the fixed moments $M_A = M_B$

Hence the fixed end moments do not in any way contribute to the shear force. The shear force diagram is same as in simply supported beam as shown in Fig. 10.12(*e*) The reader may again, verify that the deflection at centre of span is $\frac{Wl^4}{384\,EI}$, which is 20% of the deflection in a simply supported beams

Example 10.3 | *Analyse a cantilever beam propped at the end and subjected to a u.d.l. covering the entire span as shown in Fig. 10.14. Draw the shear force and bending moment diagrams.*

Step 1: To release the redundant reaction component

Propped cantilever beam is statically indeterminate by one degree. We can make the beam statically determinate by releasing one of the redundant reaction components. It can be the propped support or fixed moment at *A*. In our example support moment is released making the primary beam a simply supported beam.

Step 2: To evaluate redundant moment M_A

The given beam undergoes rotation at propped end *B* but no translation.

We can evaluate M_A by utilizing the concept that the tangential deviation at *B* from the tangent on the elastic curve at $A = 0$

or $t_{BA} = 0$, that is, the moment of the moment diagram between *A* and *B* taken about $B = 0$

We can write $$\frac{2}{3}\left(\frac{wl^2}{8}\right)l \cdot \frac{l}{2} - M_A \frac{l}{2}\frac{2}{3}l = 0$$

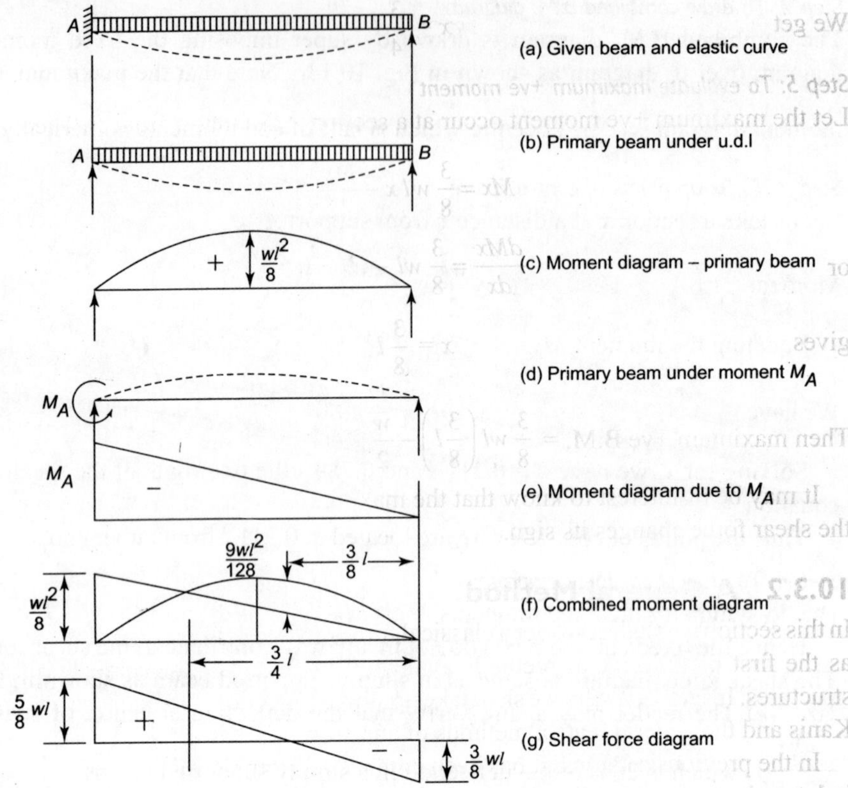

Fig. 10.14 | *Propped cantilever beam under u.d.l.*

or
$$MA = -\frac{wl^2}{8}$$

Step 3: To evaluate support reactions

The support reactions can be evaluated using equation of static equilibrium. Taking moments of all the forces about B we can write

$$M_B = R_A(l) - \frac{wl \cdot l}{2} - \frac{wl^2}{8} = 0$$

Which gives
$$R_A = \frac{wl}{2} + \frac{wl}{8} = \frac{5}{8}wl$$

and
$$R_B = w \cdot l - \frac{5}{8}wl = \frac{3}{8}wl$$

Step 4: To find point of contraflexure

Let the section be located at a distance x from support B writing down moment

$$M_x = \frac{3}{8}wlx - \frac{wx^2}{2} = 0$$

We get $$x = \frac{3}{4} l$$

Step 5: To evaluate maximum +ve moment

Let the maximum +ve moment occur at a section × distance x from B. Then

$$Mx = \frac{3}{8} w l x - \frac{\omega x^2}{2}$$

or $$\frac{dMx}{dx} = \frac{3}{8} wl - wx = 0$$

gives $$x = \frac{3}{8} l$$

Then maximum +ve B.M. $= \frac{3}{8} wl\left(\frac{3}{8} l\right) - \frac{w}{2}\left(\frac{3l}{8}\right)^2 = \frac{9}{128} wl^2$

It may be of interest to know that the maximum +ve B.M. occurs at a section the shear force changes its sign.

10.3.2 A General Method

In this section we shall consider a classical force method or a compatibility method as the first of the several methods available for the analysis of indeterminate structures. In later chapters we shall take up slope-deflection, moment distribution, Kanis and the general matrix methods of analysis.

In the previous sections, it has been emphasized that the analysis of statically indeterminate structures requires some additional conditions such as compatibility of displacements or conditions of consistent displacements.

To develop an understanding of the method of compatibility or consistent displacements, let us consider the propped cantilever beam shown in Fig. 10.15. In this the degree of indeterminancy is one. Or, alternatively, the structure has one redundant reaction. An additional equation can be obtained from deflection considerations. As a first step, the structure has to be made statically determinate by removing the redundant reaction. The choice of the redundant force is arbitrary. However, the simplicity and accuracy of the solution often depend upon the choice of the redundant. In selecting redundants it must be remembered that the structure which remains after the restraining effects of the redundants are removed must be stable. For example, in this particular case, we can choose the prop reaction as the redundants, or the moment at fixed support C, or we may even choose an internal force such as moment and introduce a hinge somewhere along the length of the beam. We shall choose the prop reaction as the redundant and the restraining action of this force is temporarily removed. The remaining structure shown in Fig. 10.15*b* is commonly referred to as the *primary structure*. The redundant force is considered to be an active force on the primary structure. Thus, the primary structure supports not only the given loads, but also the forces representing the redundants.

From now on we shall be concerned with the deflections of the primary structure. The deflections of the primary structure due to given loads and the redundant forces are considered separately. For the beam under consideration, the given loads produce displacement D_{BP} at the point of redundant reaction R_B as shown in Fig. 10.15*b*. The subscripts *BP* imply that the displacement is at *B* due to external load *P* on the primary structure. The deflection at point *B* resulting from the redundant force R_B is denoted as DBR (Fig. 10.15*c*). Again subscripts *BR* imply that the displacement is associated with the redundant reaction R_B whose value, as yet, is unknown.

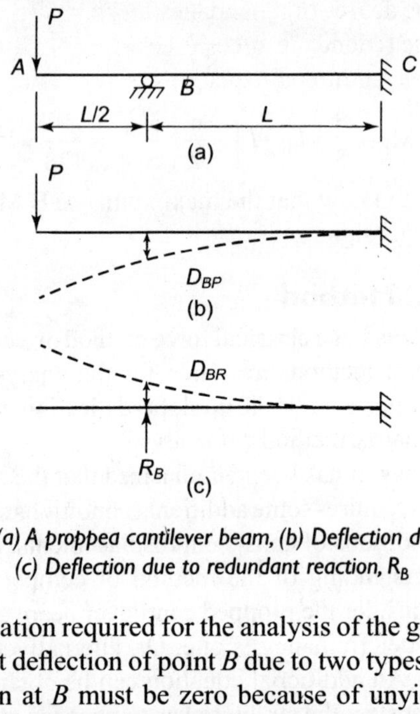

Fig. 10.15 | *(a) A proppea cantilever beam, (b) Deflection due to applied load, (c) Deflection due to redundant reaction,* R_B

The deflection equation required for the analysis of the given beam is obtained by considering the net deflection of point *B* due to two types of loading. We know that the net deflection at *B* must be zero because of unyielding roller support. Therefore, the value of R_B is determined by imposing the known condition on the displacement at point *B*.

$$D_B = D_{BP} + D_{BR} = 0 \tag{10.1}$$

The two displacements carry opposite signs in the equations as they are in opposite directions. Thus, if an upward deflection is considered positive, D_{BP} is a negative quantity. Obtaining the correct signs for forces and deflections is essential. The use of deflection and the force coordinate system will be helpful in complicated analysis.

D_{BP} and D_{BR} are deflection quantities of the statically determinate primary structure. Their value can be calculated using any of the methods discussed in Chapters 5 and 6. For example, in our problem, we have

$$D_{BP} = \frac{7}{12}\frac{PL^3}{EI} \downarrow \text{ and } D_{BR} = R_B\frac{L^3}{3EI} \uparrow$$

Therefore, $$R_B = \frac{7}{12}\frac{PL^3}{EI}\frac{3EI}{L^3} = \frac{21}{12}P$$

This method is often called the method of *consistent displacements* because the summation of displacements produced by the combined action of loads and redundants must be consistent with the given support conditions. It may also be noted that the principle of superposition has been utilised, thus restricting the validity of the analysis to the linear elastic range of the structure.

To extend the method further, consider the framed structure in Fig. 10.16. The first step is to identify the degree of redundancy and then provide releases at those points in respect of those redundant forces. The frame is statically indeterminate to the third degree. Three additional equations of deflection must, therefore, be obtained for the analysis of the frame. In the present example, the three released forces at B are selected as the redundant forces. The primary structure is attached to support A and is similar to a cantilever bent. The forces corresponding to the redundants are: shear R_1, axial force R_2 and moment R_3. These quantities, R_1, R_2 and R_3 shown in Fig. 10.16*b* are in their positive direction.

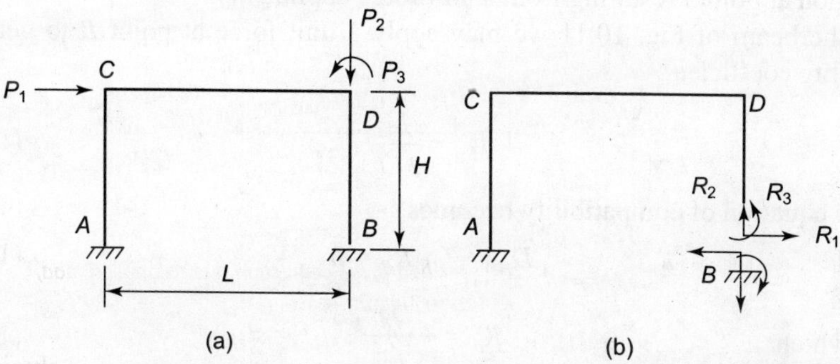

Fig. 10.16 | *(a) Portal frame, (b) Frame released at B*

The basic support or boundary condition is that the final displacements D_1, D_2 and D_3, at the release points must be zero. Displacement D_1 is the sum of four components; D_{1p} due to applied loading, D_{11} due to the force R_1, D_{13} due to the force R_2 and D_{13} due to the force R_3. The first subscript indicates correspondence to release 1, the second denotes the cause of the displacement—the loading P and the redundant forces corresponding to releases 1, 2 and 3 respectively. Since

$$D_1 = 0 \text{ we have}$$

$$D_1 = D_{1P} + D_{11} + D_{12} + D_{13} = 0$$

Similarly considering the compatibility condition for releases 2 and 3, we have

$$D_2 = D_{2P} + D_{21} + D_{22} + D_{23} = 0$$

and $$D_3 = D_{3P} + D_{31} + D_{32} + D_{33} = 0 \tag{10.2}$$

Each of the above terms represents the displacement of a statically determinate structure resulting from a specified load condition and each has its positive direction defined by the positive sense of the corresponding releases. The nine displacements D_{ij} (i = 1, 2, 3, and j=1,2, 3), are linear functions of unknown redundants R_1, R_2 and R_3.

10.4 FLEXIBILITY COEFFICIENTS

Any of the displacements D_{ij} of the primary structure is a measure of the flexibility of the structure; that is, the more flexible the structure, the higher the values of the displacements. It is convenient to define the effect of the redundant forces on the primary structure in terms of the displacements produced by unit forces corresponding to the redundants. For example, we write the displacement D_{ii} as

$$D_{ij}=f_{ij}R_j \tag{10.3}$$

where f_{ij}, the displacement at i corresponding to release i for a unit force corresponding to R_j at release j, is called the *flexibility influence coefficient.*

In general, a flexibility influence coefficient for a structure, f_{ij} is defined as the deflection at point i resulting from a unit force applied at j.

In the beam of Fig. 10.11 we may apply a unit force at point B to get the flexibility coefficient.

$$f_{BB}=\frac{L^3}{3EI} \tag{10.4}$$

The equation of compatibility becomes

$$D_{BP}+f_{BB}R_B=0 \tag{10.5}$$

from which

$$R_B=-\frac{D_{BP}}{f_{BB}} \tag{10.6}$$

The flexibility coefficients are purely functions of geometry and elastic property of primary structure. They are independent of actual loading. Thus, we can determine the flexibility coefficients for the primary structure and use them repeatedly for analysing the structure for different loading cases.

Using the concept of flexibility coefficients, Eq. 10.2 can be written as

$$\begin{aligned} D_{1P}+f_{11}R_1+f_{12}R_2+F_{13}R_3&=0\\ D_{2P}+f_{21}R_1+f_{22}R_2+f_{23}R_3&=0\\ D_{3P}+f_{31}R_1+f_{32}R_2+f_{33}R_3&=0 \end{aligned} \tag{10.7}$$

In matrix form, we write

$$\begin{bmatrix} f_{11} & f_{12} & f_{13}\\ f_{21} & f_{22} & f_{23}\\ f_{31} & f_{32} & f_{33}\end{bmatrix}\begin{bmatrix} R_1\\ R_2\\ R_3\end{bmatrix}=-\begin{bmatrix} D_{1P}\\ D_{2P}\\ D_{3P}\end{bmatrix} \tag{10.8}$$

or simply,

$$\mathbf{FR}=-\mathbf{D} \tag{10.9}$$

The values of the unknown redundants are obtained by inverting the flexibility matrix of Eq. 10.9 and pre-multiplying the negative of matrix **D** by this inverse. Thus,

$$\mathbf{R} = -\mathbf{F}^{-1}\,\mathbf{D} \tag{10.10}$$

The method of consistent displacements can be used to analyse various types of statically indeterminate problems. This procedure entails the computation of deflection quantities. Various methods for computing deflections were presented in Chapters 5 and 6. The computation method adopted for a particular structure is often a matter of personal choice, but in certain cases certain methods will have distinct advantages over others.

In the examples that follow appropriate methods have been used. It may be stressed here that some other methods of computation of deflections may be preferable. However, the emphasis in these examples is on the indeterminate aspect of analysis rather than on the comparison of different methods of computing deflections.

Example 10.4 | *It is required to determine (a) the reaction at the right hand support, (b) the fixed end moment at the left hand support and (c) the rotation at end B for the beam in Fig, 10.17.*

The given beam is statically indeterminate to the first degree. The choice of the redundant force is arbitrary, but we select R_B as the redundant one since we are required to determine this. The resulting primary structure is a cantilever. The primary structure loaded with given moment M_B is shown in Fig. 10.17*b*. Also the primary structure loaded with a unit value of redundant M_B is shown in Fig. 10.17*c*.

The condition for consistent displacement at B is written as

$$D_{BP} + f_{BB}\,R_B = 0 \tag{10.11}$$

The deflection quantities D_{BP} and f_{BB} are determined by the moment area method. From the *M/EI* diagram for an applied moment M_B in Fig. 10.17*b* we can write

$$D_{BP} = t_{BA} = \frac{M_B L^2}{2EI} \uparrow$$

From the *M/EI* diagram for the unit reaction at *B* in Fig. 10.17*c* the expression for the flexibility coefficient f_{BB} is

$$f_{BB} = t_{BA} = \frac{L^3}{3EI} \uparrow$$

The value of redundant reaction R_B from Eq. 10.6 is

$$R_B = -\frac{D_{BP}}{f_{BB}} = -\frac{3}{2}\frac{M_B}{L} \tag{10.12}$$

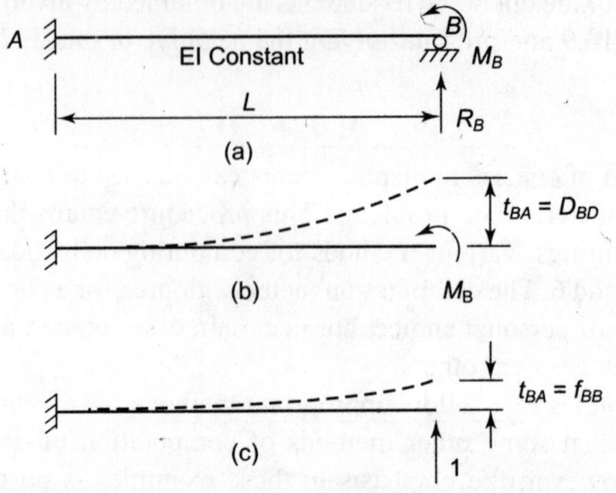

Fig. 10.17 | *(a) Beam under applied moment M_B, (b) Deflection of primary structure due to applied moment M_B, (c) Deflection of primary structure due to unit force applied along redundant force R_B*

The upward deflection is taken as positive. The negative sign for the value of R_B indicates that the reaction is downwards. After knowing R_B, the question of finding the fixed end moment at support A is only a matter of applying statics. Summing up the moments and taking the anti-clockwise moments as positive, we have

$$M_B + R_B(L) + M_A = 0$$

or

$$M_A = \frac{M_B}{2} \qquad (10.13)$$

The slope of the beam at B is obtained by the superposition of the slopes by applied moment M_B and redundant reaction R_B on the primary structure, that is

$$\theta_B = \frac{M_B L}{EI} + \frac{R_B L^2}{2EI}$$

Substituting for R_B from Eq. 10.12, we get

$$\theta_B = \frac{M_B L}{4EI} \qquad (10.14)$$

We shall solve another example of a fixed end beam for end moments which are needed quite often.

Example 10.5 | *It is required to determine the end moments for a fixed end beam subjected to concentrated load P at distance a from one end and b from the other as shown in Fig. 10.18.*

The given beam is statically indeterminate to the second degree. The beam can be made statically determinate by releasing the reaction components at one of the two fixed ends. The resulting primary structure would be a cantilever. Alternatively, the end moments may be released, reducing it to a simple beam. In this example, the end moments are considered as redundants. The elastic line of the primary structure under the given loading as well as the unit values of moments M_A and M_B are shown in Fig. 10.18 separately.

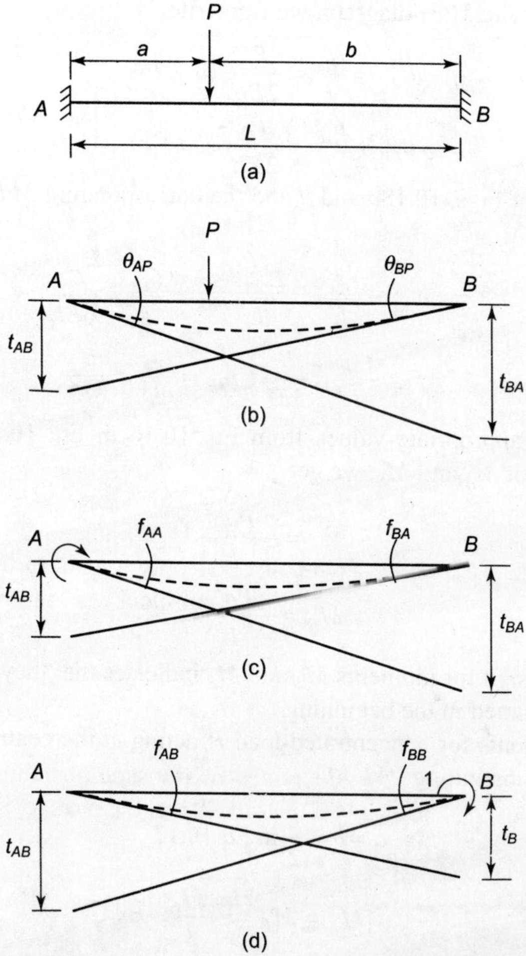

Fig. 10.18 | *(a) Fixed end beam and loading, (b) Displacement of primary structure due to applied load, (c) Displacement of primary structure due to unit moment at A, (d) Displacement of primary structure due to unit moment at B*

The two compatibility conditions which may be utilised are that the end rotations $\theta_A = \theta_B = 0$ in a fixed beam. This condition may be written as

$$\begin{aligned} \theta_A &= \theta_{AP} + f_{AA}\, M_A + f_{AB}\, M_B = 0 \\ \theta_B &= \theta_{BP} + f_{BA}\, M_A + J_{BB}\, M_B = 0 \end{aligned} \tag{10.15}$$

Writing this in matrix form,

$$\begin{Bmatrix} \theta_{AP} \\ \theta_{BP} \end{Bmatrix} + \begin{bmatrix} f_{AA} & f_{AB} \\ f_{BA} & f_{BB} \end{bmatrix} \begin{Bmatrix} M_A \\ M_B \end{Bmatrix} = \begin{Bmatrix} 0 \\ 0 \end{Bmatrix} \tag{10.15a}$$

The rotation quantities θ_{AP}, θ_{BP} and the rotation influence coefficients f_{AA}, f_{AB}, f_{BA} and f_{BB} can be obtained using the moment area method. Referring to Fig. 10.18*b*, and from the *M/EI* diagram, we can write

$$\theta_{AP} = \frac{t_{BA}}{L} = \frac{P_{ab}}{2EIL}(a + 2b)$$

$$\theta_{BP} = \frac{t_{AB}}{L} = \frac{P_{ab}}{2EIL}(2a + b)$$

Similarly, from Figs. 10.18*c* and *d* and the corresponding *M/EI* diagrams we can write

$$f_{AA} = \frac{t_{BA}}{L} = \frac{L^2}{3EI}; \; f_{BA} = \frac{t_{AB}}{L} = \frac{L^2}{6EI}$$

$$f_{BB} \frac{t_{AB}}{L} = \frac{L^2}{3EI}; \; f_{AB} = \frac{t_{BA}}{L} = \frac{L^2}{6EI} \tag{10.16}$$

Substituting the appropriate values from Eq. 10.16 in Eq. 10.15 and solving simultaneously for M_A and M_B, we get

$$M_A = -\frac{Pab^2}{L^2}$$

and

$$M_B = -\frac{Pa^2b}{L^2} \tag{10.17}$$

The negative sign for moments M_A and M_B indicates that they are opposite to the direction assumed in the beginning.

The end moments for concentrated load *P*, acting at the centre of beam may be obtained by substituting

$$a = b = \frac{L}{2} \text{ in Eq. 10.17}$$

This gives

$$M_A = M_B = -\frac{PL}{8} \tag{10.18}$$

We can also make use of Eq. 10.17 to evaluate the end moments for a fixed end beam under a uniformly distributed load extending over a distance from *a* to *b* as shown in Fig. 10.19. If an elemental load $(w \cdot dx)$ acting over a differential length (dx), is considered as concentrated we can write the end moments using Eq. 10.17 as

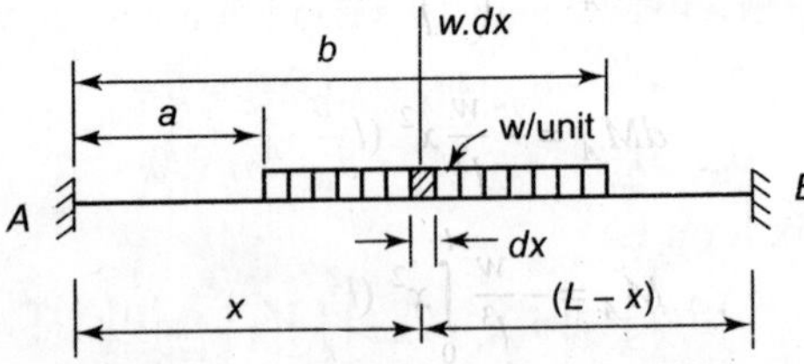

Fig. 10.19

$$dM_A = -\frac{(w \cdot dx)(x)(L-x)^2}{L^2}$$

and
$$dM_B = -\frac{(w \cdot dx)(x^2)(L-x)}{L^2}$$

The end moments developed due to the loading extending from a to b may be written as

$$M_A = -\int_a^b \frac{w(x)(L-x)^2}{L^2}\, dx \tag{10.19}$$

and
$$M_B = -\int_a^b \frac{w \cdot x^2 (L-x)}{L^2}\, dx$$

The evaluation of integrals in Eq. 10.19 gives directly the required moments. For example, for a uniformly distributed load occupying the entire span, the limits to be substituted are $a = 0$ and $b = L$. Evaluating the integral, we get

$$M_A = -\frac{wL^2}{12}$$

and
$$M_B = -\frac{wL^2}{12} \tag{10.20}$$

The reader may verify that for a load extending from the left support to the centre, the fixed end moments are

$$M_A = -\frac{11}{192} wL^2$$

and
$$M_B = -\frac{5}{192} wL^2 \tag{10.21}$$

Example 10.6 *Find end moments in a fixed beam of span l under a uniformly varying load as shown in Fig. 10.20. Draw shear force and bending moment diagrams.*

Step 1: Evaluation of fixed end moments

Consider the elemental load $\frac{wx \cdot dx}{l}$ over (dx) as a concentrated load. We can write, using equation 10.17 the elemental moment dM_A at A as

$$dM_A = -\left(\frac{w \cdot x}{l} \cdot dx\right)(x)\frac{(l-x)^2}{l^2}$$

or
$$dM_A = -\frac{w}{l^3} x^2 (l-x)^2\, dx$$

or
$$M_A = -\frac{w}{l^3}\int_0^l x^2 (l^2 - 2lx + x^2)\, dx$$

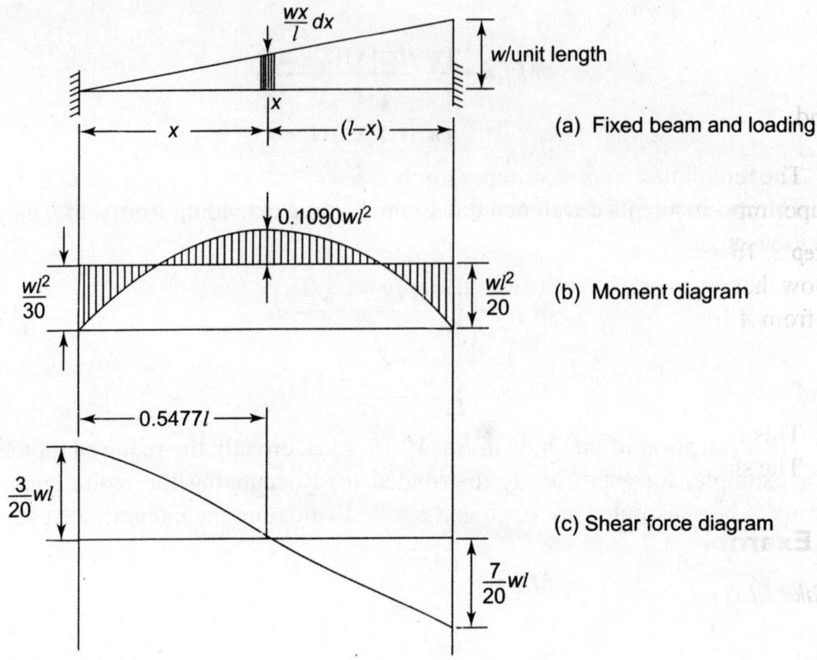

Fig. 10.20

or
$$M_A = -\frac{w}{l^3}\int_0^l (x^2 l^2 - 2lx^3 + x^4)\, dx$$

On integration and substitution of limits

We get
$$M_A = -\frac{wl^2}{30}$$

Similarly, we can write

$$dM_B = -\left(w\frac{x}{l}\,dx\right)(x^2)\left(\frac{l-x}{l^2}\right)$$

$$= -\frac{w}{l^3}x^3(l-x)\,dx$$

or
$$M_B = -\frac{w}{l^3}\int_0^l (x^3 l - x^4)\, dx$$

On evaluation of integral we get $M_B = -\dfrac{wl^2}{20}$

Step 2: To evaluate reaction components R_A and R_B

$$R_A = \frac{1}{3}\cdot\frac{wl}{2} - \left(\frac{M_B - M_A}{l}\right)$$

$$= \frac{w_l}{6} - \frac{wl}{60} = \frac{9}{60} wl \text{ or } \frac{3}{20} wl$$

and $$R_B = \frac{wl}{3} + \frac{wl}{60} = \frac{21}{60} wl \text{ or } \frac{7}{20} wl$$

The combined moment diagram is shown in Fig. 10.20*d* which is the superimposition of fixed moment diagram over μ_f diagram.

Step 3: To draw shear force diagram.

Now that R_A and R_B have been evaluated we can write shear force V_x at x, distance x from A is

$$V_x = \frac{3}{20} wl - \frac{1}{2} \cdot x \cdot \frac{wx}{l} \quad \text{or} \quad \frac{3}{20} wl - \cdot \frac{wx^2}{2l}$$

This is a second degree eqn. The S.F. is zero at $x = .5477\, l$

The shear force diagram is shown in Fig. 10.20*c*.

Example 10.7 | *Analyse the fixed beam of span l when the right end support sinks by an amount δ as shown in Fig. 10.21. Take EI is constant.*

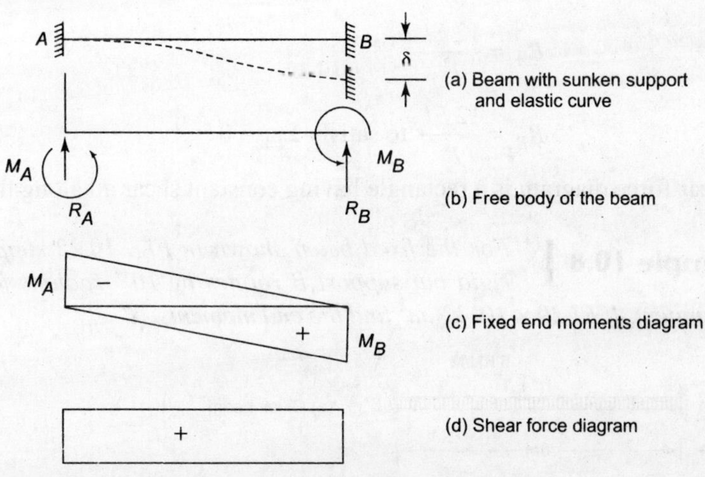

Fig. 10.21

Step 1: To find end moments M_A and M_B.

Let the support B sink by an amount δ.

The sinking of support induces fixed end moments M_A and M_B, both of them anticlockwise.

Using the beam moment sign convention the fixed moment diagram is drawn in Fig. 10.21*c*. The μ_f diagram is zero as there is no any external load. In the elastic line we notice that the change of slope on the elastic curve from A to B is zero. Using moment-area theorem 1, the area of $\frac{M}{EI}$ diagram between A and $B = 0$

or
$$-\frac{1}{2}\frac{M_A}{EI}\left(\frac{l}{2}\right)+\frac{1}{2}\frac{M_B}{EI}\left(\frac{l}{2}\right)=0$$

or
$$M_A = M_B \text{ (numerically)}$$

Again the ordinate cut at B by the tangent at A on the elastic curve is equal to t_{BA} which is same as δ. we can write

$$-M_A\left(\frac{l}{2}\right)\frac{2}{3}\frac{l}{EI}+M_B\left(\frac{l}{2}\right)\frac{l}{3EI}=\delta$$

$$-\frac{M_A l^2}{3EI}+\frac{M_B l^2}{6EI}=0$$

Gives
$$M_A = -\frac{6EI\,\delta}{l^2} \quad \text{and} \quad M_B = \frac{6EI\,\delta}{l^2}$$

Step 2: To find support reactions R_A and R_B

Using equation of equilibrium and writing summation of moments about support $B = 0$

we have
$$R_A(l) - M_A - M_B = 0$$

$$R_A = \frac{12EI}{l^3}$$

and
$$R_B = \frac{12EI}{l^3} \quad \text{to satisfy } \Sigma_{FY} = 0$$

The shear force diagram is a rectangle having constant shear all along the beam.

Example 10.8 | *For the fixed beam shown in Fig. 10.22 support A is rigid but support B rotates by 10^{-4} radians for every kN.m moment. If EI 20×10^3 kN.m^2 find the end moments.*

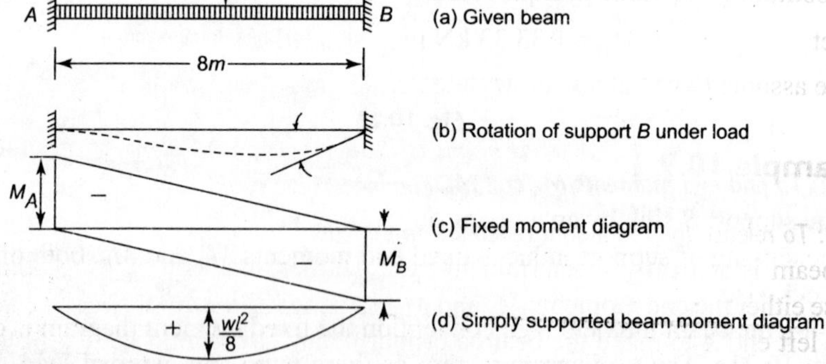

Fig. 10.22

Let M_A and M_B are the end moments and α is the angle of rotation of the support B under the load. As usual, the end moments are considered as redundant, we can evaluate the redundant moments by applying the following boundary conditions.

(i) $t_{BA} = 0$

(ii) Rotation at end $B = \alpha\, M_B$ or $\frac{t_{AB}}{l}$

Writing the above in the form of equation,

$$t_{BA} = -\frac{1}{2} M_A (l) \frac{2}{3} \frac{l}{EI} - \frac{1}{2} M_B (l) \frac{l}{3EI} + \frac{2}{3}\left(\frac{wl^2}{8}\right)\frac{l \cdot l}{2EI} = 0$$

$$\text{Rewriting } t_{BA} = -\frac{M_A l^2}{3EI} - \frac{M_B l^2}{6EI} + \frac{wl^4}{24\,EI} = 0 \tag{10.22}$$

Writing condition (ii) in equation form

$$-M_A\left(\frac{l}{2}\right)\frac{1}{3}\frac{l}{EI\,l} - M_B\left(\frac{l}{2}\right)\frac{2}{3}\frac{l}{EI\,l} + \frac{2}{3}\left(\frac{Wl^2}{8}\right)\frac{l}{2EI\,l}\cdot l - \alpha\, M_B = 0$$

or

$$-M_A\left(\frac{l}{6EI}\right) - M_B\left(\frac{l}{3EI} + \alpha\right) + \frac{wl^3}{24\,EI} = 0 \tag{10.23}$$

Multiplying Eqn (10.23) by $2l$ and subtracting from Eqn (10.22)

$$-\frac{M_B l^2}{6EI} + 2M_B\, l\left(\frac{l}{3EI} + \alpha\right) = \frac{wl^4}{24\,EI}$$

or

$$+\frac{M_B l^2}{6} + \frac{2M_B\, l}{3}(l + 3EI\,\alpha) = \frac{wl^4}{24}$$

Substituting for EI, α and l

We get $M_B = +\ 13.33$ kN

Substituting M_B value in Eqn. 10.22

We get $M_A = +\ 33.33$ kN.m

The assumed –ve values for M_A and M_B are true

Example 10.9 | *A fixed beam is loaded with a moment M as shown in Fig. 10.23. Determine end moments.*

Step 1: To release the redundant reaction components

The beam is statically indeterminate by 2 degrees. We have got the option to release either the end moments M_A and M_B or the reaction components R_A and M_A at the left end A to make the beam a primary one. The second option is chosen. The primary beam is a cantilever beam as shown in Fig. 10.23*b* with redundants R_A and M_A. The bending moment diagram on the primary beam is drawn in parts as shown in Fig. 10.23. Considering M_A as hogging.

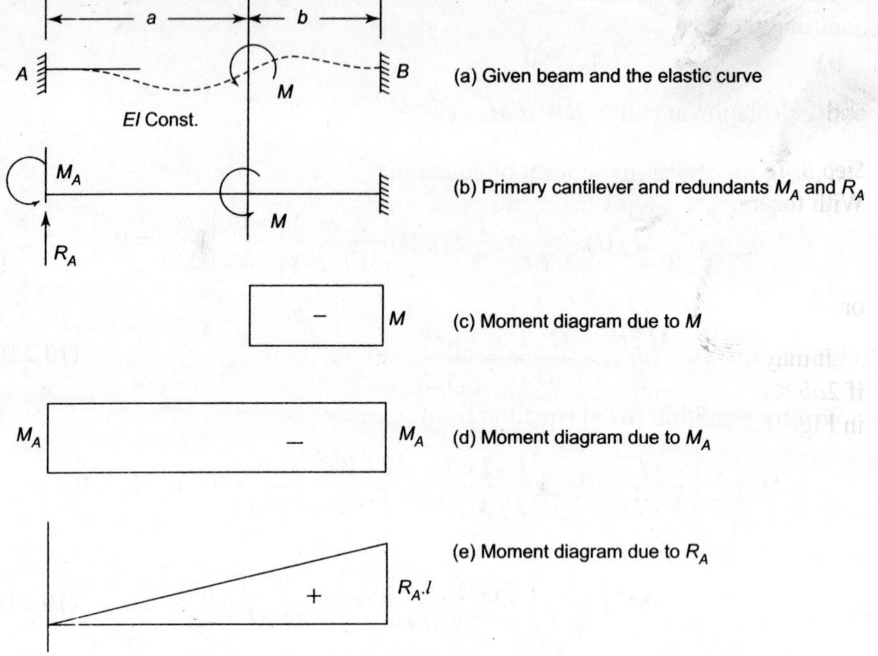

Fig. 10.23

Step 2: To evaluate redundants R_A and M_A

In the elastic curve the change of slope between ends A and B is zero and therefore the area of the moment diagrams between A and $B = 0$. writing the equation, we have

$$\frac{1}{EI}\left[-Mb - M_A\, l + \frac{1}{2} R_A \cdot l^2\right] = 0$$

or
$$\frac{R_A l^2}{2} - M_A\, l = M \cdot b \tag{10.24}$$

Again $t_{BA} = 0$ as the intercept at B from the tangent at A on the elastic curve is zero.

Writing the moment of Moment diagrams about $A = 0$ we have

$$\frac{1}{EI}\left[-M.b.\frac{b}{2} - \frac{M_A\, l^2}{2} + \frac{R_A \cdot l^3}{6}\right] = 0$$

or
$$\frac{R_A l^3}{6} - \frac{M_A\, l^2}{2} = \frac{M.b^2}{2} \tag{10.25}$$

Multiplying Eqn. (10.24) by $\frac{l}{3}$ through out and subtracting from Eqn. (10.25) we get

$$M_A = \frac{M}{l^2}\,(2ab - b^2)$$

and $$R_A = 6\,\frac{M\,ab}{l^3}$$

Step 3: To evaluate M_B

With the known redundant reaction components we can calculate.

$$M_B = R_A \cdot l - M_A - M$$

or $$M_B = \frac{M}{l^2}\,(2\,ab - a^2)$$

It may be noted that the sign of moment M_A reverse it $2ab < b^2$ and so also M_b if $2ab < a^2$. Taking $a = b = l/2$ the moment and shear force diagrams are shown in Fig. 10.24.

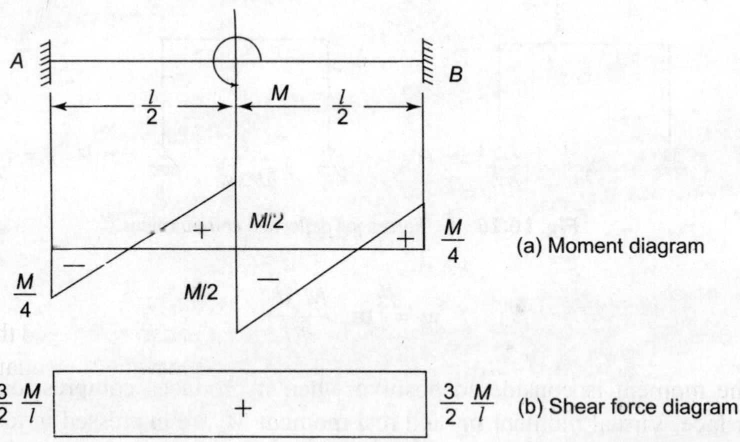

Fig. 10.24

Example 10.10 | *It is required to determine the horizontal reaction component for the frame in Fig. 10.25. The EI value for the beam is twice the value for the columns.*

The frame is statically indeterminate to the first degree. If we choose the horizontal reaction component at support D as the redundant, the frame is made into a primary structure by releasing support D from the restraint for lateral displacement. This is achieved by considering it to be placed on the roller support as shown in Fig. 10.25*b*. Redundant force R_{DX} is considered to be positive when it acts along the positive direction of X. Also indicated are the directions of x for each member which are to be used in evaluating the deflections of the structure.

We shall evaluate deflection quantities by the method of virtual work. The loadings on the primary structure to obtain each of the deflection quantities by virtual work are shown in Fig. 10.26. The internal virtual work for each of the members is obtained from the Eq. 6.51:

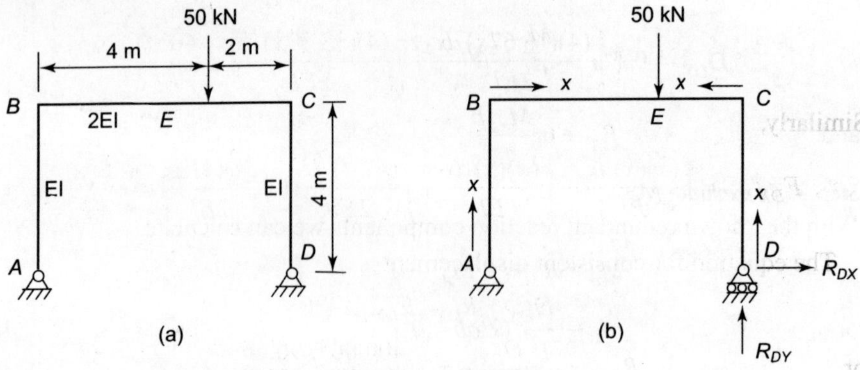

Fig. 10.25 | *(a) Frame and loading, (b) Primary structure*

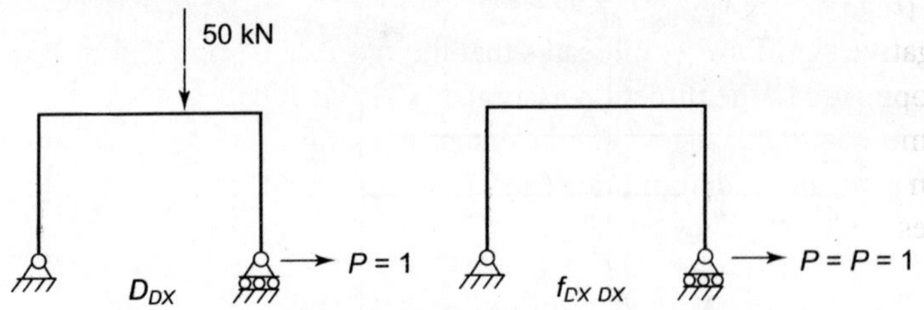

Fig. 10.26 | *Forces for deflection computations*

$$w_i = \int_0^L \mathbf{m}_x \frac{M_x d_x}{EI}$$

The moment is considered positive when it produces compression on the outer face. Virtual moment $\mathbf{m}_x$ and real moment M_x are expressed in terms of x coordinates as indicated in Fig. 10.25*b*. The values are tabulated in Table 10.1.

Table 10.1 | *Calculation for determining internal virtual work*

Section	$x = 0$ *at*	*Limits for x in m*	*Moment* $\mathbf{m}_x$	*Moment* M_x	$\int_0^L Mx\, d_x/EI$
AB	*A*	0–4	x	0	0
CD	*D*	0–4	x	0	0
BE	*B*	0–4	4	16.67x	$\int_0^L (4)\left(\frac{16.67x}{2EI}\right)dx$
CE	*C*	0–2	4	33.33x	$\int_0^L (4)\left(\frac{33.33x}{2EI}\right)dx$

Equating external virtual work to internal virtual work, we have from Eq. 6.51

$$D_{DX} = 0 + \int_0^4 \frac{(4)(16.67x)\,dx}{2EI} + \int_0^2 \frac{(4)(33.33x)\,dx}{2EI} = \frac{400.0}{EI}$$

Similarly,

$$F_{DX\,DX} = \int_0^4 \frac{(x)(x)dx}{EI} + \int_0^4 \frac{(x)(x)\,dx}{EI} + \int_0^4 \frac{(4)(4)\,dx}{2EI} + \int_0^2 \frac{(4)(4)\,dx}{2EI} = \frac{90.66}{EI}$$

The equation for consistent displacement is

$$D_{DX} + R_{DX} \cdot F_{DX\,DX} = 0$$

or $$R_{DX} = -\frac{D_{DX}}{F_{DX\,DX}} = -\frac{400.00}{EI} \Big/ \frac{90.66}{EI}$$

or $$R_{DX} = -4.412 \text{ kN}$$

The negative sign for R_{DX} indicates that the reaction component at D is inwards, that is, opposite to the direction assumed.

The method of consistent displacements can also be used to determine the bar forces in statically indeterminate truss. This aspect is illustrated by the following examples.

Example 10.11 *It is required to determine the bar forces in the steel truss of Fig. 10.27. The area of each member is 500 × 10^{-6} m^2 (500 mm^2).*

Applying the criteria for indeterminancy we find that the truss is redundant internally by two degrees. In analysing trusses with double diagonals it is convenient to select release in one of the diagonal members because the resulting primary structure will be the conventional truss.

We consider diagonal members 1–5 and 2–4 as redundants. Let the forces in the redundant bars be X_1 and X_2 as indicated in Fig. 10.27*c*.

The analysis of the truss reduces to applying an equation of compatibility to the changes in lengths of the released members. The relative displacements D_1p and D_2p corresponding to the two cut ends of the bars is shown in Fig. 10.27*b*. The displacements are always measured along the lengths of the redundant members and since the redundants are unstressed at this stage of the analysis, displacement D_1p is equal to the relative displacement of joint 4 with respect to joint 2 and D_2p is the relative displacement of joint 5 with respect to 1.

These displacements must be eliminated by the relative displacements of the cut ends of members 2–4 and 1–5 when the redundant forces are acting in the members. The desired consistency deformations are:

$$D_1p + f_{11}X_1 + f_{12}X_2 = 0$$

$$D_{2p} + f_{21}X_1 + f_{22}X_2 = 0 \qquad (10.26)$$

where f_{11} is the flexibility influence coefficient for unit tensile forces applied at the cut ends of member 4-2 and f_{22} is the flexibility influence coefficient for unit tensile forces applied at the cut ends of the members 5-1.

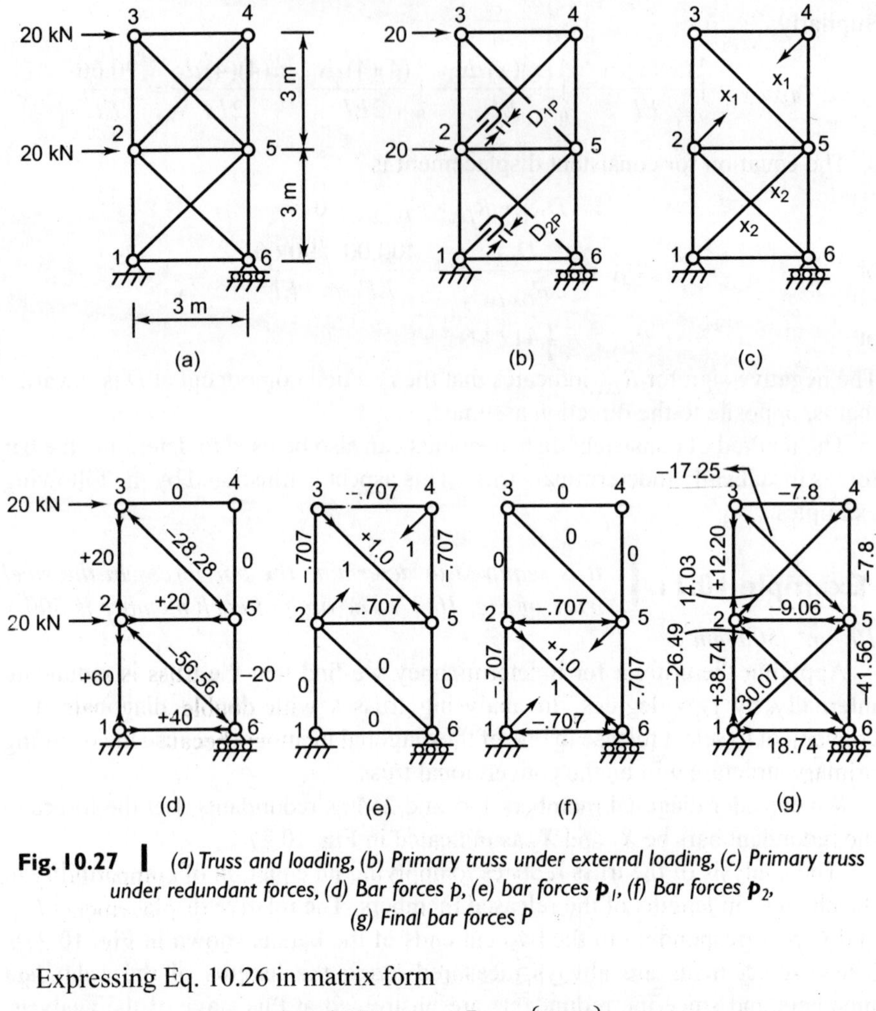

Fig. 10.27 | *(a) Truss and loading, (b) Primary truss under external loading, (c) Primary truss under redundant forces, (d) Bar forces p, (e) bar forces* $\mathbf{p}_1$*, (f) Bar forces* $\mathbf{p}_2$*, (g) Final bar forces P*

Expressing Eq. 10.26 in matrix form

$$\begin{Bmatrix} X_1 \\ X_2 \end{Bmatrix} \begin{bmatrix} f_{11} & f_{12} \\ f_{21} & f_{22} \end{bmatrix} = -\begin{Bmatrix} D_{1p} \\ D_{2p} \end{Bmatrix} \tag{10.26a}$$

The method of virtual work will be used to evaluate deflection quantities. The expression for internal virtual work in a truss is given by Eq. 6.53. The loadings on the truss for determining deflection quantities are shown in Figs. 10.27*b* and *c*. The values of bar forces resulting from each condition of loading are shown in Figs. 10.27*d*, *e* and *f* and also in Table 10.2. As before, the tensile forces in the members are considered positive. The values of D_{1p} and D_{2p} are found from the data of Table 10.2.

Table 10.2 | *Computation for truss analysis*

Member	p Kn	$\mathbf{p}_1$ kN	$\mathbf{p}_2$ kN	$\frac{L}{L_C}$	D_{1p} $p_1 p \frac{L}{L_C}$	D_{2p} $p_2 p \frac{L}{L_C}$	f_{11} $p_1 p_1 \frac{L}{L_C}$	f_{21} $p_2 p_1 \frac{L}{L_C}$	f_{12} $p_2 p_1 \frac{L}{L_C}$	f_{22} $p_1 p_2 \frac{L}{L_C}$	Final stresses in kN
1	2	3	4	5	6	7	8	9	10	11	12
1-2	+60.00	0	–0.707	1.0	0	–42.42	0			0.5	38.74
2-3	+20.00	–0.707	0	1.0	–14.14	0	0.5			0	12.20
3-4	0	–0.707	0	1.0	0	0	0.5			0	–7.80
4-5	0	–0.707	0	1.0	0	0	0.5			0	–7.80
5-6	–20.00	0	–0.707	1.0	0	14.14	0			0.5	–41.26
6-1	+40.00	0	–0.707	1.0	0	–28.28	0			0.5	18.74
2-5	+20.00	–0.707	–0.707	1.0	–14.14	–14.14	0.5	0.5	0.5	0.5	–9.06
1-5	0	0	1.0	1.414	0	0	0			1.414	30.07
2-6	–56.56	0	1.0	1.414	0	–80.80	0			1.414	–26.49
2-4	0	1.0	0	1.414	0	0	1.414			0	11.03
3-5	–28.28	1.0	0	1.414	–40.00	0	1.414			0	–1725
					Σ - 68.28	–150.70	4.828	0.5	0.5	4.828	

$$D_{1p} = \sum \frac{\mathbf{p}_1 pL}{AE} = -\frac{68.28\ L_C}{AE}$$

and
$$D_{2p} = \sum \frac{\mathbf{p}_2 pL}{AE} = -\frac{150.70\ L_C}{AE}$$

where $L_C = 3$ m and AE is constant for all members.

The same method is used to compute f_{ij}. In this case the real loading is a unit load corresponding to release i and the unit dummy load corresponds to release j, that is

$$f_{11} = \sum \mathbf{p}_1 \left(\frac{p_1 L}{AE} \right) \quad \text{for all members}$$

$$f_{12} = \sum \mathbf{p}_1 \left(\frac{p_2 L}{AE} \right)$$

also
$$f_{21} = \sum \mathbf{p}_2 \left(\frac{p_1 L}{AE} \right)$$

and
$$f_{22} = \sum \mathbf{p}_2 \left(\frac{p_2 L}{AE} \right)$$

From Table 10.2, we have

$$f_{11} = \frac{4.828\ L_c}{AE} \qquad f_{12} = \frac{0.5 L_c}{AE}$$

$$f_{21} = \frac{0.5\ L_c}{AE} \quad \text{and} \quad f_{22} = \frac{4.828\ L_c}{AE}$$

The compatibility condition is that the ends of both redundant members must match, that is, there should not be any gaps or overlaps of the members in the actual structure. Using compatibility Eq. 10.22

$$D_{1p} + f_{11}X_1 + f_{12}X_2 = 0$$

$$D_{2p} + f_{21}X_1 + f_{22}X_2 = 0$$

$$\frac{L_c}{AE}\begin{bmatrix} 4.828 & 0.500 \\ 0.500 & 4.828 \end{bmatrix}\begin{Bmatrix} X_1 \\ X_2 \end{Bmatrix} = -\frac{L_c}{AE}\begin{Bmatrix} -68.28 \\ -150.70 \end{Bmatrix}$$

and solving for X_1 and X_2 by inverting the flexibility matrix, we have

$$\begin{Bmatrix} X_1 \\ X_2 \end{Bmatrix} = \begin{bmatrix} 4.828 & 0.500 \\ 0.500 & 4.828 \end{bmatrix}^{-1}\begin{Bmatrix} -68.28 \\ -150.70 \end{Bmatrix} = \begin{bmatrix} 11.028 \\ 30.070 \end{bmatrix} \text{kN}$$

Therefore, $X_1 = 11.028$ kN (tension) and $X_2 = 30.07$ kN (tension).

The final set of forces in the truss members is obtained by adding up, for each member, the three separate effects, that is, $P = p + \mathbf{p}_1 \cdot X_1 + \mathbf{p}_2 \cdot X_2$. The final bar forces are given in Fig. 10.27g.

This procedure can also be extended to stresses produced by temperature changes in indeterminate trusses. The procedure is illustrated by the following example.

Example 10.12 | *Consider the same truss as in Example 10.11 above. Supposing that there is a temperature drop of 30°C on all the outer members, 1–2, 2–3, 3–4, 4–5 and 5–6, find the forces set up in the members due to temperature drop only. Take $\alpha = 1.0 \times 10^{-5}$ °C.*

The change in length of bars affected by temperature drop is

$$\Delta L = \alpha L \Delta T = 30 \times 10^{-5} \text{ L}$$

where $$L = 3 \text{ m}$$

To evaluate the deflections the virtual work method is used. The internal virtual work caused by the virtual forces $\mathbf{p}_1$ riding through real displacement ΔL is equated to the external virtual work caused by virtual unit load placed in the direction of redundant members, in moving through the real displacements, that is

$$1 \cdot D_{1\,\Delta L} = \Sigma\, \mathbf{p}_{1\,\Delta L} \tag{10.27}$$

$$1 \cdot D_{2\,\Delta L} = \text{X}\, \mathbf{p}_{2\,\Delta L} \tag{10.28}$$

where $\mathbf{p}_1$ and $\mathbf{p}_2$ represent forces in members due to a unit load applied separately at redundants 1 and 2 respectively. The values of $\mathbf{p}_1$ and $\mathbf{p}_2$ are tabulated (Table 10.3) for all the members affected by the temperature drop.

$$D_{1\,\Delta L} = 6.36\, L_c \times 10^{-4}$$

$$D_{2\,\Delta L} = 4.24\, L_c \times 10^{-4}$$

It may be noted that flexibility matrix **F** is the same as in Example 10.10. Therefore, the redundant forces due to temperature change are

$$\begin{Bmatrix} X_1 \\ X_2 \end{Bmatrix} = -\frac{L}{AE}\begin{bmatrix} 4.828 & 0.500 \\ 0.500 & 4.828 \end{bmatrix}^{-1} \frac{AE}{L_C}\begin{Bmatrix} -6.36 \\ -4.24 \end{Bmatrix} L_c \times 10^{-4}$$

On simplifying, we have

$$X_1 = -12.39 \text{ kN}$$

$$X_2 = -7.50 \text{ kN}$$

Table 10.3 | *Computations for temperature stresses*

Member kN	p_1 kN	p_2	ΔL	$D_{1\Delta L}$ $L_c \times 10^{-4}$	$D_{2\Delta L}$ $L_c \times 10^{-4}$
1–2	0	–0.707	–0.0003 L_c	0	2.12
2–3	–0.707	0	–0.0003 L_c	2.12	0
3–4	–0.707	0	–0.0003 L_c	2.12	0
4–5	–0.707	0	–0.0003 L_c	2.12	0
5–6	0	–0.707	–0.0003 L_c	0	2.12
				Σ 6.36	Σ 4.24

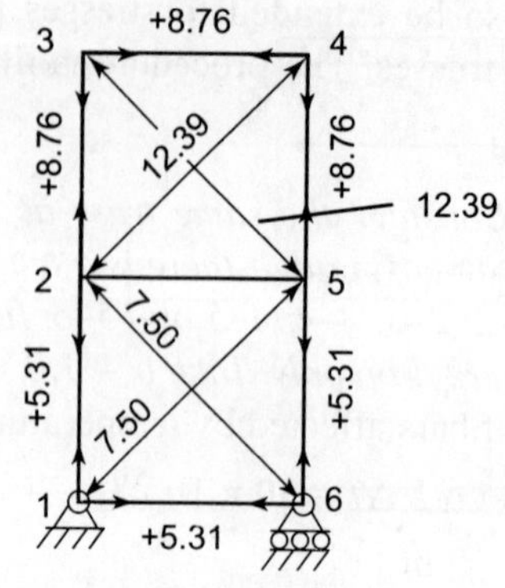

Fig. 10.28 | *Bar forces due to temperature change*

The bar forces only due to temperature drop are given by

$$P = \mathbf{p}_1 \cdot X_1 + \mathbf{p}_2 \cdot X_2$$

The results are summarised in Fig. 10.28.

The method also lends itself well to the condition of support movements and other types of initial displacements. For example, let us assume that the base of the right hand column of the portal frame in Fig. 10.16 moves by an amount Δ_1 to the right and settles down by an amount Δ_2. The expression for consistent displacement given in Eq. 10.7 now becomes

$$\begin{bmatrix} D_{1P} \\ D_{2P} \\ D_{3P} \end{bmatrix} + \begin{bmatrix} f_{11} & f_{12} & f_{13} \\ f_{21} & f_{22} & f_{23} \\ f_{31} & f_{32} & f_{33} \end{bmatrix} \begin{bmatrix} R_1 \\ R_2 \\ R_3 \end{bmatrix} = \begin{bmatrix} \Delta_1 \\ -\Delta_2 \\ 0 \end{bmatrix} \quad (10.29)$$

$$\begin{bmatrix} R_1 \\ R_2 \\ R_3 \end{bmatrix} = -\begin{bmatrix} f_{11} & f_{12} & f_{13} \\ f_{21} & f_{22} & f_{23} \\ f_{31} & f_{32} & f_{33} \end{bmatrix}^{-1} = \begin{bmatrix} D_{1P} - \Delta_1 \\ D_{2P} + \Delta_2 \\ D_{3P} \end{bmatrix} \quad (10.30)$$

Care must be exercised in such an analysis to ensure that the given settlements and computed displacements are in the same dimensions.

10.5 THEOREM OF THREE MOMENTS

A general equation based on the method of consistent displacements can be developed for continuous beams with or without support moments. We shall first discuss beams with rigid supports. The equation relates the moments at the three consecutive support points to the loading on the intermediate spans and is, therefore, referred to as *theorem of three moments*. This theorem was presented by Clapeyron in 1857 for the analysis of continuous beams. The application of the three moments equation to a continuous beam results in a set of simultaneous equations with the moments over the supports as the unknowns.

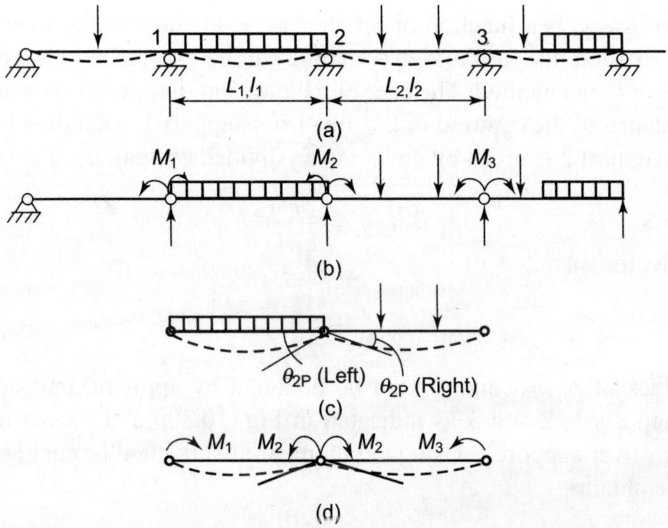

Fig. 10.29 | *(a) Continuous beam, (b) Primary structure, (c) Primary structure under external loading, (d) Primary structure under redundant moments*

Consider a continuous beam in which 1, 2 and 3 are the consecutive supports and spans 1–2 and 2–3 are arbitraily loaded as shown in Fig. 10.29*a*.

L_1 and I_1, L_2 and I_2 are the span lengths and moments of inertia corresponding to spans 1–2 and 2–3 respectively. The deflected shape of the beam under loading is given by a dotted line.

This beam can be made statically determinate by inserting hinges at the supports and considering support moments M_1, M_2 and M_3 as redundants whose values are to be determined. We now consider the deflected shape of the primary structure under given loading and the redundant moments (see Figs. 10.29*c* and *d*). The relative rotation between the segments of the beam over support 2 is represented by

$$\theta_{2P} = \theta_{2P(\text{left})} + \theta_{2P(\text{right})} \tag{10.31}$$

The rotation at 2 due to redundant moments M_1, M_2 and M_3 are related by the condition

$$\theta_{2P} + f_{21}M_1 + f_{22}M_2 + f_{23}M_3 = 0 \tag{10.32}$$

The flexibility coefficients f_{21}, f_{22}, and f_{23} are obtained by subjecting the primary structure to unit moments at supports 1, 2 and 3 and knowing the rotations at support 2.

The relative rotation θ_{2P} is obtained by considering the two segments of the beam as shown in Fig. 10.30*a*.

From Fig. 10.30*a* we see that the relative rotation at support 2 is

$$\theta_{2P} = \theta_{2P(\text{left})} + \theta_{2P(\text{right})} \tag{10.33}$$

Rotation $\theta_{2P(\text{left})}$ is a function of the transverse loading on the span 1-2. The amount of rotation due to transverse loading can be expressed in general terms by the moment area method. The area of the moment diagram is indicated by A_1 and the distance of the centroid of the area from support 1 is denoted by x_1. The rotation at support 2 is given by deviation t_{12} divided by span L_1, that is,

$$\theta_{2P(\text{left})} = \frac{A_1 x_1}{EI_1 L_1} \qquad (10.33a)$$

Similarly, for span 2–3

$$\theta_{2P(\text{right})} = \frac{A_2 x_2}{EI_2 L_2} \qquad (10.33b)$$

The values of f_{21}, f_{22} and f_{23} can be obtained by applying unit couples in turns at supports 1, 2 and 3 as indicated in Fig. 10.30*b* and then determining the rotations over support 2. Using the moment-area method or conjugate beam method, we obtain

$$f_{21} = \frac{L_1}{6EI_1}$$

$$f_{22} = f_{22(\text{left})} + f_{22\,(\text{right})} = \frac{L_1}{3EI_1} + \frac{L_2}{3EI_2}$$

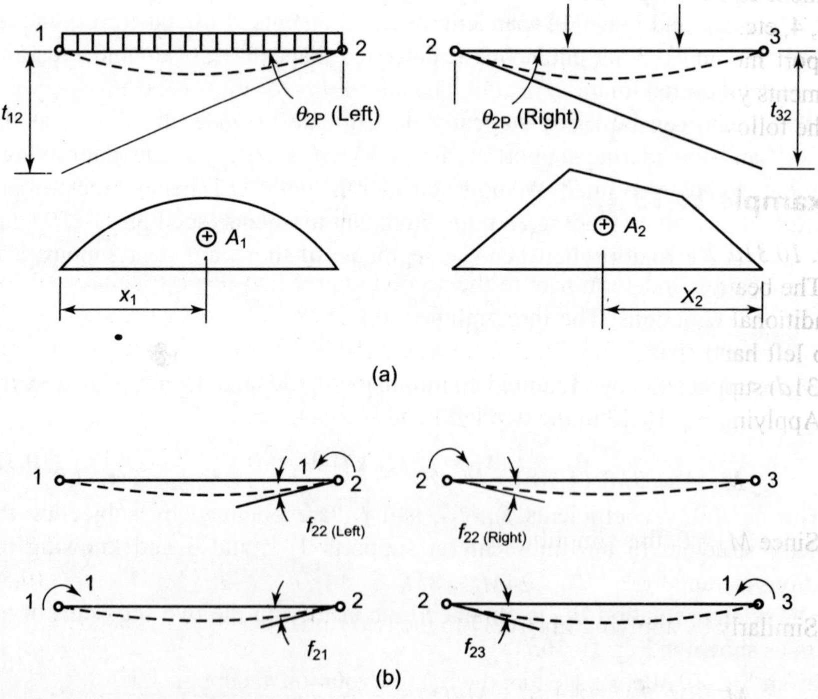

Fig. 10.30 | ***(a) Primary structure under external load and corresponding moment diagrams, (b) Primary structure under unit redundant forces***

and $$f_{23} = \frac{L_2}{6EI_2} \tag{10.34}$$

Substituting in Eq. 10.28 the values from Eqs. 10.29 and 10.30, we get

$$\frac{M_1L_1}{6EI_1} + \frac{M_2L_1}{3EI_1} + \frac{M_2L_2}{3EI_2} + \frac{M_3L_2}{6EI_2} + \frac{A_1x_1}{EI_1L_1} + \frac{A_2x_2}{EI_2L_2} = 0 \tag{10.35}$$

Rearranging Eq. 10.31, we get

$$\frac{M_1L_1}{EI_1} + 2M_2\left(\frac{L_1}{EI_1} + \frac{L_2}{EI_2}\right) + \frac{M_3L_2}{EI_2} = -\frac{6A_1x_1}{EI_1L_1} - \frac{6A_2x_2}{EI_2L_2} \tag{10.36}$$

Equation 10.36 is the general form of the three-moment equation. The moment quantities in Eq. 10.36 are positive according to the beam sign convention, that is, positive moments cause tension at the bottom fibres of the beam. If EI is constant throughout, Eq. 10.36 simplifies to

$$M_1L_1 + 2M_2\,(L_1 + L_2) + M_3L_2 = -\,6\frac{A_1x_1}{L_1} - 6\frac{A_2x_2}{L_2} \tag{10.37}$$

The three-moment equation developed above involves not only the moment over support 2 but also the moments at supports 1 and 3. In applying the three-moment equation to a particular beam, we locate the interior supports, such as 2, 3, 4, etc. successively and write as many equations as the unknown redundant support moments. A simultaneous solution of the equations for the unknown moments yields the required results. The application of this method is illustrated in the following examples.

Example 10.13 | *It is required to determine the support moments and reactions for the three-span continuous beam shown in Fig. 10.31a. EI is constant.*

The beam is indeterminate to the second degree and requires the use of two conditional equations. The three-moment equation, if used twice, once for the two left hand spans (Fig. 10.31*b*) and once for the two right hand spans (Fig. 10.31*d*) supplies the two required conditional equations.

Applying Eq. 10.37 to the two left hand spans, we have

$$M_A\,(4) + 2M_B\,(4 + 8) + M_C\,(8) = -6\frac{(60)(2)}{4} - 6\frac{(426.67)(4)}{8}$$

Since $M_A = 0$, this simplifies to

$$24M_B + 8M_C = -1460 \tag{10.38}$$

Similarly by applying Eq. 10.37 to the two right hand spans, we have

$$M_B\,(8) + 2M_C\,(8 + 6) + M_D\,(6) = -\,6\frac{(426.67)(4)}{8} - 6\frac{(180)(3)}{6}$$

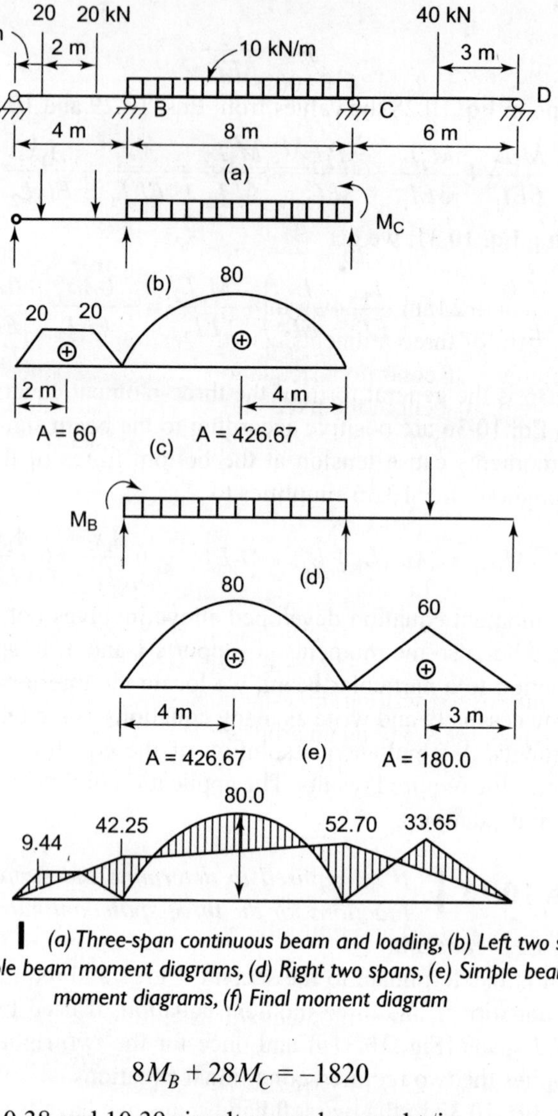

Fig. 10.31 | *(a) Three-span continuous beam and loading, (b) Left two spans, (c) Simple beam moment diagrams, (d) Right two spans, (e) Simple beam moment diagrams, (f) Final moment diagram*

or
$$8M_B + 28M_C = -1820 \tag{10.39}$$

Solving Eqs. 10.38 and 10.39 simultaneously, we obtain

$$M_B = -42.25 \text{ kN.m, and } M_C = -52.70 \text{ kN.m}$$

The reactions are determined by applying the equations of statics as follows

$$R_A = 20 + \frac{M_B}{4}$$

or
$$R_A = 20 + \frac{(-42.25)}{4} = 9.44 \text{ kN}$$

Similarly,

$$R_B = 20 + 10(4) + \frac{-M_B}{4} + \frac{M_C - M_B}{8} = 69.25 \text{ kN}$$

$$R_C = 60 + \frac{M_B - M_C}{8} - \frac{M_C}{6} = 70.09 \text{ kN}$$

Finally, $$R_D = 20 + \frac{M_C}{6} = 11.22 \text{ kN}$$

The moment diagram is shown in Fig. 10.31*f*.

The theorem of three moments can also be applied to fixed end beams. The required number of conditional equations can be obtained by considering an imaginary span adjacent to the fixed end as having an arbitrary span length with an infinite moment of inertia. This point is illustrated by solving the following example.

Example 10.14 | *It is required to determine the support moments and reactions for a continuous beam fixed at one end and having a overhang at the other as shown in Fig. 10.32a. EI is constant.*

The beam is statically indeterminate to the second degree and requires two conditional equations.

For the purpose of writing three-moment equations, an imaginary span to the left of fixed support A having an arbitrary length L' and moment of inertia $I' = \infty$ may be considered (see Fig. 10.32*b*). The three-moment equation for spans A' – A and $A - B$ can be written as

$$M'_A \left(\frac{L'}{\infty}\right) + 2M_A \left(\frac{L'}{\infty} + \frac{L_{AB}}{I}\right) + M_B \left(\frac{L_{AB}}{I}\right) = 0 - \frac{6(80)(2)}{IL_{AB}} \quad (10.40)$$

Substituting for $L_{AB} = 4$ and multiplying throughout by I, this reduces to

$$8M_A + 4M_B = -240 \quad (10.41)$$

Similarly, writing the three-moment equation for spans A–B and B–C, we have

$$M_A(4) + 2M_B(4+6) + M_C(6) = -6\frac{(80)(2)}{4} - \frac{6(80)(3)}{6} \quad (10.42)$$

We know $M_C = -40$ kN.m

Substituting this value in Eq. 10.42, we have

$$4M_A + 20M_B = -540 \quad (10.43)$$

Solving Eqs. 10.41 and 10.43 simultaneously, we get

$$M_A = -18.33 \text{ kN.m, and } M_B = -23.33 \text{ kN.m}$$

The moment diagram is shown in Fig. 10.32*c*. The reactions are evaluated from the free-body diagrams in Fig. 10.32*d* and using statics only. Therefore,

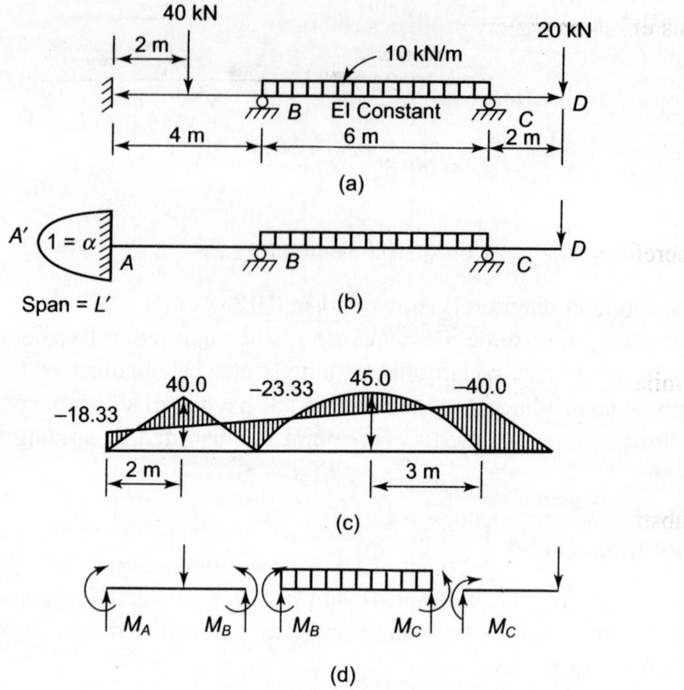

Fig. 10.32 | *(a) Beam and loading, (b) Imaginary span added to left of support A; (c) Moment diagrams, (d) Free-body diagrams of individual spans*

$$R_A = 20 + \frac{(M_B - M_A)}{L_{AB}} = 18.75 \text{ kN}$$

$$R_B = 20 + 30 + \frac{(M_A - M_B)}{L_{AB}} + \frac{(M_C - M_B)}{L_{BC}} = 48.87 \text{ kN}$$

and

$$R_C = 30 + 20 + \frac{(M_B - M_C)}{L_{BC}} = 52.78 \text{ kN}$$

The theorem of the three moments equation can be suitably modified to take into account the settlement of supports. For example, consider that supports 1, 2 and 3 of Fig. 10.33 settle downward by amounts δ_1, δ_2 and δ_3 respectively.

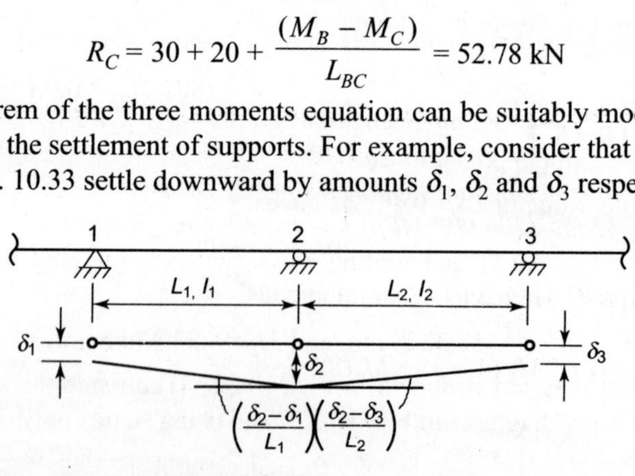

Fig. 10.33 | *Settlement of supports*

This effect separately produces rotations at 2 as

$$\theta_{2(\text{left})} = -\left(\frac{\delta_2 - \delta_1}{L_1}\right)$$

and $$\theta_{2(\text{right})} = -\left(\frac{\delta_2 - \delta_3}{L_2}\right)$$

Therefore, $\theta_{2P(\text{left})}$ (see Eq. 10.33) modifies now to

$$\theta_{2P(\text{left})} = \frac{A_1 x_1}{EI_1 L_1} - \left(\frac{\delta_2 - \delta_1}{L_1}\right) \tag{10.44}$$

Similarly, $\theta_{2P(\text{right})}$ modifies to

$$\theta_{2P(\text{right})} = \frac{A_2 x_2}{EI_2 L_2} - \left(\frac{\delta_2 - \delta_3}{L_2}\right) \tag{10.45}$$

Substituting these values in Eq. 10.32 and rearranging, we get

$$\frac{M_1 L_1}{I_1} + 2M_2\left(\frac{L_1}{I_1} + \frac{L_2}{I_2}\right) + \frac{M_3 L_2}{I_2} = -6\left(\frac{A_1 x_1}{L_1 I_1} + \frac{A_2 x_2}{L_2 I_2}\right)$$

$$+ 6E\left\{\frac{(\delta_2 - \delta_1)}{L_1} + \frac{(\delta_2 - \delta_3)}{L_2}\right\} \tag{10.46}$$

If $I_1 = I_2 = I$, Eq. 10.42 reduces to

$$M_1 L_1 + 2M_2(L_1 + L_2) + M_3 L_2 = -6\left(\frac{A_1 x_1}{L_1} + \frac{A_2 x_2}{L_2}\right)$$

$$+ 6EI\left\{\left(\frac{(\delta_2 - \delta_1)}{L_1}\right) + \left(\frac{(\delta_2 - \delta_3)}{L_2}\right)\right\} \tag{10.47}$$

We note that in Eq. 10.46 the settlements δ_1, δ_2 and δ_3 are taken as positive if downward.

Example 10.15 | *As an illustration of a typical application of the formula given in Eq. 10.47, the problem in Example 10.13 will now be solved considering that support B sinks by 10 mm under the given loading, $E = 200 \times 10^6$ kN/m² (200,000 MPa) and $I = 80 \times 10^{-6}$ m⁴ (80×10^6 mm⁴).*

Referring to Fig. 10.31 and writing down the three-moment Eq. 10.47 for spans *A-B* and *B-C*, we have

$$M_A(4) + 2M_B(4 + 8) + M_C(8) = -6\left(\frac{60 \times 2}{4} + \frac{426.67(4)}{8}\right)$$

$$+ 6 \times 16000\left(\frac{10}{4000} + \frac{10}{8000}\right)$$

This reduces to

$$24M_B + 8M_C = -\,1100 \qquad (10.48)$$

Similarly, writing down the three-moment Eq. 10.47 for spans *B-C* and *C-D*, we have

$$M_B(8) + 2M_C\,(8+6) + M_D(6) = -6\left\{\frac{426.67(4)}{8} + \frac{6(180)(3)}{6}\right\} + 6 \times 16000\left(-\frac{10}{8000}\right)$$

Writing $M_D = 0$ and simplifying, we have

$$8M_B + 28\,M_C = -1940 \qquad (10.49)$$

Solving Eqs. 10.48 and 10.49 simultaneously we obtain

$$M_B = -62.0 \text{ kN.m and } M_C = -25.1 \text{ kN.m}$$

Note that the moments have been substantially altered from the previous values.

10.6 THE METHOD OF LEAST WORK

A special form of Castigliano's second theorem is useful in the analysis of indeterminate structures. If R_i, is a reaction component, the corresponding displacement Δ_i, is zero for a rigid support and from Eq. 6.68

$$\frac{\partial U}{\partial R_i} = 0 \qquad (10.50)$$

This holds for each of the reaction components in the structure. It can be shown that Eq. 10.50 actually means that the reactions in an indeterminate structure take on values that lead to minimum strain energy level in the structure. For this reason, this approach is called *method of least work*. This method can be used to solve indeterminate beams, trusses, arches and frames. Although newer methods of analysis frequently will enable one to solve particular problems more directly than can be done using least work, the method is still in common use and preferred by some engineers.

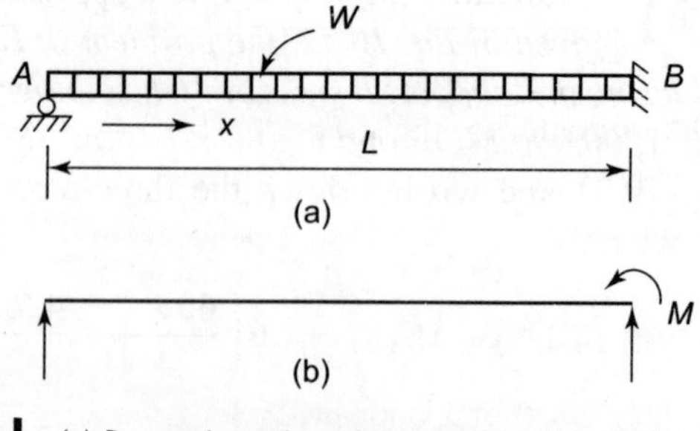

Fig. 10.34 | *(a) Propped cantilever beam and loading, (b) Primary structure under redundant moment M*

The application of the method of least work is illustrated by the following examples.

Example 10.16 | *It is required to determine the moment at the fixed end of the propped cantilever beam shown in Fig. 10.34. EI is constant.*

Let us consider moment M at the fixed end as the redundant. Then using Eq. 10.50 for the fixed end B

$$\frac{\partial U}{\partial M} = \frac{\partial}{\partial M}\int_0^L \frac{M_x dx}{2EI} = 0$$

$$\frac{\partial U}{\partial M} = \frac{1}{EI}\int_0^L M_x \frac{\partial M_x}{\partial M}\, dx = 0$$

Substituting for

$$M_x = \left(\frac{wLx}{2} - \frac{wx^2}{2} + M\frac{x}{L}\right)$$

and
$$\frac{\partial M_x}{\partial M} = +\frac{x}{L}$$

we have,

$$\frac{\partial U}{\partial M} = \frac{1}{EI}\int_0^L \left(\frac{wLx}{2} - \frac{wx^2}{2} + M\frac{x}{L}\right)\left(\frac{x}{L}\right) dx = 0$$

On evaluation, we get

$$M = -\frac{wL^2}{8}$$

The negative sign for the value of M indicates that the direction assumed for M is incorrect. Moment M produces tension at the top.

The method can also be used for frames. The following example illustrates the procedure.

Example 10.17 | *Using the method of least work, determine the horizontal reaction component for the frame shown in Fig. 10.35. Consider EI the same for all the members.*

The frame is indeterminate to the first degree. The horizontal reaction component $H_A = H_E = H$ is assumed to be a redundant reaction. The relative lateral displacement of the supports at A and E is zero. Therefore, we write

$$\frac{\partial U}{\partial H} = \frac{1}{EI}\int M_x \frac{\partial M_x}{\partial H}\, dx = 0 \qquad (10.51)$$

for the whole frame. The origin for x coordinate for each section of the frame is indicated in Fig. 10.35*b*.

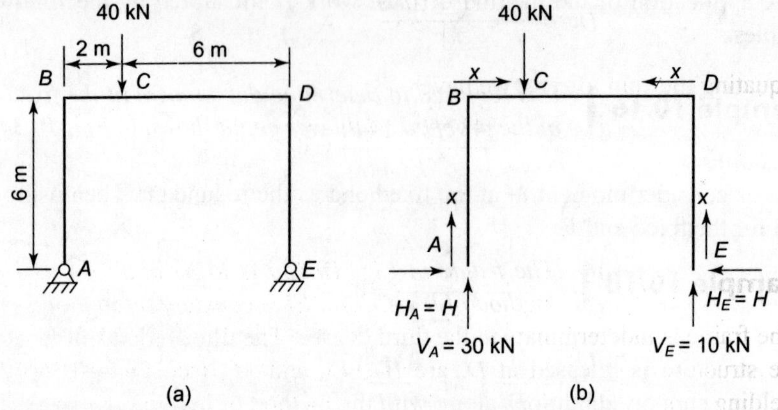

Fig. 10.35 | *(a) Frame and loading, (b) Origin for x*

The evaluation of the integral in Eq. 10.51 is carried out in Table 10.4.

Table 10.4 | *To evaluate* $\int M \frac{\partial M_x}{\partial H} \frac{dx}{EI}$ *for the frame of Fig. 10.35*

Section	$x = 0$ at	Limits for x	Moment Mx	$\frac{\partial M_x}{\partial H}$	$\int_0^L M_x \frac{\partial M_x}{\partial H} \frac{dx}{EI}$
AB	A	0–6	$-H \cdot x$	$-x$	$\frac{1}{EI}\int_0^6 H_x^2\, dx$
BC	B	0–2	30x–6H	–6	$\frac{1}{EI}\int_0^6 (30x - 6H)(-6)\, dx$
ED	E	0–6	$-H \cdot x$	$-x$	$\frac{1}{EI}\int_0^6 H_x^2\, dx$
DC	D	0–6	10x–6H	–6	$\frac{1}{EI}\int_0^6 (10x - 6H)(-6)\, dx$

Evaluating the integrals in the last column of Table 10.4, we get

$$AB = \frac{72H}{EI}$$

$$BC = \frac{72H - 360}{EI}$$

$$ED = \frac{72H}{EI}$$

and $$DC = \frac{216H - 1080}{EI}$$

Equating the sum to zero, we have

$$\frac{432H - 1440}{EI} = 0$$

or $$H = -3.33 \text{ kN}$$

Example 10.18 *The frame in Fig. 10.36a is to be analysed using the method of least work. EI is constant throughout.*

The frame is indeterminate to the third degree. The three redundant reactions, if the structure is released at D, are H_D, V_D and M_D (see Fig. 10.36*b*). The unyielding support conditions along with the method of least work give

$$\frac{\partial U}{\partial H_D} = 0, \frac{\partial U}{\partial V_D} = 0 \text{ and } \frac{\partial U}{\partial M_D} = 0 \tag{10.52}$$

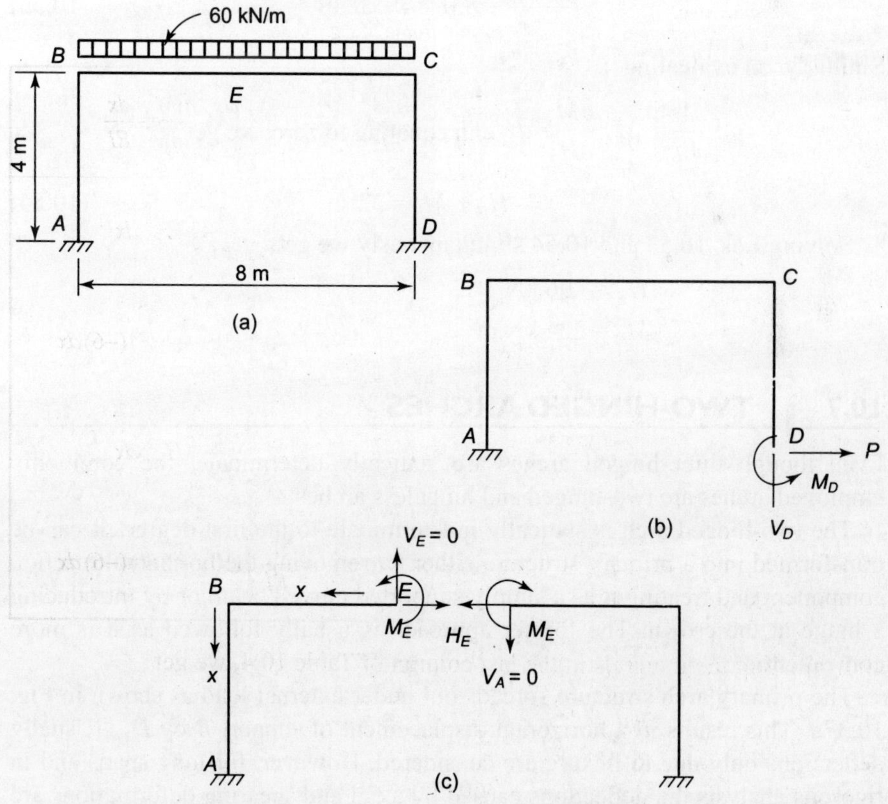

Fig. 10.36 *(a) Frame and loading, (b) Structure released at D, (c) Structure released at E*

resulting in three equations for the calculation of three unknown reaction components at D.

A simpler approach to this problem would be to release the structure at E, the mid point of girder BC, as shown in Fig. 10.36*c*. Symmetry conditions demand that $V_E = 0$ and the horizontal displacement and the rotation are also zero at E. Thus, we have two conditions

$$\frac{\partial U}{\partial H_E} = 0 \text{ and } \frac{\partial U}{\partial M_E} = 0$$

Further, it is sufficient to write the strain energy for only half of the structure. The computations are carried out for the release at E and are shown in Table 10.5.

On evaluating

$$\frac{1}{EI}\int M_x \frac{\partial M_x}{\partial H_E}\,dx \text{ and equating to zero, we get}$$

$$\frac{64H_E}{3} - 8M_E = -3840 \tag{10.53}$$

Similarly, on evaluating

$$\frac{1}{EI}\int M_x \frac{\partial M_x}{\partial H_E}\,dx \text{ and equating to zero, we get}$$

$$-H_E + M_E = 32 \tag{10.54}$$

Solving Eqs. 10.53 and 10.54 simultaneously we get

$$H_E = -96 \text{ kN}$$

$$M_E = 224 \text{ kN.m}$$

10.7 TWO-HINGED ARCHES

Even though three-hinged arches are statically determinate, the commonly employed arches are two-hinged and hingeless arches.

The two-hinged arch is statically indeterminate to the first degree. It can be transformed into a primary structure either by removing the horizontal reaction component and treating it as a simply supported curved beam or by introducing a hinge at the crown. The former approach is usually followed as it is more convenient.

The primary arch structure spreads out under external load as shown in Fig. 10.37*a*. This results in a horizontal displacement of support B by D_{1P}. Usually deflections only due to flexure are considered. However, for long spans and in rigorous analysis the deflections caused by axial and shearing deformations are also to be included.

Table 10.5 | *To evaluate* $\frac{1}{EI}\int (M_x \frac{\partial M_x}{\partial H_E} dx$ and $\frac{1}{EI}\int M_x \frac{\partial M_x}{\partial M_E\, dx}$

Section	$x = 0$ at	limits for x	Moment M_x	$\frac{\partial M_x}{\partial H_E}$	$\frac{\partial M_x}{\partial M_E}$	$\frac{1}{EI}\int_0^L M_x \frac{\partial M_x}{\partial H_E} dx$	$\frac{1}{EI}\int_0^L M_x \frac{\partial M_x}{\partial M_E} dx$
EB	*E*	0–4	$\frac{-50x^2}{2} + M_E$	0	1	$\frac{1}{EI}\int_0^4 (0)\, dx$	$\frac{1}{EI}\int_0^4 (-30x^2 + M_E)\, dx$
BA	*B*	0–4	$-480 - HE_x + M_E$	$(-x)$	1	$\frac{1}{EI}\int_0^4 (430x + H_E\, x^2 - M_E\, x)\, dx$	$\frac{1}{EI}\int_0^4 (-480x - H_E\, x + M_E)\, dx$

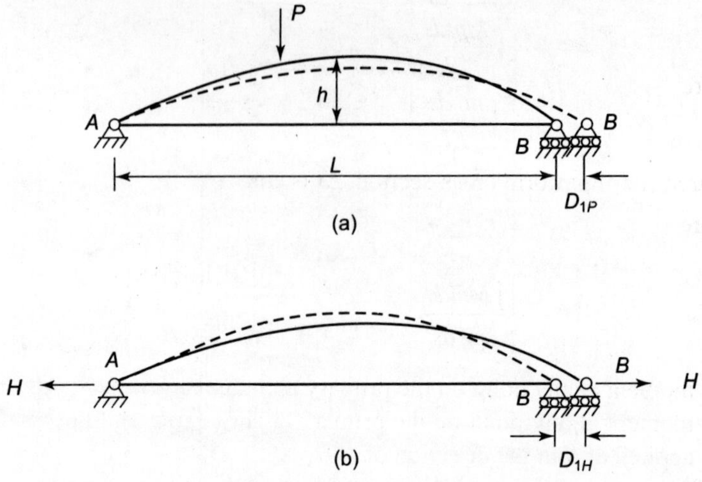

Fig. 10.37 | *(a) Primary arch under applied load, (b) Primary arch under horizontal reaction*

Since the support conditions dictate that the final displacement at support B should be zero, horizontal reaction H should be such that displacement D_{1H} caused by it must satisfy the condition

$$D_{1P} + D_{1H} = 0 \tag{10.55}$$

or

$$D_{1P} + f_{11H} = 0 \tag{10.56}$$

where f_{11} is the displacement caused by a unit force applied in the direction of H.

From Eq. 10.56

$$H = -\frac{D_{1P}}{f_{11}} \tag{10.57}$$

The problem is thus reduced to finding horizontal displacements in a primary structure caused by external loading as well as unit horizontal force.

The horizontal displacement in a curved member can be evaluated by utilising either Castigliano's second theorem or the unit load method. The only difference with respect to a straight beam is that the integration has to be carried out along the axis of the arch, that is

$$D_{1P} = \int_A^B M \frac{\partial M}{\partial H} \frac{ds}{EI}$$

$$D_{1P} = \int_A^B M \frac{mds}{EI} \tag{10.58}$$

Similarly,

$$f_{11} = \int_A^B m^2 \frac{ds}{EI} \tag{10.59}$$

Therefore, $$H = \frac{\int \frac{Mmds}{EI}}{\int \frac{m^2 ds}{EI}} \tag{10.60}$$

For an arch rib of uniform cross-section, EI is constant.

Therefore,

$$H = -\frac{\int Mmds}{\int m^2 ds} \tag{10.61}$$

where

$M =$ moment at any point on the primary arch due to given loading, and

$m =$ moment at any point on the primary arch due to a unit horizontal force applied at B in the direction of H.

The following examples will demonstrate the steps involved in evaluating the integrals in Eq. 10.61.

Example 10.19 | *A two-hinged segmental arch of span 40 m subtends an angle $2\phi = 90°$ at the centre. Find the horizontal reaction casued by a uniformly distributed load of 10 kN/m, extending from the left hand support to the centre of arch as shown in Fig. 10.38.*

$$\text{Radius } R = \frac{20}{\sin 45} = 20.28 \text{ m}$$
$$\text{Rise } h = R - R\cos 45 = 8.28 \text{ m}$$

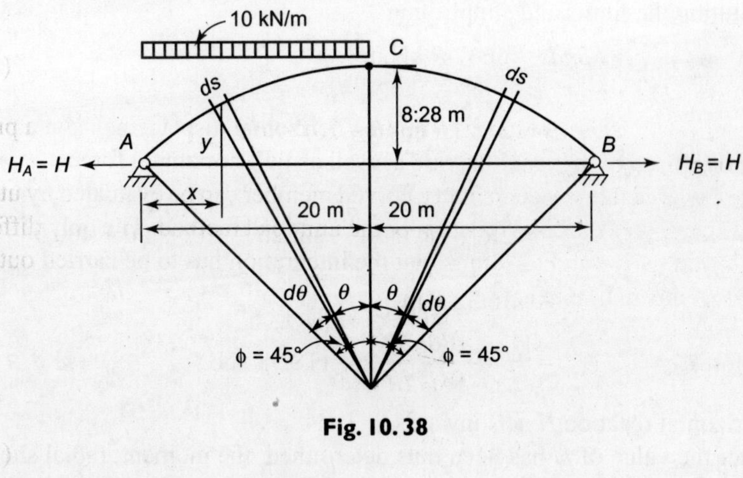

Fig. 10.38

Vertical reaction $V_A = 150$ kN

and $V_B = 50$ kN

The expression for the moment at any point in the region A to C is

$$M = V_A\,(20 - R\sin\theta) - \frac{1}{2}\,(10)(20 - R\sin\theta)^2 \text{ for } 0 \le \theta \le \frac{\pi}{4}$$
$$= 1000 + 50\,R\sin\theta - 5R^2\sin^2\theta$$

and in the region C to B is

$$M = V_B\,(20 - \sin\theta) \text{ for } 0 \le \theta \le \frac{\pi}{4}$$
$$= 1000 - 50\,R\sin\theta$$

Ordinate $\quad y = (R\cos\theta - 20)$

The moment due to unit load $m = (1)(y) = (R\cos\theta - 20)$ and $ds = Rd\theta$

$$\int Mmds = \int_0^{\pi\backslash 4} (1000 + 50\,R\sin\theta - 5\,R^2\sin^2\theta)(R\cos\theta - 20)\,Rd\theta$$
$$+ \int_0^{\pi\backslash 4} (1000 + 50\,R\sin\theta)(R\cos\theta - 20)\,Rd\theta$$
$$= R\left[1000\,R\sin\theta - \frac{50}{4}R^2\cos 2\theta - \frac{5}{3}R^3\sin^3\theta - 20{,}000\,\theta + 1000\,R\cos\theta + 50\,R^2\,\theta - 25\,R^2\sin 2\theta\right]_0^{\pi/4}$$
$$+ R\left[1000\,R\sin\theta + \frac{50}{4}R^2\cos 2\,\theta - 20{,}000\,\theta - 1000R\sin\theta\right]_0^{\pi/4}$$

Substituting the limits and simplifying

$$Mmds = 6667\,R\text{ kN.m}^3$$
$$\int m^2 ds = 2\int_0^{\pi/4} (R\cos\theta - 20)^2\,Rd\theta$$
$$= 2R\left[\frac{1}{2}R^2\theta + \frac{R^2}{4}\sin 2\theta - 40R\sin\theta + 400\,\theta\right]_0^{\pi/4}$$
$$= 56.32\,R\text{ m}^3$$

Therefore, $\quad H = \dfrac{-6667R}{56.32R} = -118.38\text{ kN}$

Horizontal reaction H acts inwards.

After the value of H has been thus determined, the moment, radial shear and normal thrust at any section can be computed as in a three-hinged arch using statics only.

Sometimes it becomes tedious to evaluate the integral, particularly when a large number of moment expressions are involved. In such cases it is easier, though less accurate, to evaluate the integral by graphical summation by dividing the arch into a number of equal parts. The example that follows illustrates the procedure.

Example 10.20 | *Solve the arch problem in Example 10.19 by evaluating the integrals by graphical summation.*

The arch is divided into eight equal parts and the parts are numbered as in Fig. 10.39. Since the arch is flat, the error in taking the divisions along the span instead of the arch axis is negligible. For any part, coordinates x and y are taken to the mid point of that part on the arch axis. The y coordinate is found from the relation

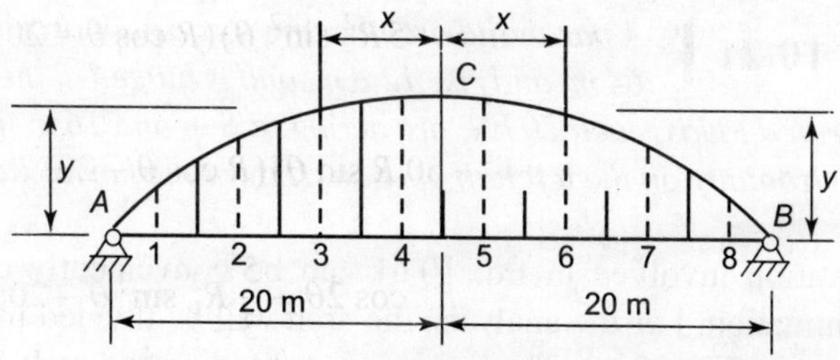

Fig. 10.39

$$(y + 20)^2 + x^2 = R^2$$

The moment from A to C is

$$M = 150\,(20 - x) - 10\,\frac{(20 - x)^2}{2}$$

and for the region C to B is

$$M = 50\,(20 - x)$$

The calculations are all shown in Table 10.6.

Table 10.6 | *Computation for evaluation of integrals in Eq. 10.61 by graphical summation*

Section	*Distance x measured from crown, m*	*Ordinate y of arch axis, m*	y^2 m^2	*Moment M kN.m*	$M \cdot y$ $kN.m^2$
1	17.5	2.21	4.88	343.75	759.69
2	12.5	5.37	28.84	843.75	4530.93
3	7.5	7.27	52.85	1093.75	7951.56
4	2.5	8.17	66.75	1093.75	8935.94
5	2.5	8.17	66.75	875.00	7148.75
6	7.5	7.27	52.85	625.00	4543.75
7	12.5	5.37	28.84	375.00	2013.75
8	17.5	2.21	4.88	125.00	276.25
			$\Sigma y^2 = 306.64$		$\Sigma M \cdot y = 36160.62$

Therefore,

$$H = \frac{-\sum Myds}{\sum y^2 ds} = \frac{-36160.62}{306.64} = -117.93 \text{ kN}$$

This value is very close to the previous value of 118.38 kN obtained by direct integration. Note that the *ds* values in the numerator and denominator cancel out.

This procedure can be followed for parabolic, elliptical or any other arch having a constant moment of inertia for the rib. The example that follows gives the details of calculations for a parabolic arch.

Example 10.21 | *A parabolic arch having a constant arch rib has a span 64 m, and rise 12.8 m, and is hinged at the two supports. Two concentrated loads, each 20 kN, are acting at 8 m and 16 m from the centre measured horizontally on the left half of span. Determine horizontal thrust at the supports.*

The integration involved in Eq. 10.61 can be conveniently carried out by graphical summation. For the analysis, the arch will be divided into eight equal parts as shown in Fig. 10.40. The error in not dividing the arch along the arch axis is negligible because of the flatness of the arch. The equation of the arch axis taking the left hand support as the origin is

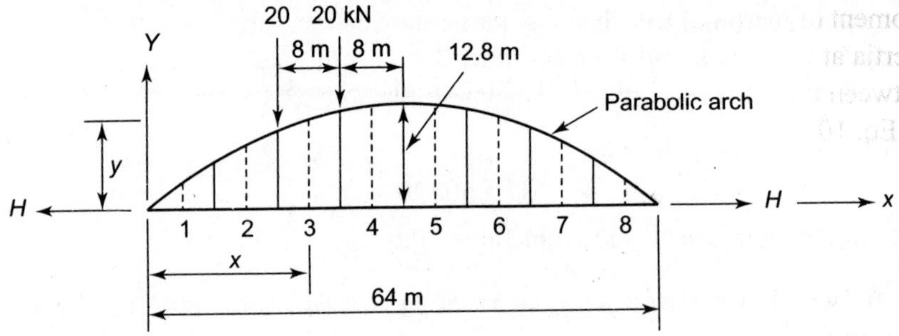

Fig. 10.40

$$y = 4h\left(\frac{x}{L} - \frac{x^2}{L^2}\right) \tag{10.62}$$

The ordinate for each part is obtained using Eq. 10.62. The entire calculations involved are shown in Table 10.7.

Therefore,

$$H = \frac{-22224}{699.2} = -31.78 \text{ kN}$$

Table 10.7 | *Computatins for evaluation of integrals in Eq. 10.61 by graphical summation*

Section	*Distance x measured from left support m*	*Ordinate y of arch axis, m*	y^2 m^2	*Moment M kN.m*	$M \cdot y$ $kN.m^2$
1	4	3.0	9.00	110.0	330.0
2	12	7.8	60.84	330.0	2574.0
3	20	11.0	121.00	470.0	5170.0
4	28	12.6	158.76	450.0	5670.0
5	36	12.6	158.76	350.0	4410.0
6	44	11.0	121.00	250.0	2750.0
7	52	7.8	60.84	150.0	1170.0
8	60	3.0	9.00	50.0	150.0
			$\Sigma y^2 = 699.2$		$\Sigma My = 22224$

Two-Hinged Parabolic Arch with a Secant Variation of Moment of Inertia

The expression for horizontal thrust H (Eq. 10.61) becomes simpler if two requirements are imposed upon the shape and proportions of the arch rib. These two requirements are: (1) the curve of the arch axis must be parabolic and (2) the moment of inertia of the rib at any particular section must be equal to moment of inertia at the crown multiplied by the secant of the angle θ, where θ is the angle between the horizontal and the tangent to the arch axis at that particular section. In Eq. 10.60

$$H = -\frac{\int \frac{Mmds}{EI}}{\int \frac{m^2 ds}{EI}}$$

the following relationships apply:

$$I = I_c \sec \theta$$

where

I_c = moment of inertia at the crown
I = moment of inertia at any other section
$m = y$ and $ds = \sec \theta \, dx$.

With these substitutions, Eq. (10.60) reduces to

$$H = -\frac{\int Mydx}{\int y^2 dx} \tag{10.63}$$

It may be noted that the integration to be carried out and the limits to be taken are along the line joining the springings and *not along the arch axis.*

The following example illustrates the point.

Example 10.22 | *A two-hinged parabolic arch of span L and central rise h, carries a load W at a distance Z = nL from the left hand support. If the moment of inertia of the arch rib varies as the secant of the slope of the arch axis, calculate horizontal reaction H.*

The arch and the loading are shown in Fig. 10.41. The arch is reduced to a primary structure by releasing the horizontal restraint at support *B*.

Then

$$M = \frac{Wx}{L}(L - Z) \text{ for } 0 \le x \le Z$$

or

$$M = \frac{WZ}{L}(L - x) \text{ for } Z \le x \le L$$

H can be evaluated using Eq. 10.63.

Taking the numerator first

$$\int_0^L My dx = \int_0^Z \frac{Wx}{L}(L - Z)\, 4h\left(\frac{x}{L} - \frac{x^2}{L^2}\right) dx$$

$$+ \int_Z^L W \frac{Z(L - x)}{L} 4h\left(\frac{x}{L} - \frac{x^2}{L^2}\right) dx$$

Integrating and substituting the limits,

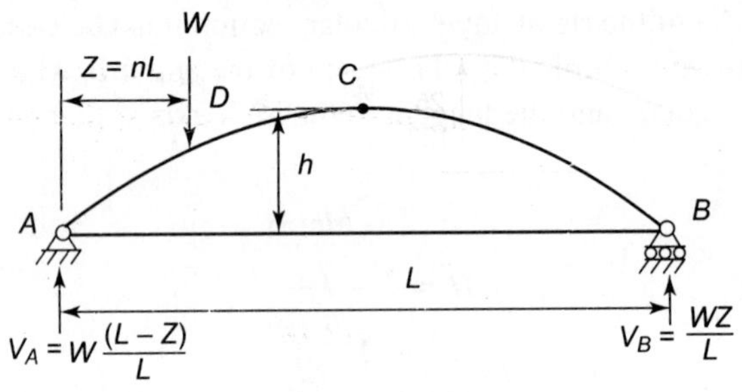

Fig. 10.41

$$\int_0^L My dx = WZ \frac{(L - Z)(L^2 + LZ - Z^2)h}{3L^2} \quad (10.64)$$

Taking the denominator next

$$\int_0^L y^2 dx = \int_0^L \left\{4h\left(\frac{x}{L} - \frac{x^2}{L^2}\right)\right\}^2 dx$$

On evaluating the integral and substituting the limits

$$\int_0^L y^2 dx = \frac{8}{15} Lh^2 \quad (10.65)$$

It may be noted that

$$\int y^2 dx = 2\int_0^L y dx \cdot \frac{y}{2}$$

$= 2 \times$ static moment of the area of the parabola about its base

This gives $\int y^2 dx = 2\left(\frac{2}{3}Lh\right)\frac{2}{5}h = \frac{8}{15}Lh^2$

Substituting the values from Eqs. 10.64 and 10.65 in Eq. 10.63

$$H = \frac{-5}{8}\frac{WZ}{hL^3}(L-Z)(L^2+LZ-Z^2) \tag{10.66}$$

Substituting

$$Z = nL$$

$$H = \frac{-5}{8}\frac{WL}{h}(n - 2n^3 + n^4) \tag{10.67}$$

The variation of H is shown plotted in Fig. 10.42.

It is seen that the value of H is dependent upon the position of the load described by distance $Z = nL$. The value of H reaches a maximum when $n = 1/2$, that is, when the load is at the crown. Substituting $n = 1/2$ in Eq. 10.67

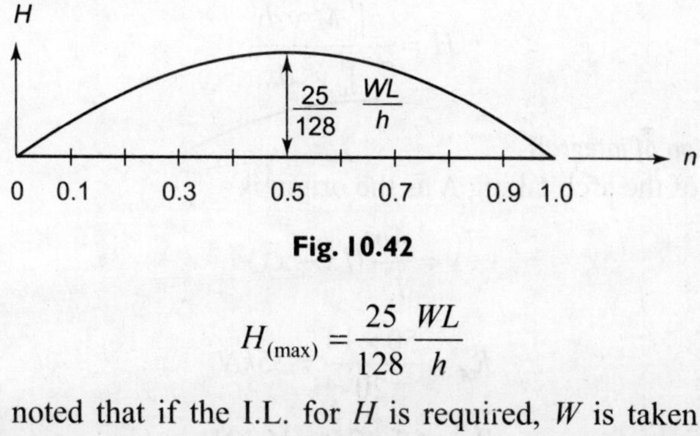

Fig. 10.42

$$H_{(\max)} = \frac{25}{128}\frac{WL}{h} \tag{10.68}$$

It may be noted that if the I.L. for H is required, W is taken as unity and values of n are substituted that correspond to the various sections for which I.L. ordinates are desired.

It there is more than one concentrated load acting on the arch, horizontal reaction H can be evaluated for individual loads and then the values superposed.

Example 10.23 | *A two-hinged parabolic arch, with I proportional to the secant of the slope of arch axis, span 20 m and rise 4m is subjected to a concentrated load of 50 kN. Placed at 6 m from the left-hand support. Calculate the horizontal thrust and normal thrust and radial shear at a section of 5 m from left support.*

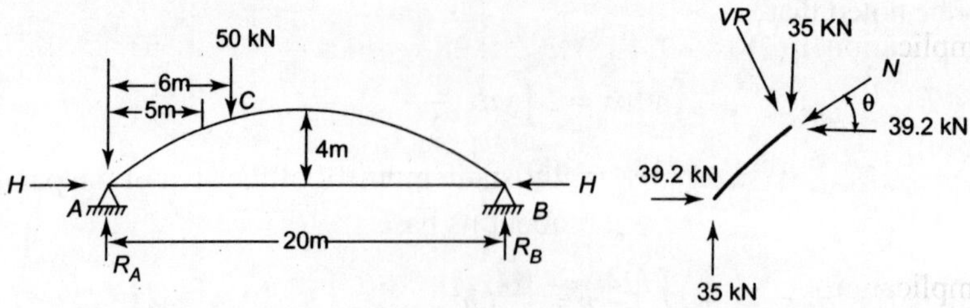

Fig. 10.43 | *A two-hinged arch*

Step 1: Calculation of H

The arch is statically indeterminate by first degree. It is rendered determinate by releasing the reaction component *H* by placing support *B* on rollers.

Under the given load the arch spreads out by an amount D_{1P}. The magnitude *H* must be such as to restore the arch to the original position. Using consistent displacement condition. We can write

$$D_{1P} + D_{1H} = 0$$

or $$H = -\frac{D_{1P}}{f_{11}} = \frac{\text{Displacement in the primary arch}}{\text{Displacement in the arch by a unit force}}$$

or writing in the form of integrals

we have $$H = -\frac{\int_0^l M\, y\, dx}{\int_0^l y^2\, dx}$$

Step 2: Evaluation of integrals

The equation of the arch taking A as the origin is

$$y = \frac{4h}{l^2}(l\,x - x^2)$$

Reaction $$R_A = \frac{50 \times 14}{20} = 35kN$$

and $$R_B = 50 - 35 = 15 \text{ kN}$$

Taking a section between *A* and *C* distance *x* from *A*

$$M_x = 35\, x$$

in the region *C* to *B*

$$M_x = 35\, x - 50\,(x - 6) = (-15\, x + 300)$$

Now $\int_0^6 M\, y\, dx + \int_6^{20} M\, y\, dx$ to be evaluated

First we take $\int_0^6 35\, x \cdot \frac{4 \times 4}{20 \times 20}(20x - x^2)\, dx$

on simplication, $\int_0^6 (28x^2 - 1.4\,x^3)\,dx = \left[28\frac{x^3}{3} - 1.4\frac{x^4}{4}\right]_0^6 = 1562.4$

Next $\int_6^{20} (-15x + 300)\frac{4 \times 4}{20 \times 20}(20x - x^2)$

on simplication, $\int_6^{20} (240x - 24x^2 + 0.6\,x^3)\,dx$

or $\left[240\frac{x^2}{2} - 24\frac{x^3}{3} + 0.6\frac{x^4}{4}\right]_6^{20} = 5123.6$

Total $\int_0^{20} My\,dx = 1562.4 + 5123.6 = 6686$

$$\int_0^{20} y^2\,dx = \frac{8}{15}(20)(4)^2 = 170.6$$

$$H = -\frac{6686}{170.6} = -\,39.20\text{ kN}$$

Step 3: Finding normal thrust and radial shear

Slope of arch at a section 5m from A = 38°42′

Normal thrust = $H\cos\theta + V\sin\theta$

= 39.2 (0.7804) + 35 (0.6239) = 52.43 kN.

Radial shear $V_r = V\cos\theta - \text{H}\sin\theta$

= 35 (0.7804) – 39.2 (0.6239) = 2.85 kN

Example 10.24 *A two-hinged parabolic arch, whose section varies such that the moment of inertia of the section is proportional to the secant of the slope of the arch axis, has a span of 100 m and a rise of 20 m. The load is transmitted to the arch by means of seven suspenders placed 12.5 m apart. Each suspender transmits a force of 50 kN. Find horizontal reaction H.*

The arch and disposition of suspenders is shown in Fig. 10.44. The suspenders 1–7, 2–6 and 3–5 are symmetrically disposed. Therefore, the value of H is evaluated for the first three forces and the results doubled. To this the value of H due to the central suspender is added. Thus, using Eq. 10.67

$$H_1 = H_7 = \frac{5}{8}\frac{(50)(100)}{20}\left\{\left(\frac{12.5}{100}\right) - 2\left(\frac{12{,}5}{100}\right)^3 + \left(\frac{12.5}{100}\right)^4\right\}$$

Similarly

$$H_2 = H_6 = 34.79\text{ kN}$$

$$H_3 = H_5 = 45.20 \text{ kN}$$

and $$H_4 = 48.83 \text{ kN}$$

Total $$H = 2\,(18.96 + 34.79 + 45.20) + 48.83$$
$$= 246.73 \text{ kN}$$

We can also make use of Eq. 10.67 to obtain horizontal reaction H caused by a uniformly distributed load spread from $Z_1 = n_1L$ to Z_2 and n_2L as shown in Fig. 10.45.

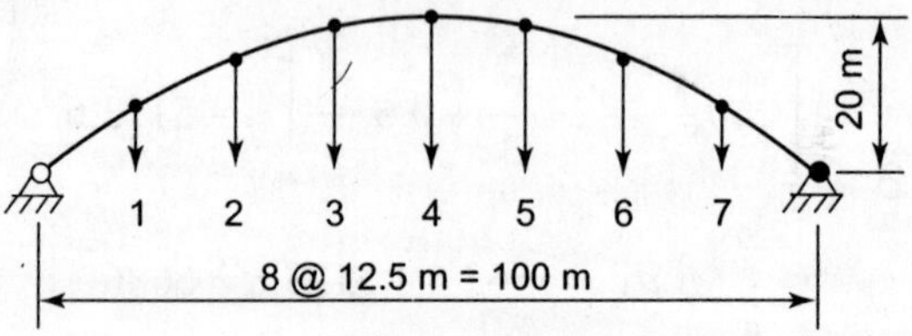

Fig. 10.44

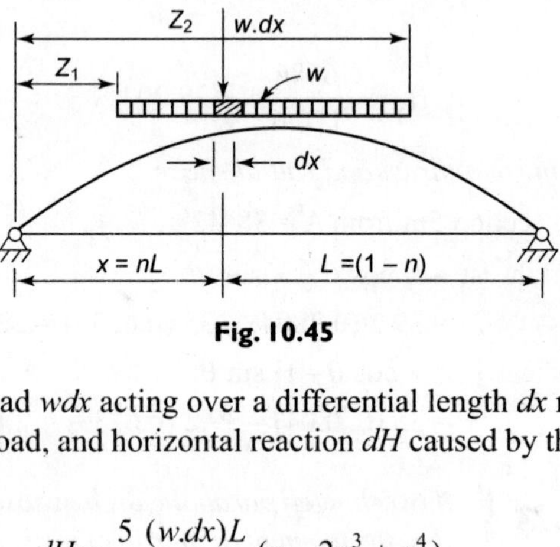

Fig. 10.45

An elemental load wdx acting over a differential length dx may be thought of as a concentrated load, and horizontal reaction dH caused by this may be written as

$$dH = \frac{5}{8}\frac{(w.dx)L}{h}(n - 2n^3 + n^4)$$

$$= \frac{5}{8}\frac{w\cdot L^2}{h}(n - 2n^3 + n^4)\,dn$$

since $dx = L\,dn$.

Hence for the load spread from Z_1 to Z_2

$$H = \int_{n_1}^{n_2} \frac{5}{8}\frac{wL^2}{h}(n - 2n^3 + n^4)\,dn$$

$$= \frac{5}{8}\frac{w\cdot L^2}{h}\left[\frac{n^2}{2} - \frac{n^2}{2} + \frac{n^5}{5}\right]_{n_1}^{n_2} \tag{10.69}$$

Taking a particular case of the whole span being loaded, substitution of limits $n_1 = 0$ and $n_2 = 1$ Eq. 10.69 gives

$$H = \frac{wL^2}{8h} \tag{10.70}$$

Again, if only one half of the span is loaded, the limits $n_1 = 0$ and $n_2 = 1/2$ when substituted in Eq. 10.69 give

$$H = \frac{wL^2}{16h} \tag{10.71}$$

Effect of Support Yielding, Rib Shortening and Temperature Changes

If the support yields under horizontal thrust an amount K per unit of horizontal reaction, then the Eq. 10.56 can be modified to

$$D_{1P} + f_{11}\, H + KH = 0$$

or

$$\int \frac{Mydx}{EI_c} + H \int \frac{y^2 dx}{EI_c} + KH = 0 \tag{10.72}$$

where I_c is the moment of inertia of the arch rib at the crown.

From Eq. 10.72 the redundant horizontal reaction H can be evaluated. Rib shortening under axial force can be accounted for by considering that the rib shortens by an amount $\frac{HL}{A_c E}$

Here A_c = area of cross-section of the rib at the crown. For a flat arch this approximation is permitted. Hence, Eq. 10.56 can be written as

$$\int_0^L \frac{Mydx}{EI_c} + \int_0^L \frac{y^2 dx}{EI_c} + \frac{HL}{A_c E} = 0 \tag{10.73}$$

Similarly, temperature change is accounted for by taking displacement in the horizontal direction

$$D_1 T = \alpha \Delta TL$$

The effect is the same as the displacement due to the horizontal yielding of supports. Then Eq. 10.56 may be written as

$$\int_0^L \frac{Mydx}{EI_c} + H \int \frac{y^2 dx}{EI_c} + \alpha \Delta TL = 0 \tag{10.74}$$

Hingeless Arches

The hingeless arch shown in Fig. 10.46*a* is indeterminate by three degrees. Therefore, we need three releases to make the structure statically determinate. Supposing that the structure is released at B, the three redundants are H_B, V_B and M_B. The three compatibility conditions that may be utilised are:

1. Horizontal displacement $\Delta_{BH} = 0$, that is

$$\int \frac{Mm_1 ds}{EI} + f_{11}H_B + f_{12}V_B + f_{13}M_B = 0$$

2. Vertical displacement $\Delta_{BV} = 0$, that is

$$\int \frac{Mm_2 ds}{EI} + f_{21}H_B + f_{22}V_B + f_{23}M_B = 0$$

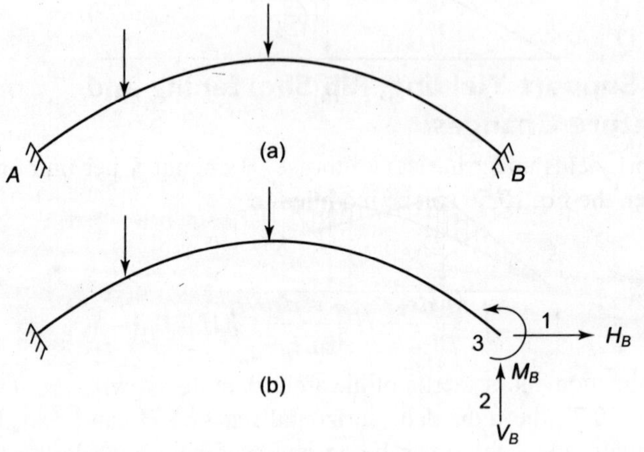

Fig. 10.46

3. Rotation $\theta_B = 0$, that is

$$\int \frac{Mm_3 ds}{EI} + f_{31}H_B + f_{32}V_B + f_{33}M_B = 0 \tag{10.75}$$

in which M is the moment at any section caused by the applied loading, and m_1, m_2 and m_3 are the moments caused by unit loads applied at the redundants.

The analysis of fixed arches can be done more conveniently either by the method of elastic centre or by column analogy. The reader is advised to consult books which deal with these methods.

Influence Lines for Two-Hinged Arches

Arch bridges are commonly employed for highway and railway bridges. Hence, influence lines for the structural forces such as horizontal thrust H, moment and shear are of paramount importance. As already pointed out, the I.L. for horizontal reaction H, can be obtained if load W is replaced by a unit load in Eq. 10.67. Therefore, for a unit load placed at a distance nL from one support

$$H = \frac{5}{8}\frac{L}{h}(n - 2n^3 + n^4) \tag{10.76}$$

It may be noted that horizontal thrust H is the same for all sections of the arch axis. The I.L. diagram for the horizontal reaction is shown in Fig. 10.47*a*.

The I.L. diagram for the moment at any section can be constructed from knowledge of the moment expression at that section. For example, the moment at any section denoted by distance *x* from one of the supports is

$$M_x = \mu_x - H \cdot y \qquad (10.77)$$

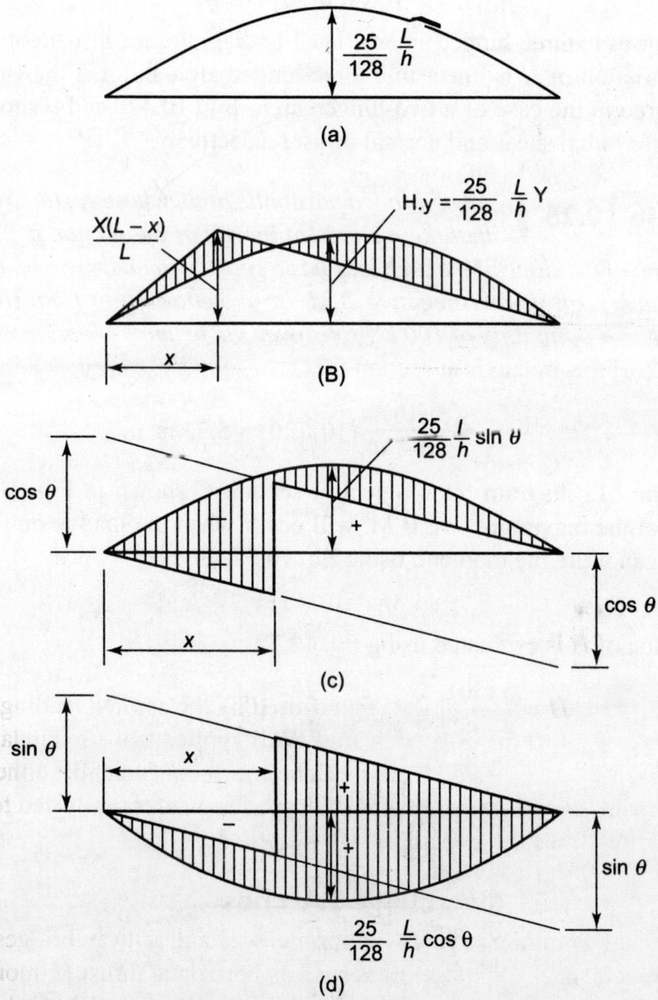

Fig. 10.47 | *(a) I.L. for horizontal reaction H, (b) I.L. for moment at section distance x from left support, (c) I.L. for radial shear, (d) I.L. for normal thrust*

where μ_x is the free bending moment, and *y* is the ordinate of the arch axis at the section under consideration.

The I.L. diagram is drawn in two separate parts representing the two terms in Eq. 10.77 and then superimposed. The net influence line ordinates are shown hatched in Fig. 10.47*b*. It may be noted that a similar procedure was followed for a three-hinged arch in Chapter 7.

Influence lines for radial shear and normal thrust are constructed in the same manner as in a three-hinged arch. The expressions for radial shear

$$V_{r} = V_A \cos\theta - H \sin\theta$$

and normal thrust

$$N = V_A \sin\theta + H\cos\theta$$

are the same as in three-hinged arches. The I.L. diagrams are also identical except that the variation of H is linear in a three-hinged arch and it is the curve of the fourth degree in the case of a two-hinged arch. Fig. 10.47c and d show the I.L. diagrams for radial shear and normal thrust respectively.

Example 10.25 | *A two-hinged parabolic arch, whose section varies such that the momeni of inertia of the section is proportionl to the secant of the slope of the arch axis has a span 30 m and rise 6 m. Determine the maximum positive and negative B.M. at a section 10 m from the left end support when a point load of 100 kN rolls over the beam.*

Ordinate of the arch axis at section C is,

$$y_c = \frac{4(6)}{30\times 30}(10)\,(20) = 5.33 \text{ m}$$

From the I.L. diagram for moment at section C shown in Fig. 10.48b it is evident that the maximum +ve B.M. will occur when the load is on the section itself. We can write the moment, using Eq. 10.77, as

$$M_C = \mu_c - H\,y$$

The value of H is evaluated using Eq. 10.72.

$$\therefore \qquad H = \frac{5}{8}\frac{(30)}{6}\left(\frac{1}{3} - \frac{2}{27} + \frac{1}{81}\right)(100) = 84.88 \text{ kN.}$$

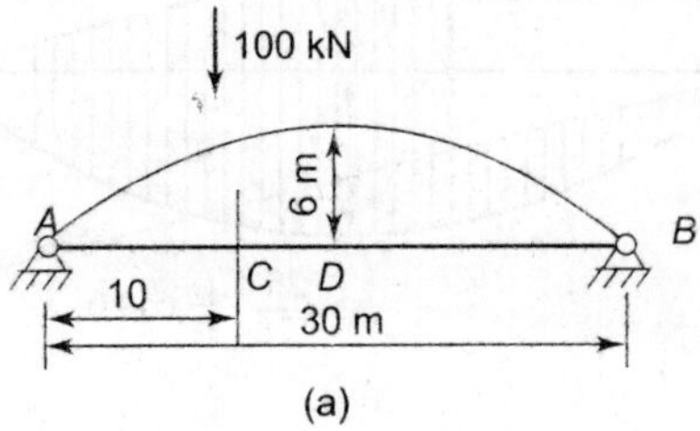

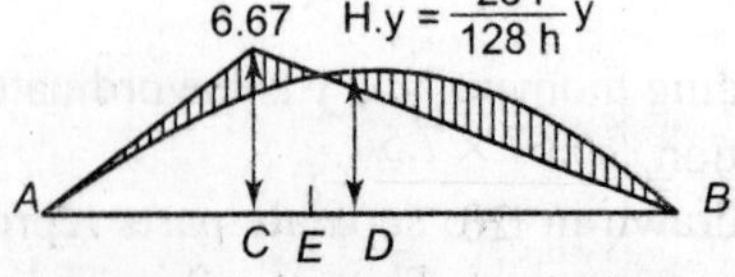

Fig. 10.48 | *(a) Two-hinged arch and the rolling load, (b) I.L. for moment at section C*

Substituting in the above equation

$$M_c = \frac{100 \times 10 \times 20}{30} - 84.88 \times 5.33$$

$$= 214.26 \text{ kN.}$$

Again from the I.L. diagram it is evident that the maximum –ve B.M. will occur when the load is in the region **E** to **B**.

Let the maximum –ve B.M. occur at a section distance x from the right hand support

$$\therefore \quad \text{Moment } M_x = \frac{Wx(l-x)}{l} - \frac{5}{8}\frac{l}{h}\left\{\frac{x}{l} - 2\left(\frac{x}{l}\right)^3 + \left(\frac{x}{l}\right)^4\right\} yx$$

For obtaining the maximum value for M_x, we get

$$\frac{dM_x}{dx} = 0$$

This results in a cubic equation in x and the value for x has to be determined by trial and error. As an alternative, the value for M_x is calculated taking values for $x = \frac{l}{8}, \frac{l}{4}, \frac{3}{8}l$, and $l/2$ from right hand support and the highest value for M_x is taken.

At $$x = \frac{l}{8}, H = \frac{5}{8}\frac{(30)}{8}\left(\frac{1}{8} - \frac{2}{64 \times 6} + \frac{1}{64 \times 64}\right) = 0.3790$$

$$x = \frac{l}{4}, H = 0.6958$$

$$x = \frac{3}{8}, H = 0.9040$$

$$x = \frac{l}{2}, H = 0.9766$$

$\therefore$ At $$x = \frac{l}{8}, M_C = \{\mu - 0.3790 \times 5.33\}\,(100)$$

$$= \left(\frac{6.67 \times 3.75}{20} - 0.3790 \times 5.33\right)(100)$$

$$= -77.0 \text{ kN.m}$$

Similarly,

at $$x = \frac{l}{4}, M_c = \left(\frac{6.67 \times 7.5}{20} - 0.6958 \times 5.33\right)(100)$$

$$= -121.0 \text{ kN.m}$$

At $\quad x = \frac{3}{8}\,l,\ M_c = 107.0$ kN.m

$x = l/2,\ M_c = 21.0$ kN.m

and the maximum –ve B.M. occurs at $x = \frac{l}{4}$ and its value is

$= 121.0$ kN.m

$\therefore$ Maxm. –ve B.M. at section $C = -121.0$ kN.m

Example 10.26 | *Draw I.L. diagrams for horizontal thrust H, and B.M. shear and thrust at D of the parabolic arch given in Fig. 10.49.*

Consider that $I_X = I_0 \sec \theta$

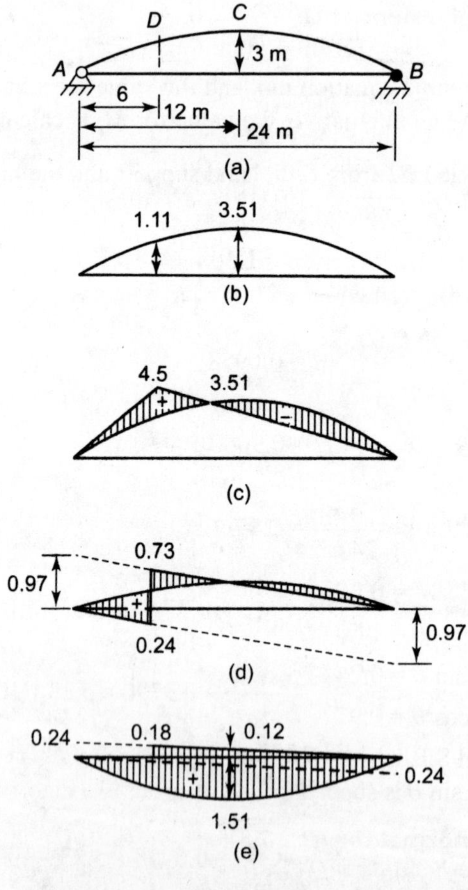

Fig. 10.49 | *(a) Parabolic arch, (b) I.L.D. for H, (c) I.L.D. for moment at section D, (d) I.L.D. for radial shear at D, (e) I.L.D. for thrust at section D*

Influence Line for H

Eqn. 10.63 gives 5

$$H = \frac{5}{8}\frac{WL}{y_c}(n - 2n^3 + n^4) \qquad \text{where } n = \frac{z}{l}$$

We can calculate H for different values of n taking $W = 1$ and plot I.L.D. for H as shown in Fig. 10.49*b*. For example

when $n = \frac{1}{4}$, $H = \frac{5}{8} \times \frac{24}{3}\left\{\left(\frac{1}{4}\right) - 2\left(\frac{1}{4}\right)^3 + \left(\frac{1}{4}\right)^4\right\}$

$$H = 1.11.$$

At $n = \frac{1}{2}$, $H = 1.56$

Influence Line for Moment at D

Moment at section D for any position of unit load is

$$M = \mu - Hy$$

The I.L.D. is drawn by the superposition of I.L. for μ and the I.L. for $H\,y$.

Ordinate y of arch at $D = \frac{4 \times 3}{24 \times 24}(6)(18) = 2.25$ m

Ordinate of the I.L. for $H\,y$ at centre $= 1.56 \times 2.25 = 3.51$

The I.L. diagram for moment at section D is shown in Fig. 10.49*c*.

Influence line for radial shear

Radial shear at $D = V_A \cos\theta - H \sin\theta$

We know $\tan\theta = \frac{dy}{dx} = \frac{4\,y_c}{l^2}(l - 2x)$

$$= \frac{4 \times 3}{24 \times 24}(24 - 12)$$

$$= 0.25$$

$\therefore$ $\theta = 14.04°$

$$\sin\theta = 0.24$$

$$\cos\theta = 0.97$$

The I.L.D. for radial S.F. is shown in Fig. 10.49*d*.

The I.L.D. for $H \sin\theta$ is superimposed over the I.L. diagram of $V_A \cos\theta$.

Influence line for normal thrust

Normal thrust at $D = V_A \sin\theta + H\cos\theta$

The I.L. diagram for normal thrust at D is shown in Fig. 10.49*e*. The diagram is drawn by superimposing I.L.D. for $H \cos\theta$ over I.L.D. for $V_A \sin\theta$. The important values are noted on the diagram.

10.8 INFLUENCE LINES FOR CONTINUOUS MEMBERS

Müller-Breslau's principle introduced in Sec. 7.12 can be used to develop influence lines for continuous members. For example, consider a two-span continuous beam as; shown in Fig. 10.50. Suppose it is required to draw the influence line for the reaction at B. According to Müller-Breslau's principle, if a unit displacement is given in the direction of reaction, the deflected shape or elastic curve gives to scale the influence line ordinates for the reaction at B. The deflected shape shown in Fig. 10.50*b* is itself the influence line for reaction R_B.

The proof for this principle can be demonstrated utilising the flexibility influence coefficients presented in Sec. 10.4. Consider the beam with the support constraint removed as in Fig. 10.50*c*. Let a unit load be placed at any point i on the beam. The deflections under the load point and at point B are indicated as f_{ii} and f_{Bi}, respectively. Now a unit load applied at B (Fig. 10.50*d*) gives deflection f_{iB} and f_{BB} at points i and B respectively.

If R_B were the reaction for a unit load at i, the compatibility condition that the final deflection at $B = 0$ gives

$$R_B f_{BB} + f_{Bi} = 0 \tag{10.78}$$

$$R_B = -\frac{f_{Bi}}{f_{BB}} = -\frac{f_{iB}}{f_{BB}} \quad \text{since } f_{Bi} = f_{iB}$$

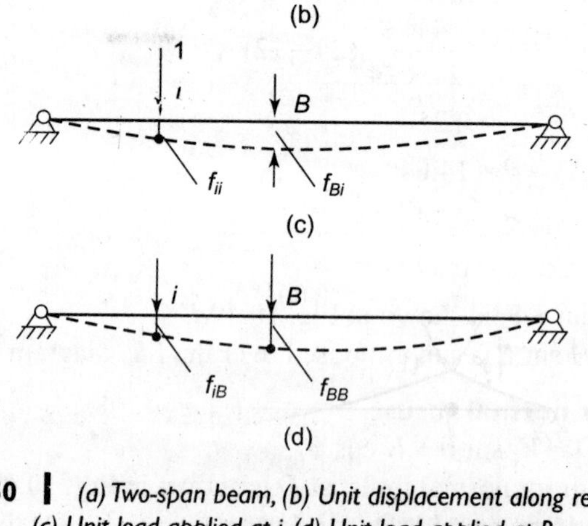

Fig. 10.50 | *(a) Two-span beam, (b) Unit displacement along reaction R_B, (c) Unit load applied at i, (d) Unit load applied at B*

The negative sign indicates that reaction R_B is upwards opposite to the direction of the unit load. From Eq. 10.78, we have

$$R_B = f_{iB} \text{ for } f_{BB} = 1 \tag{10.79}$$

This is true for any position of the unit load described by point i. Therefore, the deflected shape or the elastic line represents, to some scale, the influence line for reaction R_B and for $f_{BB} = 1$; the deflected shape gives to scale the influence line for reaction R_B

Let us proceed to establish an influence line for the moment at section i of the same two-span continuous beam (Fig. 10.51*a*). Again, according to Müller-Breslau's principle, the section has to be relieved of its moment carrying capacity by introducing a hinge as shown in Fig. 10.51*b*. The unit load applied at j deflects the beam as shown in Fig. 10.51*b*. The rotation at the hinge is denoted by f_{ij}. Suppose that a unit couple applied at i produces a rotation f_{ii} at i and deflection f_{ji} at j, then the moment at i due to the unit load at j is

$$M_i = \frac{f_{ij}}{f_{ii}} \tag{10.80}$$

or
$$M_i = \frac{f_{ji}}{f_{ii}} \text{ for } f_{ij} = f_{ji} \tag{10.81}$$

If f_{ii} were made equal to unity, the deflected shape in Fig. 10.51*c* would represent to scale the influence line for the moment at section i.

In a similar manner, the influence line for the shear at l (see Fig. 10.52*a*) can be obtained by relieving the member of its capacity to transmit shear by providing a link capable of only transmitting moment as explained in Sec. 7.12. The deflected shape of the member under a unit load placed at j is shown in Fig. 10.52*b*. Now a unit shear force is applied at i resulting in a deflected shape shown in Fig. 10.52*c*. Note that the two ends of the beam displaced laterally must have the same slope indicating the continuity of the beam in transmitting moment.

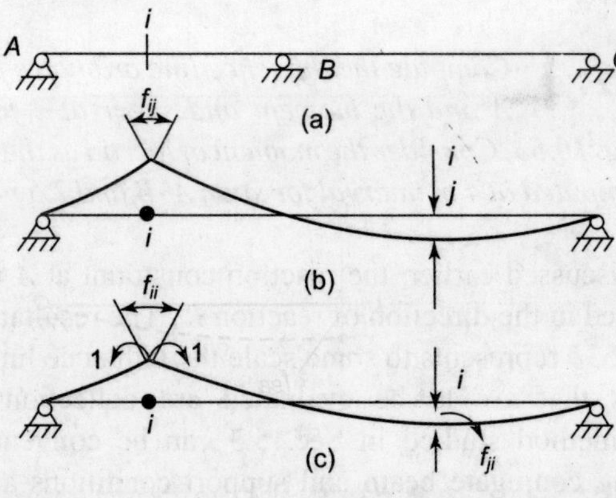

Fig. 10.51 | *(a) Two-span beam, (b) Unit load applied at j, (c) Unit couples applied at i*

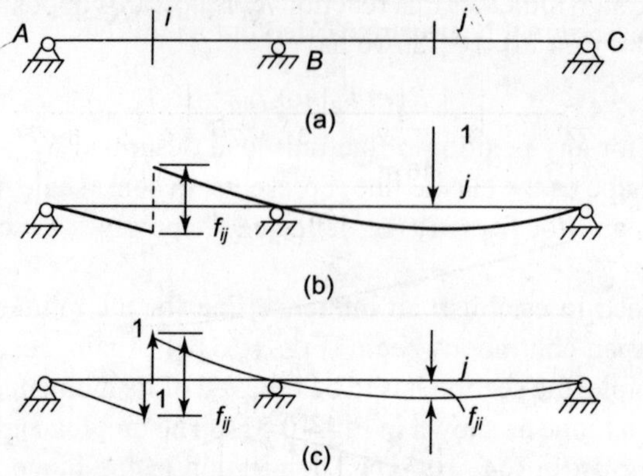

Fig. 10.52 | *(a) Beam and section i and j, (b) Deflected shape under unit load at j, (c) Deflected shape under unit shear at i*

The shear force at i due to a unit load placed at j can be expressed at

$$V_i = \frac{f_{ij}}{f_{ii}} \tag{10.82}$$

or
$$V_i = \frac{f_{ji}}{f_{ii}} \text{ since } f_{ij} = f_{ji} \tag{10.83}$$

This is true for all positions of unit load described by point j. Therefore, if displacement f_{ii}, in Fig. 10.52*c* is made equal to unity, the resulting deflected shape gives to scale the influence line for the shear at section i.

Except for simple cases, the actual calculations of influence line ordinates become quite tedious. The model analysis using brass or steel splines are profitably employed. The interested reader may consult references on model analysis of structures. Analytical solutions for simple cases are illustrated by the following examples.

Example 10.27 | *Compute the influence line ordinates for the reaction at A and the moment and shear at 2 for the continuous beam shown in Fig. 10.53. Consider the moment of inertia as the same throughout. Values may be computed at 4 m interval for span A–B and 2.5 m interval for span B–C.*

As has been discussed earlier, the reaction constraint at A is removed and a unit force is applied in the direction of reaction R_A The resultant deflected shape shown in Fig. 10.53*b* represents to some scale the influence line for reaction R_A.

The quantities that are to be evaluated are deflections; therefore, the conjugate beam method studied in Sec. 5.3 can be conveniently employed. The corresponding conjugate beam and support conditions and the conjugate beam loading are shown in Fig. 10.53*c*. The term EI is not included as it is constant throughout. The values of loading at the desired intervals are indicated.

The computations for the moments at the desired intervals, which are in fact the deflections in the original beam, are carried out as follows.

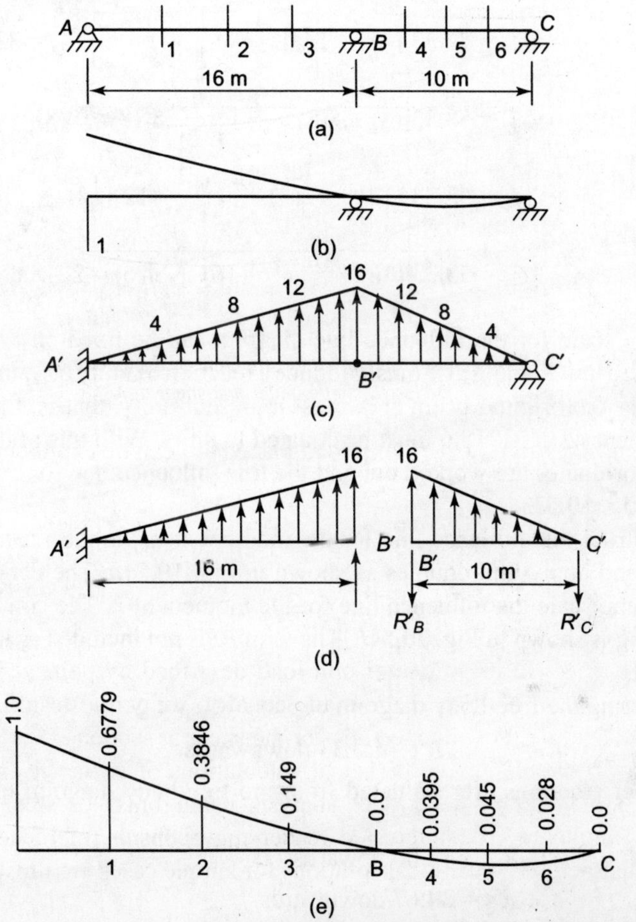

Fig. 10.53 | *(a) Continuous beam and the sections for computation of I.L. ordinates, (b) Elastic line due to a unit force applied at A, (c) Conjugate beam and loading, (d) Free-body diagrams of parts A'B' and B'C, (e) I.L. for reaction R_A*

Considering the free-body diagrams of parts $A'B'$ and $B'C'$ we have R'_B = 53.33 and R'_C = 26.67.

The moment values are

$$M_6 = -26.67\,(2.5) + \frac{1}{2}\,(4)(2.5)\left(\frac{1}{3}\right)(2.5) = -62.51$$

$$M_5 = -26.67\,(5) + \frac{1}{2}\,(8)(5)\,\frac{1}{3}\,(5) = -100.00$$

$$M_4 = -26.67\,(7.5) + \frac{1}{2}\,(12)(7.5)\,\frac{1}{3}\,7.5 = -87.53$$

$$M'_B = -26.67\,(10) + \frac{1}{2}\,(16)(10)\,\frac{1}{3}\,(10) = 0$$

$$M_3 = 53.33\,(4) + 12(4)\,\frac{4}{2} + \frac{1}{2}\,(4)(4)\,\frac{2}{3}(4) = 330.65$$

$$M_2 = 53.33\,(8) + 8(8) + \frac{8}{2} + \frac{1}{2}\,(8)(8)\,\frac{2}{3}\,(8) = 853.31$$

$$M_1 = 53.33\,(12) + 12(4)\,\frac{12}{2} + \frac{1}{2}\,(12)(12)\,\frac{2}{3}\,(12) = 1504$$

$$M'_A = 53.33\,(16) + \frac{1}{2}\,(16)(16)\,\frac{2}{3}\,(16) = 2218.6$$

Now the scale for the influence line diagram can be fixed on the following basis. Since a unit loading at A must produce a reaction of unity at A, the deflection obtained for the original beam at A must represent unity, that is, the conjugate beam moment $M'_A = 2218.6$ must be equated to unity. With this scale factor the deflection ordinates are worked out and the true influence line for reaction R_A is shown in Fig. 10.53*e*.

To construct the influence line for the moment at 2, we introduce a hinge at that point and apply unit couples as shown in Fig. 10.54*a*. The deflected shape gives to some scale the influence line for the moment at 2. The conjugate beam with loading is shown in Fig. 10.54*b*. The term EI is not included as it is constant throughout.

Considering the free-body diagram of part $B'C'$, we get

$$R'C' = 3.33 \text{ downwards}$$

The other reactions are evaluated from the free-body diagram of part $A'B'$. They are

$$R'_A = 12.00 \text{ upwards}$$

and $$R'_2 = 34.67 \text{ downwards.}$$

It may be noted that the reaction at 2′ in the conjugate beam represents the sum of shears on either side of point 2′ and hence rotation f_{22} at 2 the original beam. The moments in the conjugate beam at the desired points are

$$M_1 = 49.33,\ M_2 = 106.67,\ M3 = 41.32,\ M'_B = 0,$$
$$M_4 = -10.91,\ M_5 = -12.48,\ M_6 = -7.8,\ M'_C = 0$$

The influence line ordinates are obtained by dividing these values by 34.67 which should represent unit rotation at point 2. The ordinates thus computed result in the influence line for the moment at 2 as shown in Fig. 10.54*d*

To evaluate influence line ordinates for the shear at 2, we apply unit forces to the ends. The deflected shape and the induced reactions and couples will be as shown in Fig. 10.55*a*. The conjugate beam along with the load is shown in Fig. 10.55*b*. In this figure attention is called to moment M'_2. In Fig. 10.55*a* points D and E deflect relative to each other; E moves up and D goes down. In the conjugate beam the moment just to the right of point 2 must be different in sign and possibly in magnitude from the moment just to the left of section 2. In addition, since the tangents to the deflected beam at D and E must be parallel, the shear in the conjugate beam at 2 must be the same on either side of section 2. The imposition of moment M'_2 will satisfy these requirements.

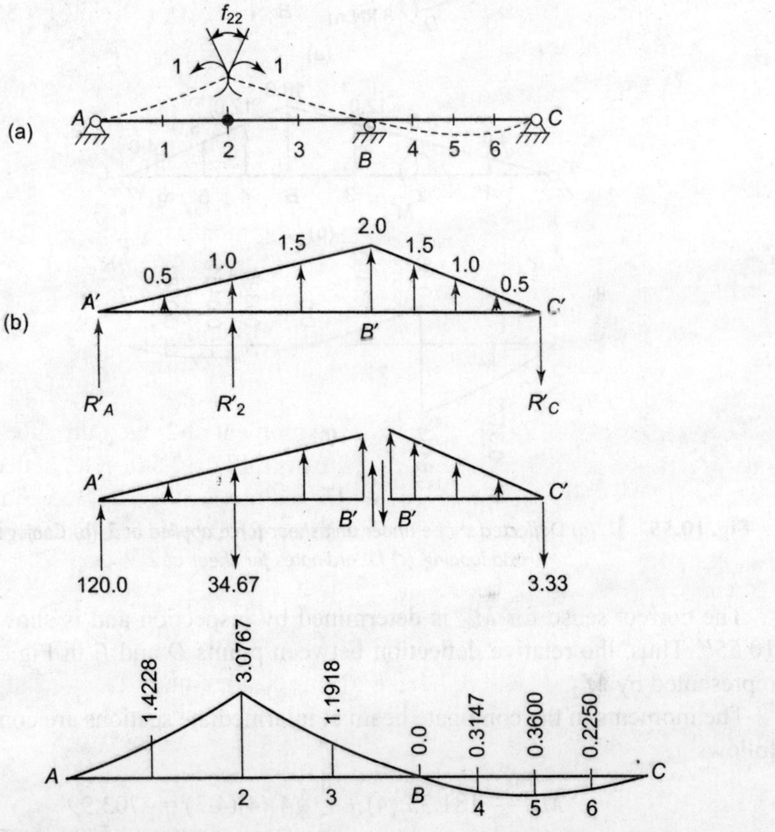

Fig. 10.54 | *(a) Deflected shape of the beam under unit couple applied at 2, (b) Conjugate beam and the loading, (c) Free-body diagrams of parts A′B′ and B′C′, (d) I.L. for moment at 2*

The values of conjugate beam reactions, R'_A and R'_C, the shear V'_B and moment M'_2 are obtained below.

Taking moments about B from the right end

$$R'_C = \frac{1}{2}\,(10)(16)(1/3) = 26.67$$

$$R'_A = \frac{1}{2}\,(26)(16) - 26.67 = 181.33$$

$$V'_B = \frac{2}{3}\,(10)(16/2) = 53.33$$

and $$M'_2 = 53.33 \times 16 + \frac{1}{2}(16)(16)\frac{2}{3}(16) = 2218.61$$

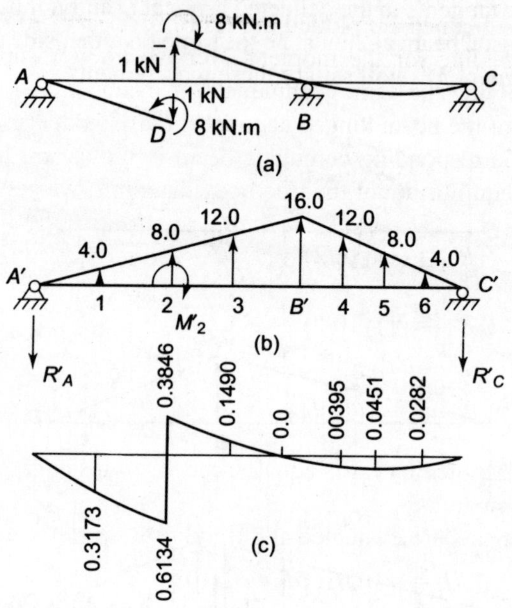

Fig. 10.55 | *(a) Deflected shape under unit shear force applied at 2, (b) Conjugate beam and loading, (c) I.L. ordinates for shear at 2*

The correct sense for M'_2 is determined by inspection and is shown in Fig. 10.55*b*. Thus, the relative deflection between points *D* and *E* in Fig. 10.55*a* is represented by M'_2.

The moments in the conjugate beam at intermediate sections are computed as follows

$$M_1 = -181.33\,(4) + \frac{1}{2}(4)(4)(4/3) = -703.99$$

$$M_{2(\text{left})} = -181.33\,(8) + \frac{1}{2}(8)(8)(8/3) = -1365.31$$

$$M_{2(\text{right})} = 2218.61 - 1365.31 = 853.3$$

$$M_3 = -181.33\,(12) + 2218.61 - \frac{1}{2}(12)(12)\frac{12}{3} = 330.65$$

$$M_4 = -26.67\,(7.5) + \frac{1}{2}(7.5)(12)(7.5/3) = -87.53$$

$$M_5 = -26.67\,(5) + \frac{1}{2}(5)(8)(5/3) = -100.00$$

$$M_6 = -26.67\,(2.5) + \frac{1}{2}(2.5)(4)(2.5/3) = -62.51$$

The above moments must be divided by the relative deflection between D and E, as represented by M'_2, in order to obtain the required ordinates. The influence line ordinates so obtained are plotted in Fig. 10.55*c*.

The influence line for the moment over the interior support can also be obtained by applying the same technique. For example, Fig. 10.56*a* gives the deflected shape of the beam hinged at support B and under unit couples applied as shown. The corresponding conjugate beam is shown in Fig. 10.56*b*. Again, considering the equilibrium of the free-body diagrams, we get

$$R'_A = (1)\,(16/2)\,\frac{1}{3} = 2.67$$

$$R'_C = (1)\,(10/2)\,\frac{1}{3} = 1.67$$

$$R'_B = \frac{1}{2}\,(26)(1) - 4.33 = 8.67$$

The rotation denoted by f_{BB} is equal to the sum of shears to the left and right of section B or reaction $R'_B = 8.67$.

Now the moments are evaluated at different sections. They are

$$M_1 = -10.01,\ M_2 = -16.03,\ M_3 = -14.04,\ M_4 = -5.49,$$
$$M_5 = -6.27 \text{ and } M_6 = -3.91$$

These moments, when divided by $R'_B = 8.67$, will directly represent the influence line ordinates for the moment over support B. The values are plotted in Fig. 10.56*c*.

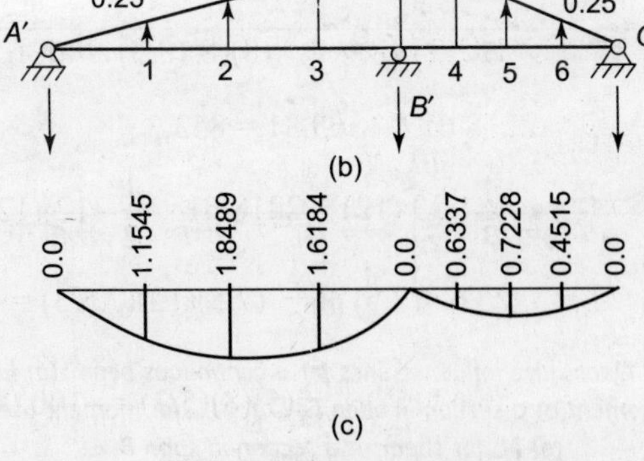

Fig. 10.56 | *(a) Deflected shape under unit couples applied at B, (b) Conjugate beam and loading, (c) I.L. ordinates for moment over support B*

The foregoing illustrations demonstrate that any desired influence line for a continuous beam can be readily computed using Müller-Breslau's principle and the conjugate beam method. Actually, when the influence line for shear is needed, it is usually easier to compute the required ordinates by statics after the influence lines for reactions have been computed. This is also probably true for moment influence lines.

10.8.1 Qualitative Influence Lines by the Müller-Breslau Principle

So far we have demonstrated that the Muller-Breslau principle is of extreme importance in the determination of quantitative influence lines. The principle also helps very much to sketch qualitative influence lines. For example, qualitative influence lines for a five-span continuous beam can be drawn with ease. To draw the influence line for reaction R_C it is enough to give a small displacement in the direction of the reaction. The deflected shape that results gives to some scale the influence line for reaction R_C. This is shown in Fig. 10.57*a*.

Next, suppose it is required to draw the qualitative influence line for the moment at the centre of span *CD*. A hinge is assumed at this point and couples are applied. The deflected shape gives the influence line for the moment in a qualitative way (Fig. 10.57*b*). I.L. for a support moment is also drawn by inserting a pin and drawing the deflected shape due to couples applied at the pin (Fig. 10.57*c*).

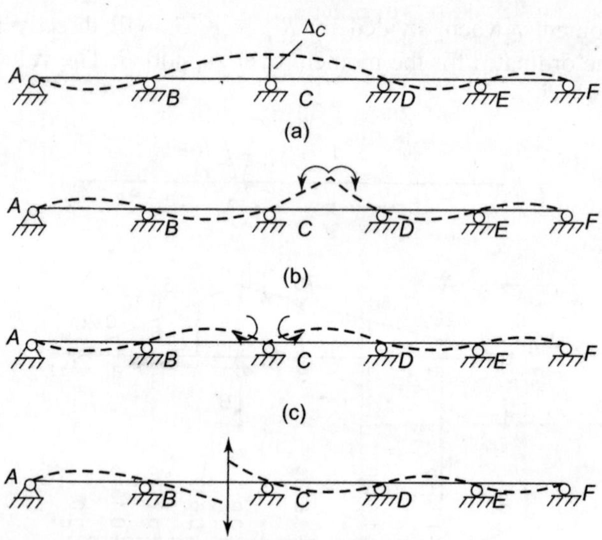

Fig. 10.57 | *Qualitative influence lines for a continuous beam: (a) I.L. for reaction R_C, (b) I.L. for moment at a section in span C–D, (c) I.L. for moment over support C, (d) I.L. for shear at a section in span B–C.*

Again suppose it is required to draw the I.L. for the shear at a section in span *BC*; the beam is cut, a roller and slide device is inserted and the ends are subjected to equal and opposite transverse forces as shown in Fig. 10.57*d*. The deflected beam and, therefore, the qualitative influence line for shear is shown in Fig. 10.57*d*.

These qualitative influence lines are useful in determining which of the spans are to be loaded with distributed live loads to get the maximum moments and shears at any section for which the qualitative influence line is drawn. For example, to obtain the maximum moment in span *C-D*, we must apply live load only on that and the alternate spans. On the other hand, to get the maximum support moment (Fig. 10.57*c*), the spans on either side of the support and only the alternate spans are to be loaded. So also, the spans that are to be loaded for obtaining maximum shear at any section can be decided by sketching the qualitative influence line for shear at the desired section.

The qualitative influence lines are particularly valuable in the analysis of building frames. For example, the I.L. for positive moment in span *B*2-*C*2 is shown in Fig. 10.58*a*. The influence line for the moment just to the right of joint *C*2 is shown in Fig. 10.58*b*.

From the qualitative influence lines for building frames it can be seen that a chequer-board loading pattern as indicated in Fig. 10.58*a* and *b* gives maximum moments. In bay *CD* a small length at levels 5, 3 and 1 should not be loaded. In addition at level 4, a short length is to be theoretically loaded. For practical analysis, bay *CD* at levels 5, 3 and 1 would be entirely loaded and at level 4 it would be unloaded. However, in practical terms, the effect of the load in any span on a member two or three bays away is negligible. Specifications, therefore, permit the analysis of large frames by the *substitute frame method*. In analysing a given floor beam and the columns above and below that floor, it is permissible to consider that all the columns are fixed at their farther ends. The ends of beams two bays away from the section under consideration are also considered to be fixed. This has the obvious effect of greatly simplifying the analysis with only a slight loss of accuracy.

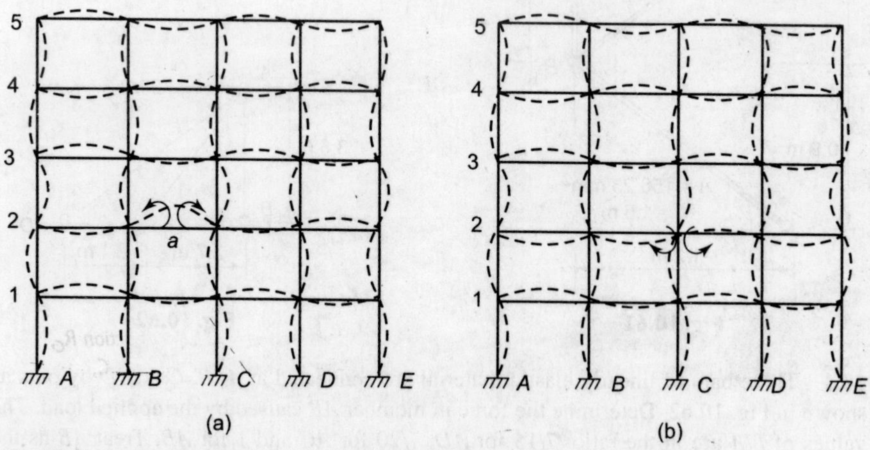

Fig. 10.58 | *Qualitative influence lines for a building frame: (a) I.L. for moment at 'a' in span B2-C2, (b) I.L. for moment just to the right of joint C2*

Problems for Practice

10.1 Using the method of consistent displacements determine for the beam given in Fig. 10.59

(a) the reaction R_b treating it as the redundant, and

(b) moment M_A treating it as the redundant for the beam shown. EI is constant.

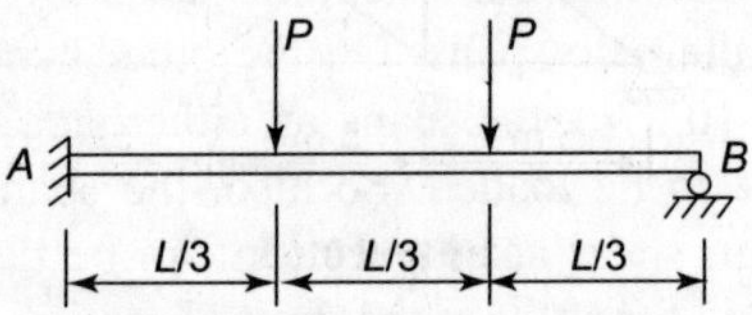

Fig. 10.59

10.2 Using the principle of consistent displacements, determine the reaction of the middle support and plot the shear and moment diagram for the two-span continuous beam shown in Fig. 10.60. Consider El constant throughout.

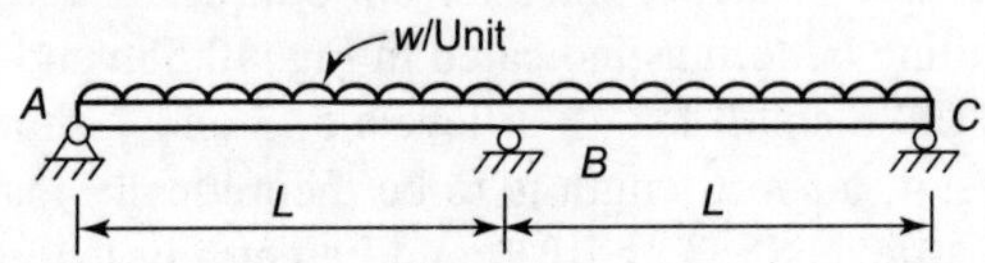

Fig. 10.60

10.3 Find the forces in pin-jointed members of the steel structure shown in Fig. 10.61 if a force of 15 kN is applied at $B.E = 200 \times 10^6$ kN/m^2 (200,000 MPa).

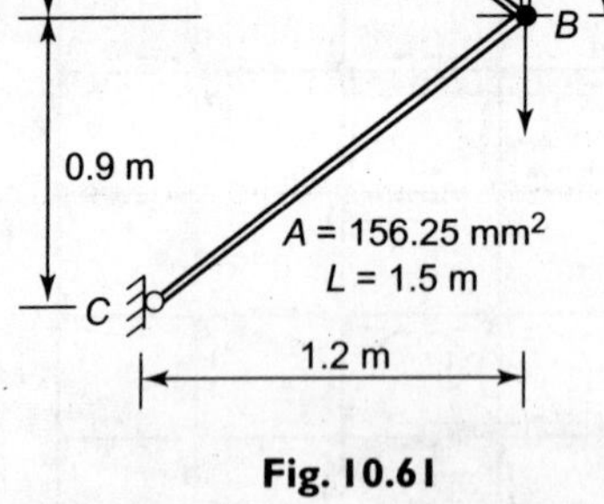

Fig. 10.61

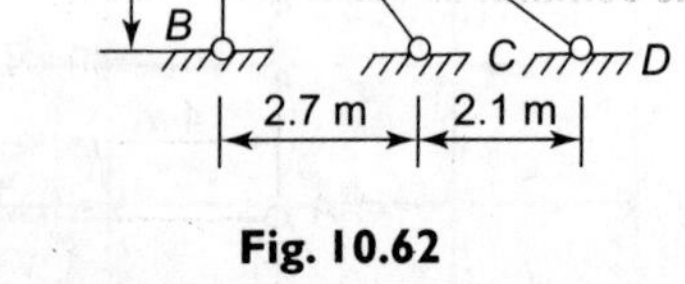

Fig. 10.62

10.4 Three bars of linearly elastic material are connected at A, B, C and D by pins as shown in Fig. 10.62. Determine the force in member AB caused by the applied load. The values of L/A are in the ratio: 7/15 for AD, 7/20 for AC and 1 for AB. Treat AB as the redundant member.

10.5 The frame in Fig. 10.63 is simply supported and loaded as shown. Compute the stresses in two diagonals BF and CE. All members are 2000 mm^2 in area. What would be the effect on these stresses of a rise in temperature of 20°C in member EF relative to other members? Take $\alpha = 12 \times 10^{-6}$/C.

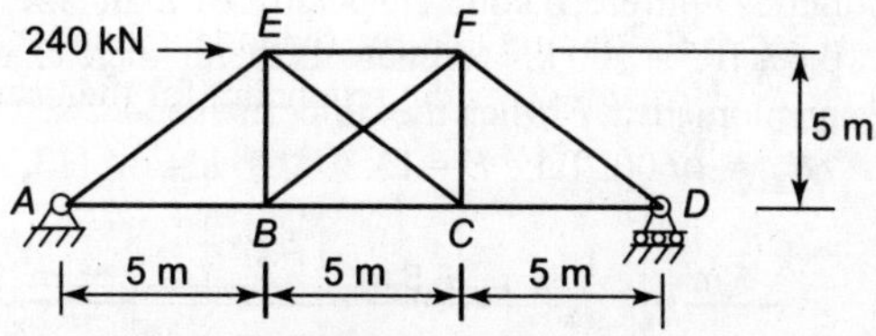

Fig. 10.63

10.6 Calculate the forces in the members of the truss shown in Fig. 10.64 if the roller support at E sinks by 1 mm. Assume A = 5000 mm^2 for all members and $E = 200 \times 10^6$ kN/m^2 (200,000 MPa). Choose force in member $AD = X_1$ and reaction at $E = X_2$ as redundants.

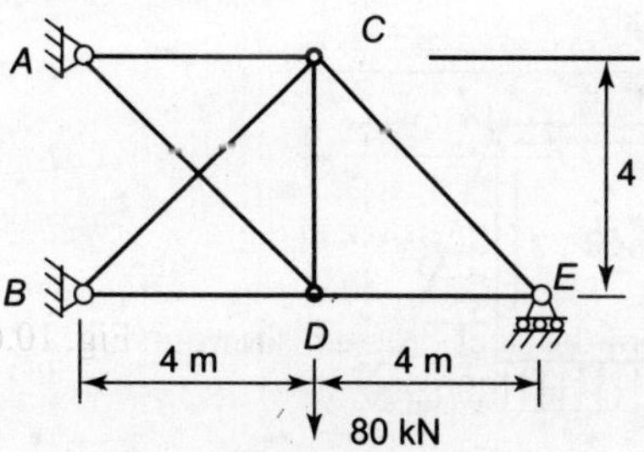

Fig. 10.64

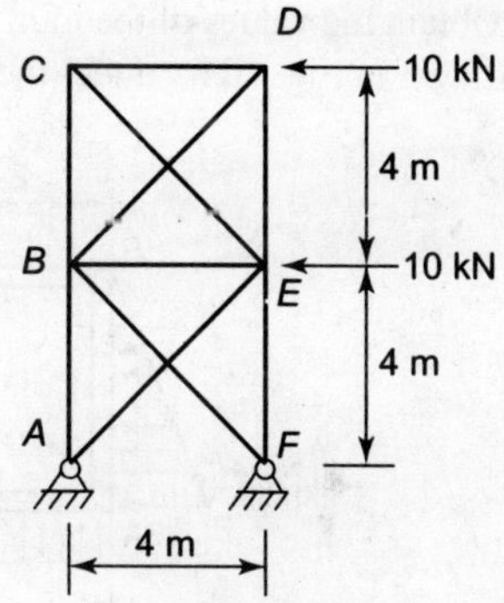

Fig. 10.65

10.7 Using the truss data in Problem 10.6 find the forces in members due to forcing of member BC which is short in length by 1 mm.

10.8 Find the forces in all members of the truss shown in Fig. 10.65. Assume L/EA to be the same for all members.

10.9 Draw the shear force and moment diagrams for a three-span continuous beam with equal spans and EI constant carrying a uniformly distributed load of intensity w/unit length. The theorem of three moment equations may be used.

10.10 Using the theorem of three moments find the support moments and reactions for the continuous beam given in Fig. 10.66. EI constant.

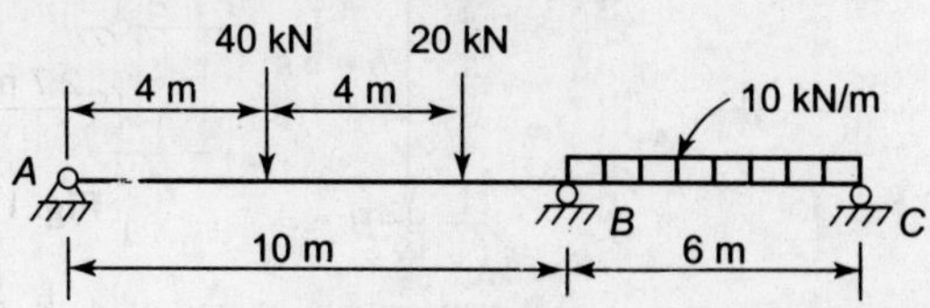

Fig. 10.66

10.11 A continuous beam of uniform section 18 m long is supported on four equally spaced elastic supports. The supports are such that they settle by 1 mm for each kN load. If the beam carries a uniformly distributed load of 30 kN/m throughout its length, obtain the reactions at the supports. $E = 200,000$ MPa and $I = 200 \times 10^{-6}$ m^4 (200×10^6 mm^4).

10.12 A continuous reinforced concrete beam *ABCD* of size 300×200 mm is loaded as shown in Fig. 10.67. If $P = 100$ kN estimate the percentage change in the support moment at *B* or *C* and span moment at *E* when the supports *B* and *C* settle through vertical distance so that $\psi_{AB} = -\psi_{CD} = 1/600$. Take $E = 13.3 \times 10^6$ kN/m^2 (13,300 MPa).

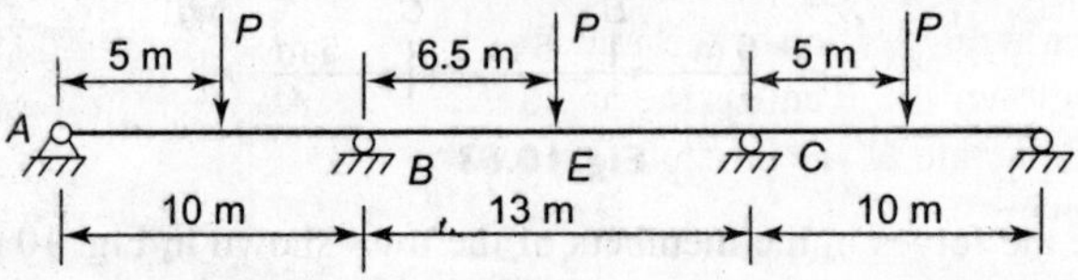

Fig. 10.67

10.13 Obtain the values of the joint moments of a frame which forms a section of a box culvert shown in Fig. 10.68 under the prescribed loading and soil pressure.

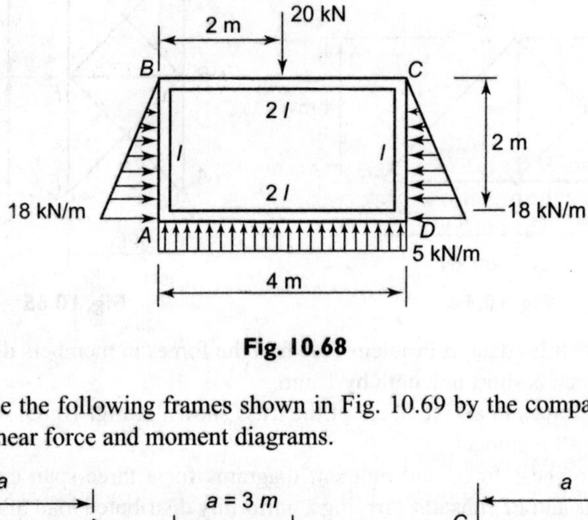

Fig. 10.68

10.14 Analyse the following frames shown in Fig. 10.69 by the compatibility method and draw the shear force and moment diagrams.

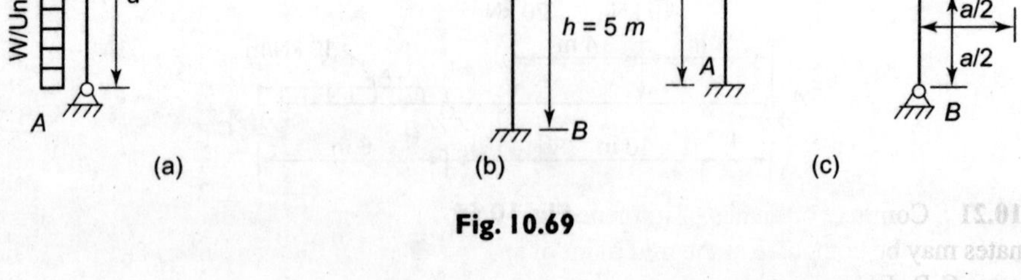

Fig. 10.69

10.15 A two-hinged circular arch of span 20 m and rise 4 m is loaded with a uniformly distributed load of 10 kN/m over the left half of the span and a concentrated load of 80 kN at the mid point of the right half of the arch. Calculate the horizontal reaction H and normal thrust N at a section just to the right of the concentrated load.

10.16 A two-hinged parabolic arch of span 60 m and rise 10 m and of constant rib cross-section carries a uniformly distributed load of 20 kN/m covering the middle one-third length of the span. Calculate the horizontal reaction at the abutments and the radialshear and normal thrust at a section just at the commencement of loading.

10.17 The axis of a two-hinged arch given in Fig. 10.70 is defined by $y = h \sin \frac{\pi x}{L}$. The moment of inertia at any section of the arch rib is equal to $I_c \sec \theta$, where I_c is the moment of inertia at the crown and θ equals the angle that a tangent to the arch axis makes with the horizontal. Calculate H, the horizontal reaction.

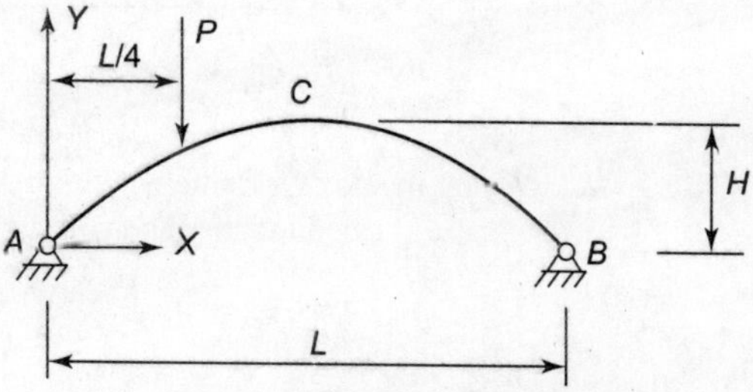

Fig. 10.70

10.18 Calculate H for load P at the centre line of the arch in Problem 10.17.

10.19 Compute the ordinates, at intervals of 2.5 m, of the influence line for the moment at A in Fig. 10.71. The moment of inertia is constant.

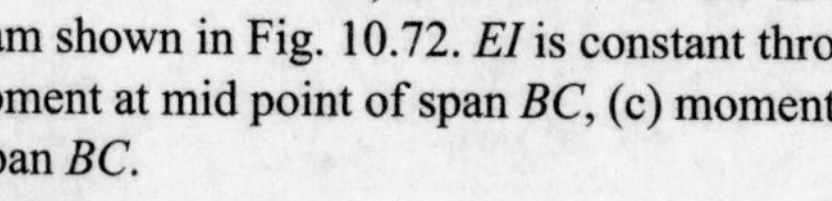

Fig. 10.71

10.20 Compute influence line ordinates, at intervals of 2.5 m for the following force components for the beam shown in Fig. 10.72. EI is constant throughout.

(a) Reaction R_A, (b) moment at mid point of span BC, (c) moment over support B and (d) shear at mid point of span BC.

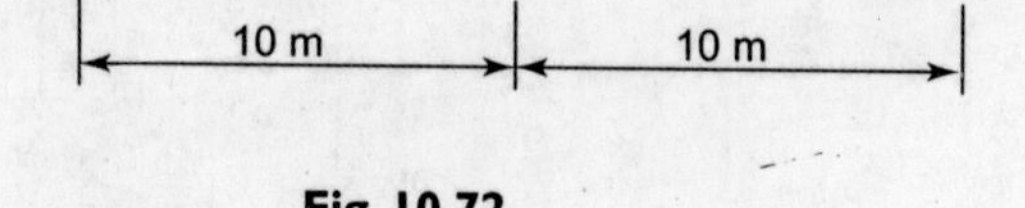

Fig. 10.72

10.21 Compute ordinates of influence line for the reaction at C in Fig. 10.73. The ordinates may be computed at the mid point of spans A-B and B-C and at intervals of 10 m in span C-D. EI is constant.

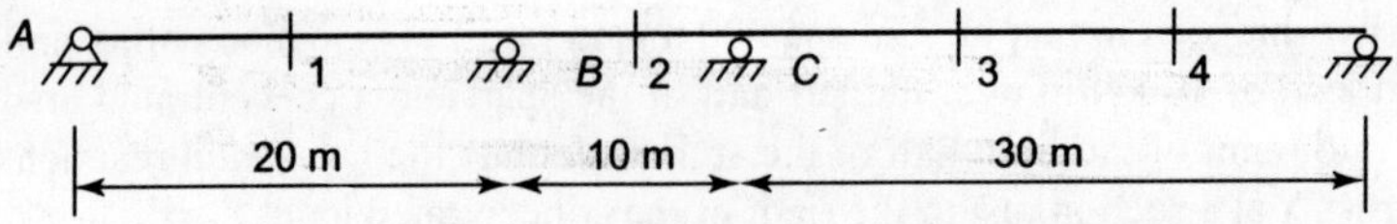

Fig. 10.73

10.22 A two-hinged parabolic arch has span $L = 30$ m and rise $h = 6$ m. Construct influence lines for (a) the horizontal reaction, (b) the moment, shear, and thrust at the one-fourth point of the span.

10.23 Construct the following influence lines for the arch in Problem 10.22: (a) thrust at crown, (b) moment and shear at crown.

11 Slope-Deflection Method

11.1 INTRODUCTION

The slope-deflection method developed by Axel Bendixen in Germany in 1914 was later presented in greater detail by G.A. Maney of the University of Minnesota in 1915. This method can be used to analyse statically indeterminate structures, composed of moment resisting members such as continuous beams and frames. The popularity of the method is lost to some extent by the advent of relaxation technique in the form of the moment distribution method and its relevance is greatly reduced by the introduction of the displacement method of matrix analysis. The method, though not preferred by engineers, is considered useful for the understanding of the relationship that exists between displacements of the joints and the forces at the ends of members.

The basic slope-deflection equation expresses the moment at the end of a member as the superposition of end moment due to external loads on the member with the ends assumed restrained and the end moments caused by the actual end rotations and displacements. In a structure composed of several members, the slope-deflection equations are applied to each member of the structure. Using appropriate equations of equilibrium of the joints along with slope-deflection equations for each member, we obtain a set of simultaneous equations with displacements as unknowns. With the displacements evaluated, the end moments can be computed using slope-deflection equations.

Before we proceed further we shall decide the sign convention for forces and displacements. As the method uses algebraic procedure, the use of correct signs is of paramount importance.

11.2 SIGN CONVENTION

It is convenient in the development and application of this method to use the following static sign convention.

1. Moment at the end of a member is considered positive when it is in the anti-clockwise direction or on a joint in the clockwise direction.

Considering a beam fixed at the ends and loaded transversely as shown in Fig. 11.1, the left hand support is subjected to a clockwise moment or the left end of the member is subjected to an anti-clockwise moment. Both of them by the above convention are positive. At the same time, the right hand support is subjected to an anti-clockwise moment or the right end of the member is subject to a clockwise moment. Again by the above convention both of them are negative.

2. Translation is considered positive when it is upward in the Y direction.
3. Angular rotation θ is considered positive when it is in the anti-clockwise direction.

To develop slope-deflection equations, let us consider a typical member in a continuous structure with its ends designated as i and j (see Fig. 11.2). The member is possibly subjected to a transverse loading as shown and is connected to other members at its ends or to the supports. When the structure undergoes deformations due to the action of applied forces, settlement of supports or any other effects, the member i-j which is a part of the structure also undergoes deformation inducing moment at its ends.

The general displacement of the member and the moments induced at the ends are shown in Fig. 11.3. All displacements and moments according to the sign convention adopted earlier are in their positive direction.

The end of moments M_{ij} and M_{ji} may be considered as caused by a combination of moments due to:

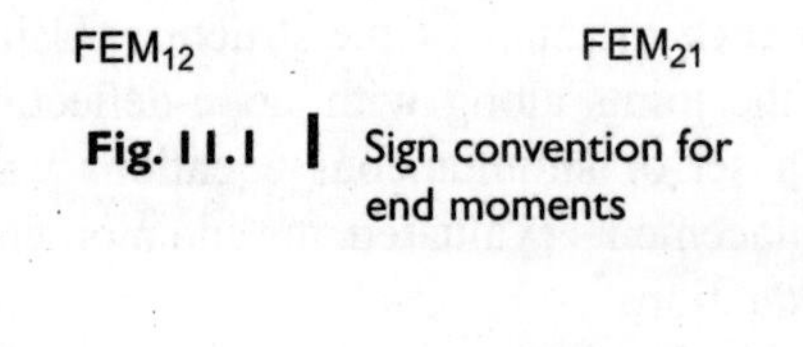

Fig. 11.1 | Sign convention for end moments

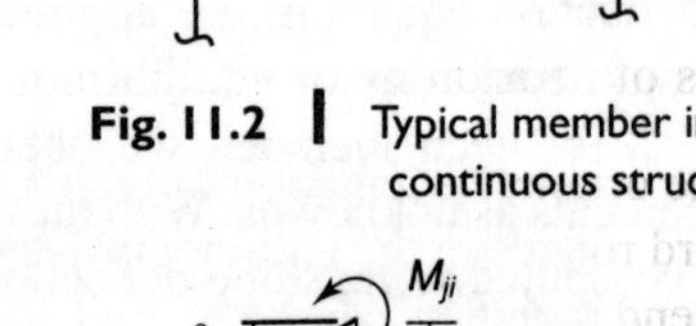

Fig. 11.2 | Typical member in a continuous structure

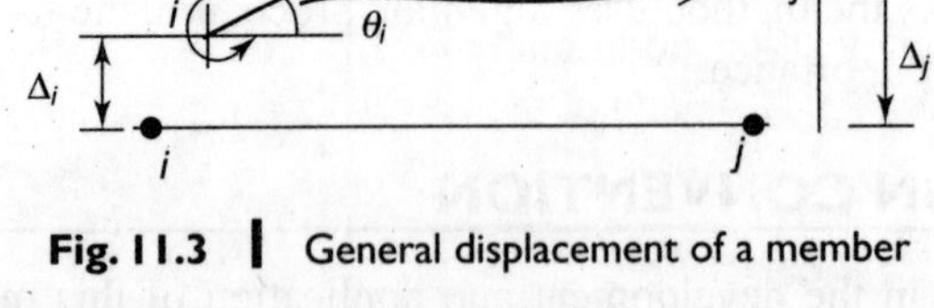

Fig. 11.3 | General displacement of a member

1. the fixed end moments developed by the transverse loading on the member;
2. moments caused by the actual rotations of the ends, θ_i and θ_j of the member; and

3. the end moments caused by relative translation of ends of the member.

At this stage we may derive expressions for fixed end moments for different loading cases and also the force displacement relationships in a member.

Fixed End Moments (FEM) The fixed end moments for a transversely loaded member are the end moments developed when the ends are fixed against rotation and translation. They may be determined by any of the standard methods of indeterminate analysis. In Examples 10.1 and 10.2 the fixed end moments are obtained by employing a general method of consistent displacements and moment-area theorems. Fixed end moments for a number of common loading cases are summarised in the Appendix C for ready reference. It may be noted that the fixed end moments are expressed in accordance with the adopted sign convention.

Force Displacement Relationships We also need information regarding the relationships that exist between the member end moments and its end displacements. They can be derived easily (one such case is presented in Example 10.1) by any of the methods discussed in Chapter 10. For ready reference they are tabulated in the Appendix D.

11.3 DEVELOPMENT OF SLOPE-DEFLECTION EQUATIONS

The end moments developed due to each of the three contributory effects are shown in Fig. 11.4*a* to *d*.

The fixed end moments for a specified loading can be obtained either from the Table in the Appendix or can be worked out independently. Next, the end moments caused by the relative translation of the member end $\Delta = \Delta j - \Delta_i$, with ends i and j fixed against rotation can be taken from the Appendix. The values of end moments are shown in Fig. 11.4*b*. It is common to express end moments in terms of rotation $\psi_{ij} = \dfrac{\Delta}{L}$, that is, the angle between the chord joining i and j and the original orientation of the member axis, ψ is considered positive when the chord rotates in the anti-clockwise direction.

The end moments caused by transverse loads and relative translation of joints are based on the fixed end condition. When the member ends are allowed to rotate to their equilibrium position, additional moments are developed. For rotations θ_i and θ_j as shown in Fig. 11.4*c* and *d*, the corresponding end moments are

$$M_{ij} = \frac{4EI\theta_i}{L} + \frac{2EI\theta_i}{L} = \frac{2EI}{L}(2\theta_i + \theta_j) \tag{11.1}$$

$$M_{ji} = \frac{2EI\theta_i}{L} + \frac{4EI\theta_j}{L} = \frac{2EI}{L}(\theta_i + 2\theta_j) \tag{11.2}$$

The true moment at each end of the member is the superposition of the zero rotation moments caused by transverse loading and relative translation, and the

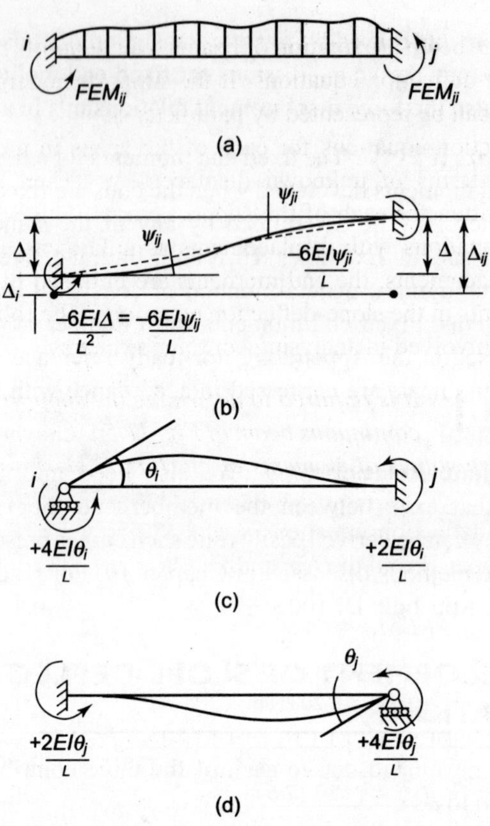

Fig. 11.4 | *(a) Fixed end moments due to transverse loading, (b) End moments due to translation of joints, (c) End moments due to rotation θ_i at end i, (d) End moments due to rotation θ_j at and j.*

moments due to the rotation of the end of the member. Thus, the true moments are

$$M_{ij} = FEM_{ij} + 2EK(2\theta_i + \theta_j - 3\ \psi_{ij}) \tag{11.3}$$

$$M_{ji} = FEM_{ji} + 2EK(\theta_i + 2\ \theta j - 3\ \psi_{ji}) \tag{11.4}$$

where

$$\psi_{ji} = \psi_{ij} = \frac{\Delta}{L} \text{ or } \frac{(\Delta_j - \Delta_i)}{L}$$

and $\quad K = \dfrac{I}{L}$ the relative stiffness

Eqs. 11.3 and 11.4 are known as slope-deflection equations.

11.4 | ANALYSIS OF CONTINUOUS BEAMS

The analysis of continuous beams by the slope-deflection method is a fairly straightforward procedure. If there are no translations of supports, the possible displacements of the beam are rotation of beams over supports represented by θ_i, and θ_j, in the slope-deflection equations. If the supports undergo displacements (translations) they can be represented by parameter ψ.

The slope-deflection equations for each of the spans in a continuous beam can be written in terms of unknown displacements. Then the equations of equilibrium are written for each of the support points. This will result in a set of simultaneous equations with displacements as unknowns. After solving the equations for displacements, the end moments are obtained by substituting the known displacements in the slope-deflection equations. The following examples illustrate the steps involved in analysing continuous beams.

Example 11.1 *It is required to determine the support moments for the continuous beam of Fig. 11.5a. Use the slope-deflection method. The relative values of moments of inertia are shown in Fig. 11.5. E is constant.*

To apply slope-deflection equations we must first determine the fixed end moments for each span. From the Appendix table, we obtain

$$FEM_{12} = \frac{100 \times 5}{8} = 62.50 \text{ kN.m}$$

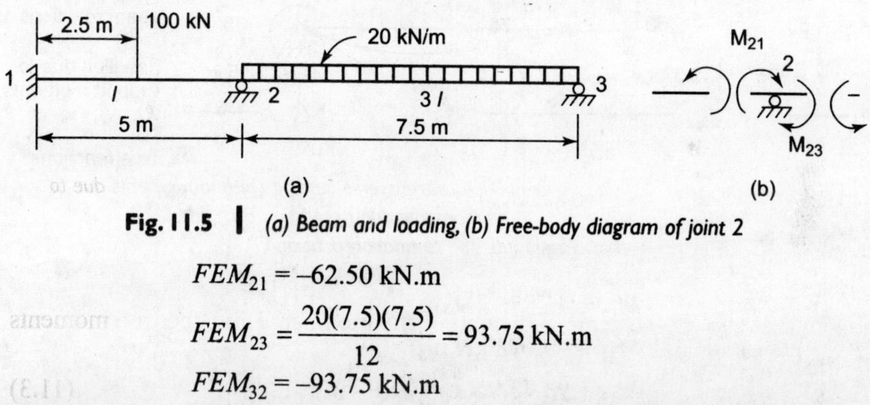

Fig. 11.5 *(a) Beam and loading, (b) Free-body diagram of joint 2*

$$FEM_{21} = -62.50 \text{ kN.m}$$

$$FEM_{23} = \frac{20(7.5)(7.5)}{12} = 93.75 \text{ kN.m}$$

$$FEM_{32} = -93.75 \text{ kN.m}$$

The supports are all rigid and no lateral translations are possible. Therefore, $\psi = 0$ for both spans. Further, for the fixed supports at the left end, $\theta_1 = 0$ and right hand support moment $M_{32} = 0$. The slope-deflection equations for the two spans can be written as

$$M_{12} = 62.50 + \frac{2EI\theta_2}{5} = 62.50 + 0.4EI\theta_2 \tag{11.5}$$

$$M_{21} = -62.50 + \frac{4EI\theta_2}{5} = -62.50 + 0.8EI\theta_2 \tag{11.6}$$

$$M_{23} = 93.75 + \frac{4(3)EI\theta_2}{7.5} + \frac{2(3)EI\theta_2}{7.5}$$
$$= 93.75 + 1.6EI\theta_2 + 0.8EI\theta_3 \tag{11.7}$$

and $$M_{32} = -93.75 + 0.8EI\theta_2 + 1.6EI\theta_3 \tag{11.8}$$

Considering the free-body diagram of a segment of the beam at joint 2 as shown in Fig. 11.5*b* we see that

$$M_{21} + M_{23} = 0 \tag{11.9}$$

Substituting the values for M_{21} and M_{23} from Eqs. 11.6 and 11.7, we get

$$31.25 + 2.4EI\theta_2 + 0.8\,EI\theta_3 = 0 \tag{11.10}$$

We can get another equation by writing $M_{32} = 0$, that is

$$-93.75 + 0.8\,EI\theta_2 + 1.6\,EI\theta_3 = 0 \tag{11.11}$$

Solving Eqs. 11.10 and 11.11 simultaneously we get

$$\theta_2 = -39.06 \text{ and } \theta_3 = 78.13$$

Substituting these values for θ_2 and θ_3 in the Eqs. 11.5, 11.6 and 11.7, we obtain

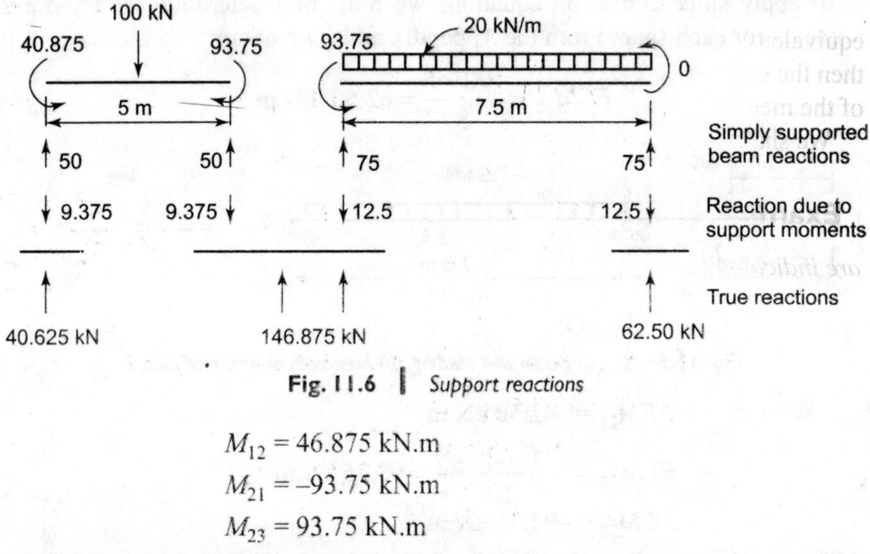

Fig. 11.6 | *Support reactions*

$$M_{12} = 46.875 \text{ kN.m}$$
$$M_{21} = -93.75 \text{ kN.m}$$
$$M_{23} = 93.75 \text{ kN.m}$$

As a check we find $M_{32} = 0$ from Eq. 11.8. Using these values the reactions may be found from the free-body diagrams of each of the spans 1–2 and 2–3 shown in Fig. 11.6. Thus, the reactions are:

$$R_1 = 40.625 \text{ kN}, R_2 = 146.875 \text{ kN and } R_3 = 62.50 \text{ kN}$$

The shear force and bending moment diagrams are shown in Fig. 11.7.

The effects of support displacements can be readily included in the slope-deflection analysis of continuous beams. From the known support settlements, we can calculate for each span, the ψ value which can be included in the slope-deflection equations.

Another condition often encountered in continuous beams is the overhanging span. Such problems can be solved by replacing the overhanging span by an

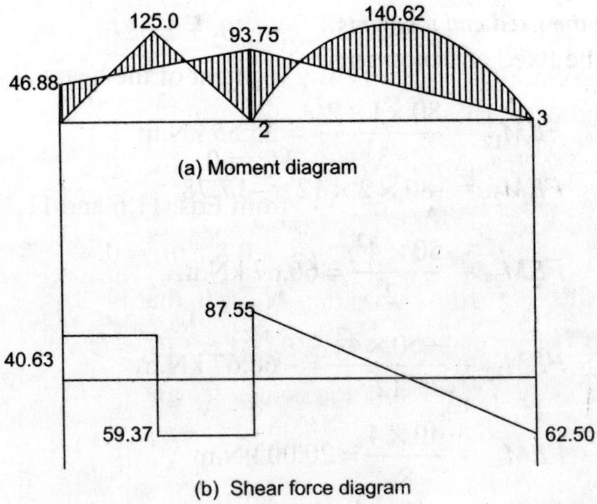

Fig. 11.7

equivalent applied moment at the support point. The procedure for analysis is then the same as before, except that there is a known value of moment at the end of the member adjacent to the overhanging span.

We shall illustrate these points by solving a few numerical examples.

Example 11.2 | *Determine the support moments for the continuous beam shown in Fig. 11.8. Relative I values for all spans are indicated on the beam. Draw the moment diagram.*

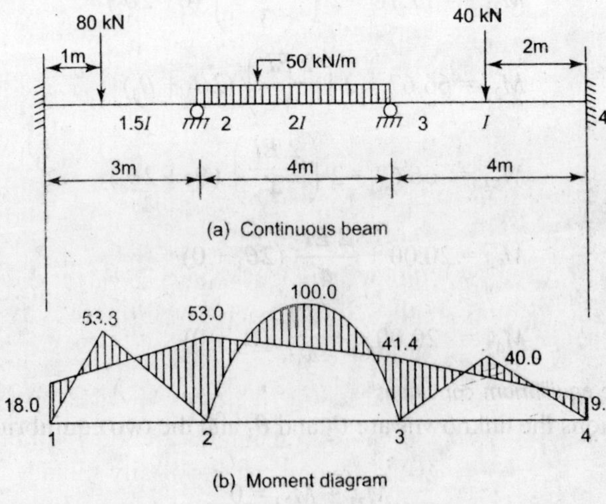

Fig. 11.8

Step 1: To fix up the fixed end moments

First we write the fixed end moments

$$FEM_{12} = \frac{80 \times 1 \times 2^2}{3^2} = 35.55 \text{ kN.m}$$

$$FEM_{21} = -80 \times 2 \times 12 = -17.78$$

$$FEM_{23} = \frac{50 \times 4^2}{12} = 66.67 \text{ kN.m}$$

$$FEM_{32} = \frac{-50 \times 4^2}{12} = -66.67 \text{ kN.m}$$

$$FEM_{34} = \frac{40 \times 4}{8} = 20.00 \text{ kN.m}$$

$$FEM_{43} = -\frac{40 \times 4}{8} = -20.00 \text{ kN.m}$$

Step 2: To write end moments

The continuous beam under goes rotation of joints at supports 2 and 3. We may designate them as θ_2 and θ_3 respectively. As the supports are rigid no translation of joints is possible.

Now we can write down slope deflection equations for moments.

$$M_{12} = 35.55 + 2\left(\frac{1.5\,EI}{3}\right)(0 + \theta_2)$$

$$M_{21} = -17.78 + 2\left(\frac{1.5\,EI}{3}\right)(0 + 2\theta_2)$$

$$M_{23} = 66.67 + 2\left(\frac{2\,EI}{4}\right)(2\theta_2 + \theta_3)$$

$$M_{32} = -66.67 + 2\left(\frac{2\,EI}{4}\right)(\theta_2 + 2\theta_3)$$

$$M_{34} = 20.00 + \frac{2\,EI}{4}(2\theta_3 + 0)$$

and

$$M_{43} = -20.00 + \frac{2\,EI}{4}(\theta_3 + 0)$$

Step 3: To write equilibrium conditions

In these equations the unknowns are θ_2 and θ_3 and the two equilibrium equations are:

$$M_{21} + M_{23} = 0 \quad \text{(a)}$$

$$M_{32} + M_{34} = 0 \quad \text{(b)}$$

Substituting the values from above we have

$$4EI\theta_2 + EI\theta_3 = -48.89$$

$$EI\theta_2 + 3EI\theta_3 = 46.67$$

Solving the equations simultaneously, we get

$$EI\theta_2 = -17.58$$

and $$EI\theta_3 = 21.42$$

Step 4: To write the end moments

Substituting these values in moment equations, we get

$$M_{12} = 35.55 - 17.58 = 17.97 \text{ kN.m}$$
$$M_{21} = -17.78 + 2(-17.58) = -52.94 \text{ kN.m}$$
$$M_{23} = 66.67 + 2(-17.58 + 21.42 = 52.93 \text{ kN.m}$$
$$M_{32} = -66.67 - 17.58 + 2(21.42) = -41.41 \text{ kN.m}$$
$$M_{34} = 20.00 + 21.42 = 41.42 \text{ kN.m}$$
$$M_{43} = -20.00 + \left(\frac{-21.42}{2}\right) = -9.29 \text{ kN.m}$$

The moment diagram is shown in Fig. 11.8*b*.

Example 11.3 *A continuous beam is shown in Fig. 11.9 during loading support 2 sinks by 10 mm, determine the support moments. E = 200 × 10⁶ kN/m² and I = 80 × 10⁻⁶ m⁴. Relative I value for each span is indicated*

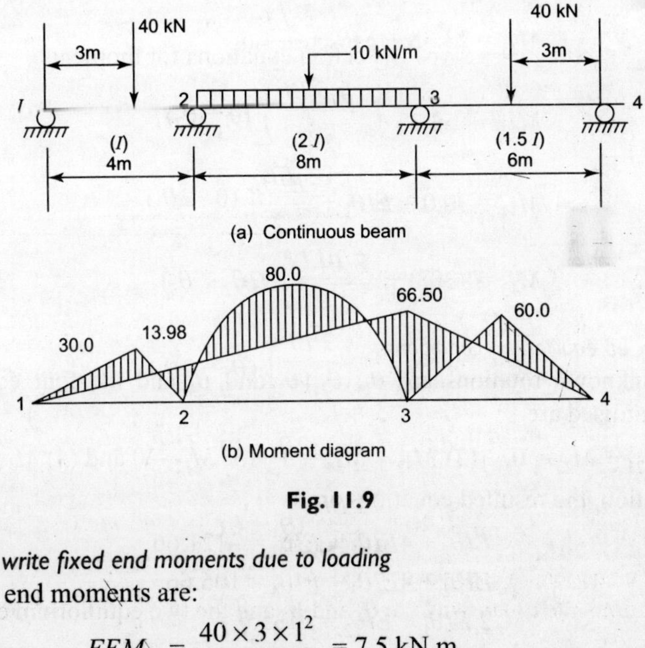

Fig. 11.9

Step 1: To write fixed end moments due to loading

The fixed end moments are:

$$FEM_{12} = \frac{40 \times 3 \times 1^2}{4^2} = 7.5 \text{ kN.m}$$

$$FEM_{21} = \frac{40 \times 3^2 \times 1}{4^2} = -22.5 \text{ kN.m}$$

$$FEM_{23} = -FEM_{32} = \frac{10 \times 8^2}{12} = 53.33 \text{ kN.m}$$

$$FEM_{32} = -53.33 \text{ kN.m}$$

$$FEM_{34} = -FEM_{43} = \frac{40 \times 6}{8} = -30.0 \text{ kN.m}$$

Step 2: To write end moments in terms of FEMs, rotation and translations

and $\psi_{12} = \frac{1}{400}, \psi_{21} = -\frac{1}{400}, \psi_{23} = +\frac{1}{800}, \psi_{32} = +\frac{1}{800}$

writing slope deflection equations for end moments

$$M_{12} = \frac{2EI}{4}(2\theta_1 + \theta_2 - 3\,\psi_{12}) + 7.5$$

$$= 7.5 + \frac{EI}{2}\left(2\theta_1 + \theta_2 + \frac{3}{4} \times 10^{-2}\right)$$

$$= 67.5 + EI\,\theta_1 + \frac{EI\theta_2}{2}$$

Similarly $M_{21} = 37.5 + \frac{EI\theta_1}{2} + EI\theta_2$

$$M_{23} = 23.33 + EI\theta_2 + \frac{EI}{2}\theta_3$$

$$M_{32} = -53.33 + \frac{EI\theta_2}{2} + EI\,\theta_3$$

$$M_{34} = 30.0 + EI\theta_3 + \frac{EI\theta_4}{2}$$

$$M_{43} = -30.0 + \frac{EI\theta_3}{2} + EI\theta_4$$

Step 3: To fix up equilibrium conditions

The four unknown rotations are θ_1, θ_2, θ_3 and θ_4 and the four equilibrium conditions utilised are

(1) $M_{21} + M_{23} = 0$, (2) $M_{32} + M_{34} = 0$ (3) $M_{12} = 0$ and (4) $M_{43} = 0$

On substitution, the resulted equations are

$$EI\theta_1 + 4EI\theta_2 + EI\theta_3 = -121.66 \quad \text{(a)}$$

$$EI\theta_2 + 4EI\theta_3 + EI\theta_4 = 106.66 \quad \text{(b)}$$

$$2\,EI\theta_1 + EI\theta_2 = -135.0 \quad \text{(c)}$$

and $EI\theta_3 + 2\,EI\theta_4 = 60.0$ (d)

Solving the above equations simultaneously the following are obtained

$EI\theta_1 = -55.68$, $EI\theta_2 = -23.65$

$EI\theta_3 = 28.66$, $EI\theta_4 = 15.67$

Step 4: To write final end moments

Substituting the above in the moment equations

$$M_{21} = 37.50 - \frac{55.68}{2} - 23.65 = -13.98 \text{ kN.m}$$

$$M_{23} = 23.33 - 23.65 + \frac{28.66}{2} = 14.01 \text{ kN.m}$$

$$M_{32} = -83.33 - \frac{23.65}{2} + 28.66 = -66.50$$

$$M_{34} = 30.00 + 28.66 + \frac{15.67}{2} = 66.50$$

The bending moment diagram is shown in Fig. 11.9 (b)

Example 11.4 | *A continuous beam is supported and loaded as shown in Fig. 11.10. During loading support 2 sinks by 10 mm. Analyse the beam for support moments and reactions. $E = 200 \times 10^6$ kN/m^2 (200,000 MPa) and $I = 100 \times 10^{-6}$ m^4 (100×10^6 mm^4) constant throughout.*

Proceeding as before we can write down fixed end moments using the table in the appendix.

$$FEM_{12} = \frac{(40)(4)}{8} = 20.0 \text{ kN.m}$$

$$FEM_{21} = -20.0 \text{ kN.m}$$

$$FEM_{23} = \frac{10(6)^2}{12} = 30.0 \text{ kN.m}$$

$$FEM_{32} = -30.0 \text{ kN.m}$$

If we consider the overhanging span 3–4 as a cantilever the fixed end moment at 3 is

$$FEM_{34} = 20(2) = 40.0 \text{ kN.m}$$

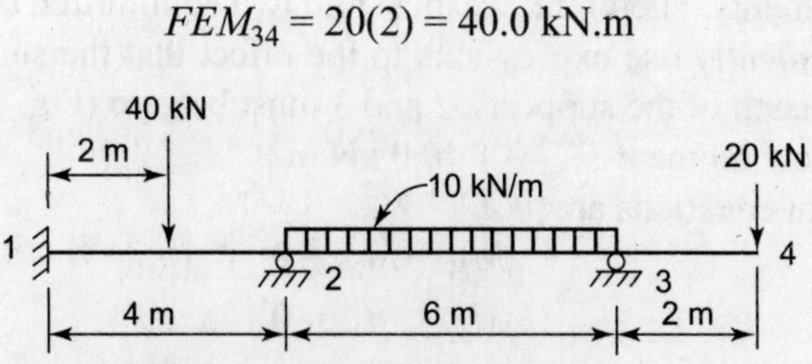

Fig. 11.10 | *Continuous beam having a sinking support*

The values of ψ for spans 1–2 and 2–3 are computed as

$$\psi_{12} = -\frac{1}{400} \text{ rad.}$$

$$\psi_{23} = +\frac{1}{600} \text{ rad.}$$

The expressions for final moments at the end of each span are written using Eqs. 11.3 and 11.4. Of special mention here is the fact that although absolute stiffnesses have been used in deriving these equations it is possible to simplify the computations by using relative values of *I*/*L*. The values of final end moments obtained by using these relative values are not affected and are correct. If, however, we wish to find out the absolute value of any unknown rotation θ or ψ, an adjustment must be made to correct for the use of relative values of I/L in writing the initial equations. Note that the relative *I*/*L* values for spans 1–2 and 2–3 may be taken as 3 and 2 respectively.

Using Eqs. 11.3 and 11.4 and remembering $\theta_1 = 0$, we can write

$$M_{12} = 20.0 + 2EK_{12}(0 + \theta_2) - \frac{6EI_{12}}{L_{12}}\psi_{12} \tag{11.12}$$

Letting $EK_{12} = 3$, and substituting numerical values for the last term, we have

$$M_{12} = 6\theta_2 + 20 + 75 \tag{11.13}$$

and

$$M_{21} = 12\,\theta_2 - 20 + 75 \tag{11.14}$$

Similarly,

$$M_{23} = 30 + 2EK_{23}(2\theta_2 + \theta_3) - \frac{6EI_{23}}{L_{23}}\psi_{23} \tag{11.15}$$

Again, letting $EK_{23} = 2$ and substituting numerical values for the last term we have

$$M_{23} = 8\theta_2 + 4\theta_3 + 30 - 33.33 \tag{11.16}$$

and

$$M_{32} = 4\theta_2 + 8\theta_3 - 30 - 33.33 \tag{11.17}$$

It is apparent that the two unknowns, θ_2 and θ_3, appear in the above expressions for various moments. Therefore, we must find two equilibrium conditions. In this case we conveniently use expressions to the effect that the sum of the internal moments over each of the supports, 2 and 3 must be zero (Fig. 11.11).

Further, we know moment $M_{34} = +\,40.0$ kN.m

The equilibrium equations are

$$M_{21} + M_{23} = 0 \tag{11.18}$$

and

$$M_{32} + M_{34} = 0 \tag{11.19}$$

Substituting for M_{21} and M_{23} from Eq. 11.14 and 11.15, we get

$$20\theta_2 + 4\theta_3 = -\,51.67 \tag{11.20}$$

Again substituting for M_{32} from Eq. 11.17 and taking $M_{34} = +\,40.0$ kN.m we get

$$4\theta_2 + 8\theta_3 = 23.33 \tag{11.21}$$

A simultaneous solution of Eqs. 11.20 and 11.21 gives

$$\theta_2 = -3.5186$$

and $$\theta_3 = 4.6755$$

Substituting back the values of θ_2 and θ_3 in Eqs. 11.13 to 11.17, we get

$$M_{12} = 73.88 \text{ kN.M}$$
$$M_{21} = 12.78 \text{ kN.M}$$
$$M_{23} = -12.78 \text{ kN.m}$$
$$M_{32} = -40.0 \text{ kN.m}$$

and $$M_{34} = +40.0 \text{ kN.m}$$

The actual values of θ_2 and θ_3 can be obtained by multiplying the relative values of θ_2 and θ_3 by $3L_{12}/EI$ or $2L_{23}/EI$. Thus,

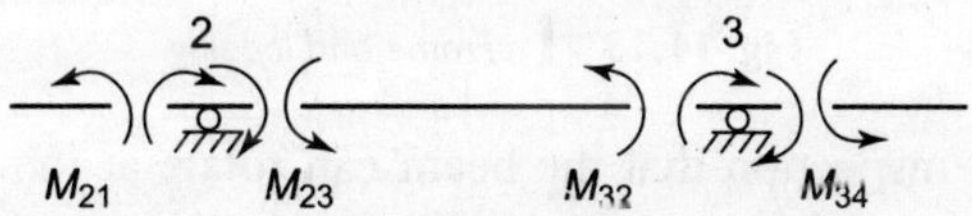

Fig. 11.11 | *Free-body diagrams of joints*

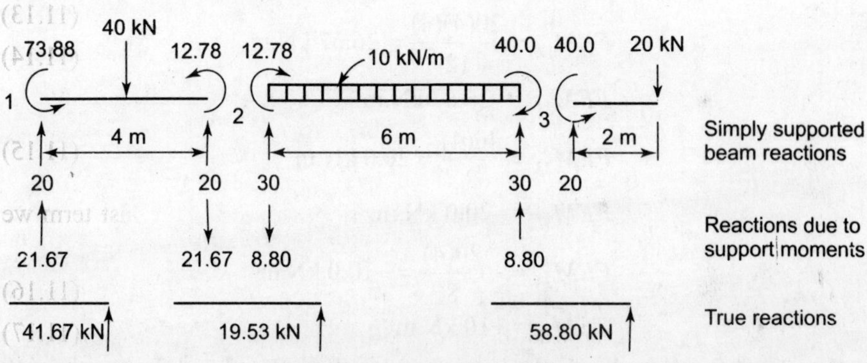

Fig. 11.12 | *Reactions*

$$\theta_2 = -2.111 \times 10^{-3} \text{ rad.}$$

and $$\theta_3 = 2.805 \times 10^{-3} \text{ rad.}$$

The negative sign for θ_2 indicates that the rotation of joint 2 is in the clockwise direction. Obviously, joint 3 rotates in the anti-clockwise direction.

The reactions may be obtained from free-body diagrams of spans 1–2 and 2–3 as shown in Fig. 11.12.

11.5 | ANALYSIS OF FRAMES WITH NO LATERAL TRANSLATION OF JOINTS

The analysis of frames in which the lateral translations of joints are restrained follows the same general procedure as for continuous beams. This aspect is illustrated in the following example.

Example 11.5 | *It is required to analyse the frame shown in Fig. 11.13 for moments at the ends of members. EI is constant for all members.*

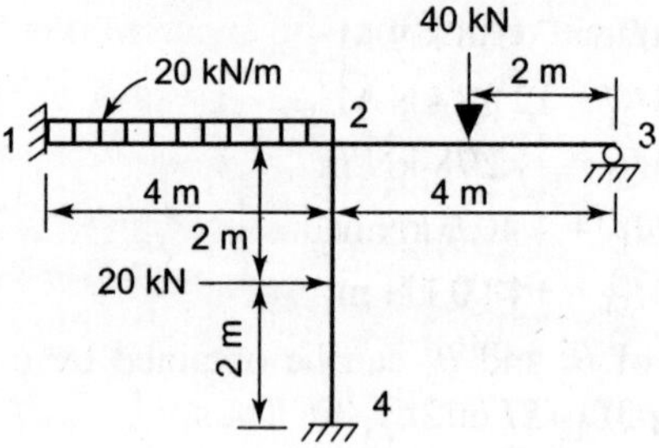

Fig. 11.13 | *Frame and loading*

It is seen from inspection that the beam can rotate at joints 2 and 3. There is no possibility of translation of any of the joints. Writing down the fixed end moments, we have

$$FEM_{12} = \frac{20(4)(4)}{12} = 26.67 \text{ kN.m}$$

$$FEM_{21} = -26.67 \text{ kN.m}$$

$$FEM_{23} = \frac{40(4)}{8} = 20.0 \text{ kN.m}$$

$$FEM_{32} = -20.0 \text{ kN.m}$$

$$FEM_{24} = -\frac{20(4)}{8} = -10.0 \text{ kN.m}$$

$$FEM_{42} = +10 \text{ kN.m}$$

Note that for the column, the bottom end is the left end and the top the right end.

Designating the rotations at joints 2 and 3 as θ_2 and θ_3 respectively, the end moments can be written as

$$M_{12} = 26.67 + 2EK\theta_2$$

Taking $EK = 1$, since only relative values are needed,

$$M_{12} = 26.67 + 2\,\theta_2$$

Similarly,

$$\left.\begin{aligned} M_{21} &= -26.67 + 4\theta_2 \\ M_{23} &= 20 + 4\theta_2 + 2\theta_3 \\ M_{32} &= -20 + 2\theta_2 + 4\theta_3 \\ M_{24} &= -10 + 4\theta_2 \\ M_{42} &= 10 + 2\theta_2 \end{aligned}\right\} \quad (11.22)$$

Now to solve for the two unknowns, θ_2 and θ_3, two conditions are used. They are

$$M_{21} + M_{23} + M_{24} = 0 \quad (11.23)$$

and $$M_{32} = 0 \text{ (support 3 is a roller support)} \quad (11.24)$$

Substituting for moment terms from the expression in Eq. 11.22, we get

$$12\,\theta_2 + 2\,\theta_3 = 16.67 \quad (11.25)$$

$$2\,\theta_2 + 4\,\theta_3 = 20.00 \quad (11.26)$$

Solving Eqs. 11.25 and 11.26 simultaneously, we get

$$\theta_2 = 0.606$$

and $$\theta_3 = 4.6968$$

Substituting back in Eq. 11.22, we have

$$M_{12} = 26.67 + 2\,(0.606) = 27.88 \text{ kN.m}$$

$$M_{21} = -26.67 + 4\,(0.606) = -24.24 \text{ kN.m}$$

$$M_{23} = 20.0 + 4(0.606) + 2(4.6968) = 31.82 \text{ kN.m}$$

$$M_{24} = -10 + 4\,(0.606) = -7.57 \text{ kN.m}$$

$$M_{32} = -20 + 2\,(0.606) + 4\,(4.6968) = 0$$

$$M_{42} = 10.0 + 2(0.606) = 11.21 \text{ kN.m}$$

The deflected shape and the moment diagram are shown in Fig. 11.14.

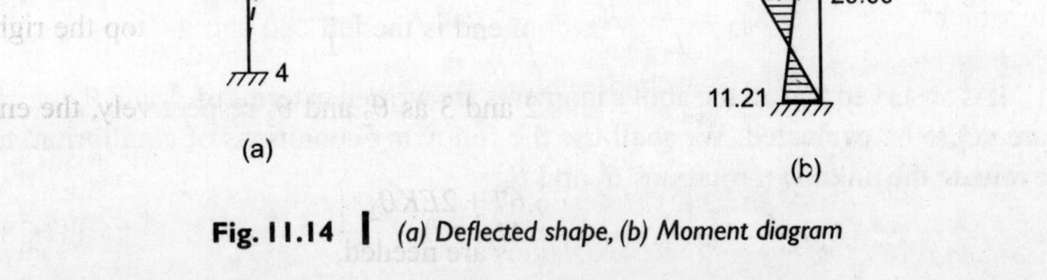

Fig. 11.14 | *(a) Deflected shape, (b) Moment diagram*

Example 11.6 | *Analyse the frame of Fig. 11.15. The relative I value for each member is indicated on the figure. E is constant.*

This is again a case of a frame where the lateral translation of joints is prevented. However, joints 2 and 3 are free to rotate. We can write down the fixed end moments as

$$FEM_{12} = \frac{40(4)}{8} = 20.0 \text{ kN.m}$$

$$FEM_{21} = -20.0 \text{ kN.m}$$

$$FEM_{23} = \frac{60(3)}{8} = 22.50 \text{ kN.m}$$

and $$FEM_{32} = -22.50 \text{ kN.m}$$

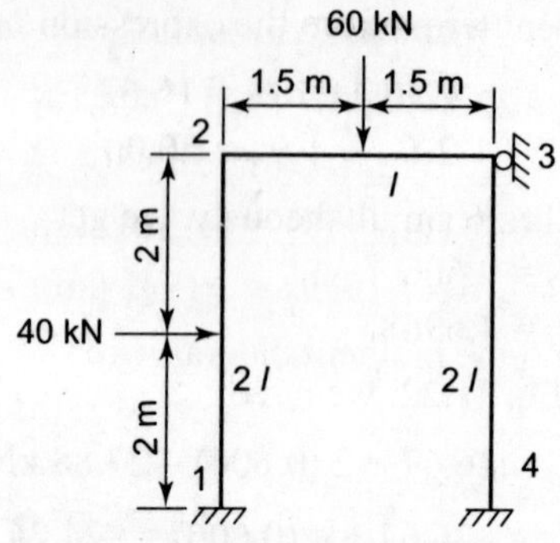

Fig. 11.15 | *Frame and loading*

Now designating rotations at joints 2 and 3 as θ_2 and θ_3 respectively, we can write down the end moments of members as

$$\left.\begin{aligned} M_{12} &= 20.0 + 2EK_{12}\,\theta_2 \\ M_{21} &= -20.0 + 4EK_{12}\,\theta_2 \\ M_{23} &= 22.50 + 4EK_{23}\,\theta_2 + 4EK_{23}\,\theta_3 \\ M_{32} &= -22.50 + 2EK_{23}\,\theta_2 + 4EK_{23}\,\theta_3 \\ M_{34} &= 0 + 4EK_{34}\,\theta_3 \\ M_{43} &= 0 + 2EK_{34}\,\theta_3 \end{aligned}\right\} \quad (11.27)$$

in which $$K_{12} = \frac{I_{12}}{L_{12}},\ K_{23} = \frac{I_{23}}{L_{23}} \text{ and } K_{34} = \frac{I_{34}}{L_{34}}$$

It is observed that all the above moments are written in terms of θ_2 and θ_3 which are yet to be evaluated. We shall use the following conditions of equilibrium to evaluate the unknown rotations θ_2 and θ_3.

$$M_{21} + M_{23} = 0 \quad (11.28)$$

and $$M_{32} + M_{34} = 0 \quad (11.29)$$

Substituting for the moment values from Eq. 11.27, we have

$$20\,E\,\theta_2 + 4\,E\,\theta_3 = -2.50 \quad (11.30)$$

and $$4\,E\,\theta_2 + 20\,E\,\theta_3 = 22.50. \quad (11.31)$$

It may be noted that the above equations were obtained taking the relative stiffness ratio as $EK_{12} = 3$, $EK_{23} = 2$ and $EK_{34} = 3$.

Solving Eqs. 11.30 and 11.31 simultaneously, we get

$$\theta_2 = -0.3646 \text{ and } \theta_3 = 1.1979$$

Substituting these values in Eq. 11.27 the end moments are evaluated as

$$M_{12} = 20 + 6(-0.3646) = 17.81 \text{ kN.m}$$
$$M_{21} = -20 + 12\,(-0.3646) = -24.38 \text{ kN.m}$$
$$M_{23} = 22.5 + 8(-0.3646) + 4(8.1979) = 24.38 \text{ kN.m}$$
$$M_{32} = -22.5 + 4(-0.3646) + 8(1.1979) = -14.38 \text{ kN.m}$$
$$M_{34} = 4(3)\,(1.1979) = 14.38 \text{ kN.m}$$
$$M_{43} = 2(3)\,(1.1979) = 7.19 \text{ kN.m}$$

The deflected shape and the moment diagram are shown in Fig. 11.16.

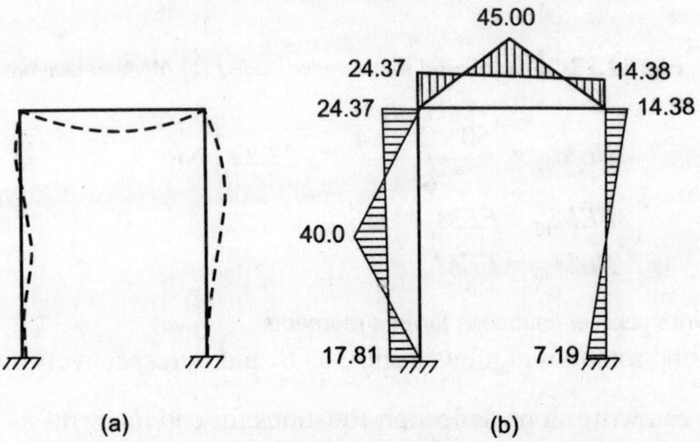

Fig. 11.16 | *(a) Deflected shape, (b) Moment diagram*

11.6 ANALYSIS OF FRAMES WITH LATERAL TRANSLATION OF JOINTS

For the analysis of frames in which the translation of joints is permitted, it is necessary to consider some other equilibrium conditions in addition to the equilibrium of joints. The following examples illustrate the steps involved in analysing such frames.

Example 11.7 | *Determine the moments at the ends of the members of the frame shown in Fig. 11.17. EI is constant for all members. Draw bending moment diagram.*

Under the loading the frame undergoes side sway as shown in the deflected shape in addition to rotations at 2 and 3

Step 1: To write fixed end moments

We can write the fixed end moments as

$$FEM_{23} = \frac{50 \times 2 \times 4^2}{6^2} = 44.44 \text{ kNm}$$

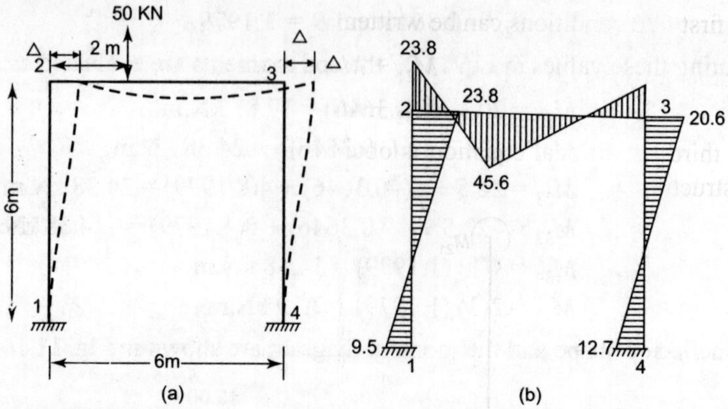

Fig. 11.17 | *(a) Portal frame under loading (b) Moment diagram*

$$FEM_{32} = \frac{50 \times 2^2 \times 4}{6^2} = -22.22 \text{ kNm}$$

$$FEM_{12} = FEM_{21} = 0$$

and $$FEM_{43} = FEM_{34} = 0$$

Step 2: To write general equations for end moments

Now assigning rotations at joints 2 and 3 as θ_2 and θ_3 respectively and $\psi_{12} = \psi_{34} = \frac{-\Delta}{6}$. We can write slope deflection equations for end moments as:

$$M_{12} = \frac{2EI}{6}\left(\theta_2 + 3\frac{\Delta}{6}\right)$$

$$M_{21} = \frac{2EI}{6}\left(2\theta_2 + 3\frac{\Delta}{6}\right)$$

$$M_{23} = \frac{2EI}{6}(2\theta_2 + \theta_3) + 44.44$$

$$M_{32} = \frac{2EI}{6}(\theta_2 + 2\theta_3) - 22.22$$

$$M_{34} = \frac{2EI}{6}\left(2\theta_3 + 3\frac{\Delta}{6}\right)$$

and $$M_{43} = \frac{2EI}{6}\left(\theta_3 + 3\frac{\Delta}{6}\right)$$

Step 3: To fix up equilibrium equations

It can be seen that the above moments are expressed in terms of θ_2, θ_3 and $\psi_{12} = \psi_{34}$. It is therefore necessary that we require three conditional equations to evaluate these unknowns.

The first two conditions can be written as

$$M_{21} + M_{23} = 0 \tag{11.32}$$

$$M_{32} + M_{34} = 0 \tag{11.33}$$

The third conditional equation is obtained by writing down $\Sigma F_H = 0$ for the entire structure

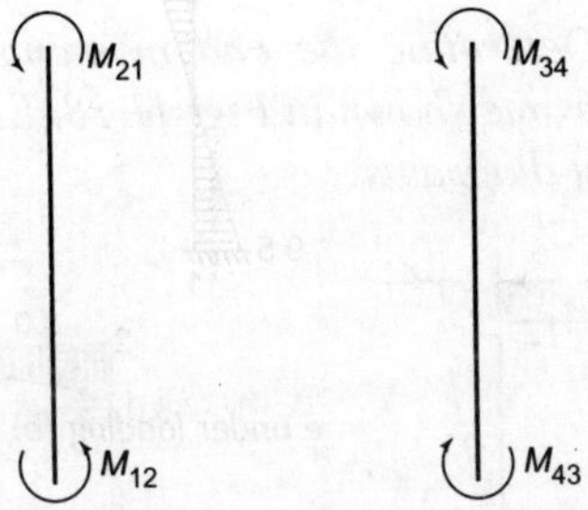

Fig. 11.17(c) | *Shear in columns*

That is:

$$\frac{M_{12} + M_{21}}{6} + \frac{M_{43} + M_{34}}{6} = 0 \tag{11.34}$$

On substitution and the values, we have

$$\frac{4EI\,\theta_2}{3} + \frac{EI\,\theta_3}{3} + \frac{EI\,\Delta}{6} = -44.44 \tag{11.35}$$

$$\frac{EI\,\theta_2}{3} + \frac{4}{3}EI\,\theta_3 + \frac{EI\,\Delta}{6} = 22.22 \tag{11.36}$$

$$EI\,\theta_2 + EI\,\theta_3 + \frac{2}{3}\,EI\,\Delta = 0 \tag{11.37}$$

Solving the above equations simultaneously, we get,

$$EI\,\theta_2 = -42.83$$
$$EI\,\theta_3 = 23.77$$

and $\quad EI\,\Delta = 28.58$

Step 4: To write final end moments

Substituting these values in moment equations we have

$$M_{12} = -\frac{4.2.83}{3} + \frac{28.58}{6} = -9.52 \text{ kN.m}$$

$$M_{21} = \frac{2}{3}(-42.83) + \frac{28.58}{6} = -23.79 \text{ kN.m}$$

$$M_{23} = 44.44 + \frac{2}{3}(-42.83) + \frac{1}{3}(23.77) = 23.81 \text{ kNm}$$

$$M_{32} = -22.22 + \frac{1}{3}(-42.83) + \frac{2}{3}(23.77) = 20.65 \text{ kNm}$$

$$M_{34} = \frac{2}{3}(23.77) + \frac{1}{6}(28.58) = 20.61 \text{ kNm}$$

$$M_{43} = 23.77 + \frac{28.58}{6} = 12.68 \text{ kNm}$$

The moment diagram is shown in Fig. 11.17*b*

Example 11.8 | *Determine the end moments of the members of the frame shown in Fig. 11.18. EI is same for all members. Draw the bending moment diagrams*

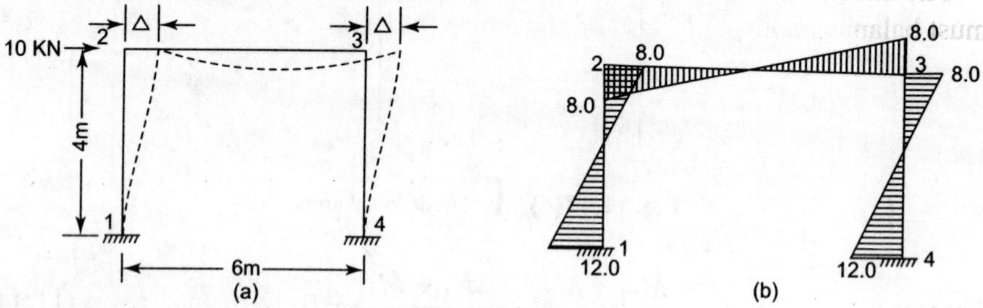

Fig. 11.18

Step 1: To identify translation and rotation of joints

The frame under goes lateral translation under the horizontal load. Joints 2 and 3 rotate by an amount θ_2 and θ_3 respectively. As the axial force is neglected in beam 2–3 the lateral translation Δ is same for both the columns. The displacements θ_2, θ_3 and Δ are to be evaluated for determining the end moments of members.

As the frame sways to the right, the angular rotation ψ_{12} and ψ_{43} are clockwise and hence taken as negative. So we have

$$\psi_{12} = \psi_{21} = \psi_{43} = \psi_{34} = -\frac{\Delta}{4}$$

Step 2: To write slope deflection equations for end moments

Writing the slope deflections we have

$$\left.\begin{aligned} M_{12} &= \frac{2EI}{4}\left(\theta_2 + \frac{3}{4}\Delta\right) \\ M_{21} &= \frac{2EI}{4}\left(2\theta_2 + \frac{3}{4}\Delta\right) \\ M_{23} &= \frac{2EI}{6}\left(2\theta_2 + \theta_3\right) \\ M_{32} &= \frac{2EI}{6}\left(2\theta_3 + 2\theta_3\right) \\ M_{34} &= \frac{2EI}{6}\left(2\theta_2 + \frac{3}{4}\Delta\right) \end{aligned}\right\} \qquad (11.38)$$

$$M_{43} = \frac{2EI}{4}\left(\theta_3 + \frac{3}{4}\Delta\right)$$

Step 3: To fix up equilibrium conditions

To evaluate θ_2, θ_3 and Δ we require three equilibrium conditions, we can write two conditions as

$$M_{21} + M_{23} = 0 \tag{11.39}$$

$$M_{32} + M_{34} = 0 \tag{11.40}$$

The third is the shear condition. The horizontal force and the shear in columns must balance making $\Sigma F_H = 0$ on the entire frame. We can write

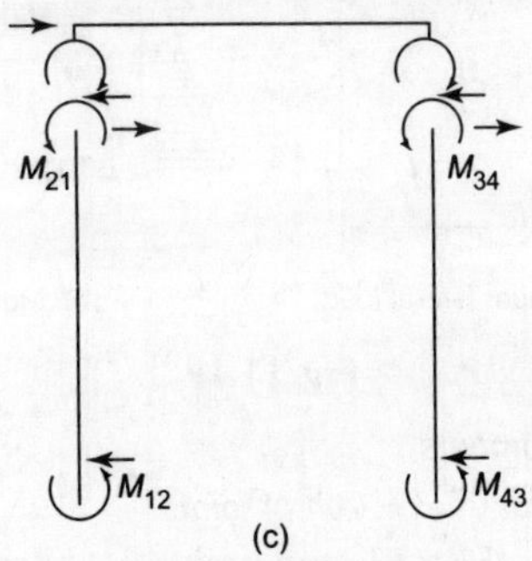

Fig. 11.18 (c)

$$\frac{M_{12} + M21}{4} + \frac{M_{43} + M_{34}}{4} = 10 \tag{11.41}$$

On substitution into equations 11.39, 11.40 and 11.41 we have

$$\frac{5}{3}EI\,\theta_2 + \frac{EI\,\theta_3}{3} + \frac{3}{8}EI\Delta = 0$$

or

$$\frac{5}{3}\theta_2 + \frac{\theta_3}{3} + \frac{3}{8}\Delta = 0 \tag{11.42}$$

$$\frac{4}{3}\theta_2 + \frac{2}{3}\theta_3 + \frac{3}{8}\Delta = 0 \tag{11.43}$$

and

$$\frac{3}{2}\theta_2 + \frac{3}{2}\theta_3 + \frac{3}{2}\Delta = 40 / EI \tag{11.44}$$

Solving the above simultaneous equations

$$\theta_2 = \theta_3 = -\frac{8}{EI} \text{ and } \Delta = \frac{128}{3}$$

Substituting back in the slope deflection equations the end moments in the members are:

$$M_{12} = 12 \text{ kNm}, \qquad M_{21} = 8 \text{ kN.m}$$

$$M_{23} = -8 \text{ kNm}, \qquad M_{32} = -8 \text{ kN.m}$$
$$M_{34} = 8 \text{ kNm}, \qquad M_{43} = 12 \text{ kN.m}$$

The moment diagram is shown in Fig. 11.18*b*.

Example 11.9 | *Determine the end moments of the members of the frame shown in Fig. 11.19. The relative EI values for each member are indicated along the members*

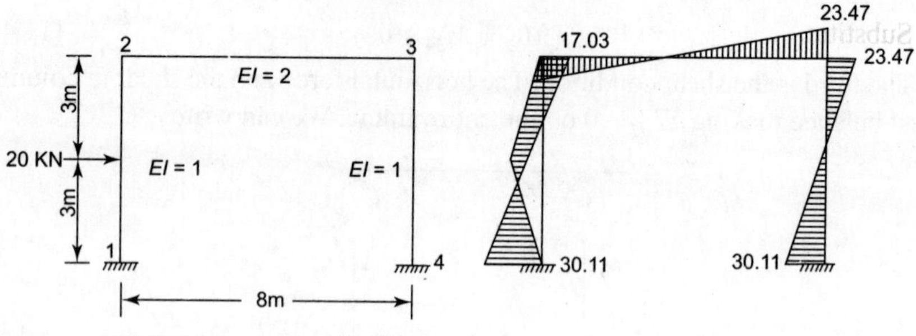

(a) Portal frame under lateral load (b) Moment diagram

Fig. 11.19

Step 1: To write fixed end moments

The fixed end moments are

$$FEM_{12} = \frac{20 \times 6}{8} = 15 \text{ kNm}$$
$$FEM_{21} = -15 \text{ kN.m}$$

Step 2: To identify translation and rotation of joints

Let θ_2, θ_3 and Δ are the displacements- θ_2 and θ_3 angular and Δ translational

We can now write the end moments in terms of displacements as

$$M_{12} = \frac{2EI}{6}(\theta_2 - 3\psi_{12}) + 15 \text{ in which } \psi_{12} = \frac{-\Delta}{6}$$

or

$$\left.\begin{aligned}
M_{12} &= \frac{EI}{3}\left(\theta_2 + 3\frac{\Delta}{6}\right) + 15 \\
M_{21} &= \frac{EI}{3}\left(2\theta_2 + 3\frac{\Delta}{6}\right) - 15 \\
M_{23} &= 2\frac{(2EI)}{8}(2\theta_2 + 2\theta_3) \\
M_{32} &= 2\frac{(2EI)}{8}(\theta_2 + 2\theta_3) \\
M_{34} &= 2\frac{2EI}{6}\left(2\theta_2 + 3\frac{\Delta}{6}\right) \\
M_{43} &= \frac{2EI}{6}\left(\theta_3 + 3\frac{\Delta}{6}\right)
\end{aligned}\right\} \tag{11.45}$$

Step 3: To fix up equilibrium equations

The three unknown displacements are determined by utilizing three equilibrium equations

(i) $M_{21} + M_{23} = 0$, (ii) $M_{32} + M_{34} = 0$

and (iii) $\Sigma F_H = 0$, i.e.$20\ H_1 - H_4 = 0$

or $$20 - \left(\frac{M_{12} + M_{21}}{6}\right) - \left(\frac{M_{34} + M_{43}}{6}\right) = 0$$

Substituting the values for the moments we have

$$\frac{5}{3}\theta_2 + \frac{\theta_3}{2} + \frac{\Delta}{6} = \frac{15}{EI}$$

$$\frac{\theta_2}{2} + \frac{5}{3}\theta_3 + \frac{\Delta}{6} = 0$$

$$\theta_2 + \theta_3 + \frac{2}{3}\Delta = \frac{120}{EI}$$

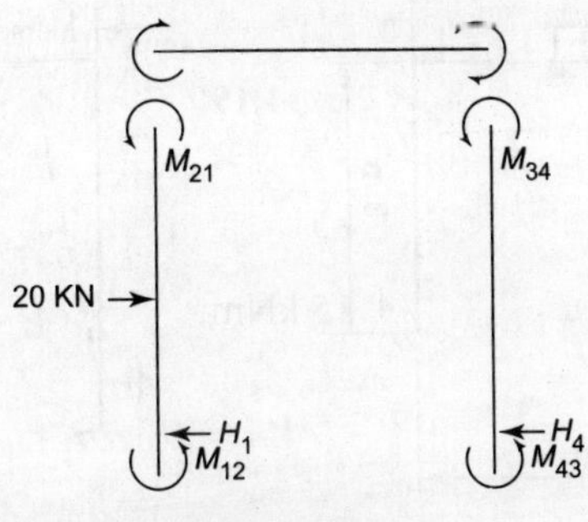

Fig. 11.19 (c)

Solving the above equations simultaneously

We have $EI\theta = -7.07$, $EI\theta_3 = -19.93$ and $EI\Delta = 220.5$

Substituting these values in moment equations

We have

$$M_{12} = -\frac{7.07}{3} + \frac{220.5}{6} + 15 = 49.39 \text{ kN.m}$$

$$M_{21} = -\frac{2}{3}(-7.07) + \frac{220.5}{6} - 15 = 17.03 \text{ kN.m}$$

$$M_{23} = -7.07 - \frac{19.93}{2} = -17.03 \text{ kN.m}$$

$$M_{32} = -\frac{7.07}{2} - 19.93 = -23.47 \text{ kN.m}$$

$$M_{34} = \frac{2}{3}(-19.93) + \frac{220.5}{6} = 23.47 \text{ kN.m}$$

$$M_{43} = -\frac{19.93}{3} + \frac{220.5}{6} = 30.11 \text{ kN.m}$$

Example 11.10 | *Using the slope-deflection method determine the end moments of the members of the frame given in Fig. 11.20a. EI is the same throughout.*

The fixed end moments for the members are written as

$$FEM_{12} = \frac{60(4)}{8} = 30.0 \text{ kN.m}$$

$$FEM_{21} = -30.0 \text{ kN.m}$$

$$FEM_{23} = \frac{40(4)(4)}{12} = 53.33 \text{ kN.m}$$

$$FEM_{32} = -53.33 \text{ kN.m}$$

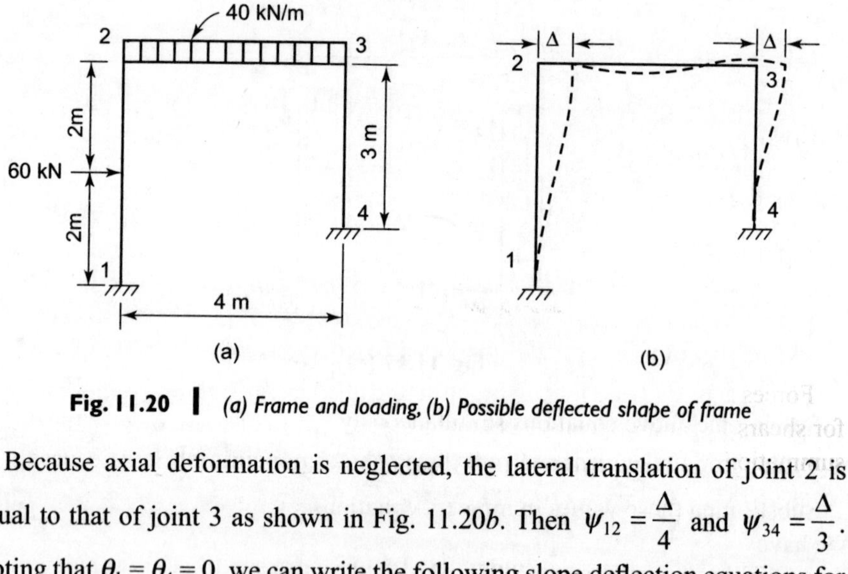

Fig. 11.20 | *(a) Frame and loading, (b) Possible deflected shape of frame*

Because axial deformation is neglected, the lateral translation of joint 2 is equal to that of joint 3 as shown in Fig. 11.20*b*. Then $\psi_{12} = \frac{\Delta}{4}$ and $\psi_{34} = \frac{\Delta}{3}$. Noting that $\theta_1 = \theta_4 = 0$, we can write the following slope deflection equations for end moments. Assigning $EK_{12} = EK_{23} = 3$ and $EK_{34} = 4$, we get

$$\left.\begin{aligned} M_{12} &= 30 + 6(\theta_2 - 3\psi_{12}) \\ M_{21} &= -30 + 6(2\theta_2 - 3\psi_{12}) \\ M_{23} &= 53.33 + 6(2\theta_2 + \theta_3 + 0) \\ M_{32} &= -53.33 + 6(\theta_2 + 2\theta_3 + 0) \\ M_{34} &= 0 + 8(2\theta_3 + 0 - 3\psi_{34}) \\ M_{43} &= 0 + 8(\theta_3 + 0 - 3\psi_{34}) \end{aligned}\right\} \quad (11.46)$$

It can be seen that all the above moments are expressed in terms of θ_2, θ_3 and $\psi_{12} = \frac{3}{4}\psi_{34}$ and, therefore, three conditional equations are required to evaluate these three unknowns. The first two conditions that can be written are

$$M_{21} + M_{23} = 0 \tag{11.47}$$

and
$$M_{32} + M_{34} = 0 \tag{11.48}$$

The third conditional equation can be obtained by considering that the summation of the forces in the horizontal direction on the entire structure is $\Sigma F_H = 0$.

The forces involved in this equilibrium equation are the 60 kN external horizontal force and the shears in the columns at the bases. Referring to Fig. 11.21, we can write

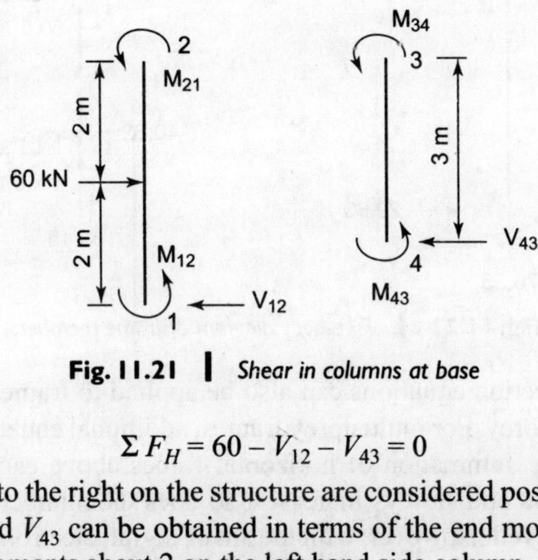

Fig. 11.21 | *Shear in columns at base*

$$\Sigma F_H = 60 - V_{12} - V_{43} = 0 \tag{11.49}$$

Forces acting to the right on the structure are considered positive. Expressions for shears V_{12} and V_{43} can be obtained in terms of the end moments. Taking the summation of moments about 2 on the left hand side column

$$M_{12} + M_{21} + 60(2) - V_{12}(4) = 0$$

or
$$V_{12} = \frac{M_{12} + M_{21}}{4} + 30 \tag{11.50}$$

Similarly, the summation of moments about 3 on the right hand side column gives

$$V_{43} = \frac{M_{34} + M_{43}}{3} \tag{11.51}$$

Simultaneously solving Eqs. 11.47, 11.48 and 11.49 after substituting for moments from Eq. 11.46, we get

$$\theta_2 = -2.005,\ \theta_3 = 1.237 \text{ and } \psi_{12} = -0.96$$

Substituting these values in Eq. 11.32, we get

$$M_{12} = 30 + 6\,(-2.0) - 6(3)\,(-0.96) = 35.28 \text{ kN.m}$$

$$M_{21} = -30 + 12\,(-2.0) - 6(3)(-0.96) = -36.70 \text{ kN.m}$$
$$M_{23} = 53.33 + 12\,(-2.0) + 6(1.237) = +\,36.70 \text{ kN.m}$$
$$M_{32} = -53.33 + 6(-2.0) + 12(1.237) = -50.50 \text{ kN.m}$$
$$M_{34} = 0 + 16(1.237) - 24(4/3)(-0.96) = 50.50 \text{ kN.m}$$
$$M_{43} = 0 + 8(1.237) - 24\,(4/3)\,(-\,0.96) = 40.62. \text{ kN.m.}$$

The free-body diagrams of the members of the frame are shown in Fig. 11.22. As a check we notice that the sum of shear in columns and the external horizontal load satisfy the condition $\Sigma F_H = 0$.

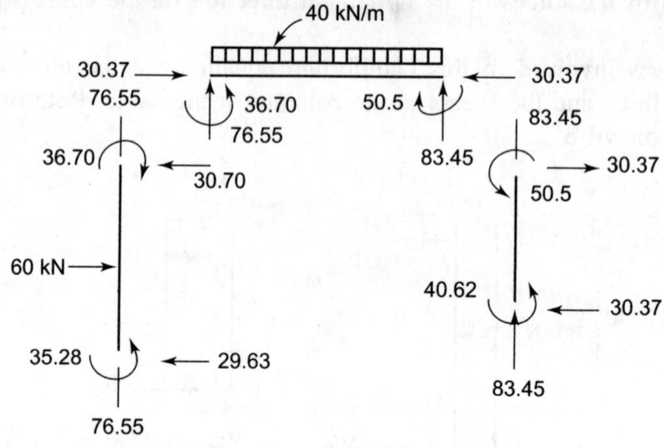

Fig. 11.22 | *Free-body diagram of frame members*

The slope-deflection equations can also be applied to frames with more than one bay and one storey. For multistorey frames, additional equations are obtained by considering the summation of horizontal forces above each floor level. As the number of bays and storeys increases, so does the number of simultaneous equations to be solved. However, with the aid of digital electronic computers, the solving of the large number of simultaneous equations poses no problem.

Another type of frame sometimes encountered in practice is the one with columns that are inclined instead of being vertical. The analysis of such a frame is illustrated in the example using slope-deflection equations.

Example 11.11 | *The frame of Fig. 11.23a is to be analysed by the slope-deflection method. EI is same for all members.*

The given frame has three degrees of freedom. Two rotations, one at each of joints 2 and 3, and one lateral translation of the frame. The rotations at 2 and 3 are denoted as usual by θ_2 and θ_3 and the lateral horizontal displacement is denoted by Δ. The general displacement of the frame is shown in Fig. 11.23*b* .

Referring to Fig. 11.23*b* we can write

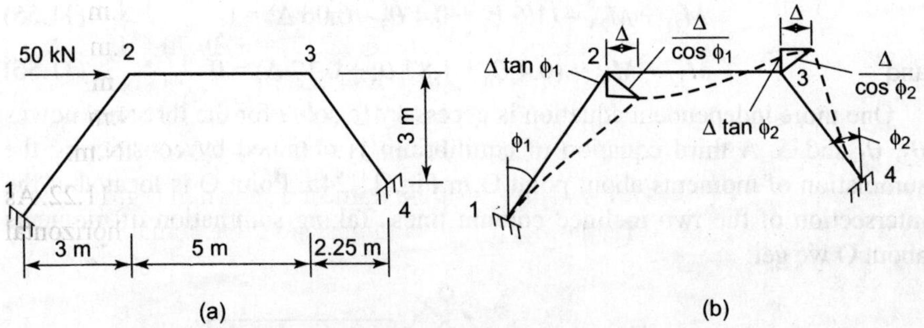

Fig. 11.23 | *(a) Frame and loading, (b) Deflected shape of frame*

$$\left.\begin{aligned}\psi_{12} &= \frac{-\Delta}{L_{12}\cos\Phi_1}\\ \psi_{43} &= \frac{-\Delta}{L_{43}\cos\Phi_2}\\ \psi_{23} &= \frac{\Delta}{L_{23}}(\tan\Phi_1 + \tan\Phi_2)\end{aligned}\right\} \quad (11.52)$$

For the member length and slopes of columns in the frame, we obtain

$$\left.\begin{aligned}\psi_{12} &= \frac{-\Delta}{(5)(0.8)} = -0.25\Delta\\ \psi_{43} &= \frac{-\Delta}{(3.75)(0.8)} = -0.3333\Delta\\ \text{and} \quad \psi_{23} &= \frac{\Delta}{5}(0.75 + 0.75) = 0.30\Delta\end{aligned}\right\} \quad (11.53)$$

In this example the fixed end moments for all the members of the frame are zero. Noting $\theta_1 = \theta_4 = 0$. we can write the following slope-deflection equations,

$$M_{12} = 2EK_{12}\,(0 + \theta_2 - 3\psi_{12})$$

Taking $EK_{12} = \frac{1}{5}$, we get

$$\left.\begin{aligned}M_{12} &= (0.4\theta_2 + 0.3\Delta)\\ M_{21} &= (0.8\theta_2 + 0.3\Delta)\\ M_{23} &= (0.8\theta_2 + 0.4\theta_3 - 0.36\Delta)\\ M_{32} &= (0.4\theta_2 + 0.8\theta_3 - 0.36\Delta)\\ M_{34} &= (1.07\theta_3 + 0.53\Delta)\\ M_{43} &= (0.53\theta_3 + 0.53\Delta)\end{aligned}\right\} \quad (11.54)$$

Then taking the summation of the moments at joints 2 and 3, we obtain

$$M_{21} + M_{23} = (1.6\,\theta_2 + 0.4\,\theta_3 - 0.06\,\Delta) = 0 \qquad (11.55)$$

and
$$M_{32} + M_{34} = (0.4\,\theta_2 + 1.87\,\theta_3 + 0.17\,\Delta) = 0 \qquad (11.56)$$

One more independent equation is necessary to solve for the three unknowns θ_2, θ_3 and Δ. A third equation of equilibrium is obtained by considering the summation of moments about point O in Fig. 11.24*a*. Point O is located at the intersection of the two inclined column lines. Taking summation of moments about O we get:

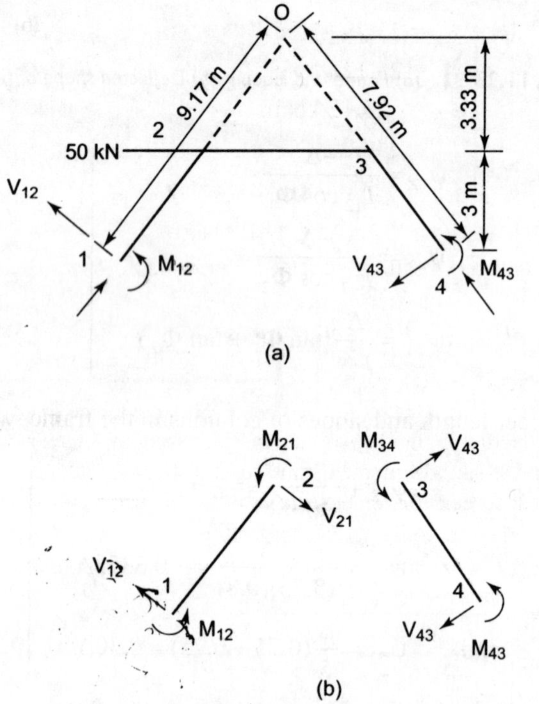

Fig 11.24 | *(a) Free-body diagram of the structure, (b) Moment and shear in columns*

$$\Sigma M_o = M_{12} + M_{43} - V_{12}\,(9.17) - V_{43}(7.92) + 50(3.33) = 0 \qquad (11.57)$$

The shears V_{12} and V_{43} can be expressed in terms of moments at the ends of members 1-2 and 3-4 respectively. Referring to Fig. 11.24*b*

$$V_{12} = \frac{M_{12} + M_{21}}{L_{12}} \qquad (11.58)$$

$$V_{43} = \frac{M_{34} + M_{43}}{L_{34}} \qquad (11.59)$$

After expressing the moments as in Eq. 11.54 and substituting in Eq. 11.57, and on simplification, we get

$$(1.8\,\theta_2 + 2.84\,\theta_3 + 2.52\,\Delta) = 166.67 \qquad (11.60)$$

We have thus three unknowns θ_2, θ_3 and Δ, and three conditional equations given by Eqs. 11.55, 11.56 and 11.60. On solving these three equations simultaneously, we get

$$\theta_2 = 4.59$$
$$\theta_3 = -7.62$$
and
$$\Delta = 71.51$$

Substituting these values in Eq. 11.54, we get the end moments as

$$M_{12} = 23.29 \text{ kN.m}$$
$$M_{21} = 25.12 \text{ kN.m}$$
$$M_{23} = -25.12 \text{ kN.m}$$
$$M_{32} = -30.00 \text{ kN.m}$$
$$M_{34} = +30.00 \text{ kN.m}$$
$$M_{43} = +34.07 \text{ kN.m}$$

The values of the remaining reaction components can be obtained from the free-body diagrams of the members.

Problems for Practice

Use the slope-deflection method in solving the following problems.

11.1 Determine the support moments and reactions for the beam shown in Fig. 11.25. Construct the shear force and moment diagrams for the beam. *EI* is constant.

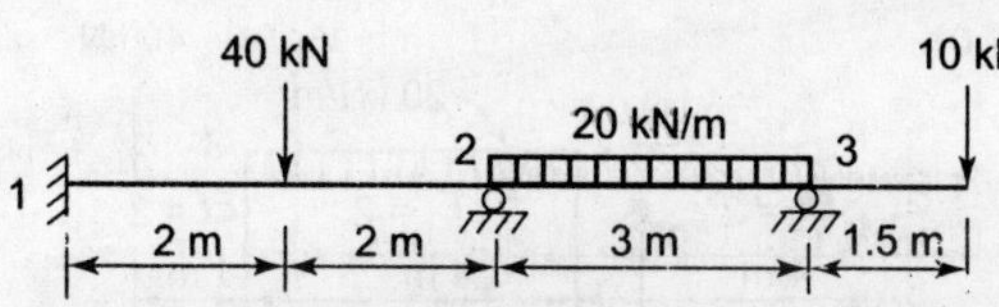

Fig. 11.25

11.2 Determine the support moments of the continuous beam shown in Fig. 11.26. *EI* is constant.

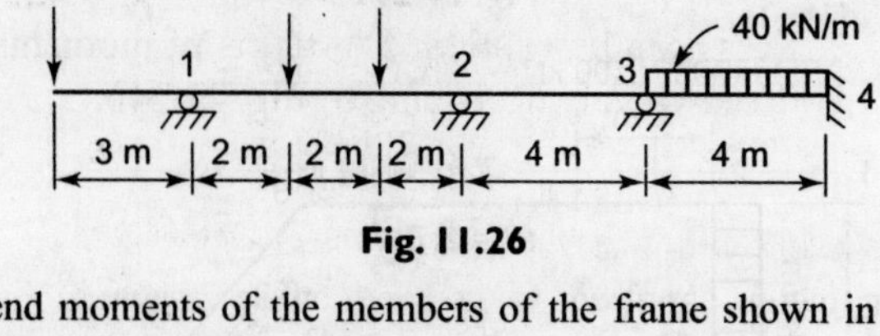

Fig. 11.26

11.3 Find.the end moments of the members of the frame shown in Fig. 11.27. El is constant.

11.4 Determine the end moments of the members of the framed structure shown in Fig. 11.28 Sketch the deflected shape and draw the moment diagram on the tension face of the members.

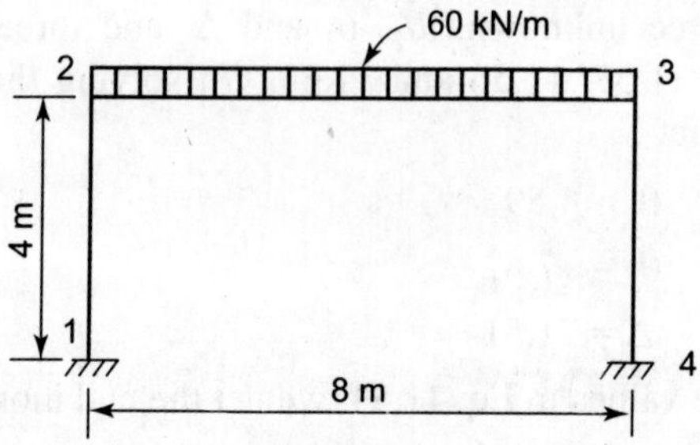

Fig. 11.27

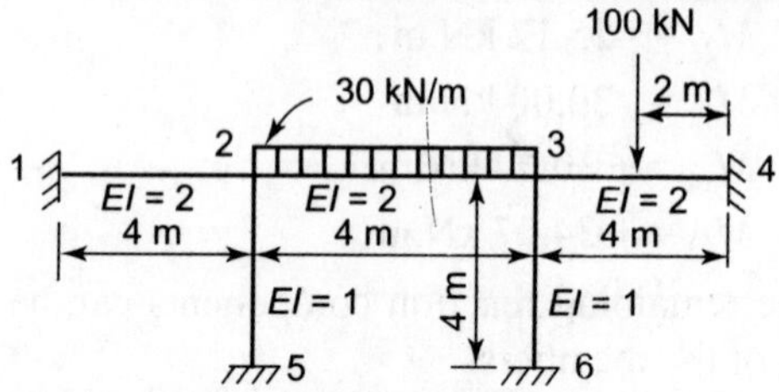

Fig. 11.28

11.5 Determine the end moments in all the members of the framed structure shown in Fig. 11.29. *EI* value for each member is indicated along the members.

11.6 Determine the end moments in all the members of the frame shown in Fig. 11.30. *EI* values are indicated along the members.

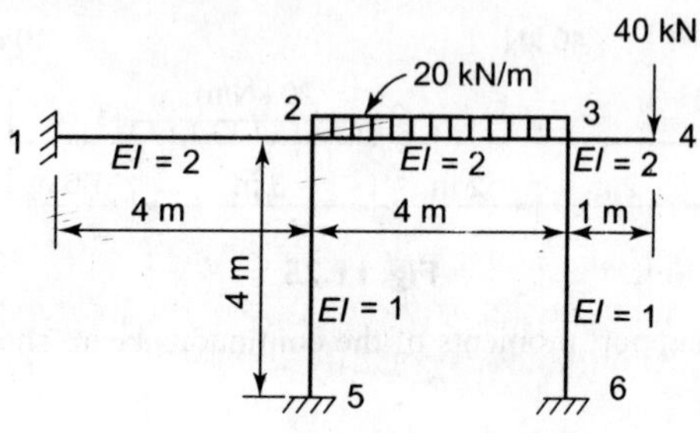

Fig. 11.29

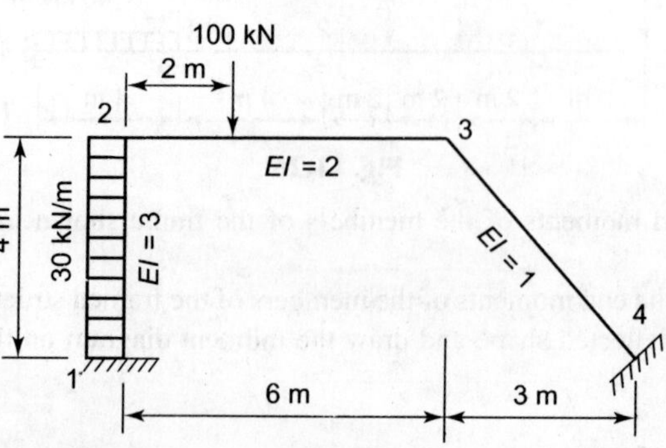

Fig. 11.30

12 Moment Distribution Method

12.1 INTRODUCTION

The moment distribution method, also known as the Hardy Cross method, provides a convenient means of analysing statically indeterminate beams and frames by simple hand calculations. This is basically an iterative process. The procedure, in general, involves artificially restraining temporarily all the joints against rotation and writing down the fixed end moments for all the members. The joints are then released one by one in succession. At each released joint the unbalanced moments are distributed to all the ends of the members meeting at that joint. A certain fraction of these distributed moments are carried over to the far end of members. The released joint is again restrained temporarily before proceeding to the next joint. The same set of operations are carried out at each joint till all the joints are completed. This completes one cycle of operations. The process is iterated for a number of cycles till the values obtained are within the desired accuracy.

The moment distribution method is also a displacement method of analysis. However, this method does not involve solving any equations. This method is highly popular among engineers as the calculations involved are minimum and are free from solving simultaneous equations if the frames do not undergo lateral translations.

Sign Convention The sign convention followed in the development of this method is the same as the one followed for the slope-deflection method. Reference may be made to Sec. 11.2.

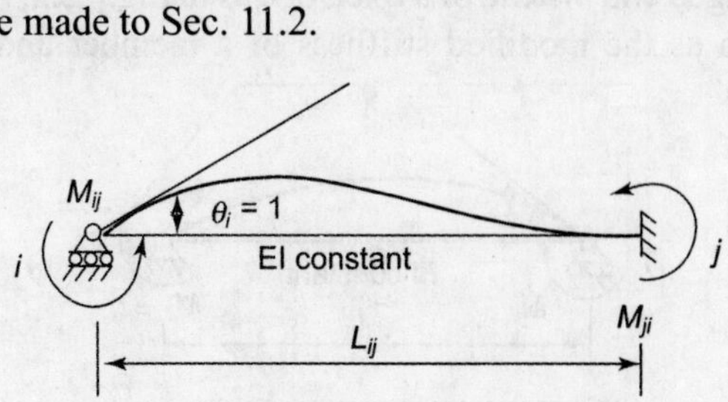

Fig. 12.1 | *Absolute stiffness of a member when the far end is fixed*

Before developing the moment distribution method of analysis, it is necessary to define certain terms employed in this method. They are presented and discussed below.

12.1.1 Absolute and Relative Stiffness of Members

The absolute stiffness of a member can be defined as the moment required to produce a unit rotation at the simply supported end, *i*, while the farther end, *j*, is fully restrained. For the beam in Fig. 12.1, the moment M_{ij} thus represents the absolute stiffness of member *i-j*. From the Appendix table, the moment

$$M_{ij} = \frac{4EI}{L} \tag{12.1}$$

considering θ_i as unity.

The moment at the farther fixed end is equal to

$$M_{ji} = \frac{2EI}{L} \tag{12.2}$$

The ratio *I*/*L* in Eqs. 12.1 and 12.2 is referred to as the relative stiffness and is denoted by letter K. For prismatic member *i-j*, the relative stiffness

$$\frac{I_{ij}}{L_{ij}} = K_{ij} = K_{ji}.$$

12.1.2 Carry Over Factor (C.O.F.)

If a moment M_{ij} is applied at end *i* of the member in Fig. 12.1, a specified amount of moment, M_{ji}, is generated at the farther restrained end. The *carry over factor* (C.O.F.) is defined as the factor by which the moment at simply supported end *i*, M_{ij}, is multiplied to get the moment carried over to the other end, that is, M_{ji} or

$$M_{ji} = C_{ij} M_{ij} \tag{12.3}$$

C_{ij} is the carry over factor. From Eqs. 12.1 and 12.2, we can write

$$M_{ji} = \left(\frac{1}{2}\right) M_{ij} \tag{12.4}$$

Thus, for a prismatic member the carry over factor is always $\left(+\frac{1}{2}\right)$. If the farther end is a hinged end instead of a fixed one as in Fig. 12.2, the corresponding stiffness is known as the modified stiffness of a member and is equal to (see Appendix D)

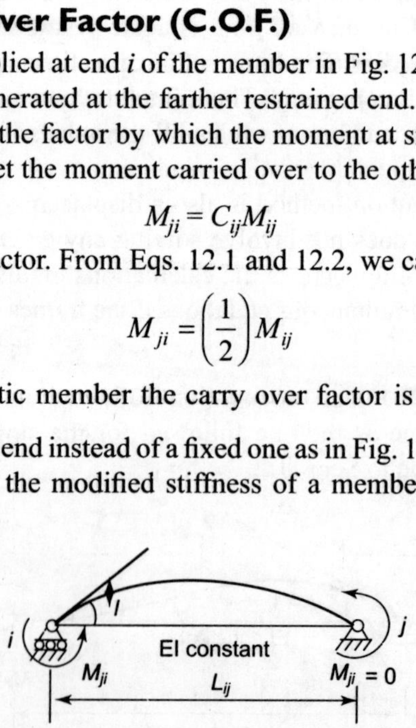

Fig. 12.2 | *Absolute stiffness of a member when far end is hinged*

$$M_{ij} = \frac{3EI}{L} \tag{12.5}$$

and the relative stiffness

$$K' = \frac{3}{4}K \tag{12.6}$$

Obviously, the moment carried over to the farther hinged end, $M_{ji} = 0$.

12.1.3 Distribution Factor (D.F.)

Consider a joint in a structure where two or more members meet. If an external moment M is applied to such a joint, the joint undergoes a rotation θ as shown in Fig. 12.3*a*. Since all the members meeting at this joint undergo the same rotation θ, the applied moment M is distributed to each of the ends of the members according to their relative stiffness values. The factor by which the applied moment is multiplied to obtain the end moment of any member is known as the *distribution factor* (D.F.).

Consider the free-body diagram of joint 1 in Fig. 12.3*b*. For equilibrium of joint 1, we have

$$M_{12} + M_{13} + M_{14} + M_{15} - M = 0 \tag{12.7}$$

In writing the equilibrium Eq. 12.7, clockwise moments on a joint are considered positive. The end moments of members can be written in terms of the angle of rotation θ as

$$M_{12} = 4EK_{12}\theta,\ M_{13} = 4\ EK_{13}\theta,\ M_{14} = 4EK_{14}\theta$$

and $$M_{15} = 4EK_{15}\theta \tag{12.8}$$

Substituting these values in Eq. 12.7, we get

$$M = 4E\theta\,(K_{12} + K_{13} + K_{14} + K_{15})$$

or $$M = 4E\theta\Sigma K \tag{12.9}$$

or $$\theta = \frac{M}{4E\Sigma K} \tag{12.10}$$

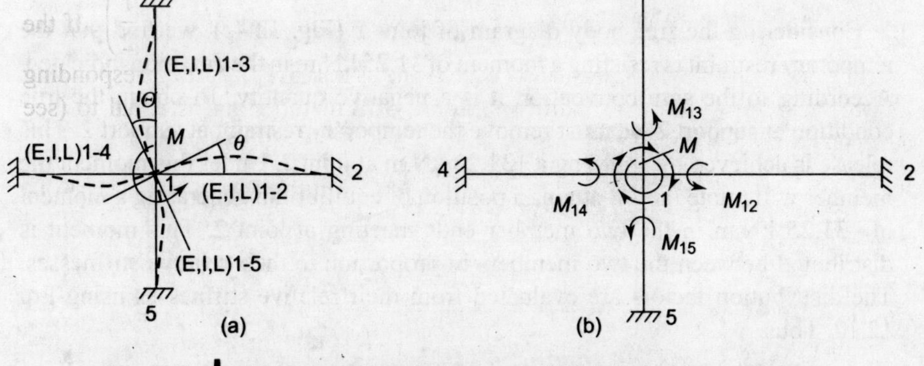

Fig. 12.3 | *(a) Moment M applied at joint 1, (b) Free-body diagram of joint 1*

where ΣK denotes the sum of the relative stiffnesses of all the members meeting at joint 1.

Substituting the value for θ from Eq. 12.10 in the first of Eq. 12.8, we get

$$M_{12} = 4EK_{12}\frac{M}{4E\Sigma K} = \frac{K_{12}}{\Sigma K}M \tag{12.11}$$

Ratio $\dfrac{K_{12}}{\Sigma K}$ indicates the fraction by which the applied moment M is to be multiplied to get the moment resisted by member 1–2. This ratio by definition is the distribution factor. The distribution factor for any member *i-j* is defined in general as

$$r_{ij} = \frac{K_{ij}}{\Sigma K} \tag{12.12}$$

The distribution factor for a member is thus equal to the stiffness (or relative stiffness) of the member divided by the sum of stiffnesses (or relative stiffnesses) of all the members meeting at the joint.

12.2 DEVELOPMENT OF METHOD

The basic idea underlying the moment distribution method may be illustrated by considering the analysis of a continuous beam shown in Fig. 12.4*a*. The beam is fixed at its ends but is free to rotate over support 2. Now, we begin the analysis by temporarily restraining the beam against rotation over support as shown in Fig. 12.4*b*. The fixed end moments in each span caused by the transverse loads are evaluated using the Appendix Table, and following the sign convention adopted, we have

$$FEM_{12} = \frac{100 \times 5}{8} = 62.50 \text{ kN.m}$$

$$FEM_{21} = -62.50 \text{ kN.m}$$

$$FEM_{23} = \frac{20(7.5)^2}{12} = 97.75 \text{ kN.m}$$

$$FEM_{32} = -93.75 \text{ kN.m}$$

Considering the free-body diagram of joint 2 (Fig. 12.4*c*), we find that the temporary restraint is resisting a moment of 31.25 kN.m in the direction indicated. According to the sign convention it is a negative quantity. To obtain the true condition at support 2 we must remove the temporary restraint at support 2. This release is achieved by applying a +31.25 kN.m at joint 2. Under this moment the member will rotate until it attains a position of equilibrium generating a moment of –31.25 kN.m in the two member ends meeting at joint 2. This moment is distributed between the two members in proportion to their relative stiffnesses. The distribution factors are evaluated from their relative stiffnesses using Eq. 12.12. Thus

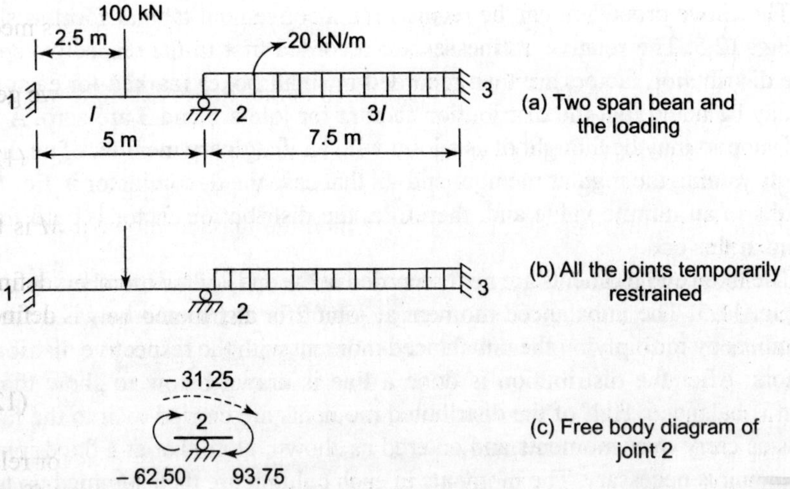

Fig. 12.4 *(a) Two span beam and the loading, (b) All the joints temporarily restrained, (c) Free body diagram of joint 2*

$$r_{21} = \frac{I/5}{(I/5 + 3I/7.5)} = \frac{1}{3}$$

And
$$r_{23} = \frac{3I/7.5}{(I/5 + 3I/7.5)} = \frac{2}{3}$$

Therefore, the moment developed at the end of each member is

$$M_{21} = \frac{1}{3}(-31.25) = -10.42 \text{ kN.m}$$

$$M_{23} = \frac{2}{3}(-31.25) = -20.83 \text{ kN.m}$$

Half of these moments are carried to their farther ends as carry over moments, that is,

$$M_{12} = (1/2)(-10.42) = -5.21 \text{ kN.m}$$

$$M_{32} = (1/2)(-20.83) = -10.42 \text{ kN.m}$$

The beam is now in its true position under the given loading. The true moment at each end of the member is obtained by adding algebraically the fixed end moments and the moments caused by the release of joint 2.

Thus, the true moments are

$$M_{12} = 62.50 - 5.21 = 57.29 \text{ kN.m}$$

$$M_{21} = -62.50 - 10.42 = -72.92 \text{ kN.m}$$

$$M_{23} = +93.75 - 20.83 = 72.92 \text{ kN.m}$$

and
$$M_{32} = -93.75 - 10.42 = -104.17 \text{ kN.m}$$

The above procedure can be recorded in a convenient tabular form as shown in Fig. 12.5. The relative stiffnesses are recorded first in the respective spaces. The distribution factors are then recorded in small boxes marked for each joint. It may be noted that the distribution factors for joints 1 and 3 are zero. A fixed end support may be thought of as a joint with an imaginary member of negligible length joining the regular member end. In that case the denominator in Eq. 12.12 results in an infinite value and, therefore, the distribution factor is zero for the beam at this end.

The fixed end moments are next recorded at the ends of the members as shown in Fig. 12.5. The unbalanced moment at joint 2 is distributed between the two members by multiplying the unbalanced moment with the respective distribution factors. After the distribution is done a line is drawn below to show that the joint is balanced. Half of the distributed moments are carried over to the farther ends as carry over moments and entered as shown. Note that at a fixed joint no balancing is necessary. The moments in each column are then summed up to get the final or true moments.

As an extension of the procedure consider now a variation in the support condition at the right end of the beam as shown in Fig. 12.6*a*. This is a simply supported end. It is now possible for both joint 2 and joint 3 to rotate.

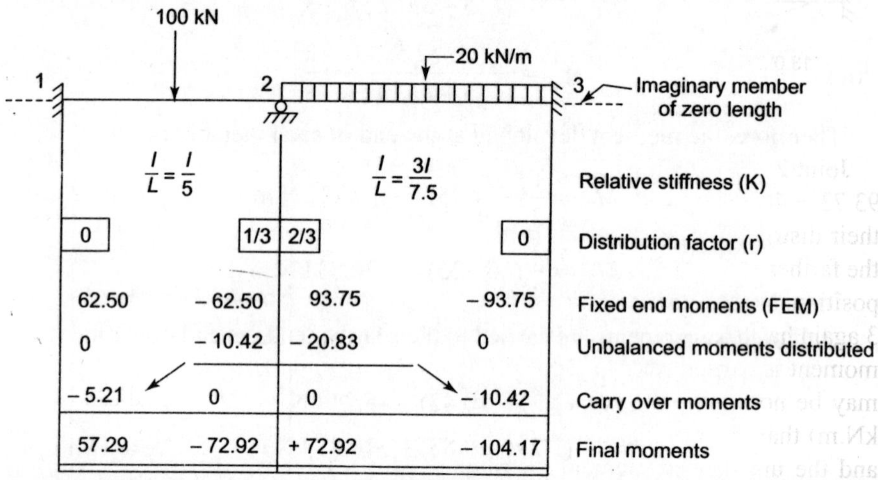

Fig. 12.5

The analysis is started by considering that both joints 2 and 3 are restrained temporarily as shown in Fig. 12.6*b*. The fixed end moments are the same as in the previous example and are shown entered in their respective ends. The two temporarily restrained joints are to be released one by one in turn. In releasing joint 3, we first notice that the joint is actually a simply supported end and the final moment must be equal to zero. Therefore, the unbalanced moment at joint 3 is balanced as shown in Fig. 12.6*c*. Half of the balanced moment is carried to joint 2 as the carry over moment. Joint 3 is again restrained temporarily in the rotated position as shown in Fig. 12.6*d*.

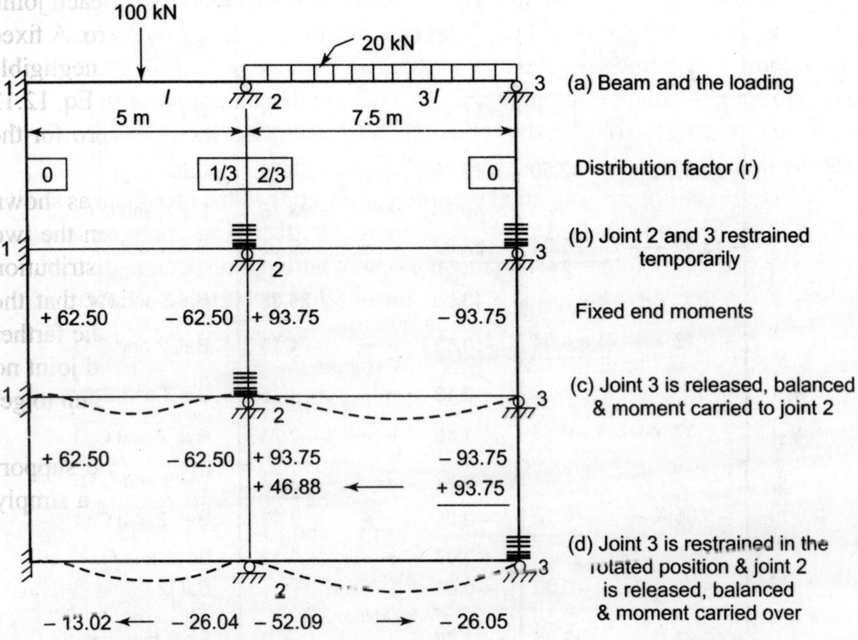

Fig. 12.6

Joint 2 is then released and the unbalanced moment –78.13 kN.m (–62.50 + 93.73 + 46.88) is distributed to the two members, 2-1 and 2-3 in proportion to their distribution factors. Half of these distributed moments are carried over to the farther ends. At this stage joint 2 is again temporarily restrained in the rotated position. No balancing is necessary at joint 1 which is a fixed end. However, joint 3 again has to be released by balancing the moment to zero. Half of the balanced moment is carried over to joint 2. The joint 2 again is unbalanced. However, it may be noted that the unbalanced moment at joint 2 is much smaller (–13.03 kN.m) than the original unbalanced moment (–78.13 kN.m). Joint 2 is released and the unbalanced moment distributed. This set of operations is repeated in a cyclic order until the unbalanced moments are within the desired degree of accuracy.

The whole procedure can be condensed and performed in a tabular form where the values are recorded in a compact form. The procedure involved is shown in the table of Fig. 12.7. The iteration is stopped when the unbalanced moments become very small and can be neglected. The final step before summing the moments is the balancing and distribution of the moments at the joints. Thus, the final moments are in equilibrium.

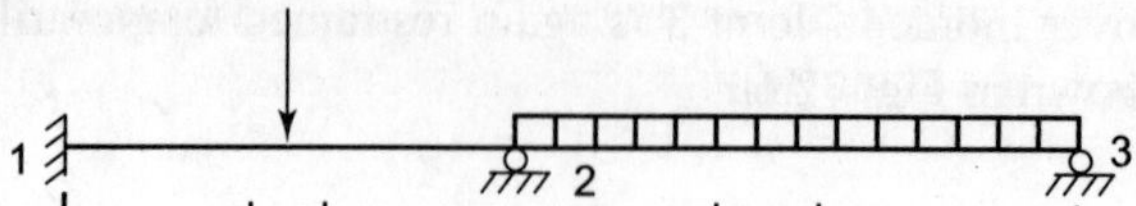

$\frac{I}{L}=\frac{I}{5}$		$\frac{I}{L}=\frac{I}{2.5}$		Rel. stiff
0	1/3	2/3	0	D.F
62.50	− 62.50	93.75	− 93.75	FEM
		46.88 ←	+ 93.75	Bal.3 and C.O
− 13.02 ←	− 26.04	− 52.09 →	− 26.05	Bal. 2 and C.O
		+ 13.03 ←	+ 26.05	Bal.3 and C.O
− 2.17 ←	− 4.34	− 8.69 →	− 4.35	Bal. 2 and C.O
		+ 2.18 ←	+ 4.35	Bal.3 and C.O
− 0.37 ←	− 0.73	− 1.45 →	− 0.73	Bal. 2 and C.O
		+ 0.37 ←	+ 0.73	Bal.3 and C.O
− 0.06 ←	− 0.12	− 0.25 →	− 0.13	Bal. 2 and C.O
		+ 0.07 ←	+ 0.13	Bal.3 and C.O
	− 0.02	− 0.05		Bal. 2
+ 46.87	− 93.75	+ 93.75	0	Final moments

Fig. 12.7

The convergence in the above solution has been rather slow owing to the fact that joint 3 is a hinged end and continuously throws back sizeable carry over moments to joint 2.

In a structure where one end is simply supported, such as the right hand support of the beam in Fig. 12.7, a considerable amount of work can be saved by using the modified or reduced stiffness factor for right span 2-3. According to Eq. 12.6 the modified stiffness factor is three-fourth of the stiffness factor of a beam whose farther end is fixed.

We shall revise the distribution factors using the modified stiffness factor for span 2-3. The distribution factors are worked out as shown in Fig. 12.8. The joints are temporarily restrained as earlier and the fixed end moments are written. Again, as a first step, we release joint 3 and allow the member to rotate and develop a moment of + 93.75 kN.m and carry over half of it to support 2. At this point we leave joint 3 free of any restraint so that it can rotate freely and hence develop no moment. Next, we move on to joint 2. The joint is released so that the unbalanced moment (–78.13 kN.m) is distributed to the two members meeting at that joint as shown in Fig. 12.8. Then, there is the usual carry over to joint 1 and no carry over moment to joint 3 since the modified stiffness for span 2-3 is used considering that the farther end is hinged. Thus a considerable amount of computation is avoided. The summed up moments are seen to be the same as in the previous method.

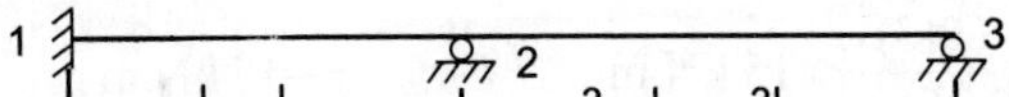

1	2	2	3	
$\frac{I}{L}=\frac{I}{5}$		$\frac{3}{4}\frac{I}{2.5}=\frac{3I}{10}$		Rel. stiff.
0	0.4	0.6	0	D.F
+ 62.50	− 62.50	+ 93.75	+ 93.75	FEM
		+ 46.88 ←	+ 93.75	Bal. 3 and C.O
− 15.63 ←	−31.25	− 46.88		Bal. 2 and C.O
+ 46.87	−93.75	+ 93.75	0	Final moments

Fig. 12.8

Structures with overhanging members can be solved by replacing the overhanging span with an equivalent applied moment at the adjacent support point. The procedure is then similar to the case of a simply supported end but with a known moment at that end. We shall make this point clear by working out the following example.

Example 12.1 | *It is required to determine the support moments for the continuous beam shown in Fig. 12.9 by the moment distribution method. EI is the same throughout.*

First, we can replace the overhanging span at the right end with an equivalent moment of +10.0 kN.m. The moment is positive according to our sign convention because the 10 kN force tends to rotate the joint in a clockwise direction. This moment is entered along with the fixed end moments in Fig. 12.9.

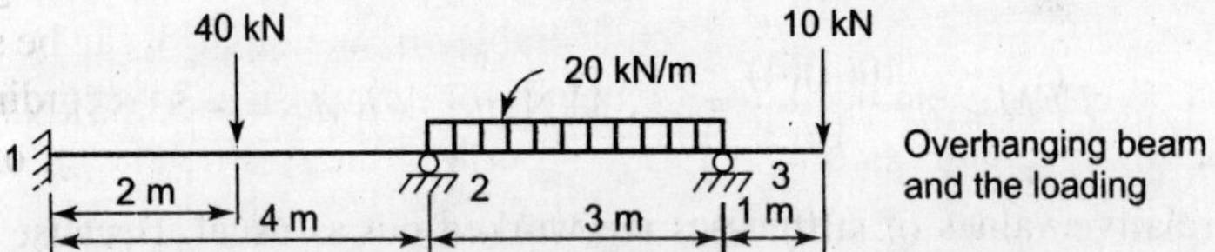

1	2	2	3	3	
$\frac{I}{L}=\frac{I}{4}$		$\frac{3}{4}\frac{I}{L}\frac{I}{4}$			Rel. stiff
0	0.5	0.5	0		D.F
+ 20.00	−20.00	+ 15.00	− 15.00	+ 10.00	FEM.
		+ 2.50 —	+ 5.00		Bal. 3 and C.O
+ 0.63 ←	+ 1.25	+ 1.25			Ba. 2, distr. C.O
+ 20.63	−18.75	+18.75	− 10.00	+ 10.00	Final moments

Fig. 12.9

The fixed end moments are

$$FEM_{12} = \frac{4(40)}{8} = 20 \text{ kN.m} \quad FEM_{21} = -20 \text{ kN.m}$$

$$FEM_{23} = \frac{60(3)^2}{12} = 15 \text{ kN.m} \quad FEM_{32} = -15 \text{ kN.m}$$

Joint 2 is released so that the final moment is kept at 10.0 kN.m and the joint is then treated similar to a simply supported end. Note that the modified stiffness factor 3/4 is used for span 2-3. Half of the balanced moment at 3 is carried over to joint 2. Joint 2 is released and the unbalanced moment is distributed. Half of the distributed moment is carried over to fixed end –1 as usual. The final moments are obtained by summing up the moments in each column entry. As a check, we see that the moments at support 2 sum up to zero.

Consider another example of a continuous beam of three spans with one end fixed and the other with a overhang.

Example 12.2 | *It is required to determine the support moments for the continuous beam of Fig. 12.10. El is the same throughout.*

As in the previous example, the overhanging end can be replaced by applying a concentrated moment –150.00 kN.m at the left of support 1. This is entered in the same row as the fixed end moments in other spans in Fig. 12.10. The fixed end moments are written down using the Appendix table.

$$FEM_{12} = \frac{80(2)(4)^2}{(6)^2} + \frac{80(4)(2)^2}{(6)^2} = 106.66 \text{ kN.m}$$

$$FEM_{21} = -106.66 \text{ kN.m}$$

$$FEM_{23} = FEM_{32} = 0$$

$$FEM_{34} = \frac{40(4)(4)}{12} = 53.33 \text{ kN.m}; \quad FEM_{43} = -53.33 \text{kN.m}$$

The relative values of stiffnesses are worked out as usual. Because support 1 can be considered as simply supported with a definite moment, the reduced stiffness value is taken for span 1-2.

As shown in Fig. 12.10, the first step in moment distribution is to balance the simply supported end at 1. An equivalent moment of the overhanging span is included in balancing the joint. After carrying over half the balancing moment to joint 2, the moment distribution analysis follows the usual procedure to completion. The entire analysis is shown entered in the moment distribution table of Fig. 12.10.

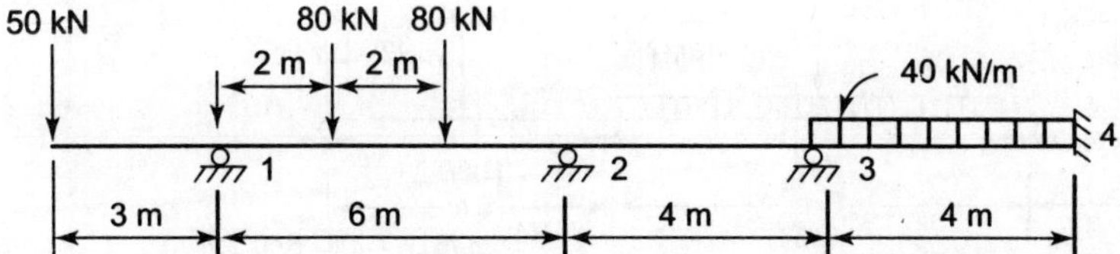

							Beam and the loading
	$\frac{3}{4}\frac{I}{6}=\frac{I}{8}$		$\frac{I}{4}$		$\frac{I}{4}$		Rel. stiff
	1.0	1/3	2/3	1/2	1/2	0	D.F
– 150.00	+ 106.66	– 106.66	0.0	0.0	+ 53.33	– 53.33	FEM
	+ 43.34	+ 21.67					Bal. 1 and C.O
0	0	+ 28.33	+ 56.66	– 26.67	– 26.66		Dist.
			13.34	+ 28.33		– 13.33	C.O
		+ 4.45	+ 8.89	+ 14.77	– 14.16		Dist.
			– 7.09	+ 4.45		– 7.08	C.O
		+ 2.36	+ 4.73	– 2.23	– 2.22		Dist.
			– 1.12	+ 2.37		– 1.11	C.O
		+ 0.37	+ 0.75	– 1.19	– 1.18		Dist.
			– 0.60	+ 0.38		– 0.59	C.O
		+ 0.20	+ 0.40	– 0.19	– 0.19		Dist.
			– 0.10	+ 0.20		– 0.10	C.O
		– 0.03	– 0.07	– 0.1	– 0.1		Dist.
– 150.00	+ 150.00	– 49.25	+ 49.25	– 8.82	+ 8.82	– 75.54	Final moments

Fig. 12.10

In the discussion of moment distribution above, none of the joints was considered to have translated in a direction transverse to the axis of the member. Where such transverse joint translations are possible, for example, as in the case of settlement of supports or elastic supports, they can be taken into account while writing the fixed end moments. We shall illustrate the procedure by solving the following examples.

Example 12.3 | *Determine the support moments for the continuous beam shown in Fig. 12.11. E is constant and I values are as indicated on the beam.*

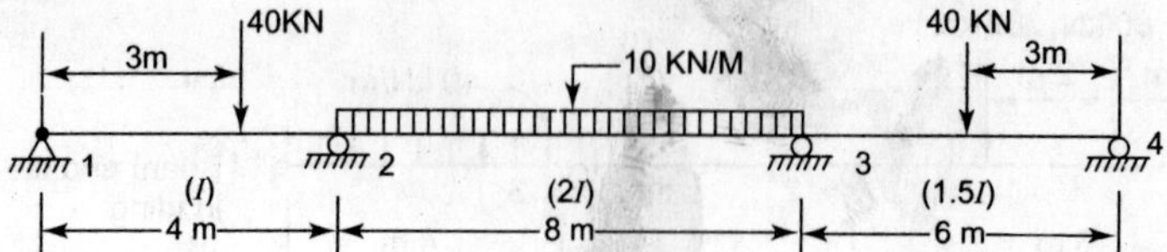

	1	2	2	3	3	4	
	$I/4$		$2I/8$		$1.5I/6$		Rel, Stiff
	$\frac{3}{4}\, I/4$		$I/4$		$\frac{3}{4}\left(\frac{1.5I}{6}\right)$		Modified Rel Stiff.
	0	$\frac{3}{7}$	$\frac{4}{7}$	$\frac{4}{7}$	$\frac{3}{7}$	0	Dist Fact
	+ 7.50	–22.50	+53.33	–53.33	+30.00	–30.00	FEM Clue to loading
	–7.50	–3.75			+15.00	+30.00	Due to Simply supported ends
	0	–26.25	+53.33	–53.33	+45.00	0	Moments for distribution
		–11.61	–15.47	+4.76	+3.57		Dist.
			+2.38	–7.74			C.O
		–1.02	–1.36	+ 4.42	+ 3.32		Dist.
			+ 2.21	–0.68			C.O
		–0.95	–1.26	+ 0.39	+ 0.29		Dist.
			+ 0.20	–0.63			C.O
		–0.09	–0.11	+ 0.36	+ 0.27		Dist.
			+0.18	–0.06			C.O
		–0.08	–0.10	+0.03	+0.03		Dist.
	0	–40.00	+40.00	–52.48	+52.48	0	

Fig. 12.11

Example 12.4 | *In the continuous beam in Example 12.3, the support 2 sinks by 10 mm under the loading. Determine the support moments due to combined loading and sinking of support take $E = 200 \times 10^6\ kN/m^2$ and $I = 80 \times 10^{-6}\ m^4$.*

In the Example 12.3 above. The support moments are worked out due to the loading on the beam. Now we can work out separately the support moments as a result of sinking of support 2. The combined moments can be obtained by adding them algebraically. The results are shown entered in the Table that follows.

0	$\frac{3}{7}$	$\frac{4}{7}$	$\frac{4}{7}$	$\frac{3}{7}$	0	D.F.
+ 60.0	+ 60.0	– 15.0	– 15.0			FEM Due to settlement of supp.
– 60.0	– 30.0					Bal. & c.o.
0	+ 30.0	– 15.0	– 15.0			Moments for Pist
	– 6.43	– 8.57	+ 8.57	+ 6.43		Dist
		+ 4.29	+ 4.29			C.O
	– 1.84	– 2.45	+ 2.45	+ 1.84		Dist
		+ 1.23	– 1.23			C.O
	– 0.53	– 0.70	+ 0.70	+ 0.53		Dist
		+ 0.35	– 0.35			C.O
	– 0.15	– 0.20	+ 0.20	+ 0.15		Dist
		+ 0.10	– 0.10			C.O
	– 0.04	– 0.06	+ 0.06	+ 0.04		Dist
0	+ 21.01	– 21.01	– 8.99	+ 8.99	0	Support Moments
0	+ 40.00	+ 40.00	+ 52.48	+ 52.48	0	Moments Previous
0	– 18.99	+ 18.99	– 61.47	+ 61.47	0	Final Moments

Fig. 12.12

Example 12.5 *Let us consider the continuous beam of Fig. 12.13. Under the loading support 2 sinks by 10 mm. Determine the support moments using the moment distribution method. $E = 200 \times 10^6$ kN/m^2 (200,000 MPa) and $I = 100 \times 10^6$ m^4 (100×10^6 mm^4).*

The example was previously solved by the slope-deflection method in Example 11.4. Note *EI* is the same throughout.

The fixed end moments are written, as usual, after restraining all the joints temporarily. The fixed end moments are

$$FEM_{12} = \frac{40 \times 4}{8} = 20.00 \text{ kN.m}; \quad FEM_{21} = -20.00 \text{ kN.m}$$

$$FEM_{23} = \frac{10 \times 6 \times 6}{12} = 30.00 \text{ kN.m}; \quad FEM_{32} = -30.00 \text{ kN.m}$$

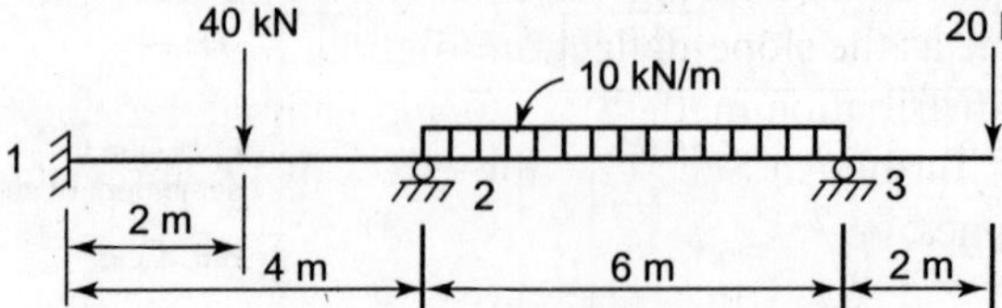

	1	2 (span 1–2)	2 (span 2–3)	3 (span 2–3)	3 (overhang)
Rel. stiff	$\frac{I}{4}$		$\frac{3}{4}\frac{I}{6}=\frac{I}{8}$		
D.F	0	2/3	1/3	1.0	
FEM due to loading	+ 20.00	– 20.00	+ 30.00	– 30.00	+ 40.00
FEM due to translation of support 2	+ 75.00	– 75.00	+ 33.33	– 33.33	
Total FEM	+ 95.00	+ 75.00	+ 3.33	– 63.33	+ 40.00
Bal. 3 and C.O			+ 11.67 ←	+ 23.33	
Bal. 2, distr. and C.O	– 21.12 ←	– 42.23	– 21.11		
Final moments	+ 73.88	+ 12.77	– 12.77	– 40.00	+ 40.00

Fig. 12.13

The overhanging end is replaced by applying a moment of + 40.00 kN.m at joint 3 and the support shall be treated as simply supported for all purposes hereafter. The effect of the translation of supports can now be included by considering the fixed end moments caused by the translation of joints in the temporarily restrained condition. The fixed end moments due to the translation of support 2 are (see Appendix table).

$$FEM_{12} = \frac{6EI\Delta}{L^2} = \frac{6(200\times10^6)(100\times10^{-6})}{(4)^2}\left(\frac{10}{1000}\right)$$

$$= 75.00 \text{ kN.m}$$

$$FEM_{21} = 75.00 \text{ kN.m}$$

$$FEM_{23} = \frac{-6(200\times10^6)(100\times10^{-6})}{(6)^2}\left(\frac{10}{1000}\right)$$

$$= -33.33 \text{ kN.m}$$

$$FEM_{32} = -33.33 \text{ kN.m}$$

These fixed end moments are to be added algebraically to the fixed end moments caused by transverse loads. For clarity, the fixed enu moments caused by transverse loads are entered first in one row and the fixed end moments due to the translation of the support are entered in the next row as shown in Fig. 12.13. The algebraic sum of the fixed end moments are entered in the third row. The procedure from now on is the same as in the previous examples. The final moments are summed up at the end taking the total fixed end moments. The results obtained are the same as the slope-deflection solution. Notice the extreme simplicity of the moment distribution method in dealing with the translation of supports. We shall discuss further in Sec. 12.4 the effect of the translation of joints with reference to frames.

12.3 ANALYSIS OF FRAMES WITH NO LATERAL TRANSLATION OF JOINTS

Moment distribution for the analysis of frames in which the joint translations are prevented follows the same general procedure as for continuous beams. It is quite usual that in frames more than two members meet at a joint. Care must be taken in such cases to include the stiffness of all members meeting at any joint while evaluating the distribution factors. Additional consideration must be given to the recording of computations. For a single bay, single storey frame, it is convenient to spread out the legs of the frame so that the frame lies along a straight line. The entries are then made as on a continuous beam. Another convenient method of recording calculations is to enter the values on a sketch of the framed structure. Both these methods are illustrated in the following examples.

Example 12.6 *The end moments of the members of the portal frame of Fig. 12.14 are to be obtained using the moment distribution method. The relative values of EI are shown along the members.*

The frame is prevented from undergoing lateral translation. The calculations are entered on the opened up frame as shown in Fig. 12.15*a*. The procedure followed is the same as for a continuous beam. The entries are self-explanatory. The final summed up values give a check as regards the correctness of the calculations. The calculations can also be recorded on a sketch of the frame as shown in Fig. 12.15*b*.

Example 12.7 *Using the moment distribution method, determine the end moments of the members of the frame of Fig. 12.16 and draw the moment diagram. EI is the same throughout.*

The support condition at end 1 prevents the frame from undergoing lateral translation. As pointed out, the procedure for the analysis of this frame is the same as that for continuous beams without translation of supports. The fixed end moments due to external loading in the temporarily restrained joints are

$$FEM_{12} = \frac{30(4)^2}{12} = 40.00 \text{ kN.m}, \quad FEM_{21} = -40.00 \text{ kN.m}$$

$$FEM_{23} = \frac{(100)(4)}{8} = 50.00 \text{ kN.m}, \quad FEM_{32} = -50.00 \text{ kN.m}$$

$$FEM_{34} = \frac{20(3)^2}{12} = 15.00 \text{ kN.m}, \quad FEM_{43} = -15.00 \text{ kN.m}$$

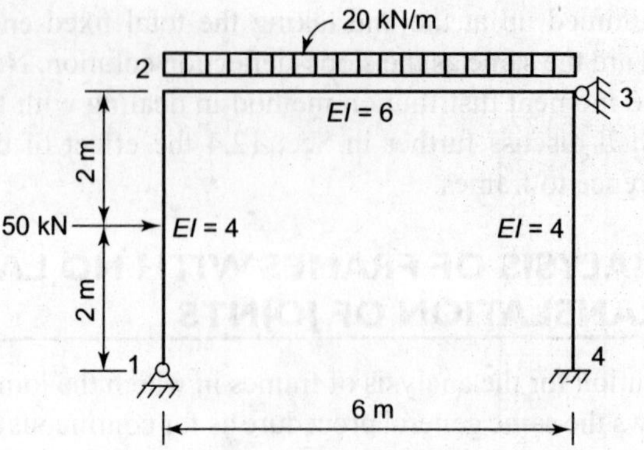

Fig. 12.14 | *Frame and loading*

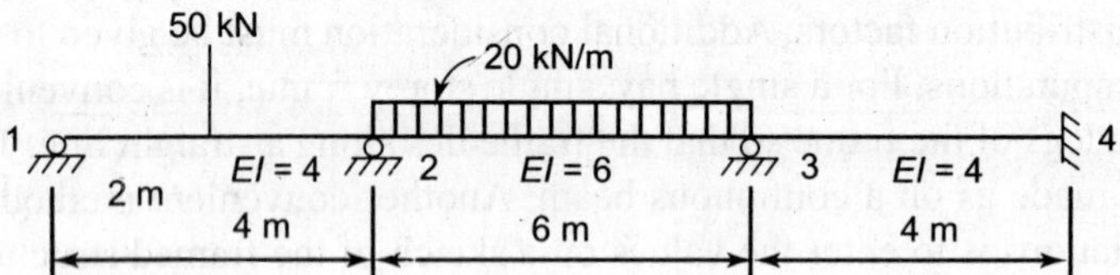

1	2		3		4	
$\frac{3}{4}\frac{EI}{L} = \frac{3}{4}$		$\frac{EI}{L} = 1$		$\frac{EI}{L} = 1$		Rel. stiff
1.0	3/7	4/7	1/2	1/2	0	D.F
+ 25.00	– 25.00	+ 60.00	– 60.00	0.0	0.0	FEM
– 25.00 →	– 12.50					Bal. 1 and C.O
	– 9.64	– 12.86	+ 30.00	+ 30.00		Dist.
		+ 15.00	– 6.43		+ 15.00	C.O
	– 6.43	– 8.57	+ 3.22	+ 3.21		Dist.
		+ 1.61	– 4.29		+ 1.61	C.O
	– 0.69	– 0.92	+ 2.15	+ 2.14		Dist.
		+ 1.08	– 0.46		+ 1.07	C.O
	– 0.46	– 0.62	+ 0.23	+ 0.23		Dist.
		+ 0.12	– 0.31		+ 0.12	C.O
	– 0.05	– 0.07	+ 0.16	+ 0.16		Dist.
0	– 54.77	+ 54.77	– 35.73	+ 35.74	+ 17.80	Final moments

Fig. 12.15(a) | *Frame opened out and distribution carried out*

The calculations are performed on a sketch of the frame as shown in Fig. 12.17. For the girder of the frame, the calculations are entered as in a continuous beam. For columns, however, the values are entered below on a column line as shown in Fig. 12.17. This arrangement is convenient for single storey frames of any number of bays. For frames of more than one storey, the arrangement which is slightly different but more convenient is shown later in Sec. 12.6.

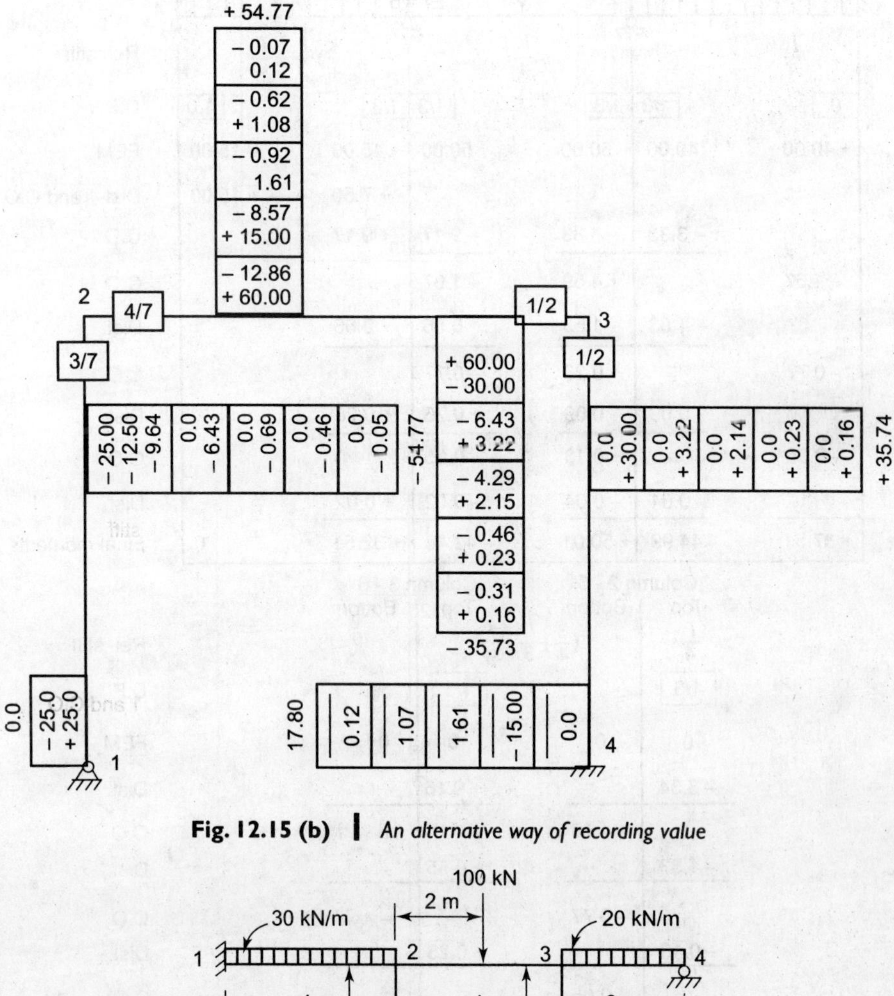

Fig. 12.15 (b) | *An alternative way of recording value*

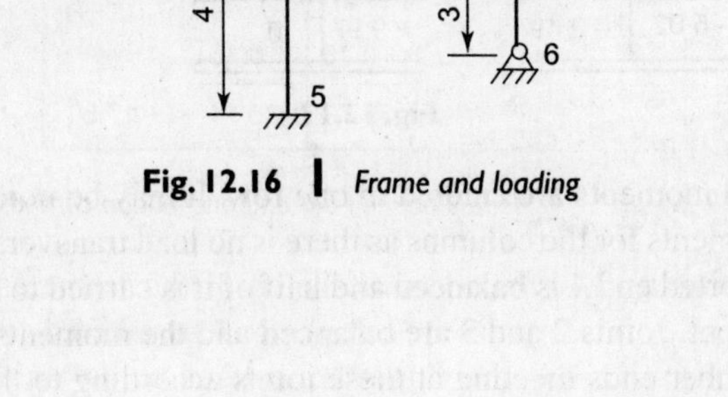

Fig. 12.16 | *Frame and loading*

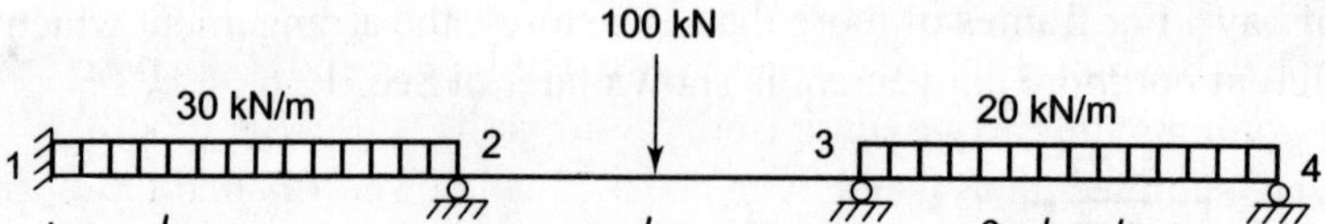

End 1	End 2 (1-2)	End 2 (2-3)	End 3 (2-3)	End 3 (3-4)	End 4	
$\frac{I}{4}$		$\frac{I}{4}$		$(\frac{3}{4})\frac{I}{3}=\frac{I}{4}$		Rel. stiff
0	1/3	1/3	1/3	1/3	1.0	D.F
+ 40.00	– 40.00	50.00	50.00	+ 15.00	– 15.00	FEM
				+ 7.50	← + 15.00	Dist.4 and C.O
	– 3.33	– 3.33	– 9.17	+ 9.17		C.O
– 1.67		+ 4.59	– 1.67			C.O
	– 1.53	– 1.53	+ 0.56	+ 0.56		Dist.
– 0.77		+ 0.28	– 0.77			C.O
	– 0.09	– 0.09	+ 0.26	+ 0.26		Dist.
– 0.05		+ 0.13	– 0.05			C.O
	– 0.04	– 0.04	– 0.02	+ 0.02		Dist.
+ 37.51	– 44.99	+ 50.01	– 42.48	+ 32.51	0	Final moments

Column 2 - 5 Top	Column 2 - 5 Bottom	Column 3 - 6 Top	Column 3 - 6 Bottom	
$\frac{I}{4}$	$(\frac{3}{4})\frac{I}{3}=\frac{I}{4}$			Rel. stiff
1/3		1/3		D.F
0	0	0	0	FEM
– 3.34	–	+ 9.16	–	Dist.
	– 1.67			C.O
– 1.53	–	+ 0.55	–	Dist.
	– 0.77	–	–	C.O
– 0.10		+ 0.25		Dist.
	– 0.05			C.O
– 0.05		+ 0.01		Dist.
– 5.02	– 2.49	+ 9.97	0	Final moments

Fig. 12.17

The fixed end moments are entered in one row. It may be noted that there are no fixed end moments for the columns as there is no load transverse to them. Next the simply supported end 4 is balanced and half of it is carried to support 3 as the carry over moment. Joints 2 and 3 are balanced and the moments are distributed to the three member ends meeting at these joints according to their distribution factors. The distributed moment for the column tops are recorded below under 'column top'. Next the moments are carried over to the farther ends as carry over moments. For columns the carry over is from 'column top' to 'column bottom'. The moments are summed up as usual. As a check it can be verified that the sum of the moments at any joint must be zero. It may be noted that the moments at the bottom of columns are either equal to zero as in the case of hinged supports or equal to half the moment at the top as in the case of fixed bases. The moment diagram is shown in Fig. 12.18 drawn on the tension face of the members.

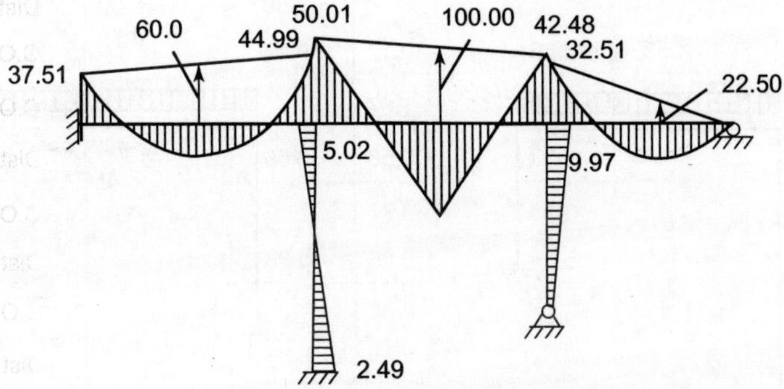

Fig. 12.18 | *Moment diagram*

12.4 | ANALYSIS OF FRAMES WITH LATERAL TRANSLATION OF JOINTS

So far we have considered frames in which the joints are not allowed to translate laterally. However, in frames, the translation of some joints is commcn due to forces acting in the lateral direction as in Fig. 12.19*a* or due to asymmetrical forces as in Fig.12.19*b* or due to asymmetry in the make-up of the frame even though the load is symmetrical as in Figs. 12.19*c* and *d*.

In frames undergoing lateral translation, the analysis is carried out in two stages. In the first stage, the frame is prevented from undergoing any lateral translation by applying an artificial joint restraint as shown in Fig. 12.20. The procedure is then similar to the one adopted for frames without sway.

The value of artificial restraining force X is obtained by first evaluating the shear at the bases of columns. Then from the equilibrium condition, $\Sigma F_H = 0$, the value of X can be evaluated. At this stage the end moments obtained are true only when restraining force X is acting. To achieve the true condition of the structure, the frame has to be analysed again by applying a force equal and opposite to artificial restraining force X. The member end moments resulting from this condition of loading will be combined with the moments obtained from the earlier restrained condition to obtain the true values of moments in the frame.

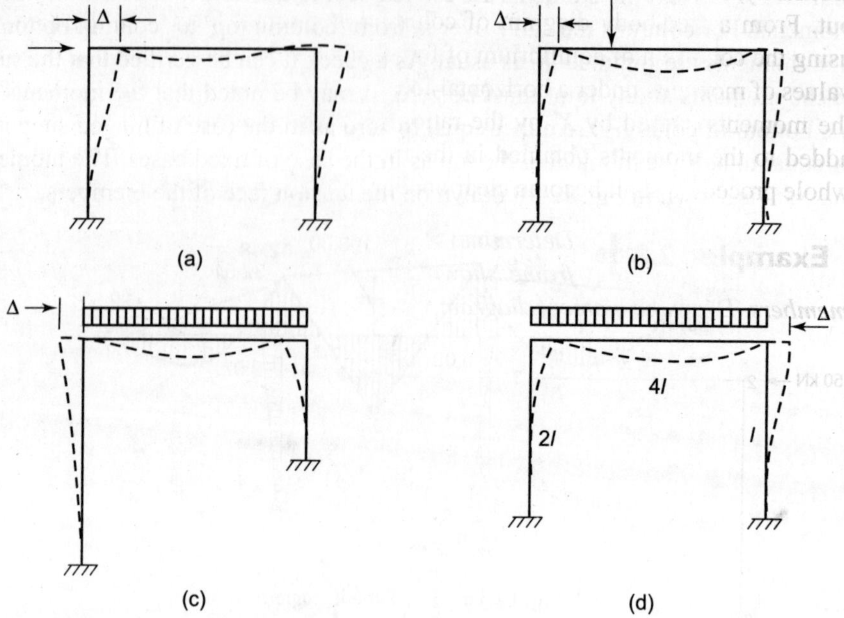

Fig. 12.19 | *Lateral translations due to: (a) Lateral loading, (b) unsymmetrical loading, (c) Unequal column heights, (d) Unequal column stiffnesses*

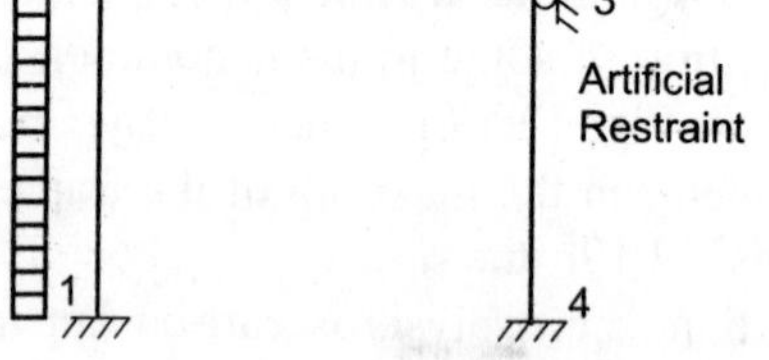

Fig. 12.20 | *Frame restrained from lateral translation*

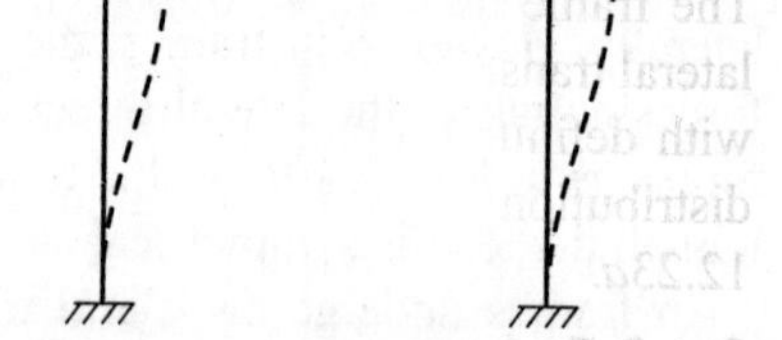

Fig. 12.21 | *Frame under an arbitrary lateral force*

The moments in the members of the frame due to application of the force $(-X)$ are obtained in an indirect manner. The frame is assumed to be subjected to an arbitrary loading say X', as shown in Fig. 12.21. If only translations are allowed restraining the rotations temporarily, the frame deflects laterally by an amount Δ. Lateral translation Δ is the same for both joints, if the axial deformation in the beam is neglected. The fixed end moments for this condition can be written using the Appendix table. It is not necessary to know the true value of Δ. We can arbitrarily fix moments in columns on the basis that the joints translate equally without undergoing rotation. It is good practice to assume the moments that lie in the range of moments we are working with. With the fixed end moments chosen arbitrarily, but following definite proportions, the moment distribution is worked out. From a free body diagram of columns, horizontal force X is worked out using the condition of equilibrium of forces in the horizontal direction. The true values of moments under a horizontal force ($-X$) can be obtained by multiplying the moments caused by X' by the ratio, X/X'. The moments thus obtained are added to the moments obtained in the first stage of moment distribution. The whole procedure shall become clear once we work out some examples.

Example 12.8 *Determine the end moments of the members of the frame shown in Fig. 12.22a. EI is same for all the members. Draw the moment diagram.*

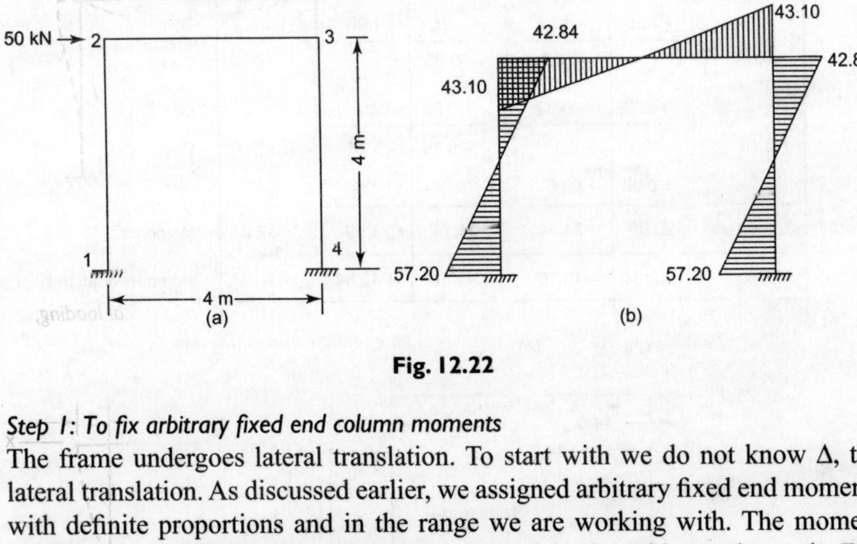

Fig. 12.22

Step 1: To fix arbitrary fixed end column moments

The frame undergoes lateral translation. To start with we do not know Δ, the lateral translation. As discussed earlier, we assigned arbitrary fixed end moments with definite proportions and in the range we are working with. The moment distribution is carried out as usual and entered in the table as shown in Fig. 12.23*a*.

Step 2: To determine column shears

From the free-body diagram of columns Fig. 12.23b.

$$H_1 = \frac{M_1 + M_2}{4} = \frac{32.03 + 23.99}{4} = 14.00$$

$$H_2 = \frac{M_3 + M_4}{4} = \frac{32.00 + 23.99}{4} = 14.00$$

$$H_1 + H_2 = 28.00 \text{ kN}.$$

As the given load is 50 kN, the moments are multiplied by a ratio 50/28. The final adjusted moments are entered in the table bending moment diagram is shown in Fig. 12.15*b*.

1	2		3		4	
$\frac{I}{4}$		$\frac{I}{4}$		$\frac{I}{4}$		Rel. Stiffness
0	$\frac{1}{2}$	$\frac{1}{2}$	$\frac{1}{2}$	$\frac{1}{2}$	0	D.F
+ 40.0	+ 40.00		0	+ 40.0	+ 40.0	FEM – Arbitrary
	– 20.0	– 20.0	– 20.0	– 20.0	+ 20.0	Dist
– 10.0		– 10.0	– 10.0		– 10.0	C.O
	+ 5.0	+ 5.0	+ 5.0	+ 5.0		Bal
+ 2.5		+ 2.5	+ 2.5		+ 2.5	C.O
	– 1.25	– 1.25	+ 1.25	– 1.25		Dist
– 0.63		– 0.63	– 0.63		– 0.63	C.O
	+ 0.32	+ 0.32	+ 0.32	+ 0.32		Dist
+ 0.16	– 0.16	– 0.16	– 0.16	– 0.16	+ 0.16	C.O
	+ 0.08	+ 0.08	+ 0.08	+ 0.08		Dist
32.03	23.99	– 24.14	– 24.14	+ 23.99	+ 32.03	Moments
57.20	42.84	– 43.10	– 43.10	+ 42.84		Moments adjusted

(a)

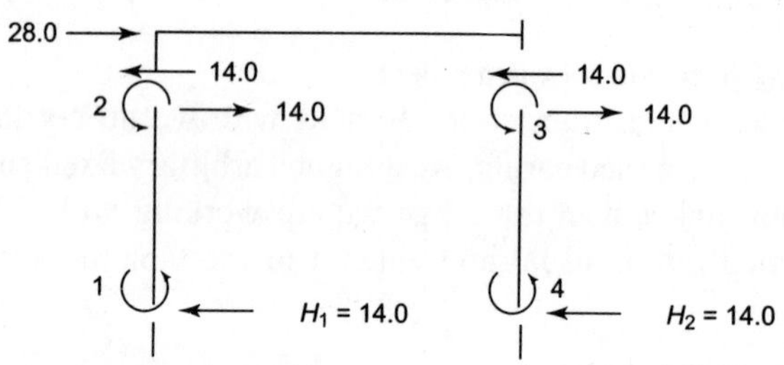

(b)

Fig. 12.23

Example 12.9 | *Determine the end moments of the members of the frame shown in Fig. 12.24a. EI values are indicated along the members.*

Step 1: To fix up end moments in a restrained structure

As a first step, the joint 3 is artificially restrained from undergoing lateral translation. The moment distribution is worked out in the table that follows.

The artificial restraining force '*X*' is worked out considering the free-body diagrams of the columns as shown in Fig. 12.24*b*.

From the free-body diagram of columns in Fig. 12.24*b*, we can arrive at the column shears by applying $\Sigma F_H = 0$

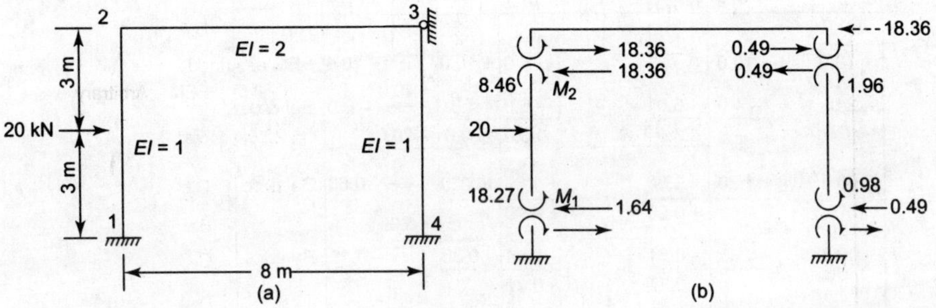

Fig. 12.24

1	2		3		4	
$\frac{I}{6}$		$\frac{2I}{8}$		$\frac{I}{6}$		Rel . Stiff
0	0.4	0.6	0.6	0.4	0	D.F.
+ 15.0	− 15.0	0	0	0	0	FEM
	+ 6.0	+ 9.0				Bal
+ 3.0			+ 4.5			C.O
			− 2.7	− 1.8		Bal
		− 1.35			− 0.9	C.O
	+ 0.54	+ 0.81				Bal
+ 0.27			+ 0.41			C.O
			− 0.25	− 0.16		Bal
		− 0.13			− 0.08	C.O
	+ 0.05	+ 0.08				Bal
+ 18.27	− 8.41	+ 8.41	+ 1.96	− 1.96	− 0.98	Moments sway prevented
+ 31.12	+ 25.51	− 25.51	− 25.51	+ 25.51	+ 31.12	Moments due to sway
+ 49.39	+ 17.10	− 17.10	− 23.55	+ 23.55	+ 30.14	Final Moments

Fig. 12.25

We get $X = 18.85$ kN.

The frame is now permitted to sway. The FEM caused due to sway are assumed appropriately and the moment distribution is carried out as follows.

1	2		3		4	
O	0.4	0.6	0.4	0.6	O	D.F
+ 10.0	+ 10.0	0	0	+ 10.0	+ 10.0	FEM assumed
– 2.0	– 4.0	– 6.0	– 6.0	– 4.0	– 2.0	Bal & co
		0 3.0	0 3.0			
0.60	+ 1.20	+ 1.80	+ 1.80	+ 1.20	0.60	Bal & co
		+ 0.90	+ 0.90			
– 0.18	– 0.36	– 0.54	– 0.54	– 0.36	– 0.18	Bal & co
		– 0.27	– 0.27			
+ 0.06	+ 0.11	+ 0.16	+ 0.16	+ 0.11	+ 0.06	Bal & co
+ 8.48	+ 6.95	– 6.95	– 6.95	+ 6.95	+ 8.48	Moments due to FEM assumed

Fig. 12.26

The shear in columns and hence the horizontal fore

$$X' = \frac{8,48 + 6.95}{6} + \frac{6.95 + 8.48}{6} = 5.14 \text{ kN}$$

The ratio $$\frac{X}{X'} = \frac{18.85}{5.14} = 3.67$$

The moments in the table above are multiplied by the factor 3.67 and added algebraically to the moments obtained by presenting sway. The find moments are shown entered in Table Fig. 12.25.

Example 12.10 | *Determine the end moments of the members of the frame shown in Fig. 12.27a. E is constant and relative I values are indicated on the frame.*

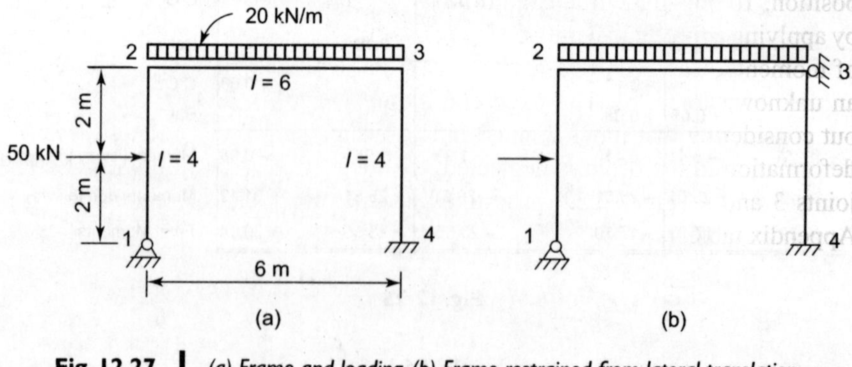

Fig. 12.27 | *(a) Frame and loading, (b) Frame restrained from lateral translation*

As a first step we artificially restrain joint 3 from undergoing lateral translation as in Fig. 12.27*b*. The moment distribution for this restrained condition is already worked out in Example 12.6. We shall take the values from the table of Fig. 12.15. The artificial restraining force X is worked out considering the free-body diagram of the columns as in Fig. 12.28. Summing the moments about joint 2, we have

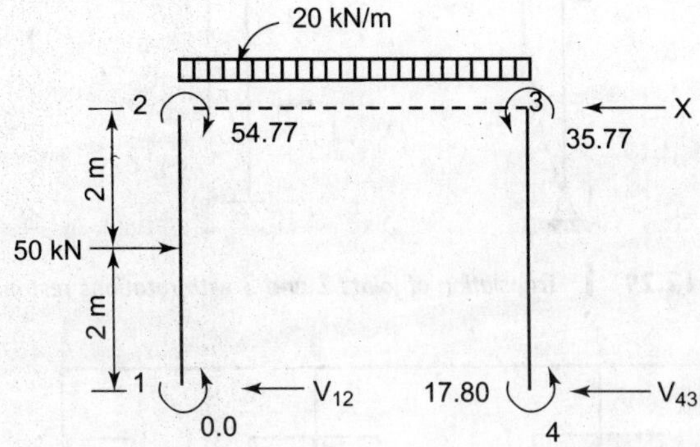

Fig. 12.28 | *Free-body diagram of columns*

$$V_{12} = \frac{50(2) - 54.77}{4} = 11.31 \text{ kN}$$

Similarly, summing the moments about joint 3, we get

$$V_{43} = \frac{35.74 + 17.80}{4} = 13.39 \text{ kN}$$

Applying equilibrium condition $\Sigma F_H = 0$ for the entire structure

$$50 - 11.31 - 13.39 - X = 0$$

or $$X = 25.3 \text{ kN}$$

This is the force the artificial restraint exerts on the frame to prevent lateral translation and the moments in the first distribution are true for this constrained position. To obtain the true condition, the artificial constraint has to be removed by applying a force equal but opposite to force X. This needs a second distribution of moments. However, this has to be worked out in an indirect manner. Apply an unknown force, X'. The fixed end moments in the columns can be worked out considering that joints 2 and 3 only translate and do not rotate. If the axial deformation in the beam is neglected, the translations (Δ) are the same at both joints 3 and 4 (Fig. 12.29). The fixed end moments can be written using the Appendix table.

$$FEM_{12} = 0, \; FEM_{21} = \frac{3EI\Delta}{L^2} = \frac{(3)(4)}{(4)^2} E\Delta = 0.75\, E\Delta$$

$$FEM_{43} = FEM_{34} = \frac{6EI\Delta}{L^2} = \frac{(6)(4)}{(4)^2} E\Delta = 1.5\, E\Delta$$

Letting $E\Delta = 20$

$$FEM_{21} = 15.0 \text{ kN.m}$$
$$FEM_{34} = FEM_{43} = 30.0 \text{ kN.m}$$

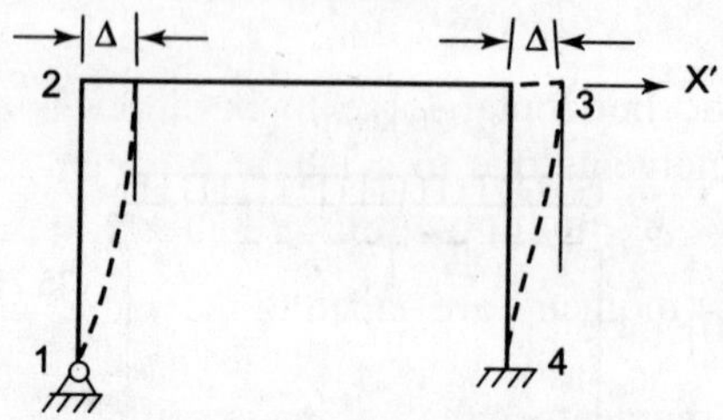

Fig. 12.29 | *Translation of joints 2 and 3 with rotations restrained*

1	2		3		4	
1.0	3/7	4/7	1/2	1/2	1/2	D.F
0.0	+ 15.00	0.0	0.0	+ 30.00	+ 30.00	FEM
	− 6.43	− 8.57	− 15.00	− 15.00		Dist.
		− 7.50	− 4.29		− 7.5	C.O
	+ 3.21	+ 4.29	+ 2.15	+ 2.14		Dist.
		+ 1.08	+ 2.15		+ 1.07	C.O
	− 0.46	− 0.62	− 1.08	− 1.07		Dist.
		− 0.54	− 0.32		− 0.54	C.O
	+ 0.23	+ 0.31	+ 0.16	+ 0.16		Dist.
		+ 0.08	+ 0.16		+ 0.08	C.O
	− 0.03	− 0.05	− 0.08	− 0.08		Dist.
0.0	+ 11.52	− 11.52	− 16.15	+ 18.15	+ 23.11	Final moments

Fig. 12.30

The fixed moments are entered, as usual, in the table of Fig. 12.30. The distribution factors are the same as in Example 12.6. The moment distribution is carried out as usual and the final moments are shown in the last row of the table. The lateral force X' which produced these moments can be evaluated by considering the free-body diagram of the columns as in Fig. 12.31. The summation of the moments about joint 2 gives,

$$V_{12} = \frac{11.52}{4} = 2.88 \text{ kN}$$

Similarly, the summation of the moments about joint 3 gives

$$V_{43} = \frac{16.15 + 23.11}{4} = 9.82 \text{ kN}$$

From considerations of equilibrium of horizontal forces $\Sigma F_H = 0$, we have

$$X' = 2.88 + 9.82 = 12.70 \text{ kN}$$

The true value of the horizontal force to be applied to the frame is 25.30 kN. Therefore, the moments due to a lateral force of 25.30 kN are obtained by proportion, that is, by multiplying the moments in Fig. 12.30 by a factor $\frac{25.30}{12.70} = 1.992$. These moments are added to the values of the moments in Fig. 12.15. Thus, the true moments in the frame are

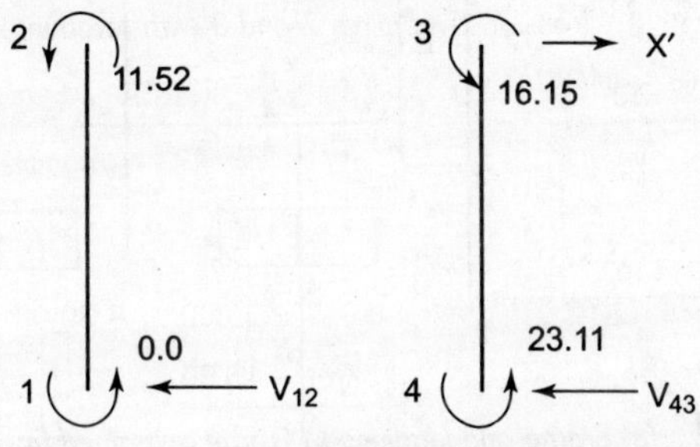

Fig. 12.31 | *Free-body diagram of columns*

$$M_{12} = 0$$
$$M_{21} = -54.77 + 11.52 \times 1.992 = -31.82$$
$$M_{32} = -35.73 - 16.15 \times 1.992 = -67.90 = -M_{34}$$
$$M_{43} = +17.80 + 23.11 \times 1.992 = 63.84$$

As a check it is seen that the sum of the shears in the columns is equal to the external lateral force (Fig. 12.32).

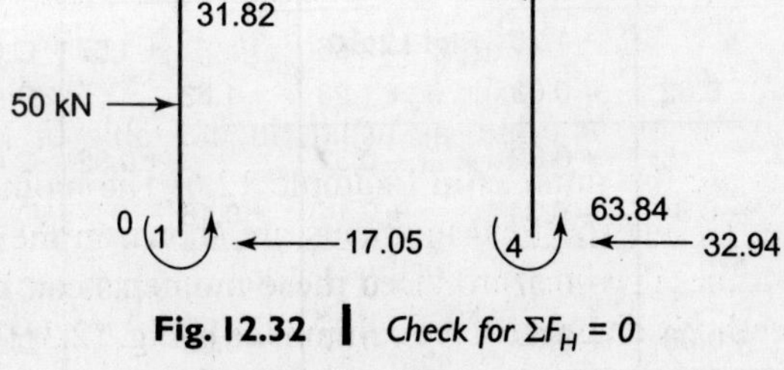

Fig. 12.32 | *Check for $\Sigma F_H = 0$*

Example 12.11 | *Determine the end moments of the members of the frame of Fig. 12.33a. EI is constant for all members.*

As earlier, the frame is artificially restrained temporarily by applying an unknown lateral force at joint 3 as shown in Fig. 12.33*b*. For the restrained condition, the moment distribution is carried out in the table of Fig. 12.34 in the usual manner. We shall find out the artificial restraining force by considering the shear force in the columns and the equilibrium of horizontal forces (Fig. 12.35).

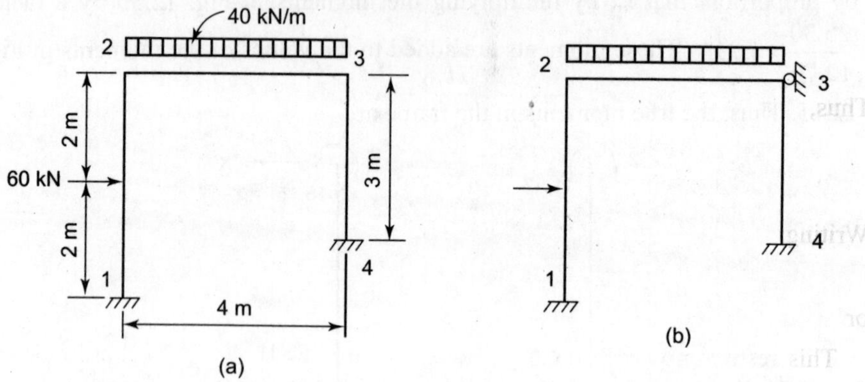

Fig.12.33 | *(a) Frame and loading, (b) Frame restrained from translation*

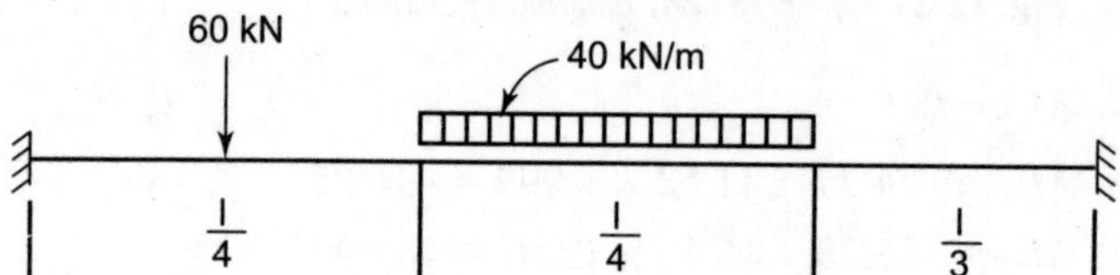

	$\frac{I}{4}$		$\frac{I}{4}$		$\frac{I}{3}$	Rel. stiff
0	1/2	1/2	3/7	4/7	0	Dist.
+ 30.00	– 30.00	+ 53.33	– 53.33	0	0	FEM
	– 11.67	– 11.66	+ 22.86	+ 30.47		Dist.
– 5.84		+ 11.43	– 5.83		+ 15.24	C.O
	– 5.72	– 5.71	+ 2.50	+ 3.33		Dist.
– 2.86		+ 1.25	– 2.86		+ 1.67	C.O
	– 0.62	– 0.63	+ 1.23	+ 1.63		Dist.
– 0.31		+ 0.62	– 0.32		+ 0.82	C.O
	– 0.31	– 0.31	+ 0.14	+ 0.18		Dist.
– 0.16		+ 0.07	– 0.16		+ 0.09	C.O
	– 0.03	– 0.04	+ 0.07	+ 0.09		Dist.
+ 20.83	– 48.35	+ 48.35	– 35.70	+ 35.70	– 17.82	Final moments

Fig. 12.34

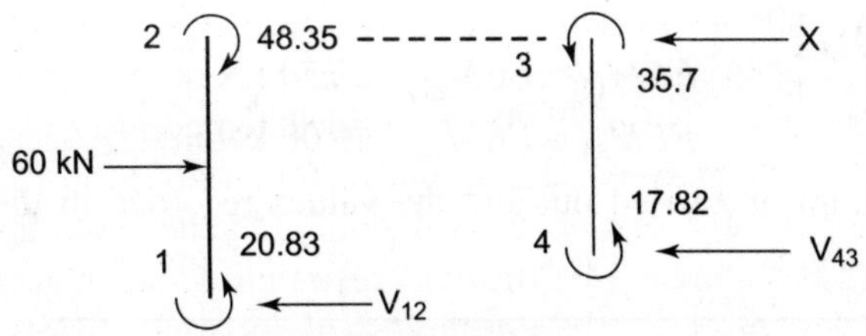

Fig. 12.35 | *Free-body diagram of columns*

Thus, $$V_{12} = \frac{60(2) + 20.83 - 48.35}{4} = 23.12 \text{ kN}$$

$$V_{43} = \frac{17.82 + 35.7}{3} = 17.84 \text{ kN}$$

Writing, $\Sigma F_H = 0$, we have

$$60 - 23.12 - 17.84 - X = 0$$

or $$X = 19.04 \text{ kN}$$

This restraining force was not there in the original structure. Therefore, a force of $(-X)$ has to be applied and the moments corresponding to this force $(-X)$, have to be worked out. This is done in an indirect way. Let the frame undergo a translation, Δ, without undergoing rotation under lateral force X' as shown in Fig. 12.36.

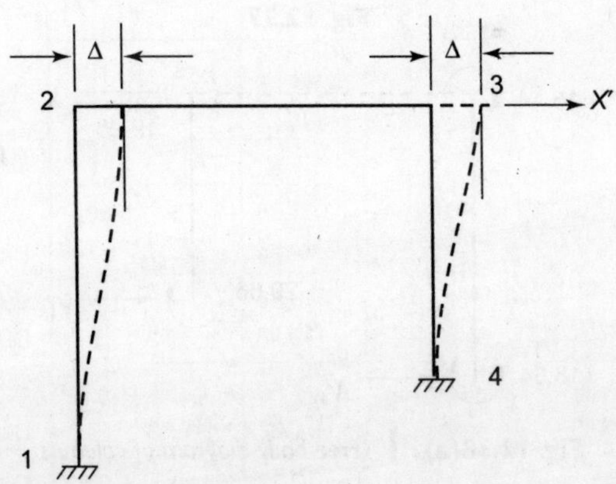

Fig. 12.36 | *Frame undergoing translation without rotation*

The fixed end moments are

$$FEM_{12} = FEM_{21} = \frac{6EI\Delta}{(4)^2} = \frac{3}{8}EI\Delta$$

$$FEM_{34} = FEM_{43} = \frac{6EI\Delta}{(3)^2} = \frac{2}{3}EI\Delta$$

Letting $EI\,\Delta = 60$

$$FEM_{12} = FEM_{21} = 22.50 \text{ kN.m}$$

$$FEM_{34} = FEM_{43} = 40.00 \text{ kN.m}$$

The distribution is carried out and the values recorded in the table of Fig. 12.37.

0	1/2	1/2	3/7	4/7	0	D.F
+ 22.50	+ 22.50	0	0	+ 40.00	+ 40.00	FEM – Arbitrary
	– 11.25	– 11.25	– 17.14	– 22.86		Dist
– 5.63		– 8.57	– 5.63		– 11.43	C.O
	+ 4.29	+ 4.28	+ 2.41	+ 3.22		Dist
+ 2.15		+ 1.21	+ 2.14		+ 1.61	C.O
	– 0.60	– 0.61	– 0.92	–1.22		Dist
– 030		– 0.46	– 0.31		– 0.61	C.O
	+ 0.23	+ 0.23	+ 0.13	+ 0.18		Dist
+ 0.12		+ 0.07	+ 0.12		+ 0.09	C.O
	– 0.03	– 0.04	– 0.05	–0.07		Dist
+ 18/84	+ 15.14	– 15.14	– 19.25	+ 19.25	+ 29.66	Final Moments

Fig. 12.37

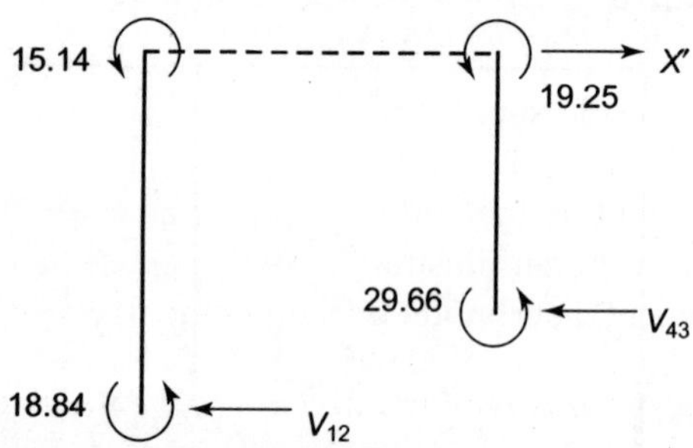

Fig. 12.38(a) | *Free-body diagram of columns*

Columns shears are evaluated considering the free body diagrams of the columns (Fig. 12.38*a*)

$$V_{12} = \frac{15.14 + 18.84}{4} = 8.50$$

$$V_{43} = \frac{29.66 + 19.25}{3} = 16.30$$

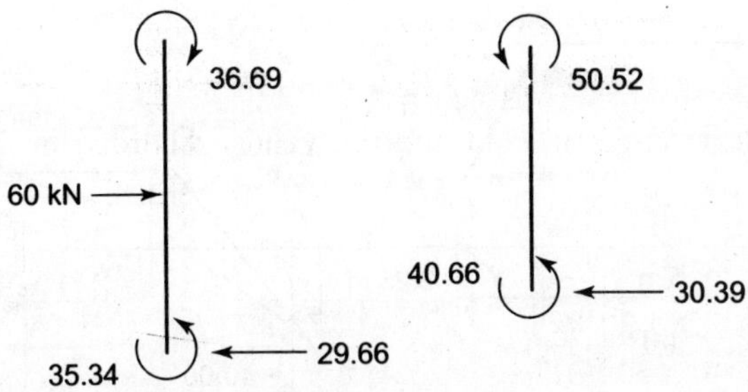

Fig. 12.38(b) | *Check for $\Sigma F_H = 0$*

$$X' = 8.50 + 16.30 = 24.80 \text{ kN}$$

The moments in the table of Fig. 12.30 are to be multiplied by the ratio

$$\frac{X}{X'} = \frac{19.04}{24.80} = 0.77$$

The true final moments are

$$M_{12} = 20.83 + 18.84(0.77) = 35.34 \text{ kN.m}$$
$$M_{21} = -48.35 + 15.14(0.77) = -36.69 \text{ kN.m}$$
$$M_{23} = +36.69 \text{ kN.m}$$
$$M_{32} = -35.70 - 19.25(0.77) = -50.52 \text{ kN.m}$$
$$M_{34} = +50.52 \text{ kN.m}$$
$$M_{43} = 17.82 + 29.66(0.77) = 40.66 \text{ kN.m}$$

The final check that the sum of the shears in the columns must balance the lateral force is satisfied (Fig. 12.38*b*).

The moment distribution method can be conveniently employed for frames with inclined columns. As an illustration, the frame in Example 11.11 is again analysed by the moment distribution method in the following example.

Example 12.12 | *It is required to analyse the frame in Fig. 12.39a using the moment distribution method. EI is the same throughout.*

The lateral translation of joints 2 and 3 under external force is shown in Fig. 12.39*b*. If axial deformations are neglected, lateral translation Δ is the same for both joints. The amount of translation transverse to the columns and the beam is indicated in Fig. 12.39*b*. The fixed end moments caused by joint translations only restraining joint rotations are

$$FEM_{12} = FEM_{21} = \frac{6EI}{L_{12}^2} \frac{\Delta}{\cos \phi_1} = \frac{6}{5 \times 5} \frac{EI\Delta}{0.8}$$

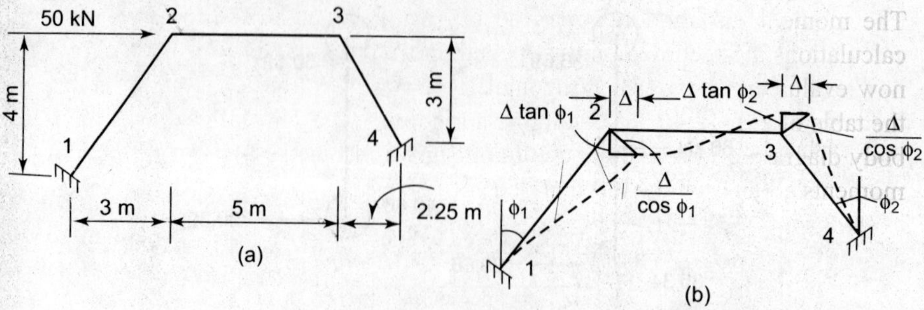

Fig. 12.39 | *(a) Frame and loading, (b) Lateral translation of joints*

1	2		3		4	
$\frac{I}{5}$		$\frac{I}{5}$		$\frac{I}{3.75}$		Rel. stiff
0	1/2	1/2	3/7	4/7	0	D.F
+ 30.00	+ 30.00	– 36.00	– 36.00	+ 53.33	+ 53.33	FEM
	+ 3.00	+ 3.00	– 7.43	– 9.90		Dist
+ 1.50		– 3.72	+ 1.50		– 4.95	C.O
	+ 1.86	+ 1.86	– 0.64	– 0.88		Dist
+ 0.93		– 0.32	+ 0.93		– 0.43	C.O
	+.0.16	+ 0.16	– 0.40	– 0.53		Dist
+ 0.08		– 0.20	– 0.08		– 0.05	C.O
	+ 0.10	+ 0.10	– 0.03	– 0.05		Dist
+ 32.51	+ 35.12	– 35.12	– 41.99	+ 41.99	– 47.68	Final Moments

Fig. 12.40

Letting $EI\Delta = 100$

$$FEM_{12} = FEM_{21} = \frac{6 \times 100}{20} = +30.0 \text{ kN.m}$$

$$FEM_{23} = FEM_{32} = -\frac{6EI}{L_{23}^2}(\Delta \tan \phi_1 + \Delta \tan \phi_2)$$

$$= -\frac{6}{5^2}(1.5)EI\Delta = -36.00 \text{ kN.m}$$

$$FEM_{34} = FEM_{43} = \frac{6EI\Delta}{L_{34}^2 \cos \phi_2} = \frac{6\,EI\,\Delta}{3.75^2 \times 0.8} = 53.33 \text{ kN.m}$$

The moment distribution is carried out using the general procedure and the calculations are recorded on the opened up frame shown in Fig. 12.40. We shall now evaluate the external horizontal force which caused the final moments in the table of Fig. 12.40. To obtain the horizontal force we shall consider the free-body diagram of the beam and column shown in Fig. 12.41. The summation of moments about 2 on the left inclined column gives

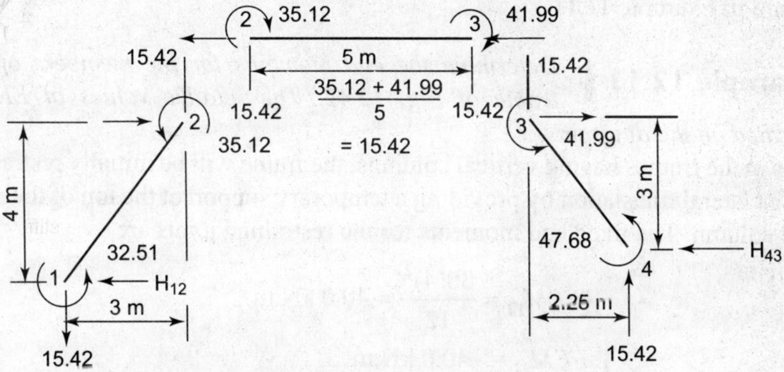

Fig. 12.41 | *Free-body diagram of frame members*

$$\Sigma M_2 = -H_{12}(4) + 15.42(3) + 32.51 + 35.12 = 0$$

or $$H_{12} = 28.47 \text{ kN}$$

Similarly, for the right hand column, taking moments about 3 and equating $\Sigma M_3 = 0$, we have

$$-H_{43}(3) + 15.42(2.25) + 47.68 + 41.99 = 0$$

or $$H_{43} = 41.46 \text{ kN}$$

The resultant external lateral force is $X' = 69.93$ kN from left to right. But the frame was actually subjected to an external force of 50.0 kN only. Therefore, the final moments in the table of Fig. 12.42 are to be multiplied by a factor $\frac{50}{69.93} = 0.715$. The true moments are

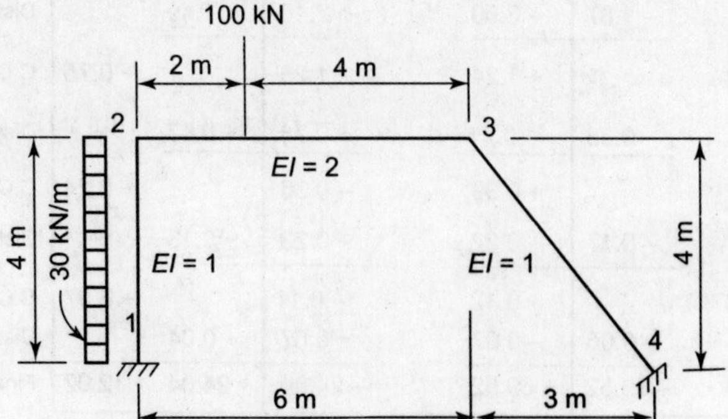

Fig. 12.42 | *Frame and loading*

$$M_{12} = 32.51 \times 0.715 = 23.24 \text{ kN.m}$$
$$M_{21} = -M_{23} = 35.12(0.715) = 25.11 \text{ kN.m}$$
$$M_{34} = -M_{32} = 41.99(0.715) = 30.92 \text{ kN.m}$$
$$M_{43} = 47.68(0.715) = 34.09 \text{ kN.m}$$

These values tally well with the values obtained by the slope-deflection method in Example 11.11.

Example 12.13 | *Determine the end moments for the members of the frame of Fig. 12.42. The relative values of EI are indicated on the diagram.*

As in the frames having vertical columns, the frame will be initially restrained against lateral translation by providing a temporary support at the top of the right hand column. The fixed end moments for the restrained joints are

$$FEM_{12} = \frac{30(4)^2}{12} = 40.0 \text{ kN.m}$$

$$FEM_{21} = -40.0 \text{ kN.m}$$

$$FEM_{23} = \frac{100(2)(4)^2}{36} = 88.89 \text{ kN.m}$$

$\frac{I}{L} = \frac{I}{4}$		$\frac{2I}{6} = \frac{I}{3}$		$\frac{I}{L} = \frac{I}{5}$		Rel. stiff
0	3/7	4/7	5/8	3/8	0	D.F
– 40.00	+ 40.00	+ 88.89	– 44.44	0	0	FEM
	– 20.95	– 27.94	+ 27.78	– 16.67		Dist
– 10.48		+ 13.89	– 13.97		+ 8.34	C.O
	– 5.95	– 7.94	+ 8.73	+ 5.24		Dist
– 2.98		+ 4.37	– 3.97		+ 2.62	C.O
	– 1.87	– 2.50	+ 2.48	+ 1.49		Dist
– 0.94		+ 1.24	– 1.25		+ 0.75	C.O
	– 0.53	– 0.71	+ 0.78	+ 0.47		Dist
– 0.27		+ 0.39	– 0.36		+ 0.24	C.O
	– 0.17	– 0.22	+ 0.23	+ 0.13		Dist
– 0.09		+ 0.12	– 0.11		+ 0.07	C.O
	– 0.05	– 0.07	+ 0.07	+ 0.04		Dist
+ 25.24	– 69.52	+ 69.52	– 24.04	+ 24.04	+ 12.02	Final Moments

Fig. 12.43

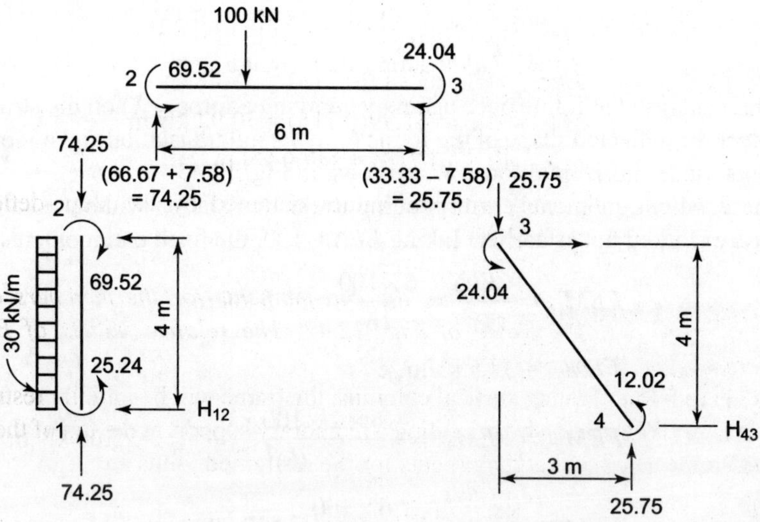

Fig. 12.44 | *Free-body diagram of frame members*

$$FEM_{32} = \frac{100(4)(2)^2}{36} = -44.44 \text{ kN.m}$$

$$FEM_{34} = FEM_{43} = 0$$

The moment distribution is carried out in the table of Fig. 12.43. Next, the free-body diagrams of the columns and girder are shown in Fig. 12.44 indicating all the forces acting on them. From the free-body diagram of the left hand column, summing up moments about joint 2, we have

$$\Sigma M_2 = -H_{12}(4) + 120(2) + 25.24 - 69.52 = 0$$

or $\quad H_{12} = 48.93$ kN

Similarly, summing up moments about joint 3 on the right hand column, we get

$$\Sigma M_3 = -H_{43}(4) + 25.75(3) + 12.02 + 24.04 = 0$$

or $\quad H_{43} = 28.33$ kN

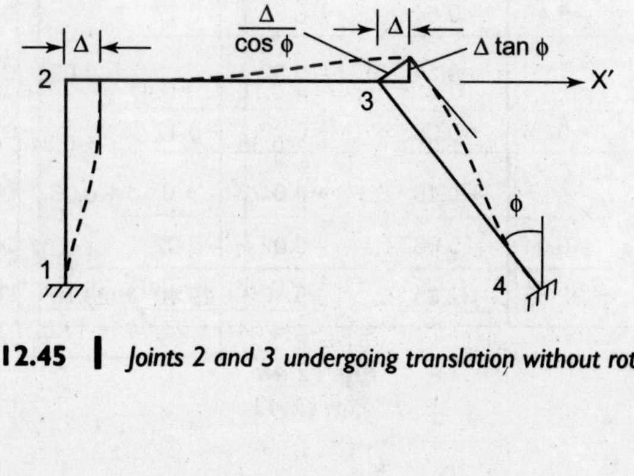

Fig. 12.45 | *Joints 2 and 3 undergoing translation without rotation*

Total artificial restraining force $X = 120 - (48.93 + 28.33)$

or $X = 42.74$ kN acting from right to left.

We shall analyse the frame once again by applying a force $(-X)$ on the structure. Consider the deflected shape of the frame (only lateral translation and no rotation of joints) under horizontal force X' as shown in Fig. 12.45.

The fixed end moments corresponding to this translation without rotation of joints are worked out as earlier. Taking $EI\,\Delta = 100$, the fixed end moments are

$$FEM_{12} = \frac{6EI\Delta}{(4)^2} = \frac{6 \times 100}{16} = 37.5 \text{ kN.m}$$

$$FEM_{21} = 37.5 \text{ kN.m}$$

$$FEM_{23} = FEM_{32} = -\frac{6 \times 2 \times 100 \times 0.75}{(6)^2} = -25.00 \text{ kN.m}$$

$$FEM_{34} = FEM_{43} = \frac{6 \times 100}{(5)^2\,(0.8)} = 30.00 \text{ kN.m}$$

The moment distribution is carried out in the table of Fig. 12.46. The free-body diagrams of the members of the frame are shown in Fig. 12.47. Summing up moments about joint 2 on the left hand column, we have

$$\Sigma M_2 = -H_{12}(4) + 34.95 + 32.35 = 0$$

or $H_{12} = 16.83$ kN

0	3/7	4/7	5/8	3/8	0	D.F
+ 37.50	+ 37.50	− 25.00	− 25.00	+ 30.00	+ 30.00	FEM
	− 5.36	− 7.14	− 3.13	− 1.87		Dist
− 2.68		− 1.57	− 3.57		− 0.94	C.O
	+ 0.67	+ 0.90	+ 2.23	+ 1.34		Dist
+ 0.34		+ 1.12	+ 0.45		+ 0.67	C.O
	− 0.48	− 0.64	− 0.28	− 0.17		Dist
− 0.24		− 0.14	− 0.32		− 0.09	C.O
	+ 0.06	+ 0.08	+ 0.20	+ 0.12		Dist
+ 0.03		+ 0.10	+ 0.04		+ 0.06	C.O
	− 0.04	− 0.06	− 0.02	− 0.02		Dist
+ 34.95	+ 32.35	− 32.35	− 29.40	+ 29.40	+ 29.70	Final Moments

Fig. 12.46

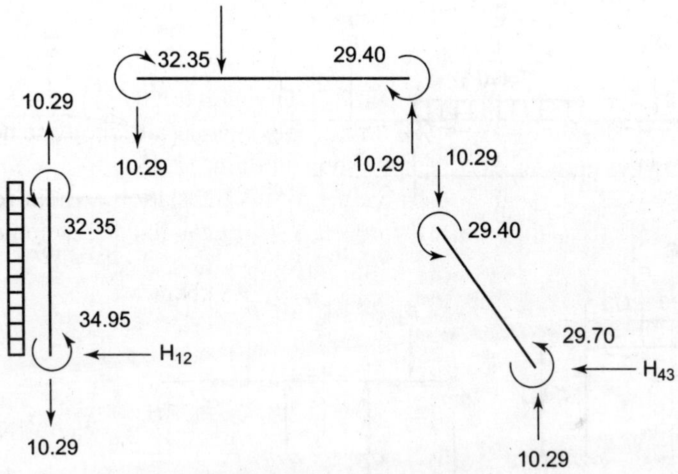

Fig. 12.47 | *Free-body diagram of frame members*

Similarly, summing up moments about joints 3 on the right hand column, we have

$$\Sigma M_3 = -H_{43}(4) + 10.29(3) + 29.7 + 29.4 = 0$$

or $\quad H_{43} = 22.49$ kN

Therefore, lateral force $X' = 16.83 + 22.49 = 39.32$ kN acting from left to right.

The moments in Fig. 12.46 are to be multiplied by a factor, $\dfrac{42.74}{39.32} = 1.087$, to get the true moments for the lateral force of $(-X)$ or 42.74 kN acting from left to right. Therefore, the final and the true moments by superposition are

$$M_{12} = 25.24 + 34.95(1.087) = 63.23 \text{ kN.m}$$
$$M_{21} = -69.52 + 32.35(1.0987) = -34.36 \text{ kN.m}$$
$$M_{23} = +34.36 \text{ kN.m}$$
$$M_{32} = -24.04 - 29.4(1.087) = -56.00 \text{ kN.m}$$
$$M_{34} = +56.00 \text{ kN.m}$$
$$M_{43} = 12.02 + 29.7(1.087) = 44.30 \text{ kN.m}$$

12.5 | SYMMETRICAL FRAMES

Symmetric Loading In symmetrical structures symmetrically loaded as in Fig. 12.41*a*, symmetrical joints rotate by the same amount but in the opposite direction.

Making use of this fact the stiffness of member AB in Fig. 12.48*b* is established.

The moment developed at end A can be written as

$$M_{AB} = 2E\frac{I}{L}(2\theta_A + \theta_B)$$

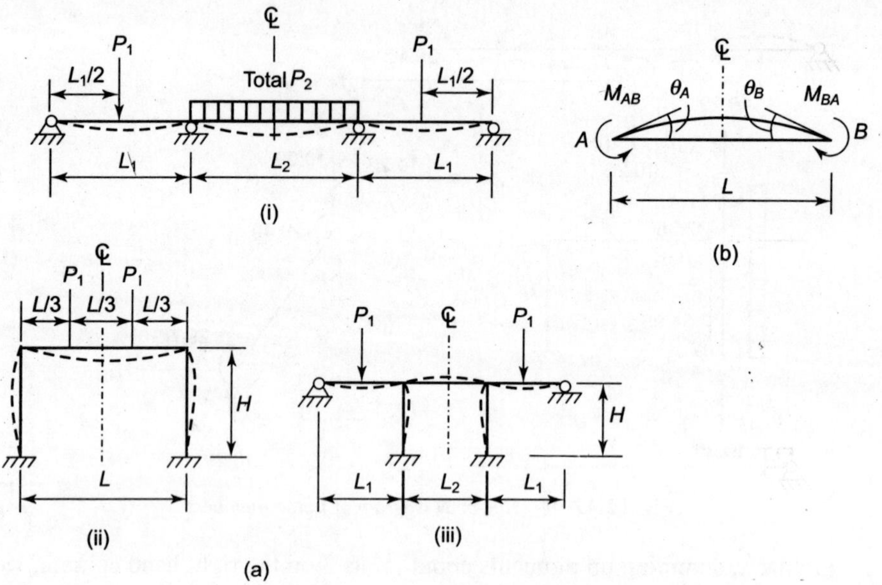

Fig. 12.48 | *(a) Symmetrical structure under symmetrical loading, (b) Symmetrically deflected beam*

But due to symmetry $\theta_A = -\theta_B$
Therefore,

$$M_{AB} = \frac{2EI}{L}\theta_A$$

or absolute stiffness $K' = \dfrac{2EI}{L}$

Comparing this with the stiffness value in Eq. 12.1, we have

$$K' = \frac{1}{2}K$$

Thus, if the distribution factors at the ends of the members common to each half of the structure are adjusted, the distribution procedure need only be carried out on one half and there will be no carry-over moments across the axis of symmetry. This would considerably reduce computational work.

Skew-Symmetric Loading Now consider symmetrical structures under a skew-symmetrical loading as shown in Fig. 12.49 in which symmetrical joints rotate the same amount but, in this case, in the same direction.

Consider the beam in Fig. 12.50 subject to end moments M_{AB} and M_{BA}. There is a point of contraflexure at mid span, that is

$$\theta_A = \theta_B = \theta$$

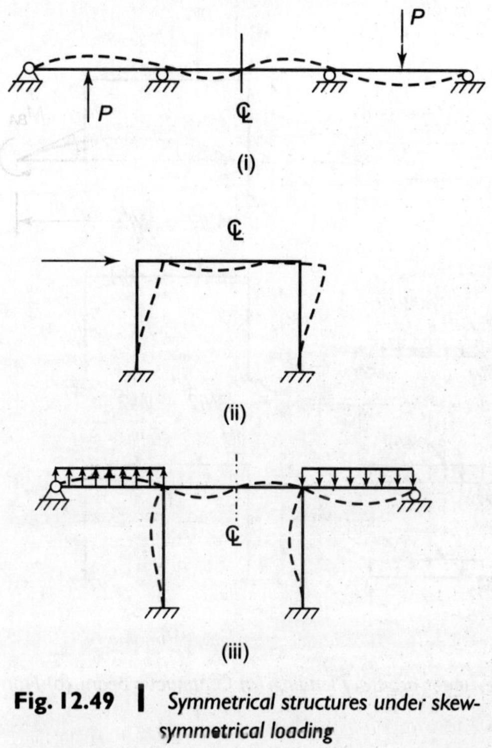

Fig. 12.49 | *Symmetrical structures under skew-symmetrical loading*

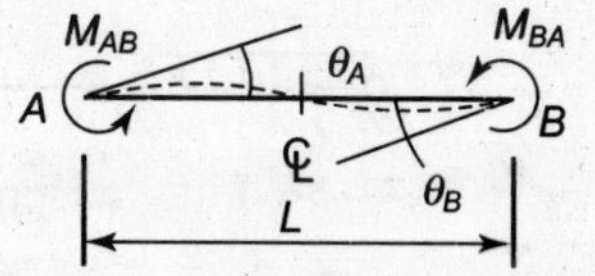

Fig. 12.50

Writing the moment at end A in terms of rotations, we have

$$M_{AB} = \frac{2EI}{L}(2\theta_A + \theta_B)$$

or

$$M_{AB} = \frac{6EI\theta}{L}$$

Thus, the absolute stiffness of member AB is

$$K' = \frac{6EI}{L} = \frac{3}{2}K$$

Or, alternatively, consider one half as a pin-ended member, $\frac{L}{2}$ long, then $K' = \frac{3}{4}\frac{I}{L/2} = \frac{3}{2}K$ as above. Again, if the distribution factors at the ends of the common members are adjusted, there is no carry over across the axis of symmetry and only half of the structure need be analysed.

Any Loading

Since any load system can be broken into two systems, one symmetrical and the other skew-symmetrical, these devices are very useful. Two examples are shown in Fig. 12.51.

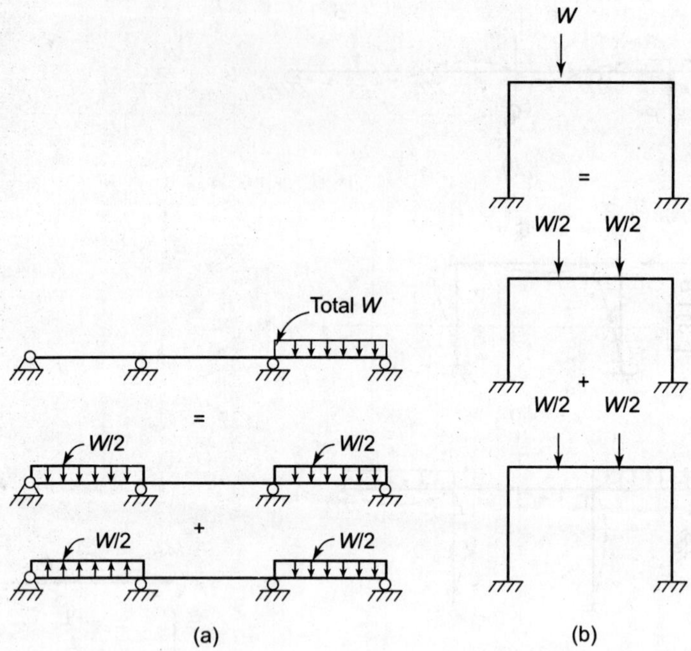

Fig. 12.51 | *Symmetrical structure, under arbitrary loading; (a) Continuous beam, (b) Frame*

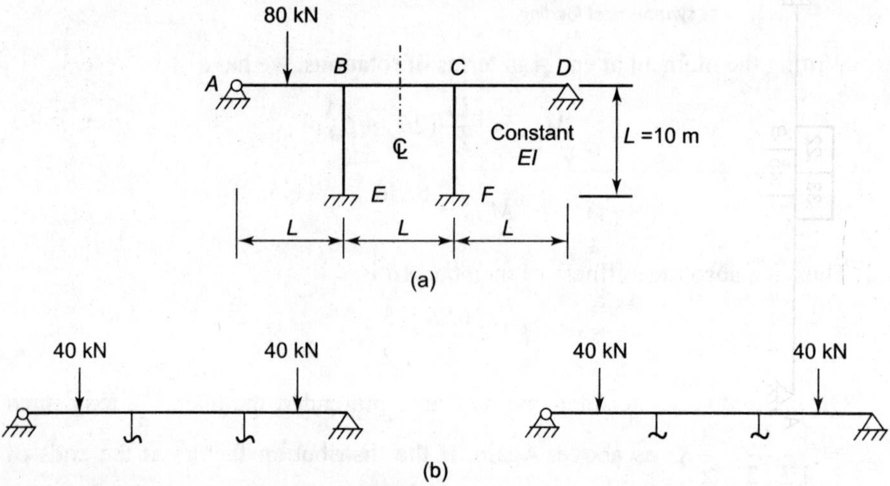

Fig. 12.52 | *(a) Frame and loading, (b) Symmetrical and skew-symmetric loading*

We shall illustrate the procedure by solving an example.

Example 12.14 | *Find the bending moment in the symmetrical frame shown in Fig. 12.52a replacing the loading by equivalent symmetrical and anti-symmetrical loading.*

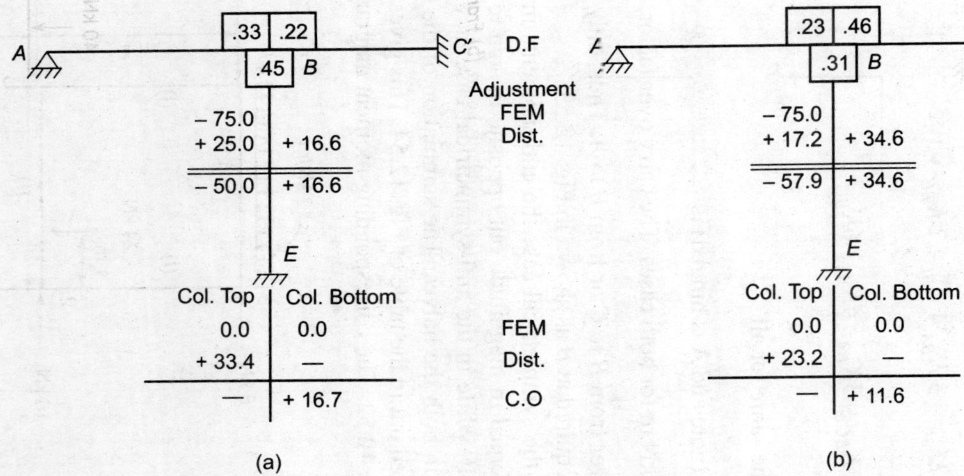

Fig. 12.53 | *(a) Symmetrical loading, (b) Skew-symmetrical loading*

End	*BA*	*BE*	*BC*	*EB*	*CD*	*CF*	*CB*	*FC*
Symmetric Loading	– 50.0	+ 33.4	+ 16.6	+ 16.7	+ 50.0	– 33.4	– 16.6	– 16.7
Antisymmetric Loading	– 57.8	+ 23.2	+ 34.6	+ 11.6	– 57.8	+ 23.2	+ 34.6	+ 11.6
Final End Moments for the frame	– 107.8	+ 56.6	+ 51.2	+ 28.3	– 7.8	– 10.2	+ 18.0	– 5.1

Fig. 12.54 | *Results of analysis*

The given loading on the symmetric structure is replaced by a symmetrical and an anti-symmetrical loading as indicated in Fig. 12.52*b*.

With symmetry or anti-symmetry of loading, the moment distribution need be carried out for only one-half of the frame. Figure 12.53*a* and *b* deal with the symmetrical and anti-symmetrical cases respectively. The relative stiffnesses for the members meeting at *B* are calculated on the basis

$$K'_{BA} : K'_{BE} : K'_{BC} = 3\ K_{BA}\ 4\ K_{BE} : 2\ K_{BC}$$

for the symmetrical case, and

$$K'_{BA} : K'_{BE} : K_{BC} = 3K_{BA} : 4K_{BE} : 6K_{BC}$$

for the anti-symmetrical case,

where $K = \frac{I}{L}$. In this case *K* is the same for all.

The FEM at B in member BA (note end A is hinged) is

$(FEM_{BA} = \frac{1}{2} FEM_{AB}) = -75.0$ kN.m for both cases. The carry over factor, C_{BE} = 0.5. No moments are carried over from *B* to *C* or from *B* to *A*. Thus, only one cycle of moment distribution is required at *B* as shown in Fig. 12.53.

It is important to note that in the symmetrical case the end moments in the right hand half of the frame are equal in magnitude but opposite in sign to the end moments in the left hand half, while in the anti-symmetrical case they are equal and also of the same sign in both the halves. The summation of the end moments in the two cases is carried out in the table of Fig. 12.54. This gives the end moments of the frame in Fig. 12.52. The corresponding moment diagram is shown in Fig. 12.55.

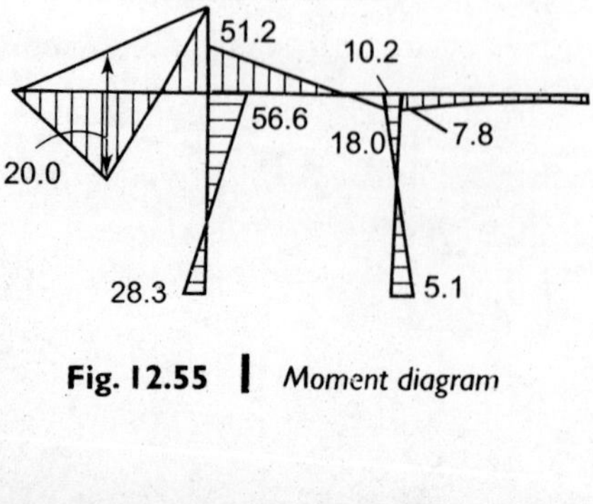

Fig. 12.55 | *Moment diagram*

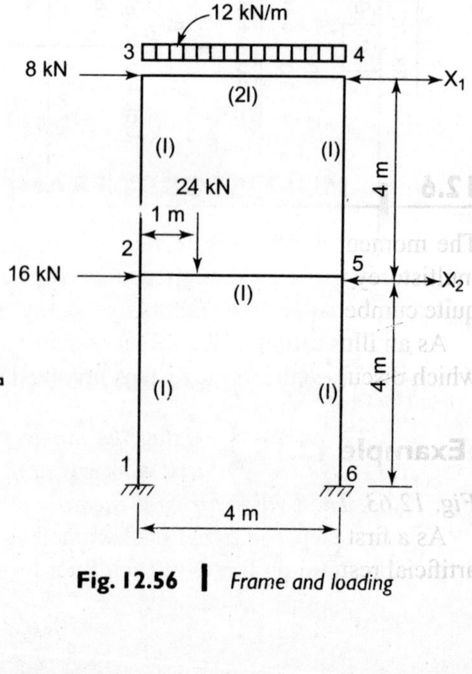

Fig. 12.56 | *Frame and loading*

COL. Top	COL. Bottom	Beam Left	Beam Right	Col. Top	Col. Bottom	
0	1/3	2/3	2/3	0	1/3	D.F
		– 16.00	+ 16.00			FEM
	– 5.33	+ 10.67	– 10.67		+ 5/33	Dist
	– 2.25	+ 5.34	– 5.34		+ 0.75	C.O
	– 1.03	– 2.06	+ 3.06		+ 1.53	Dist
	+ 0.32	+ 1.53	– 1.03		– 0.07	C.O
	– 0.62	– 1.23	+ 0.73		+ 0.37	Dist
	+ 0.10	+ 0.37	– 0.62		– 0.18	C.O
	– 0.16	– 0.31	+ 0.53		+ 0.27	Dist
	– 8.97	+ 8.97	– 8.00		+ 8.00	C.O

COL. Top	COL. Bottom	Beam Left	Beam Right	Col. Top	Col. Bottom	
1/3	1/3	1/3	1/3	1/3	1/3	D.F
		+ 13.50	– 4.50			FEM
– 4.50	– 4.50	– 4.50	+ 1.50	+ 1.50	– 1.50	Dist
– 2.67		+ 0.75	– 2.25	+ 2.67		C.O
+ 0.64	+ 0.64	+ 0.64	– 0.14	– 0.14	– 0.14	Dist
– 0.52		– 0.07	+ 0.32	+ 0.77		C.O
+ 0.20	+ 0.20	+ 0.20	– 0.36	– 0.36	– 0.36	Dist
– 0.31		– 0.18	+ 0.10	+ 0.19		C.O
+ 0.16	+ 0.16	+ 0.16	– 0.10	– 0.10	– 0.10	Dist
– 7.00	– 3.50	+ 10.50	– 5.43	+ 4.53	+ 0.90	C.O

+ 0.45

Fig. 12.57

12.6 MULTISTOREY FRAMES

The moment distribution technique can be extended to include the analysis of multistorey frames, although manual computation for such structures can become quite cumbersome if the number of storeys is large.

As an illustration of the moment distribution approach, an example is solved which discusses the various steps involved in the analysis.

Example 12.15 *Using the moment distribution method, determine the end moments of all the members of the frame shown in Fig. 12.63. The I value for each member is indicated on the frame.*

As a first step, the frame is restrained against lateral translation by providing artificial restraining forces at each floor level.

The distribution factors are evaluated using Eq. 12.12. The values are shown recorded at all the joints in Fig. 12.57. The moment distribution is carried out as usual. It may be noted that the carry over moments are always from column to column and beam to beam only. The moments obtained from the first stage of moment distribution are true only if lateral translation is prevented by the restraining forces acting at each floor level. First the value of the restraining forces is determined. In multistoreyed frames they are best obtained by finding the shears in the columns at the bases. The column shears are worked out using the free-body diagrams of columns as shown in Fig. 12.58. The summation of the horizontal forces in the top storey gives

$$8 + 3.99 - 3.13 + X_1 = 0$$

The forces acting from left to right are considered positive.

This gives $X_1 = -8.86$ kN.

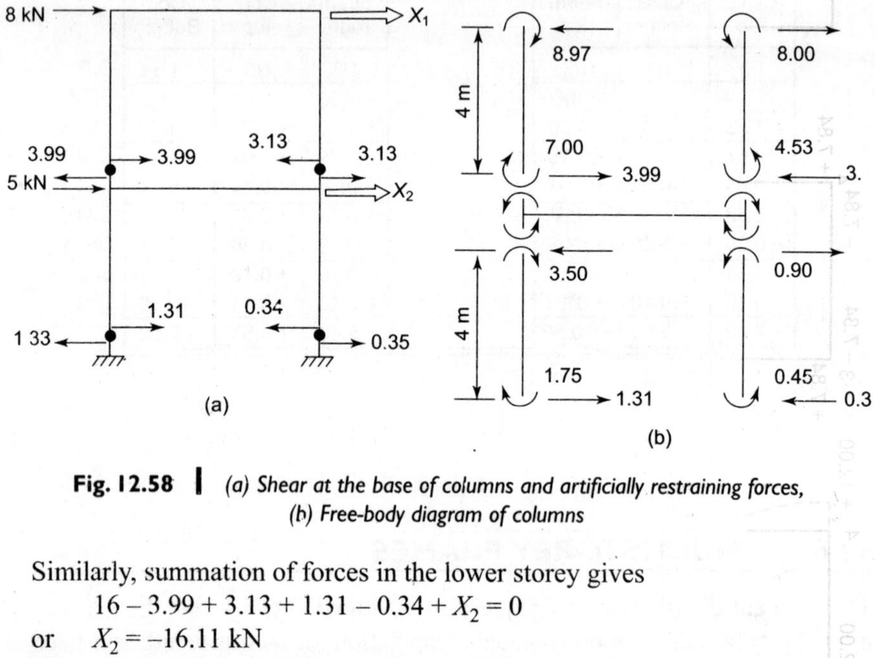

Fig. 12.58 | *(a) Shear at the base of columns and artificially restraining forces, (b) Free-body diagram of columns*

Similarly, summation of forces in the lower storey gives

$$16 - 3.99 + 3.13 + 1.31 - 0.34 + X_2 = 0$$

or $\quad X_2 = -16.11$ kN

The negative sign for forces X_1 and X_2 indicates that the restraining forces were acting from right to left.

The next step is to apply forces equal but opposite in direction to the restraining forces and work out moments induced in the ends of the members. As we have seen earlier, this procedure is to be carried out in an indirect manner.

A horizontal force of an unknown magnitude is assumed to act at the top level and at the same time holding the lower storey by an another unknown force. Under these forces, the frame lurches to the right as shown in Fig. 12.59*a*. The magnitude and sense of these forces can be determined later. The resulting fixed end moments due to translation but without rotation are shown in Fig. 12.59*a*.

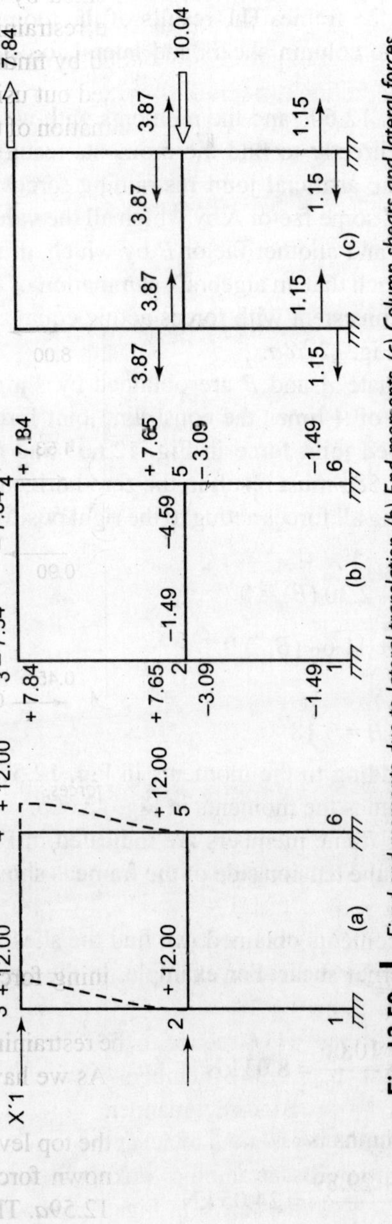

Fig. 12.59 | Fixed end moments due to lateral translation, (b) End moments, (c) Column shears and lateral forces

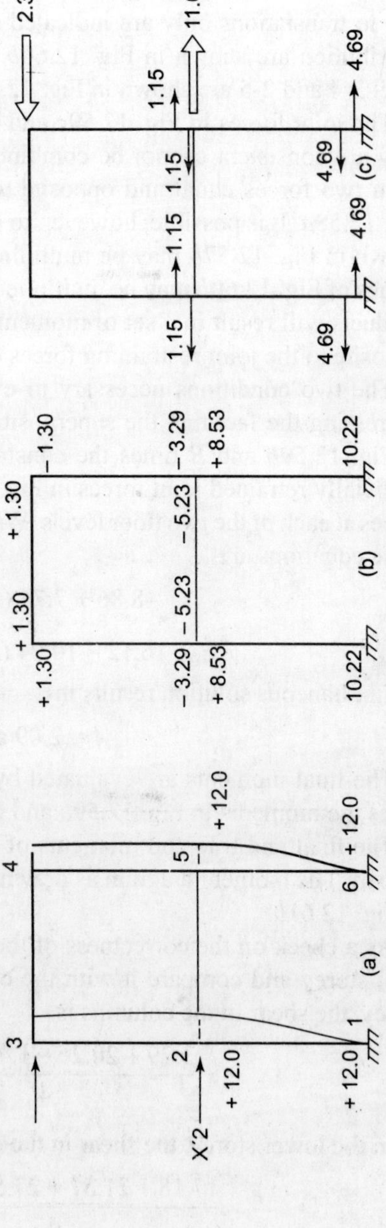

Fig. 12.60 | (a) Fixed end moments due to lateral translation, (b) End moments, (c) Column shears and lateral forces

Moments are balanced and distributed in the usual manner. The results of moment distribution are shown in Fig. 12.59*b*. The column shears and lateral forces at floor levels 3-4 and 2-5 are shown in Fig. 12.59*c*.

By another combination of horizontal forces acting at levels 3-4 and 2-5, the frame is next forced to translate as shown in Fig. 12.60*a*. The fixed end moments due to translations only are indicated on the frame. The results of the moment distribution are shown in Fig. 12.60*b*. The column shears and lateral forces at level 3-4 and 2-5 are shown in Fig. 12.60*c*.

The joint forces in Fig. 12.59*c* and Fig. 12.60*c* and the moments with which they are consistent cannot be combined directly to find the moments resulting from two forces equal and opposite to the artificial joint restraining forces of Fig. 12.58*a*. It is possible, however, to find some factor A by which all the values shown in Fig. 12.57*b* may be multiplied, and another factor *B* by which all the values of Fig. 12.60*b* may be multiplied, such that an algebraic summation of the products will result in a set of moments consistent with forces acting equal and opposite to the joint restraining forces of Fig. 12.58*a*.

The two conditions necessary to evaluate *A* and *B* are obtained by simply expressing the fact that the superposition of *A* times the constraint joint forces in Fig. 12.59*b* and *B* times the constrained joint force in Fig. 12.60*b* and the artificially retrained joint forces in Fig. 12.58*a* must result in the zero horizontal forces at each of the two floor levels. Writing all forces acting to the right positive, these equations are

$$-8.86 + 7.74\,(A) - 2.30\,(B) = 0$$

and

$$-16.12 - 10.04\,(A) + 11.68\,(B) = 0.$$

A simultaneous solution results in

$$A = 2.09 \text{ and } B = 3.18$$

The final moments are evaluated by adding to the moments in Fig. 12.57 A times the moments in Fig. 12.59*b* and B times the moments in Fig. 12.60*b*.

The final and true end moments of the frame members are indicated in Fig. 12.61*a*. The moment diagram as drawn on the tension side of the frame is shown in Fig. 12.61*b*.

As a check on the correctness of the moments obtained, we find the shear in each storey and compare it with the external shear. For example, in the upper storey, the shear in the columns is

$$\frac{3.29 + 20.26 - 1.47 + 10.06}{4} = 8.03 \text{ kN}$$

In the lower storey, the shear in the columns is

$$\frac{17.18 + 21.57 + 27.56 + 29.89}{4} = 24.05 \text{ kN}$$

They agree with the external shear.

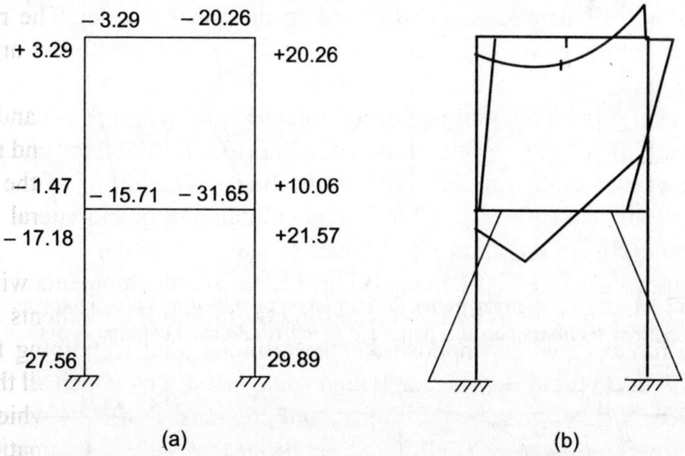

Fig. 12.61 | *(a) Results of analysis, (b) Moment diagram*

Thus the moment distribution method for frames undergoing lateral translations involves repeated distribution of moments and the solving of simultaneous equations.

12.7 | NO-SHEAR MOMENT DISTRIBUTION

In the analysis of frames subjected to lateral loading, moment distribution has to be carried out, which results in solving of simultaneous equations. In the no-shear moment distribution, the side sway is allowed to occur freely during the moment distribution, that is, no change in the forces acting at the floor level takes place when the joints are allowed to rotate, thus the shear in the columns is not changed during the distribution. The method was originally developed for symmetrical one bay multistorey frames supporting anti-symmetrical loading. The method is also known as *cantilever moment distribution*.

The adjusted end rotational stiffnesses and fixed end moments required for the no-shear moment distribution are discussed below.

Consider a symmetrical frame loaded anti-symmetrically as in Fig. 12.62*a*. The deflected shape of the column is shown in Fig. 12.62*b* in which the end B is allowed to sway by Δ without rotation of joint when rotation θ occurs at A due to moment applied at that end.

The moments at ends A and B for the member can be obtained as the summation of moments due to independent deformations shown in Fig. 12.62*c* and *d*.

$$M_A = \frac{6EI\phi}{l} - \frac{4E\theta}{l} \tag{12.13}$$

and

$$M_B = \frac{6EI\phi}{l} - \frac{2EI\theta}{l} \tag{12.14}$$

M_A should be equal to $-M_B$ for no shear in columns AB.

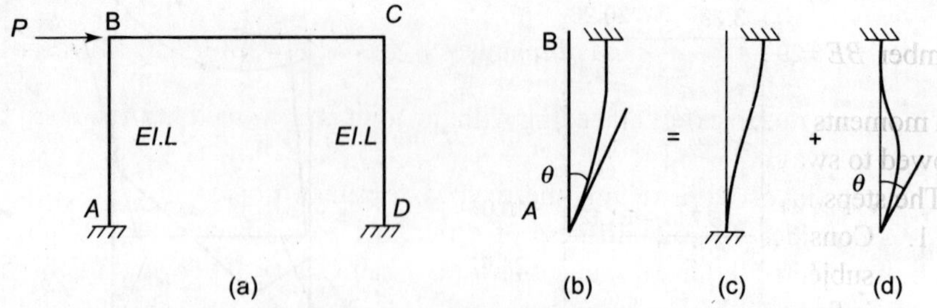

Fig. 12.62 | *(a) Symmetrical frame, (b) Translation permitted at top and rotation imposed at bottom, (c) Translation only permitted at top, (d) Rotation imposed at bottom*

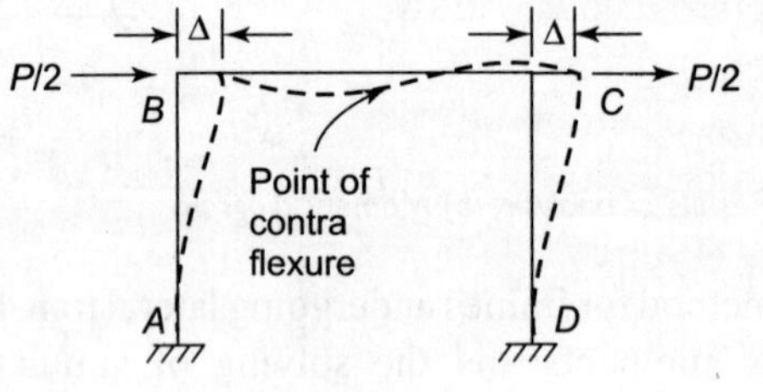

Fig. 12.63

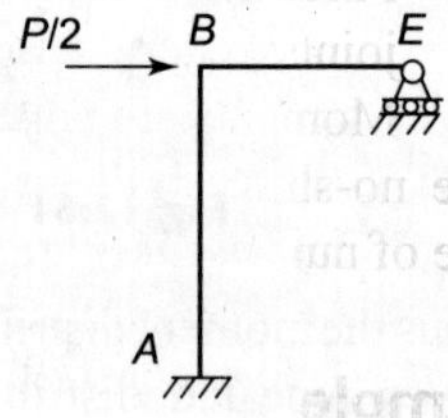

Fig. 12.64

Equating $M_A = -M_B$

$$\frac{6EI\phi}{l} - \frac{4EI\theta}{l} = -\frac{6EI\phi}{l} + \frac{4EI\theta}{l}$$

$$\phi = \frac{\theta}{2} \tag{12.15}$$

Substituting in Eqs. 12.13 and 12.14

Moment at $$A = -\frac{EI}{l} \tag{12.16}$$

Moment at $$B = \frac{EI}{l} \tag{12.17}$$

Thus the stiffness for member $AB = \dfrac{EI}{l}$ and the carry over factor $= -1$.

The values for stiffness and C.O.F. can also be obtained by the moment distribution method. If the horizontal load is split into two equal forces as shown in Fig. 12.63 the frame will be subjected to anti-symmetrical loading. Under the loading the joints B and C translate horizontally by the same amount, and there is a contraflexure point at the mid-point of BC. It is therefore sufficient to consider the frame in Fig. 12.64 which will have the same end moments as in the left half of the original frame.

Our task is thus to carry out the moment distribution at joint B with the translation allowed to take place freely. The relative and rotational stiffness of member $BE = \frac{3EI}{l/2} = \frac{6EI}{l}$ and of member $BA = \frac{EI}{l}$ as earlier. The FEM are the end moments due to external loading with the joint *B* prevented from rotation but allowed to sway.

The steps involved in solving these types of frames are:

1. Consider that the stiffness of a member parallel to axis of symmetry subjected to lateral translation is EI/l and C.O.F. is –1 when the far end is fixed.
2. Consider that the stiffness of a member perpendicular to the axis of symmetry is $6\,EI/l$ and the C.O.F. is zero.
3. Fixed end moments are determined due to sway only without rotation of joints under lateral loading.
4. Moments are balanced for one half of the frame only.

The no-shear moment distribution method is best explained by solving a couple of numerical examples.

Example 12.16 | *Using the no-shear moment distribution, obtain the bending moment diagram for the frame shown in Fig. 12.65. Consider that all the members have the same value of EI.*

The frame under anti-symmetric loading is shown in Fig. 12.65*b* and the analysis has to be carried out for the frame shown in Fig. 12.65*c*.

Stiffness of member $AB = \frac{EI}{4}$

$$BE = \frac{6EI}{6}$$

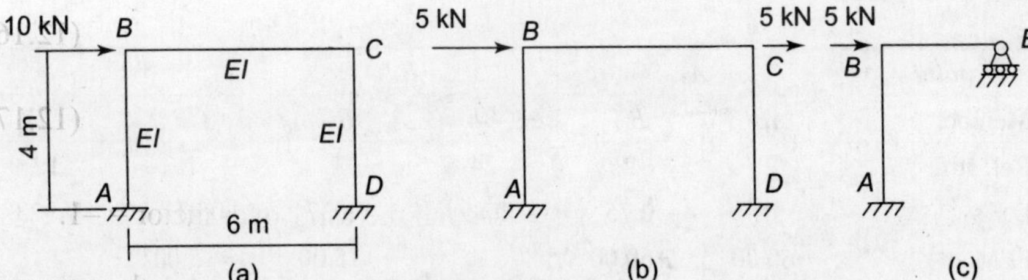

Fig. 12.65 | *(a) Frame under lateral load, (b) Frame under antisymmetric loading, (c) Frame for analysis*

The moment distribution for the frame is shown below. It may be noted that the fixed end moments are obtained by allowing lateral translation freely but restraining end rotations. The fixed end moment $FEM_{AB} = FEM_{BA}$ and using equilibrium equation $FEM_{AB}/2 = 5$ which gives $FEM_{AB} = FEM_{BA} = 10.0$ kN.m

Joint	*A*	*B*	
Member	*AB*	*BA*	*BE*
Rel. stiff	1/4	1/4	1
D.F.s	0	0.2	0.8
FEM	+10	+10	0
Bal. & C.O.	+2	–2	–8
Final moments	+12	+8	–8

The B.M. diagram is shown in Fig. 12.66.

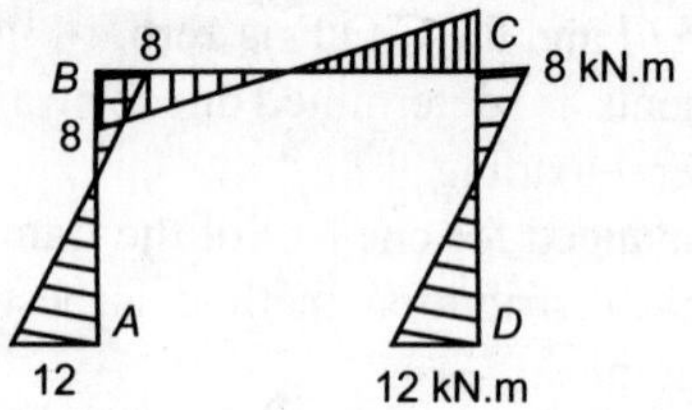

Fig. 12.66

Example 12.17 | *Using no-shear moment distribution, analyse the symmetrical two-storey frame loaded as shown in Fig. 12.60. Draw the B.M. diagram. The values encircled are the relative stiffness, I/l, of members.*

The moment distribution is carried out at joints A, B and C for half of the frame shown in Fig. 12.67b. The relative end-rotational stiffnesses are $6I/l$ for the beams and I/l for the columns. The C.O.Fs are $C_{BA} = C_{BC} = C_{CB} = -1$. The fixed end moments in the columns are calculated using equilibrium equations in each floor level.

Joint	*A*		*B*		*C*	
Member	*AB*	*BA*	*BE*	*BC*	*CB*	*CD*
Rel. stiff	2	2	24	1	1	12
D.F.s	0	0.75	0.89	0.037	0.077	0.923
FEM	+60.00	+60.00		+15.00	+15.00	
Bal.		–5.55	–66.75	–2.78	–1.16	–13.84
C.O.	+5.55			+1.16	+2.78	
Bal.		–0.09	–1.03	–0.04	–0.21	–2.56
C.O.	+0.09			+2.14	+0.04	
Bal.		–2.02	–0.19	–0.01	0.0	–0.04
Final moments	+65.64	+54.34	–67.97	+13.54	+16.44	–6.44

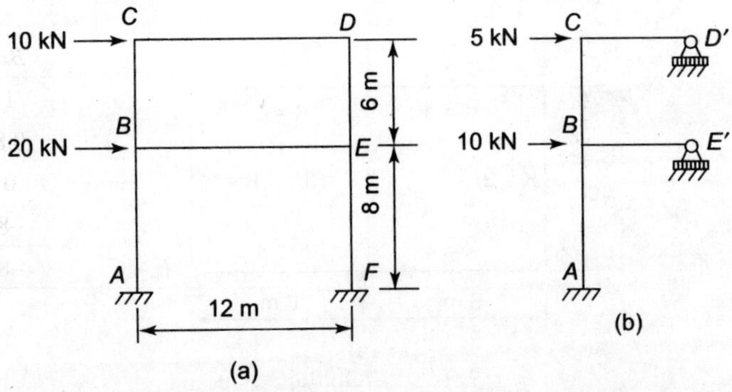

**Fig. 12.67 | ** *(a) Frame and the loading (b) Frame for analysis*

The moment diagram is shown in Fig. 12.68.

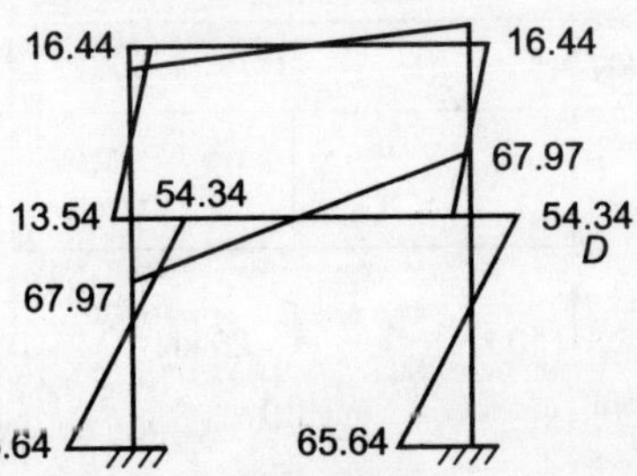

Fig. 12.68

Example 12.18 | *Using no-shear moment distribution method, analyse the vierendel girder in Fig. 12.70. The relative values of $K = I/l$ for the members are indicated on the girder.*

The frame is symmetrical about mid height. The symmetrical frame under anti-symmetric loading is shown in Fig. 12.69*b*. Therefore, the analysis needs to be carried out for one half of the frame. The relative stiffness values and the carry-over factors are shown in the table that follows. The fixed end moments are calculated using equilibrium conditions for each bay.

Joint	*B*		*C*			*D*	
Member	*BA*	*BC*	*CB*	*CF*	*CD*	*DC*	*DE*
Rel. stiff	18	4	4	18	4	4	18
Dist.Factory	0.82	0.18	0.15	0.70	0.15	0.18	0.82
C.O.F.	0	–1	–1	0	–1	–1	0
FEM	0	+30.0	+30.0		–30.0	–30.0	0
Bal.	–24.6	–5.4	0	0	0	+5.4	+24.6
C.O.		0	+5.4		–5.4		
Final moments	–24.6	+24.6	+35.4	0	–35.4	–24.6	+24.6

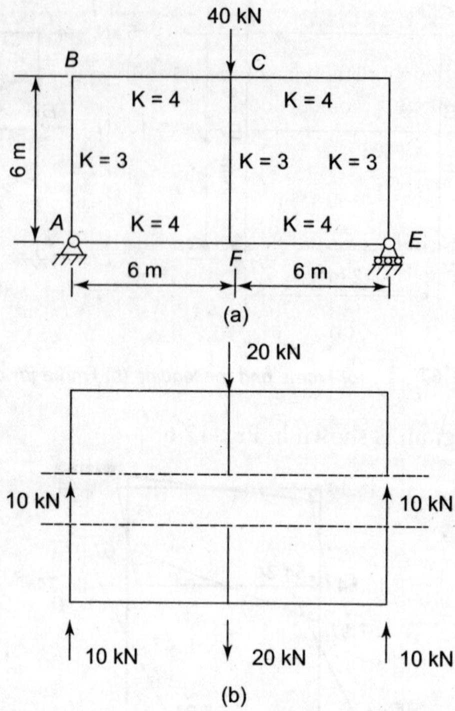

Fig. 12.69 | *(a) Vierendel girder and the loading, (b) Symmetrical frame under anti-symmetrical loading*

The bending moment diagram is shown in Fig. 12.70.

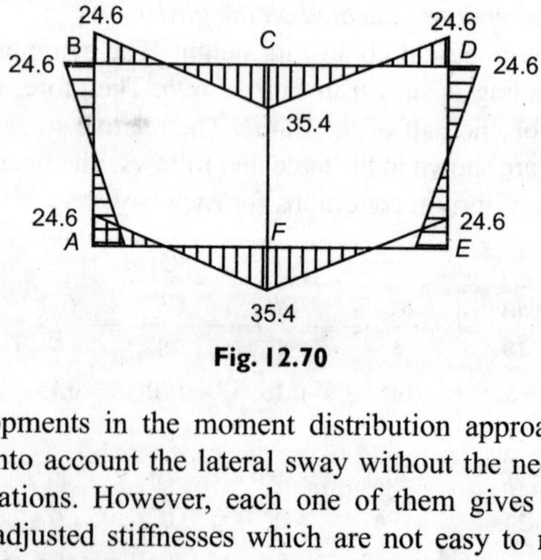

Fig. 12.70

Several developments in the moment distribution approach have made it possible to take into account the lateral sway without the necessity of solving simultaneous equations. However, each one of them gives a set of rules in working out the adjusted stiffnesses which are not easy to remember. Kani's method which is dealt with in chapter 13 may be more convenient to apply in such cases.

Problems for Practice

Use the moment distribution method in solving the following problems.

12.1 Determine the support moments of the beam shown in Fig. 12.71. *EI* is constant throughout.

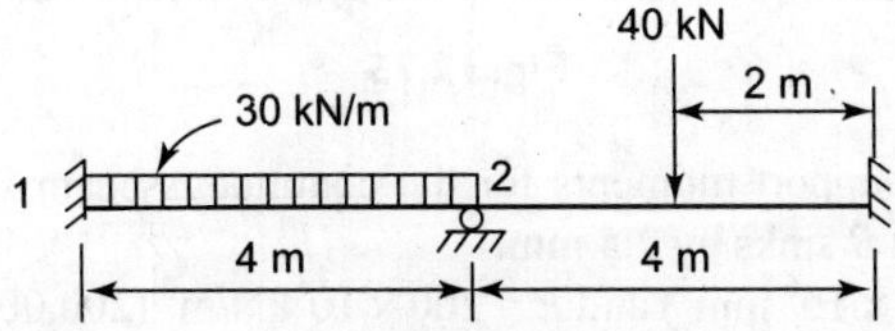

Fig. 12.71

12.2 Determine the moment over the central support and sketch the shear force and moment diagrams for the beam shown in Fig. 12.72. *E* is constant and *I* values are indicated on the beam.

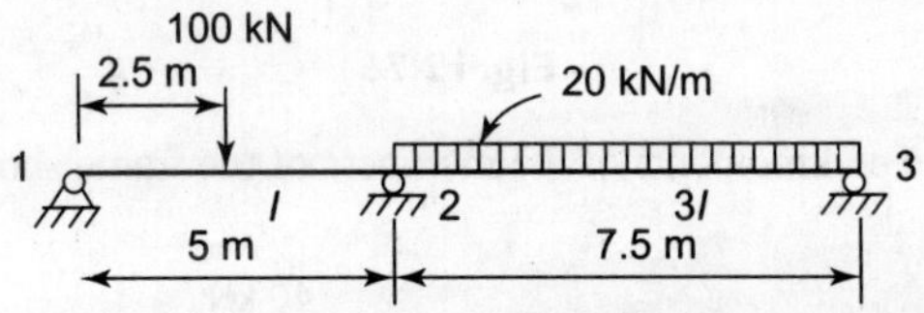

Fig. 12.72

12.3 Determine the support moments and sketch the moment diagram for the beam shown in Fig. 12.73. *EI* is constant.

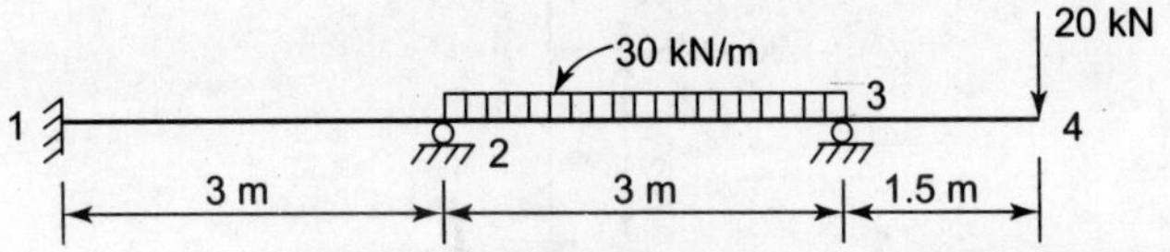

Fig. 12.73

12.4 Determine the support moments for the beam shown in Fig. 12.74. EI is constant.

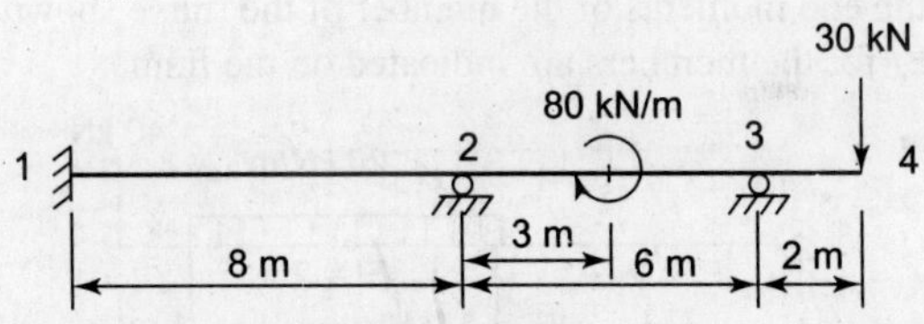

Fig. 12.74

12.5 Determine the support moments for the beam shown in Fig. 12.75. *E* is constant and *I* values are as indicated on the beam.

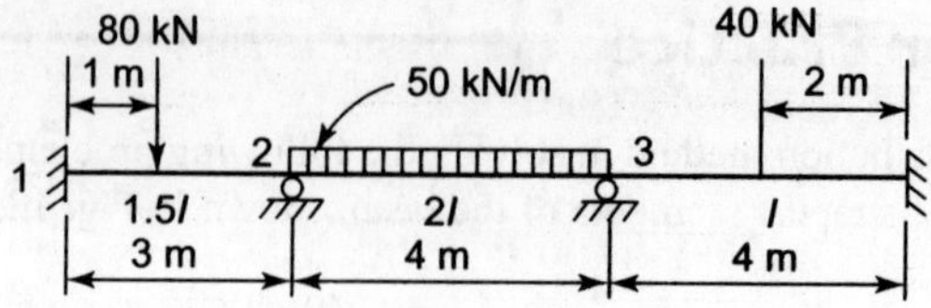

Fig. 12.75

12.6 Determine the support moments for the continuous beam shown in Fig. 12.76. Under the load support B sinks by 2.5 mm.
$I = 350 \times 10^{-6}$ m^4 (350×10^6 mm^4) and $E = 200 \times 10^6$kN/m^2 (200,000 MPa) for all members.

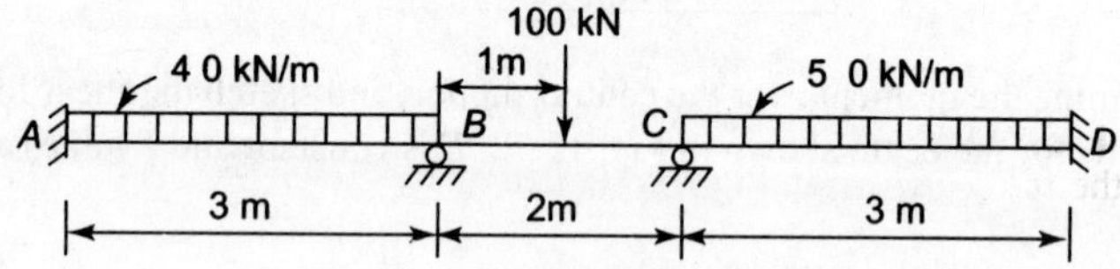

Fig. 12.76

12.7 Determine the end moments of the members of the frame shown in Fig. 12.77. *EI* is constant throughout.

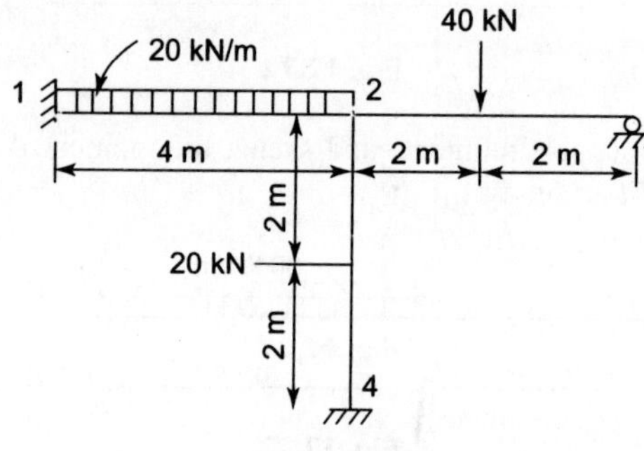

Fig. 12.77

12.8 Determine the end moments of the member of the frame shown in Fig. 12.78. The relative values of *EI* for the members are indicated on the frame.

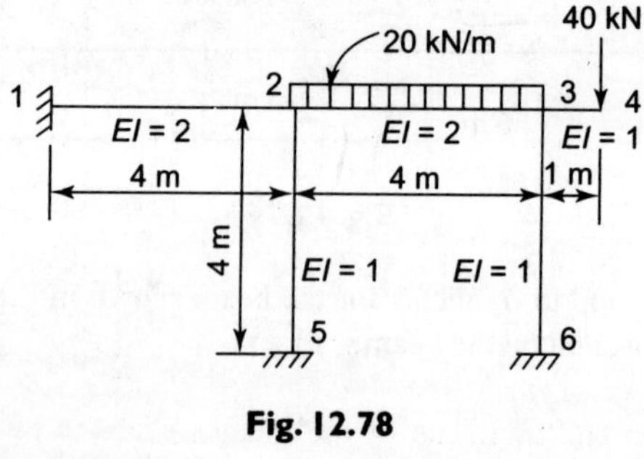

Fig. 12.78

12.9 Determine the moments at the ends of the members of the frame shown in Fig. 12.79. *EI* values are indicated on the frame.

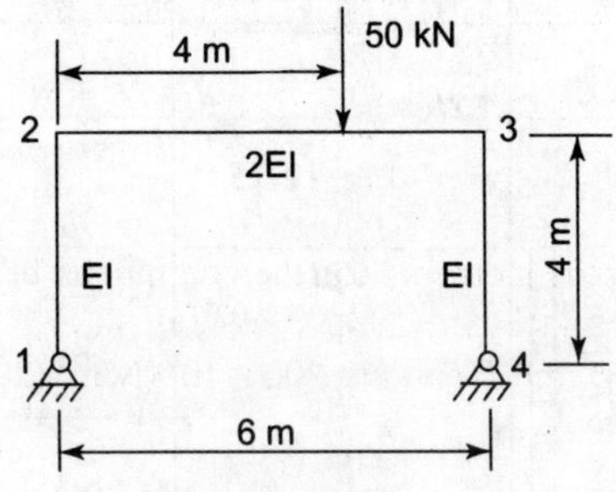

Fig. 12.79

12.10 For the frame shown in Fig. 12.80, determine the end moments of members. *EI* is constant.

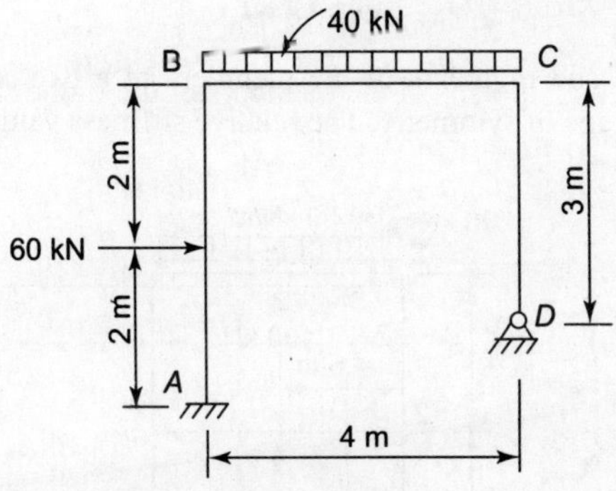

Fig. 12.80

12.11 Determine the end moments of the members of the frame shown in Fig. 12.81. *EI* is constant.

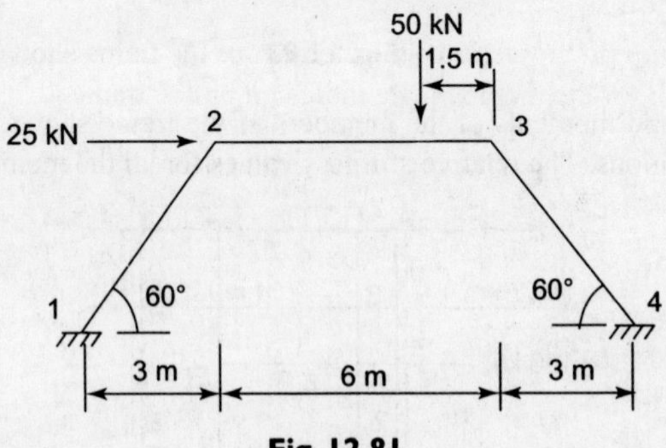

Fig. 12.81

12.12 Determine the end moments of the members of the two-storey frame shown in Fig. 12.82. *E* is constant and relative *I* values are indicated on the members of the frame.

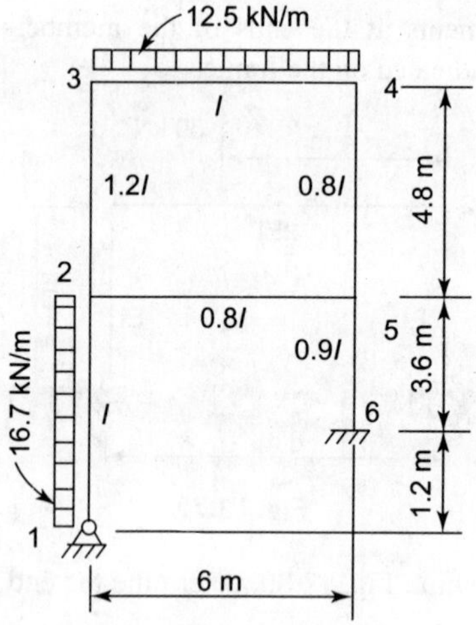

Fig. 12.82

12.13 Determine the end moments of the members of a two-storey frame shown in Fig.12.83. Take advantage of symmetry. The relative stiffness values for all the members are indicated.

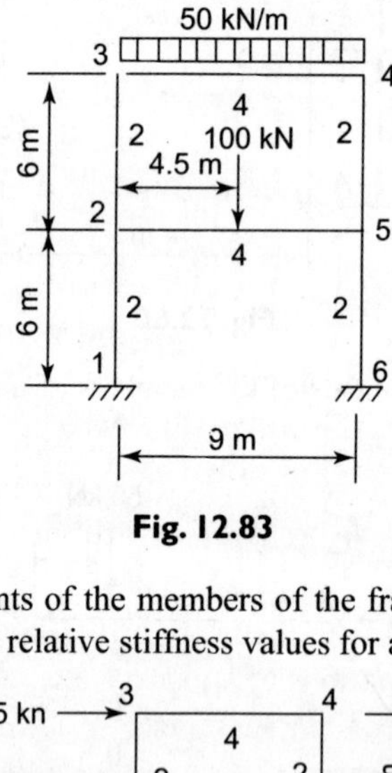

Fig. 12.83

12.14 Solve the end moments of the members of the frame shown in Fig. 12.84. Use anti-symmetric relations. The relative stiffness values for all the members are indicated

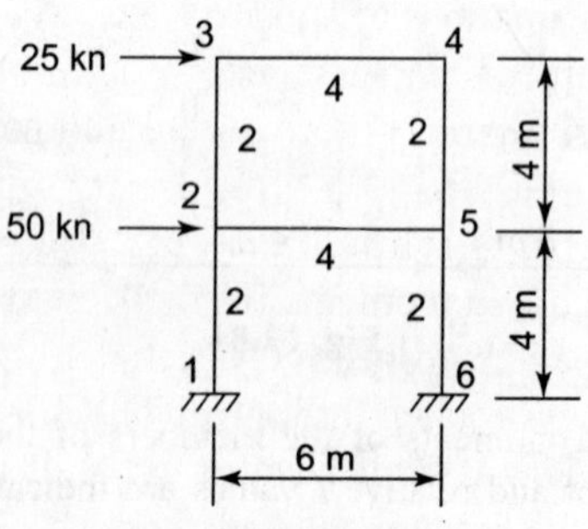

Fig. 12.84

13 Kani's Method

13.1 INTRODUCTION

This is an iteration method. This method was developed by Gasper Kani of Germany in 1947. The method is an excellent extension of the slope-deflection method. It has the simplicity of moment distribution. Since the method has been recognised as one which is very useful, it is discussed in some detail in this chapter.

13.2 BASIC CONCEPT

13.2.1 Members without Translation of Joints

Let AB represent a beam in a frame or a continuous structure under transverse loading as shown in Fig. 13.1*a*. *A* general deflected shape of the member under the loading is shown in Fig. 13.1*b*. For the time being it is assumed that the joints do not translate and only ends *A* and *B* undergo rotations θ_A and θ_B respectively. Let M_{AB} and M_{BA} represent the end moments of beam *AB*.

Sign Convention

We follow the same sign convention as is followed in the slope-deflection method, that is,

1. anti-clockwise end moments are positive, and
2. anti-clockwise rotations are positive.

The actual end moments in member *AB* may be thought of as moments developed due to a superimposition of the following three components of deformation:

1. The member *AB*, to start with, is regarded as completely restrained or fixed. The fixed end moments for this condition are written as M^F_{AB} and M^F_{BA} at ends *A* and *B* respectively (Fig. 13.1*c*).
2. Only the end *A* is rotated through an angle θ_A inducing a moment $2M'_{AB}$ at end *A* and M'_{AB} at farther end *B* which is fixed (Fig. 13.1*d*) Moment M'_{AB} is called the *rotation moment* at end *A*.

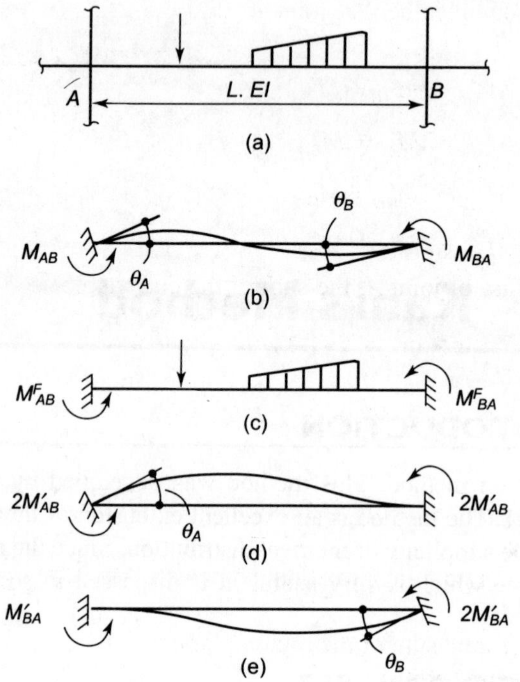

Fig. 13.1 | *(a) Beam in a continuous structure, (b) Deflected shape—ends undergo rotation only, (c) Fixed end moments, (d) Only end A rotates by θ_A, (e) Only end B rotates by θ_B.*

3. Next end A is considered as fixed in the rotated condition and only end B is rotated through an angle θ_B which induces a moment $2M'_{BA}$ at B and moment M'_{BA} at end A (Fig. 13.1*e*). The moment M'_{BA} is called the rotation moment at end B.

Thus, the final moments M_{AB} and M_{BA} can be expressed as the superposition of the three moments, that is

$$M^F_{AB} = M^F_{AB} + 2M'_{AB} + M'_{BA}$$

and
$$M_{BA} = M^F_{BA} + 2M'_{BA} + M'_{AB} \tag{13.1}$$

For member AB, when we refer to the final moment M_{AB} at A, end A may be referred to as the near end and B as the far end. Similarly, when we refer to moment M_{BA} at B, end B may be referred to as the near end and end A as the far end. Therefore, the relationship in Eq. 13.1 may be stated as follows: the true moment at the near end of a member is the algebraic sum of (a) the fixed end moment at the near end due to applied loading, (b) twice the rotation moment of the near end and (c) the rotation moment of the far end.

Figure 13.2 shows a multi-storeyed frame. If no translation of joints occur, Eq. 13.1 is applicable to all the members.

Consider various members at joint A. End moments at A for the members meeting at A are given by

$$M_{AB} = M^F_{AB} + 2M'_{AB} + M'_{BA}$$

$$M_{AC} = M^F_{AC} + 2M'_{AC} + M'_{CA} \tag{13.2}$$

$$M_{AD} = M^F_{AD} + 2M'_{AD} + M'_{DA}$$

$$M_{AE} = M^F_{AE} + 2M'_{AE} + M'_{EA}$$

For the equilibrium of joint A, the sum of the end moments at A must be zero, that is

$$\Sigma M_{AB} = 0 \tag{13.3}$$

or

$$\Sigma M^F_{AB} + 2\,\Sigma M'_{AB} + \Sigma M'_{BA} \tag{13.4}$$

where

ΣM^F_{AB} = algebraic sum of fixed end moments at A of all members meeting at A.

$\Sigma M'_{AB}$ = algebraic sum of rotation moments at A of all the members meeting at A.

$\Sigma M'_{BA}$ = algebraic sum of the rotation moments of far ends of the members meeting at A.

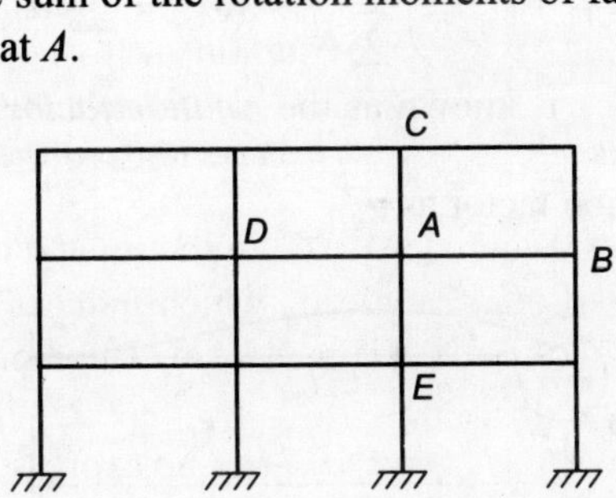

Fig. 13.2 | *A building frame*

From Eq. 13.4, we have

$$\sum M'_{AB} = \left(-\frac{1}{2}\right)\left(\sum M^F_{AB} + \sum M'_{BA}\right) \tag{13.5}$$

or

$$\sum M'_{AB} = \left(-\frac{1}{2}\right)\left(M^F_A + \sum M'_{BA}\right) \tag{13.6}$$

where $M^F_A = \Sigma M^F_{AB}$ = sum of fixed end moments at joint A.

From the moment-rotation relationship given in the Appendix we have for the beam in Fig. 13.3,

$$2M'_{AB} = \frac{4\,E I_{AB}\,\theta_A}{L_{AB}} = 4\,E K_{AB}\,\theta_A \tag{13.7}$$

where

$$K_{AB} = \frac{I_{AB}}{L_{AB}} \text{ the relative stiffness of member } AB.$$

Therefore,

$$M'_{AB} = 2E\, K_{AB}\, \theta_A \tag{13.8}$$

At joint A all the members undergo the same rotation θ_A. Assuming E is the same for all

$$\sum M'_{AB} = 2\, E\theta_A \sum K_{AB} \tag{13.9}$$

Dividing Eq. 13.8 by Eq. 13.9. we have M_{AB}

$$\frac{M'_{AB}}{\sum M'_{AB}} = \frac{K_{AB}}{\sum K_{AB}} \tag{13.10}$$

or
$$M'_{AB} = \frac{K_{AB}}{\sum K_{AB}} \sum M'_{AB} \tag{13.11}$$

Substituting for ΣM_{AB} from Eq. 13.6

$$M'_{AB} = \left(-\frac{1}{2}\right)\frac{K_{AB}}{\sum K_{AB}}\left(M_A^F + \sum M'_{BA}\right) \tag{13.12}$$

The ratio $\left(-\frac{1}{2}\right)\frac{K_{AB}}{\sum K_{AB}}$ is known as the *rotation factor* for the member AB at joint A. Denoting rotation factor as u_{AB}

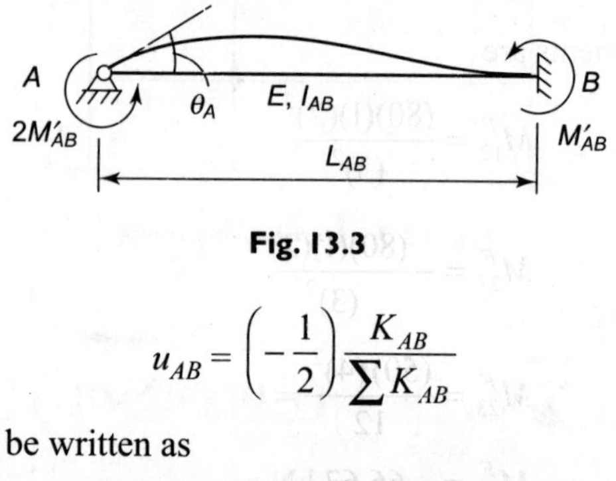

Fig. 13.3

$$u_{AB} = \left(-\frac{1}{2}\right)\frac{K_{AB}}{\sum K_{AB}} \tag{13.13}$$

Eq. 13.12 can now be written as

$$M'_{AB} = u_{AB}\,(M^F_A + \Sigma M'_{BA}) \tag{13.14}$$

In this equation the summation of fixed end moments M^F_A is a known quantity. To start with, the far end rotation moments M'_{BA} are not known and hence they may be taken as zero.

By a similar approximation, the rotation moments at other joints are also determined. With the approximate values of rotation moments computed, it is possible to determine again a more correct value of the rotation moment at A for member A_B using Eq. 13.14.

The process mentioned above is iterated till the desired accurate values of the rotation moments are obtained. After attaining the desired degree of accuracy in the values of the rotation moments, the final moments can be computed using Eq. 13.2.

Some Important Points

1. The sum of the rotation factors at a joint is $\left(-\frac{1}{2}\right)$
2. If an end of a member is fixed, the rotation at that end being zero, the rotation moment is also zero.
3. If an end of a member is hinged or pinned, it is convenient to consider it as fixed and take the relative stiffness as $\left(\frac{3}{4}\right)\frac{I}{L}$
4. The following examples illustrate the procedure involved.

Example 13.1 | *Determine the support moments for the continuous beam of Fig. 13.4. The relative I values are indicated along the member in each span. E is constant.*

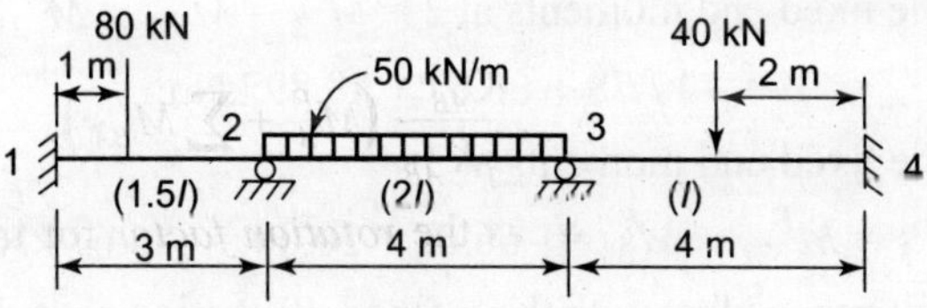

Fig.13.4 | *Beam and loading*

The fixed end moments are

$$M_{12}^F = \frac{(80)(1)(2)^2}{(3)^2} = 35.56 \text{ kN.m}$$

$$M_{21}^F = -\frac{(80)(1)(2)^2}{(3)^2} = -17.78 \text{ kN.m}$$

$$M_{23}^F = \frac{(50)(4)^2}{12} = 66.67 \text{ kN.m}$$

$$M_{32}^F = -66.67 \text{ kN.m}$$

$$M_{34}^F = \frac{40(4)}{8} = 20 \text{ kN.m}$$

$$M^F{}_{43} = -20 \text{ kN.m}$$

Next we evaluate the rotation factors at joints 2 and 3.

Joint	Members	Rel. stiffness K	ΣK	Rotation factor $u = \left(-\frac{1}{2}\right)\frac{K}{\sum K}$
2	2–1	$1.5I/3$	I	$\left(-\frac{1}{4}\right)$
	2–3	$2I/4$		$\left(-\frac{1}{4}\right)$
3	3–2	$2I/4$	$\frac{3}{4}I$	$\left(-\frac{1}{3}\right)$
	3–4	$I/4$		$\left(-\frac{1}{6}\right)$

The sum of the fixed end moments at 2 = $M^F{}_2 = M^F{}_{21} + M^F{}_{23}$

$$= -17.78 + 66.67 + 48.89 \text{ kN.m}$$

The sum of the fixed end moments at 3

$$M^F{}_3 = M^F{}_{32} + M^F_{34} = -66.67 + 20.00 = -46.67 \text{ kN.m}$$

The scheme for proceeding with the method of rotation contributions is shown in Fig. 13.5. The beam line is drawn and joints 2 and 3 are marked by two squares one inside the other. The sum of the fixed end moments at each joint are entered in the inner squares.

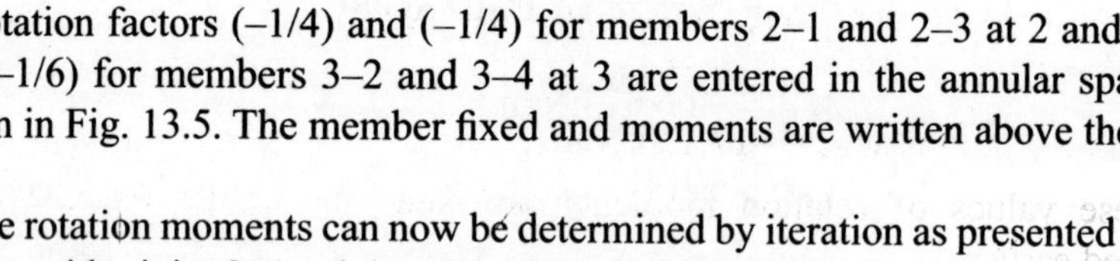

Fig. 13.5

Rotation factors (–1/4) and (–1/4) for members 2–1 and 2–3 at 2 and (–1/3) and (–1/6) for members 3–2 and 3–4 at 3 are entered in the annular spaces as shown in Fig. 13.5. The member fixed and moments are written above the beam line.

The rotation moments can now be determined by iteration as presented below. First consider joint 2. Applying Eq. 13.14 to this joint, we have

$$M'_{21} = u_{21}\left(M_2^F + \sum M'_{12}\right)$$

and

$$M'_{23} = u_{23}\left(M_2^F + \sum M'_{12}\right) \quad (13.15)$$

in which

M^F_2 = sum of fixed end moments at 2 = 48.89 kN.m

$\Sigma M'_{12}$ = rotation moments of far ends of members meeting at joint 2.

We know $\quad M'_{12} = 0$ since end 1 is fixed end

$M'_{32} = 0$ assumed to start with.

Substituting these values in Eq. 13.15

$$M'_{21} = (-1/4)\,(48.89 + 0) = -12.22 \text{ kN.m}$$

and

$$M'_{23} = (-1/4)\,(48.89 + 0) = -12.22 \text{ kN.m}$$

These rotation moments are entered below the beam line at the appropriate places as shown in Fig. 13.6.

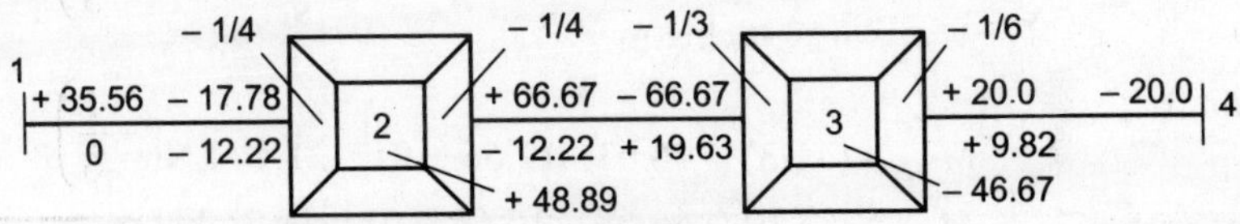

Fig. 13.6

Now consider joint 3. Rotation moments M'_{32} and M'_{34} will be determined using Eq. 13.14

$$M'_{32} = u_{32}(-46.67 - 12.22 + 0)$$
$$= (-1/3)\,(-58.89) = +19.63 \text{ kN.m}$$

and

$$M'_{34} = (-1/6)\,(-58.89) = +9.82 \text{ kN.m}$$

These rotation moments are shown entered in Fig. 13.6 in the appropriate places.

This completes one cycle. The procedure is repeated starting again from joint 2. More accurate rotation moments at joint 2 can be obtained by taking the approximate values obtained in the first cycle. For example, considering joint 2:

The sum of the rotation moments at the far ends

at 1 = 0

at 3 = + 19.63 kN.m

Therefore,

$$M'_{21} = (-1/4)\,(48.89 + 0 + 19.63)$$
$$= -17.13 \text{ kN.m}$$
$$M'_{23} = -17.13 \text{ kN.m}$$

These values of rotation moments supersede the values (–12.22 kN.m) obtained earlier.

Now consider joint 3. New values of rotation moments are determined as explained earlier. Therefore,

$$M'_{32} = (-1/3)\,(-46.67 - 17.13) = +21.27 \text{ kN.m}$$

and

$$M'_{34} = (-1/6)\,(-46.67 - 17.13) = +10.63 \text{ kN.m}$$

These values replace the previous values of +19.63 kN.m and +9.82 kN.m. The new values are shown in Fig. 13.7 under the previous values shown struck off. This completes the second cycle.

Proceeding again, starting from joint 2

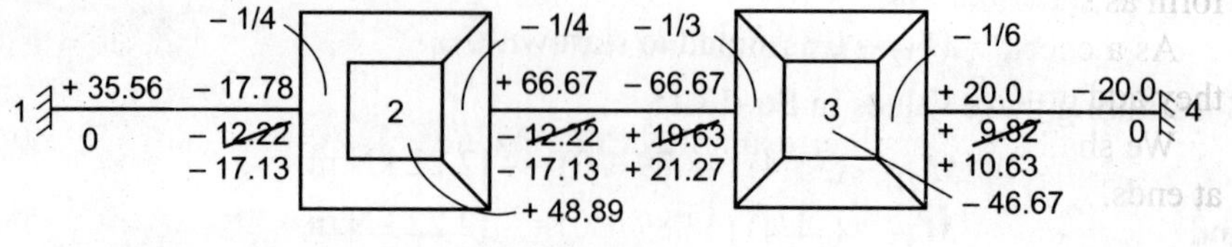

Fig. 13.7

$$M'_{21} = (-1/4)\,(48.89 + 0 + 21.27) = -17.66 \text{ kN.m}$$
$$M'_{23} = -17.66 \text{ kN.m}$$

Joint 3

$$M'_{32} = (-1/3)\,(-46.67 - 17.66 + 0) = 21.44 \text{ kN.m}$$
$$M'_{34} = (-1/6)\,(-46.67 - 17.66 + 0) = 10.72 \text{ kN.m}$$

These values are entered in Fig. 13.8 striking the previous values. Proceeding on to the fourth cycle, we have at joint 2

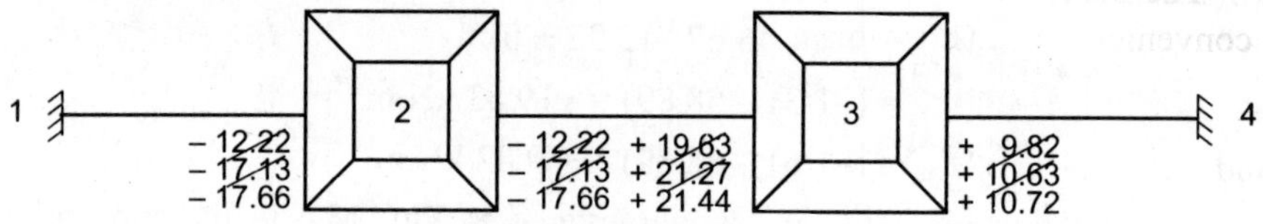

Fig. 13.8

$$M'_{21} + (-1/4)\,(48.89 + 0 + 21.44) = -17.58 \text{ kN.M}$$
$$M'_{23} = -17.58 \text{ kN.m}$$

At joint 3

$$M'_{32} = (-1/3)\,(-46.67 - 17.58 + 0) = 21.42 \text{ kN.m}$$
$$M'_{34} = (\ 1/6)\,(-64.25) = 10.71 \text{ kN.m}$$

These values now replace the previous values. The previous values are struck off and the new values are entered as shown in Fig. 13.9.

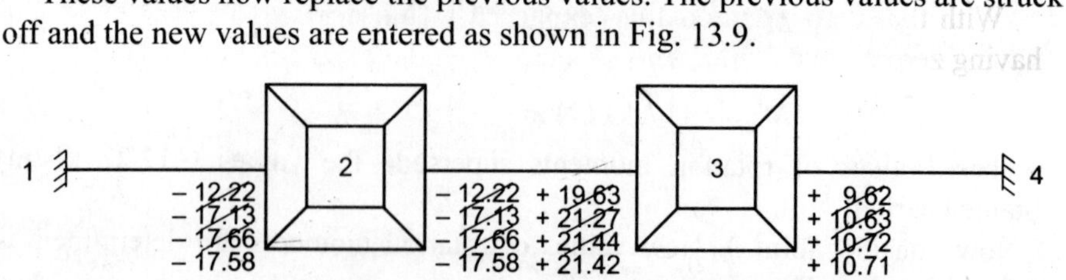

Fig. 13.9

At this stage it is seen that the maximum difference between the immediately previous and present values is 0.08 at joint 2 and 0.02 at joint 3. If we are satisfied that this difference is within acceptable limits, the iteration process can be stopped.

Now we have the acceptable values of rotation moments so that the final moments can be determined using Eq. 13.1. This is accomplished in a tabular form as shown in Fig. 13.10.

As a check we can sum up the moments over joints 2 and 3 and see whether they add up to zero.

We shall solve another example of a continuous beam with hinged supports at ends.

1		2		3	4
+ 35.56	− 17.78	+ 66.67	− 66.67	+ 20.00	− 20.00
0	− 17.58	− 17.58	+ 21.42	+ 10.71	0
0	− 17.58	− 17.58	+ 21.42	+ 10.71	+ 10.71
− 17.58	0	+ 21.42	− 17.58	0	+ 10.71
+ 17.98	− 52.94	+ 52.93	− 41.41	+ 41.42	− 9.29

Fig. 13.10 | *Computation of end moments*

Example 13.2 | *Determine the end moments of the continuous beam shown in Fig. 13.11. The relative values of I for each span are indicated along the members. E is constant.*

The beam has hinged and roller supports at 1 and 4 respectively. It will be convenient to consider that

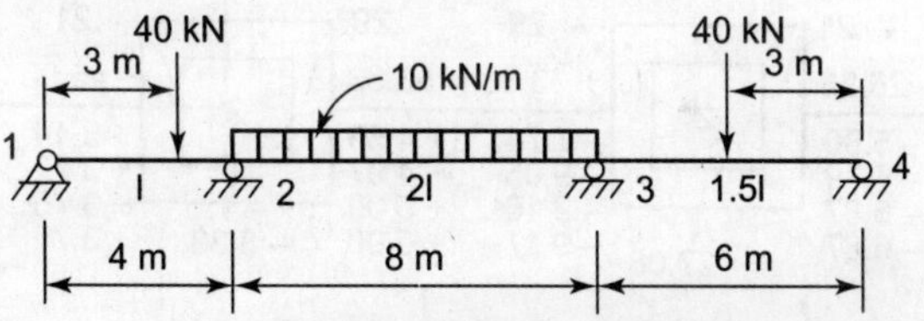

Fig. 13.11 | *Beam and loading*

1. the ends are fixed and the relative stiffness of spans 1-2 and 3-4 are taken as $(3/4)K_{12}$ and $(3/4)K_{34}$ respectively, and
2. the fixed end moments, M^F_{21} and M^F_{34} are modified as $(M^F_{21} - \frac{1}{2} M^F_{12})$ and $(M^F_{34} - \frac{1}{2} M^F_{34})$ respectively.

With these modifications the hinged or roller ends are assumed to be fixed having zero fixed end moments. We can write the fixed end moments as

$$M^F_{12} = \frac{40(3)(1)^2}{(4)^2} = 7.50 \text{ kN.m}$$

$$M^F_{21} = -\frac{40(3)^2(1)}{(4)^2} = -22.50 \text{ kN.m}$$

$$M^F_{23} = -M^F_{32} = \frac{10(8)^2}{12} = 53.33 \text{ kN.m}$$

$$M_{34}^F = -M_{43}^F = \frac{40(6)}{8} = 30.00 \text{ kN.m}$$

The moments are adjusted to account for the hinged or roller supports at 1 and 4. The adjusted moments are:

$$M^F_{21} = -22.50 - \frac{1}{2}(7.50) = -26.25 \text{ kN.m}$$

$$M^F_{34} = +30.00 - \frac{1}{2}(-30.00) = 45.00 \text{ kN.m}$$

$$M_{12}^F = -M_{43}^F = 0$$

The rotation factors are

Joint	Member	Rel. Stiff. K	ΣK	Rotation factor $u = (-1/2)\, K/\Sigma K$
2	2–1	$(3/4)(I/4)$	$7/16I$	–0.21
	2–3	$(2I/8)$		–0.29
3	3–2	$(2I/8)$	$7/16I$	–0.29
	3–4	$(3/4)(1.5I/6)$		–0.21

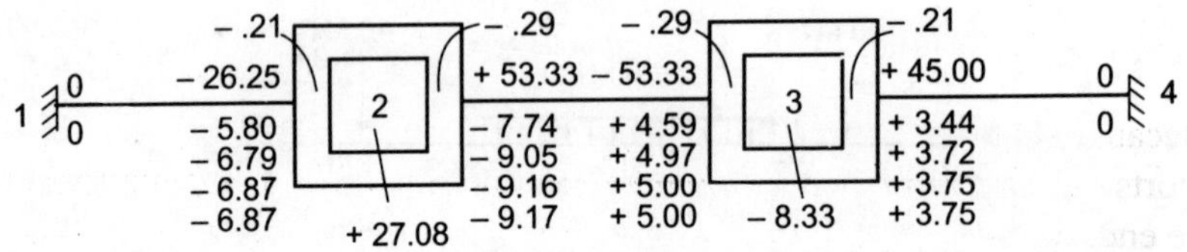

Fig. 13.12

The fixed end moments and the rotation factors are shown entered in Fig. 13.12.

Cycle 1

Joint 2

To start with the rotation moments of the far ends are assumed to be zero. Therefore,

$$M'_{21} = (-0.21)(27.08 + 0 + 0) = -5.80 \text{ kN.m}$$

$$M'_{23} = (-0.29)(27.08) = -7.74 \text{ kN.m}$$

Joint 3

$$M'_{32} = (-0.29)(-8.33 - 7.74 + 0) = 4.59 \text{ kN.m}$$

$$M'_{34} = (-0.21)(-8.33 - 7.74 + 0) = 3.44 \text{ kN.m}$$

These rotation moments are shown entered in Fig. 13.12. The rotation moments at 1 and 4 are noted as zero since they have been replaced by fixed ends.

Cycle 2

Joint 2

More accurate rotation moments can be obtained by taking the previously evaluated values of rotation moments. For example, at joint 2

$$M'_{21} = (-0.21)\,(27.08 + 0 + 4.59) = -6.79 \text{ kN.m}$$
$$M'_{23} = (-0.29)\,(27.08 + 0 + 4.59) + -9.05 \text{ kN.m}$$

Joint 3

$$M'_{32} = (-0.29)\,(-8.33 - 9.04 + 0) = 4.97 \text{ kN.m}$$
$$M'_{34} = (-0.21)\,(-8.33 - 9.05 + 0) = 3.72 \text{ kN.m}$$

The procedure is continued and the values up to four cycles are shown entered in Fig. 13.12.

Now that the rotation moments are known, the final moments are easily evaluated using Eq. 13.1. The computations are conveniently done in a tabular form as shown in Fig. 13.13.

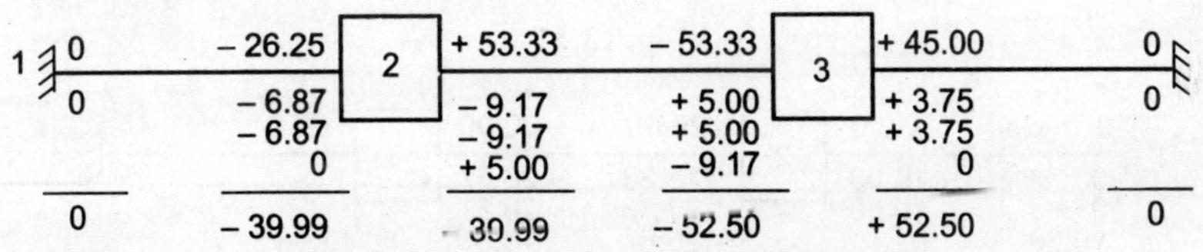

Fig.13.13 | *Computation of end moments*

Because supports 1 and 4 are actually simply supports, the moments at those supports must be zero. Hence, moment computations need not be carried out at those ends.

The overhanging ends in a continuous member should be considered as having a stiffness equal to zero. The rotation moments are computed in the usual manner. This is illustrated in the following example.

Example 13.3 | *Find the support moments for the continuous beam shown in Fig. 13.14. EI is constant.*

The fixed end moments using the Appendix table are

$$M^F_{12} = M^F_{21} = 0$$

$$M^F_{23} = -\frac{80(3)(3)}{6 \times 6} = -20.0 \text{ kN.m}$$

$$M^F_{32} = -20.00 \text{ kN.m}$$

$$M^F_{34} = 30\,(2) = 60.0 \text{ kN.m}$$

The rotation factors are:

$$u_{21} = -0.21,\; u_{23} = -0.29 \text{ and } u_{32} = (-1/2).$$

These values are shown entered in Fig. 13.15. Proceeding in the usual manner the rotation moments are evaluated. The rotation moments for five cycles are shown.

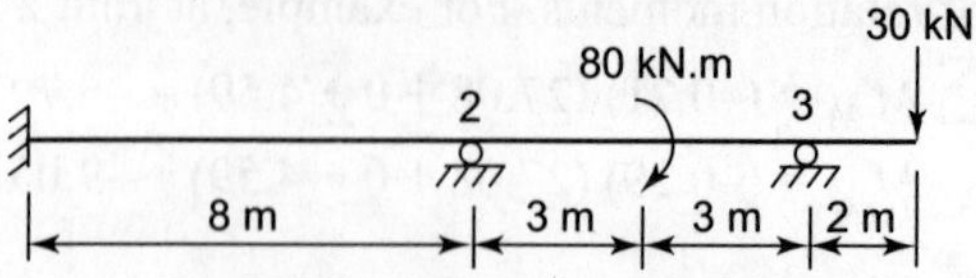

Fig. 13.14 | *Beam and loading*

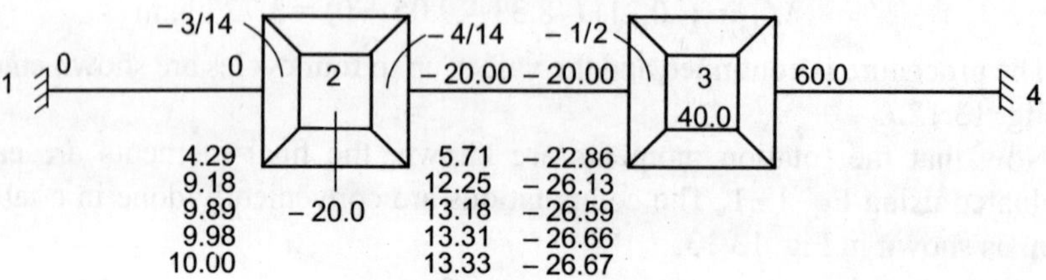

Fig. 13.15

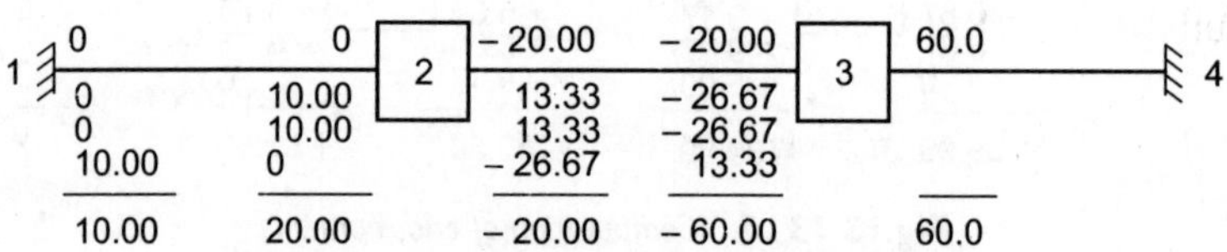

Fig. 13.16 | *Compulation of end moments*

The final end moments are worked out in the Table of Fig. 13.16.

13.2.2 Members with Translatory Joints

Figure 13.17 shows a member *AB* in a frame which has undergone lateral displacements at *A* and *B* so that the relative displacement is $\Delta = (\Delta_B - \Delta_A)$. It may be noted that the ends are restrained from rotation. The fixed end moments corresponding to this displacement are

$$M''_{AB} = M''_{BA} = \frac{6\,EI\,\Delta}{L^2} \tag{13.16}$$

When the translation of joints occurs along with rotations, the true end mpments are given by

$$M_{AB} = M^F_{AB} + 2\,M'_{AB} + M'_{BA} + M''_{AB}$$

$$M_{BA} = M^F_{BA} + 2\,M'_{BA} + M'_{AB} + M''_{BA} \tag{13.17}$$

The quantity $M''_{AB} = M''_{BA}$ is known as the *displacement moment* of member *AB*.

If A happens to be a joint where two or more members meet (Fig. 13.2), then for the condition of equilibrium of joint *A*, we have

$$\sum M_{AB} = 0 \tag{13.18}$$

that is, $$\sum M^F_{AB} + 2\sum M'_{AB} + \sum M'_{BA} + \sum M''_{AB} = 0 \tag{13.19}$$

Therefore, $$\sum M'_{AB} = (-1/2)\,(\sum M^F_{AB} + \sum M'_{BA} + \sum M''_{AB}) \tag{13.20}$$

Again from Eq. 13.11

$$M'_{AB} = \frac{K_{AB}}{\Sigma K_{AB}} \Sigma M'_{AB}$$

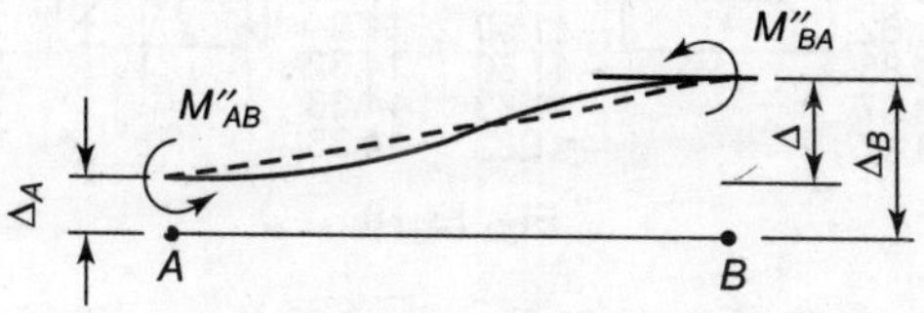

Fig. 13.17

Substituting for $\Sigma M'_{AB}$ from Eq. 13.20 we have

$$M'_{AB} = \frac{K_{AB}}{\Sigma K_{BA}} \left(-1/2\,(\Sigma M^F_{AB} + \Sigma M'_{BA} + \Sigma M''_{AB}\right) \tag{13.21}$$

or $$M'_{AB} = u_{AB}\left(M^F_A + \Sigma M'_{BA} + \Sigma M''_{AB}\right) \tag{13.22}$$

In a similar way, we can write

$$M'_{BA} = u_{AB}\left(M^F_B + \sum M'_{AB} + \sum M''_{BA}\right) \tag{13.23}$$

Using these relationships, rotation moments can be determined by the iterative procedure followed earlier. If lateral displacements are known, the displacement moments can be determined from Eq. 13.16. If lateral displacements are unknown, then additional equations are to be used. This aspect is discussed later in Sec. 13.4.

Example 13.4 | *Consider the same beam and loading as in Example 13.2. Under the load, support 2 sinks by 10 mm. Determine the end moments. $E = 200 \times 10^6$ kN/m^2 (200,000 MPa) and $I = 80 \times 10^6$ mm^4).*

The fixed end moments after taking into account the simply supported conditions at the ends were worked out earlier as

$$M^F{}_{12} = 0 \text{ and } M^F{}_{21} = -26.25 \text{ kN.m}$$
$$M^F{}_{34} = +\,45.00 \text{ kN.m and } M^F_{43} = 0$$

The fixed end moments due to the known settlement of support are to be added to the above moments to arrive at net fixed end moments. However, for clarity, the fixed end moments due to transverse loading are entered in the first row and moments due to translation of joints in the second row in Fig. 13.18. The rotation factors are taken from Example 13.2 and shown entered in Fig. 13.18.

The procedure, after this step, is the same as was followed in earlier examples. The rotation moments for five cycles are shown recorded in Fig. 13.18. The final moments are obtained using Eq. 13.17. The computations are shown tabulated in Fig. 13.19.

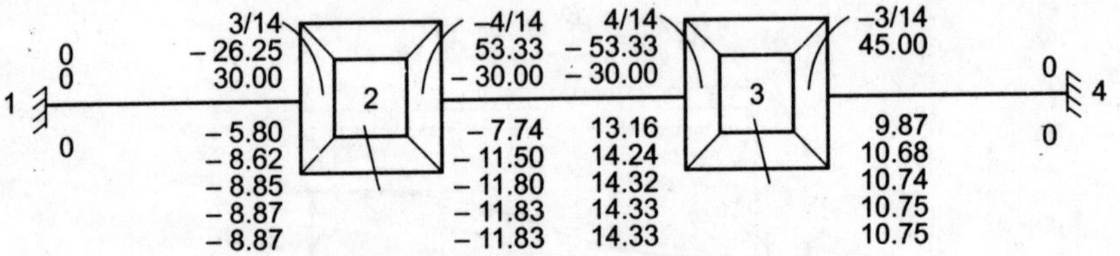

Fig. 13.18

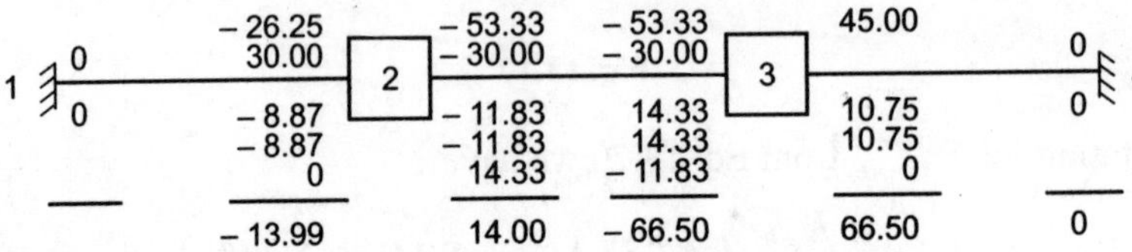

Fig. 13.19 | Computations of end moments

13.3 FRAMES WITHOUT LATERAL TRANSLATION OF JOINTS

The frames in which lateral translations are prevented are analysed in the same way as continuous beams. The lateral sway is prevented either due to support conditions or due to the symmetry of the frame and loading. The procedure is illustrated by solving the following example.

Example 13.5 | *Determine the end moments of the members of the frame of Fig. 13.20. The relative values of I are indicated on the Figure.*

Due to symmetry of frame and loading, the frame does not undergo lateral translation. The fixed end moments are

$$M_{23}^F = -M_{32}^F = \frac{40(6)(6)}{12} = 120.0 \text{ kN.m}$$

$$M_{12}^F = M_{21}^F = M_{34}^F = M_{43}^F = 0$$

Rotation factors:

Joint	Member	Rel. Stiff K	$\sum K$	Rot. Factor (u)
2	2–1	$I/3$		–0.2
	2–3	$3I/6$	$5/6\ I$	–0.3
3	3–2	$3I/6$	$5/6\ I$	–0.3
	3–4	$I/3$		–0.2

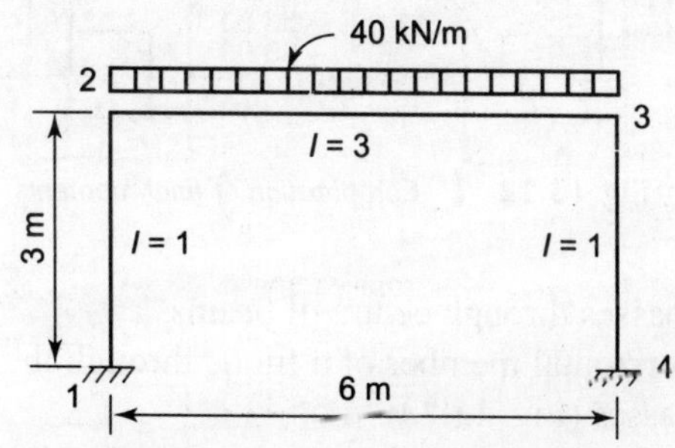

Fig. 13.20 | *Frame and loading*

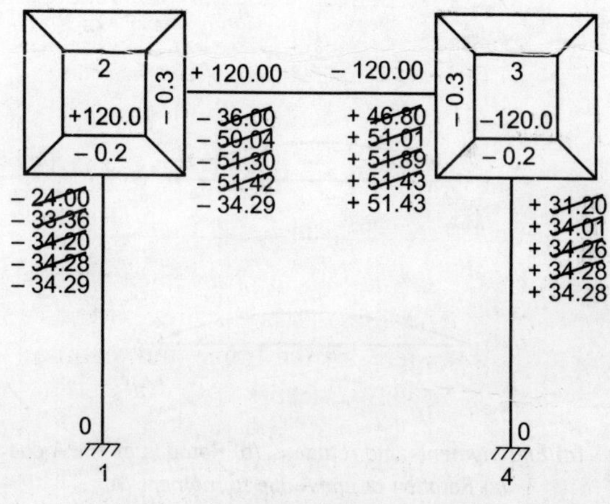

Fig. 13.21

Now the rotation moments are worked out in the same manner as was done in the previous examples. These are shown entered in Fig. 13.21 up to five cycles.

The final moments are computed and shown in Fig. 13.22.

13.3.1 Symmetrical Frames Under Symmetrical Loading

Considerable computational work can be saved if we make use of symmetry of frames and loading. Two cases of symmetry arise, namely, frames in which the axis of symmetry passes through the centre line of the beams, and frames with the axis of symmetry passing through the column line.

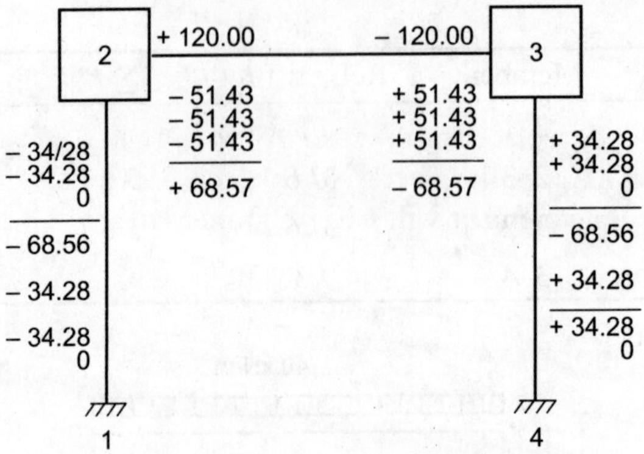

Fig. 13.22 | *Computation of final moments*

Case I

Axis of symmetry passes through centre of beams.

Let *AB* be any horizontal member of a frame through the centre of which the axis of symmetry passes (Fig. 13.23).

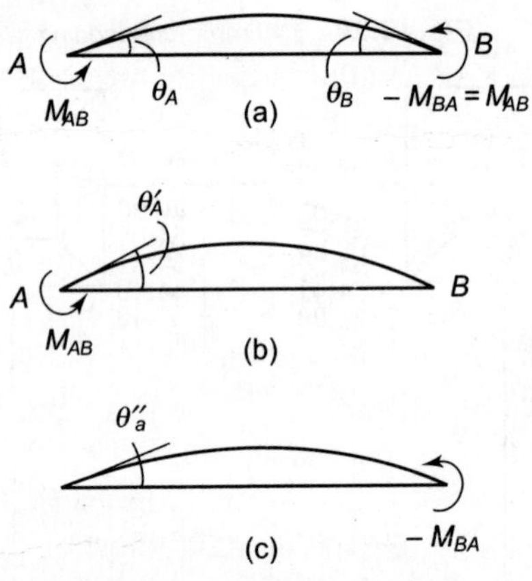

Fig. 13.23 | *(a) End moments and rotations, (b) Rotation at end A due to moment M_{AB}. (c) Rotation at end A due to moment M_{BA}*

Let M_{AB} and M_{BA} be the end moments. Due to symmetry of deformation, M_{AB} and M_{BA} are numerically equal but are opposite in sense.

The slope at $A = \theta_A$ is obtained as a superposition of the rotations due to M_{AB} and $-M_{BA}$ as shown in Figs. 13.23*b* and *c* respectively. Therefore,

$$\theta_A = \theta'_A + \theta''_A$$

From the moment rotation relationships given in Appendix *D*

$$\theta'_A = \frac{M_{AB}L}{3\,EI} \text{ and } \theta''_A = -\frac{M_{BA}L}{6\,EI} = \frac{M_{AB}L}{6\,EI}$$

Therefore, $$\theta_A = \theta'_A + \theta''_A = \frac{M_{AB}L}{2\,EI}$$

Let this member be replaced by member AB' whose end A will undergo rotation θ_A due to moment M_{AB} applied at end A while end B' is being B' restrained (Fig. 13.24). The substitute member will have the same value of I as for the original member.

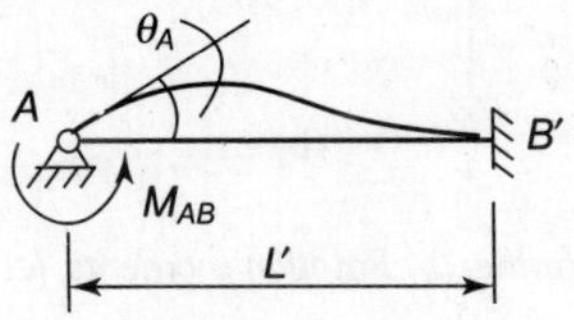

Fig. 13.24

For such a beam the force displacement relationship is

$$\theta_A = \frac{M_{AB}L}{4EI}$$

where L' is the length of the substitute member. Hence for the equality of rotations between original member AB and the substitute member AB'

$$\theta_A = \frac{M_{AB}L}{2EI} = \frac{M_{AB}L'}{4EI} \quad \text{or} \quad \frac{I}{L} = \frac{2I}{L'}$$

or $$K = 2K'$$

or $$K' = \frac{K}{2} \qquad (13.24)$$

Thus, if K is the relative stiffness of original member AB, this member can be replaced by substitute member AB' having relative stiffness $K/2$. With this substitute member, the analysis then needs to be carried out for only one half of the frame considering the line of symmetry as the fixed end. The following examples illustrate the steps involved.

Example 13.6 | *We shall analyse the same frame as in Example 13.5 (Fig. 13.20) taking advantage of symmetry of the frame and loading.*

Since the axis of symmetry passes through the middle of the beam 2-3, only one half of the frame need be considered. The substitute frame is shown in Fig. 13.25*a*.

The rotation factors at joint 2 are

$$u_{21} = -\frac{4}{14} \quad \text{and} \quad u'_{23} = \frac{-3}{14}$$

The true rotation moments are obtained in the first distribution only. They are shown entered in Fig. 13.25*b*. The final moments are worked out in Fig. 13.25*c*.

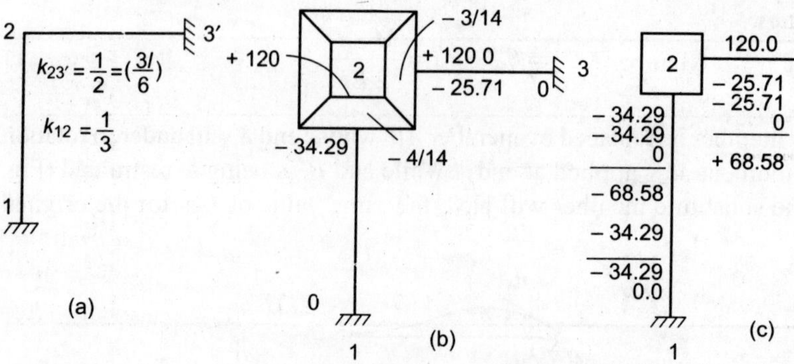

Fig. 13.25 | *(a) Substitue frame, (b) Rotation moments, (c) Computation of end moments*

Example 13.7 | *Analyse the frame of Fig. 13.26 for end moments taking advantage of symmetry of the frame and loading.*

The fixed end moments are

$$M_{25}^F = \frac{30(2)(4)^2}{(6)^2} + \frac{30(2)^2(4)}{(6)^2} = 40.0 \text{ kN.m}$$

$$M_{34}^F = \frac{20(6)^2}{12} = 60.0 \text{ kN.m}$$

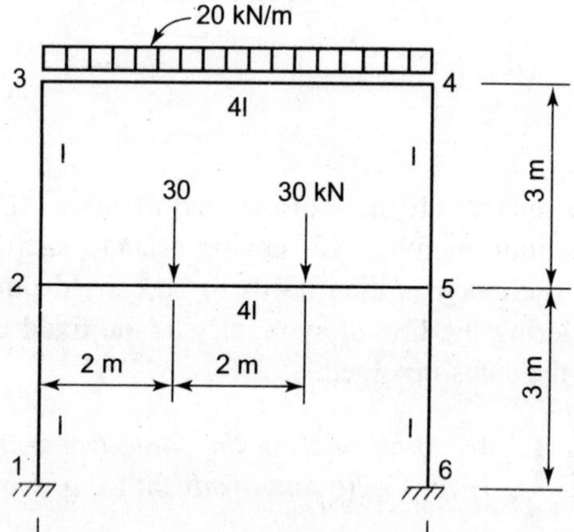

Fig. 13.26 | *Frame and loading*

The substitute frame is shown in Fig. 13.27*a*.

For the substitute frame the relative stiffnesses and the rotation factors are worked out. The rotation moments are evaluated at joints 2 and 3, These are shown in Fig. 13.27*b*.

Rotation factors:

Joint	Member K	Rel. Stiff K	$\sum K$	Rot. Factor $(-1/2)K/\sum K$
	2–1	$I/3$		–1/6
2	2–3	$I/3$	I	–1/6
	2–5	$I/3$		–1/6
3	3–2	$I/3$		–1/4
	3–4	$I/3$	$2/3I$	–1/4

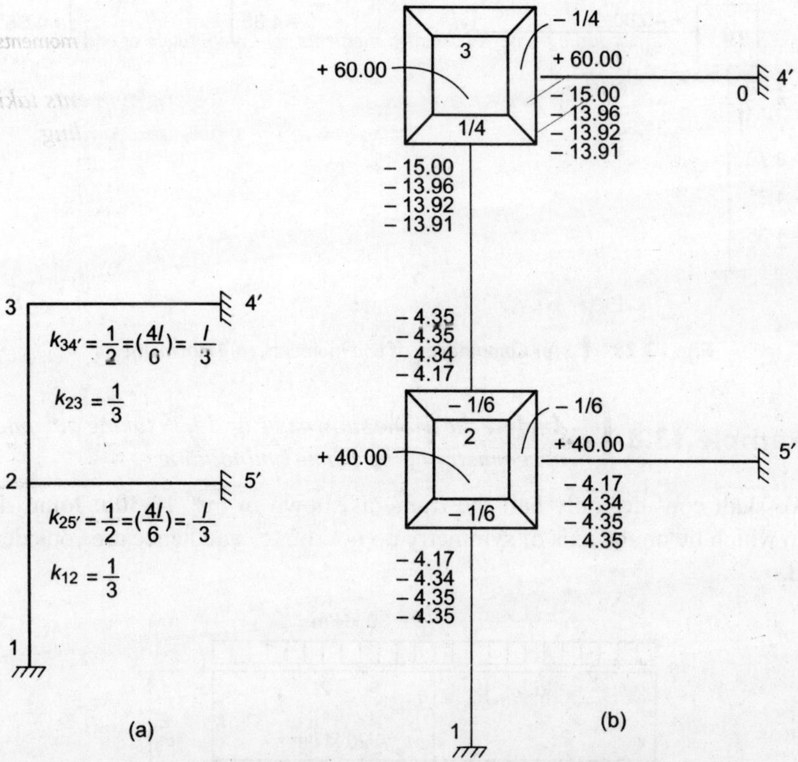

Fig. 13.27 | *(a) Substitute frame, (b) Rotation moments*

The final moments are computed, as shown in Fig. 13.28*a*. Figure shows the end moments of the members of the frame.

Case II

The axis of symmetry passes through the column. This case occurs when the number of bays is an even number. Due to symmetry of the loading and frame, the joints on the axis of symmetry will not rotate. Hence, it is sufficient if half the frame is analysed. The following example illustrates the procedure.

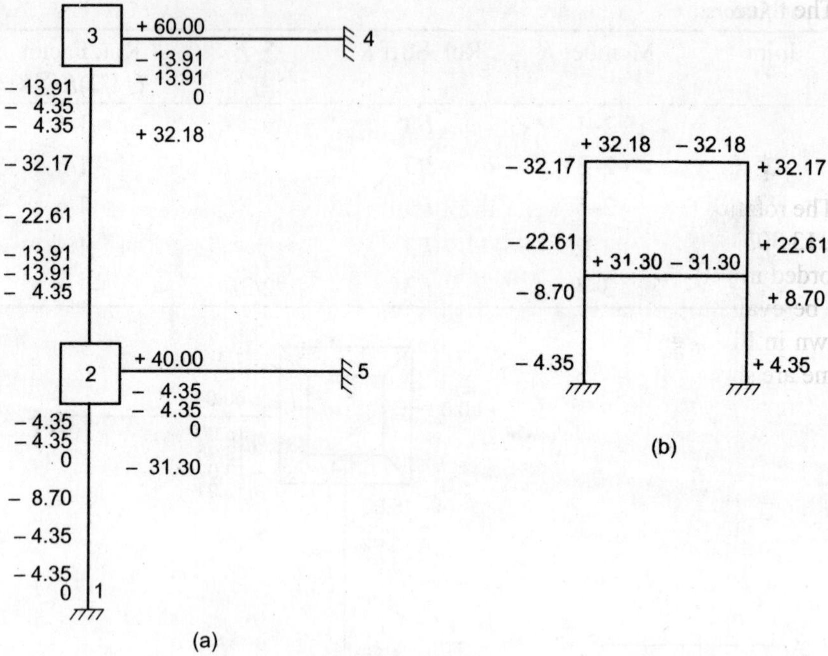

Fig. 13.28 | *(a) Computation of end moments, (b) Final moments*

Example 13.8 | *Analyse the frame shown in Fig. 13.29 taking advantage of symmetry of the frame and loading.*

We shall consider only half the frame as shown in Fig. 13.30*a*. Joints 4, 5 and 6 which lie on the axis of symmetry do not rotate" and hence are considered fixed.

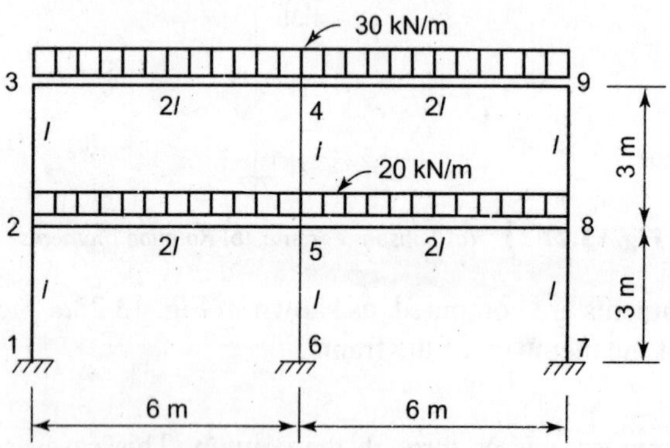

Fig. 13.29 | *Frame and loading*

The fixed end moments are

$$M_{34}^F = -M_{43}^F = \frac{30\times6\times6}{12} = 90.0 \text{ kN.m}$$

$$M_{25}^F = -M_{52}^F = \frac{20\times6\times6}{12} = 60.0 \text{ kN.m}$$

The rotation factors at joints 2 and 3 are worked out as usual and are shown in Fig. 13.30*b*. The rotation moments are evaluated up to three cycles and are shown recorded in Fig. 13.30*b*. The rotation moments being known, the final moments can be evaluated using Eq. 13.1. The computations for final end moments are shown in Fig. 13.31*a*. The final end moments of all the members in the given frame are shown in Fig. 13.31*b*.

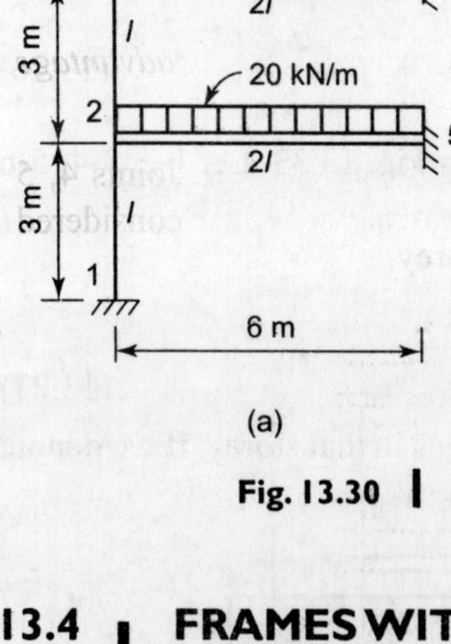

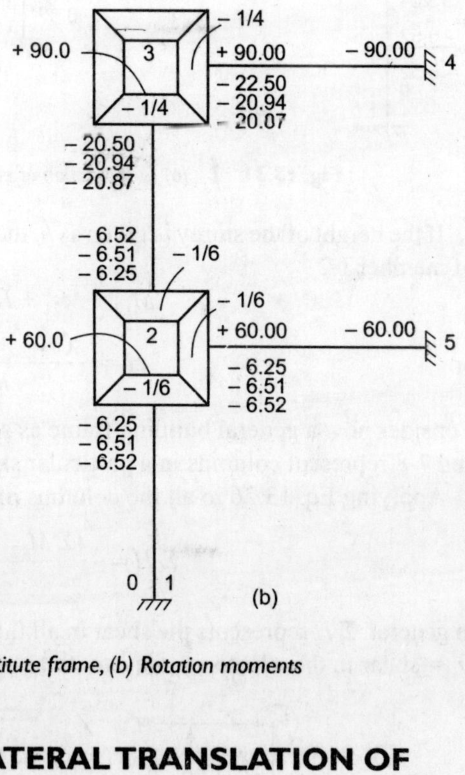

Fig. 13.30 | *(a) Substitute frame, (b) Rotation moments*

13.4 FRAMES WITH LATERAL TRANSLATION OF JOINTS

13.4.1 Vertical Loading

Let 1-2 represent a vertical member in any storey of a multi storeyed frame (Fig. 13.32). M_{12} and M_{21} are the end moments at 1 and 2. Let the horizontal force exerted by the frame on column 1-2 be H.

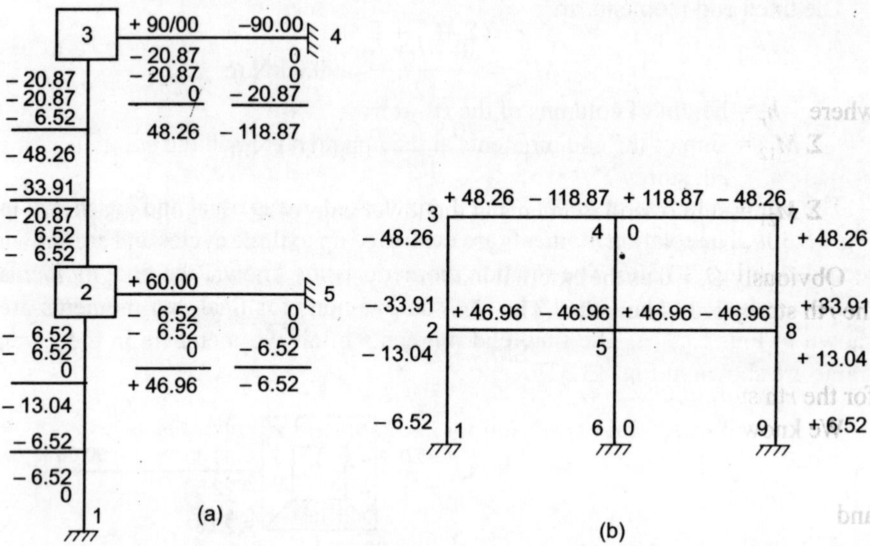

Fig. 13.31 | *(a) Computation of end moments, (b) Final moment*

If the height of the storey is given as h, then from the equilibrium consideration of member I-2

$$M_{12} + M_{21} + H(h) = 0 \tag{13.25}$$

or

$$H = -\frac{(M_{12} + M_{21})}{h} \tag{13.26}$$

Consider now a general building frame as shown in Fig. 13.33. Let 1-2, 3-4, 5-6 and 7-8 represent columns in a particular storey.

Applying Eq. 13.26 to all the columns of the storey

$$\Sigma H = -\frac{(\Sigma M_{12} + \Sigma M_{21})}{h} \tag{13.27}$$

In general, ΣH represents the shear in all the columns in that storey. If we denote Q_r = shear in the rth storey, we can write

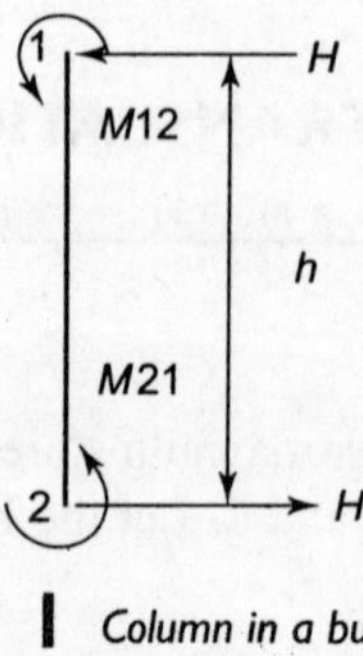

Fig. 13.32 | *Column in a building frame*

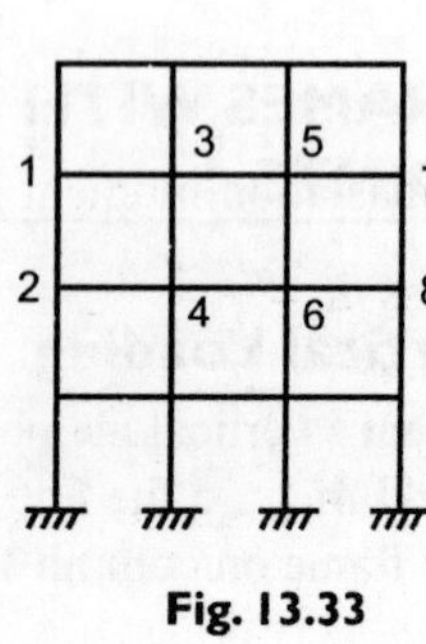

Fig. 13.33

$$Q_r = -\frac{(\Sigma M_{12} + \Sigma M_{21})}{h_r} \tag{13.28}$$

where h_r = height of columns of the rth storey;

ΣM_{12} = sum of the end moments at the upper ends of all the columns in the rth storey;

ΣM_{21} = sum of end moments at the lower ends of all the columns in the rth storey.

Obviously $Q_r = 0$ as the external loading is vertical. Further, all the columns of the rth storey are of height h_r. Therefore, we have from Eq. 13.28

$$\Sigma M_{12} + \Sigma M_{21} = 0 \tag{13.29}$$

for the rth storey.

We know, the general expression for end moments for member 1-2 is

$$M_{12} = M^F{}_{12} + 2M'_{12} + M'_{21} + M''_{12}$$

and

$$M_{21} = M^F{}_{21} + 2\,M'_{21} + M'_{12} + M''_{21} \tag{13.30}$$

For a column which is vertical $M^F{}_{12} = M^F{}_{21} = 0$ since the loading on the frame is vertical. For any prismatic member it may be noted that

$$M''_{12} = M''_{21}$$

Therefore,

$$\Sigma M_{12} + \Sigma M_{21} = 3\,\Sigma M'_{12} + 3\Sigma M'_{21} + 2\Sigma M''_{12} \tag{13.31}$$

From Eq. 13.29 $\quad \Sigma M_{12} + \Sigma M_{21} = 0$

Hence, $\quad 3\,\Sigma M'_{12} + 3\Sigma M'_{21} + 2\Sigma M''_{12} = 0$

or

$$\Sigma M''_{12} = -\frac{3}{2}(\Sigma M'_{12} + \Sigma M'_{21}) \tag{13.32}$$

Equation 13.32 gives a relation between the rotation and translation moments.

We know that the relative lateral displacement Δ is the same for all the columns in any one storey. For any column, the translation moment is

$$M''_{12} = \frac{6EI\Delta}{h^2} = \frac{6EI\varphi}{h}$$

Where

$$\varphi = \frac{\Delta}{h}$$

Thus, the translation moment of a column in a storey is proportional to the relative stiffness $K = \frac{I}{h}$.

Therefore,

$$\frac{M''_{12}}{\Sigma M''_{12}} = \frac{K_{12}}{\Sigma K_{12}}$$

or

$$M''_{12} = \frac{K_{12}}{\Sigma K_{12}}\Sigma M''_{12} \tag{13.33}$$

Substituting for $\Sigma M''_{12}$ from Eq. 13.32 we can write Eq. 13.33 as

$$M''_{12} = \frac{K_{12}}{\Sigma K_{12}}(-3/2)\,(\Sigma M'_{12} + \Sigma M'_{21})$$

Or
$$M''_{12} = v_{12}\,(\Sigma M'_{12} + \Sigma M'_{21}) \tag{13.34}$$

in which $v_{12} = (-3/2)\, K/\Sigma K_{12}$ is called the *displacement factor* or *translation factor* of member 1-2.

It may be noted that in Eq. 13.34, $(\Sigma M'_{12} + \Sigma M'_{21})$ sum of the rotation moments at the top and bottom ends of all the columns of the storey under consideration.

ΣK_{12} = sum of the relative stiffnesses of all the columns in the storey under consideration

Obviously, the sum of the translation factors of all the columns of a storey will be equal to (–3/2). Summing up, the various relationships obtained earlier are

$$M'_{12} = u_{12}\,(M_1^F + \Sigma M'_{21} + \Sigma M''_{12})$$

$$M'_{21} = u_{21}\,(M_2^F + \Sigma M'_{12} + \Sigma M''_{21}) \tag{13.35}$$

$$M''_{12} = M''_{21} = v_{12}\,(\Sigma M'_{12} + \Sigma M'_{21}) \tag{13.36}$$

$$M_{12} = M_{12}^F + 2\,M'_{12} + M'_{21} + M''$$

$$M_{21} = M_{21}^F + 2\,M'_{21} + M'_{12} + M''_{21} \tag{13.37}$$

By applying Eqs. 13.35 and 13.36, the rotation and translation moments may be determined by iteration for all the storey in turn. Once the acceptable values of rotation and translation moments are known, the final moments may be determined by Eq. 13.37.

13.4.2 Horizontal Loading

For a frame subjected to horizontal loading, the storey shear, $\Sigma H = Q_r$. This is shown in Fig. 13.34 by making a cut through all the columns. If all the columns of the storey are of height h_r, we can write the equilibrium equation $\Sigma M = 0$, that is

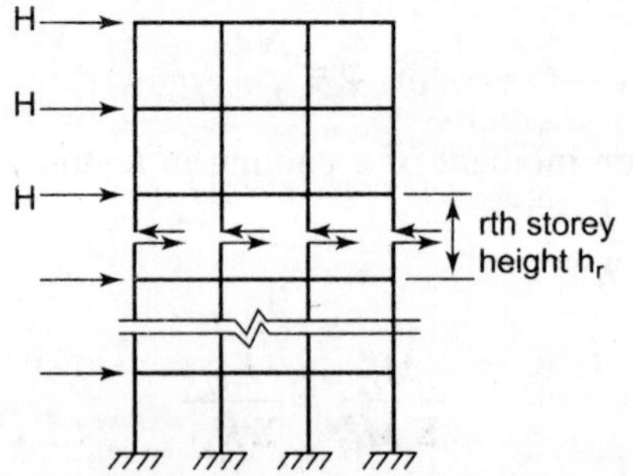

Fig. 13.34 | *Shear in columns*

$$\Sigma M_{12} + \Sigma M_{21} + Q_r\,(h_r) = 0 \tag{13.38}$$

or
$$Q_r\, h_r = -(\Sigma M_{12} + \Sigma M_{21}) \tag{13.39}$$

Writing column moments using Eq. 13.30 and knowing $M_{12}^F = M_{21}^F = 0$ for columns, we have

$$Q_r\, h_r = -\sum_r \{3\,(M'_{12} + M'_{21}) + 2\,M''_{12}\} \tag{13.40}$$

Summation $\sum_r$ is for all the columns in the rth storey

or
$$\frac{Q_r\, h_r}{3} = -\sum_r \left\{(M'_{12} + M'_{21}) + \frac{2}{3}M''_{12}\right\} \tag{13.41}$$

This gives

$$\Sigma M''_{12} = -\frac{3}{2}\left\{\frac{Q_r\, h_r}{3} + \sum_r (M'_{12} + M'_{21})\right\} \tag{13.42}$$

or
$$\Sigma M''_{12} = -\frac{3}{2}\left\{M_r^F + \sum_r (M'_{12} + M'_{21})\right\} \tag{13.43}$$

Here $M^F{}_r = Q_r\, h_r/3$ is known as the *storey moment*. This is positive when Q acts from right to left. From Eq. 13.43 we can write

$$M''_{12} = \frac{K_{12}}{\sum_r K_{12}}(-3/2)\left\{M_r^F + \sum_r (M'_{12} + M'_{21})\right\} \tag{13.44}$$

or
$$M''_{12} = v_{12}\left\{M_r^F + \sum_r (M'_{12} + M'_{21})\right\} \tag{13.45}$$

The analysis of a multi storey building frame with horizontal loading differs from that of a frame with vertical loading only by the fact that in performing the basic operation for the determination of the translation moments, the sum of the rotation moments of all member ends of the storey must also contain storey moment M_r^F.

We shall illustrate various points by solving a few numerical examples. First we consider frames under vertical loading only.

Example 13.9 *Determine the end moments of the members of the frame shown in Fig. 13.35. Relative I values are indicated along the members. E is constant.*

The fixed end moments are

$$M_{23}^F = \frac{33.75(2)(4)^2}{(6)^2} = 33.0 \text{ kN.m}$$

$$M_{32}^F = -\frac{33.75(2)^2(4)}{(6)^2} = -15.0 \text{ kN.m}$$

The rotation factors are computed in the usual way.

$$u_{21} = u_{23} = -\frac{1}{4} \text{ and } u_{32} = u_{34} = -\frac{1}{4}$$

This displacement factor for each column is (–3/4) as the two columns have the same relative stiffness.

The fixed end moments, and the rotation factors are shown entered in Fig. 13.36 as in the previous examples. The translation factors are written by the side of columns in small boxes as shown. Now we can proceed with the iteration process. The various computations follow this order: first joints 2 and 3 and then storey columns 1-2 and 3-4.

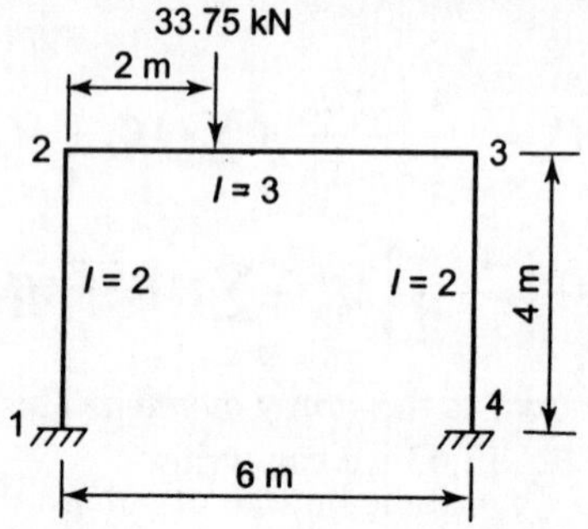

Fig. 13.35 | *Frame and loading*

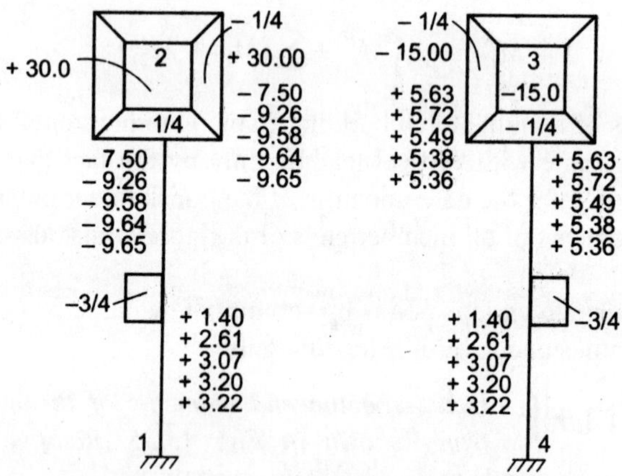

Fig. 13.36 | *Rotation and translation moments*

Cycle I

Joint 2

Assuming all far end rotation and translation moments to be zero, the following first approximations are obtained for the near end rotation moments.

$$M'_{23} = (-1/4)\,(+30.0) = -7.50 \text{ kN.m}$$

$$M'_{21} = -7.50 \text{ kN.m}$$

Joint 3

Sum of fixed end moments	$= -15.00$	kN.m
Rotation moment at 2	$= -7.50$	kN.m
at 4	$= 0$	(fixed end)
Displacement moments of column 1–2	$= 0$	(assumed)
column 3–4	$= 0$	(assumed)
Total	-22.50	kN.m

Therefore,

$$M'_{32} = (-1/4)\,(-22.50) = +5.63 \text{ kN.m}$$

and

$$M'_{34} = (-1/4)\,(-22.50) = +5.63 \text{ kN.m}$$

Storey I

(There is only one storey in the present case.)

Rotation moments at the top of column 1–2	$= -7.50$ kN.m
column 3–4	$= +5.63$ kN.m
at the bottom of column	$= 0$
Total	-1.87 kN.m

Therefore,

$$M''_{12} = (-3/4)\,(-1.87) = +1.40 \text{ kN.m}$$

$$M''_{34} = (-3/4)(-1.87) = +1.40 \text{ kN.m}$$

The rotation moments are entered for the beam ends as earlier and for the columns at their ends (see Fig. 13.36). Note that the translation moments are entered along the columns at the mid height of the storey. This completes the first cycle.

Cycle 2

Joint 2

Sum of fixed end moments	$= +30.0$	kN.m
Rotation moments at 1	$= 0$	(fixed end)
at 3	$= +5.63$	kN.m
Translation moments of column 1–2	$= +1.40$	kN.m
Total	37.03	kN.m

$$M'_{23} = (-1/4)\,(37.03) = -9.26 \text{ kN.m}$$

$$M'_{21} = (-1/4)\,(37.03) = -9.26 \text{ kN.m}$$

Joint 3

Sum of fixed end moments	$= -15.00$ kN.m
Rotation moments: at 2	$= -9.26$ kN.m
at 4	$= 0$
Translation moments of column 3-4	$= +1.40$ kN.m
Total	-22.86 kN.m

$$M'_{32} = (-1/4)(-22.86) = 5.72 \text{ kN.m}$$
$$M' = (-1/4)(-22.86) = 5.72 \text{ kN.m.}$$

Storey I

Rotation moments at the top of column 1–2 = –9.26 kN.m
column 3–4 = 5.72 kN.m
Rotation moment at the bottom of the columns = 0

Total –3.54 kN.m

$$M''_{12} = M''_{34} = (-3/4)(-3.54) = 2.61 \text{ kN.m}$$

This completes the second cycle. In a similar manner the computations were carried out up to five cycles, the rotation and translation moments in successive cycles are entered in Fig. 13.36.

Once the rotation moments and translation moments are known with the desired accuracy, the final moments can be computed using Eq. 13.37. These computations are shown in Fig. 13.37*a* and the final moments are shown in Fig. 13.37*b*.

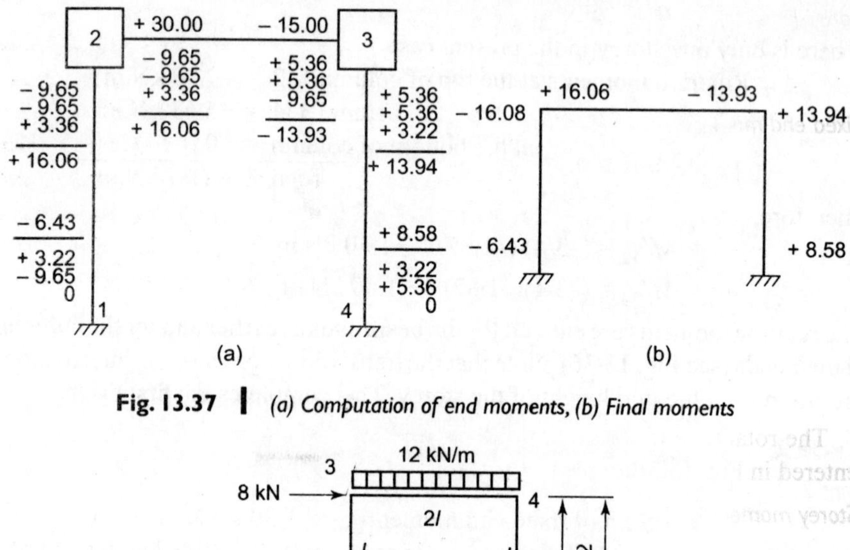

Fig. 13.37 | *(a) Computation of end moments, (b) Final moments*

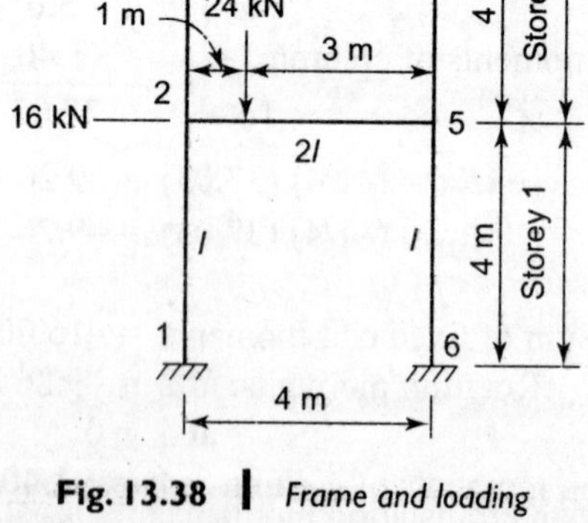

Fig. 13.38 | *Frame and loading*

Example 13.10 *We shall consider a two storey frame subjected to both vertical and lateral loading. The details of the frame and loading are given in Fig. 13.38.*

Rotation factors

Joint	Members	*Rel. K*	$\sum K$	*u*
2	2-1	*I*/4		–1/8
	2-3	*I*/4	*I*	–1/8
	2-5	*I*/2		–1/4
3	3-2	*I*/4	(3/4)*I*	–1/6
	3-4	*I*/2		–1/3

Translation factors

In each storey there are only two columns and both of them have the same relative stiffness. Therefore, the translation factor for each column

$$= \frac{1}{2}\left(-\frac{3}{2}\right) = \left(-\frac{3}{4}\right)$$

Fixed end moments

$$M^F_{34} = -M^F_{43} = \frac{12(4)^2}{12} = +16.00 \text{ kN.m}$$

$$M^F_{25} = \frac{24(1)(3)^2}{(4)^2} = +13.50 \text{ kN.m}$$

$$M^F_{52} = \frac{24(1)^2(3)}{(4)^2} = +4.50 \text{ kN.m}$$

The rotation and translation factors as well as fixed end moments are shown entered in Fig. 13.39.

Storey moments

Storey 2 $Q_2 = -8$ kN, $\quad M_r^F = -\dfrac{8 \times 4}{3} = -10.67$ kN.m

Storey 1 $Q_1 = (-8) + (-16) = -24$ kN, $\quad M_r^F = -\dfrac{24 \times 4}{3} = -32.00$ kN.m

The storey moments are recorded in small rectangular blocks at the mid height of each storey to the left of the first column line.

The iterations are carried out in the following order: first joints 2-5-4 and 3 and then storey 2 and 1.

Cycle I

Joint 2

The rotation moments and translation moments are initially assumed to be zero.

Sum of joint moments		= 13.50 kN.m
Sum of rotation moments at ends:	1	= 0 (fixed end)
	3	= 0 (assumed)
	5	= 0 (assumed)
Sum of translation moments of		
column 2-3 above		= 0 (assumed)
column 2-1 below		= 0 (assumed)
Total		+13.50 kN.m

Using Eq. 13.35, $M'_{21} = M'_{23} = (-1/8)\ (13.5) = -1.69$ kN.m

$M'_{25} = (-1/4)\ (13.5) = -3.38$ kN.m

Joint 5

Sum of joint moments		= –4.50 kN.m
Sum of rotation moments at far ends:		
	2	= – 3.38 kN.m
	4	= 0 (assumed)
	6	= 0 (fixed end)
Sum of translation moments of:		
column 5-4 above		= 0
column 5-6 below		= 0
Total		–7.88 kN.m

Therefore, $M'_{56} = M'_{54} = (-1/8)\ (-7.88) = 0.99$ kN.m

$M'_{52} = (-1/4)\ (-7.88) = 1.97$ kN.m

Joint 4

Sum of joint moments		= –16.00 kN.m
Sum of rotation moments at far ends:	3	= 0
	5	= +0.99 kN.m
Sum of translation moment of column	4-5	= 0
Total		–15.01 kN.m

$M'_{45} = (-1/6)\ (-15.01) = 2.50$ kN.m

$M'_{43} = (-1.3)\ (-15.01) = 5.00$ kN.m

Joint 3

Sum of joint moments		= +16.00 kN.m
Sum of rotation moments at far ends:	2	= –1.69 kN.m
	4	= +5.00 kN.m
Sum of translation moments of column	2-3	= 0
Total		19.31 kN.m

$M'_{32} = (-1/6)\ (19.31) = -3.22$ kN.m

$M'_{34} = (-1/3)\ (19.31) = -6.44$ kN.m

Having considered all the joints for rotation moments, now we shall proceed to evaluate the translation moments—storey 2 first and then storey 1.

Storey 2

Storey moment		= –10.67 kN.m
Rotation moments at ends of columns	2-3	= –3.22 kN.m (top)
		+2.50 kN.m (bottom)
	and 4-5	= –1.69 kN.m (top)
		+0.99 kN.m (bottom)
Total		–12.09 kN.m

Therefore, using Eq. 13.36

$$M''_{23} = M''_{45} = (-\,^3/4)\,(12.09) = 9.07 \text{ kN.m}$$

Storey 1

Storey moment		= –32.00 kN.m
Rotation moments from column ends:	1-2	= –1.69 kN.m (top)
		0 (bottom)
	5-6	= 0.99 kN.m (top)
		0 (bottom)
Total		–32.7 kN.m

$$M'12 = M''56 = (-3/4)\,(-32.7)$$
$$= +\,24.53 \text{ kN.m}$$

The rotation moments and translation moments are entered in Fig. 13.39. This completes the first cycle of iteration.

Cycle 2

Improved values of rotation moments and translation moments can be obtained by taking the values obtained in the first cycle. Consider,

Joint 2

Sum of joint moments		= + 13.50 kN.m
Sum of rotation moments at far ends:	1	= 0
	2	= –3.22 kN.m
	5	=+ 1.97 kN.m
Sum of translation moments: columns 2-3 above		= + 9.07 kN.m
column 2-1 below		= + 24.53 kN.m
Total		–45.85 kN.m

$$M'_{21} = M'_{23} = (-1/8)\,(45.85) = -5.73 \text{ kN.m}$$
$$M'_{25} = (-1/4)\,(45.85) = -11.46 \text{ kN.m}$$

Joint 5

Sum of joint moments		= –4.50 kN.m
Sum of rotation moments at far ends:	2	= –11.46 kN.m
	4	= + 2.50 kN.m
	6	= 0.
Sum of translation moments: column 5-4 above		= + 9.07 kN.m
column 5-6 below		= + 24.53 kN.m
Total		+20.14 kN.m

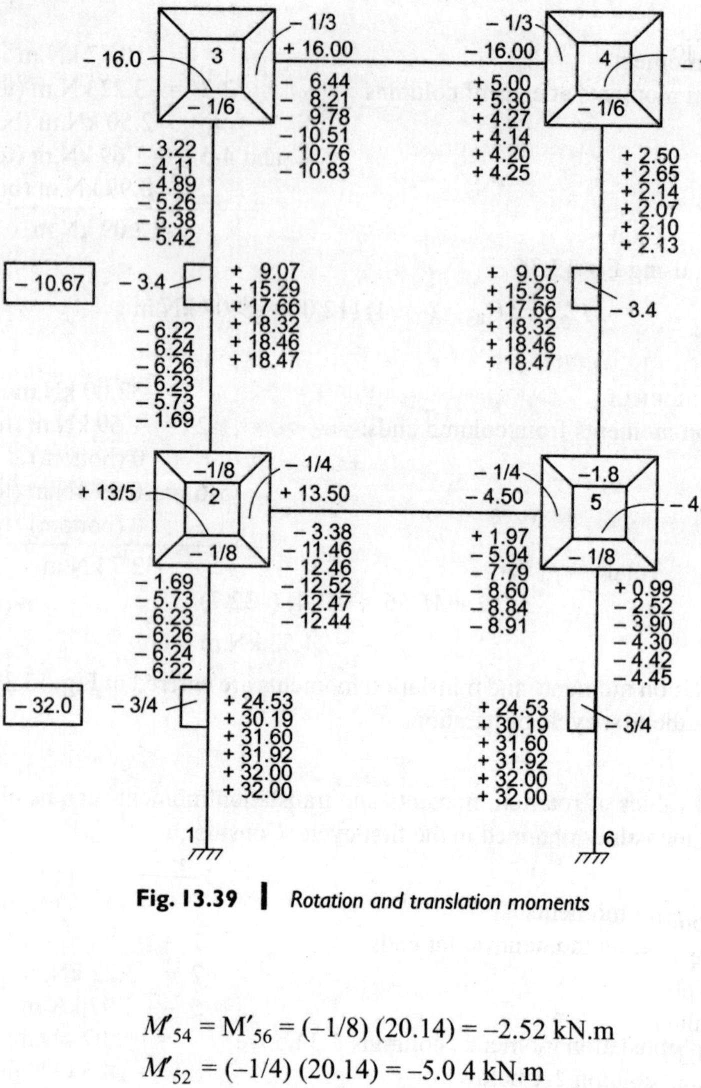

Fig. 13.39 | *Rotation and translation moments*

$$M'_{54} = M'_{56} = (-1/8)\,(20.14) = -2.52 \text{ kN.m}$$
$$M'_{52} = (-1/4)\,(20.14) = -5.0\,4 \text{ kN.m}$$

Joint 4

Sum of joint moments	= –16.00 kN.m
Rotation moments at far ends: 3	= –6.44 kN.m
5	= –2.52 kN.m
Translation moments of column 4-5 below	= + 9.07 kN.m
Total	–15.89 kN.m

$$M'_{43} = (1/3)\,(-15.89) = +5.30 \text{ kN.m}$$
$$M'_{45} = (-1/6)\,(-15.89) = +2.65 \text{ kN.m}$$

Joint 3

Sum of joint moments = + 16.00 kN.m

Rotation moments at far ends: 4 = + 5.30 kN.m

2 = –5.73 kN.m

Translation moment: column 2-3 below = + 9.07 kN.m

Total +24.64 kN.m

$$M'_{34} = (-1/3)(24.64) = -8.21 \text{ kN.m}$$

$$M'_{32} = (-1/6)(24.64) = -4.11 \text{ kN.m}$$

Storey 2

Storey moment = –10.67 kN.m

Rotation moments from column ends:

column 2-3 = –4.11 kN.m (top)

= –5.73 kN.m (bottom)

and column 4-5 = +2.65 kN.m (top)

–2.52 kN.m (bottom);

Total –20.38 kN.m

$$M''_{23} = M''_{54} = (-3/4)(-20.38) = +15.29 \text{ kN.m}$$

Storey 1

Storey moment = –32.00 kN.m

Rotation moments of column ends:

column 1-2 = –5.73 kN.m

column 5-6 = –2.52 kN.m

Total –40.25 kN.m

$$M''_{12} = M''_{56} = (-3/4)(-40.25) = 30.19 \text{ kN.m}$$

This completes the second cycle of iteration. In a similar manner the subsequent cycles are carried out taking each time the improved rotation and translation moments obtained in the immediately previous cycle.

The values for the rotation and translation moments for six cycles are shown in Fig. 13.39. The values in the sixth cycle are taken as acceptable and the values in the previous cycles are ignored.

Once the rotation and translation moments are known the final end moments can be computed using Eq. 13.37. The computations have been done on the outline of the frame as shown in Fig. 13.40a.

The final moments are shown in Fig. 13.40*b*. It may be noted that all the joints satisfy the equilibrium condition, $\Sigma M = 0$, with only slight errors, if any due to rounding off of values.

As an additional check we can find the shear in the columns in each storey and compare it with the external shear on the structure. Shear in the columns in storey 2 is

$$(1/4)(1.41 + 0.61 + 18.28 + 11.70) = 8.0 \text{ kN}$$

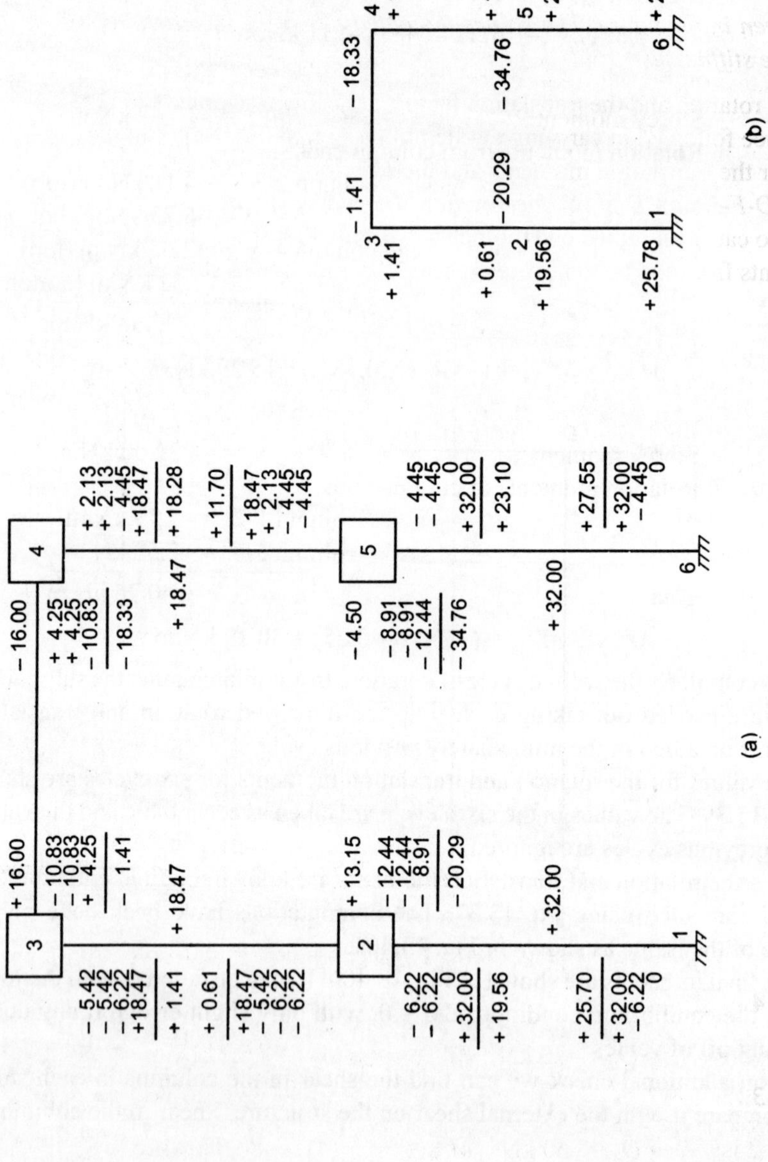

Fig. 13.40 | *(a) Computation of end moments, (b) Final moments*

and in storey 1

$$(1/4)\ (19.56 + 25.78 + 23.10 + 27.55) = 24.0 \text{ kN}$$

Both of them satisfy external shear and thus act as a check for the accuracy of results.

Example 13.11 | *A four-storeyed three-bay building frame under lateral loading is shown in Fig. 13.41. The data of the frame are given in the figure. The values indicated along the members represent their relative stiffnesses.*

The rotation and the translation factors are shown recorded in Fig. 13.42. The sequence followed in carrying out iterations is: first the storeys in the order, 4-3-2-1, for the translation moments and then the joints in the order, *B-H-N-Q*, *C-G-M-P*, *D-F-L* and *E-K* for the rotation moments. Since there are no transverse loads to cause any fixed end moments, we start with the evaluation of translation moments from the known storey moment as follows. The storey moments are:

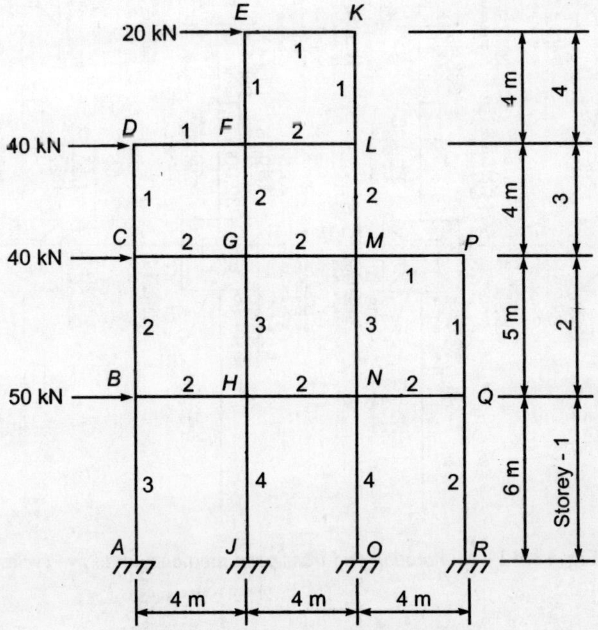

Fig. 13.41 | *Frame and loading*

Storey 4

$$Q_4 = -20 \text{ kN} \quad M^F{}_r = -\frac{20}{3}(4) = -26.7 \text{ kN.m}$$

Storey 3

$$Q_3 = -60 \text{ kN} \quad M^F{}_r = -\frac{60}{3}(4) = -80.0 \text{ kN.m}$$

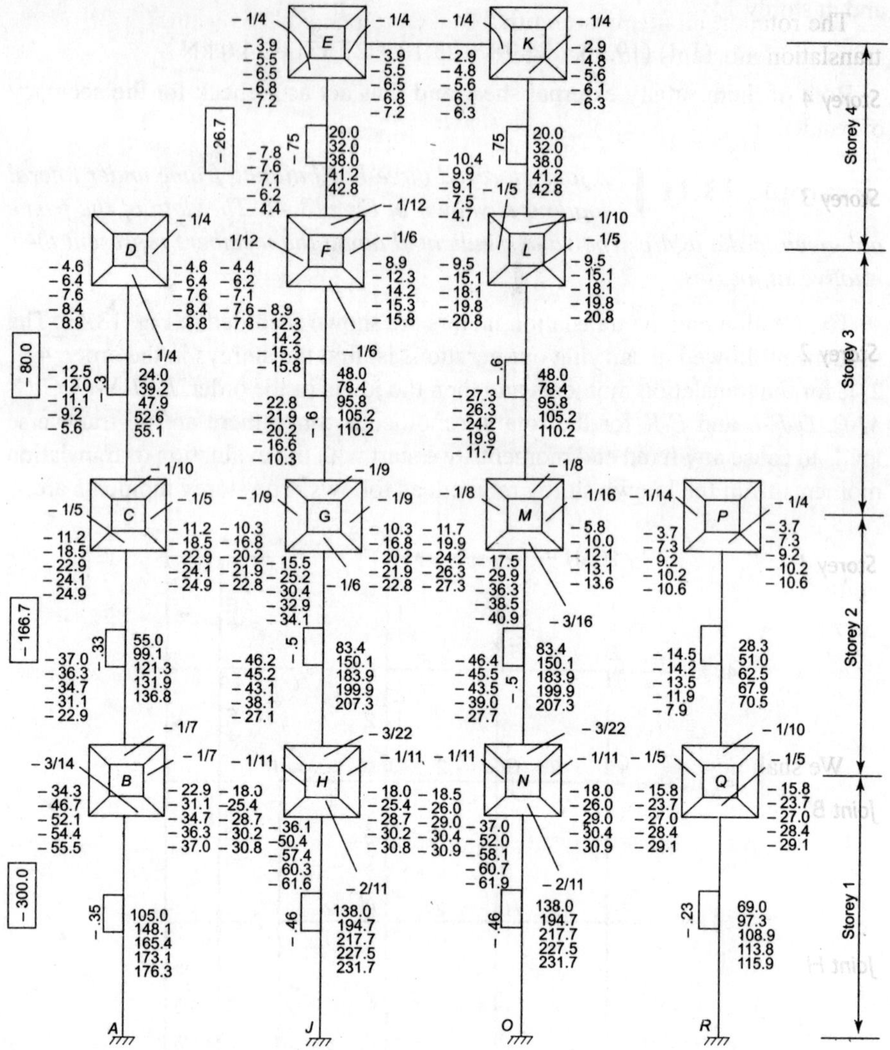

Fig. 13.42 | *Rotation and translation moments up to five cycles*

Storey 2

$$Q_2 = -100 \text{ kN} \quad M^F{}_r = -\frac{100}{3}(5) = -166.7 \text{ kN.m}$$

Storey 1

$$Q_1 = -150 \text{ kN} \quad M^F{}_r = -\frac{150}{3}(6) = -300 \text{ kN.m}$$

These storey moments are shown entered in small boxes at the mid height of each storey to the left of the first column line.

The rotation moments are initially taken to be zero and using Eq. 13.45 the translation moments are calculated.

Storey 4

$$M''_{EF} = M''_{KL} = (-3/4)\,(26.7) = +\,20.0$$

Storey 3

$$M''_{DC} = (-0.3)\,(-80.0) = +\,24.0$$

$$M''_{FG} = M''_{LM} = (-0.96)\,(-80.0) = +\,48.0$$

Storey 2

$$M''_{CB} = (-0.33)\,(-166.67) = +\,55.0$$

$$M''_{GH} = M''_{MN} = (-0.5)\,(-166.7) = +\,83.4$$

$$M''_{PQ} = (-0.17)\,(-166.7) = +\,28.3$$

Storey 1

$$M''_{BA} = (-0.35)\,(-300.0) = +\,105.0$$

$$M''_{HJ} = M''_{NO} = (-0.46)\,(-300.0) = 138.0$$

$$M''_{QR} = (-0.23)\,(-300.0) = 69.0$$

We shall now proceed to determine the rotation moments.

Joint B

$$M'_{BH} = M'_{BC} = (-1/7)\,(55.0 + 105) = -22.9$$

$$M'_{BA} = (-3/14)\,(55.0 + 105.0) = -34.3.$$

Joint H

$$M'_{HB} = M'_{HN} = (-1/11)\,(83.4 + 138.0 - 22.9) = -18.0$$

$$M'_{HG} = (-3/22)\,(83.4 + 138.0 - 22.9) = -27.1$$

$$M'_{HJ} = (-2/11)\,(83.4 + 138.0 - 22.9) = -36.1$$

Joint N

$$M'_{NH} = M'_{NQ} = (-1/11)\,(83.4 + 138.0 - 18.0) = -18.5$$

$$M'_{NO} = (-2/11)\,(83.4 + 138.0 - 18.0) = -37.0$$

$$M'_{NM} = (-1.5/11)(83.4 + 138.0 - 18.0) = -27.7$$

Joint Q

$$M'_{QN} = M'_{QR} = (-1/5)\,(28.3 + 69.0 - 18.5) = -15.8$$

$$M'_{QP} = (-1/10)\ (28.3 + 69.0 - 18.5) = -7.9$$

Joint C

$$M'_{CG} = M'_{CB} = (-1/5)\ (24.0 + 22.0 - 22.9) = -11.2$$

$$M'_{CD} = (-1/10)\ (24.0 + 55.0 - 22.9) = -5.6$$

Joint G

$$M'_{GC} = M'_{GF} = M'_{GM} = (-1/9)\ (48.0 + 83.4 - 11.2 - 27.1) = -10.3$$

$$M'_{GH} = (-1/6)\ (93.1) = -15.5$$

Joint M

$$M'_{MG} = M'_{ML} = (-1/8)(48.0 + 83.4 - 10.3 - 27.7) = -11.7$$

$$M'_{MP} = (-1/16)\ (93.4) = -5.8$$

$$M'_{MN} = (-3/16)\ (93.4) = -17.5$$

Joint P

$$M'_{PM} = M'_{PQ} = (-1/4)\ (28.3 - 5.8 - 7.9) = -3.7$$

Joint D

$$M'_{DC} = M'_{DF} = (-1/4)\ (24.0 - 5.6) = -4.6$$

Joint F

$$M'_{FL} = M'_{FG} = (-1/6)\ (20.0 + 48.0 - 4.6 - 10.3) = -8.9$$

$$M'_{FD} = M'_{FE} = (-1/12)\ (5.3.1) = -4.4$$

Joint L

$$M'_{LF} = M'_{LM} = (-1/5)\ (20.0 + 48.0 - 8.9 - 11.7) = -9.5$$

$$M'_{LK} = (-1/10)\ (20.0 + 48.0 - 8.9 - 11.7) = -4.7$$

Joint E

$$M'_{EK} = M'_{EF} = (-1/4)\ (20.0 - 4.4) = -3.9$$

Joint K

$$M'_{KE} = M'_{KL} = (-1/4)\ (20.0 - 3.9 - 4.7) = -2.9$$

This completes one cycle. The rotation and translation moments have been shown entered in Fig. 13.42 as the first row in each entry.

The second cycle again starts with the determination of translation moments. Again using Eq. 13.36, the improved values of translation moments are:

Storey 4

$$M''_{EF} = M''_{KL} = (-3/4)\ (-26.7 - 3.9 - 4.4 - 2.9 - 4.7) = 32.0$$

Storey 3

$$M''_{DC} = (-0.3)(-80.0 - 4.6 - 5.6 - 8.9 - 10.3 - 9.5 - 11.7) = 39.2$$

$$M''_{FG} = M''_{LM} = (-0.6)(130.6) = 78.4$$

Storey 2

$$M''_{CB} = (-0.33)(-166.7 - 11.2 - 22.9 - 15.5 - 27.1 - 17.5 - 27.7 - 3.7 - 7.9)$$
$$= 99.1$$

$$M''_{GH} = M''_{MN} = (-0.5)(-300.2) = 150.1$$

$$M''_{PQ} = (-0.17)(-300.2) = 51.0$$

Storey 1

$$M''_{BA} = (-0.35)(-300.0 - 34.3 - 36.1 - 37.0 - 15.8) = 148.1$$

$$M''_{HJ} = M''_{NO} = (-0.46)(-423.2) = 194.7$$

$$M'_{QR} = (-0.23)(-423.2) = 97.3$$

Having determined the translation moments we shall proceed to determine the rotation moments.

Joint B

$$M'_{BH} = M'_{BC} = (-1/7)(-11.2 - 18.0 + 99.1 + 148.1) = -31.1$$

$$M'_{BA} = (-3/14)(218.0) = -46.7$$

Joint H

$$M'_{HB} = M'_{HN} = (-1/11)(-31.1 - 15.5 - 18.5 + 0 + 150.1 + 194.7) = -25.4$$

$$M'_{HG} = (-3/2)(279.7) = -38.1$$

$$M'_{HJ} = (-2/11)(279.7) = -50.9$$

Joint N

$$M'_{NH} = M'_{NQ} = (-1/11)(150.1 + 194.7 - 25.4 - 17.5 - 15.8) = -26.0$$

$$M'_{NO} = (-2/11)(286.1) = -52.0$$

$$M'_{NM} = (-1.5/11)(286.1) = -39.0$$

Joint G

$$M'_{QN} = M'_{QR} = (-1/5)(51.0 + 97.3 - 26.0 - 3.7) = -23.7$$

$$M'_{QP} = (-1/10)(118.6) = -11.9$$

Joint C

$$M'_{CG} = M'_{CB} = (-1/5)(39.2 + 99.1 - 10.3 - 31.1 - 4.6) = -18.5$$

$$M'_{CD} = (-1/10)(92.3) = -9.2$$

Joint G

$$M'_{GC} = M'_{GF} = M'_{GM} = (-1/9)\,(78.4 + 150.1 - 18.5 - 8.9 - 11.7 - 38.1) = -16.8$$

$$M'_{GH} = (-1/6)\,(151.3) = -25.2$$

Joint M

$$M'_{MG} = M'_{ML} = (-1/8)\,(78.4 + 150.1 - 16.8 - 9.5 - 3.7 - 39.0) = -19.9$$

$$M'_{MP} = (-1/16)\,(159.5) = -10.0$$

$$M'_{MN} = (-3/16)\,(159.5) = -29.9$$

Joint P

$$M'_{PM} = M'_{PQ} = (-1/4)\,(51.0 - 10.0 - 11.9) = -7.3$$

Joint D

$$M'_{DC} = M'_{DF} = (-1/4)\,(39.2 - 9.2 - 4.4) = -6.4$$

Joint F

$$M'_{FL} = M'_{FG} = (-1/6)\,(32.0 + 78.4 - 6.4 - 3.9 - 9.5 - 16.8) = -12.3$$

$$M'_{FD} = M'_{FE} = (-1/12)\,(73.8) = -6.2$$

Joint L

$$M'_{LF} = M'_{LM} = (-1/5)\,(32.0 + 78.4 - 12.3 - 2.9 - 19.9) = -15.1$$

$$M'_{LK} = (-1/10)\,(75.3) = -7.5$$

Joint E

$$M'_{EK} = M'_{EF} = (-1/4)\,(32.0 - 6.2 - 3.9) = -5.5$$

Joint K

$$M'_{KE} = M'_{KL} = (-1/4)\,(32.0 - 5.5 - 7.5) = -4.8$$

This completes the second cycle. In a similar manner computations are carried out in the subsequent cycles of iteration. The values of the rotation and translation moments up to five cycles have been shown entered in Fig. 13.42.

The values obtained in the fifth cycle are taken as acceptable and all the values in the previous cycles are ignored. The fifth cycle values of rotation and translation moments are shown separately in Fig. 13.43 for reference while computing the final end moments.

The final moments are worked out as usual using Eq. 13.37 and the results are shown entered in Fig. 13.44.

As a check we evaluate the shear in each storey and compare it with the external shear.

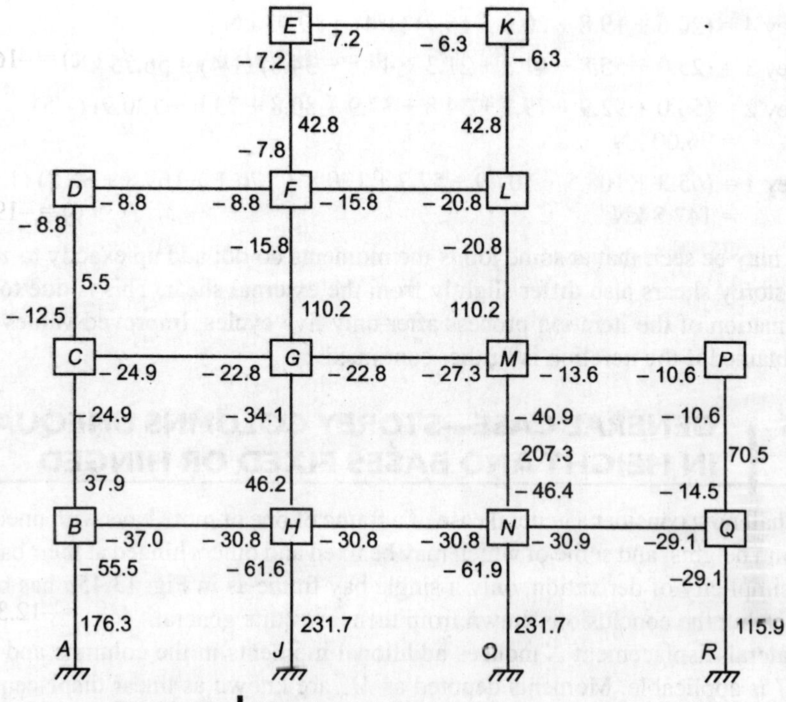

Fig. 13.43 | *Rotation and translation moments in fifth cycle*

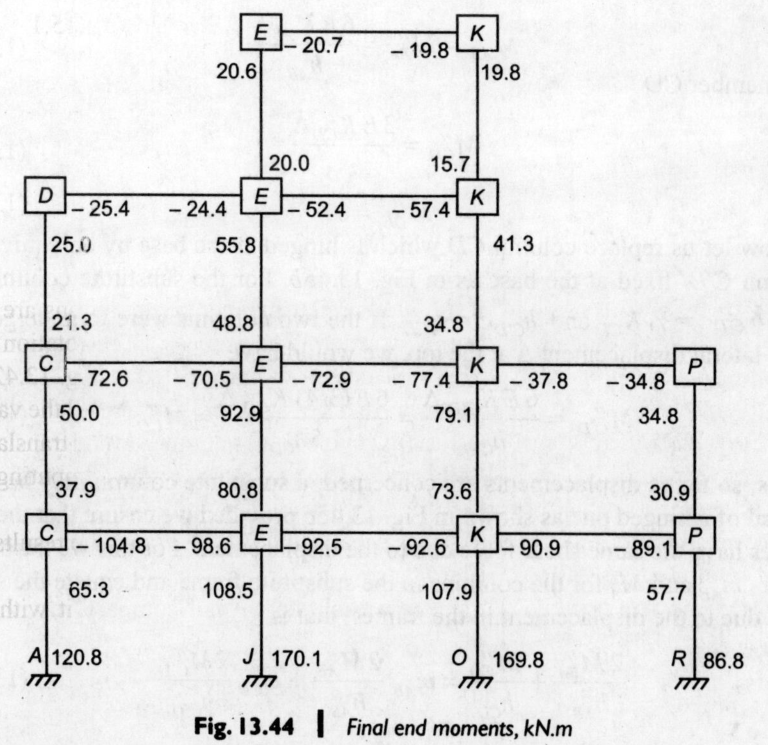

Fig. 13.44 | *Final end moments, kN.m*

Storey 4 = (20.6 + 19.8 + 20.0 + 15.7) (1/4) = 19.03 kN

Storey 3 = (25.0 + 55.8 + 41.3 + 21.3 + 48.8 + 34.8) (1/4) = 56.75 kN

Storey 2 = (50.0 + 92.9 + 79.1 + 34.8 + 37.9 + 80.8 + 73.6 + 130.9) (1/5)
= 96.00 kN

Storey 1 = (65.3 + 108.5 + 107.9 + 57.7 + 120.8 + 170.1 + 169.8 + 86.8) (1/6)
= 147.8 kN

It may be seen that at some joints the moments do not add up exactly to zero. The storey shears also differ slightly from the external shear. This is due to the termination of the iteration process after only five cycles. Improved values can be obtained if the iteration is further continued.

13.5 GENERAL CASE—STOREY COLUMNS UNEQUAL IN HEIGHT AND BASES FIXED OR HINGED

We shall now consider a general case of a frame of one or more bays with unequal column heights, and some of which may be fixed and others hinged at their bases. For simplicity of derivation, only a single bay frame as in Fig. 13.45*a* has been chosen but the conclusions drawn from them are quite general.

Lateral displacement Δ induces additional moments in the columns and Eq. 13.37 is applicable. Moments denoted as M'' are known as linear displacement moments and are as follows

$$M''_{AB} = M''_{BA} = \frac{6EK_{AB}\Delta}{h_{AB}} \tag{13.46}$$

For member CD

$$M''_{CD} = \frac{3EK_{CD}\Delta}{h_{CD}} \tag{13.47}$$

and

$$M''_{DC} = 0 \tag{13.48}$$

Now let us replace column *CD* which is hinged at the base by an equivalent column *C′D′* fixed at the base as in Fig. 13.45*b*. For the substitute column we take $K_{C'D'}$ = ¾ K_{CD} and $h_{C'D'}$. = h_{CD}. If the two columns were to undergo the same lateral displacement Δ at the top, we would have

$$M''_{C'D'} = \frac{6EK_{C'D'}\Delta}{h_{C'D'}} = \frac{6E(3/4)K_{CD}\Delta}{1.5h_{CD}} = M''_{CD} \tag{13.49}$$

that is, so far as displacements are concerned, a substitute column may be used instead of a hinged one as shown in Fig. 13.45*c* provided we ensure that the two frames have the same shear force due to the displacement. For this we introduce factors m_{AB} and M_A for the columns in the substitute frame and equate the shear force due to the displacement in the frames, that is

$$\frac{2M''_{BA}}{h_{AB}} + \frac{M''_{CD}}{h_{CD}} = m_{AB}\frac{2M''_{AB}}{h_{AB}} + m_{CD}\frac{2M''_{C'D'}}{h_{C'D'}} \tag{13.50}$$

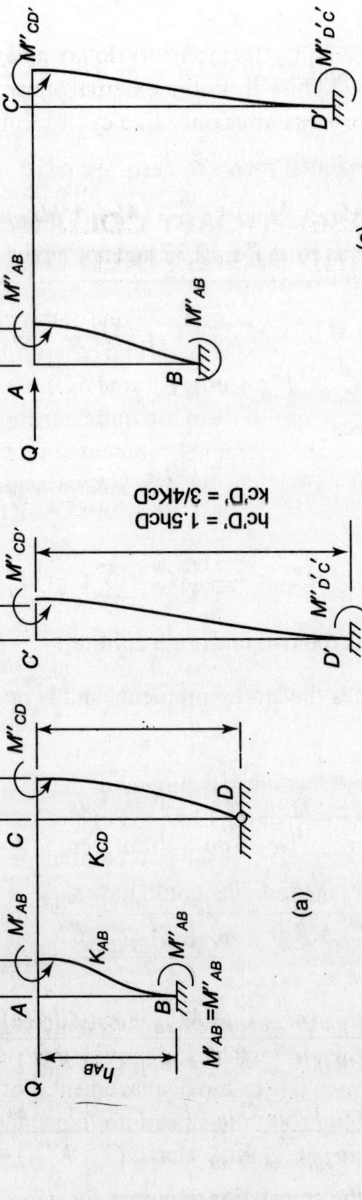

Fig. 13.45 | (a) Frame with unequal column heights and one leg fixed and the other hinged at the base, (b) Substitute column, (c) Frame with substitute column

Which gives $m_{AB} = 1$ for the column fixed at the base

and $$m_{CD} = \frac{M''_{CD}}{M''_{C'D'}} \frac{h_{C'D'}}{2h_{CD}} = 3/4 \tag{13.51}$$

for the hinged column.

We are now in a position to deal with the substitute frame by choosing storey height $h_r = H_{AB}$ and writing

$$C_{AB} = \frac{h_r}{h_{AB}} \text{ and } C_{C'D'} = \frac{h_r}{h_{C'D'}} \text{ or } C'_{CD} = \frac{h_r}{h'_{CD}} \tag{13.52}$$

for convenience of notation.

Now summing up the horizontal forces to zero. We have

$$Q\,h_r + C_{AB}\,(M_{AB} + M_{BA}) + C'_{CD}\,(M_{CD} + M_{DC}) = 0 \tag{13.53}$$

Substituting for the moments from Eq. 13.37 and noting the fixed end moments = 0 we have

$$Q\,h_r + 3\,[C_{AB}\,(M'_{AB} + M'_{BA}) + C'_{CD}\,(M'_{C'D'} + M'_{D'C'})]$$
$$+ 2(m_{AB}\,C_{AB}\,M''_{AB} + m_{CD}\,C'_{C'D'}M''_{C'D'}) = 0 \tag{13.54}$$

This after transformation gives

$$m_{AB}\,C_{AB}M''_{AB} + m_{CD}\,C'_{CD}\,M''_{CD} = -\frac{3}{2}\left\{\frac{Q\,h_r}{3} + \sum C_{ik}\,(M'_{ik} + M'_{ki}\right\} \tag{13.55}$$

or $$m_{AB}\,C_{AB}M''_{AB} + m_{CD}\,C'_{CD}\,M''_{CD} = -\frac{3}{2}\left\{M_r^F + \sum C_{ik}\,(M'_{ik} + M'_{ki}\right\} \tag{13.56}$$

(here *i-k* in general represents the two ends of a column)

The terms $M_r^F = \dfrac{Q\,h_r}{3}$ defines the storey moments and is positive when Q acts from right to left. We know

$$\frac{M''_{AB}}{M''_{C'D'}} = \frac{K_{AB}}{h_{AB}}\frac{h'_{CD}}{K'_{CD}} = \frac{K_{AB}}{K'_{CD}}\frac{C_{AB}}{C'_{CD}} \tag{13.57}$$

or $$\frac{m_{AB}\,C_{BA}M''_{AB}}{m_{CD}\,C'_{CD}M''_{C'D'}} = \frac{m_{AB}}{m_{CD}}\frac{C^2{}_{AB}}{C^2{}_{CD}} = \frac{K_{AB}}{K'_{CD}} \tag{13.58}$$

which gives

$$M''_{A'B'} = \frac{C_{AB}\,K_{AB}\,(m_{AB}\,C_{AB}\,M''_{AB} + m_{CD}C'_{CD}M''_{C'D'})}{(m_{AB}\,C^2_{AB}\,K_{AB} + m_{CD}\,C^2_{CD}\,K'_{CD})} \tag{13.59}$$

$$M''_{C'D'} = \frac{C'_{CD}\,K'_{CD}\,(m_{AB}\,C_{AB}\,M''_{AB} + m_{CD}\,C'_{CD}M''_{C'D'})}{(m_{AB}\,C^2_{AB}\,K_{AB} + m_{CD}\,C'^2_{CD}\,K'_{CD})} \tag{13.60}$$

With the help of Eq. 13.56 the translation moment for any column *i-k* may be written as

$$M''_{ik} = \left(-\frac{3}{2}\right)\frac{C_{ik}k_{ik}}{\sum m_{ik}\, C_{ik}^2\, K_{ik}}\left\{M_r^F + \sum C_{ik}\,(M'_{ik} + M'_{ki})\right\} \quad (13.61)$$

$$M''_{ik} = v_{ik}\left\{M_r^F + \sum C_{ik}\,(M'_{ik} + M'_{ki})\right\} \quad (13.62)$$

$$v_{ik} = \left(-\frac{3}{2}\right)\frac{C_{ik}\,k_{ik}}{\sum m_{ik}\, C_{ik}^2\, K_{ik}} \quad (13.63)$$

which we know is the translation factor for column *i-k*. A control on the calculations v_{ik} is given by

$$\sum m_{ik}\, C_{ik}\, V_{ik} = -\frac{3}{2} \quad (13.64)$$

These derivations are perfectly general in nature and can be extended to different cases as described below.

Case 1

All the columns fixed at base.

For this condition

$$M_{ik} = 1 \text{ for all columns}$$

and
$$v_{ik} = \left(-\frac{3}{2}\right)\frac{C_{ik}\, K_{ik}}{\sum C_{ik}^2\, K_{ik}} \quad (13.65)$$

and the control is
$$\sum C_{ik}\, v_{ik} = -\frac{3}{2} \quad (3.66)$$

Case 2

All the columns hinged at base.

For this condition

$$m_{ik} = \frac{3}{2} \text{ for all columns}$$

and
$$v_{ik} = -\frac{2\, C_{ik}\, K_{ik}}{\sum C_{ik}^2\, K_{ik}} \quad (13.67)$$

and the control is $\Sigma C_{ik}\, v_{ik} = -2$

The example that follows illustrates the steps involved.

Example 13.12 | *It is required to analyse the frame in Fig. 13.46 for the end moments. E is constant and the relative values of I are indicated along the members.*

The same frame is once solved in Example 12.6 using the moment distribution method. This gives a good comparison of the two methods.

The hinged column is replaced by a column fixed at the base having $h'_{12} = 1.5\, h_{12} = 1.5(4) = 6.0$ m and $K'_{12} = (3/4)\, K_{12} = (3/4)\,(I) = 3/4$. If we choose frame height $h_r = h_{34} = 4$ m, $C'_{12} = 4/6 = 2/3$ and $C_{34} = 4/4 = 1.0$.

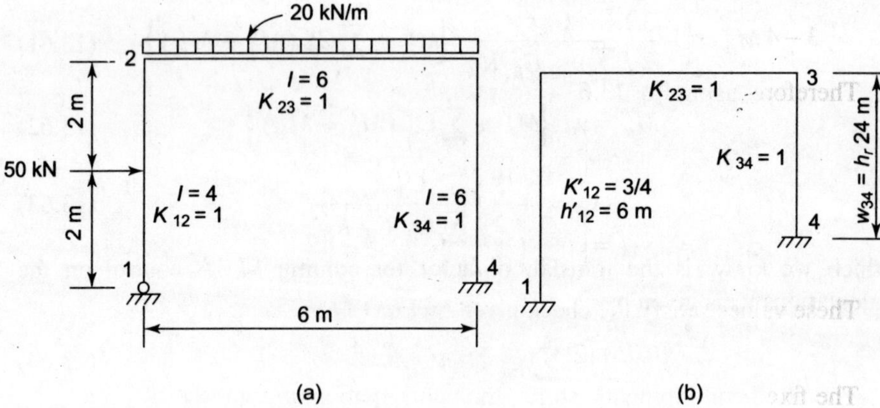

Fig. 13.46 | *(a) Frame and loading, (b) Frame with substitute column*

Fixed end moments

The fixed end moments are calculated as usual. However, the fixed end moment M^F_{21} is modified as $(M^F_{21} - 1/2\ M^F_{12})$ to take into account the hinged condition at support 1.

Storey moment

$$\text{Lateral force } Q_r = \left(25 + \frac{37.5}{4}\right) = 34.375 \text{ kN}$$

and the storey moment

$$M_r^F = \frac{Q_r\,(h_r)}{3} = -\frac{34.375\,(4)}{3} = -45.83 \text{ kN.m}$$

Rotation factors

Joint 2
$$u_{21} = (-½)\ \frac{(3/4)}{(1+3/4)} = -0.21$$

$$u_{23} = (-1/2)\frac{1}{(1+3/4)} = -0.29$$

Joint 3
$$u_{32} = (-1/2)\frac{1}{(1+1)} = -0.25$$

$$u_{34} = (-1/2)\frac{1}{(1+1)} = -0.25$$

Transaction factors

$$1-2: m_{12} = 0.75,\ C'_{12} = \frac{2}{3},\ K'_{12} = \frac{3}{4},\ C'_{12}\ K'_{12} = 0.5,\ m_{12}\ C'^2_{12}\ K'_{12} = 0.25$$

$$3-4: m_{34} = 1.0, C_{34} = 1, K_{34} = 1, C_{34}\ K_{34} = 1.0, m_{34}\ C_{34}^2\ K_{34} = 1.0$$

Therefore, using Eq. 13.63

$$v_{12} = \left(-\frac{3}{2}\right)\frac{0.5}{(0.25+1.0)} = -0.6$$

$$v_{34} = \left(-\frac{3}{2}\right)\frac{1.0}{(0.25+1.0)} = -1.2$$

These values satisfy the check given by Eq. 13.66, that is.

$$(0.75)\ (2/3)\ (-0.6) + 1 \times 1\ (-1.2) = -1.5$$

The fixed end moments, storey moment, rotation and translation factors are shown entered in Fig. 13.47a.

The iterations have been carried out in the usual manner. To start with, the rotation and translation moments are considered to be zero.

Cycle 1

Rotation moments

Joint 2

$$M'_{23} = -0.29\ (+22.5 + 0 + 0) = -6.53$$

$$M'_{21} = -0.21\ (22.5) = 4.73$$

Joint 3

$$M'_{32} = M'_{34} = -0.25\ (-60 - 6.53 + 0) = +16.63$$

Translation moments

$$M''_{12} = -0.6\ [-45.83 + (2/3)\ (-4.73) + (16.63)] = +19.41$$

$$M''_{34} = -1.2\ (-32.3) = +38.82$$

This completes the first cycle of computations and the values are shown entered in Fig. 13.47a.

Cycle 2

Rotation moments

Joint 2

$$M'_{23} = -0.29\ (+22.5 + 16.63 + 19.41) = -16.98$$

$$M'_{21} = -0.21\ (58.54) = -12.29$$

Joint 3

$$M'_{32} = M'_{34} = -0.25\ (-60 - 19.68 + 38.82) = +9.54$$

Translation moments

$$M''_{12} = -0.6\ [-45.83 + 2/3(-12.29) + 9.54] = +26.69$$

$$M''_{34} = -1.2\ (44.48) = 53.58$$

This completes the second cycle. In a similar manner, further iterations are carried out each time improving the values of the previous cycles. The rotation and translation moments up to five cycles have been shown entered in Fig. 13.47*a*. The final moments are computed using Eq. 13.37 and the computations are shown entered in Fig. 13.47*b*.

The final moments compare very well with the values obtained (shown in brackets) in Example 12.6 solved by the moment distribution method. The reader can easily judge the versatility of Kani's method and the simplicity of computations.

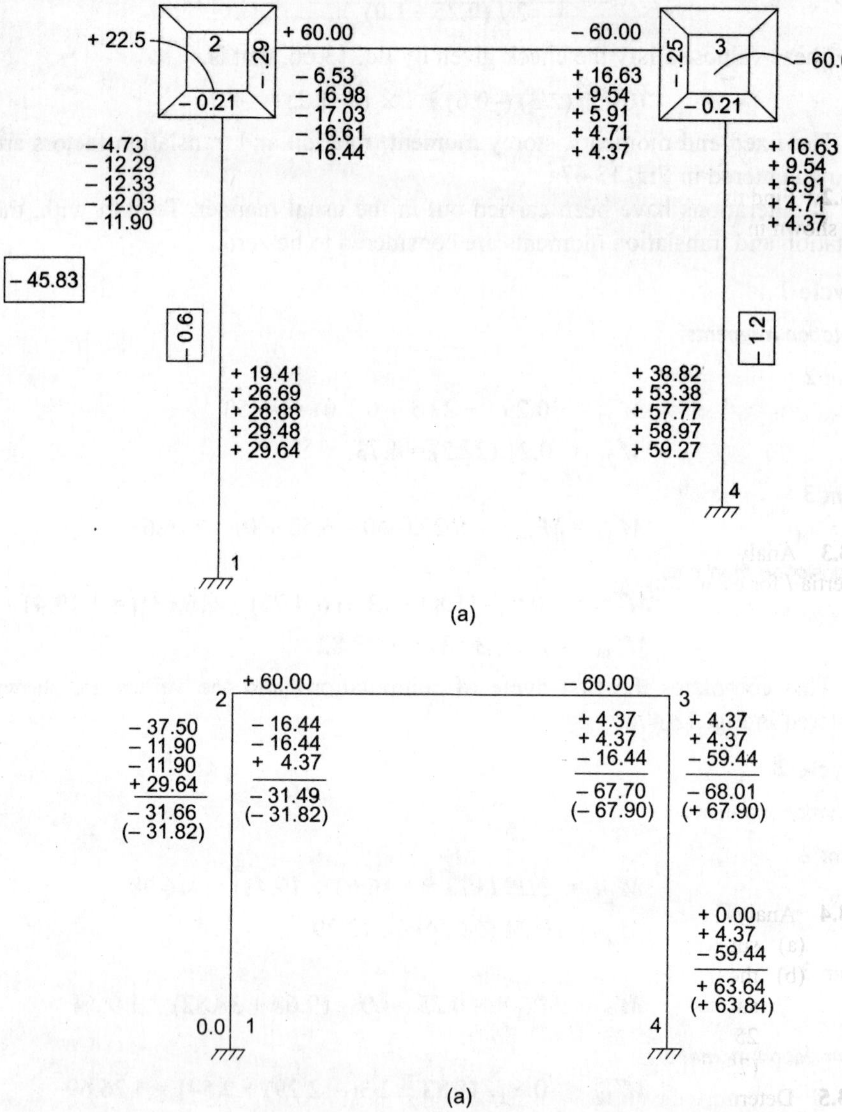

Fig. 13.47 | *Results of analysis: (a) Rotation and translation moments, (b) Computation of final end moments*

Problems for Practice

Use Kani's method in solving the following problems.

13.1 Analyse the continuous beam loaded as shown in Fig. 13.48 and sketch shear and moment diagrams. EI is constant.

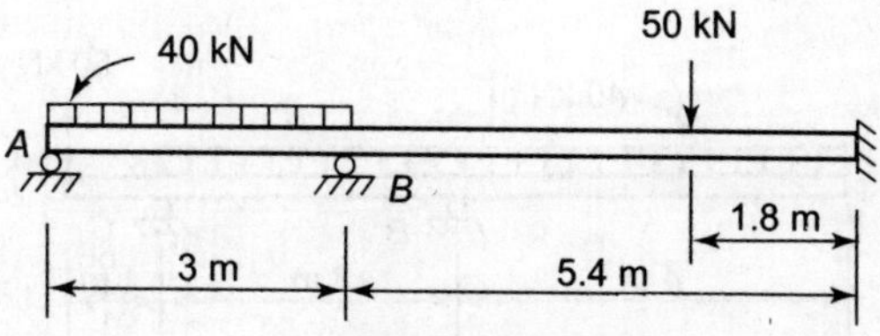

Fig. 13.48

13.2 Find the moments at all the supports and reactions for the continuous beam loaded as shown in Fig. 13.49. Flexural rigidity EI is constant.

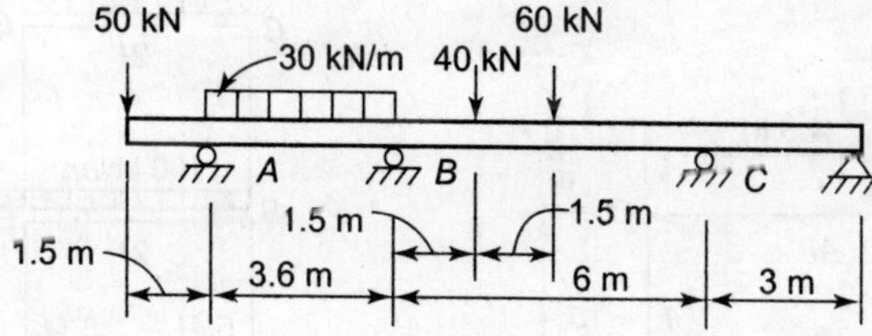

Fig. 13.49

13.3 Analyse the continuous beam given in Fig. 13.50 for support moments. Moment of inertia I for each span is indicated. E is constant.

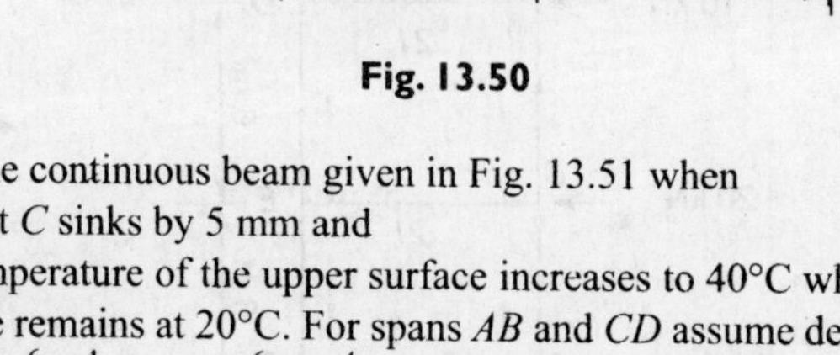

Fig. 13.50

13.4 Analyse the continuous beam given in Fig. 13.51 when

(a) support C sinks by 5 mm and

(b) the temperature of the upper surface increases to 40°C while that of the lower surface remains at 20°C. For spans AB and CD assume depth $h = 200$ mm, $I = 25 \times 10^{-6}$ m^4 (25×10^6 mm^4); for span BC, $h = 300$ mm and $I = 75 \times 10^{-6}$ m^4 (75×10^6 mm^4). $E = 210 \times 10^6$ kN/m^2 (210,000 MPa), $\alpha_t = 12 \times 10^{-6}$ per °C.

13.5 Determine the moments at supports if support B yields by 10 mm under the given loading for the beam shown in Fig. 13.52. $E = 204 \times 10^6$ kN/m^2 (204,000 MPa) and $I = 30 \times 10^{-6}$m^4(30×10^6 mm^4).

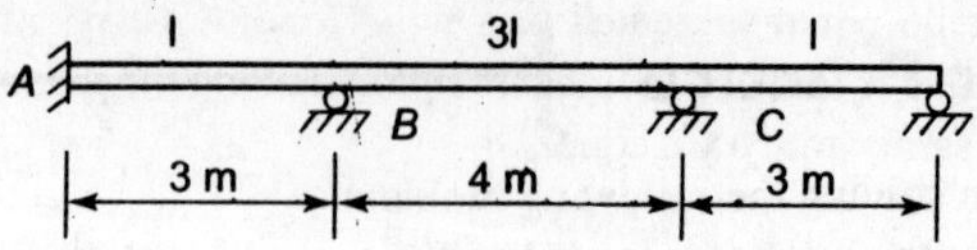

Fig. 13.51

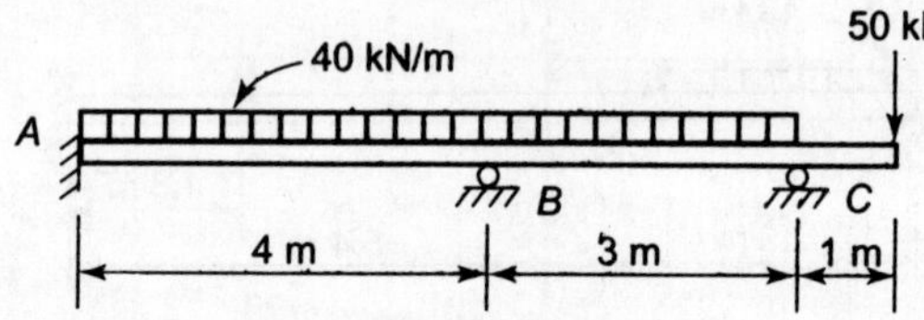

Fig. 13.52

13.6 Analyse the frame shown in Fig. 13.53 for end moments of members.

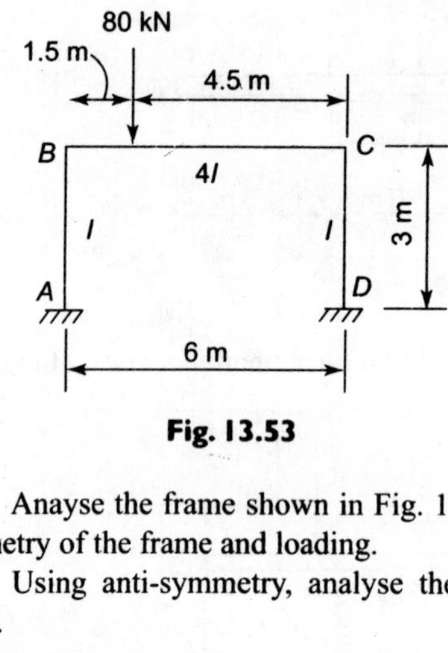

Fig. 13.53

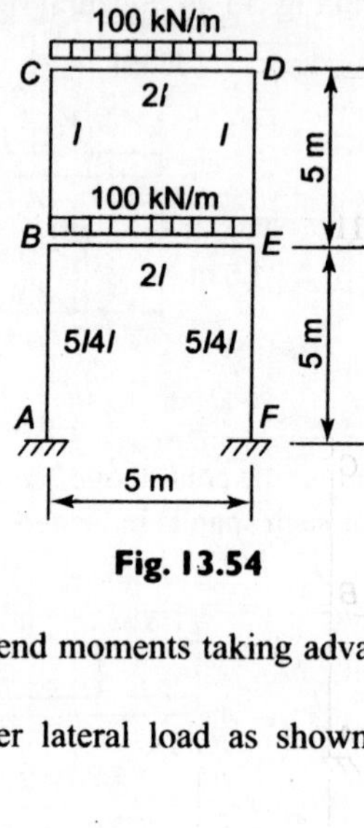

Fig. 13.54

13.7 Anayse the frame shown in Fig. 13.54 for the end moments taking advantage of symmetry of the frame and loading.

13.8 Using anti-symmetry, analyse the frame under lateral load as shown in Fig. 13.55.

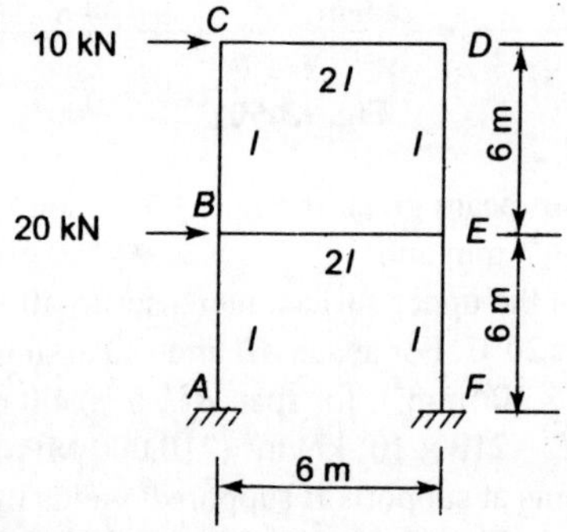

Fig. 13.55

13.9 Analyse the frame when subjected to a wind load of 10 kN/m as shown in Fig. 13.56. The stiffness value K for each member is marked along the members. The K values of the columns in the same storey are equal.

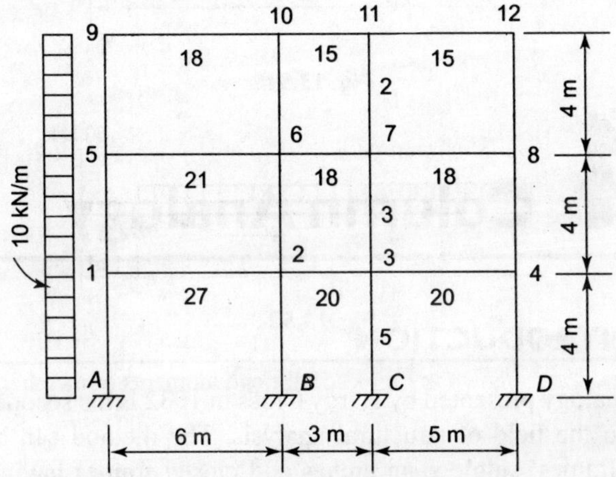

Fig. 13.56

13.10, 13.11 Analyse the frames shown for end moments.

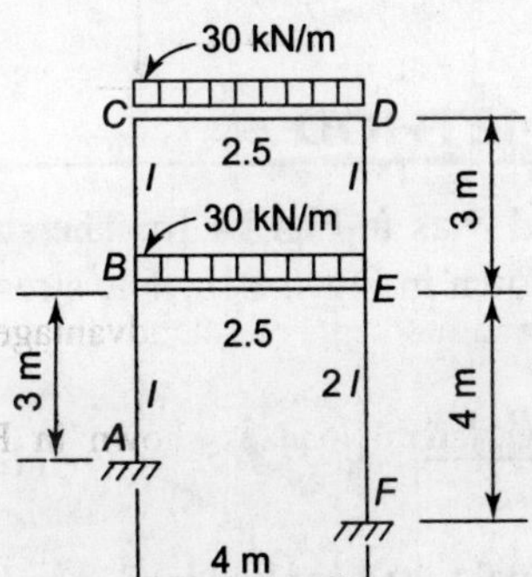

Fig. 13.57

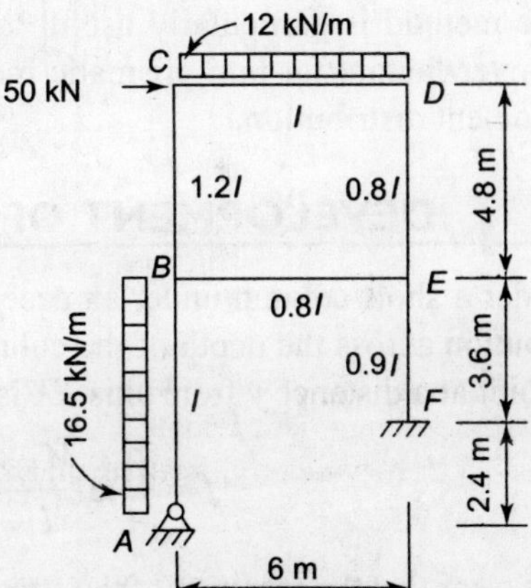

Fig. 13.58

14 Column Analogy

14.1 INTRODUCTION

The column analogy presented by Hardy Cross in 1932 is his second outstanding contribution to the field of structural analysis. The method can be applied to fixed beams, frames, single span arches and closed frames having degrees of indeterminancy not more than three. The anology pertains to the identities between the moments in a statically indeterminate structure and the stresses produced in an eccentrically loaded short column.

The method is particularly useful to determine the fixed end moments and carry-over factors for non-prismatic members which are necessary in carrying out moment distribution.

14.2 DEVELOPMENT OF THE METHOD

Consider a short column under an eccentric load P as in Fig. 14.1*a*. The stress distribution across the depth of the column is shown in Fig. 14.1*b*. The stress at any point at a distance y from axis XX is

$$f = \frac{P}{A} \pm \frac{M_y x}{I_y} = \frac{P}{A} \pm \frac{P \cdot e \cdot x}{I_y} \tag{14.1}$$

It is clear that the resultant of the stresses is equal to the applied load P and the centroid of the resultant coincides with the line of action of P.

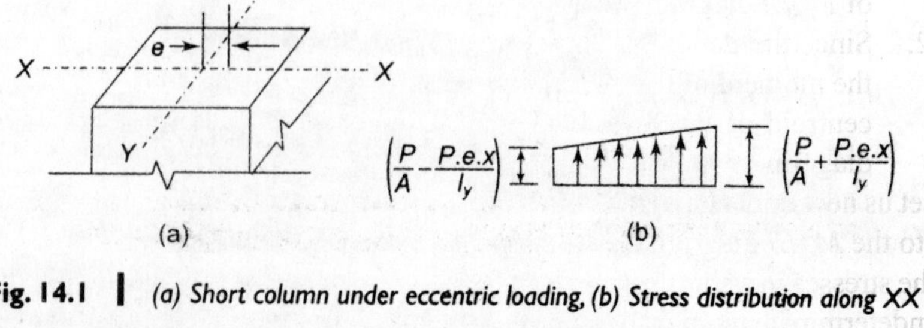

Fig. 14.1 | *(a) Short column under eccentric loading, (b) Stress distribution along XX*

Next consider a beam fixed at the ends and subjected to a concentrated load P as shown in Fig. 14.2. Suppose it is required to determine the end moments at A and B. The beam is made statically determinate by releasing the restraining moments M_A and M_B. The simply supported moment M_s diagram is shown in Fig. 14.2*b*. The bending moment diagram due to unknown end moments M_i is shown in Fig. 14.2*d*.

Now consider that the beam is replaced by an analogous column whose width is equal to $1/EI$ and depth same as the length of the beam along the axis. The loading on the column is the M_s diagram acting downward and the upward stress distribution across the depth of the column is the M_i diagram. From the moment area theorems the following, can be stated.

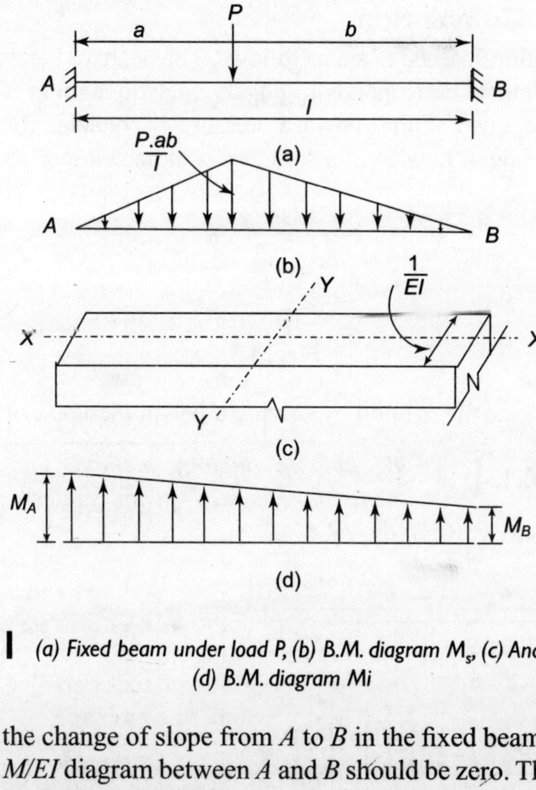

Fig. 14.2 | *(a) Fixed beam under load P, (b) B.M. diagram M_s, (c) Analogous column, (d) B.M. diagram Mi*

1. Since the change of slope from A to B in the fixed beam is zero, the area of the M/EI diagram between A and B should be zero. Therefore, the area of M_s/EI diagram should be equal to the $-M_i/EI$ diagram.
2. Since the deflection of the tangent at A to the tangent at end B is zero, the moment of the M/EI diagram should be equal to zero. Therefore, the centroid of the M_s/EI diagram should coincide with that of the M_i/EI diagram.

Let us now understand the column analogy. The load on the analogous column due to the M_s/EI diagram is analogous to stress distribution in the short column. As the stresses in a short column can be evaluated from the known applied load P, the indeterminate moment M_i can be evaluated from the known M_s/EI diagram. If the total area of the M_s/EI diagram, represented by N_i is applied to the analogous column at a point corresponding to the centroid of the diagram, the stress at any point on the analogous column will be

$$f = \frac{N}{A} \pm \frac{N \cdot e \cdot x}{I_y} \tag{14.2}$$

Hence the inderterminate moment M_i at any section is given by

$$M_i = \frac{N}{A} \pm \frac{N \cdot e \cdot x}{I_y} \tag{14.3}$$

The moment at any section in the beam is given by $M = M_s - M_i$

14.2.1 Sign Convention

The sign convention shall be taken as follows. The positive bending moment M_s, causing tension inside corresponds to positive elastic loading N. For the stress diagram of analogous column, upward pressure is considered as positive. The domain of the inside or outside of a structure is indicated in Fig. 14.3.

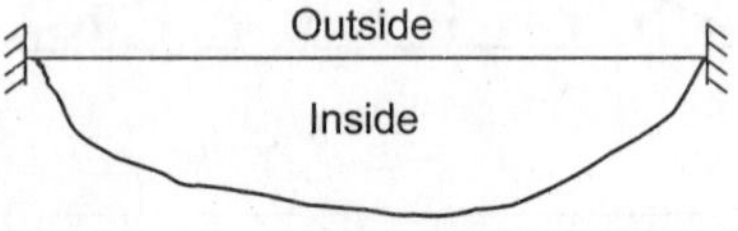

Fig. 14.3

We shall illustrate the method by solving a few numerical examples.

Example 14.1 | *Using column analogy method, find the fixed end moments for a fixed beam subjected to uniformly distributed load as shown in fig. 14.4.*

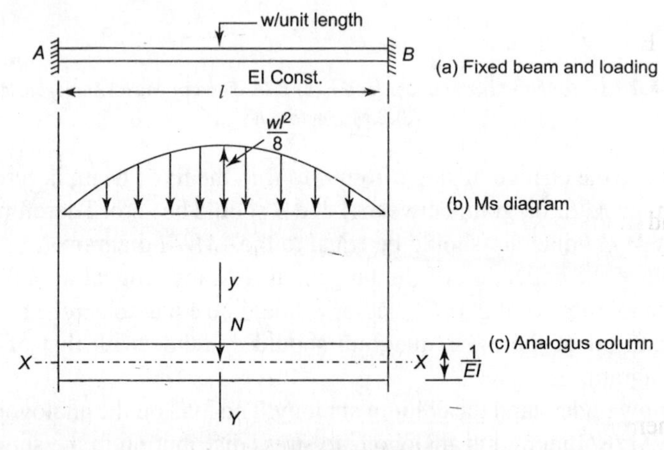

Fig. 14.4

Consider support moments M_A and M_B are redundant and are released to make the beam a simply supported. The analogous column is shown in Fig. 14.4c

Area of the column $= \dfrac{l}{EI}$

Axial load $N = \dfrac{2}{3}\dfrac{wl^2}{8}\cdot\dfrac{l}{EI} = \dfrac{wl^2}{12EI}$

The load is axial without eccentricity. Hence

$$M_{iA} = M_s - \frac{N}{A}$$

$$= 0 - \frac{wl^3}{12EI}\cdot\frac{EI}{l} = -\frac{wl^2}{12}$$

$$M_{iB} = -\frac{wl^2}{12}$$

Example 14.2 *Using the column analogy method, determine the fixed end moments for the beam shown in Fig. 14.5 draw the moment diagram.*

The beam is statically indeterminate by two degrees. We can release the redundants R_B and M_B and consider it as a cantilever beam.

The beam, the redundants, the moment diagram and the analogous column are shown in Fig. 14.5.

$$\text{Area of the column} = \frac{l}{EI}$$

$$\text{Axial load} = \frac{1}{3}\left(-\frac{wl^2}{8}\right)\frac{l}{2EI} = -\frac{wl^2}{48EI}$$

$$\text{Eccentricity } e = \frac{3}{8}l$$

$$\text{and } I_{yy} = \frac{1}{12}\left(\frac{I}{EI}\right)l^3 = \frac{l^3}{12EI}$$

$$\text{End moment } M_{iA} = -\frac{wl^2}{8} - \left(-\frac{wl^3}{48EI}\cdot\frac{EI}{l} - \frac{wl^3}{48EI}\cdot\frac{3l}{8}\cdot\frac{l}{2}\cdot\frac{12EI}{l^3}\right)$$

$$= -\frac{wl^2}{8} - \left(-\frac{wl^2}{48} - 3\frac{wl^2}{64}\right) = -\frac{11}{192}wl^2$$

$$\text{End moment } M_{iB} = 0 - \left\{-\left(\frac{wl^3}{48EI}\cdot\frac{EI}{l}\right) - \frac{wl^3}{48EI}\left(\frac{3l}{8}\right)\left(\frac{-l}{2}\right)\left(\frac{12EI}{l^3}\right)\right\} = \frac{-5}{192}wl^2$$

The bending moment diagram is shown in Fig. 14 5c

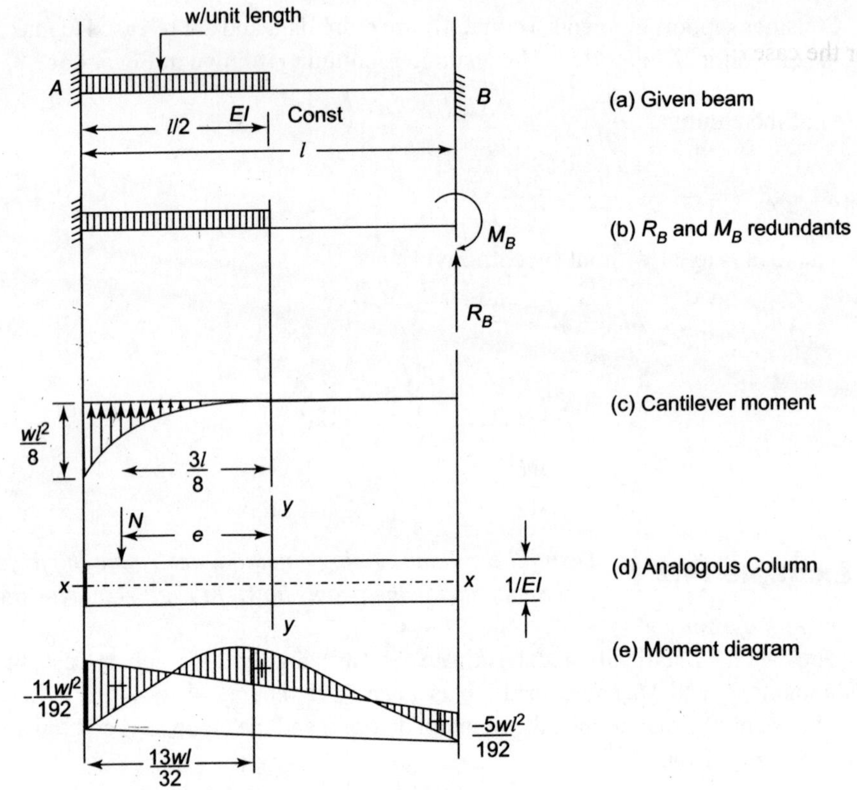

Fig. 14.5

Example 14.3 | *Using column analogy method, analyse the prismatic beam fixed at the ends as shown in Fig. 14.6 for the end moments under the given loading.*

Consider M_A and M_B as redundants. The M_s diagram for a simple beam and the analogous column are shown in Fig. 14.6*b* and *c*.

$$\text{Indeterminate moment } M_{iA} = \frac{N}{A} + \frac{N \cdot e \cdot x}{I_y}$$

$$\text{in which } A = \frac{l}{EI} \text{ and } I_y = \frac{l^3}{12EI}$$

$$\text{Load } N = \frac{1}{2}\frac{Pab}{l}\frac{l}{EI} = \frac{Pab}{2EI}$$

$$\text{Eccentrictiy } e = \frac{l}{2} - \frac{l+a}{3} = \frac{l-2a}{6}$$

$$\text{and } \frac{N}{A} = \frac{Pab}{2l}$$

for the case $a < b$, $x_A = +\dfrac{l}{2}$.

$$\frac{M_y}{I_y} x_A = \frac{Pab}{2EI} \frac{(l-2a)}{6} \frac{l}{2} \frac{12\,EI}{l^3}$$

$$= \frac{Pab}{2l^2}(l-2a)$$

$$M_{iA} = \frac{Pab}{2l} + \frac{Pab}{2l^2}(l-2a)$$

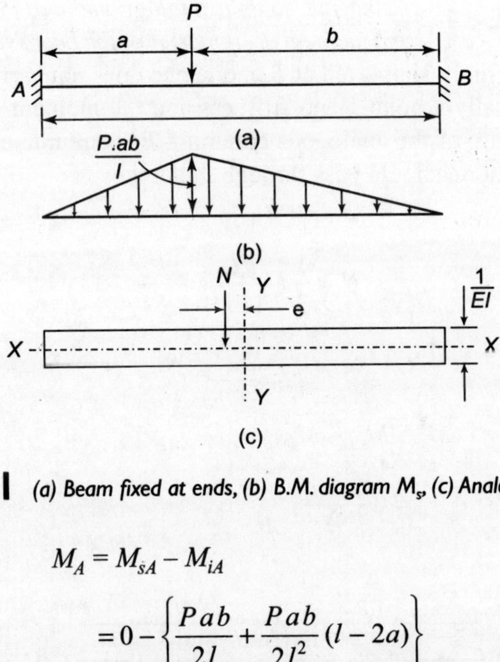

Fig. 14.6 | *(a) Beam fixed at ends, (b) B.M. diagram M_s, (c) Analogous column*

$$M_A = M_{sA} - M_{iA}$$

$$= 0 - \left\{ \frac{Pab}{2l} + \frac{Pab}{2l^2}(l-2a) \right\}$$

Simplifying, $\quad M_A = -\dfrac{Pab^2}{2l^2}$

For the end B, $\quad x_B = -l/2$

$$M_B = M_{sB} - M_{iB} = 0 - \left\{ \frac{Pab}{2l} + \frac{Pab}{2l^2}(l-2a) \right\}$$

$$= -\frac{Pab}{2l} + \frac{Pab}{2l^2}(a+b-2a)$$

$$= -\frac{Pab}{2l} + \frac{Pab}{2l^2}(b-a)$$

$$= -\frac{Pab}{2l^2}(l - b + a)$$

$$= -\frac{Pa^2 b}{l^2}$$

$$\therefore \quad M_A = -\frac{Pab^2}{l^2}$$

and $$M_B = -\frac{Pa^2 b}{l^2}$$

Example 14.4 *Using the column analogy method, determine the fixed end moment at support A of the beam shown in Fig. 14.7.*

The beam is simply supported at B and hence does not carry any moment. A hinge is theoretically a point of no stiffness and the moment of inertia is zero. Therefore the width of the analogous column $1/EI$ is infinite and the YY axis of the analogous column should pass through the hinge.

Area of the column $$A = \propto$$

Axial load $$N = \frac{1}{2} l \frac{pl}{4EI} = \frac{pl^2}{8\,EI}$$

$$I_y = \frac{1}{3}\frac{l^3}{EI}$$

$$\therefore \quad M_A = M_{sA} - M_{iA}$$

$$= 0 - \frac{Ne}{I_x} y$$

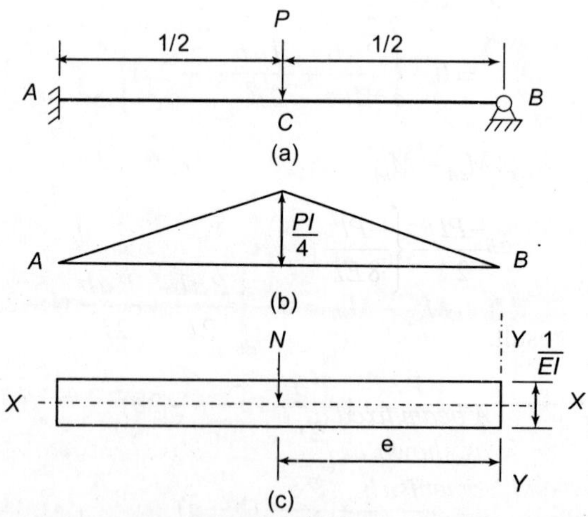

Fig. 14.7

$$= 0 - \frac{Pl^2}{8EI}\left(\frac{l}{2}\right)\frac{3EI}{l^3}$$

$$= -\frac{3}{16}pl$$

Alternatively we can release the reaction component at B and consider it as a cantilever beam (Fig. 14.8).

Area $\qquad A = \infty$

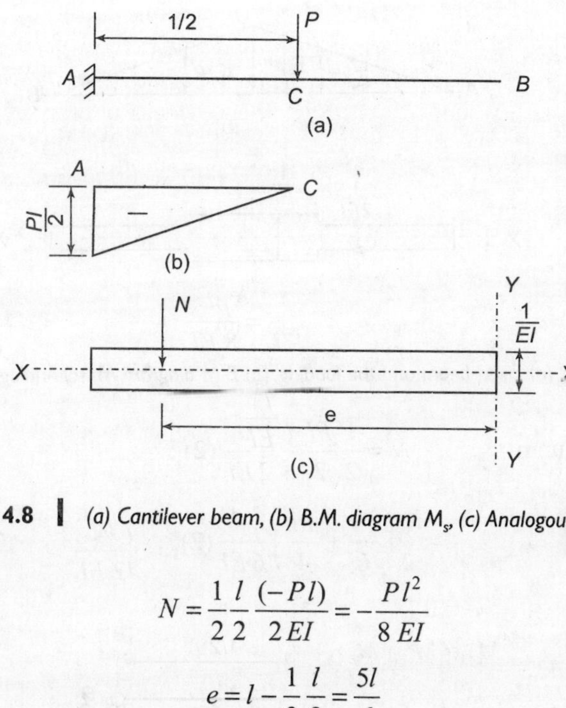

Fig. 14.8 | *(a) Cantilever beam, (b) B.M. diagram M_s, (c) Analogous column*

Load $\qquad N = \frac{1}{2}\frac{l}{2}\frac{(-Pl)}{2EI} = -\frac{Pl^2}{8EI}$

$$e = l - \frac{1}{3}\frac{l}{2} = \frac{5l}{6}$$

$$M_A = M_{sA} - M_{iA}$$

$$= \frac{-Pl}{2} - \left\{\frac{-Pl^2}{8EI}\left(\frac{5}{l}\right)(l)\frac{3EI}{l^3}\right\} = -\frac{Pl}{2} + \frac{5}{16}Pl = -\frac{3}{16}Pl$$

same as the earlier result.

Example 14.5 | *A beam fixed at the ends has varying moment of inertia as shown in Fig. 14.9. Using column analogy method determine the fixed end moments.*

The beam is reduced to a simply supported beam by removing the restraining moments at A and B.

The 'bending moment diagram M_s is shown in Fig. 14.9*b*. The analogous column with the width changing in steps is shown in Fig. 14.9*c*.

Area of analogous column $A = \frac{2 \times 1 \times l}{2\,EI\,3} + \frac{l}{3}\frac{1}{2\,EI} = \frac{2}{3}\frac{l}{EI}$

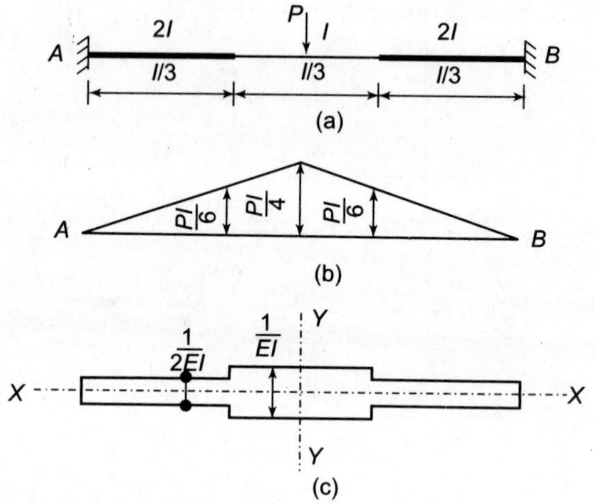

Fig. 14.9 | *(a) Fixed beam and the loading, (b) B.M diagram M_s (c) Analogous column*

Load on the column $\qquad N = \frac{1}{2}\frac{Pl}{2}\frac{l}{3}\frac{1}{2\,EI}(2)$

$$= \frac{1}{2}\left(\frac{Pl}{6} + \frac{Pl}{4}\right)\frac{l}{6\,EI}(2) = \frac{7\,Pl^2}{72\,EI}$$

Eccentricity $\qquad e = 0$

$$M_A = M_s - M_i$$

$$= 0 - \left[\frac{N}{A} + \frac{Ney}{I_y}\right] = 0 - \frac{7}{72}\frac{Pl^2}{EI}\frac{3\,EI}{2l} = -\frac{21}{144}Pl$$

$$M_B = -\frac{21}{144}Pl$$

Example 14.6 | *Obtain fixed end moments, stiffness factors and carry over factors for the ends A and B of the beam shown in Fig. 14.10. Use the column analogy method.*

Step 1: To fix up column area and load

The end moments M_A and M_B are considered as redundants. The moment diagram is drawn in parts; first for u.d.l and then for point load as shown. The analogous column is shown in Fig. 14.10*d*.

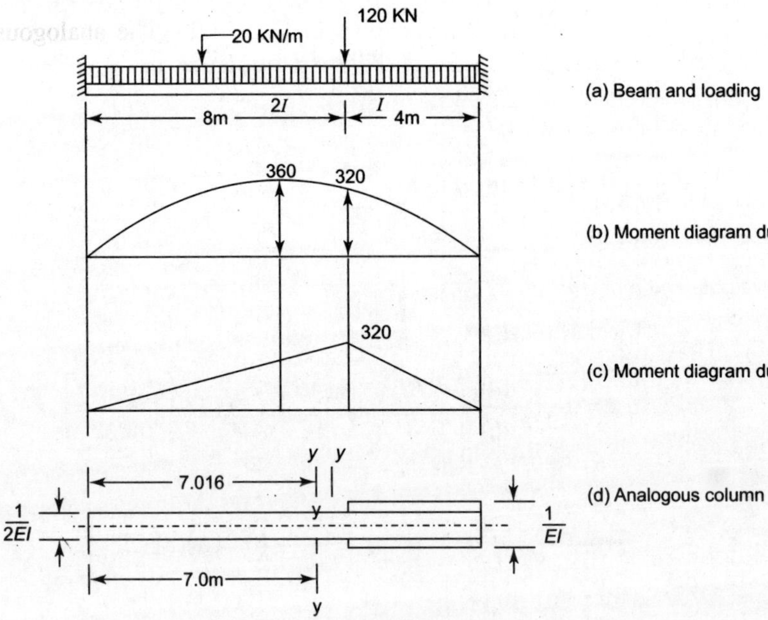

Fig. 14.10

$$\text{Area of the column} = \frac{8}{2EI} + \frac{4}{EI} = \frac{8}{EI}$$

$$\text{Column load } N = \frac{2}{3}\frac{(360)(12)}{2\,EI} + \frac{2}{3}\frac{(320)(4)}{2EI} + \frac{1}{2}\frac{(320)(8)}{2EI} + \frac{1}{2}\frac{(320)(4)}{EI}$$

$$= \frac{3146.7}{EI}$$

Step 2: To evaluate M.I. of column

The position of YY axis is determined by taking moments of the column areas about A

$$\bar{x} = \frac{\dfrac{12}{2EI}(6) + \dfrac{4}{2EI}(10)}{8\,EI} = 7.0 \text{ m}$$

$$I_{yy} = \frac{1}{3}\left(\frac{1}{2\,EI}\right)(7)^3 + \frac{1}{3}\left(\frac{1}{EI}\right)(5)^3 - \frac{1}{3}\left(\frac{1}{2EI}\right)(1)^3 = \frac{98.67}{EI} \text{ m}^4$$

Step 3: To fix up eccentricity of column load

Next, The resultant of the column load from end A is obtained by taking moments about A as.

$$\bar{x} = \frac{\dfrac{1440 \times 6}{EI} + \dfrac{\dfrac{426.7}{EI}(8+1.5)}{} + \dfrac{640}{EI}\left(\dfrac{16}{3}\right) + \dfrac{640}{EI}(9.33)}{\dfrac{3146.7}{EI}}$$

$$= \frac{22078.18}{3146.7} = 7.016 \text{ m}$$

or $\quad e = 0.016$ m

Step 4: To evaluate end moments

Moment $\quad M_A = M_s - M_{iA}$

$$= 0 - \left[\frac{3146.7}{EI}\left(\frac{EI}{8}\right) + \frac{3146.7}{EI}(-0.016)\frac{(7)(EI)}{98.67}\right]$$

$$= -389.77 \text{ kNm}$$

and $\quad M_B = 0 - \left[\frac{3146.7}{EI}\left(\frac{EI}{8}\right) + \frac{3146.7}{EI}(0.016)(5)\,(EI)\right]$

$$= -395.89 \text{ kN.m}$$

To obtain stiffness and carry over factor apply a unit load at A for $N = 1$

$$M_{iA} = \frac{EI}{8} + 1\frac{(7)(7)(EI)}{98.67} = 0.6216\ EI$$

$$M_{iB} = \frac{EI}{8} - 1\frac{(7)(5)(EI)}{98.67} = -0.2297\ EI$$

C.O.F. $\quad \dfrac{M_{iB}}{M_{iA}} = \dfrac{0.2297\ EI}{0.6216\ EI} = -0.3695$

Again apply a unit load for N at B

$$M_{iB} = \frac{EI}{8} + \frac{(1)(5)(5)\ EI}{98.67} = 0.3783\ EI$$

$$M_{iA} = \frac{EI}{8}(1)\frac{(1)(5)(7)\ EI}{98.67} = -0.2297$$

C.O.F. $\quad = \dfrac{M_{iA}}{M_{iB}} = \dfrac{-0.2297}{0.3783} = -0.6087$

We can make a check $C_{AB}\,K_A = C_{BA}\,K_B$ which is satisfied.

$$(-0.3695)\ (0.6216) = (-0.3783)(0.6072)$$

14.2.2 Stiffness and Carry-over Factors

The column analogy method is specially useful for determining stiffness and carry-over factors for non-prismatic members.

Let us first evaluate the stiffness and carry-over factors for a prismatic member. Let a moment M_A be applied at the simply supported end A to give a unit rotation at that end while the farther end B is fixed as shown in Fig. 14.11*a*.

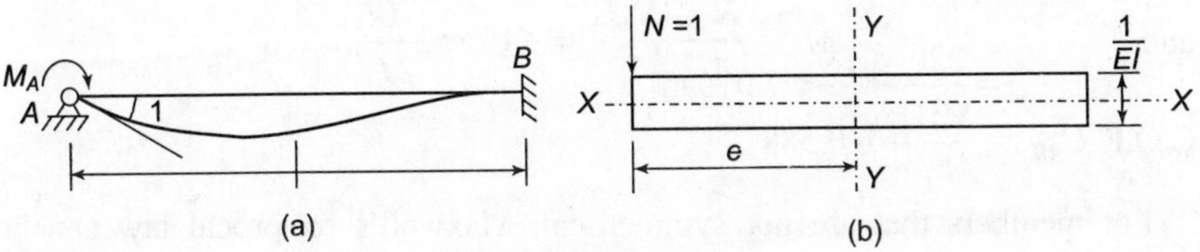

Fig. 14.11

The analogous column is shown in Fig. 14.11*b*. Change of slope in the beam from A to $B = 1$. Therefore, the area of the M/EI diagram between A and B is also $= 1$. Accordingly the load on the analogous column $N = 1$. The load N has to be located at A along axis XX as the location of it elsewhere implies a moment of it about A and hence falsely indicates deflection at A.

Now $$A = \frac{l}{EI}, N = 1, e = l/2, I_y = \frac{l^3}{12\,EI}$$

$$M_A = \frac{N}{A} + \frac{Nex}{I_y} = \frac{EI}{l} + \frac{(l/2)}{l^3/12EI}$$

$$= \frac{EI}{l} + \frac{3\,EI}{l} = \frac{4\,EI}{l} \tag{14.4}$$

And $$M_B = \frac{N}{A} - \frac{Nex}{I_y} = \frac{E\,I}{l} - \frac{3\,E\,I}{l} = -\frac{2EI}{l} \tag{14.5}$$

Therefore stiffness of member at end $A = \frac{4\,EI}{l}$ and carry-over factor from A to B

$$C_{AB} = \frac{M_B}{M_A} = -\frac{1}{2} \tag{14.6}$$

Now let us evaluate the stiffness and carry-over factors for a non-prismatic member. This is best explained by solving a numerical example.

Example 14.7 | *Compute the stiffness at end A of the member AB and carry-over factor from A to B for the fixed beam in Example 14.5.*

From the previous example $A = \frac{2}{3}\frac{l}{EI}$ and $N = 1$ as shown in Fig 14.12

$$e = l/2, y = l/2$$

$$I_y = \frac{1}{12}\left(\frac{1}{2EI}\right)l^3 + \frac{1}{12}\left(\frac{1}{2EI}\right)\left(\frac{l}{3}\right)^3 = \frac{7l^3}{162\,EI}$$

$$M_A = \frac{N}{A} + \frac{Nex}{l_y} = \frac{3\,EI}{2l} + 1\,(l/2)\,(l/2)\,\frac{162\,EI}{7l^3} = 7.2857\,\frac{EI}{l}$$

and $$M_B = \frac{N}{A} - \frac{Nex}{l_y} = -4.2857\,\frac{EI}{l}$$

$$\text{C.O.F. } C_{AB} = \frac{M_B}{M_A} = -0.588$$

For members that are not symmetrical, Maxwell's reciprocal law can be utilised in checking the stiffness and carry-over factors obtained at the two ends. The moment M_A applied at A to induce a unit rotation at A produces a moment M_B at B which is equal to the moment M_A developed at A due to a moment M_B applied at B to induce a unit rotation at B. The relationship may be expressed

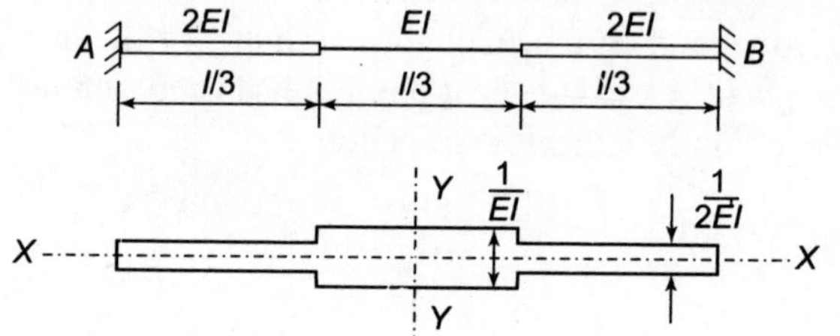

Fig. 14.12

$$C_{AB}\,K_A = C_{BA}\,K_B \tag{14.7}$$

This is made clear in Fig. 14.13.

Here

K_A = Stiffness of member AB at end A

K_B = Stiffness of member BA at end B

C_{AB} = Carry-over factor from A to B

A_{BA} = Carry-over factor from B to A.

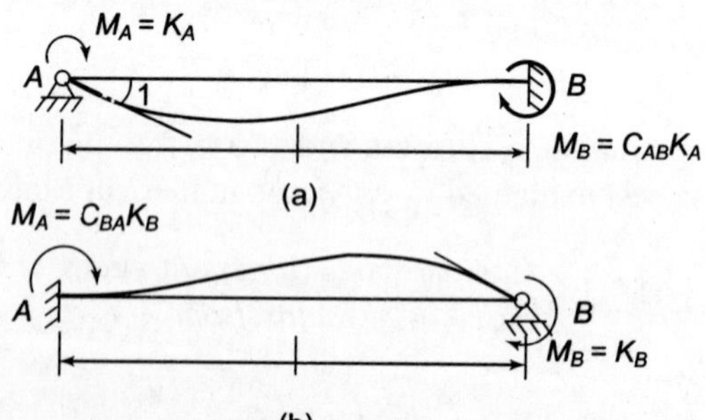

Fig. 14.13

Example 14.8 | *Determine stiffness factors and carry over factors for the fixed ends A and B of the shown in Fig. 14.14. Check the relation $C_{AB}\,K_A = C_{BA}\,K_B$*

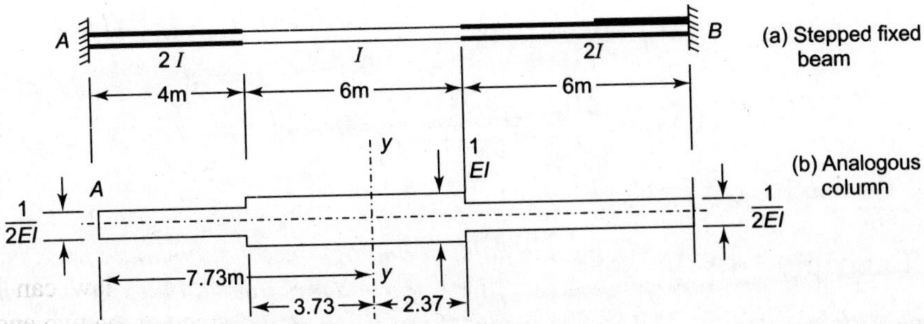

Fig. 14.14

Step 1: To fix up analogous column and area

The analogous column is drawn as shown (Fig. 14.14*b*)

$$\text{Area of the analogous column} = \frac{4}{2EI} + \frac{6}{EI} + \frac{6}{2EI} = \frac{11}{EI}$$

Step 2: To evaluate M.I. of the column

The axis YY is determined by taking moments of the area about A

$$\bar{x} = \frac{\frac{4}{2EI}(2) + \frac{6}{EI}(7) + \frac{6}{2EI}(13)}{\frac{11}{EI}} = 7.73 \text{ m}$$

Then

$$I_{yy} = \frac{1}{12}\left(\frac{1}{2}EI\right)(4)^3 + \frac{4}{2EI}(5.73)^2 + \frac{1}{3}\left(\frac{1}{EI}\right)(3.73)^3$$

$$+ \frac{1}{3}\left(\frac{1}{EI}\right)(2.27)^3 + \frac{1}{12}\left(\frac{1}{2EI}\right)(6)^3 + \frac{6}{2EI}(5.27)^2 = \frac{190.80}{EI} \text{ m}^4$$

Step 3: To evaluate end moments

Now applying a load $N = 1$ at A

$$M_{iA} = \frac{EI}{11} + (1)(7.73)(7.73)\frac{EI}{190.80}$$

$$= (0.09 + 0.3132)\, EI$$

$$= 0.4031\, EI$$

$$M_{iB} = \frac{EI}{11} - (1)(7.73)\frac{(8.27)(EI)}{190.80} = -0.2450\, EI$$

$$\text{C.O.F.} = \frac{M_{iB}}{M_{iA}} = -\frac{-0.2450}{0.4031}\frac{EI}{EJ} = -0.607$$

Similarly applying a load $N = 1$ at B

$$M_{iB} = \frac{EI}{11} + (1)(8.27)\frac{(8.27)(EI)}{190.80} = 0.467\, EI$$

$$M_{iA} = \frac{EI}{11} - (1)(8.27)\frac{(7.73)(EI)}{190.80} = -0.2450\,EI$$

$$\text{C.O.F.} = \frac{M_{iA}}{M_{iB}} = \frac{-0.2450EI}{0.4670EI} = 0.5246$$

The check $C_{AB}\,K_A = C_{BA}\,K_B$ satisfies.

Example 14.9 | *Using the column analogy method, compute fixed end moments, carry-over factors and stiffnesses for the beam shown in Fig. 14.15. The depth of the beam is varying but the width is constant and is equal to 500 mm.*

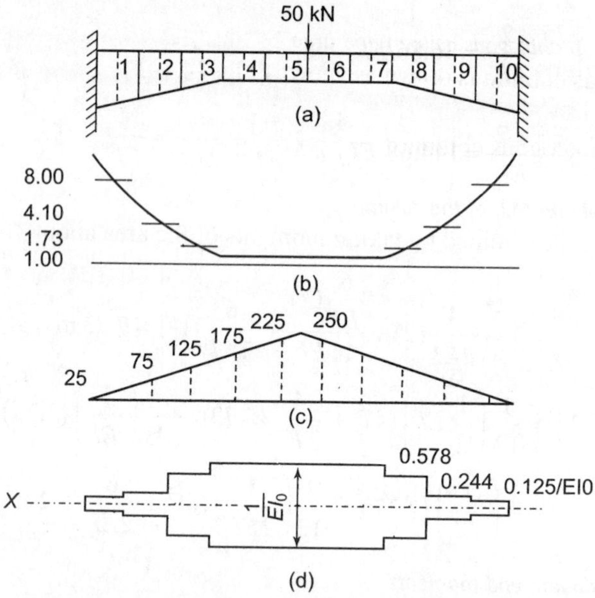

Fig. 14.15 | *(a) Non-prismatic beam and the loading, (b) Moment of inertia in steps, (c) Moment M_s ordinates, (d) Analogous column width in steps*

The beam is divided into ten parts each of 2 m length for computing I_x and A. Greater accuracy can be obtained by dividing the beam into more parts.

Let I_o be the moment of inertia in the uniform depth region. The moments of inertia at the mid width of other sections are obtained as follows.

$$I_{(1)} = \left(\frac{1000}{500}\right)^3 I_o = 8.00 I_o$$

$$I_{(2)} = \left(\frac{800}{500}\right)^3 I_o = 4.10\,I_o$$

$$I_{(3)} = \left(\frac{600}{500}\right)^3 I_o = 1.73\,I_o$$

The moment diagram M_S is shown in Fig. 13.15*c*. The moment ordinates at the mid width of each section are calculated.

The Fig. 14.15*d* shows the analogous column for the haunched beam. The width of analogous column for each of the sections is arrived at by calculating $1/EI$ using a appropriate value for I. The width of the anologous column for half of the column depth is indicated in Fig. 14.15*d*. The values of A and N for each of the sections for one half are tabulated below.

Section	N	A
1.	$\frac{25 \times 2}{8.0\, EI_a} = 6.25/EI_o$	$\frac{2 \times 0.125}{EI_o} = 0.25/EI_o$
2.	$\frac{75 \times 2}{4.01\, EI_o} = 36.86/EI_o$	$\frac{2 \times 0.244}{EI_o} = 0.488/EI_o$
3.	$\frac{125 \times 2}{1.73\, EI_o} = 144.68/EI_o$	$\frac{2 \times 0.578}{EI_o} = 1.156/EI_o$
4.	$\frac{175 \times 2}{1.0\, EI_o} = 350.00/EI_o$	$\frac{2 \times 1}{EI_o} = 2.00/EI_o$
5.	$\frac{225 \times 2}{1.0\, EI_o} = \frac{450.00/EI_o}{\Sigma 978.51/EI_o}$	$\frac{2 \times 1}{EI_o} = \frac{2.00/EI_o}{\Sigma 5.894/EI_o}$

$$N = \frac{2 \times 987.51}{EI_o} = \frac{1975.02}{EI_o}$$

$$A = \frac{2 \times 5.89}{EI_o} = \frac{11.78}{EI_o}$$

$$I = \frac{1}{12\, EI_o}\{0.125(20)^3 + 0.199\,(16)^3 + 0.234(12)^3 + 0.422\,(8)^3\} = \frac{175.65}{EI_o}$$

$$e = 0$$

$$\therefore \quad M_i = \frac{N}{A} = \frac{1975.02}{11.78} = 167.66$$

$$M_A = M_B = 0 - M_i = -167.66 \text{ kN.m}$$

Stiffness and C.O.F.

Take $N = 1$ and $e = 10$ m.

$$M_i = \frac{N}{A} \pm \frac{Nex}{I_y}$$

$$Mi_A = \frac{1.0}{11.78/EI_o} + \frac{1\times10\times10}{175.65/EI_o} = 0.65.43\ EI_o$$

$$M_{iB} = \frac{1.0}{11.78/EI_o} - \frac{1\times10\times10}{175.65/EI_o} = -0.484\ EI_o$$

$$M_A = 0 - M_{iA} = -\,0.6543\ EI_o$$

$$M_B = 0 - (-\,0.484\ EI_o) = +\,0.484\ EI_o$$

$$C_{AB} = C_{BA} = -\frac{0.484}{0.654} = -0.74$$

$$K_A = K_B = 0.6543\ EI_o$$

Example 14.10 | *Determine the stiffness at ends A and B and carry-over factors from A to B and B to A for the unsymmetrical haunched beam of constant width b shown in Fig. 14.16.*

The analogous column is shown in Fig. 14.16*b*, the beam is divided into 4 elements. For each element, the value of $1/EI$ and hence the width of the analogous column is calculated. The values are tabulated in the table that follows. The depth of the beam in the prismatic length is taken as $d_o = 0.6$ m and moment of inertia is taken as I_o.

Element	Length	d	$\left(\frac{do}{d}\right)^3$	Distance x from YY axis of beam	Area	ax	$ax^2 + i_x$
1	1 m	1.1 m	0.1622	4.0 m	$\frac{0.1622}{EI_o}$	$\frac{0.6488}{EI_o}$	$\frac{2.5952+0}{EI_o}$
2	1 m	0.9 m	0.2963	3.0 m	$\frac{0.2963}{EI_o}$	$\frac{0.8889}{EI_o}$	$\frac{2.6667+0}{EI_o}$
3	1 m	0.7 m	0.6297	2.0 m	$\frac{0.6297}{EI_o}$	$\frac{1.2594}{EI_o}$	$\frac{2.5188+0}{EI_o}$
4	6 m	0.6 m	1.00	–1.5 m	$\frac{6.0}{EI_o}$	$\frac{-9.00}{EI_o}$	$\frac{13.50+18}{EI_o}$
					$\sum\frac{7.0882}{EI_o}$	$\frac{-6.2029}{EI_o}$	$\frac{39.2807}{EI_o}$

$$\bar{x} = -\frac{6.2029}{7.0882} = -0.875$$

$$I_{GG} = \frac{39.2807}{EI_o} - \frac{7.0882}{EI_o}(0.875)^2 = \frac{33.8537}{EI_o}$$

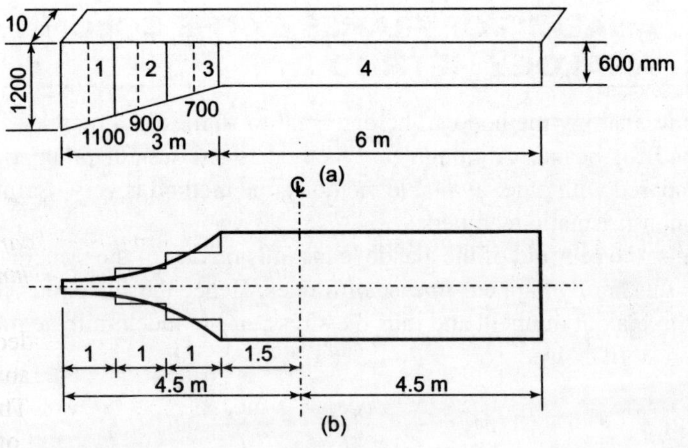

Fig. 14.16

Applying a unit elastic load $N = 1$ at end A

$$M_{iA} = \frac{EI_o}{7.0882} + 1\frac{(5.375)(5.375)}{33.8537/EI_o} = 0.9943\ EI_o$$

$$M_{iB} = \frac{EI_o}{7.0882} - 1\frac{(5.375)(3.625)}{33.8537/EI_o} = -0.4355\ EI_o$$

$$M_A = 0 - M_{iA} = -0.9943\ EI_o$$

$$M_B = 0 - M_{iB} = -0.4355\ EI_o$$

$$\text{C.O.F. } C_{AB} = \frac{-0.4355}{0.9943} = -0.438$$

Applying a unit elastic load $N = 1$ at end B

$$M_{iA} = \frac{EI_o}{7.0882} - 1\frac{(4.5 - 0.875)(4.5 + 0.875)}{33.8537/EI_o} = -0.4355\ EI_o$$

$$M_{iB} = \frac{EI_o}{7.0882} + 1\frac{(4.5 - 0.875)(4.5 - 0.875)}{33.8537/EI_o} = 0.529\ EI_o$$

$$M_A = 0.4355\ EI_o$$

$$M_B = -0.529\ EI_o$$

$$\text{C.O.F. } C_{BA} = -\frac{0.4355}{0.5299} = -0.823$$

Checking by the reciprocal theorem

$$C_{AB}\,K_A = C_{BA}\,K_B$$
$$C_{AB}\,K_A = -0.438\,(-0.9943)\,EI_o = 0.4353\,EI_o$$
$$CB_A\,K_B = -0.823\,(-0.529)\,EI_o = 0.4355\,EI_o$$

14.3 ANALYSIS OF FRAMES BY THE COLUMN ANALOGY METHOD

The column analogy method can be extended to frames in a manner similar to that applied for beams. Although this method is tedious for prismatic frames when compared with other available methods, the method is very useful when it comes to non-prismatic members.

In beams, the centroid of the elastic loads falls on one of the principal axes of the beam; this hardly happens in case of frames. Hence the analogous column is subjected to biaxial moment and thus the stress, or the indeterminate moment at any section is given by

$$M_i = \frac{N}{A} \pm \frac{N e_x\, y}{I_y} \pm \frac{N e_y\, x}{I_x} \tag{14.8}$$

Consider a portal frame fixed at the base as in Fig. 14.17*a*. In the analogous column the width of each member is $1/EI$. The frame is symmetrical about the *YY* axis. The *XX* axis passes through the centroid of the analogous column as shown.

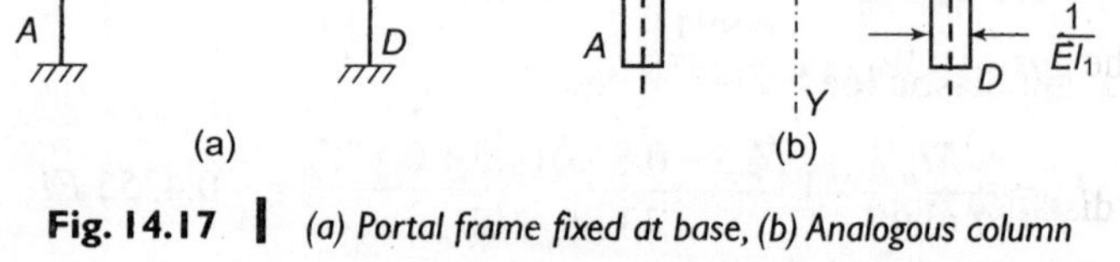

Fig. 14.17 | *(a) Portal frame fixed at base, (b) Analogous column*

Next consider the same frame but hinged at the feet of the base. As discussed earlier, the hinge offers no resistance to rotation, the flexural rigidity $EI = 0$. The width $1/EI$ of the analogous column becomes infinite. Consequently the centroid of the column area lies on the line joining hinges and forms the *XX* axis. The axis *YY* lies on the line of symmetry as shown in Fig. 14.18*b*.

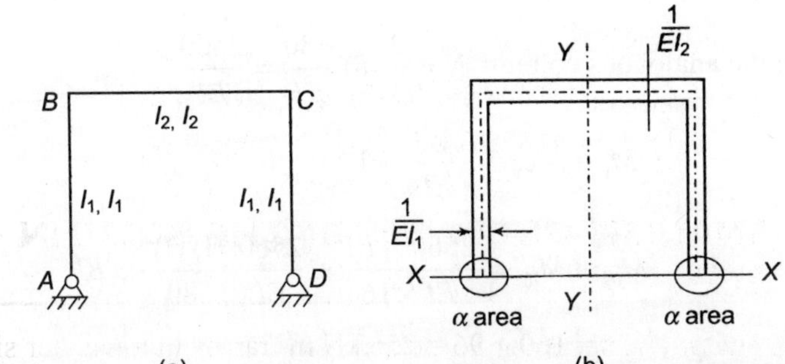

(a) (b)

Fig. 14.18 | *(a) Portal frame hinged at base, (b) Analogous column with axes*

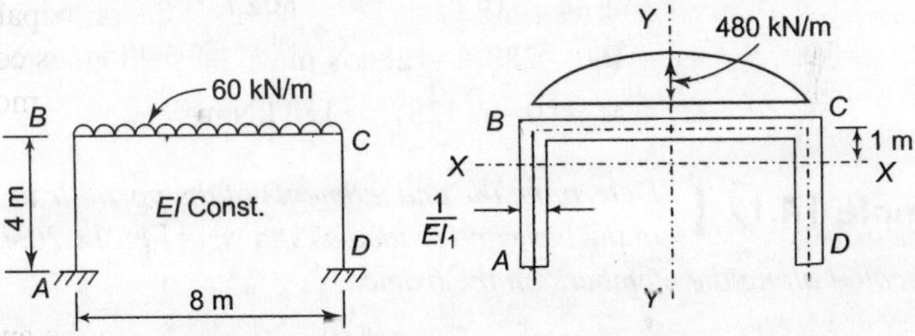

(a) (b)

Fig. 14.19 | *(a) Frame and the loading, (b) Analogous column and M_s diagram*

The following worked out examples will make the procedure clear.

Example 14.11 | *Using column analogy method determine the moments at A, B, C and D of the frame shown in Fig. 14.19.*

The moment constraints at A and D and the horizontal reaction component at D are released to make the frame statically determinate. The moment diagram M_s for the released structure is a parabola.

Area of the analogous columns $A = \dfrac{2 \times 4}{EI} + \dfrac{8}{EI} = \dfrac{16}{EI}$

Centroid distance from top face $\bar{y} = \dfrac{2 \times 4}{EI} \times \dfrac{2 \times EI}{16} = 1$ m

$$I_{xx} = I_{BC} - A_{\bar{y}}^{2}$$

$$= \frac{1}{3}\left(\frac{1}{EI}\right) 4^3 \times 2 - \frac{16}{EI}(1)^2 = \frac{80}{3\,EI}$$

C.G. of moment diagram M_s lies on the YY axis

$$e_x = 0 \quad e_y = 1 \text{ m}$$

Load on the analogous column $N = \frac{2}{3}(8)\frac{480}{EI} = \frac{2560}{EI}$

$$M_x = N\,e_y = \frac{2560}{EI} \times 1$$

$$M_{iB} = M_{iC} = \frac{2560 \times EI}{EI \times 16} + \frac{2560}{EI}\,\frac{(1)\,(1)^3}{80}\,3\,EI$$

$$= 160 + 96 = 256 \text{ kN.m}$$

End moments $M_B = M_C = 0 - 256 = -256$ kN.m.

Similarly $M_{iB} = M_{iD} = \frac{2560\,EI}{16\,EI} - \frac{2560 \times 3\,EI\,(3)}{80\,EI}$

$$= 160 - 288 = -128 \text{ kN.m}$$

$$M_A = M_D = 0 - (-128) = +128 \text{ kN.m}$$

Example 14.12 | *Determine the end moments of the portal frame hinged at the base and loaded as shown in Fig. 14.20. EI values are indicated along the members on the frame.*

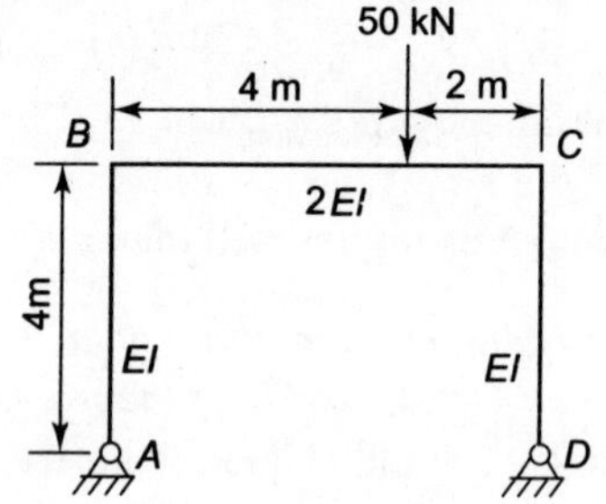

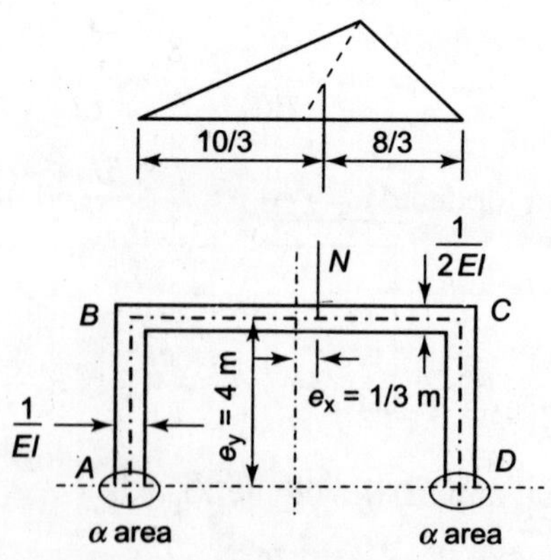

Fig. 14.20

Hinged support points represent an infinite area of the analogous column. Therefore the XX axis passes through hinge points.

Column load $$N = \frac{1}{2}(6)\frac{(66.67)}{2\,EI} = \frac{100}{EI}$$

$$e_x = \frac{1}{3}\text{ m} \quad e_y = 4\text{ m}$$

$$M_x = N\,e_y = \frac{100}{EI}(4) = \frac{400}{EI}$$

$$I_{xx} = \frac{1}{3}\left(\frac{1}{EI}\right)(4)^3\,(2) + \frac{6}{2\,EI}(4)^2 = \frac{272}{3\,EI}$$

$$M_{iA} = M_{iD} = 0 \text{ since } A = \propto \text{ and } I_{YY} = \propto$$

$$Mi_B = 0 + \frac{400}{EI}\frac{(4)}{272}3\,EI = 17.65$$

$$M_{iC} = 17.65$$

$$\therefore \quad M_B = 0 - 17.65 = -17.65 \text{ kN.m (tension outside)}$$

$$M_C = 0 - 17.65 = -17.65 \text{ kN.m (tension outside)}$$

Example 14.13 | *Determine the moments at A, B, C and D of the frame shown in Fig. 14.21.*

The degree of indeterminancy of the frame is 3. The frame is made statically determinate by releasing the three reaction components at D. The resulting frame is a cantilever bent for which the M_s diagram is shown. As the moment M_s causes tension outside, the axial load N is negative.

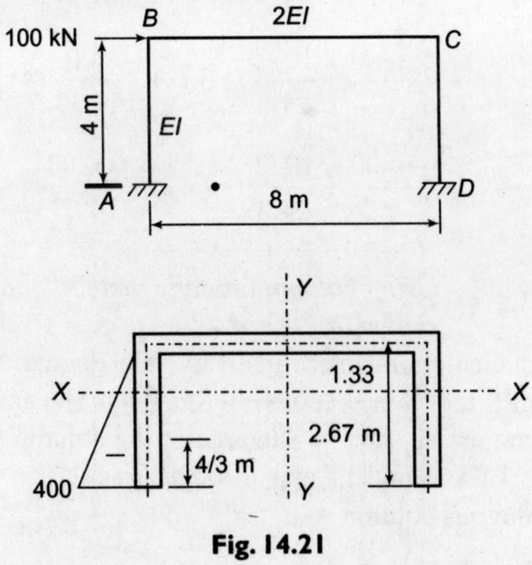

Fig. 14.21

Column load $$N = -\frac{1}{2}(400)\frac{4}{EI} = -\frac{800}{EI}$$

Column area $A = \dfrac{2 \times 4}{EI} + \dfrac{8}{2\,EI} = \dfrac{12}{EI}$

Taking moment about top of beam

$$\bar{y} = \frac{8}{EI}\frac{(2)EI}{12} = 1.33 \text{ m}$$

$$I_{xx} = \frac{1}{3}\left(\frac{1}{EI}\right)(1.33^3 + 2.67^3)(2) + \frac{8}{2\,EI}(1.33)^2 = \frac{21.3}{EI}$$

$$I_{yy} = \frac{1}{12}\left(\frac{1}{2\,EI}\right)(8)^3 + 2\left(\frac{4}{EI}\right)4^2 = \frac{149.33}{EI}$$

$$e_x = -4.0 \text{ m}$$

$$e_y = 1.34 \text{ m}$$

$$M_x = N\,e_y = \frac{800}{EI}(-1.34) = \frac{1072}{EI}$$

$$M_y = N\,e_x = -\frac{800}{EI}(-4) = \frac{3200}{EI}$$

Moment $M_A = M_{sA} - M_{iA}$

$$= -400 - \left\{\frac{-800}{12} + \frac{1072}{EI}\frac{EI}{21.3}(-2.67) + \frac{3200}{EI}\frac{EI}{149.33}(-4)\right\}$$

$$= -113.24 \text{ kN.m}$$

Similarly $M_B = 0 - \left\{\dfrac{-800}{12} + \dfrac{1072}{21.3}(1.33) + \dfrac{3200}{149.33}(-4)\right\} = 85.46 \text{ kN.m}$

$$M_C = 0 - \left\{\frac{-800}{12} + \frac{1072}{21.3}(1.33) + \frac{3200}{149.33}(4)\right\} = 85.98 \text{ kN.m}$$

$$M_x = 0 - \left\{\frac{-800}{12} + \frac{1072}{21.3}(-2.67) + \frac{3200}{149.33}(4)\right\} = 115.32 \text{ kN.m}$$

Example 14.14 | *Using column analogy method analyse the portal bent shown in Fig. 14.22a*

The frame is statically in determinate by one degree. On removal of the reaction constant at C the frame becomes a free bent. The analogous column and the moment diagram are as shown. The area of the column at hinged support C is ∞ and hence the XX axis and YY axis through C as shown.

Area of the analogous column = ∞

Column load = $N_1 + N_2$

$$= \frac{1}{2}(-30)(3)\frac{1}{2\,EI} + \frac{(-30)(4)}{EI}$$

$$= -22.5/EI + 120/EI$$

$$= -\frac{142.5}{EI}$$

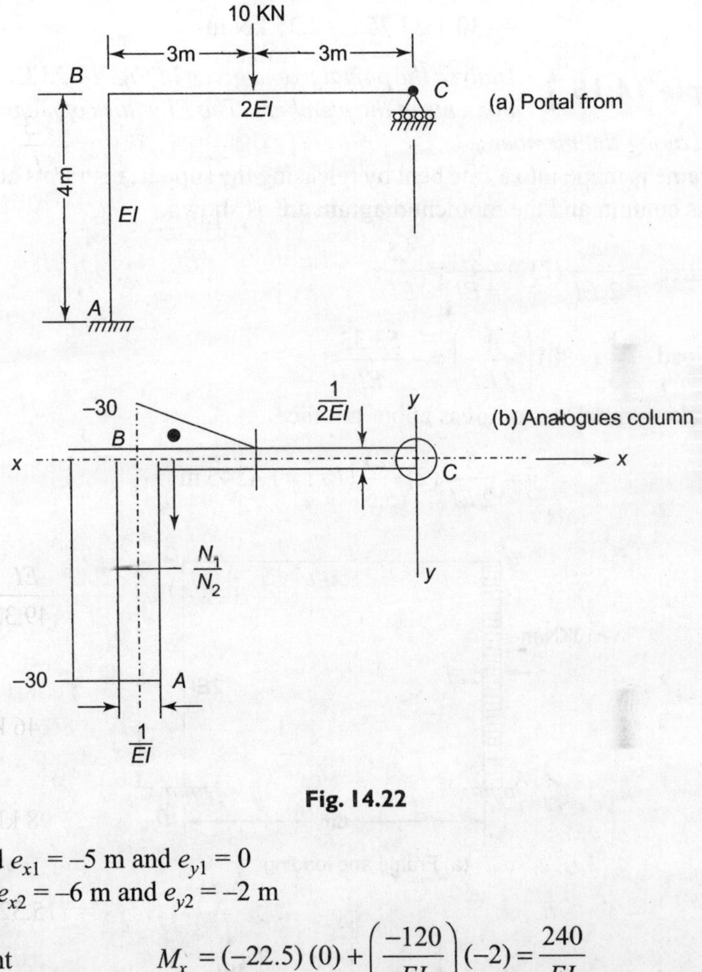

Fig. 14.22

We find $e_{x1} = -5$ m and $e_{y1} = 0$
and $e_{x2} = -6$ m and $e_{y2} = -2$ m

Moment $\quad M_x = (-22.5)\,(0) + \left(\frac{-120}{EI}\right)(-2) = \frac{240}{EI}$

$$M_y = \frac{(-22.5)}{EI}\;(-5) + \left(\frac{-120}{EI}\right)(-6) = \frac{832.5}{EI}$$

$$I_{xx} = \frac{1}{3}\left(\frac{1}{EI}\right)(4)^3 = \frac{21.33}{EI}$$

$$I_{yy} = \frac{1}{3}\left(\frac{1}{2\,EI}\right)(6)^3 + 4\left(\frac{1}{EI}\right)(6)^2\;\frac{180}{EI}$$

Now moment $\quad M_A = -30 - \left[0 + \frac{240}{EI}\,\frac{(EI)}{21.33}\,(-4) + \frac{832.5}{EI}\,\frac{EI}{180}\,(-6)\right]$

$$= -30 - [(45) + (-27.75)] = 42.75 \text{ kN.m}$$

and
$$M_B = -30\left[0 + \frac{240}{EI}\frac{EI}{21.33}(0) + \frac{832.5}{EI}\frac{(EI)}{180}(-6)\right]$$
$$= -30 + 27.75 = -2.25 \text{ kN.m}$$

Example 14.15 | *Analyse the portal frame given in Fig. 14.23 for the end moments of the members. The EI values of members are indicated along the members.*

The frame is made into a free bent by releasing the support restraints at D. The analogous column and the moment diagram are as shown.

$$\text{Column area} = \frac{4}{2\,EI}(2) + \frac{6}{4\,EI} = \frac{5.5}{EI}$$

$$\text{Column load} = \frac{1}{3}(-80)\left(\frac{4}{2\,EI}\right) = -\frac{53.33}{EI}$$

Now taking moments of areas about top face

$$\bar{y} = \frac{4}{2\,EI}(2)\frac{(2)}{5.5}(EI) = 1.4545 \text{ m}$$

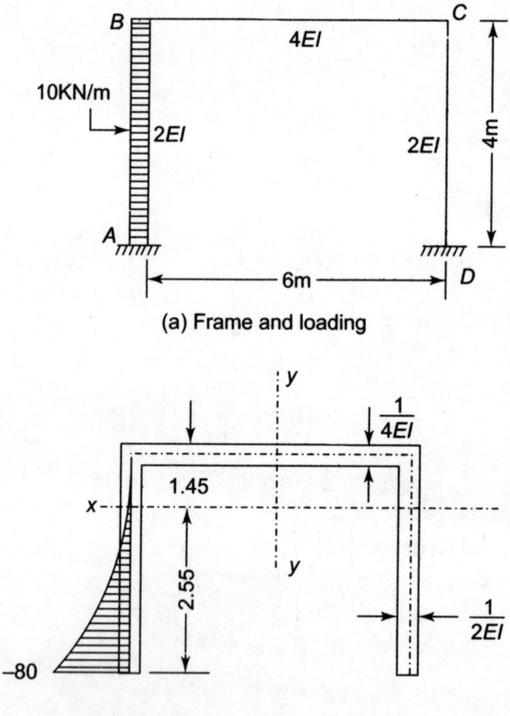

(a) Frame and loading

(b) Analogous column

Fig. 14.23

$$I_{xx} = \frac{1}{3}\left(\frac{1}{2\,EI}\right)(4)^3(2) - \frac{5.5}{EI}(1.4545)^2 = \frac{9.69}{EI}$$

$$I_{yy} = \frac{1}{12}\left(\frac{1}{4\,EI}\right)(6)^3 + 2\left(\frac{4}{2\,EI}\right)(3)^2 = \frac{40.5}{EI}$$

$$M_A = M_s - \left[\frac{N}{A} + \frac{N\,e_y}{I_{xx}}(y) + \frac{N\,e_x}{I_{yy}}(x)\right]$$

Substituting,
$$M_A = 10 - \left[-\frac{53.33\,(EI)}{EI\quad 5.5} + \left(\frac{-53.33}{EI}\right)\left(\frac{-1.5}{9.69}\right)(EI)(-2.55)\right.$$
$$\left. + \left(\frac{-53.33}{EI}\right)\left(\frac{(-3)}{40.5}\right)(EI)\,(-3)\right]$$
$$= -80 - (-9.70 - 21.75 - 11.85)$$
$$= -80 + 43.30 = -36.7$$

$$M_B = 0 - \left[-\frac{53.33\,(EI)}{EI\quad 5.5} + \left(\frac{-53.33}{EI}\right)\left(\frac{-1.55}{9.69}\right)(EI)\,(1.45)\right.$$
$$\left. + \left(\frac{-53.33}{EI}\right)\left(\frac{(-3)}{40.5}\right)(EI)\,(-3)\right]$$
$$= 0 - (-9.70 + 12.37 - 11.85)$$
$$= +\,9.18 \text{ kN.m}$$
$$M_C = 0 - (-9.70 + 12.37 + 11.85) = -14.52 \text{ kN.m}$$
$$M_D = 0 - (-9.70 - 21.75 + 11.85) = +\ 19.6 \text{ kN.m}$$

14.3.1 Closed Frames

Column analogy method can also be extended to closed frames. The following example will illustrate the procedure.

Example 14.16 | *Using the column analogy method, determine the end moments and draw the bending moment diagram for the closed frame shown in Fig. 14.24.*

We can release the structure say at C just to the right of load and make it determinate. The analogous column is shown in Fig. 14.24. The bending moment diagram M_s for the released structure is shown in Fig. 14.25. The loads on the analogous column and their centroids are shown in Fig. 14.25.

$$N_1 = \frac{1}{2}\left(-\frac{300}{2\,EI_o}\right)(3) = -\frac{225}{EI_o} \qquad e_{x_1} = -\,2 \text{ m}, e_{y_1} = -\,1.29 \text{ m}$$

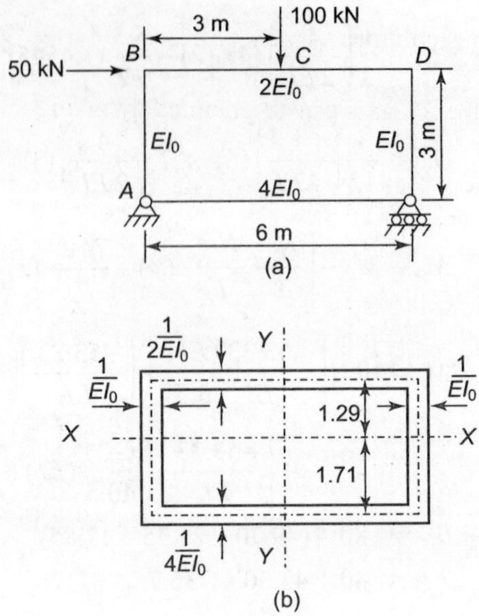

Fig. 14.24

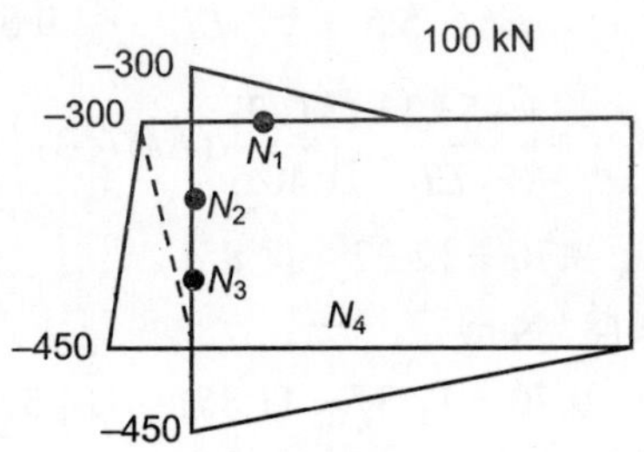

Fig. 14.25

$$N_2 = \frac{1}{2}\left(\frac{-300}{EI_o}\right)(3) = -\frac{-450}{EI_o} \qquad e_{x_2} = -3\text{ m}, e_{y_2} = -0.29\text{ m}$$

$$N_3 = \frac{1}{2}\left(\frac{-450}{EI_o}\right)(3) = -\frac{-675}{EI_o} \qquad e_{x_3} = -3\text{ m}, e_{y_3} = +0.71\text{ m}$$

$$N_4 = \frac{1}{2}\left(\frac{-450}{4EI_o}\right)(6) = \frac{-337.5}{EI_o} \qquad e_{x_4} = -1\text{ m}, e_{y_4} = +0.71\text{ m}$$

$$\Sigma N = -16875/EI_o$$

Each of the loads N_1, N_2, N_3 and N_4 are equal to the area of the M_s, diagram multiplied by the appropriate $1/EI$ of the column. These loads act at points on the centre line of the analogous column, the point of application of each being aligned with the centroid of the area of the appropriate M_s diagram.

Area of the analogous columns $A = 2\left(\dfrac{3}{EI_o}\right) + \dfrac{6}{2EI_o} + \dfrac{6}{4EI_o} = \dfrac{21}{2EI_o}$

The location of the XX axis can be obtained by taking moment about the top face of the column.

$$\bar{y} = 2\left(\frac{3}{EI_o}\right)\left(\frac{3}{2}\right) + \left(\frac{6}{4EI_o}\right)(3) = 1.29 \text{ m.}$$

$$I_{xx} = \frac{6}{2EI_o}(1.29)^2 + \frac{6}{4EI_o}(1.71)^2 + 2\left(\frac{1}{3EI_o}\right)(1.29^3 + 1.71^3)$$

$$= 4.99 + 4.39 + 4.77 = \frac{14.15}{EI_o} \text{ m}^4$$

$$I_{yy} = \frac{1}{12}\left(\frac{1}{2EI_o}\right)(6)^3 + \frac{1}{12}\left(\frac{1}{4EI_o}\right)(6)^3 + 2\left(\frac{3}{EI_o}\right)(3)^2$$

$$= \frac{9}{EI_o} + \frac{4.5}{EI_o} + \frac{54}{EI_o} = \frac{67.5}{EI_o} \text{ m}^4$$

Let us evaluate the moments M_x and M_y

$$M_x = \sum N\, e_y = N_1\, e_{y1} + N_2\, e_{y2} + N_3\, e_{y3} + N_4\, e_{y4}$$

$$= \frac{1}{EI_o}\{(-225)(-129) + (-450)(-0.29) + (-675)(0.71) + (-337.5)(1.71)\}$$

$$= \frac{-635.63}{EI_o} \text{ kN.m}$$

$$M_Y = \sum N\, e_x = N_1\, e_{x1} + N_2\, e_{x2} + N_3\, e_{x3} + N_4\, e_{x4}$$

$$= \frac{1}{EI_c}\{(-225)(-2) + (-45)(-3) + (-675)(-3) + (-337.5)(-1)\}$$

$$= \frac{4162.5}{EI_o} \text{ kN.m}$$

Substituting these values in

$$M_A = M_{sA} - M_{iA}$$

$$= -450 - \left\{\frac{-1687}{10.5} - \frac{535-63}{14.15}(1.71) + \frac{4162.5}{67.5}(-3)\right\}$$

$$= -450 + 422.65 = -27.35 \text{ kN.m}$$

$$M_B = -300 - \left\{-160.67 - \frac{635.63}{14.15}(-1.29) + \frac{4162.5}{67.5}(-3)\right\}$$

$$= -300 + 287.9 = -12.1 \text{ kN.m}$$

$$M_c = 0 - \left\{-160.67 - \frac{635.63}{14.15}(-1.29)\right\}$$

$$= 160.67 - 57.77 = 102.9 \text{ kN.m}$$

$$M_D = 0 - \left\{-160.67 - \frac{635.63}{14.15}(-1.29) + \frac{4162.5}{67.5}(+3)\right\}$$

$$= 0 - 82.9 = -82.9 \text{ kN.m}$$

$$M_E = 0 - \left\{-160.67 - \frac{635.63}{14.15}(1.71) + \frac{4162.5}{67.5}(+3)\right\}$$

$$= 0 + 52.66 = 52.66 \text{ kN.m}$$

The bending moment diagram is shown in Fig. 14.26.

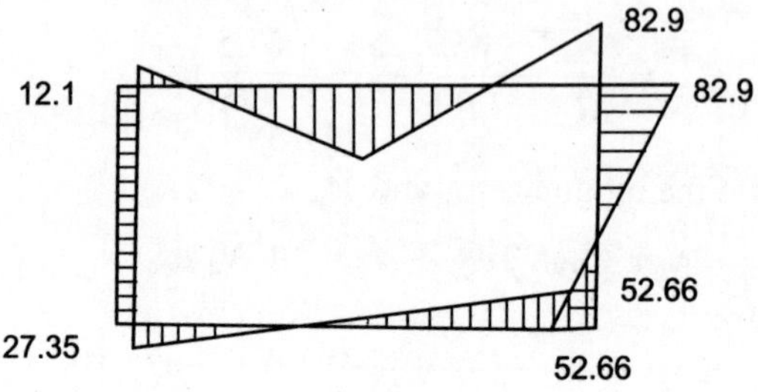

Fig. 14.26

14.4 GABLE FRAMES

Symmetrical gable frames can be analysed using the column analogy method. However, while analysing a gable frame it is necessary to determine the moment of inertia of the line area about a centroidal axis which is at an angle θ with the direction of the axis XX. Consider a line area as shown in Fig. 14.27.

$$I_x = 2\int_0^{l/2} b\cdot dx\,(x\sin\theta)^2$$

$$= 2\int_0^{l/2} b\cdot \sin^2\theta\, x^2\, dx$$

$$= 2b\sin^2\theta\left[\frac{x^3}{3}\right]_0^{l/2} = \frac{bl^3}{12}\sin^2\theta$$

Similarly $$I_y = \frac{bl^3}{12}\cos^2\theta$$

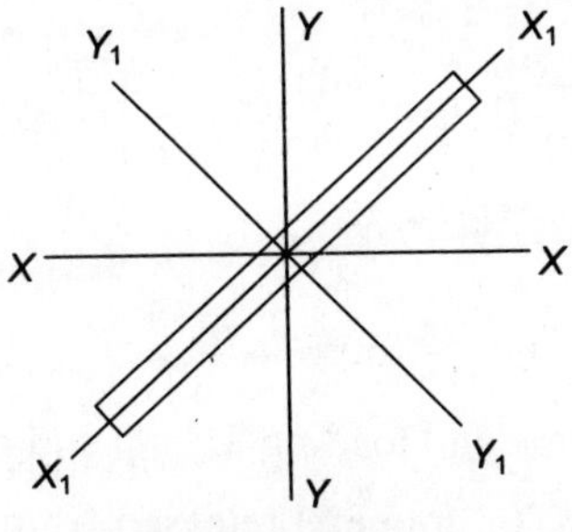

Fig. 14.27

Example 14.17 | *Analyse the gable frame shown in Fig. 14.28 by column analogy method.*

The analogous column is shown in Fig. 14.28. The properties of the analogous column are:

$$A = \left(\frac{4}{EI}\right)(2) + \frac{6.32}{2\,EI}(2) = \frac{14.32}{EI}\ \text{m}^2$$

Centroid distance $\overline{y}$ from base

$$\overline{y} = \frac{\dfrac{8(2)}{EI} + \dfrac{6.32}{2\,EI}(2)(5)}{14.32/EI} = 3.32\ \text{m}$$

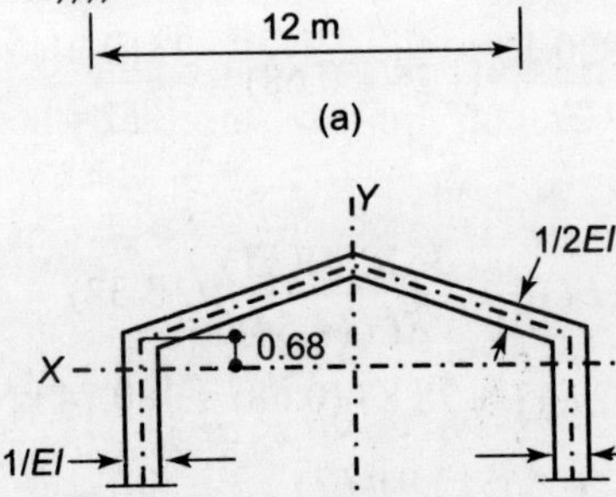

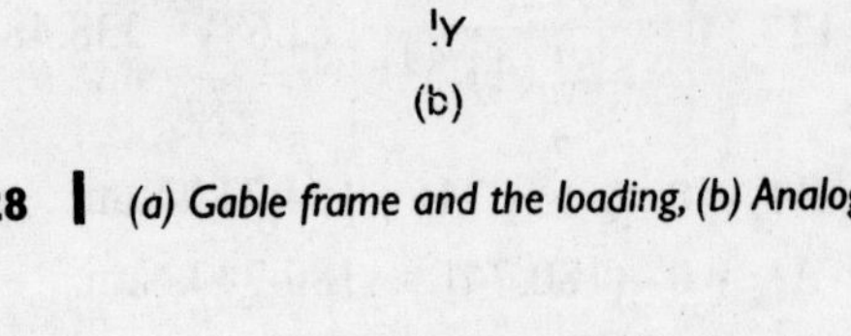

Fig. 14.28 | *(a) Gable frame and the loading, (b) Analogous column*

$$I_{xx} = 2\left[\frac{1}{12}\left(\frac{1}{EI}\right)(4)^3 + \frac{4}{EI}(1.32)^2 + \frac{1}{12}\left(\frac{1}{2EI}\right)(6.32)^3(0.1) + \frac{6.32}{2EI}(1.68)^2\right]$$

$$= \frac{44.54}{EI}\ \text{m}^4$$

Due to symmetry of frame and loading $M_y = 0$ and hence I_{yy} is not required.

Load on Analogous Column The frame is released from the restraining moments at A and E and the horizontal reaction component at E making it a statically determinate frame as shown in Fig. 14.29*a*. The M_s diagram is shown plotted on the inclined member in Fig. 14.29*b*.

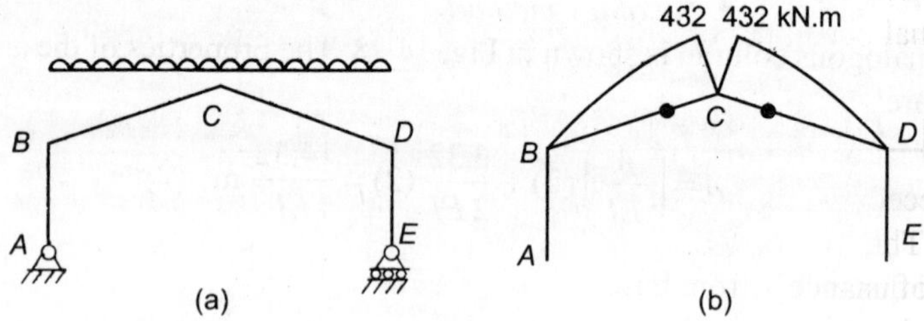

Fig. 14.29

Maxm. $M_s = \frac{wl^2}{8} = \frac{24 \times 12^2}{8} = 432$ kN.m

Column load $N = 2\left\{\frac{2}{3}\left(\frac{6.32}{2EI}\right)(432)\right\} = \frac{1820.16}{EI}$ kN

Column area $A = \frac{4}{EI} + \frac{4}{EI} + \frac{6.32}{2EI} \times 2 = \frac{14.32}{EI}$ m^2

Moment $M_x = \frac{1820.16}{EI}(1.25 + 0.68) = \frac{3512.91}{EI}$

$M_y = 0$

$$M_{iA} = \frac{1820.16}{EI(14.32)}(EI) + = \frac{3512.91(EI)}{EI\ (44.54)}(-3.32) = -134.74 \text{ kN.m}$$

$$M_{iB} = 127.11 + 78.87\ (0.68) = 180.74 \text{ kN.m}$$

$$M_{iC} = 127.11 + \frac{3512.91(EI)}{EI\ (44.54)}(2.68) = 338.48 \text{ kN.m}$$

The moments are

$$M_A = 0 - (-134.74) = 134.74 \text{ kN.m}$$

$$M_B = 0 - (180.74) = -180.74 \text{ kN.m}$$

$$M_C = 432 - 338.48 = 93.51 \text{ kN.m}$$

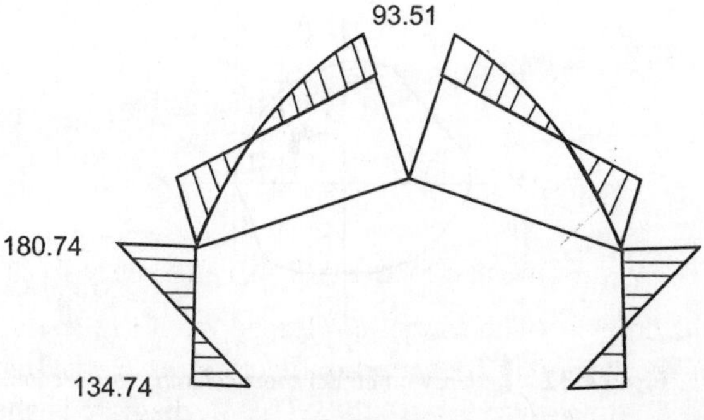

Fig. 14.30

The final bending moment diagram is indicated in Fig. 14.30.

14.5 ANALYSIS OF UNSYMMETRICAL FRAMES

In the preceding section, we required to know the principal axes of the analogous column. This presented no problem when the frame was symmetrical. In the analysis of unsymmetrical frames it is necessary to use the principal moments of inertia and product moments of inertia in the solution. The typical frames and the axes to be considered are shown in Fig 14.31.

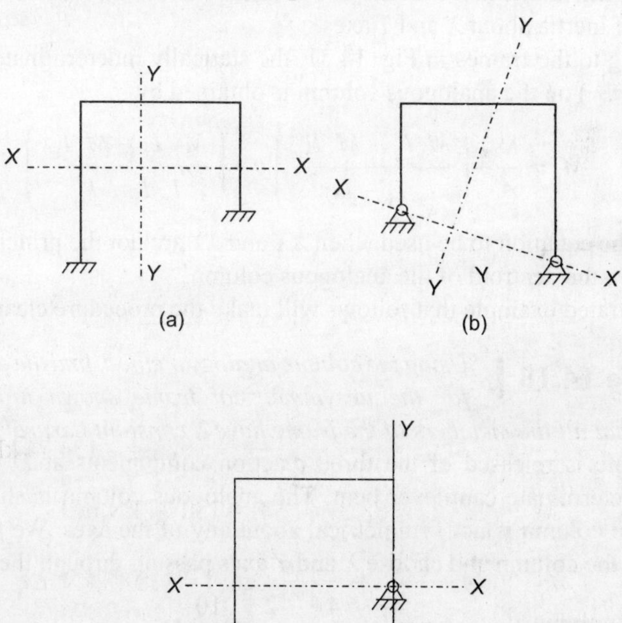

Fig. 14.31 | *Unsymmetrical frames and the axes*

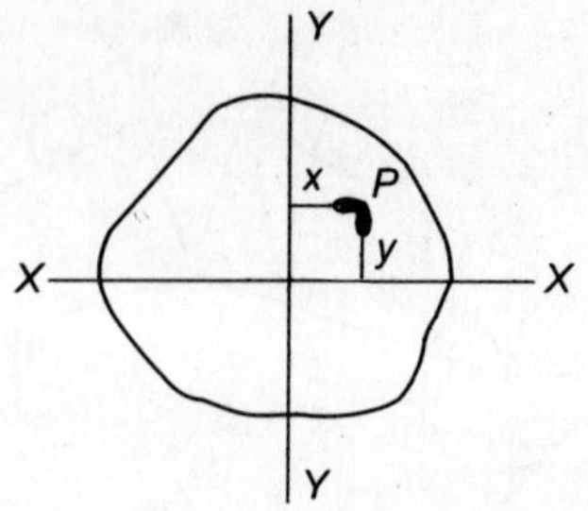

Fig. 14.32 | Unsymmetrical short column cross-section

Consider an area as shown in Fig. 14.32 subjected to a tensile force (+ve) P at a point (x, y) with respect to any coordinate axes X and Y through the centroid of the section.

The eccentric force P can be replaced by a force N at the centroid and by two moments $M_x = P \cdot y$ and $M_y = P \cdot x$. The normal stress at any point is given by

$$f = \frac{P}{A} + \left(\frac{M_x I_y - M_y I_{xy}}{I_x I_y - I_{xy}^2} \right) y + \left(\frac{M_x I_y - M_x I_{xy}}{I_x I_y - I_{xy}^2} \right) \tag{14.9}$$

where A is the area of cross-section.

I_x, I_y are the moments of inertia about x and y axes and I_{xy}. is the product of moments of inertia about X and Y axes.

Referring to the frames in Fig. 14.31, the statically indeterminate moment at any point (x, y) on the analogous column is obtained by

$$M_i = \frac{N}{A} + \left(\frac{M_x I_y - M_y I_{xy}}{I_x I_y - Ixy^2} \right) y + \left(\frac{M_x I_y - M_x I_{xy}}{I_x I_y - I_{xy}^2} \right) \tag{14.10}$$

This is the equation to be used when XX and YY are not the principal axes but pass through the centroid of the analogous column.

An illustrated example that follows will make the procedure clear.

Example 14.18 | *Using the column analogy method, find the end moments for the unsymmetrical frame shown in Fig. 14.33. Consider that all the members of the frame have a constant flexural rigidity EI.*

The frame is released of the three reaction components at D making it a statically determinate cantilever bent. The analogous column is shown in Fig. 14.33*b*. The column is not symmetrical about any of the axes. We find first the centroid of the column and choose X and Y axes passing through the centroid.

Area of the column $\quad A = \frac{4}{EI} + \frac{4}{EI} + \frac{2}{EI} = \frac{10}{EI} \text{ m}^2$

Distance of centroid from top $\bar{y} = \dfrac{\frac{4}{EI}(2) + \frac{2}{EI}(1)}{10/EI} = 1.0 \text{ m}$

Distance of centroid from left hand column line

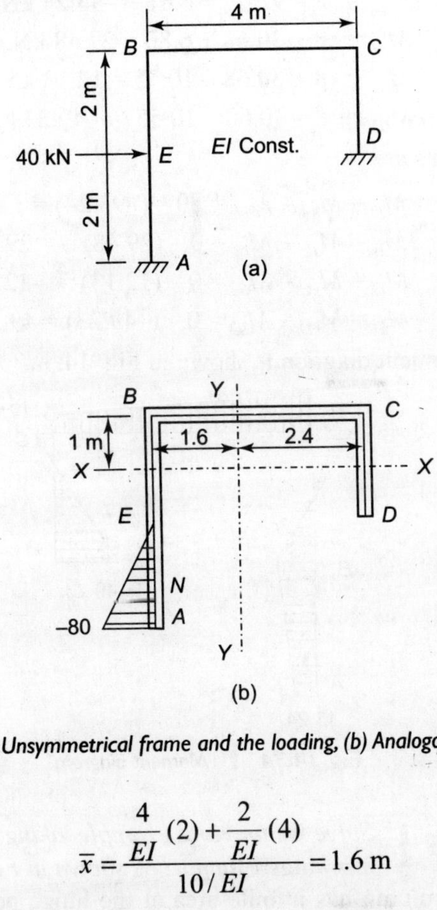

Fig. 14.33 | *(a) Unsymmetrical frame and the loading, (b) Analogous column and the axes*

$$\bar{x} = \frac{\frac{4}{EI}(2) + \frac{2}{EI}(4)}{10/EI} = 1.6 \text{ m}$$

$$I_{xx} = \frac{1}{12}\left(\frac{1}{EI}\right)(4)^3 + \frac{4}{EI}(1)^2 + \frac{1}{2}\left(\frac{2}{EI}\right)(2)^3 + \frac{4}{EI}(1)^2 = \frac{7.17}{EI} \text{ m}^2$$

$$I_{yy} = \left(\frac{4}{EI}\right)(1.6)^2 + \frac{2}{EI}(2.4)^2 + \frac{1}{12}\left(\frac{1}{EI}\right)(4)^3 + \frac{4}{EI}(0.4)^2 = \frac{27.73}{EI} \text{ m}^4$$

$$I_{xy} = \frac{4}{EI}(-1)(-1.6) + \frac{2}{EI}(2.4)(0) + \frac{4}{EI}(0.4)(1) = \frac{8.0}{EI} \text{ m}^4$$

Loading on the column $N = \frac{1}{2}(-80)\frac{2}{EI} = \frac{-80}{EI}$ kN.

Moment $M_x = N e_y = \frac{-80}{EI}\left(-\frac{7}{3}\right) = \frac{186.67}{EI}$ kN.m

$$M_y = N e_x = \frac{-80}{EI}(1-6) = \frac{128}{EI} \text{ kN.m}$$

Using Equation 14.8 and substituting the values as worked out earlier

$$M_{iA} = -8 - 92.05 + 6.81 = -93.24 \text{ kN.m}$$
$$M_{iB} = -8 + 30.68 + 6.80 = 29.48 \text{ kN.m}$$
$$M_{iC} = -8 + 30.68 - 10.55 = 12.13 \text{ kN.m}$$
$$M_{iD} = -8 - 30.68 - 10.55 = -49.23 \text{ kN.m}$$

The end moments are

$$M_A = M_{sA} - M_{iA} = 80 - (-93.24) = 13.24 \text{ kN.m}$$
$$M_B = M_{sB} - M_{iB} = 0 - (29.48) = -29.48 \text{ kN.m}$$
$$M_C = M_{sC} - M_{iC} = 0 - (12.13) = -12.13 \text{ kN.m}$$
$$M_D = M_{sD} - M_{iD} = 0 - (-49.23) = 49.23 \text{ kN.m}$$

The bending moment diagram is shown in Fig. 14.34.

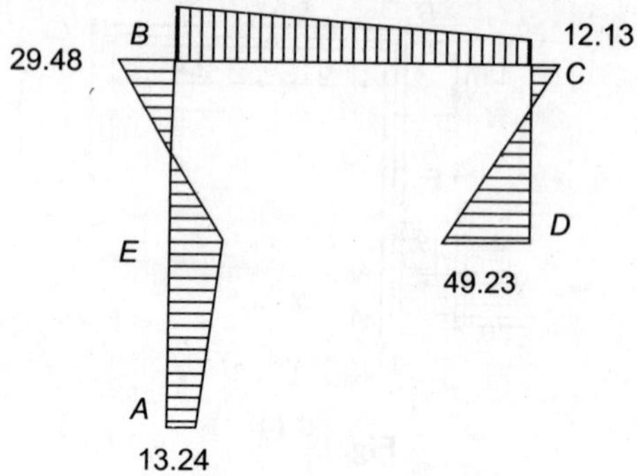

Fig. 14.34 | *Moment diagram*

Example 14.19 | *Solve the previous example taking that the right column has hinged support as shown in Fig 14.35a.*

The analogous column has infinite area at the hinge point. The centroid of the column is located at the hinge point. We choose two coordinate axes X and Y passing through the hinge point which are not principal axes.

Area of the column $A = \alpha$

$$I_{xx} = \frac{1}{3}\left(\frac{1}{EI}\right)(2)^3 + \frac{1}{12}\left(\frac{1}{EI}\right)(4)^3 + 4\left(\frac{1}{EI}\right)(2)^2 = \frac{24}{EI}\text{ m}^4$$

$$I_{yy} = 4\left(\frac{1}{EI}\right)(4)^2 + \frac{1}{3}\left(\frac{1}{EI}\right)(4)^3 + = \frac{85.33}{EI}\text{ m}^4$$

$$I_{xy} = 4\left(\frac{1}{EI}\right)(2) + (-2) = \frac{-16}{EI}\text{ m}^4$$

$N = \dfrac{-80}{EI}$ kN, as earlier.

$$M_x = N e_y = \frac{-80}{EI}(-4/3) = \frac{106.67}{EI}\text{ kN.m}$$

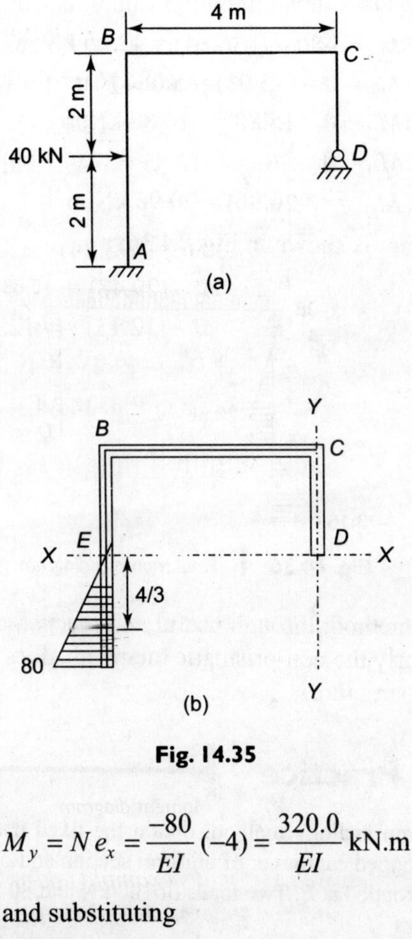

Fig. 14.35

$$M_y = N\,e_x = \frac{-80}{EI}(-4) = \frac{320.0}{EI}\ \text{kN.m}$$

Using Equation 14.8 and substituting

$$M_{iA} = 0 + \left(\frac{106.67 \times 85.33 + 320 \times 16}{24 \times 85.33 - 256}\right)(-2) + \frac{(320 \times 24 + 106.67 \times 16)}{24 \times 85.33 - 256}(-4)$$

$$= 0 + 7.94\,(-2) + 5.24\,(-4)$$

$$= -36.84\ \text{kN.m}$$

$$M_{iB} = 0 + 7.94\,(2) + 5.24\,(-4)$$

$$= -5.08\ \text{kN.m}$$

$$M_{iC} = 0 + 7.94\,(2) + 5.24\,(0)$$

$$= +\,15.88\ \text{kN.m}$$

$$M_{iD} = 0$$

$$M_{iE} = 0 + 7.94\,(0) + 5.24\,(-4)$$

$$= -20.96\ \text{kN.m}$$

The final moments are

$$M_A = -80 - (-36.84) = 43.16 \text{ kN.m}$$
$$M_B = 0 - (-5.08) = 5.08 \text{ kN.m}$$
$$M_C = 0 - 15.88 = -15.88 \text{ kN.m}$$
$$M_D = 0$$
$$M_E = -(-20.96) = 20.96 \text{ kN.m}$$

The final B.M. diagram is shown in Fig. 14.36.

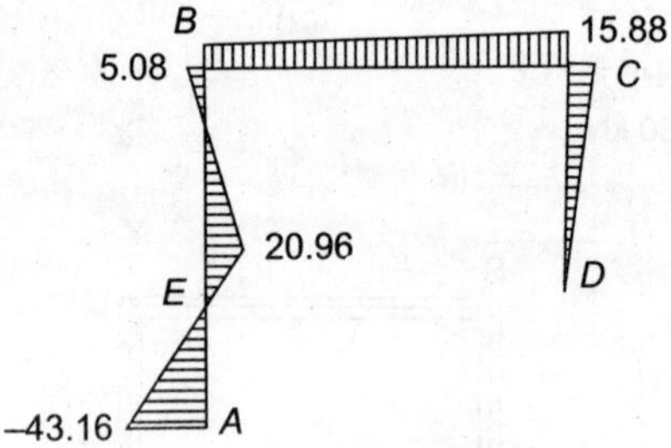

Fig. 14.36 | *Final moment diagram*

Column analogy method, through useful and mechanical in analysing beams and frames, particularly the non-prismatic members, does not throw any light on the behaviour of the structure.

Problems for Practice

14.1 Using the column analogy method, obtain the fixed end moment and draw the B.M. diagram for a propped cantilever of uniform section and of length 12 m. The beam is fixed at end *A* and propped at *B*. Two loads of 100 kN and 80 kN are placed at 4 m and 8 m respectively from *A*.

14.2, 14.3 Using the column anology method, determine the fixed end moments and draw the B.M. diagram for the beams shown in Figs. 14.37 and 14.38.

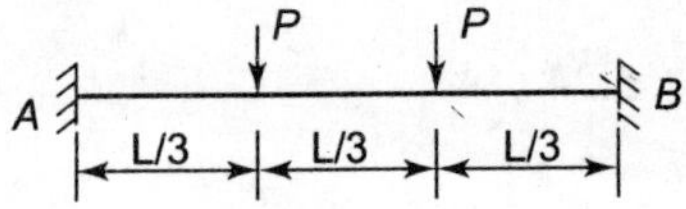

Fig. 14.37

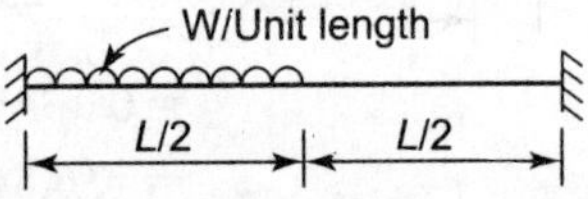

Fig. 14.38

14.4, 14.5 Obtain fixed end moments, stiffness factors and carry-over factors for the ends *A* and *B* of the beams shown in Figs. 14.39 and 14.40. Use the column anology method.

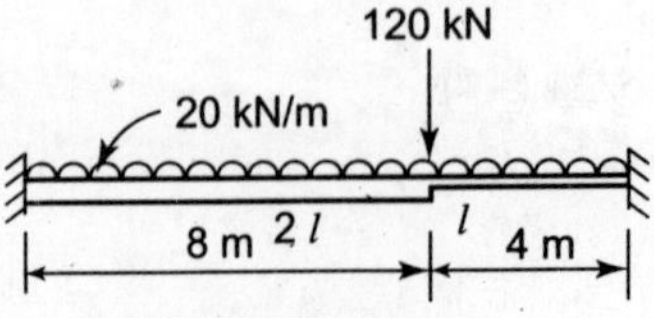

Fig. 14.39

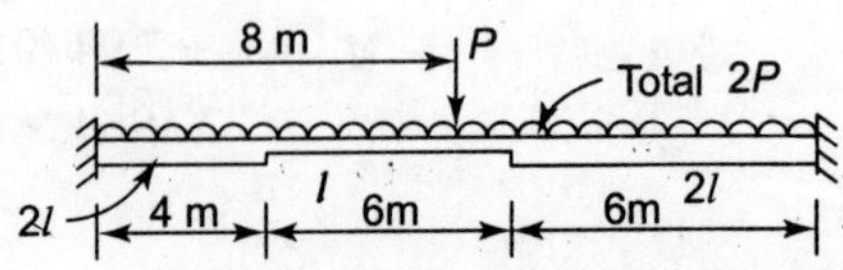

Fig. 14.40

14.6, 14.7, 14.8 Using the column analogy method analyse the portal frames shown in Fig. 14.41, 14.42 and 14.43.

14.9 Analyse the frame shown in Fig. 14.44 using column analogy method.

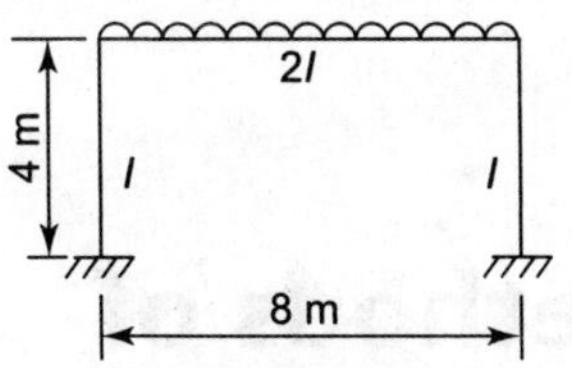

Fig. 14.41

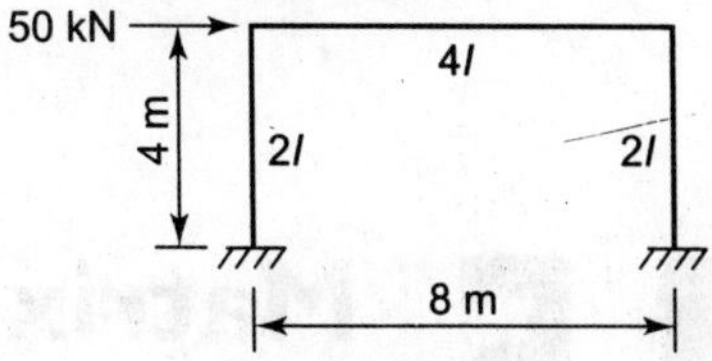

Fig. 14.42

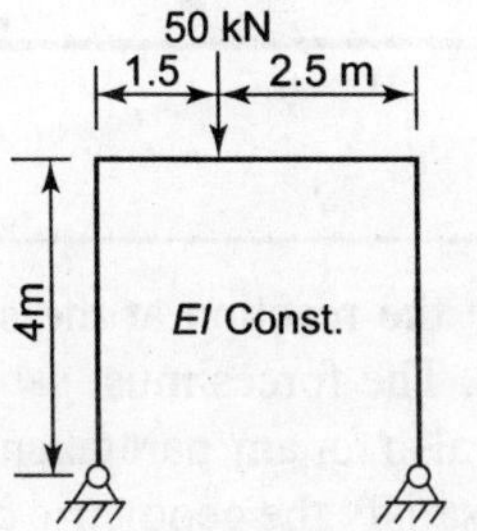

Fig. 14.43

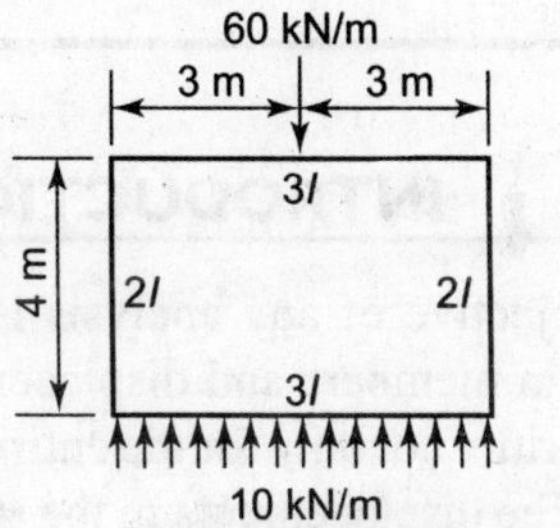

Fig. 14.44

14.10, 14.11, 14.12, 14.13 Obtain the rotational stiffness and the carry-over factors for the beams shown in Figs. 14.45, 14.46, 14.47 and 14.48.

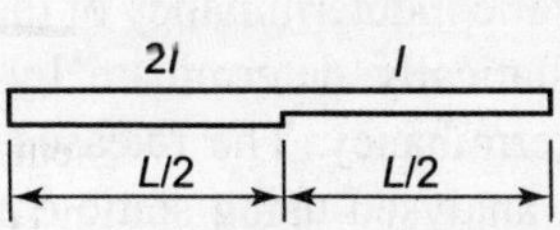

Fig. 14.45

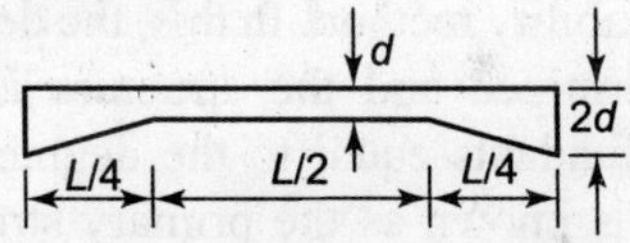

Fig. 14.46

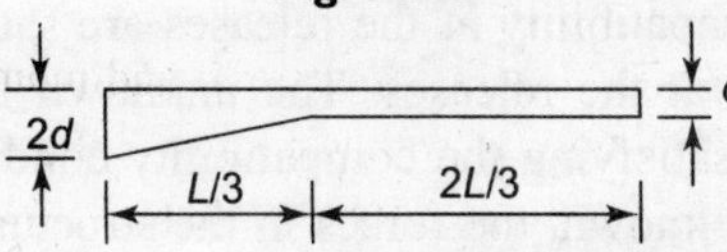

Fig. 14.47

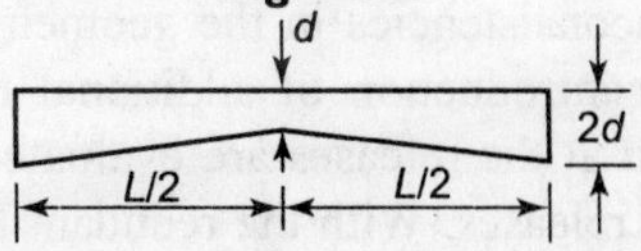

Fig. 14.48

15 Matrix Methods of Structural Analysis

15.1 INTRODUCTION

The objective of any analysis is to determine the reaction at the supports, the forces in members and displacement of joints. The forces must satify the static equilibrium not only for the entire structure but also for any part taken out as a free body. The displacements in the structure must satisfy the geometric continuity of the structure and be compatible with support conditions.

15.1.1 Methods of Analysis

In the analysis, two general methods are adopted. The first is the force method or the flexibility method. In this, the degree of static indeterminancy of the structure is determined and the structure is made statically determinate by releasing the redundants equal to the degree of indeterminancy. The released structure, which is known as the primary structure, is analysed using static equations of equilibrium and the displacements in the direction of released are determined. The inconsistencies in the geometric compatibility at the releases are satisfied by the introduction of additional forces at the releases. The unknown forces applied at the releases are evaluated by satisfying the compatibility conditions on the releases. With the redudant forces known, the forces in the structure are determined by the superimposition of the forces in the released structure and the forces due to redudant forces.

The second approach is the displacement method or the stiffness method. In this, the structure is restrained from undergoing displacements at the joints. The restraining forces are determined at the joints. The number of artificial restraints added to make the structure kinematically determinate is equal to the degree of freedom of the structure. The restrained structure, however, does not satisfy the equilibrium of forces at the joints. Displacements are then allowed to take place at the joints until the artificial restraining forces vanish. The displacements are evaluated by satisfying the equilibrium conditions of the joints. With the joint displacements known, the forces on the structure are determined by the superposition of the forces in the restrained structure and the forces due to displacements at the joints.

Either the force method or the displacement method can be used for the analysis of a given structure. In the force method, the unknowns are the forces required at the releases to satisfy the geometric compatibility. The analysis results in a number of simultaneous equations equal to the number of releases. On the other hand the unknowns in the displacement method are the displacements at the joints. The analysis results in a number of simultaneous equations equal to the number of independent displacements. The possible number of independent displacements represents a different type of indeterminancy known as kinematic indeterminancy.

15.1.2 Kinematic Indeterminancy of a Structure

There are two types of indeterminancies that may be used to describe a structural system; (1) static indeterminancy and (2) kinematic indeterminancy. Static indeterminancy, discussed in Chapter 10, refers to the number of redundant forces that are to be released to transform the structure into a statically determinate and yet a stable structure. The second type of indeterminancy in a structural system, kinematic indeterminancy, refers to the number of independent components of joint displacements with respect to a specified set of axes.

Any joint in space will have six independent components of displacements known as degrees of freedom (d.o.f), three translations and three rotations. A joint in a plane frame will have three degrees of freedom, two translations and one rotation. A plane truss joint naturally will have two degrees of freedom, both translations. A few illustrations presented below will make the point clear.

Consider a continuous beam *ABC* as shown in Fig. 15.1. At end *A* the beam is prevented from undergoing any rotation or translation. The roller supports at *B* and *C* prevent any translation in the vertical direction. It may be noted that a roller support is capable of taking up either the upward or downward reaction.

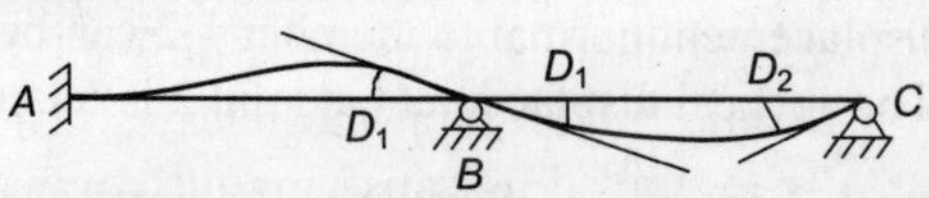

Fig. 15.1

If we neglect axial deformations in the beams, there will be no horizontal displacement at supports *B* or *C*. Therefore, the only unknown displacements are the rotations at *B* and *C* denoted by D_1 and D_2 respectively. These displacements are independent of one another as either can be given an arbitrary value by the application of appropriate forces.

The number of independent joint displacements or degrees of freedom in a structure is called the degree of kinematic indeterminancy. Therefore, in the continuous beam referrred to above, the degree of kinematic indeterminancy is two.

Consider an example of a plane frame shown in Fig. 15.2. The fixed supports at *C* and *D* prevent translations and rotations. If the axial forces are neglected joints *A* and *B* undergo only rotations and no translations as shown in Fig. 15.2.

As these rotations are independent of one another the kinematic indeterminancy of the frame is two.

As another example, consider a plane frame with inclined legs as shown in Fig. 15.3.

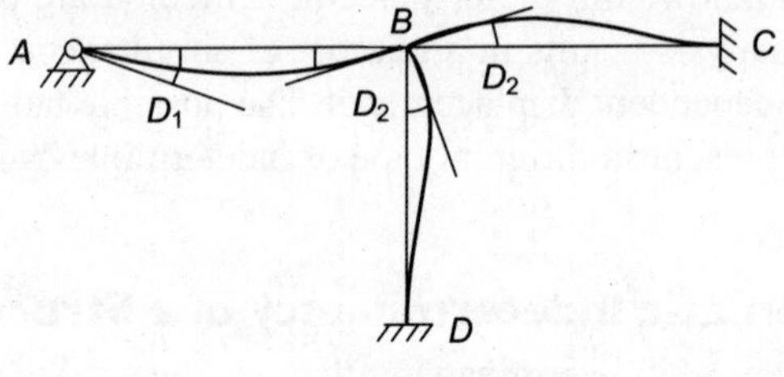

Fig. 15.2

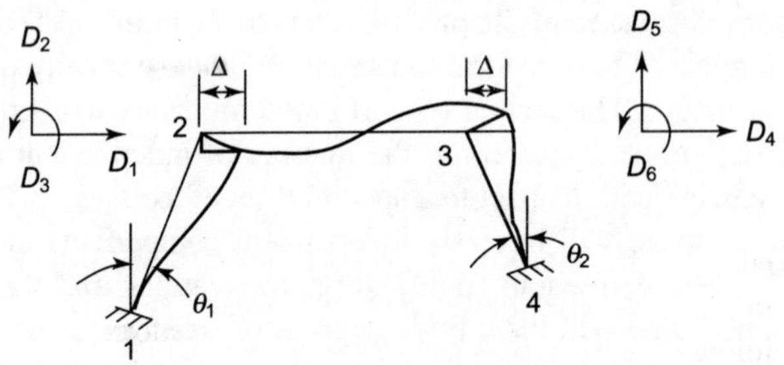

Fig. 15.3

The frame has fixed supports at 1 and 4. Joints 2 and 3 apparently have three degrees of freedom each, two translations and one rotation. However, if axial deformation in member 2-3 is neglected, the horizontal displacement A at joint 3 is equal to the horizontal displacement at joint 2. Further, the vertical displacement at joint 2 or the displacement normal to member 1-2 can be related to horizontal displacement Δ. The vertical displacement at joint 2 is $\Delta \tan \theta_1$ and displacement normal to member 1-2 is $\frac{\Delta}{\cos \theta_1}$. Similarly, the vertical displacement at joint 3 is $\Delta \tan \theta_2$ and displacement normal to member 4-3 is $\frac{\Delta}{\cos \theta_2}$. Thus, there are only three independent displacements, rotations at 2 and 3, and lateral displacement which is same at both the joints. Therefore, the degree of kinematic indeterminancy for the frame is 3. However, if the axial deformations are taken into account, all the four translational displacements are independent and the kinematic indeterminancy of the structure is 6.

Consider an example of a pin-jointed truss with the forces acting at the joints only as shown in Fig. 15.4.

The members undergo axial deformations only and remain straight. The deformations in the structure are completely defined if the components of translations along two orthogonal axes are determined for each joint. Thus each joint has two degrees of freedom. The pin-jointed frame in Fig. 15.4 is kinematically indeterminate by two degrees.

Consider now a space frame as shown in Fig. 15.5. All the four columns are fixed and hence no displacements take place. The four joints A, B, C and D in space have six degrees of freedom each—three translation and three rotations. The kinematic indeterminancy of the structure is $4 \times 6 = 24$.

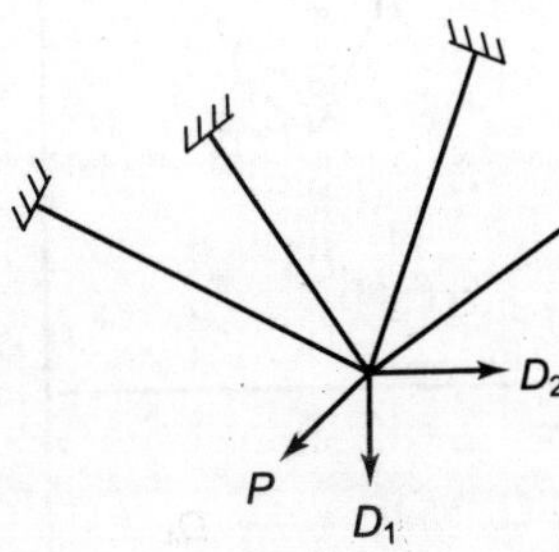

Fig. 15.4

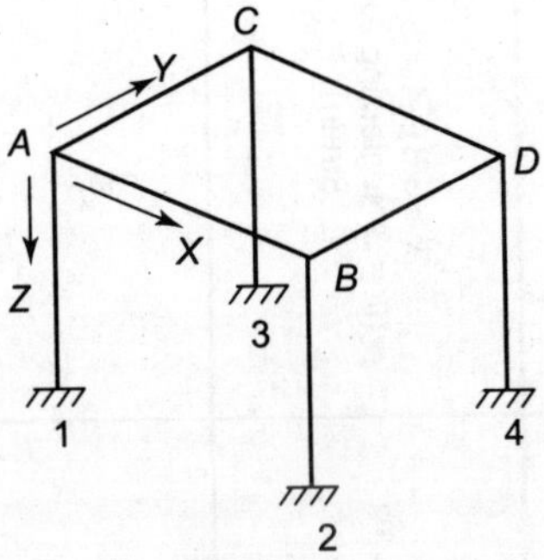

Fig. 15.5

If the axial deformations are not accounted for, the four columns remain unchanged in their lengths and hence the vertical displacements in the Z direction at joints A, B, C and D vanish reducing the kinematic indeterminancy by four. Further, the displacements in the X direction at joints A and B are equal. So also the displacements in the X direction at C and D are equal. Similarly the displacements in the Y direction at A and C, and B and D are equal. The kinematic indeterminancy is further reduced by four. Therefore the kinematic indeterminancy of the structure neglecting axial deformations is 16.

To determine the kinematic indeterminancy of a structural system, consider a plane frame or truss having J joints, let C be the number of displacements constrained giving rise to reaction components, then the kinematic indeterminancy of a frame is given by

$$I_K = NJ - C$$

in which I_K is the kinematic indeterminancy and N the number of degrees of freedom at the joint.

Examples illustrating the degree of kinematic indeterminancy of various plane structures are given in Fig. 15.6. It may be noted that the restrained degrees of freedom are indicated in broken lines.

It is emphasized here that the kinematic indeterminancy should not be confused with static indeterminancy. For instance, the frame in Fig. 15.2 is statically indeterminate by five degrees. If the fixed support at C is replaced by a hinge, the degree of static indeterminancy is reduced by one. This, however, introduced one independent displacement at C and hence increases the kinematic indeterminancy by one.

Structure	Degree of Freedom Restrained and Unrestrained	Degree of Kinematic Indeterminancy
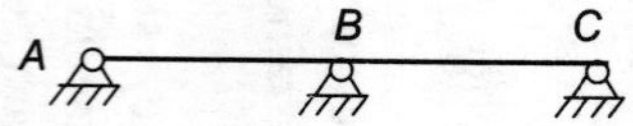 (a)	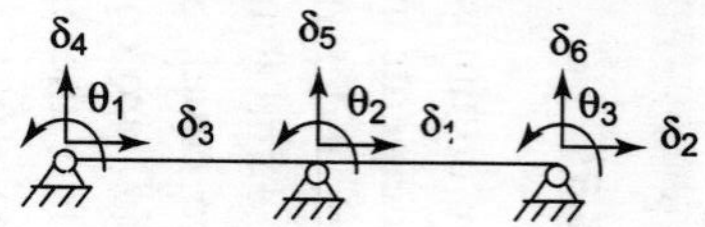	$Ik = 3(3) - 4 = 5$ ($Ik = 3$ Neglecting Axial Strain)
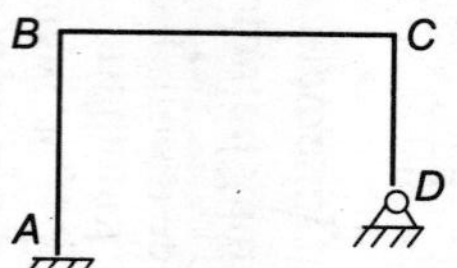 (b)	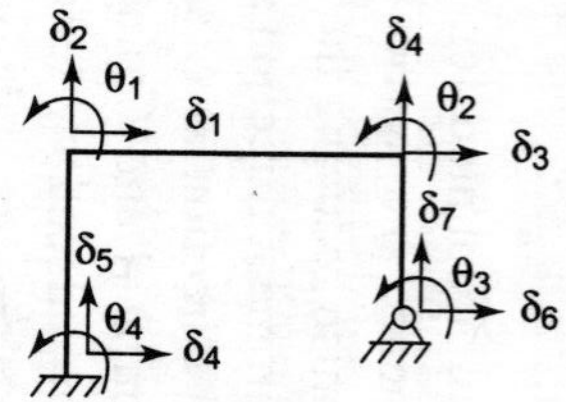	$Ik = 4(3) - 5 = 7$ ($Ik = 4$ Neglecting Axial Strain)
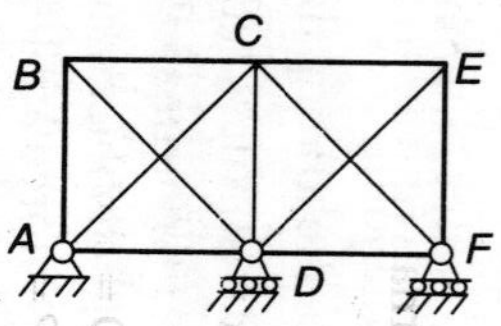 (c)	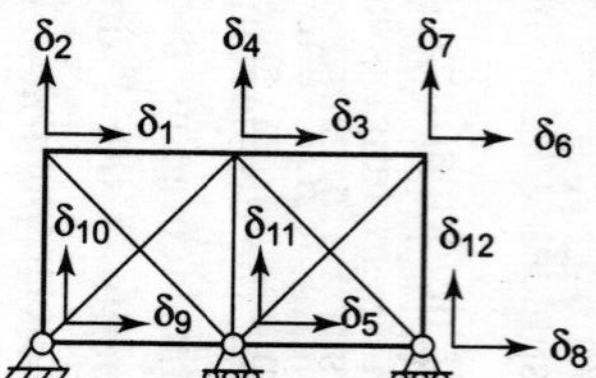	$Ik = 2(6) - 4 = 8$

Fig. 15.6 | *Examples of kinematic indeterminancy*

In general the introduction of a release decreases the statical indetermiancy and increases the kinematic indeterminancy. For this reason the displacement method of analysis is more suitable for structures having a higher degree of indeterminancy.

Before we embark upon any of the above methods we shall consider the necessary preliminaries that are essential to the development of methods.

15.2 STIFFNESS AND FLEXIBILITY COEFFICIENTS

In Section 10.4 we discussed the force displacement relationship in terms of flexibility influence coefficients. In this section we shall formalise the procedures of relating the forces and displacements in a structure in terms *of flexibility and stiffness coefficients.* These coefficients are characteristic of a structure relating the forces and displacement at its coordinates.

The flexibility coefficients characterise the behaviour of the structure by specifying the displacement response to the applied forces at the coordinates. On the other hand, the stiffness coefficients specify the forces required to produce the given displacements at the coordinates.

15.2.1 Structure with a Single Coordinate

Consider a simple example of a cantilever beam in Fig. 15.7*a* with a single coordinate indicated for force displacement measurements. The deformation of the structure may be expressed as

$$D = fP \tag{15.1}$$

in which D = deformation at coordinate point 1

f = flexibility coefficient which is defined as the displacement at coordinate 1 caused by a unit force at 1

P = load applied at coordinate 1

Using the moment area method, we find for the beam of Fig. 15.7*a*

$$f = \frac{L^3}{3EI} \tag{15.2}$$

An alternative way to relating the force and displacement at coordinate 1 is

$$P = kD \tag{15.3}$$

in which k = stiffness coefficient which is defined as the force required at coordinate 1 to produce a unit displacement at 1

D = displacement at point 1

P = force applied at point 1

The value of k for the beam of Fig. 15.7*a* is

$$k = \frac{3EI}{L^3} \tag{15.4}$$

Again for a beam of Fig. 15.7*b*, we have

$$f = \frac{L}{EI} \text{ and } k = \frac{EI}{L} \tag{15.5}$$

We find from a comparison of the values of f and k, one is the inverse of the other, or

$$fk = 1 \tag{15.6}$$

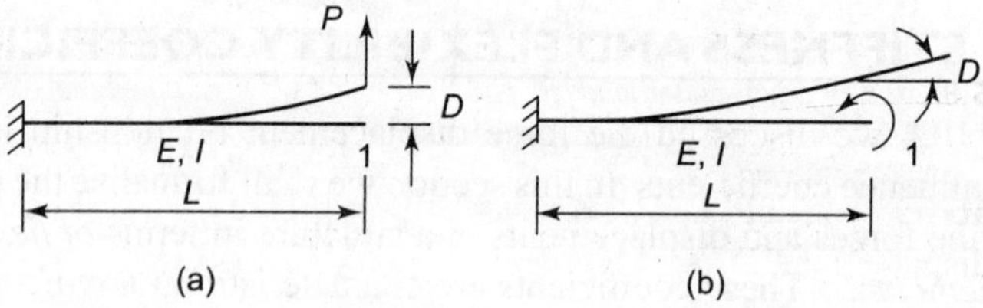

Fig. 15.7 | *(a) Cantilever beam under load P, (b) Displacement due to moment*

15.2.2 Structure with Two Coordinates

We now extend the concept of stiffness and flexibility matrices to a structure having two coordinates.

Flexibility Matrix Figure 15.8*a* again shows a cantilever beam with two coordinates. Let us relate the forces and the corresponding displacements through flexibility coefficients.

To do this, we apply the superposition of forces as follows: First we apply a unit force at coordinate 1 only (Fig. 15.8*b*) and designate the displacements at 1 and 2 as f_{11} and f_{12} respectively. Next, we apply a unit force at 2 only (Fig. 15.8*c*) and designate the displacements at 1 and 2 as f_{12} and f_{22} respectively.

The displacements D_1 and D_2 due to forces P_1 and P_2 acting simultaneously are

$$D_1 = f_{11}P_1 + f_{12}P_2$$

And
$$D_2 = f_{21}P_1 + f_{22}P_2 \tag{15.7}$$

This can be written in the form of a matrix as

$$\begin{Bmatrix} D_1 \\ D_2 \end{Bmatrix} = \begin{bmatrix} f_{11} & f_{12} \\ f_{21} & f_{22} \end{bmatrix} \begin{Bmatrix} P_1 \\ P_2 \end{Bmatrix} \tag{15.8}$$

or simply

$$\mathbf{D} = \mathbf{fP} \tag{15.9}$$

The matrix **f** is the flexibility matrix for the structure of Fig. 15.8. It may be noted that the elements of the first column of this matrix are generated by applying a unit force at 1 only and the elements of the second column by applying a unit force at 2 only. The elements of the flexibility matrix for the structure is

$$\mathbf{f} = \begin{bmatrix} f_{11} & f_{12} \\ f_{21} & f_{22} \end{bmatrix} = \frac{L}{EI} \begin{bmatrix} L^2/3 & L/2 \\ L/2 & 1 \end{bmatrix} \tag{15.10}$$

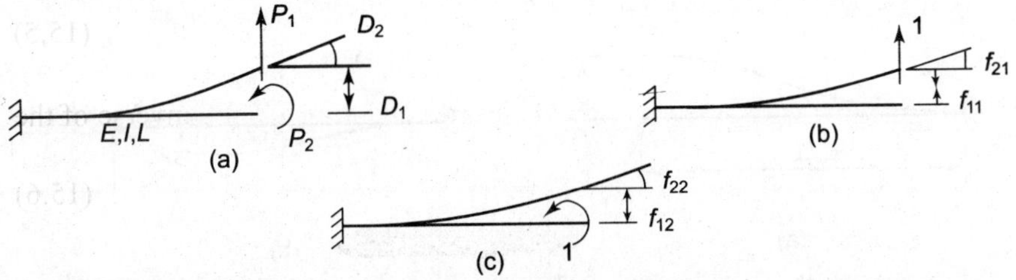

Fig. 15.8 | *(a) Beam under loads P_1 and P_2, (b) Unit load at coordinate 1 only, (c) Unit load at coordinate 2 only*

Stiffness Matrix We shall now relate displacements and forces in an alternative way so that displacement information at the coordinates can be easily transferred into forces. To do this we apply a superposition of displacements as shown in Fig. 15.9. First we apply a unit displacement at 1 only (Fig. 15.9*a*) and designate the required forces at 1 and 2 as k_{11} and k_{21} respectively. The first subscript is the coordinate where the force is measured and the second subscript is the coordinate where the unit displacement is applied. Similarly, in Fig. 15.9*b* forces k_{12} and k_{22} are required to cause a unit displacement at 2 only. The forces required to produce displacements D_1 and D_2 simultaneously are obtained by a superposition of the results obtained in Figs. 15.9*a* and *b*. This yields

$$\begin{aligned} P_1 &= k_{11}D_1 + k_{12}D_2 \\ P_2 &= k_{21}D_1 + k_{22}D_2 \end{aligned} \tag{15.11}$$

Expressing this in matrix form, we have

$$\begin{Bmatrix} P_1 \\ P_2 \end{Bmatrix} = \begin{bmatrix} k_{11} & k_{12} \\ k_{21} & k_{22} \end{bmatrix} \begin{Bmatrix} D_1 \\ D_2 \end{Bmatrix} \tag{15.12}$$

or simply

$$\mathbf{P} = \mathbf{kD} \tag{15.13}$$

The stiffness matrix **k** has the following elements for the structure of Fig. 15.9

$$\mathbf{k} = \begin{bmatrix} k_{11} & k_{12} \\ k_{21} & k_{22} \end{bmatrix} = \frac{EI}{L}\begin{bmatrix} 12/L^2 & -6/L \\ -6/L & 4 \end{bmatrix} \tag{15.14}$$

The reciprocal relation on the basis of Eqs. 15.10 and 15.14 is not apparent because in Eq. 15.10 $f_{11} = L^3/3EI$ and in Eq. 15.14 $k_{11} = {}^{12EI}/_{L^3}$. However, in cases where more than one coordinate is considered, the reciprocity between stiffness and flexibility matrices exists in matrix form. This is shown to be true by the operation in Eq. 15.15 in which the flexibility and stiffness matrix of Eqs. 15.10 and 15.14 are multiplied to yield the identity matrix.

$$\frac{L}{EI}\begin{bmatrix} L^2/3 & L/2 \\ L/2 & 1 \end{bmatrix}\frac{EI}{L}\begin{bmatrix} 12/L^2 & -6/L \\ -6/L & 4 \end{bmatrix} = \begin{bmatrix} 1 & 0 \\ 0 & 1 \end{bmatrix} \tag{15.15}$$

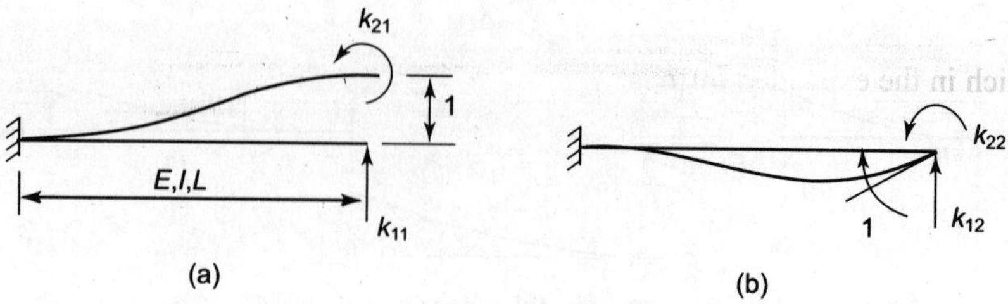

Fig. 15.9 | *(a) Unit displacement at coordinate 1 only, (b) Unit displacement at coordinate 2 only*

Equation 15.15 is written in compact notation as

$$\mathbf{fk} = \mathbf{I} \tag{15.16}$$

The procedure can be extended to structures with more than two coordinates and can be generalised as follows.

15.2.3 Flexibility and Stiffness Matrices in *n* Coordinates

Consider a linear elastic structure with n coordinates. To generate the elements of column 1 of the flexibility matrix **f** we apply a unit force at coordinate 1 only and compute displacements at all the coordinates f_{i1} ($i = 1, 2, ..., n$). This will give the elements in column 1. To generate, again say, column n of matrix **f**, we apply a unit force at coordinate n only and compute displacements f_{in} ($i = 1, 2, ..., n$). The values of these displacements form the elements in the nth column of matrix **f**. In general, to generate the elements in the jth column, apply a unit force at coordinate j only and compute the displacements fij ($i = 1, 2 ..., n$). The values of these displacements form the elements of the jth column of the matrix **f**. Thus, it is seen that the complete flexibility matrix **f** will have n rows and n columns forming a square matrix $n \times n$.

Similarly, to generate the elements in column 1 of matrix **k**, we impose a unit displacement at coordinate 1 only and compute forces needed at all the coordinates k_{i1} ($i = 1, 2, ..., n$) to hold the structure in that configuration. These forces form the elements of the first column of the stiffness matrix k. To generate, say, column n of matrix k, we impose a unit displacement at coordinate n only and compute forces k_{in}(i = 1,2, ..., n) needed to hold the structure with no displacements at other coordinate. In general, to generate the elements of column j ($j = 1, 2, ..., n$), we impose a unit displacement at j only and compute forces k_{ij}($i = 1, 2, ..., n$) needed to hold the structure with no displacements at other coordinates. These forces form the elements of the jth column of matrix k. Thus, a stiffness matrix for n coordinates will have n rows and n columns of elements, forming a square matrix $n \times n$.

15.2.4 Force Displacement Relations

Using matrices **f** and **k**, the force and displacement vectors at the coordinates are related by

$$\mathbf{D} = \mathbf{fP} \tag{15.17}$$

which in the expanded form is

$$D_1 = \sum_{j=1}^{n} f_{ij}\, P_j = f_{11}P_1 + f_{12}P_2 \ldots f_{1n}P_n$$

$$D_2 = \sum_{j=1}^{n} f_{2j}\, \boldsymbol{Pj}$$

$$\vdots \qquad \vdots \qquad \vdots$$

$$D_n = \sum_{j=1}^{n} f_{nj}\, \boldsymbol{P}_j \tag{15.18}$$

that is, the *i*th element of vector *D* in Eq. 15.18 is equal to the sum of the products of $f_{ij}(j = 1, 2, ..., n)$ in row *i* of **f** and the corresponding elements *Pj* of **P**.

In terms of the stiffness matrix, the force displacement relationship is

$$\mathbf{P} = \mathbf{kD} \tag{15.19}$$

which in the expanded form is

$$P_1 = \sum_{j=1}^{n} k_{ij}\, D_j = k_{11}D_1 + k_{12}D_2 \ldots k_{in}D_n$$

$$P_2 = \sum_{j=1}^{n} k_{2j}\, \boldsymbol{Dj}$$

$$\vdots \qquad \vdots \qquad \vdots$$

$$P_n = \sum_{j=1}^{n} k_{nj}\, \boldsymbol{D}_j \tag{15.20}$$

A number of examples will help to reinforce the concept of the stiffness and flexibility matrices.

Example 15.1 | *Generate the flexibility matrix [f] for cordinates 1 and 2 of the beam shown in Fig. 15.10*

To generate flexibility matrix [*f*] we apply a unit force at coordinate 1 and find displacements corresponding to coordinates 1 and 2. These displacements form the elements f_{11} and f_{21} of the flexibility matrix. The required displacements are obtained by using conjugate beam method.

From the above

$$R'_A = \frac{3}{8}\frac{L}{EI} \quad \text{or} \quad f_{11} = \frac{3}{8}\frac{L}{EI}$$

$$M'_C = \frac{L^2}{12EI} \quad \text{or} \quad f_{21} = \frac{L^2}{12EI}$$

Again applying a unit force at coordinate 2 and using conjugate beam method

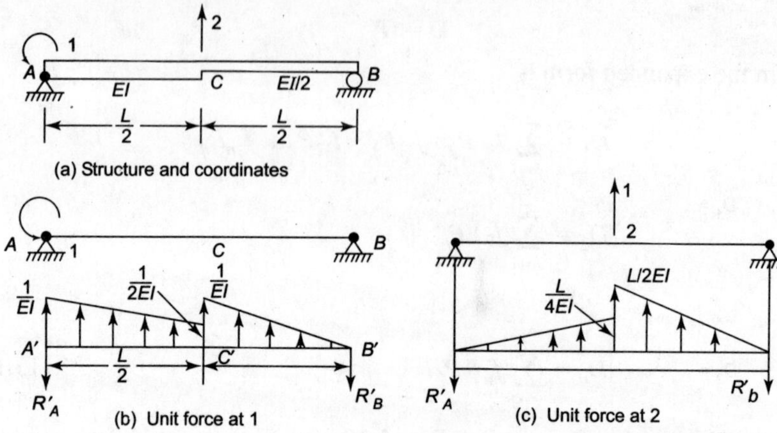

Fig. 15.10

$$R'_A = \frac{L^2}{12EI} \text{ or } f_{12} = \frac{L^2}{12EI}$$

and
$$M'_C = \frac{L^3}{32EI} \text{ or } f_{22} = \frac{L^3}{32EI}$$

The flexibility matrix

$$[f] = \frac{L}{EI}\begin{bmatrix} \frac{3}{8} & \frac{L}{12} \\ \frac{L}{12} & \frac{L^2}{32} \end{bmatrix}$$

Example 15.2 | *Generate the stiffness matrix [k] for the structure with the coordinates as shown in Fig. 15.11*

The stiffness matrix [k] can be obtained by imposing a unit displacement at each of the coordinates one at a time and computing the forces required to hold the structure in the deflected configuration.

First, we apply a unit displacement $D_1 = 1$ at coordinate 1 and work out the forces required to hold the structure in that configuration. The elements in the first column of the matrix [k] are as shown in Fig. 15.11. The elements in the second column of the matrix [k] are obtained by imposing a unit displacement $D_2 = 1$ at coordinate 2 and working out the forces. The elements are shown in figure. Similarly, the elements in third and fourth columns of the matrix [k] are obtained by imposing displacements $D_3 = 1$ and $D_4 = 1$ in turns. The resulting elements in the third and fourth quadrants are also shown. The complete stiffness matrix.

$$[k] = \begin{bmatrix} 8 & & & \\ 0 & 4 & Sym & \\ 4 & 2 & 12 & \\ -12 & 6 & -6 & 36 \end{bmatrix}$$

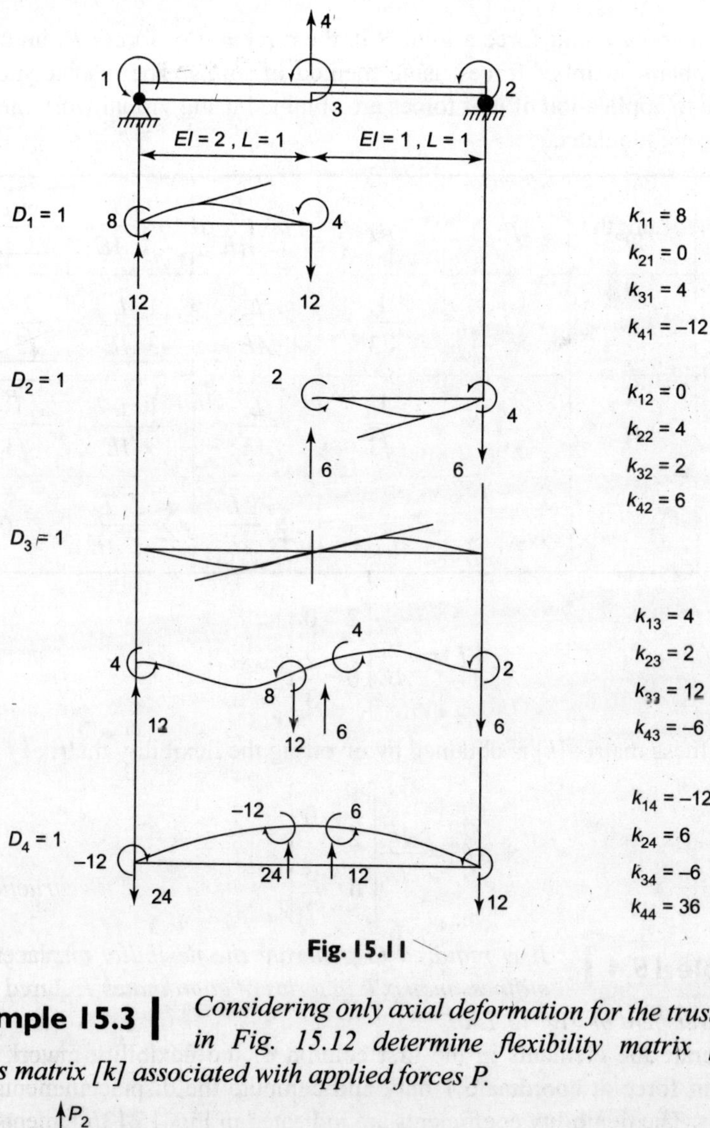

Fig. 15.11

Example 15.3 | *Considering only axial deformation for the truss shown in Fig. 15.12 determine flexibility matrix [f] and stiffness matrix [k] associated with applied forces P.*

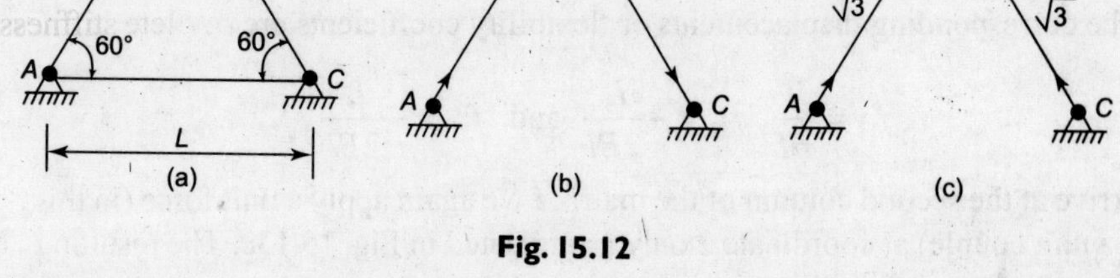

Fig. 15.12

First, we apply a unit force a joint B in the direction of forces P_1 and P_2 in turns and obtain member forces using method of joints. The displacement of joint B due to application of unit forces are obtained using virtual work method. The results are tabulated.

Member	Length	p_1	p_2	$p_1^2 \frac{L}{AE}$	$p_2^2 \frac{L}{AE}$	$\frac{p_1 p_2 L}{AE}$
AB	L	1	$\frac{1}{\sqrt{3}}$	$\frac{L}{AE}$	$\frac{1}{3}\frac{L}{AE}$	$\frac{1}{\sqrt{3}}\frac{L}{AE}$
BC	L	−1	$\frac{1}{\sqrt{3}}$	$\frac{L}{AE}$	$\frac{1}{3}\frac{L}{AE}$	$-\frac{1}{\sqrt{3}}\frac{L}{AE}$
				$\sum \frac{2L}{AE}$	$\sum \frac{2}{3}\frac{L}{AE}$	0

$$[f] = \frac{L}{AE}\begin{bmatrix} 2 & 0 \\ 0 & \frac{2}{3} \end{bmatrix}$$

The stiffness matrix $[k]$ is obtained by inverting the flexibility matrix $[f]$ as

$$[k] = \frac{AE}{L}\begin{bmatrix} \frac{1}{2} & 0 \\ 0 & \frac{3}{2} \end{bmatrix}$$

Example 15.4 | *It is required to generate the flexibility matrix* **f** *and stiffness matrix* **k** *in terms of coordinates 1, 2 and 3 for the cantilever bent of Fig. 15.13a.*

To generate the elements in the first column of the flexibility matrix **f** we apply a unit force at coordinate 1 only and compute the displacements at the coordinates. The flexibility coefficients are indicated in Fig. 15.13*b*.

The displacements correspond to the translation at coordinate 1 and rotations at coordinates 2 and 3. Displacements only due to bending are considered. Any method such as the moment area or virtual work method can be used in the computation of displacements.

The corresponding displacements or flexibility coefficients are

$$f_{11} = \frac{L^3}{EI},\ f_{21} = -\frac{L^2}{2EI} \text{ and } f_{31} = -\frac{L^2}{2EI}$$

To arrive at the second column of the matrix **f** we again apply a unit force (in this case a unit couple) at coordinate 2 only as indicated in Fig. 15.13*c*. The resulting displacements at the coordinates give

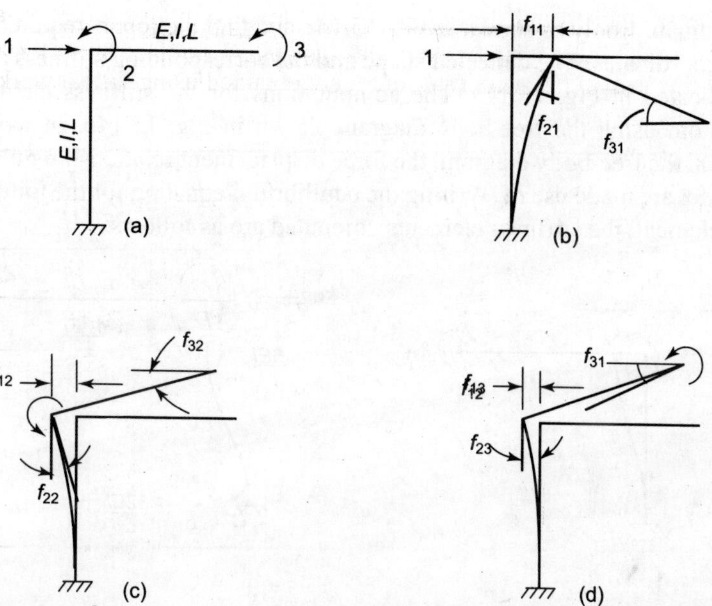

Fig. 15.13 | *(a) Frame and coordinates, (b) Unit load applied at coordinate 1 only, (c) Unit load applied at coordinate 2 only, (d) Unit load applied at coordinate 3 only*

$$f_{12} = -\frac{L^2}{2EI},\ f_{22} = \frac{L}{EI} \text{ and } f_{32} = \frac{L}{EI} \tag{15.21}$$

Lastly, a unit couple is applied at only coordinate 3 (Fig. 15.13*d)* and the elements in the third column of matrix **f** are determined. The flexibility coefficients are

$$f_{13} = -\frac{L^2}{2EI},\ f_{23} = \frac{L}{EI} \text{ and } f_{33} = \frac{2L}{EI} \tag{15.22}$$

The complete flexibility matrix **f** is

$$\mathbf{f} = \frac{L}{EI}\begin{bmatrix} \frac{L^2}{3} & -\frac{L}{2} & -\frac{L}{2} \\ -\frac{L}{2} & 1 & 1 \\ -\frac{L}{2} & 1 & 2 \end{bmatrix} \tag{15.23}$$

It is seen that flexibility matrix **f** is a square matrix and is symmetric, **that is,** $f_{ij} = f_{ji}$.

The stiffness matrix **k** can likewise be determined by imposing a unit displacement at one coordinate at a time and computing the forces required at the coordinates to hold the structure in that configuration. For example, to generate the first column of the stiffness matrix **k** we impose a unit displacement at coordinate 1 only as shown in Fig. 15.14*a* and find the forces required at each of the coordinates. The deflected shape and the corresponding stiffness elements are indicated in Fig. 15.14*a*. The computations for the stiffness elements are carried out using the free-body diagram shown in Fig. 15.14*b*. In writing the forces on the free-body diagram, the force displacement relationship given in the Appendix are made use of. Writing the equilibrium equation for the joint and the beam element, the stiffness elements computed are as follows.

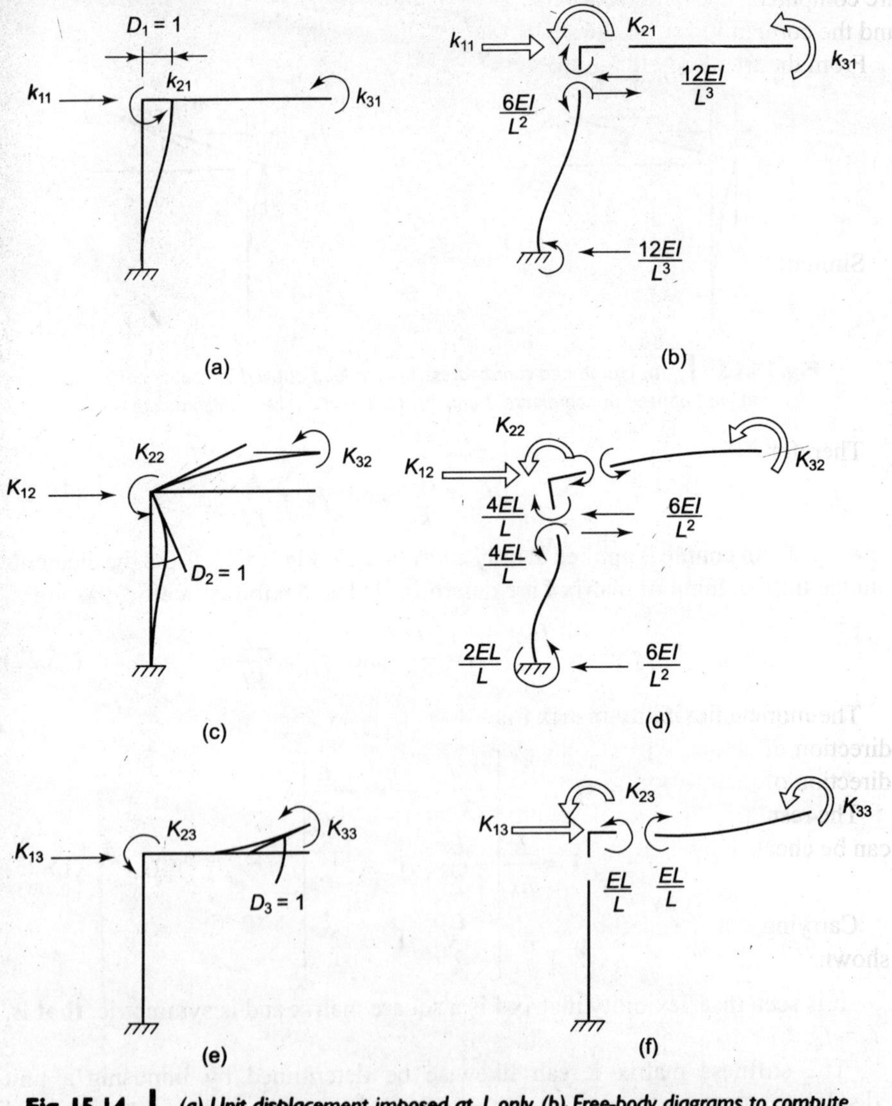

Fig. 15.14 | *(a) Unit displacement imposed at 1 only, (b) Free-body diagrams to compute k_{11}, k_{21} and k_{31}, (c) Unit displacement imposed at coordinate 2, (d) Free-body diagram to compute k_{12}, k_{22} and k_{32}. (e) Unit displacement imposed at coordinate 3, (f) Free-body diagrams to compute k_{13}, k_{23} and k_{33}*

$$k_{11} - 12\frac{EI}{L^3} = 0,\ k_{21} - \frac{6EI}{L^2} = 0,\ k_{31} + 0 = 0$$

or $$k_{11} = \frac{12EI}{L^3},\ k_{21} = \frac{6EI}{L^2},\ k_{31} = 0 \qquad (15.24)$$

To generate the stiffness elements in the second column of matrix **k**, unit displacement is imposed at coordinate 2 and the forces needed at all the coordinates are computed. The deflected shape of the cantilever bent is shown in Fig. 15.14*c* and the computations for the stiffness elements are shown in Fig. 15.14*d*.

From the free-body diagram of Fig. 15.14*d* we have

$$k_{12} - \frac{6EI}{L^2} = 0,\ k_{22} - \frac{5EI}{L} = 0,\ k_{32} + \frac{EI}{L} = 0$$

$$k_{12} = \frac{6EI}{L^2},\ k_{22} = \frac{5EI}{L},\ k_{32} = -\frac{EI}{L} \qquad (15.25)$$

Similarly, from Fig. 15.14*e* and *f*, we have

$$k_{13} + 0 = 0,\ k_{23} + \frac{EI}{L} = 0,\ k_{33} - \frac{EI}{L} = 0$$

$$k_{13} = 0,\ k_{23} = -\frac{EI}{L},\ k_{33} = \frac{EI}{L} \qquad (15.26)$$

Therefore, the complete stiffness matrix **k** is

$$\mathbf{k} = \frac{EI}{L}\begin{bmatrix} \frac{12}{L^2} & \frac{6}{L} & 0 \\ \frac{6}{L} & 5 & -1 \\ 0 & -1 & 1 \end{bmatrix} \qquad (15.27)$$

The moments and shears shown in the free-body diagrams are in their actual direction of action, whereas the stiffness elements k_{ij} are shown in the positive direction of their action.

The results of **f** and **k** referred to the same structure with identical coordinates can be checked by applying the condition

$$\mathbf{fk} = \mathbf{I} \qquad (15.28)$$

Carrying out the multiplication of **f** and **k** results in the identity matrix as shown.

$$\frac{L}{EI}\begin{bmatrix} \frac{L^2}{3} & -\frac{L}{2} & -\frac{L}{2} \\ -\frac{L}{2} & 1 & 1 \\ -\frac{L}{2} & 1 & 2 \end{bmatrix} \frac{EI}{L}\begin{bmatrix} \frac{12}{L^2} & \frac{6}{L} & 0 \\ \frac{6}{L} & 5 & -1 \\ 0 & -1 & 1 \end{bmatrix} = \begin{bmatrix} 1 & 0 & 0 \\ 0 & 1 & 0 \\ 0 & 0 & 1 \end{bmatrix} \qquad (15.29)$$

Example 15.5 | *It is required to generate stiffness matrix* **k** *for the frame of Fig. 15.15a.*

A unit displacement is imposed at coordinate 1 as in Fig. 15.15*b* and the forces required at all the coordinates to hold the structure in that configuration determined. The forces on the free-body diagram of Fig. 15.15*c* gives

$$K_{11} - 15 = 0 \qquad K_{21} - 6 = 0 \qquad K_{31} - 3 = 0$$
$$K_{11} = 15 \qquad K_{21} = 6 \qquad K_{31} = 3 \tag{15.30}$$

The second column of matrix **k** is obtained by imposing a unit displacement at coordinate 2 only as in Fig. 15.15*d* and computing the forces required at the coordinates to hold the structure in that configuration. From the free-body diagram of Fig. 15.15*e* we have

$$K_{12} - 6 = 0 \qquad K_{22} - 4 - 4 = 0 \qquad K_{32} - 2 = 0$$
$$K_{12} = 6 \qquad K_{22} = 8 \qquad K_{32} = 2 \tag{15.31}$$

Similarly, to obtain the third column elements of matrix **k** we impose a unit displacement at coordinate 3 only as in Fig. 15.15*f* and the forces needed to hold the structure in that configuration are worked out. From the free-body diagram of Fig. 15.15*g* we have

$$k_{13} = 3 \quad k_{23} = 2 \quad k_{33} = 7 \tag{15.32}$$

The complete stiffness matrix is

$$\mathbf{k} = \begin{bmatrix} 15 & 6 & 3 \\ 6 & 8 & 2 \\ 3 & 2 & 7 \end{bmatrix} \tag{15.33}$$

The generation of flexibility matrix **f** for the frame of Fig. 15.15*a* is more complicated than stiffness matrix **k**. For example, to generate the elements in column one of matrix **f**, the solution of the frame which is two times redundant is necessary. We shall see later that flexibility matrix **f** can be obtained by inverting stiffness matrix **k** so that in the present case we can use **k** to find **f**.

Example 15.6 | *We shall now consider another example where* **f** *can be generated more easily than* **k**. *Figure 15.16a shows a cantilever beam with three coordinates.*

The flexibility elements of matrix **f** can be computed by applying a unit force at each of the coordinates in turn and then computing the displacements at the coordinates.

The flexibility matrix **f** is

$$\mathbf{f} = \frac{L^3}{6EI} \begin{bmatrix} 2 & 5 & 8 \\ 5 & 16 & 28 \\ 8 & 28 & 54 \end{bmatrix} \tag{15.34}$$

The generation of stiffness matrix **k** for the beam of Fig. 15.16*a* is more involved and requires the solution of an indeterminate beam. For example, to generate the first column of matrix **k** we need to compute forces k_{11}, k_{21} and k_{31} (see Fig. 15.17) which requires the solution of a three times redundant structure.

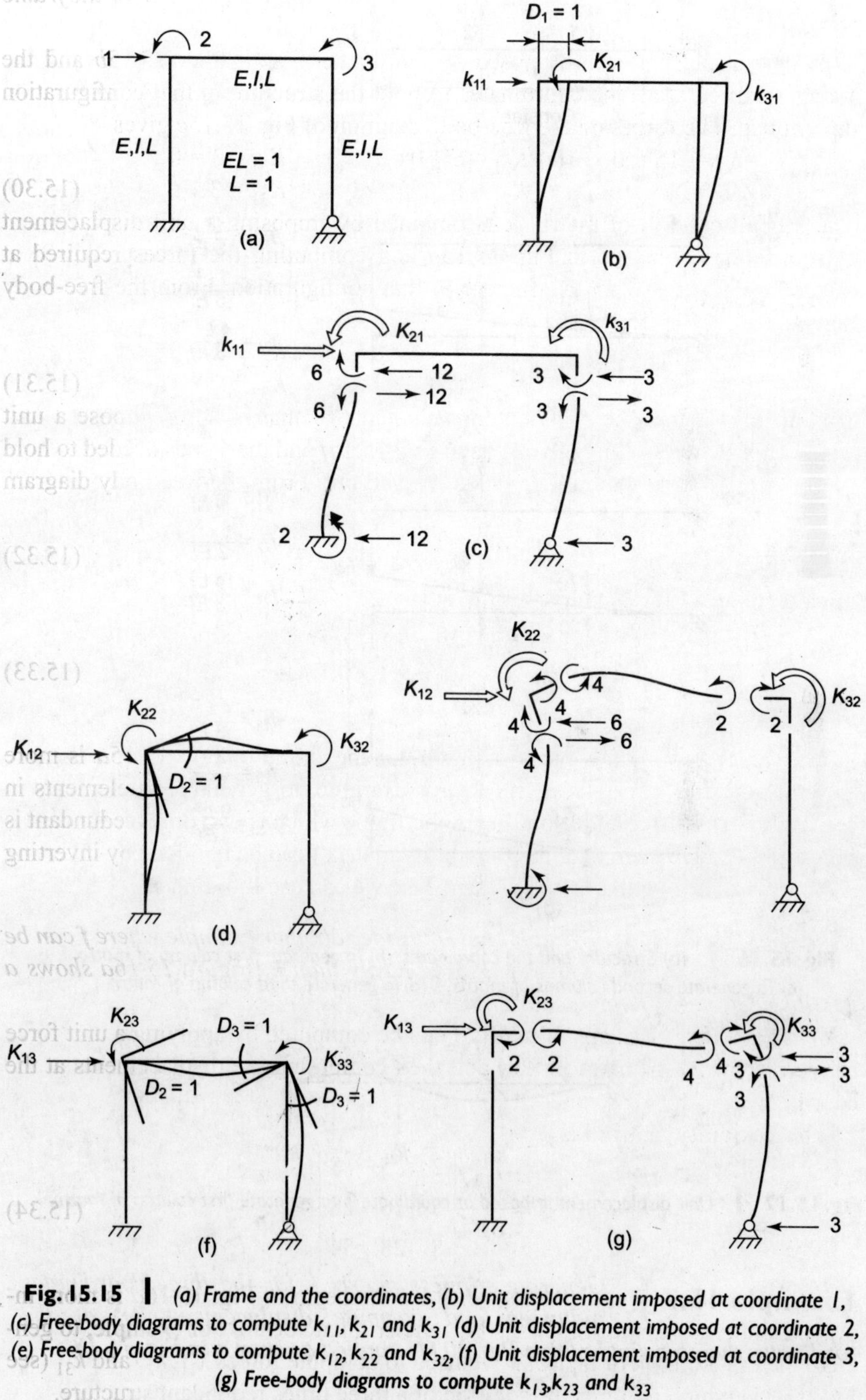

Fig. 15.15 | *(a) Frame and the coordinates, (b) Unit displacement imposed at coordinate 1, (c) Free-body diagrams to compute k_{11}, k_{21} and k_{31} (d) Unit displacement imposed at coordinate 2, (e) Free-body diagrams to compute k_{12}, k_{22} and k_{32}, (f) Unit displacement imposed at coordinate 3, (g) Free-body diagrams to compute k_{13}, k_{23} and k_{33}*

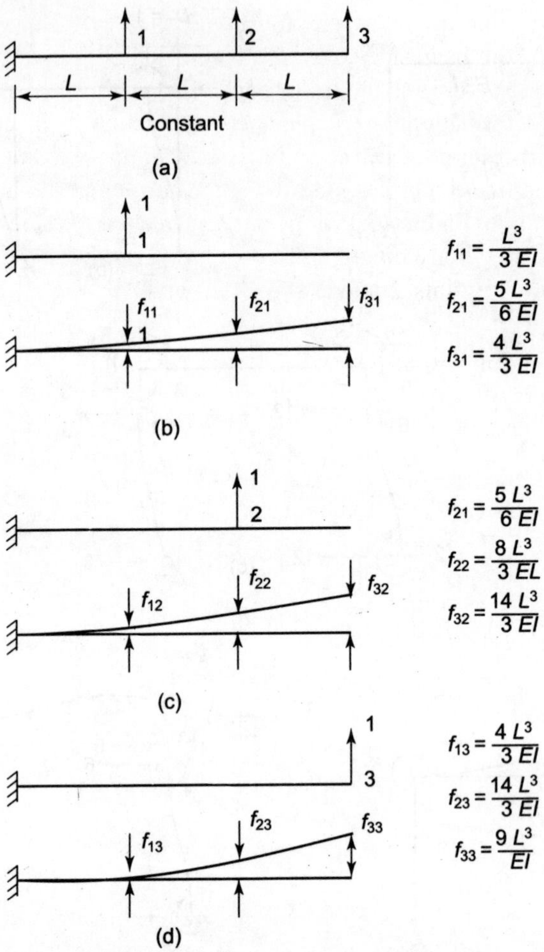

Fig. 15.16 | *(a) Structure and the coordinates, (b) To generate first column of matrix* **f**, *(c) To generate second columns of matrix, f, (d) To generate third column of matrix f*

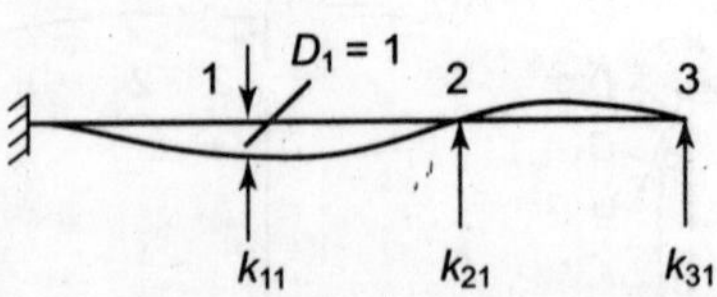

Fig. 15.17 | *Unit displacement imposed at coordinate 1 to generate first column of matrix k*

Example 15.7 | *Generate stiffness matrix* **k** *for the three coordinates indicated and compute displacements D_i at the coordinates due to a single force $P = 30$ kN applied as shown in Fig. 15.18.*

The procedure to be followed in determining the elements of the stiffness matrix **k** is the same as in the previous Example 15.5. However, due to the inclination of the left hand side column, attention should be paid to the geometry of the displaced structure and the resolution of forces into axial and lateral forces. Only bending deformations are considered, neglecting axial deformations. To generate the first column of stiffness matrix **k**, we impose a unit displacement at coordinate 1 and find the forces required at the coordinates to hold the structure in that configuration. Figure 15.19*a* shows the deflected form of the structure and Fig. 15.19*b* the free-body diagram for computing stiffness elements. Consider now equilibrium of joints 2 and 3 shown separately in Fig. 15.19*c*. The total vertical component of $\left(\frac{9EI}{L^3}+\frac{9EI}{L^3}\right)$ is replaced by a horizontal component $\left(\frac{13.5EI}{L^3}\right)$ and an axial componentas $\left(\frac{22.5EI}{L^3}\right)$ shown in Fig. 15.19*c*. Writing the equations of equilibrium

$$k_{11}-\frac{25.5\,EI}{L^3}-\frac{12EI}{L^3}=0 \qquad k_{21}-\frac{7.5\,EI}{L^2}-\frac{4.5EI}{L^2}=0$$

$$k_{11}=\frac{37.5\,EI}{L^3} \qquad k_{21}=\frac{3\,EI}{L^2}$$

$$k_{31}+\frac{4.5\,EI}{L^2}-\frac{6EI}{L^2}=0 \qquad k_{31}=\frac{1.5\,EI}{L^2} \tag{15.35}$$

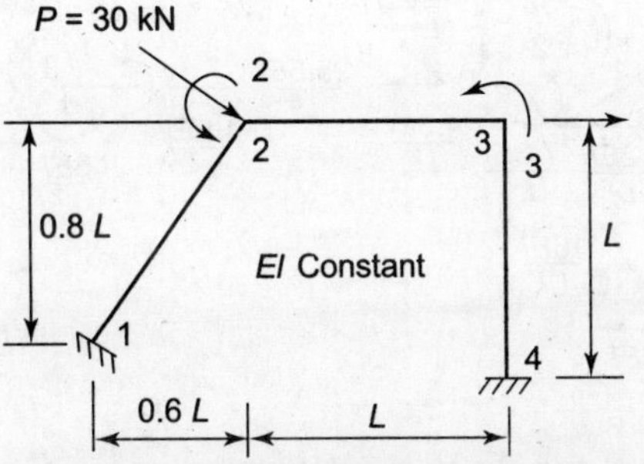

Fig. 15.18 | *Structure and coordinates*

The elements of column 2 of stiffness matrix **k** are obtained by imposing a unit displacement at coordinate 2 only and computing the forces required to hold the structure in that configuration (Fig. 15.19*d* and *e*). Again, resolving the forces at coordinate 2 along the axis of the inclined column and in the horizontal direction, we have from equilibrium conditions

$$k_{12}-\frac{3\,EI}{L^2}=0 \quad k_{22}-\frac{4EI}{L}-\frac{4EI}{L}=0 \quad k_{32}+\frac{2\,EI}{L}=0$$

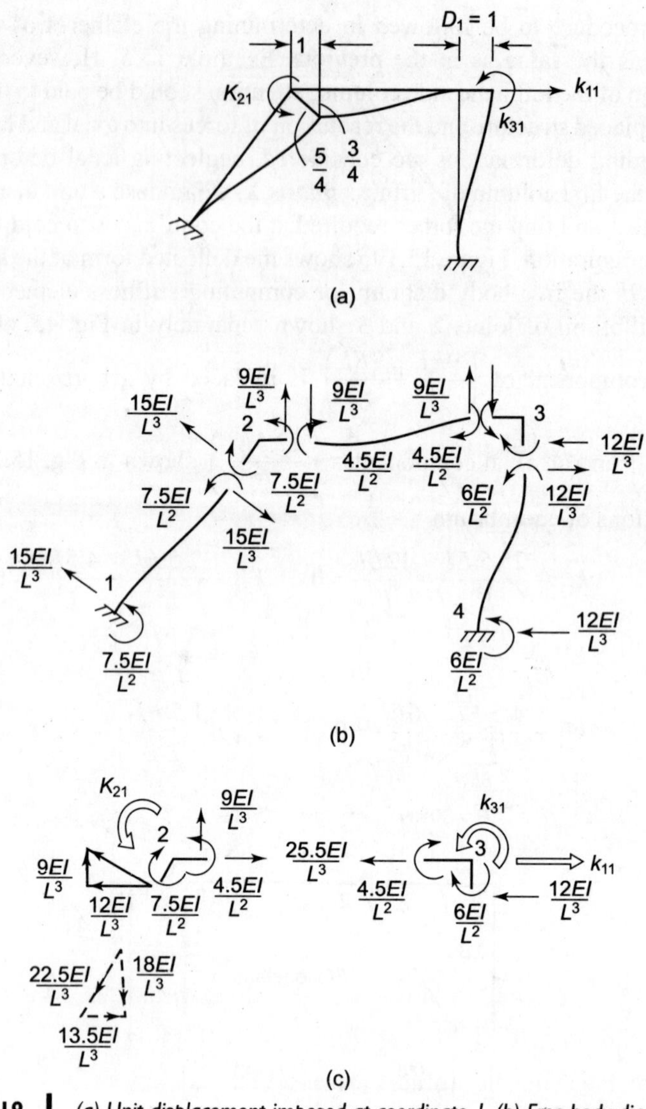

Fig. 15.19 | *(a) Unit displacement imposed at coordinate 1, (b) Free-body diagrams to compute k_{11}, k_{21} and k_{31}, (c) Equilibrium of joints 2 and 3* **(contd.)**

$$k_{12} = \frac{3EI}{L^2} \qquad k_{22} = \frac{8EI}{L} \qquad k_{32} = -\frac{2EI}{L} \tag{15.36}$$

In a similar manner the elements of the third column of matrix **k** can be obtained by imposing a unit displacement at coordinate 3 only and computing the forces required at the coordinates. The values obtained are

$$k_{13} = \frac{1.5EI}{L^2}, \qquad k_{23} = -\frac{2EI}{L}, \qquad k_{33} = \frac{8EI}{L} \tag{15.37}$$

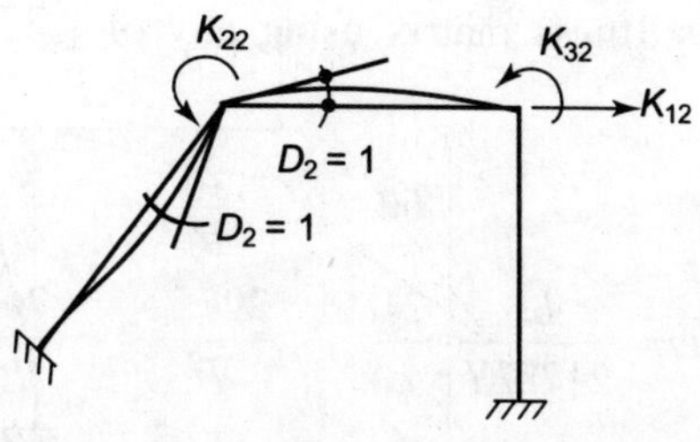

(d)

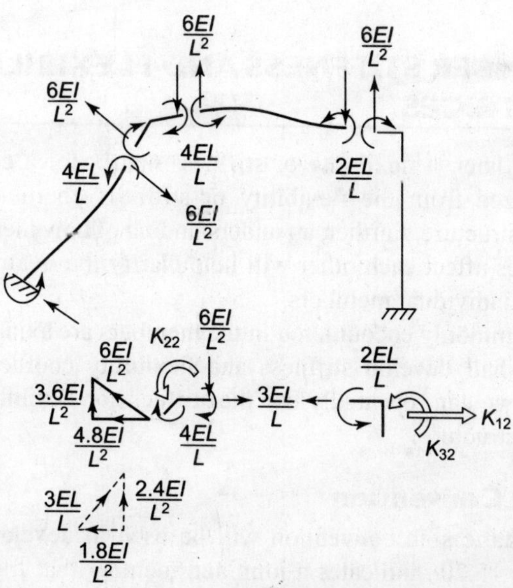

(e)

Fig. 15.19 | ***(Contd.)*** *(d) Unit displacement imposed at coordinate 2, (e) Free-body diagrams to compute k_{12}, k_{22} and k_{32}*

Therefore, the complete stiffness matrix is

$$\mathbf{k} = \frac{EI}{L}\begin{bmatrix} \frac{37.5}{L^2} & \frac{3.0}{L} & \frac{1.5}{L} \\ \frac{3.0}{L} & 8 & -2 \\ \frac{1.5}{L} & -2 & 8 \end{bmatrix} \tag{15.38}$$

Now the displacements due to the given load can be found using Eq. 15.19

$$\mathbf{P} = \mathbf{kD}$$

or

$$\mathbf{D} = \mathbf{k}^{-1}\mathbf{P}$$

Inverting the stiffness matrix using any of the standard methods, we can write

$$D = \frac{L^3}{2178EI}\begin{bmatrix} 60 & \dfrac{21}{L} & \dfrac{6}{L} \\ \dfrac{21}{L} & \dfrac{297.75}{L^2} & -\dfrac{70.5}{L^2} \\ \dfrac{6}{L} & -\dfrac{70.5}{L^2} & \dfrac{291}{L^2} \end{bmatrix}\begin{Bmatrix} 37.5 \\ 0 \\ 0 \end{Bmatrix} \tag{15.39}$$

15.3 MEMBER STIFFNESS AND FLEXIBILITY MATRICES

As we shall see later, a flexibility or stiffness matrix for a complete structure can be syntherized from the flexibility or stiffness coefficients of members constituting the structure. Further, an understanding of how member deformation or member forces affect each other will help clarify the treatment of structures assembled from individual members.

The forces commonly encountered in the members are axial, bending, torsion and shear. We shall develop stiffness and flexibility coefficients for a beam element so that we can repeatedly use the stiffness or flexibility matrix for the members in the structure.

15.3.1 Sign Convention

The following static sign convention will be used in developing the member matrices. Figure 15.20*a* indicates a joint and member that frames into it along with the internal forces all in their positive sense. Figure 15.20*b* indicates a set of external joint forces and Fig. 15.20*c* gives the displacements all in their positive sense.

It may be noted that the moment M_{12} at the end 1 of member 1-2 is counter-clockwise while the same moment on the joint is clockwise. In ternal forces acting on either end of a member are positive in the positive direction of coordinate axes. Couples are included in this convention by using the right hand screw rule.

The positive directions of internal member end forces, external joint forces and nodal displacements coincide.

We shall generate a member stiffness matrix for a simple beam element shown in Fig. 15.21*a*. Figure 15.21*b* gives the beam element in its general deformed state.

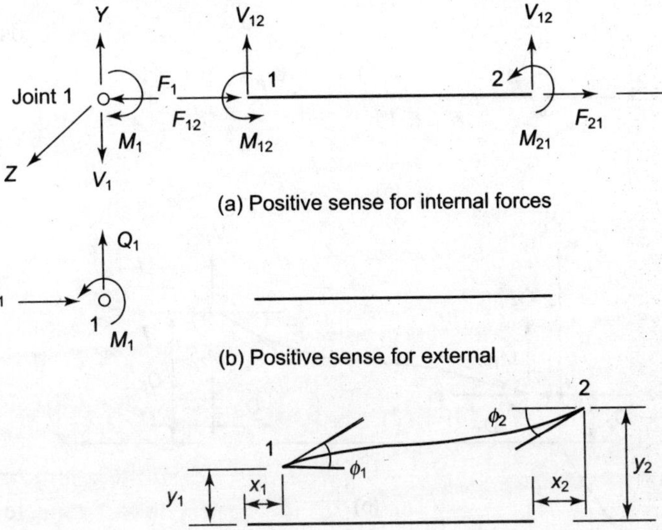

(a) Positive sense for internal forces

(b) Positive sense for external

(c) Positive sense for displacements

Fig. 15.20 | *Illustration of sign convention (a) Positive sense for internal forces, (b) Positive sense for external forces, (c) Positive sense for displacements*

The stiffness matrix to be developed is of the form

$$\begin{Bmatrix} F_{12} \\ V_{12} \\ M_{12} \\ F_{21} \\ V_{21} \\ M_{21} \end{Bmatrix} = \begin{bmatrix} k_{11} & k_{12} & k_{13} & k_{14} & k_{15} & k_{16} \\ k_{21} & k_{22} & k_{23} & k_{24} & k_{25} & k_{26} \\ k_{31} & k_{32} & k_{33} & k_{34} & k_{35} & k_{36} \\ k_{41} & k_{42} & k_{43} & k_{44} & k_{45} & k_{46} \\ k_{51} & k_{52} & k_{53} & k_{54} & k_{55} & k_{56} \\ k_{61} & k_{62} & k_{63} & k_{64} & k_{65} & k_{66} \end{bmatrix} \begin{Bmatrix} D_1 \\ D_2 \\ \phi_1 \\ D_4 \\ D_5 \\ \phi_2 \end{Bmatrix} \tag{15.40}$$

$$\mathbf{P} = \mathbf{k} \quad \mathbf{D} \tag{15.41}$$

Although there are several ways in which the stiffness influence coefficients can be developed, we shall develop the elements from the basic definition of the stiffness element k_{ij}. The beam element has six displacement directions corresponding to the six degrees of freedom for the member ends as defined in Fig. 15.21*b*. The stiffness matrix **k** is generated by imposing a unit displacement at each of the degrees of freedom in turn and computing the forces required at all the coordinate points. Fig. 15.22 indicates the unit displacement given to each degree of freedom (d.o.f) with the corresponding forces and couples required to impose the displacement.

The vectors are shown in the assumed positive direction and the minus sign indicates that the force actually acts opposite to the direction indicated. The member stiffness matrix may be written directly from the information

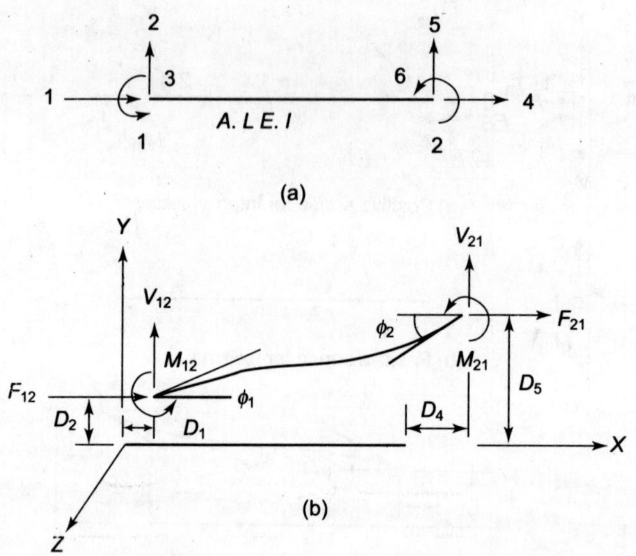

Fig. 15.21 | *(a) Beam element with coordinates, (b) General deformation of a beam element*

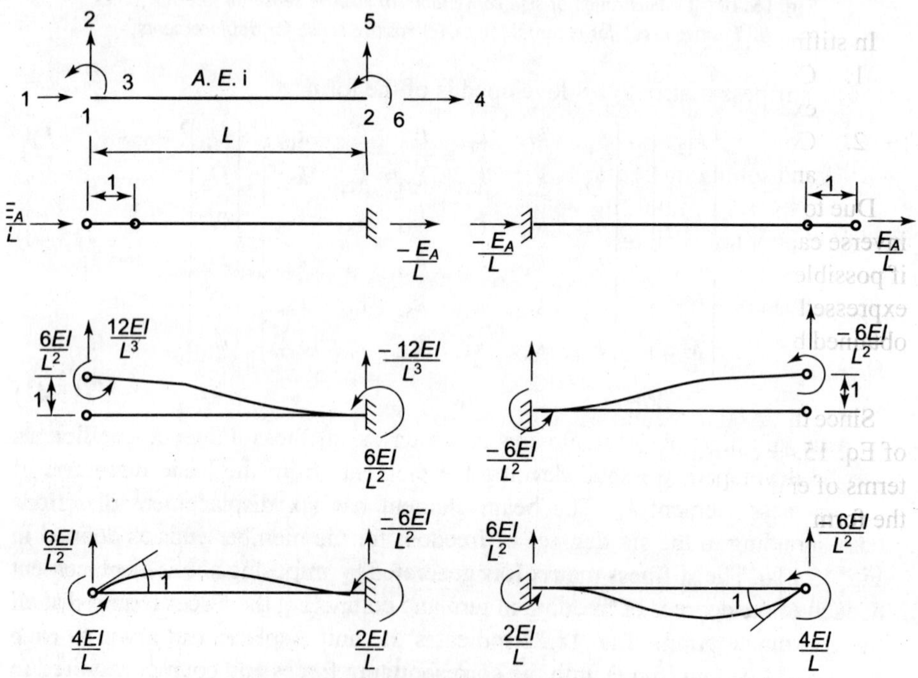

Fig. 15.22 | *Member stiffness influence coefficients*

displayed in Fig. 15.22. Writing the end forces in terms of the stiffness matrix and displacement vector

$$\qquad\qquad 1 \qquad\quad 2 \qquad\quad 3 \qquad\quad 4 \qquad\quad 5 \qquad\quad 6$$

$$\begin{Bmatrix} F_{12} \\ V_{12} \\ M_{12} \\ F_{21} \\ V_{21} \\ M_{21} \end{Bmatrix} = \begin{bmatrix} \frac{EA}{L} & 0 & 0 & -\frac{EA}{L} & 0 & 0 \\ 0 & \frac{12EI}{L^3} & \frac{6EI}{L^2} & 0 & -\frac{12EI}{L^3} & \frac{6EI}{L^2} \\ 0 & \frac{6EI}{L^2} & \frac{4EI}{L} & 0 & -\frac{EI}{L^2} & \frac{2EI}{L} \\ -\frac{EA}{L} & 0 & 0 & \frac{EA}{L} & 0 & 0 \\ 0 & -\frac{12EI}{L^3} & -\frac{6EI}{L^2} & 0 & \frac{12EI}{L^3} & -\frac{6EI}{L^2} \\ 0 & \frac{6EI}{L^2} & \frac{2EI}{L} & 0 & -\frac{6EI}{L^2} & \frac{4EI}{L} \end{bmatrix} \begin{Bmatrix} D_1 \\ D_2 \\ \phi_1 \\ D_4 \\ D_5 \\ \phi_2 \end{Bmatrix} \qquad (15.42a)$$

$$\mathbf{P} = \mathbf{k} \quad \mathbf{D} \qquad (15.42b)$$

In stiffness matrix **k** the following can be noticed:

1. Column 4 is identical to column 1 and column 5 identical to column 2 except for their signs being reversed.
2. Column 6 can be obtained by multiplying elements of column 5 by *(-L)* and subtracting column 3 from it.

Due to these identities, the value of the determinant is zero and, therefore, the inverse cannot be obtained. For this reason and also because it is often desirable, if possible, to work with a reduced size of matrix **k**, the stiffness relation can be expressed in terms of a 3 × 3 matrix. The reduced size of the **k** matrix can be obtained by considering

$$F_{12} = -F_{21},\ V_{12} = -V_{21} \text{ and } M_{21} = M_{12} - V_{21}(L) \qquad (15.43)$$

Since the axial force and shear are constant along the beam and the relationship of Eq. 15.43 can serve as a supplemental equation for determining the shear in terms of end moments, it is sufficient to rewrite the relationship of Eq. 15.42a in the form

$$\begin{Bmatrix} F_{12} \\ M_{12} \\ M_{21} \end{Bmatrix} = \begin{bmatrix} \frac{EA}{L} & 0 & 0 & -\frac{EA}{L} & 0 & 0 \\ 0 & \frac{6EA}{L^2} & \frac{4EA}{L} & 0 & -\frac{EI}{L^2} & \frac{2EI}{L} \\ 0 & \frac{6EA}{L^2} & \frac{2EI}{L} & 0 & -\frac{6EI}{L^2} & \frac{4EI}{L} \end{bmatrix} \begin{Bmatrix} D_1 \\ D_2 \\ \phi_1 \\ D_4 \\ D_5 \\ \phi_2 \end{Bmatrix} \qquad (15.44)$$

The deformations due to axial strains at the coordinates 1 and 4 can be sufficiently described by the extension as

$$e_{12} = -(D_1 - D_4) \tag{15.45}$$

The deflections of a member due to bending can be adequately described by rotations θ_1 and θ_2 as shown in Fig. 15.20 where end rotations are referred to with respect to the chord joining the ends rather than the coordinate axes, that is

$$\phi_1 = \theta_1 + \frac{(D_5 - D_2)}{L} \tag{15.46}$$

and

$$\phi_2 = \theta_2 + \frac{(D_5 - D_2)}{L} \tag{15.47}$$

Therefore,

$$F = -\frac{EA}{L}(D_1 - D_4) = \frac{EA}{L}(e) \tag{15.48}$$

Using force displacement relationships given in Appendix D and superimposing the effects, we have

$$M_{12} = \frac{6EID_2}{L^2} - \frac{6EID_5}{L^2} + \frac{4EI\phi_1}{L} + \frac{4EI\phi_2}{L}$$

$$= \frac{6EI}{L^2}(D_2 - D_5) + \frac{4EI}{L}\left\{\theta_1 + \frac{(D_5 - D_2)}{L}\right\} + \frac{2EI}{L}\left\{\theta_2 + \frac{(D_5 - D_2)}{L}\right\}$$

$$= \frac{4EI\theta_1}{L} + \frac{2EI\theta_2}{L} \tag{15.49}$$

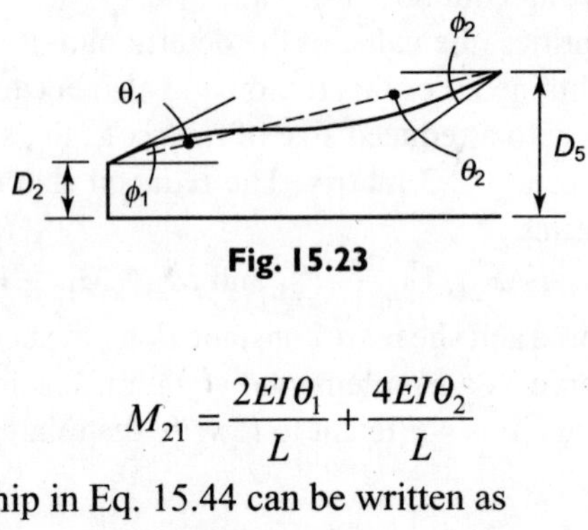

Fig. 15.23

Similarly,

$$M_{21} = \frac{2EI\theta_1}{L} + \frac{4EI\theta_2}{L} \tag{15.50}$$

Hence, the relationship in Eq. 15.44 can be written as

$$\begin{Bmatrix} F \\ M_{12} \\ M_{21} \end{Bmatrix} = \begin{bmatrix} \frac{EA}{L} & 0 & 0 \\ 0 & \frac{4EA}{L} & \frac{2EA}{L} \\ 0 & \frac{2EA}{L} & \frac{4EA}{L} \end{bmatrix} \begin{Bmatrix} e \\ \theta_1 \\ \theta_2 \end{Bmatrix} \tag{15.51}$$

$$\mathbf{P} = \mathbf{k} \quad \mathbf{D} \tag{15.52}$$

In this form stiffness matrix **k** is expressed as a 3 × 3 matrix. It may be noted that the axial force is uncoupled from the moments, that is, the axial effect does not influence the moment.

If only bending deformations are predominant so that the axial effect can be neglected, matrix **k** can be written as

$$\mathbf{k} = \frac{2EI}{L}\begin{bmatrix} 2 & 1 \\ 1 & 2 \end{bmatrix} \tag{15.53}$$

The flexibility matrix for the member can be obtained by inverting the stiffness matrix. Thus,

$$f = \frac{L}{6EI}\begin{bmatrix} 2 & -1 \\ -1 & 2 \end{bmatrix} \tag{15.54}$$

The stiffness and flexibility of a member due to twist can be developed in a manner similar to that used for axial loading. Torsional force like an axial force is uncoupled and does not influence the other forces. For a circular member, the deformation resulting from twist is shown in Fig. 15.24.

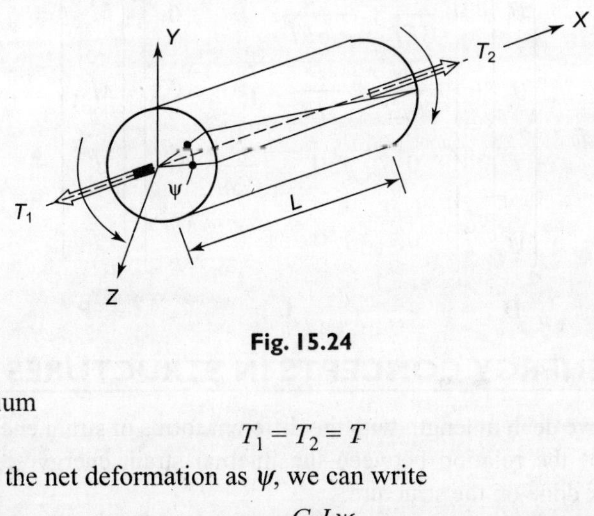

Fig. 15.24

For equilibrium

$$T_1 = T_2 = T$$

Denoting the net deformation as ψ, we can write

$$T = \frac{GJ\psi}{L}$$

where

J = polar moment of inertia

and G = modulus of elasticity in shear.

The stiffness relationship for the member can be written as

$$T = k\psi \tag{15.55}$$

where

$k = \dfrac{GJ}{L}$, stiffness of a circular member subjected to twisting moment.

The flexibility of the member due to twist is

$$f = \frac{L}{GJ} \tag{15.56}$$

Thus, to summaries, for a member subjected to end moments, axial and twist forces, the general stiffness matrix can be written as

$$\begin{Bmatrix} M_{12} \\ M_{21} \\ F \\ T \end{Bmatrix} = \begin{bmatrix} \frac{4EI}{L} & \frac{2EI}{L} & 0 & 0 \\ \frac{2EI}{L} & \frac{4EI}{L} & 0 & 0 \\ 0 & 0 & \frac{EA}{L} & 0 \\ 0 & 0 & 0 & \frac{GJ}{L} \end{bmatrix} \begin{Bmatrix} \theta_1 \\ \theta_2 \\ e \\ \psi \end{Bmatrix} \tag{15.57}$$

$$\mathbf{P} = \mathbf{k} \quad \mathbf{D} \tag{15.58}$$

The corresponding flexibility matrix can be written as

$$\begin{Bmatrix} \theta_1 \\ \theta_2 \\ e \\ \psi \end{Bmatrix} = \begin{bmatrix} \frac{L}{3EI} & -\frac{L}{6EI} & 0 & 0 \\ -\frac{L}{6EI} & \frac{L}{3EI} & 0 & 0 \\ 0 & 0 & \frac{L}{EA} & 0 \\ 0 & 0 & 0 & \frac{GJ}{L} \end{bmatrix} \begin{Bmatrix} M_{12} \\ M_{21} \\ F \\ T \end{Bmatrix} \tag{15.59}$$

$$\mathbf{D} = \quad \mathbf{f} \quad \mathbf{P} \tag{15.60}$$

15.4 ENERGY CONCEPTS IN STRUCTURES

In Chapter 6 we dealt at length with the different forms of strain energy in elastic structures and the relation between the internal strain energy stored and the external work done on the structure.

Using tools of matrix algebra, we shall now derive some important concepts relating to strain energy and properties of stiffness and flexibility matrices of structures.

15.4.1 Symmetry Property of the Stiffness and Flexibility Matrices

It has been pointed out in Section 6.6 that the strain energy depends only on the final deflected shape of the structure and is independent of the order of loading. Making use of this property we can prove that the stiffness and flexibility matrices are symmetrical. Let us show it by means of a simple example.

The beam in Fig. 15.25 is being acted upon by two systems of forces P_i and Pj causing displacements D_i, and D_j. For the purpose of clarity only two loads are shown.

Since the sequence of loading has no bearing on the final value of strain energy U, the same strain energy will be obtained in the following two sequences of loading in which the forces are gradually applied.

In loading sequence I, consider that only load P_i is gradually applied first. The displacement caused by this load is indicated in Fig. 15.26*a*. The work done by P_i can be written as

$$\frac{1}{2}(P_i)(f_{ii}P_i) = \frac{1}{2}f_{ii}P_i^2$$

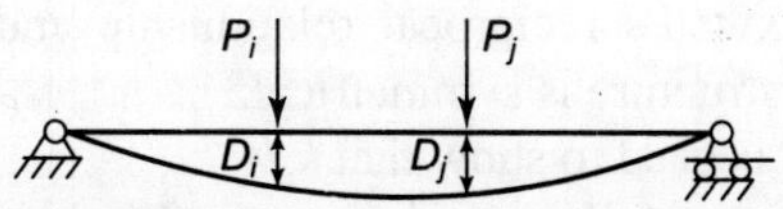

Fig. 15.25 | *Deflections due to loads P_i and P_j*

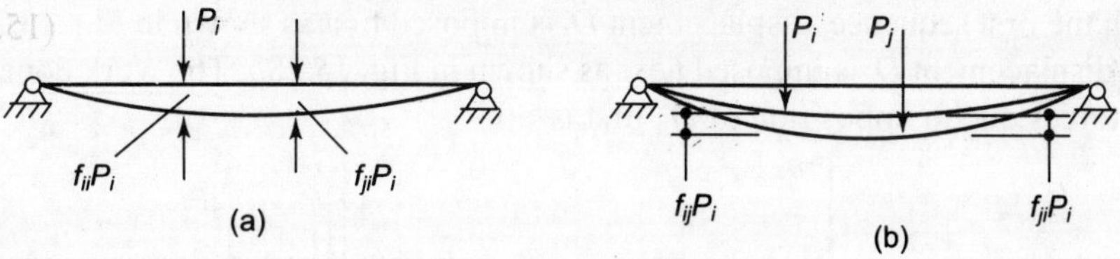

Fig. 15.26 | *(a) Load P_i is applied first; deflections due to load p_i, (b) Load P_j is applied next; deflections due to loads P_i and P_j*

Next apply P_j gradually which results in additional displacements shown in Fig. 15.26*b*. The corresponding work done by forces P_i and P_j during this operation is

$$f_{ii}P_iP_j + \frac{1}{2}f_{jj}P_j^2$$

It may be noted that there is no coefficient (1/2) in the first term because P_i rides in full through the displacement caused by P_j. The total work done or strain energy in loading sequence I is

$$U_I = \frac{1}{2}f_{ii}P_i^2 + f_{ij}P_iP_j + \frac{1}{2}f_{ji}P_j^2 \tag{15.61}$$

Next consider loading sequence II in which P_j is applied gradually first and then P_i. The displacement caused due to loading P_j alone is shown in Fig. 15.27*a* and under P_j and P_i together is shown in Fig. 15.27*b*.

The work done when P_i is applied first is

$$\frac{1}{2}(P_j)(f_{jj}P_j) = \frac{1}{2}f_{ij}P_j^2$$

The work done when P_i is applied gradually next is

$$f_{ji}P_iP_j + \frac{1}{2}f_{ii}P_i^2$$

The total work done or strain energy stored in loading sequence II is

$$U_{II} = \frac{1}{2}f_{ii}P_i^2 + f_{ji}P_iP_j + \frac{1}{2}f_{jj}P_j^2 \quad (15.62)$$

Since $U_I = U_{II}$ we find from Eqs. 14.61 and 14.62

$$f_{ij} = f_{ji} \quad (15.63)$$

This is known as Maxwell's reciprocal relationship and it indicates that the flexibility matrix for a structure is symmetrical.

A similar procedure is used to show that $k_{ij} = k_{ji}$.

We use the following two displacement sequences so that the displacements D_i and D_j are imposed to result in forces P_i and P_j.

In the first sequence, displacement D_i is imposed first as shown in Fig. 15.28*a* and displacement D_j is imposed next as shown in Fig. 15.28*b*. The work done by the forces due to imposition of D_i, first is

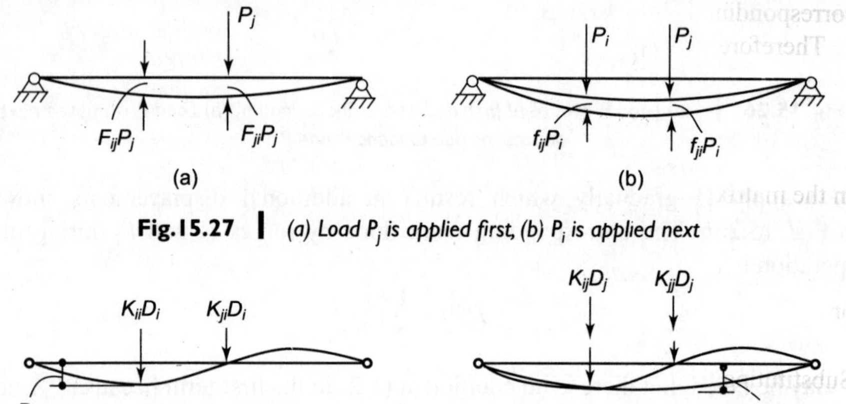

Fig.15.27 | *(a) Load P_j is applied first, (b) P_i is applied next*

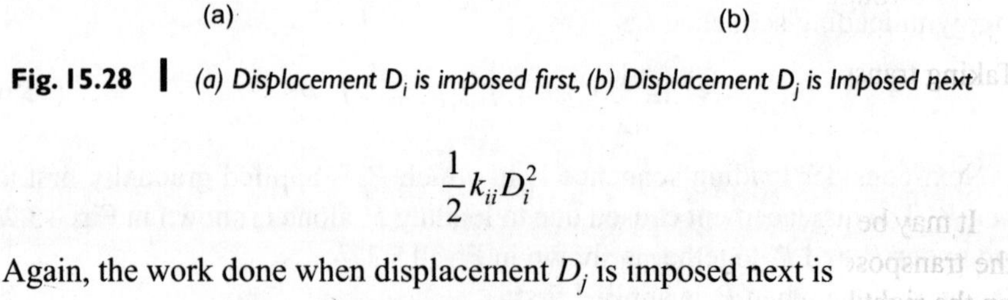

Fig. 15.28 | *(a) Displacement D_i is imposed first, (b) Displacement D_j is Imposed next*

$$\frac{1}{2}k_{ii}D_i^2$$

Again, the work done when displacement D_j is imposed next is

$$k_{ij}D_iD_j + \frac{1}{2}k_{jj}D_j^2$$

It may be noted that the force $k_{ji}D_i$ rides in full through displacement D_j. The total work done in sequence *I* is

$$U_I = \frac{1}{2}k_{ii}D_i^2 + k_{ji}D_iD_j + \frac{1}{2}k_{jj}D_j^2 \tag{15.64}$$

Similarly, in sequence II we impose displacement D_j first keeping $D_i = 0$ and then impose displacement D_i holding D_j at its value. This yields

$$U_{II} = \frac{1}{2}k_{ii}D_i^2 + k_{ij}D_iD_j + \frac{1}{2}k_{jj}D_j^2 \tag{15.65}$$

Since $U_I = U_{II}$, we find from Eqs. 15.64 and 15.65

$$kij = kji \tag{15.66}$$

which indicates that the stiffness matrix of a structure is symmetrical.

15.4.2 Strain Energy in Terms of Stiffness and Flexibility Matrices

We studied in Section 6.1 that when a linear elastic structure is acted upon by a number of forces $P_i(i = 1, 2, ..., n)$ applied gradually, the strain energy U, stored in the structure is equal to the work done by these forces in moving through the corresponding D_i $(i = 1, 2, ..., n)$.

Therefore,

$$U = W_e = \frac{1}{2}\sum_{i=1}^{n} P_iD_i \tag{15.67}$$

In the matrix notation

$$U = \frac{1}{2}\mathbf{P}^\mathbf{T}\mathbf{D} \tag{15.68}$$

or

$$U = \frac{1}{2}\mathbf{D}^\mathbf{T}\mathbf{P} \tag{15.69}$$

Substituting for P from Eq. 15.19

$$U = \frac{1}{2}\mathbf{D}^\mathbf{T}\mathbf{k}\,\mathbf{D} \tag{15.70}$$

Taking transpose on either side, we get

$$U = \frac{1}{2}\mathbf{D}^\mathbf{T}\,\mathbf{k}^\mathbf{T}\,\mathbf{D} \tag{15.71}$$

It may be noted that the left hand side of Eq. 15.70 remains unchanged becuase the transpose of a scalar quantity does not alter its value. The operation carried on the right hand side is according to the reversal law of transpose. We conclude from Equations 15.70 and 15.71 that

$$\mathbf{k}^\mathrm{T} = \mathbf{k} \tag{15.72}$$

Again consider Eq. 15.68. Substituting for **D** from Eq. 15.17, we have

$$U = \frac{1}{2}\mathbf{P}^\mathbf{T}\,\mathbf{f}\,\mathbf{P} \tag{15.73}$$

Taking transpose on either side results

$$U = \frac{1}{2}\mathbf{P}^{\mathrm{T}}\,\mathbf{f}^{\mathrm{T}}\,\mathbf{P} \tag{15.74}$$

From Equations 15.73 and 15.74, we conclude

$$\mathbf{f}^{\mathrm{T}} = \mathbf{f} \tag{15.75}$$

15.4.3 Stiffness and Flexibility Coefficients in Terms of Strain Energy

Stiffness Coefficients Let us consider Equation 15.70

$$U = \frac{1}{2}\mathbf{D}^{\mathrm{T}}\,\mathbf{k}\,\mathbf{D}$$

In this equation, each element of k_{ij} multiplies D_i of the left row vector and D_j of the right column vector, then all the products are added and the sum multiplied by 1/2.

If we take a partial derivative of strain energy U with respect to any displacement D_i in Eq. 15.70, then on the right hand side of the equation, only the terms in which D_i appear will contribute to this partial derivative. From Eq. 15.70 it is seen that the terms in which D_i appear are associated with l row and l column of the stiffness element k_{ij} Taking the symmetric property $k_{ij} = k_{ji}$ the sum of these terms is given by

$$\frac{1}{2}(2k_{l1}D_lD_1 + 2k_{l2}D_lD_2 + \cdots + k_{ll}D^2{}_l + \cdots 2k_{ln}D_lD_n) \tag{15.76}$$

Taking a partial derivative with respect to D_i this sum becomes

$$k_{l1}D_1 + k_{l2}D_2 + \cdots + k_{ll} + D_l + \cdots + k_{ln}\,D_n = \sum_{j=1}^{n} k_{lj}D_j \tag{15.77}$$

or

$$\frac{\partial U}{\partial D_l} = \sum_{j=1}^{n} k_{lj}D_j \tag{15.78}$$

The right hand side of Eq. 15.78 is equal to force P_l at coordinate l (see Equation 15.20), hence

$$\frac{\partial U}{\partial D_l} = P_l \tag{15.79}$$

This, we are familiar, is Castigliano's first theorem. If we now take the partial derivative with respect to any D_s in Eq. 15.77, we have

$$\frac{\partial^2 U}{\partial D_l\,\partial D_s} = k_{1s} \qquad \begin{array}{l} l = 1, 2, \ldots, n \\ s = 1, 2, \ldots, n \end{array} \tag{15.80}$$

Hence, in general, cross stiffness coefficient k_{ls} is equal to the second partial derivative of the strain energy with respect to displacements at l and s.

Direct stiffness coefficient k_{ll} is obtained by taking the partial derivative of Eq. 15.77 with respect to D_l, that is

$$\frac{\partial^2 U}{\partial D_l^2} = k_{ll} \tag{15.81}$$

Flexibility Coefficients Starting with Eq. 15.73

$$U = \frac{1}{2}\mathbf{P}^{\mathrm{T}}\mathbf{f}\mathbf{P}$$

and proceeding as earlier, but taking a partial derivative of Equation 15.73 with respect to any force P_l (with all the P_i considered to be independent) and making use of the reciprocal relationship $f_{ij} = f_{ji}$ we obtain

$$\frac{\partial U}{\partial P_l} = \sum_{j=1}^{n} f_{lj} P_j \tag{15.82}$$

The right hand side of Eq. 15.82 is equal to displacement D_l at coordinate l (see Eq. 15.18), hence

$$\frac{\partial U}{\partial P_l} = D_l \tag{15.83}$$

If we take the partial derivative with respect to any force P_s in Eq. 15.82, we have

$$\frac{\partial^2 U}{\partial P_l \partial P_s} = f_{ls}$$

$$l = 1, 2, \ldots, n;\ s = 1, 2, \ldots n \tag{15.84}$$

The partial derivative of Equation 15.82 with respect to P_l gives

$$\frac{\partial^2 U}{\partial P^2} = f_{ll} \tag{15.85}$$

15.5 MAXWELL'S AND BETTI'S RECIPROCAL DEFLECTIONS

Maxwell-Betti's reciprocal relationships were developed in Section 6.6. However, it is of interest to establish the relationship using the matrix relationship as follows.

Consider the two identical elastic structures of Fig. 15.29*a* and *b*. The applied forces are arbitrary in the two structures and we designate them as systems I and II respectively. Let us fix the coordinates for systems I and II as shown in Fig. 15.29*c*.

Let **f** be the flexibility matrix corresponding to the coordinates. For convenience of reference we designate the forces and displacements in Fig. 15.29*a* as $\mathbf{P}_{\mathrm{I}}$ and $\mathbf{D}_{\mathrm{I}}$ and in Fig. 15.29*b* as $\mathbf{P}_{\mathrm{II}}$ and $\mathbf{D}_{\mathrm{II}}$ respectively. For system **I** in Fig. 15.29*a* we can write

$$\mathbf{D}_{\mathrm{I}} = \mathbf{f}\,\mathbf{P}_{\mathrm{I}} \tag{15.86}$$

and for system **II** we write

$$\mathbf{D}_{\text{II}} = \mathbf{f}\,\mathbf{P}_{\text{II}} \tag{15.87}$$

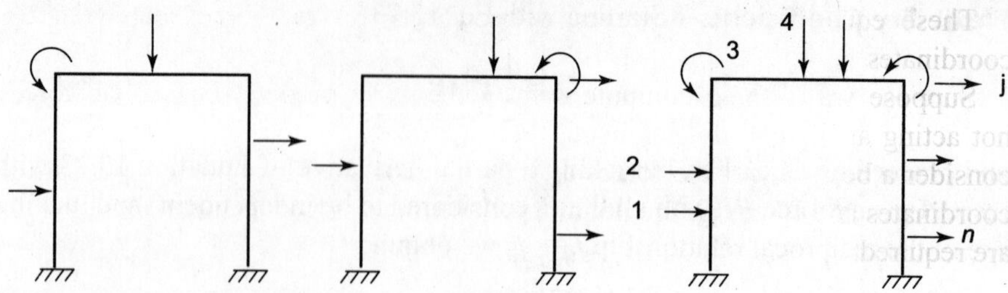

(a) (b) (c)

Fig. 15.29 | *(a) Loading system I, (b) Loading system II, (c) Loading systems I and II and structure coordinates*

If we now compute the product $\mathbf{P}^{\text{T}}_{\text{I}}\,\mathbf{D}_{\text{II}}$, (the forces in system I and the displacements in system II) substituting for $\mathbf{D}_{\text{II}}$ from Eq. 15.87, we have

$$\mathbf{P}^{\text{T}}_{\text{I}}\,\mathbf{f}\mathbf{P}_{\text{II}} \tag{15.88}$$

Again, finding product $\mathbf{P}^{\text{T}}_{\text{II}}\,\mathbf{D}_{\text{I}}$ (the forces in system II and the displacements in system I) and substituting for D_I from Eq. 15.86, we get

$$\mathbf{P}^{\text{T}}_{\text{II}}\,\mathbf{f}\mathbf{P}_{\text{I}} \tag{15.89}$$

This is a scalar quantity and the transposition of it does not effect the value. Taking the transpose of Eq. 15.89 it can be written as

$$\mathbf{P}^{\text{T}}_{\text{I}}\,\mathbf{f}\mathbf{P}_{\text{II}} \tag{15.90}$$

This is identical with Eq. 15.88. Hence, we conclude

$$\mathbf{P}^{\text{T}}_{\text{I}}\,\mathbf{D}_{\text{I}} = \mathbf{P}^{\text{T}}_{\text{II}}\,\mathbf{D}_{\text{I}} \tag{15.91}$$

The relationship expressed by Eq. 15.91 is known as Betti's law. This may be stated as: *For a linear elastic structure subjected to two different force systems I and II, the work done by the forces in system I acting through the corresponding displacements (virtual) in system II, is equal to the work done by the forces in system II acting through the displacements (virtual) in system I.*

15.5.1 Application of Betti's Law

Betti's law can be made use of to deal with structures in which forces are not acting at the coordinates.

As an example consider the beam element with the coordinates defined as shown in Fig. 15.30. Considering only bending deformations, the stiffness matrix k is

$$\mathbf{k} = \frac{EI}{L}\begin{bmatrix} 4 & 2 \\ 2 & 4 \end{bmatrix}$$

The force displacement relationship is

$$\mathbf{P} = \mathbf{kD}$$

or
$$\mathbf{D} = \mathbf{k}^{-1}\mathbf{P}$$

These equations are, however, defined only for the forces acting at the coordinates

Suppose we wish to compute displacements D_1 and D_2 caused by forces not acting at the coordinates using the same stiffness matrix. As an example, consider a beam under the forces as shown in Fig. 15.31a. Forces P^f_i are at the coordinates and forces Qi are not at the coordinates. Displacements D_i $(i = 1, 2)$ are required.

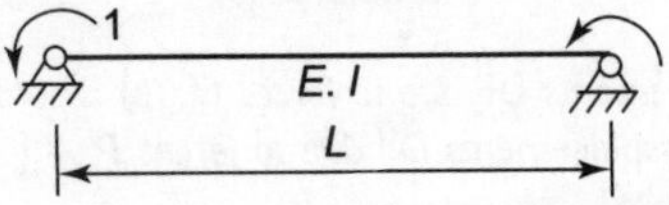

Fig. 15.30 | *Beam element and coordinates*

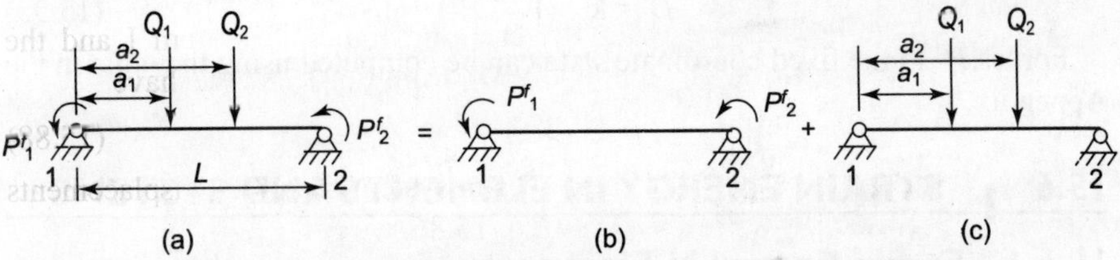

Fig.15.31 | *(a) Forces and coordinates, (b) Forces at coordinates, (c) Forces not at coordinates*

The forces on the beam can be separated into forces acting at the coordinates (Fig. 15.31*b*) and forces not acting at the coordinates (Fig. 15.31*c*). The displacements due to forces at coordinates pose no problem and consideration has to be only given to forces not at the coordinates.

In Fig. 15.32*a* we apply a superposition of displacements to the structure. In Fig. 15.32*b* we represent the fixed coordinate state in which no displacements are permitted at the coordinates, whereas in Fig. 15.32*c* forces P_1 and P_2 are such as to produce, displacements corresponding to the ones in Fig. 15.32*a*. The forces at the coordinates are shown in their positive direction. Since the forces at the coordinates in Fig. 15.32*a* are zero, it, therefore, follows that the sum of corresponding forces in Fig. 15.32*a* and 15.32*c* must be zero. Thus,

$$P^o_i + P_i = 0 \tag{15.92}$$

or
$$P_i = -P^o_i \tag{15.93}$$

It may be remembered that forces P^o_i in Fig. 15.32*b* are fixed end moments. Forces P, in Fig. 15.32*c* are equal in magnitude and opposite in sign to the forces at coordinates P^o_i in the fixed coordinate state. Thus, any distributed load or concentrated load applied at other than coordinate points can be replaced by equivalent joint forces given by Equation 15.93.

To obtain displacements D_i due to the loads in Fig. 15.32*a* we add the results in Fig. 15.32*b*, 15.32*b* and *c* which are respectively

$$\mathbf{k}^{-1}\mathbf{P}^f, 0, -\mathbf{k}^{-1}\mathbf{P}^o \tag{15.94}$$

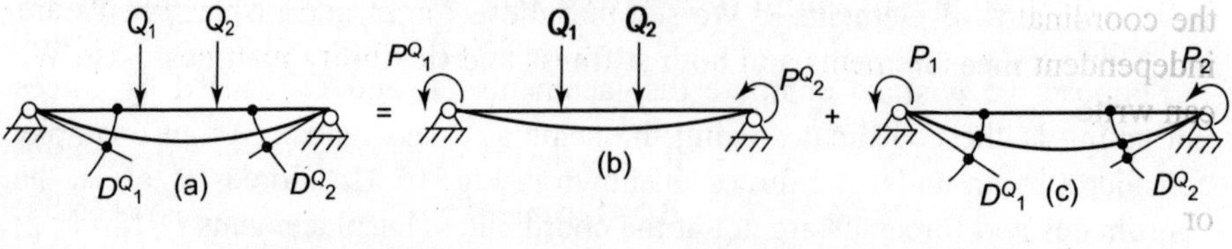

Fig. 15.32 | *(a) Displacements D^Q due to forces Q_i, (b) Displacements $D^Q = 0$ in the fixed coordinate state, (c) Displacements D^Q due to forces $P^f = \{-P\}^o$ at the coordinates*

Final displacements *D* are, therefore,

$$D = \mathbf{k}^{-1}\,(\mathbf{P}^{\mathrm{f}} - P^o) \tag{15.95}$$

Forces P^o in the fixed coordinate state can be computed using the table in the Appendix.

15.6 | STRAIN ENERGY IN ELEMENTS AND SYSTEMS

15.6.1 Strain Energy in Elements

In Section 15.4 we showed that the strain energy in a structure can be expressed in terms of flexibility or stiffness matrices when the forces are applied at the coordinates. Similar expressions can be written for strain energy in any element when the forces are applied at only the coordinates. The only requirement for expressing the strain energy in terms of flexibility or stiffness matrices is that these matrices should exist for the coordinates defined for the element.

Consider the element of Fig. 15.33 with forces p_i and displacements d_i at the two coordinates as shown.

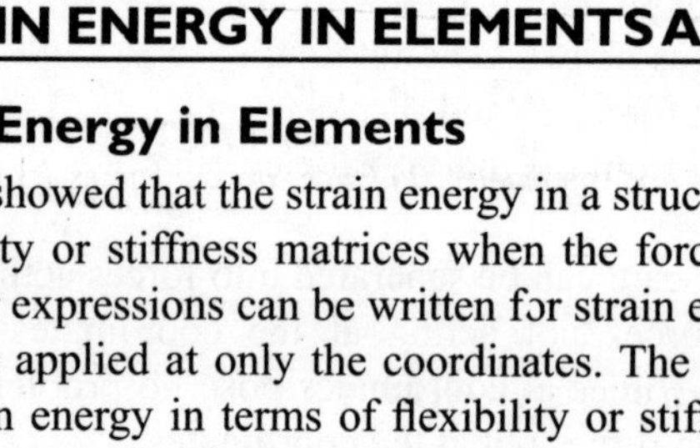

Fig. 15.33 | *Element forces and displacements at coordinates*

We shall identify the element as *s*. The strain energy U_s in this element is equal to the work done by forces p_i in going through corresponding displacements d_i, that is

$$U_s = \frac{1}{2}\mathbf{p}_s^{\mathrm{T}}\mathbf{d}_s \tag{15.96}$$

or
$$U_s = \frac{1}{2}\mathbf{d}_s^{\mathrm{T}}\mathbf{p}_s \tag{15.97}$$

in which $\mathbf{p}_s$ and $\mathbf{d}_s$ are respectively the vectors of forces and displacements at the coordinates of elements s. We see that these forces and displacements are independent measurements and both stiffness and flexibility matrices exist. We can write

$$\mathbf{p}_s = \mathbf{k}_s\,\mathbf{d}_s \tag{15.98}$$

or
$$\mathbf{d}_s = \mathbf{f}_s\,\mathbf{p}_s \tag{15.99}$$

Substituting for $\mathbf{p}_s$ from Equation 15.98 into Equation 15.97, we get

$$U_s = \frac{1}{2}\mathbf{d}_s^{\mathrm{T}}\mathbf{k}_s\mathbf{d}_s \tag{15.100}$$

Again substituting for $\mathbf{d}_s$ from Equation 15.99 into Equation 15.96, we get

$$U_s = \frac{1}{2}\mathbf{p}_s^{\mathrm{T}}\mathbf{f}_s\mathbf{p}_s \tag{15.101}$$

15.6.2 Strain Energy in a System in Terms of Strain Energy in the Elements

Consider the structure of Fig. 15.34 with forces P_i producing displacements D_i at the coordinates as shown. The total strain energy in the structure can be expressed in terms of forces P_i and displacements D_i, as in Equations 15.68 or 15.69 or strain energy from the individual elements as

$$U = \sum_{s=1}^{m} U_s \tag{15.102}$$

in which U_s is the strain energy in element s and m is the total number of elements (m = 6 in this case).

Figure 15.35 shows the internal forces and corresponding displacements at the ends of each element in the structure of Fig. 15.34. Two coordinates are defined for each element to identify forces and displacement as in Fig. 15.35. Applied forces P_i are not included in the figure for clarity.

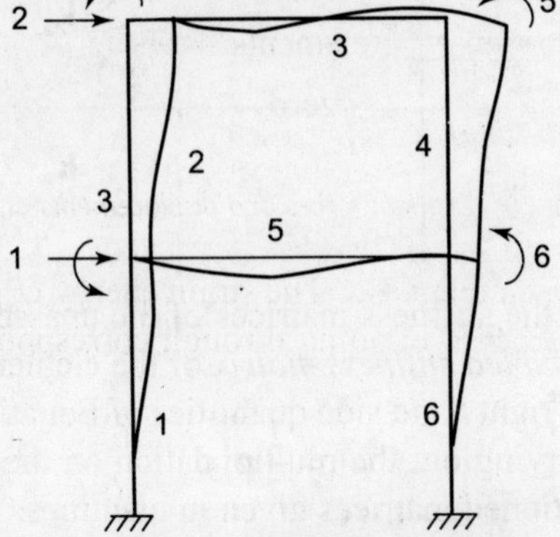

Fig. 15.34 | *Coordinates and forces P_i*

The strain energy U_s, in any element s can be expressed in terms of Equations 15.96 or 15.97. The total strain energy in the structure can be written as

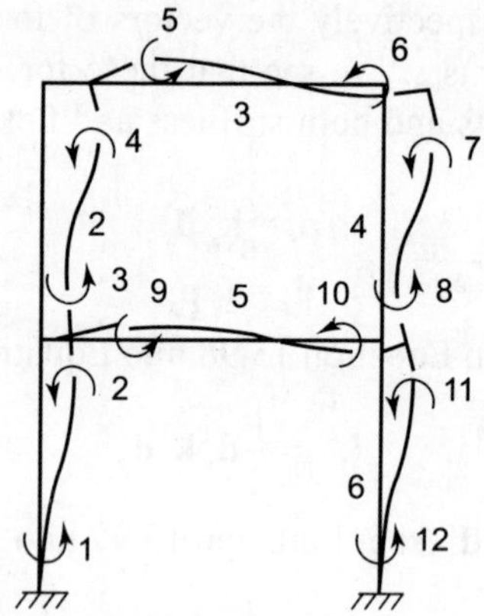

Fig. 15.35 | *Internal forces P_i at element coordinates in the structure*

$$U = \sum_{s=1}^{m} U_s = \frac{1}{2}\sum_{s=1}^{m} \mathbf{d}_s^{\mathrm{T}}\mathbf{p}_s \tag{15.103}$$

or

$$U = \frac{1}{2}\sum_{s=1}^{n} \mathbf{d}_s^{\mathrm{T}}\mathbf{k}_s\mathbf{d}_s \tag{15.104}$$

Equation 15.104 can also be written in the form

$$U = \frac{1}{2}\mathbf{d}^{\mathrm{T}}\mathbf{k}\mathbf{d} \tag{15.105}$$

in which

$$d = \begin{Bmatrix} \mathbf{d}_1 \\ \mathbf{d}_1 \\ \vdots \\ \mathbf{d}_s \\ \vdots \\ \mathbf{d}_m \end{Bmatrix} \text{ and } \mathbf{k} = \begin{bmatrix} \mathbf{k}_1 & & & \text{All other elements zero} \\ & \mathbf{k}_2 & & \\ & \vdots & & \\ \text{All other elements zero} & & \mathbf{k}_s & \\ & & \vdots & \\ & & & \mathbf{k}_m \end{bmatrix} \tag{15.106}$$

Martrix **k** contains the stiffness matrices of the unassembled elements and is referred to as the *uncoupled stiffness matrix* of the elements.

The equality of the right hand side quantities of Equations 15.104 and 15.105 can be verified by carrying out the multiplication on the right hand side of Eq. 15.105 using the partitioned matrices given in identities.

The total strain energy U, in the structure in Fig. 15.31 can also be expressed in terms of the flexibility matrix **f** of the elements, that is

$$U = \frac{1}{2}\sum_{s=1}^{m} \mathbf{p}_s^{\mathrm{T}} \mathbf{d}_s \tag{15.107}$$

or

$$U = \frac{1}{2}\sum_{s=1}^{m} \mathbf{p}_s^{\mathrm{T}} \mathbf{f}_s \, \mathbf{p}_s \tag{15.108}$$

or

$$U = \frac{1}{2}\mathbf{p}^{\mathrm{T}} \mathbf{f} \, \mathbf{p} \tag{15.109}$$

in which

$$\mathbf{p} = \begin{Bmatrix} \mathbf{p}_1 \\ \mathbf{p}_1 \\ \vdots \\ \mathbf{p}_s \\ \vdots \\ \mathbf{p}_m \end{Bmatrix} \text{ and } \mathbf{f} = \begin{bmatrix} \mathbf{f}_1 & & & & \text{All other elements zero} \\ & \mathbf{f}_2 & & & \\ & & \vdots & & \\ \text{All other elements zero} & & & \mathbf{f}_s & \\ & & & & \vdots \\ & & & & \mathbf{f}_m \end{bmatrix} \tag{15.110}$$

The equality of expressions on the right hand side of Equations 15.108 and 15.09 can be verified.

Equations 15.105 and 15.109 are very important in the development of stiffness and flexibility methods of structural analysis which are discussed in Chapters 17 and 18. The reader is advised to make an effort to thoroughly understand the physical significance of these equations.

Problems for Practice

15.1 Compute the 3 × 3 flexibility matrix f considering axial and flexural deformations for the beam shown in Fig. 15.36.

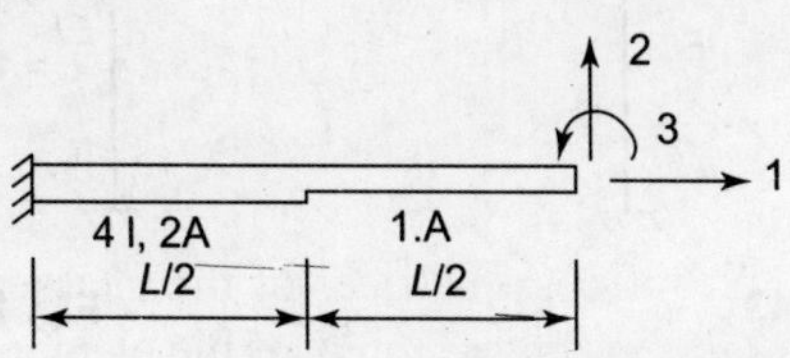

Fig. 15.36

15.2, 15.3 Considering only bending deformation, determine the flexibility matrix f and stiffness matrix **k** associated with the actions shown in Figs. 15.37 and 15.38.

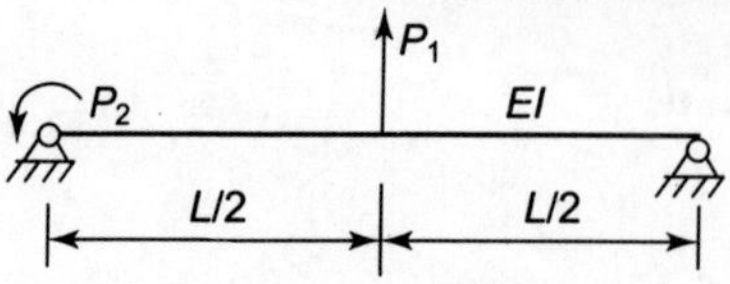

Fig. 15.37

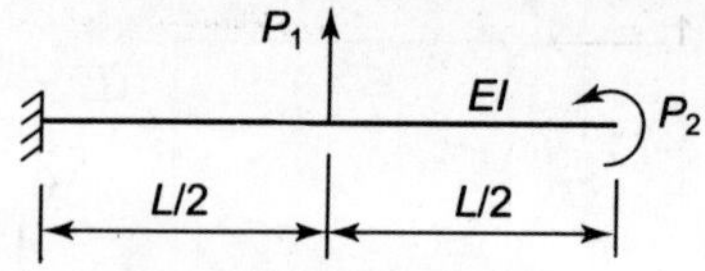

Fig. 15.38

15.4, 15.5, 15.6, 15.7, 15.8 Generate the stiffness matrix **k** for each of the following structures with coordinates as shown in Figs. 15.39, 15.40, 15.41, 15.42 and 15.43.

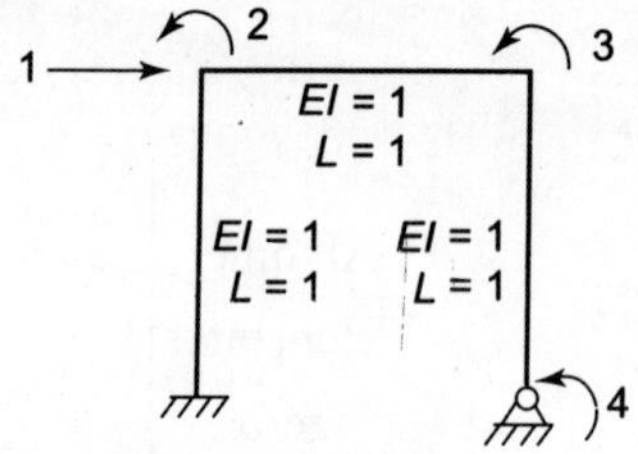

Fig. 15.39

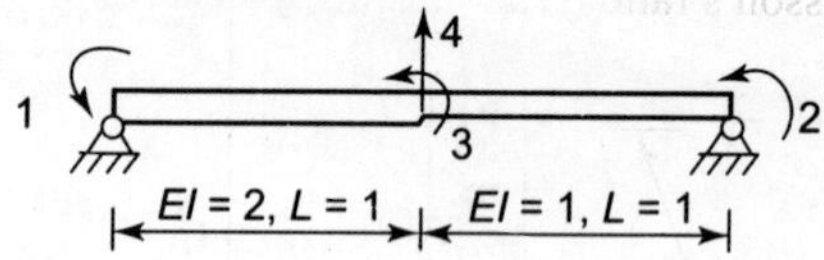

Fig. 15.40

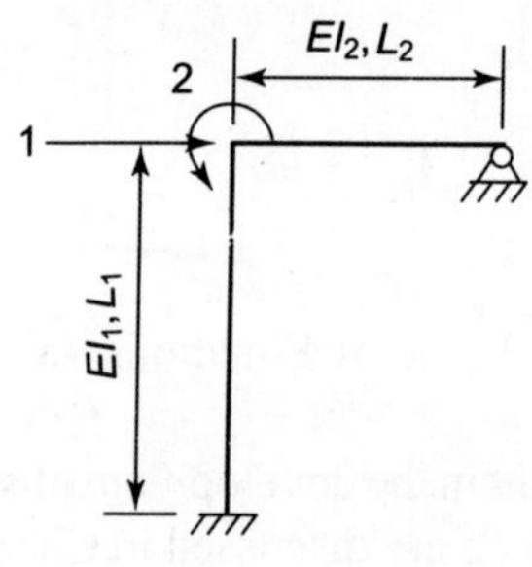

Fig. 15.41

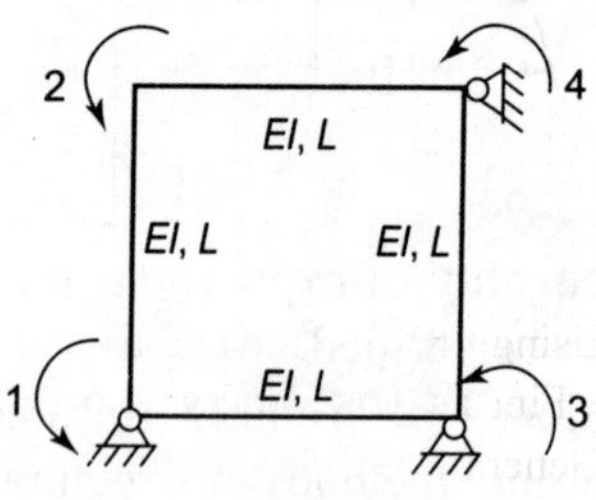

Fig. 15.42

15.9, 15.10 Generate the flexibility matrix f for each of the structures with the coordinates shown in Figs. 15.44 and 15.45.

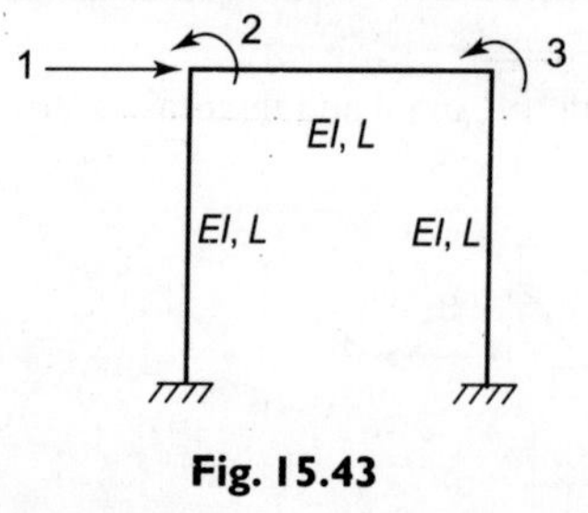

Fig. 15.43

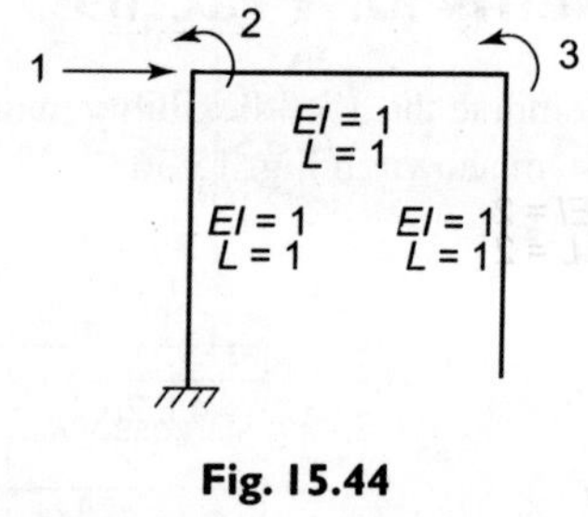

Fig. 15.44

15.11 Generate the flexibility matrix **f** for the stepped beam in Problem 15.5. Check your answer by applying **f k** = **I**.

15.12 Considering only axial deformations for the trusses shown in Figs. 15.46 determine flexibility matrix **f** and stiffness matrix **k** associated with applied forces *P*.

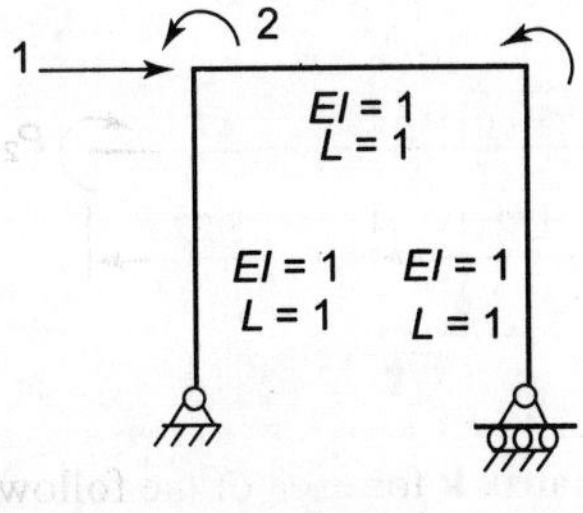

Fig. 15.45

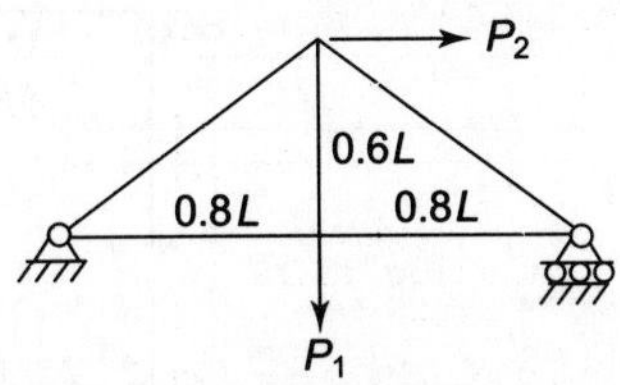

Fig. 15.46

15.13 Calculate flexibility matrix **f** for coordinates 1, 2 and 3 of the pipe bend shown in Fig. 15.47. Consider only the effects of flexural and torsional deformations. Assume Poisson's ratio of 0.5 leading to $G = E/3$.

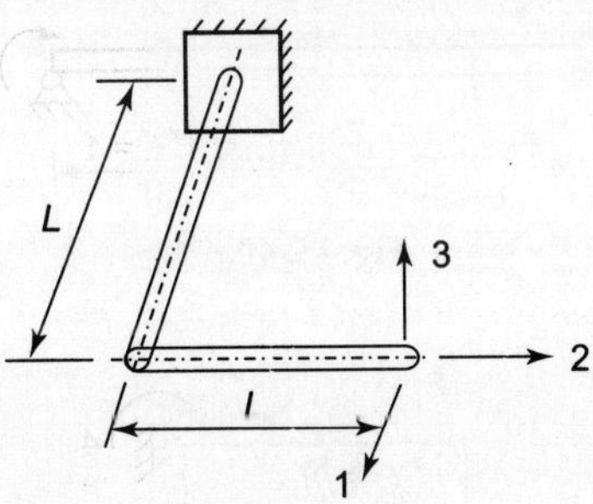

Fig. 15.47

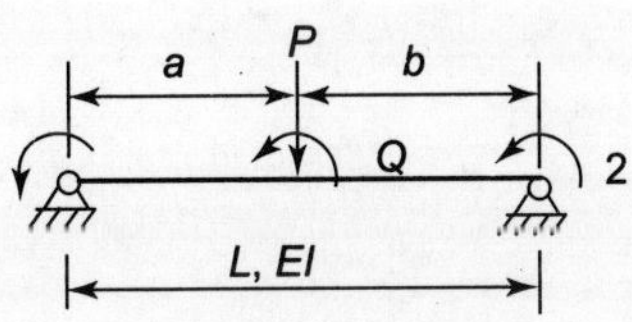

Fig. 15.48

15.14 Using the coordinates as shown in Fig. 15.48 generate flexibility matrix **f** for structure. Find displacements D_1 and D_2 for the leading as shown.

15.15 Generate stiffness matrix k for the four coordinates shown in Fig. 15.49.

15.16 Generate the stiffness matrix k for the structure with the coordinates shown in Fig. 15.49.

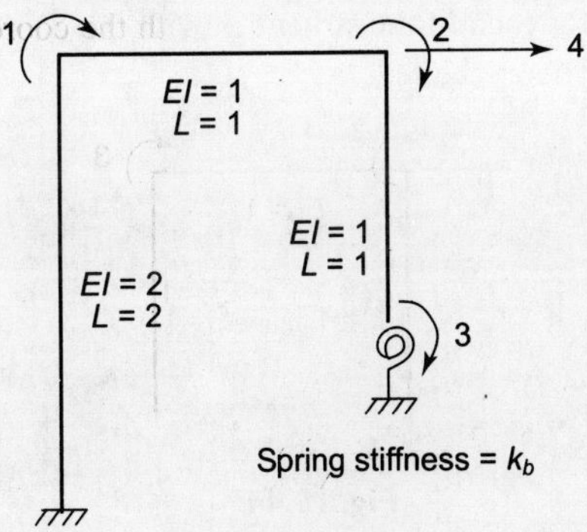

Fig. 15.49

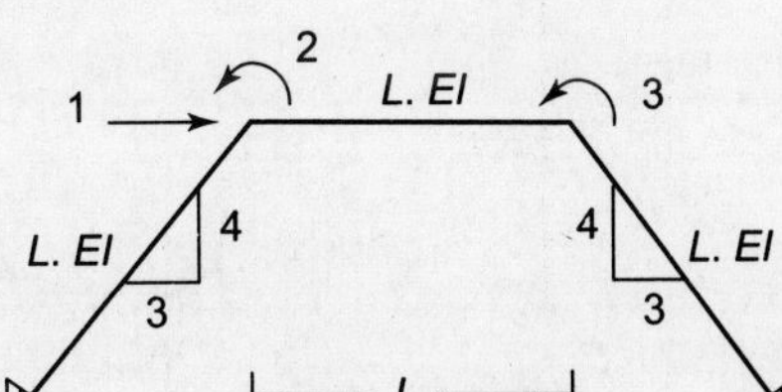

Fig. 15.50

15.17 Write the stiffness matrix corresponding to the coordinates 1 and 2 of the frame shown in Fig. 15.51. EI is constant.

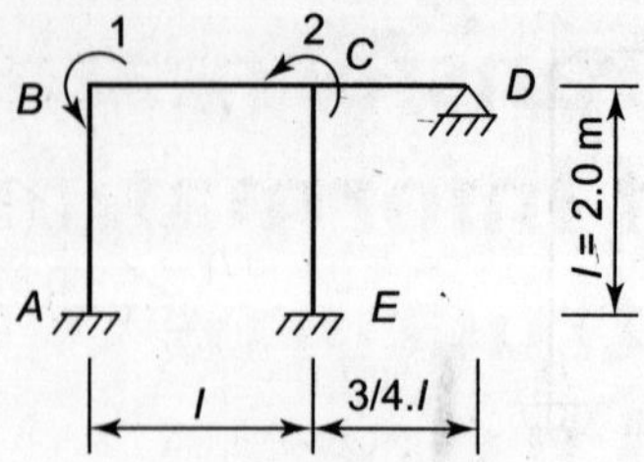

Fig. 15.51

16 Transformation of Information in Structures through Matrices

16.1 TRANSFORMATION OF SYSTEM FORCES TO ELEMENT FORCES

A common objective in the analysis of structures is finding the internal forces resulting from external forces. We relate the external forces $\mathbf{P}_i$ at the system coordinates to the forces $\mathbf{p}_i$ defined at the element coordinates by a matrix $\mathbf{A}$ by the expression

$$\mathbf{P}_i = \mathbf{A}\mathbf{P}_i \qquad (16.1)$$

Equation 16.1 transforms system forces $\mathbf{P}_i$ to element-forces $\mathbf{p}_i$ and constitutes an equation of equilibrium. Matrix $\mathbf{A}$ is known as the *force transformation matrix* and can be easily generated for a determinate structure.

Let us now show how forces $\mathbf{P}_i$ at the coordinates of a structure are transformed to forces $\mathbf{p}_i$ at the coordinates of the elements in the structure. This requires that we generate matrix $\mathbf{A}$ of Eq. 16.1 for the structure. As an example, we consider the statically determinate frame of Fig. 16.1.

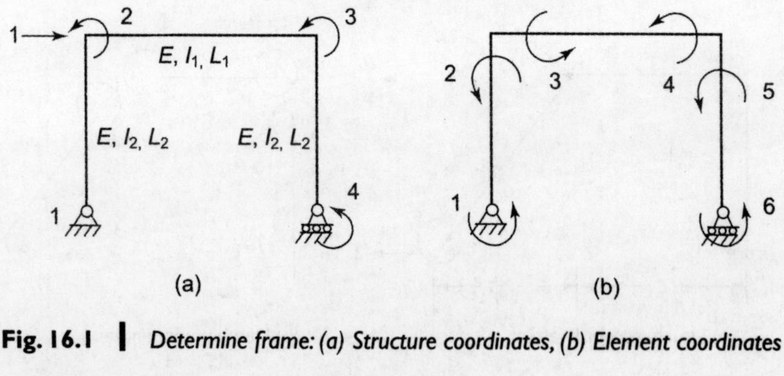

Fig. 16.1 | *Determine frame: (a) Structure coordinates, (b) Element coordinates*

The structure and element coordinates are indicated in Figs. 16.1*a* and *b* respectively. One can generate matrix $\mathbf{A}$ in Eq. 16.1 by assigning arbitrary values to $\mathbf{P}_i$ (i = 1, 2, 3 and 4) and computing the element forces at the coordinates from the equations of equilibrium. However, an alternative approach, is to generate the elements of matrix **A** column by column as follows. For example, to generate elements in column *j*, we apply a unit force at coordinate *j* only and compute the element forces P_i ($i = 1, 2, ..., 6$) at the element coordinates. To generate column 1 of matrix **A** we apply $P_1 = 1$ and $P_i = 0$ for $i \neq 1$ and compute internal forces p_i from equilibrium considerations. In the present example, we get

$$p_1 = 0, p_2 = L, p_3 = -L, p_4 = p_5 = p_6 = 0.$$

These forces form the first column of **A**.

Applying next a unit force only at 2 in Fig. 16.1*a* and computing the internal forces, we have

$$p_1 = 0, p_2 = 0, p_3 = 1, p_4 = p_5 = p_6 = 0$$

This will form the second column of matrix **A**.

Proceeding in a similar manner, the application of a unit force at coordinate 3 gives

$$P_1 = P_2 = P_3 = 0, P_4 = 1, P_5 = P_6 = 0.$$

Following in a similar manner, the elements in the fourth column are obtained. The complete transformation matrix **A** is given as

$$\mathbf{A} = \begin{bmatrix} 0 & 0 & 0 & 0 \\ L & 0 & 0 & 0 \\ -L & 1 & 0 & 0 \\ 0 & 0 & 1 & 1 \\ 0 & 0 & 0 & -1 \\ 0 & 0 & 0 & 1 \end{bmatrix} \tag{16.2}$$

Example 16.1 | *Generate the force transformation matrix [A] for the structure and the coordinates shown in Fig. 16.2.*

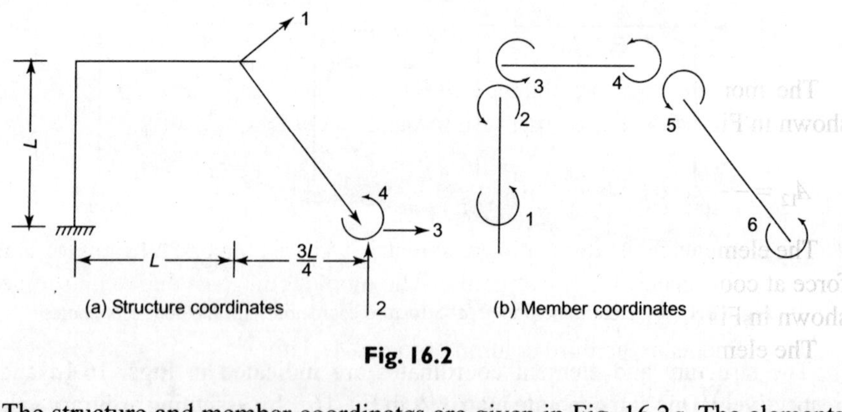

Fig. 16.2

The structure and member coordinates are given in Fig. 16.2*a*. The elements in the first column of the transformation matrix [A] are obtained by applying a unit force at the structure coordinate 1 and finding the forces at the member coordinates. Since the structure is statically determinate this presents no difficulty. The moment diagram obtained due to unit force applied at coordinate 1 is shown in Fig. 16.3*a*. The forces at the element coordinates are as shown in Fig. 16.3*b*. The elements are:

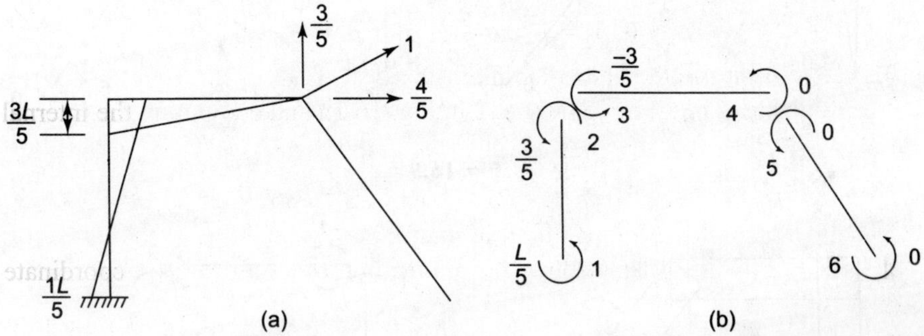

Fig. 16.3

$$A_{11} = +\frac{L}{5}, A_{21} = +\frac{3}{5}, A_{31} = -\frac{3}{5}L, A_{41} = 0, A_{51} = 0, A_{61} = 0$$

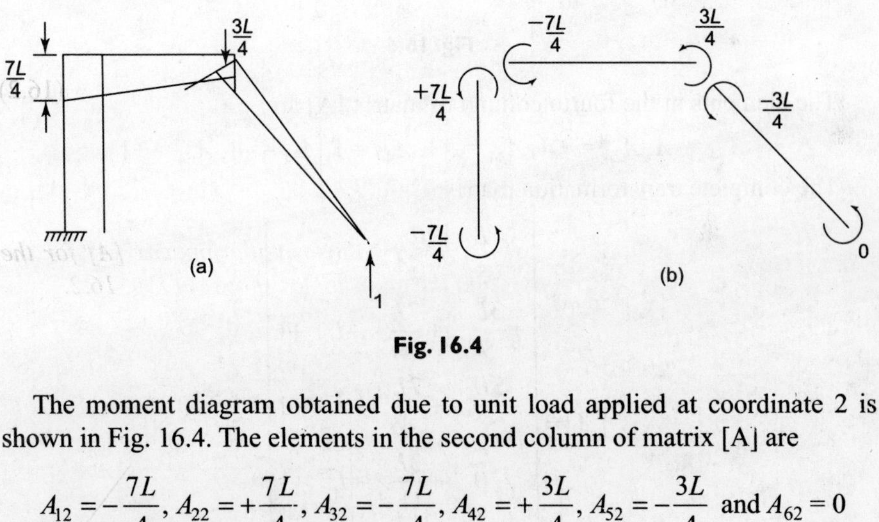

Fig. 16.4

The moment diagram obtained due to unit load applied at coordinate 2 is shown in Fig. 16.4. The elements in the second column of matrix [A] are

$$A_{12} = -\frac{7L}{4}, A_{22} = +\frac{7L}{4}, A_{32} = -\frac{7L}{4}, A_{42} = +\frac{3L}{4}, A_{52} = -\frac{3L}{4} \text{ and } A_{62} = 0$$

The elements in the third column of matrix [A] are obtained by applying a unit force at coordinates 3 in the structure. The moment diagram due to unit force is shown in Fig. 16.5.

The elements in the third column of matrix [A] are

$$A_{13} = 0, A_{23} = +L, A_{33} = -L, A_{43} = +L, A_{53} = -L, A_{63} = 0$$

Lastly, applying a unit force at structure coordinate 4 the moment diagram obtained is shown in Fig. 16.6.

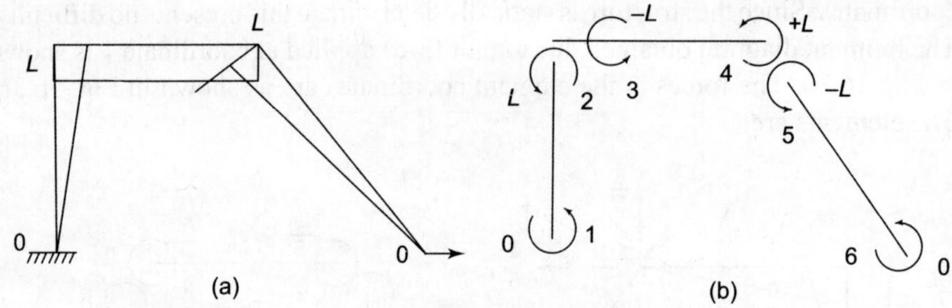

Fig. 16.5

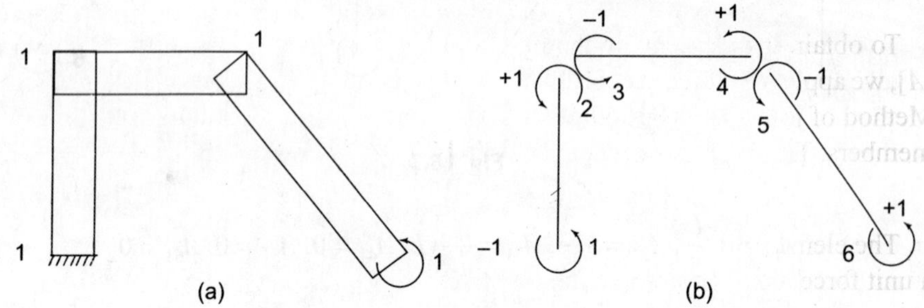

Fig. 16.6

The elements in the fourth column of matrix [A] are:

$$A_{14} = -1, A_{24} = +1, A_{34} = -1, A_{44} = 1, A_{54} = -1, A_{64} = +1$$

The complete transformation matrix:

$$[A] = \begin{bmatrix} \frac{L}{5} & -\frac{7}{4}L & 0 & -1 \\ +\frac{3L}{5} & +\frac{7L}{4} & +L & +1 \\ -\frac{3L}{5} & +\frac{7L}{4} & -L & -1 \\ 0 & +\frac{3L}{4} & +L & +1 \\ 0 & -\frac{3L}{4} & -L & -1 \\ 0 & 0 & 0 & +1 \end{bmatrix}$$

Example 16.2 | *Generate the force transformation matrix [A] relating the member forces to the external forces along the coordinates for the truss shown in Fig. 16.7. Find the forces in members due to applied load*

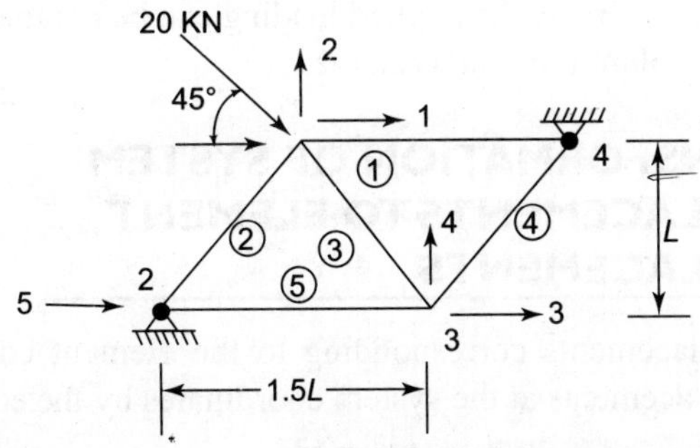

Fig. 16.7

To obtain the elements in the first column of the force transformation matrix [A], we apply unit force at coordinate 1 and work out the forces in all the members. Method of joints and method of sections are utilized in calculating the forces in members. The elements of the matrix are:

$$A_{11} = -1.0, A_{21} = A_{31} = A_{41} = A_{51} = 0$$

The elements n the second column of the matrix [A] are obtained by applying a unit force along coordinate 2. They are:

$$A_{12} = 0.25, A_{22} = 0.85, A_{32} = -0.43, A_{42} = +0.43 \text{ and } A_{52} = -0.51$$

Similarly, the elements in the third column of matrix [A] are obtained by applying a unit force at coordinate 3 and finding the forces in members of the truss. The elements are:

$$A_{13} = -0.66, A_{23} = -0.55, A_{33} = +0.55, A_{43} = -0.55 \text{ and } A_{53} = 0.33$$

The elements in the fourth column of matrix are:

$$A_{14} = 0.5, A_{24} = 0.42, A_{34} = -0.42, A_{44} = -0.83 \text{ and } A_{54} = -0.25$$

The fifth column elements are:

$$A_{15} = -0.66, A_{25} = -0.55, A_{35} = 0.55, A_{45} = -0.55 \text{ and } A_{55} = -0.67$$

The complete transformation matrix [A] is

$$[A] = \begin{bmatrix} -1.00 & 0.25 & -0.66 & 0.50 & -0.66 \\ 0 & 0.85 & -0.55 & 0.42 & -0.55 \\ 0 & -0.43 & 0.55 & -0.42 & 0.55 \\ 0 & 0.43 & -0.55 & -0.83 & -0.55 \\ 0 & -0.51 & 0.33 & -0.25 & -0.67 \end{bmatrix}$$

$$\text{or } \{p\} = [A] \begin{Bmatrix} 14.14 \\ -14.14 \\ 0 \\ 0 \\ 0 \end{Bmatrix} = \begin{Bmatrix} -17.68 \\ -12.0 \\ +6.00 \\ -6.00 \\ +7.00 \end{Bmatrix}$$

The forces in members by the applied loading can be obtained by multiplying [*A*] by the loading column matrix as above.

16.2 TRANSFORMATION OF SYSTEM DISPLACEMENTS TO ELEMENT DISPLACEMENTS

The element displacements corresponding to the element coordinates can be related to the displacements at the system coordinates by the equation

$$\mathbf{d}_i = \mathbf{B}\,\mathbf{D}_i \tag{16.3}$$

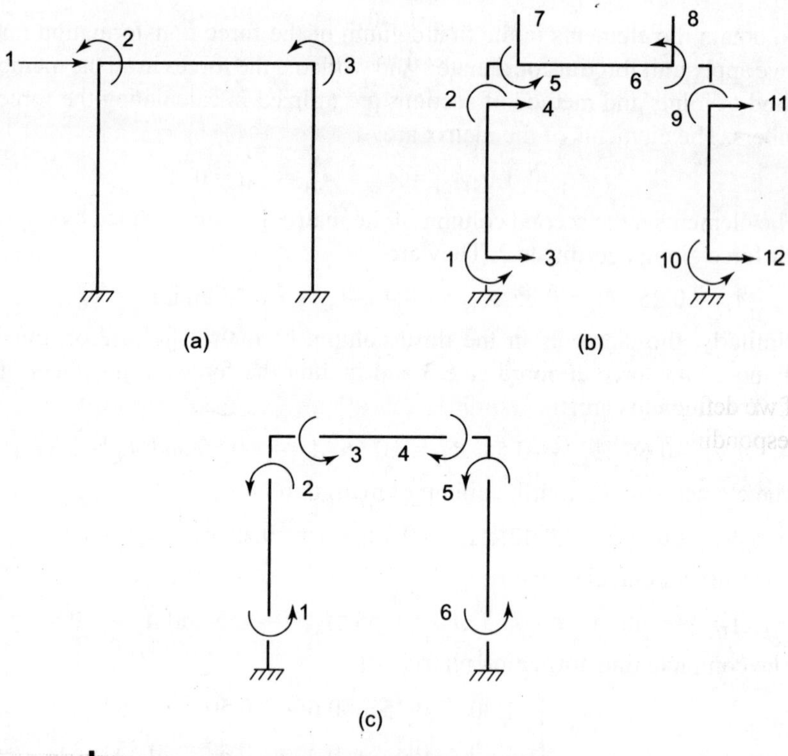

Fig. 16.8 | *(a) Structure and structure coordinates, (b) Elements and element coordinates, (c) Alternate element coordinates*

in which **B** is the *displacement transformation matrix* and $\mathbf{d_i}$ and $\mathbf{D_i}$ are the displacements at element and system coordinates respectively. Equation 16.3 constitutes an equation of displacement compatibility which ensures the continuity of the structure.

Let us generate **B** for the frame of Fig. 16.8 in which the structure coordinates are defined in Fig. 16.8a and element coordinates in Fig. 16.8*b*. One way to generate matrix **B** in Eq. 16.3 is to apply arbitrary displacements $D_i (i = 1, 2, 3)$ and use conditions of compatibility to find corresponding displacements $d_i (i = 1, 2, ..., 12)$.

An alternate approach is to apply a unit displacement at system coordinate j only and compute the displacements at the element coordinates. These displacements d_i, form the element of the jth column of matrix [**B**]. For example to generate column 2 of matrix **B** we apply a unit displacement at structure coordinate 2 only and compute displacements at the element coordinates. This results in $d_2 = 1$ and $d_5 = 1$ and all the other element displacements zero. Following this procedure for other structure coordinate displacements of Fig. 16.8a, we obtain

$$\mathbf{B} = \begin{bmatrix} 0 & 0 & 0 \\ 0 & 1 & 0 \\ 0 & 0 & 0 \\ 1 & 0 & 0 \\ \hline 0 & 1 & 0 \\ 0 & 0 & 1 \\ 0 & 0 & 0 \\ 0 & 0 & 0 \\ \hline 0 & 0 & 1 \\ 0 & 0 & 0 \\ 1 & 0 & 0 \\ 0 & 0 & 0 \end{bmatrix} \tag{16.4}$$

If we define the element coordinates as in Fig. 16.8c, deleting the coordinates corresponding to transverse forces, matrix B reduces to

$$\mathbf{B} = \begin{bmatrix} \frac{1}{L} & 0 & 0 \\ \frac{1}{L} & 1 & 0 \\ \hline 0 & 1 & 0 \\ 0 & 0 & 1 \\ \hline \frac{1}{L} & 0 & 1 \\ \frac{1}{L} & 0 & 0 \end{bmatrix} \tag{16.5}$$

It may be noted that for this coordinate system the rotations are to be taken relative to a chord connecting the ends of the member.

Example 16.3 | *Generate displacement transformation matrix [B] for the structure with structure and element coordinates shown in Fig. 16.9. For the same structure, prove the contra gradient law.*

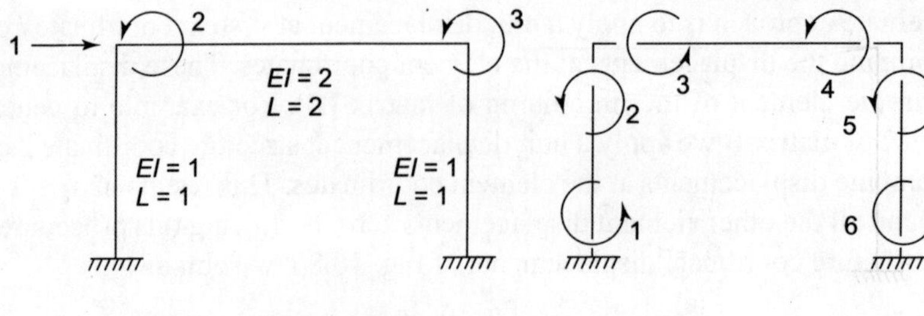

(a) Structure coordinates (b) Element coordinates

Fig. 16.9

We can generate displacement transformation matrix $[B]$ by imposing unit displacement at each of the structure coordinates in turns and working out the displacements at member coordinates.

A unit displacement $D_1 = 1$ is given to the structure. The displacements at the member coordinates are worked out as shown in Fig. 16.10 *a*, *b*. The elements in the first column of the matrix $[B]$ are:

$$B_{11} = \frac{1}{L}, B_{21} = \frac{1}{L}, B_{31} = B_{41} = 0, B_{51} = \frac{1}{L} \text{ and } B_{61} = \frac{1}{L}$$

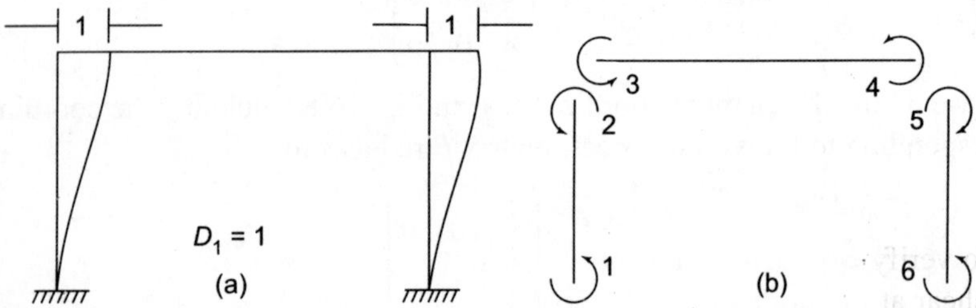

Fig. 16.10

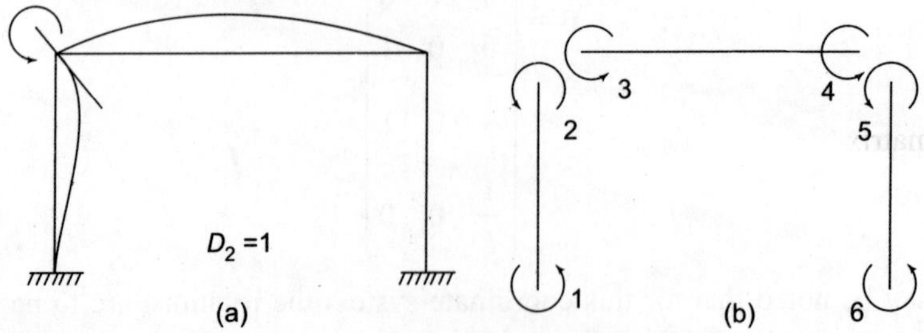

Fig. 16.11

The displacement $D_2 = 1$ imposed at the structure is as shown in Fig. 6.11*a*. The displacements at the element coordinates are.

$$B_{12} = 0, B_{22} = 1, B_{32} = 1, B_{42} = B_{52} = B_{62} = 0$$

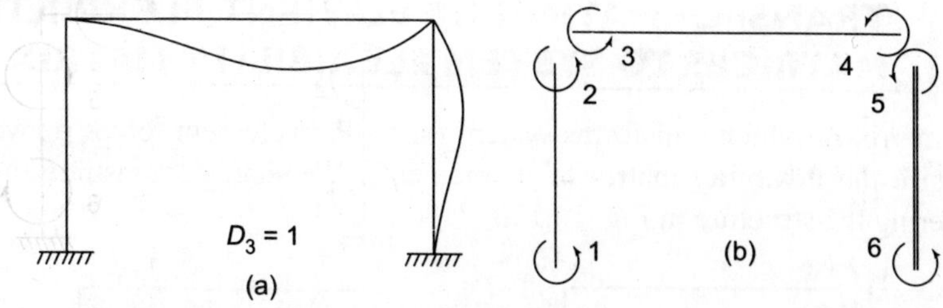

Fig. 16.12

Similarly, the elements in the third column are:

$$B_{13} = 0, B_{23} = 0, B_{33} = 0, B_{43} = 1, B_{53} = 1 \text{ and } B_{63} = 0$$

The complete transformation matrix $[B]$ is

$$[B] = \begin{bmatrix} \frac{1}{L} & 0 & 0 \\ \frac{1}{L} & 1 & 0 \\ 0 & 1 & 0 \\ 0 & 0 & 1 \\ \frac{1}{L} & 0 & 1 \\ \frac{1}{L} & 0 & 0 \end{bmatrix}$$

To verify contra gradient law $P = B^T p$, we write the equation of equilibrium for shear at top portion of the columns and moment about two joints (structure coordinates 2 and 3) as

$$P_1 = p_1 + p_2 + p_5 + p_6$$

$$P_2 = p_2 + p_3$$

and $$P_3 = p_4 + p_5$$

In matrix form we have,

$$\{P\} = [B]^T \{p\}$$

$$\begin{Bmatrix} P_1 \\ P_2 \\ P_3 \end{Bmatrix} = \begin{bmatrix} \frac{1}{L} & \frac{1}{L} & 0 & 0 & \frac{1}{L} & \frac{1}{L} \\ 0 & 1 & 1 & 0 & 0 & 0 \\ 0 & 0 & 0 & 1 & 1 & 0 \end{bmatrix} \begin{Bmatrix} p_1 \\ p_2 \\ p_3 \\ p_4 \\ p_5 \\ p_6 \end{Bmatrix}$$

16.3 TRANSFORMATION OF ELEMENT FLEXIBILITY MATRICES TO SYSTEM FLEXIBILITY MATRIX

Using matrix **A**. which transforms system forces $\mathbf{P_i}$ to element forces $\mathbf{p}_i$, we can synthesise the flexibility matrix of its elements. We shall demonstrate this by considering the structure in Fig. 16.13*a*.

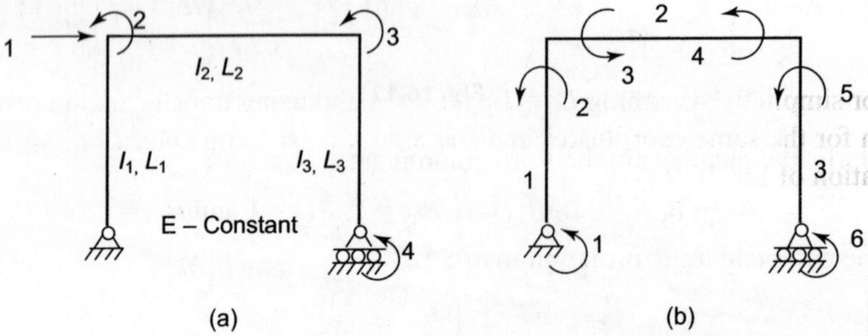

Fig. 16.13 *(a) Structure and structure coordinates, (b) Element and element coordinates*

Let the structure undergo deformations $D_i(i = 1, 2, 3, 4)$ as a result of forces $P(i = 1, 2, 3, 4)$ gradually applied at the coordinates. The total strain energy, U, in the structure is equal to the work done by forces P_i in moving through displacements D_i, at the coordinates. This can be written in terms of flexibility matrix **F** of the system (*see* Eq. 15.73).

$$U = \frac{1}{2}\,\mathbf{P^T F P} \tag{16.6}$$

This strain energy can also be expressed as the sum of the strain energies, U_s, in the elements (*see* Eq. 15.109)

$$U = \frac{1}{2}\,\mathbf{p^T\ f p} \tag{16.7}$$

Substituting $\quad \mathbf{p = A\,P}$

And $\quad \mathbf{p^T = P^T\,A^T}$

in Eq. 16.7, and comparing with Eq. 16.6, we have

$$\mathbf{P^T F P = P^T\,A^T\,f\,A\,P} \tag{16.8}$$

Since forces P_i are independent, both flexibility matrix **F** and matrix product $\mathbf{A^T f\ A}$ are symmetrical matrices. That they are identical can be easily proved from Equation 16.8. Therefore,

$$\mathbf{F = A^T f A} \tag{16.9}$$

Hence the system flexibility matrix **F** is synthesised from the element flexibility matrices $\mathbf{f_s}$ which are recorded as sub-matrices in **f** as in Equation 15.110.

For the structure in Fig. 16.13*a*

$$\mathbf{f} = \begin{bmatrix} \dfrac{L_1}{6EI_1}\begin{bmatrix} 2 & -1 \\ -1 & 2 \end{bmatrix} & & \\ & \dfrac{L_2}{6EI_2}\begin{bmatrix} 2 & -1 \\ -1 & 2 \end{bmatrix} & \\ & & \dfrac{L_3}{6EI_3}\begin{bmatrix} 2 & -1 \\ -1 & 2 \end{bmatrix} \end{bmatrix} \tag{16.10}$$

For simplicity, assuming $L_1 = L_2 = L_3 = L$ and using transformation matrix **A** given for the same coordinates and frame in Eq. 16.2 and **f** from Eq. 16.10, the operation of Eq. 16.9 gives

$$\mathbf{f} = \frac{L}{6EI}\begin{bmatrix} 4 & -2 & 1 & 1 \\ -2 & 2 & -1 & -1 \\ 1 & -1 & 1 & 2 \\ 1 & -1 & 2 & 4 \end{bmatrix} \tag{16.11}$$

Equation 16.9 can also be derived from virtual work. Consider virtual displacement $\overline{\mathbf{D}}_i$ at system coordinates (Fig. 16.8*a*). The compatible element displacements **d** can be obtained from matrix **B** of Eq. 16.5. Equating the internal and external virtual works, we get

$$\mathbf{p}^T\overline{\mathbf{D}} = \mathbf{P}^T\overline{\mathbf{d}} \tag{16.12}$$

Substituting for

$$\overline{\mathbf{D}} = \mathbf{FP}$$

$$\overline{\mathbf{d}} = \mathbf{f\,p}$$

$$\mathbf{p}^T = (\mathbf{A\,P})^T = \mathbf{P}^T\,\mathbf{A}^T$$

in Equation 16.12, we get

$$\mathbf{P}^T\,\mathbf{FP} = \mathbf{P}^T\,\mathbf{A}^T\,\mathbf{f\,A\,P} \tag{16.13}$$

This is same as Equation 16.8, therefore,

$$\mathbf{F} = \mathbf{A}^T\mathbf{f\,A} \tag{16.14}$$

Example 16.4 | *Using the relationship*

$$[F] = [A]^T\,[f]\,[A]$$

generate the flexibility matrix [F] for the structure shown in Fig. 16.14

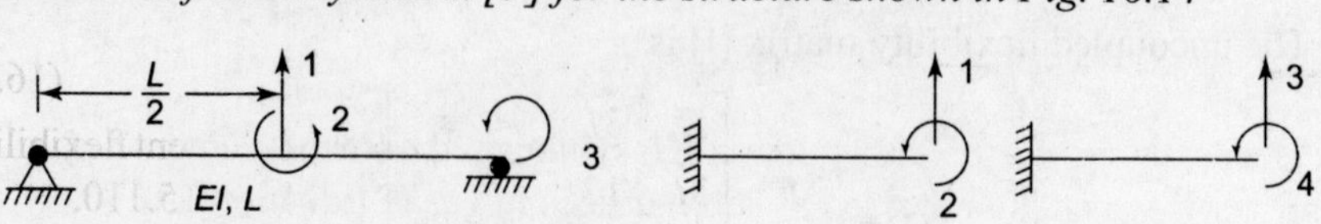

(a) Structure coordinates (b) Elements coordinates

Fig. 16.14

First, let us establish the force transformation matrix $[A]$. Unit forces are applied in turns at the structure coordinates and the forces at the member coordinates of the elements are worked out. The elements in the first column of the matrix $[A]$ are

$$A_{11} = -\frac{1}{2},\ A_{21} = -\frac{L}{4},\ A_{31} = -\frac{1}{2} = \text{and } A_{41} = 0$$

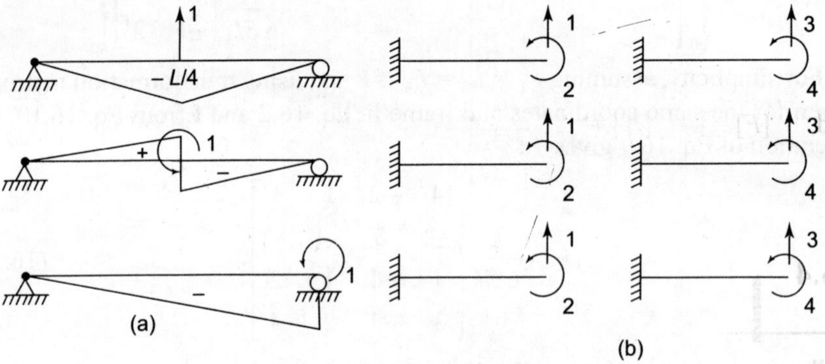

Fig. 16.15

The elements in the second column of the matrix $[A]$ are :

$$A_{12} = -\frac{1}{L},\ A_{22} = -\frac{1}{2},\ A_{32} = \frac{1}{L} = \text{ and } A_{42} = 0$$

The elements of the third column are,

$$A_{13} = -\frac{1}{L},\ A_{23} = \frac{1}{2},\ A_{33} = -\frac{1}{L} = \text{ and } A_{43} = 0$$

The complete force transformation matrix [A] is

$$[A] = \begin{bmatrix} \frac{1}{2} & -\frac{1}{L} & -\frac{1}{L} \\ -\frac{L}{4} & -\frac{1}{2} & \frac{1}{2} \\ -\frac{1}{2} & \frac{1}{L} & -\frac{1}{L} \\ 0 & 0 & 1 \end{bmatrix}$$

The uncoupled flexibility matrix [f] is

$$[f] = \frac{1}{24\,EI}\begin{bmatrix} L^2 & 3L & & \\ 3L & 12 & & \\ & & L^2 & 3L \\ & & 3L & 12 \end{bmatrix}$$

The product $$[A]^T [f] = \frac{L}{24\,EI}\begin{bmatrix} -\frac{L^2}{4} & -\frac{3L}{2} & -\frac{L^2}{2} & -\frac{3}{2}L \\ \frac{L}{2} & 3 & -L & -3 \\ \frac{L}{2} & 3 & 2L & 9 \end{bmatrix}$$

and $$[F] = [A]^T [f]\,[A] = \frac{L}{24\,EI}\begin{bmatrix} \frac{L^2}{2} & 0 & -\frac{3}{2}L \\ 0 & 2 & -1 \\ -\frac{3}{2}L & -1 & 8 \end{bmatrix}$$

16.4 TRANSFORMATION OF ELEMENT STIFFNESS MATRICES TO SYSTEM STIFFNESS MATRIX

Following a development similar to that of Section 16.3 we can use displacement transformation matrix **B** to synthesize the stiffness matrix **K** of a system from the stiffness matrices of its element $\mathbf{k}_s$.

Consider the frame of Fig. 16.8*a* under force P_i at the system coordinates. The strain energy of the system in terms of stiffness matrix **K** can be written as (Eq. 15.70)

$$U = \frac{1}{2}\mathbf{D}^{\mathbf{T}}\,\mathbf{K}\mathbf{D}$$

The same strain energy can also be written as the sum of strain energies U_s in the elements of the system (Eq. 15.70) that is

$$U = \frac{1}{2}\mathbf{d}^{\mathbf{T}}\,\mathbf{k}\mathbf{d} \tag{16.15}$$

Substituting for

$$\mathbf{d} = \mathbf{B}\mathbf{D}$$

And $$\mathbf{d}^{\mathbf{T}} = \mathbf{D}^{\mathbf{T}}\,\mathbf{B}^{\mathbf{T}}$$

in Eq. 16.15 and comparing with Eq. 16.14, we have

$$\mathbf{D}^{\mathbf{T}}\,\mathbf{K}\mathbf{D} = \mathbf{D}^{\mathbf{T}}\,\mathbf{B}^{\mathbf{T}}\mathbf{k}\mathbf{B}\mathbf{D} \tag{16.16}$$

Since D_i are independent (otherwise **k** does not exist) and both **K** and matrix product $\mathbf{B}^{\mathbf{T}}\,\mathbf{k}\mathbf{B}$ are symmetical matrices, we can see that they are also identical, Hence,

$$\mathbf{K} = \mathbf{B}^{\mathbf{T}}\,\mathbf{k}\mathbf{B} \tag{16.17}$$

For the structure of Fig. 16.8*a* and element coordinates as in Fig. 16.8*c*, we can write

$$\mathbf{k} = \begin{bmatrix} \mathbf{k}_1 & & \\ & \mathbf{k}_2 & \\ & & \mathbf{k}_3 \end{bmatrix} \tag{16.18}$$

in which

$$\mathbf{k}_s = \frac{EI}{L}\begin{bmatrix} 4 & 2 \\ 2 & 4 \end{bmatrix}$$

$$s = 1, 2, 3$$

On performing the operation of Eq. 16.17 taking the values of matrix **B** from Eq. 16.5 and **k** from the foregoing, we get

$$\mathbf{K} = \frac{EI}{L}\begin{bmatrix} \frac{24}{L^2} & \frac{6}{L} & \frac{6}{L} \\ \frac{6}{L} & 8 & 2 \\ \frac{6}{L} & 2 & 8 \end{bmatrix} \tag{16.19}$$

Equation 16.17 can also be derived from virtual work. Considering virtual forces $\overline{\mathbf{P}}_\mathbf{i}$ at the system coordinates and the corresponding internal forces $\mathbf{P}_\mathbf{i}$ which are in equilibrium with forces **Pi**, we can write from the virtual work relationship

$$\mathbf{D}^\mathbf{T}\overline{\mathbf{P}} = \mathbf{d}^\mathbf{T}\,\overline{\mathbf{p}} \tag{16.20}$$

Substituting for

$$\overline{\mathbf{P}} = \mathbf{KD}$$

$$\mathbf{d} = \mathbf{BD} \quad \text{or} \quad \mathbf{d}^\mathbf{T} = \mathbf{D}^\mathbf{T}\mathbf{B}^\mathbf{T}$$

and

$$\overline{\mathbf{p}} = \mathbf{kd} \quad \text{or} \quad \mathbf{p} = \mathbf{kBD}$$

in Eq. 16.20, we get

$$\mathbf{D}^\mathbf{T}\mathbf{KD} = \mathbf{D}^\mathbf{T}\mathbf{B}^\mathbf{T}\mathbf{kBD} \tag{16.21}$$

This is same as Eq. 16.16. Therefore,

$$\mathbf{K} = \mathbf{B}^\mathbf{T}\,\mathbf{kB} \tag{16.22}$$

16.5 TRANSFORMATION OF FORCES AND DISPLACEMENTS IN GENERAL

The transformation of forces and displacements from a set of independent coordinates of the system to dependent coordinates of the elements are given in Sections 16.1 and 16.2. When the same element coordinates are used to identify element forces p_1, and displacements d_i there then exists a relationship between the transformations of forces and displacements.

To show the nature of relationship, we consider the frame of Fig. 16.8a. From the conditions of compatibility, we have (*see* Eq. 16.3)

$$\mathbf{d} = \mathbf{BD}$$

and the total strain energy stored in the structure is (*see* Eq. 15.69)

$$U = \frac{1}{2}\mathbf{D}^{\mathbf{T}}\mathbf{P}$$

The same strain energy can also be obtained by summing up the strain energies in the elements, that is

$$U = \frac{1}{2}\sum_{s=1}^{m}\mathbf{d}_s^{\mathbf{T}}\mathbf{P}_s = \frac{1}{2}\mathbf{d}^{\mathbf{T}}\mathbf{p}$$

Equating the strain energy obtained from both the approaches

$$\mathbf{D}^{\mathbf{T}}\mathbf{P} = \mathbf{d}^{\mathbf{T}}\mathbf{p} \tag{16.23}$$

Substituting for $\mathbf{d}^{\mathbf{T}} = \mathbf{D}^{\mathbf{T}}\,\mathbf{B}^{\mathbf{T}}$, we get

$$\mathbf{D}^{\mathbf{T}}\,\mathbf{P} = \mathbf{D}^{\mathbf{T}}\mathbf{B}^{\mathbf{T}}\mathbf{p} \tag{16.24}$$

or

$$\mathbf{D}^{\mathbf{T}}\,(\mathbf{P} - \mathbf{B}^{\mathbf{T}}\mathbf{p}) = 0 \tag{16.25}$$

Since the displacements **D** are independent and can be assigned arbitrary values, it follows that the expression inside the parenthesis must be zero. Thus, we have

$$\mathbf{P} = \mathbf{B}^{\mathbf{T}}\mathbf{p} \tag{16.26}$$

Forces *P* are the generalised forces in the *D* coordinates.

Equations 16.3 and 16.26 show the relationship that exists between the transformation of displacements and forces from element coordinates *d* (which identify d_i and p_i) to generalised coordinates *D* (which identify D_i and *Pi*).

If, instead of starting with Eq. 16.3, the transformation of displacements, we start with the transformation of forces (Eq. 16.1)

$$\mathbf{p} = \mathbf{AP}$$

Again equating the total energy stored in the structure obtained from external forces and their displacements and from the elements forces and their displacements, we have

$$\mathbf{P}^{\mathbf{T}}\mathbf{D} = \mathbf{p}^{\mathbf{T}}\mathbf{d} \tag{16.27}$$

Substituting for

$$\mathbf{p}^{\mathbf{T}} = \mathbf{P}^{\mathbf{T}}\,\mathbf{A}^{\mathbf{T}}$$

we have

$$\mathbf{P}^{\mathbf{T}}\,\mathbf{D} = \mathbf{P}^{\mathbf{T}}\,\mathbf{A}^{\mathbf{T}}\,\mathbf{d} \tag{16.28}$$

or

$$\mathbf{P}^{\mathbf{T}}(\mathbf{D} - \mathbf{A}^{\mathbf{T}}\mathbf{d}) = 0 \tag{16.29}$$

Since forces P_i are independent and can be assigned arbitrary values, the expression inside the parenthesis must vanish to satisfy the above relationship, that is

$$\mathbf{D} = \mathbf{A}^T\mathbf{d} \tag{16.30}$$

Equations 16.3 and 16.26, and Equations 16.1 and 16.30 constitute *contragradient laws.*

To verify Equation 16.26 for the structure of Fig. 16.8*a* and element coordinates (Fig. 16.8*b*) we write the equations of equilibrium for shear across top portion of the frame and moment about two joints (structure coordinates 2 and 3) as follows:

$$P_i = p_4 + p_{11} \qquad P_2 = P_2 + P_5 \qquad P_3 = P_6 + P_9$$

In matrix form, we have

$$\begin{Bmatrix} P_1 \\ P_2 \\ P_3 \end{Bmatrix} = \underbrace{\begin{bmatrix} 0 & 0 & 0 & 1 & 0 & 0 & 0 & 0 & 0 & 0 & 1 & 0 \\ 0 & 1 & 0 & 0 & 1 & 0 & 0 & 0 & 0 & 0 & 0 & 0 \\ 0 & 0 & 0 & 0 & 0 & 1 & 0 & 0 & 1 & 0 & 0 & 0 \end{bmatrix}}_{\mathbf{B}^T} \begin{Bmatrix} p_1 \\ p_2 \\ p_3 \end{Bmatrix} \tag{16.31}$$

The above 3 × 12 matrix is the transpose of matrix **B** obtained earlier (*see* Eq. 16.4) from conditions of compatibility.

16.6 TRANSFORMATION OF INFORMATION FROM MEMBER COORDINATES TO STRUCTURE COORDINATES AND VICE VERSA

For convenience we fix element coordinates coincident with member coordinates called *local coordinates.* It often becomes necessary to transform information from member coordinates to structure coordinates called *global coordinates* and vice versa.

Consider an axial force member in a truss arbitrarily oriented as shown in Fig. 16.16*a*. The orientation of local coordinates are represented by X' and Y' and structure coordinates by X and Y. The member stiffness matrix with reference to local coordinates is (see information displayed in Fig. 15.19 and neglect flexural stiffness)

$$\mathbf{k}' = \begin{bmatrix} \dfrac{EA}{L} & 0 & -\dfrac{EA}{L} & 0 \\ 0 & 0 & 0 & 0 \\ -\dfrac{EA}{L} & 0 & \dfrac{EA}{L} & 0 \\ 0 & 0 & 0 & 0 \end{bmatrix} \tag{16.32}$$

Let the force components be p_1 and p_2 at one end and p_3 and p_4 at the other end along the structure coordinates. If we designate the force components along member coordinates as $p'_1 p'_2$ at one end and p'_3 and p'_4 at the other, we can relate them as

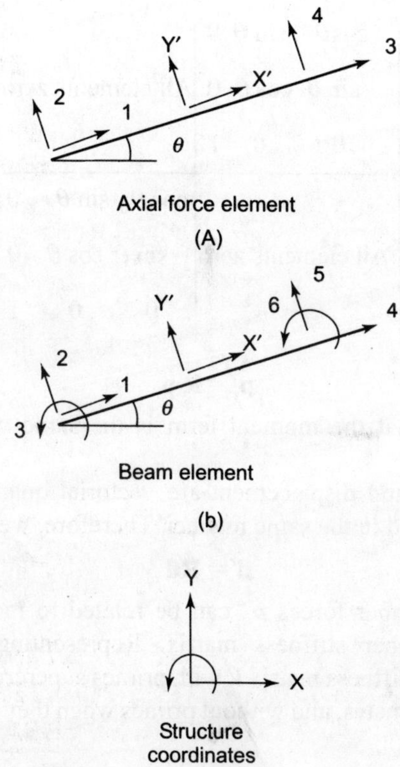

Fig. 16.16

$$p'_1 = p_1 \cos\theta + p_2 \sin\theta \tag{16.33}$$

$$p'_2 = -p_1 \sin\theta + p_2 \cos\theta$$

Writing Equation 16.33 in the matrix form, we have

$$\begin{Bmatrix} p'_1 \\ p'_2 \\ \hline p'_3 \\ p'_4 \end{Bmatrix} = \left[\begin{array}{cc|cc} \cos\theta & \sin\theta & 0 & 0 \\ \sin\theta & \cos\theta & 0 & 0 \\ \hline 0 & 0 & \cos\theta & \sin\theta \\ 0 & 0 & -\sin\theta & \cos\theta \end{array}\right] \begin{Bmatrix} p_1 \\ p_2 \\ p_3 \\ p_4 \end{Bmatrix} \tag{16.34a}$$

or
$$\mathbf{p}' = \mathbf{R}\mathbf{p} \tag{16.34b}$$

where **R** is called the *rotation transformation matrix*.

Consider next a beam element subjected to axial and bending forces and oriented in an arbitrary manner as shown in Fig. 16.16*b*. The member stiffness matrix with repeat to local coordinates is the same as given in Eq. 15.42. Again denoting force components along structure coordinates as p_i and along member coordinates as p'_i they can be related by

$$\begin{Bmatrix} p_1' \\ p_2' \\ p_3' \\ \hline p_4' \\ p_5' \\ p_6' \end{Bmatrix} = \left[\begin{array}{ccc|ccc} \cos\theta & \sin\theta & 0 & & & \\ -\sin\theta & \cos\theta & 0 & & \text{All elements zero} & \\ 0 & 0 & 1 & & & \\ \hline & & & \cos\theta & \sin\theta & 0 \\ & \text{All elements zero} & & -\sin\theta & \cos\theta & 0 \\ & & & 0 & 0 & 1 \end{array}\right] \begin{Bmatrix} p_1 \\ p_2 \\ p_3 \\ \hline p_4 \\ p_5 \\ p_6 \end{Bmatrix} \quad (16.35a)$$

or

$$\mathbf{p}' = \mathbf{R}\,\mathbf{p} \quad (16.35b)$$

It may be noted that the moment term is the same whatever may be the orientation of axes.

Since both forces and displacement are vectorial quantities, displacements can also be transformed in the same manner. Therefore, we can write

$$\mathbf{d}' = \mathbf{R}\mathbf{d} \quad (16.36)$$

We know that member forces p'_i can be related to member displacements d'_i through the member stiffness matrix. Representing member forces p_i displacements d_i and stiffness matrix **k** with primes supercribed for identification in terms of local coordinates, and without primes when they are in terms of global coordinates, we have

$$\mathbf{p}' = \mathbf{k}'\,\mathbf{d}' \quad (16.37)$$

Substituting for **p**′ from Eq. 16.35*b*, we have

$$\mathbf{R}\mathbf{p} = \mathbf{k}'\mathbf{d}' \quad (16.38)$$

Replacing **d**′ = **Rd** from Eq. 16.36

$$\mathbf{R}\mathbf{p} = \mathbf{k}'\,\mathbf{R}\mathbf{d} \quad (16.39)$$

Pre-multiplying by $\mathbf{R}^{-1}$ on both sides, we get

$$\mathbf{R}^{-1}\,\mathbf{R}\,\mathbf{p} = \mathbf{R}^{-1}\,\mathbf{k}'\,\mathbf{R}\mathbf{d} \quad (16.40)$$

or

$$\mathbf{p} = \mathbf{R}^{-1}\,\mathbf{k}'\,\mathbf{R}\mathbf{d} \quad (16.41)$$

Since **p** and **d** are member end forces and displacements in the structure coordinate system, we may interpret $\mathbf{R}^{-1}\,\mathbf{k}'\,\mathbf{R}$ as the member stiffness matrix in the structure coordinate system. Therefore,

$$\mathbf{k} = \mathbf{R}^{-1}\mathbf{k}'\,\mathbf{R} \quad (16.42)$$

or

$$\mathbf{k} = \mathbf{R}^{\mathrm{T}}\,\mathbf{k}'\,\mathbf{R} \quad (16.43)$$

since $\mathbf{R}^{-1} = \mathbf{R}^{\mathrm{T}}$ which can be easily verified.

Therefore, Eq. 16.41 can be written as

$$\mathbf{p} = \mathbf{k}\,\mathbf{d} \quad (16.44)$$

Hence, if we find stiffness matrix $\mathbf{k}'$ referred to local coordinates, it is a simple matter to transform it to global coordinates by using Eq. 16.43.

In a similar manner, by making use of rotation matrix $\mathbf{R}$ we can write for displacements as

$$\mathbf{d}' = \mathbf{R}\mathbf{d} \tag{16.45}$$

Substituting for

$$\mathbf{d}' = \mathbf{f}'\mathbf{p}$$

Eq. 16.45 can be written as

$$\mathbf{R}\mathbf{d} = \mathbf{f}'\mathbf{p} \tag{16.46}$$

where $\mathbf{f}'$ is the flexibility matrix in terms of local coordinates. Pre-multi plying both sides of Equation 16.46 by $\mathbf{R}^{-1}$

$$\mathbf{R}^{-1}\,\mathbf{R}\mathbf{d} = \mathbf{R}^{-1}\,\mathbf{f}'\mathbf{p}' \tag{16.47}$$

Replacing $\mathbf{p}' = \mathbf{R}\,\mathbf{p}$ and $\mathbf{R}^{-1} = \mathbf{R}^{\mathrm{T}}$ in Eq. 15.47, we have

$$\mathbf{d} = \mathbf{R}^{\mathrm{T}}\,\mathbf{f}'\mathbf{R}\,\mathbf{p} \tag{16.48}$$

$$\mathbf{d} = \mathbf{f}\,\mathbf{p} \tag{16.49}$$

in which

$$\mathbf{f} = \mathbf{R}^{\mathrm{T}}\,\mathbf{f}'\,\mathbf{R} \tag{16.50}$$

Matrix $\mathbf{f}$ represents the flexibility matrix of a member in a global coordinates.

Hence, from the flexibility matrix of a member in its local coordinates, the flexibility matrix in terms of global coordinates can be obtained from the operation of Eq. 16.50.

Problems for Practice

16.1 The space truss is pinned at all joints and supported as shown in Fig. 16.17. Compute the transformation matrix of order (3 × 3) relating the member forces to the load applied along coordinates 1, 2 and 3.

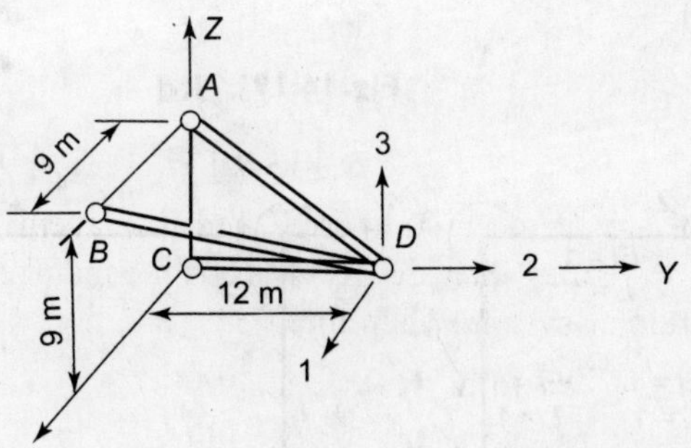

Fig. 16.17

16.2 Generate the displacement transformation matrix B for the structure with structure and element coordinates as shown in Fig. 16.18.

16.3 Generate the transformation matrix **B** for the structure and element coordinates shown in Fig. 16.19.

16.4 Using the relationship

$$\mathbf{F} = \sum \mathbf{A}_s^T \, \mathbf{f}_a \, \mathbf{A}_s$$

or

$$\mathbf{F} = \mathbf{A}^T \mathbf{f} \mathbf{A}$$

generate the flexibility matrix **F** for the structures shown in Figs. 16.20 and 16.21.

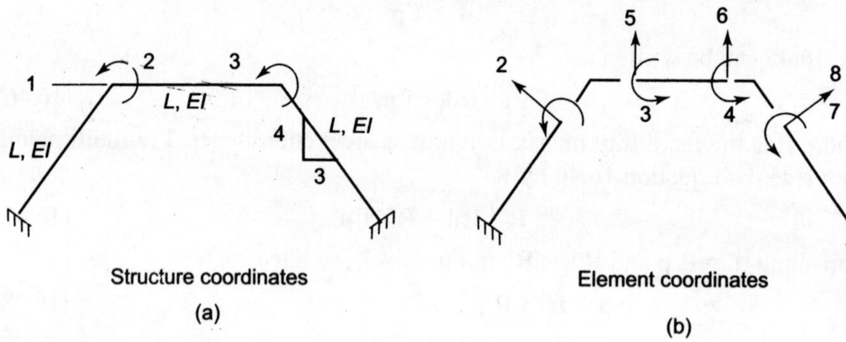

Fig. 16.18

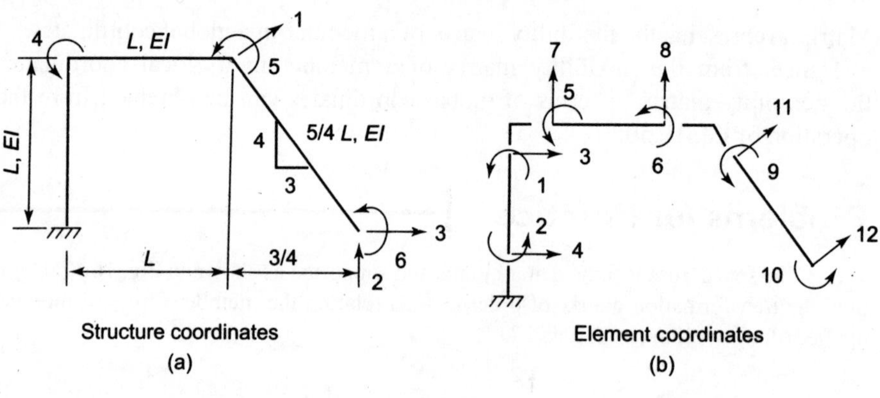

Fig. 16.19

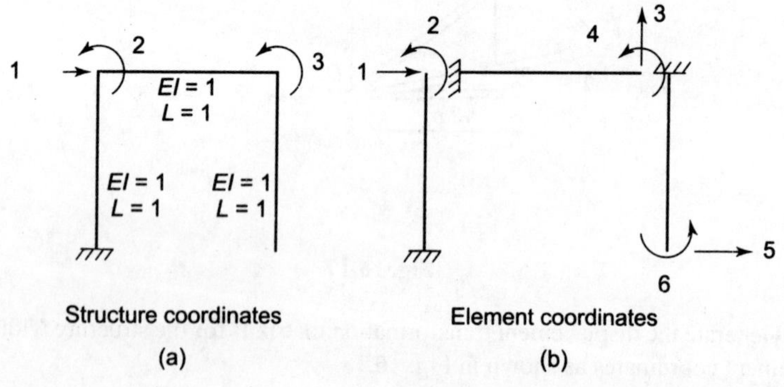

Fig. 16.20

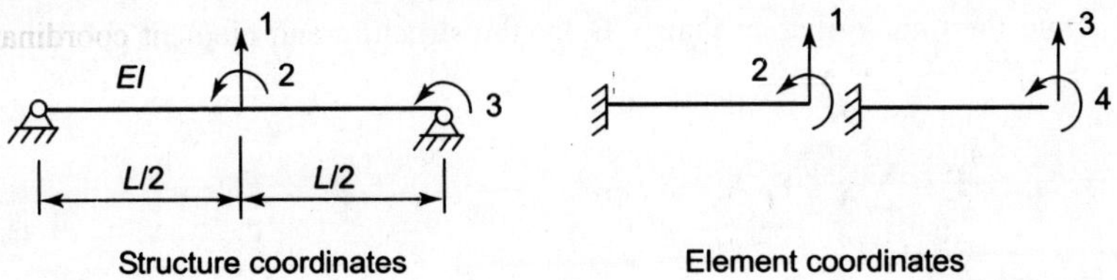
1
2
3
EI
L/2
L/2
Structure coordinates
1
2
3
4
Element coordinates

Fig. 16.21

17 Flexibility or Force Method of Analysis

17.1 INTRODUCTION

The flexibility method of matrix analysis is basically a consistent displacement method cast in a matrix form. This method can be used for the analysis of statically determinate as well as indeterminate structures. There is, however, no advantage in employing the matrix approach for determinate structures as they can easily be solved using only equations of equilibrium. In the analysis of indeterminate structures the procedures outlined in Section 10.3 are followed.

17.1.1 Flexibility Method—Steps to be Followed

1. As a first step the degree of static indeterminancy of the structure is determined. The structure is then reduced to a statically determinate one by releasing the redundant forces equal to the degree of indeterminancy, say n. The releases can be either external or internal or a combination of both. Care has to be exercised in selecting the releases so that the released structure is not only statically determinate and stable but also convenient for evaluation of displacements at the released coordinates.
2. The displacements of the primary structure under given loading are determined at all releases in the direction of releases. Any of the standard methods may be adopted for determining the displacements which may be translations or rotations or both. Sometimes one may refer to standard tables for obtaining the above displacements. These displacements D_{1p}, D_{2p}, ... D_{np} form the displacement vector {**Dp**}.
3. The next step consists of determining the displacements at the released coordinates due to unit values of the redundants applied one by one in turn at all the coordinates. These displacements form the elements of the flexibility matrix [**F**] for the structure.
4. The values of the redundant forces necessary to ensure geometric continuity or compatibility of the structure are determined by the relation

$$\{\mathbf{Dp}\} + [\mathbf{F}]\,(\mathbf{X_R}) = 0 \qquad (17.1)$$

where $\underset{n \times 1}{\mathbf{D_p}}$ = Displacements at the releases due to applied loading on the primary structure

$\underset{n \times n}{\mathbf{F}}$ = Flexibility matrix, the elements of which are displacements due to unit values of redundants on the primary structure.

$\underset{n \times 1}{\mathbf{X_R}}$ = Redundant forces on the released structure which are to be determined.

From Equation 17.1 the unknown redundant forces are obtained as

$$\{\mathbf{X_R}\} = [\mathbf{F}]^{-1}\{-\mathbf{Dp}\} \tag{17.2}$$

5. The final forces in the structure are obtained by superimposing the effects of external loading and the redundant forces on the released structure. The force in any member '*i*' can be expressed as

$$P_i = P_{si} + \{f_{i1}X_i + f_{i2}X_2 \ldots f_{in}X_n\} \tag{17.3}$$

Expressing in matrix form, the forces in all the members, m

$$\underset{m\times1}{\{\mathbf{p}\}} = \underset{m\times1}{\{\mathbf{p_s}\}} + \underset{m\times n}{[\mathbf{F}]}\ \underset{n\times1}{\{\mathbf{X_r}\}}$$

With this brief description, the application of the force or flexibility method will now be illustrated by examples.

17.1.2 Sign Convention

Any consistent sign convention can be adopted. The sign convention adopted in this matrix method of analysis is the stress sign convention as the static sign convention cannot be adopted for moment at a joint or section. The redundant forces and displacements in the direction of the coordinates are considered positive.

Example 17.1 *Fig. 17.1a shows a continuous beam ABC fixed at end A and supported on rollers at B and C. EI is the same throughout. Determine the reactions and moments over supports.*

The beam is statically indeterminate by two degrees and therefore two redundant forces are to be released to make the beam statically determinate. The following alternatives exist for the release of redundants: (1) M_A and M_B, (2) M_A and R_B, (3) R_B and R_C and (4) M_A and R_C. In this example the moments at A and B are considered as redundants. The released structure and the structure coordinates are shown in Fig. 17.1*b*. It may be noted that the released structure comprises two simple beams AB and BC.

The displacements D_1 and D_2 in the primary structure are obtained using moment area theorems.

$$D_{1P} = +\frac{100(8)^2}{16\,EI} = +\frac{400}{EI}$$

$$D_{2P} = +\frac{100(8)^2}{16\,EI} + \frac{60(8)^2}{16\,EI} = +\frac{640}{EI}$$

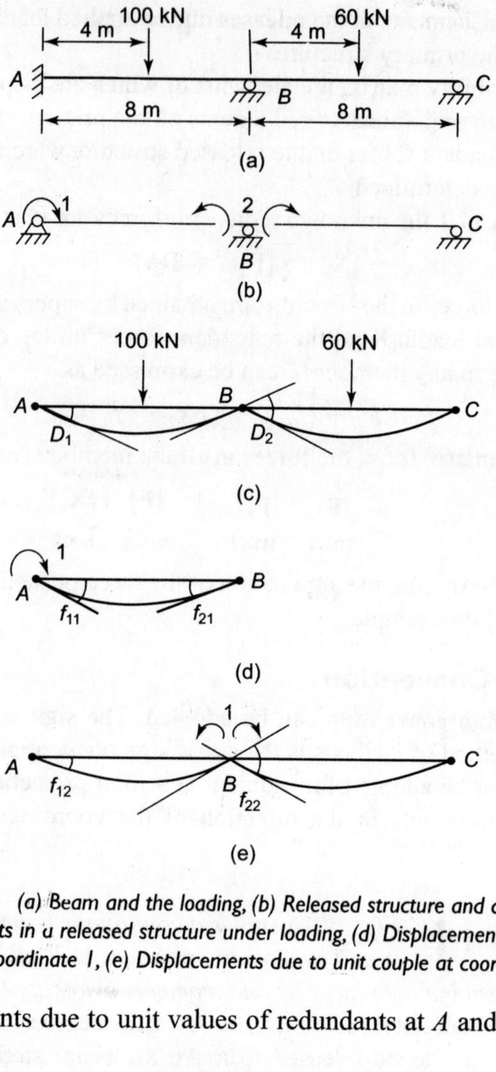

Fig. 17.1 | *(a) Beam and the loading, (b) Released structure and coordinates, (c) Displacements in a released structure under loading, (d) Displacements due to unit couple at coordinate 1, (e) Displacements due to unit couple at coordinate 2*

The displacements due to unit values of redundants at A and B are shown in Fig. 17.1d and e.

$$f_{11} = \frac{8}{3EI}$$

$$f_{21} = \frac{4}{3EI}$$

$$f_{22} = \frac{8}{3EI} + \frac{8}{3EI} = \frac{16}{3EI}$$

The flexibility matrix [**F**] for the coordinates chosen is

$$[\mathbf{F}] = \begin{bmatrix} f_{11} & f_{12} \\ f_{21} & f_{22} \end{bmatrix} \text{ or } [\mathbf{F}] = \frac{1}{3EI}\begin{bmatrix} 8 & 4 \\ 4 & 16 \end{bmatrix}.$$

The redundant forces are obtained by using Equation 17.2

$$\{\mathbf{X_R}\} = [\mathbf{F}]^{-1}\{-\mathbf{D_p}\}$$

$$[\mathbf{F}]^{-1} = \frac{3EI}{112}\begin{bmatrix} 16 & -4 \\ -4 & 8 \end{bmatrix}$$

∴ $$\{\mathbf{X_R}\} = \frac{3EI}{112}\begin{bmatrix} 16 & -4 \\ -4 & 8 \end{bmatrix}\frac{1}{EI}\begin{Bmatrix} -400 \\ -640 \end{Bmatrix}$$

Solving, $$\{\mathbf{X_R}\} = \begin{Bmatrix} -102.86 \\ -94.29 \end{Bmatrix}$$

∴ $$X_1 = -102.86 \text{ kN.m}$$

and $$X_2 = -94.29 \text{ kN.m}$$

According to our sign convention the moments at A and B cause tension at the top.

The reactions can be obtained by using statics. The vertical reactions at A, B and C are:

$$R_A = 51.07 \text{ kN}, R_B = 90.72 \text{ kN and } R_C = 18.21 \text{ kN}$$

Example 17.2 | *Using the flexibility method analyse the continuous beam shown in Fig. 17.2. The value of EI for each span is as indicated.*

The beam is statically indeterminate by two degrees. Internal moments at A and B are considered as redundants. The released structure consists of two simple beams AB and BC.

$$D_1 = \frac{wl^3}{24EI} = \frac{10(10)^3}{24EI} = \frac{416.17}{EI}$$

$$D_2 = \frac{wl^3}{24EI} = \frac{wl^2}{16(3EI)} = \frac{450}{EI}$$

The elements of the flexibility matrix are obtained by applying a unit value of the redundants as shown in Figs. 17.2*d* and *e*.

$$f_{11} = \frac{10}{3EI}$$

$$f_{21} = \frac{5}{3EI}$$

$$f_{22} = \frac{10}{3EI} + \frac{10}{9EI} = \frac{40}{9EI}$$

$$f_{21} = \frac{5}{3}EI$$

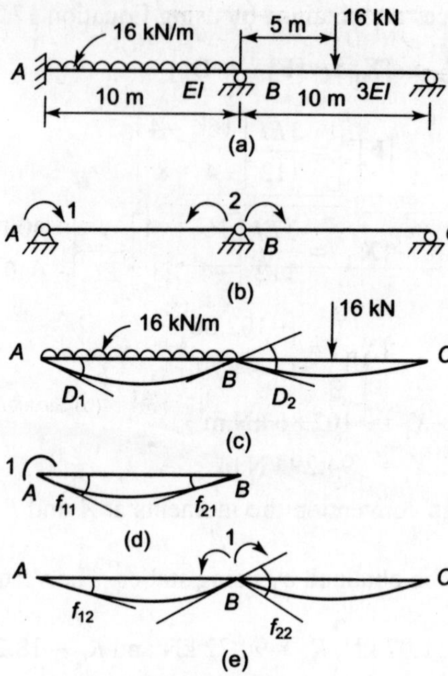

Fig. 17.2 | *(a) Beam and the loading, (b) Released structure and coordinates, (c) Displacements in a released structure under loading, (d) Displacements under unit couple at coordinate 1, (e) Displacements under unit couple at coordinates 2*

$$\therefore \qquad [\mathbf{F}] = \frac{1}{3EI}\begin{bmatrix} 10 & 5 \\ 5 & \dfrac{40}{3} \end{bmatrix}$$

$$[\mathbf{F}]^{-1} = \frac{3EI}{108.33}\begin{bmatrix} \dfrac{40}{3} & -5 \\ -5 & 10 \end{bmatrix}$$

The redundant vector

$$\{\mathbf{X_R}\} = \frac{3EI}{108.33}\begin{bmatrix} \dfrac{40}{3} & -5 \\ -5 & 10 \end{bmatrix}\frac{1}{EI}\begin{Bmatrix} -416.67 \\ -450 \end{Bmatrix}$$

$$X_1 = -91.54 \text{ kN.m}$$

and $$X_2 = -66.92 \text{ kN.m}$$

The reactions and moments at other sections can be worked out using equations of statics.

Example 17.3 | *Analyse the beam given in Fig. 17.3 for the central deflection and rotation at the ends.*

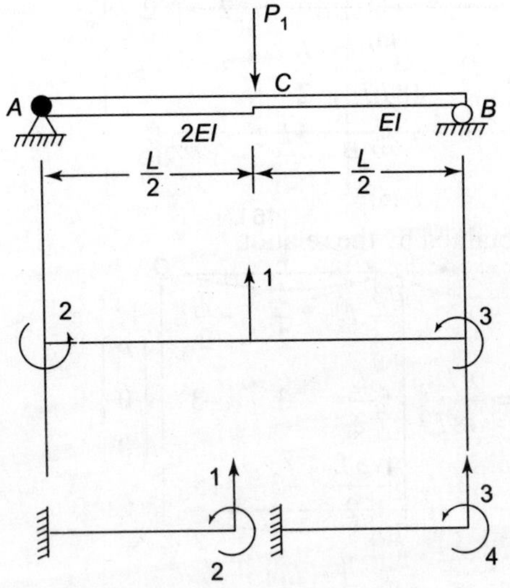

Fig. 17.3

Let us generate the flexibility matrix $[F]$ by utilising the relation:

$$[F] = [A]^T [f] [A]$$

The transformation matrix $[A]$ is obtained by applying a unit force at the structure coordinates in turns and finding out the forces at the member coordinates the resulting matrix is:

$$[A] = \begin{bmatrix} +\dfrac{1}{2} & -\dfrac{1}{L} & -\dfrac{1}{L} \\ -\dfrac{L}{4} & \dfrac{1}{2} & \dfrac{1}{2} \\ -\dfrac{1}{2} & -\dfrac{1}{L} & -\dfrac{1}{L} \\ 0 & 0 & +1 \end{bmatrix}$$

The uncoupled flexibility matrix $[f]$ is obtained by applying a unit force at the member coordinates one by one in turns and finding the displacements using standard formulae. The resulting matrix is

$$[f] = \frac{L}{48\,EI}\begin{bmatrix} L^2 & 3L & 0 & 0 \\ 3L & 12 & 0 & 0 \\ 0 & 0 & 2L^2 & 6L \\ 0 & 0 & 6L & 24 \end{bmatrix}$$

Now the flexibility matrix is $[F]$ is obtained as

$$[F]=[A]^T\,[f][A]=\frac{L}{48EI}\begin{bmatrix}\frac{3}{4}L^2 & \frac{L}{2} & -\frac{5}{2}L\\ \frac{L}{2} & 3 & -3\\ -\frac{5L}{2} & -3 & 15\end{bmatrix}$$

Next: The displacements are obtained by the relation

$$\{D\}=[F]\{P\}=\frac{L}{48EI}\begin{bmatrix}\frac{3}{4}L^2 & \frac{L}{2} & -\frac{5}{2}L\\ \frac{L}{2} & 3 & -3\\ -\frac{5L}{2} & -3 & 15\end{bmatrix}\begin{Bmatrix}P_1\\0\\0\end{Bmatrix}$$

$$\begin{Bmatrix}\Delta_C\\ \theta_A\\ \theta_B\end{Bmatrix}=\frac{P_1L^2}{EI}\begin{Bmatrix}-\frac{L}{64}\\ -\frac{1}{24}\\ \frac{5}{96}\end{Bmatrix}$$

Example 17.4 | *Analyse the beam in Example 17.3 for central deflection and end rotations if the beam is subjected to a u.d.l. from A to B of intensity w/unit length.*

The flexibility matrix $[F]$ developed in Example 7.3 is used. The displacements are calculated using the relation

$$\{D\} = [F]\,\{P\}$$

The forces $\{P\}$ are the equivalent joint loads as shown in Fig. 17.4.

or $$\{D\}=\frac{L}{48EI}\begin{bmatrix}\frac{3}{4}L^2 & \frac{L}{2} & -\frac{5}{2}L\\ \frac{L}{2} & 3 & -3\\ -\frac{5L}{2} & -3 & 15\end{bmatrix}\begin{Bmatrix}\frac{wL}{2}\\ \frac{wL^2}{48}\\ -\frac{wl^2}{48}\end{Bmatrix}$$

gives

$$\{D\} = \begin{Bmatrix} \frac{7}{768}\frac{wL^4}{EI} \\ \frac{wL^3}{128\,EI} \\ -\frac{13}{384}\frac{wl^3}{EI} \end{Bmatrix}$$

Hence,
$$\begin{Bmatrix} \Delta_C \\ \theta_A \\ \theta_B \end{Bmatrix} = \frac{wL^3}{EI}\begin{Bmatrix} \frac{7}{768}L \\ \frac{1}{128} \\ -\frac{13}{384} \end{Bmatrix}$$

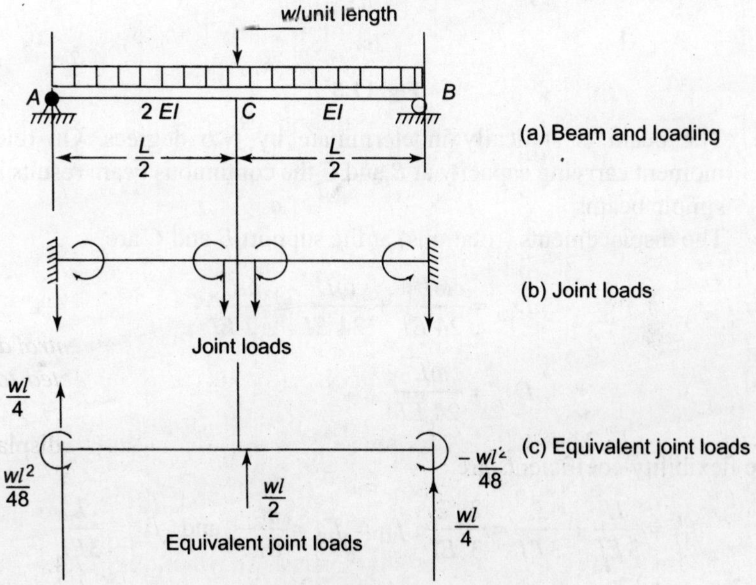

Fig. 17.4

Example 17.5 | *Analyse the prismatic two span continuous beam shown in Fig. 17.5 a due to the following effects:*

(a) A uniform load w/unit length over the entire beam.

(b) Concentrated load P at the centre of span BC

(c) Support settlement of Δ at point B.

(d) Temperature variation ΔT between the top and bottom of beam: the depth of the beam is d and coefficient of thermal expansion is α.

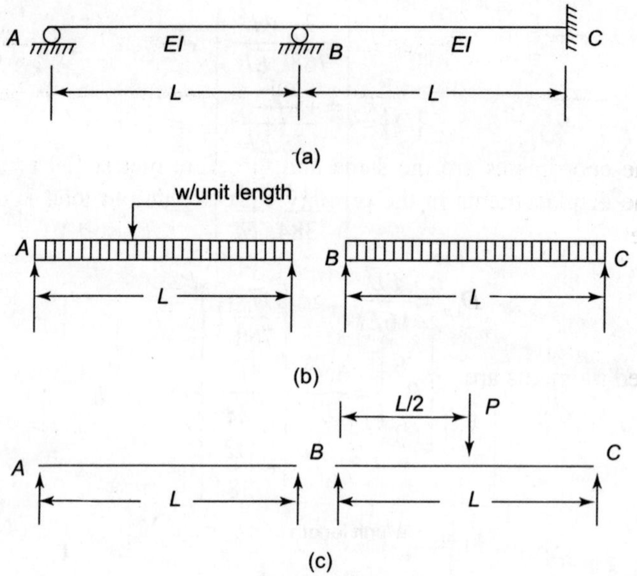

Fig. 17.5

(a) The beam is statically indeterminate by two degrees. On releasing moment carrying capacity at B and C the continuous beam results in two simple beams.

The displacements (rotations) at the support B and C are

$$D_{1p} = \frac{\omega L^3}{24\,EI} + \frac{\omega L^3}{24\,EI} = \frac{\omega L^3}{12\,EI}$$

$$D_{2p} = \frac{\omega L^3}{24\,EI}$$

The flexibility coefficients are:

$$f_{11} = \frac{L}{3\,EI} + \frac{L}{3\,EI} = \frac{2}{3}\frac{L}{EI},\; f_{21} = f_{12} = \frac{L}{6\,EI} \text{ and } f_{22} = \frac{L}{3EI}$$

Hence $$[F] = \frac{L}{6\,EI}\begin{bmatrix} 4 & 1 \\ 1 & 2 \end{bmatrix}$$

and $$[F]^{-1} = \frac{6EI}{7L}\begin{bmatrix} 2 & -1 \\ -1 & 4 \end{bmatrix}$$

The released moments are, say X_1 and X_2 at B and C are:

$$\begin{Bmatrix} X_1 \\ X_2 \end{Bmatrix} = \frac{6\,EI}{7\,L}\begin{bmatrix} 2 & -1 \\ -1 & 4 \end{bmatrix}\frac{\omega L^2}{EI}\begin{bmatrix} -\dfrac{1}{12} \\ -\dfrac{1}{24} \end{bmatrix}$$

gives $$X_1 = -\frac{3}{28}\omega l^2$$

$$X_2 = +\frac{\omega l^2}{14}$$

(b) The coordinates are the same and therefore matrix [*F*] remains same. The displacements in the primary structure due to load *P* on span *BC* are:

$$D_{1p} = \frac{PL^2}{16EI} \text{ and } D_{2p} = \frac{PL^2}{16EI}$$

The released moments are

$$\begin{Bmatrix} X_1 \\ X_2 \end{Bmatrix} = \frac{6}{7}\frac{EI}{L}\begin{bmatrix} 2 & -1 \\ -1 & 4 \end{bmatrix}\frac{PL^2}{16EI}\begin{Bmatrix} -1 \\ -1 \end{Bmatrix}$$

gives $$X_1 = -\frac{3}{6}PL \text{ and } X_2 = -\frac{9}{56}PL$$

(c) The settlement of support *B* gives the displacements in the primary structure as

$$D_{1P} = -\frac{\Delta}{L} - \frac{\Delta}{L} = -2\frac{\Delta}{L}$$

$$D_{2P} = \frac{\Delta}{L}$$

We get $$\begin{Bmatrix} X_1 \\ X_2 \end{Bmatrix} = \frac{6}{7}\frac{EI}{L}\begin{bmatrix} 2 & -1 \\ -1 & 4 \end{bmatrix}\begin{Bmatrix} \frac{2\Delta}{L} \\ -\frac{\Delta}{L} \end{Bmatrix}$$

Gives $$X_1 = \frac{30}{7}\frac{EI\Delta}{L^2} \text{ and } X_2 = -\frac{36}{7}\frac{EI\Delta}{L^2}$$

(d) If we consider the variation of temperature Δ^T over the depth of the beam '*d*' the higher temperature being at the top the primary beams deflect uploads.

Let the rotation over a unit distance be $d\theta = \alpha\frac{\Delta Td}{d} = \alpha\Delta^T$

Rotation at *B* from both the sides $= -\alpha\Delta T\left(\frac{L}{2} + \frac{L}{2}\right) = -\alpha\Delta TL$

Rotation at $C = -\frac{\alpha\Delta TL}{2}$

Therefore, $$\begin{Bmatrix} X_1 \\ X_2 \end{Bmatrix} = \frac{6}{7}\frac{EI}{L}\begin{bmatrix} 2 & -1 \\ -1 & 4 \end{bmatrix}\begin{Bmatrix} \alpha\ \Delta TL \\ \alpha\ \frac{\Delta TL}{2} \end{Bmatrix}$$

Gives $$X_1 = \frac{9}{7}\alpha\,\Delta TL$$

and $$X_2 = \frac{6}{7}\alpha\,\Delta TL$$

Example 17.6 *Analyse the frame shown in Fig. 17.6 and draw the B.M. diagram. Consider only flexural deformations and take El as constant throughout.*

The frame is statically indeterminate by two degrees. The vertical and horizontal reaction components at C are considered as redundants. The frame is reduced to a cantilever bent as shown in Fig. 27.6b on which the positive directions of coordinates are indicated.

The unit load method is adopted to determine displacements D_{IP} and D_{2P} under the given loading

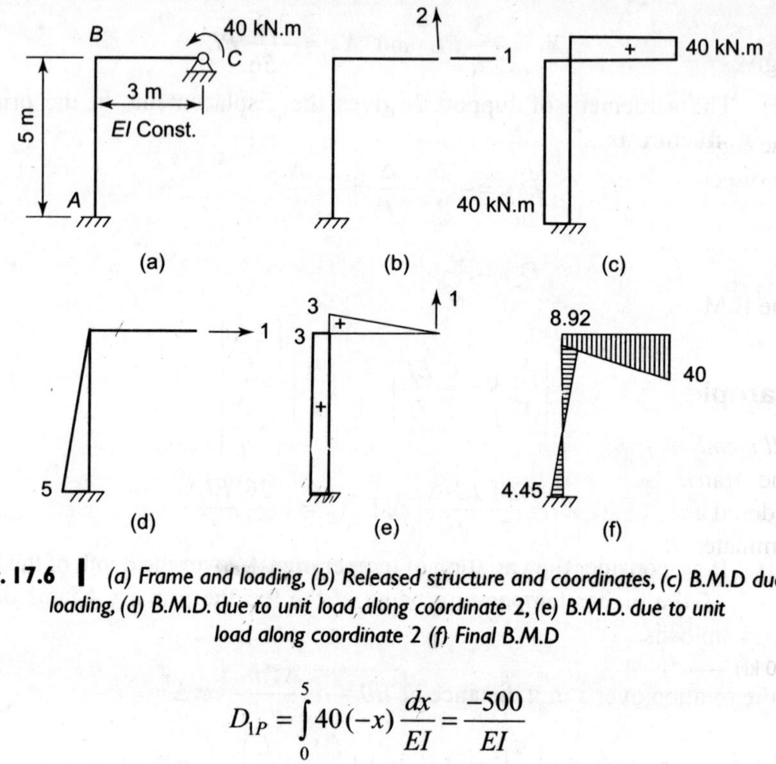

Fig. 17.6 | *(a) Frame and loading, (b) Released structure and coordinates, (c) B.M.D due to loading, (d) B.M.D. due to unit load along coordinate 2, (e) B.M.D. due to unit load along coordinate 2 (f) Final B.M.D*

$$D_{1P} = \int_0^5 40(-x)\frac{dx}{EI} = \frac{-500}{EI}$$

$$D_{2P} = \int_0^3 40(x)\frac{dx}{EI} = \int_0^5 40(3)\,dx/EI\,\frac{780}{EI}$$

The elements of the flexibility matrix are determined using the unit load method again. From Fig. 17.6d and e.

$$f_{11} = \frac{125}{3\,EI}$$

$$f_{12} = f_{21} = \frac{-37.5}{EI} \quad \therefore [\mathbf{F}] = \frac{1}{EI}\begin{bmatrix} \frac{125}{3} & -37.5 \\ -37.5 & 54 \end{bmatrix}$$

$$f_{22} = \frac{54}{EI} \quad \text{and} \quad [\mathbf{F}]^{-1} = \frac{EI}{843.75}\begin{bmatrix} 54 & 37.5 \\ 37.5 & \frac{125}{3} \end{bmatrix}$$

The redundants vector

$$\{\mathbf{X_R}\} = \frac{EI}{843.75}\begin{bmatrix} 54 & 37.5 \\ 37.5 & \frac{125}{3} \end{bmatrix}\frac{1}{EI}\begin{Bmatrix} 500 \\ -780 \end{Bmatrix}$$

This gives $X_1 = -2.67$ kN

$X_2 = -16.30$ kN

The negative signs for the reactions indicate that these components are opposite to the direction of the coordinates. The moments are determined using statics

$$M_A = 40 + 2.67\,(5) - 16.3\,(3) = 4.45 \text{ kN.m}$$

$$M_B = 40 - 16.3\,(3) = 8.92 \text{ kN.m}$$

The B.M. diagram is shown in Fig. 17.6*f*.

Example 17.7 *Using the flexibility method, analyse the pin-jointed frame in Fig. 17.7. The cross-sectional areas A and E for all members is the same.*

The frame is internally redundant by one degree. The member *AC* is considered as redundant. The removal of member *AC* makes the frame statically determinate.

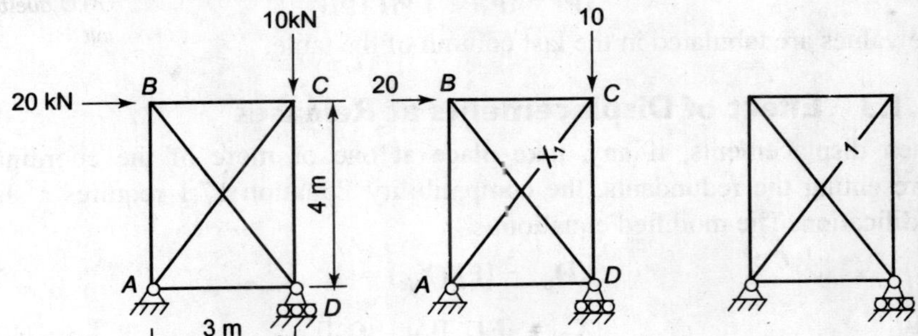

Fig. 17.7

The displacement in direction AC is determined using the unit load method. The displacement calculations are shown in tabular form. Tension is taken as positive and compression negative.

Displacement in the direction of the redundant in the released structure is

Member	l in l_m	p_s kN	p_i kN	$p_s p_i l$	$p_i^2 l$	$p = p_s + X_1 \{p_i\}$
AB	4	26.67	–0.8	–85.33	2.56	14.82
BC	3	0	–0.6	0	1.08	–8.88
CD	4	–10.0	–0.8	320.0	2.56	–21.85
AD	3	20.0	–0.6	–36.0	1.08	11.11
BD	5	–33.33	1	–166.7	5.00	–18.52
AC	5	0	1	0	5.00	14.81
				Σ –256.03	Σ 17.28	

$$D_p = \sum_{i=1}^{6} \frac{p_s\, p_i\, l}{AE} = \frac{-256.03}{AE}$$

Flexibility coefficient

$$f_{11} = \frac{\Sigma\, p_i^2 l}{AE} = \frac{17.28}{AE}$$

The compatibility condition can be written as

$$D_p + \mathrm{f}_{11} X_1 = 0$$

$$\therefore \quad X_1 = f_{11}^{-1} - D_p = \frac{256.03}{17.28} = 14.82$$

The final forces in the members are obtained by the superposition of the forces due to external loading and the redundant force

$$P_i = P_{si} + X_1\, (p_{1i})$$

Expressing in matrix form

$$\{\mathbf{p}\} = \{\mathbf{p_s}\} + \{\mathbf{X_1}\}\, \{\mathbf{p_1}\} \tag{17.4}$$

The values are tabulated in the last column of the table.

17.1.3 Effect of Displacements at Releases

When displacements, if any, take place at one or more of the coordinates representing the redundants, the compatibility Equation 17.1 requires a slight modification. The modified equation is

$$\{\mathbf{D_p}\} + [\mathrm{F}]\, (\mathbf{X_R}) = |\Delta| \tag{17.5}$$

and

$$|\mathbf{X_R}| = [\mathbf{F}]^{-1}\, [|\Delta| - |\mathbf{D_P}|] \tag{17.6}$$

in which $|\Delta|$ represents the displacements at the coordinates.

The displacements at the coordinates may be translations and rotations of supports, temperature effects and lack of fit in truss members.

Example 17.8 | *Analyse the continuous beam in Example 17.1 if the beam undergoes settlement of supports B and C by $\frac{300}{EI}$ and $\frac{200}{EI}$ respectively. The beam was analysed earlier taking support moments M_A anu M_B as redundants.*

From the earlier examples, the displacements at coordinates 1 and 2 under the given loading were

$$D_{1P} = \frac{400}{EI} \quad \text{and} \quad D_{2P} = \frac{640}{EI}$$

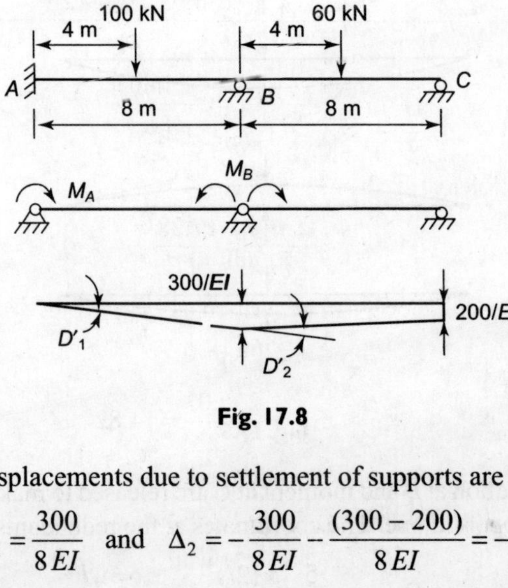

Fig. 17.8

Additional displacements due to settlement of supports are

$$\Delta_1 = \frac{300}{8EI} \quad \text{and} \quad \Delta_2 = -\frac{300}{8EI} - \frac{(300-200)}{8EI} = -\frac{50}{EI}$$

The combined displacement at coordinate $1 = \frac{400}{EI} + \frac{300}{8EI} = \frac{437.5}{EI}$

and at coordinates $2 = \frac{640}{EI} - \frac{50}{EI} = \frac{590}{EI}$

The flexibility matrix **[F]** and the inverse $[\mathbf{F}]^{-1}$ are the same as earlier. The redundants are evaluated using the Equation 17.2.

$$\{\mathbf{X_R}\} = \frac{3EI}{112}\begin{bmatrix} 16 & -4 \\ -4 & 8 \end{bmatrix} \frac{1}{EI} \begin{Bmatrix} -437.5 \\ -590.0 \end{Bmatrix}$$

Solving $$\{\mathbf{X_R}\} = \begin{Bmatrix} -124.28 \\ -79.55 \end{Bmatrix} \text{kN.m}$$

$\therefore \qquad X_1 = -124.28$ kN.m

$X_2 = -79.55$ kN.m

It may be noted that the magnitude of the moment at fixed end A is increased and the moment over support B is decreased due to settlement of supports.

Example 17.9 | *Analyse the continuous beam shown in Fig. 17.9 under the given loading. Support B sinks by 1/100 and support C rotates by 0.004 radians in the anti-clockwise direction.*

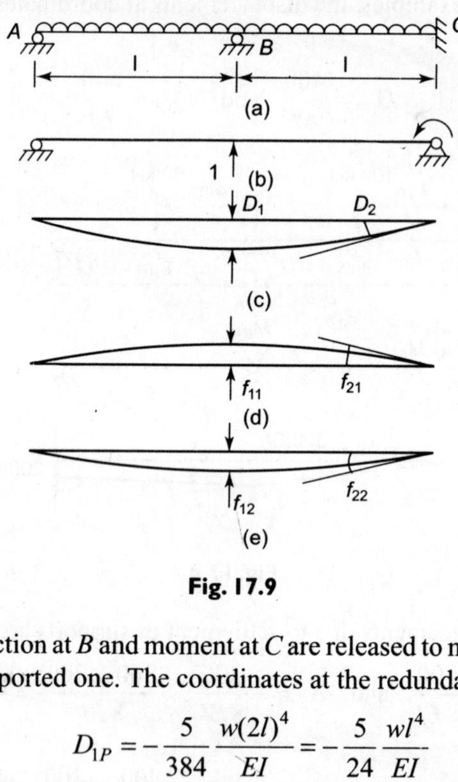

Fig. 17.9

The vertical reaction at B and moment at C are released to make the continuous beam a simply supported one. The coordinates at the redundants are indicated.

Now $$D_{1P} = -\frac{5}{384}\frac{w(2l)^4}{EI} = -\frac{5}{24}\frac{wl^4}{EI}$$

$$D_{2P} = +\frac{w(2l)^3}{24\,EI} = +\frac{wl^3}{3\,EI}$$

Given $$\Delta_i = \frac{-l}{100} \text{ and } \Delta_2 = +0.004$$

The negative sign for Δ_1 is assigned as the displacement is opposite to the direction of coordinate 1.

The elements of the flexibility matrix are obtained by applying unit values of redundants at the coordinates one after the other as shown in Fig. 17.9*d* and *e*.

$$f_{11} = \frac{(2l)^3}{48\,EI} = \frac{l^3}{6\,EI}$$

$$f_{21} = f_{12} - \frac{l^2}{4EI}$$

$$f_{22} = \frac{2l}{3EI}$$

$$[\mathbf{F}] = \frac{l}{12EI}\begin{bmatrix} 2l^2 & -3l \\ -3l & 8 \end{bmatrix}$$

$$[\mathbf{F}]^{-1} = \frac{12}{7}\frac{EI}{l^3}\begin{bmatrix} 8 & 3l \\ 3l & 2l^2 \end{bmatrix}$$

Using the compability equation

$$\{\mathbf{X}\} = -[\mathbf{F}]^{-1}[\{\Delta\} + [\{\mathbf{D}\}]$$

$$= \frac{12\,EI}{7l^3}\begin{bmatrix} 8 & 3l \\ 3l & 2l^2 \end{bmatrix}\begin{Bmatrix} +\dfrac{l}{100} & +\dfrac{5wl^4}{24EI} \\ -0.004 & -w\dfrac{l^3}{3EI} \end{Bmatrix}$$

The moments and reactions can be calculated using statics.

Example 17.10 | *Analyse the pin-jointed frame in Example 17.7 if the member AC is too long by $\frac{50}{AE}$ to fit.*

The displacement D_{1p} and the flexibility coefficient f_{11} are same as in Example 17.7. Since the member has to be compressed to fit in $\Delta = -{}^{50}/_{AE}$ the modified compatibility condition can be written as

$$D_{1P} + f_{11}(X_1) = \Delta$$

$$\therefore \qquad X_1 = (f_{11})^{-1}\{\Delta - D_{1p}\}$$

Substituting the values with proper sign

$$X_1 = \frac{AE}{17.28}\left\{\frac{-50}{AE} + \frac{256.03}{AE}\right\} = 11.92 \text{ kN (tension)}$$

The reduced value of tension in member AC is due to pre-compressor of the member before fitting. The revised values in other members car be calculated using the revised value for X_1.

17.2 | GENERALISED METHOD OF ANALYSIS

Up to this point the flexibility method has been carried out on the lines of the consistent displacement method with the resulting equations cast in matrix form. This approach, though useful in the beginning, does not lend itself to the generalised procedure for the use of computers.

It may be pointed out that only simple examples have been worked out in the following sections for a clear illustration of the method. The reader may be inclined to think that the application of matrix analysis for relatively simply problems is cumbersome. However, the method is a powerful tool in the solution of more complex problems especially with the use of computer techniques. Matrix methods are at their best when applied to complex structures which are difficult to solve by traditional methods.

17.3 STATICALLY DETERMINATE STRUCTURES

In this approach, coordinates on the structure are selected at all degrees of freedom. Additional coordinates can be included at points where external forces are applied and also where displacement measurements are desired.

The structure is then broken up into a number of individual elements such that the structure coordinates occur only at their ends. The element coordinates are then selected for which the flexibility matrix exists so that we can write $\mathbf{d_s} = \mathbf{f_s}$ $\mathbf{p_s}$ for each element and $\mathbf{d} = \mathbf{f}\,\mathbf{p}$ for all elements in an uncoupled state. A relation between the external forces and element forces is established (Eq. 16.1).

$$\mathbf{p} = \mathbf{AP}$$

This can be easily generated in the case of statically determinate structures. From the element flexibility matrices, the system flexibility matrix is synthesised using Equation 16.9.

$$\mathbf{F} = \mathbf{A^T f\, A}$$

After evaluating flexibility matrix **F**, structure displacements are computed to the set of given external forces using Eq. 15.17.

$$\mathbf{D} = \mathbf{FP}$$

The internal forces are evaluated by the relation

$$\mathbf{p} = \mathbf{AP}$$

and the internal displacements

$$\mathbf{d} = \mathbf{f\,p} = \mathbf{f\,A\,P}$$

The preceding analysis will now be illustrated by the following examples.

Example 17.11 | *For the truss of Fig. 17.10a it is required to determine the vertical and horizontal displacements of joint B and the horizontal displacement of joint C.*

This is a statically determinate truss with three degrees of freedom; two at joint *B* and one at joint *C*. We can easily establish the relation between external and internal forces by using the method of joints, which yields the relation

$$\begin{Bmatrix} p_1 \\ p_2 \\ p_3 \end{Bmatrix} = \begin{bmatrix} 0.36 & -0.48 & 1 \\ 0.80 & 0.60 & 0 \\ -0.60 & 0.80 & 0 \end{bmatrix} \begin{Bmatrix} P_1 \\ P_2 \\ P_3 \end{Bmatrix}$$

Or $\qquad \mathbf{p} = \mathbf{A} \quad \mathbf{P}$

The flexibility of an element taking only axial effect is

$$\mathbf{f}_s = \frac{L}{AE}$$

Therefore, the flexibility matrix **f** of unassembled elements is

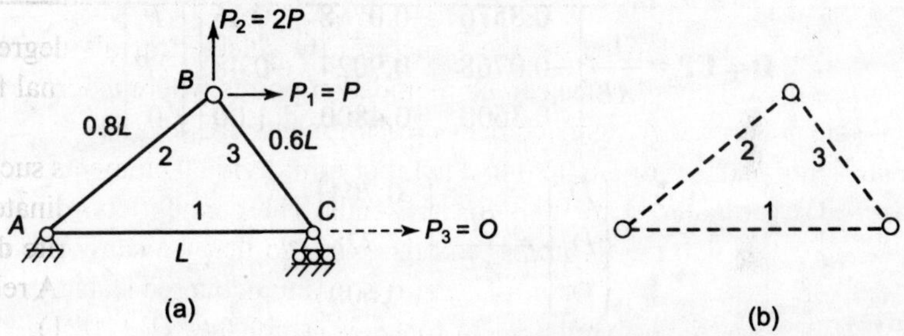

Fig. 17.10 | *(a) Truss and loading, (b) Truss elements*

$$\mathbf{f} = \frac{L}{AE} \begin{bmatrix} 1 & 0 & 0 \\ 0 & 0.8 & 0 \\ 0 & 0 & 0.6 \end{bmatrix}$$

Hence displacements

$$\mathbf{d} = \mathbf{f}\,\mathbf{p} = \mathbf{f}\,\mathbf{A}\,\mathbf{P}$$

$$\begin{Bmatrix} d_1 \\ d_2 \\ d_3 \end{Bmatrix} = \frac{L}{AE} \begin{bmatrix} 1 & 0 & 0 \\ 0 & 0.8 & 0 \\ 0 & 0 & 0.6 \end{bmatrix} \begin{bmatrix} 0.36 & -0.48 & 1 \\ 0.80 & 0.60 & 0 \\ -0.60 & 0.80 & 0 \end{bmatrix} \begin{Bmatrix} P_1 \\ P_2 \\ P_3 \end{Bmatrix}$$

or
$$\begin{Bmatrix} d_1 \\ d_2 \\ d_3 \end{Bmatrix} = \frac{L}{AE} \begin{bmatrix} 0.36 & -0.48 & 1 \\ 0.64 & 0.48 & 0 \\ -0.36 & 0.48 & 0 \end{bmatrix} \begin{Bmatrix} P_1 = P \\ P_2 = 2P \\ P_3 = 0 \end{Bmatrix}$$

or
$$\begin{Bmatrix} d_1 \\ d_2 \\ d_3 \end{Bmatrix} = \frac{PL}{AE} \begin{Bmatrix} -0.60 \\ 1.60 \\ 0.60 \end{Bmatrix}$$

The flexibility matrix **F** of the entire truss can be obtained using Eq. 16.9.

$$\mathbf{F} = \mathbf{A}^{\mathrm{T}}\,\mathbf{F}\,\mathbf{A} = \begin{bmatrix} 0.36 & 0.80 & -0.60 \\ -0.48 & 0.60 & 0.80 \\ 1.00 & 0 & 0 \end{bmatrix} \frac{L}{AE} \begin{bmatrix} 1 & 0 & 0 \\ 0 & 0.8 & 0 \\ 0 & 0 & 0.6 \end{bmatrix} \begin{bmatrix} 0.36 & -0.48 & 1 \\ 0.80 & 0.60 & 0 \\ -0.60 & 0.80 & 0 \end{bmatrix}$$

This gives

$$\mathbf{F} = \frac{L}{AE} \begin{bmatrix} 0.8576 & -0.0768 & 0.36 \\ -0.0768 & 0.9024 & -0.48 \\ 0.3600 & -0.4800 & 1.00 \end{bmatrix}$$

Hence the displacement vector D is given by

$$\mathbf{D} = \mathbf{F}\mathbf{P} = \frac{L}{AE} \begin{bmatrix} 0.8576 & -0.0768 & 0.36 \\ -0.0768 & 0.9024 & -0.48 \\ 0.3600 & -0.4800 & 1.00 \end{bmatrix} \begin{Bmatrix} P \\ 2P \\ 0 \end{Bmatrix}$$

$$\begin{Bmatrix} D_1 \\ D_2 \\ D_3 \end{Bmatrix} = \frac{PL}{AE} \begin{Bmatrix} 0.704 \\ 1.728 \\ -0.600 \end{Bmatrix} m$$

Example 17.12 | *It is required to determine the horizontal and vertical displacements of joint B and C, and the horizontal displacement component of joint D for the truss illustrated in Fig. 17.11a.*

The structure has five degrees of freedom and coordinates are indicated in Fig. 17.11*b* at all the degrees of freedom. The transformation matrix **A** relating the bar forces to applied forces is obtained by the method of joints. This results in

$$\mathbf{p} = \begin{bmatrix} 0 & 0 & 0 & 0 & 1 \\ 0 & 1 & 0 & 0 & 0 \\ -1 & 0 & 0 & 0 & 0 \\ -1 & 0 & -1 & -1 & 0 \\ \sqrt{2} & 0 & \sqrt{2} & 0 & 0 \end{bmatrix} \begin{Bmatrix} P_1 \\ P_2 \\ P_3 \\ P_4 \\ 0 \end{Bmatrix}$$

As before, the displacements **d** are obtained by the relation

$$\mathbf{d} = \mathbf{f}\mathbf{p} = \frac{L}{AE} \begin{bmatrix} 1 & 0 & 1 & 0 & 0 \\ 0 & 1 & 0 & 0 & 0 \\ 0 & 0 & 1 & 0 & 0 \\ 0 & 0 & 0 & 1 & 0 \\ 0 & 0 & 0 & 0 & \sqrt{2} \end{bmatrix} \mathbf{p}$$

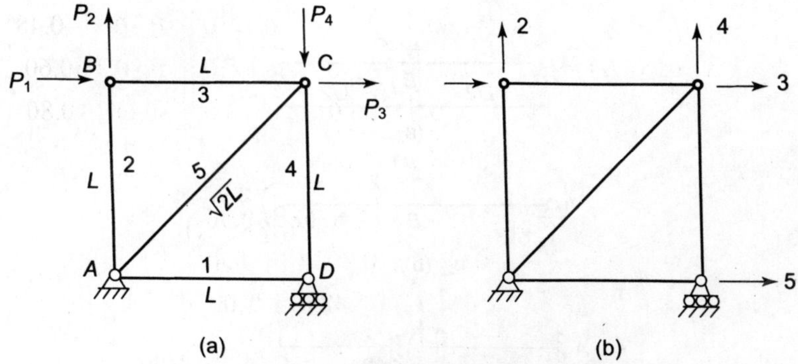

Fig. 17.11 | *(a) Truss and loading, (b) System coordinates*

The flexibility matrix **F** is synthesised using Eq. 16.9.

$$\mathbf{F} = \mathbf{A}^{\mathrm{T}}\,\mathbf{f}\mathbf{A} = \begin{bmatrix} 0 & 1 & -1 & -1 & \sqrt{2} \\ 0 & 1 & 0 & 0 & 0 \\ 0 & 0 & 0 & -1 & \sqrt{2} \\ 0 & 0 & 0 & -1 & 0 \\ 1 & 0 & 0 & 0 & 0 \end{bmatrix} \frac{L}{AE} \begin{bmatrix} 1 & 0 & 0 & 0 & 0 \\ 0 & 1 & 0 & 0 & 0 \\ 0 & 0 & 1 & 0 & 0 \\ 0 & 0 & 0 & 1 & 0 \\ 0 & 0 & 0 & 0 & \sqrt{2} \end{bmatrix}$$

$$\times \begin{bmatrix} 0 & 0 & 0 & 0 & 1 \\ 0 & 1 & 0 & 0 & 0 \\ -1 & 0 & 0 & 0 & 0 \\ -1 & 0 & -1 & -1 & 0 \\ \sqrt{2} & 0 & \sqrt{2} & 0 & 0 \end{bmatrix} = \frac{L}{AE} \begin{bmatrix} 4.828 & 0 & 3.828 & 1 & 0 \\ 0 & 1 & 0 & 0 & 0 \\ 3.828 & 0 & 3.828 & 1 & 0 \\ 1 & 0 & 1 & 1 & 0 \\ 0 & 0 & 0 & 0 & 1 \end{bmatrix}$$

Hence,

$$\mathbf{D} = \mathbf{f}\mathbf{P} = \frac{L}{AE} \begin{bmatrix} 4.828P_1 + 3.828P_3 + P_4 \\ P_2 \\ 3.828P_1 + 3.828P_3 + P_4 \\ P_1 + P_3 + P_4 \\ 0 \end{bmatrix}$$

The preceding analysis can be easily extended to statically determinate beams and frames. This is illustrated by the following examples.

Example 17.13 | *For the simply supported beam loaded as shown in Fig. 17.12a, it is required to determine the vertical displacement at B and the rotations at B and C.*

The structure coordinates are indicated in Fig. 17.12*b*. The structure is broken into two elements, *AB* and *BC*, and the coordinates for each element are indicated in Fig. 17.12*c*.

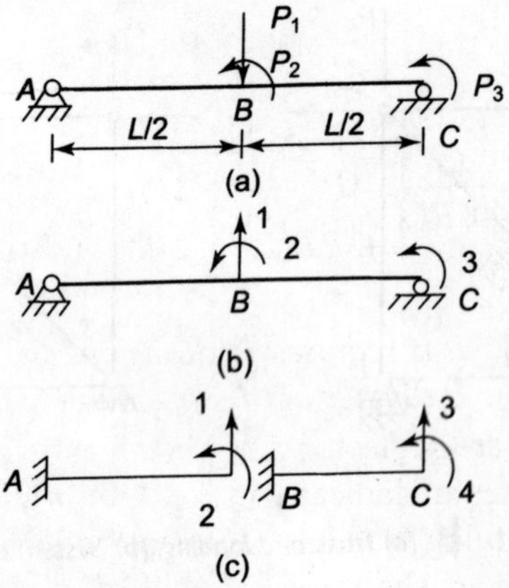

Fig. 17.12 | *(a) Beam and loading, (b) Structure coordinates, (c) Elements and element coordinates*

Since the beam is statically determinate, we can relate without much effort internal forces p of the elements to applied forces P.

$$\mathbf{P} = \mathbf{AP}$$

$$\begin{Bmatrix} p_1 \\ p_2 \\ p_3 \\ p_4 \end{Bmatrix} = \begin{bmatrix} -\frac{1}{2} & -\frac{1}{L} & -\frac{1}{L} \\ +\frac{L}{4} & +\frac{1}{2} & +\frac{1}{2} \\ +\frac{1}{2} & -\frac{1}{L} & -\frac{1}{L} \\ 0 & 0 & +1 \end{bmatrix} \begin{Bmatrix} P_1 \\ P_2 \\ P_3 \end{Bmatrix}$$

We can get the displacements of the elements by the relation

$$\mathbf{d} = \mathbf{f\,P}$$

The uncoupled flexibility matrix f is

$$\mathbf{f} = \frac{L}{24\,EI}\left[\begin{array}{cc|cc} L^2 & 3L & 0 & 0 \\ 3L & 12 & 0 & 0 \\ \hline 0 & 0 & L^2 & 3L \\ 0 & 0 & 3L & 12 \end{array}\right]$$

The flexibility matrix F is synthesised using Eq. 16.9.

$$\mathbf{F} = \mathbf{A^T\,f\,A} = \frac{L}{48\,EI}\begin{bmatrix} L^2 & 0 & 3L \\ 0 & 4 & -2 \\ 3L & -2 & 16 \end{bmatrix}$$

The displacement vector D is given by

$$\mathbf{D} = \mathbf{F}\,\mathbf{P}$$

$$\mathbf{D} = \frac{L}{48EI}\begin{bmatrix} -P_1L^2 & + & 3P_3L & \\ 4P_2 & - & 2P_3 & \\ -3P_1L & - & 2P_2 & + & 16P_3 \end{bmatrix} m$$

Example 17.14 | *It is required to determine the displacements at the co-ordinates of the structure for the frame of Fig. 17.13a.*

Axial deformations are neglected. Only bending deformations are considered. The structure coordinates are indicated in Fig. 17.13*b* and the element coordinates in Fig. 17.13*c*. The varying load has been replaced by equivalent joint loads as given by Eq. 15.93 and are shown in Fig. 17.13*d*.

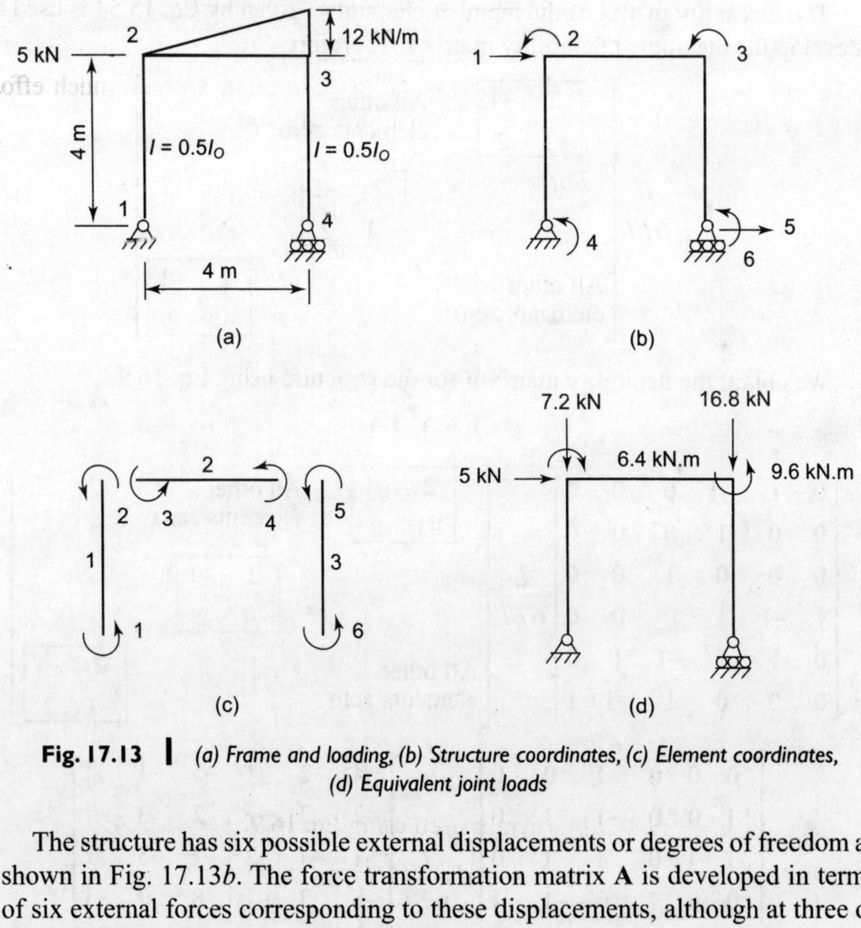

Fig. 17.13 | *(a) Frame and loading, (b) Structure coordinates, (c) Element coordinates, (d) Equivalent joint loads*

The structure has six possible external displacements or degrees of freedom as shown in Fig. 17.13*b*. The force transformation matrix **A** is developed in terms of six external forces corresponding to these displacements, although at three of the coordinates the forces have zero values.

Since the frame is statically determinate, from statics alone we construct the **A** matrix.

$$\begin{array}{ccccc} \mathbf{P} & = & \mathbf{A} & & \mathbf{P} \end{array}$$

$$\begin{Bmatrix} p_1 \\ p_2 \\ p_3 \\ p_4 \\ p_5 \\ p_6 \end{Bmatrix} = \begin{bmatrix} 0 & 0 & 0 & 1 & 0 & 0 \\ 1 & 0 & 0 & -1 & 1 & 0 \\ -1 & 1 & 0 & 1 & 1 & 0 \\ 0 & 0 & 1 & 0 & -1 & 1 \\ 0 & 0 & 0 & 0 & 1 & -1 \\ 0 & 0 & 0 & 0 & 0 & 1 \end{bmatrix} \begin{Bmatrix} P_1L \\ P_2 \\ P_3 \\ P_4 \\ P_5L \\ P_6 \end{Bmatrix}$$

The linear force quantities are multiplied by a factor L in order to maintain the nondimensional characteristic nature of matrix **A**. In this case $L = 4$ m.

The flexibility matrix of the member element as given by Eq. 15.54 is used to develop the uncoupled flexibility matrix **f**. Therefore,

$$\mathbf{f} = \frac{L}{6EI}\begin{bmatrix} \begin{array}{|cc|}\hline 2 & -1 \\ -1 & 2 \\ \hline \end{array} & & \text{All other elements zero} \\ & \begin{array}{|cc|}\hline 2 & -1 \\ -1 & 2 \\ \hline \end{array} & \\ \text{All other elements zero} & & \begin{array}{|cc|}\hline 2 & -1 \\ -1 & 2 \\ \hline \end{array} \end{bmatrix}$$

We obtain the flexibility matrix **F** for the structure using Eq. 16.9.

$$\mathbf{F} = \mathbf{A}^{\mathrm{T}}\,\mathbf{f}\,\mathbf{A}$$

$$\begin{bmatrix} 0 & 1 & -1 & 0 & 0 & 0 \\ 0 & 0 & 1 & 0 & 0 & 0 \\ 0 & 0 & 0 & 1 & 0 & 0 \\ 1 & -1 & 1 & 0 & 0 & 0 \\ 0 & 1 & 1 & -1 & 1 & 0 \\ 0 & 0 & 0 & 1 & -1 & 1 \end{bmatrix} \frac{L}{6EI}\begin{bmatrix} \begin{array}{|cc|}\hline 2 & -1 \\ -1 & 2 \\ \hline \end{array} & & \text{All other elements zero} \\ & \begin{array}{|cc|}\hline 2 & -1 \\ -1 & 2 \\ \hline \end{array} & \\ \text{All other elements zero} & & \begin{array}{|cc|}\hline 2 & -1 \\ -1 & 2 \\ \hline \end{array} \end{bmatrix}$$

$$\begin{bmatrix} 0 & 0 & 0 & 1 & 0 & 0 \\ 1 & 0 & 0 & -1 & 1 & 0 \\ -1 & 1 & 0 & 1 & 1 & 0 \\ 0 & 0 & 1 & 0 & -1 & 1 \\ 0 & 0 & 0 & 0 & 1 & -1 \\ 0 & 0 & 0 & 0 & 0 & 1 \end{bmatrix} = \frac{L}{6EI}\begin{bmatrix} 4 & -2 & 1 & -5 & -1 & 1 \\ -2 & 2 & -1 & 2 & 3 & -1 \\ 1 & -1 & 2 & -1 & -3 & 2 \\ -5 & 2 & -1 & 8 & 0 & -1 \\ -1 & 3 & -3 & 0 & 10 & -6 \\ 1 & -1 & 2 & -1 & -6 & 8 \end{bmatrix}$$

The external loads are related to external displacements in the form.

$$\begin{Bmatrix} D_1/L \\ D_2 \\ D_3 \\ D_4 \\ D_5/L \\ D_6 \end{Bmatrix} = \frac{L}{6EI}\begin{bmatrix} 4 & -2 & 1 & -5 & -1 & 1 \\ -2 & 2 & -1 & 2 & 3 & -1 \\ 1 & -1 & 2 & -1 & -3 & 2 \\ -5 & 2 & -1 & 8 & 0 & -1 \\ -1 & 3 & -3 & 0 & 10 & -6 \\ 1 & -1 & 2 & -1 & -6 & 8 \end{bmatrix}\begin{Bmatrix} P_1L \\ P_2 \\ P_3 \\ P_4 \\ P_5L \\ P_6 \end{Bmatrix}$$

For the loading shown in Fig. 17.10*d* the load vector *P* and the resulting deflections are

$$\begin{Bmatrix} P_1L \\ P_2 \\ P_3 \\ P_4 \\ P_5 \\ P_6 \end{Bmatrix} = \begin{vmatrix} 20.0 \\ -6.4 \\ 9.6 \\ 0 \\ 0 \\ 0 \end{vmatrix} \text{ and } \begin{Bmatrix} D_1L \\ D_2 \\ D_3 \\ D_4 \\ D_5/L \\ D_6 \end{Bmatrix} = \frac{1}{EI}\begin{Bmatrix} 68.27 \\ -41.60 \\ 30.40 \\ -81.60 \\ -45.33 \\ 30.40 \end{Bmatrix} \text{ radians}$$

If displacements are desired only at the points of application of external loads, matrix **A** need be developed only for those external loads. This would have resulted in a smaller **F** matrix. However, the displacements at other points cannot be obtained. In the present example, if only displacements corresponding to given external loads P_1, P_2 and P_3 were to be determined, the **F** matrix would have been a 3×3 matrix.

17.3.1 Computer Programme for Statically Determinate Structures

The same general procedure can be used for the analysis of any type of statically determinate structural system. When performing these analyses, it is relatively a simply task to compute the initial matrices [**A**] and [**f**]. However, the computations required to generate [**f**], {**p**}, and {**D**} are very tedious and prone to error. Computer programme can be easily developed for the matrix operations on a computer.

17.3.2 Flow Chart

The flow chart given in Fig. 17.14 enables one to perform a flexibility analysis of a statically determinate structure. The input data for the programme consists of the equilibrium matrix [**A**], the member flexibility matrix [**f**] and the joint load matrix {**P**}. The output consists of the structure flexibility matrix [**F**], the member end forces matrix {**p**} and the joint displacement matrix {**D**}.

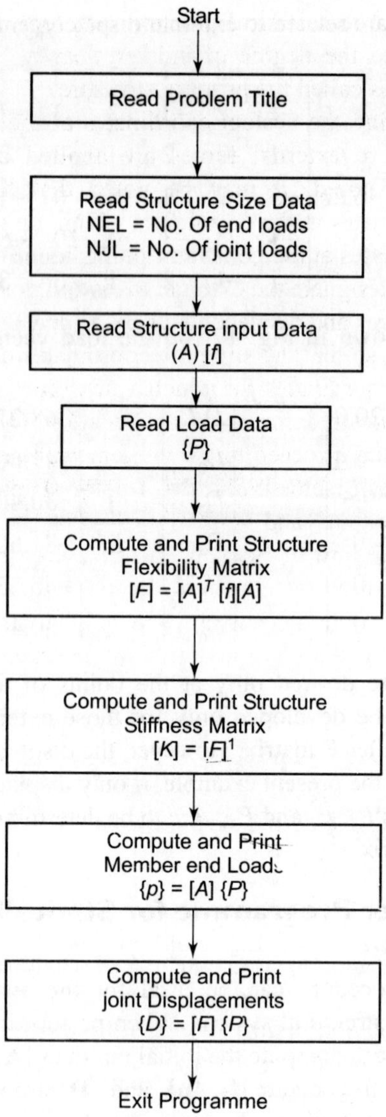

Fig. 17.14 | *Flow chart for flexibility analysis of statically determinate structures*

17.4 | STATICALLY INDETERMINATE STRUCTURES

For statically determinate structures, matrix **A** can be generated by conditions of equilibrium. However, for statically indeterminate structures, the conditions of equilibrium alone are not sufficient to generate **A**.

We shall now show how the flexibility method is used to analyse statically indeterminate structures. To determine matrix A, we reduce the statically indeterminate structure to a statically determinate one by removing a number of constraints equal to the degree of indeterminancy. This reduced statically determinate structure is called the primary structure.

Next, we shall define the system coordinates at the points of interest, that is, at the points where external forces are applied or where displacement measurements are required. It may be noted that the system coordinates should include coordinates at the constraints which are removed in the primary structure. The coordinates at the constraint points identify the redundant forces. For identification we designate the external loads and displacements by $\mathbf{P}_i$ and $\mathbf{D}_i$ and the redundant forces and displacements by $\mathbf{X}_r$ and $\mathbf{D}_r$ respectively.

We select elements so that the structure coordinates occur only at their ends. Next we fix element coordinates for which a flexibility matrix exists and takes into account all the desired forms of deformation. We number the element coordinates in sequence, proceeding from element to element so that we can write $\mathbf{d}_s = \mathbf{f}_s\, \mathbf{p}_s$ for each element and $\mathbf{d} = \mathbf{f}\, \mathbf{p}$ for all elements before they are coupled to form the system.

Following the procedure of Section 16.1, we construct the transformation matrix $\mathbf{A}$ which transforms system forces $\mathbf{P}_i$ and redundants $\mathbf{X}_r$ on the primary structure to element forces $\mathbf{p}$. To distinguish the external forces $\mathbf{P}_i$ and redudant forces $\mathbf{X}_r$ and the corresponding transformation matrix $\mathbf{A}$, we write Eq. 16.1 in the partitional form as

$$\mathbf{p} = [\mathbf{A}_l\ \mathbf{A}_r]\begin{Bmatrix} \mathbf{P}_l \\ \mathbf{X}_r \end{Bmatrix} \tag{17.7}$$

or

$$\mathbf{p} = \mathbf{A}_l\, \mathbf{p}_l + \mathbf{A}_r\, \mathbf{X}_r \tag{17.8}$$

in which

$\mathbf{P}_l$ = external load vector

$\mathbf{X}_r$ = redundant forces vector

$\mathbf{A}_l$ = transformation matrix connecting the external forces to the element forces

$\mathbf{A}_r$ = transformation matrix connecting the redundant forces to the element forces.

At this point, it may be noted that redundant forces **X***r* are not known and hence **p** cannot be computed. However, flexibility matrix **F** can be synthesized using Eq. 16.9. This equation can be written in the partitioned form as

$$\mathbf{F} = \begin{bmatrix} \mathbf{A}_l^T \\ \hline \mathbf{A}_r^T \end{bmatrix} \mathbf{F}[\mathbf{A}_l : \mathbf{A}_r] \tag{17.9}$$

or

$$\mathbf{F} = \left[\begin{array}{c|c} \mathbf{F}_{ll} & \mathbf{F}_{lr} \\ \hline \mathbf{F}_{rl} & \mathbf{F}_{rr} \end{array}\right] \tag{17.10}$$

Expanding the submatrices, we get

$$\mathbf{F}_{ll} = \mathbf{A}_l^T \mathbf{f} \mathbf{A}_l$$

$$\mathbf{F}_{lr} = \mathbf{A}_l^T \mathbf{f} \mathbf{A}_r \quad (17.11)$$

$$\mathbf{F}_{rl} = \mathbf{A}_r^T \mathbf{f} \mathbf{A}_l$$

$$\mathbf{F}_{rr} = \mathbf{A}_r^T \mathbf{f} \mathbf{A}_r$$

The submatrices of Eq. 17.11 are simply the deflection influence coefficients of the primary structure for forces $\mathbf{P}_l$ and $\mathbf{X}_r$.

We can write down the force displacement relationship to the primary structure as

$$\mathbf{D} = \mathbf{FP}$$

or in the partitioned form

Unknown displacements at load points — Known forces (loads)

$$\begin{Bmatrix} D_l \\ \hline D_r \end{Bmatrix} = \begin{bmatrix} \mathbf{F}_{ll} & \mathbf{F}_{lr} \\ \hline \mathbf{F}_{rl} & \mathbf{F}_{rr} \end{bmatrix} \begin{Bmatrix} \mathbf{P}_l \\ \hline \mathbf{X}_r \end{Bmatrix}$$

known displacements (specified) at redundants — Unknown forces (redundants)

(17.12)

In this equation the redundant force vector $\mathbf{X}_r$ which depends on applied forces $\mathbf{P}_l$ must take such values as to restore the displacements $\mathbf{D}_r$ at the constraint points to their specified values in the original structure under the action of forces $\mathbf{P}_l$. This is equivalent to saying that the values of $\mathbf{X}_r$ must be such as to satisfy the conditions of compatibility.

Considering that there are no displacements at the points of constraints in the original unreduced structure, we substitute $D_r = 0$ in Eq. 17.12 and write it into two separate matrix equations as

$$\mathbf{D}_l = \mathbf{F}_{ll}\,\mathbf{P}_i + \mathbf{F}_{lr}\,\mathbf{X}r \quad (17.13)$$

$$\mathbf{0} = \mathbf{F}_{rl}\,\mathbf{P}_l + \mathbf{F}_{rr}\,\mathbf{X}_r \quad (17.14)$$

From Eq. 17.14 we solve for redundant forces $\mathbf{X}_r$ in terms of the applied forces

$$\mathbf{X}_r = -\,\mathbf{F}_{rr}^{-1}\,\mathbf{F}_{rl}\,\mathbf{P}_l \quad (17.15)$$

Substituting for $\mathbf{X}_r$ in Eq. 17.13 we get

$$\mathbf{D}_l = \mathbf{F}_l\,\mathbf{P}_l \quad (17.16)$$

in which $$\mathbf{F}_l = \mathbf{F}_{ll} - \mathbf{F}_{ir}\,\mathbf{F}_{rr}^{-1}\,\mathbf{F}_{rl} \quad (17.17)$$

$\mathbf{F}_l$ is known as *reduced flexibility matrix* corresponding to the coordinates at which the applied forces exist.

The member end forces can now be computed by superimposing the effects of external loads $\{\mathbf{P}_l\}$ and the redundant forces $\{\mathbf{X}_r\}$ on the reduced structure

$$\{\mathbf{p}\} = \{\mathbf{p}_l\} + \{\mathbf{p}_r\} = [\mathbf{A}_l]\,\{\mathbf{P}_l\} + [\mathbf{A}_r]\,\{\mathbf{X}_r\} \tag{17.18}$$

The internal forces P can be obtained by substituting for $\mathbf{X}_r$ in Eq. 17.8. This gives

$$\mathbf{p} = (\mathbf{A}_l - \mathbf{A}_r\,\mathbf{F}^{-1}{}_{rr}\,\mathbf{F}_{rl})\,\mathbf{P}_l \tag{17.19}$$

The displacements $\mathbf{d}$ at the element coordinates can be obtained in terms of applied forces $\mathbf{P}_l$ by substituting for $\mathbf{p}$ from Eq. 17.19, that is

$$\mathbf{d} = (\mathbf{f}A_l - \mathbf{A}_r\,\mathrm{F}^{-1}{}_{rr}\,\mathbf{F}_{rl})\,\mathbf{P}_l \tag{17.20}$$

The procedure is illustrated through the simple examples that follow.

Example 17.15 *It is required to determine the bar forces and displacements at coordinates 1 and 2 for the three-bar truss loaded as shown in Fig. 17.15a.*

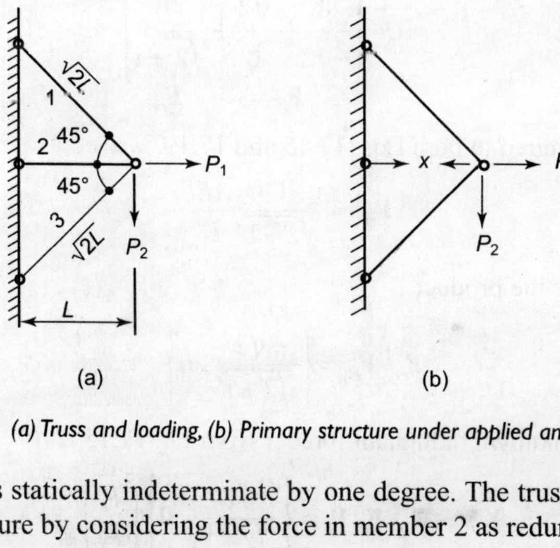

Fig. 17.15 *(a) Truss and loading, (b) Primary structure under applied and redundant forces*

The truss is statically indeterminate by one degree. The truss is reduced to a primary structure by considering the force in member 2 as redundant.

This yields

$$\mathbf{p} = \mathbf{AP} = \begin{Bmatrix} p_1 \\ p_2 \\ p_3 \end{Bmatrix} = \frac{1}{\sqrt{2}} \underset{\mathbf{A}_l \qquad\quad \mathbf{A}_r}{\left[\begin{array}{cc:c} 1 & 0 & -1 \\ 0 & 0 & \sqrt{2} \\ 1 & -1 & -1 \end{array}\right]} \begin{Bmatrix} P_1 \\ P_2 \\ X \end{Bmatrix}$$

The flexibility matrix of each bar $\mathbf{f}_s = \dfrac{L}{AE}$. The flexibility matrix $\mathbf{f}$ of the uncoupled members can be written as

$$\mathbf{f} = \frac{L}{AE}\begin{bmatrix} \sqrt{2} & 0 & 0 \\ 0 & 1 & 0 \\ 0 & 0 & \sqrt{2} \end{bmatrix}$$

Now we generate matrix **F** by carrying out the operation of Eq. 16.9.

$$\mathbf{F} = \mathbf{A}^T\, \mathbf{f}\, \mathbf{A} = \frac{1}{\sqrt{2}}\begin{bmatrix} 1 & 0 & 1 \\ 1 & 0 & -1 \\ \hline -1 & \sqrt{2} & -1 \end{bmatrix} \frac{L}{AE}\begin{Bmatrix} \sqrt{2} & 0 & 0 \\ 0 & 1 & 0 \\ 0 & 0 & \sqrt{2} \end{Bmatrix}$$

$$\frac{1}{\sqrt{2}}\begin{Bmatrix} 1 & 1 & -1 \\ 0 & 0 & \sqrt{2} \\ 1 & -1 & -1 \end{Bmatrix}$$

$$\mathbf{F}_{ll} \qquad \mathbf{F}_{lr}$$

$$\mathbf{F} = \frac{1}{AE}\begin{Bmatrix} \sqrt{2} & 0 & -\sqrt{2} \\ 0 & \sqrt{2} & 0 \\ \hline -\sqrt{2} & 0 & \sqrt{2}+1 \end{Bmatrix}$$

$$\mathbf{F}_{rl} \qquad \mathbf{F}_{rr}$$

Since $F^{-1}{}_{rr}$ is required in both Eqs. 17.15 and 17.19, we see

$$\mathbf{F}_{rr}^{-1} = \frac{1}{\sqrt{2}+1}\,\frac{AE}{L}$$

Next we evaluate the product

$$\mathbf{F}_{rr}^{-1}\,\mathbf{F}_{rl} = \begin{Bmatrix} \dfrac{-\sqrt{2}}{\sqrt{2}+1} & 0 \end{Bmatrix}$$

Therefore, the unknown redundant force X from Eq. 17.15 is

$$\mathbf{X} = -\,\mathbf{F}_{rr}^{-1}\,\mathbf{F}_{rl}\,\mathbf{P}_l = \begin{Bmatrix} \dfrac{\sqrt{2}}{\sqrt{2}+1} & 0 \end{Bmatrix}\begin{pmatrix} P_1 \\ P_2 \end{pmatrix}$$

Flexibility matrix $\mathbf{F}_l$ corresponding to external forces P_l is given by Eq. 17.17.

$$\mathbf{F}_l = \mathbf{F}_{ll} - \mathbf{F}_{lr}\,\mathbf{F}_{rr}^{-1}\,\mathbf{F}_{rl}$$

$$= \frac{L}{AE}\begin{bmatrix} \sqrt{2} & 0 \\ 0 & \sqrt{2} \end{bmatrix} - \frac{L}{AE}\begin{Bmatrix} -\sqrt{2} \\ 0 \end{Bmatrix}\begin{Bmatrix} \dfrac{-\sqrt{2}}{\sqrt{2}+1} & 0 \end{Bmatrix}$$

$$\mathbf{F}_l = \frac{L}{AE}\begin{bmatrix} \dfrac{\sqrt{2}}{\sqrt{2}+1} & 0 \\ 0 & \sqrt{2} \end{bmatrix}$$

Then $$\mathbf{D}_l = \mathbf{F}_l\,\mathbf{P}_l$$

$$= \frac{L}{AE}\begin{bmatrix} \frac{\sqrt{2}}{\sqrt{2}+1} & 0 \\ 0 & \sqrt{2} \end{bmatrix} \begin{Bmatrix} P_1 \\ P_2 \end{Bmatrix} = \frac{\sqrt{2}L}{AE}\begin{bmatrix} \frac{P_1}{\sqrt{2}+1} \\ P_2 \end{bmatrix}$$

The bar forces are obtained from Eq. 17.18.

$$\mathbf{p} = (\mathbf{A}_l - \mathbf{A}_r\,\mathbf{F}_{rr}^{-1}\,\mathbf{F}_{rl})\,\mathbf{P}_l$$

$$= \left(\frac{1}{\sqrt{2}}\begin{bmatrix} 1 & 1 \\ 0 & 0 \\ 1 & -1 \end{bmatrix} - \frac{1}{\sqrt{2}}\begin{bmatrix} -1 \\ \sqrt{2} \\ -1 \end{bmatrix} \begin{Bmatrix} \frac{-\sqrt{2}}{\sqrt{2}+1} & 0 \end{Bmatrix} \right) \begin{Bmatrix} P_1 \\ P_2 \end{Bmatrix}$$

$$= \frac{1}{\sqrt{2}}\begin{bmatrix} \frac{1}{1+\sqrt{2}} & 1 \\ \frac{2}{\sqrt{2}+1} & 0 \\ \frac{1}{\sqrt{2}+1} & -1 \end{bmatrix} \begin{Bmatrix} P_1 \\ P_2 \end{Bmatrix}$$

Example 17.16 *Generate the 2×2 matrix $\mathbf{F}_l$ and find the internal forces in terms of $\mathbf{P}_l$ for the continuous beam shown in Fig. 17.16a. EI is the same throughout.*

The structure is statically indeterminate to the second degree. The structure can be reduced to a primary structure by removing the two interior supports or by removing the moments constraints over supports. For our illustration, the support moments are chosen as redundants. Using equations of equilibrium, we can construct the following matrix **A**.

$$\mathbf{A} = \left[\begin{array}{cc|cc} 0 & 0 & 0 & 0 \\ -L/4 & 0 & 1/2 & 0 \\ L/4 & 0 & -1/2 & 0 \\ 0 & 0 & 1 & 0 \\ 0 & 0 & -1 & 0 \\ 0 & -L/4 & +1/2 & +1/2 \\ 0 & +L/4 & -1/2 & -1/2 \\ 0 & 0 & 0 & +1 \\ 0 & 0 & 0 & -1 \\ 0 & 0 & 0 & 0 \end{array}\right]$$

$$\qquad\qquad \mathbf{A}_l \qquad\qquad\qquad \mathbf{A}_r$$

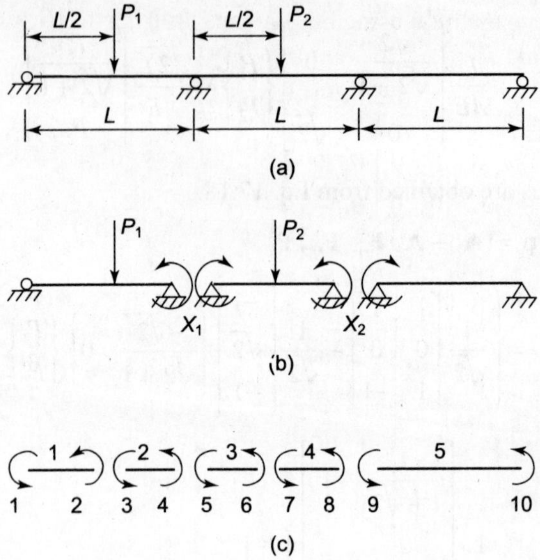

Fig. 17.16 | *(a) Continuous beam and loading, (b) Primary structure with system coordinates, (c) Elements and element coordinates*

The element flexibility coefficients are

$$\mathbf{f}_1 = \mathbf{f}_2 = \mathbf{f}_3 = \mathbf{f}_4 = \frac{L}{12\,EI}\begin{bmatrix} 2 & -1 \\ -1 & 2 \end{bmatrix}$$

and
$$\mathbf{f}_5 = \frac{L}{12\,EI}\begin{bmatrix} 4 & -2 \\ -2 & 4 \end{bmatrix}$$

The uncoupled flexibility matrix **f** of the structure is

$$\mathbf{f} \times \frac{L}{12\,EI} \begin{bmatrix} 2 & -1 & & & & & & & & \\ -1 & 2 & & & & & & & & \\ & & 2 & -1 & & & & & & \\ & & -1 & 2 & & & & & & \\ & & & & 2 & -1 & & & & \\ & & & & -1 & 2 & & & & \\ & & & & & & 2 & -1 & & \\ & & & & & & -1 & 2 & & \\ & & & & & & & & 4 & -2 \\ & & & & & & & & -2 & 4 \end{bmatrix}$$

All other elements zero

Now we generate matrix **F** using Eq. 16.9 and utilising the partitioned matrices as in Eqs. 17.10 and 17.11.

$$\mathbf{F}_{ll} = \mathrm{A}_l^T \; f \; \mathbf{A}_l$$

$$= \frac{L}{4}\left[\begin{array}{cc|cc|cc|cc|cc} 0 & -1 & 1 & 0 & 0 & 0 & 0 & 0 & 0 & 0 \\ 0 & 0 & 0 & 0 & 0 & -1 & 1 & 0 & 0 & 0 \end{array}\right]$$

$$\times \frac{L}{12EI}\begin{bmatrix} 2 & -1 & & & & & & & & \\ -1 & 2 & & & & & & & & \\ & & 2 & -1 & & & & & & \\ & & -1 & 2 & & & & & & \\ & & & & 2 & -1 & & & & \\ & & & & -1 & 2 & & & & \\ & & & & & & 2 & -1 & & \\ & & & & & & -1 & 2 & & \\ & & & & & & & & 4 & -2 \\ & & & & & & & & -2 & 4 \end{bmatrix}$$

(All other elements zero)

$$\times \frac{L}{4}\left[\begin{array}{cc} 0 & 0 \\ -1 & 0 \\ \hline 1 & 0 \\ 0 & 0 \\ \hline 0 & 0 \\ 0 & -1 \\ \hline 0 & 1 \\ 0 & 0 \\ \hline 0 & 0 \\ 0 & 0 \end{array}\right] = \frac{L^3}{192\,EI}\begin{bmatrix} 4 & 0 \\ 0 & 4 \end{bmatrix}$$

Similarly, proceeding with other partitioned matrices

$$\mathbf{F}_{rr} = \frac{L}{48\,EI}\begin{bmatrix} 32 & 8 \\ 8 & 32 \end{bmatrix}$$

$$\mathbf{F}_{lr} = \frac{L^2}{96\,EI}\begin{bmatrix} -6 & 0 \\ -6 & -6 \end{bmatrix}$$

$$\mathbf{F}_{rl} = \frac{L^2}{96\,EI}\begin{bmatrix} -6 & -6 \\ 0 & -6 \end{bmatrix}$$

These sub-matrices will be utilised to evaluate the required quantities. We first evaluate

$$\mathbf{F}_{rr}^{-1} = \frac{2}{5}\frac{EI}{L}\begin{bmatrix} 4 & -1 \\ -1 & 4 \end{bmatrix}$$

and
$$\mathbf{F}_{rr}^{-1}\,\mathbf{F}_{rl} = \frac{L}{40}\begin{bmatrix} -4 & -3 \\ 1 & -3 \end{bmatrix}$$

Then
$$X_r = \frac{L}{40}\begin{bmatrix} -4 & -3 \\ 1 & -3 \end{bmatrix}\begin{Bmatrix} P_1 \\ P_2 \end{Bmatrix}$$

or
$$X_r = \frac{L}{40}\begin{bmatrix} -4P_1 & -3P_2 \\ P_1 & -3P_2 \end{bmatrix}$$

and
$$\mathbf{F}_{lr}\,\mathbf{F}_{rr}^{-1}\,\mathbf{F}_{rl} = \frac{L^3}{640\,EI}\begin{bmatrix} 4 & 3 \\ 3 & 6 \end{bmatrix}$$

The reduced flexibility matrix $\mathbf{F}_l$ of order 2×2 can be obtained using Eq. 17.17 Thus,

$$\mathbf{F}_l = \frac{L^3}{1920}\begin{bmatrix} 28 & -9 \\ -9 & 22 \end{bmatrix}$$

The internal forces are evaluated using Eq. 17.17.

$$\mathbf{p} = \mathbf{A}_l\,\mathbf{P}_l + \mathbf{A}_r\mathbf{X}_r$$

Substituting for $\mathbf{X}_r$, we have

$$\mathbf{p} = (\mathbf{A}_l - \mathbf{A}_r\,\mathbf{F}_{rr}^{-1}\,\mathbf{F}_{rl})\,\mathbf{P}_l$$

$$\begin{Bmatrix} p_1 \\ p_2 \\ p_3 \\ p_4 \\ p_5 \\ p_6 \\ p_7 \\ p_8 \\ p_9 \\ p_{10} \end{Bmatrix} = \left\{ \frac{L}{4}\begin{bmatrix} 0 & 0 \\ -1 & 0 \\ 1 & 0 \\ 0 & 0 \\ 0 & 0 \\ 0 & -1 \\ 0 & 1 \\ 0 & 0 \\ 0 & 0 \\ 0 & 0 \end{bmatrix} - \frac{1}{2}\begin{bmatrix} 0 & 0 \\ 1 & 0 \\ -1 & 0 \\ 2 & 0 \\ -1 & 20 \\ 1 & 1 \\ -1 & -1 \\ 0 & 2 \\ 0 & -2 \\ 0 & 0 \end{bmatrix}\frac{L}{40}\begin{bmatrix} -4 & -3 \\ 1 & -3 \end{bmatrix} \right\} P_1$$

or

$$\begin{Bmatrix} p_1 \\ p_2 \\ p_3 \\ p_4 \\ p_5 \\ p_6 \\ p_7 \\ p_8 \\ p_9 \\ p_{10} \end{Bmatrix} = \frac{L}{80}\begin{bmatrix} 0 & 0 \\ -16 & 3 \\ 16 & -3 \\ 8 & 6 \\ -8 & -6 \\ 3 & -14 \\ -3 & 14 \\ -2 & 6 \\ +2 & -6 \\ 0 & 0 \end{bmatrix}\begin{Bmatrix} P_1 \\ P_2 \end{Bmatrix}$$

Example 17.17 *Generate the flexibility matrix $\mathbf{F}_l$ in terms of $\mathbf{P}_l$ for the structure shown in Fig. 17.17a. Find the external displacements and the member forces for the coordinates shown.*

The structure is indeterminate by two degrees, It is reduced to a primary structure by removing two reaction constraints at the left hand support. The primary structure is shown in Fig. 17.17*b* and the element coordinates in Fig. 17.17*c*.The element coordinates for element 1 are different from those for elements 2 and 3. This is taken only to show the facility available. One can as well choose, and in fact, it is preferable to have the same type of coordinates for all the elements. Matrix A is constructed in the usual manner in the partitioned form, that is

$$\mathbf{A} = \left[\begin{array}{cc|cc} 0 & 0 & 0 & -1 \\ 0 & 0 & -1 & -L \\ \hline 1 & 0 & 1 & L \\ 0 & 1 & 0 & -L \\ \hline 0 & 0 & 0 & L \\ 0 & 0 & 0 & 0 \end{array}\right]$$
$$\qquad\quad \mathbf{A}_l \qquad \mathbf{A}_r$$

The element flexibility matrix can be written as

$$\mathbf{f}_1 = \frac{L}{6}\begin{bmatrix} 2L^2 & -3L \\ -3L & 6 \end{bmatrix}, \mathbf{f}_2 = \frac{L}{12}\begin{bmatrix} 2 & -1 \\ -1 & 2 \end{bmatrix}$$

and

$$\mathbf{f}_3 = \frac{L}{6}\begin{bmatrix} 2 & -1 \\ -1 & 2 \end{bmatrix}$$

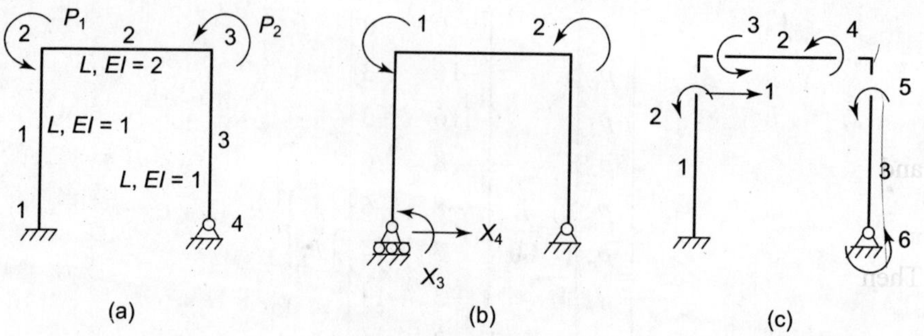

Fig. 17.17 | *(a) Structure and loading, (b) Primary structure and system coordinates, (c) Elements and coordinates*

The uncoupled flexibility matrix **f** can be written as

$$\mathbf{f} = \frac{L}{12}\begin{bmatrix} 4L^2 & -6L & & & & \\ -6L & 12 & & & & \\ & & 2 & -1 & & \\ & & -1 & 2 & & \\ & & & & 4 & -2 \\ & & & & -2 & 4 \end{bmatrix}$$

The flexibility matrix **F** for the structure is obtained by peforming the operation of Eq. 16.9.

$$\mathbf{F} = \mathbf{A}^{\mathrm{T}}\,\mathbf{f}\,\mathbf{A}$$

$$\mathbf{F} = \begin{bmatrix} 0 & 0 & 1 & 0 & 0 & 0 \\ 0 & 0 & 0 & 1 & 0 & 0 \\ 0 & -1 & 1 & 0 & 0 & 0 \\ -1 & -L & +L & -L & +L & 0 \end{bmatrix} \frac{L}{12}\begin{bmatrix} 4L^2 & -6L & & & & \\ -6L & 12 & & & & \\ & & 2 & -1 & & \\ & & -1 & 2 & & \\ & & & & 4 & -2 \\ & & & & -2 & 4 \end{bmatrix}$$

$$\begin{bmatrix} 0 & 0 & 0 & -1 \\ 0 & 0 & -1 & -L \\ 1 & 0 & 1 & +L \\ 0 & 1 & 0 & -L \\ 0 & 0 & 0 & +L \\ 0 & 0 & 0 & 0 \end{bmatrix} = \frac{L}{12}\left[\begin{array}{cc|cc} 2 & -1 & 2 & 3L \\ -1 & 2 & -1 & -3L \\ \hline 2 & -1 & 14 & 9L \\ 3L & -3L & 9L & 14L^2 \end{array}\right]$$

We now have,

$$\mathbf{F}_{rr}^{-1} = \frac{12}{115L^3}\begin{bmatrix} 14L^2 & -9L \\ -9L & 14 \end{bmatrix}$$

and $$\mathbf{F}_{rr}^{-1}\,\mathbf{F}_{rl} = \frac{1}{115L^3}\begin{bmatrix} L^2 & 13L^2 \\ 24L & -33L \end{bmatrix}$$

Then $$X_r = \frac{1}{115\,L^2}\begin{bmatrix} L^2 & 13L^2 \\ 24\,L & -33\,L \end{bmatrix}\begin{Bmatrix} P_1 \\ P_2 \end{Bmatrix} \text{ and}$$

$$\mathbf{F}_{lr}\,\mathbf{F}_{rr}^{-1}\,\mathbf{F}_{rl} = \frac{L}{12\times 115}\begin{bmatrix} 74 & -73 \\ -73 & 86 \end{bmatrix}$$

Therefore, the reduced flexibility matrix corresponding to forces $\mathbf{P}_l$ is

$$\mathbf{F}_l = \frac{L}{12}\begin{bmatrix} 2 & -1 \\ -1 & 2 \end{bmatrix} - \frac{L}{12\times 115}\begin{bmatrix} 74 & -73 \\ -73 & 86 \end{bmatrix} = \frac{L}{12\times 115}\begin{bmatrix} 156 & -42 \\ -42 & 114 \end{bmatrix}$$

The external displacements are obtained from the relation of Eq. 17.16

$$\mathbf{D}_l = \mathbf{F}_l\,\mathbf{P}_l$$

$$= \frac{L}{230}\begin{bmatrix} 26 & -7 \\ -7 & 24 \end{bmatrix}\begin{Bmatrix} P_1 \\ P_2 \end{Bmatrix}$$

The internal forces are obtained using Eq. 17.19

$$\mathbf{p} = \left\{\begin{bmatrix} 0 & 0 \\ 0 & 0 \\ 1 & 0 \\ 0 & 1 \\ 0 & 0 \\ 0 & 0 \end{bmatrix} - \begin{bmatrix} 0 & -1 \\ -1 & -L \\ 1 & L \\ 0 & -L \\ 0 & L \\ 0 & 0 \end{bmatrix}\frac{1}{115\,L^2}\begin{bmatrix} L^2 & 13L^2 \\ 24\,L & -33\,L \end{bmatrix}\right\}\begin{Bmatrix} P_1 \\ P_2 \end{Bmatrix}$$

or $$\mathbf{p} = \frac{1}{115}\begin{bmatrix} 24/L & -33/L \\ 25 & -20 \\ 90 & +20 \\ +20 & 82 \\ -24 & +33 \\ 0 & 0 \end{bmatrix}\begin{Bmatrix} P_1 \\ P_2 \end{Bmatrix}$$

17.4.1 Computer Programme for Statically Indeterminate Structures

We have seen that the general analysis of a statically indeterminate structure is reduced to a set of simple matrix multiplications, starting with matrices [$\mathbf{A}_l$], [$\mathbf{A}_r$], [**f**] and [**P**]. The computations can be carried out manually to generate the initial matrices [$\mathbf{A}_l$], [$\mathbf{A}_r$] and [**f**] but the computations of [**A**] and [**F**] are very tedious and time-consuming. The computer is much better suited to perform these types of operations.

17.4.2 Flow Chart

Fig. 17.18 shows a flow chart for a computer programme which uses the analysis described earlier. The flow chart is drawn such that it can be used both for statically determinate and indeterminate structures. If the number of redundants is zero it will branch out for the analysis of statically determinate structures. On the other hand if the redundants number one or more the structure is statically indeterminant and the matrix [**A**] will be computed in the programme. The input matrices for this case will be [$\mathbf{A}_1$] and [$\mathbf{A}_2$]. The steps outlined in the flow chart are obvious.

17.5 TEMPERATURE STRESSES, LACK OF FIT, SUPPORT SETTLEMENTS, ETC.

When a statically determinate structure is subjected to a change in temperature, the element displacements d_i will take place with no resulting internal forces p_r. However, in statically indeterminate structures, thermal change will result in internal forces p_i. We shall now discuss how these forces are computed by the flexibility method. Let us designate

Δ_s = displacements due to temperature at the coordinates of the element before it is connected to the structure,

Δ_i = displacements at the coordinates of the elements uncoupled.

If in addition to the change in temperature, each element s is subjected to forces p_s at the coordinates, then an additional displacement $f_s\,p_s$ will take place at the coordinates. The total displacement d_s for any element, s, is then

$$d_s = f_x\,p_s + \Delta_s \tag{17.21}$$

For all the elements in an unconnected state, we can write

$$\mathbf{d} = \mathbf{f}\,\mathbf{p} + \Delta \tag{17.22}$$

From Eq. 16.1 we can write

$$\mathbf{p} = \mathbf{AP}$$

and also from Eq. 16.30, we have

$$\mathbf{D} = \mathbf{A}^{\mathrm{T}}\mathbf{d}$$

and partitioning the matrices of Eq. 16.30 to distinguish between the coordinates at the applied forces and redundants we can write

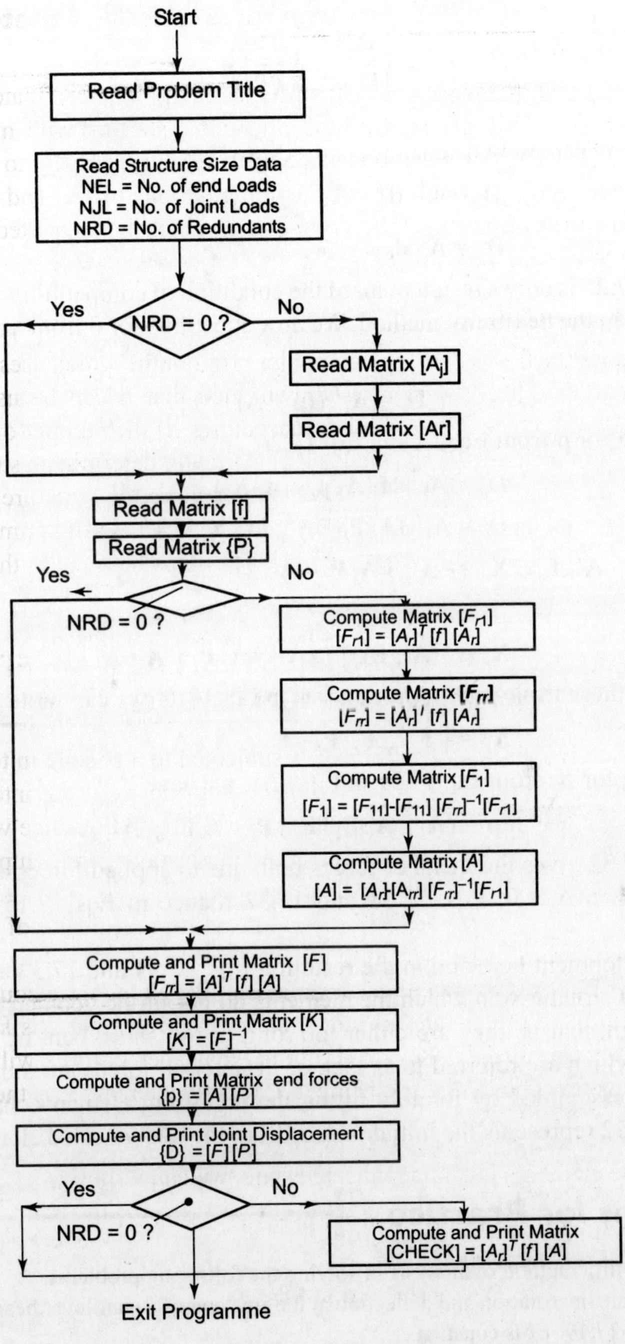

Fig. 17.18 | *Flow chart for the flexibility analysis for statically indeterminate structures*

$$\left\{\frac{\mathbf{D}_l}{\mathbf{D}_r}\right\} = \left\{\frac{A_l^T}{\mathbf{A}_r^T}\right\}\mathbf{d} \tag{17.23}$$

Writing the matrix in two separate matrices, we have

$$\mathbf{D}_l = \mathbf{A}^{\mathrm{T}}{}_l\,\mathbf{d} \tag{17.24}$$

and
$$\mathbf{D}_r = \mathbf{A}_r{}^{\mathrm{T}}\mathbf{d} \tag{17.25}$$

Equation 17.25 is only a restatement of the condition of compatibility which must be satisfied in the flexibility method. We now substitute for d from Eq. 17.22 into Eq. 17.25

$$\mathbf{D}_r = \mathbf{A}_r{}^T(\mathbf{fp} + \Delta) = 0 \tag{17.26}$$

Substituting for **p** from Eq. 17.8 in Eq. 17.26

$$D_r = \mathbf{A}_r{}^{\mathrm{T}}\ \{\mathbf{f}\,(\mathbf{A}_l\,\mathbf{p}_l + \mathbf{A}_r\mathbf{X}_r) + \Delta\} = 0 \tag{17.27}$$

or
$$D_r = \mathbf{A}_r{}^T\,\mathbf{fA}_l\,\mathbf{P}_l + \mathbf{A}^T{}_r\,\mathbf{fA}_r\,\mathbf{X}_r + \mathbf{A}^T{}_r\,\Delta = 0 \tag{17.28}$$

or
$$\mathbf{A}^T{}_r\,\mathbf{f}\,\mathbf{A}_r\,\mathbf{X}_r = -\mathbf{A}^T{}_r\,\mathbf{f}\,\mathbf{A}_l\,\mathbf{P}_l - \mathbf{A}_r\,\Delta \tag{17.29}$$

Therefore,

$$\mathbf{X}_r = -(\mathbf{A}^T{}_r\,\mathbf{fA}_r)^{-1}(\mathbf{A}^T{}_r\,\mathbf{f}\,\mathbf{A}_l\,\mathbf{P}_l + \mathbf{A}^T{}_r\,\Delta) \tag{17.30}$$

Identifying these triple matrix products as in Eq. 17.11 we can write

$$\mathbf{X}_r = -\,\mathbf{F}^{-1}{}_{rr}\,\mathbf{F}_{rl}\,\mathbf{P}_l - \mathbf{F}^{-1}{}_{rr}\,\mathbf{A}^T{}_r\,\Delta \tag{17.31}$$

Substituting for $\mathbf{X}_r$ from Eq. 17.31 in Eq. 17.8, we have

$$\mathbf{p} = (\mathbf{A}_l - \mathbf{A}_r\,\mathbf{F}^{-1}_{rr}\,\mathbf{F}_{rl})\,\mathbf{P}_l - \mathbf{A}_r\mathbf{F}^{-1}{}_{rr}\,\mathbf{A}^T{}_r\,\Delta \tag{17.32}$$

Equation 17.32 gives the member forces both due to applied forces and thermal changes. When $\Delta = 0$, Eqs. 17.31 and 17.32 reduce to Eqs 17.15 and 17.19 respectively.

The development here and in the resulting Eqs. 17.31 and 17.32 also applies to redundant structures in which the elements do not fit the geometry called for by the design, that is, they are either too long or too short, bent or twisted. In such cases which are referred to as lack of fit, internal forces are induced when structure is assembled by forcibly fitting the nonfitting elements. In this case, Δ in Eq. 17.22 represents the initial lack of fit in the unconnected elements.

Problems for Practice

Use the flexibility method of analysis in solving the following problems.

17.1 Calculate the rotation and deflection at the free end of a cantilever beam loaded as shown in Fig. 17.19. *EI* is constant.

17.2 Determine the nodal displacements for the pin-connected trusses shown in Figs. 17.20*a* and *b*. The displacement coordinates are indicated.

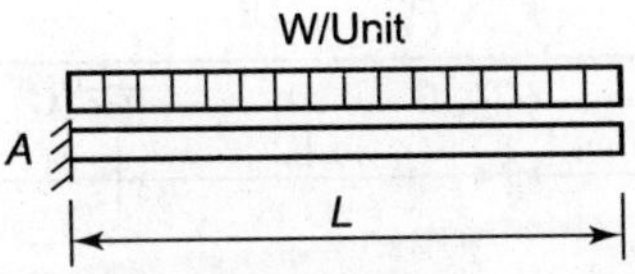

Fig. 17.19

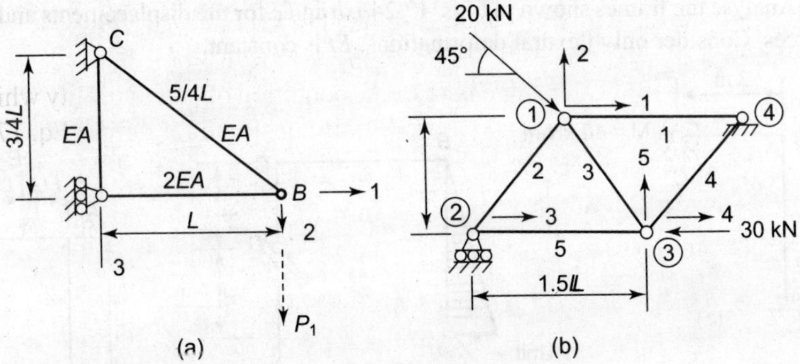

Fig. 17.20

17.3 Analyse the beams shown in Figs. 17.21*a* and *b* for reaction components at *A* and *B* and rotations at *B*.

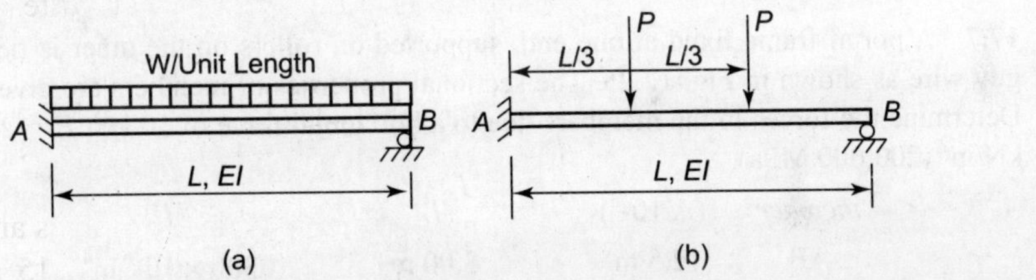

Fig. 17.21

17.4 For the structure shown in Fig. 17.22 choose as redudants (a) R_B and R_C, (b) M_A and M_B. By inspection of **F** state which of the choices is the most desirable. Using the best system, solve for redundants and reactions and draw final shear and moment diagrams.

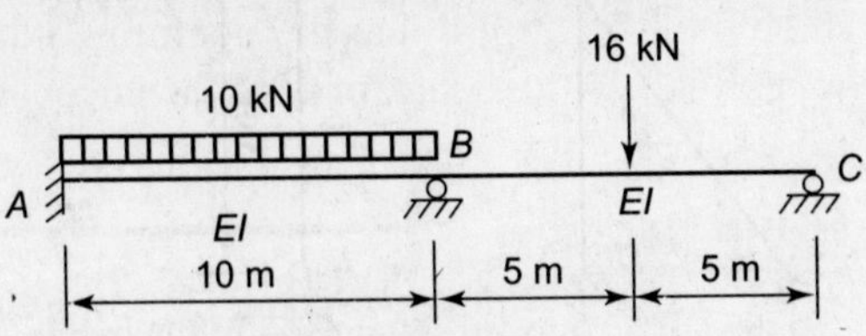

Fig. 17.22

17.5 Generate the 2 × 2 matrix $\mathbf{F}_l$ and find the internal forces in terms of P_l for the continuous beam shown in Fig. 17.23. *EI* is the same throughout.

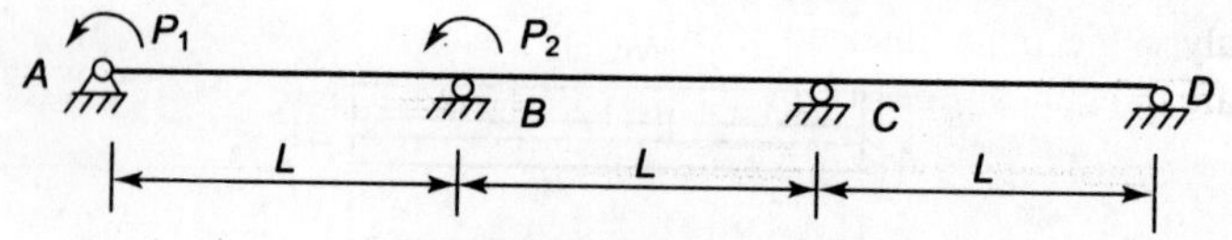

Fig. 17.23

17.6 Analyse the frames shown in Figs. 17.24*a*, *b* and *c* for the displacements and member forces. Consider only flexural deformations. *EI* is constant.

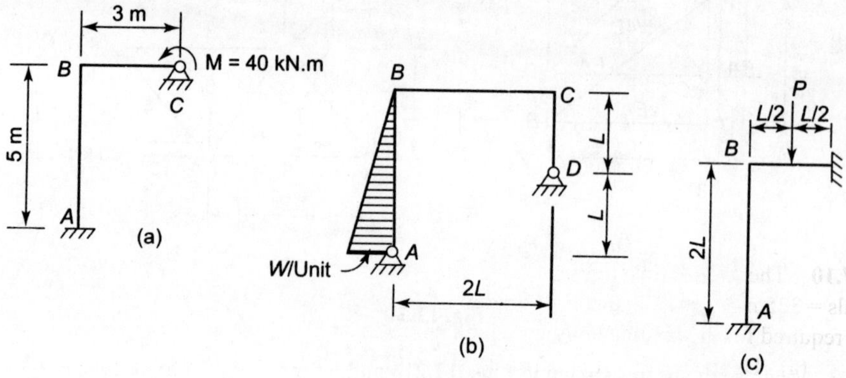

Fig. 17.24

17.7 A portal frame fixed at one end, supported on rollers on the other is tied with a guy wire as shown in Fig. 17.25. The sectional properties of members are given below. Determine the forces in the members due to a horizontal force of 50 kN. $E = 200 \times 10^6$ kN/m^2 (200,000 MPa).

member	$(A.10^{-3})$	L	I
AB	12.5 m^2	3.00 m	300×10^{-6} m^4
BC	12.5 m^2	3.00 m	200×10^{-6} m^4
CD	12.5 m^2	3.00 m	200×10^{-4} m^4
CE	3.125 m^2	3.75 m	0

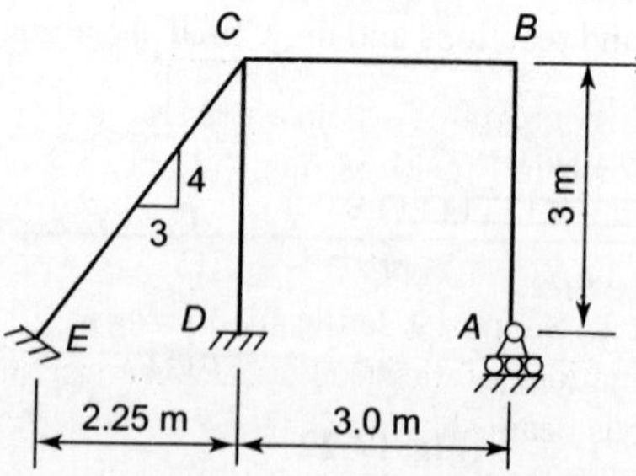

Fig. 17.25

17.8 Member *AB* of the portal frame in Prob. 17.7 is subjected/ to a temperature drop of 30°C. Determine forces in members, $\alpha_c = 8.25 \times 10^{-6}$°C.

17.9 Analyse the pin-connected truss shown in Fig. 17.26 for applied load *P*. All members are of equal stiffness *EA*.

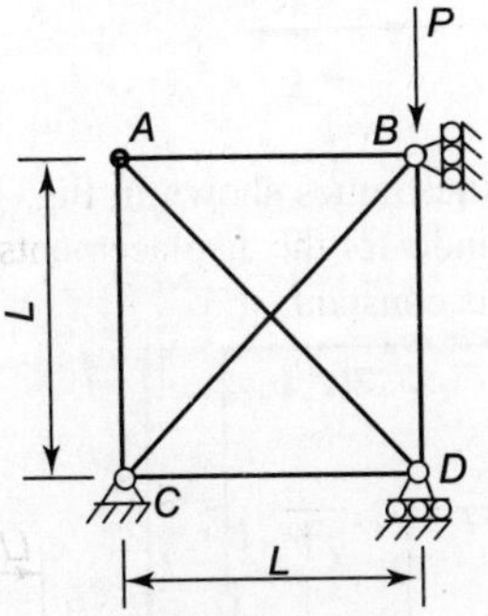

(a) Structure

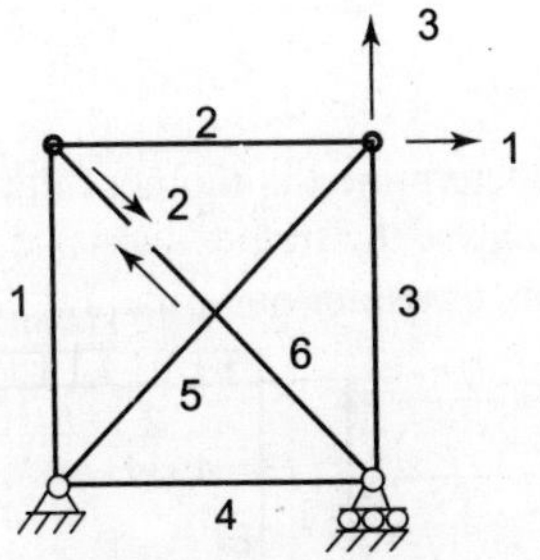

(b) Numbering System

Fig. 17.26

17.10 The truss shown in Fig. 17.27 has top and bottom chords area = 650 mm^2, verticals = 325 mm^2 and diagonals = 975 mm , E = 200 kN/mm^2 (200,000 MPa). An analysis is required for the loading and fabrication defects described below.

(a) Define an approriate primary structure.
(b) Obtain flexibility matrix *FI*.
(c) Analyse the structure for a concentrated load of 50 kN applied downwards at joint 4.
(d) For the unloaded truss, find the bar forces resulting from a fabrication error in which bars 1-4 and 3-5 were fabricated 8.33 mm longer than required.

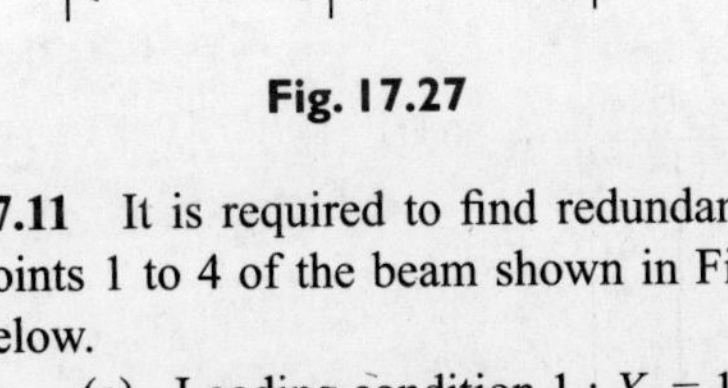

Fig. 17.27

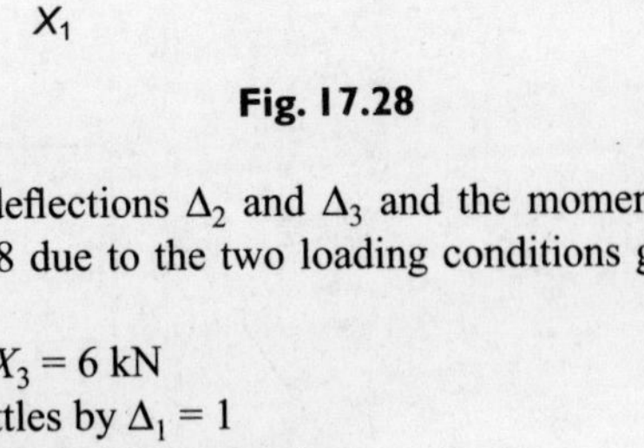

Fig. 17.28

17.11 It is required to find redundant X_1, deflections Δ_2 and Δ_3 and the moments at points 1 to 4 of the beam shown in Fig 17.28 due to the two loading conditions given below.

(a) Loading condition 1 : X_2 = 10 kN, X_3 = 6 kN
(b) Loading condition 2 : Support 1 settles by Δ_1 = 1

Consider only flexural deformations of the structure. *EI* is constant.

17.12 Analyse the continuous beam shown in Fig. 17.29 due to:

(a) A uniform load w/unit length over the entire beam.
(b) Concentrated load P at the centre of span BC.
(c) Support settlement Δ at point B.

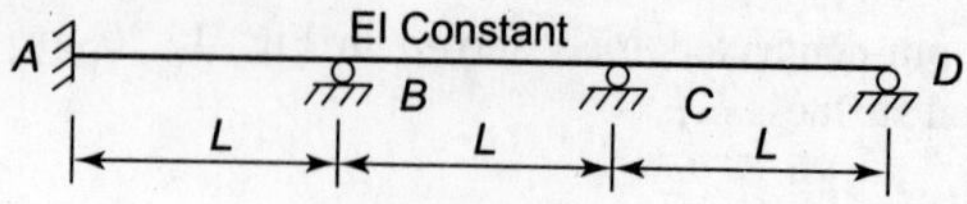

Fig. 17.29

17.13 Determine the moment in the members of the frames shown in Fig. 17.30*a* and *b*.

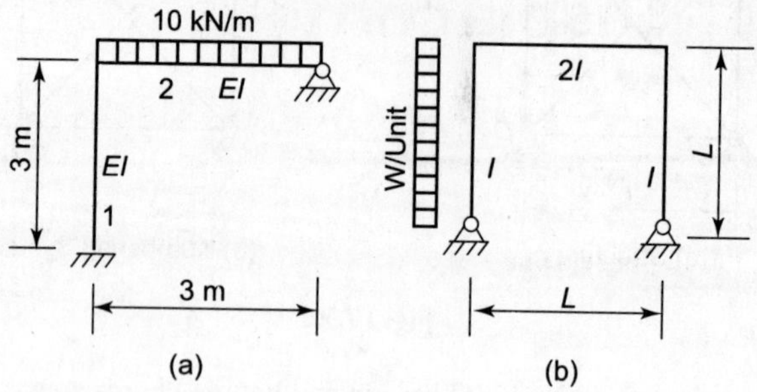

Fig. 17.30

18 Stiffness or Displacement Method of Analysis

18.1 INTRODUCTION

The stiffness or displacement method in structural analysis is analogous to the flexibility method; whereas in the latter the forces were unknowns, here the displacements are unknowns. Some of the basic principles and equations applying to the stiffness method have already been developed in Chapters 15 and 16 by the generalisation of existing methods and well-known principles. In the matrix formulation of the stiffness method there is no need to distinguish between a statically determinate structure and an indeterminate one, since the steps are identical in both cases.

Of the two methods, the matrix displacement method of analysis is commonly preferred particularly when the degree of static indeterminancy is high. The stiffness method aims at solving for unknown joint equations at the joints. The steps involved in the displacement method of analysis are presented in the following section.

18.1.1 Stiffness Method—Steps to be Followed

1. As a first step the degree of freedom or the kinematic indeterminancy of the structure is determined. The coordinates for the structure are established identifying the location and direction of joint displacements. Restraining forces are applied at the coordinates to prevent joint displacements. In this method there is no choice exercised in the selection of joint displacements unlike the redundants in the flexibility method. This is a favourable point for the adoption of the displacement method.
2. The restraining forces are determined as a sum of fixed end forces for the members meeting at a joint. The fixed end forces are obtained with the aid of standard tables (Table in Appendix).
3. In the next step, the forces required to hold the restrained structure with a unit displacement at one of the coordinates only, and with zero displacements at all other coordinates, are determined. This is done for all other coordinates one by one and the forces required are determined. These forces form the elements of the stiffness matrix **[K]**.

4. The values of displacements {**D**} necessary to ensure the equilibrium of the joints are determined using the relation

$$\{\mathbf{P}\} + [\mathbf{K}]\{\mathbf{D}\} = 0 \tag{18.1}$$

in which

{**P**} = restraining forces at the joints,
{**K**} = stiffness matrix corresponding to the coordinates, and
{**D**} = unknown displacements at the coordinates.

Displacements {**D**} are obtained by solving Equation 18.1.

$$\{\mathbf{D}\} = [\mathbf{K}]^{-1}\{-\mathbf{P}\} \tag{18.2}$$

5. Finally the forces in the given structure are obtained by adding the forces on the restrained structure and the forces caused by the joint displacements found as above.

The examples that follow will make the procedure clear.

Example 18.1 | *Using the displacement method, analyse the continuous beam in example 17.1 solved by the flexibility method.*

The degree of kinematic indeterminacy is two because of the two independent rotations at *B* and *C*. The coordinates 1 and 2 are chosen at the displacements. The restraining forces at the joints which are equal to the sum of end forces, are calculated. As earlier, they are considered positive when their directions accord with those of the coordinates.

Therefore, to obtain the restraining forces, it is sufficient to add the end forces at each joint as indicated following the static sign convention used in other methods.

Any external couples or forces applied at the joints require equal and opposite restraining forces.

The forces in the present case are

$$\{\mathbf{P}\} = \begin{Bmatrix} -40 \\ -60 \end{Bmatrix} \text{kN.m}$$

The elements of the stiffness matrix are obtained by determining the forces required to hold the beam with $D_1 = 1$ and $D_2 = 0$ and again with $D_1 = 0$ and $D_2 = 1$ as in Fig. 18.1*d* and *e*.

The elements of the stiffness matrix are

$$k_{11} = \frac{4EI}{8} + \frac{4EI}{8} = EI$$

$$k_{21} = \frac{2EI}{8} + \frac{EI}{4}$$

$$k_{12} = \frac{2EI}{8} + \frac{EI}{4}$$

$$k_{22} = \frac{4EI}{8} + \frac{EI}{2}$$

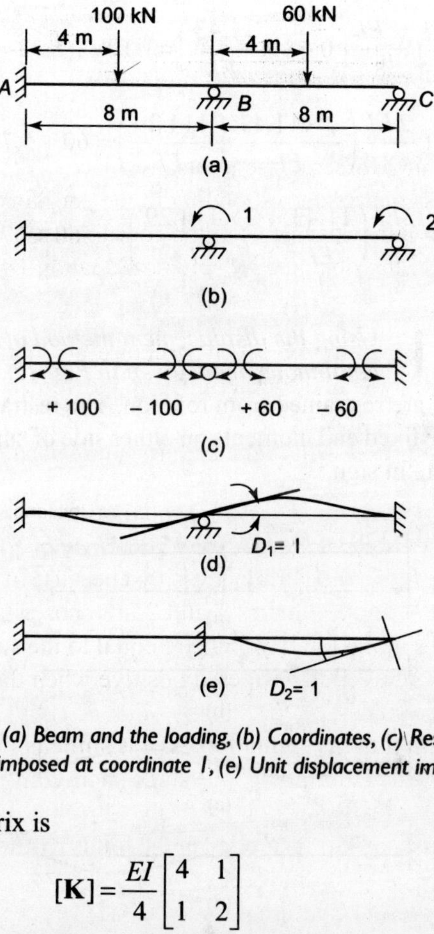

Fig. 18.1 | *(a) Beam and the loading, (b) Coordinates, (c) Restraining forces, (d) Unit displacement imposed at coordinate 1, (e) Unit displacement imposed at coordinate 2*

The stiffness matrix is

$$[\mathbf{K}] = \frac{EI}{4}\begin{bmatrix} 4 & 1 \\ 1 & 2 \end{bmatrix}$$

$$[\mathbf{K}]^{-1} = \frac{4}{7\,EI}\begin{bmatrix} 2 & -1 \\ -1 & 4 \end{bmatrix}$$

Substituting in Equation 18.2,

$$|\mathbf{D}| = \frac{4}{7\,EI}\begin{bmatrix} 2 & -1 \\ -1 & 4 \end{bmatrix}\begin{Bmatrix} +40 \\ +60 \end{Bmatrix}$$

$$D_1 = \frac{11.43}{EI}$$

$$D_2 = \frac{114.29}{EI}$$

The final moments are calculated using slope deflection equations.

$$M_{AB} = +100 + \frac{2\,EI}{8}\left(0 + \frac{11.43}{EI}\right) = 100 + 2.86 = 102.86$$

$$M_{BA} = -100 + \frac{2EI}{8}\left(0 + 1 \times \frac{11.43}{EI}\right) = -100 + 5.72 = -94.28$$

$$M_{BC} = +60 + \frac{2EI}{8}\left(\frac{2 \times 11.43}{EI} + \frac{114.29}{EI}\right) = 60 + 5.72 + 28.57 = 94.29$$

$$M_{CB} = -60 + \frac{2EI}{8}\left(\frac{11.43}{EI} + \frac{2 \times 114.29}{EI}\right) = -60 + 57.15 = 0$$

Example 18.2 | *Using the displacement method of analysis analyse the continuous beam given in Fig. 18.2a.*

Supports B and C are restrained from rotation. The restraining forces are,

$P_1 = 0$ as the fixed end moments on either side of support B are equal but opposite in sign

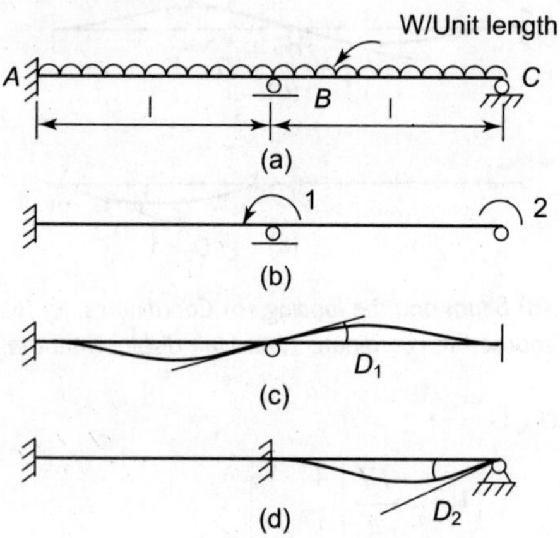

Fig. 18.2

$$P_2 = \frac{-wl^2}{12}$$

The stiffness matrix

$$[\mathbf{K}] = \frac{EI}{l}\begin{bmatrix} 8 & 2 \\ 2 & 4 \end{bmatrix}$$

$$[\mathbf{K}]^{-1} = \frac{l}{28EI}\begin{bmatrix} 4 & -2 \\ -2 & 8 \end{bmatrix}$$

and
$$D = \frac{l}{28EI}\begin{bmatrix} 4 & -2 \\ -2 & 8 \end{bmatrix}\begin{Bmatrix} 0 \\ \dfrac{wl^2}{12} \end{Bmatrix}$$

$$\therefore \qquad D_1 = \frac{wl^3}{168\,EI}$$

and
$$D_2 = \frac{wl^3}{42\,EI}$$

The end moments of the members are determined using slope deflection equations. The moment values are

$$M_{AB} = \frac{wl^2}{12} + \frac{2\,EI}{l}\left(0 - \frac{wl^3}{168\,EI}\right) = \frac{wl^2}{12} - \frac{wl^2}{84} = \frac{wl^2}{14}$$

$$M_{BA} = \frac{-wl^2}{12} + \frac{2\,EI}{l}\left(0 - \frac{2 \times wl^3}{168\,EI}\right) = -\frac{wl^2}{12} - \frac{wl^2}{42} = \frac{-3}{28}wl^2$$

$$M_{BC} = \frac{+wl^2}{12} + \frac{2\,EI}{l}\left(-\frac{2wl^3}{168\,EI} + \frac{wl^3}{42\,EI}\right) = \frac{3}{28}wl^2$$

$$M_{CB} = -\frac{wl^2}{12} + \frac{2\,EI}{l}\left(-\frac{wl^3}{168\,EI} + \frac{2 \times wl^3}{42\,EI}\right) = 0$$

18.1.2 Effect of Support Displacements, Temperature Changes, etc.

The effect of support displacements, temperature changes, truss members being too long or short to fit in can be readily incorporated in the stiffness method of analysis. A convenient procedure is to consider all such effects on the restrained structure and add the resulting joint forces to the joint forces caused by the loads. The combined restraining forces Pc can be written as

$$\{\mathbf{P}_C\} = \{\mathbf{P}_1\} + \{\mathbf{P}_t\} + \{\mathbf{P}_r\}$$

where $\{\mathbf{P}_1\}$ = Restraining forces due to applied loads
$\{\mathbf{P}_t\}$ = Restraining forces due to temperature change
$\{\mathbf{P}_r\}$ = Restraining forces due to displacement of supports.

Equation 18.2 is modified as

$$\{\mathbf{D}\} = [\mathbf{K}^{-1}]\,\{-\mathbf{P}_C\} \qquad (18.3)$$

The procedure is illustrated by the following examples.

Example 18.3 | *Using the displacement method, analyse the continuous beam given in Fig. 18.3. Consider that under the given loading the support B sinks by* $\frac{300}{EI}$ *and support C by* $\frac{200}{EI}$.

The fixed end moments produced by the translation of joint are added to the fixed end moment caused by external loading, The forces at the joints are

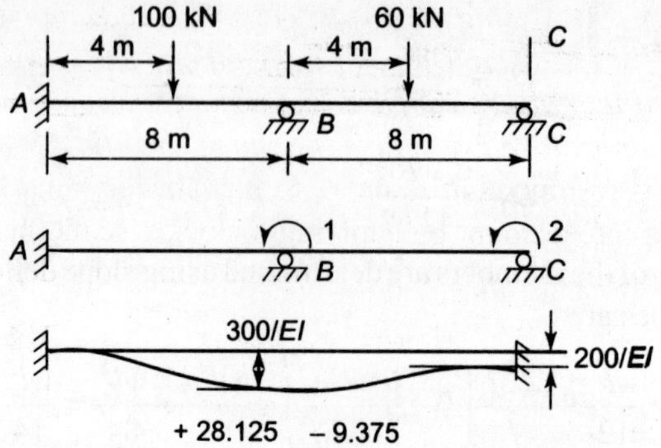

Fig. 18.3

$$\{\mathbf{P}\} = \begin{Bmatrix} -40 + 18.75 \\ -60 - \ 9.38 \end{Bmatrix} = \begin{Bmatrix} -21.25 \\ -69.38 \end{Bmatrix}$$

The stiffness matrix for the same coordinates as in example 18.2 is

$$[\mathbf{K}] = \frac{EI}{8}\begin{bmatrix} 8 & 2 \\ 2 & 4 \end{bmatrix}$$

$$[\mathbf{K}]^{-1} = \frac{2}{7\,EI}\begin{bmatrix} 4 & -2 \\ -2 & 8 \end{bmatrix}$$

The displacements are

$$\{\mathbf{D}\} = \frac{2}{7\,EI}\begin{bmatrix} 4 & -2 \\ -2 & 4 \end{bmatrix}\begin{Bmatrix} 21.25 \\ 69.38 \end{Bmatrix}$$

$$\therefore \qquad D_1 = \frac{-15.36}{EI} \text{ and } D_2 = \frac{146.44}{EI}$$

The end moments for the members are calculated using slope deflection equations:

$$M_{AB} = 100 + 28.13 + \frac{2\,EI}{8}\left(0 - \frac{15.36}{EI}\right) = +124.29 \text{ kN.m}$$

$$M_{BA} = -100 + 28.13 + \frac{2EI}{8}\left(0 - \frac{2 \times 15.36}{EI}\right) = -79.55 \text{ kN.m}$$

$$M_{BC} = 60 - 9.38 + \frac{2\,EI}{8}\left(\frac{-2 \times 15.36}{EI} + \frac{146.44}{EI}\right) = 79.55 \text{ kN.m}$$

$$M_{CB} = -60 - 9.38 + \frac{2\,EI}{8}\left(\frac{2 \times 146.44}{EI} - \frac{15.76}{EI}\right) = 0$$

Example 18.4 *Using the displacement method, analyse the continuous beam with overhang on one end as shown in Fig. 18.4, for conditions: (a) that all the supports are rigid and (b) support B sinks by 10 mm under the loading. Take $E = 200 \times 10^6$ N/mm^2 and $I = 100 \times 10^3$ mm^4.*

(a) The beam is restrained at B and C to prevent rotations. The effect of the overhang is taken into account by applying a clockwise moment of 40 kN.m at joint C.

The load vector $$\{\mathbf{P}\} = \begin{Bmatrix} -20+30=10 \\ -30+40=10 \end{Bmatrix}$$

Stiffness matrix $$[\mathbf{K}] = \frac{EI}{3}\begin{bmatrix} 5 & 1 \\ 1 & 2 \end{bmatrix}$$

$$[\mathbf{K}]^{-1} = \frac{1}{3EI}\begin{bmatrix} 2 & -1 \\ -1 & 5 \end{bmatrix}$$

$\therefore$ $$[\mathbf{D}] = \frac{1}{3EI}\begin{bmatrix} 2 & -1 \\ -1 & 5 \end{bmatrix}\begin{Bmatrix} -10 \\ -10 \end{Bmatrix}$$

which gives $$D_1 = \frac{-10}{3EI} \text{ and } D_2 = \frac{-40}{3EI}$$

The end moments are

$$M_{AB} = +20 + \frac{2EI}{4}\left(0 - \frac{10}{3EI}\right) = +18.33 \text{ kN.m}$$

$$M_{BA} = -20 + \frac{2EI}{4}\left(-2 \times -\frac{10}{3EI} + 0\right) = -23.33 \text{ kN.m}$$

$$M_{BC} = +30 + \frac{2EI}{6}\left(\frac{-2 \times 10}{EI} + \frac{40}{3EI}\right) = +23.33 \text{ kN.m}$$

$$M_{CB} = -30 + \frac{2EI}{6}\left(-2 \times \frac{40}{3EI} + \frac{10}{3EI}\right) = -40.0 \text{ kN.m}$$

(b) The settlement of support B causes fixed end moments. The fixed end moments are

$$FEM_{AB} = FEM_{BA} = 6 \times \frac{200 \times 100}{4 \times 4} \times \frac{10}{1000} = +75.0 \text{ kN.m}$$

$$FEM_{BC} = FEM_{CB} = -6 \times \frac{200 \times 100}{6 \times 6} \times \frac{10}{1000} = -33.33 \text{ kN.m}$$

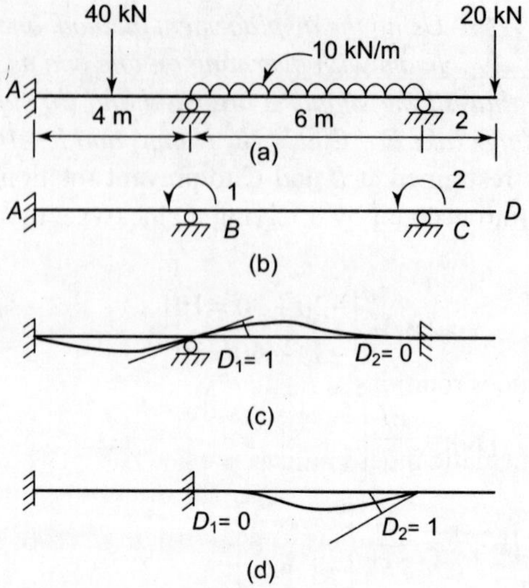

Fig.18.4

The restraining forces at the joints are

$$\{\mathbf{P}\} = \begin{Bmatrix} 10 + 75 - 33.33 & = 51.67 \\ 10 - 33.33 & = -23.33 \end{Bmatrix}$$

The stiffness matrix $[\mathbf{K}]$ and the inverse $[\mathbf{K}]^{-1}$ remains the same

Hence $$[\mathbf{D}] = \frac{1}{3EI}\begin{bmatrix} 2 & -1 \\ -1 & 5 \end{bmatrix}\begin{Bmatrix} -51.67 \\ +27.33 \end{Bmatrix}$$

This gives $$D_1 = -\frac{42.22}{EI} \quad \text{and} \quad D_2 = +\frac{56.11}{EI}$$

The resulting final moments are

$$M_{AB} = +20 + 75 + \frac{2EI}{4}\left(0 - \frac{42.22}{EI}\right) = 73.89 \text{ kN.m}$$

$$M_{BA} = -20 + 75 + \frac{2EI}{4}\left(-2 \times \frac{42.22}{EI} + 0\right) = 12.78 \text{ kN.m}$$

$$M_{BC} = +30 - 33.33 + \frac{2EI}{6}\left(-2 \times \frac{42.22}{EI} + \frac{56.11}{EI}\right) = 12.78 \text{ kN.m}$$

$$M_{CB} = -30 - 33.33 + \frac{2EI}{6}\left(\frac{2 \times 56.11}{EI} - \frac{42.22}{EI}\right) = -40 \text{ kN.m}$$

Example 18.5 | *Using the stiffness method of analysis obtain the moments at the ends of members for the portal frame shown in Fig. 18.5.*

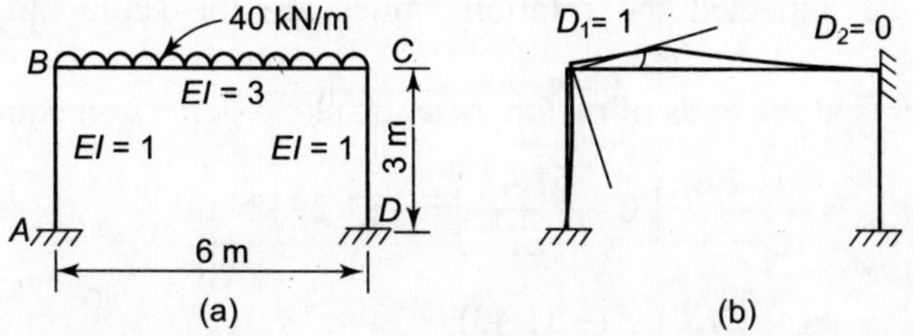

Fig. 18.5

The frame undergoes rotations at B and C without translation due to symmetry of frame and loading.

The degree of kinematic indeterminacy is two. The coordinates at B and C are denoted as 1 and 2.

The elements in the first column of the stiffness matrix [**K**] are obtained by setting $D_1 = 1$ and $D_2 = 0$ and the second column by setting $D_1 = 0$ and $D_2 = 1$.

$$k_{11} = \frac{4EI}{3} + \frac{4\times 3EI}{6} = \frac{10}{3}EI$$

$$k_{21} = \frac{2\times 3EI}{6} = EI$$

$$k_{12} = k_{21} = EI$$

and $$k_{22} = k_{11} = \frac{10}{3}EI$$

The restraining forces required to prevent rotations at B and C are

$$\{\mathbf{P}\} = \begin{Bmatrix} +120 \text{ kN.m} \\ -120 \text{ kN.m} \end{Bmatrix}$$

The stiffness matrix for the coordinates is

$$[\mathbf{K}] = \frac{EI}{3}\begin{bmatrix} 10 & 3 \\ 3 & 10 \end{bmatrix}$$

$$[\mathbf{K}]^{-1} = \frac{3}{91EI}\begin{bmatrix} 10 & -3 \\ -3 & 10 \end{bmatrix}$$

The displacements are

$$\{\mathbf{D}\} = \frac{3}{91EI}\begin{bmatrix} 10 & -3 \\ -3 & 10 \end{bmatrix}\begin{Bmatrix} -120 \\ +120 \end{Bmatrix}$$

which gives $$D_1 = \frac{-51.43}{EI}$$

and $$D_2 = \frac{51.43}{EI}$$

As could be expected the rotation values are the same but opposite in direction.

The moments at the ends of members using slope deflection equations are,

$$M_{AB} = 0 + \frac{2EI}{3}\left(0 - \frac{51.43}{EI}\right) = -34.29 \text{ kN.m}$$

$$M_{BA} = 0 + \frac{2EI}{3}\left(2\frac{(-51.43)}{EI} + 0\right) = -68.58 \text{ kN.m}$$

$$M_{BC} = +120 + \frac{2\times 3EI}{6}\left\{2\frac{(-51.43)}{6EI} + \frac{51.43}{EI}\right\} = 68.57 \text{ kN.m}$$

$$M_{CB} = -120 + \frac{2\times 3EI}{6}\left\{\frac{2\times 51.43}{EI} - \frac{51.43}{EI}\right\} = -68.5 \text{ kN.m}$$

Example 18.6 | *Using the displacement method, analyse the frame shown in Fig. 18.6 for end moments of members AB and BC.*

The kinematic indeterminancy of the frame is two. The frame undergoes independent rotations at B and C. The displacements are denoted as D_1 and D_2 at coordinates 1 and 2 as indicated in Fig. 18.7*b*. The joint loads are:

$$\{\mathbf{P}\} = \begin{Bmatrix} 7.5 \\ -7.5 \end{Bmatrix}$$

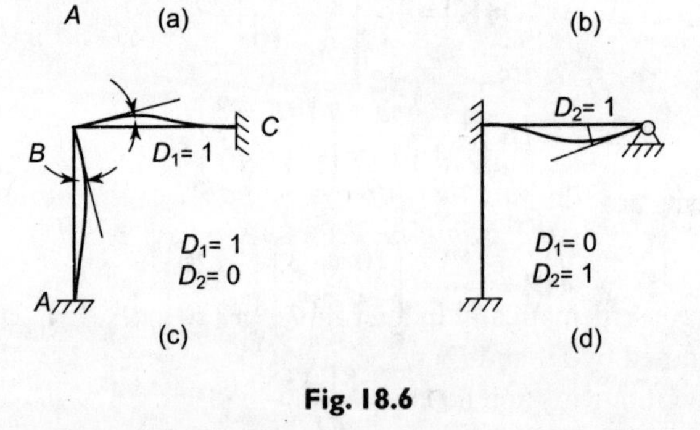

Fig. 18.6

The elements of the stiffness matrix are obtained by setting $D_1 = 1$ and $D_2 = 0$ and, $D_1 = 0$ and $D_2 = 1$ in turn and determining the forces required to hold the structure in the displaced position as shown in Fig. 18.6*c* and *d*.

The elements of the stiffness matrix are:

$$k_{11} = \frac{4EI}{3} + \frac{4EI}{3} = \frac{8}{3}EI \qquad k_{21} = \frac{2EI}{3}$$

$$k_{21} = \frac{2EI}{3} \qquad k_{22} = \frac{4EI}{3}$$

The stiffness matrix

$$[\mathbf{K}] = \frac{EI}{3}\begin{bmatrix} 8 & 2 \\ 2 & 4 \end{bmatrix}$$

$$[\mathbf{K}]^{-1} = \frac{3}{28EI}\begin{bmatrix} 4 & -2 \\ -2 & 8 \end{bmatrix}$$

$$\therefore \qquad \{\mathbf{D}\} = [\mathbf{K}]^{-1}\{-\mathbf{P}\} = \frac{3}{28EI}\begin{bmatrix} 4 & -2 \\ -2 & 8 \end{bmatrix}\begin{Bmatrix} -7.5 \\ +7.5 \end{Bmatrix}$$

$$D_1 = -\frac{4.82}{EI}$$

$$D_2 = +\frac{8.04}{EI}$$

The moments at the ends of the members are obtained using slope deflection equations.

$$M_{AB} = 0 + \frac{2EI}{3}\left(0 - \frac{4.82}{EI}\right) = -3.21 \text{ kN.m}$$

$$M_{BA} = 0 + \frac{2EI}{3}\left(\frac{2(-4.82)}{EI} + 0\right) = -6.43 \text{ kN.m}$$

$$M_{BC} = +7.5 + \frac{2EI}{3}\left\{2\frac{(-4.82)}{EI} + \frac{8.04}{EI}\right\} = 6.43$$

$$M_{CB} = -7.5 + \frac{2EI}{3}\left\{2\frac{(8.04)}{EI} - \frac{4.82}{EI}\right\} = 0$$

Example 18.7 | *Using the stiffness method, analyse for end moments of the frame in Fig. 18.7 which is same as the one in Example 11.3.*

The structure is kinematically indeterminate by two degrees. The rotations at *B* and *C* are denoted by D_1 and D_2 respectively.

The fixed end moments and hence the joint loads are shown in Fig. 18.7*b*.

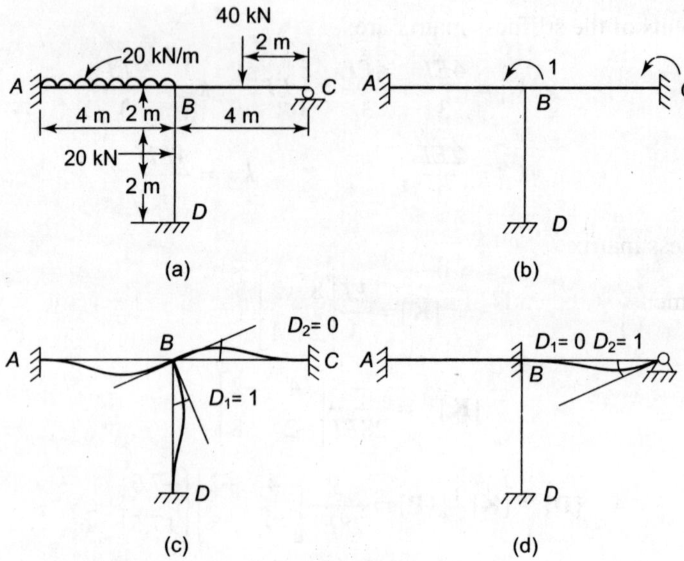

Fig. 18.7

The load vector is

$$\{\mathbf{P}\} = \begin{Bmatrix} -\dfrac{50}{3} \\ -20 \end{Bmatrix}$$

The elements of the stiffness matrix are

$$k_{11} = \frac{4EI}{L} + \frac{4EI}{L} + \frac{4EI}{L} = \frac{12EI}{L}$$

$$k_{21} = \frac{2EI}{L}$$

$$k_{12} = \frac{2EI}{L}$$

$$k_{22} = \frac{4EI}{L}$$

The stiffness matrix,

$$[\mathbf{K}] = \frac{2EI}{L}\begin{bmatrix} 6 & 1 \\ 1 & 2 \end{bmatrix}$$

$$[\mathbf{K}]^{-1} = \frac{L}{22\,EI}\begin{bmatrix} 2 & -1 \\ -1 & 6 \end{bmatrix}$$

$$\{\mathbf{D}\} = \frac{L}{22\,EI}\begin{bmatrix} 2 & -1 \\ -1 & 6 \end{bmatrix}\begin{Bmatrix} +\dfrac{50}{3} \\ +20 \end{Bmatrix}$$

$$D_1 = \frac{2.42}{EI}$$

$$D_2 = \frac{18.79}{EI}$$

The moments at the ends of members are obtained by using slope deflection equations.

$$M_{AB} = \frac{+80}{3} + \frac{2\,EI}{4}\left(0 + \frac{2.42}{EI}\right) = 27.88 \text{ kN.m}$$

$$M_{BA} = \frac{-80}{3} + \frac{2\,EI}{4}\left(\frac{2 \times 2.42}{EI} + 0\right) = -24.25 \text{ kN.m}$$

$$M_{BC} = +20 + \frac{2\,EI}{4}\left(\frac{2 \times 2.42}{EI} + \frac{18.79}{EI}\right) = 31.80 \text{ kN.m}$$

$$M_{CB} = -20 + \frac{2\,EI}{4}\left(\frac{2 \times 18.79}{EI} + \frac{2.42}{EI}\right) = 0 \text{ kN.m}$$

$$M_{BD} = -10 + \frac{2\,EI}{4}\left(\frac{2 \times 2.42}{EI} + 0\right) = -7.58 \text{ kN.m}$$

$$M_{DB} = +10 + \frac{2\,EI}{4}\left(0 + \frac{2.42}{EI}\right) = 11.21 \text{ kN.m}$$

The results tally with the values obtained by the slope deflection method.

Example 18.8 | *Using the stiffness method of analysis, determine the moments at the ends of members for the frame as in Example 12.7.*

Neglecting axial deformations, the structure is kinematically indeterminate by three degrees. The restrained structure and the coordinates are indicated in Fig. 18.8*b*. It may be noted that the lateral translation at 1 is same for joints 2 and 3, as the axial strain in member 2-3 is neglected.

The restraining forces are

$$\{\mathbf{P}\} = \begin{Bmatrix} -30.00 \\ +23.23 \\ -53.33 \end{Bmatrix} \text{ kN.m}$$

The stiffness matrix is generated by giving unit displacement at each coordinate one by one and determining the forces required to hold the structure with zero displacements at the other coordinates as shown in Fig. 18.8*c*, *d* and *e*. The elements as shown in Fig. 18.8*c*, *d* and *e* of the stiffness matrix are:

$$k_{11} = \frac{12EI}{4^3} + \frac{12EI}{3^3} = \frac{91EI}{144};\ k_{22} = \frac{4EI}{4} + \frac{4EI}{4} = 2EI;\ k_{33} = \frac{4EI}{4} + \frac{4EI}{3} = \frac{7}{3}EI$$

$$k_{21} = \frac{6EI}{4^2} = \frac{3}{8}EI;\quad k_{32} = \frac{2EI}{4^2} = \frac{3}{8}EI;\ k_{13} = \frac{12EI}{3^3} = \frac{4}{9}EI$$

$$k_{31} = \frac{6EI}{3^2} = \frac{2}{3}EI;\quad k_{32} = \frac{2EI}{4} = \frac{EI}{2};\ k_{23} = \frac{2EI}{4} = \frac{EI}{2}$$

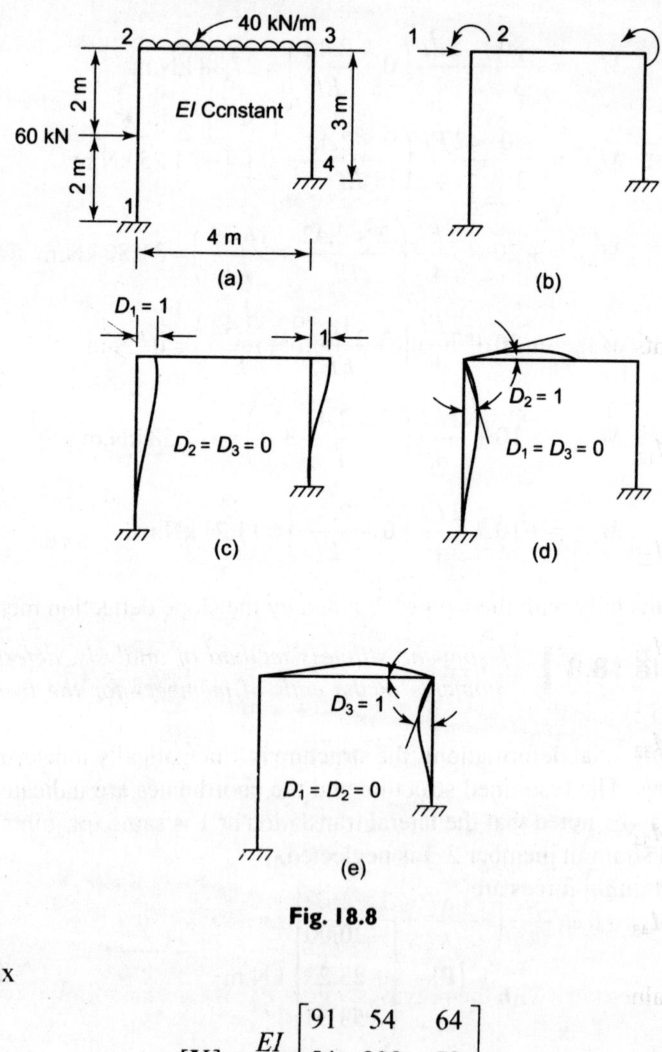

Fig. 18.8

The matrix

$$[\mathbf{K}] = \frac{EI}{144}\begin{bmatrix} 91 & 54 & 64 \\ 54 & 288 & 72 \\ 96 & 72 & 336 \end{bmatrix}$$

$$[\mathbf{K}]^{-1} = \frac{144 \times 10^{-3}}{EI}\begin{bmatrix} 16.80 & -2.06 & -4.36 \\ -2.06 & 3.92 & -0.25 \\ -4.36 & -0.25 & -4.28 \end{bmatrix}$$

The displacements vector

$$\{\mathbf{D}\} = [\mathrm{K}]^{-1}\,\{-\mathbf{P}\}$$

or

$$\{\mathbf{D}\} = \frac{144 \times 10^{-3}}{EI}\begin{bmatrix} 16.80 & -2.06 & -4.36 \\ -2.06 & 3.92 & -0.25 \\ -4.36 & -0.25 & 4.28 \end{bmatrix}\begin{bmatrix} +30.00 \\ -23.33 \\ +53.33 \end{bmatrix}$$

Solving

$$D_1 = \frac{46.01}{EI}$$

$$D_2 = \frac{-23.98}{EI}$$

$$D_3 = \frac{14.87}{EI}$$

The moments at the ends of members are obtained using the slope deflection equations.

$$M_{12} = +30 + \frac{2EI}{4}\left(0 - \frac{23.98}{EI} + \frac{3 \times 46.01}{4EI}\right) = +35.26 \text{ kN.m}$$

$$M_{21} = -30 + \frac{2EI}{4}\left(\frac{-2 \times 23.98}{EI} + \frac{3 \times 46.01}{4EI}\right) = -36.73 \text{ kN.m}$$

$$M_{23} = +53.33 + \frac{2EI}{4}\left(-2 \times \frac{23.98}{EI} + \frac{14.87}{4EI}\right) = +36.79 \text{ kN.m}$$

$$M_{32} = -53.33 + \frac{2EI}{4}\left(2 \times \frac{14.87}{EI} - \frac{23.98}{EI}\right) = +50.45 \text{ kN.m}$$

$$M_{34} = 0 + \frac{2EI}{3}\left(2 \times \frac{14.87}{EI} + \frac{3 \times 46.01}{EI}\right) = +50.50 \text{ kN.m}$$

$$M_{43} = 0 + \frac{2EI}{3}\left(\frac{14.87}{EI} + \frac{3 \times 46.01}{EI}\right) = 40.58 \text{ kN.m}$$

These values tally with the previous values except for the minor rounding-off errors.

It may be mentioned that the lateral translations, though positive, give rotations which are negative for substitution in slope deflection equations.

18.2 DEVELOPMENT OF STIFFNESS MATRIX FOR A PIN-JOINTED STRUCTURE

Consider a pin-jointed structure as in Fig. 18.9*a*. The structure is kinematically indeterminate by two degrees as the displacement at A will have two components along X and Y directions. To prevent joint displacement at A, a restraining force equal to but opposite to external force P has to be applied. The components of the restraining force along the coordinate axes are

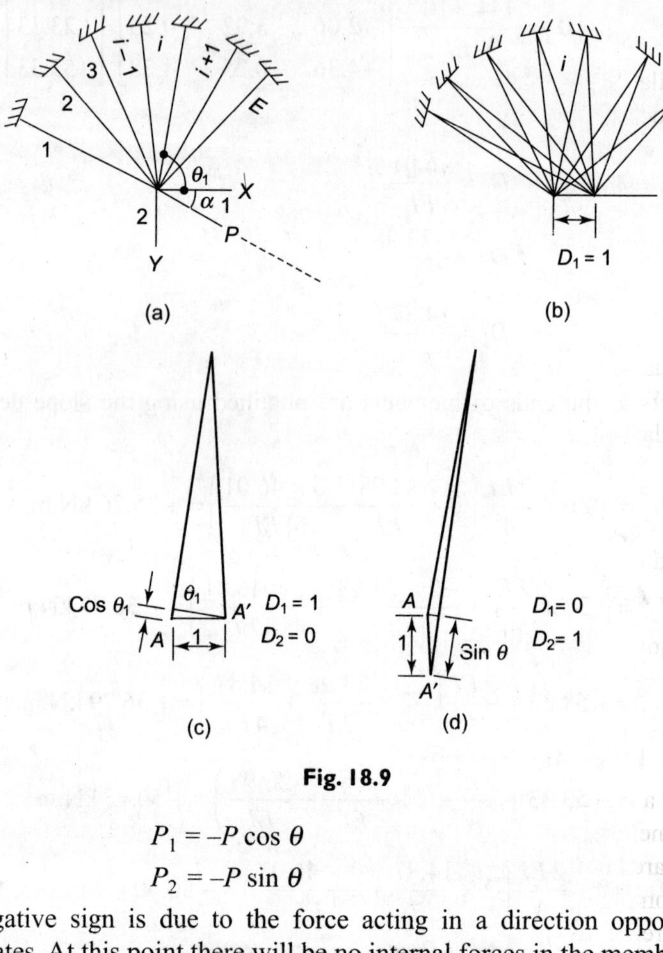

Fig. 18.9

$$P_1 = -P \cos \theta$$

$$P_2 = -P \sin \theta$$

The negative sign is due to the force acting in a direction opposite to the coordinates. At this point there will be no internal forces in the members.

Now let us determine the forces required in the members to hold the structure in displaced position such that $D_1 = 1$ and $D_2 = 0$ as in Fig. 18.9*b*.

The member 'i' as shown in Fig. 18.9*c* shortens by an amount cos θ_i and produces a compressive force $a_i\, E_i \cos \theta_i / l_i$, in which a_i, l_i and E_i refer to area of cross-section, length and Young's modulus for member 'i' The components of this force along X and Y directions are,

$$\frac{a_i E_i}{l_i} \cos^2 \theta_1 \text{ and } \frac{a_i E_i}{l_i} \sin \theta_i \cos \theta_i$$

The forces required to hold all the bars in the displaced position are,

$$k_{11} = \sum_{i=1}^{m} \frac{a_i E_i}{l_i} \cos^2 \theta_i \tag{18.4a}$$

And

$$k_{21} = \sum_{i=1}^{m} \frac{a_i E_i}{l_i} \sin \theta_i \cos \theta_i \tag{18.4b}$$

By a similar argument, the forces required to hold the joint in the displaced position such that $D_1 = 0$ and $D_2 = 1$ as in Fig. 18.19*a* are

$$k_{12} = \sum_{i=1}^{m} \left(\frac{a_i E_i}{l_i} \right) \sin \theta_i \cos \theta_i$$

$$k_{22} = \sum_{i=1}^{m} \left(\frac{a_i E_i}{l_i} \right) \sin^2 \theta_1$$

In the actual structure the joint A undergoes translations D_1 along X direction and D_2 along Y direction under the given loading without any restraining forces. The statical relationship can be expressed as

$$P_1 + k_{11} D_1 + k_{12} D_2 = 0$$
$$P_2 + k_{21} D_1 + k_{22} D_2 = 0$$

Expressed in matrix form

$$\{\mathbf{P}\} + [\mathbf{K}]\ \{\mathbf{D}\} = 0$$

The unknown displacements are determined from

$$\{\mathbf{D}\} = [\mathbf{K}]^{-1} \{-\mathbf{P}\}$$

18.2.1 Member Forces

Consider a typical member AB connecting joints A and B in a plane truss. The force in member AB can be calculated if the displacements at the two ends of the member are known.

The components of the displacements at A along X and Y coordinates are D_{AX} and D_{AY} respectively. Similarly D_{BX} and D_{BY} are the displacement components along X and Y respectively at B. It is clear that the shortening of the member at A due to joint displacement at A is $(D_{AX} \cos \theta_{AB} + D_{AY} \sin \theta_{AB})$. Similarly the elongation of the member at joint B due to joint displacement at B is $(D_{RX} \cos \theta_{AB} + D_{BY} \sin \theta_{AB})$

Therefore, the net elongation of member AB is

$$\{D_{BX} - D_{AX}) \cos \theta_{AB} + (D_{BY} - D_{AY}) \sin \theta_{AB}\} \tag{18.5}$$

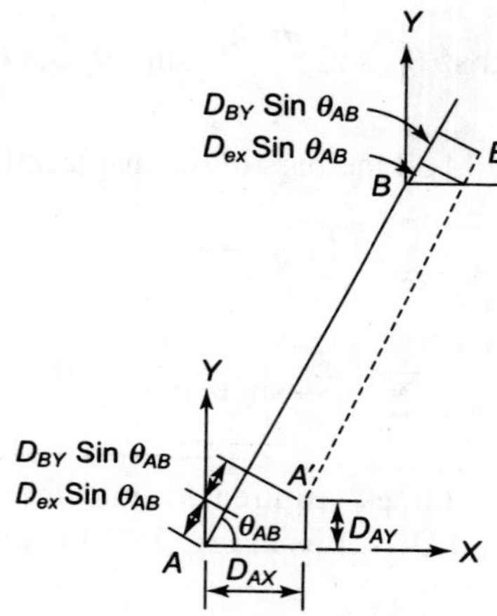

Fig. 18.10

Consequently the force in member AB is given by

$$P_{AB} = \frac{AE}{L}\{(D_{BX} - D_{AX})\cos\theta_{AB} + (D_{BY} - S_{AY})\sin\theta_{AB}\} \qquad (18.6)$$

If the member is reckoned as BA instead of AB, then the inclination of the member should be measured at end B and the equation for the force in member BA can be written as

$$P_{BA} = \frac{AE}{L}\{(D_{BX} - D_{AX})\cos\theta_{BA} + (D_{AY} - S_{BY})\sin\theta_{BA}\}$$

The angle of inclination of the member at any end is always measured counter-clockwise from the positive direction of the X axis. Also it may be remembered that the force in members P_{AB} and P_{BA} are found to be equal since the member carries the same force from one joint to the other.

A few examples will make the procedure clear.

Example 18.9 | *Using the stiffness method determine the displacements at the joint B of a pin-jointed frame shown in Fig. 18.11a. Also calculate the forces in members AB and BC due to the given loading. The values of area of cross-section are indicated. Take E = 2×10^5 N/mm^2*

Joint B has two degrees of freedom; displacements in the direction of the coordinates X and Y.

Let D_1 and D_2 be the displacements in the direction of the coordinates 1 and 2 due to applied load. The restraining forces are:

$$P_1 = 0 \text{ and } P_2 = +\,10 \text{ kN}.$$

The stiffness matrix with reference to the chosen coordinates may be developed by diving a unit displacement at coordinates 1 and 2 sucessively. The necessary computations for the evaluation of stiffness elements have been listed in the table that follows.

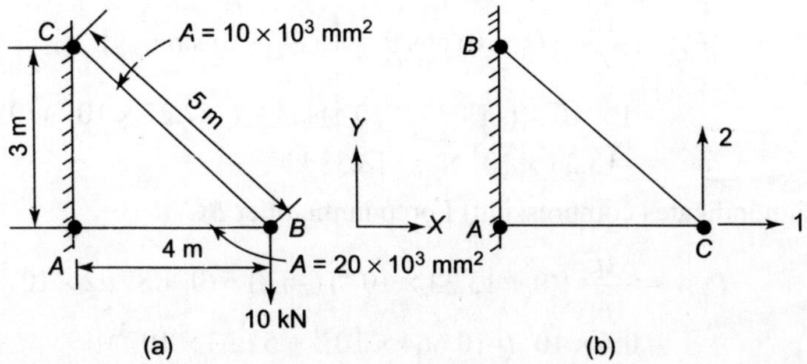

Fig. 18.11

Table 18.1

Member	$\frac{AE}{L}$	θ	cos θ	sin θ	$\frac{AE}{L}\cos^2\theta$	$\frac{AE}{L}\sin\theta^2$	$\frac{AE}{L}\sin\theta\cos\theta$
1	2	3	4	5	6	7	8
BA	1×10^6	180	–1.0	0	1×10^6	0	0
BC	0.4×10^6	143.3	–0.8	0.6	0.256×10^6	0.144×10^6	-0.192×10^6
					$\Sigma\ 1.256 \times 10^6$	0.144×10^6	-0.192×10^6

From the Table $k_{11} = 1.256 \times 10^6$

$$k_{21} = -0.192 \times 10^6$$

$$k_{22} = 0.144 \times 10^6$$

The stiffness matrix

$$[\mathbf{K}] = 10^3 \begin{bmatrix} 1256 & -192 \\ -192 & 144 \end{bmatrix}$$

$$[\mathbf{K}]^{-1} = \frac{1}{144 \times 10^6}\begin{bmatrix} 144 & +192 \\ +192 & 1256 \end{bmatrix}$$

$$\therefore \qquad D = \frac{1}{144 \times 10^6}\begin{bmatrix} 144 & 192 \\ 192 & 1256 \end{bmatrix}\begin{Bmatrix} 0 \\ -10 \times 10^3 \end{Bmatrix}$$

Solving,

$$D_1 = 13.33 \times 10^{-3}$$

and $D_2 = -87.22 \times 10^{-3}$

The member forces may be determined using Eqns 18.5 and 18.6. Force in member AB

$$P_{AB} = \frac{AE}{L}\{(D_1 - 0)\cos\theta_{AB} + (D_2 - 0)\sin\theta_{AB}\}$$
$$= 1\times10^6\{(-13.33\times10^{-3})(1) + (-87.22\times10^{-3})(0)\}$$
$$= -13.33\times10^3\text{ N} = -13.33\text{ kN}.$$

(Minus sign indicates compression) Force in member BC

$$P_{BC} = \frac{AE}{L}\{(0 + 13.33\times10^{-3})(-0.8) + (0. + 87.22\times10^{-3})(0.6)\}$$
$$= 0.4\times10^6(-10.664\times10^{-3} + 52.33\times10^{-3})$$
$$= +16.66\times10^3\text{ N} = +16.66\text{ kN}.$$

Example 18.10 | *Using the stiffness matrix method determine the displacements of the joint A of the pin-jointed plane frame shown in Fig. 18.12. Also determine the bar forces for the given loading.*

The degree of freedom or the kinematic indeterminancy is two as the joint A can undergo displacements in X and Y directions.

Let the coordinates be as shown in Fig. 18.12*b* to determine displacements D_1 and D_2 at joint A. The necessary computations for the evaluation of stiffness elements have been shown in the Table that follows.

Table 18.2

Member	$\frac{AE}{L}$	θ	$\cos\theta$	$\sin\theta$	$\frac{AE}{L}\cos^2\theta$	$\frac{AE}{L}\sin^2\theta$	$\frac{AE}{L}\sin\theta\cos\theta$
1	2	3	4	5	6	7	8
AF	$\frac{AE}{1.414L}$	45°	0.707	0.707	$0.3535\frac{AE}{L}$	$0.3535\frac{AE}{L}$	$0.3535\frac{AE}{L}$
AE	$\frac{AE}{1.154L}$	60°	0.500	0.866	$0.2166\frac{AE}{L}$	$0.6500\frac{AE}{L}$	$0.3752\frac{AE}{L}$
AD	$\frac{AE}{L}$	90°	0	1.000	0	$1.0000\frac{AE}{L}$	0
AC	$\frac{AE}{1.154L}$	120°	–0.500	0.866	$0.2166\frac{AE}{L}$	$0.6500\frac{AE}{L}$	$-0.3752\frac{AE}{L}$
AB	$\frac{AE}{1.1414L}$	135°	–0.707	0.707	$0.3535\frac{AE}{L}$	$0.3535\frac{AE}{L}$	$0.3752\frac{AE}{L}$
					$\Sigma\ 1.1402\frac{AE}{L}$	$3.007\frac{AE}{L}$	0

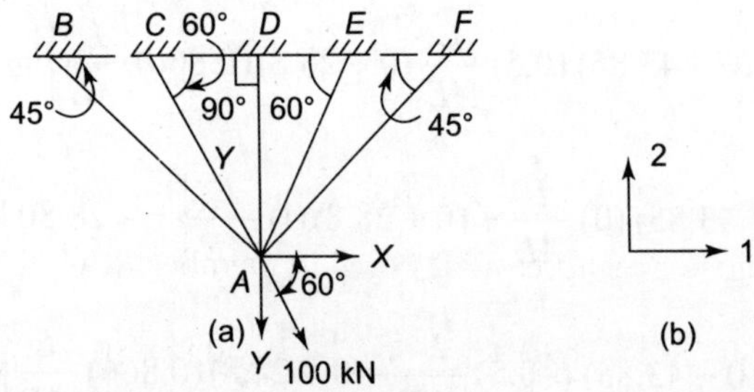

Fig. 18.12

The Elements of the matrix [**K**] are

$$k_{11} = 1.1402\ \frac{AE}{L}$$

$$k_{21} = 0$$

$$k_{22} = 3.007\ \frac{AE}{L}$$

Matrix
$$[\mathbf{K}] = \frac{AE}{L}\begin{bmatrix} 1.1402 & 0 \\ 0 & 3.007 \end{bmatrix}$$

$$[\mathbf{K}]^{-1} = \frac{L}{3.4286\ AE}\begin{bmatrix} 3.0070 & 0 \\ 0 & 1.1402 \end{bmatrix}$$

The restraining forces

$$\{\mathbf{P}\} = \begin{Bmatrix} -50.0\ \text{kN} \\ +86.6\ \text{kN} \end{Bmatrix}$$

$$\{\mathbf{D}\} = \frac{L}{3.4286\ AE}\begin{bmatrix} 3.0070 & 0 \\ 0 & 1.1402 \end{bmatrix}\begin{Bmatrix} 50.0 \\ +86.6 \end{Bmatrix}$$

Soving

$$D_1 = +43.85\ \frac{L}{AE}$$

$$D_2 = -28.8\ \frac{L}{AE}$$

The member forces may be obtained using Eqns 18.5 and 18.6.

$$P_{AF} = \frac{AE}{1.414\,L}\{(0 - D_1)\cos\theta + (0 - D_2)\sin\theta\}$$

$$= \frac{AE}{1.414\,L}\left\{(043.85)\frac{(0.707)L}{AE} - 28{,}8\,(0.707)\frac{L}{AE}\right\} = -7.525\ \text{kN}.$$

$$P_{AE} = \frac{AE}{1.154L}\left\{(0-43.85)\,(0.5)\frac{L}{AE}\,(0+28.8)(0.866)\frac{L}{AE}\right\} = +2.61\text{ kN.}$$

$$P_{AD} = \frac{AE}{L}\left\{(0-43.85)\,(0)\frac{L}{AE}+(0+28.8)(1)\frac{L}{AE}\right\} = +28.80\text{ kN.}$$

$$P_{AC} = \frac{AE}{1.541L}\left\{(0-43.85)\,(-0.5)\frac{L}{AE}+(0+28.8)(0.866)\frac{L}{AE}\right\} = +40.61\text{ kN.}$$

$$P_{AB} = \frac{AE}{1.414L}\left\{(0-43.85)\,(-0.707)\frac{L}{AE}+(0+28.8)(0.707)\frac{L}{AE}\right\} = +36.33\text{ kN.}$$

Example 18.11 | *Using the stiffness method determine the displacements at joint A and bar forces under loads P_1 and P_2 for the pin-jointed frame shown in Fig. 18.13. AE is the same for all members.*

The stiffness elements have been worked out in Table 18.3.

Table 18.3

Member	$\frac{AE}{L}$	θ	*cos θ*	*sin θ*	$\frac{AE}{L}\cos^2\theta$	$\frac{AE}{L}\sin^2\theta$	$\frac{AE}{L}\sin\theta\cos\theta$
1	*2*	*3*	*4*	*5*	*6*	*7*	*8*
AB	$0.707\frac{AE}{L}$	135°	–0.707	0.707	$0.3534\frac{AE}{L}$	$0.3534\frac{AE}{L}$	$-0.3534\frac{AE}{L}$
AC	$\frac{AE}{L}$	180°	–1.00	0	$1.000\frac{AE}{L}$	0	0
AD	$0.707\frac{AE}{L}$	225°	–0.700	–0.700	$0.3534\frac{AE}{L}$	$0.3534\frac{AE}{L}$	$+0.3534\frac{AE}{L}$
					$\Sigma\ 1.7068\frac{AE}{L}$	$0.7068\frac{AE}{L}$	0

The stiffness elements are

$$k_{11} = 1.7068\frac{AE}{L}$$

$$k_{21} = 0$$

$$k_{22} = 0.7068\frac{AE}{L}$$

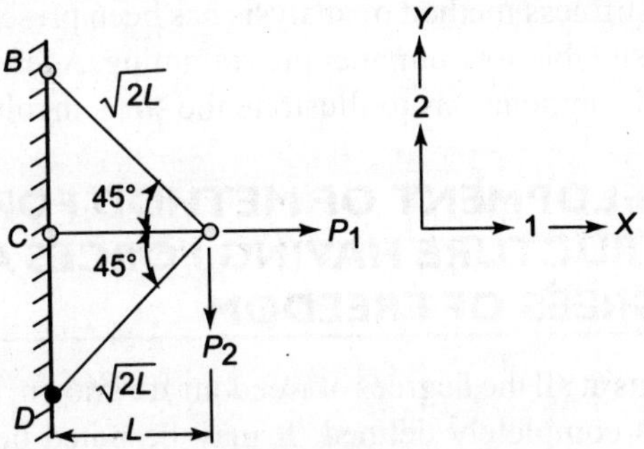

Fig. 18.13

$$[\mathbf{K}] = \frac{AE}{L}\begin{bmatrix} 1.7068 & 0 \\ 0 & 0.7068 \end{bmatrix}$$

$$[\mathbf{K}]^{-1} = \frac{L}{1.2071\,AE}\begin{bmatrix} 0.7068 & 0 \\ 0 & 1.7068 \end{bmatrix}$$

$$[\mathbf{D}] = \frac{1}{1.2071\,AE}\begin{bmatrix} 0.707 & 0 \\ 0 & 1.707 \end{bmatrix}\begin{Bmatrix} P_1 \\ -P_2 \end{Bmatrix}$$

$$D_1 = 0.5857\,P_1\,\frac{L}{AE}$$

$$D_2 = -1.414\,P_2\,\frac{L}{AE}$$

The bar forces are

$$P_{AB} = \frac{AE}{1.414\,L}\left\{\left(0 - 0.5857 P_1 \frac{L}{AE}\right)(-0.707) + \left(0 + 1.414\,P_2\,\frac{L}{AE}\right)0.707\right\}$$

$$= +\,0.2929\,P_1 + 0.707\,P_2$$

$$P_{AC} = \frac{AE}{L}\left\{\left(0 - 0.5857 P_1 \frac{L}{AE}\right)(-1.0) + \left(0 + 1.414\,P_2\,\frac{L}{AE}\right)(0)\right\} = +\,0.5857\,P_1$$

$$P_{AD} = \frac{AE}{1.414\,L}\left\{\left(0 - 0.5857\,\frac{P_1 L}{AE}\right)(0.707) + \left(\frac{0 + 1.414\,P_2 L}{AE}\right)(-0.707)\right\}$$

$$= +\,0.292\,P_1 - 0.707\,P_2$$

Up to this point, the stiffness matrix [K] has been developed direct for the structure at the required coordinates. A few simple examples have been solved to bring out the procedure involved. This procedure, however, is not suitable for structures having high degree of kinematic indeterminancy and also it does not lend itself to computer programming.

A generalised stiffness method of analysis has been presented in the following sections which is suitable for computer programming. A few examples have been solved using hand computations to illustrate the steps involved in the process.

18.3 DEVELOPMENT OF METHOD FOR A STRUCTURE HAVING FORCES AT ALL DEGREES OF FREEDOM

If the displacements at all the degrees of freedom are known, then the deformation of the structure is completely defined. It may be noted here that the stiffness matrix to be inverted is of order $n \times n$, where n represents the degree of freedom of the structure. If forces exist at all the degrees of freedom (n in number), the structure is said to be kinematically indeterminate by n degrees and the kinematic deficiency is zero. However, if forces are applied at m ($m < n$) degrees of freedom only, the displacements associated with the applied loads only cannot fully describe the deformation of the structure. Such a structure is kinematically deficient by $(n - m)$ degrees.

We shall first develop the method for a structure having forces applied at all the degrees of freedom and next for a general case considering that forces are only applied at some of the degrees of freedom.

If the forces exist at all the degrees of freedom, the kinematic deficiency is zero. The displacements associated with the applied forces completely describe the deformation of the structure.

We can proceed with the analysis by defining the system coordinates at all the nodal points and numbering them: first, the nodal points that undergo displacements, and then, the nodal points that are restrained from undergoing displacements. We must ensure that the displacements are independent and hence corresponding stiffness matrix **K** exists so that we can write Eq. 18.1.

Next, we select elements so that the ends of the elements coincide with system coordinates. We fix for each element s, element coordinats for which a stiffness matrix exists, so that we can write

$$\mathbf{p_s} = \mathbf{k_s}\, \mathbf{d_s}$$

or

$$\mathbf{p} = \mathbf{k}\, \mathbf{d}$$

for all unassembled elements.

We next generate displacement transformation matrix B using the procedure of Section 16.2 which ensures element-structure compatibility, that is

$$\mathbf{d} = \mathbf{BD}$$

Structure stiffness matrix **K** is synthesised from the element stiffness matrices using Eq. 16.17

$$\mathbf{K} = \mathbf{B^T}\, \mathbf{k}\, \mathbf{B}$$

From the known stiffness matrix, the nodal displacements are obtained from Eq. 18.2. From the known nodal displacements, we can write the displacements of the elements as

$$\mathbf{d} = \mathbf{BD}$$

or
$$\mathbf{d} = \mathbf{BK}^{-1}\,\mathbf{P} \tag{18.7}$$

The internal stresses can be written as

$$\mathbf{p} = \mathbf{kd}$$

or
$$\mathbf{p} = \mathbf{kBD} \tag{18.8}$$

or
$$\mathbf{p} = \mathbf{kBK}^{-1}\,\mathbf{P} \tag{18.9}$$

We shall demonstrate the complete procedure in the following simple examples.

Example 18.12 | *Two loads are applied at joint A of the trussed structure shown in Fig. 18.14a. Determine the displacements and internal forces of the members of the truss by the stiffness method. The cross-section is the same for all members.*

On inspection, we can notice that the structure has two degrees of freedom. These are identified and represented by coordinates 1 and 2. The elements are identified and numbered. Transformation matrix **B** can be constructed by imposing unit displacements along coordinates 1 and 2, in turn, and evaluating the displacements of elements (*see* Fig. 18.14*a* and *c*). The resulting matrix relationship is

$$\underset{\mathbf{d}}{\begin{Bmatrix} d_1 \\ d_2 \\ d_3 \\ d_4 \\ d_5 \end{Bmatrix}} = \underset{\mathbf{B}}{\begin{bmatrix} 1/\sqrt{2} & 1/\sqrt{2} \\ 1/2 & \sqrt{3}/2 \\ 0 & 1 \\ -1/2 & \sqrt{3}/2 \\ -1\sqrt{2} & 1/\sqrt{2} \end{bmatrix}} \underset{\mathbf{D}}{\begin{Bmatrix} D_1 \\ D_2 \end{Bmatrix}}$$

Next the stiffness matrix for the structure can be developed using Eq. 16.17.

$$\mathbf{K} = \mathbf{B}^{\mathrm{T}}\,\mathbf{kB}$$

where **k** is the unassembled stiffnesses of all members. For the truss structure, we can write matrix **k** as

$$\mathbf{k} = \frac{AE}{L}\begin{bmatrix} 1/\sqrt{2} & & & & \\ & 3/\sqrt{2} & & & \\ & & 1 & & \\ & & & 3/\sqrt{2} & \\ & & & & 1/\sqrt{2} \end{bmatrix}$$

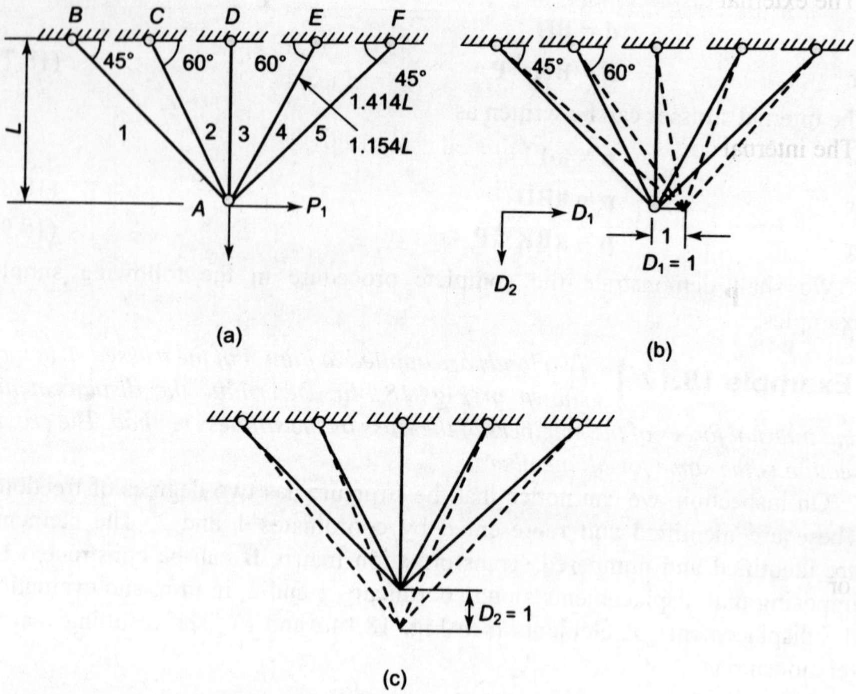

Fig. 18.14 | *(a) Structure and loading, (b) Unit displacement imposed along coordinate 1, (c) Unit displacement imposed along coordinate 2*

and

$$\mathbf{K} = \begin{bmatrix} 1/\sqrt{2} & 1/2 & 0 & -1/2 & -1/2 \\ 1/\sqrt{2} & \sqrt{3}/2 & 1 & \sqrt{3}/2 & -1/\sqrt{2} \end{bmatrix}$$

$$\frac{AE}{L}\begin{bmatrix} 1/\sqrt{2} & & & & \\ & 3/\sqrt{2} & & & \\ & & 1 & & \\ & & & 3/\sqrt{2} & \\ & & & & 1/\sqrt{2} \end{bmatrix}\begin{bmatrix} 1/\sqrt{2} & 1/\sqrt{2} \\ 1/2 & \sqrt{3}/2 \\ 0 & 1 \\ -1/2 & \sqrt{3}/2 \\ -1/\sqrt{2} & 1/\sqrt{2} \end{bmatrix}$$

or

$$\mathbf{K} = \frac{AE}{L}\begin{bmatrix} 1.140 & 0 \\ 0 & 3.006 \end{bmatrix}$$

$$\mathbf{K}^{-1} = \frac{L}{3.83\,AE}\begin{bmatrix} 3.006 & 0 \\ 0 & 1.140 \end{bmatrix}$$

The external displacements are

$$\mathbf{D} = \frac{L}{3.83\,AE}\begin{bmatrix} 3.006 & 0 \\ 0 & 1.140 \end{bmatrix}\begin{Bmatrix} P_1 \\ P_2 \end{Bmatrix},$$

The internal stresses can be evaluated using Eq. 18.8.

$$\mathbf{P} = \frac{AE}{L}\begin{bmatrix} 1/\sqrt{2} & & & & \\ & 3/\sqrt{2} & & & \\ & & 1 & & \\ & & & 3/\sqrt{2} & \\ & & & & 1/\sqrt{2} \end{bmatrix}\begin{bmatrix} 1/\sqrt{2} & 1/\sqrt{2} \\ 1/2 & \sqrt{3}/2 \\ 0 & 1 \\ -1/2 & \sqrt{3}/2 \\ -1/\sqrt{2} & 1/\sqrt{2} \end{bmatrix}$$

$$\frac{L}{3.43\,AE}\begin{bmatrix} 3.006 & 0 \\ 0 & 1.140 \end{bmatrix}\begin{Bmatrix} P_1 \\ P_2 \end{Bmatrix}$$

or

$$\mathbf{P} = \begin{bmatrix} 0.4382\,P_1 & +0.1662\,P_2 \\ 0.3795\,P_1 & +0.2493\,P_2 \\ 0 & +0.3324\,P_2 \\ -0.3795\,P_1 & +0.2493\,P_2 \\ -0.4389\,P_1 & +0.1662\,P_2 \end{bmatrix}$$

Example 18.13 | *Using the stiffness method, calculate the end deflectio.. and rotation of a cantilever beam loaded uniformly as shown in Fig. 18.15a. EI is constant.*

The structure has two degrees of freedom, one rotation and the other translation at the free end.

Displacements transformation matrix **B** is

$$\mathbf{B} = \begin{bmatrix} -\dfrac{1}{4} & 0 \\ -\dfrac{1}{4} & 1 \end{bmatrix}$$

Member stiffness matrix $\mathbf{k} = \dfrac{EI}{L}\begin{bmatrix} 4 & 2 \\ 2 & 4 \end{bmatrix}$

Structure stiffness matrix **K** is

$$\mathbf{K} = \mathbf{B}^T\mathbf{kB}$$

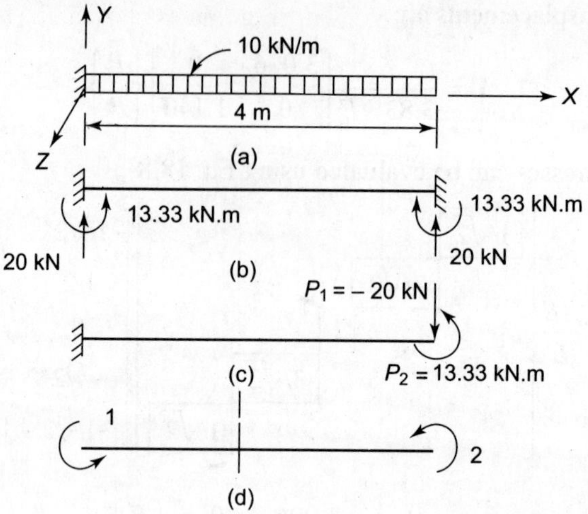

Fig. 18.15 | *(a) Cantilever beam and loading, (b) Fixed coordinate state-forces P* at coordinates, (c) Forces at coordinates (–P*), (d) Member coordinates*

$$= \frac{EI}{4}\begin{bmatrix} 3/4 & -3/2 \\ -3/2 & 4 \end{bmatrix}$$

Then $$\mathbf{D} = \mathbf{K}^{-1}\,\mathbf{P}$$

$$= \frac{16}{3EI}\begin{bmatrix} 4 & 3/2 \\ 3/2 & 3/4 \end{bmatrix}\begin{Bmatrix} -20.00 \\ +13.33 \end{Bmatrix} = \frac{16}{3EI}\begin{Bmatrix} -60 \\ -20 \end{Bmatrix}$$

The member forces are given by Eq. 18.4

$$\mathbf{p} = \frac{EI}{2}\begin{bmatrix} 2 & 1 \\ 1 & 2 \end{bmatrix}\begin{bmatrix} -1/4 & 0 \\ -1/4 & 1 \end{bmatrix}\frac{16}{3EI}\begin{Bmatrix} -60 \\ -20 \end{Bmatrix} = \frac{4}{3}\begin{Bmatrix} +50 \\ +10 \end{Bmatrix} \text{kN.m}$$

Final internal forces p must include the fixed end moments shown in Fig. 18.15*b*.

or

$$\mathbf{p}^f = \frac{4}{3}\begin{Bmatrix} +50 \\ +10 \end{Bmatrix} + \begin{Bmatrix} 40/3 \\ -40/3 \end{Bmatrix} = \begin{Bmatrix} 80 \\ 0 \end{Bmatrix} \text{kN.m}$$

Example 18.14 | *Using the stiffness method of analysis, analyse the frame of Fig. 18.16a for displacements at the coordinates of the structure and the internal forces in the members.*

The structure has two degrees of freedom. They are the rotations at coordinates 1 and 2. The deformations corresponding to these coordinates completely describe the deflected shape of the structure. We choose the element coordinates as shown in Fig. 18.16*c*. For the element coordinates chosen, the stiffness matrix of the elements is

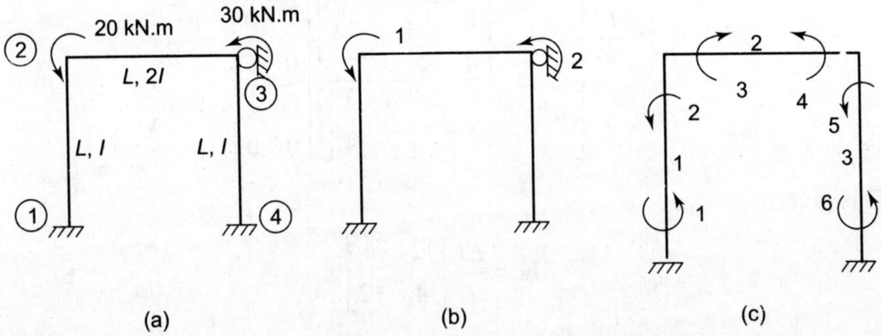

Fig. 18.16 | *(a) Structure and loading, (b) Structure coordinates, (c) Element coordinates*

$$\mathbf{k}_s = \frac{EI}{L}\begin{bmatrix} 4 & 2 \\ 2 & 4 \end{bmatrix}$$

so that

$$\mathbf{P}_s = \mathbf{k}_s\, \mathbf{d}_s$$

Therefore we can write matrix **k**, the uncoupled matrix of the structure as

$$\mathbf{k} = \frac{EI}{L}\begin{bmatrix} 4 & 2 & & & & \\ 2 & 4 & & & & \\ & & 8 & 4 & & \\ & & 4 & 8 & & \\ & & & & 4 & 2 \\ & & & & 2 & 4 \end{bmatrix}$$

Next, we construct displacement transformation matrix **B**

$$\mathbf{B} = \begin{Bmatrix} 0 & 0 \\ 1 & 0 \\ \hline 1 & 0 \\ 0 & 1 \\ \hline 0 & 1 \\ 0 & 0 \end{Bmatrix}$$

Structure matrix **K** is synthesised using Eq. 16.17

$$\mathbf{K} = \begin{bmatrix} 0 & 1 & 1 & 0 & 0 & 0 \\ 0 & 0 & 0 & 1 & 1 & 0 \end{bmatrix}$$

$$\times \frac{EI}{L}\begin{bmatrix} 4 & 2 & & & & \\ 2 & 4 & & & & \\ & & 8 & 4 & & \\ & & 4 & 8 & & \\ & & & & 4 & 2 \\ & & & & 2 & 4 \end{bmatrix}\begin{bmatrix} 0 & 0 \\ 1 & 0 \\ 1 & 0 \\ 0 & 1 \\ 0 & 1 \\ 0 & 0 \end{bmatrix}$$

This gives

$$\mathbf{K} = \frac{EI}{L}\begin{bmatrix} 12 & 4 \\ 4 & 12 \end{bmatrix}$$

or

$$\mathbf{K}^{-1} = \frac{L}{128\,EI}\begin{bmatrix} 12 & -4 \\ -4 & 12 \end{bmatrix}$$

The displacement at the coordinates are

$$\mathbf{D} = \mathbf{K}^{-1}\,\mathbf{P}$$

$$= \frac{L}{128\,EI}\begin{bmatrix} 12 & -4 \\ -4 & 12 \end{bmatrix}\begin{Bmatrix} 20 \\ 30 \end{Bmatrix}$$

$$\mathbf{D} = \frac{L}{EI}\begin{Bmatrix} 15/16 \\ 35/16 \end{Bmatrix}$$

We can evaluate the member forces using Eq. 18.8

$$\mathbf{p} = \frac{1}{128}\begin{bmatrix} 24 & -8 \\ 48 & -16 \\ 80 & 16 \\ 16 & 80 \\ -16 & 48 \\ -8 & 24 \end{bmatrix}\begin{Bmatrix} 20 \\ 30 \end{Bmatrix} = \frac{1}{8}\begin{Bmatrix} 15 \\ 30 \\ 130 \\ 170 \\ 70 \\ 35 \end{Bmatrix} \text{kN.m}$$

The results satisfy the conditions of equilibrium at the joints where the external forces are applied.

Example 18.15 | *If the structure of Example 18.14 has applied loads at the coordinates as well as along members as shown in Fig. 18.17a, establish the displacements at the coordinates and internal stresses in members.*

The structure coordinates and element coordinates are the same as in the previous example.

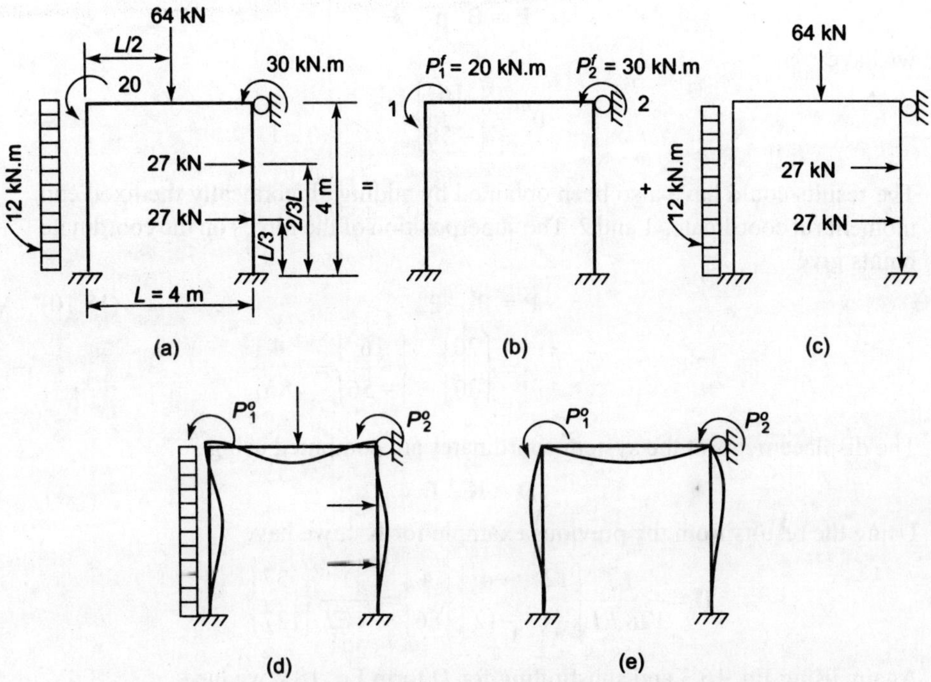

Fig. 18.17 | *(a) Structure and loading, (b) Forces P^f at coordinates, (c) Forces not at the coordinates, (d) Fixed coordinate state, (e) Forces at the coordinates ($-P^f$)*

It is convenient to separate the forces at the joints and forces not at the joints as in Fig. 18.176 and *c* respectively. Fixed coordinate forces $\mathbf{P}^o$ are computed from the corresponding $\mathbf{p}^o{}_s$ at the element coordinates using the Appendix table. From Eq, 16.26, we can write

$$\mathbf{P}^o = \mathbf{B}^T \mathbf{p}^o$$

In the present example, we have

$$\begin{Bmatrix} p_1^o \\ p_2^o \end{Bmatrix} = \begin{Bmatrix} +16 \\ -16 \end{Bmatrix}, \begin{Bmatrix} p^{o3} \\ p_4^o \end{Bmatrix} = \begin{Bmatrix} +32 \\ -32 \end{Bmatrix}, \begin{Bmatrix} p^o 5 \\ p^o 6 \end{Bmatrix} = \begin{Bmatrix} -24 \\ +24 \end{Bmatrix}$$

or

$$\mathbf{p}^o = \begin{Bmatrix} 16 \\ -16 \\ \hline +32 \\ -32 \\ \hline -24 \\ +24 \end{Bmatrix}$$

Matrix **B** is the same as in the earlier example. Performing the operation of Eq. 16.26

$$\mathbf{P} = \mathbf{B}^T \mathbf{p}^o$$

we have

$$\mathbf{p}^o = \begin{Bmatrix} 16 \\ -56 \end{Bmatrix}$$

The results could have also been obtained by adding algebraically the fixed end moments at coordinates 1 and 2. The superposition of the forces on the coordinate points give

$$\mathbf{P} = \mathbf{P}^f - \mathbf{P}^o \tag{18.10}$$

$$= \begin{Bmatrix} 20 \\ 30 \end{Bmatrix} - \begin{Bmatrix} 16 \\ -56 \end{Bmatrix} = \begin{Bmatrix} 4 \\ 86 \end{Bmatrix}$$

The displacements at the system coordinates are computed using

$$\mathbf{D} = \mathbf{K}^{-1}\,\mathbf{P}$$

Using the results from the previous example for $\mathbf{K}^{-1}$, we have

$$\mathbf{D} = \frac{L}{128\,EI}\begin{bmatrix} 12 & -4 \\ -4 & -12 \end{bmatrix}\begin{Bmatrix} 4 \\ 86 \end{Bmatrix} = \frac{1}{4\,EI}\begin{Bmatrix} 37 \\ 127 \end{Bmatrix}$$

Again, using Eq. 16.3 and substituting for D form Eq. 18.2 we have

$$\mathbf{d} = \mathbf{B}\mathbf{K}^{-1}\,(\mathbf{p}^f - \mathbf{P}^o) \tag{18.11}$$

and

$$\mathbf{p} = \mathbf{k}\mathbf{d} = \mathbf{k}\mathbf{B}\mathbf{K}^{-1}\,(\mathbf{P}^f - \mathbf{P}^o) \tag{18.12}$$

The superposition of forces and displacements in Figs. 18.17*b* and *e* (forces at coordinates + forces opposite to the fixed coordinate state) give the corresponding forces and displacements at any point on the structure as desired. Therefore, the net forces at the element coordinates are

$$\mathbf{P}^f = \mathbf{p}^o + \mathbf{k}\mathbf{B}\mathbf{K}^{-1}\,(\mathbf{P}^f - \mathbf{P}^o) \tag{18.13}$$

In the example, we have

$$\mathbf{p}^f = \begin{Bmatrix} +16 \\ -16 \\ +32 \\ -32 \\ -24 \\ +24 \end{Bmatrix} + \frac{1}{128}\begin{bmatrix} 24 & -8 \\ 48 & -16 \\ 80 & 16 \\ 16 & 80 \\ -16 & 48 \\ -8 & 24 \end{bmatrix}\begin{Bmatrix} 4 \\ 86 \end{Bmatrix} = \frac{1}{16}\begin{Bmatrix} 96 \\ -404 \\ 724 \\ 356 \\ 124 \\ 630 \end{Bmatrix}\ \text{kN.m}$$

These results satisfy the conditions of equilibrium at the structure coordinates 1 and 2.

With forces $\mathbf{p}^f$ known, each element can be analysed as a statically determinate member to compute the displacements and internal forces at any point in the structure.

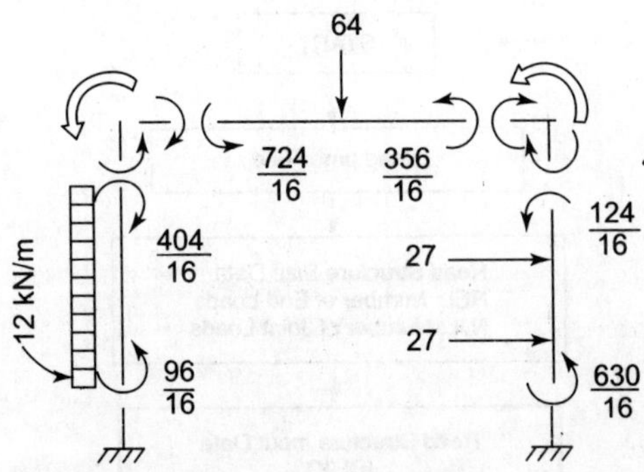

Fig. 18.18 | *Free-body diagrams of elements*

The free-body diagrams shown in Fig. 18.18 give an idea about the forces.

18.3.1 Computer Programme for the Stiffness Analysis of Kinematically Determinate System

The analysis of kinematically determinate structures by the stiffness method has been reduced to a set of matrix opeations. The flow chart is shown in Fig. 18.19. The input data consists of the matrices **[B]**,{**k**}, and {**P**}. The output consists of the matrices [**K**], [**F**], {**D**}, and {**p**}.

18.4 DEVELOPMENT OF METHOD FOR A GENERAL CASE

Generally in a structure, every node will have a force and a corresponding displacement, one of which is known. For example, specified loads are applied at all degrees of freedom (some of them can be zero), hence the forces are known and the displacements are unknown; on the other hand, at the support points displacements are zero or specified but the reactive forces are unknown. We can collect *n* coordinates with *unknown displacements and known forces* (loads) and the remaining *m* coordinates corresponding to *known displacements {at supports) and unknown forces* (*reactions)* and partition the matrix equation as

Known forces (loads) — Unknown displacement at load point

$$\left\{\begin{array}{c}\mathbf{P}_1\\ \hline \mathbf{P}_2\end{array}\right\} = \begin{array}{c} \\ 1\\ 0\end{array}\begin{array}{c}\quad 1 \qquad\quad 2\\ \left[\begin{array}{c|c} \underset{n\times n}{K_{11}} & \underset{n\times m}{K_{12}} \\ \hline \underset{m\times n}{K_{21}} & \underset{m\times m}{K_{22}}\end{array}\right]\end{array} \left\{\begin{array}{c}\mathbf{D}_1\\ \hline \mathbf{D}_2\end{array}\right\} \tag{18.14}$$

Unknown forces (reactions) — Known (zero or specified) displacements at supports

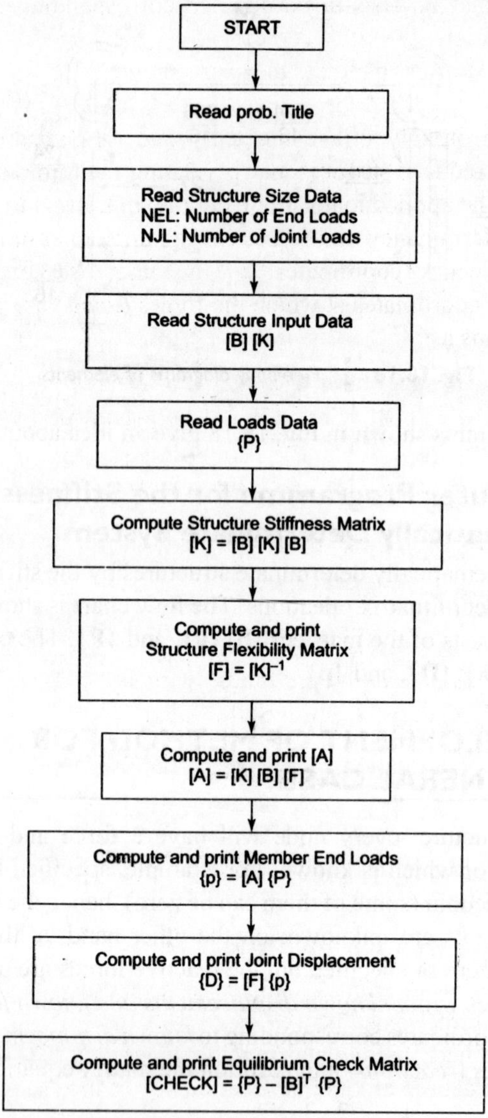

Fig. 18.19 | *Flow chart for stiffness analysis of kinematically determinate structure*

Writing Equation 18.14 in the expanded form

$$\mathbf{P}_1 = \mathbf{K}_{11}\,\mathbf{D}_1 + \mathbf{K}_{12}\,\mathbf{D}_2 \qquad (18.15)$$

and

$$\mathbf{P}_2 = \mathbf{K}_{21}\,\mathbf{D}_1 + \mathbf{K}_{22}\,\mathbf{D}_2 \qquad (18.16)$$

The solution is obtained in two steps. First *n* equations are solved for unknown displacements $\mathbf{D}_1$. From Eq 18.15,

$$\mathbf{D}_1 = \mathbf{K}^{-1}_{11}\,(\mathbf{P}_1 - \mathbf{K}_{12}\,\mathbf{D}_2) \qquad (18.17)$$

It may be noted that $\mathbf{K}_{11}$ has to be inverted corresponding to a solution of n simultaneous equations.

Displacements $\mathbf{D}_1$ thus found are then substituted in Eq. 18.16 to solve for the unknown forces.

We can also profitably utilise the technique of partitioning the stiffness matrix to obtain a reduced stiffness matrix relating the forces and displacements corresponding to the applied loads. For example, in a structure having n degrees of kinematic indeterminancy and forces (loads) applied at m of the coordinates only, we can collect m coordinates at which the loads are applied and the remaining $(n - m)$ coordinates at which the forces (loads) are zero, and partition the matrix equations as

$$\begin{matrix}\text{Applied forces known (load)} \\ \text{Applied forces (zero)}\end{matrix}\left\{\begin{array}{c}\mathbf{P}_1 \\ \hline \mathbf{P}\end{array}\right\} = \left[\begin{array}{c|c}\underset{m\times m}{\mathbf{K}_{11}} & \underset{m\times(n-m)}{\mathbf{K}_{11}} \\ \hline \underset{(n-m)(m)}{\mathbf{K}_{21}} & \underset{(n-m)(n-m)}{\mathbf{K}_{22}}\end{array}\right]\left\{\begin{array}{c}\mathbf{D}_1 \\ \mathbf{D}_2\end{array}\right\}\begin{matrix}\text{Displacements corresponding to applied forces (unknown)} \\ \text{Displacements corresponding to zero applied forces (unknown)}\end{matrix} \tag{18.18}$$

Writing Equation 18.18 in the expanded form

$$\mathbf{P}_1 = \mathbf{K}_{11}\mathbf{D}_1 + \mathbf{K}_{12}\mathbf{D}_2 \tag{18.19}$$

and

$$0 = \mathbf{K}_{21}\,\mathbf{D}_1 + \mathbf{K}_{22}\,\mathbf{D}_2 \tag{18.20}$$

As a first step we evaluate D_2 by pre-multiplying Eq. 18.20 by $\mathbf{K}_{22}^{-1}$

$$\mathbf{D}_2 = -\,\mathbf{K}_{22}^{-1}\,\mathbf{K}_{21}\,\mathbf{D}_1 \tag{18.21}$$

Substituting for $\mathbf{D}_2$ in Equation 18.19 and rearranging, we get

$$\mathbf{P}_1 = (\mathbf{K}_{11} - \mathbf{K}_{12}\mathbf{K}^{-1}{}_{22}\,\mathbf{K}_{21})\mathbf{D}_1 \tag{18.22}$$

or

$$\mathbf{P}_1 = \mathbf{K}_1\,\mathbf{D}_1 \tag{18.23}$$

where

$$\mathbf{K}_1 = \mathbf{K}_{11} - \mathbf{K}_{12}\mathbf{K}^{-1}{}_{22}\,\mathbf{K}_{21} \tag{18.24}$$

$\mathbf{K}_1$ is known as *reduced stiffness matrix* corresponding to the coordinates at which the applied forces exist.

It may be noted here that the matrix to be inverted is $\mathbf{K}_{22}$, the order of which corresponds to the degree of kinematic deficiency of the structure.

With these basic concepts, the steps necessary for the formulation of the stiffness method are as follows.

1. Identify the nodal points (joints) and number them, first the joints that undergo displacements, and then the joints that are restrained from undergoing displacements.
2. Select elements so that the ends of the elements coincide with the structure coordinates.
3. Write the stiffness matrix for each element using element coordinates to only account for the desired energy forms.
4. Generate transformation matrix **B** using the procedure of Section 16.2, which ensures element-structure compatibility.

5. Synthesise structure stiffness matrix **K** using Eq. 16.17.
6. Write the matrix equation in the partitioned form (Eq. 18.14 or 18.18) and solve for unknown displacements $\mathbf{D}_1$ or $\mathbf{D}_2$ and unknown reactions P_2.
7. The element displacements can be written in the form

$$\mathbf{d} = \mathbf{BD}$$

and the element (internal) forces $\mathbf{p} = \mathbf{kd}$.

This concludes the analysis by the stiffness method. We shall demonstrate the complete procedure in terms of simple examples.

Example 18.16 | *Figure 18.20a shows a frame subjected to lateral force P_1. Find reduced stiffness matrix $\mathbf{K}_l$ and corresponding displacements $\mathbf{D}_l$ and the internal stresses and displacements in elements.*

Neglecting axial deformations the structure has three degrees of freedom (two rotations and one translation) while external force P_1 is applied at coordinate 1 only. Therefore, the kinematic deficiency is two. The element coordinates are chosen as indicated in Fig. 18.20*c*. Transformation matrix **B** which ensures compatibility by relating element displacements **d** to system displacements **D** is obtained from Eq. 16.3

$$\mathbf{d} = \mathbf{BD}$$

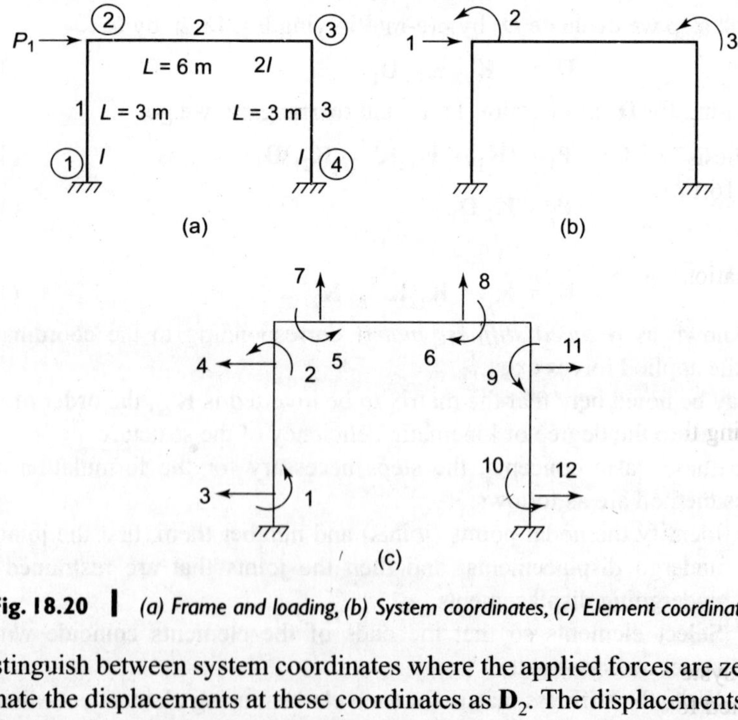

Fig. 18.20 | *(a) Frame and loading, (b) System coordinates, (c) Element coordinates*

To distinguish between system coordinates where the applied forces are zero, we designate the displacements at these coordinates as $\mathbf{D}_2$. The displacements at the coordinates where forces P_1 are applied are denoted by $\mathbf{D}_1$. Using this notation, Eq. 16.3 can be written in the partitioned form as

$$d = \left[\mathbf{B}_1 \,\vdots\, \mathbf{B}_2\right]\begin{Bmatrix}\mathbf{D}_1\\ \hline \mathbf{D}_2\end{Bmatrix} \tag{18.25}$$

At this stage it may be noted that displacements D_1 are not known. Writing Eq. 18.25 in expanded form

$$\mathbf{d} = \mathbf{B}_1\mathbf{D}_1 + \mathbf{B}_2\mathbf{D}_2 \tag{18.26}$$

We construct the transformation matrix **B** and write

$$\begin{Bmatrix} d_1\\ d_2\\ d_3\\ d_4\\ \hline d_5\\ d_6\\ d_7\\ d_8\\ \hline d_9\\ d_{10}\\ d_{11}\\ d_{12}\end{Bmatrix} = \left[\begin{array}{c|cc} 0 & 0 & 0\\ 0 & 1 & 0\\ 0 & 0 & 0\\ -1 & 0 & 0\\ \hline 0 & 1 & 0\\ 0 & 0 & 1\\ 0 & 0 & 0\\ 0 & 0 & 0\\ \hline 0 & 0 & 1\\ 0 & 0 & 0\\ 1 & 0 & 0\\ 0 & 0 & 0 \end{array}\right]\begin{Bmatrix}\mathbf{D}_{11}\\ \hline \mathbf{D}_{21}\\ \mathbf{D}_{22}\end{Bmatrix} \tag{18.27}$$

$$\quad\quad\uparrow \quad\quad \uparrow$$

To synthesise stiffness matrix [**K**] from the stiffness matrices of elements, we use Eq. 16.17

$$\mathbf{K} = \mathbf{B}^{\mathrm{T}}\,\mathbf{kB}$$

The equation can be written in the partitioned form as

$$\left[\begin{array}{c|c}\mathbf{K}_{11} & \mathbf{K}_{12}\\ \hline \mathbf{K}_{21} & \mathbf{K}_{22}\end{array}\right] = \left[\begin{array}{c}\mathbf{B}_1^T\\ \hline \mathbf{B}_2^T\end{array}\right]\mathbf{kB}_1 \,\vdots\, \mathbf{B}_2 \tag{18.28}$$

Expanding the right hand side of Eq. 18.28,

$$\left.\begin{aligned}\mathbf{K}_{11} &= \mathbf{B}_1^T\ \mathbf{kB}_1\\ \mathbf{K}_{12} &= \mathbf{B}_1^T\ \mathbf{kB}_2\\ \mathbf{K}_{21} &= \mathbf{B}_2^T\ \mathbf{kB}_1\\ \mathbf{K}_{22} &= \mathbf{B}_2^T\ \mathbf{kB}_2\end{aligned}\right\} \tag{18.29}$$

For the system coordinates we write the force displacement relationship in the partitioned form

$$\begin{Bmatrix} \mathbf{P}_1 \\ \mathbf{0} \end{Bmatrix} = \left[\begin{array}{c|c} \mathbf{K}_{11} & \mathbf{K}_{12} \\ \hline \mathbf{K}_{21} & \mathbf{K}_{22} \end{array}\right] \begin{Bmatrix} \mathbf{D}_1 \\ \mathbf{D}_2 \end{Bmatrix} \tag{18.30}$$

This is the same as Eq. 18.18 and all the relationships developed there from can be made use of.

Thus,

$$\mathbf{D}_2 = -\mathbf{K}_{22}\,\mathbf{K}_{21}\,\mathbf{D}_1 \tag{18.31}$$

and the reduced stiffness matrix is

$$\mathbf{K}_1 = \mathbf{K}_{11} - \mathbf{K}_{12}\,\mathbf{K}_{22}\,\mathbf{K}_{21} \tag{18.32}$$

We can write

$$\mathbf{P}_1 = \mathbf{K}_1\,\mathbf{D}_1 \tag{18.33}$$

or

$$\mathbf{D}_1 = \mathbf{K}_1^{-1}\,\mathbf{P}_1 \tag{18.34}$$

The element displacements can be written from Eq. 18.26 as

$$\mathbf{d} = (\mathbf{B}_1 - \mathbf{B}_2\,\mathbf{K}^{-1}{}_{22}\,\mathbf{K}_{21})\mathbf{D}_1 \tag{18.35}$$

and internal forces $\mathbf{p} = \mathbf{kd} = \mathbf{k}(\mathbf{B}_1 - \mathbf{B}_2\,\mathbf{K}^{-1}{}_{22}\,\mathbf{K}_{21})\mathbf{D}_1$ (18.36)

To complete the solution for the frame of Fig. 18.19 we can write

$$\mathbf{k} = \begin{bmatrix} \boxed{\mathbf{k}_1} & & \\ & \boxed{\mathbf{k}_2} & \\ & & \boxed{\mathbf{k}_3} \end{bmatrix}$$

For the coordinates of the elements given in Fig. 18.19*c*,

$$\mathbf{k}_1 = \mathbf{k}_2 = \mathbf{k}_3 = \frac{EI}{L}\begin{bmatrix} 4 & 2 & 6/L & -6/L \\ 2 & 4 & 6/L & -6/L \\ 6/L & 6/L & 12/L^2 & -12/L^2 \\ -6/L & -6/L & -12/L^2 & 12/L^2 \end{bmatrix}$$

It may be noted that we could have taken only the end moments for each member. In that case the displacements at the ends (rotations θ) must b with reference to a chord joining the ends of the member. Now w generate the stiffness matrix **K** using Eq. 16.17.

$$\underset{(3\times3)}{\mathbf{K}} = \underset{(3\times12)}{\mathbf{B}^T}\;\underset{(12\times12)}{\mathbf{k}}\;\underset{(12\times3)}{\mathbf{B}} = \frac{EI}{L}\left[\begin{array}{c|cc} 24/L^2 & 6/L & 6/L \\ \hline 6/L & 8 & 2 \\ 6/L & 2 & 8 \end{array}\right]$$

Since $\mathbf{K}_{22}^{-1}$ is required in Eqs. 18.31 and 18.32, we evaluate this first

$$\mathbf{K}_{22}^{-1} = \frac{L}{60\,EI}\begin{bmatrix} 8 & -2 \\ -2 & 8 \end{bmatrix}$$

Next, we evaluate the product

$$\mathbf{K}_{22}^{-1}\,\mathbf{K}_{21} = \frac{L}{60\,EI}\begin{bmatrix} 8 & -2 \\ -2 & 8 \end{bmatrix}\frac{EL}{L}\begin{Bmatrix} 6/L \\ 6/L \end{Bmatrix} = \frac{1}{5L}\begin{Bmatrix} 3 \\ 3 \end{Bmatrix}$$

Substituting this in Eq. 18.32

$$\mathbf{K}_1 = \mathbf{K}_{11} - \mathbf{K}_{12}\,\mathbf{K}_{22}^{-1}\,\mathbf{K}_{21} = \frac{EI}{L}\left(\frac{24}{L^2}\right) - \frac{EI}{L}\begin{Bmatrix} \frac{6}{L} & \frac{6}{L} \end{Bmatrix}\frac{1}{5L}\begin{Bmatrix} 3 \\ 3 \end{Bmatrix}$$

$$\mathbf{K}_1 = 16.8\frac{EI}{L^3}$$

Therefore,

$$D_1 = \frac{P_1 L^3}{16.8\,EI}$$

since

$$D_1 = \mathbf{K}_1^{-1} P_1$$

Internal displacements d are obtained using Eq. 18.35.

$$\begin{pmatrix} d_1 \\ d_2 \\ d_3 \\ d_4 \\ \hline d_5 \\ d_6 \\ d_7 \\ d_8 \\ \hline d_9 \\ d_{10} \\ d_{11} \\ d_{12} \end{pmatrix} = \begin{pmatrix} 0 \\ 0 \\ 0 \\ -1 \\ \hline 0 \\ 0 \\ 0 \\ 0 \\ \hline 0 \\ 0 \\ 1 \\ 0 \end{pmatrix} - \begin{pmatrix} 0 \\ 3/5L \\ 0 \\ 0 \\ \hline 3/5L \\ 3/5L \\ 0 \\ 0 \\ \hline 3/5L \\ 0 \\ 0 \\ 0 \end{pmatrix}\mathbf{D}_1 \quad \text{or} \quad \mathbf{d} = \begin{Bmatrix} 0 \\ -3/5L \\ 0 \\ -1 \\ -3/5L \\ -3/5L \\ 0 \\ 0 \\ \hline -3/5L \\ 0 \\ 1 \\ 0 \end{Bmatrix}\mathbf{D}_1$$

The internal stresses are obtained using Eq. 18.36.

$$\mathbf{p} = \frac{EI}{L}\begin{Bmatrix} 24/5L \\ 18/5L \\ 42/5L^2 \\ -42/5L^2 \\ \hdashline -18/5L \\ -18/5L \\ -36/5L^2 \\ +36/5L^2 \\ \hdashline 18/5L \\ 24/5L \\ 42/5L^2 \\ -42/5L^2 \end{Bmatrix}\mathbf{D}_1$$

Example 18.17 | *Construct reduced stiffnes $\mathbf{K}_I$ for the frame of Fig. 18.21a from the stiffness matrices of its elements. The structure and element coordinates are indicated. Compute the internal forces in the elements. EI is constant.*

This structure has three degrees of freedom. They are indicated by the coordinates of the structure. If the external load is resolved into two components so that the 37.5 kN force in the direction of coordinate 1 and the other component of 18 kN is along the inclined leg of the frame which does not cause any moments if axial deformations are neglected. Thus, external forces do not exist at two of the three structure coordinates. Transformation matrix **B** is constructed in the partitioned form as shown.

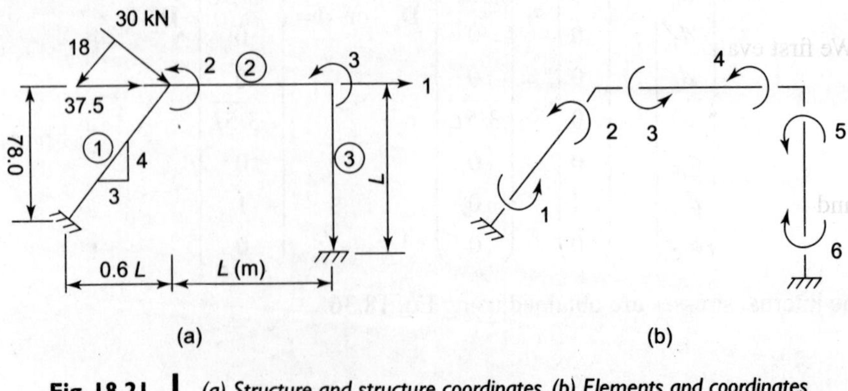

Fig. 18.21 | *(a) Structure and structure coordinates, (b) Elements and coordinates*

$$\mathbf{B} = \begin{array}{c} \\ \\ \\ \\ \\ \\ \end{array}\overset{\begin{array}{cc}\mathbf{B}_1 \quad & \quad \mathbf{B}_2\end{array}}{\left[\begin{array}{c:cc} 5/4L & 0 & 0 \\ 5/4L & 1 & 0 \\ -3/4L & 1 & 0 \\ -3/4L & 0 & 1 \\ 1/L & 0 & 1 \\ 1/L & 0 & 0 \end{array}\right]}$$

Element stiffness matrix $\mathbf{k}_i = \mathbf{k}_2 = \mathbf{k}_3 = \dfrac{EI}{L}\begin{bmatrix} 4 & 2 \\ 2 & 4 \end{bmatrix}$

The uncoupled stiffness matrix $\mathbf{k}$ is

$$\mathbf{k} = \frac{EI}{L}\begin{bmatrix} 4 & 2 & & & & \\ 2 & 4 & & & & \\ & & 4 & 2 & & \\ & & 2 & 4 & & \\ & & & & 4 & 2 \\ & & & & 2 & 4 \end{bmatrix}$$

Stiffness matrix $\mathbf{K}$ of the structure is obtained using Eq. 16.17 in the partitioned form as

$$\underset{3\times 3}{\mathbf{K}} = \underset{(3\times 12)}{\mathbf{B}^T}\ \underset{(12\times 12)}{\mathbf{k}}\ \underset{(12\times 3)}{\mathbf{B}} = \frac{EI}{L}\left[\begin{array}{c:cc} \dfrac{37.5}{2} & \dfrac{3.0}{L} & \dfrac{1.5}{L} \\ \hdashline \dfrac{3.0}{L} & 8 & 2 \\ \dfrac{1.5}{L} & 2 & 8 \end{array}\right]$$

We first evaluate

$$\mathbf{K}_{22}^{-1} = \frac{L}{EI}\frac{1}{60}\begin{bmatrix} 8 & -2 \\ -2 & 8 \end{bmatrix}$$

and

$$\mathbf{K}_{22}^{-1}\,\mathbf{k}_{21} = \frac{L}{60\,EI}\begin{bmatrix} 8 & -2 \\ -2 & 8 \end{bmatrix}\frac{EI}{L}\begin{Bmatrix} \dfrac{3.0}{L} \\ \dfrac{1.5}{L} \end{Bmatrix} = \frac{1}{60\,L}\begin{Bmatrix} 21 \\ 6 \end{Bmatrix}$$

Now, we can construct the reduced stiffness matrix using Eq 18.32.

$$\mathbf{K}_1 = \frac{EI}{L}\left(\frac{37.5}{L^2}\right) - \frac{EI}{L}\begin{Bmatrix} \frac{3.0}{L} & \frac{1.5}{L} \end{Bmatrix} \frac{1}{60L}\begin{Bmatrix} 21 \\ 6 \end{Bmatrix}$$

or
$$\mathbf{K}_1 = 36.3\frac{EI}{L^3}$$

Displacement
$$D_1 = \frac{L^3}{36.3\,EI}(37.5) = 1.033\frac{L^3}{EI}$$

Internal displacements **d** can be obtained using Eq. 18.35.

$$\mathbf{d} = \begin{Bmatrix} 5/4L \\ 5/4L \\ -3/4L \\ -3/4L \\ 1/L \\ 1/L \end{Bmatrix} - \begin{bmatrix} 0 & 0 \\ 1 & 0 \\ 1 & 0 \\ 0 & 1 \\ 0 & 1 \\ 0 & 0 \end{bmatrix} \frac{1}{60L}\begin{Bmatrix} 21 \\ 6 \end{Bmatrix}\left(1.033\frac{L^3}{EI}\right)$$

or
$$\mathbf{d} = \frac{1}{L}\begin{Bmatrix} 1.25 \\ 0.90 \\ -1.10 \\ -0.85 \\ 0.90 \\ 1.00 \end{Bmatrix}\left(1.03\frac{L^3}{EI}\right) = \frac{L^2}{EI}\begin{Bmatrix} 1.29 \\ 0.93 \\ \hline -1.14 \\ -0.88 \\ \hline 1.93 \\ 0.03 \end{Bmatrix}$$

Internal forces **p** are obtained as

$$\underset{(6\times1)}{\mathbf{p}} = \underset{(6\times6)}{\mathbf{k}}\ \underset{(6\times1)}{\mathbf{d}}$$

or
$$\mathbf{p} = \begin{Bmatrix} 6.96\,L \\ 6.30\,L \\ \hline -6.30\,L \\ -5.80\,L \\ \hline 5.80\,L \\ 5.98\,L \end{Bmatrix} \text{kN.m}$$

The application of Eq. 18.14 is illustrated later in Examples 18.18, 18.19 and 18.20.

18.4.1 Computer Programme for the Stiffness Analysis of Kinematicaliy Indeterminate Structures

At this point, the reader might have noticed that there are a number of similarities between the equations which have been developed in this chapter for the stiffness analysis of kinematically indeterminate structures, and those which were developed in Chapter 17 for the flexibility analysis of statically indeterminate structures. This comparison shows that the form of the equations is same in each step of the analysis. The primary difference is that if an equation represents an equilibrium condition in one of the procedures, the corresponding equation in another procedure represents a geometry condition. The computer programmes in the two types of analyses are similar. The input data consists of the matrices $[\mathbf{B}_1]$, $[\mathbf{B}_2]$, $[\mathbf{K}]$ and $\{\mathbf{P}\}$. The computations can be carried out to generate matrices $[\mathbf{B}]$ and $[\mathbf{K}]$. A flow chart is given in Fig. 18.22 which traces the steps involved in solving the kinematically indeterminate structures using stiffness analysis. The flow chart is drawn to analyse the kinematically determinate structure also, so as to be in line with the flow chart given in Fig. 17.15 for the analysis of statically determinate and indeterminate structures. The reader will not miss the striking similarity between the two flow charts.

18.4.2 Temperature Stresses, Lack of Fit, Support Settlements, etc.

Let us designate by Δ the thermal displacements at the coordinates of the elements before they are connected to form the system. The solution for final internal forces $\mathbf{p}^f$ is given by the relation (Eq. 18.13).

$$\mathbf{p}^f = \mathbf{p}^\circ + \mathbf{kBK}^{-1}\{\mathbf{P}^f - \mathbf{P}^o)$$

or

$$\mathbf{p}^f = \mathbf{p}^o + \mathbf{kBK}^{-1}(\mathbf{P}^f - \mathbf{B}^{\mathrm{T}}\mathbf{p}^o) \tag{18.37}$$

applies here too, except that forces $\mathbf{p}^o$ in the fixed coordinate state are due to thermal changes or lack of fit, whereas in Section 18.3 they were caused by forces not at the coordinates.

In the fixed coordinate state the element forces are

$$\mathbf{p}^o = -\mathbf{k}\Delta \tag{18.38}$$

The minus sign in this case is needed because forces $\mathbf{p}^o$ must be applied in a direction opposite to displacements Δ.

If the structure is subjected only to thermal changes or lack of fit represented by displacements Δ of the unassembled elements then we set $\mathbf{p}^f = 0$ in the Eq. 18.37 that is

$$\mathbf{p}^f = (\mathbf{kBK}^{-1}\,\mathbf{B}^T - \mathbf{I})\,\mathbf{k}\,\Delta \tag{18.39}$$

If there are forces $\mathbf{P}^f$ at the coordinates, then

$$\mathbf{p}^f = \mathbf{kBK}^{-1}\,\mathbf{P}^f + (\mathbf{kBK}^{-1}\,\mathbf{B}^{\mathrm{T}} - \mathbf{I})\,\mathbf{k}\,\Delta \tag{18.40}$$

This aspect has been illustrated in Example 18.20.

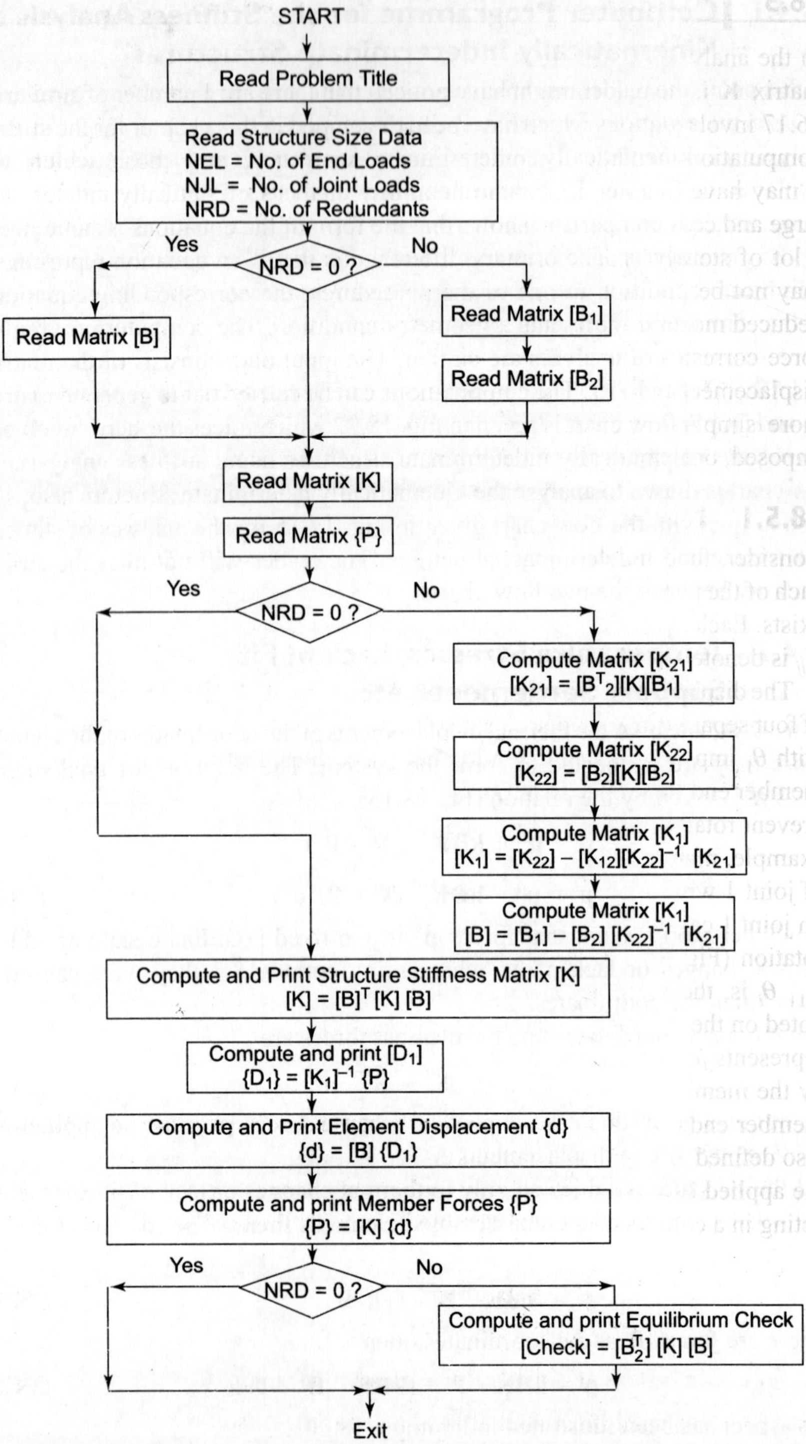

Fig. 18.22 | *Flow chart for stiffness analysis of kinematically indeterminate structure*

18.5 DIRECT STIFFNESS METHOD

In the analysis of structures by the stiffness method, the formation of stiffness matrix **K** is a major step in the process. The matrix **K** was achieved by Eq. 16.17 involving a triple matrix product. The numerical work involved in manual computations tends to become voluminous even for a simple structure. Also, it may have been noticed that uncoupled stiffness matrix **k** is sometimes very large and contains a large number of zero terms so that when a computer is used a lot of storage space is wasted. For this reason the transformation procedure may not be the best way of assembling structure matrix **K**. This matrix can be deduced more easily by noting the fact that any stiffness element K_{ij} is the nodal force corresponding to degree of freedom i caused by the imposition of a unit displacement corresponding to degree of freedom j. The same result is, therefore, more simply obtained if the forces caused by the displacements as they are imposed, one at a time, on the restrained structure are computed and assembled.

18.5.1 Nodal Stiffness of a Continuous Beam

Consider, for example, a three-span continuous beam with moments applied at each of the joints as shown in Fig. 17.23a. At each support one degree of freedom exists. Each support point can be considered as a nodal point and the stiffness of K_{ij} is denoted as nodal stiffness.

The displacements shown in Fig. 17.23*b* can be considered as the superposition of four separate cases. Figure 18.23*c* shows the displacement shape of the structure with θ_1 imposed and all other degrees of freedom locked against rotations. The member end forces needed to accomplish these displacements (to cause θ_1 and to prevent rotation at the locked joints) are indicated in Fig. 18.23c. At joint 1, for example, moment K_{11} is the moment acting on joint 1 caused by a unit rotation of joint 1 with joints 2, 3 and 4 fixed against rotation. K_{11} is the moment acting on joint 1 caused by a unit rotation of joint 2 with joints 1, 3 and 4 fixed against rotation (Fig. 18.23*d*). The θ terms represent the actual rotation of the joints. $K_{11}\ \theta_i$ is, therefore, the actual moment at joint 1. The joint moments, K_{ij}, are noted on the figures as totals for both members framing into a joint and thus, K_{ij} represents *joint stiffness*. The totals are equal to the moment acting on the joint by the members. It may be noticed that the counter-clockwise moments on the member ends which are positive, result in clockwise moments acting on joints, also defined as positive. The total internal moment on any joint must be equal to the applied external moment. If external moments are defined as positive when acting in a counter-clockwise direction, we may express the joint equilibrium as

$$\begin{bmatrix} K_{11} & K_{12} & K_{13} & K_{14} \\ K_{21} & K_{22} & K_{23} & K_{24} \\ K_{31} & K_{32} & K_{33} & K_{34} \\ K_{41} & K_{42} & K_{43} & K_{44} \end{bmatrix} \begin{Bmatrix} \theta_1 \\ \theta_2 \\ \theta_3 \\ \theta_4 \end{Bmatrix} = \begin{Bmatrix} M_1 \\ M_2 \\ M_3 \\ M_4 \end{Bmatrix} \tag{18.41}$$

or

$$\mathbf{K} \qquad \mathbf{D} = \mathbf{P}$$

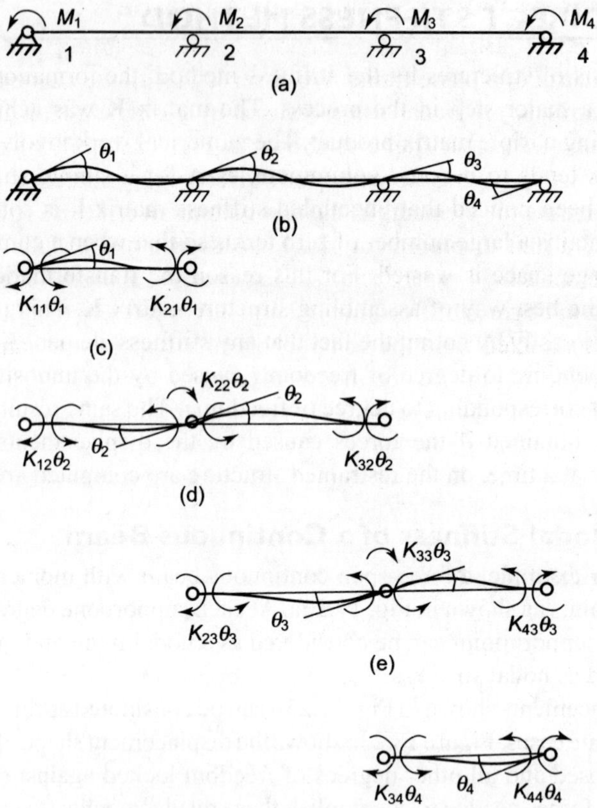

Fig. 18.23 | *(a) Continuous beam and applied moments, (b) Displacements at coordinates, (c) 81 imposed at coordinate 1, (d) 82 imposed at coordinate 2, (e) 62 imposed at coordinate 3, (f) 64 imposed at coordinate 4*

In this example $K_{13}, = K_{31} = 0, K_{14} = K_{41} = 0, K_{24} = K_{42} = 0.$

For the continuous beam, the joint or nodal stiffness values, K_{ij} are easily computed from a knowledge of the member stiffness values. Therefore,

$$K_{11} = \frac{4\,EI}{L_{12}} \quad K_{12} = \frac{2EI}{L_{12}} \quad K_{13} = 0 \quad K_{14} = 0$$

$$K_{12} = \frac{2\,EI}{L_{12}} \quad K_{22} = \left(\frac{4\,EI}{L_{12}} + \frac{4\,EI}{L_{23}}\right) \quad K_{23} = \frac{2EI}{L_{23}} \quad K_{24} = 0$$

$$K_{31} = 0 \quad K_{32} = \frac{2\,EI}{L_{32}} \quad K_{33} = \left(\frac{4\,EI}{L_{32}} + \frac{4\,EI}{L_{34}}\right) \quad K_{34} = \frac{2EI}{L_{34}}$$

$$K_{41} = 0 \quad K_{42} = 0 \quad K_{43} = \frac{2EI}{L_{34}} \quad K_{44} = \frac{4\,EI}{L_{34}}$$

18.5.2 Joint or Nodal Stiffness or a Frame

A similar procedure, as that of a continuous beam, can be applied to assemble the stiffness matrix of a plane frame. We observe that all the elements in a given column of matrix **K** are nodal forces caused by a single nodal displacement. Consider a part of a planar frame as shown in Fig. 18.24*a*. For simplicity, the structure is shown as having three degrees of freedom numbered as 1, 2 and 3. If we wish to compute the elements in the first column of structure stiffness matrix **K**, we impose a unit displacement along the degree of freedom 1 with other nodal displacements held at zero values. Elements K_{11},K_{21} and K_{31} are the forces corresponding to degrees of freedom 1, 2 and 3 respectively (Fig. 18.24*b*). This can also be constructed from the components of the members framing into the joint. For example, K_{11} can be obtained as the sum of the terms k_{11} for each of the three members as shown in Fig. 18.24*c*. Thus, the problem has been reduced to that of computing appropriate member stiffnesses and assembling the structure stiffness elements from the member stiffness elements. To summarise, each term of the structure stiffness matrix can be computed directly by examining the member ends at each node and adding the stiffness computed for each member. This is the feature which is the origin of the term *direct stiffness approach.*

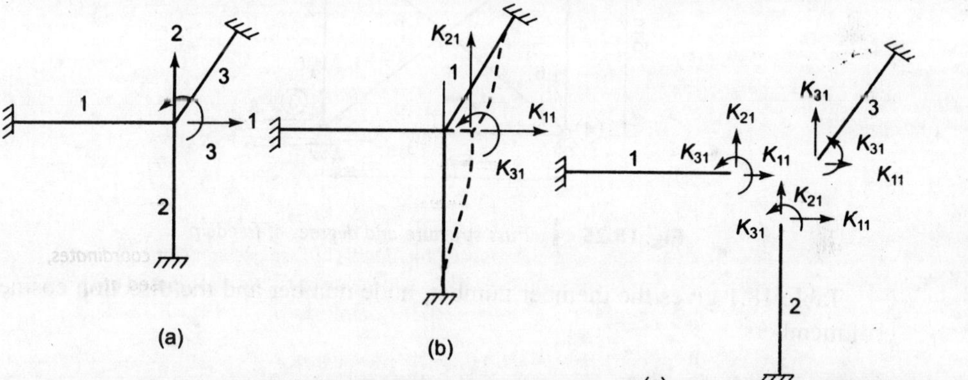

Fig. 18.24 | *(a) Frame and degrees of freedom (structure coordinates), (b) Unit displacement imposed along coordinate 1, (c) Stiffness coefficient of elements along structure coordinates*

18.5.3 Member Stiffness Matrix in the Structure Coordinate System

In the computation of nodal stiffnesses, it is necessary that the member stiffness matrix must be computed in terms of coordinate directions established for the structure. Difficulties arise when the member orientation differs from the structure coordinate directions. In such cases it is convenient first to establish member stiffness coefficients in relation to its local coordinates. The member stiffnesses are then transferred to the structure coordinate system as discussed in Section 16.6. Then nodal stiffness K_{ij} is simply the sum of corresponding member stiffnesses k_{ij}. It is essential that each node of the structure is carefully labelled, and that the nodal numbering of each element corresponds to that of the structure. We shall illustrate the steps involved by solving a few examples.

Example 18.18 *Construct the direct stiffness matrix* **K** *for the truss of Fig. 18.25.*

$$E = 200 \times 10^6 \text{ kN/m}^2 \text{ (200,000 MPa)}$$

$$A = 2500 \times 10^{-6} \text{m}^2 \text{ (2500 mm}^2\text{)}$$

At each joint, the truss has only two degrees of freedom. The degrees of freedom are numbered, first the unrestrained and then the restrained degrees of freedom. The positive member senses are shown in Fig. 18.25. It may be noticed that they have been chosen so that the far end of the member has a higher joint number than the one at the near end. This is a useful convention that allows **k** to be formed in the arrangement that is added into stiffness matrix **K**.

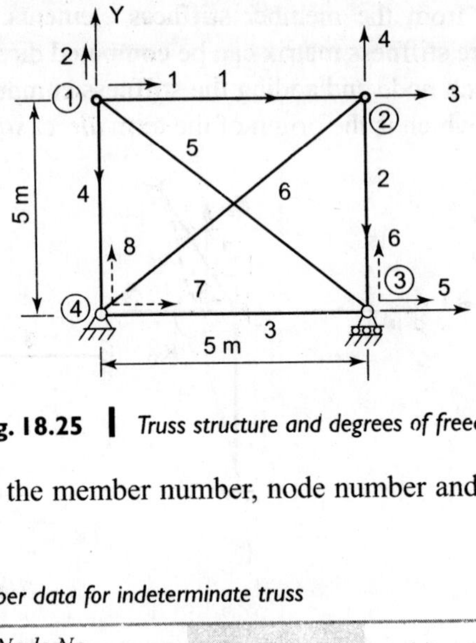

Fig. 18.25 | *Truss structure and degrees of freedom*

Table 18.1 gives the member number, node number and the direction cosine of members.

Table 18.1 | *Member data for indeterminate truss*

Member	*Node No.* *Near*	*Far*	θ	$\sin\theta$	$\cos\theta$
1	1	2	0°	0	1.0
2	2	3	270°	–1.0	0
3	3	4	180°	0	–1.0
4	1	4	270°	–1.0	0
5	1	3	315°	–0.707	0.707
6	2	4	225°	–0.707	–0.707

The origin for each member is chosen at the lesser joint number end and the X axis is directed towards the joint having a higher joint number. The member stiffness matrix denoted by **k′** with reference to local coordinate axes is (Eq. 16.32):

$$\mathbf{k}' = \begin{bmatrix} EA/L & 0 & -EA/L & 0 \\ 0 & 0 & 0 & 0 \\ -EA/L & 0 & EA/L & 0 \\ 0 & 0 & 0 & 0 \end{bmatrix}$$

The rotation matrix for the plane truss member is (Eq. 16.34*a*)

$$\mathbf{R} = \begin{bmatrix} \cos\theta & \sin\theta & 0 & 0 \\ -\sin\theta & \cos\theta & 0 & 0 \\ 0 & 0 & \cos\theta & \sin\theta \\ 0 & 0 & -\sin\theta & \cos\theta \end{bmatrix}$$

Member stiffness matrix for members 1, 2, 3 and 4 is

$$\mathbf{k}'_{1,2,3,4} = \begin{bmatrix} 1 & 0 & -1 & 0 \\ 0 & 0 & 0 & 0 \\ -1 & 0 & 1 & 0 \\ 0 & 0 & 0 & 0 \end{bmatrix} \times 10^5 \text{ kN/m}$$

and for members 5 and 6 is

$$\mathbf{k}'_{5,6} = \begin{bmatrix} 0.707 & 0 & -0.707 & 0 \\ 0 & 0 & 0 & 0 \\ -0.707 & 0 & 0.707 & 0 \\ 0 & 0 & 0 & 0 \end{bmatrix} \times 10^5 \text{ kN/m}$$

The rotation matrices for members are

$$\mathbf{R}_1 = \begin{bmatrix} 1 & 0 & 0 & 0 \\ 0 & 0 & 0 & 0 \\ 0 & 0 & 1 & 0 \\ 0 & 0 & 0 & 0 \end{bmatrix}, \mathbf{R}_2 = \mathbf{R}_4 = \begin{bmatrix} 0 & -1 & 0 & 0 \\ 1 & 0 & 0 & 0 \\ 0 & 0 & 0 & -1 \\ 0 & 0 & 1 & 0 \end{bmatrix}$$

$$\mathbf{R}_3 = \begin{bmatrix} -1 & 0 & 0 & 0 \\ 0 & -1 & 0 & 0 \\ 0 & 0 & -1 & 0 \\ 0 & 0 & 0 & -1 \end{bmatrix}, \mathbf{R}_5 = \begin{bmatrix} 0.707 & -0.707 & 0 & 0 \\ 0.707 & 0.707 & 0 & 0 \\ 0 & 0 & 0.707 & -0.707 \\ 0 & 0 & 0.707 & 0.707 \end{bmatrix}$$

$$\mathbf{R}_6 = \begin{bmatrix} -0.707 & -0.707 & 0 & 0 \\ 0.707 & -0.707 & 0 & 0 \\ 0 & 0 & -0.707 & -0.707 \\ 0 & 0 & 0.707 & -0.707 \end{bmatrix} \tag{18.42}$$

Now we have for all members, 1 to 6, the stiffness matrices in their local coordinates and the transformation matrices related to structure coordinates. The member stiffness matrix is transformed to the structure coordinate system using Eq. 16.43.

$$\mathbf{k} = \mathbf{R}^{\mathbf{T}}\mathbf{k}'\mathbf{R}$$

where each one of them is a 4×4 array. Taking member 5 as an example,

$$\mathbf{k}_5 = \begin{bmatrix} 0.707 & 0.707 & 0 & 0 \\ -0.707 & 0.707 & 0 & 0 \\ 0 & 0 & 0.707 & 0.707 \\ 0 & 0 & -0.707 & 0.707 \end{bmatrix} \begin{bmatrix} 0.707 & 0 & -0.707 & 0 \\ 0 & 0 & 0 & 0 \\ -0.707 & 0 & 0.707 & 0 \\ 0 & 0 & 0 & 0 \end{bmatrix} \times 10^5$$

$$\begin{bmatrix} 0.707 & -0.707 & 0 & 0 \\ 0.707 & 0.707 & 0 & 0 \\ 0 & 0 & 0.707 & -0.707 \\ 0 & 0 & 0.707 & 0.707 \end{bmatrix}$$

$$= \begin{array}{c} \\ 1 \\ 2 \\ 3 \\ 4 \end{array} \begin{array}{c} \begin{array}{cccc} 1 & 2 & 5 & 6 \end{array} \\ \begin{bmatrix} 0.353 & -0.353 & -0.353 & 0.353 \\ -0.353 & 0.353 & 0.353 & -0.353 \\ -0.353 & 0.353 & 0.353 & -0.353 \\ 0.353 & -0.353 & -0.353 & 0.353 \end{bmatrix} \end{array} \times 10^5 \text{ kN/m}$$

Similarly, we get the member stiffness matrices in the structure coordinates system for the remaining members.

Member 1

$$\mathbf{k}_1 = \begin{array}{c} \begin{array}{cccc} 1 & 2 & 3 & 4 \end{array} \\ \begin{bmatrix} 1 & 0 & -1 & 0 \\ 0 & 0 & 0 & 0 \\ -1 & 0 & 1 & 0 \\ 0 & 0 & 0 & 0 \end{bmatrix} \end{array} \begin{array}{c} \\ 1 \\ 2 \\ 3 \\ 4 \end{array} \times 10^5 \text{ kN/m}$$

Member 2

$$\mathbf{k}_2 = \begin{array}{c} \begin{matrix} 3 & 4 & 5 & 6 \end{matrix} \\ \begin{bmatrix} 0 & 0 & 0 & 0 \\ 0 & 1 & 0 & -1 \\ 0 & 0 & 0 & 0 \\ 0 & -1 & 0 & 1 \end{bmatrix} \end{array} \begin{matrix} \\ 3 \\ 4 \\ 5 \\ 6 \end{matrix} \times 10^5 \text{ kN/m}$$

Member 3

$$\mathbf{k}_3 = \begin{array}{c} \begin{matrix} 5 & 6 & 7 & 8 \end{matrix} \\ \begin{bmatrix} 1 & 0 & -1 & 0 \\ 0 & 0 & 0 & 0 \\ -1 & 0 & 1 & 0 \\ 0 & 0 & 0 & 0 \end{bmatrix} \end{array} \begin{matrix} \\ 5 \\ 6 \\ 7 \\ 8 \end{matrix} \times 10^5 \text{ kN/m}$$

Member 4

$$\mathbf{k}_4 = \begin{array}{c} \begin{matrix} 1 & 2 & 7 & 8 \end{matrix} \\ \begin{bmatrix} 0 & 0 & 0 & 0 \\ 0 & 1 & 0 & -1 \\ 0 & 0 & 0 & 0 \\ 0 & -1 & 0 & 1 \end{bmatrix} \end{array} \begin{matrix} \\ 1 \\ 2 \\ 7 \\ 8 \end{matrix} \times 10^5 \text{ kN/m}$$

Member 6

$$\mathbf{k}_6 = \begin{array}{c} \begin{matrix} 3 & 4 & 7 & 8 \end{matrix} \\ \begin{bmatrix} 0.353 & 0.353 & -0.353 & -0.353 \\ 0.353 & 0.353 & -0.353 & -0.353 \\ -0.353 & 0.353 & 0.353 & 0.353 \\ -0.353 & -0.353 & 0.353 & 0.353 \end{bmatrix} \end{array} \begin{matrix} \\ 1 \\ 2 \\ 7 \\ 8 \end{matrix} \times 10^5 \text{ kN/m}$$

While this step is simple, it becomes tedious when the computations are done by hand. A simple computer programme can be utilised to do the job.

The stiffness matrix for the unrestrained structure is constructed by inserting the elements from the member stiffness matrix into the correct position (like numbered) in the structure stiffness matrix. The resulting 8x8 array is shown in Eq. 18.43. The 8×8 array is necessary since there are eight degrees of freedom for the unrestrained structure. The array is partitioned so that the first five rows and columns correspond to the actual degrees of freedom with the structure restrained at its supports.

$$
\mathbf{K} = \begin{array}{c} \\ 1 \\ 2 \\ 3 \\ 4 \\ 5 \\ 6 \\ 7 \\ 8 \end{array}
\begin{array}{c} \begin{array}{ccccc:ccc} \quad 1 & \quad 2 & \quad 3 & \quad 4 & \quad 5 & \quad 6 & \quad 7 & \quad 8 \end{array} \\
\left[\begin{array}{ccccc:ccc}
1.353 & -0.353 & -1.000 & 0 & -0.353 & 0.353 & 0 & 0 \\
0 & 1.353 & 0 & 0 & 0.353 & -0.353 & 0 & -1.000 \\
-1.000 & 0 & 1.353 & 0.353 & 0 & 0 & -0.353 & -0.353 \\
0 & 0 & 0.353 & 1.353 & 0 & -1.000 & -0.353 & -0.353 \\
-0.353 & 0.353 & 0 & 0 & 1.353 & -0.353 & -1.000 & 0 \\
\hdashline
0.353 & -0.353 & 0 & -1.000 & -0.353 & 1.353 & 0 & 0 \\
0 & 0 & -0.353 & -0.353 & -1.000 & 0 & 1.353 & 0.353 \\
0 & -1.000 & -0.353 & -0.353 & 0 & 0 & 1.353 & 1.353
\end{array}\right] \end{array}
\tag{18.43}
$$

Up to this point the analysis is independent of loading.

If loading were given at any or all the degrees of freedom, the upper left portion of stiffness matrix $\mathbf{K}_{11}$ can be inverted to obtain displacements $\mathbf{D}_1$ by the relation

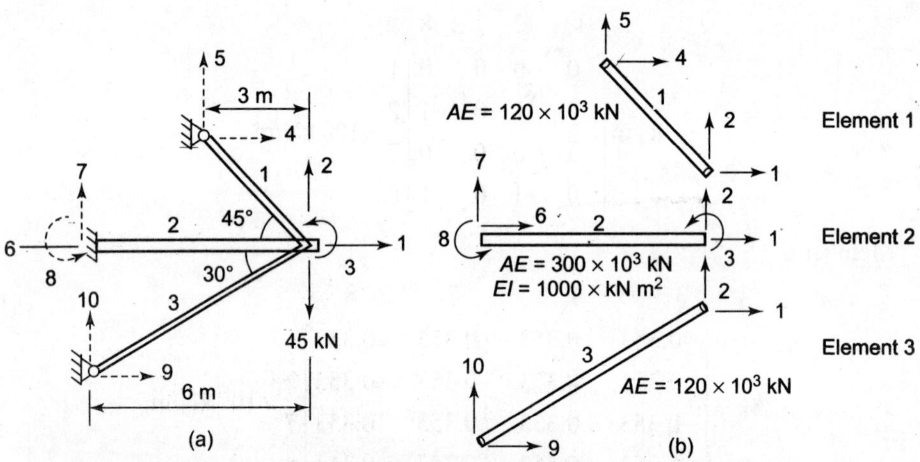

Fig. 18.26 | *(a) Structure and structure coordinates, (b) Elements and coordinates*

$$\mathbf{D}_1 = \mathbf{K}_{11}^{-1}\mathbf{P}_1$$

since $\qquad \mathbf{D}_2 = 0 \qquad$ in Eq. 18.17

Now that the displacements of joints have been found we can find member forces $\mathbf{p}' = \mathbf{k}'\mathbf{R}\mathbf{d}$.

It may be noted that the matrix to be inverted is of the order 5×5 and hence a computer programme would be required for this purpose.

Example 18.19 | *Assemble the structure stiffness matrix and determine the nodal displacements and member forces due to an* *applied load of 45 kN as shown in Fig. 18.26a. Members 1 and 3 are axial force members and member 2 is subject to flexural and axial deformations.*

We shall first write out the element stiffness matrices, member by member, taking care to adhere to the established numbering shown in Fig. 18.26*a*.

We can write the member stiffness matrix for axial force members 1 and 3 using Eq. 16.43, and for beam element 2, using Eq. 15.42 directly since the local coordinates for the members coincide with structure coordinates. Using the structure properties in Fig. 18.26*b* and substituting values for k' and R in Eq. 16.43, we obtain

$$\mathbf{k}_1 = \frac{120\times 10^3}{3\sqrt{2}} \begin{array}{c} \\ 1 \\ 2 \\ 4 \\ 5 \end{array} \begin{array}{c} \begin{array}{cccc} 1 & 2 & 4 & 5 \end{array} \\ \begin{bmatrix} +\frac{1}{2} & -\frac{1}{2} & -\frac{1}{2} & +\frac{1}{2} \\ -\frac{1}{2} & +\frac{1}{2} & +\frac{1}{2} & -\frac{1}{2} \\ -\frac{1}{2} & +\frac{1}{2} & +\frac{1}{2} & -\frac{1}{2} \\ +\frac{1}{2} & -\frac{1}{2} & -\frac{1}{2} & +\frac{1}{2} \end{bmatrix} \end{array} \text{kN/m}$$

$$\mathbf{k}_3 = \frac{120\times 10^3}{(6)2\sqrt{3}} \begin{array}{c} \\ 1 \\ 2 \\ 9 \\ 10 \end{array} \begin{array}{c} \begin{array}{cccc} 1 & 2 & 9 & 10 \end{array} \\ \begin{bmatrix} \frac{3}{4} & \frac{\sqrt{3}}{4} & -\frac{3}{4} & -\frac{\sqrt{3}}{4} \\ \frac{\sqrt{3}}{4} & \frac{1}{4} & -\frac{\sqrt{3}}{4} & -\frac{1}{4} \\ -\frac{3}{4} & -\frac{\sqrt{3}}{4} & \frac{3}{4} & \frac{\sqrt{3}}{4} \\ -\frac{\sqrt{3}}{4} & -\frac{1}{4} & \frac{\sqrt{3}}{4} & \frac{1}{4} \end{bmatrix} \end{array} \text{kN/m}$$

and

$$\mathbf{k}_2 = \begin{array}{c} \\ 6 \\ 7 \\ 8 \\ 1 \\ 2 \\ 3 \end{array} \begin{array}{c} \begin{array}{cccccc} 6 & 7 & 8 & 1 & 2 & 3 \end{array} \\ \begin{bmatrix} 50.0 & 0 & 0 & -50.0 & 0 & 0 \\ 0 & 0.0556 & 0.1667 & 0 & -0.0556 & 0.1667 \\ 0 & 0.1667 & 0.6667 & 0 & -0.1667 & 0.3333 \\ -50.0 & 0 & 0 & 50.0 & 0 & 0 \\ 0 & -0.0556 & -0.1667 & 0 & 0.0556 & -0.1667 \\ 0 & 0.1667 & 0.3333 & 0 & -0.1667 & 0.6667 \end{bmatrix} \end{array} \times 10^3 \text{ kN/m}$$

Assembling these element stiffnesses into the location of the (10 × 10) structure stiffness matrix as indicated by the numbering of rows and columns, we obtain the matrix **K** (Eq. 18.44).

The physical meaning of these numbers should be clearly understood; element *Kn-* for instance, is more than four times the value of K_{22}, that is, the horizontal force required to horizontally stretch the structure by a specified amount is four times as much as a vertical force at the same point causing the same amount of vertical displacement. It may be verified that

K =

$$
\begin{array}{c} \\ 1\\2\\3\\4\\5\\6\\7\\8\\9\\10 \end{array}
\begin{array}{c}
\begin{array}{cccccccccc} 1 & 2 & 3 & 4 & 5 & 6 & 7 & 8 & 9 & 10 \end{array} \\
\left[\begin{array}{cccccccccc}
77.13 & & & & & \text{Sym.} & & & & \\
-6.64 & 18.53 & & & & & & & & \\
0 & -0.1667 & 0.6667 & & & & & & & \\
-14.14 & +14.14 & & +14.14 & & & & & & \\
14.14 & -14.14 & & -14.14 & +14.14 & & & & & \\
-50.00 & 0 & 0 & & & 50.00 & & & & \\
0 & -0.0556 & 0.1667 & & & 0 & 0.0556 & & & \\
0 & -0.1667 & 0.333 & & & 0 & 0.1667 & 0.6667 & & \\
-12.99 & -7.50 & & & & & & & 12.99 & \\
-7.50 & -4.33 & & & & & & & 7.50 & 4.33
\end{array}\right]
\end{array}
\times 10^3 \text{ kN/m} \qquad (18.44)
$$

the contribution of beam member (element 2) k_{22} is only a small fraction of the total value of stiffness k_{22} indicating that this stiffness is mainly due to inclined truss members.

Now writing the force displacement relationship of the structure in the partitioned form, we get

$$
\left[\begin{array}{c} 0 \\ -45 \\ 0 \\ \hline X_4 \\ X_5 \\ Y_6 \\ X_7 \\ X_8 \\ X_9 \\ X_{10} \end{array}\right]
=
\begin{array}{c} \\ 1\\2\\3\\4\\5\\6\\7\\8\\9\\10 \end{array}
\begin{array}{c}
\begin{array}{ccc|ccccccc} 1 & 2 & 3 & 4 & 5 & 6 & 7 & 8 & 9 & 10 \end{array} \\
\left[\begin{array}{ccc|ccccccc}
77.13 & & & & & & & & & \\
-6.64 & 18.53 & & & & \text{sym} & & & & \\
0 & -0.1667 & 0.667 & & & & & & & \\ \hline
-14.14 & +14.14 & & 14.14 & & & & & & \\
14.14 & -14.14 & & -14.14 & +14.14 & & & & & \\
-50.00 & 0 & 0 & & & 50.00 & & & & \\
0 & -0.0556 & 0.1667 & & & 0 & 0.0556 & & & \\
0 & -0.1667 & 0.3333 & & & 0 & 0.1667 & 0.6667 & & \\
-12.99 & -7.50 & & & & & & & 12.99 & \\
-7.50 & -4.33 & & & & & & & 7.50 & 4.33
\end{array}\right]
\end{array}
\times 10^3
\left[\begin{array}{c} D_1 \\ D_2 \\ D_3 \\ \hline 0 \\ 0 \\ 0 \\ 0 \\ 0 \\ 0 \\ 0 \end{array}\right]
\qquad (18.45)
$$

The given values of applied loads P_1 to P_3 and the specified support displacements D_4 to D_{10} (all zero) are now inserted into the force displacement relationship (Eq. 18.14). Equations 18.15 and 18.16 can now be solved to yield the matrices of unknown displacements and reactions. The displacements are

$$\begin{Bmatrix} D_1 \\ D_2 \\ D_3 \end{Bmatrix} = (10^{-3}) \begin{bmatrix} 0.0134 & 0.0048 & 0.0012 \\ 0.0048 & 0.0558 & 0.0140 \\ 0.0012 & 0.0140 & 1.5035 \end{bmatrix} \begin{Bmatrix} 0 \\ -45 \\ 0 \end{Bmatrix} \text{ m.}$$

Or

$$\begin{Bmatrix} D_1 \\ D_2 \\ D_3 \end{Bmatrix} = \begin{Bmatrix} -0.2160 \\ -2.5110 \\ -0.0006 \end{Bmatrix} \begin{matrix} \text{mm} \\ \text{mm} \\ \text{rad.} \end{matrix}$$

The reactions are

$$\begin{Bmatrix} X_4 \\ X_5 \\ X_6 \\ X_7 \\ X_8 \\ X_9 \\ X_{10} \end{Bmatrix} = \begin{Bmatrix} -32.4500 \\ 32.4500 \\ 10.8000 \\ 0.1395 \\ 0.4184 \\ 21.6383 \\ 12.4926 \end{Bmatrix} \text{ All kN, except } X_8 \text{ in kN.m}$$

It may be noted that these reactions satisfy the equilibrium conditions. The element forces for elements 1, 2 and 3 can now be found by applying $\mathbf{p}_s = \mathbf{k}_s \mathbf{d}_s$ for each member. For example, for member 1

$$\mathbf{p}_1 = \mathbf{k}_1 \, \mathbf{d}_1$$

$$\begin{Bmatrix} p_1' \\ p_2' \\ p_4' \\ p_5' \end{Bmatrix} = \begin{bmatrix} 14.14 & -14.14 & -14.14 & 14.14 \\ -14.14 & 14.14 & 14.14 & -14.14 \\ -14.14 & 14.14 & 14.14 & -14.14 \\ 14.14 & -14.14 & -14.14 & 14.14 \end{bmatrix} \begin{Bmatrix} -0.216 \\ -2.511 \\ 0 \\ 0 \end{Bmatrix} = \begin{Bmatrix} +32.45 \\ -32.45 \\ -32.45 \\ +32.45 \end{Bmatrix} \text{kN}$$

It may be noted that only the horizontal and vertical components of the bar forces are obtained. The bar force which is needed for design purposes can be determined by simple statics as

$$p = \frac{32.45}{\sqrt{2}} + \frac{32.45}{\sqrt{2}} = 45.92 \text{ kN}$$

Example 18.20 | *The structure in Example 18.19 is subjected to two different loading conditions:*

1. *A uniform load $w = 15$ kN/m on the horizontal beam;*
2. *An axial shortening of 2.5×10^{-3} m of the inclined bracing member 3.*

For each case compute nodal forces and nodal displacements.

We computed the fixed end forces for both members due to the two different loading conditions. In the loading condition 1, the beam has only the fixed end forces as shown in Fig. 18.27*b*. In loading condition 2, the shortening of member 3 leads to axial fixed end forces which are proportional to the member stiffness. The fixed end forces along the direction of member 3 is

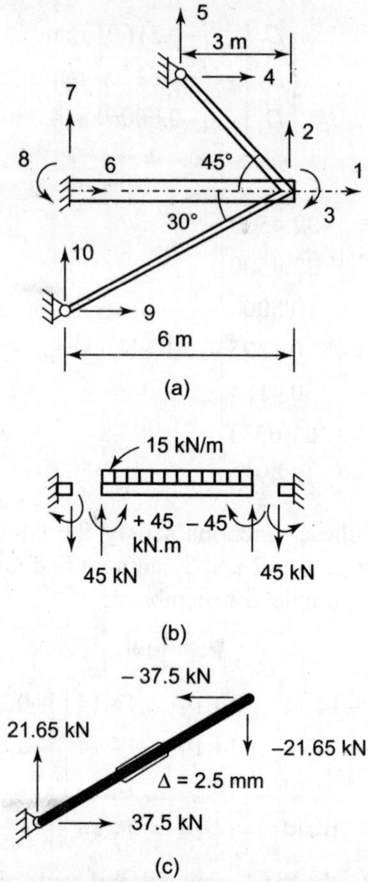

Fig. 18.27 | *(a) Structure, (b) Loading condition 1—fixed end forces, (c) Loading condition 2—fixed end forces*

$$\frac{AE\Delta}{l} = \frac{120 \times 10^3 \times 2.5 \times 10^3}{6(2\sqrt{3})} = 43.30 \text{ kN (tension)};$$

and this fixed end force has to be expressed in terms of its components parallel to the nodal axes as shown in Fig. 18.27*c*.

The fixed end forces acting on the joints for both loading conditions are now assembled in fixed end force matrix $\mathbf{P}^f$, adhering to the nodal numbering system shown in Fig. 18.27*a*.

$$
P^f = \begin{array}{c} \\ 1 \\ 2 \\ 3 \\ 4 \\ 5 \\ 6 \\ 7 \\ 8 \\ 9 \\ 10 \end{array}
\begin{array}{c} \begin{array}{cc} 1 & \quad 2 \end{array} \\
\begin{bmatrix} 0 & -37.50 \\ -45 & -21.65 \\ +45 & 0 \\ 0 & 0 \\ 0 & 0 \\ 0 & 0 \\ -45 & 0 \\ -45 & 0 \\ 0 & 37.50 \\ 0 & 21.65 \end{bmatrix} \end{array}
$$

The other matrices for this structure were already calculated and presented in Example 18.19. Using the relation, $\mathbf{D} = \mathbf{K}^{-1}\,\mathbf{P}^f$ and taking the values for $\mathbf{K}^{-1}$ from the previous example, the displacements in the two loading cases are

$$
\begin{bmatrix} D_{1,1} & D_{1,2} \\ D_{2,1} & D_{2,2} \\ D_{3,1} & D_{3,2} \end{bmatrix} = (10^{-3}) \begin{bmatrix} 0.0134 & 0.0048 & 0.0012 \\ 0.0048 & 0.0558 & 0.0140 \\ 0.0012 & 0.0140 & 1.5035 \end{bmatrix} \begin{bmatrix} 0 & -37.5 \\ -45 & -21.65 \\ 45 & 0 \end{bmatrix}
$$

$$
= \begin{bmatrix} -0.1620 & -0.6064 \\ -1.8810 & -0.1388 \\ +0.0670 & -0.00035 \end{bmatrix} \begin{array}{l} \text{mm} \\ \text{mm} \\ \text{rad.} \end{array}
$$

The reactions are

$$
\begin{bmatrix} X_{4,1} & X_{4,2} \\ X_{5,1} & X_{5,2} \\ X_{6,1} & X_{6,2} \\ X_{7,1} & X_{7,2} \\ X_{8,1} & X_{8,2} \\ X_{9,1} & X_{9,2} \\ X_{10,1} & X_{10,2} \end{bmatrix} = \begin{bmatrix} -24.31 & -11.05 \\ 24.31 & 11.05 \\ 8.1 & 30.32 \\ 0.1158 & 0.0766 \\ 0.3359 & 0.2313 \\ 16.21 & -19.21 \\ 9.36 & -11.09 \end{bmatrix} \text{ All kN, except } X_8 \text{ in kN.m}
$$

Member forces can now be calculated using Eq. 18.13.

18.6 ANALYSIS BY TRIDIAGONALIZATION OF STIFFNESS MATRIX

Consider the building frame in Fig. 18.28a which is three bay wide and say 20 storey high. At each floor level, neglecting axial deformations, there exist five degrees of freedom, four joint rotations and the lateral translation. We have stiffness matrix **K** of order 100×100 which needs to be inverted. Most computers do not have this capacity.

We shall now discuss a step-by-step procedure which reduces the size of the matrix, so that even small capacity computers can be employed.

18.6.1 Numbering of Coordinates

First consider the continuous beam in Fig. 18.28*b*. For the coordinates as shown, the resulting siffness matrix is a band matrix of the form shown in Eq. 18.46. Each coordinate j (except for coordinates 1 and n which are coupled only to single coordinates) is coupled to two coordinates, the coordinate $(j - 1)$ that proceeds it and the coordinate $(j + 1)$ that follows. If the numbering is not in sequence then we would not obtain this band matrix.

$$\mathbf{K} = \begin{bmatrix} K_{11} & K_{12} & & & & & \\ K_{21} & K_{22} & K_{23} & & & & \\ & K_{32} & K_{33} & K_{34} & & & \\ & & \cdots\cdots & & & & \\ & & & \cdots\cdots & & & \\ & & & K_{j,j-1} & K_{jj} & K_{j,j+1} & \\ & & & & \cdots\cdots & & \\ & & & & & \cdots\cdots & \\ & & & & & K_{n,n-1} & K_{nm} \end{bmatrix} \tag{18.46}$$

Similar considerations apply to larger systems. For example the resulting stiffness matrix for a building frame such as the one in Fig. 18.28*a* is

$$\mathbf{K} = \begin{bmatrix} \mathbf{K}_{11} & \mathbf{K}_{12} & & & & & \\ \mathbf{K}_{21} & \mathbf{K}_{22} & \mathbf{K}_{23} & & & & \\ & \mathbf{K}_{32} & \mathbf{K}_{33} & \mathbf{K}_{34} & & & \\ & & \cdots\cdots & & & & \\ & & & \cdots\cdots & & & \\ & & & \mathbf{K}_{j,j-1} & \mathbf{K}_{jj} & \mathbf{K}_{j,j+1} & \\ & & & & \cdots\cdots & & \\ & & & & & \cdots\cdots & \\ & & & & & \mathbf{K}_{n,n-1} & \mathbf{K}_{nm} \end{bmatrix} \tag{18.47}$$

The above two stiffness are similar except that in the latter case, each element $\mathbf{K}_{jj}$ corresponds to the stiffness matrix of level j and matrices $\mathbf{K}_{j,j-1}$ and $\mathbf{K}_{j,j+1}$ are the coupling matrices corresponding to level $(j - 1)$ and $(j+1)$ respectively. Each element is a submatrix, the order of which is equal to the degrees of freedom at each floor level. The sequence of numbering must be maintained to develop the tridiagonal band matrix.

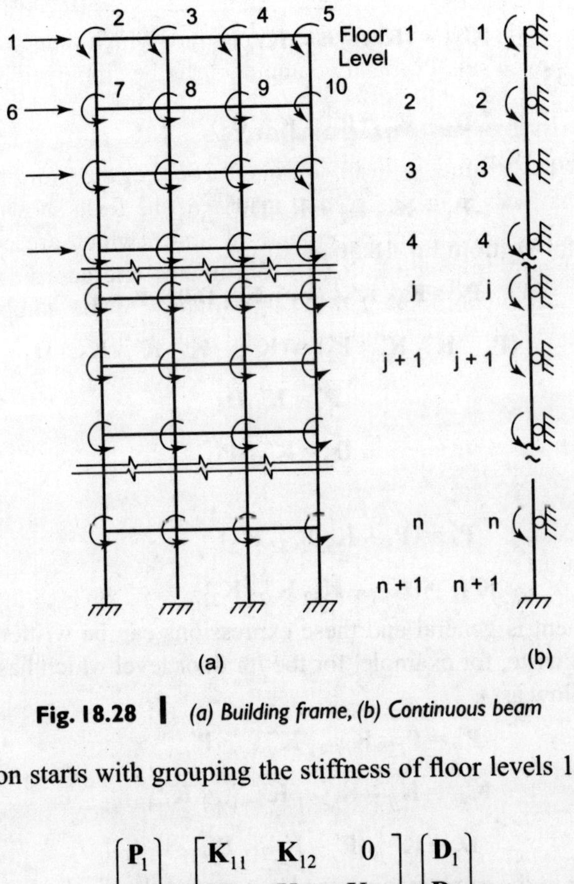

Fig. 18.28 | *(a) Building frame, (b) Continuous beam*

The solution starts with grouping the stiffness of floor levels 1, 2 and 3, that is

$$\begin{Bmatrix} \mathbf{P}_1 \\ \mathbf{P}_2 \\ \mathbf{P}_3 \end{Bmatrix} = \begin{bmatrix} \mathbf{K}_{11} & \mathbf{K}_{12} & 0 \\ \mathbf{K}_{21} & \mathbf{K}_{22} & \mathbf{K}_{23} \\ 0 & \mathbf{K}_{32} & \mathbf{K}_{33} \end{bmatrix} \begin{Bmatrix} \mathbf{D}_1 \\ \mathbf{D}_2 \\ \mathbf{D}_3 \end{Bmatrix} \tag{18.48}$$

This yields

$$\mathbf{P}_1 = \mathbf{K}_{11}\,\mathbf{D}_1 + \mathbf{K}_{12}\,\mathbf{D}_2 \tag{18.49}$$

Therefore,

$$\mathbf{D}_1 = \mathbf{K}_{11}^{-1}(\mathbf{P}_1 - \mathbf{K}_{12}\mathbf{D}_2) \tag{18.50}$$

Again

$$\mathbf{P}_2 = \mathbf{K}_{21}\,\mathbf{D}_1 + \mathbf{K}_{22}\,\mathbf{D}_2 + \mathbf{K}_{23}\,\mathbf{D}_3 \tag{18.51}$$

Substituting for $\mathbf{D}_1$ from Eq. 18.50 into Eq. 18.51

$$\mathbf{P}_2 = \mathbf{K}_{21}\,\mathbf{K}_{11}^{-1}\,\mathbf{P}_1 - \mathbf{K}_{21}\,\mathbf{K}_{11}^{-1}\,\mathbf{K}_{12}\,\mathbf{D}_2 + \mathbf{K}_{22}\,\mathbf{D}_2 + \mathbf{K}_{23}\,\mathbf{D}_3 \tag{18.52}$$

or $$(\mathbf{P}_2 - \mathbf{K}_{21}\,\mathbf{K}_{11}^{-1}\,\mathbf{P}_1) = (\mathbf{K}_{22} - \mathbf{K}_{21}\,\mathbf{K}_{11}^{-1}\,\mathbf{K}_{12})\mathbf{D}_2 + \mathbf{K}_{23}\,\mathbf{D}_3$$

or $$\mathbf{P}'_2 = \mathbf{K}'_2\,\mathbf{D}_2 + \mathbf{K}_{23}\,\mathbf{D}_3 \tag{18.53}$$

in which

$$\mathbf{P}'_2 = (\mathbf{P}_2 - \mathbf{K}_{21}\,\mathbf{K}_{11}^{-1}\,\mathbf{P}_1) \tag{18.54}$$

and

$$\mathbf{K}'_{22} = (\mathbf{K}_{22} - \mathbf{K}_{21}\,\mathbf{K}_{11}^{-1}\,\mathbf{K}_{12}) \tag{18.55}$$

From Eq. 18.53

$$\mathbf{D}_2 = \mathbf{K}'_{22}\,(\mathbf{P}'_2 - \mathbf{K}_{23}\,\mathbf{D}_3) \tag{18.56}$$

Again from Eq. 18.48

$$\mathbf{P}_3 = \mathbf{K}_{32}\,\mathbf{D}_2 + \mathbf{K}_{33}\mathbf{D}_3 \tag{18.57}$$

Substituting for D_2 from Eq. 18.56

$$\mathbf{P}_3 = \mathbf{K}_{32}\,\mathbf{K}'^{-1}_{22}\,(\mathbf{P}'_2 - \mathbf{K}_{23}\,\mathbf{D}_3) + \mathbf{K}_{33}\,\mathbf{D}_3 \tag{18.58}$$

or

$$(\mathbf{P}_3 - \mathbf{K}_{32}\,\mathbf{K}'^{-1}_{22}\,\mathbf{P}'_2) = (\mathbf{K}_{33} - \mathbf{K}_{32}\,\mathbf{K}'^{-1}_{22}\,\mathbf{K}_{23})\,\mathbf{D}_3$$

$$\mathbf{P}'_3 = \mathbf{K}'_{33}\mathbf{D}_3 \tag{18.59}$$

or

$$\mathbf{D}_3 = \mathbf{K}'^{-1}_{33}\,\mathbf{P}'_3 \tag{18.60}$$

in which

$$\mathbf{P}'_3 = (\mathbf{P}_3 - \mathbf{K}_{32}\mathbf{K}'^{-1}_{22}\,\mathbf{P}'_2) \tag{18.61}$$

and

$$\mathbf{K}'_{33} = (\mathbf{K}_{33} - \mathbf{K}_{32}\,\mathbf{K}'^{-1}_{22}\,\mathbf{K}_{23}) \tag{18.62}$$

The treatment is general and these expressions can be written for any floor level. We can write, for example, for the jth floor level which has a floor above and a floor below as

$$\mathbf{P}'_j = \mathbf{P}_j - \mathbf{K}_{j,j-1}\,\mathbf{K}'^{-1}_{j-1,j-1}\mathbf{P}'_{j-1} \tag{18.63}$$

$$\mathbf{K}'_{jj} = \mathbf{K}_{jj} - \mathbf{K}_{j,j-1}\,\mathbf{K}'^{-1}_{j-1,j-1}\,\mathbf{K}_{j+1} \tag{18.64}$$

$$\mathbf{D}_j = \mathbf{K}'^{-1}_{jj}\,(\mathbf{P}'_j - \mathbf{K}_{j,j+1}\,\mathbf{D}_{j+1}) \tag{18.65}$$

Proceeding in the same manner for $j = n$ we get the displacement at the floor immediately above the foundation

$$\mathbf{D}_n = \mathbf{K}'^{-1}_{nn}\,(\mathbf{P}'_n - \mathbf{K}_{n,n+1}\,\mathbf{D}_{n+1}) \tag{18.66}$$

For fixity at the base, that is, $(n + 1)$ level, $\mathbf{D}_n + 1$ is a null matrix leading to

$$\mathbf{D}_n = \mathbf{K}'^{-1}_{nn}\,\mathbf{P}'_n \tag{18.67}$$

Likewise, for a hinged base condition, we get for $j = n + 1$

$$\mathbf{D}_{n+1} = \mathbf{K}'^{-1}_{n+1,n+1}\,\mathbf{P}'_{n+1} \tag{18.68}$$

in which

$$\mathbf{K}'_{n+1,n+1} = \mathbf{K}_{n+1,n+1} - \mathbf{K}_{n+1,n}\,\mathbf{K}'^{-1}_{n,n}\,\mathbf{K}_{n,n+1} \tag{18.69}$$

and

$$\mathbf{P}'_{n+1} = \mathbf{P}_{n+1} - \mathbf{K}_{n+1,n}\,\mathbf{K}'^{-1}_{n,n}\,\mathbf{P}'_n \tag{18.70}$$

Once displacements $\mathbf{D}_n$ (Eqn 18.67) or $\mathbf{D}_{n+1}$ (Eq. 18.68) have been determined, the remaining displacements are obtained by successive back substitution till topmost floor level displacements $\mathbf{D}_1$ are obtained. The displacements being known, the member end forces are obtained by using the force-displacement relationships or by the slope-deflection equations.

The advantage of this approach is obvious. The largest matrix to be inverted at any time is $m \times m$, where m is the number of displacements in the floor level under consideration.

The computational steps involved for a three-storeyed building frame fixed at the base may be summarised as follows:

1. Set up the stiffness matrix which takes the form of a tridiagonal matrix

$$\mathbf{K} = \begin{bmatrix} \mathbf{K}_{11} & \mathbf{K}_{12} & \\ \mathbf{K}_{21} & \mathbf{K}_{22} & \mathbf{K}_{23} \\ & \mathbf{K}_{32} & \mathbf{K}_{33} \end{bmatrix}$$

2. Compute

(a) $\mathbf{K}_{11}^{-1}$

(b) $\mathbf{K}_{21}\mathbf{K}_{11}^{-1}$

(c) $\mathbf{K}_{21}\mathbf{K}_{11}^{-1}\mathbf{P}_1$

3. Calculate

$$\mathbf{P}'_2 = \mathbf{P}_2 - \mathbf{K}_{21}\,\mathbf{K}_{11}^{-1}\,\mathbf{P}_1$$

Then $\quad \mathbf{K}'_{22} = \mathbf{K}_{22} - \mathbf{K}_{21}\,\mathbf{K}'^{-1}_{11}\mathbf{K}_{12}$

4. Compute

(a) $\mathbf{K}'^{-1}_{22}$

(b) $\mathbf{K}_{32}\mathbf{K}_{22}^{-1}$

(c) $\mathbf{K}_{32}\mathbf{K}_{22}^{-1}\mathbf{P}'_1$

5. Calculate

$$\mathbf{P}'_3 = \mathbf{P}_3 - \mathbf{K}_{32}\,\mathbf{K}_{22}^{-1}\,\mathbf{P}'_2$$

and $\quad \mathbf{K}'_{33} = \mathbf{K}_{33} - \mathbf{K}_{32}\,\mathbf{K}'^{-1}_{22}\,\mathbf{K}_{23}$

6. Compute

$$\mathbf{K}'^{-1}_{33}$$

7. Evaluate

$$\mathbf{D}_3 = \mathbf{K}'^{-1}_{33}\,\mathbf{P}'_3$$

$$\mathbf{D}_2 = \mathbf{K}'^{-1}_{22}\,(\mathbf{P}'_2 - \mathbf{K}_{23}\,\mathbf{D}_3)$$

and $\quad \mathbf{D}_1 = \mathbf{K}'^{-1}_{11}\,(\mathbf{P}_1 - \mathbf{K}_{12}\,\mathbf{D}_2)$

The entire procedure involved will be further made clear by solving the following examples.

Example 18.21 | *Consider the single bay four storey building frame shown in Fig. 18.29a. Using the step-by-step or storey-by-storey procedure developed above, calculate displacements* $\mathbf{D}_1$ *through* $\mathbf{D}_4$ *and member end forces for the loading indicated. Consider only bending deformations.*

The frame has three degrees of freedom at each floor level, one lateral translation and two joint rotations. They are numbered floor by floor in a particular sequence starting from the top as shown in Fig. 18.29*b*.

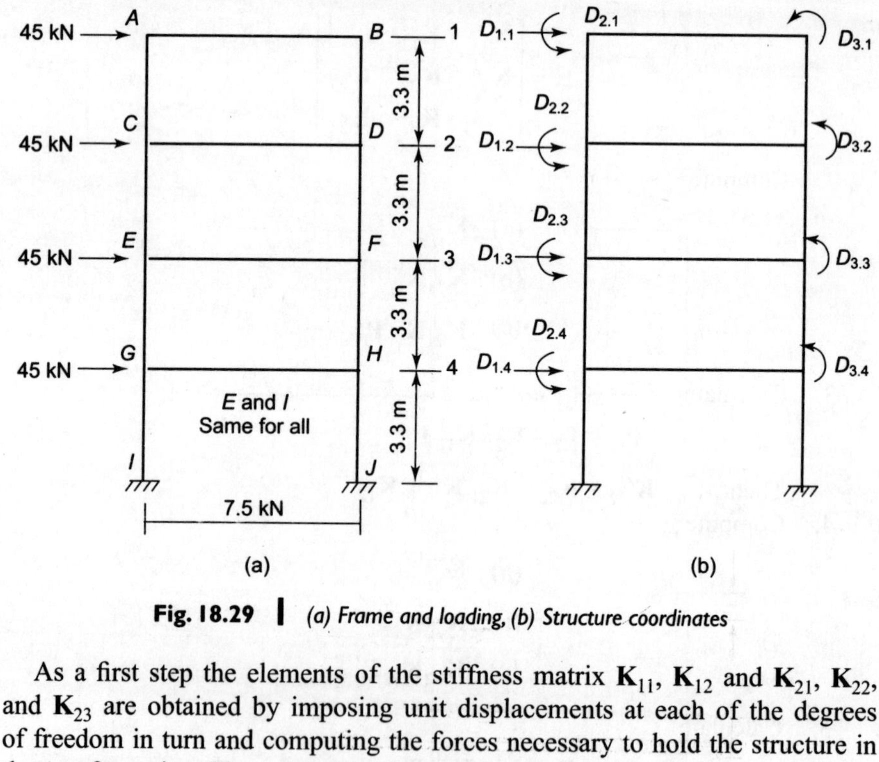

Fig. 18.29 | *(a) Frame and loading, (b) Structure coordinates*

As a first step the elements of the stiffness matrix $\mathbf{K}_{11}$, $\mathbf{K}_{12}$ and $\mathbf{K}_{21}$, $\mathbf{K}_{22}$, and $\mathbf{K}_{23}$ are obtained by imposing unit displacements at each of the degrees of freedom in turn and computing the forces necessary to hold the structure in that configuration. The values are shown computed in Fig. 18.30. The complete stiffness matrix K is shown in Eq. 18.71.

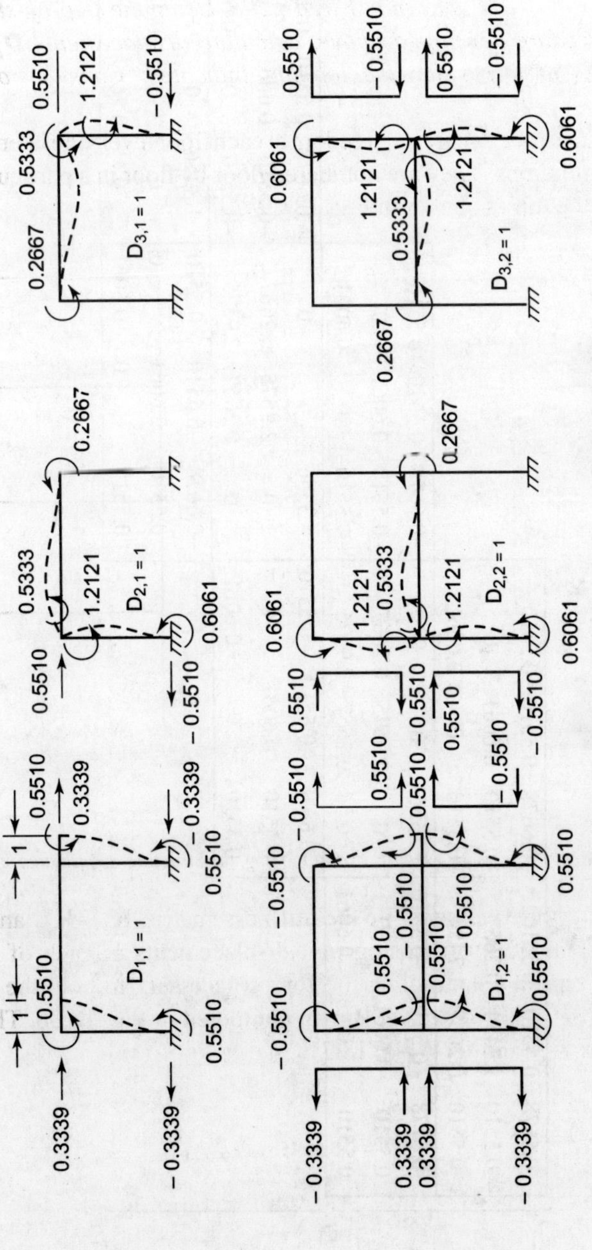

Fig. 18.30 | *Development of stiffness matrices* $\mathbf{K}_{11}$, $\mathbf{K}_{12}$ *and* $\mathbf{K}_{21}$, $\mathbf{K}_{22}$ *and* $\mathbf{K}_{33}$

$$
\underset{\mathbf{P}}{\begin{Bmatrix} 45 \\ 0 \\ 0 \\ \hline 45 \\ 0 \\ 0 \\ \hline 45 \\ 0 \\ 0 \\ \hline 45 \\ 0 \\ 0 \end{Bmatrix}} = EI \underset{\mathbf{K}}{\left\{\begin{array}{ccc|ccc|ccc|ccc}
0.6678 & 0.5510 & 0.5510 & -0.6678 & 0.5510 & 0.5510 & & & & & & \\
0.5510 & 1.7454 & 0.2667 & -0.5510 & 0.6061 & 0 & & & & & & \\
0.5510 & 0.2667 & 1.7454 & -0.5510 & 0 & 0.6061 & & & & & & \\
\hline
-0.6778 & -0.5510 & -0.5510 & 1.3356 & 0 & 0 & -0.6678 & 0.5510 & 0.5510 & & & \\
0.5510 & 0.6061 & 0 & 0 & 2.9575 & 0.2667 & -0.5510 & 0.6061 & 0 & & & \\
0.5510 & 0 & 0.6061 & 0 & 0.2667 & 2.9575 & -0.5510 & 0 & 0.6061 & & & \\
\hline
 & & & -0.6678 & -0.5510 & -0.5510 & 1.3356 & 0 & 0 & -0.6678 & 0.5510 & 0.5510 \\
 & & & 0.5510 & 0.6061 & 0 & 0 & 2.9575 & 0.2667 & -0.5510 & 0.6061 & 0 \\
 & & & 0.5510 & 0 & 0.6061 & 0 & 0.2667 & 2.9575 & -0.5510 & 0 & 0.6061 \\
\hline
 & & & & & & -0.6678 & -0.5510 & -0.5510 & 1.3356 & 0 & 0 \\
 & & & & & & 0.5510 & 0.6061 & 0 & 0 & 2.9575 & 0.2667 \\
 & & & & & & 0.5510 & 0 & 0.6061 & 0 & 0.2667 & 2.9575
\end{array}\right\}} \underset{\mathbf{D}}{\begin{Bmatrix} \mathbf{D}_{1,1} \\ \mathbf{D}_{2,1} \\ \mathbf{D}_{3,1} \\ \hline \mathbf{D}_{1,2} \\ \mathbf{D}_{2,2} \\ \mathbf{D}_{3,2} \\ \hline \mathbf{D}_{1,3} \\ \mathbf{D}_{2,3} \\ \mathbf{D}_{3,3} \\ \hline \mathbf{D}_{1,4} \\ \mathbf{D}_{2,4} \\ \mathbf{D}_{3,4} \end{Bmatrix}} \tag{18.71}
$$

First, we evaluate

$$\mathbf{K}_{11}^{-1} = \frac{1}{EI}\begin{bmatrix} 2.7296 & -0.7475 & -0.7475 \\ -0.7475 & 0.7908 & 0.1151 \\ -0.7475 & 0.1151 & 0.7908 \end{bmatrix}$$

Next

$$\mathbf{K}_{21}\mathbf{K}_{11}^{-1}\mathbf{P}_1 = \begin{bmatrix} -0.6678 & -0.5510 & -0.5510 \\ 0.5510 & 0.6061 & 0 \\ 0.5510 & 0 & 0.6061 \end{bmatrix}\begin{bmatrix} 2.7296 & -0.7475 & -0.7475 \\ -0.7475 & 0.7908 & 0.1151 \\ -0.7475 & 0.1151 & 0.7908 \end{bmatrix}$$

$$\begin{Bmatrix} 45 \\ 0 \\ 0 \end{Bmatrix} = \begin{Bmatrix} -45.00 \\ 47.32 \\ 47.32 \end{Bmatrix}$$

Then,

$$\mathbf{P}_2' = \mathbf{P}_2 - \mathbf{K}_{21}\mathbf{K}_{11}^{-1}\mathbf{P}_1$$

$$= \begin{Bmatrix} 45 \\ 0 \\ 0 \end{Bmatrix} - \begin{Bmatrix} -45.00 \\ 47.32 \\ 47.32 \end{Bmatrix} = \begin{Bmatrix} 90.00 \\ -47.32 \\ -47.32 \end{Bmatrix} \begin{matrix} \text{kN} \\ \text{kN.m} \\ \text{kN.m} \end{matrix}$$

We now evaluate

$$\mathbf{K'}_{22} = \mathbf{K}_{22} - \mathbf{K}_{21}\,\mathbf{K}_{11}^{-1}\,\mathbf{K}_{12}$$

$$\mathbf{K'}_{22} = EI\begin{bmatrix} 0.6678 & 0.5510 & 0.5510 \\ 0.5510 & 2.3372 & -0.1053 \\ 0.5510 & -0.1053 & -2.3372 \end{bmatrix}$$

$$\mathbf{K'}_{22}^{-1} = \frac{1}{EI}\begin{bmatrix} 2.5267 & -0.6238 & -0.6238 \\ -0.6238 & 0.5827 & 0.1733 \\ -0.6238 & 0.1733 & 0.5827 \end{bmatrix}$$

$$\mathbf{K}_{32}\mathbf{K'}_{22}^{-1} = \begin{bmatrix} -1.0000 & 0 & 0 \\ 1.0142 & 0.0095 & -0.2386 \\ 1.0142 & -0.2386 & 0.0095 \end{bmatrix}$$

and

$$\mathbf{K}_{32}\mathbf{K'}_{22}^{-1}\mathbf{P'}_2 = \begin{Bmatrix} -90.00 \\ +102.12 \\ +102.12 \end{Bmatrix}$$

$$\mathbf{K}_{32}\mathbf{K}'^{-1}_{22}\mathbf{K}_{23} = \frac{1}{EI}\begin{bmatrix} 0.6678 & -0.5510 & -0.5510 \\ -0.5510 & 0.5646 & 0.4142 \\ -0.5510 & 0.4142 & 0.5646 \end{bmatrix}$$

Now, we have

$$\mathbf{P}'_3 = \mathbf{P}_3 - \mathbf{K}_{32}\mathbf{K}'^{-1}_{22}\mathbf{P}'_2$$

$$= \begin{Bmatrix} 45.00 \\ 0 \\ 0 \end{Bmatrix} - \begin{Bmatrix} -90.00 \\ 102.12 \\ 102.12 \end{Bmatrix} = \begin{Bmatrix} 135.00 \\ 102.12 \\ 102.12 \end{Bmatrix}$$

and

$$\mathbf{K}'_{33} = \mathbf{K}_{33} - \mathbf{K}_{32}\mathbf{K}'^{-1}_{22}\mathbf{K}_{23}$$

$$= EI\begin{bmatrix} 0.6678 & 0.5510 & 0.5510 \\ 0.5510 & 2.3929 & -0.1475 \\ 0.5510 & -0.1475 & 22.3929 \end{bmatrix}$$

$$\mathbf{K}'^{-1}_{33} = \frac{1}{EI}\begin{bmatrix} 2.5159 & -0.6172 & 0.6172 \\ -0.6172 & 0.5709 & 0.1773 \\ -0.6172 & 0.1773 & 0.5709 \end{bmatrix}$$

$$\mathbf{K}_{43}\,\mathbf{K}'^{-1}_{33} = \begin{bmatrix} -1.0000 & 0 & 0 \\ 1.0122 & 0.0059 & -0.2326 \\ 1.0122 & -0.2326 & 0.0059 \end{bmatrix}$$

$$\mathbf{K}_{43}\,\mathbf{K}'^{-1}_{33}\mathbf{P}'_3 = \begin{Bmatrix} -135.00 \\ 159.80 \\ 159.80 \end{Bmatrix} \begin{matrix} \text{kN} \\ \text{kN.m} \\ \text{kN.m} \end{matrix}$$

$$\mathbf{P}'_4 = \begin{Bmatrix} 45.00 \\ 0 \\ 0 \end{Bmatrix} - \begin{Bmatrix} -135.00 \\ 159.80 \\ 159.80 \end{Bmatrix} = \begin{Bmatrix} 180.00 \\ -159.80 \\ -159.80 \end{Bmatrix} \begin{matrix} \text{kN} \\ \text{kN.m} \\ \text{kN.m} \end{matrix}$$

$$\mathbf{K}'_{44} = \mathbf{K}_{44} - \mathbf{K}_{43}\mathbf{K}'^{-1}_{33}\mathbf{K}_{34}$$

$$= EI\begin{bmatrix} 0.6678 & 0.5510 & 0.5510 \\ 0.5510 & 2.3962 & -0.1500 \\ 0.5510 & -0.1500 & 2.3962 \end{bmatrix}$$

and

$$\mathbf{K}'^{-1}_{44} = \frac{1}{EI}\begin{bmatrix} 2.5158 & -0.6171 & -0.6171 \\ -0.6171 & 0.5704 & 0.1776 \\ -0.6171 & 0.1776 & 0.5704 \end{bmatrix}$$

Now, we can evaluate the displacements by back substitution. For example

$$\mathbf{D}_4 = \mathbf{K}'_{44}\,\mathbf{P}'_4$$

or

$$\begin{Bmatrix} \mathbf{D}_{1,4} \\ \mathbf{D}_{2,4} \\ \mathbf{D}_{3,4} \end{Bmatrix} = \frac{1}{EI}\begin{Bmatrix} 650.11 \\ -230.66 \\ -230.66 \end{Bmatrix} \begin{matrix} \text{m} \\ \text{rad} \\ \text{rad} \end{matrix}$$

Next we evaluate

$$\underset{(3\times3)}{\mathbf{K}_{34}}\ \underset{(3\times1)}{\mathbf{D}_4} = \begin{Bmatrix} -688.33 \\ -498.01 \\ -498.01 \end{Bmatrix}$$

and

$$\mathbf{P}'_3 - \mathbf{K}_{34}\mathbf{D}_4 = \begin{Bmatrix} 823.33 \\ 395.89 \\ 395.89 \end{Bmatrix}$$

then

$$\mathbf{D}_3 = \mathbf{K}'^{-1}_{33}\,(\mathbf{P}'_3 - \mathbf{K}_{34}\,\mathbf{D}_4)$$

or

$$\begin{Bmatrix} \mathbf{D}_{1,3} \\ \mathbf{D}_{2,3} \\ \mathbf{D}_{3,3} \end{Bmatrix} = \frac{1}{EI}\begin{Bmatrix} 1582.73 \\ -211.95 \\ -211.95 \end{Bmatrix} \begin{matrix} \text{m} \\ \text{rad} \\ \text{rad} \end{matrix}$$

Proceeding again,

$$\underset{(3\times3)}{\mathbf{K}_{23}}\ \underset{(3\times1)}{\mathbf{D}_3} = \begin{Bmatrix} -1290.51 \\ -1000.54 \\ -1000.54 \end{Bmatrix}$$

and

$$\mathbf{P}'_2 - \mathbf{K}_{23}\mathbf{D}_3 = \begin{Bmatrix} 1380.50 \\ 953.22 \\ 953.22 \end{Bmatrix}$$

then

$$\mathbf{D}_2 = \mathbf{K}'^{-1}_{22}\,(\mathbf{P}'_2 - \mathbf{K}_{23}\,\mathbf{D}_3)$$

or

$$\begin{Bmatrix} \mathbf{D}_{1,2} \\ \mathbf{D}_{2,2} \\ \mathbf{D}_{3,2} \end{Bmatrix} = \frac{1}{EI}\begin{Bmatrix} 2298.87 \\ -140.52 \\ -140.52 \end{Bmatrix} \begin{matrix} \text{m} \\ \text{rad} \\ \text{rad} \end{matrix}$$

Proceeding in a similar way

$$\mathbf{K}_{12}\mathbf{D}_2 = \begin{Bmatrix} 2298.87 \\ -140.52 \\ -140.52 \end{Bmatrix}$$

$$\mathbf{P}_1 - \mathbf{K}_{12}\,\mathbf{D}_2 = \begin{Bmatrix} 1735.04 \\ 1351.85 \\ 1351.85 \end{Bmatrix}$$

Then

$$\mathbf{D}_1 = \mathbf{K}_{11}^{-1} - (\mathbf{P}_1 - \mathbf{K}_{12}\,\mathbf{D}_2).$$

$$\begin{Bmatrix} \mathbf{D}_{1,1} \\ \mathbf{D}_{2,1} \\ \mathbf{D}_{3,1} \end{Bmatrix} = \frac{1}{EI}\begin{Bmatrix} 2716.69 \\ -72.30 \\ -72.30 \end{Bmatrix} \begin{matrix} \text{m} \\ \text{rad} \\ \text{rad} \end{matrix}$$

The displacements are indicated in Fig. 18.31*a*.

It may be noted that the rotation $\mathbf{D}_{2,i} = \mathbf{D}_{3,i}$ at all the floor levels for i = 1,2,3 and 4.

Moments

The end moments of the elements can be found using slope-deflection equations. Starting from the top storey.

$$M_{AB} = \frac{1}{7.5}\,(4\,D_{2,1} + 2\,D_{3,1}) = \frac{6}{7.5}\,D_{3,1} = -\,57.84 \text{ kN.m}$$

$$M_{AC} = \frac{1}{3.3}\,(4\,D_{2,1} + 2\,D_{2,2}) = \frac{6}{3.3^2}\,(D_{1,1} - D_{1,2}) = +57.40 \text{ kN.m}$$

Proceeding on similar lines

$$M_{CD} = \frac{6}{7.5}\,(-140.52) = -112.42 \text{ kN.m}$$

$$M_{CA} = \frac{1}{3.3}\,\{4(-140.52) + 2(-72.30)\} + \frac{6}{3.3^2}\,(2716.69 - 2298.87)$$

$$= 16.05 \text{ kN.m}$$

$$M_{CE} = \frac{1}{3.3}\,\{4(-140.52) + 2(-211.95)\} + \frac{6}{3.3^2}\,(2298.87 - 1582.73)$$

$$= 95.79 \text{ kN.m}$$

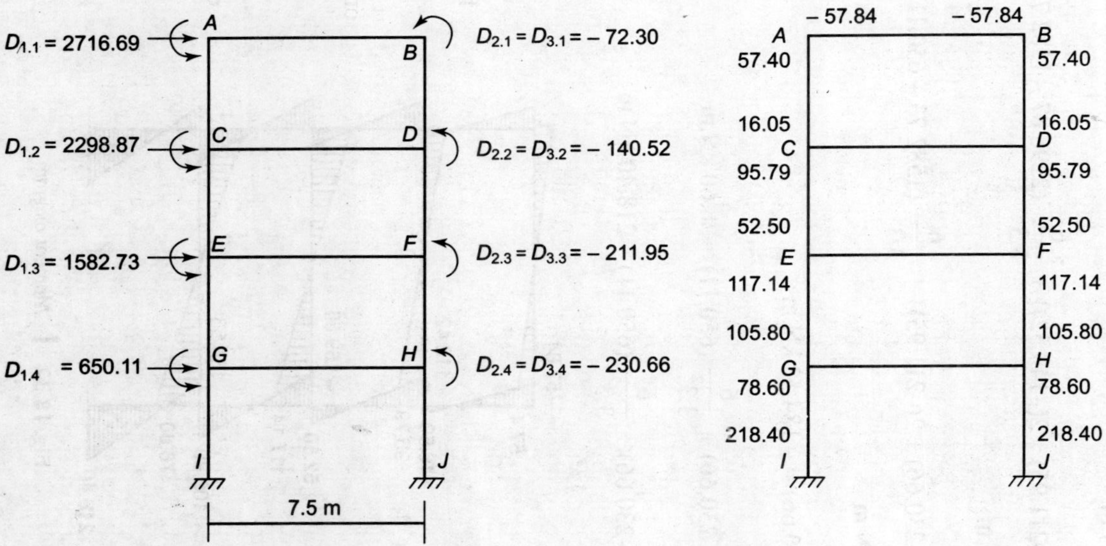

Fig. 18.31 | *Results of analysis: (a) Displacements at all degrees of freedom, (b) Final end moments*

$$M_{EF} = \frac{6}{7.5}(-211.95) = -169.56 \text{ kN.m}$$

$$M_{EG} = \frac{1}{3.3}\{4(-211.95) + 2(-230.66)\} + \frac{6}{3.3^2}(1582.73 - 650.11)$$
$$= 117.14 \text{ kN.m}$$

$$M_{EC} = \frac{1}{3.3}\{4(-211.95) + 2(-140.52)\} + \frac{6}{3.3^2}(2298.87 - 1582.73)$$
$$= 52.50 \text{ kN.m}$$

$$M_{GE} = \frac{1}{3.3}\{4(-230.66) + 2(-211.95)\} + \frac{6}{3.3^2}(1582.73 - 650.11)$$
$$= 105.80 \text{ kN.m}$$

$$M_{GH} = \frac{6}{7.5}(-230.66) = -184.53 \text{ kN.m}$$

$$M_{GI} = \frac{1}{3.3}(4)(-230.66) + \frac{6}{3.3^2}(650.11) = 78.60 \text{ kN.m}$$

$$M_{IG} = \frac{1}{3.3}(2)(-230.66) + \frac{6}{3.3^2}(650.11) = 218.40 \text{ kN.m}$$

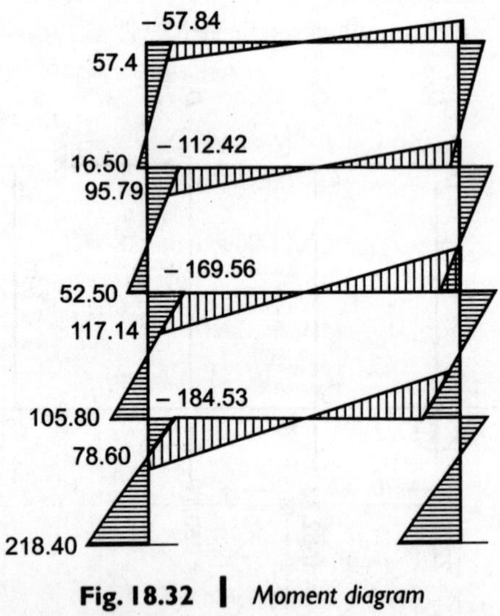

Fig. 18.32 | *Moment diagram*

The results satisfy the equilibrium of the joints with only a rounding off of error, if any. The end moments in the columns satisfy the external shear. The final moment values are shown in Fig. 18.31*b*. The moment diagram drawn on the tension side of members is shown in Fig. 18.32. The computations have been carried in long hand to show in detail the various steps involved. However, the method is well suited for a complete matrix formulation of the problem and for carrying out computations through electronic digital computers. For a complete discussion of this and other efficient methods best suited for matrix formulation, the reader should refer to more advanced text books.

18.7 COMPARISON OF FLEXIBILITY AND STIFFNESS METHODS

The flexibility and stiffness methods of structural analysis are quite similar in many respects, especially in the fomulation of the problem. For this reason the choice of one method or the other is primarily a matter of computational convenience.

In the flexibility method there are several alternatives as to redundants, and the choice of redundants has a significant effect on the nature and amount of computational effort required. In the stiffness method, on the other hand, there is no choice of unknowns since the structure can be restrained in a definite manner; thus, the method of analysis follows a rather set procedure. However, there are both advantages and disadvantages in both approaches and when carrying out the analysis by hand computations, the method that produces fewer unknowns generally involves the least amount of computations. For example, the inversion of a flexibility or stiffness matrix depends upon the number of unknowns involved. For a structure that has numerous redundants but very few joint displacements as in Fig. 18.33*a*, the stiffness method will be preferred. The flexibility method needs an inversion of a 7×7 matrix, whereas the stiffness method needs an inversion of a 2×2 matrix. When there are fewer redundants in a structure than the number of

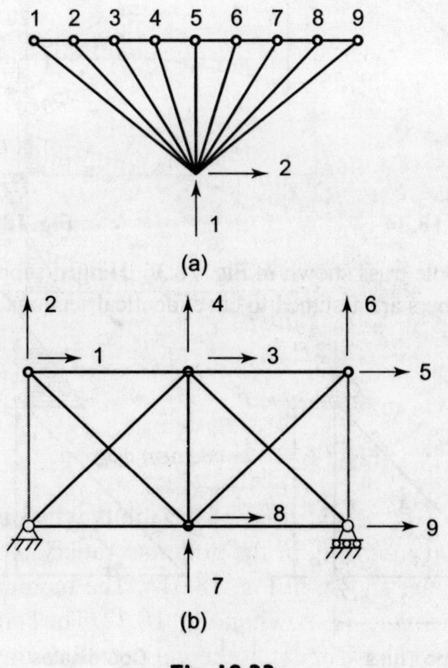

Fig. 18.33

joint displacements, as in Fig. 18.33*b*, the flexibility method is preferred. Since the structure is redundant to the second degree, the flexibility method requires an inversion of a 2×2 matrix. On the other hand, the stiffness method requires the inversion of a 9×9 matrix in order to compute displacements.

The order of the matrix to be inverted is obviously important for manual computations. However, if an electronic digitial computer is to be used to execute the analysis, the manner in which the required set of equations is formulated becomes the important factor in selecting the method of analysis. To have an effective computer programme, the computations required to develop the equations for the analysis should be general and repetitive and not unique to any particular problem. In this respect the stiffness approach is preferred. In the stiffness method there is no actual choice involved as far as the required structure is concerned, and hence a general programme can be written that will solve all classes of problems.

Problems for Practice

Use the stiffness method in solving the following problems.

18.1, 18.2 Analyse the plane three-member trusses shown in Fig. 18.34 and 18.35 due to the applied load. All members have identical axial stiffness *AE*.

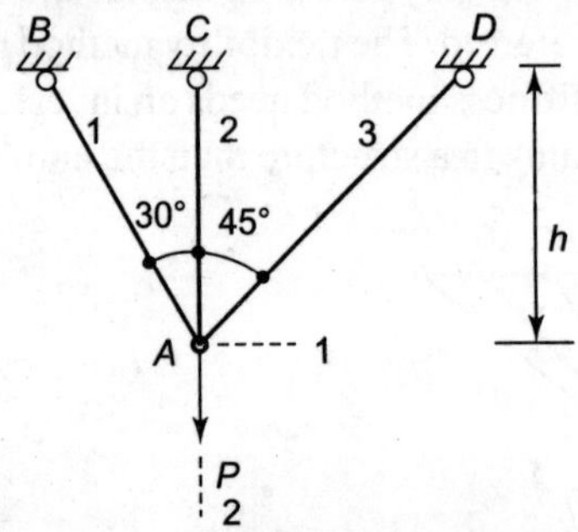

Fig. 18.34

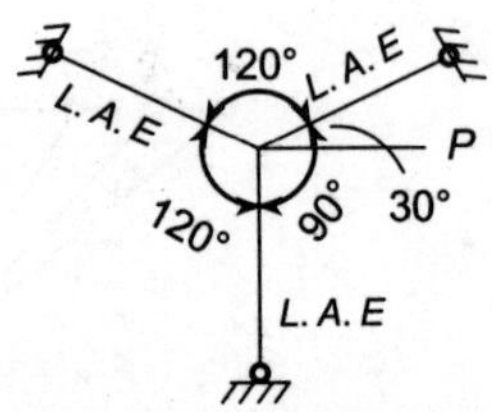

Fig. 18.35

18.3 Analyse the simple truss shown in Fig. 18.36. Hinged supports are provided at A and B and all the members are assumed to have identical sectional properties.

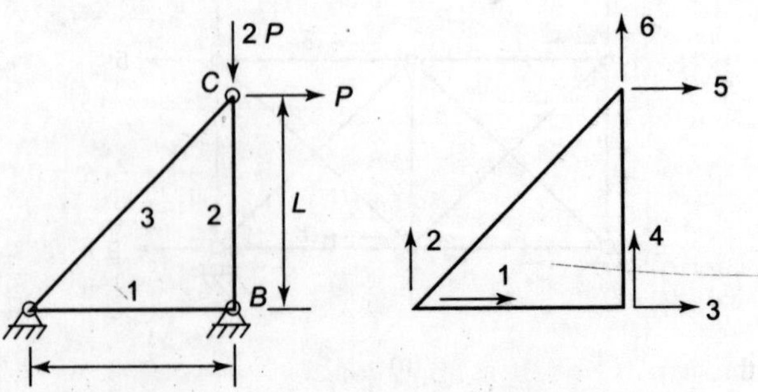

Fig. 18.36

18.4 Analyse the plane truss given in Fig. 18.37 for the loads shown, assuming all members to have the same axial stiffness *AE*.

18.5 Analyse the beam shown in Fig. 18.38 assuming flexural rigidity *EI* constant for the span.

18.6 Analyse the beam shown in Fig 18.39 taking points A, B and C as joints and assuming constant *EI*.

18.7 Find the displacements over supports and member and forces of the beam given in Fig. 18.40.

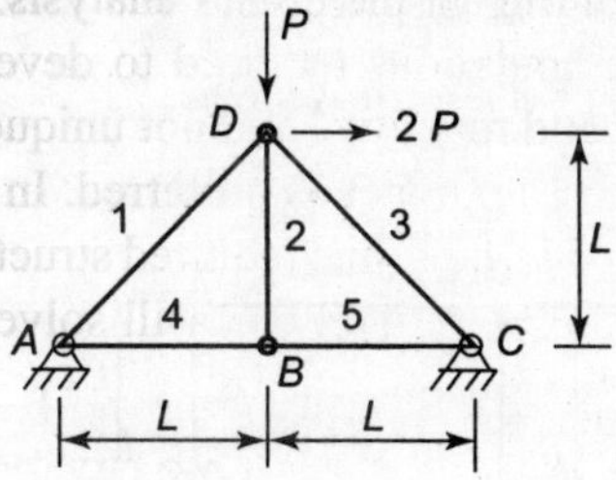

Fig. 18.37

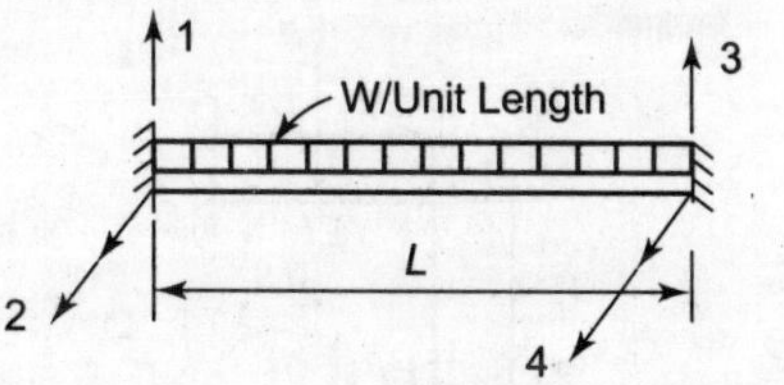

Fig. 18.38

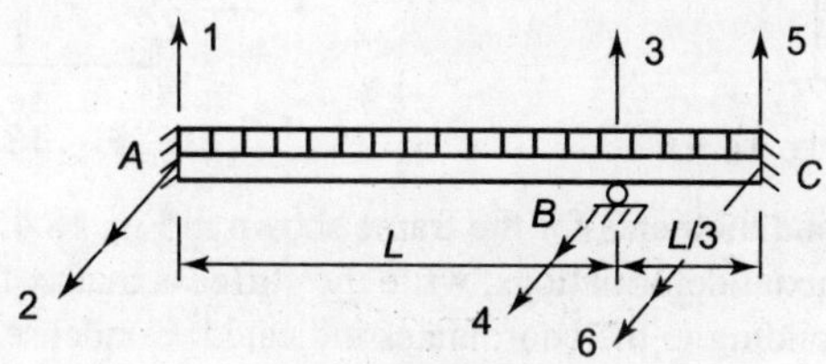

Fig. 18.39

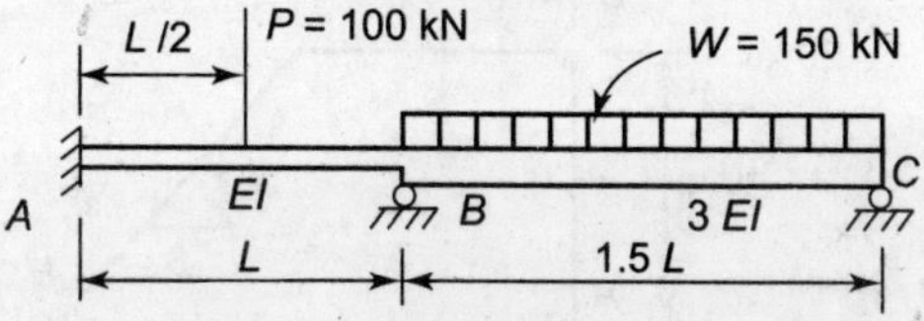

Fig. 18.40

18.8 Analyse the beam shown in Fig. 18.41 taking points *A*, *B*, *C* and *D* as joints.

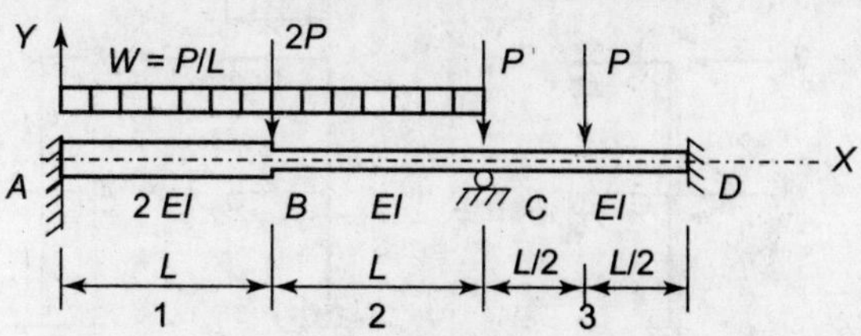

Fig. 18.41

18.9 Find the support moment at *B* for the beam shown in Fig. 18.42. *EI* is constant.

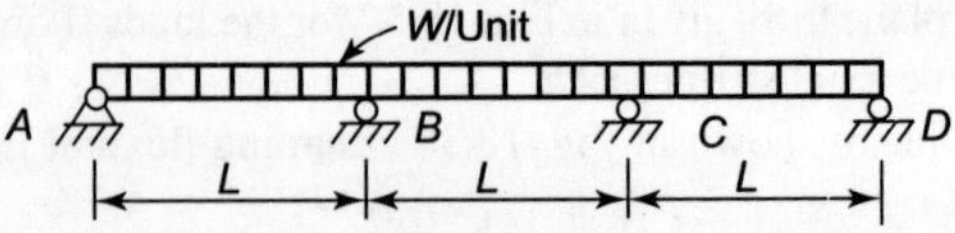

Fig. 18.42

18.10 Analysis the structure shown in Fig. 18.43 taking rotation at joint B as the only degree of freedom. Draw the moment diagram and evaluate all reactions.

18.11 Analyse the rigid frame shown in Fig. 18.44 for displacements and members forces. Neglect axial deformations. *EI* is constant.

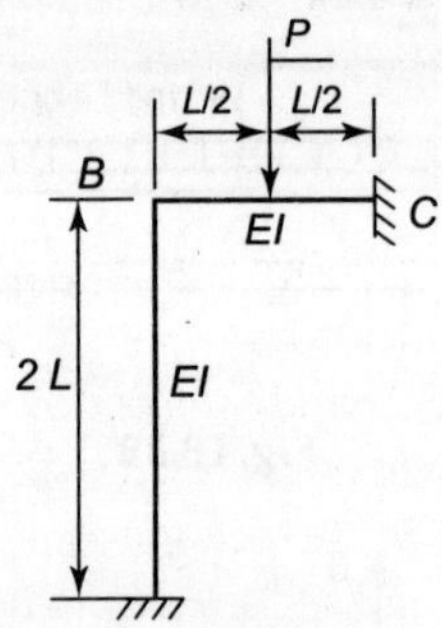

Fig. 18.43

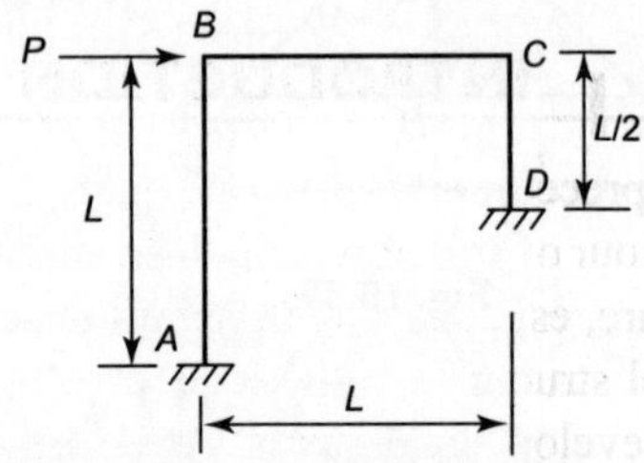

Fig. 18.44

18.12 Obtain the end moments for the frame shown in Fig. 18.45. *EI* is constant.

18.13 Neglecting axial deformations, write the stiffness matrix for the frame shown in Fig. 10.46*a* corresponding to the coordinates indicated. Condense this matrix to find the stiffness matrix corresponding to the coordinates in Fig. 18.46*b*. *EI* is constant.

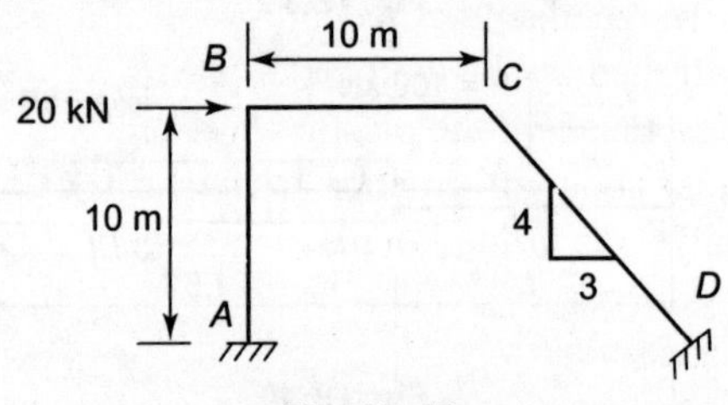

Fig. 18.45

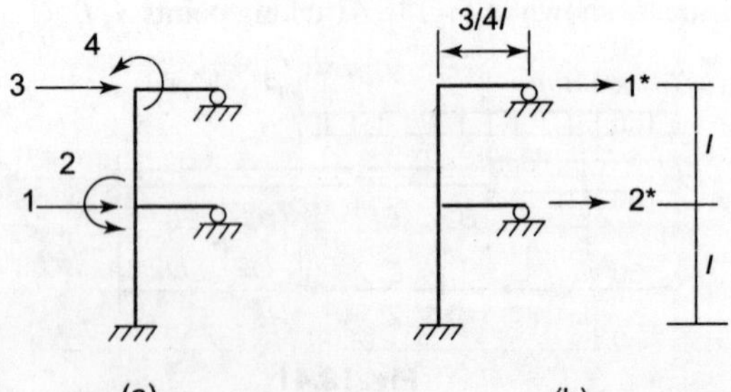

Fig. 18.46

19 Plastic Analysis of Steel Structures

19.1 INTRODUCTION

In the preceding chapters the analysis has been carried out on the basis of elastic behaviour of structures. Such an analysis is useful to study the performance of the structure, especially with regard to serviceability under working load. However, in steel structures if the load is increased, some of the sections in the structure may develop yield stress. Any further increase in load causes the structure to undergo elasto-plastic deformations and some of the sections may develop a fully plastic condition at which a sufficient number of plastic hinges are formed transforming the structure into a mechanism. The mechanism would collapse without noticeable additional loading. A study of the mechanism of failure and knowledge of the load causing the mechanism is necessary to determine the load factor. A structure is designed so that its collapse load is equal to or higher than the working load multiplied by the load factor specified.

Design of structures based on the plastic approach is being increasingly used and adopted in various codes of practice particularly for steel structures. The present outline of analysis is limited to plastic analysis of steel structures only.

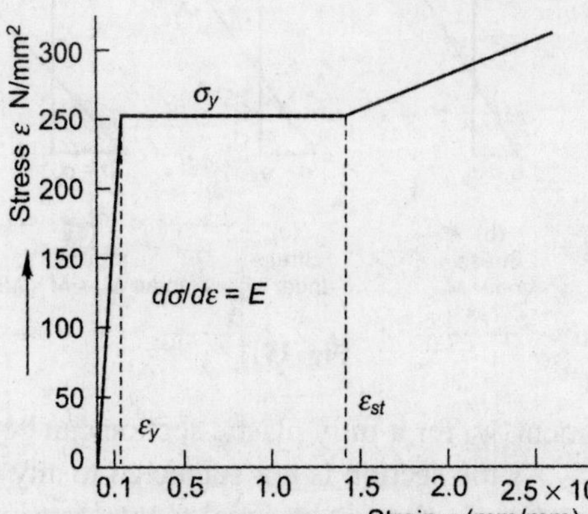

Fig. 19.1 | *Idealised stress-strain curve*

19.2 STRESS-STRAIN CURVE

An idealised stress-strain relationship for structural steel is shown in Fig. 19.1. It may be noted that the large amount of plastic deformation (12-14 times the elastic deformation) is useful for the section to develop a plastic hinge. The large reserve strength available in the strain-hardening region is not utilised in the design of structures.

We shall now consider the principles of plastic analysis as applicable to simple and continuous beams and frames.

19.3 PLASTIC MOMENT

Consider a beam cross-section symmetrical about the plane of bending subjected to a moment M under a working load as shown in Fig. 19.2*a*. The stresses developed are in the elastic region (Fig. 19.2b)

The stress distributions across the depth of the beam under different moment levels are shown in Fig. 19.2*b*, *c*, *d* and *e* in which

M = Moment corresponding to working load under which the stresses are within the proportional limit

M_y = Moment at which the section develops yield stress

and M_p = Moment at which the entire section is under yield stress.

When the moment is increased to M_y the stress variation continues to be linear, the maximum stress at the extreme top fibre reaching yield stress σ_y as in Fig. 19.2*c*. With a further increases in the moment $M_y < M < M_p$ the bottom fibre stress also reaches the yield stress while the yield stress penetrates into the inner fibres at the top as in Fig. 19.2*d*. As the bending moment is increased to M_p the yield will penetrate until the two zones of yield meet; the cross-section at this stage is said to be fully plastic and the moment M_p is known as a plastic moment.

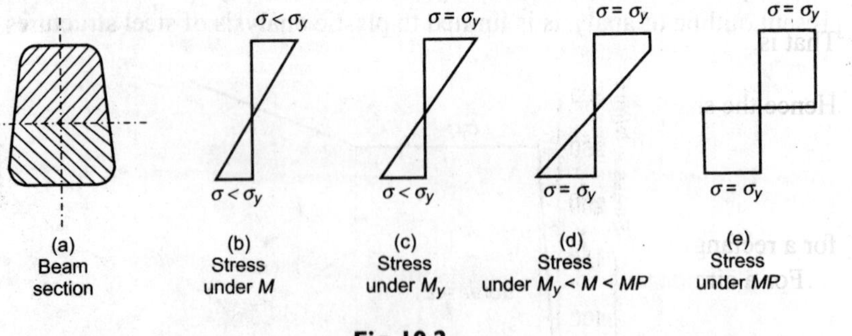

Fig. 19.2

The value of moment M_p for a fully plastic section can be calculated in terms of the yield stress σ_y. As the section is not subjected to any external axial force the total compression $C = A_C\,\sigma_y$ must be equal to total tension $T = A_t\sigma_y$ in which A_c and A_t, are the areas of compression and tensile zones in the cross section.

Equating,

$$A_c\,\sigma_y = A_t\,\sigma_y \tag{19.1}$$

$$\therefore \quad A_c = A_t = \frac{A}{2} \tag{19.2}$$

where A is the area of cross-section.

That is, the neutral axis divides the area of cross-section into two equal parts and the resulting compression and tension $\frac{A\sigma_y}{2}$ form a couple equal to the plastic moment M_p.

$$\therefore \quad M_p = \frac{A}{2}\sigma_y(y_c + y_t) = \sigma_y Z_p \tag{19.3}$$

where y_c and y_t are the distances of centroids of compression and tension areas from N.A. of the plastic section and Z_p is known as the plastic modulus of section.

The maximum moment which a section can carry when the stress first reaches yield value is $M_y = \sigma_y Z_e$, where Z_e is the elastic modulus of the section.

19.3.1 Plastic Modulus, Shape Factor

The ratio $S = \frac{Z_p}{Z_e}$ is known as the shape factor which depends on the shape of the cross-section; it is always greater than unity.

For a rectangular cross-section having breadth b and depth d the modulus of section in elastic analysis is $Z_e = \frac{bd^2}{6}$. The corresponding modulus of section in plastic analysis is obtained by taking the static moment of the compression and tensile areas about the neutral axis as shown in Fig. 19.3*a*.

That is,

$$Z_p = b\left(\frac{d}{2}\right)\left(\frac{d}{4}\right)(2) = \frac{bd^2}{4} \tag{19.4}$$

Hence the shape factor

$$S = \frac{Z_p}{Z_e} = \frac{bd^2}{4} \div \frac{bd^2}{6} = 1.5 \tag{19.5}$$

for a rectangular section.

For a circular section the elastic modulus is

$$Z_e = \frac{\pi}{32}d^3$$

The modulus of section in the plastic analysis can be obtained again by taking the static moment of the tensile and compression areas about the neutral axis as shown in Fig 19.3*b*.

$$Z_p = \frac{\pi}{8}d^2 \frac{2d}{3\pi}(2)$$

$$= \frac{d^3}{6} \tag{19.6}$$

$\therefore$ Shape factor: $$S = \frac{Z_p}{Z_e}$$

$$= \frac{d^3}{6} \div \frac{\pi}{32}d^3 = 1.7 \tag{19.7}$$

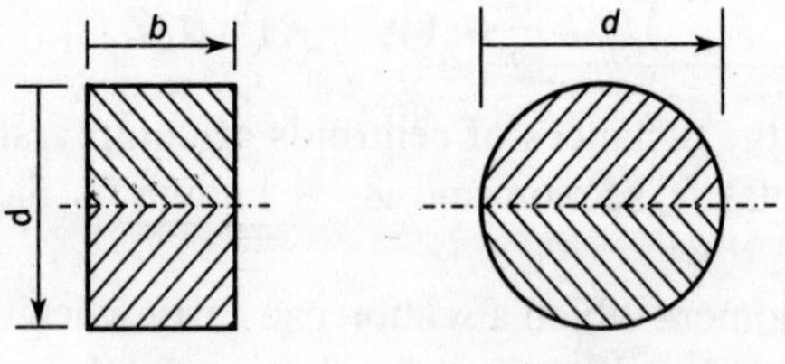

Fig. 19.3

For rolled steel I beams and other built-up I sections the value of the shape factor varies from 1.14 to 1.18.

19.3.2 Load Factor

The ratio of the load causing collapse to the working load is called the load factor. The load factor is dependent upon the shape of the section as the working load is dependent upon the I and Z values and the collapse load is dependent upon the shape of the section.

Considering a rectangular beam of breadth b and depth d, the moment of resistance under working load $M = \sigma_b \frac{bd^2}{6}$, where σ_b is the allowable stress in bending = $\sigma_y/1.5$.

For collapse load the moment of resistance $M_p = \sigma_y \frac{bd^2}{4}$.

Therefore, the load factor = $\frac{M_p}{M} = \sigma_y \left(\frac{bd^2}{4}\right) \bigg/ \frac{\sigma_y}{1.5}\frac{bd^2}{6} = 1.5 \times 1.5 = 2.25$

19.3.3 Mechanism of Failure

(i) Simple Beam

Consider a simply supported beam under a concentrated load W at the centre. (Fig. 19.4.)

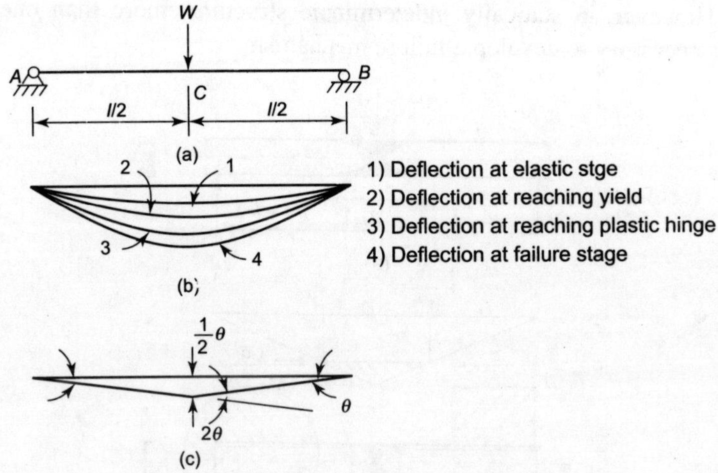

Fig. 19.4 | *(a) Beam and the loading, (b) Deflection curves under increasing load, (c) Rotations and deflections at collapse (4-3)*

As the load W is increased gradually, the mid span section develops yield stress σ_y. As the load is further increased the yield stress penetrates deeper and at a moment M_p a plastic section is developed at the centre. The formation of a hinge under the load point can constitute a mechanism of failure. Without any addition of load the plastic hinge undergoes increasing rotation and so also the deflection in the beam, causing collapse. The load Wu at the collapse stage is related to plastic moment M_p by statics as

$$M_p = \frac{M_u l}{4} \quad \text{or} \quad M_u = \frac{4M_p}{l}$$

Distribution of the Plastic Hinge In the discussions above, the plastic hinge has been assumed to be formed at a point and all the rotation occurred at that point. In reality, the hinge extends over a length of member that is dependent on the loading and the geometry. For example, in the rectangular beam, the hinge length is equal to one-third of the span. For a wide-flanged beam with a shape factor 1.14, the hinge length is $L/8$. In other words, the hinge length ΔL is the length of the beam over which the moment is greater than M_y. In the analysis however, the plastic hinge is considered as a point at which all the plastic rotations occur. (Fig. 19.5.)

The deflections and rotations resulting in a collapse mechanism are shown in Fig. 19.6. The increase in deflection during collapse is due to rotation of the hinge at the centre without a corresponding change in the curvature in the two halves which remain straight.

The plastic analysis is of no advantage for determinate beams and frames as they would develop collapse mechanism soon after formation of the first plastic hinge. However, in statically indeterminate structures more than one plastic hinge is necessary to develop a failure mechanism.

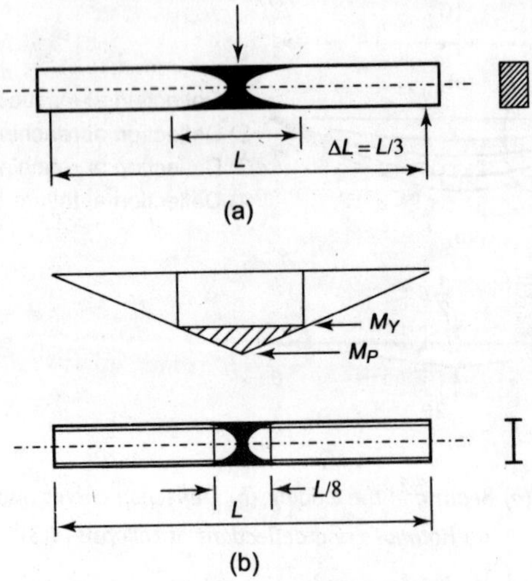

Fig. 19.5 | *Plastic hinge length in (a) rectangular beam, (b) I section*

Virtual Displacements and Virtual Work The principle of virtual displacements is useful in expressing the equilibrium condition. It may be stated that if a system of forces in equilibrium is subjected to virtual displacement, the work done by the external forces equals the work done by the internal forces.

If the external work is called W_E and the internal work is called W_I, this principle may be expressed in the form

$$W_E = W_I \tag{19.8}$$

The collapse load of the beam can be calculated by equating the virtual work done by the external load and internal forces during virtual moment of the collapse mechanism as shown in Fig. 19.6.

In the virtual movement of failure mechanism there is rotation θ at the ends and 2 θ at centre. The downward displacement under load point W_u is $\left(\frac{l}{2}\theta\right)$

Virtual work done by load W_u is $W_E = W_u\left(\frac{l}{2}\right)\theta$

Internal virtual work done by the plastic moment M_p at the hinge, $W_I = M_p$ (2θ). No work is done at the ends as no moment or plastic hinge exists there. Equating external virtual work to internal virtual work we get

$$W_u\left(\frac{l}{2}\right)\theta = 2M_p\,\theta$$

$$W_u = \frac{4M_p}{l}$$

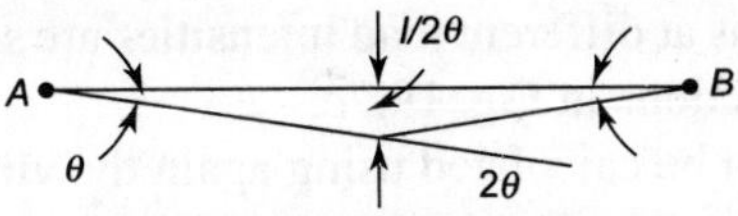

Fig. 19.6 | *Virtual rotations and deflections in a failure mechanism*

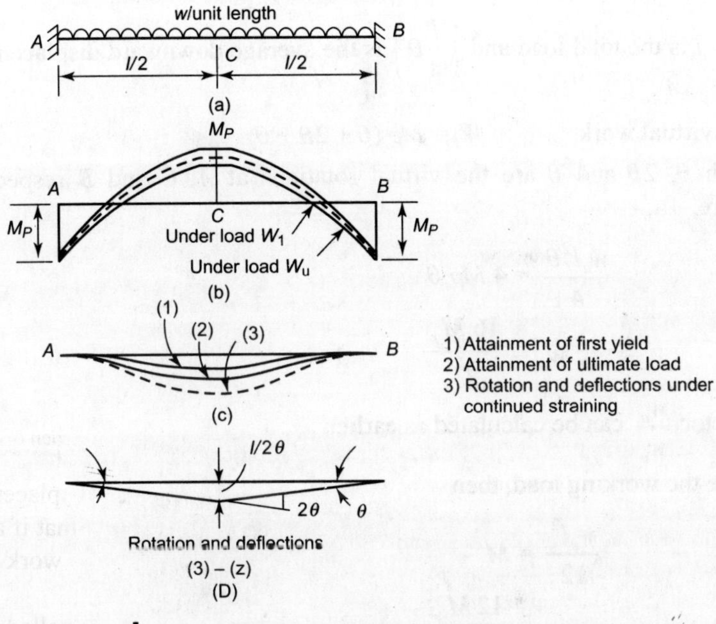

Fig. 19.7 | *(a) Fixed beam, (b) Moment diagrams under loads w_l and w_u (c) Deflections under w_l and w_u and (d) Failure mechanism*

(ii) Fixed Beam

Consider a beam fixed at both the ends and subjected to a uniformly distributed load w as shown in Fig. 19.7. The fixed end moments are $M_A = M_B = \dfrac{-wl^2}{12}$ and $M_c = \dfrac{wl^2}{24}$ in the elastic analysis. If the u.d.1. is increased to w_1 the ends develop fully plastic moment M_p and plastic hinges are formed at A and B.

Although the ends develop plastic moment M_p and the plastic hinges form, the beam will not fail but act as a simple beam to carry further load. If w_1 is further increased, the moments at supports will remain constant at M_p and the moments in the beam will increase, as it would be in a simple beam. In this process the plastic hinges at the ends freely rotate and the increase in deflections will be the same as in a simply supported beam. At a load w_u the beam develops plastic moment M_p at the centre and a third plastic hinge is formed, which constitute a failure mechanism.

The moment diagrams at different load intensities are shown in Fig. 19.7*b* and the corresponding deflections in Fig. 19.7*c*.

The collapse load can be calculated using again the virtual work equations.

External virtual work $\quad W_E = (w_u l)\left(\frac{l}{4}\theta\right)$

where $w_u l$ is the total load and $\left(\frac{l}{4}\theta\right)$ is the average downward displacement of the load.

Internal virtual work $\quad W_I = Mp\,(\theta + 2\theta + \theta)$

in which θ, 2θ and θ are the virtual rotations at A, C and B respectively. Therefore,

$$\frac{w_u l^2 \theta}{4} = 4\,Mp\,\theta$$

$$w_u = \frac{16\,M_p}{l^2}$$

Load factor $\frac{w_u}{w_e}$ can be calculated as earlier.

Let w_e be the working load, then

$$\frac{w_e l^2}{12} = M$$

$$w_e = \frac{12\,M}{l^2}$$

Considering a factor of safety 1.5 to the yield stress, we can write

$$w_y = \frac{12\,M_y}{l^2}$$

$$\text{Ratio } \frac{w_u}{w_y} = \left(\frac{16\,M_p}{l^2}\right)\left(\frac{l^2}{12\,M_y}\right)$$

or

$$\frac{w_u}{w_e} = (1.5)\,\frac{16\,M_p}{12\,M_y}$$

For a rectangular section $\frac{M_p}{M_y} = 1.5$

$$\therefore \quad \frac{w_u}{w_e} = 1.5\left(\frac{16}{12}\right)(1.5) = 3.0$$

This clearly indicates that the design of a fixed beam on the basis of elastic theory is conservative.

(iii) Propped Cantilever Beam

Consider a propped cantilever beam of span l and loaded centrally by a concentrated load W as shown in Fig. 19.8a. The elastic moment diagram is shown in Fig. 19.8b.

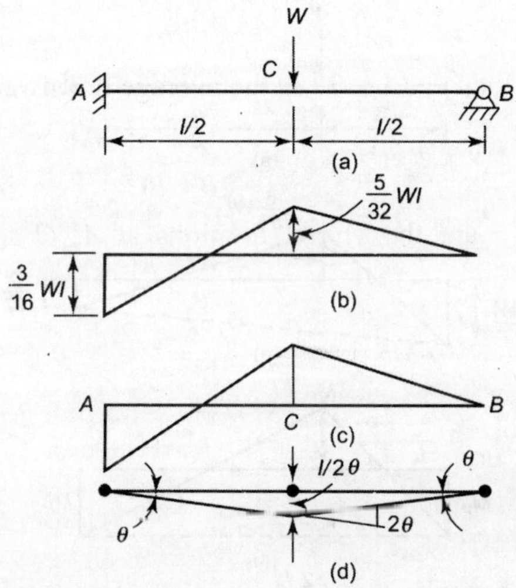

Fig. 19.8 | *(a) Propped cantilever beam and the loading, (b) Moment diagram under working load, (c) M_p at A and C, (d) Deflections under failure mechanism*

The moment at the fixed end is larger than the one under the load point. If the load is increased, the moment at the fixed end reaches M_p and a plastic hinge is formed.

After a plastic hinge is formed at A the beam will act as though it is a simply supported beam having a plastic hinge at A and a real hinge at B. An increase in load W will not increase the moment at the fixed end but will increase the moment in the beam as in a simply supported beam. Eventually the moment under the load point reaches M_p and a failure mechanism will form with two plastic hinges and one real hinge.

The load W_u at which the beam develops a collapse mechanism is determined using virtual work equations as earlier.

$$W_u\left(\frac{l}{2}\theta\right) = M_p\,(\theta) + M_p\,(2\,\theta)$$

where θ = rotation of the hinge at left end and 2 θ = rotation of hinge under load point

(iv) Fixed Beam Under Unsymmetrical Loading

Consider a fixed beam under load W positioned at $l/3$ from the left hand support as shown in Fig. 19.9a. The elastic moment diagram is shown in Fig. 19.9b. The plastic moment will develop first at end A. As the load is increased the next section which will develop plastic moment M_p is under load point. Then the beam between C and B is a cantilever. The third hinge will form at support B, forming a failure mechanism. The collapse load W_u can be calculated using virtual work equations as earlier.

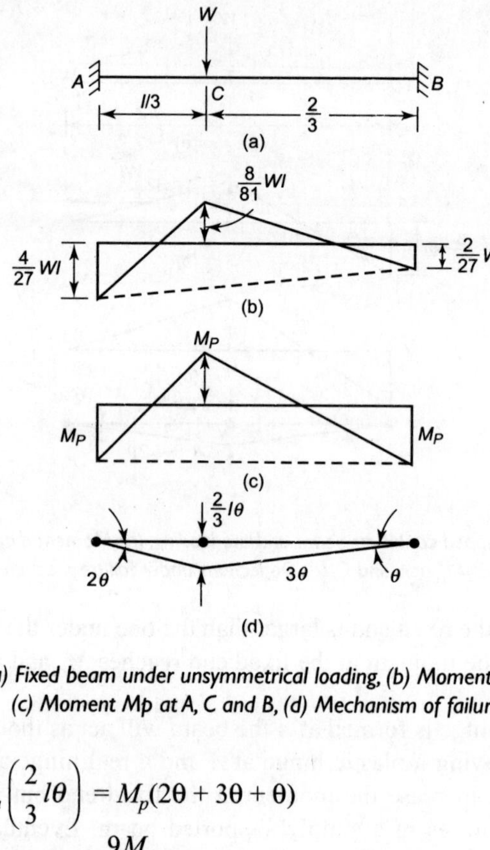

Fig. 19.9 | *(a) Fixed beam under unsymmetrical loading, (b) Moment under working load, (c) Moment Mp at A, C and B, (d) Mechanism of failure*

$$W_u\left(\frac{2}{3}l\theta\right) = M_p(2\theta + 3\theta + \theta)$$

$$W_u \frac{9M_p}{l}$$

(v) Propped Cantilever Under U.D.L.

Consider a propped cantilever beam subjected to a uniformly distributed load w/ unit length as shown in Fig. 19.10*a*.

The elastic moment diagram is shown in Fig. 19.10*b*. The beam has to develop a plastic hinge at fixed end A and another in the span to form a collapse mechanism. The final moment diagram before collapse is as shown in Fig. 19.10*c*. But in this case, the hinge along the span is not at mid point but at a point C where the effective moment is M_p. Let this point be at a distance x from the simply supported end.

$$R_B = \frac{wl}{2} \frac{M_p}{l}$$

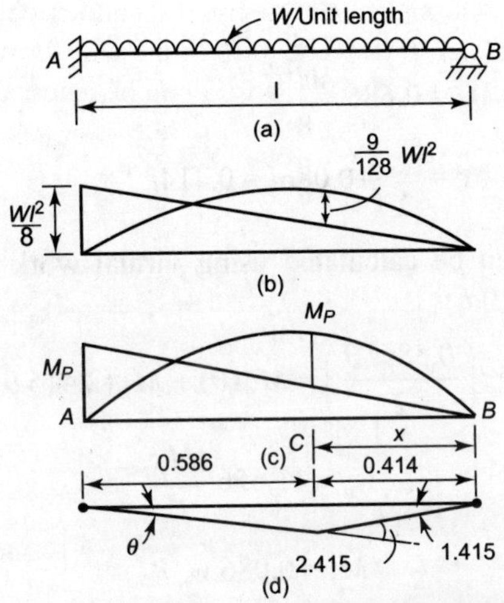

Fig. 19.10 | *(a) Propped cantilever beam, (b) Elastic moment diagram, (c) Location of plastic moment, (d) Failure mechanism*

Shear at $C = R_B - w\,x = 0$ since maximum moment is assumed to occur at C.

Therefore
$$\frac{wl}{2} - \frac{M_p}{l} - wx = 0$$

$$x = \frac{l}{2} - \frac{M_p}{wl}$$

B.M. at C,
$$M_C = R_B x - \frac{wx^2}{2}$$

Substituting for R_B and x

$$M_C = \left(\frac{wl}{2} - \frac{M_p}{l}\right)\left(\frac{l}{2} - \frac{M_p}{wl}\right) - \frac{w}{2}\left(\frac{l}{2} - \frac{M_p}{wl}\right)^2$$

Simplifying

$$M_C = \frac{wl^2}{8} - \frac{M_p}{2} + \frac{M_p^2}{2wl^2}$$

For formation of plastic hinge at C this moment M_C must be equal to M_p.
Therefore

$$M_p = \frac{wl^2}{8} - \frac{M_p}{2} + \frac{M_p^2}{2wl^2}$$

or
$$\frac{M_p^2}{2wl^2} - \frac{3}{2}M_p + \frac{wl^2}{8} = 0$$

Solving,

$$M_p = 0.686\ \frac{wl^2}{8}$$

and $$x = \frac{l}{2} - 0.086l = 0.414l$$

The collapse load can be calculated using virtual work equations as earlier. Referring to Fig. 19.10*d*,

$$w_u l\left(\frac{0.586l\theta}{2}\right) = M_p(\theta) + M_p(2.415\,\theta)$$

$$w_u = 11.656\ \frac{M_p}{l^2}$$

and $$M_p = 0.086\ w_u\ l^2$$

The reader may note the collapse load $w_u = 12\ \frac{M_p}{l^2}$ if the plastic hinge in the span is assumed to form at centre of span. This information is very useful in the design of continuous beams by plastic theory.

Example 19.1 | *Calculate the collapse load W_u for a proposed cantilever loaded as shown in Fig. 19.11. Take the plastic moment capacity of the beam as M_p.*

The beam will develop a mechanism of failure by the formation of a plastic hinge at the fixed end and the second one under one of the two concentrated loads. The location of the second plastic hinge is not obvious; a trial and error procedure is necessary.

First the collapse load is calculated considering that a hinge is formed under load 0.8 W_u as shown in Fig. 19.11*b*. Using the virtual work equation we have

$$W_u\left(\frac{l}{3}\theta\right) + 0.8\,W_u\left(\frac{2}{3}l\theta\right) = M_p(\theta + 3\theta)$$

Solving

$$W_u = 4.62\frac{M_p}{l}$$

If the second hinge is assumed to be formed under load W_u, the collapse load W_u can be calculated from Fig. 19.11*d* and *e* as follows.

$$W_u\left(\frac{2}{3}l\theta\right) + 0.8W_u\left(\frac{l}{3}\theta\right) = M_p(2\theta + 3\theta)$$

Solving,

$$W_u = 5.36\frac{M_p}{l}$$

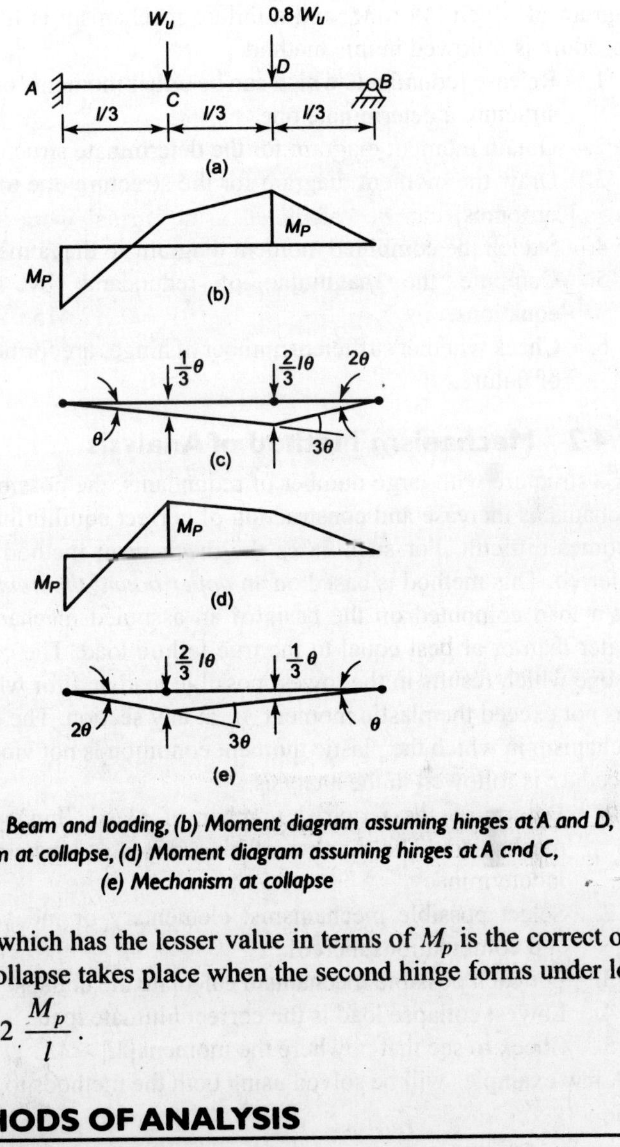

Fig. 19.11 | *(a) Beam and loading, (b) Moment diagram assuming hinges at A and D, (c) Mechanism at collapse, (d) Moment diagram assuming hinges at A and C, (e) Mechanism at collapse*

The value of W_u which has the lesser value in terms of M_p is the correct one. For the beam, the collapse takes place when the second hinge forms under load 0.8 W_u and $W_u = 4.62 \dfrac{M_p}{l}$.

19.4 | METHODS OF ANALYSIS

The two methods of analysis which are followed in the plastic analysis are:

1. Statical Method of Analysis
2. Mechanism Method of Analysis or Kinematic Method of Analysis

19.4.1 Statical Method of Analysis

The statical method is based on the *lower bound theorem*. The theorem states that a load computed on the basis of the assumed equilibrium moment diagram in which the moments are not greater than M_p is less than or at best equal to the true ultimate load. The objective of this method is to find an equilibrium moment diagram in which $|M| < M_p$ and a failure mechanism is formed. The following procedure is followed in this method.

1. Release redundants which can be either moments or forces and make the structure a determinate one.
2. Obtain moment diagram for the determinate structure.
3. Draw the moment diagram for the structure due to redundant moments or forces.
4. Sketch the combined moment diagram so that a mechanism is formed.
5. Compute the magnitude of redundants by solving equilibrium equations.
6. Check whether sufficient number of hinges are formed for the mechanism of failure.

19.4.2 Mechanism Method of Analysis

For a structure with large number of redundants, the possible number of failure mechanisms increase and construction of correct equilibrium moment diagrams becomes difficult. For such cases the mechanism method of analysis may be preferred. This method is based on an *upper bound theorem*. The theorem states that a load computed on the basis of an assumed mechanism will always be greater than or ar best equal to the true failure load. The correct mechanism is the one which results in the lowest possible load and for which the moment $|M|$ does not exceed the plastic moment M_p at any section. The objective is to find a mechanism in which the plastic moment condition is not violated. The following procedure is followed in the analysis.

1. Determine the required number of plastic hinges necessary for the mechanism. The number of hinges is $n + 1$ where n is the degree of indeterminacy.
2. Select possible mechanisms: elementary or independent mechanisms and combinations thereof.
3. For each possible mechanism calculate the collapse load.
4. Lowest collapse load is the correct ultimate load.
5. Check to see that nowhere the moment $|M| > M_p$.

A few examples will be solved using both the methods to make the procedure clear.

(i) Continuous Beams

Example 19.2 | *A two span continuous beam of uniform cross-section is fixed at end A and simply supported at B and C, as shown in Fig. 19.12. The loading in each span is shown. Determine the collapse load in terms of plastic moment Mp.*

Statical Method The structure is statically indeterminate by two degrees. Moments at A and B are identified as redundants. The moment diagram is drawn taking AB and BC as two simple beams. The redundant moment diagram is superimposed over the simple beam moments.

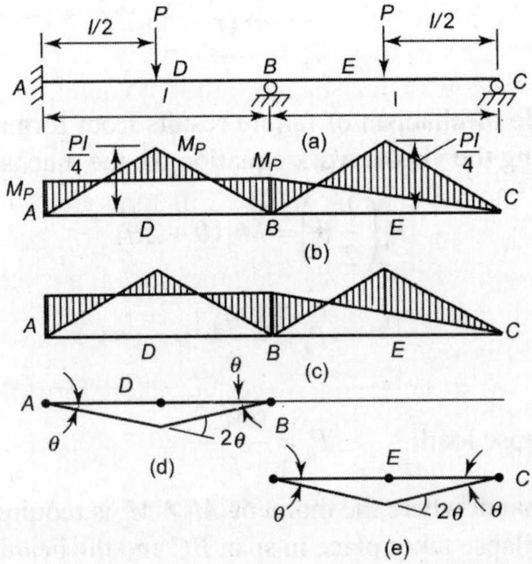

Fig. 19.12 | *(a) A two-span continuous beam, (b) Moment diagram with hinges at A, D and B, (c) Moment diagram with hinges at B and E, (d) and (e) Failure mechanisms*

A failure mechanism requires three plastic hinges. Consider the hinges at A, B and D as shown in Fig. 19.12b. From the moment diagram we can write

$$\frac{Pl}{4} = 2M_p$$

$$P_u = \frac{8M_p}{l}$$

In this, the moment at E, $M_E > M_p$. Hence this is not the correct mechanism. Next we try the moment diagram which causes collapse of the beam BC by forming hinges at B and E as in Fig. 19.12c.

$$\frac{Pl}{4} = M_p + \frac{M_p}{2}$$

$$P_u = \frac{6M_p}{l}$$

Obviously the moment at D, $M_D < M_p$

Therefore the true collapse load is $P_u = \dfrac{6M_p}{l}$

Mechanism Method It is assumed that the beam AB will collapse by the formation of plastic hinges at A, B and D. The mechanism is shown in Fig. 19.12c.

Equating virtual work done by the external and internal forces

$$P_u\left(\frac{l}{2}\theta\right) = M_p(\theta + 2\theta + \theta)$$

$$\therefore \qquad P_u = \frac{8M_p}{l}$$

Another possible mechanism of failure results from formation of hinges at B and E. Again writing the virtual work equation for the mechanism,

$$P_u\left(\frac{l}{2}\theta\right) = M_p(\theta + 2\theta)$$

$$\therefore \qquad P_u = \frac{6M_p}{l}$$

$\therefore$ The least collapse load $\qquad P_u = \dfrac{6M_p}{l}$

A moment check that nowhere the moment $|M| > M_p$ is required. However, in the present case the collapse takes place in span BC and the beam AB is intact and is redundant. The exact magnitude of the moment in the redundant portion of the beam, is not of interest. There are some simple methods available for obtaining the possible moment diagram when the structure is partially redundant at failure. The reader may refer to *Plastic Design of Steel Frames* by Lynn S. Beedle for those methods.

(ii) Rectangular Portal Frames

The plastic analysis of fixed and continuous beams was carried out either by the statical method or by the mechanism method. There was only one distribution of moment in each span and identification of failure mechanism was relatively easy. In case of frames, there exist several possible failure mechanisms. The total number of independent or elementary mechanisms is equal to $(n - m)$ where n is the number of possible plastic hinges and m is the degree of indeterminacy of the frame. Besides the elementary mechanism, combined mechanisms may also form. Each possible mechanism results in a particular failure load, only the lowest of which is correct. In a frame, it is usually convenient to make the analysis by the kinematic method. In the examples that follow only this method of analysis has been followed.

Example 19.3 | *Determine the collapse load for the portal frame shown in Fig. 19.13.*

The frame is indeterminate by three degrees. The number of hinges necessary for total collapse is 4. The number of independent of elementary mechanisms is 2; that is the number of possible hinges (5) minus the degree of indeterminacy.

The two independent mechanisms are (i) the beam mechanism and (ii) the sway mechanism as shown in Fig. 19.13*b* and *c*. The third one shown in Fig. 19.13*d* is the combined mechanism in which the hinge at B is eliminated.

Virtual work equations for each of these mechanisms give

Mechanism 1 $\qquad P\Delta = M_p(\theta) + 2M_p(2\theta) + M_p(\theta)$

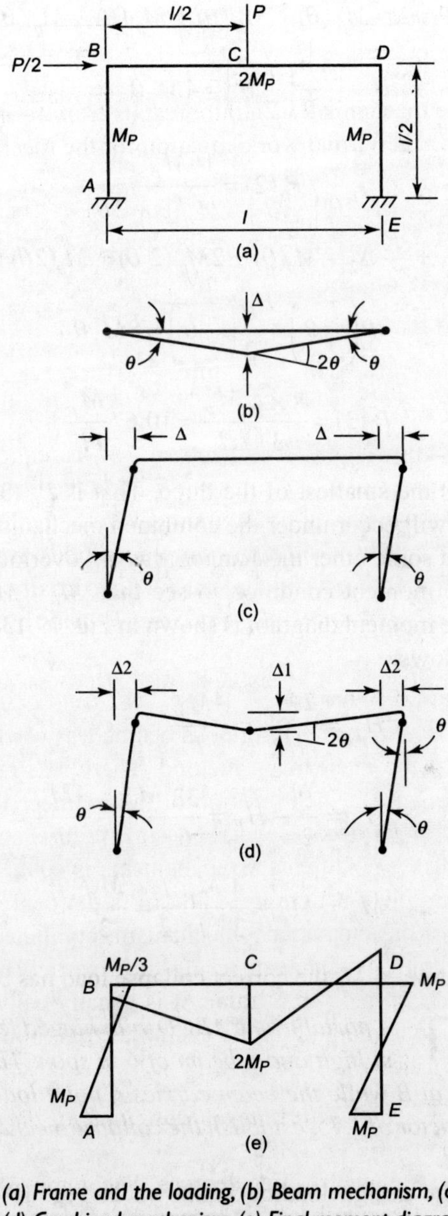

Fig. 19.13 *(a) Frame and the loading, (b) Beam mechanism, (c) Sway mechanism, (d) Combined mechanism, (e) Final moment diagram*

$$P\left(\frac{l}{2}\theta\right) = 6M_p\theta$$

$$P_u(1) = \frac{12M_p}{l}$$

Mechanism 2 $\quad P\Delta/2 = M_p(\theta) + M_p(\theta) + M_p(\theta) + M_p(\theta)$

$$\frac{P}{2}\left(\frac{l}{2}\theta\right) = 4M_p\theta$$

$$P_u(2) = \frac{16M_p}{l}$$

Mechanism 3 $\quad P\Delta_1 + \frac{P}{2}\Delta_2 = M_p(\theta) + 2M_p(2\theta) + M_p(2\theta) + M_p(\theta)$

$$P\left(\frac{l}{2}\theta\right) + \frac{P}{2}\left(\frac{l}{2}\theta\right) = 8M_p\theta$$

$$P_u(3) = \frac{8\times 4}{3}\frac{M_p}{l} = 10.67\frac{M_p}{l}$$

The collapse load is the smallest of the three. That is $P_u(3) = 10.67\ M_p/l$ and failure of the frame will occur under the combined mechanism.

To make sure that some other mechanism was not overlooked, it is necessary to check the plastic moment condition to see that $|M| < M_p$ anywhere on the frame. The complete moment diagram is shown in Fig. 19.13*e*. The moment at *B* is determined as follows:

$$H_E = \frac{2M_p}{l/2} = \frac{4M_p}{l}$$

$$H_A = \frac{P}{2} - H_E = \frac{32}{6}\frac{M_p}{l} - \frac{4M_p}{l} = \frac{4}{3}\frac{M_p}{l}$$

$$M_B = M_p - \frac{4}{3}\frac{M_p}{l}\frac{l}{2} = \frac{M_p}{3}$$

Since the moment $M_B < M_p$ the correct collapse load has been obtained.

Example 19.4 *A portal frame ABCD with hinged feet has stanchions 4 m high and a beam of 6 m span. There is a horizontal point load of 40 kN at B while the beam carries a point load of 120 kN at mid span. Using a load factor of 1.75, establish the collapse mechanism and calculate the collapse moment.*

The frame and the design loading is shown in Fig. 19.14*a*. Using virtual equations in each of the mechanisms, we get,

Beam Mechanism $\quad 210(\Delta) = M_p(\theta) + M_p(2\theta) + M_p(\theta)$

$$210 \times 3\theta = 4M_p\theta$$

$$\therefore \quad M_p = 157.5 \text{ kN.m}$$

Sway Mechanism $\quad 70(\Delta) = M_p(\theta) + M_p(\theta)$

$$70(4\theta) = 2M_p\theta$$

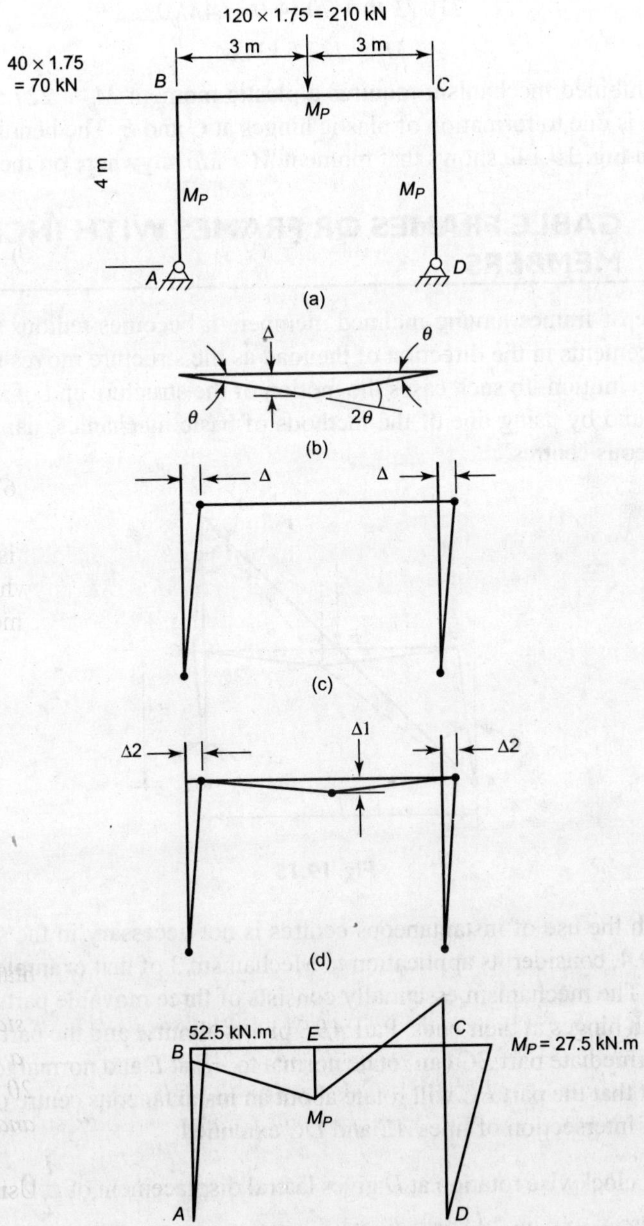

Fig. 19.14 | *(a) Frame and the loading, (b) Beam mechanism, (c) Sway mechanism, (d) Combined mechanism, (e) Final moment diagram*

$\therefore$ $$M_p = 140.0 \text{ kN.m}$$

Combined Mechanism $\quad 210\,(\Delta_1) + 70\,(\Delta_2) = M_p\,(2\,\theta) + M_p\,(2\,\theta)$

$$210\,(3\,\theta) + 70\,(4\,\theta) = 4M_p\theta$$

$$\therefore \qquad M_p = 227.5 \text{ kN.M}$$

The combined mechanism requires a plastic moment M_p = 227.5 kN.m and the failure is due to formation of plastic hinges at *C* and *E*. The bending moment diagram in Fig. 19.14*e* shows that moment *M* > *Mp* anywhere on the frame.

19.5 GABLE FRAMES OR FRAMES WITH INCLINED MEMBERS

In the case of frames having inclined members it becomes tedious to compute the displacements in the direction of the load as the structure moves through the mechanism motion. In such cases the motion of the structure and of its elements may be found by using one of the methods of basic mechanics, namely that of "instantaneous centres".

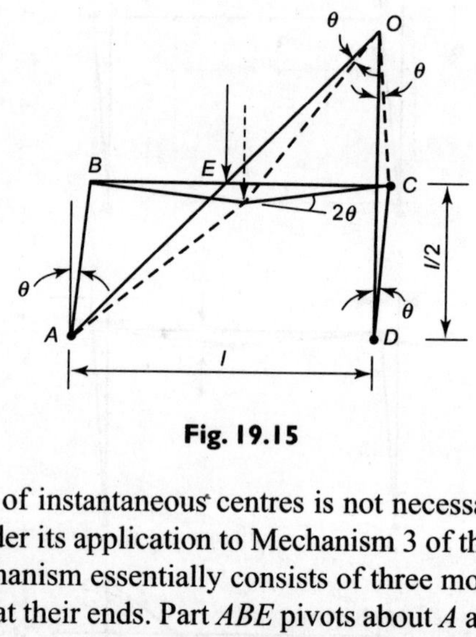

Fig. 19.15

Although the use of instantaneous centres is not necessary in the solution of example 19.4, consider its application to Mechanism 3 of that example shown in Fig. 19.15. The mechanism essentially consists of three movable parts *ABE*, *EC* and *CD* with hinges at their ends. Part *ABE* pivots about *A* and the part *CD* about *D*. The intermediate part *EC* can rotate normal to *AE* at *E* and normal to *DC* at *C*. It is evident that the part *EC* will rotate about an instantaneous centre ι which is the point of intersection of lines *AE* and *DC* extended.

A virtual clockwise rotation at *D* gives lateral displacement of $\frac{l}{2}\theta$ at *C*. Since the length *CO* is equal to $\frac{l}{2}$, the rotation of *EC* about *O* is = θ. The total rotation at *C* is 2 θ and that at *E* is also 2 θ by geometry. The vertical motion of the load at *E* = $\frac{l}{2}\,\theta$ as before. These angular rotations and translations are the same as in Example 19.4 and the virtual work equations give the same results.

The use of "instantaneous centres" is more appropriate for gable frames. Consider a global frame shown in Fig. 19.16*a*. One of the trial mechanisms is shown in Fig. 19.16*b*. The mechanism results in the break up of the structure into three parts: *ABC*, *CDE* and *EF* with hinges at their ends. As discussed earlier part *CDE* rotates about the instantaneous centre *O*. The instantaneous centre is located at the intersection of the lines *AC* and *FE* extended. Let θ_1 be the rotation at hinge point *F*, then by geometry

$$\theta_2 = \frac{\theta_1}{4} \text{ and } \theta_3 = 3\,\theta_2 = \frac{3}{4}\theta_1$$

Equating external virtual work to internal virtual work in a virtual displacement as shown in Fig. 19.16*b*,

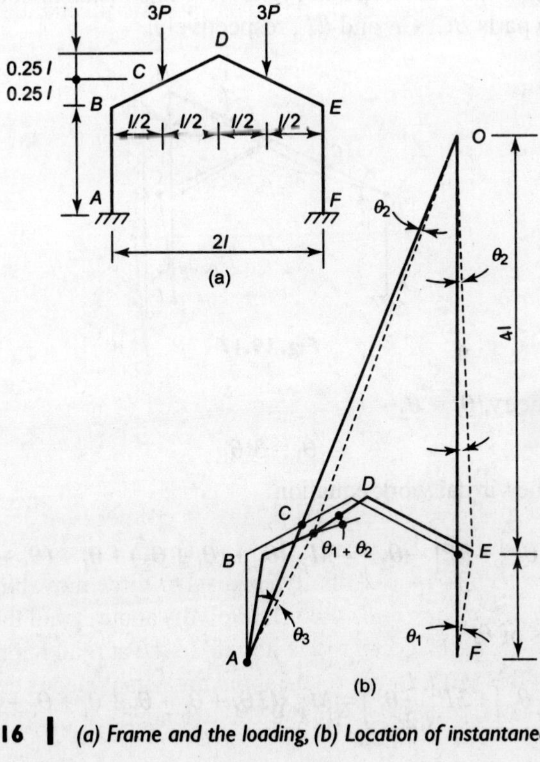

Fig. 19.16 | *(a) Frame and the loading, (b) Location of instantaneous centre*

$$P(l\theta_3) + 3P\left(\frac{l}{2}\theta_3\right) + 3P\left(\frac{l}{2}\theta_2\right)$$

$$= \underset{\text{at A}}{M_p(\theta_3)} + \underset{\text{at F}}{M_p(\theta_1)} + \underset{\text{at C}}{M_p(\theta_2 + \theta_3)} + \underset{\text{at E}}{M_p(\theta_1 + \theta_2)}$$

Writing in terms of θ_1

$$P\left(\frac{3}{4}l\theta_3\right)+3P\left(\frac{3}{4}\frac{l}{2}\theta_1\right)+3P\left(\frac{1}{4}\frac{l}{2}\theta\right)$$
$$= M_p\left(\frac{3}{4}\theta_1\right)+M_p(\theta_1)+M_p(\theta_1)+M_p\left(\frac{5}{4}\theta_1\right)$$

$$\frac{9}{4}Pl\theta_1 = 4M_p\theta_1$$

$$P_{u(1)} = \frac{4\times 4}{9}\frac{M_p}{l} = 1.777\frac{M_p}{l}$$

A second mechanism is shown in Fig. 19.17. The instantaneous centres are *B*, *O* and *F* for the parts *BC*, *CE* and *EF*, respectively.

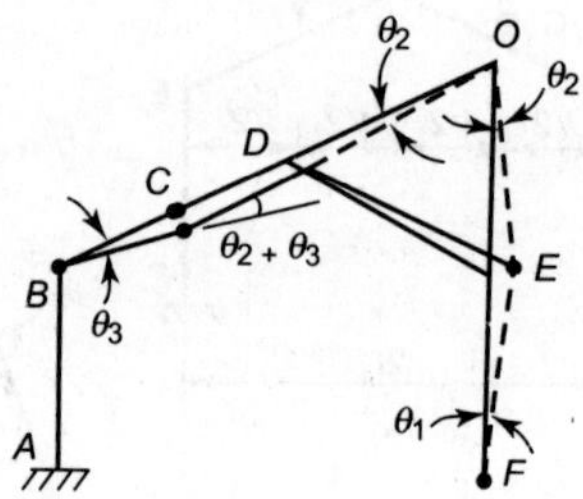

Fig. 19.17

In this, by geometry, $\theta_1 = \theta_2$

$$\theta_3 = 3\,\theta_1$$

Writing down the virtual work equation

$$3P\left(\frac{l}{2}\theta_3\right)+3P\left(\frac{l}{2}\theta_2\right) = M_p\{\underset{\text{at }B}{\theta_3}+\underset{\text{at }E}{(\theta_1+\theta_2)}+\underset{\text{at }F}{\theta_1}+\underset{\text{at }C}{(\theta_1+3\theta_1)}\}$$

Writing in terms of θ_1

$$3P\left(\frac{l}{2}\theta_1\right)+3P\left(\frac{l}{2}\theta_1\right) = M_p\{(3\theta_1+\theta_1+\theta_1+\theta_1+\theta_1+3\theta_1)\}$$

$$6pl\theta_1 = 10M_p\theta_1$$

$$P_{u(2)} = \frac{5}{3}\frac{M_p}{l}$$

It is necessary to make a check that the moment *M* anywhere does not exceed M_p.

19.6 TWO BAY PORTAL FRAME

We shall now take up a two bay portal frame and determine the collapse load by the elementary mechanism and combined mechanisms.

Example 19.5 *Consider a two bay portal frame as shown in Fig. 19.18. Determine the collapse load. The moment M_p values are specified along the members.*

Beam mechanisms for beams 4-5-6 and 6-7-8 and sway mechanisms are considered as shown in Fig. 19.18*b*, *c* and *d* and the collapse loads are determined.

Beam 4-5-6 $\quad 4P(l\theta) = M_p(\theta) + 2M_p(2\theta) + 2M_p(\theta)$

$$4Pl\theta = 7M_p\theta$$

$$P_u(1) = 1.75\frac{M_p}{l}$$

Beam 6-7-8 $\quad 5P\left(\frac{3}{2}l\theta\right) = 3M_p(\theta) + 3M_p(2\theta) + 2M_p(\theta)$

$$\frac{15}{2}Pl\theta = 11M_p\theta$$

$$P_{u(2)} = 1.47\frac{M_p}{l}$$

Sway Mechanism:

$$3P(2l\theta) = M_p(\theta+\theta) + 2M_p(\theta+\theta) + 2M_p(\theta+\theta)$$

$$6Pl\theta = 10M_p\theta$$

$$P_{u(3)} = 1.67\frac{M_p}{l}$$

Combining sway and beam mechanism for beam 6-7-8 as in Fig. 19.18*e* the virtual work equation is,

$$3P(2l\theta) + 5P\left(\frac{3}{2}l\theta\right) = M_p(\theta+\theta) + 2M_p(\theta+\theta) + 2M_p(\theta+2\theta) + 3M_p(2\theta)$$

$$13.5Pl\theta = 18M_p\theta$$

$$P_{u(4)} = \frac{18}{13.5}\frac{M_p}{l} = 1.33\frac{M_p}{l}$$

Combining sway and beams 4-5-6 and 6-7-8 as in Fig. 19.18*f* the virtual work equation is

$$4P(\Delta_1) + 5P(\Delta_2) + 3P(\Delta_3) = M_p(\theta) + 2M_p(\theta) + 2M_p(\theta) + 2Mp(2\theta + 2\theta)$$
$$+ 3M_p(2\theta) + 2M_p(2\theta)$$

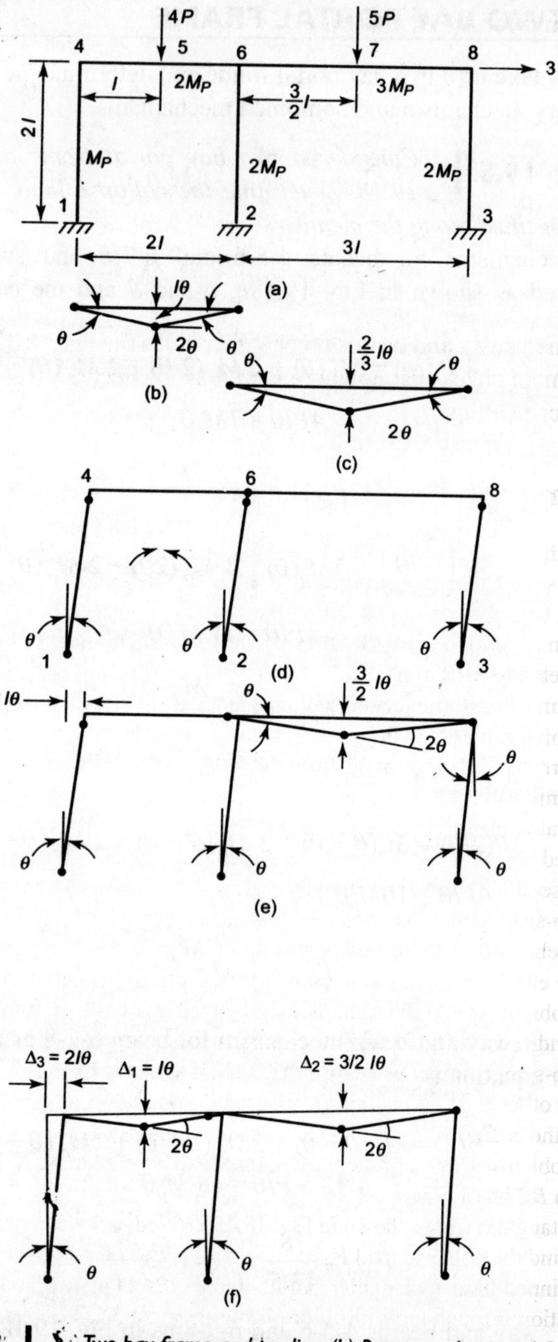

Fig. 19.18 | (a) Two-bay frame and loading, (b) Beam mechanism 4-5-6, (c) Beam mechanism 6-7-8, (d) Sway mechanism, (e) Combined mechanism, (f) Beam and sway mechanism combined

Substituting for Δ_1, Δ_2 and Δ_3,

$$4P(l\theta) + 5P\left(\frac{3}{2}l\theta\right) + 3\,P(2l\theta) = 23M_p\theta$$

$$\frac{35}{2}\,Pl\theta = 23\,M_p\theta$$

$$P_{u(5)} = \frac{23 \times 2}{35}\frac{M_p}{l} = 1.314\frac{M_p}{l}$$

This combined sway and beams mechanism gives the least value for $P_u = 1.3142\ M_p/l$. A moment check that nowhere is $M > M_p$, is necessary to ensure that this is the correct critical load.

Problems for Practice

19.1 Find the shape factor for the following sections:

(a) A square section having side 'a' placed with one of its diagonals vertical.

(b) A tubular section with outer diameter equal to twice the inner diameter.

(c) An I section having compression and tension flanges 250 × 15 mm each and a web 500 × 10 mm.

19.2 A beam of uniform section having span 1 and plastic moment M_p is fixed at one end and simply supported at the other. What is the maximum concentrated load W that the beam can carry if the load is at $l/3$ from the fixed end?

19.3 Determine the collapse load for a propped cantilever beam 15 MB 250 @ 373 N/m if a concentrated load W is acting at mid span.

19.4 A fixed beam 8 m span carries a u.d.l. on the left half of the span. If the plastic moment of the section is 120 kN.m find the value of the collapse load.

19.5 A two-span continuous beam having span AB = 6 m and BC = 8 m is subjected to central concentrated loads of 60 kN and 80 kN respectively. If the beam is simply supported at the ends calculate plastic moment required for the beam.

19.6 In problem 19.5, if the span BC is subjected to a u.d.l. of 100 kN instead of 80 kN point load and end C is fixed find the plastic moment M_p required for the beam.

19.7 A two-span continuous beam ABC each of span l is fixed at end A and simply sup ported at the other end C. Find the collapse load if it is subjected to u.d.l. of w/unit length. Take it that the beam is uniform and has plastic moment M_p.

19.8 In problem 19.7 determine the collapse load w/unit length if the span AB has 2 M_p and the span BC has 1.5 M_p.

19.9 A rectangular frame shown in Fig. 19.19 is fixed at the column bases and is loaded as shown. Find the collapse load P_u.

19.10 A pinned-base rectangular portal frame $ABCD$ of height L and span $3L$ is of uniform section throughout with fully plastic moment M_p. The frame is subjected to a horizontal load P at the top left column together with a vertical load P at a distance L from right end of the beam. Find the value of P which would cause collapse.

19.11 Find the required value of M_p for the fixed-base pitched roof portal frame shown in Fig. 19.20.

19.12 Determine the collapse load of the two bay portal frame shown in Fig. 19.21.

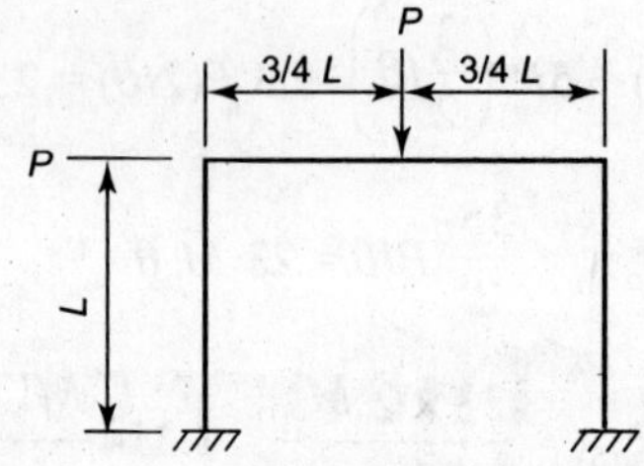

Fig. 19.19

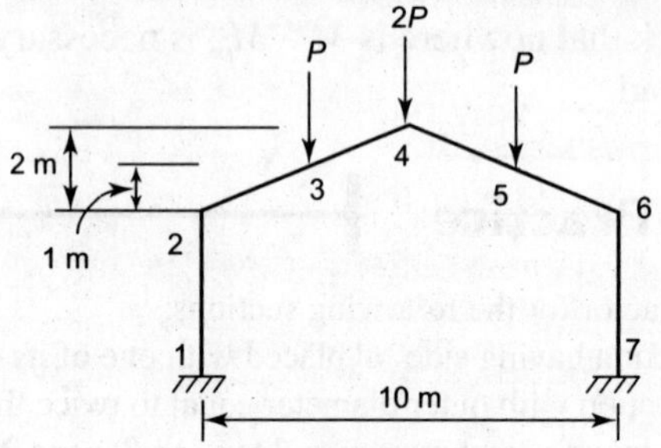

Fig. 19.20

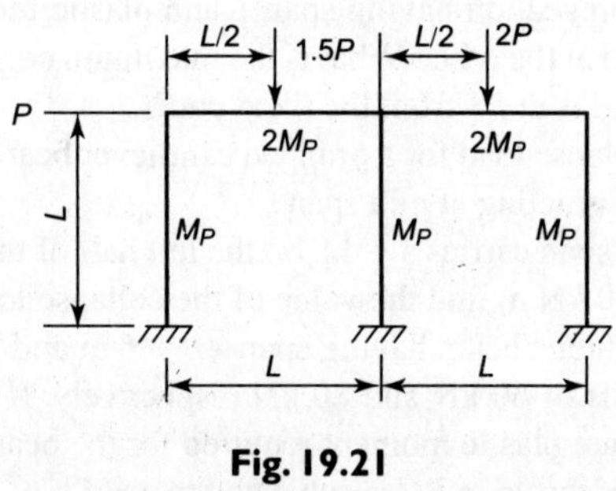

Fig. 19.21

Appendix A Theory of Vectors and Matrices

A.1 VECTORS

In structural analysis forces and displacements can be expressed conveniently by vectors. The use of vector algebra simplifies certain calculations particularly in space members, trusses and frames. Only an elementary theory of vectors is presented here. For greater details the reader should refer to books dealing with vector algebra. A force can be represented vectorially as

$$\mathbf{F} = a_1\mathbf{i} + a_2\mathbf{j} + a_3\mathbf{k} \tag{A.1}$$

where a_1, a_2 and a_3 are the scalar components along coordinate axes **X**, **Y** and **Z** respectively, and **i**, **j** and **k** are the basic unit vectors directed along X, Y and Z axes respectively as shown in Fig. A.1.

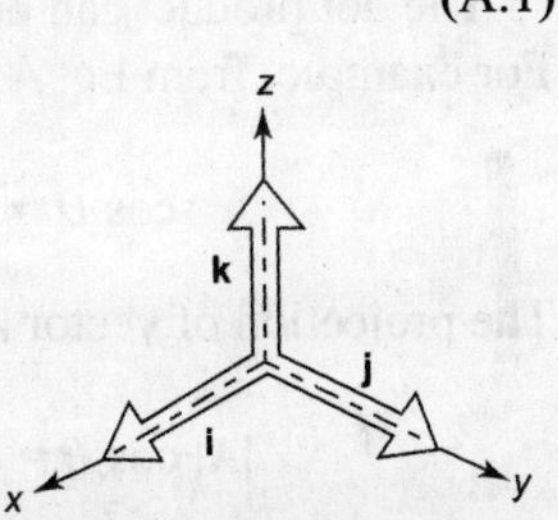

Fig. A.1 *Basic unit vectors*

A.1.1 Addition of Vectors

The addition of vectors follows the law of parallelogram of forces. Let $\mathbf{A} = (a_1\mathbf{i} + a_2\mathbf{j} + a_3\mathbf{k})$ and $\mathbf{B} = (b_1\mathbf{i} + b_2\mathbf{j} + b_3\mathbf{k})$ be the two vectors to be added.

Then,
$$\mathbf{C} = \mathbf{A} + \mathbf{B} = (c_1\mathbf{i} + c_2\mathbf{j} + c_3\mathbf{k}) \tag{A.2}$$

in which
$$c_1 = (a_1 + b_1),\ c_2 = (a_2 + b_2),\ c_3 = (a_3 + b_3)$$

that is, the components of vector **A** are added to the components of vector **B**. In the same way more than two vectors can be added by adding the corresponding components of the elements of the vectors. The only necessary condition is that they must all have the same dimension and expressed in terms of the same coordinate system.

Subtraction is also carried out similarly. For example, **A** – **B** is obtained by subtracting the components of vector B from the corresponding components of vector **A**.

Addition is commutative

$$\mathbf{A} + \mathbf{B} = \mathbf{B} + \mathbf{A} \tag{A.3}$$

Addition is associative

$$\mathbf{A} + (\mathbf{B} + \mathbf{C}) = (\mathbf{A} + \mathbf{B}) + \mathbf{C} \tag{A.4}$$

A.1.2 Product of Vectors

A.1.2.1 Dot Product

The scalar or dot product of two vectors $\mathbf{A} = (a_1\text{i} + a_2\mathbf{j} + a_3\mathbf{k})$ and $B = \{b_1\mathbf{i} + b_2\mathbf{j} + b_3\mathbf{k})$ results in a scalar quantity. This is represented as

$$\mathbf{A} \cdot \mathbf{B} = |\,\mathbf{A}\,|\,|\,\mathbf{B}\,| \cos\theta \tag{A.5}$$

This product $\mathbf{A} \cdot \mathbf{B} = a_1b_1 \cdot \text{i} + a_2b_2\mathbf{j} \cdot \mathbf{j} + a_3b_3k \cdot k$

$$= a_1b_1 + a_2b_2 + a_3b_3 \tag{A.6}$$

since $\mathbf{i} \cdot \mathbf{i} = 1, \mathbf{j} \cdot \mathbf{j} = 1$ and $\mathbf{k} \cdot \mathbf{k} = 1$

In mechanics, this simple operation is useful in many ways. If vectors **A** and **B** represent force and displacement respectively, the product results in work. Work is a scalar quantity.

From the definition it follows readily that the dot product is commutative and is distributive with respect to vector addition.

Thus, $\mathbf{A} \cdot \mathbf{B} = \mathbf{B} \cdot \mathbf{A}$ or $\mathbf{A}(\mathbf{B} + \mathbf{C}) = \mathbf{A} \cdot \mathbf{B} + \mathbf{A} \cdot \mathbf{C}$ (A.7)

The dot product can be used to determine the angle between the two vectors. For example, from Eq. A.5

$$\cos\theta = \frac{\mathbf{A} \cdot \mathbf{B}}{|\,\mathbf{A}\,|\,|\,\mathbf{B}\,|}$$

The projection of vector **A** along an axis represented by **B** is equal to

$$|\mathbf{A}| \cos\theta = \frac{a_1b_1 + a_2b_2 + a_3b_3}{|\,\mathbf{B}\,|} \tag{A.8}$$

The dot product can also be employed to define a plane passing through a given point P and perpendicular to a vector N.

Let $Q\,(x, y, z)$ be an arbitrary point on a plane and **A** and **B** the position vectors to points P and Q as shown in Fig. A.2.

The displacement vector $\mathbf{PQ} = \mathbf{B} - \mathbf{A}$

The equation of the plane is obtained by solving

$$(\mathbf{B} - \mathbf{A}) \cdot \mathbf{N} = 0 \tag{A.9}$$

Example A.1 | *What is the rectangular component of the 500 N force shown in Fig. A.3 along the diagonal from B to A.*

Unit vector along 500 N force

$$\mathbf{n} = \frac{5\mathbf{i} + 10\mathbf{j} + 8\mathbf{k}}{\sqrt{5^2 + 10^2 + 8^2}} = \frac{5\mathbf{i} + 10\mathbf{j} + 8\mathbf{k}}{\sqrt{189}}$$

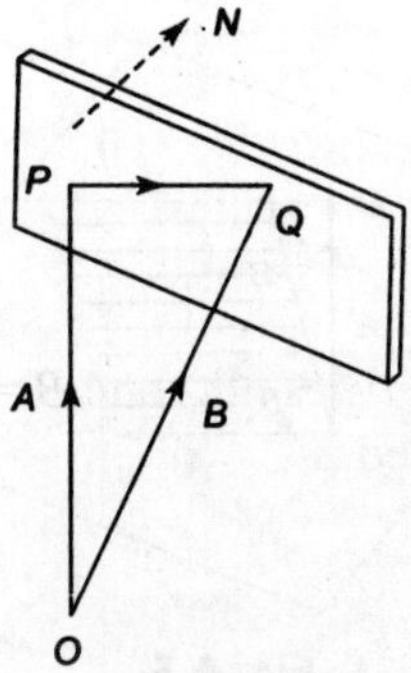

Fig. A.2

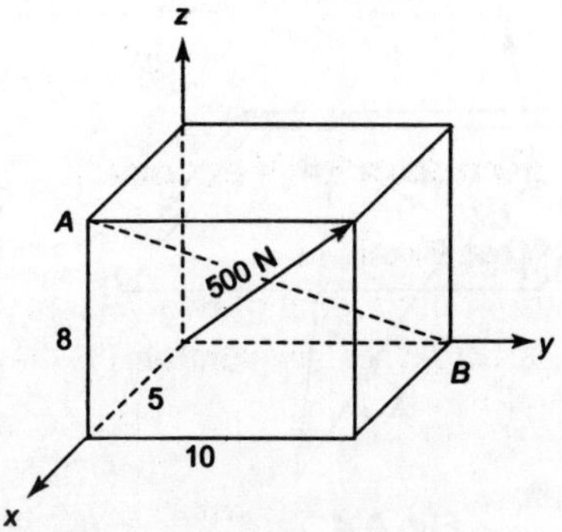

Fig. A.3

Expressing 500 N force in the vector form, we get

$$\mathbf{F} = \frac{2500\mathbf{i}}{\sqrt{189}} + \frac{5000\mathbf{j}}{\sqrt{189}} + \frac{4000\mathbf{k}}{\sqrt{189}}$$

Displacement vector $\mathbf{BA} = 5\,\mathbf{i} - 10\,\mathbf{j} + 8\,\mathbf{k}$

Unit vector along **BA** is $\mathbf{n} = \dfrac{5\mathbf{i} - 10\,\mathbf{j} + 8\mathbf{k}}{\sqrt{5^2 + 10^2 + 8^2}} = \dfrac{5\mathbf{i} - 10\,\mathbf{j} + 8\mathbf{k}}{\sqrt{189}}$

The component of 500 N force along **BA** is obtained by

$$\mathbf{F} \cdot \mathbf{n} = -29.07 \text{ N}$$

Example A.2 | *Define the plane which passes through point B and is perpendicular to axis A given that A = (2, 3, 5) and B = (8, 1, –2).*

Let $Q(x, y, z)$ be any point on the plane (Fig. A.4). Displacement vector $\mathbf{QB} = (8 - x)\,\mathbf{i} + (1 - y)\mathbf{j} + (-2 - z)\mathbf{k}$

The equation for the plane can be obtained by using Eq. A.9

$$\mathbf{QB} \cdot \mathbf{A} = 0$$

Substituting for **QB** and **A** and simplifying we get

$$2x + 3y + 5z - 9.$$

A.1.2.2 Cross Product

This is also known as *vector product* since it results in a vector. One such result may be the moment of the force. For the vector (having possibly different dimensions) shown in Fig. A.5 as **A** and **B** the operation is defined as

$$\mathbf{A} \times \mathbf{B} = \mathbf{C} \tag{A.10}$$

where **C** has a magnitude that is given as

$$|\mathbf{C}| = |\mathbf{A}|\,|\mathbf{B}| \sin\theta \tag{A.11}$$

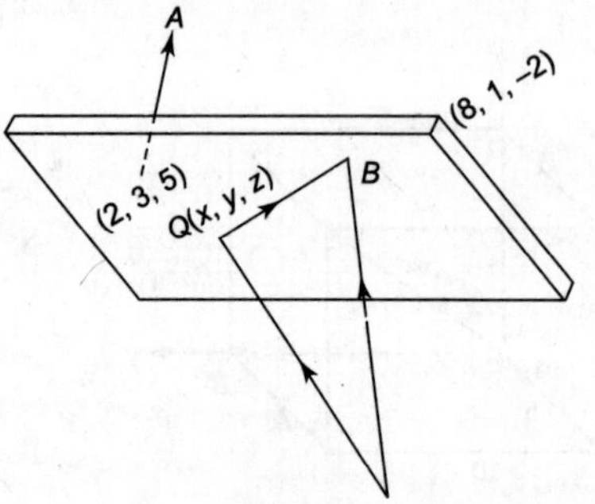

Fig. A.4

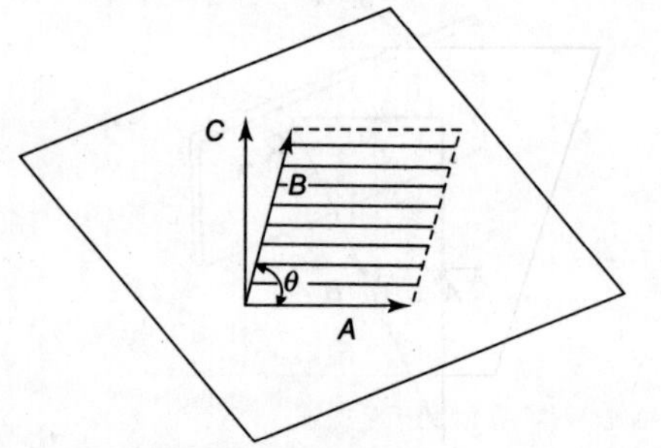

Fig. A.5

which is equal to the area of the parallelogram formed by **A** and **B**. Angle θ is the smaller angle between the two vectors **A** and **B**. Vector **C** has a direction normal to the plane formed by **A** and **B**.

The commutative law does not hold good. From the definition of the cross product

$$(\mathbf{A} \times \mathbf{B}) = -(\mathbf{B} \times \mathbf{A}) \tag{A.12}$$

In a right-handed Cartesian coordinate system

$$\mathbf{A} \times \mathbf{B} = \mathbf{i}\begin{vmatrix} a_2 & a_3 \\ b_2 & b_3 \end{vmatrix} + \mathbf{j}\begin{vmatrix} a_3 & a_1 \\ b_3 & b_1 \end{vmatrix} + \mathbf{k}\begin{vmatrix} a_1 & a_2 \\ b_1 & b_2 \end{vmatrix} \tag{A. 13}$$

where a_1, a_2, a_3 and b_1, b_2, b_3 are the scalar components of **A** and **B** respectively.

For convenience in memorising, it is useful to note that Eq. A. 13 can be interpreted as the expansion of the determinant, that is

$$\mathbf{A} \times \mathbf{B} = \begin{vmatrix} \mathbf{i} & \mathbf{j} & \mathbf{k} \\ a_1 & a_2 & a_3 \\ b_1 & b_2 & b_3 \end{vmatrix} \tag{A.14}$$

The application of cross-product occurs in the definition of moment. The moment of a force **F** about an axis through a given point O is

$$\mathbf{M} = \mathbf{r} \times \mathbf{F} \tag{A.15}$$

where r is a position vector from point O to any point A on the line of action of force F as shown in Fig. A.6. Moment vector **M** is perpendicular to the plane defined by **r** and **F**.

It is easy to see that the magnitude of moment **M** is

$$\mathbf{r} \times \mathbf{F} = |\mathbf{r}| \cdot |\mathbf{F}| \sin \theta = F \cdot d \tag{A.16}$$

a result we are familiar with.

We can easily see that an arbitrary selection of point A anywhere along the line of action of force **F** has no effect on the value of **M**. This in effect only stipulates that **F** is a transmissible vector in the computation of **M**.

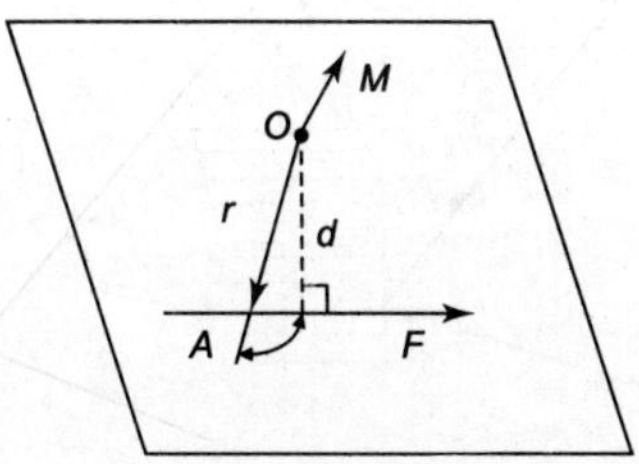

Fig. A.6

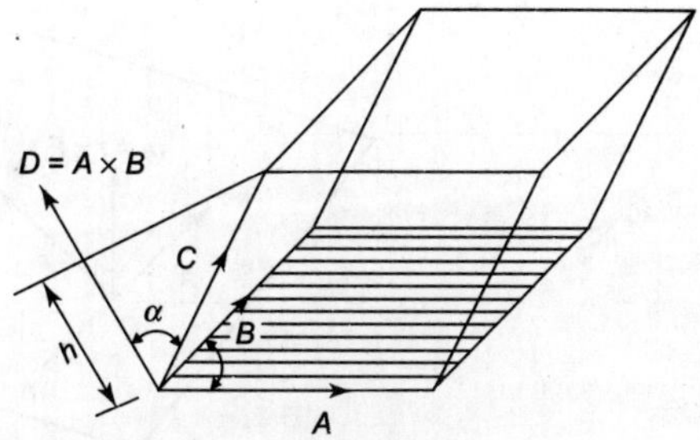

Fig. A.7

A.1.2.3 The Scalar Triple Produpt

Another useful quantity is the *scalar triple product*, which for a set of vectors A, B and C is defined as

$$(\mathbf{A} \times \mathbf{B}) \cdot \mathbf{C} = S \tag{A.17}$$

where S is a scalar quantity. A simple geometric meaning can be associated with this operation. In Fig. A.7 we have shown **A, B** and **C** as an arbitrary set of vectors.

Let $$\mathbf{A} \times \mathbf{B} = \mathbf{D}$$

D is a vector perpendicular to the plane defined by **A** and **B** and is equal in magnitude to the shaded area of the parallelogram. The scalar product of $\mathbf{D} \cdot \mathbf{C}$ results in

$$\begin{aligned} \mathbf{D}\,\mathbf{C} &= D \cdot C \cos \alpha \\ &= D \cdot h \end{aligned} \tag{A.18}$$

in which D is the base area and h the normal height of the parallel-piped. Thus, the scalar triple product represents the volume of the parallel-piped formed by the concurrent vectors of the triple product.

The computation of the scalar triple product is a straightforward process. It takes the form

$$(\mathbf{A} \times \mathbf{B}) \cdot C = \begin{vmatrix} a_1 & a_2 & a_3 \\ b_1 & b_2 & b_3 \\ c_1 & c_2 & c_3 \end{vmatrix} \tag{A.19}$$

This can be written as

$$(\mathbf{A} \times \mathbf{B}) \cdot \mathbf{C} = -(\mathbf{B} \times \mathbf{A}) \cdot \mathbf{C} \tag{A.20}$$

$$\mathbf{A} \cdot (\mathrm{B} \times \mathrm{C}) = (\mathbf{A} \times \mathbf{B}) \cdot \mathrm{C}$$

One important application of the triple scalar product in mechanics is the moment of a force along an axis passing through a point.

In Fig. A.8 let **A** represent an arbitrary vector along axis *aa* and **r** the position vector of a point **B** on the line of the action of force **F**.

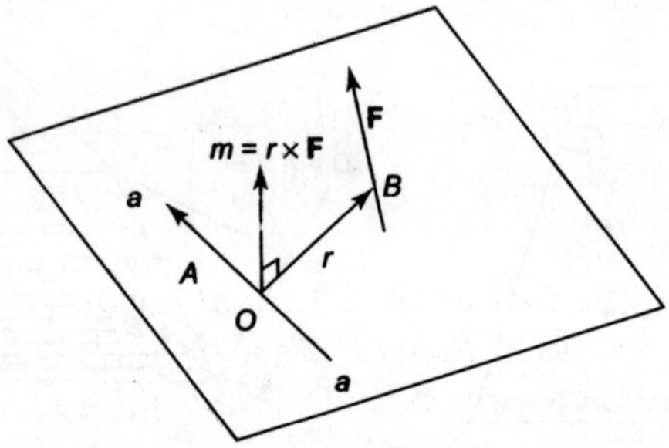

Fig. A.8

Then,

$$\mathbf{M} = (\mathbf{r} \times \mathbf{F}) \times A \tag{A.21}$$

where **M** is the moment of force **F** along axis aa passing through *O*. It is evident that if axis *aa* lies in the same plane as **F** and *O*, moment **M** is equal to zero. Consequently, **M** becomes maximum when axis *aa* lies normal to the plane defined by *O* and **F**.

A.1.2.4 Vector Triple Product

Another operation involving three vectors is the *vector triple product* defined by vectors **A**, **B** and **C** as **A** × (**B** × **C**). The vector triple product is a vector quantity and will appear often in the study of dynamics. The triple vector product can be carried out by using the dot product as

$$\mathbf{A} \times (\mathbf{B} \times \mathbf{C}) = \mathbf{B}(\mathbf{A} \cdot \mathbf{C}) - \mathbf{C}(\mathbf{A} \cdot \mathbf{B}) \tag{A.22}$$

Example A.3 | *A force* $\boldsymbol{F} = (10\boldsymbol{i} + 16\boldsymbol{j})$ *kN goes through the origin of the coordinate system. What is the moment of this force* $\boldsymbol{F}$ *about an axis going through points 1 and 2 with position vectors* $\boldsymbol{r}_1 = 6\boldsymbol{i} + 3\boldsymbol{k}$ *and* $\boldsymbol{r}_2 = 16\boldsymbol{j} - 4\boldsymbol{k}$?

To compute this we can take the moment of **F** about either point 1 or 2 and then find the component of this vector along the direction of the displacement vector between 1 and 2. Using point 1

$$\mathbf{M}_{12} = \{(-\mathbf{r}_1) \times \mathbf{F}\} \cdot \mathbf{n}_{12}$$

where $\mathbf{n}_{12}$ is the unit vector along the direction between points 1 and 2.

This is in the form of the triple scalar product. Accordingly, we can use the determinant approach (see Sec. A.2). for the calculation, once the required vectors have been determined.

$$\mathbf{n}_{12} = \frac{\mathbf{r}_2 - \mathbf{r}_1}{|\mathbf{r}_2 - \mathbf{r}_1|} = \left[\frac{-6\mathbf{i} + 16\mathbf{j} - 7\mathbf{k}}{\sqrt{36 + 256 + 49}}\right] = 0.3249\mathbf{i} + 0.8664\mathbf{j} - 0.3790\mathbf{k}$$

We then have for $\mathbf{M}_{12}$

$$\begin{vmatrix} -6 & 0 & -3 \\ 10 & 16 & 0 \\ -.3249 & +.8664 & -.3790 \end{vmatrix} = -5.20 \text{ kN.m}$$

Example A.4 | *The coordinate of ends of vectors from the origin to the points A, B and C are A = (1, –1, 2) B = (3, –2, 0) and C =(–4, 1, 2). Find a vector N perpendicular to plane ABC.*

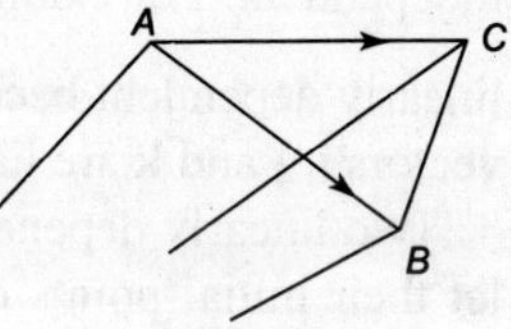

Fig. A.9

In Fig. A.9 let A, B and C be the points, the coordinates of which are given. The displacement vectors **AB** and **AC** are

$$\mathbf{AB} = (2\mathbf{i} - \mathbf{j} - 2\mathbf{k})$$
$$\mathbf{AC} = (-5\mathbf{i} + 2\mathbf{j})$$

The cross product **AB** × **AC** gives a vector N normal to the plane defined by ABC. Carrying out the cross product in the determinant form

$$\begin{vmatrix} \mathbf{i} & \mathbf{j} & \mathbf{k} \\ 2 & -1 & -2 \\ -5 & +2 & 0 \end{vmatrix}$$

We get the vector perpendicular to the plane ABC as

$$4\mathbf{i} + 10\mathbf{j} - \mathbf{k}$$

Example A.5 | *Find the volume of the tetrahedron that has the following vertices*

A(4, 7, –1), B(3, 3, 3), C(1, –1, 2) and D(2, 1, 0)

Let ABC form the corners of the base and D the apex. The displacement vectors **AB**, **AC** and **AD** are

$$\mathbf{AB} = -\mathbf{i} - 4\mathbf{j} + 4\mathbf{k}$$
$$\mathbf{AC} = -3\mathbf{i} - 8\mathbf{j} + 3\mathbf{k}$$
$$\mathbf{AD} = -2\mathbf{i} - 6\mathbf{j} + \mathbf{k}$$

Volume V of the parallel-piped having **AB**, **AC** and **AD** as the three adjacent edges is obtained from the scalar triple product. That is

$$V = (\mathbf{AB} \times \mathbf{AC})\,\mathbf{AD}$$

Evaluating in the determinant form

$$\begin{vmatrix} -1 & -4 & +4 \\ -3 & -8 & +3 \\ -2 & -6 & 1 \end{vmatrix} = 10$$

$$\text{Volume of tetrahedron} = \frac{1}{6}(10) = 1.67 \text{ units}^3$$

A.1.2.5 Orthogonal Vectors

Vectros **A** and **B** are said to be orthogonal if, and only if, their scalar is ..... zero.

A.1.2.6 Linear Dependence of Vectors

A set of m vectors $\mathbf{A}_1$, $\mathbf{A}_2$, ..., $\mathbf{A}_m$ are said to be linearly dependent if at least one of the them can be represented as a linear combination of the other $m - 1$ vectors, that is, as the sum of those vectors each multiplied by a constant. If none of the vectors can be represented in this fashion, they are said to be linearly independent. For example, vectors $\mathbf{A} = \mathbf{i} + 2\mathbf{j} + \mathbf{k}$, $\mathbf{B} = 3\mathbf{k}$ and $\mathbf{C} = 2\mathbf{i} + 4\mathbf{j}$ are linearly dependent because $6\mathbf{A} - 2\mathbf{B} - 3\mathbf{C} = 0$, that is $\mathbf{A} = \frac{1}{3}\mathbf{B} + \frac{1}{2}\mathbf{C}$. The unit vectors **i**, **j** and **k** are linearly independent.

Two linearly dependent non-zero vectors **A** and **B** are collinear, that is, if we let their initial points coincide they lie along the same line. Consequently, we have $\mathbf{A} \times \mathbf{B} = 0$ from the definition of cross product. This yields that two vectors **A** and **B** are linearly dependent if, and only if, their vector product is the zero vector.

Three linearly dependent non-zero vectors **A**, **B** and **C** are co-planar, that is, if we allow their initial points to coincide, they lie in the same plane. By interpreting the triple scalar product, we can state that the three vectors are linearly dependent if, and only if, their triple scalar product is zero.

A.2 DETERMINANTS

Determinants arise in the solution of simultaneous linear algebraic equations. We shall briefly discuss the elements of determinants. Consider the system of simultaneous equations

$$\begin{aligned} a_{11}x_1 + a_{12}x_2 + a_{13}x_3 &= b_1 \\ a_{21}x_1 + a_{22}x_2 + a_{23}x_3 &= b_2 \\ a_{31}x_1 + a_{32}x_2 + a_{33}x_3 &= b_3 \end{aligned} \tag{A.23}$$

Solving for x_1, x_2 and x_3 we get

$$x_1 = \frac{b_1(a_{22}a_{23} - a_{33}a_2) - b_{32}(a_{12}a_{33} - a_{13}a_{32}) + b_3(a_{12}a_{23} - a_{13}a_{22})}{a_{11}(a_{22}a_{33} - a_{23}a_{32}) - a_{21}(a_{12}a_{33} - a_{13}a_{32}) + a_{31}(a_{12}a_{23} - a_{13}a_{22})}$$

$$x_2 = \frac{b_1(a_{23}a_{31} - a_{21}a_{33}) - b_2(a_{13}a_{31} - a_{11}a_{33}) + b_3(a_{13}a_{21} - a_{11}a_{23})}{a_{11}(a_{22}a_{33} - a_{23}a_{32}) - a_{21}(a_{12}a_{33} - a_{13}a_{32}) + a_{31}(a_{12}a_{23} - a_{13}a_{22})}$$

$$x_3 = \frac{b_1(a_{21}a_{32} - a_{22}a_{31}) - b_2(a_{11}a_{32} - a_{12}a_{31}) + b_3(a_{11}a_{22} - a_{12}a_{21})}{a_{11}(a_{22}a_{33} - a_{23}a_{32}) - a_{21}(a_{12}a_{33} - a_{13}a_{32}) + a_{31}(a_{12}a_{23} - a_{13}a_{22})} \tag{A.23}$$

The expression in the denominators of x_1, x_2 and x_3 is written in the form

$$|A| = \begin{vmatrix} a_{11} & a_{12} & a_{13} \\ a_{21} & a_{22} & a_{23} \\ a_{31} & a_{32} & a_{33} \end{vmatrix} \tag{A.25}$$

and is called a determinant of order 3, since it has three rows and three columns

Note that the expression in Eq. (A.25) can also be written as

$$a_{11}\begin{vmatrix} a_{22} & a_{23} \\ a_{32} & a_{33} \end{vmatrix} - a_{12}\begin{vmatrix} a_{21} & a_{23} \\ a_{31} & a_{33} \end{vmatrix} + a_{13}\begin{vmatrix} a_{21} & a_{22} \\ a_{31} & a_{32} \end{vmatrix} \tag{A.26}$$

Thus, a determinant of order 3 is defined in terms of a determinant of order 2. Similarly, determinants of order 4 can be defined in terms of determinants of order 3. In general, we may define a determinant of order n in terms of determinants of order $(n-1)$.

A.2.1 Minors and Cofactors

The minor of an element a_{ij} of a determinant of size n is the determinant of order $(n-1)$ obtained by deleting the ith row and jth column of the original determinant. This is denoted as M_{ij}. Thus, there would be n first minors, each of order $(n-1)$.

The cofactor of determinant $|A|$, corresponding to the element a_{ij}, is designated as C_{ij} and is

$$C_{ij} = (-1)^{i+j} M_{ij} \tag{A.27}$$

Because of this, cofactors are sometimes referred to as signed minors. For example in the third order determinant

$$\begin{vmatrix} a_{11} & a_{12} & a_{13} \\ a_{21} & a_{22} & a_{23} \\ a_{31} & a_{32} & a_{33} \end{vmatrix}$$

$$M_{32} = \begin{vmatrix} a_{11} & a_{13} \\ a_{21} & a_{23} \end{vmatrix} = \begin{vmatrix} a_{11} & a_{13} \\ a_{21} & a_{23} \end{vmatrix} = (a_{11}a_{23} - a_{13}a_{21}) \tag{A.28}$$

and

$$C_{32} = (-1)^{3+2} M_{32} = -a_{11}a_{23} + a_{13}a_{21} \tag{A.29}$$

A.2.2 Evaluation of Determinants by Cofactors–Laplace Expansion

The numerical value of an $n \times n$ determinant $|A|$ can be found either from

$$|A| = \sum_{j=1}^{n} a_{ij} C_{ij} \tag{A.30}$$

for any value of i, or

$$|A| = \sum_{i=1}^{n} a_{ij} C_{ij} \tag{A.31}$$

for any value of j. Equation A.30 describes the expansion of the determinant about a row. Equation A.31 describes the expansion of the determinant about a column.

Example A.6 | *Using Laplace expansion, evaluate the determinant*

$$|A| = \begin{vmatrix} 1 & 3 & 1 & -2 \\ 2 & 0 & -2 & -4 \\ -1 & 1 & 0 & 1 \\ 2 & 5 & 3 & 6 \end{vmatrix}$$

The determinant will be expanded about the third row. Equation A.30 with $i = 3$ and $j = 1, 2, 3, 4$ is applicable

$$|A| = a_{31}C_{31} + a_{32}C_{32} + a_{33}C_{33} + a_{34}C_{34}$$
$$= (-1)(+M_{31}) + (1)(-M_{32}) + (0)(+M_{33}) + (1)(-M_{34})$$

Now

$$M_{31} = \begin{vmatrix} 3 & 1 & -2 \\ 0 & -2 & -4 \\ 5 & 3 & 6 \end{vmatrix} \quad M_{32} = \begin{vmatrix} 1 & 1 & -2 \\ 2 & -2 & -4 \\ 2 & 3 & 6 \end{vmatrix} \quad M_{34} = \begin{vmatrix} 3 & 1 & 1 \\ 0 & 2 & -2 \\ 5 & 2 & 3 \end{vmatrix}$$

Expansion of M_{31} about the first column leaves only a 2×2 determinant. We shall use Eq. A.31 with $j = 1$ and $i = 1, 2, 3$ to evaluate M_{31}. Thus,

$$M_{31} = 3[(-2)(6)-(-4)(3)] + 0 + 5[(1)(-4) - (-2)(-2)] = -40$$

Similarly expanding M_{32} about the first column,

$$M_{32} = 1[(-2)(6)-(-4)(3)] - 2[(1)(6) - (-2)(3)] + 2[(1)(-4) - (-2)(-2)] = -40$$

Expanding M_{34} about the second column

$$M_{34} = -(3)[(2)(3) - (-2)(2)] + 0 - (5)[(1)(-2) - (1)(2)] = -10$$

Now $$|A| = (-1)(-40) + (1)(40) + (1)(10) = 90$$

A.2.3 Properties of Determinants

The following properties of determinants play an important part in matrix algebra:

1. The value of a determinant is not altered when its rows and columns are interchanged, that is, transposed.
2. If all the elements in one row (or column) of a determinant are zero, the determinant is zero.
3. Interchanging of two adjacent rows or columns of a determinant only alters the sign of its determinant.
4. If all the elements in a row or a coiumn of a determinant are multiplied by a factor k, the value of the determinant is k times the value of the given determinant.
5. If each element of a row (or a column) of a determinant is expressed as a binomial, the determinant can be written as the sum of two determinants.

6. If corresponding elements of two rows or columns of a determinant are proportional, the value of the determinant is zero.
7. The value of a determinant is left unchanged if the elements of a row (or column) are altered by adding to them any constant multiple of the corresponding elements in any other row (or column).
8. The value of a triangular determinant is equal to the product of the diagonal elements.

The evaluation of a higher order determinant by Laplace expansion is every time consuming. There are many special methods that are suitable for large determinants. Only one method which is known as pivotal condensation or Gauss' method is presented. This method is based on triangularisation of the determinant using No. 7 from the above properties of the determinants. Then by virtue of No. 8, the determinant will be the product of the elements on the main diagonal.

This is illustrated for a determinant of order 3, but the application of the technique to an nth order determinant will be apparent.

Example A.7 | *Consider a determinant*

$$|A| = \begin{bmatrix} 1 & -2 & 1 \\ -2 & 2 & 3 \\ 1 & 3 & 3 \end{bmatrix}$$

Multiply the first row by +2 and add it to the second row. The resulting determinant value does not change (Property 7). That is,

$$|A| = \begin{bmatrix} 1 & -2 & 1 \\ 0 & -2 & 5 \\ 1 & 3 & 3 \end{bmatrix}$$

Now subtracting first row from the third

$$|A| = \begin{bmatrix} 1 & -2 & 1 \\ 0 & -2 & 5 \\ 1 & 3 & 2 \end{bmatrix}$$

Multiplying second row by 5/2 and adding to the third

$$|A| = \begin{bmatrix} 1 & -2 & 1 \\ 0 & -2 & 5 \\ 0 & 0 & 14.5 \end{bmatrix}$$

Then $$|A| = (1)(-2)(14.5) = -29.$$

A.3 MATRICES

A matrix is defined as an array of elements arranged in rows and columns. The general representation of an $m \times n$ matrix **A** is

$$\mathbf{A} = \begin{bmatrix} a_{11} & a_{12} & a_{13} & \cdots & a_{1n} \\ a_{21} & a_{22} & a_{23} & \cdots & a_{2n} \\ a_{31} & a_{32} & a_{33} & \cdots & a_{3n} \\ \vdots & \vdots & \vdots & \cdots & \vdots \\ a_{m1} & a_{m2} & a_{m3} & \cdots & a_{mn} \end{bmatrix} \tag{A.32}$$

An element of matrix **A** in the *i*th row and *j*th column is represented by the notation a_{ij}. An exmple of a matrix notation are the coefficients in a set of simultaneous equations

$$\begin{aligned} 2x_1 - 3x_2 + x_3 &= 4 \\ -x_1 + 2x_2 + x_3 &= 1 \\ 3x_1 + x_2 + 2x_3 &= -3 \end{aligned} \tag{A.33}$$

represented by a matrix

$$\mathbf{A} = \begin{bmatrix} +2 & -3 & +1 \\ -1 & +2 & +1 \\ +3 & +1 & +2 \end{bmatrix} \tag{A.34}$$

where $a_{11} = +2, a_{12} = -3$ and so on.

A.3.1 Types of Matrices

A.3.1.1 Row Matrix

A matrix,

$$\{a_1\ a_2\ \dots\ a_n\} \tag{A.35}$$

having only one row is called *row matrix* or row vector.

A.3.1.2 Column Matrix

A matrix,

$$\begin{Bmatrix} b_1 \\ b_2 \\ \vdots \\ b_m \end{Bmatrix} \tag{A.36}$$

having only one column is called *column matrix* or column vector. Two matrices **A** and **B** of $(m \times n)$ are said to be equal if, and only if, the corresponding elements are equal, that is $a_{ij} = b_{ij}$ for all $\boldsymbol{i} = 1, 2, \dots, n$ and $\boldsymbol{j} = 1, 2, \dots, m$. Then **A** = **B**.

A.3.1.3 Square Matrix

A matrix having the same number of rows and columns is called *square matrix* and the number of rows or columns is called its order. The diagonal containing the elements $a_{11}, a_{21}, ..., a_{nn}$ of the square matrix is called the principal diagonal. Square matrices are of special importance in structural analysis.

A.3.1.4 Diagonal Matrix

A square matrix **A** whose elements above and below the principal diagonal are all zero, that is $a_{ij} = 0$ for all $i \neq j$, is called *diagonal matrix*. For example,

$$\begin{bmatrix} 3 & 0 & 0 \\ 0 & 1 & 0 \\ 0 & 0 & -4 \end{bmatrix} \text{ and } \begin{bmatrix} 2 & 0 & 0 \\ 0 & -3 & 0 \\ 0 & 0 & 5 \end{bmatrix} \tag{A.37}$$

are diagonal matrices.

A.3.1.5 Unit Matrix

A diagonal matrix whose elements in the principal diagonal are all unity and zero elsewhere is called *unit matrix* or identity matrix. For example,

$$\mathbf{I} = \begin{bmatrix} 1 & 0 & 0 & \cdots & 0 \\ 0 & 1 & 0 & \cdots & 0 \\ 0 & 0 & 1 & \cdots & 0 \\ \hline 0 & 0 & 0 & 0 & 1 \end{bmatrix} \tag{A.38}$$

is a unit matrix.

A.3.1.6 Null Matrix

A matrix whose elements are all zero is called *null matrix* or zero matrix

A.3.1.7 Real Matrix

If all the elements of a matrix are real, the matrix is called *real matrix*.

A.3.1.8 Transpose of a Matrix

Interchanging the rows and columns of a matrix **A** results in *transpose matrix* and is denoted as $\mathbf{A}^T$. If **A** is a $m \times n$ matrix then $\mathbf{A}^T = n \times m$ matrix. For example,

$$\mathbf{A} = \begin{bmatrix} 2 & 3 & 6 \\ 4 & 1 & 0 \end{bmatrix} \text{ and } \mathbf{A}^T = \begin{bmatrix} 2 & 4 \\ 3 & 1 \\ 6 & 0 \end{bmatrix}$$

Likewise, after transposition, a column vector becomes a row vector 2

$$\mathbf{A} = \begin{Bmatrix} 2 \\ -1 \\ 3 \end{Bmatrix} \text{ and } \mathbf{A}^T = \{2 \ -1 \ 3\}$$

The fundamental properties of transposition are

$$(\mathbf{A}^T)^T = \mathbf{A} \quad \text{(A.39a)}$$

$$(\mathbf{A} \pm \mathbf{B})^T = \mathbf{A}^T \pm \mathbf{B}^T \quad \text{(A.39b)}$$

$$(\mathbf{A}\,\mathbf{B})^T = \mathbf{B}^T\mathbf{A}^T \quad \text{(A.39c)}$$

$$(\mathbf{K}\,\mathbf{A})^T = \mathbf{K}\mathbf{A}^T \quad \text{(A.39d)}$$

$$\mathbf{A}^T = \mathbf{A} \text{ if and only if } \mathbf{A} \text{ is symmetric} \quad \text{(A.39e)}$$

$$\mathbf{A}^T\mathbf{A} = \mathbf{A}\,\mathbf{A}^T = \mathbf{B} \quad (\mathbf{B} \text{ is symmetric}) \quad \text{(A.39f)}$$

$$\mathbf{A} + \mathbf{A}^T = \mathbf{B} \quad (\mathbf{B} \text{ is symmetric}) \quad \text{(A.39g)}$$

$$\mathbf{A} - \mathbf{A}^T = \mathbf{B} \quad (\mathbf{B} \text{ is skew-symmetric}) \quad \text{(A.39h)}$$

A.3.1.9 Symmetric and Skew Symmetric Matrices

A real square matrix **A** is said to be symmetric if it is equal to its transpose, that is,

$$\mathbf{A}^T = \mathbf{A}, \text{ that is, } a_{ij} = a_{ji}\ (i, j = 1, 2, ..., n) \quad \text{(A.40)}$$

A real square matrix **A** is said to be skew-symmetric if

$$\mathbf{A}^T = \mathbf{A}, \text{ that is, } a_{ij} = -a_{ji}\ (i, j = 1, 2, ..., n)$$

which implies that the elements in the principal diagonal of a skew-symmetric matrix are all zero.

A.3.1.10 Triangular Matrices (Upper and Lower)

A square matrix whose elements above the principal diagonal (or below the principal diagonal) are all zero is called *triangular matrix.* For example, the matrices **U** and **L**,

$$\mathbf{U} = \begin{bmatrix} 0 & -6 & 1 \\ 0 & 2 & 3 \\ 0 & 0 & 4 \end{bmatrix} \mathbf{L} = \begin{bmatrix} 1 & 0 & 0 \\ -2 & 3 & 0 \\ 5 & 0 & 2 \end{bmatrix} \quad \text{(A.41)}$$

are called the upper and lower triangular matrices respectively.

A.3.1.11 Band Matrix

A matrix whose non-zero elements are located on or near the principal diagonal such that $a_{ij} = 0$ for $|\,i - j\,| > b$ is known as *band matrix.* In case the bandwidth $b = 3$, the band matrix is often referred to as a tridiagonal (triple diagonal) matrix. Band matrices are commonly encountered in structural analysis.

A.3.1.12 Orthogonal Matrix

This is a square matrix whose inverse is equal to its transpose. That is

$$\mathbf{A}^{-1} = \mathbf{A}^T \quad \text{(A.42)}$$

Rotation matrix **R** between rectangular coordinate frame is orthogonal, that is $\mathbf{R}^T = \mathbf{R}^{-1}$ or $\mathbf{R}\,\mathbf{R}^T = \mathbf{I}$. This is frequently employed in matrix analysis of structures.

A.4 MATRIX OPERATIONS

A.4.1 Addition and Subtraction of Matrices

The addition of matrix **A** and matrix **B** results in another matrix **C**, the elements of which are equal to the sum of the corresponding elements of **A** and **B**. The addition can be written as

$$\mathbf{A} + \mathbf{B} = \mathbf{C} \tag{A.43}$$

or $$a_{ij} + b_{ij} = c_{ij} \text{ for each } i \text{ and } j.$$

Similarly, the difference of matrices **A** and **B** is another matrix **D**. That is

$$\mathbf{A} - \mathbf{B} = \mathbf{D} \tag{A.44}$$

or $$a_{ij} - b_{ij} = d_{ij} \text{ for each } i \text{ and } j.$$

The only necessary condition for addition or subtraction is that matrices must all have the same dimension, that is, the same number of rows and columns. Such matrices are said to be *conformable for addition.*

Example A.8 *Let,*

$$\mathbf{A} = \begin{bmatrix} -4 & 6 & 3 \\ 0 & 1 & 2 \end{bmatrix} \text{ and } \mathbf{B} = \begin{bmatrix} 5 & -1 & 2 \\ 3 & 1 & 0 \end{bmatrix}$$

Then

$$\mathbf{A} + \mathbf{B} = \begin{bmatrix} 1 & 5 & 5 \\ 3 & 2 & 2 \end{bmatrix} \text{ and } \mathbf{A} - \mathbf{B} = \begin{bmatrix} -9 & 7 & 1 \\ -3 & 0 & 2 \end{bmatrix}$$

The addition is commutative and also associative.
That is,

$$\mathbf{A} + \mathbf{B} = \mathbf{B} + \mathbf{A} \tag{A.45}$$

and $$(\mathbf{A} + \mathbf{B}) + \mathbf{C} = \mathbf{A} + (\mathbf{B} + \mathbf{C}) \tag{A.46}$$

A.4.2 Matrix Multiplication

The product of matrix **A** and scalar quantity C is defined as matrix $C \cdot \mathbf{A}$ or $\mathbf{A} \cdot C$ and is obtained by multiplying each element of **A** by scalar C. Thus

$$C \cdot \mathbf{A} = \mathbf{A} \cdot C = \begin{bmatrix} c \cdot a_{11} & c \cdot a_{12} & \cdots & c \cdot a_{1n} \\ c \cdot a_{21} & c \cdot a_{22} & \cdots & c \cdot a_{2n} \\ \cdots & \cdots & \cdots & \cdots \\ \cdots & \cdots & \cdots & \cdots \\ c \cdot a_{m1} & c \cdot a_{m2} & \cdots & c \cdot a_{mn} \end{bmatrix} \tag{A.47}$$

The product of two matrices **A** and **B** results in another matrix **C**. This can be written as

$$\mathbf{A}\,\mathbf{B} = \mathbf{C}$$

where the elements of resulting matrix **C** are by definition

$$C_{ij} = \sum_{k=1}^{n} a_{ik} b_{kj}$$

or $$C_{ij} = a_{i1}\, b_{1j}, + a_{i2}, b_{2}, - a_{in}\, b_{nj} \tag{A.48}$$

that is, the ijth element of the resultant matrix **C** is the sum of the multiplication of the elements in the ith row of matrix **A** with the elements of jth column of matrix **B**. Thus, the process of matrix multiplication is conveniently referred to as the multiplication of rows into columns.

In the multiplication of **A** and **B**, **A** is called *pre-multiplier* and **B** is called *post-multiplier.* Such a distinction must be kept in mind since matrix multiplication is in general non-commutative, that is,

$$\mathbf{AB} \neq \mathbf{BA} \tag{A.49}$$

Two matrices can be multiplied only when the matrices are *conformable for multiplication,* that is, the number of columns of pre-multiplier is equal to the number of rows of the post-multiplier. Thus, if **A** is $m \times n$, then **B** must be $n \times p$ so that the product matrix **C** will be $m \times p$

$$\underset{m \times n}{\mathbf{A}} \quad \underset{n \times p}{\mathbf{B}} = \underset{m \times p}{\mathbf{C}} \tag{A.50}$$

Figure A. 10 represents schematically the matrix multiplication.

Figure A. 10 shows an arrangement of matrices **A** and **B** and their product **C** = **AB** which is convenient for numerical work. The point of interest is that each element $\mathbf{C}_{ij}$ of the product matrix occupies the intersection of the ith row of the pre-multiplier matrix and jth column of post-multiplier which are used for computing C_{ij}.

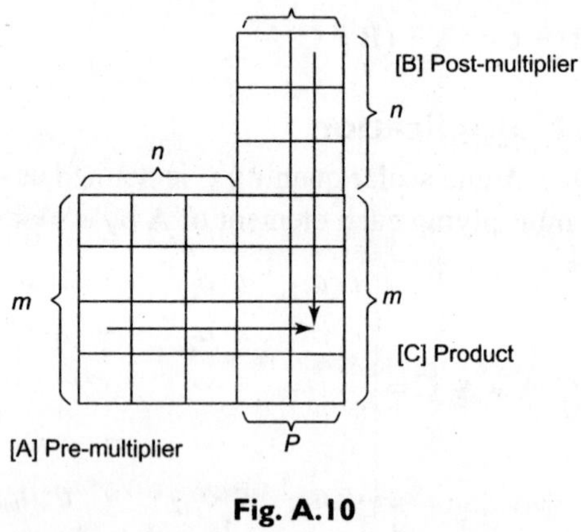

Fig. A.10

Example A.9 *Let,*

$$\mathbf{A} = \begin{bmatrix} 3 & 2 & -1 \\ 0 & 4 & 6 \end{bmatrix} \text{ and } \mathbf{B} = \begin{bmatrix} 1 & 0 & 2 \\ 5 & 3 & 1 \\ 6 & 4 & 2 \end{bmatrix}$$

We are to determine product **AB**.

As a first step it is desirable to write the two matrices with their dimensions. The inner dimension 3 × 3

$$\underset{2\times 3}{\mathbf{A}}\ \underset{3\times 3}{\mathbf{B}} = \underset{2\times 3}{\mathbf{C}}$$

indicates conformability of matrix multiplication and the outer 2 × 3 indicates the dimension of the resulting matrix. Thus,

$$\mathbf{C} = \begin{bmatrix} 3 & 2 & -1 \\ 0 & 4 & 6 \end{bmatrix}\begin{bmatrix} 1 & 0 & 2 \\ 5 & 3 & 1 \\ 6 & 4 & 2 \end{bmatrix} = \begin{bmatrix} 7 & 2 & 6 \\ 56 & 36 & 16 \end{bmatrix}$$

Sum 3 6 5 Sum 63 38 22

$$C_{11} = (3)(1) + (2)(5) + (-1)(6) = 7$$
$$C_{12} = (3)(0) + (2)(3) + (-1)(4) = 2$$
$$C_{13} = (3)(2) + (2)(1) + (-1)(2) = 6$$
$$C_{21} = 0(1) + (4)(5) + (6)(6) = 56$$
$$C_{22} = 0 + (4)(3) + (6)(4) = 36$$
$$C_{23} = 0 + (4)(1) + (6)(2) = 16$$

Numerical work can be checked by the use of the sums of the elements of the columns. In example A.9 we have,

$$(3)(1) + (6)(5) + (5)(6) = 63$$
$$(3)(0) + (6)(3) + (5)(4) = 38$$
$$(5)(2) + (6)(1) + (5)(2) = 22$$

which checks with the sum of elements of the columns in product **C**. The reader will realise that for this numerical example product **B A** is not possible. In matrix multiplication it may happen that **AB** = 0, even though **A** ≠ 0 or **B** ≠ 0. For example, the product of

$$\mathbf{A} = \begin{bmatrix} 1 & 1 \\ 2 & 2 \end{bmatrix} \text{ and } \mathbf{B} = \begin{bmatrix} -1 & 1 \\ 1 & -1 \end{bmatrix} \text{ results in } \mathbf{C} = \begin{bmatrix} 0 & 0 \\ 0 & 0 \end{bmatrix}$$

It may be shown that in this example **BA** ≠ 0.

Furthermore, if **AB** = **AC** it does not necessarily imply that **B** = **C**. For example,

$$\mathbf{A} = \begin{bmatrix} 0 & -3 \\ 0 & 1 \end{bmatrix}, \mathbf{B} = \begin{bmatrix} 1 & 2 \\ -4 & 3 \end{bmatrix}, \mathbf{C} = \begin{bmatrix} 5 & 8 \\ -4 & 3 \end{bmatrix}$$

Then $\mathbf{AB} = \mathbf{AC} = \begin{bmatrix} +12 & -9 \\ -4 & +3 \end{bmatrix}$, yet **B** ≠ **C**

Multiplication of three or more matrices can exist provided they are conformable in sequence as shown

$$\underset{m\times n}{\mathbf{A}}\quad \underset{n\times p}{\mathbf{B}}\quad \underset{p\times q}{\mathbf{C}}\quad \underset{q\times s}{\mathbf{D}} = \underset{m\times s}{\mathbf{E}} \tag{A. 51}$$

The associative and distributive laws are valid in matrix multiplication

$$(\mathbf{AB})\mathbf{C} = \mathbf{A}(\mathbf{BC}) \tag{A.52}$$

$$\mathbf{A}(\mathbf{BC}) = \mathbf{AB} + \mathbf{AC} \tag{A.53}$$

A.4.3 Partitioned Matrix

It is often necessary to partition a matrix into a number of smaller matrices known as sub-matrices. For example a 3 × 4 matrix can be partitioned as

$$\mathbf{A} = \left[\begin{array}{c|ccc} a_{11} & a_{12} & a_{13} & a_{14} \\ a_{21} & a_{22} & a_{23} & a_{24} \\ \hline a_{31} & a_{32} & a_{33} & a_{34} \end{array}\right] = \left[\begin{array}{c|c} \mathbf{A}_{11} & \mathbf{A}_{12} \\ \hline \mathbf{A}_{21} & \mathbf{A}_{22} \end{array}\right] \tag{A.54}$$

where

$$\mathbf{A}_{11} = \begin{bmatrix} a_{11} \\ a_{21} \end{bmatrix} \qquad \mathbf{A}_{12} = \begin{bmatrix} a_{12} & a_{13} & a_{14} \\ a_{22} & a_{23} & a_{24} \end{bmatrix}$$

$$\mathbf{A}_{21} = [a_{31}] \qquad \mathbf{A}_{22} = [a_{32} \quad a_{33} \quad a_{34}]$$

Two matrices **A** and **B** which are of the same size can be added or subtracted by adding or subtracting their sub-matrices, treating them as elements, provided they are partitioned into the same number and size of submatrices. Thus, if **A** and **B** are partitioned as above, their sum **C** can be written as

$$\mathbf{C} = \mathbf{A} + \mathbf{B} = \left[\begin{array}{c|c} \mathbf{A}_{11} + \mathbf{B}_{11} & \mathbf{A}_{12} + \mathbf{B}_{12} \\ \hline \mathbf{A}_{21} + \mathbf{B}_{21} & \mathbf{A}_{22} + \mathbf{B}_{22} \end{array}\right] \tag{A.55}$$

The multiplication of two matrices can be performed in terms of their submatrices, treating them as elements, provided the original matrices and their partitioned matrices are conformable for multiplication.

Example A.10 | *Given two matrices*

$$\mathbf{A} = \left[\begin{array}{cc|c} -1 & 0 & 4 \\ 0 & 2 & 5 \\ \hline 4 & 5 & 1 \end{array}\right] \text{ and } \mathbf{B} = \left[\begin{array}{cc|cc} 4 & 2 & 3 & 1 \\ 2 & 1 & 0 & 3 \\ \hline 5 & 2 & 1 & 0 \end{array}\right]$$

The product $\mathbf{AB} = \mathbf{C} = \begin{bmatrix} 16 & 6 & 1 & -1 \\ 29 & 12 & 5 & 6 \\ 31 & 15 & 13 & 19 \end{bmatrix}$

The two matrices are partitioned keeping in mind the conformity principle for multiplication. The product **AB** in terms of their sub-matrices is

$$\begin{bmatrix} \mathbf{A}_{11} & \mathbf{A}_{12} \\ \mathbf{A}_{21} & \mathbf{A}_{22} \end{bmatrix} \begin{bmatrix} \mathbf{B}_{11} & \mathbf{B}_{12} \\ \mathbf{B}_{21} & \mathbf{B}_{22} \end{bmatrix}$$

where

$$\mathbf{A}_{11} = \begin{bmatrix} -1 & 0 \\ 0 & 2 \end{bmatrix} \qquad \mathbf{B}_{11} = \begin{bmatrix} 4 & 2 \\ 2 & 1 \end{bmatrix}$$

$$\mathbf{A}_{12} = \begin{bmatrix} 4 \\ 5 \end{bmatrix} \qquad \mathbf{B}_{12} = \begin{bmatrix} 3 & 1 \\ 0 & 3 \end{bmatrix}$$

$$\mathbf{A}_{21} = [4 \quad 5] \qquad \mathbf{B}_{21} = [5 \quad 2]$$

$$\mathbf{A}_{22} = [1] \qquad \mathbf{B}_{21} = [1 \quad 0]$$

Performing the required multiplication we obtain

$$\mathbf{C}_{11} = \mathbf{A}_{11}\mathbf{B}_{11} + \mathbf{A}_{12}\mathbf{B}_{21}$$

$$= \begin{bmatrix} -1 & 0 \\ 0 & 2 \end{bmatrix} \begin{bmatrix} 4 & 2 \\ 2 & 1 \end{bmatrix} + \begin{bmatrix} 4 \\ 5 \end{bmatrix} [5 \quad 2]$$

$$= \begin{bmatrix} -4 & -2 \\ 4 & 2 \end{bmatrix} + \begin{bmatrix} 20 & 8 \\ 25 & 10 \end{bmatrix} = \begin{bmatrix} 16 & 6 \\ 29 & 12 \end{bmatrix}$$

$$\mathbf{C}_{12} = \mathbf{A}_{11}\mathbf{B}_{12} + \mathbf{A}_{12}\mathbf{B}_{22}$$

$$= \begin{bmatrix} -1 & 0 \\ 0 & 2 \end{bmatrix} \begin{bmatrix} 3 & 1 \\ 0 & 3 \end{bmatrix} + \begin{bmatrix} 4 \\ 5 \end{bmatrix} [1 \quad 0]$$

$$= \begin{bmatrix} -3 & -1 \\ 0 & 6 \end{bmatrix} + \begin{bmatrix} 4 & 0 \\ 5 & 0 \end{bmatrix} = \begin{bmatrix} 1 & -1 \\ 5 & 6 \end{bmatrix}$$

$$\mathbf{C}_{21} = \mathbf{A}_{21}\mathbf{B}_{11} + \mathbf{A}_{22}\mathbf{B}_{21}$$

$$= [4 \quad 5] \begin{bmatrix} 4 & 2 \\ 2 & 1 \end{bmatrix} + [1][5 \quad 2]$$

$$= [26 \ 13] + [5 \quad 2] = [31 \ 15]$$

$$\mathbf{C}_{22} = \mathbf{A}_{21}\mathbf{B}_{12} + \mathbf{A}_{22}\mathbf{B}_{22}$$

$$= [4 \quad 5] \begin{bmatrix} 3 & 1 \\ 0 & 3 \end{bmatrix} + [1][1 \quad 0]$$

$$= [12 \ 19] + [1 \quad 0] = [13 \ 19]$$

Therefore,

$$\mathbf{C} = \begin{bmatrix} \begin{bmatrix} 16 & 6 \\ 29 & 12 \\ [31 & 15] \end{bmatrix} & \begin{bmatrix} 1 & -1 \\ 5 & 6 \\ [31 & 15] \end{bmatrix} \end{bmatrix} = \begin{bmatrix} 16 & 6 & 1 & -1 \\ 29 & 12 & 5 & 6 \\ 31 & 15 & 13 & 19 \end{bmatrix}$$

which is same as the previous result.

A.4.4 Adjoint Matrix

Let **A** be a square matrix of order n and C_{ij} be the signed minors or cofactors, then

$$\mathbf{C} = \begin{vmatrix} C_{11} & C_{12} & \cdots & C_{1n} \\ C_{21} & C_{22} & \cdots & C_{2n} \\ \hline \\ \hline C_{n1} & C_{n2} & \cdots & C_{nn} \end{vmatrix} \tag{A.56}$$

The transpose of this matrix, by definition is *adjoint of matrix* **A**. Thus,

$$\text{Adj}\,\mathbf{A} = \begin{vmatrix} C_{11} & C_{12} & \cdots & C_{n1} \\ C_{21} & C_{22} & \cdots & C_{n2} \\ \hline \\ \hline C_{1n} & C_{2n} & \cdots & C_{nn} \end{vmatrix} \tag{A.57}$$

The use of adjoint matrices is found in the inversion of matrices dealt with in Sec. A.5.

A.4.4.1 Singular Matrices and Rank of a Matrix

If the determinant of a square matrix vanishes, that is, if the rows (or columns) are nearly dependent, the matrix is said to be *singular.* For example, the matrix

$$\mathbf{A} = \begin{bmatrix} 1 & -2 & -1 \\ -3 & 3 & 0 \\ 2 & 2 & 4 \end{bmatrix}$$

is singular because determinant $\mathbf{A} = 0$. It is seen that the elements in the third column are obtained by adding columns 1 and 2.

The maximum number of linearly independent rows or columns of matrix **A** is called *rank.* The rank of a matrix is defined as being the order of its largest non-singular minor. The rank, therefore, of every square matrix with a non-vanishing minors is the same as its order.

The rank of the transpose of a matrix is same as that of the original matrix.

A.5 INVERSE OF A MATRIX

In an ordinary algebraic equation such as $ax = b$ the unknown x can be obtained by dividing both sides by a that is, $x = b/a$. However, in matrix equation $\mathbf{AX} = \mathbf{B}$, where $\mathbf{A}$ is a $n \times n$ non-singular matrix, $\mathbf{X}$ and $\mathbf{B}$ are $n \times 1$ column vectors, the solution is obtained by pre-multiplying both sides by inverse or reciprocal matrix $\mathbf{A}^{-1}$ of $\mathbf{A}$, that is,

$$\mathbf{A}^{-1}\,\mathbf{AX} = \mathbf{A}^{-1}\mathbf{B} \tag{A.58}$$

or

$$\mathbf{IX} = \mathbf{A}^{-1}\mathbf{B} \tag{A.59}$$

or

$$\mathbf{X} = \mathbf{A}^{-1}\mathbf{B} \tag{A.60}$$

The inverse matrix, then, is the matrix which when pre-multiplied or post-multiplied by matrix $\mathbf{A}$, results in an identity matrix I, that is,

$$\mathbf{AA}^{-1} = \mathbf{A}^{-1}\,\mathbf{A} = \mathbf{I} \tag{A.61}$$

The multiplication here is commutative.

The only necessary and sufficient condition for the existence of the inverse is that the original matrix be a non-singular square matrix. If a matrix has an inverse, the inverse is unique and is referred to as proper inverse.

Inversion in matrix algebra is analogous to division in ordinary algebra. However, inversion in matrix is perhaps the most time consuming procedure in the handling of matrices if carried by hand calculations. However, the availability of electronic digital computers has changed the situation completely. Now there are several methods available for the inversion of a matrix. However, only three of them are described in the following sections.

A.5.1 Inversion of a Matrix by Adjoint

The inverse of a matrix $\mathbf{A}$ is obtained by dividing the adjoint matrix, Adj $\mathbf{A}$, by the determinant of the original matrix. Thus

$$\mathbf{A}^{-1} = \left[\frac{\text{Adj}\,\mathbf{A}}{|\mathbf{A}|}\right] \tag{A.62}$$

The procedure is illustrated by a numerical example.

Example A.11 *Given a square matrix*

$$\mathbf{A} = \begin{bmatrix} -3 & 6 & -11 \\ 3 & -4 & 6 \\ 4 & -8 & 13 \end{bmatrix}$$

Determine the inverse of matrix A.

First let us evaluate the co-factors and hence

$$C_{ij} = \begin{bmatrix} \begin{vmatrix} -4 & 6 \\ -8 & 13 \end{vmatrix} & -\begin{vmatrix} 3 & 6 \\ 4 & 13 \end{vmatrix} & \begin{vmatrix} 3 & -4 \\ 4 & -8 \end{vmatrix} \\ \begin{vmatrix} 6 & -11 \\ -8 & 13 \end{vmatrix} & \begin{vmatrix} -3 & -11 \\ 4 & 13 \end{vmatrix} & \begin{vmatrix} -3 & 6 \\ 4 & -8 \end{vmatrix} \\ \begin{vmatrix} 6 & -11 \\ -4 & 6 \end{vmatrix} & -\begin{vmatrix} -3 & -11 \\ 3 & 6 \end{vmatrix} & \begin{vmatrix} -3 & 6 \\ 3 & -4 \end{vmatrix} \end{bmatrix}$$

$$= \begin{bmatrix} -4 & -15 & -8 \\ 10 & 5 & 0 \\ -8 & -15 & -6 \end{bmatrix}$$

Next, by transposing **C** above the adjoint is obtained. Thus,

$$\text{Adj } \mathbf{A} = \begin{bmatrix} -4 & 10 & -8 \\ -15 & 5 & -15 \\ -8 & 0 & -6 \end{bmatrix}$$

Evaluate the determinant of the original matrix using Eq. A.30

$$|A| = \sum_{j=1}^{n} a_{ij} c_{ij}$$

Operating on the second row .

$$|A| = 3(10) - 4(5) + 0 = 10$$

Therefore,

$$\mathbf{A}^{-1} = \frac{\text{Adj } \mathbf{A}}{|A|} = \begin{bmatrix} -0.4 & 1.0 & -0.8 \\ -1.5 & 0.5 & -1.5 \\ -0.8 & 0 & -0.6 \end{bmatrix}$$

It can be verified that

$$\mathbf{AA}^{-1} = \mathbf{I}$$

A.5.2 Inversion of Matrix by Gauss-Jordan Method

When the size of matrix is larger than 4×4, inversion by Eq. A.62 becomes very cumbersome. Many additional methods have been developed for inverting large matrices. One method which is most commonly used is the Gauss-Jordan or complete elimination method. This is by far the quickest method for the inversion of a matrix on a computer.

As an example, consider the matrix

$$\mathbf{A} = \begin{bmatrix} 2 & 4 & 3 \\ 1 & 3 & 4 \\ -1 & 3 & 6 \end{bmatrix} \tag{A. 63}$$

As a first step, a rectangular matrix is formed by augmenting the given matrix with an identity matrix as shown

$$\begin{bmatrix} 2 & 4 & 3 & 1 & 0 & 0 \\ 1 & 3 & 4 & 0 & 1 & 0 \\ -1 & 3 & 6 & 0 & 0 & 1 \end{bmatrix} \tag{A.64}$$

The Gauss-Jordan elimination process is applied to the rectangular matrix reducing the left part of the matrix to an identity matrix with the right part attaining the elements denoted by b_{ij}. The resulting matrix is of the form

$$\begin{bmatrix} 1 & 0 & 0 & b_{11} & b_{12} & b_{13} \\ 0 & 1 & 0 & b_{21} & b_{22} & b_{23} \\ 0 & 0 & 1 & b_{31} & b_{32} & b_{33} \end{bmatrix} \tag{A.65}$$

The inversion of **A** is

$$\mathbf{A}^{-1} = \mathbf{B} = \begin{bmatrix} b_{11} & b_{12} & b_{13} \\ b_{21} & b_{22} & b_{23} \\ b_{31} & b_{32} & b_{33} \end{bmatrix} \tag{A.66}$$

Example A.12 | *Let us apply the procedure to invert matrix A given in Eq. A.63. The augmented matrix is given in Eq. A.64.*

$$\begin{bmatrix} 1 & 2 & \frac{3}{2} & \frac{1}{2} & 0 & 0 \\ 0 & 1 & \frac{5}{2} & -\frac{1}{2} & 1 & 0 \\ 0 & 5 & \frac{15}{2} & \frac{1}{2} & 0 & 1 \end{bmatrix}$$

Divide first row by $a_{11} = 2$.
Multiply first row by $a_{21} = 1$ and subtract from second row.
Multiply first row by $a_{31} = -1$ and subtract from third row.

After this, the first row and the first column are untouched for some time. The process on the second row is repeated.

$$\begin{bmatrix} 1 & 2 & \frac{3}{2} & \frac{1}{2} & 0 & 0 \\ 0 & 1 & \frac{5}{2} & -\frac{1}{2} & 1 & 0 \\ 0 & 0 & -5 & 3 & -5 & 1 \end{bmatrix}$$

First row left untouched
Divide second row by $a_{22} = 1$.
Multiply second row by $a_{32} = 5$ and subtract from third row.

What we have followed so far is the Gaussian elimination.

$$\begin{bmatrix} 1 & 0 & -\frac{7}{2} & \frac{3}{2} & -2 & 0 \\ 0 & 1 & \frac{5}{2} & -\frac{1}{2} & 1 & 0 \\ 0 & 0 & 1 & -\frac{3}{5} & 1 & -\frac{1}{5} \end{bmatrix}$$

Multiply second row by $a_{12} = 2$ and subtract from first row. This eliminates a_{12}
Second row is left untouched
Divide third row by $a_{33} = -5$

$$\begin{bmatrix} 1 & 0 & 0 & -\frac{3}{5} & \frac{3}{2} & -\frac{7}{10} \\ 0 & 1 & 0 & 1 & -\frac{3}{2} & \frac{1}{2} \\ 0 & 0 & 1 & -\frac{3}{5} & 1 & -\frac{1}{5} \end{bmatrix}$$

Multiply third row by $a_{13} = -7/2$ and subtract from first row. This eliminates a_{13}.
Multiply third row by $a_{23} = 5/2$ and subtract from second row. This eliminates a_{23}.
This eliminates a_{23}.

Hence the inverse matrix is

$$\mathbf{A}^{-1} = \begin{bmatrix} -\frac{3}{5} & \frac{3}{2} & -\frac{7}{10} \\ 1 & -\frac{3}{2} & \frac{1}{2} \\ -\frac{3}{5} & 1 & -\frac{1}{5} \end{bmatrix}$$

The result can be verified by the relationship

$$\mathbf{A}^{-1}\mathbf{A} = \mathbf{I}$$

$$\begin{bmatrix} -\frac{3}{5} & \frac{3}{2} & -\frac{7}{10} \\ 1 & -\frac{3}{2} & \frac{1}{2} \\ -\frac{3}{5} & 1 & -\frac{1}{5} \end{bmatrix} \begin{bmatrix} 2 & 4 & 3 \\ 1 & 3 & 4 \\ -1 & 3 & 6 \end{bmatrix} = \begin{bmatrix} 1 & 0 & 0 \\ 0 & 1 & 0 \\ 0 & 0 & 1 \end{bmatrix}$$

A.5.3 Matrix Inversion by Partitioning

Suppose we have a computer and a programme for inversion which can handle matrices of order $\leq n$. To invert a matrix of a higher order, say $m > n$, we can partition the matrix and generate its inversion in terms of inverses on submatrices of order $\leq n$.

Let $\mathbf{A}$ be an $m \times m$ square matrix which is partitioned into four sub-rnatrices.

$$\mathbf{A} = \begin{array}{c} p \\ \\ q \end{array}\left[\begin{array}{c|c} \mathbf{A}_{11}^{p} & \mathbf{A}_{12}^{q} \\ (p \times p) & (p \times q) \\ \hline \mathbf{A}_{21} & \mathbf{A}_{22} \\ (q \times p) & (q \times q) \end{array}\right] \tag{A.67}$$

Here p and q are smaller than n, and p can be equal to q. Attention must be paid to the fact that at least one of the sub-matrices on the main diagonal, that is, $\mathbf{A}_{11}$ or $\mathbf{A}_{12}$ should be a non-singular square matrix. In other words, an inverse matrix should exist for at least to one of them.

The following procedure is developed on the basis that $\mathbf{A}_{11}$ is non-singular. A similar procedure can be developed if $\mathbf{A}_{22}$ rather than $\mathbf{A}_{11}$ is non-singular. Let the inverted matrix of $\mathbf{A}$ be $\mathbf{B}$ and is partitioned likewise, that is,

$$\mathbf{A}^{-1} = \begin{array}{c} p \\ q \end{array}\left[\begin{array}{c|c} \mathbf{B}_{11}^{n} & \mathbf{B}_{12}^{q} \\ \hline \mathbf{B}_{21} & \mathbf{B}_{22} \end{array}\right] \tag{A.68}$$

Then from the definition of inverse

$$\left[\begin{array}{c|c} \mathbf{A}_{11} & \mathbf{A}_{12} \\ \hline \mathbf{A}_{21} & \mathbf{A}_{22} \end{array}\right]\left[\begin{array}{c|c} \mathbf{B}_{11} & \mathbf{B}_{12} \\ \hline \mathbf{B}_{21} & \mathbf{B}_{22} \end{array}\right] = \left[\begin{array}{c|c} \mathbf{I} & 0 \\ \hline 0 & \mathbf{I} \end{array}\right] \tag{A.69}$$

The following equations can be written

$$\mathbf{A}_{11}\,\mathbf{B}_{11} + \mathbf{A}_{12}\,\mathbf{B}_{21} = \mathbf{I} \tag{A.70}$$

$$\mathbf{A}_{11}\,\mathbf{B}_{12} + \mathbf{A}_{12}\,\mathbf{B}_{22} = 0 \tag{A.71}$$

$$\mathbf{A}_{21}\,\mathbf{B}_{11} + \mathbf{A}_{22}\,\mathbf{B}_{21} = 0 \tag{A.72}$$

$$\mathbf{A}_{21}\,\mathbf{B}_{12} + \mathbf{A}_{22}\,\mathbf{B}_{22} = \mathbf{I} \tag{A.73}$$

Using these four equations and remembering that the inverse is defined only for a square matrix, we can solve for the four submatrices $\mathbf{B}_{ij}$. For example, the pre-multiplication of each term in Eq. A.71 by $\mathbf{A}_{11}^{-1}$ gives

$$\mathbf{B}_{12} = -\mathbf{A}_{11}^{-1}\,\mathbf{A}_{12}\mathbf{B}_{22} \tag{A.74}$$

Substituting this in Eq. A.73,

$$(\mathbf{A}_{22} - \mathbf{A}_{21}\mathbf{A}_{11}^{-1}\mathbf{A}_{12})\mathbf{B}_{22} = \mathrm{I} \tag{A.75}$$

The matrix in parenthesis is square and of order q, the same as $\mathbf{B}_{22}$ and $\mathbf{I}$ in Eq. A.73. Therefore, we can pre-multiply by its inverse and write

$$\mathbf{B}_{22} = (\mathbf{A}_{22} - \mathbf{A}_{21}\mathbf{A}_{11}^{-1}\mathbf{A}_{12})^{-1} \tag{A.76}$$

where

$$\mathbf{L} = (\mathbf{A}_{22} - \mathbf{A}_{21}\mathbf{A}_{11}^{-1}\mathbf{A}_{12}) \tag{A.77}$$

From Eq. A.74, $\qquad \mathbf{B}_{12} = -\mathbf{A}_{11}^{-1}\mathbf{A}_{12}\mathrm{L}^{-1}$ (A.78)

From Eqs. A.70, we have

$$\mathbf{B}_{11} = \mathbf{A}_{11}^{-1} - \mathbf{A}_{11}^{-1}\,\mathbf{A}_{12}\mathbf{B}_{21} \tag{A.79}$$

Again from Eq. A.72 and A.79 we can write

$$\mathbf{A}_{21}\mathbf{A}_{11}^{-1} - \mathbf{A}_{21}\mathbf{A}_{11}^{-1}\mathbf{A}_{12}\mathbf{B}_{21} + \mathbf{A}_{22}\mathbf{B}_{21} = 0 \quad \text{(A.80)}$$

or

$$\mathbf{A}_{21}\mathbf{A}_{11}^{-1} + (\mathbf{A}_{22}\mathbf{A}_{21}\mathbf{A}_{11}^{-1}\mathbf{A}_{12})\mathbf{B}_{21} = 0 \quad \text{(A.81)}$$

or

$$\mathbf{L}\mathbf{B}_{21} = -\mathbf{A}_{21}\mathbf{A}_{11}^{-1} \quad \text{(A.82)}$$

or

$$\mathbf{B}_{21} = -\mathbf{L}^{-1}\mathbf{A}_{21}\,\mathbf{A}_{11}^{-1} \quad \text{(A.83)}$$

The complete inverse is given by

$$\mathbf{A}^{-1} = \mathbf{B} = \left[\begin{array}{c|c} \mathbf{A}_{11}^{-1} - \mathbf{A}_{11}^{-1}\mathbf{A}_{12}\mathbf{B}_{21} & -\mathbf{A}_{11}^{-1}\mathbf{A}_{12}\mathbf{L}^{-1} \\ \hline -\mathbf{L}^{-1}\mathbf{A}_{21}\mathbf{A}_{11}^{-1} & \mathbf{L}^{-1} \end{array}\right] \quad \text{(A.84)}$$

The required steps are as follows:
Compute

1. $\mathbf{A}_{11}^{-1}$
2. $\mathbf{L} = \mathbf{A}_{22} - \mathbf{A}_{21}\mathbf{A}_{11}^{-1}\,\mathbf{A}_{12}$
3. $\mathbf{B}_{22} = \mathbf{L}^{-1}$
4. $\mathbf{B}_{21} = -\mathbf{L}^{-1}\,\mathbf{A}_{21}\mathbf{A}_{11}^{-1}$
5. $\mathbf{B}_{12} = -\,\mathbf{A}_{11}^{-1}\,\mathbf{A}_{12}\mathbf{L}^{-1}$
6. $\mathbf{B}_{11} = \mathbf{A}_{11}^{-1}\,\mathbf{A}_{11}^{-1}\,\mathbf{A}_{12}\,\mathbf{B}_{21}$ (A.85)

The above steps indicate that the inversion of a matrix of order n requires the inversion of two matrices $\mathbf{A}_{11}$ and $\mathbf{L}$ of order p and q respectively, where $p + q = n$.

Example A.13 | *Invert the following matrix by partitioning.*

$$A = \left[\begin{array}{cc|c} 6 & 1 & -5 \\ -2 & -5 & 4 \\ \hline -3 & 3 & -1 \end{array}\right]$$

We find,

$$\mathbf{A}_{11}^{-1} = \frac{1}{(-28)}\begin{bmatrix} -5 & -1 \\ 2 & 6 \end{bmatrix}$$

$$\mathbf{L} = [-1] - [-3 \;\; 3]\frac{1}{(-28)}\begin{bmatrix} -5 & -1 \\ 2 & 6 \end{bmatrix}\begin{bmatrix} -5 \\ 4 \end{bmatrix} = -\frac{49}{28}$$

$$\mathbf{B}_{22} = \mathbf{L}^{-1} = -\frac{28}{49} = -\frac{4}{7}$$

$$\mathbf{B}_{21} = \left(-\frac{28}{49}\right)[-3 \;\; 3]\frac{1}{(-28)}\begin{bmatrix} -5 & 1 \\ 2 & 6 \end{bmatrix} = -\frac{1}{7}[3 \;\; 3]$$

$$\mathbf{B}_{11} = -\left(\frac{1}{28}\right)\begin{bmatrix}-5 & 1\\ 2 & 6\end{bmatrix} + \frac{1}{28}\begin{bmatrix}-5 & 1\\ 2 & 6\end{bmatrix}\begin{bmatrix}-5\\ 4\end{bmatrix}\frac{1}{7}[-3\ \ -3] = -\frac{1}{7}\begin{bmatrix}1 & 2\\ 2 & 3\end{bmatrix}$$

$$\mathbf{B}_{12} = -\left(\frac{1}{28}\right)\begin{bmatrix}-5 & -1\\ 2 & 6\end{bmatrix}\begin{bmatrix}-5\\ 4\end{bmatrix}\left(-\frac{28}{29}\right) = -\frac{1}{7}\begin{bmatrix}3\\ 2\end{bmatrix}$$

Therefore, the inverse matrix is

$$\mathbf{A}^{-1} = \mathbf{B} = -\frac{1}{7}\begin{bmatrix}1 & 2 & 3\\ 2 & 3 & 2\\ 3 & 3 & 4\end{bmatrix}$$

A.5.4 Properties of the Inverse

There are many useful properties of the inverse. Some of the important ones are mentioned below.

1. The inverse of a non-singular square matrix is unique. If

$$\mathbf{AB} = \mathbf{I} \text{ and } \mathbf{AC} = \mathbf{I} \tag{A.86}$$

then $\mathbf{B} = \mathbf{C} = \mathbf{A}^{-1}$

2. The inverse of an inverse is the original matrix, that is

$$(\mathbf{A}^{-1})^{-1} = \mathbf{A} \tag{A.87}$$

3. The inverse of a product of matrices is equal to the product of their individual inverses in reverse order, that is,

$$[\mathbf{ABCD}]^{-1} = [\mathbf{D}]^{-1}[\mathbf{C}]^{-1}[\mathbf{B}]^{-1}[\mathbf{A}]^{-1} \tag{A.88}$$

4. The inverse of the transpose is equal to the transpose of the inverse, that is.

$$(\mathbf{A}^{T})^{-1} = (\mathbf{A}^{-1})^{-T} \tag{A89}$$

5. If K is non-zero scalar, than

$$(K\mathbf{A})^{-1} = \frac{1}{K}\mathbf{A}^{-1} \tag{A.89}$$

6. The inverse of a diagonal matrix is also a diagonal matrix.

$$\mathbf{A} = \begin{bmatrix}a_{11} & & & \\ & a_{22} & & \\ & & a_{33} & \\ & & & a_{nn}\end{bmatrix} \quad \mathbf{A}^{-1} = \begin{bmatrix}1/a_{11} & & & \\ & 1/a_{22} & & \\ & & 1/a_{33} & \\ & & & 1/a_{nn}\end{bmatrix} \tag{A.90}$$

7. The inverse of an upper (or lower) triangular matrix is also a upper (or lower) triangular matrix. The inverse exists only if all the diagonal elements of a triangular matrix are non-zero.

Problems for Practice

A.1 Find the work done by force **F** = 4**i** + 2**j** + 3**k** acting on a particle if the particle is displaced from point *A*(l, 2, 0) to point *B*(2, –1, 3).

A.2 Calculate the area of triangle *ABC* where *A*, *B*, and *C* are points (1, 2, 4), (3, 1, – 2) and (4, 3, 1) respectively. Also find the angles at *A*, *B* and *C*.

A.3 Show that vectors *A*(2, –1, 1), *B*(1, –3, –5) and *C*(3, –4, –4) form a right triangle.

A.4 Find the volume of a tetrahedron with **A**, **B**, **C** as adjacent edges where **A** = **i** + 2**k**, **B** = 4**i** + 6**j** + 2**k** and **C** = 3**i** + 3**j** – 6**k** with respect to right-handed cartesian coordinates.

A.5 Making use of the cross product give the unit vector *n* normal to the inclined surface *ABC* given in Fig. A.11.

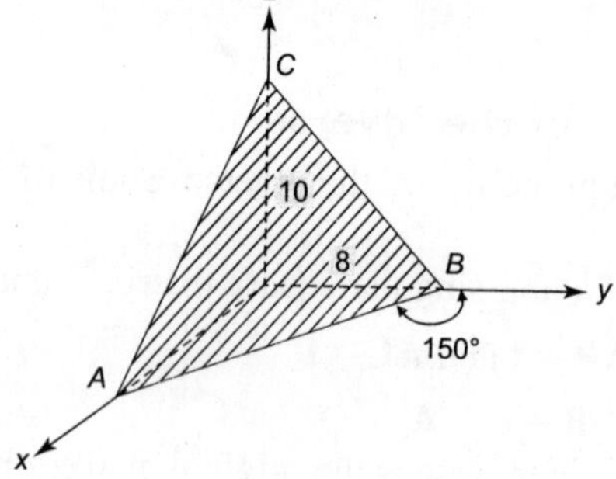

Fig. A.11

A.6 Evaluate the following determinants

(i) $\begin{vmatrix} 2 & 3 & 1 & -2 \\ 2 & 0 & -2 & -4 \\ -1 & 1 & 0 & 1 \\ 2 & 5 & 3 & 6 \end{vmatrix}$ (ii) $\begin{vmatrix} b^2+c^2 & a^2 & a^2 \\ b^2 & a^2+c^2 & b^2 \\ c^2 & c^2 & a^2+b^2 \end{vmatrix}$

A.7 Using the scalar triple product find the area projected on to the plane N from the surface *ABC* given in Fig. A.12. Plane N is infinite and is normal to vector **r** = 50**i** + 40**j** + 30**k**.

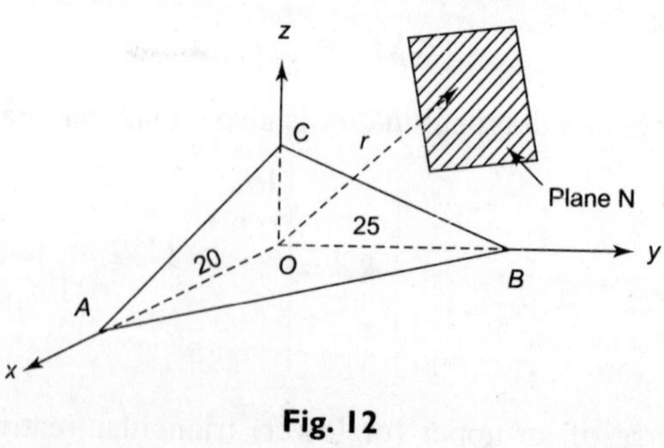

Fig. 12

A.8 A force **F** acts at position (3, 2, 0). It is in *xy* plane and is inclined at 30° from *x* axis with a sense directed away from the origin. What is the moment of this force about an axis going through the points (6, 2, 5) and (0, –2, 3)?

A.9 Find the moment of 1 kN force shown in Fig. A.13 about points A and B.

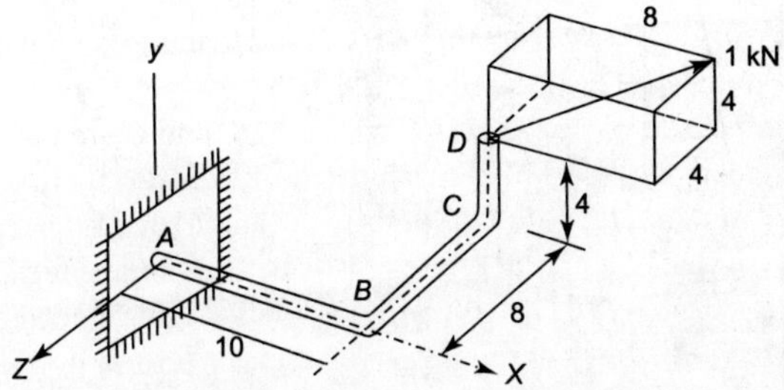

Fig. A.13

A.10 Prove the following relationship

(a) $(\mathbf{AB})^T = \mathbf{B}^T\mathbf{A}^T$; (b) $(\mathbf{A}^T)^T = \mathrm{A}$; (c) $(\mathbf{A} + \mathbf{B})^T = \mathbf{A}^T + \mathbf{B}^T$;

(d) $(\mathbf{AC})^{-1} = \mathbf{C}^{-1}\mathbf{A}^{-1}$; (e) $\mathbf{AC}(\mathbf{AC})^{-1} = \mathbf{I}$; (f) $(\mathbf{A}^2)^1 = (\mathbf{A}-^1)^2$

A.11 Given

$$\mathbf{A} = \begin{bmatrix} 3 & 2 & -4 \\ 1 & 3 & 2 \\ 2 & -1 & 2 \end{bmatrix} \text{ and } \mathbf{B} = \begin{bmatrix} 1 & -2 & 3 \\ 3 & 1 & 0 \\ 0 & 4 & 2 \end{bmatrix}$$

Compute $\mathbf{C} = \mathbf{AB}$ and $\mathbf{C} = \mathbf{BA}$

A.12 Find the product **AB** for the following square matrices.

$$\mathbf{A} = \begin{bmatrix} 4 & -4 & -8 \\ -2 & 6 & 8 \\ 2 & -4 & -6 \end{bmatrix} \quad \mathbf{B} = \begin{bmatrix} -2 & 4 & 8 \\ 2 & -4 & -8 \\ -2 & 4 & 8 \end{bmatrix}$$

A.13 Invert the following matrices solving at least one by both the adjoint and Gauss-Jordan elimination methods.

(a)

$$\mathbf{A} = \begin{bmatrix} 3 & -1 & 1 \\ -15 & 6 & -5 \\ 5 & -2 & 2 \end{bmatrix}$$

(b)

$$\mathbf{B} = \begin{bmatrix} -3 & 6 & -11 \\ 3 & -4 & 6 \\ 4 & -8 & 13 \end{bmatrix}$$

(c)

$$\mathbf{R} = \begin{bmatrix} \cos\theta & \sin\theta & 0 \\ -\sin\theta & \cos\theta & 0 \\ 0 & 0 & 1 \end{bmatrix}$$

A.14 Invert the following matrix by partitioning.

$$\begin{bmatrix} 2 & 1 & 3 & 1 \\ 1 & 4 & -2 & -1 \\ 3 & -1 & -4 & 3 \\ 1 & 2 & 3 & 1 \end{bmatrix}$$

Appendix B

Product Integrals $\int_0^L m_i m_j \, ds$

mi / mj	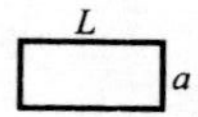	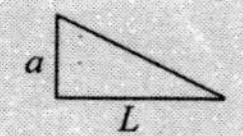	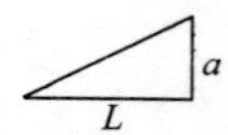	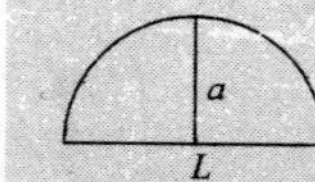	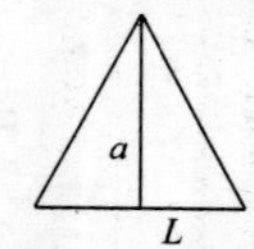	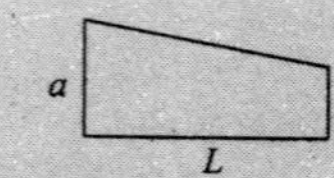
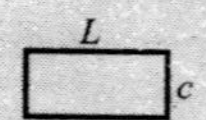	Lac	$\frac{1}{2}Lac$	$\frac{1}{2}Lac$	$\frac{2}{3}Lac$	$\frac{1}{2}Lac$	$\frac{1}{2}L(2a+b)c$
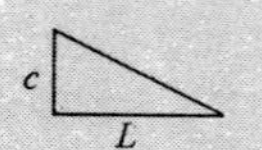	$\frac{1}{2}Lac$	$\frac{1}{3}Lac$	$\frac{1}{6}Lac$	$\frac{1}{3}Lac$	$\frac{1}{4}Lac$	$\frac{1}{6}L(2a+b)c$
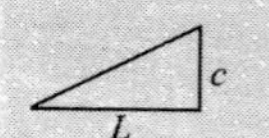	$\frac{1}{2}Lac$	$\frac{1}{6}Lac$	$\frac{1}{3}Lac$	$\frac{1}{3}Lac$	$\frac{1}{4}Lac$	$\frac{1}{6}L(a+2b)c$
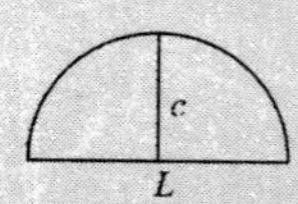	$\frac{2}{3}Lac$	$\frac{1}{3}Lac$	$\frac{1}{3}Lac$	$\frac{8}{15}Lac$	$\frac{5}{12}Lac$	$\frac{1}{3}L(a+b)c$
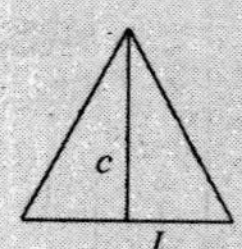	$\frac{1}{2}Lac$	$\frac{1}{4}Lac$	$\frac{1}{4}Lac$	$\frac{5}{12}Lac$	$\frac{1}{3}Lac$	$\frac{1}{4}L(a+b)c$
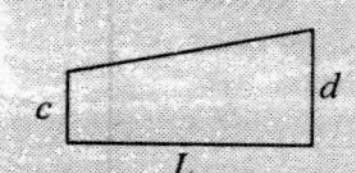	$\frac{1}{2}La(c+d)$	$\frac{1}{6}La(2c+d)$	$\frac{1}{6}La(c+d)$	$\frac{1}{3}La(c+d)$	$\frac{1}{4}La(c+d)$	$\frac{L}{6}[a(2c+d)+b(2d+c)]$

Appendix C Fixed End Moments in A Prismatic Beam

Type of Loading	Fixed End Moment FEM_{AB}	Fixed End Moment FEM_{BA}
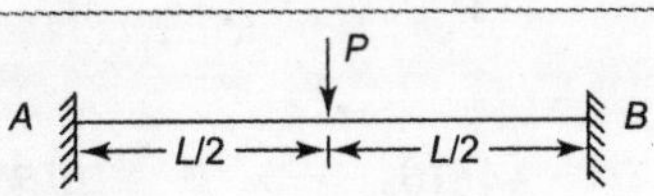	$\frac{+PL}{8}$	$\frac{-PL}{8}$
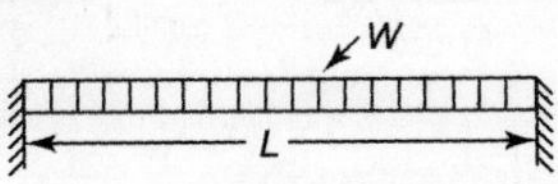	$\frac{+WL^2}{12}$	$\frac{-WL^2}{12}$
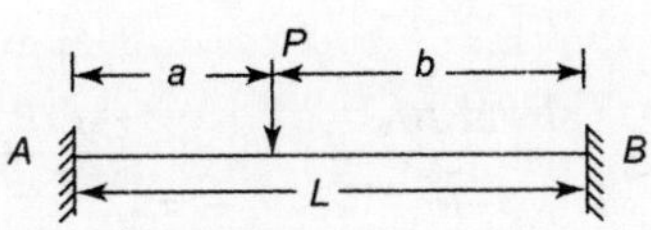	$\frac{+Pab^2}{L^2}$	$\frac{-Pa^2b}{L^2}$
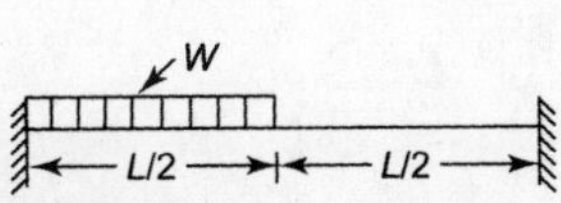	$\frac{+11}{192}WL^2$	$\frac{-5}{192}WL^2$
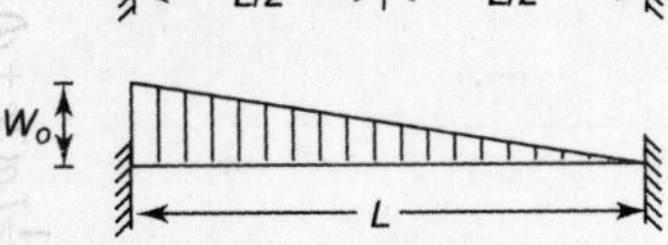	$\frac{+WoL^2}{20}$	$\frac{-WoL^2}{30}$
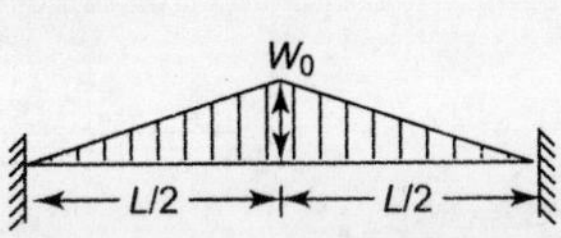	$\frac{+5}{96}WoL^2$	$\frac{-5}{96}WoL^2$
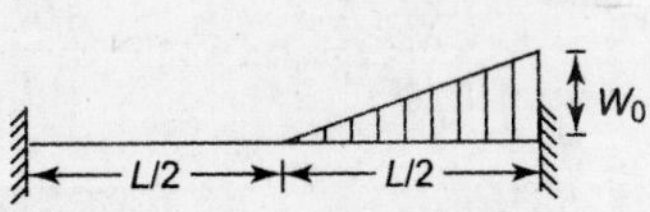	$\frac{+7}{960}WoL^2$	$\frac{-23}{960}WoL^2$
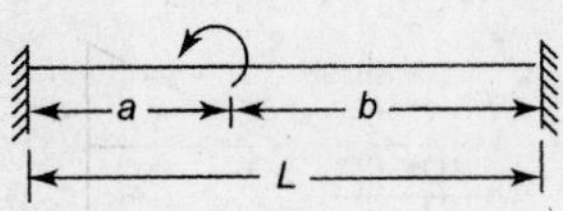	$\frac{+M}{L^2}b(b-2a)$	$\frac{+M}{L^2}a(2b-a)$

Appendix D Force Displacement Relationship in a Prismatic Member

Type of Displacement	End Moment	
	M_{AB}	M_{BA}
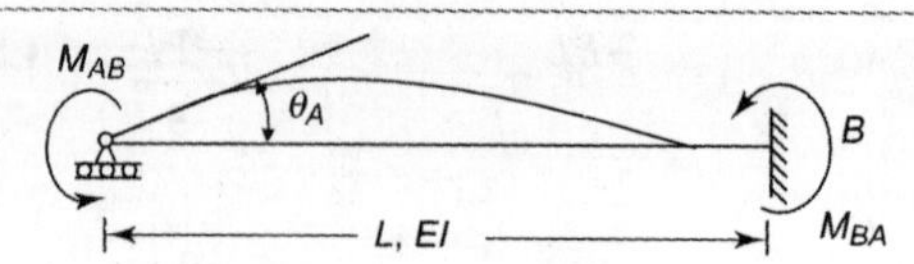	$\dfrac{+4EI\theta_A}{L}$	$\dfrac{+2EI\theta_A}{L}$
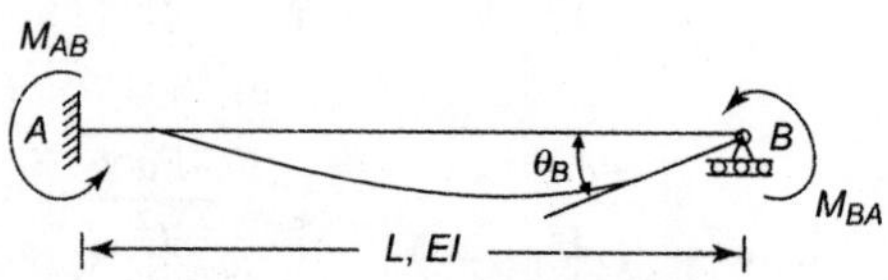	$\dfrac{+2EI\theta_B}{L}$	$\dfrac{+4EI\theta_B}{L}$
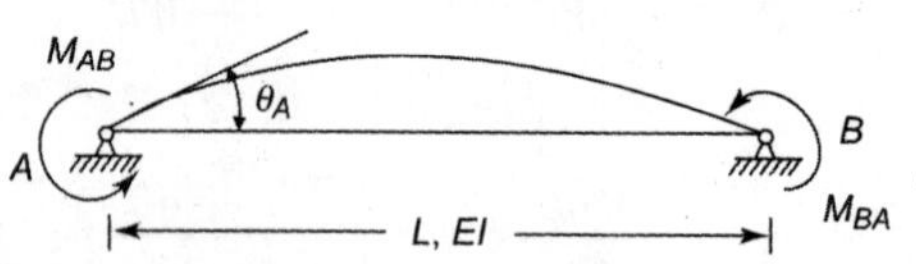	$\dfrac{+3EI\theta_A}{L}$	0
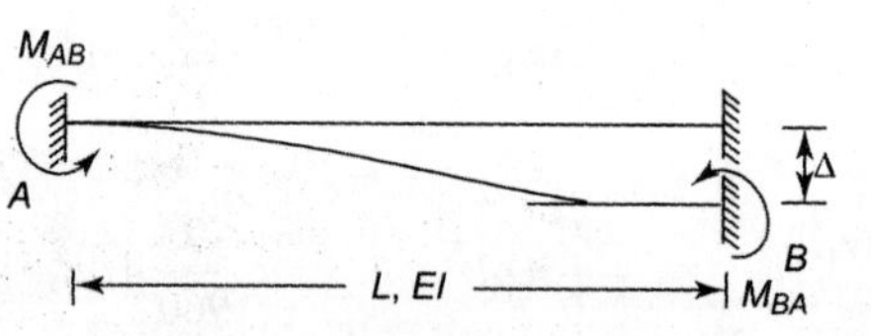	$\dfrac{+6EI\Delta}{L^2}$	$\dfrac{+6EI\Delta}{L^2}$
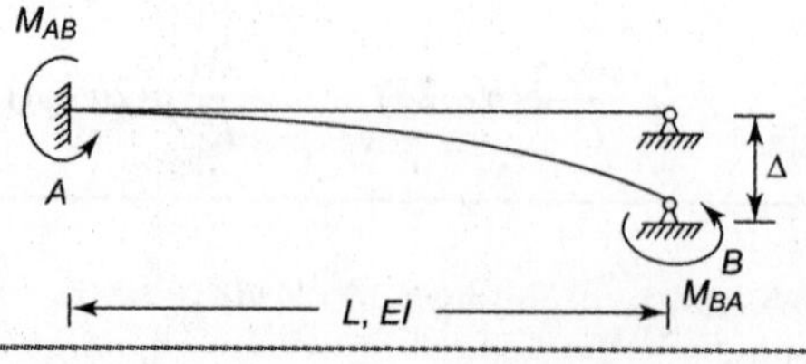	$\dfrac{+3EI\Delta}{L^2}$	0

Objective Type Questions

CHAPTER 2

2.1 On a free-body diagram,
(a) only external and internal forces are indicated
(b) external, internal as well as reaction components are shown
(c) only internal forces need be shown
(d) only reaction components and external forces are indicated.

2.2 The principle of super position is valid only if the material is
(a) elastic (b) stress-strain relationship is linear
(c) plastic (d) elasto-plastic

2.3 In a plane structure, the equilibrium equations are
(a) $\Sigma F_x = 0, \Sigma F_y = 0$ (b) $\Sigma F_x = 0, \Sigma F_y = 0, \Sigma F_z = 0$
(c) $\Sigma M_x = 0, \Sigma My = 0, \Sigma M_z = 0$ (d) $\Sigma F_x = 0, \Sigma F_y = 0, \Sigma M_z = 0$

2.4 A closed funicular polygon of forces acting on a body indicates
(a) $\Sigma F_x = 0, \Sigma F_y = 0$ and $\Sigma M_z = 0$
(b) only $\Sigma F_x = 0$, and $\Sigma F_y = 0$
(c) the body dies not rotate
(d) no relevance with regard to forces

2.5 In a cable stretched between two level supports the horizontal tension in the cable is
(a) same through out (b) maximum at the supporting towers
(c) maximum at the centre of cable
(d) not predictable

2.6 In a cable subjected to v.d.l. the tension in the cable is
(a) maximum at the centre
(b) uniform through out
(c) maximum at the supporting towers
(d) not predictable

2.7 A three-hinged arch is
(a) statically indeterminate because of central hinge
(b) determinate if the springings are at the same level
(c) statically determinate
(d) statically determinate or indeterminate depending upon loading

2.8 Arches are not subjected to bending moment if the
(a) arch is parabolic and symmetrically loaded
(b) arch is parabolic and springings are at the same level

(c) arch is parabolic and subjected to v.d.l all through
(d) thrust line does not coincide with the arch axis

2.9 If a cable stretched between two towers at the same level has span l and rise h, the horizontal tension in the cable due to v.d.l. w/unit length is
(a) $\frac{wl^2}{8h}$ (b) $\frac{wl^2}{12h}$ (c) $\frac{wl^2}{16h}$ (d) $\frac{wl^2}{4h}$

2.10 Masonry arches resist the external loads by
(a) bending moment and radial shear
(b) thrust only
(c) normal thrust, radial shear and moment
(d) normal thrust and shear

2.11 Line of thrust in an arch follows the
(a) line joining the hinge points (b) Funicular polygon of forces
(c) axis of the arch (d) line joining the springings

CHAPTER 3

3.1 In a truss, the joints are considered to be
(a) rigid (b) pin joints (c) bolted joints (d) welded

3.2 A panel in a truss means the space between any two
(a) members (b) joints
(c) lower chord joints (d) lower chord and upper chord joints

3.3 A member in a truss is subjected to only
(a) axial tension (b) axial compression
(c) axial compression or tension (d) axial force and moment

3.4 Tension coefficient of a member indicates
(a) tension in the member (b) force in a member
(c) force per unit displacement (d) force per unit length of member

3.5 The tension coefficient method is based on only
(a) equilibrium condition (b) method of joints and equilibrium
(c) method of joints and sections (d) method of sections and equilibrium

3.6 The pitch of a truss is the ratio of
(a) height to half span (b) height to span
(c) height of truss to panel length (d) average height of truss to span

3.7 If $2j > m + r$ in a plane truss, the truss is
(a) redundant (b) determinate (c) stable (d) unstable

3.8 If $2j < m + r$ in a plane truss, the truss is
(a) redundant (b) determinate (c) stable (d) unstable

3.9 The forces in collinear members at a joint of a plane truss are equal if
(a) the joint is not loaded
(b) joint is not loaded and has only three members
(c) loaded
(d) loaded and has only three members

3.10 A truss and the loading is shown in Fig.3.1. The force in member 2 – 4 is
(a) 8.66 KN tension (b) 8.66 KN compression
(c) 22.5 KN compression (d) 22.5 KN tension

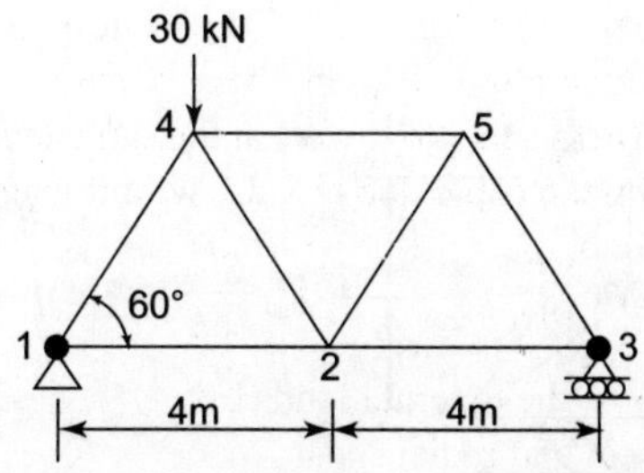

Fig : 3.1

3.11 The force in member 2.5 in the truss in Fig.3.1 is
(a) 22.5 KN compression (b) 7.5 KN tension
(c) 8.66 KN compression (d) 8.66 KN tension

3.12 The force in member 4 – 5 in the above truss is
(a) 8.66 KN compression (b) 7.5 KN tension
(c) 7.5 KN compression (d) 8.66 KN tension

CHAPTER 4

4.1 In a space truss if $3j < m + r$, the truss is
(a) unstable (b) stable (c) determinate (d) indeterminate

4.2 In a space truss if $3j > m + r$, the truss is
(a) unstable (b) stable (c) determinate (d) indeterminate

4.3 The force in the non-coplanar member at the joint of a space truss is zero if the joint is
(a) not loaded (b) not loaded and has only three members
(c) loaded (d) loaded and has only four members

4.4) If all the members, except one, are coplanar at the joint of a truss, the force in the force in the non-coplanar member is zero if the
(a) joint is not loaded
(b) joint loaded in the coplanar
(c) joint load normal to the coplanar
(d) joint load not normal to the coplanar

4.5 The method of tension coefficients was developed by
(a) clapeyron (b) hardy cross (c) williot molr (d) south well

4.6 At a joint in a space truss if all the bar forces, except two, are zero, the two bars also have no bar forces if
(a) joint is loaded
(b) the two bars collinear
(c) joint is not loaded
(d) joint is not loaded and bars are not collinear

4.7 A space truss and the loading on it is given in Fig.4.1. The coordinates of the joints are indicated. The equilibrium equations using tension coefficients are given. Indicate right or wrong against each

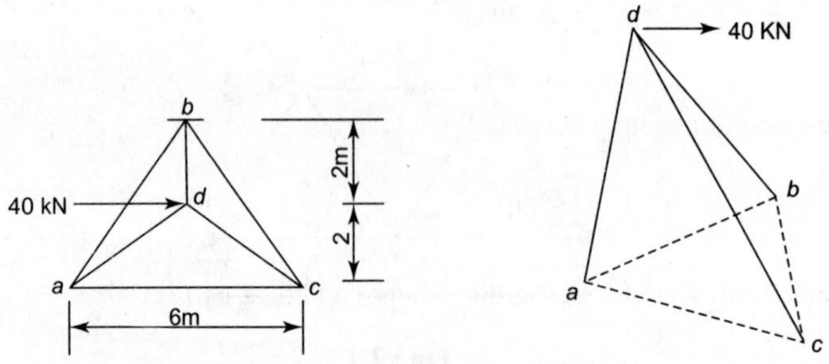

Coordinates: $a(-3, -6, 2)$, $b(0, -6, -2)$, $c(3, -6, 2)$, $d(0, 0, 0)$

Fig. 4.1

(a) $t_{da}(-3) + t_{db}(0) + t_{dc}(3) + 40 = 0$ right/wrong
(b) $t_{da}(-4) + t_{db}(-6) + t_{dc}(-6) + 0 = 0$ right/wrong
(c) $t_{da}(2) + t_{db}(-2) + t_{dc}(2) + 0 = 0$ right/wrong

CHAPTER – 5

5.1 The moment area method is valid for
(a) single span beams (b) continuous beams
(c) frames (d) no limitations

5.2 At the free end of a beam
(a) the slope and deflection are zero(b) the moment and shear are zero
(c) the moment is zero but not shear(d) the shear is zero but not moment

5.3 Moment area method yields
(a) only deflection at a section (b) only slope at a section
(c) slopes and deflections (d) elastic curve

5.4 The slope of the cantilever beam of space l at the free end due to load concentrated load at the free end is

(a) $\dfrac{Pl^2}{EI}$ (b) $\dfrac{Pl^2}{2EI}$ (c) $\dfrac{Pl^2}{3EI}$ (d) $\dfrac{Pl^2}{4EI}$

5.5 The slopes at the ends of a simply supported beam of span l under a u.d.l w/unit length is

(a) $\dfrac{wl^3}{8EI}$ (b) $\dfrac{wl^3}{16EI}$ (c) $\dfrac{wl^3}{24EI}$ (d) $\dfrac{wl^3}{48EI}$

5.6 The ratio of deflection at centre of a fixed beam and a simply supported beam under a concentrated load p at the centre of the space is
(a) 0.2 (b) 0.25 (c) 0.5 (d) 0.75

5.7 The ratio of deflections of the beam above under a u.d.l is
(a) 0.1 (b) 0.2 (c) 0.25 (d) 0.5

5.8 In a cantilever beam, a moment M applied at the free end yields slope at the free end

(a) $\dfrac{Ml}{4EI}$ (b) $\dfrac{Ml}{3EI}$ (c) $\dfrac{Ml}{2EI}$ (d) $\dfrac{Ml}{EI}$

5.9 The deflection of a cantilever beam at the free end due to a moment M applied at that end is

(a) $\frac{Ml^2}{EI}$ (b) $\frac{Ml^2}{2EI}$ (c) $\frac{Ml^2}{3EI}$ (d) $\frac{Ml^2}{4EI}$

5.10 The slope of a simply supported beam at the end under a moment M at that end is

(a) $\frac{Ml}{2EI}$ (b) $\frac{Ml}{3EI}$ (c) $\frac{Ml}{4EI}$ (d) $\frac{Ml}{6EI}$

5.11 The deflection centre of a simply supported beam under a moment M at the end is

(a) $\frac{Ml^2}{8EI}$ (b) $\frac{Ml^2}{16EI}$ (c) $\frac{Ml^2}{24EI}$ (d) $\frac{Ml^2}{48EI}$

5.12 In a conjugate beam the free end of a real beam will have

(a) a free end (b) a fixed end (c) hinged end (d) none of the above

5.13 In a conjugate beam the intermediate hinge in the original beam will be

(a) a hinge in the conjugate beam also
(b) an intermediate support
(c) with a hinge but with a moment applied
(d) moment without hinge

5.14 The deflection at the free end of a overhanging beam under load p at centre of span is

(a) $\frac{Pla^2}{16EI}$ (b) $\frac{Pl^2a}{16EI}$ (c) $\frac{Pa^3l}{16EI}$ (d) $\frac{Pl^3a}{16EI}$

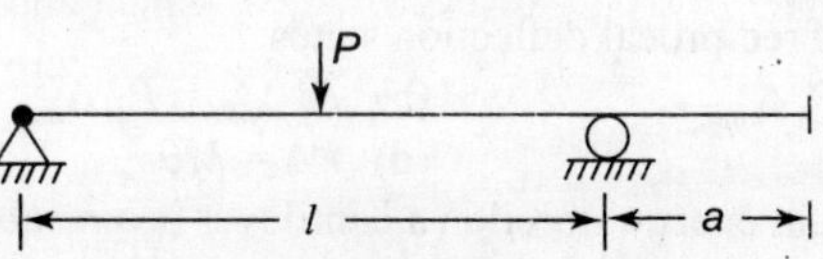

Fig. 4.2

5.15 The slope at the free end in the above beam is

(a) $\frac{Pl^2a}{8EI}$ (b) $\frac{Pl^2a}{16EI}$ (c) $\frac{Pl^2}{16EI}$ (d) $\frac{Pl^2}{24EI}$

CHAPTER – 6

6.1 The strain energy stored in a circular rod of length l and axial rigidity AE due to a pull P is

(a) $\frac{P^2l}{AE}$ (b) $\frac{P^2l}{2AE}$ (c) $\frac{P^2l}{3AE}$ (d) $\frac{P^2l}{4AE}$

6.2 Two uniform steel rods A and B of lengths l and $2l$ having diameters d and $2d$ are subjected to tensile force P and $2P$ respectively, which of the following statement is correct

(a) elongation in A is twice the elongation in B
(b) strain energy in A is half the strain energy in B
(c) strain energy in A and B are same
(d) strain energy in A is twice the strain energy in B

6.3 In a cantilever beam of space l and flexural rigidity EI the total strain energy under a concentrated W is
(a) $\frac{W^2l^3}{6EI}$ (b) $\frac{W^2l^4}{3EI}$ (c) $\frac{W^2l^4}{8EI}$ (d) $\frac{W^2l^4}{6EI}$

6.4 Two beams, one of simply supported of space l and the other a cantilever of space $l/2$, are subjected to a concentrated P at centre of span of simply supported and at free end in a cantilever beam. Choose the correct statement in the following :
(a) The strain energy stored in both the beams is same
(b) The strain energy in S.S. beam is twice the strain energy in cantilever beam
(c) The strain energy in S.S. beam is half the strain energy in cantilever beam
(d) The strain energy in cantilever is four times the strain energy is S.S. beam

6.5 The shear strain energy stored in a rectangular beam of length l under shear force V and shear rigidity GA is equal to
(a) $\frac{V^2l}{2GA}$ (b) $\frac{1.2V^2l}{2GA}$ (c) $\frac{V^2l}{GA}$ (d) $\frac{1.5V^2l}{2GA}$

6.6 The elastic strain energy stored in a beam of flexural rigidity EI and length l subjected to pure moment M is
(a) $\frac{Ml}{EI}$ (b) $\frac{Ml^2}{2EI}$ (c) $\frac{M^2l}{2EI}$ (d) $\frac{M^2l}{4EI}$

6.7 Maxwell-Bettis theorem states that
(a) $P_m\Delta_{mn} = P_n\Delta_{nm}$ (b) $P_n\Delta_{mn} = P_m\delta_{nm}$
(c) $\Delta_{mn} = \Delta_{nm}$ (d) $P\Delta = M\theta$

6.8 Maxwell law of reciprocal deflection states
(a) $P_m\Delta_{mn} = P_n\Delta_{nm}$ (b) $P_n\Delta_{mn} = P_m\Delta_{nm}$
(c) $\Delta_{mn} = \Delta_{nm}$ (d) $P\Delta = M\theta$

6.9 The ratio of strain energy stored in a cantilever beam subjected to a concentrated load at the free end and at centre of span is
(a) 2 (b) 2 (c) 4 (d) 8

6.10 The displacements caused by a load P acting at points m and n are shown in Fig.6.1. As per Maxwell's reciprocal deflections

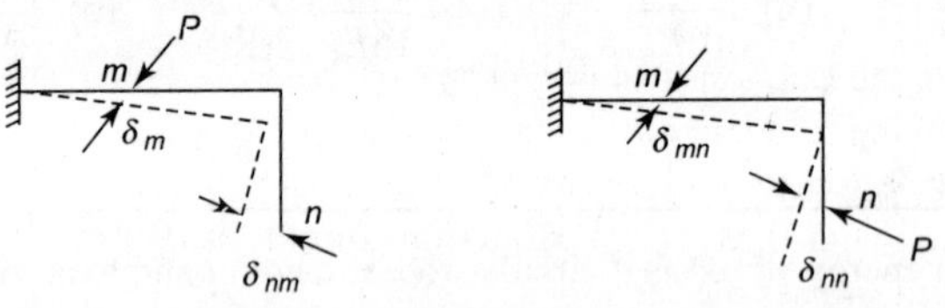

Fig. 6.1

(a) $\delta_{mm} = \delta_{nm}$ (b) $\delta_{nn} = \delta_{mn}$ (c) $\delta_{nm} = \delta_{nn}$ (d) $\delta_{nm} = \delta_{mn}$

6.11 The displacements caused in a beam subjected to a unit load and a unit couple are shown in Fig. 6.2. Then the as per Maxwell-Bettis theorem the relationship is

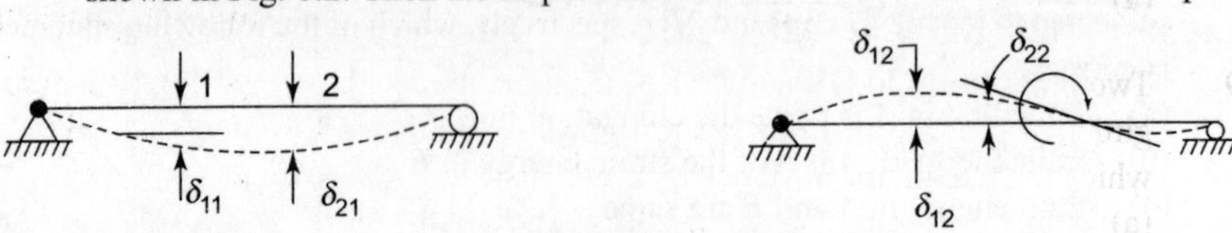

Fig. 6.2

(a) $\delta_{12} = \delta_{21}$ (b) $\delta_{11} = \delta_{22}$ (c) $\delta_{11} = \delta_{12}$ (d) $\delta_{21} = \delta_{22}$

CHAPTER – 7

7.1 The influence diagram for maximum bending moment in a simply supported beam is

(a) rectangular (b) triangular (c) parabolic (d) irregular

7.2 The influence line for structural function is used for obtaining maximum value due to

(a) single point load only (b) uniformly distributed load only
(c) several point loads (d) all the above

7.3 The ordinate of I.L.D for moment at $\frac{1}{4}$ span of a simply supported beam is $\frac{3}{16}l$. If a u.d.l., w/unit length occupies the whole span the moment at $\frac{1}{4}$ span is

(a) $\frac{wl^2}{16}$ (b) $\frac{wl^2}{32}$ (c) $\frac{3wl^2}{32}$ (d) $\frac{3wl^2}{16}$

7.4 The maximum B.M. at a section under rolling u.d.l. shorter than span occurs when the moving load

(a) just reaches the section
(b) just leaves the section
(c) load occupies centre of span
(d) when the section divides the span and load in the same ratio

7.5 The maximum +ve shear force at a section under a moving u.d.l shorter than the span occurs when the load

(a) just reaches the section
(b) the tail end just leaves the section
(c) load occupies centre of span
(d) section divides the span and the load in the same ratio

7.6 The maximum B.M. under a particular load among several moving loads occurs when that load is

(a) at centre of span
(b) when the load and the resultant of loads are at equidistant from the middle of span
(c) when the heaviest load is at centre of span
(d) when the resultant of the loads is at centre of span

7.7 A u.d.l of length 10 m passes over a simple beam of span 25 m from left to right. For a maximum B.M. at a section 10 m from left support the part length of the load that must pass the section is

(a) 4 m (b) 5 m (c) 6 m (d) 7 m

7.8 A u.d.l., w/unit length covering a length $l/2$ passes over a simply supported beam of span l. The maximum B.M. at the left quarter span is equal to

(a) $\frac{9wl^2}{128}$ (b) $\frac{7wl^2}{128}$ (c) $\frac{5wl^2}{128}$ (d) $\frac{3wl^2}{128}$

7.9 Two point loads 30 KN and 20 KN separated by 5 m crosses a simply supported girder of span 16 m from left to right the 30 KN load leading. The section at which the maximum B.M. occurs is at a distance from left hand support

(a) 8 m (b) 9 m (c) 10 m (d) 11 m

7.10 The area of the I.L.D. for the reaction of a simply supported beam of span l is
(a) $l/8$ (b) $l/2$ (c) $l/6$ (d) $l/4$

7.11 A simply supported beam is traversed by two concentrated loads 80 kN and 40 kN at 3 m apart with 80 kN load leading. The maximum moment under 80 kN load occurs when the load is at
(a) mid span (b) 1 m from centre
(c) 1.5 m from centre (d) 3 m from centre

7.12 Muller-Breslan's principal for influence line is applicable for only
(a) simple beams (b) continuous beams
(c) trusses and frames (d) all the above

7.13 In a three-hinged arch the B.M. caused by horizontal thrust at the hinge point is equal to
(a) $H.l/4$ (b) Zero (c) $H.l/8$ (d) $H.l/6$

7.14 The I.L.D for a force in a truss member is shown in Fig.7.1. If a u.d.l. of intensity 10 kN/m longer than span traverses the maximum tension in the member is

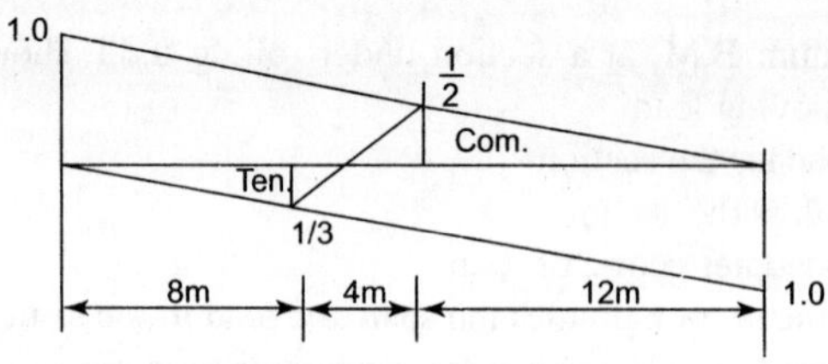

Fig. 7.1

(a) 20 kN (b) 16 kN (c) 12 kN (d) 8 kN

7.15 The I.L.D. for force in a truss member is given in Fig. 7.1. If a u.d. load of 10 kN/m longer than the span passes over the maximum compression in the member is
(a) 20 kN (b) 24 kN (c) 36 kN (d) 48 kN

CHAPTER – 8

8.1 Under a uniformly distributed load the cable takes the shape of a
(a) parabola (b) circular (c) catenary (d) funicular polygon

8.2 The length of a parabolic cable of span l and dip d is
(a) $l+\frac{d^2}{3l}$ (b) $l+\frac{4}{3}\frac{d^2}{l}$ (c) $l+\frac{8}{3}\frac{d^2}{l}$ (d) $l+\frac{16}{3}\frac{d^2}{l}$

8.3 A cable of span l and dip d is subjected to a v.d.l., w/unit length along the span. The horizontal tension H in the cable is
(a) $\frac{wl^2}{4d}$ (b) $\frac{wl^2}{8d}$ (c) $\frac{wl^2}{12d}$ (d) $\frac{wl^2}{16d}$

8.4 A cable of a suspension bridge of span 100 m is hung from towers which are 10 m and 5 m respectively above the lowest point of the cable. The ratio of the horizontal length of the cable from the higher and lower towers to the lowest point of the cable $\frac{l_1}{l_2}$ is
(a) $\left(\frac{10}{5}\right)^{\frac{1}{2}}$ (b) $\left(\frac{5}{10}\right)^{\frac{1}{2}}$ (c) $\left(\frac{10}{5}\right)$ (d) $\left(\frac{5}{10}\right)^2$

8.5 The horizontal length of the cable from the higher end to the lowest point of the cable is
(a) 50 m (b) 55 m (c) 57.5 m (d) 58.58 m

8.6 The total length of the above cable from end to end is
(a) 104 m (b) 103 m (c) 102 m (d) 101 m

8.7 In a three-hinged stiffening girder, the maximum +ve moment due to a rolling concentrated load occurs at a section distance x from one end. The value of x is
(a) $0.211\,l$ (b) $0.25\,l$ (c) $0.234\,l$ (d) $\frac{3}{8}l$

8.8 The maximum – ve moment due to a concentrated moving load on a three-hinged stiffening girder occurs at a section the distance of which from one end is
(a) $0.211\,l$ (b) $0.25\,l$ (c) $0.234\,l$ (d) $\frac{3}{8}l$

8.9 In a symmetrical three-hinged stiffening girder the influence line ordinate for H at centre of span is
(a) $\frac{l}{16yc}$ (b) $\frac{l}{12yc}$ (c) $\frac{l}{8yc}$ (d) $\frac{l}{4yc}$

8.10 In a two-hinged stiffening girder the maximum +ve or –ve moment due to a moving concentrated load W is
(a) $\frac{Wl}{4}$ (b) $\frac{Wl}{8}$ (c) $\frac{Wl}{16}$ (d) $\frac{Wl}{24}$

8.11 The maximum +ve or –ve moment in a two-hinged stiffening girder due to a v.d.l. longer than the span crosses over is
(a) $\frac{Wl^2}{32}$ (b) $\frac{Wl^2}{16}$ (c) $\frac{Wl^2}{12}$ (d) $\frac{Wl^2}{8}$

8.12 A suspension cable having 50 m span and dip 4 m is stiffened by a three-hinged stiffening girder. A concentrated load 100 kN is placed at 8 m from the left end. The equivalent v.d.l. we on the cable is
(a) 2.0 kN/m (b) 1.28 kN/m (c) 6.25 kN/m (d) 12.5 kN/m

8.13 A cable of a suspension bridge has a span 40 m and dip 5 m. A three-hinged stiffening girder is used. The maximum + ve moment at a section 10 m from left end as a 50 kN load rolls over is
(a) 312.5 kN.m (b) 300 kN.m (c) 187.5 kN.m (d) 625.0 kN.m

CHAPTER 9

9.1 A redundant truss shown in Fig. 9.1 is analysed using approximation that the compression diagonal members do not carry any force. Then the force in member 2-7 will be

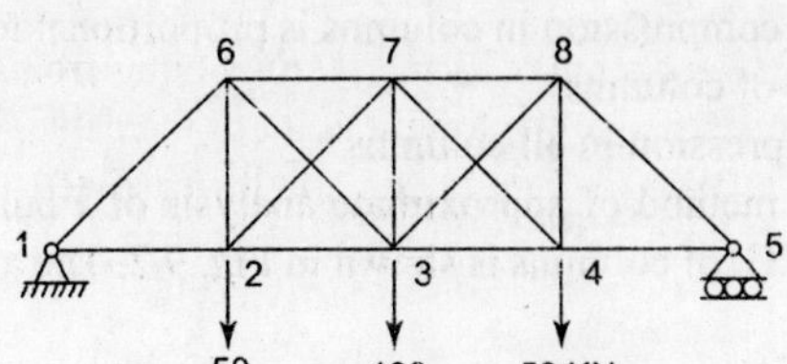

Fig. 9.1

(a) 0 (b) $50\sqrt{2}$ KN (c) 100 KN (d) 150 KN

9.2 The force in member 3-c in the above truss is

(a) 100 KN (b) 50 KN (c) $50\sqrt{2}$ KN (d) 25 KN

9.3 The truss in Fig. 9.1 is analysed using the approximation that the shear in panel 2-3 will be shared equally by the diagonal members the force in member 2-7 is

(a) $25\sqrt{2}$ compression (b) $25\sqrt{2}$ tension

(c) $50\sqrt{2}$ compression (d) $50\sqrt{2}$ tension

9.4 The force in member 6-3 in the above truss is

(a) $50\sqrt{2}$ compression (b) $50\sqrt{2}$ tension

(c) $25\sqrt{2}$ compression (d) $25\sqrt{2}$ tension

9.5 A portal frame of span 6 m and height 4 m is fixed at the base. It is subjected to a v.d.l. of 10 KN/m over the entire span. Assuming the hinge points at 0.2 l from the ends on the beam and hinges at 1/3 height of columns, the moment at the top of columns is

(a) 30 kN.m (b) 28.8 kN.m (c) 4.5 kN.m (d) 22.5 kN.m

9.6 In the portal frame above the moment at the foot of the columns is

(a) 45 kN.m (b) 30 kN.m (c) 28.8 kN.m (d) 14.4 kN.m

9.7 In the approximate analysis of building frames under lateral loads the points of contra flexure in beams and columns are assumed at

(a) $\frac{l}{2}$ for beams $\frac{2}{3}h$ from base for columns

(b) $\frac{1}{10}l$ for beams and $\frac{h}{2}$ for columns

(c) $\frac{l}{2}$ for beams $\frac{1}{3}h$ for columns

(d) $\frac{l}{2}$ for beams $\frac{h}{2}$ for columns

9.8 In the approximate analysis of building frames using portal method the shear in the frame is shared

(a) equally by all the columns

(b) the interior columns take twice as much as the exterior columns

(c) only interior columns take the shear

(d) the exterior columns take twice as much as the interior columns.

9.9 In the approximate analysis of building frames using cantilenar method, the axial force in the columns is assumed to the

(a) equal tension in all columns

(b) tension in windward columns and compression in leeward columns

(c) tension or compression in columns is proportional to the distance of column from C.G. of columns

(d) equal compression in all columns

9.10 In a cantilenar method of approximate analysis of a building frame the column lines and the C.G. of columns is shown in Fig. 9.2. The axial force in column 1-4 is

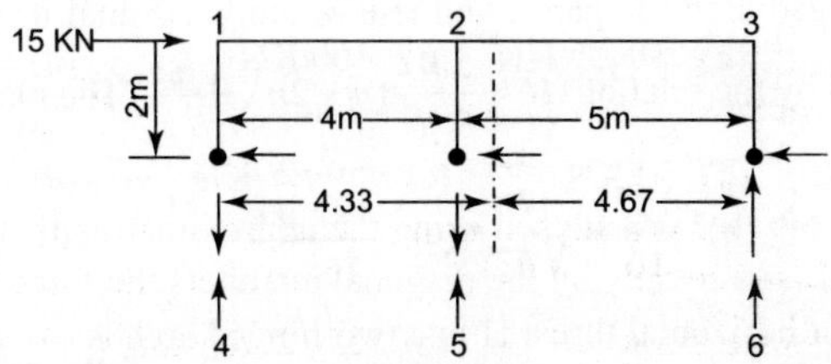

Fig. 9.2

(a) 3.20 kN tension (b) 3.20 compression
(c) 3.45 compression (d) 0.24 kN tension

9.11 The force in column 3-6 in the above building frame is
(a) 3.20 kN compression (b) 3.45 tension
(c) 3.45 compression (d) 0.24 kN compression

CHAPTER – 10

10.1 A fixed beam loaded transversely is statically indeterminate by
(a) 3 degrees (b) 2 degrees (c) 1 degree (d) no indeterminacy

10.2 A portal frame fixed at the base is indeterminate by
(a) 3 degrees (b) 2 degrees (c) 1 degree (d) no indeterminacy

10.3 A three span continuous beam fixed at one end and hinged at the other end is statically indeterminate by
(a) 4 degrees (b) 3 degrees (c) 2 degrees (d) 1 degree

10.4 A two-storey single bay portal frame fixed at the base having a hinge in each of the beams is indeterminate by
(a) 4 degrees (b) 3 degrees (c) 2 degrees (d) 1 degree

10.5 In a portal frame, one column is fixed at the base and the other a hinge on rollers. If there is a hinge in the beam the frame is
(a) statically indeterminate by one degree
(b) statically determinate
(c) statically determinate and stable
(d) unstable

10.6 Consistent displacements or compatibility condition means
(a) displacements caused by the redundant forces
(b) displacements caused by the forces other than redundant forces
(c) displacements caused by redundant and applied forces
(d) displacements caused by redundant and applied forces satisfying the boundary conditions

10.7 The theorem of three moments cannot be applied to
(a) single span fixed beams
(b) continuous beams with over hangs
(c) trusses and frames
(d) continuous beams with sinking supports and rotating joints

10.8 In a two-hinged arch, the I.L. ordinate at centre of span for horizontal thrust H is

(a) $\frac{25}{128}\frac{l}{h}$ (b) $\frac{15}{128}\frac{l}{h}$ (c) $\frac{1}{16}\frac{l}{h}$ (d) $\frac{1}{8}\frac{l}{h}$

10.9 In a two-hinged arch of span l, and rise h, the horizontal thrust under a rolling load is given by the relation $H = \frac{5}{8}\frac{Wl}{h}(n - 2n^3 + n^4)$. The maximum value of H is

(a) $\frac{125}{16}\frac{Wl}{h^2}$ (b) $\frac{125}{16}W\sqrt{\frac{l}{h}}$ (c) $\frac{125}{16}\frac{Wl}{h}$ (d) $\frac{Wl}{16h}$

10.10 The I.L.D. for horizontal thrust H in a two-hinged arch is
(a) parabola (b) tringle
(c) curve of third degree (d) curve of fourth degree

10.11 The radial shear in a parabolic two-hinged arch when the load is at the crown is
(a) $V_r = V_A + \cos\theta + H\sin\theta$
(b) $V_r = V_A - \cos\theta - H\sin\theta$
(c) $V_r = V_A + \sin\theta + H\cos\theta$
(d) $V_r = V_A - \sin\theta - H\cos\theta$

10.12 Muller-Breslau principal can be utilized for
(a) drawing influence lines for force quantities qualitatively
(b) drawing influence lines for forces quantitatively
(c) drawing influence lines for slopes and deflections
(d) drawing influence lines for cables and arches

CHAPTER 11

11.1 The propped end of a cantilever beam of span l settles by an amount δ, the rotation of the propped end is

(a) $\frac{\delta}{l}$ (b) $\frac{1.5\delta}{l}$ (c) $\frac{2\delta}{l}$ (d) zero

11.2 The propped end of a cantilever beam of span l settles by an amount δ without rotation the moment at the fixed end is

(a) $\frac{3EI\delta}{l^2}$ (b) $\frac{4EI\delta}{l^2}$ (c) $\frac{6EI\delta}{l^2}$ (d) $\frac{8EI\delta}{l^2}$

11.3 A moment M is applied at the propped end of a cantilever beam of span l and flexural rigidity EI. The moment at the fixed end will be
(a) 2 M (b) M (c) $M/2$ (d) $M/3$

11.4 In a fixed beam AB of span l, one of the supports undergoes a rotation θ then the shear in beam is

(a) $\frac{EI\theta}{l^2}$ (b) $\frac{2EI\theta}{l^2}$ (c) $\frac{4EI\theta}{l^2}$ (d) $\frac{6EI\theta}{l^2}$

11.5 A fixed beam AB of span l is subjected to a moment M at the centre of span. The fixed end moments are
(a) M (b) $M/2$ (c) $M/4$ (d) $M/8$

11.6 In a fixed beam AB of span l under a v.d.l w/unit length, the support B rotates by an amount θ. The moment at end A is

(a) $\left(\frac{wl^2}{12} + \frac{2EI\theta}{l}\right)$ (b) $\left(\frac{wl^2}{12} + \frac{4EI\theta}{l}\right)$

(c) $\left(\frac{wl^2}{12} - \frac{2EI\theta}{l}\right)$ (d) $\left(\frac{wl^2}{12} - \frac{4EI\theta}{l}\right)$

11.7 In the beam above the moment at end B is

(a) $\left(\frac{wl^2}{12}+\frac{2EI\theta}{l}\right)$ (b) $\left(-\frac{wl^2}{12}+\frac{4EI\theta}{l}\right)$

(c) $\left(\frac{wl^2}{12}+\frac{2EI\theta}{l}\right)$ (d) $\left(\frac{wl^2}{12}+\frac{4EI\theta}{l}\right)$

11.8 In a fixed beam AB of span l, the support B undergoes a unit rotation and a unit translation. The moment at end B is

(a) $\left(\frac{2EI}{l}+\frac{6EI}{l^2}\right)$ (b) $\left(\frac{4EI}{l}+\frac{6EI}{l^2}\right)$

(c) $\left(\frac{2EI}{l}-\frac{6EI}{l^2}\right)$ (d) $\left(-\frac{4EI}{l}+\frac{6EI}{l^2}\right)$

11.9 In the above beam the beam at end A is

(a) $\left(\frac{4EI}{l}+\frac{6EI}{l^2}\right)$ (b) $\left(-\frac{4EI}{l}+\frac{6EI}{l^2}\right)$

(c) $\left(\frac{2EI}{l}+\frac{6EI}{l^2}\right)$ (d) $\left(\frac{2EI}{l}-\frac{6EI}{l^2}\right)$

11.10 In a two span continuous beam shown in Fig. 11.1 the support B settles by an amount δ. Then

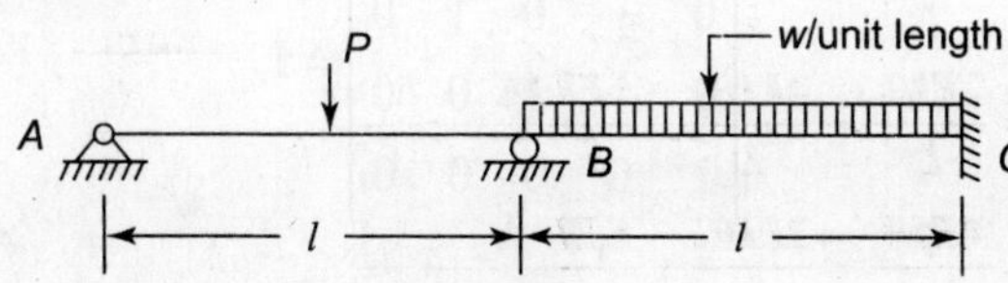

Fig. 11.1

(a) the moment at B increases
(b) the moments at B and C decrease
(c) the moment at B increase and moment at C decrease
(d) the moment at B decrease and moment at C increase

11.11 In an analysis of a portal frame, the moments at the column ends are given in Fig. 11.2. The shear in the left column at the base is

(a) 60 kN (b) 30 kN (c) 29.66 kN (d) 36.69 kN

11.12 In Fig. 11.2, the end moments of columns are given; the shear at the base of the right column is

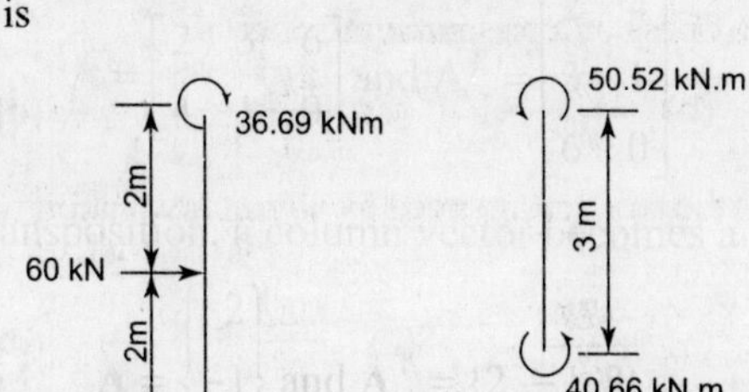

Fig. 11.2

(a) 0 kN (b) 30.39 kN (c) 40.66/3 kN (d) 30 kN

11.13 A member 1-2 in a framed structure undergoes deformation as shown in Fig. 11.3. The moment M_{12} at end 1 can be written as

(a) $M_{12} = \frac{4EI\theta_1}{l} - \frac{2EI\theta_2}{l} + \frac{6EI}{l}\frac{(\Delta_2 - \Delta_1)}{l}$

(b) $M_{12} = -\frac{4EI\theta_1}{l} - \frac{2EI\theta_2}{l} + \frac{6EI}{l}\frac{(\Delta_2 - \Delta_1)}{l}$

(c) $M_{12} = -\frac{4EI\theta_1}{l} + \frac{2EI\theta_2}{l}$

(d) $M_{12} = \frac{4EI\theta_1}{l} + \frac{2EI\theta_2}{l} - \frac{6EI}{l}\frac{(\Delta_2 - \Delta_1)}{l}$

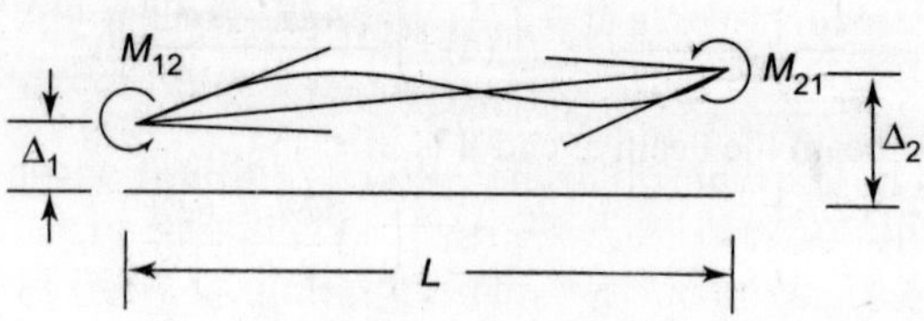

Fig. 11.3

11.14 From the Fig.11.3, one can write the end moment M_{21} as

(a) $M_{21} = \frac{4EI\theta_1}{L} + \frac{2EI\theta_2}{L} - \frac{6EI}{L}\frac{(\Delta_2 - \Delta_1)}{L}$

(b) $M_{21} = \frac{2EI\theta_1}{L} + \frac{4EI\theta_2}{L} - \frac{6EI}{L}\frac{(\Delta_2 - \Delta_1)}{L}$

(c) $M_{21} = \frac{4EI\theta_1}{L} + \frac{2EI\theta_2}{L} + \frac{6EI}{L}\frac{(\Delta_2 - \Delta_1)}{L}$

(d) $M_{21} = \frac{2EI\theta_1}{L} + \frac{4EI\theta_2}{L} + \frac{6EI}{L}\frac{(\Delta_2 - \Delta_1)}{L}$

CHAPTER 12

12.1 The stiffness of a prismatic beam of length l and flexural rigidity EI is

(a) $\frac{EI}{l}$ (b) $\frac{2EI}{l}$ (c) $\frac{3EI}{l}$ (d) $\frac{4EI}{l}$

12.2 The modified stiffness of a prismatic member is

(a) $\frac{EI}{L}$ (b) $\frac{3EI}{l}$ (c) $\frac{4EI}{l}$ (d) $\frac{6EI}{l}$

12.3 In a fixed beam, the moment induced by w/unit translation without rotation at one end is

(a) $\frac{2EI}{l^2}$ (b) $\frac{4EI}{l^2}$ (c) $\frac{6EI}{l^2}$ (d) $\frac{8EI}{l^2}$

12.4 If a moment M is applied at the propped end of a cantilever beam, the moment induced at fixed end is

(a) M (b) $M/2$ (c) $M/3$ (d) $M/4$

12.5 The carry over factor for a prismatic beam is
(a) 0 (b) 1/4 (c) 1/2 (d) 1

12.6 Moment distribution method
(a) gives only approximate results
(b) not suitable for non prismatic members
(c) highly useful for continuous beams and frames
(d) can be used for solving plane and space trusses

12.7 A two-span continuous beam is shown in Fig. 12.1 under the load the support 2 sinks by 10 mm. The flexural rigidity $EI = 20 \times 10^3$ kN m^2. The fixed end moment to be considered for moment distribution $M^F_{12} = M^F_{21}$ is

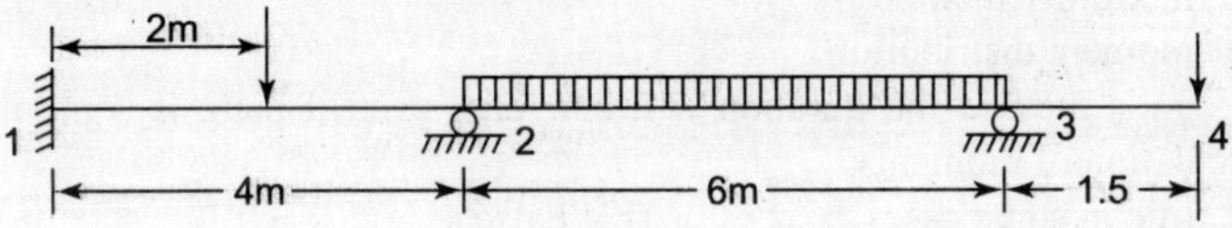

Fig. 12.1

(a) 50 kN.m (b) 75 kN.m (c) 100 kN.m (d) 150 kN.m

12.8 In the continuous beam in Fig.12.1 the fixed end moments $M^F_{23} = M^F_{32}$ to ue considered for moment distribution are
(a) 100/3 kN.m (b) 50 KN.m
(c) 75 kN.m (d) 100 kN.m

12.9 In the continuous beam in Fig. 12.1 support 1 settles by 10 mm and yields ratio 10^{-3} radians under the load. The fixed end moment to be considered for moment distribution M^F_{12} is
(a) 75 kN.m (b) 95 kN.m (c) 100 kN.m (d) 150 kN.m

12.10 In the continuous beam in Fig. 12.1 the fixed end moment M^F_{21} is
(a) 50 kN.m (b) 75 kN.m (c) 85 kN.m (d) 100 kN.m

12.11 In an analysis of a portal frame, with sway the following column end moments are obtained by arresting lateral sway with a horizontal force 'x' at the top of the column. The resulting shear at the top of the column 1–2, V_{21} is (see Fig. 12.2)

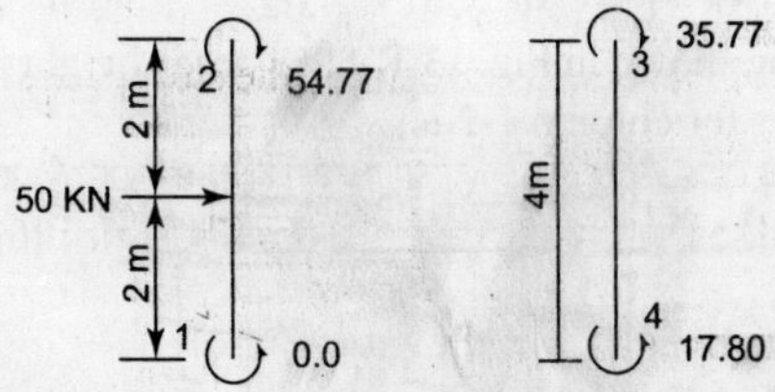

Fig. 12.2

(a) 50 kN (b) 0 (c) 38.69 kN (d) 25

12.12 In Fig. 12.2 the shear at the top of column 3–4, V_{43} is
(a) 13.39 kN (b) 0.0 (c) 17.8 kN (d) 8.94 kN

12.13 In the analysis of the frame in the above the force 'x' applied to present sway in the analysis is
(a) 50 kN (b) 38.69 kN (c) 13.39 kN (d) 24.70 kN

CHAPTER 13

13.1 In Kani's method the sign convention followed for end moments is based on
(a) static sign convention
(b) beam sign convention
(c) beam convention for beams and static sign convention for columns
(d) no uniform convention

13.2 Kani's method is based on
(a) method of consistent displacements
(b) slope deflection method
(c) flexibility method
(d) moment distribution

13.3 In Kani's method the iteration will converge even if there is some mistake in calculation of moments
(a) true in some cases (b) False
(c) true (d) true for beams only

13.4 The sum of rotation factors at a joint is
(a) 1 (b) 1/2 (c) –1/2 (d) –3/4

13.5 The sum of displacement or translation factors due to sway in a storey of a frame is equal to
(a) –1/4 (b) –1/2 (c) –1.0 (d) –3/2

13.6 In a member fixed at the end the rotation factor is
(a) $\frac{4EI}{L}$ (b) $\frac{2EI}{L}$ (c) 0 (d) $\frac{6EI}{L}$

13.7 If an end of a member is hinged it can be considered as fixed, if the relative stiffness is taken as
(a) $\frac{I}{L}$ (b) $\frac{I}{2L}$ (c) $\frac{I}{1.5L}$ (d) $\frac{3}{4}\frac{I}{L}$

13.8 In a portal frame hinged at the base and of columns equal height the translation factor is equal to
(a) –1/2 (b) –1 (c) –3/2 (d) –2

13.9 In the portal frame shown in Fig. 13.1 if the storey height is taken as 6 m the sway distribution factor for column 1-2 is

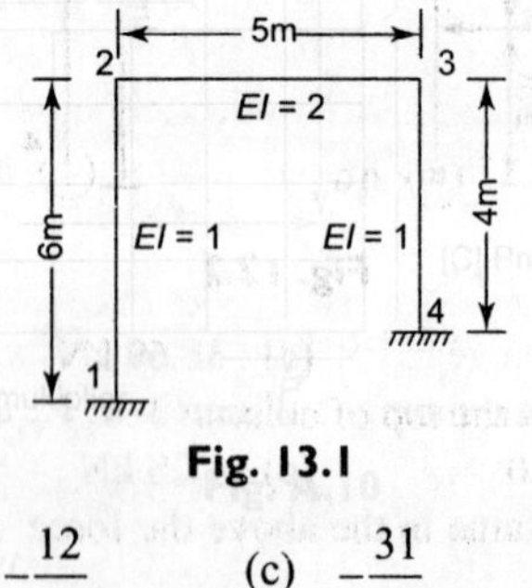

Fig. 13.1

(a) $\frac{31}{65}$ (b) $-\frac{12}{35}$ (c) $-\frac{31}{35}$ (d) $-\frac{81}{70}$

13.10 In the portal frame shown in Fig. 13.1 the sway distribution factor for column 3–4 is

(a) $-\dfrac{12}{35}$ (b) $-\dfrac{31}{35}$ (c) $-\dfrac{27}{35}$ (d) $\dfrac{81}{70}$

13.11 In a multi-storey building frame the storey height of rth storey is h_r and storey shear is Q, then the storey moment is equal to

(a) $\dfrac{Qh_r}{4}$ (b) $\dfrac{Qh_r}{3}$ (c) $\dfrac{Qh_r}{2}$ (d) Qh_r

13.12 The general expression for end moment in a member 1-2 in a frame with translatory joint is

(a) $M_{12}^F + 2M'_{12} + M'_{21} + M''_{12}$

(b) $M_{12}^F + 2M'_{12} - M'_{21} - M''_{12}$

(c) $M_{12}^F - 2M'_{12} - M'_{21} + M''_{12}$

(d) $M_{12}^F + 2M'_{12} + M'_{21} - M''_{12}$

CHAPTER 14

14.1 Column analogy method is applicable to
(a) determinate structures
(b) continuous beams
(c) fixed beams and single storey frames
(d) multi-storey frames

14.2 The area of an analogous column of a propped cantilever beam of length L and flexural rigidity EI is equal to

(a) $\dfrac{L}{EI}$ (b) $\dfrac{2L}{EI}$ (c) $\dfrac{3L}{EI}$ (d) α

14.3 The moment of inertia of an analogous column of a propped cantilever beam as above is

(a) $\dfrac{L^3}{2EI}$ (b) $\dfrac{L^3}{3EI}$ (c) $\dfrac{L^3}{4EI}$ (d) $\dfrac{L^3}{12EI}$

14.4 The moment of inertia of an analogous column of a fixed beam of length L and flexural rigidity EI is

(a) $\dfrac{L^3}{3EI}$ (b) $\dfrac{L^3}{4EI}$ (c) $\dfrac{L^3}{12EI}$ (d) $\dfrac{L^3}{48EI}$

14.5 The M.I of a beam of rectangular cross section varies I. If the width of the beam is b and span l the area of an analogous column will be

(a) $\dfrac{3L}{EI}$ (b) $\dfrac{4.5L}{EI}$ (c) $\dfrac{6L}{EI}$ (d) $\dfrac{7.5L}{EI}$

14.6 A fixed beam AB has a hinge in the span at a distance from A and b from B. If EI is flexural rigidity, then the moment of inertia of the analogous column is

(a) $\left(\dfrac{a^3+b^3}{3EI}\right)$ (b) $\left(\dfrac{a^3+b^3}{2EI}\right)$ (c) $\left(\dfrac{a^3+b^3}{4EI}\right)$ (d) $\left(\dfrac{a^3+b^3}{6EI}\right)$

14.7 The stiffness K_A and K_B of a non prismatic beam element are : $\dfrac{5EI}{L}$ and $\dfrac{6EI}{L}$ respectively. If the c.o.f $C_{AB} = 0.45$ the c.o.f C_{BA} will be

(a) 0.6 (b) 0.45 (c) 0.375 (d) 0.2875

14.8 In a non-prismatic beam element AB the c.o.f C_{AB} and C_{BA} are 0.607 and 0.525 respectively. If the stiffness $K_A = 0.4031\ EI$ then the stiffness K_B will be

(a) 0.2450 (b) 0.3350 (c) 0.3716 (d) 0.4670

14.9 On an analogous column of a propped cantilever beam AB of span L and flexural rigidity EI, a load $N = 1$ is placed on the column at A, then the moment at end A, the fixed end is equal to

(a) $\frac{EI}{L}$ (b) $\frac{2EI}{L}$ (c) $\frac{3EI}{L}$ (d) $\frac{4EI}{L}$

14.10 In a non prismatic beam AB the relationship between c.o.f and stiffness factors is

(a) $K_A = K_B$ (b) $C_{AB} = C_{BA}$ (c) $C_{AB}K_A = C_{BA}K_B$ (d) $C_{AB}K_B = C_{BA}K_A$

CHAPTER 15

15.1 State True or False the following statements

(a) In matrix analysis of structures the static indeterminancy or kinematic indeterminancy is one and the same

(b) In force method of analysis the unknown are the forces at the releases to satisfy the geometric compatibility

(c) In the stiffness method of analysis the unknowns are the displacements at the joints independent of equilibrium equation

(d) The displacement method of analysis is more suitable for structures having a higher degree of static indeterminancy

(e) The flexibility coefficient f_{ij} is equal to stiffness coefficient K_{ji}

(f) The stiffness matrix $[\mathbf{K}] : [\mathbf{F}]^{-1}$

15.2 For a cantilever beam the structure coordinates are shown in Fig. 15.1. The flexibility coefficient are

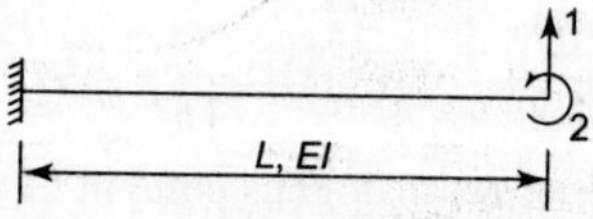

Fig. 15.1

(i) $f_{11} =$ (a) $\frac{L^3}{2EI}$ (b) $\frac{L^3}{3EI}$ (c) $\frac{L^3}{4EI}$ (d) $\frac{L^3}{6EI}$

(ii) $f_{12} =$ (a) $\frac{L^2}{2EI}$ (b) $\frac{L^2}{3EI}$ (c) $\frac{L^2}{4EI}$ (d) $\frac{L^2}{6EI}$

(iii) $f_{22} =$ (a) $\frac{L}{EI}$ (b) $\frac{L}{2EI}$ (c) $\frac{L}{3EI}$ (d) $\frac{L}{4EI}$

15.3 In a simply supported beam the structure coordinates are as shown in Fig. 15.2. The flexibility coefficients are

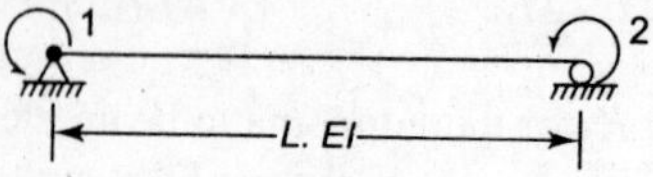

Fig. 15.2

(i) $f_{11} =$ (a) $\frac{L}{3EI}$ (b) $\frac{L}{4EI}$ (c) $\frac{L}{2EI}$ (d) $\frac{L}{EI}$

(ii) $f_{12} = f_{21} =$ (a) $\frac{L}{2EI}$ (b) $\frac{L}{3EI}$ (c) $\frac{L}{4EI}$ (d) $\frac{L}{6EI}$

15.4 The structure coordinates are as indicated on a simply supported beam in Fig. 15.3. The flexibility coefficients are

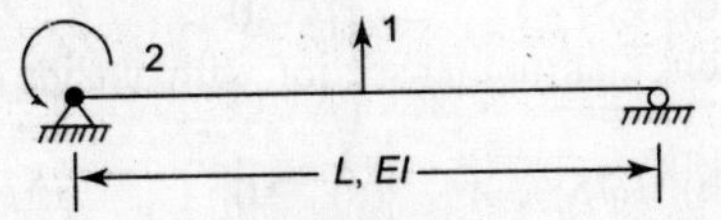

Fig. 15.3

(i) $f_{11} =$ (a) $\frac{L^3}{12EI}$ (b) $\frac{L^3}{16EI}$ (c) $\frac{L^3}{24EI}$ (d) $\frac{L^3}{48EI}$

(ii) $f_{12} = f_{21} =$ (a) $\frac{L^2}{4EI}$ (b) $\frac{L^2}{8EI}$ (c) $\frac{L^2}{16EI}$ (d) $\frac{L^2}{24EI}$

15.5 The flexibility coefficients for the structure shown in Fig. 15.4 are

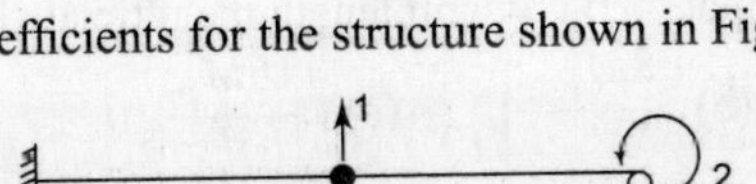

Fig. 15.4

(i) $f_{11} =$ (a) $\frac{L^3}{EI}$ (b) $\frac{L^3}{2EI}$ (c) $\frac{L^3}{3EI}$ (d) $\frac{L^3}{4EI}$

(i) $f_{22} =$ (a) $\frac{L}{EI}$ (b) $\frac{2L}{3EI}$ (c) $\frac{L}{3EI}$ (d) $\frac{3L}{4EI}$

(ii) $f_{12} = f_{21} =$ (a) $\frac{L^2}{3EI}$ (b) $\frac{-L^2}{3EI}$ (c) $\frac{L^2}{4EI}$ (d) $\frac{-L^2}{4EI}$

15.6 which of the following relationship is correct

(a) $[F] = [A]^T[f][A]$ (b) $[F] = [A][f][A]$

(c) $[F] = [A][f][A]^T$ (d) $[F] = [B][f][B]$

15.7 State True or False the following relationships

(a) $[f][K] = [I]$ (b) $[K]^{-1} = [f]$

(c) $U = \frac{1}{2}\{P\}^T[D]$ (d) $[K] = [F]^T$

15.8 If stiffness matrix $[k] = \frac{2EI}{L}\begin{bmatrix} 2 & 1 \\ 1 & 2 \end{bmatrix}$ the flexibility matrix $[f]$ is

(a) $\frac{L}{2EI}\begin{bmatrix} 2 & -1 \\ -1 & 2 \end{bmatrix}$ (b) $\frac{L}{6EI}\begin{bmatrix} 2 & 1 \\ 1 & 2 \end{bmatrix}$

(c) $\frac{L}{6EI}\begin{bmatrix} -2 & 1 \\ 1 & -2 \end{bmatrix}$ (d) $\frac{L}{6EI}\begin{bmatrix} 2 & -1 \\ -1 & 2 \end{bmatrix}$

CHAPTER 19

19.1 The shape factor for a square section having side 'a' placed with one of its diagonal vertical is

(a) 1.0 (b) 1.5 (c) 2.0 (d) 3.0

19.2 Shape factor for a circular tubular section with outside diameter equal to twice the inner diameter is

(a) 1.58 (b) 1.75 (c) 2.0 (d) 2.5

19.3 For a steel rolled beam section the shape factor varies from

(a) 1.0–1.5 (b) 1.5–2.0 (c) 1.14–1.18 (d) 1.25–1.5

19.4 Load factor for a steel rectangular beam is

(a) 1.0 (b) 2.25 (c) 1.15 (d) 1.5

19.5 In a simply supported beam centrally loaded the ultimate load W_u is

(a) $\frac{M_p}{l}$ (b) $\frac{2M_p}{l}$ (c) $\frac{3M_p}{l}$ (d) $\frac{4M_p}{l}$

19.6 In a fixed beam under u.d.l. w/unit length the ultimate load W_u is

(a) $\frac{4M_p}{l^2}$ (b) $\frac{8M_p}{l^2}$ (c) $\frac{16M_p}{l^2}$ (d) $\frac{20M_p}{l^2}$

19.7 In a fixed beam under u.d.l. the load factor is

(a) 1.0 (b) 2.0 (c) 3.0 (d) 4.0

19.8 In a fixed beam under a concentrated load W at 1/3 span the ultimate load W_u is

(a) $\frac{3M_p}{l}$ (b) $\frac{6M_p}{l}$ (c) $\frac{9M_p}{l}$ (d) $\frac{12M_p}{l}$

19.9 In a propped cantilever beam under central load W the ultimate load W_u is

(a) $\frac{12M_p}{l}$ (b) $\frac{9M_p}{l}$ (c) $\frac{6M_p}{l}$ (d) $\frac{3M_p}{l}$

19.10 A propped cantilever beam is under a concentrated load W at 1/3 span point from fixed end. The maximum load the beam can carry is W_u equal to

(a) $\frac{7.5M_p}{l}$ (b) $\frac{5M_p}{l}$ (c) $\frac{6M_p}{l}$ (d) $\frac{9M_p}{l}$

19.11 Number of plastic hinges necessary for collapse of the structure if the degree of indeterminancy of the structure n is

(a) n (b) $n+3$ (c) $n+2$ (d) $n+1$

19.12 A portal frame is shown in Fig. 19.1. In a beam mechanism the ultimate load P_u is

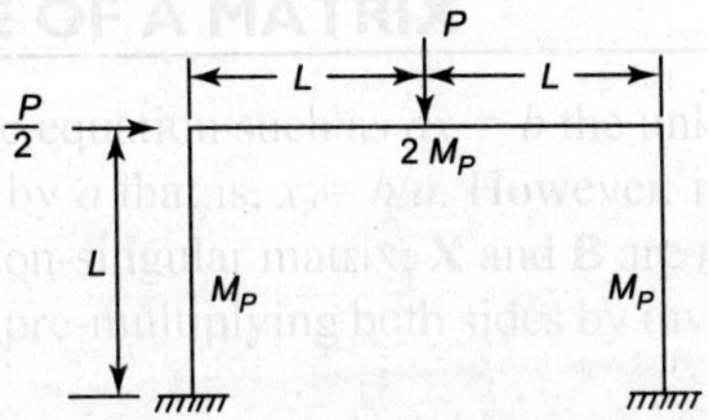

Fig. 19.1

(a) $\frac{4M_p}{L}$ (b) $\frac{6M_p}{L}$ (c) $\frac{7.5M_p}{L}$ (d) $\frac{9M_p}{L}$

19.13 For the frame in Fig. 19.1 the ultimate load in sway mechanism is

(a) $\frac{4M_p}{L}$ (b) $\frac{6M_p}{L}$ (c) $\frac{8M_p}{L}$ (d) $\frac{12M_p}{L}$

19.14 For the frame in Fig. 19.1 the ultimate load in combined mechanism is

(a) $\frac{4.5M_p}{L}$ (b) $\frac{16}{3}\frac{M_p}{L}$ (c) $\frac{7.5M_p}{L}$ (d) $\frac{9M_p}{L}$

Answers

ANSWERS TO PROBLEMS FOR PRACTICE

Chapter 2

2.2 $y_D = 7.07$ m; $y_E = 5.29$ m
$T_{AC} = 100.95$ kN; $T_{CD} = 86.34$, $T_{DE} = 87.64$ kN, $T_{EB} = 98.79$ kN
2.3 $H_A = 5.0$ kN, $V_A = 7.5$ kN, $V_B = 5.0$ kN, $T_{max} = 9.01$ kN
2.4 (a) $y_D = 4.09$ m; (b) 261.07 kN; (c) 407.42 kN; (d) 409.24 kN
2.5 $H = 1080$ kN, $V_A = 720$ kN, $V_B = 360$ kN, $T_{A(max)} = 1298$ kN
2.6 (a) $T_{BC(max)} = 194.81$ kN; (b) $T_{AB} = 257.15$ kN
2.7 $H = 108.89$ KN, $V_A = 67.67$ kN, $N_D = 103.63$ (comp) kN,
$V_{D(r)} = -33.52$ kN, $M_D = 112.02$ kN.m
2.8 $H = 136.0$ kN, $V_A = 85.6$ kN, $V_B = 74.4$ kN,
$M = -60.3$ kN.m, $N = 159.18$ kN, $V_{(r)} = 21.65$ kN
2.15 $H = 10.67$ kN (to the left), $V_{(left)} = 21.34$ kN (upwards) $V_{(right)} = 18.66$ kN
2.16 $R_{(left)} = 31.82$ kN (normal to rollers), $H_{(right)} = 2.5$ kN (to the right)
$V_{(right)} = 27.5$ kN (upwards)

Chapter 3

3.1 (a) Stable, statically indeterminate, degree of indeterminacy = 1
(b) Stable, statically indeterminate, degree of indeterminacy = 2
(c) Stable, statically determinate, complex truss
(d) Unstable
(e) Stable, statically indeterminate, degree of indeterminacy = 2
(f) Unstable
(g) Unstable
(h) Unstable
3.2 1–2 = 12.5 kN, 2–3 = 75.0 kN, 3–4 = 25.0 kN, 5–6 = 0 kN, 6–7 = –12.5 kN,
7–8 = –75.0 kN, 8–9 = –25.0 kN,
1–6 = – 17.68 kN, 2–7 =. –88.38 kN, 3–8 = 70.71 kN, 4–9 = 35.36 kN,
1–5 = 12.50 kN, 2–6 = 62.50 kN, 3–7 = –50.0 kN, 4–8 = – 25.0 kN
3.3 1–2 = 22.99 kN, 2–3 = 1.44 kN, 4–5 = – 1.55 kN, 1–5 = –24.64 kN,
5–2= –21.55 kN, 2–4 = 21.55 kN, 4–3 = – 21.55 kN

3.4 1–2 = –10.0 kN, 2–3= –40.0 kN, 3–4= –10.0 kN, 4–5 = 42.42 kN, 5–3 = 14.14 kN, 5–2 = 28.28 kN, 5–6 = 20.00 kN, 6–1 = 14.14 kN, 6–2 = 14.14 kN

3.5 1–2 = –15.0 kN, 2–3 = 0 kN, 2–4 = 9.6 kN, 2–6 = –9.6 kN, 4–5 = –37.5 kN, 4–7 = 6.0 kN

3.6 2–3 = –76.03 kN, 9–3 = –50.26 kN, 9–16 = 42.62 kN, 15–16 = 75.00 kN, 2–9 = 56.25 kN, 9–15 = –6.25 kN

3.7 1–2 = 180.0 kN, 8–4= 5–10 = 0

Chapter 4

4.1 $R_{aY} = -10.0$ kN, $R_{bY} = 10.0$ kN, $R_{cY} = 0$ kN, $R_{aZ} = -8.66$ kN, $R_{bZ} = 8.66$ kN, $R_{cX} = -10.0$ kN, $P_{ab} = -5.0$ kN, $P_{bc} = 10.0$ kN, $P_{ca} = -10.0$ kN, $P_{ed} = -10.0$ kN, $P_{eh} = -10.0$ kN, $P_{ea} = 14.14$ kN, All others = 0

4.2 $R_{aY} = -11.54$ kN, $R_{aZ} = -2.50$ kN, $R_{bY} = 5.77$ kN, $R_{bZ} = 7.50$ kN, $R_{cY} = 5.77$ kN, $R_{cX} = -8.67$ kN, $P_{ab} = -4.33$ kN, $P_{bc} = 8.67$ kN, $P_{ca} = -2.90$ kN, $P_{ed} = -5.77$ kN, $P_{eb} = -5.77$ kN, $P_{da} = 5.77$ kN, $P_{dc} = -8.15$ kN, $P_{ea} = 8.15$ kN, All others = 0

4.3 $R_{zY} = 32.0$ kN, $R_{3Y} = -28.0$ kN, $R_{4Y} = 6.0$ kN, $R_{3Z} = -20.0$ kN, $R_{4Z} = 20.0$ kN, $R_{4X} = 20.0$ kN, $P_{21} = -37.4$ kN, $P_{24} = 17.2$ kN, $P_{23} = 4.3$ kN, $P_{31} = 32.7$ kN, $P_{34} = -36.6$ kN

4.6 $R_{2Z} = 0$, $R_{1X} = -133.33$ kN, $R_{3X} = 33.33 =$ kN

Chapter 5

5.2 $\Delta = 0.723$ mm (downwards)
$\theta = 0.00289$ (clockwise)

5.3 $\Delta = \dfrac{Pa^3}{3EI}$

5.4 $\Delta_E = \dfrac{4Pa^3}{3EI}$

5.5 $\Delta = 15.94$ mm

5.6 $\Delta_{\text{centre}} = 8.16$ mm (downward)
$\Delta_{\text{free end}} = 5.24$ mm (upwards)

5.7 $\Delta A = \dfrac{2PL^3}{3EI}$

5.8 $\Delta_5 = 4.23$ mm

5.9 $\Delta_{\text{centre}} = 52.0$ mm (taking eight equal parts of 0.75 m each)

5.10 $\Delta_B = \dfrac{14PL^3}{3EI}$ (to the right), $\Delta_D = \dfrac{31PL^3}{6EI}$ (to the right),

CHAPTER 6

6.1 $\Delta = \dfrac{3PL^3}{16EI}$ $\theta = \dfrac{5PL^2}{16EI}$

6.2 $\Delta_{cen} = \frac{1040}{EI}$ (downwards) $\Delta_{end} = \frac{350}{EI}$ (upwards)

$\theta_{supp.} = \frac{360}{EI}$

6.3 $\Delta_A = \frac{5PL^3}{16EI}$ (downwards)

6.4 $\Delta_C = 17.5$ mm

6.5 $\Delta_C = \frac{PR^2}{EI}\left(L + \frac{\pi R}{4}\right)$

6.6 $\Delta_{AH} = \frac{13}{192}\frac{PL^3}{EI}$ (to the right)

6.7 $\Delta_c = 41.76$ mm

6.8 $\Delta_{max} = \frac{\alpha \Delta T L^2}{8d}$

Chapter 7

7.1 Shear at A:I.L ordinates, left end –0.2, left of supp. –0.0, right of supp.–1.0, right hand supp. – 0.0 right end– – 0.3. Shear at *B*: left end –0.2, left of *B*–0.3, right of *B*–0.7, right supp. –0.0, right end – 0.3.

7.2 Moment at *A*: –0.0, *B*–4.0, *C*– –6.0, *D* –0.0, *E* – –4.0
Shear at *B*: right of *B*–1.0, *C* – 1.0, *D*–0.0, *E*– –0.67. Moment at *B*: *A* –0.0, *B*–0.0, *C*– –2.0, *D* – 0.0, *E* – 4/3

7.3 Reaction at 1:1–1.0, 3–0.0, 4– –0.3, 6–0.0
Shear at 2:1–0.0, left of 2— – 0.6, right of 2–0.4, 3–0.0, 4– –0.3, 6–0.0
Moment at 2:1–0.0, 2–3.6, 3–0.0, 4– –2.7 6–0.0

7.4 Reaction at *A*: *A*–1.0, *B*–0.0, *C* –1/3
Moment at *B*:panel point in span next to supp. *B*–0.0, *B* – –0.75, *C*– –3.0
Shear left of *B*:*A*–0.0 panel point in span next of *B*— –12/13.5 to the right of panel point – 1.5/13.5, *B*–0.50 *C* — –1/3

7.5 Moment at *A*:*B*–6.0, *E*–0.0, *D*– –6.0, and *C*–0.0

7.6 Member *BC*: *A*–0.0, *B*–0.72, *C*–0.87, *E*–0.0 (comp.)
Member *HC*: *A*–0.0, *B*–0.29, *C*–0.58, *E*–0.0 (–Comp, +Ten.)

7.7 Reaction 1:1–1.0, 2–0.0, 3 – 0.25, 7–0.0
Moment at 2:2–0.0, 3 – 2.50, 7–0.0
Member 5–6:3–0.0, 6–0.75, 7–0.0
Member 5–12:3–0.0, 5–0.71, 6 – 0.35 (comp.), 7–0.0
Member 11–12:1–0.0, 5 – 1.0 (comp.), 7–0.0

7.8 Max. shear at *C*: 40.63 kN or –15.62kN, max. moment at *C*.–175. 13 kN.m

7.9 Max. shear–84.0 kN or –48.0 kN, max. moment –688.0 kN.m

7.10 Max. shear –80.0 or –80.0 kN, max. moment +875.0 or –800.0 kN.m

7.11 Max. shear next to supp. –220.0 or +220.0 kN
Max moment –580.8 kN.m under 160 kN load at 5.28 m from supp.

7.12 Max shear –260.0 or –260.0 kN
Max. moment –517.1 kN.m under 180 kN load at 4.09 m from supp.

7.13 Max. force in members: CD–562.5, CH–96.0, GH–612.0 kN (comp.)
7.14 Max. shear–276.0 or –156.0 kN, max. moment –9080 kN.m
7.15 Max. shear at C–157.5 or –78.75 kN, max. moment at C–1437.2 kN.m, absolute max. shear next to supp. –275.6 kN, absolute max. moment under interior 140 kN load at 11.25 m from supp. –1503.7 kN.m

Chapter 8

8.1 $t = 22.09°C$
8.2 (i) $L = 103.8$ m, (ii) $H = 171.52$ kN, (iii) $T_A = 203.0$ kN, $T_B = 194.36$ kN
8.3 $V_{40} = 4.0$ kN, $M_{40} = -320.0$ kN.m
8.5 (i) $M_{max} = +240.5$ kN.m, –156.25 kN.m
$V_{max} = +25$kN., (ii) $T_{max} = 201.9$ kN
8.6 $A = 2951$ mm^2, $M_{max} = +577.2$ kN.m
–375.0 kN.m
8.7 $V_{25} = 27.03$ kN, $M_{25} = 1881$ kN.m
$T_{max} = 740.73$ kN.
8.8 (i) $H = 12.0$ kN, $M_{max} = 90.0$ kN.m
(iii) $M_{max} = +120$ kN.m at ends
$V_{max} = 5.0$ kN. const, al through

Chapter 9

9.1 $M_B = -2.31\,P$ and $M_c = -0.46P$ assuming hinge points 3 m from B and 2 m from C.
9.2 $M_B = M_C = -64.8$ kN.m, hinges assumed at 0.1 L from B and C.
9.3 (a) 1–7 = 5–9 = 62.49 kN, 2–8 = 4–8 = 31.24 kN
(b) 1–7 = 31.24 kN, 6–2 = 31.24 kN, 2–8 = 15.63 kN, 3–7 = –15.63 kN
9.4 (a) Windward col. $N = 9.17$ kN(ten), $M = -45.0$ kN.m, $V = 15.0$ kN
Leeward col. $N = -9.17$ kN (comp), $M = 45.0$ kN.m, $V = 15.0$ kN
hinges assumed 3 m above base; (c) 1–2 = –16.53 kN
9.5 $P_{CL} = -30.0$ kN, $P_{LD} = 10.0$ kN, $P_{BM} = P_{ML} = 28.29$ kN,
$P_{LN} = P_{EN} = -28.29$ kN. $P_{CM} = P_{MN} = P_{ND} = 0$
9.6 Reactions at base:
Vindward col. $N = 7.5$ kN (ten), $M = 200.0$ kN.m, $V = 70.0$ kN
Leeward col. $N = -7.5$ kN(comp), $M = 120.0$ kN.m, $V = 30.0$ kN
9.7 Beam end moments upper row–4.0 kN.m, lower row–22.0 kN.m
Column moments upper storey – 4.0, 8.0, 8.0 and 4.0 kN.m
lower storey – 18.0, 36.0, 36.0 and 18.0 kN.m
9.8 Beam moments upper row –3.3, 5.43, 5.43 and 3.3 kN.m
lower row –11.49, 30.45, 30.45 and 11.49 kN.m
Column moments upper storey – 3.3, 8.72, 8.72 and 3.3 kN.m
lower storey –11.49, 30.45. 30.45 and 11.49 kN.m

Chapter 10

10.1 $R_B = \frac{2}{3}P,\ M_A = -\frac{PL}{3}$
10.2 $R_B = \frac{5}{4}wL$

10.3 $P_{AB} = 5.91$, $P_{BC} = -5.91$ (comp.), $P_{BD} = 7.90$ kN
10.4 $P_{AB} = 58.63$ kN (ten.)
10.5 $P_{EC} = -56.6$ kN, $P_{BF} = 0$, $P_{EC} = -42.5$ (comp.) kN, $P_{BF} = -14.1$ kN
10.6 $X_1 = 65.3$ kN (ten), $X_2 = 3.48$ kN downwards
10.7 $P_{BC} = 73.34$ kN (ten), $X_2 = 19.65$ kN upwards
10.8 $P_{AB} = -19.5$ kN, $P_{BC} = -3.8$ kN, $P_{CD} = -3.8$ kN, $P_{DE} = 6.2$ kN
$P_{EF} = 20.6$, $P_{BE} = -3.3$ kN, $P_{CE} = 5.4$ kN, $P_{DB} = -8.8$ kN
$P_{BF} = 13.4$ kN, $P_{EA} = -14.9$ kN
10.9 Interior supp. moments $-\dfrac{wL^2}{10}$, Interior reaction $-\dfrac{11wL}{10}$
10.10 $M_B = -76.88$ kN.m, $R_A = 20.30$ kN, $R_B = 82.51$ kN, $R_C = 14.52$ kN
10.11 Interior reaction –178.62 kN
10.12 Interior support moments decreased by 0.595%, span moments in creased by 0.65%
10.13 $M_{BC} = M_{CB} = -6.77$ kN.m. $M_{AD} = M_{AB} = -4.57$ kN.m
10.14 (a) $H_A = -\dfrac{3}{7}wa$, $V_A = \dfrac{3}{28}wa$, (b) $V_A = 14.4$ kN
(c) $V_B = 1.26\,P$(upwards), $H_B = 0.41\,P$ (to the right)
10.15 $H = 115.77$ kN (graphical summation from eight equal parts)
10.16 $H = 447.88$ kN (graphical summation from eight equal parts)
10.17 $H = \sqrt{2}\dfrac{PL}{\pi^2 h}$
10.18 $H = \dfrac{2PL}{\pi^2 h}$
10.19 1–1.64, 2–1.87, 3–1.17 kN.m
10.20

	A	1	2	3	B	4	5	6
(a)	1.000	0.692	0.406	0.168	0	–0.082	–0.094	–0.059
(b)	0	–0.293	–0.468	–0.410	0	0.841	2.032	0.956
(c)	0	0.590	0.940	0.820	0	0.820	0.940	0.590
(d)	0	0.059	0.094	0.082	0	0.168	0.410 } 0.590	0.308

10.21 1 – 0.30, 2 – 0.50, C – 1.0, 3 – 1.31, 4 – 0.85
10.22 $H = \dfrac{125}{128}$ when unit load is at crown, moment quarter span point = 2.49, centre –0.64.

Chapter 11

11.1 $M_{12} = 21.25$ kN.m, $M_{21} = -17.50$ kN.m, $M_{23} = 17.50$ kN.m
$M_{32} = -15.00$ kN.m
11.2 $M_{12} = 150.0$ kN.m, $M_{21} = -49.25$ kN.m, $M_{32} = -8.82$ kN.m.
$M_{43} = -75.60$ kN.m.
11.3 $M_{12} = -128.0$ kN.m, $M_{21} = -256.0$ kN.m, $M_{32} = -256.0$ kN.m.
$M_{43} = 128.0$ kN.m.
11.4 $M_{12} = -23.09$ kN.m, $M_{21} = -46.18$ kN.m, $M_{23} = 69.27$ kN.m,
$M_{25} = -23.09$ kN.m, $M_{32} = -85.24$ kN.m, $M_{34} = -70.15$ kN.m,
$M_{36} = 12.73$ kN.m, $M_{43} = -37.28$ kN.m.

11.5 $M_{12} = -4.76$ kN.m, $M_{21} = -9.52$ kN.m, $M_{23} = 14.29$ kN.m,
$M_{25} = -4.76$ kN.m, $M_{32} = -37.15$ kN.m, $M_{36} = -2.86$ kN.m,
$M_{34} = 40.00$ kN.m, $M_{52} = -2.38$ kN.m. $M_{63} = -1.43$ kN.m.

11.6 $M_{12} = 63.42$ kN.m. $M_{21} = -34.17$ kN.m, $M_{32} = -56.20$ kN.m,
$M_{43} = 44.50$ kN.m.

Chapter 12

12.1 $M_{12} = 45.00$ kN.m, $M_{21} = -30.00$ kN.m, $M_{23} = 30.00$ kN.m, $M_{32} = -15.00$ kN.m.

12.2 $M_{21} = -109.4$ kN.m, $M_{23} = 109.4$ kN.m.

12.3 $M_{12} = -5.30$ kN.m. $M_{23} = 10.70$ kN.m, $M_{34} = 30.00$ kN.m.

12.4 $M_{12} = 10.0$ kN.m, $M_{21} = -M_{23} = 20.0$ kN.m, $M_{32} = -60.0$ kN.m.

12.5 $M_{12} = 18.00$ kN.m. $M_{23} = 52.90$ kN.m, $M_{34} = 41.40$ kN.m. $M_{43} = -9.30$ kN.m.

12.6 $M_{AB} = 45.02$ kN.m. $M_{BC} = 11.59$ kN.m, $M_{CD} = 40.94$ kN.m,
$M_{DC} = -35.78$ kN.m.

12.7 $M_{12} = 27.92$ kN.m, $M_{21} = -24.26$ kN.m, $M_{23} = 31.83$ kN.m,
$M_{24} = -7.56$ kN.m, $M_{42} = 11.22$ kN.m.

12.8 See 11.6

12.9 $M_{21} = -17.60$ kN.m. $M_{23} = 17.60$ kN.m, $M_{32} = -17.59$ kN.m.

12.10 $M_{AB} = 60.50$ kN.m, $M_{BA} = -19.70$ kN.m, $M_{BC} = 19.70$ kN.m.
$M_{CB} = -59.40$ kN.m, $M_{CD} = 59.40$ kN.m.

Chapter 13

13.1 $M_{BA} = -30.64$ kN.m, $M_{EC} = 30.64$ kN.m, $M_{CB} = -34.68$ kN.m

13.2 $M_{AB} = 75.0$ kN.m, $M_{BA} = -57.0$ kN.m, $M_{CB} = -44.8$ kN.m.

13.3 $M_{BC} = 13.04$ kN.m, $M_{CD} = 15.90$ kN.m, $M_{DC} = -52.06$ kN.m.

13.4 (a) $M_{AB} = -3.77$ kN.m, $M_{BC} = -7.54$ kN.m, $M_{CD} = -11.75$ kN.m

13.5 $M_{AB} = 83.98$ kN.m. $M_{BC} = 14.99$ kN.m, $M_{CB} = -50.00$ kN.m.

13.6 $M_{AB} = -9.53$ kN.m, $M_{BA} = -24.26$ kN.m, $M_{CB} = 20.60$ kN.m.
$M_{DC} = 13.0$ kN.m.

13.7 $M_{CD} = -128.0$ kN.m, $M_{BF} = -162.8$ kN.m, $M_{BC} = 106.3$ kN.m.
$M_{BA} = 56.4$ kN.m, $M_{AB} = 28.2$ kN.m, $M_{DC} = 128.0$ kN.m,
$M_{EB} = 162.8$ kN.m, $M_{CB} = -106.3$ kN.m.

13.8 $M_{AB} = M_{FE} = 49.4$ kN.m, $M_{BA} = M_{EF} = 40.6$ kN.m.
$M_{BE} = M_{EB} = -52.7$ kN.m. $M_{BC} = M_{ED} = 12.0$ kN.m,
$M_{CD} = M_{DC} = -18.0$ kN.m.

13.10 $M_{BA} = -10.3$ kN.m, $M_{EF} = 13.7$ kN.m, $M_{DE} = 23.1$ kN.m,
$M_{CB} = -22.9$ kN.m.

CHAPTER 14

14.1 $M_A = -364.38$ kN.m

14.2 $M_A = -\frac{2}{9}\ Pl,\ M_B = -\frac{2}{9}\ Pl$

14.3 $M_A = -\frac{11}{192}\ wl^2,\ M_B = -\frac{5}{192}\ wl^2$

14.4 $M_A = -389.77$ kN.m. $M_B = -395.89$ kN.m

14.5 $M_A = -7.19$ P kN.m, $M_B = -3.36$ P kN.m
Stiffness factors $M_{iA} = 0.4186\,EI$, $M_{iB} = -0.2615EI$,
C.O.Fs = –0.6247, –0.5610

14.6 $M_A = M_D = 17.78$ kN.m. $M_B = M_C = -35.56$ kN.m.

14.7 $M_A = -56.75$ kN.m. $M_B = -M_C = 42.86$ kN.m.

14.8 $M_B = -14.06$ kN.m, $M_C = -14.06$ kN.m.

14.9 $M_A = -13.12$ kN.m, $M_B = -24.38$ kN.m.
$M_c = -24.38$ kN.m, $M_D = -13.12$ kN.m

14.10 Stiffness factors $M_{iA} = 8.21\,\dfrac{EI}{l}$, $M_{iB} = 4.84\,\dfrac{EI}{l}$
CO. $F_s = -0.4356, -0.74$

Chapter 15

15.1

$$\mathbf{f} = \begin{bmatrix} \dfrac{3L}{4AE} & & \\ 0 & \dfrac{11L^3}{96EI} & \text{sym.} \\ 0 & \dfrac{7L^2}{32EI} & \dfrac{5L^2}{8EI} \end{bmatrix}$$

15.2

$$\mathbf{f} = \frac{L}{EI}\begin{bmatrix} \dfrac{1}{3} & \dfrac{L}{16} \\ \dfrac{L}{16} & \dfrac{L^2}{48} \end{bmatrix}$$

15.3

$$\mathbf{f} = \frac{L}{EI}\begin{bmatrix} \dfrac{L^2}{24} & \dfrac{L}{8} \\ \dfrac{L}{8} & 1 \end{bmatrix}$$

15.4

$$\mathbf{k} = \begin{bmatrix} 24 & & & \\ 6 & 8 & & \text{sym.} \\ 6 & 2 & 8 & \\ 6 & 0 & 2 & 4 \end{bmatrix}$$

15.5

$$\mathbf{k} = \begin{bmatrix} 8 & & & \\ 0 & 4 & \text{sym.} & \\ 4 & 2 & 12 & \\ -12 & 6 & -6 & 36 \end{bmatrix}$$

15.6

$$\mathbf{k} = \begin{bmatrix} \dfrac{12EI_1}{L_1^3} & \dfrac{6EI_1}{L_1^2} \\ \dfrac{6EI_1}{L_1^2} & \left(\dfrac{4EI_1}{L_1} + \dfrac{3EI_2}{L_2}\right) \end{bmatrix}$$

15.7

$$\mathbf{k} = \begin{bmatrix} 8 & & & \\ 2 & 8 & \text{sym.} & \\ 2 & 0 & 8 & \\ 0 & 2 & 2 & 8 \end{bmatrix}$$

15.8

$$\mathbf{f} = \begin{bmatrix} \dfrac{24}{L^2} & \text{sym.} & \\ \dfrac{6}{L} & 8 & \\ \dfrac{6}{L} & 2 & 8 \end{bmatrix}$$

15.9

$$\mathbf{f} = \begin{bmatrix} \dfrac{1}{3} & & \\ \dfrac{-1}{2} & 1 & \text{sym.} \\ \dfrac{-1}{2} & 2 & 2 \end{bmatrix}$$

15.10

$$\mathbf{f} = \frac{L}{EI} \begin{bmatrix} \dfrac{2L}{3} & & \text{sym.} \\ \dfrac{-1L}{3} & \dfrac{1}{3} & \\ \dfrac{1L}{6} & \dfrac{-1}{6} & \dfrac{1}{3} \end{bmatrix}$$

15.11

$$\mathbf{f} = \begin{bmatrix} \dfrac{3}{8} & & & \\ \dfrac{-1}{4} & \dfrac{5}{8} & \text{sym.} & \\ 0 & \dfrac{-1}{8} & \dfrac{1}{8} & \\ \dfrac{1}{6} & \dfrac{-5}{24} & \dfrac{1}{24} & \dfrac{1}{8} \end{bmatrix}$$

15.13

$$\mathbf{f} = \frac{L^3}{6EI}\begin{bmatrix} 8 & & \text{sym.} \\ -3 & 2 & \\ 0 & 0 & 13 \end{bmatrix}$$

15.14

$$\mathbf{f} = \frac{L}{6EI}\begin{bmatrix} 2 & -1 \\ -1 & 2 \end{bmatrix}$$

15.15

$$\mathbf{k} = \begin{bmatrix} 8 & & & \\ 2 & 8 & \text{sym.} & \\ 0 & 2 & (4+k_b) & \\ 3 & 6 & 6 & 15 \end{bmatrix}$$

15.16

$$\mathbf{k} = \frac{EI}{2L^3}\begin{bmatrix} 129 & \text{sym.} & \\ -3L & 16L^2 & \\ -3L & 4L^2 & 16L^2 \end{bmatrix}$$

15.17

$$\mathbf{k} = \frac{EI}{L}\begin{bmatrix} 2 & 8 \\ 2 & 12 \end{bmatrix}$$

Chapter 16

16.1

$$\mathbf{A} = \begin{matrix} AD \\ BD \\ CD \end{matrix}\begin{bmatrix} \frac{5}{3} & 0 & \frac{-5}{3} \\ \frac{-\sqrt{34}}{3} & 0 & 0 \\ 0 & \frac{4}{3} & -\frac{4}{3} \end{bmatrix}$$

16.2

$$\mathbf{B} = \begin{matrix} 1 \\ 2 \\ 3 \\ 4 \\ 5 \\ 6 \\ 7 \\ 8 \end{matrix}\begin{bmatrix} 0 & 1 & 0 \\ \frac{-5}{4} & 0 & 0 \\ \hline 0 & 1 & 0 \\ 0 & 0 & 1 \\ \frac{-3}{4} & 0 & 0 \\ \frac{3}{4} & 0 & 0 \\ \hline 0 & 0 & 1 \\ \frac{5}{4} & 0 & 0 \end{bmatrix}$$

16.3

$$\mathbf{B} = \begin{array}{r} 1\\2\\3\\4\\5\\6\\7\\8\\9\\10\\11\\12 \end{array}\begin{bmatrix} & & & 1 & & \\ & & & & & \\ \frac{4}{5} & \frac{-12}{25} & \frac{9}{25} & & & \\ & & & & & \\ & & & 1 & & \\ & & & & 1 & \\ & & & & & \\ \frac{8}{5} & \frac{16}{25} & \frac{-12}{25} & & & \\ & & & & 1 & \\ & & & & & 1 \\ 1 & & & & & \\ & \frac{3}{5} & \frac{4}{5} & & & \end{bmatrix}$$

Elements not recorded are zero

16.4 See 15.10

Chapter 17

17.1

$$\begin{Bmatrix} D_B \\ \theta_B \end{Bmatrix} = \frac{WL^3}{EI}\begin{Bmatrix} \frac{-L}{8} \\ \frac{-1}{6} \end{Bmatrix}$$

17.2

(a) $$\begin{Bmatrix} D_1 \\ D_2 \\ D_3 \end{Bmatrix} = \frac{P_1 L}{EI}\begin{Bmatrix} -0.67 \\ 4.36 \\ 0 \end{Bmatrix}$$

(b) $$\begin{Bmatrix} D_1 \\ D_2 \\ D_3 \\ D_4 \\ D_5 \end{Bmatrix} = \frac{L}{AE}\begin{Bmatrix} -3.49 \\ -12.50 \\ -33.33 \\ -37.70 \\ -5.13 \end{Bmatrix}$$

17.3

(a) $$\begin{Bmatrix} R_A \\ R_B \\ \theta_B \end{Bmatrix} = \frac{WL}{8}\begin{Bmatrix} 5 \\ 3 \\ \frac{L^2}{6EI} \end{Bmatrix}$$

$$\text{(b)}\quad \begin{Bmatrix} R_A \\ R_B \\ \theta_B \end{Bmatrix} = \frac{P}{3} \begin{Bmatrix} 4 \\ 2 \\ \dfrac{L^2}{6EI} \end{Bmatrix}$$

17.4 Second choice is most desired

$$\begin{Bmatrix} M_{AB} \\ M_{BA} \end{Bmatrix} = \begin{Bmatrix} -91.3 \\ -66.92 \end{Bmatrix} \text{kN.m and} \begin{bmatrix} R_A \\ R_B \\ R_C \end{bmatrix} = \begin{bmatrix} 52.46 \\ 62.24 \\ 1.3 \end{bmatrix} \text{kN}$$

17.6

$$\text{(b)}\quad H_D = \frac{104}{345} wL$$

$$\text{(c)}\quad \theta_3 = \frac{PL^2}{48EI}$$

17.9

$$\begin{Bmatrix} X_1 \\ X_2 \end{Bmatrix} = \begin{pmatrix} -0.262\,P \\ 0 \end{pmatrix} \text{and} \begin{bmatrix} p_1 \\ p_2 \\ p_3 \\ p_4 \\ p_5 \\ p_6 \end{bmatrix} = \begin{bmatrix} 0 \\ 0 \\ -0.738P \\ 0 \\ -0.731P \\ 0 \end{bmatrix}$$

17.10 (a) 1–4 and 3–6 may be removed to make it a primary structure

$$\text{(b)}\quad \mathbf{F}_1 = \frac{1}{EI} \begin{bmatrix} 24.78 & 3.32 \\ 3.32 & 24.78 \end{bmatrix}$$

$$\begin{matrix}\text{(c)} \\ \text{(d)}\end{matrix}\quad \begin{Bmatrix} 1-2 \\ 2-4 \\ 1-3 \\ 2-3 \\ 1-4 \\ 3-4 \end{Bmatrix} = \begin{bmatrix} -8.07, & 35.6 \\ -10.75, & 47.4 \\ 22.60, & 47.4 \\ 13.45, & -59.3 \\ -28.22, & -59.3 \\ -16.13, & 71.2 \end{bmatrix}$$

17.11

$$\mathbf{F} = \frac{L^3}{6EI} \begin{bmatrix} 54 & & \text{sym.} \\ 28 & 16 & \\ 8 & 5 & 2 \end{bmatrix} \quad X_1 = \left\{ 1.85\text{ kN}, \ -0.11\frac{EI}{L^3} \right\}$$

$$\begin{bmatrix} D_{21} & D_{22} \\ D_{31} & D_{32} \end{bmatrix} = \begin{bmatrix} 1.016 \dfrac{L^3}{EI} & 0.519 \\ 0.70 \dfrac{L^3}{EI} & 0.148 \end{bmatrix}$$

17.12

$$\begin{Bmatrix} M_{AB} \\ M_{BC} \\ M_{CD} \end{Bmatrix} = \begin{bmatrix} 0.0863wL^2 & \frac{-5}{112}PL & 5.54\frac{EI\Delta}{L^2} \\ 0.0774wL^2 & \frac{19}{224}PL & -5.08\frac{EI\Delta}{L^2} \\ 0.1059wL^2 & \frac{111}{1568}PL & 2.77\frac{EI\Delta}{L^2} \end{bmatrix}$$

Chapter 18

18.1

$$\begin{Bmatrix} p_1 \\ p_2 \\ p_3 \end{Bmatrix} = \begin{Bmatrix} 0.365P \\ 0.500P \\ 0.259P \end{Bmatrix}$$

18.2

$$\begin{matrix} \text{left incl.} \\ \text{right incl.} \\ \text{vert.} \end{matrix} \begin{Bmatrix} 0.5774P \\ 0.5774P \\ 0.0 \end{Bmatrix}$$

18.3

$$\begin{Bmatrix} D_5 \\ D_6 \end{Bmatrix} = \frac{PL}{AE}\begin{Bmatrix} 5.828 \\ -3.000 \end{Bmatrix}, \quad \begin{Bmatrix} p_1 \\ p_2 \\ p_3 \end{Bmatrix} = \begin{Bmatrix} 0 \\ -3.000P \\ 1.414P \end{Bmatrix}, \quad \begin{Bmatrix} R_1 \\ R_2 \\ R_3 \\ R_4 \end{Bmatrix} = \begin{Bmatrix} -P \\ -P \\ 0 \\ 3P \end{Bmatrix}$$

18.4

$$\begin{Bmatrix} p_1 \\ p_2 \\ p_3 \\ p_4 \\ p_5 \end{Bmatrix} = \begin{Bmatrix} 0.707P \\ 0.0 \\ -2.121P \\ 0.0 \\ 0.0 \end{Bmatrix}$$

18.7

$$\begin{bmatrix} \theta_B \\ \theta_C \end{bmatrix} = \frac{L}{EI}\begin{bmatrix} -1.5625 \\ 3.125 \end{bmatrix} \text{ rad. and } \begin{bmatrix} M_{AB} \\ M_{BC} \end{bmatrix} - \begin{bmatrix} 9.38 \\ 18.75 \end{bmatrix} \text{ kN.m}$$

18.10

$$\theta_B = \frac{-PL^2}{48EI}, \quad \begin{Bmatrix} M_{AB} \\ M_{BA} \\ M_{BC} \\ M_{CB} \end{Bmatrix} = \frac{PL}{48}\begin{Bmatrix} -1 \\ -2 \\ 2 \\ -8 \end{Bmatrix}$$

18.11

$$\begin{Bmatrix} \theta_B \\ \theta_C \\ \Delta \end{Bmatrix} = \frac{PL^2}{EI} \begin{Bmatrix} -0.0044 \\ 0.0329 \\ 0.0168L \end{Bmatrix}, \quad \begin{Bmatrix} M_{AB} \\ M_{BA} \\ M_{CD} \\ M_{DC} \end{Bmatrix} = \begin{Bmatrix} 0.0921 \\ 0.0833 \\ 0.1400 \\ 0.2716 \end{Bmatrix} PL$$

18.12

$$\begin{Bmatrix} M_{AB} \\ M_{BC} \\ M_{CD} \\ M_{DA} \end{Bmatrix} = \begin{Bmatrix} 40.2 \\ -37.5 \\ 34.6 \\ 34.5 \end{Bmatrix} \text{kN.m}$$

18.13

$$\mathbf{k} = \frac{EI}{L} \begin{bmatrix} \frac{24}{L^2} & & & \\ 0 & 12 & \text{sym.} & \\ \frac{-12}{L^2} & \frac{-6}{L^2} & \frac{12}{L^2} & \\ \frac{-6}{L} & 2 & \frac{6}{L} & 8 \end{bmatrix}$$

$$\mathbf{k}^* = \frac{EI}{L^3} \begin{bmatrix} 19.304 & -8.087 \\ -8.087 & 5.743 \end{bmatrix}$$

Chapter 19

19.1 (a) 2.0, (b) 1.58, (c) 1.13

19.2 $W_u = 7.5 \dfrac{M_P}{l}$

19.3 $W_u = 6.0 \dfrac{M_P}{l}$

19.4 $W_u = 53.33$ kN/m

19.5 $M_P = 106.67$ kN.m

19.6 $M_P = 60.0$ kN.m

19.7 $w_u = 11.67 \dfrac{M_P}{l^2}$

19.8 $w_u = 17.5 \dfrac{M_P}{l^2}$

19.9 $P_u = 3.43 \dfrac{M_P}{l}$

19.10 $P_u = 1.5 \dfrac{M_P}{l}$

ANSWERS TO OBJECTIVE TYPE QUESTIONS

Ch.2 :	(1) b	(2) b	(3) d	(4) b	(5) a
	(6) c	(7) c	(8) c	(9) a	(10) b
	(11) b				
Ch.3 :	(1) b	(2) d	(3) c	(4) d	(5) b
	(6) b	(7) d	(8) a	(9) b	(10) b
	(11) b	(12) a			
Ch.4 :	(1) d	(2) a	(3) a	(4) a	(5) d
	(6) d	(7) a			
Ch.5 :	(1) a	(2) b	(3) c	(4) b	(5) c
	(6) b	(7) b	(8) d	(9) b	(10) b
	(11) b	(12) b	(13) b	(14) b	(15) c
Ch.6 :	(1) b	(2) b	(3) a	(4) c	(5) b
	(6) c	(7) a	(8) c	(9) d	(10) d
	(11) a				
Ch.7 :	(1) b	(2) d	(3) c	(4) d	(5) b
	(6) b	(7) c	(8) a	(9) d	(10) b
	(11) b	(12) d	(13) a	(14) b	(15) c
Ch.8 :	(1) a	(2) c	(3) b	(4) a	(5) d
	(6) b	(7) a	(8) b	(9) d	(10) b
	(11) a	(12) b	(13) c		
Ch.9 :	(1) a	(2) c	(3) a	(4) d	(5) b
	(6) d	(7) d	(8) b	(9) c	(10) a
	(11) c				
Ch.10 :	(1) b	(2) a	(3) b	(4) a	(5) d
	(6) d	(7) c	(8) a	(9) c	(10) d
	(11) b	(12) a			
Ch.11 :	(1) b	(2) c	(3) c	(4) d	(5) c
	(6) a	(7) b	(8) b	(9) c	(10) d
	(11) c	(12) b	(13) d	(14) b	
Ch.12 :	(1) d	(2) b	(3) c	(4) b	(5) c
	(6) c	(7) b	(8) a	(9) b	(10) c
	(11) c	(12) a	(13) d		
Ch.13 :	(1) a	(2) d	(3) c	(4) c	(5) d
	(6) c	(7) d	(8) d	(9) b	(10) c
	(11) b	(12) a			
Ch.14 :	(1) c	(2) d	(3) b	(4) c	(5) b
	(6) a	(7) c	(8) d	(9) b	(10) c
Ch.15 :	(1)	(a) F	(b) T	(c) F	(d) T
		(e) F	(f) T		
	(2)	(i) b	(ii) a	(iii) a	
	(3)	(i) a	(ii) d		
	(4)	(i) d	(ii) c		
	(5)	(i) c	(ii) b	(iii) b	
Ch.19 :	(1) c	(2) a	(3) c	(4) b	(5) d
	(6) c	(7) c	(8) b	(9) c	(10 a
	(11) d	(12) b	(13) c	(14) b	

Bibliography

BEAUFAIT, F.W., ROWAN, W.H., Jr., HOADLEY P.G. and HACKETT R.M., Computer Methods of Structural Analysis, Prentice-Hall, Inc. Englewood Cliffs; New Jersey, 1970.

BORG. S.F. and GENNARO, J.J., Advanced Structural Analysis, D. van Nostrand Company, Inc., 1959.

DAYARATNAM, P. Advanced Structural Analysis, Tata McGraw-Hill Publishing Co. Ltd., New Delhi, 1978.

GERE, J.I. and WEAVER, W., Jr., Analysis of Framed Structures, D. van Nostrand Co., Inc., Princeton, New Jersey, 1965.

GERSTLE KURI, H., Basic Structural Analysis, Prentice-Hall, Inc., Englewood Cliffs, New Jersey, 1974.

GHALI, A. and NEVILLE, A.M., Structural Analysis, Intext Educational Publishers, San Francisco, 3 Edn. 1989.

HALL, A.S. and WOODHEAD, R.W., Frame Analysis, Second edn., John Wiley & Sons, Inc., New York, 1967.

JACK C. McCORMC, Structural Analysis Harper and Row Publishers, New York, 1984.

JOHN F. FLEMING, Computer Analysis of Framed Structures, McGraw-Hill, International Edition

JUNNARKAR, S.B., Mechanics of Structure Vol. II, Charotar Publishing House, Anand 1989.

KARDESTUNCER, H., Elementary Matrix Analysis of Structures, McGraw-Hill Book Co., New York, 1974.

KINNEY, J.S., Indeterminate Structural Analysis, Addison-Wesley Publishng Co., Inc., Reading Massachusettes, 1957.

LAURSEN, H.I., Matrix Analysis of Structures, McGraw-Hill Book Co., New York, 1966.

LAURSEN, H.I., Structural Analysis, McGraw-Hill Book Co., New York, 3 Edn. 1988.

LIVESLEY, R.K., Matrix Methods of Structural Analysis, Pergaman Press, London, 1964.

MALLICK, S.K. and RANGASWAMY, K.S., Introduction to Matrix Analy¬sis of Structures, Khanna Publishers, New Delhi, 1971. MARSHALL, W.T. and NELSON, H.M., Structures, Sir Isaac Pitman and Sons Ltd., London, 1969.

MARTIN, H.C., Introduction to Matrix Methods of Structural Analysis, McGraw-Hill Book Co., New York, 1966.

McMINN, S.J., Matrices for Structural Analysis, John Wiley & Sons, Inc., New Yofk, 1962.

MEEK, J.L., Matrix Structural Analysis, McGraw-Hill Book Co., New York,, 1971.

MORICE, P.B., Matrix for Structural Analysis, Pergaman Press, London, 1964.

NEGI, L.S. and JANGID, R.S., Structural Analysis, Tata McGraw-Hill, 1997.

NORIS, C.H. and WILBUR, J.B., Elementary Structural Analysis, McGraw-Hill Book Co. New York, 1960.

PANDIT, G.S., GUPTA, S.P., Structural Analysis-A Matrix Approach. Tata McGraw-Hill Publishing Co. Ltd., New Delhi, 1981.

PESTEL., E.C. and LECKIE, F.A., Matrix Methods in Elasto Mechanics, McGraw-Hill Book Co., New York, 1963.

POPOV, E.P., Introduction to Mechanics of Solids, Prentice-Hall of India Ltd., New Delhi, 1973.

PRAKASA RAO, D.S., Strength of Materials, University Press (India) Limited, 1999.

RUBINSTEIN, M.F., Matrix Computer Analysis of Structures, Prentice-Hall Inc., Englewood Cliffs, N.J., 1966.

SINHA, N.C., and GAYEN, P.K., Advanced Theory of Structures, Dhanpat Rai & Sons, New Delhi 1990.

THADANI, B.N., Modern Methods in Structural Analysis, Asia Publishing House, New York, 1963.

TIMOSHENKO, S. and YOUNG, D.H., Theory of Structures, Second edn., McGraw-Hill Book Co., New York, 1965..

WANG, C.K., Matrix Methods of Structural Analysis, International Text¬book Co., 1970.

WHITE, RICHARD N., GERGELY, PETER and SEXSMITH, ROBERT G., Structural Engineering, combined edn., John Wiley & Sons, Inc., 1972.

WILLEMS, NICHOLAS and LUCAS, WILLIAM, M. Jr., Matrix Analysis for Structural Engineers, Prentice-Hall, Inc. Énglewood Cliffs, N.J., 1968.

Index

9.8 In the approximate analysis of building frames using portal method the shear in the frame is shared

(a) equally by all the columns

(b) the interior columns take twice as much as the exterior columns

(c) only interior columns take the shear

(d) the exterior columns take twice as much as the interior columns

9.9 In the approximate analysis of building frames using cantilever method the axial force in the columns is assumed to be

(a) equal tension in all columns

(b) tension in windward columns and compression in leeward columns

(c) tension or compression in columns is proportional to the distance of column from C.G. of columns

(d) equal compression in all columns

9.10 In a cantilever method of approximate analysis of a building frame the column lines and the C.G. of columns is shown in Fig. 9.2. The axial force in column [illegible] is